Principles of condensed matter physics

Now in paperback, this book provides an overview of the physics of condensed matter systems. Assuming a familiarity with the basics of quantum mechanics and statistical mechanics, the book establishes a general framework for describing condensed phases of matter, based on symmetries and conservation laws. It explores the role of spatial dimensionality and microscopic interactions in determining the nature of phase transitions, as well as discussing the structure and properties of materials with different symmetries. Particular attention is given to critical phenomena and renormalization group methods. The properties of liquids, liquid crystals, quasicrystals, crystalline solids, magnetically ordered systems and amorphous solids are investigated in terms of their symmetry, generalized rigidity, hydrodynamics and topological defect structure. Written in a clear and pedagogic style, the book contains problems at the end of each chapter and extensive real-world examples.

In addition to serving as a course text, this book is an essential reference for students and researchers in physics, applied physics, chemistry, materials science and engineering, who are interested in modern condensed matter physics.

Principles of condensed matter physics

P. M. CHAIKIN
Princeton University

T. C. LUBENSKY
University of Pennsylvania

CAMBRIDGE
UNIVERSITY PRESS

CAMBRIDGE
UNIVERSITY PRESS

314 to 321, 3rd Floor, Plot No.3, Splendor Forum, Jasola District Centre, New Delhi 110025, India

Cambridge University Press is part of the University of Cambridge.

It furthers the University's mission by disseminating knowledge in the pursuit of education, learning and research at the highest international levels of excellence.

www.cambridge.org
Information on this title: www.cambridge.org/9780521794503

First published 1995
Reprint 1997 (with corrections), 2000
First paperback edition (with corrections), 2000
First South Asia edition 2004
Reprint 2007, 2008, 2012, 2017, 2018, 2025

This South Asia edition is based on P. M. Chaikin and T. C. Lubensky / Principles of Condensed Matter Physics / 9780521794503 / 1995

Printed in India by Manipal Technologies Limited, Manipal

A catalogue record for this publication is available from the British Library

ISBN 978-8-175-96025-1 Paperback

To Amy, David, Ellen, Paula, Diana, and Valerie

Contents

Preface

The use and understanding of matter in its condensed (liquid or solid) state have gone hand in hand with the advances of civilization and technology since the first use of primitive tools. So important has the control of condensed matter been to man that historical ages – the Stone Age, the Bronze Age, the Iron Age – have often been named after the material dominating the technology of the time. Serious scientific study of condensed matter began shortly after the Newtonian revolution. By the end of the nineteenth century, the foundations of our understanding of the macroscopic properties of matter were firmly in place. Thermodynamics, hydrodynamics and elasticity together provided an essentially complete description of the static and dynamic properties of gases, liquids and solids at length scales long compared to molecular lengths. These theories remain valid today. By the early and mid-twentieth century, new ideas, most notably quantum mechanics and new experimental probes, such as scattering and optical spectroscopy, had been introduced. These established the atomic nature of matter and opened the door for investigations and understanding of condensed matter at the microscopic level. The study of quantum properties of solids began in the 1920s and continues today in what we might term "conventional solid state physics". This field includes accomplishments ranging from electronic band theory, which explains metals, insulators and semiconductors, to the theory of superconductivity and the quantum Hall effect. The fundamental problems of how to treat the effects of the strong Coulomb interaction in many electron systems and the effects of lattice disorder remain only partially resolved to this day.

The second half of the twentieth century has seen a new set of paradigms introduced into physics, originating in condensed matter and spreading to other areas. The idea is to span length scales, to see what remains as an observer steps back from the microscopics of a system and then keeps stepping back. X-ray, neutron and light scattering have become powerful probes of structure from microscopic to near macroscopic length scales. The study of critical phenomena has led to the notions of scaling and universality and has spawned the renormalization group, which shows how identical behavior at long length scales can arise from widely different microscopic interactions. At the same time, the concepts of broken symmetry and order parameters have emerged as unifying theoretical concepts applicable not only to condensed matter physics but also to particle physics and even to cosmology. These theoretical advances have provided a framework for describing condensed matter phases: liquid crystals, superfluid helium, incommensurate crystals, quasicrystals and systems in one and two dimensions, as well as classical fluids and regular periodic solids.

In spite of these unifying advances, conventional solid state physics and "soft" condensed matter physics (which includes the study of many statistical problems such as critical phenomena as well as the study of soft material such as fluids and liquid crystals) have very much remained distinct fields. The present book grew out of the conviction that graduate programs in physics should offer a course in the broad subject of condensed matter physics, a course that would prepare students to begin research in any area of this vast, yet still expanding, field. Our experience was that students learned either conventional solid state physics or soft condensed matter physics, and that each group remained blissfully ignorant of the other. We therefore developed, and began to teach, a one-year course in condensed matter physics at the University of Pennsylvania.

The first semester of this course was designed to establish a general framework, based on concepts of symmetry, for approaching condensed phases, from high-temperature fluids to low-temperature quantum crystals. It included an overview of the great variety of condensed systems found in nature and a description of their symmetry in terms of order parameters. It then discussed phase transitions, elasticity, hydrodynamics and topological defect structure in terms of these order parameters. It revisited many of the problems of the nineteenth century from a modern viewpoint. The second semester treated subjects normally associated with conventional solid state physics and many-body theory: normal Fermi liquids, electrons, phonons, magnetism and superconductivity. However, these topics were taught within the general framework established during the first semester. None of the concepts in the first semester involved quantum mechanics in an essential way, whereas those in the second semester did. We, therefore, in our own minds, referred to the first semester as "$\hbar = 0$" and the second semester as "$\hbar \neq 0$". The first semester also dealt much more extensively with "soft" systems, such as liquid crystals or microemulsions, and we sometimes referred to the first semester as "soft" condensed matter physics and to the second semester as "hard" condensed matter physics. The concepts to be covered in the first semester were, however, quite general and applied to both "soft" and "hard" systems. We have each taught the full year course described above many times to second-year graduate students at both the University of Pennsylvania and Princeton University.

The present book evolved from notes prepared for the first semester of the course. While there are several excellent texts dealing with $\hbar \neq 0$ solid state physics and with many-body physics, we have been unable to find a text dealing with $\hbar = 0$, or soft condensed matter physics, to recommend to our students or colleagues. Different aspects of this subject are available in the research literature and in several, sometimes material-specific, books. We, and others, have long felt that there is an acute need for a text on modern aspects of condensed matter physics, one that would present a unified picture of structures other than periodic solids, that would treat broken symmetry, critical phenomena and the renormalization group, and that would explore the role of fluctuations and topological defects in determining the existence of order and the nature of phase transitions. This book is an attempt to address this need.

What do you need to make use of this book? Some knowledge of quantum mechanics would be helpful but not essential. Statistical mechanics is an important prerequisite and is used throughout the book. (Although Chapter 3 provides a review of statistical mechanics, it is intended as a refresher to define notation rather than as a substitute for prior exposure.) A course in solid state physics would be helpful, but again not absolutely essential. If you are a field theorist, the book should make nice bedtime reading and introduce you to some really interesting relevant physics. The book is meant as a first course in condensed matter physics for second-year graduate students regardless of their field of specialization. It relies more on a general background of physical understanding and mathematical tools appropriate to that level than it does on any specific previous course.

Though originally intended as a text for graduate courses, the book should also serve as a reference text for researchers in condensed matter physics, materials science, chemistry, engineering and applied physics. We have attempted to cover each subject as completely as possible, beginning with simple ideas and ending with advanced concepts. Thus, for example, we present mean-field theory in a variety of guises, beginning with Bragg-Williams theory, but also including variational and field-theoretic approaches; or we cover descriptive aspects of topological defects and more advanced concepts like lattice duality transformations. Parts of the book could be, and some have been, used in more elementary courses such as undergraduate solid state physics, statistical mechanics or materials science. At the other extreme, in many scientific arguments with colleagues and competitors, we have found the notes for this text an invaluable resource in proving either our point or theirs.

The text as it stands is suitable for a full year graduate course, although we have never taught it as such. Chapters 1–6 establish the fundamentals. They introduce the systems to be studied, present experimental and theoretical tools, set up mean-field theories and show how they break down, investigate critical phenomena, and discuss symmetry breaking and the resulting generalized elasticity. We usually teach all of Chapters 1–6 and parts of the remaining four chapters (usually all of Chapter 9 on topological defects and bits and pieces of the other three chapters). On occasion, we have, instead of including all of Chapter 9, taught Chapters 7 and 8 on dynamical processes and hydrodynamics, followed by parts of Chapter 10 on domain walls, kinks and solitons. When the whole year sequence was taught, we sometimes taught Chapters 1–6 and Chapter 9 in the first semester, followed by parts of Chapters 7 and 8 in the second semester, before moving on to many-body physics. Though we have generally taught this text as a part of the full year sequence discussed above, we believe it can serve as an excellent text book for a second semester of statistical mechanics, as a secondary text for a course in many-body physics, and as a stand alone text for condensed matter physics.

Each chapter concludes with a set of problems. We have tried to include problems at all levels of difficulty. Many problems are, however, challenging, even for seasoned professionals. Where possible, we have tried to provide answers or answer clues for the more difficult problems.

Astonishingly, none of our friends or colleagues has tried to dissuade us from completing this book. Some have had a direct or indirect influence on both the content and style of this book. The pedagogical and research approaches of Phil Anderson, P.G. de Gennes, Bert Halperin, Paul Martin and David Nelson can be seen throughout the book. We received constant encouragement and help from Shlomo Alexander, Mark Azbel, Daniel Fisher, Gary Grest, Scott Milner, Burt Ovrut, Phil Pincus, David Pine, Jacques Prost, Cyrus Safinya, David Weitz and Tom Witten. We are extremely grateful to Phil Nelson, who read every word of the manuscript, corrected many misprints and made numerous suggestions for improving the text, to Tetsuji Tokihiro for using preliminary notes as a basis for a course at the University of Tokyo and for providing a lengthy list of corrections, to Ray Goldstein, Mark Robbins, and Holger Stark for pointing out errors right up to the publication date, and to numerous students and postdocs at Penn and Princeton who read and commented on various versions of the notes leading to the final manuscript. We are also grateful to Chris Henley and Josh Socolar for using the unfinished manuscript in courses they taught at Cornell and Duke University, respectively. Finally, we are grateful to Exxon Research and Engineering Co. for providing a friendly environment where many discussions about this book took place. We owe particular thanks to Jodi Forlizzi for most of the artwork.

1

Overview

1.1 Condensed matter physics

Imagine that we knew all of the fundamental laws of nature, understood them completely, and could identify all of the elementary particles. Would we be able to explain all physical phenomena with this knowledge? We could do a good job of predicting how a single particle moves in an applied potential, and we could equally well predict the motion of two interacting particles (by separating center of mass and interparticle coordinates). But there are only a few problems involving three particles that we could solve exactly. The phenomena we commonly observe involve not two or three but of order 10^{27} particles (e.g., in a liter of water); there is little hope of finding an analytical solution for the motion of all of these particles. Moreover, it is not clear that such a solution, even if it existed, would be useful. We cannot possibly observe the motion of each of 10^{27} particles. We can, however, observe macroscopic variables, such as particle density, momentum density, or magnetization, and measure their fluctuations and response to external fields. It is these observables that characterize and distinguish the many different thermodynamically stable phases of matter: liquids flow, solids are rigid; some matter is transparent, other matter is colored; there are insulators, metals and semiconductors, and so on.

Condensed matter physics provides a framework for describing and determining what happens to large groups of particles when they interact via presumably well-known forces. Nature provides us with an almost unlimited variety of many-body systems, from dilute gases and quantum solids to living cells and quark-gluon plasmas. Collections of even the simplest atoms exist in a number of different states. Helium, for example, can be found not only in gaseous, liquid, and solid phases but also as a non-viscous superfluid at low temperatures. Condensed matter physics is the study of all of these many-body states of matter. Its paradigms can and do provide insight into fields as diverse as biology and particle physics.

Indeed, many of the seminal ideas of modern theories of fundamental interactions, such as broken symmetry, had their origins in condensed matter physics.

Condensed matter physics deals with many-body interacting systems. However, it builds on, and in turn contributes to, other fields. It requires a knowledge of the fundamental force laws between atoms and molecules and the properties of small groups of these particles; it thus builds on atomic and molecular physics as well as on classical and quantum mechanics. Since it focuses on macroscopic properties rather than trajectories of individual particles, condensed matter physics requires an understanding of how things behave under different averaging processes; it builds on statistical mechanics and thermodynamics. Because most of the macroscopic variables of interest vary slowly in space, their statistical mechanics can be described by continuum field theories of the type first introduced in particle physics; modern condensed matter physics thus builds on quantum field theory.

Probably the most important unifying concept to emerge from the study of condensed matter physics is that macroscopic properties are governed by conservation laws and broken symmetries. In a system of particles, particle number, energy, and momentum are conserved. At high temperatures, all such systems are disordered, uncorrelated, uniform and isotropic. The probability of finding a particle at a given point in space is independent of the position of that point in space and independent of whether there is another particle nearby. This high-temperature state has the full rotational and translational symmetry of free space. The low-frequency dynamical properties of this state are controlled entirely by hydrodynamical equations, which in turn are determined by conservation laws. As temperature is lowered, new thermodynamically stable states condense. These states have progressively lower symmetry. For example, a periodic crystal is invariant with respect to only a discrete set of translations rather than to the continuum of translations that leave the high-temperature state unchanged. Associated with each broken symmetry are distortions, defects, and dynamical modes that provide paths to restore the symmetry of the original high-temperature state. The properties of each broken-symmetry phase are largely controlled by these distortions, defects, and modes. A crystalline solid for example can be sheared. The energy of shear distortions is determined by an elastic constant, which is a particular rigidity associated with broken translational symmetry. There are shear sound modes in crystals not found in the high-temperature isotropic phase. Finally, there are various defects that interrupt an otherwise ideal crystal structure.

Conservation laws and broken symmetries are equally important in classical and quantum systems. Their consequences, when expressed in the appropriate language, are to a considerable degree independent of whether the underlying particle dynamics is classical or quantum mechanical. Thus, general truths about all of nature's phases can be obtained by studying classical rather than quantum systems. This book will explore condensed matter physics in a largely classical context. Many of its ideas, however, apply quite generally.

1.2 An example - H_2O

1 *Gaseous and liquid states*

To see how some of these ideas work, let us consider our experience with a rather common material – water. Although the water molecule is not the physicist's ideal (argon would probably be closer to ideal because of its filled atomic shell, spherically symmetric shape, and isotropic interparticle interactions), our experience with the phase transitions and different states of water is more extensive. At high temperature, water is steam or water vapor. Its kinetic energy dominates over its potential energy, and, as a result, it exists in a state that is isotropic and homogeneous and that fills any volume allowed it. This gaseous or fluid phase has complete translational and rotational symmetry. There is equal probability of finding a molecule anywhere in the containing volume. The density is uniform. There are very few correlations between the positions of the molecules. If the gas were ideal, then the pointlike particles would completely ignore the presence of each other.

If we look at this gas, the water vapor in the atmosphere, we do not see it. In order for something that has no direct absorptions at the optical frequency to be seen, it must scatter light. That means there must be a mismatch in the refractive index over some distance. In most cases, the refractive index is directly proportional to the density. Since the density of the gas is uniform, there are no index variations, and there is no scattering. Of course, there will always be fluctuations in the density, but, to be seen, they must have a length scale comparable to the wavelength of light.

Now let us lower the temperature, i.e., the average kinetic energy. As the potential energy becomes more important, specific intermolecular interactions come into play. For neutral water molecules, the dominant interaction is the dipole-dipole interaction, which for particular configurations is attractive. At short distances, comparable to the charge separation in the dipoles, the individual charges attract each other more strongly than the dipole approximation would predict. This stronger, more orientationally-dependent interaction, is called hydrogen bonding. Attraction tends to enhance density fluctuations: each molecule would prefer to spend most of its time in a region where there are other molecules rather than in one where there are none. This clustering leads not only to a lower energy but also to a lower entropy. As temperature is lowered, density fluctuations brought about by clustering grow in amplitude and persist for longer times. The larger fluctuations take longer to develop and longer to decay. Increased size dictates a slower dynamics. Density is still uniform but only when averaged over large regions of space or over long intervals of time. The end result of these attractive interactions is the formation of another fluid phase, a liquid phase (water) whose density is greater than that of the gas phase. The principal physical quantity distinguishing the liquid and gas phases is their density.

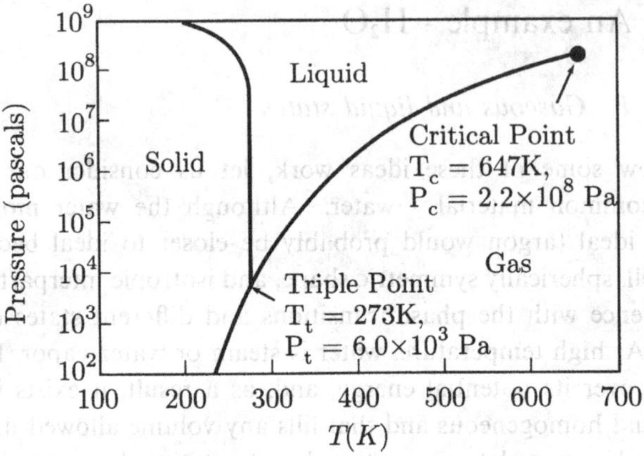

Fig. 1.2.1. The phase diagram for water.

2 *The liquid-gas phase transition*

Now suppose we have a closed container of water vapor at a density of 0.322 g/cc at high temperature. As the temperature is lowered, density fluctuations continue to grow and live longer. The system now no longer looks homogeneous: there are regions with greater and lesser density. As the size of these regions approaches the wavelength of visible light, scattering increases dramatically – the system looks "milky" (milk has droplets of fat whose diameter is of order one micron (1μ) and comparable to the wavelength of visible light $\sim 0.5\mu$ – that is why it looks "milky"). This is the phenomenon of critical opalescence and critical slowing down (the divergence of length and time scales). Finally, the size and size range of fluctuations become so large that some high- or low-density regions span the container. They also live long enough that the denser regions fall and the lighter ones rise in the gravitational field. The denser regions coalesce on the bottom, and the dense liquid and less-dense gas phases separate. Once again each of the phases is homogeneous and nonscattering. The only evidence we have that the two phases differ is that they are separated by a meniscus, made visible by the difference in the index of refraction of the two phases.

This most common condensation or phase transition from a gas to a liquid is different from most of the other phase transitions we will encounter. The symmetry of the two phases is the same, and there is no loss of symmetry in going to the low-temperature phase (both gas and liquid are fluid phases). This is reflected in the fact that in the phase diagram shown in Fig. 1.2.1, it is possible, by going around the critical point, to go from the gas phase to the liquid phase without traversing any phase boundary.

The scenario in the above paragraph resulted from a special choice of density and does not correspond to our usual experience with water condensation. When water is not at the critical density (0.322 g/cc) in a closed container, something else happens. As temperature is lowered (at pressures below the critical pressure), there is a discontinuous change in the thermodynamically stable state as the gas-liquid phase boundary is crossed. Consider now the gas phase at some temperature. Its average density is homogeneous and uniform. There will, however, be rare fluctuations creating droplets of the higher-density liquid phase. As temperature is lowered, the number and size of these droplets will grow, but none will become very large nor persist for a very long time. When the temperature is lowered beyond the gas-liquid phase boundary, the sample does not homogeneously and instantaneously change to the higher-density liquid phase. Rather, droplets of the liquid phase, already present by virtue of fluctuations in the gas phase, will grow larger and persist for longer times. Long before the average size of these droplets diverges, a few droplets will grow to be very large, most often nucleating on a dust particle or a salt molecule. They become large enough that, rather than decaying, they grow with time and absorb surrounding droplets and gas molecules as they grow. Their size is determined by kinetics, by how fast molecules can diffuse to their outer surface and be incorporated into their masses. As the size of these dense droplets becomes comparable to or larger than the wavelength of visible light, they scatter light strongly. This is what is responsible for the milky whiteness of clouds (Fig. 1.2.2), which are suspended droplets of water. This is not critical opalescence, but its effect is similar. The growth of droplets at the discontinuous gas-liquid transition is more rapid than the growth of fluctuations at the critical point. This is one of the characteristic differences between discontinuous, or first-order, transitions and continuous, or second-order, transitions.

If we apply pressure to the gas, its density changes, i.e., it is compressible. At the critical point, the liquid and gas phases with different densities are in equilibrium. Pressure can cause transformation of some volume of gas into the denser liquid phase with no cost in energy: a small pressure change leads to a large density change. There is a divergent rate of change of density with pressure, i.e., a divergent compressibility. Most of the continuous transitions that we will study are signaled by the divergence at a critical temperature of a quantity usually referred to as a susceptibility. The compressibility at the liquid-gas transition is an example of such a susceptibility. The diverging compressibility at the liquid-gas transition can literally be seen via critical opalescence. The diverging size and slowing down of fluctuations are just another manifestation of the same phenomenon that produces a diverging compressibility.

3 Spatial correlations in the liquid state

The liquid state is different from the gaseous state, if not by symmetry then by other properties: density and compressibility, for example. Less obvious is that the particles in the liquid are much more correlated. The distance between particles is now set by the trade-off between the repulsive and attractive parts

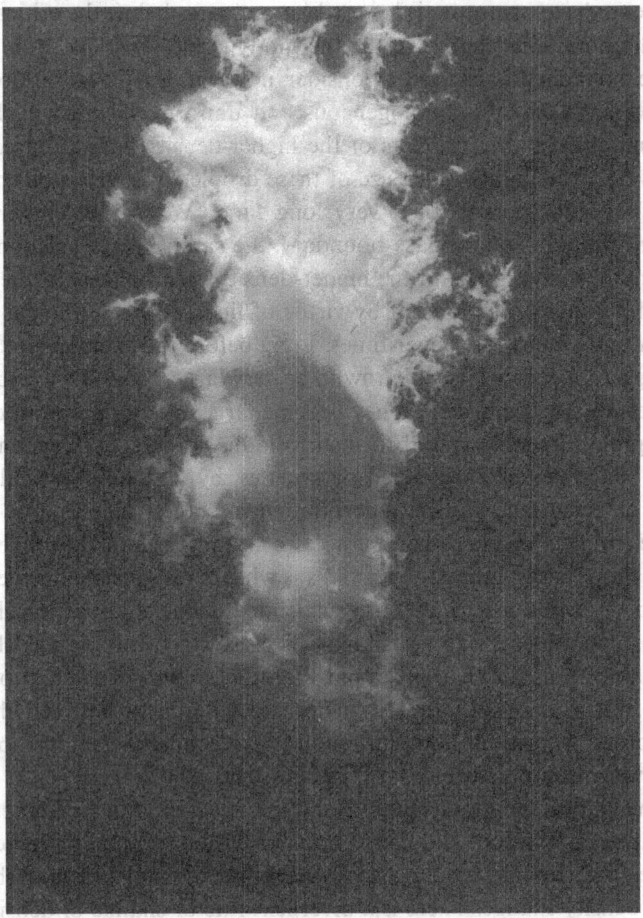

Fig. 1.2.2. Clouds are droplets of water or ice with characteristic size 1–10μ. The inhomogeneous density on the scale of the wavelength of visible light ($\sim 0.5\mu$) is responsible for the strong multiple scattering and white or milky appearance. It is similar to the phenomenon of critical opalescence observed in second-order phase transitions.

of the interparticle interactions. Although the density is uniform, the correlation between the positions of neighboring atoms is strong. If there is a particle at one point, there is no chance that another will sit on top of it and a good chance that another will be a particle-diameter away.

We now proceed to cool the system further. The desire of attractive interactions to bring particles close together has largely been satisfied by the formation of the high-density nearly incompressible liquid. The local packing of molecules is determined predominantly by the repulsive interaction, which prevents atoms from overlapping. When particles are pushed together, the electronic energy increases very rapidly because particles with the same spin cannot occupy the same point in space (exclusion principle) and because electrons with any spin repel each other via the direct Coulomb interaction. In a simpler liquid, say liquid argon, the repulsive interaction would be well described as a hard-wall potential at twice the atomic radius. Such a hard-sphere model gives us the essence of the liquid and solid physics of many systems. Attraction wants to bring atoms together. Hard-sphere repulsion leads to a discrete set of local configurations that take maximum advantage of attractive interactions. Atoms want to form triangles and then tetrahedra and then fill the triangular faces of the tetrahedra to form larger clusters. Two things prevent this. First, thermal energy keeps atoms from packing too tightly in the liquid phase. Secondly, the local algorithm for packing atoms as densely as possible by making tetrahedra from all exposed triangular faces and so on cannot be continued indefinitely without the introduction of voids that are disfavored by the attractive interaction: it is impossible to fill space by packing tetrahedra or icosahedra. There is a sort of frustration arising from the inability of the system to satisfy simultaneously local packing rules and global packing constraints. This process, however, paints a reasonably good picture of the structure of simple liquids and their atomic correlations. The strong correlations – local order – become increasingly more important as temperature is decreased. In order to see correlations at this intermolecular length scale, we have to probe with X-rays or neutrons which can probe this characteristic distance.

Liquid water behaves in much the same way as liquid argon, but the complex shape of water molecules and the complicated interactions between them lead to interesting differences between argon and water. The oxygen in a water molecule bonds its two hydrogens at an angle of 105° and arranges its four other electrons in two lone-pair bonds. To keep out of each other's way, the four bonds point toward the vertices of a tetrahedron. The liquid gains attractive energy by pointing the negative lone pairs toward the positive hydrogen atoms (this is an alternative description of the hydrogen bonding that is responsible for the structures of water and ice as well as much of biology). The water molecules try to form chains or clumps in which oxygens are tetrahedrally arranged but in which the twisted dumbbell molecules at the same time do not overlap. Liquid water gets its condensation energy from these directional bonds. Correlations again build up in response to these geometrical constraints. An X-ray scattering study of water has been analyzed to show the density of molecules around a molecule located at the origin. In Fig. 1.2.3, we see that the density is depressed near the central molecule, increases in a shell of order a molecular distance away, and then oscillates and decays to the uniform density at fairly short distances. Note also that correlations increase significantly as the water is chilled.

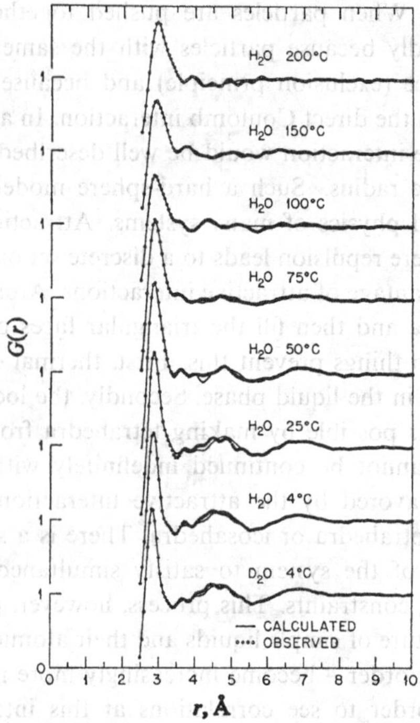

Fig. 1.2.3. The radial distribution function for liquid water is the probability distribution for water molecules surrounding a water molecule. There is an excluded region close to the central molecule, then an increased density for close neighbors, then an oscillating decrease in correlations to the average density at distances of a couple of molecular diameters. [A.H. Narton, W.D. Danford, and H.A. Levy, *Disc. Faraday Soc.*, **43**, 97 (1967).]

4 *Ice – crystallized water*

Our experience tells us that, at some point on cooling, water takes on a different form – ice. Ice is a solid, and the first thing we notice about it is that it does not flow like water. A solid is rigid, it resists shear. But there is a more fundamental difference between ice and water. The molecules in ice are arranged in a uniform repetitive way on a periodic lattice. The crystal structure of ice is illustrated in Fig. 1.2.4. It consists of layers of rippled hexagons in which neighboring atoms do not touch but in which the preferred local tetrahedral arrangement of oxygens is almost maintained. Although we cannot see the periodic lattice directly with our eyes, we can easily see some of its consequences. The hexagonal planar structure is responsible for the faceting planes and six-fold rotation symmetry of the ice crystals we find as snowflakes.

In the far simpler case of argon, the structure of the solid phase is determined

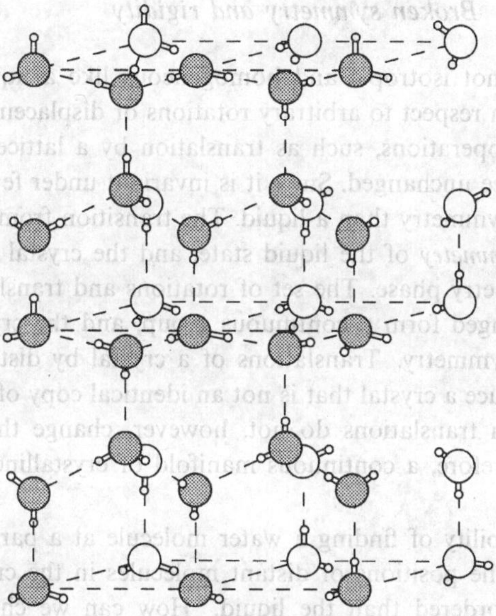

Fig. 1.2.4. Crystal structure of common ice. Note the directionality of the hydrogen bonds and the approximate tetrahedral coordination of each oxygen atom. The structure is a hexagonal "wurzite" form.

by the hard-sphere constraint at short distances and by the attractive interaction at somewhat larger distances. The attraction prefers as many close neighbors as possible and favors the densest periodic packing of spheres consistent with the hard-sphere repulsion. This is the FCC (face-centered cubic) structure with cubic symmetry consisting of hexagonal planes of close-packed spheres stacked on top of each other.

In a liquid, there are substantial local correlations in the positions of particles. Hard-sphere repulsion prevents two molecules from overlapping so that there will be no molecules within a molecular diameter of a given molecule. There will certainly be several molecules about a diameter away and, as a result, a density of molecules greater than the average. By about four to six diameters away, however, the density of particles will differ little from the average (Fig. 1.2.3). Knowledge of the position of one molecule gives essentially no information about the positions of far away molecules. The situation in a crystal is quite different. Molecules occupy, on average, sites on a periodic lattice. The position of one molecule (to specify an origin) and maybe one other (to specify a direction) will determine the positions of all other molecules out to infinity (or at least to the end of the crystallite).

5 Broken symmetry and rigidity

A crystal is not isotropic and homogeneous like a liquid. Rather than being invariant with respect to arbitrary rotations or displacements, it is invariant only under those operations, such as translation by a lattice spacing, that leave the periodic lattice unchanged. Since it is invariant under fewer operations, a crystal has a lower symmetry than a liquid. The transition from the liquid to the crystal *breaks the symmetry* of the liquid state, and the crystal is often referred to as a broken-symmetry phase. The set of rotations and translations leaving the liquid phase unchanged form a continuous group, and the crystal state has a broken continuous symmetry. Translations of a crystal by distances less than a lattice spacing produce a crystal that is not an identical copy of the untranslated lattice. Such uniform translations do not, however, change the energy of the crystal. There is, therefore, a continuous manifold of crystalline ground states with the same energy.

The probability of finding a water molecule at a particular position in space depends on the positions of distant molecules in the crystal. The crystal looks much more ordered than the liquid. How can we characterize the difference between the two states? Average density does not provide a good characterization, although the average density of liquid and crystal phases do differ (ice has a lower density than water at 0°C; most solids have a slightly higher density than the liquids with which they are in equilibrium). There is long-range order in the crystal associated with its periodic density. Molecules in a crystal are situated on the set of periodically arranged mathematical points called a lattice. We can tell whether there is long-range periodic order in the same way we test for periodicity in anything – by taking a Fourier transform and looking for discrete peaks in its spectrum. Scattering waves from a crystal is the experimental way of taking the spatial Fourier transform because the matrix element, $\langle \mathbf{k} \mid \text{sample} \mid \mathbf{k}' \rangle$, between incident and scattered plane waves $\mid \mathbf{k} \rangle$ and $\mid \mathbf{k}' \rangle$ is just the Fourier transform of the sample perturbation evaluated at $\mathbf{k} - \mathbf{k}'$. So it is the existence of a discrete spatial Fourier spectrum that distinguishes a crystal from a liquid, i.e. the existence of "Bragg spots" in the scattering spectrum.

The Fourier spectrum or scattering pattern does not change when the sample is displaced as a whole; it is only sensitive to the relative positions of molecules. (It is interferences of waves scattered from the molecules at their various positions that add up to give the Bragg spots.) The molecules are held in their positions by interactions with their neighbors; but even in the solid, individual molecules and groups of molecules are subjected to thermal (or quantum) fluctuations that lead to instantaneous configurations in which molecules are not arranged on an ideal periodic lattice. We can obtain an estimate of the magnitude of molecular displacements at finite temperature by considering the ice to be an elastic medium and using the equipartition theorem. An ideal crystal consists of periodically repeated unit cells with a particular size and shape. Distortions of the unit cell are described by strains x/a, which are displacements of one part of the cell relative to another by a distance x divided by the characteristic dimension a of the unit cell. The fact that there is a continuum of strains determined by a continuous variable x is intimately associated with the fact that a continuous symmetry is broken in going from the liquid to the crystal state. The stress, or force per unit area, required to produce such a strain is Gx/a, where G is an elastic modulus (for either shear or compression) which provides a measure of the rigidity of the crystal phase. The force on a unit cell associated with a stress is thus $-a^2Gx/a \equiv -kx$, where $k = Ga$ is an effective harmonic spring constant. The equipartition theorem states that the average potential energy at temperature T of a harmonic oscillator with spring constant k is $k\langle x^2\rangle/2 = k_BT/2$ or that the mean-square displacement is $\langle x^2\rangle = k_BT/k$, where k_B is the Boltzmann constant. Thus, the mean-square displacement in a crystal is inversely proportional to an elastic modulus: $\langle x^2\rangle \propto k_BT/Ga$. If any of the elastic moduli are zero, then a mean-square displacement will diverge. Once a random displacement is comparable in size to a lattice constant, the periodic order and the discrete peaks in the Fourier spectrum are destroyed. Therefore, *the rigidity is a necessary condition for the existence of the periodicity.* We will find in general that, associated with each phase transition to a state with a broken continuous symmetry, there will be a new rigidity or elastic constant preventing thermal fluctuations from destroying the new state.

The density in a high-temperature gas or liquid is uniform, and the probability of finding a molecule is independent of position in space. In a crystal, there is a higher probability of finding a molecule at one point than at another. How was the higher probability point chosen? There was nothing in the original problem favoring one point over another. There must be a mechanism to restore this lost or broken symmetry. It is found in the long-wavelength excitations of the system. The energies involved in distorting the periodic system depend on the relative displacement of neighboring molecules. We might expect that the dynamical modes of such a system are elastic waves. Conventional compressional-sound modes exist in water as well as in ice; shear sound modes, on the other hand, exist in ice but not in water. These modes have frequencies ω that vary linearly with wave vector q: $\omega = cq$, where c is the sound velocity and $q = 2\pi/\lambda$, where λ is the wavelength. In the long-wavelength limit, the frequency or energy of the mode approaches zero. There is no restoring force against a long-wavelength displacement. Mechanically, this is a result of the fact that at long wavelengths we can get a reasonable displacement of a molecule with just an infinitesimal change in each bond length over a large number of bonds. Physically, it is the consequence of the fact that a uniform translation of the system does not cost any energy. We can find the origin for the ice lattice with equal probability anywhere in space, but once we have located it, the rest of the molecular positions are fixed. The appearance of a hydrodynamic (that is long-wavelength) mode with zero frequency is another general feature of every transition yielding a broken continuous symmetry.

6 Dislocations – topological defects

There is another property of ice with which you may be familiar (depending on where you live): it flows. The rigidity, which we discussed above, is a measure of the resistance of ice to deformation. If applied forces or stresses are sufficiently weak, ice will respond by distorting or straining in a time-independent way. (The shear modulus is defined in the limit of zero stress.) When the stress is released, ice returns to its initial undistorted form. However, as stresses increase, ice will eventually flow. It will distort continuously with time, and when the stress is removed it will not return to its original shape. It has undergone plastic deformation. The most dramatic effects associated with this flow are seen in glaciers.

We might imagine that ice could flow if all of the bonds between its hexagonal planes were broken so that they could slide over one another. A quick estimation tells us that a "yield" stress of essentially the shear modulus would be required to do this. But it has been found that there is considerable flow or "creep" well below this value, often five orders of magnitude below. Perhaps not all of the bonds in the plane have to be broken in order for the ice to flow. Imagine that we cut half of the bonds in the plane, move them over one lattice to the left and reattach them as depicted in Fig. 1.2.5. We pay the price of a line of cut bonds and some strain energy, but since everything matches up far away from the line, the energy cost is finite. Now we can move this line defect or "dislocation" quite readily since it means breaking a *line* of bonds and remaking them one site over. Each time we do this the whole crystal on top moves a little in the direction of the dislocation motion. This edge dislocation "glides" easily in the plane and allows the ice to shear above and below the plane.

The motion of dislocations is what allows for creep and dynamic recrystallization in ice glaciers. The pinning of dislocations and dissipation associated with dislocation motion are responsible for most of the mechanical properties of crystalline solids. What makes dislocations possible is a combination of the periodicity of the ideal crystalline state and the elasticity of that state. A displacement of the ideal crystal by one lattice spacing leads to an identical crystal. It is thus possible to cut a crystal along a half plane, displace the crystal above that plane by one lattice spacing, and "glue" it to the undisplaced crystal below that plane. Far from the edge of the half plane, there is a slightly strained but otherwise perfect crystal. This construction yields a dislocation whose existence is determined by the nature and topology of the manifold of displacements that leave the energy of the solid unchanged. It is a *topological* defect. Like rigidity, topological defects are a general feature of broken continuous symmetries.

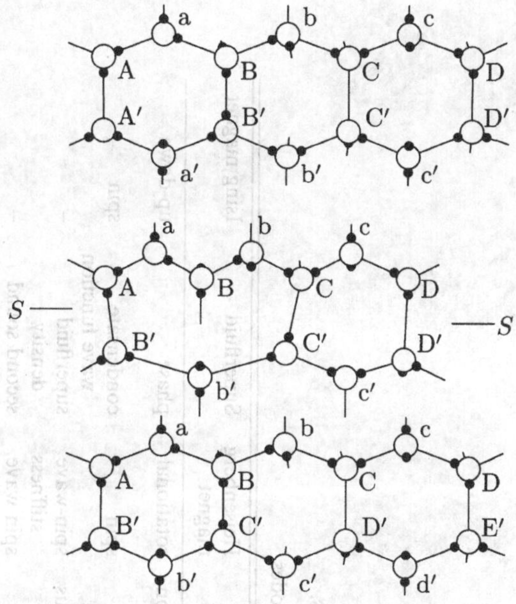

Fig. 1.2.5. An edge dislocation in ice. The dislocation motion is more complex than in simpler materials since it also leaves defects in the hydrogen bonding which must be relaxed by diffusion. [J.C. Poirier, *Creep of Crystals* (Cambridge University Press, 1985).]

7 *Universality of the water example*

Water is a part of our everyday experience. Many of its properties discussed here, when properly interpreted, are universal. They are found throughout the realm of condensed matter physics. At high temperature, kinetic energy dominates over potential energy, and equilibrium phases of matter are isotropic and homogeneous. As temperature is lowered there are phase transitions to more strongly correlated states. These transitions can be continuous (like the liquid-gas transition at the critical density) or discontinuous (like the boiling of water). At continuous transitions, characteristic lengths, susceptibilities, and relaxation times diverge. At discontinuous transitions, there is the phenomenon of nucleation. At sufficiently low temperatures, when potential energy is truly dominant over kinetic energy, equilibrium states (like ice) will in general have a lower symmetry than the high-temperature phase. If the broken symmetry is continuous, then the broken-symmetry phase is characterized by a rigidity (like the elastic modulus of ice), low-frequency dynamical modes (shear sound waves), and topological defects (dislocations). Table 1.2.1 lists properties of some broken-symmetry phases.

The water-ice transition provides an example of a transition in which a continuous symmetry is broken. There are transitions in which a discrete symmetry is

Table 1.2.1. *Properties of some representative broken-symmetry phases.*

The Ising magnet has a broken discrete symmetry and no new rigidity or modes.

	Fluid	Nematic	Smectic-*A*	Crystal	Heisenberg magnet	Superfluid	Ising magnet
Broken symmetry	none	rotational	1D translation	3D translation	rotational	phase	up-down
New order	none	orientational	1D periodic density	3D periodic density	spin	condensate wave function	spin
Rigidity	none	rotational elastic constant	layer modulus	shear modulus	spin-wave stiffness	superfluid density	–
New modes	none	diffusive orientational	second sound – undulation	shear sound	spin wave	second sound	–
Defects	none	disclinations, hedgehogs	dislocations	dislocations	hedgehog	vortices	domain walls

broken. The most familiar of these is the transition to the ferromagnetic state of an Ising model. Spins in an Ising model can point only up or down so that spin-flip is the only nontrivial symmetry operation that leaves the high-temperature paramagnetic state with randomly aligned spins unchanged. There are only two equivalent low-temperature ground states: that with all spins up and that with all spins down, and there are no low-energy excitations taking the system from one ground state to another. The elementary excitations are domain walls separating up spin from down spin regions. There are no low-energy excitations characterized by a rigidity or low-frequency hydrodynamic modes as there are in states with a broken continuous symmetry. Thus, the water example does not provide a good description of low-temperature broken-discrete-symmetry systems. It continues to provide, however, a remarkably correct description of many of the properties of the high-temperature phase of these systems.

Crystalline ice breaks both the translational and rotational symmetry of the fluid water phase. It has a very low symmetry. Physical systems with symmetry intermediate between the highest symmetry fluid and the lowest symmetry crystal phases also occur in nature. Liquid crystalline mesophases successively break the symmetries of the fluid phase. An isotropic fluid is invariant with respect to arbitrary translations in any direction and with respect to rotations about any axis. The nematic phase is invariant with respect to arbitrary translations but only with respect to arbitrary rotations about a single preferred axis: it is uniaxial. The smectic-*A* phase is uniaxial and breaks translational symmetry along a single direction. Discotic phases break translational symmetry along two directions and are invariant only with respect to discrete rotations. These phases and others will be explored throughout this book. They are reviewed in Table 1.2.2.

8 *Fluctuations and spatial dimension*

Water is an example from our real three-dimensional world. There are many materials and systems, however, that behave as though they were either one- or two-dimensional rather than three-dimensional. Furthermore, theoretical models can be formulated in any spatial dimension, and it is quite instructive to do so. As spatial dimension is increased, fluctuations become less and less important. Above a critical dimension d_c, fluctuations become so unimportant that mean-field theory, a simple approximation scheme, provides an analytically correct description of continuous phase transitions and essentially numerically correct descriptions of both low- and high-temperature phases. As dimension is reduced, fluctuations become increasingly important. They are quite important at continuous transitions in three dimensions even though mean-field theory continues to provide a very good qualitative description.

Below three dimensions, fluctuations become so violent that they can destroy the ordered state and finite-temperature phase transitions. In one dimension, fluctuations destroy all long-range order and phase transitions. This is essentially

Table 1.2.2. *Sequence of phases with decreasing symmetry from the highest-symmetry isotropic fluid to the lowest-symmetry crystalline solid.*

Phase	Invariances	Order
Isotropic	all translations and rotations	none
Nematic	all rotations about **n** axis, rotations by $\pi \perp$ to **n**, all translations	uniaxial orientational
Biaxial nematic	rotations by π about **n** and $\perp$ to **n**, all translations	biaxial orientational
Smectic-*A*	same rotational invariances as nematic, all translations $\perp$ to **n** and translations by lattice vector $\parallel$ to **n**	uniaxial orientational 1D periodic density
Smectic-*C*	same translational as smectic-*A*, rotation by π about **n** and $\perp$ **n**	biaxial orientational 1D periodic density
Hexatic	same translational as smectic-*A*, rotations by $2\pi/6$ about **n** and by $\pi \perp$ to **n**	six-fold orientational 1D periodic density
Discotic	same rotational as hexatic, all translations along **n** and by lattice spacings $\perp$ to **n**	six-fold orientational 2D periodic density
Crystal	discrete rotations in all directions, translations by lattice vectors in all directions	discrete orientational 3D periodic density

a problem of connectivity. The only way one end of a one-dimensional system knows what is going on at the other end is via information transmitted directly along the chain. For an infinitely long system, any fluctuation cuts the flow of information and hence the order. Since there are always fluctuations at any finite temperature, a one-dimensional system cannot be ordered except at zero temperature. In two-dimensional systems, there are many paths that can connect one part of the system to the others. Fluctuations are strong enough to destroy long-range order in systems with broken continuous symmetry but not necessarily strong enough to destroy phase transitions. Fluctuations in two dimensions do not destroy long-range order in systems with a broken discrete symmetry.

9 Overview of book

This book presents an overview of condensed matter physics that follows the water example just discussed. Properties of condensed matter systems depend on their symmetry and structure, which in turn depend on the nature of their constituent particles. We, therefore, begin in Chapter 2 by exploring the different symmetry phases and structures that occur in nature. We will find that these structures depend on the nature and shape of particles from which they are composed. For example, spherical particles usually form close-packed solids, whereas rigid bar-like molecules form anisotropic crystals and liquid crystals. We are interested in averaged quantities such as susceptibilities and two-point correlation functions that can be measured experimentally. These are best described by the machinery of thermodynamics and statistical mechanics, which we review in Chapter 3. Phase transitions, such as those from water to ice or from liquid to gas, are among the most striking phenomena we observe. These are studied first in mean-field theory in Chapter 4. Mean-field theory provides a qualitatively correct description of most phase transitions. In Chapter 5, we study how fluctuations modify mean-field theory. We show how fluctuations lead to a breakdown of mean-field theory below an upper critical dimension d_c. Each phase with a broken continuous symmetry is characterized by a generalized elasticity characterized by rigidities such as the shear modulus of ice. Static properties of broken-continuous-symmetry systems, from simple spin models to defected crystals, are explored in Chapter 6. Emphasis is placed on thermal fluctuations of elastic modes, which become increasingly violent as spatial dimension is lowered, until finally, at the lower critical dimension d_L, they destroy order altogether. The long-wavelength, low-frequency dynamics of a given phase is determined by its conservation laws and the nature of its broken continuous symmetries. The dynamics of liquid water is controlled by its five conservation laws (mass, energy, and three components of momentum), whereas the dynamics of ice is controlled by its shear and bulk rigidities in addition to its conservation laws. In Chapter 7 we develop a general language for describing dynamical phenomena, which is a generalization to dynamics of the time-independent statistical mechanics of Chapter 3. Then, in Chapter 8, we set up a general formalism for determining the hydrodynamics of any broken-symmetry system and derive the hydrodynamic equations for a number of particular systems. The dynamical correlation functions predicted by these equations reduce in the zero-frequency limit to those obtained statically in Chapters 4 to 6. Associated with each broken symmetry are defects whose presence tends to restore the high-temperature disordered state. In ice, these defects are topological dislocations. In systems with a broken discrete symmetry, they are domain walls. Chapter 9 is devoted to topological defects and Chapter 10 discusses domain walls. Both chapters begin with a description of the nature of these defects and then investigate their energy and the nature of interactions among them.

1.3 Energies and potentials

1 Energy scales

The characteristic distance on an atomic scale is the angstrom (10^{-8} cm). The diameter of electron orbits, the size of atoms, and the distance between atoms in condensed systems (e.g. solids) are all of this order. At the angstrom length scale, electrostatic energies are of order

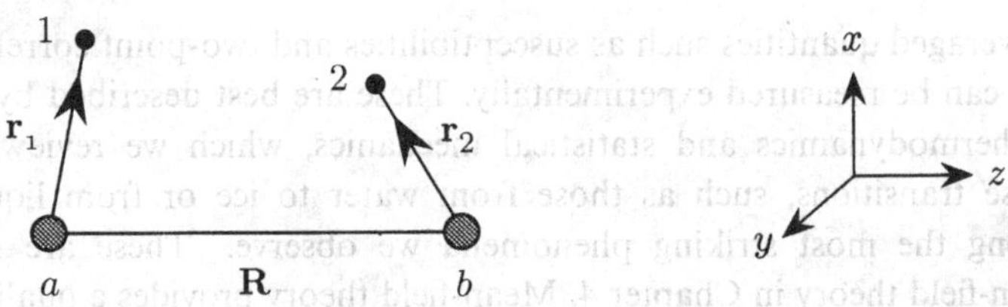

Fig. 1.3.1. Two hydrogen atoms with protons labeled a and b and electrons labeled 1 and 2.

$$e^2/(1\,\text{Å}) \sim 2.3\times10^{-11}\text{erg},\quad 14\,\text{eV}\quad \text{or } 160{,}000\,\text{K}. \tag{1.3.1}$$

The kinetic energy associated with localizing an electron in a box of side one angstrom is

$$\hbar^2(1/\text{Å})^2/2m \sim 6.1\times10^{-12}\ \text{erg},\quad 3.8\ \text{eV or } 44{,}000\,\text{K}. \tag{1.3.2}$$

The above two energies are comparable (of course that is why atoms are of this size) and are both much larger than room temperature 300 K $\sim$ 0.025 eV. Thus, if we had a large number of ions with equal numbers of charges of opposite signs, we could form a very stable ionic salt like NaCl with a binding energy of several eV per atom. Similarly, it is possible to approximate the binding energy of a metal by allowing some subset of the total number of electrons to have wavefunctions extended over the entire solid rather than localized on an atomic site. This lowers the kinetic energy by several eV per electron. Interparticle forces are in essence determined by the above effects: Coulomb attraction between opposite charges and the lowering of kinetic energy by delocalization of quantum wavefunctions to reduce kinetic energy. The implementation of these effects is, however, not trivial when many atoms or many electrons are present.

2 Van der Waals attraction

We now want to look at an example of the effective interaction between two neutral atoms as a function of the separation of their nuclei. The easiest system we can imagine is two hydrogen atoms. We adopt the coordinate system shown in Fig. 1.3.1, labeling the protons with letters and the electrons with numbers. Proton a is located at the origin ($\mathbf{R}_a = 0$), and proton b is located at position $\mathbf{R}_b = \mathbf{R} = R\mathbf{e}_z$. Electron 1 is at position $\mathbf{x}_1 = \mathbf{r}_1$ and electron 2 is at $\mathbf{x}_2 = \mathbf{R} + \mathbf{r}_2$.

Since we have presumably solved Schrödinger's equation for the eigenstates and energies of individual hydrogen atoms, we separate the Hamiltonian $\mathscr{H}$ into a part, $\mathscr{H}_0$, for two separate atoms and a remainder, $\mathscr{H}'$, treated as a perturbation:

$$\mathscr{H} = \mathscr{H}_0 + \mathscr{H}'. \tag{1.3.3}$$

$\mathscr{H}_0$ contains the kinetic energy of the two electrons and the interaction of each electron with its nucleus:

$$\mathscr{H}_0 = \frac{\hbar^2}{2m}[\nabla_1^2 + \nabla_2^2] - \frac{e^2}{r_1} - \frac{e^2}{r_2}, \tag{1.3.4}$$

where m is the electron mass and $r_{1,2} = |\mathbf{r}_{1,2}|$. The Hamiltonian $\mathscr{H}'$ includes the Coulomb attraction between each proton and the opposite electron and the Coulomb repulsion between the protons and between electrons:

$$\mathscr{H}' = \frac{e^2}{R} + \frac{e^2}{r_{12}} - \frac{e^2}{r_{1b}} - \frac{e^2}{r_{2a}}, \tag{1.3.5}$$

where $R = |\mathbf{R}|$ is the separation between protons, $r_{12} =| \mathbf{R} + \mathbf{r}_2 - \mathbf{r}_1 |$ is the separation between electrons, and $r_{1b} =| \mathbf{r}_1 - \mathbf{R} |$ and $r_{2a} =| \mathbf{R} + \mathbf{r}_2 |$ are the separations between electrons and opposite nuclei.

The solution to $\mathscr{H}_0$ can simply be obtained by using a product of hydrogenic wavefunctions on the separate atoms. Let $\phi_n(\mathbf{r})$ be the eigenfunction of the hydrogen atom with energy

$$E_n = -\frac{me^4}{2\hbar^2 n^2}. \tag{1.3.6}$$

Then, the eigenfunctions of $\mathscr{H}_0$ are

$$\Psi_p(1,2) = \phi_n(\mathbf{r}_1)\phi_m(\mathbf{r}_2) \tag{1.3.7}$$

with

$$\mathscr{H}_0\Psi_p(1,2) = (E_n + E_m)\Psi_p(1,2), \tag{1.3.8}$$

where the index p stands for the pair (n,m) and where the shorthand convention $1 \equiv \mathbf{r}_1$, $2 = \mathbf{r}_2$ is understood. In general, we will have to take into account the indistinguishability of the electrons, but first we will treat the case where atoms are separated by a distance considerably larger than an atomic radius, a_0 $(= \hbar^2/me^2 \sim 0.53\,\text{Å})$, so that the individual wavefunctions do not overlap. In this large separation, $R \gg a_0$ limit, we can approximate the perturbation Hamiltonian as a dipole-dipole interaction,

$$\mathscr{H}' \sim \frac{e^2}{R^3}(x_1x_2 + y_1y_2 - 2z_1z_2). \tag{1.3.9}$$

We will now consider only the ground state wavefunction since excited states have energies considerably above conventional temperatures. To lowest order in perturbation theory, the expectation values of $\mathbf{r}_1$ and $\mathbf{r}_2$ are zero in the ground state. In second order perturbation theory, the energy shift is obtained by taking the squared matrix elements to excited states, dividing by the excitation energies, and summing over all excited states:

$$\Delta E = \sum_p \frac{| \mathscr{H}'_{0p} |^2}{E_0 - E_p}, \tag{1.3.10}$$

where

$$\begin{aligned}\mathscr{H}'_{0p} &= \langle \Psi_p(1,2) \mid \mathscr{H}' \mid \Psi_0(1,2)\rangle \\ &= \frac{e^2}{R^3}[\langle \phi_n(1)|x_1|\phi_0(1)\rangle\langle \phi_m(2)|x_2|\phi_0(2)\rangle + ...].\end{aligned} \tag{1.3.11}$$

The required atomic matrix elements are the "dipole" matrix elements. In atomic physics, it is conventional to denote the sum

$$\alpha = \sum_p \frac{e^2|x_{0p}|^2}{E_p - E_0} \tag{1.3.12}$$

as the atomic polarizability (the macroscopic static dielectric constant of a collection of these atoms with density n is just $\epsilon = 1 + 4\pi n\alpha$). Since we are starting with the ground state, second order perturbation theory always reduces the energy, and we are left from Eq. (1.3.9) with a net attraction between atoms of the form

$$\Delta E \sim -\frac{\alpha_1\alpha_2}{R^6} . \tag{1.3.13}$$

The same equation is applicable for the interaction between any pair of neutral atoms or molecules at large separation. Thus, we have found that there is always an attractive interaction between neutral particles that dies off as R^{-6} at large separation and that is proportional to the product of atomic polarizabilities. This interaction is conventionally called the *Van der Waals* attraction. Between neutral atoms at atomic distances it is of order 10^{-2} eV.

3 Molecular hydrogen – the Heitler-London approach

If we let the protons move closer together in our original two hydrogen atom problem, the Pauli exclusion principle requires that the two-electron wavefunction be properly antisymmetrized with respect to interchange of electrons:

$$\Psi(\mathbf{r}_1, \mathbf{s}_1; \mathbf{r}_2, \mathbf{s}_2) = -\Psi(\mathbf{r}_2, \mathbf{s}_2; \mathbf{r}_1, \mathbf{s}_1), \tag{1.3.14}$$

where we now explicitly include the spins $\mathbf{s}_1$ and $\mathbf{s}_2$. Since there is no spin-orbit coupling in our Hamiltonian, we can separate the spin and position variables:

$$\Psi(\mathbf{r}_1, \mathbf{s}_1; \mathbf{r}_2, \mathbf{s}_2) = \psi(\mathbf{r}_1, \mathbf{r}_2)\chi(\mathbf{s}_1, \mathbf{s}_2) . \tag{1.3.15}$$

The spatial wavefunction can then be taken as a combination of products of hydrogenic wavefunctions of the individual electrons. The Pauli principle requires that the spin function be antisymmetric if the spatial function is symmetric under interchange of electrons and vice versa. Ψ thus has a spin singlet (s) and a spin triplet (t) part:

$$\begin{aligned} \Psi_s(1,2) &= N_s[\phi_n(\mathbf{r}_1)\phi_m(\mathbf{r}_2) + \phi_m(\mathbf{r}_1)\phi_n(\mathbf{r}_2)]\chi_s(\mathbf{s}_1, \mathbf{s}_2) \\ \Psi_t(1,2) &= N_t[\phi_n(\mathbf{r}_1)\phi_m(\mathbf{r}_2) - \phi_m(\mathbf{r}_1)\phi_n(\mathbf{r}_2)]\chi_t(\mathbf{s}_1, \mathbf{s}_2), \end{aligned} \tag{1.3.16}$$

where N_s and N_t are normalization constants and χ_s and χ_t are, respectively, the singlet and triplet spin wavefunctions.

The particular form we have chosen for the wavefunctions is due to Heitler and London. It does not allow two electrons to occupy the same site and, therefore, has a built in correlation that reduces the Coulomb repulsion between electrons, thereby anticipating the effects of $\mathscr{H}'$. Our Hamiltonian is independent of the electron spins. Thus, the only way the spin variables affect the energies of our problem is in the determination of the symmetry of the spatial wavefunctions. We will, therefore, suppress them in what follows and deal only with the 1s hydrogen wavefunctions, $\phi_\alpha(i) = \phi_0(\mathbf{x}_i - \mathbf{R}_\alpha)$ with $\alpha = a, b$ and $i = 1, 2$. The triplet and single wavefunctions are therefore

$$\psi_{s,t} = N_{s,t}[\phi_a(1)\phi_b(2) \pm \phi_b(1)\phi_a(2)] \,. \tag{1.3.17}$$

The normalization constant can be obtained by requiring $\int \Psi^*\Psi dx_1 dx_2 = 1$. The result is

$$N_{s,t}^2 = \frac{1}{2}[1 \pm \beta^2]^{-1}, \tag{1.3.18}$$

where

$$\beta = \int \rho_{ab}(1) d^3x_1 \tag{1.3.19}$$

is the overlap integral and

$$\rho_{ab}(1) = \phi_a^*(1)\phi_b(1) \tag{1.3.20}$$

is the overlap charge density. $\rho_{ab}(1)$ is a strong function of the separation of the atoms since it represents the extent to which the unperturbed wavefunctions occupy the same point in space. The charge densities,

$$\begin{aligned} \rho_a(1) &= \phi_a^*(1)\phi_a(1), \\ \rho_b(2) &= \phi_b^*(2)\phi_b(2), \end{aligned} \tag{1.3.21}$$

with $\int \rho_a(1) d^3x_1 = 1$ of the unperturbed atomic orbitals will also be of some use in what follows.

Lowest order perturbation theory makes a nonzero contribution to the energy shift, which is evaluated from $\int \Psi^* \mathscr{H}' \Psi d^3r_1 d^3r_2$. The result can be expressed as

$$\Delta E_{s,t} = \frac{Q \pm J}{[1 \pm \beta^2]}, \tag{1.3.22}$$

where Q represents the interaction between the time average charge cloud on separate unperturbed atoms and J is the *exchange integral,* which appears as a result of the symmetry of the spatial wavefunction under interchange of electrons. The precise forms of Q and J follow from the the perturbation Hamiltonian,

$$\mathscr{H}' = \frac{e^2}{R} + \frac{e^2}{r_{12}} - \frac{e^2}{r_{a2}} - \frac{e^2}{r_{b1}}. \tag{1.3.23}$$

The quantity $Q = Q_1 + Q_2$ is then the sum of the Coulomb repulsions between electron clouds and between protons:

$$Q_1 = \int \rho_a(1) \frac{e^2}{r_{12}} \rho_b(2) d^3x_1 d^3x_2 + \frac{e^2}{R}, \tag{1.3.24}$$

and the Coulomb attraction of each electron with the opposite proton:

$$Q_2 = -2 \int \rho_b(2) \frac{e^2}{r_{a2}} \rho_a(1) d^3x_1 d^3x_2 \rightarrow -2 \int \rho_b(2) \frac{e^2}{r_{a2}} d^3x_2 \,. \tag{1.3.25}$$

The exchange integral $J = J_1 + J_2$ is the sum of the Coulomb repulsion between overlap charges,

$$J_1 = \int \rho_{ab}(1)\frac{e^2}{r_{12}}\rho_{ab}(2)d^3x_1d^3x_2, \tag{1.3.26}$$

and the attraction between overlap charge densities and the protons:

$$\begin{aligned} J_2 &= -2\int \rho_{ab}(1)\frac{e^2}{r_{b1}}\rho_{ab}(2)d^3x_1d^3x_2 \\ &\rightarrow -2\beta\int \rho_{ab}(1)\frac{e^2}{r_{b1}}d^3x_1. \end{aligned} \tag{1.3.27}$$

The normalization of the electron wavefunctions and Eq. (1.3.20) were used to obtain the final forms of Eqs. (1.3.25) and (1.3.27).

For large separation, this formulation actually predicts incorrect results. There is no overlap charge density between widely separated hydrogen atoms, so that J_2 is zero. In this case, the first term in Q would dominate and predict a repulsion between hydrogen atoms, whereas we have seen that the correct result is that there should be Van der Waals attraction. When there is a small overlap, the second term in J dominates. There is a repulsion of the small overlap region with itself, but an attraction of the overlap density with both protons (with their full charge) is stronger. Note that the net result is attractive only when the system is in the singlet state, as seen in the sign associated with the spin configuration in Eq. (1.3.18). This is called the *bonding orbital.* If the electron spins are aligned (triplet), then the effect of J is repulsive. This is called the *antibonding orbital.* The binding energy of the hydrogen molecule comes mostly from the buildup of the overlap charge density between the protons. To see this explicitly, notice that the probability of finding an electron in the center of the bond changes in going from the bonding to the antibonding configuration, as shown in Fig. 1.3.2.

The energies of the bonding and antibonding states for molecular hydrogen within the Heitler-London approximation are shown in Fig. 1.3.3. Qualitatively the binding energy and its dependence on internuclear separation are correct, although in detail the actual binding is greater than the calculation gives.

4 Hard-sphere repulsion

The interparticle interaction has the same qualitative form for a vast number of different atoms and molecules. There is an attractive interaction, which at large distances approaches the Van der Waals behavior, $1/R^6$. There is a minimum energy on the atomic scale followed by an abrupt increase in repulsion. The equilibrium separation is at the bottom of a highly anharmonic potential. The sharp rise in the potential at short distances for the bonding state is the result of the rapid increase in the first term in the exchange integral as the overlap charge densities increase. This "hard-sphere" repulsion is common to the interaction of most atoms and molecules at short distances. In other more realistic models, it results from a combination of the repulsion of the overlapping electron densities,

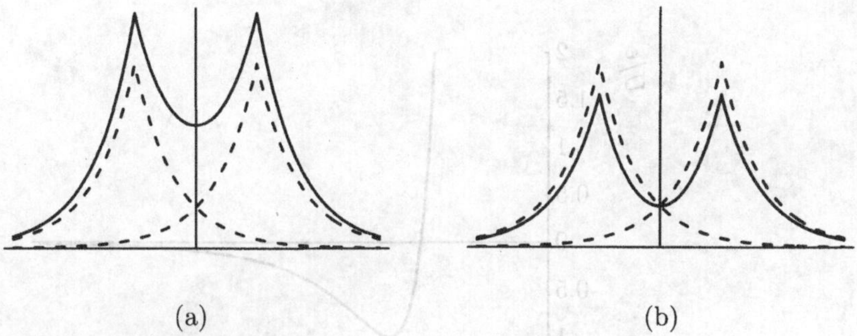

Fig. 1.3.2. Schematic representation of (a) bonding and (b) antibonding charge densities for the hydrogen molecule. The bonding charge density shown as the solid curve in (a) is $\rho = \rho_a(1) + \rho_b(1) + 2\beta\rho_{ab}(1)$. The antibonding charge density shown as the solid curve in (b) is $\rho = \rho_a(1) + \rho_b(1) - 2\beta\rho_{ab}(1)$. The individual charge densities $\rho_a(1)$ and $\rho_b(2)$ centered, respectively, at nuclei (1) and (2) are the dashed curves in both (a) and (b).

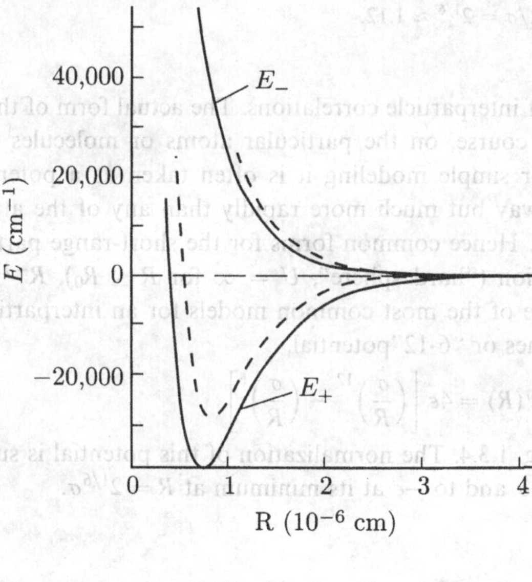

Fig. 1.3.3. The energies E_- and E_+ of antibonding and bonding orbitals calculated using Heitler-London theory and the observed energy of the hydrogen molecule as a function of separation R of hydrogen atoms.

the exclusion principle limiting the volume which each electron can occupy (and hence increasing the kinetic energy), and the Coulombic repulsion of the nuclei.

In many of the systems that we will study, strong short-range repulsion plays a particularly important role in determining local and later global structures and

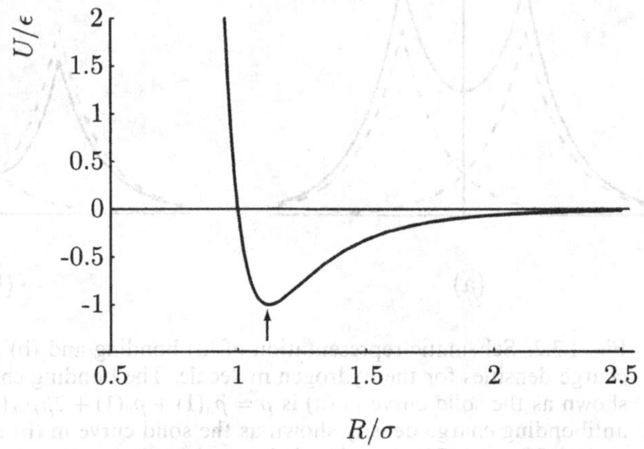

Fig. 1.3.4. The Lennard-Jones potential [Eq. (1.3.28)] showing the R^{-12} repulsive core and the R^{-6} attractive tail. The potential passes through zero at $R/\sigma = 1.0$ and has a minimum indicated by the arrow at $R/\sigma = 2^{1/6} \approx 1.12$.

the nature of interparticle correlations. The actual form of the repulsion potential depends, of course, on the particular atoms or molecules under consideration. However, for simple modeling it is often taken as a potential that varies in a convenient way but much more rapidly than any of the attractive potentials in the problem. Hence common forms for the short-range part of the potential are a step function ("hard sphere", $U = \infty$ for $R \leq R_0$), R^{-12} or $\exp(R_0/R)$. For example, one of the most common models for an interparticle interaction is the Lennard-Jones or "6-12"potential,

$$U(R) = 4\epsilon \left[\left(\frac{\sigma}{R}\right)^{12} - \left(\frac{\sigma}{R}\right)^{6} \right] \tag{1.3.28}$$

shown in Fig. 1.3.4. The normalization of this potential is such that it is equal to zero at $R = \sigma$ and to $-\epsilon$ at its minimum at $R = 2^{1/6}\sigma$.

5 Exchange interaction and magnetism

One of the most interesting results from the study of molecular hydrogen is the presence of a spin-dependent interaction of a magnitude given by electrostatic forces. The exchange integral J is of order $3-4$ eV in the case of hydrogen, and it is comparable in other systems. As we have seen, it arises from the requirement that the symmetry of the orbital part of the wavefunction be complementary to that of the spin part to make the total wavefunction antisymmetric. Nonetheless, it is an energy associated with the spin arrangement, so that flipping a spin requires an energy change of order $2J$. For the simple case we have treated, the spin interaction alone can, therefore, be written in the form

$$\mathscr{H}_{\text{spin}} = -2J\boldsymbol{\sigma}_1 \cdot \boldsymbol{\sigma}_2 \tag{1.3.29}$$

(where $\boldsymbol{\sigma}_i = \mathbf{s}_i/\hbar$ is the unitless Pauli spin operator), which will be useful later in a less restrictive context when we study magnetism. In fact, most magnetism found in nature is due to this exchange interaction. The interaction of the elementary magnetic dipoles associated with electron spins is orders of magnitude too small to explain magnetism at temperatures comparable to room temperature. The *dipole interaction* between Bohr magnetons at angstrom distances is

$$\mu_B^2/(1\text{Å})^3 \sim 10^{-4}\,\text{eV} \sim 1\text{K}. \tag{1.3.30}$$

However, since the dipolar interaction is long range (it dies out only as $1/R^3$, whereas the exchange interaction decays exponentially as it depends on electron overlap), it may have important macroscopic effects, for example in the form of demagnetization fields.

The exchange interaction between electrons in unfilled atomic shells due to the formation of bonds is negative, i.e. it favors the formation of a singlet. Ferromagnetic exchange, favoring parallel spin alignment, is also possible, but usually results from the existence of degenerate levels on the same atom or molecule. This can be seen by noting that in the expression for J, Eqs. (1.3.26) and (1.3.27), we may choose ϕ_a and ϕ_b as degenerate states on the *same* atom. In that case, the second or bonding J_2 term in the expression for J does not exist since the electron-nucleus Coulomb interaction has already been accounted for in the solution for the single electron wavefunctions. What is left, therefore, is only the repulsive first term in J. This is the primary argument leading to *"Hund's rule"* in atomic physics, which says that the minimum energy is achieved by maximizing the spin alignment in degenerate orbitals. To use this ferromagnetic interaction between spins on neighboring atoms, there must be an additional interatomic "bonding" that takes advantage of the intra-atomic degeneracy.

6 *The hydrogen molecule, molecular orbitals, and bands in metals*

There is a different approximate treatment of the hydrogen atom that leads to an understanding of the role that delocalization energy (kinetic energy lowering) plays in interparticle interactions and later for the formation of energy bands in solids. The Heitler-London approach starts from the viewpoint that the dominant interaction is the Coulomb repulsion of the electrons. It, therefore, uses a wavefunction in which the electrons are correlated apart, never sharing the same atom. In the *molecular-orbital* approximation, the one-electron–two-proton problem (H_2^+ molecular hydrogen ion) is solved first and the Coulomb repulsion is treated as a perturbation (if at all).

The one-electron Hamiltonian is taken as

$$\mathscr{H}_{0,\text{mo}} = -\frac{\hbar^2\nabla_1^2}{2m} - \frac{e^2}{r_{a1}} - \frac{e^2}{r_{b1}} \tag{1.3.31}$$

and the perturbation (Coulomb interaction between electrons) as

$$\mathscr{H}_{12} = \frac{e^2}{r_{12}}. \tag{1.3.32}$$

The actual solution of the one-electron problem is no longer trivial, but an approximate wavefunction can be constructed from the atomic wavefunctions ϕ_a and ϕ_b centered at protons a and b:

$$\phi_\pm = 2^{-1/2}[\phi_a(\mathbf{x}) \pm \phi_b(\mathbf{x})]. \tag{1.3.33}$$

These wavefunctions are called *"linear combination of atomic orbitals"* (LCAO) and are sketched in Fig. 1.3.5. For ground state wavefunctions, the LCAOs of Eq. (1.3.33) have energies

$$E_\pm = E_0 \pm t_{\text{mo}}. \tag{1.3.34}$$

The one-electron energies are split from the single-atom case by $2t_{\text{mo}}$, a fraction of the localization energy ($\sim \hbar^2/2mR^2$) or, alternatively, by the attraction of the additional charge density in the middle to the protons, $t_{\text{mo}} = \int \rho_{ab}(1)(e^2/r_{1b})dr_1$, which is similar to the second term in the exchange integral. We now treat these levels as we would orbitals of an atom. If we have two electrons, we just put them with opposite spins in the lowest energy state. The result is that we have lowered the energy by t_{mo} per electron as compared to two isolated hydrogen atoms. However, we lose a great deal of energy because of the Coulomb repulsion. The two-electron wavefunction, which is properly symmetrized, has one electron on each site only half the time, and both electrons on the same site half the time. This is the price that is payed for having an uncorrelated wavefunction,

$$\begin{aligned}\Psi(1,2) &= \phi_+(1)\phi_+(2) \\ &= \frac{1}{2}[\phi_a(1)+\phi_b(1)][\phi_a(2)+\phi_b(2)] \\ &= \frac{1}{2}[\phi_a(1)\phi_a(2)+\phi_a(1)\phi_b(2)+\phi_b(1)\phi_a(2)+\phi_b(1)\phi_b(2)].\end{aligned} \tag{1.3.35}$$

The same process can be attempted for four atoms, as depicted in Fig. 1.3.6, or for an infinite number of atoms. First we find the one-electron energies and then we fill them up according to the exclusion principle.

If we started with one electron per atom, or anything but an even number of electrons per atom, the net result is a lowering of the energy. If the number of levels (equal to the number of atoms we start with) is very large, as for delocalized electrons in a metal, the spectrum becomes almost continuous. For one electron per atom, the binding energy per electron or per atom is then of the order of $t_{\text{mo}}/2$.

In practice, the Heitler-London approximation yields better results for small molecules such as H_2 but not as good results for large molecules, where the cor-

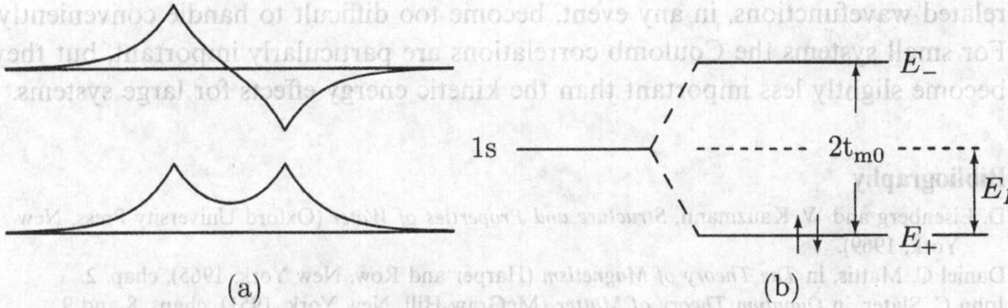

Fig. 1.3.5. (a) The + (top) and − (top) wavefunctions of Eq. (1.3.33). (b) Energy levels for separated hydrogen atoms and for the hydrogen molecule. The binding energy E_B per electron of the hydrogen molecule is one-half the difference between the energies of the separated atoms and the ground state of the molecule in which the lowest energy state is occupied by a spin up and a spin down electron.

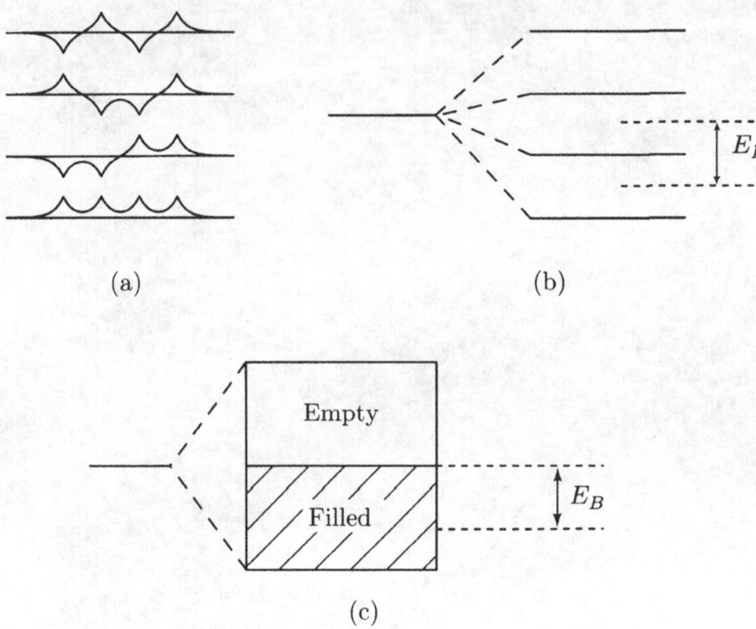

Fig. 1.3.6. (a) Wavefunctions and (b) energy levels of a four-hydrogen-atom molecule. Each atom contributes one electron to the molecule. The energy levels of the separated atoms split into four levels corresponding to the wavefunctions as shown. In the ground state, each of the two lowest energy states is occupied by two electrons of opposite spin giving rise to a binding energy E_B per electron. (c) Energy bands in a solid, showing occupied and empty states in the ground state and the binding energy per electron.

related wavefunctions, in any event, become too difficult to handle conveniently. For small systems the Coulomb correlations are particularly important, but they become slightly less important than the kinetic energy effects for large systems.

Bibliography

D. Eisenberg and W. Kauzmann, *Structure and Properties of Water* (Oxford University Press, New York, 1969).

Daniel C. Mattis, in *The Theory of Magnetism* (Harper and Row, New York, 1965), chap. 2.

John C. Slater, in *Quantum Theory of Matter* (McGraw-Hill, New York, 1951), chaps. 8 and 9.

References

A.H. Narton, W.D. Danford, and H.A. Levy, *Disc. Faraday Soc.*, **43**, 97 (1967).

J.C. Poirier, *Creep of Crystals* (Cambridge University Press, 1985).

2

Structure and scattering

Large collections of particles can condense into an almost limitless variety of equilibrium and nonequilibrium structures. These structures can be characterized by the average positions of the particles and by the interparticle spatial correlations. Periodic solids, with their regular arrangements of particles, are more ordered and have lower symmetry than fluids with their random arrangements of particles in thermal motion. There are a number of equilibrium thermodynamic phases that have higher symmetry than periodic solids but lower symmetry than fluids. Typically interacting particles at low density and/or high temperature form a gaseous phase characterized by minimal interparticle correlations. As temperature is lowered or density increased, a liquid with strong local correlations but with the same symmetries as a gas can form. Upon further cooling, various lower-symmetry phases may form. At the lowest temperatures, the equilibrium phase of most systems of particles is a highly ordered low-symmetry crystalline solid. Nonequilibrium structures such as aggregates can have unusual symmetries not found in equilibrium structures.

In this chapter, we will investigate some of the prevalent structures found in nature and develop a language to describe their order and symmetry. We will also study how these structures can be probed with current experimental methods. Though tools such as scanning force and tunneling microscopes can now provide direct images of charge and particle density, at least near surfaces, most information about bulk structure, especially at the angstrom scale, is obtained via scattering of neutrons, electrons, or photons. In this chapter, we will focus on elastic or quasi-elastic scattering in which changes in the energy of scattered particles are not probed. We will consider inelastic scattering in detail in Chapter 7.

2.1 Elementary scattering theory – Bragg's law

The easiest example of scattering yielding structural information is that of Bragg scattering of a wave from a set of partially reflecting equally spaced parallel planes. An incident wave will be diffracted by the set of planes, its intensity being modulated by constructive or destructive interference. For an infinite set of such planes (with infinitesimal reflection coefficient) the only surviving reflection is one for which there is constructive interference between waves reflected by each set of

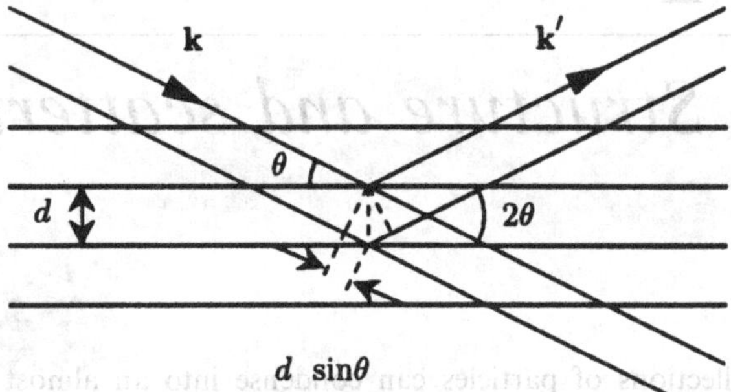

Fig. 2.1.1. Scattering from parallel planes showing the origin of Bragg's law. The planes are separated by a distance d. The incident wave vector is $\mathbf{k}$ and the scattered wave vector is $\mathbf{k}'$. The magnitude of both $\mathbf{k}$ and $\mathbf{k}'$ is $2\pi/\lambda$, and the path difference between waves partially reflected from successive planes is $2d\sin\theta$.

neighboring planes. Thus, the difference in path length between waves reflected from adjacent planes separated by a distance d must be an integral multiple of the wavelength λ as illustrated in Fig. 2.1.1. This leads to *Bragg's law*,

$$2d\sin\theta = n\lambda, \tag{2.1.1}$$

where n is an integer and θ is defined in Fig. 2.1.1. (Note that the angle between incident and scattered particles is 2θ.) As we shall see, a more sophisticated analysis gives the same result, but most discussions of scattering phenomena tend to center around the simple description given above. That is, the scattered intensity at angle 2θ reflects a fluctuation or inhomogeneity of the system with periodicity $\lambda/(2\sin\theta)$.

In a slightly more advanced approach, the quantum mechanical transition rate between plane wave states of scattered particles is calculated. Let $|\mathbf{k}\rangle$ and $|\mathbf{k}'\rangle$ be the incident (incoming) and final (outgoing) plane wave states of the scattered particle with respective momenta $\hbar\mathbf{k}$ and $\hbar\mathbf{k}'$. If the scattered particle interacts with the scattering medium via a potential U (and the interaction is sufficiently weak that only lowest order scattering need be considered for the entire sample), then by Fermi's golden rule, the transition rate between $|\mathbf{k}\rangle$ and $|\mathbf{k}'\rangle$ is proportional to the square of the matrix element,

$$M_{\mathbf{k},\mathbf{k}'} = \langle\mathbf{k}|U|\mathbf{k}'\rangle = \int d^dx e^{-i\mathbf{k}\cdot\mathbf{x}}U(\mathbf{x})e^{i\mathbf{k}'\cdot\mathbf{x}}, \tag{2.1.2}$$

where $U(\mathbf{x})$ is the scattering potential in the coordinate representation of the scattered particle. We use here the unnormalized wavefunction $\langle\mathbf{x}|\mathbf{k}\rangle = e^{i\mathbf{k}\cdot\mathbf{x}}$ for the scattered particle. We also treat the $\mathbf{x}$ as a vector in a d-dimensional space. For most physical systems, d is two, three, or possibly one. It is, however, useful to imagine generalizations of d to other dimensions. The differential cross-section $d^2\sigma/d\Omega$ per unit solid angle of the final wave vector $\mathbf{k}'$ is

$$\frac{d^2\sigma}{d\Omega} \sim \frac{2\pi}{\hbar}|M_{\mathbf{k},\mathbf{k}'}|^2. \tag{2.1.3}$$

Eq. (2.1.3) represents a static cross-section obtained experimentally by integrating over all possible energy transfers to the medium. In practice, this integration is naturally accomplished by X-ray diffraction but not by neutron diffraction. In this and the next several chapters, we will be interested only in static rather than dynamic phenomena, and we will use Eq. (2.1.3) for the scattering cross-section. In Chapter 7, we will see how the static approximation is derived from a full dynamical description.

In multiparticle systems, the scattering potential is the sum of terms arising from each of the individual atoms in the system:

$$U(\mathbf{x}) = \sum_\alpha U_\alpha(\mathbf{x} - \mathbf{x}_\alpha), \tag{2.1.4}$$

where $\mathbf{x}_\alpha$ is the position of the atom arbitrarily labeled α. The matrix element in $d^2\sigma/d\Omega$ then has the form

$$\langle\mathbf{k}|U|\mathbf{k}'\rangle = \sum_\alpha \int e^{-i\mathbf{k}\cdot\mathbf{x}} U_\alpha(\mathbf{x} - \mathbf{x}_\alpha) e^{i\mathbf{k}'\cdot\mathbf{x}} d^d x. \tag{2.1.5}$$

This can be placed in a more convenient form by taking $\mathbf{R}_\alpha = \mathbf{x} - \mathbf{x}_\alpha$ so that the scattering "form factor" (corresponding to the magnitude and direction of the scattering from each individual atom) appears multiplicatively times a factor with information about the atomic positions:

$$\begin{aligned}\langle\mathbf{k}|U|\mathbf{k}'\rangle &= \sum_\alpha \int e^{-i\mathbf{k}\cdot(\mathbf{x}_\alpha+\mathbf{R}_\alpha)} U_\alpha(\mathbf{R}_\alpha) e^{i\mathbf{k}'\cdot(\mathbf{x}_\alpha+\mathbf{R}_\alpha)} d^d R_\alpha \\ &= \sum_\alpha [\int e^{-i\mathbf{q}\cdot\mathbf{R}_\alpha} U_\alpha(\mathbf{R}_\alpha) d^d R_\alpha] e^{-i\mathbf{q}\cdot\mathbf{x}_\alpha} \\ &= \sum_\alpha U_\alpha(\mathbf{q}) e^{-i\mathbf{q}\cdot\mathbf{x}_\alpha}.\end{aligned} \tag{2.1.6}$$

Here the scattering wave vector is $\mathbf{q} \equiv \mathbf{k} - \mathbf{k}'$, and $U_\alpha(\mathbf{q})$ is the *atomic form factor* or Fourier transform of the atomic potential. The momentum transferred from the scattered particle to the sample is $\hbar\mathbf{q}$. (An alternative convention in which $\hbar\mathbf{q} = \hbar\mathbf{k}' - \hbar\mathbf{k}$ is the momentum gained by the scattered particle is often used, but the present convention is preferable for inelastic scattering.) The differential cross-section is proportional to the matrix element [Eq. (2.1.6)] squared:

$$|\langle\mathbf{k}|U|\mathbf{k}'\rangle|^2 = \sum_{\alpha,\alpha'} U_\alpha(\mathbf{q}) U^*_{\alpha'}(\mathbf{q}) e^{-i\mathbf{q}\cdot\mathbf{x}_\alpha} e^{i\mathbf{q}\cdot\mathbf{x}_{\alpha'}}. \tag{2.1.7}$$

Eq. (2.1.7) expresses the scattering matrix element for a particular configuration, specified by the position vectors $\mathbf{x}_\alpha$, of atoms in the sample. A typical scattering geometry illustrating $\mathbf{k}$, $\mathbf{k}'$, and $\mathbf{q}$ is shown in Fig. 2.1.2.

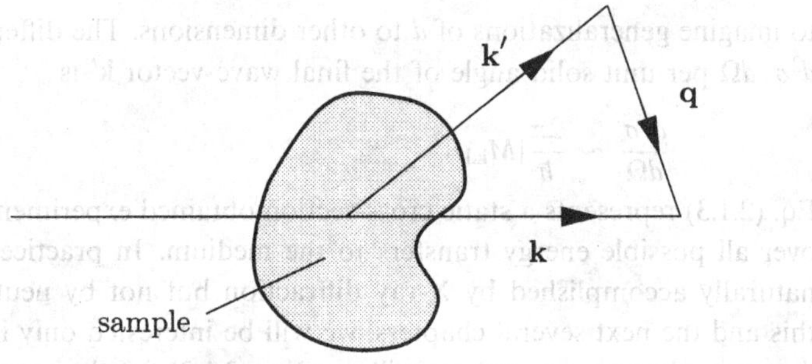

Fig. 2.1.2. Typical scattering geometry showing the incident, final, and scattering wave vectors **k**, **k**′, and **q** = **k** − **k**′.

If the positions of the atoms are rigidly fixed, as they would be in a classical system at absolute zero, then Eqs. (2.1.7) and (2.1.3) correctly give the cross-section. In real materials, particles move about, probing large regions of phase space determined by the rules of statistical mechanics, and some ensemble average of the ideal cross-section is required. If our detector accepts all particles scattered by a certain wave vector independent of their energy change (effectively integrating over frequency), then each scattering event takes a snapshot of the sample. Since data are taken over a period of time that is long compared to thermal equilibration times, the different snapshots correspond to a time average over many sample configurations. Assuming that time averaging and averages over all allowed configurations (ensemble averages – denoted by angular brackets $\langle\,\rangle$) are equivalent (i.e., that the system is ergodic) we have the *static* or *quasi-elastic* limit.

If the atoms are identical, then the form factor, $|U_\alpha(\mathbf{q})|^2$, in Eq. (2.1.7) comes outside the sum, and the cross-section for scattering from a statistical system becomes

$$(d^2\sigma/d\Omega) \sim |U_\alpha(\mathbf{q})|^2\, I(\mathbf{q}), \tag{2.1.8}$$

where the function

$$I(\mathbf{q}) = \Big\langle \sum_{\alpha,\alpha'} e^{-i\mathbf{q}\cdot(\mathbf{x}_\alpha - \mathbf{x}_{\alpha'})}\Big\rangle \tag{2.1.9}$$

depends only on the positions of the atoms in the scattering medium and not on the nature of the interaction between atoms and the scattering probe. We will call $I(\mathbf{q})$ the *structure function.* For a system of N atoms, $I(\mathbf{q})$ contains a sum of N^2 complex numbers with phases determined by the positions of all N particles. If the relative positions of the atoms are random (as for an ideal gas) then the only terms that do not average to zero are those with $\alpha = \alpha'$ for which $\sum_{\alpha,\alpha'} \to \sum_\alpha$. In this case, $I(\mathbf{q})$ increases linearly with N (rather than with N^2), i.e., $I(\mathbf{q})$ is extensive. For fluid phases, where relative positions are not random for some close neighbor particles, $I(\mathbf{q})$ remains extensive. An intensive version of the structure function (independent of N) is obtained by dividing $I(\mathbf{q})$ by N or V. The resulting function,

$$S(\mathbf{q}) = N^{-1}I(\mathbf{q}) \quad \text{or} \quad S(\mathbf{q}) = V^{-1}I(\mathbf{q}), \tag{2.1.10}$$

is called the *structure factor*. The first definition is more commonly used for classical fluids, whereas the second is used more often for quantum fluids. They clearly differ from each other only by a factor of the particle density $n = N/V$. We will generally employ the second convention in this book unless otherwise specified. However, most experimental data are presented in the first (dimensionless) form.

In this section, we have seen that the differential cross-section for scattering of plane-wave states provides direct information about the spatial structure of many-particle systems. It is important to understand both the generality and the limitations of the formulae we have derived. First, the derivation applies to any plane-wave states. Thus it applies to the scattering of quantum particles such as neutrons, photons (light), and electrons, provided the quasi-elastic approximation applies. Secondly, the scattering formula [Eq. (2.1.8)] was derived using the Fermi golden rule that results when only the lowest order term in an expansion in the scattering potential is retained. It is an approximation that is valid only so long as *single scattering* events dominate the cross-section. This is true if the interaction potential is small. If the scattering potential is large, *multiple scattering* events become important, and interpretation of scattering cross-sections becomes much more difficult. Scattering as a tool to probe structure is most useful when the scattering particles interact only weakly with the medium under study. Quantitatively this says that the scattering mean-free-path of the probe particle should be much larger than the thickness of the sample being probed.

2.2 Photons, neutrons, or electrons

From Bragg's law [Eq. (2.1.1)] it is clear that the wavelength of the particle to be scattered must be smaller than twice the nearest neighbor distance, i.e., $\lambda < 2d$. The materials with which we usually work have interparticle spacing on the angstrom scale, and we must, therefore, consider what energies the scattered particles must have to correspond to these wavelengths and what potentials these particles scatter from.

For photons, the dispersion relation relating energy ϵ to wave number $k = 2\pi/\lambda$ is

$$\epsilon = \hbar\omega = \hbar ck = hc/\lambda. \tag{2.2.1}$$

Visible light has $\epsilon \sim 1$ eV and $\lambda = 0.4 - 0.7 \times 10^4$ Å, which is suitable for probing structure on the scale of a micron. Scattering in this case is from variations in the dielectric constant or index of refraction. Probing structure at the angstrom scale requires photons with energy $\sim 10^4$ eV, and scattering is from variations in the dielectric constant caused by variations in the electronic charge density. Typically, X-rays in this range can penetrate up to a millimeter of matter and thus can provide "bulk" information.

Electrons with mass m_e have a dispersion relation

$$\epsilon = \frac{\hbar^2 k^2}{2m_e} = \frac{h^2}{2m_e \lambda^2}. \tag{2.2.2}$$

A wavelength of $\lambda = 1$Å corresponds to an energy $\epsilon \sim 100$ eV. Scattering is from the electrostatic potential, which is often large. Unless thin ($\sim$ 1 micron thick) samples are used, there is a problem with multiple scattering.

Neutrons (with mass m_n) have a similar dispersion but with a much larger mass, $\epsilon = \hbar^2 k^2/2m_n = h^2/(2m_n\lambda^2)$. For $\lambda = 1$Å, $\epsilon = 0.1$eV ~ 400 K. This means that room temperature or "thermal" neutrons have the correct energy to probe angstrom scales. Scattering is from nuclear forces or from electron spins (since the neutron itself has a spin).

Heavier charged particles (e.g. ions) tend to interact too strongly with the electrostatic potentials and are, therefore, used more frequently as probes of surface structure to avoid multiple scattering.

The energies of typical excitations (e.g. the average kinetic energy of a particle in a fluid) in condensed matter systems are of order a fraction of an eV. This is much less than the energy of X-rays but of the same order as the energy of neutrons needed to probe structure at the angstrom scale. At the moment, it is difficult to detect energy changes of 0.1 eV in a 10^4 eV photon. For this reason, X-ray scattering detects all photons scattered in a given direction regardless of energy change. Thus, X-rays scatter quasi-elastically and measure the static structure factor $S(\mathbf{q})$. Changes in neutron energy of order 0.1 eV are easy to detect, and neutrons are used extensively to study dynamical excitations in condensed systems. Presently, X-ray scattering is a more useful probe for determining the static structure of most materials. Within the past decade, laser light scattering has come into its own (via time correlation spectroscopy) for studying the dynamics of relatively slow processes ($\sim$ 1–10^{-6} s) at micron length scales.

2.3 The density operator and its correlation functions

The structure function clearly contains information about the average relative positions of atoms. We will now show that it is in fact a Fourier transform of a correlation function of the density of particles. The number density operator specifying the number of particles per unit volume at position $\mathbf{x}$ [$= (x, y, z)$ in three dimensions] in space is defined as

$$n(\mathbf{x}) \equiv \sum_\alpha \delta(\mathbf{x} - \mathbf{x}_\alpha). \tag{2.3.1}$$

In quantum systems, $\mathbf{x}_\alpha$ is the position operator for particle α; in classical systems, it is the dynamical variable specifying the position of particle α. In either case, $n(\mathbf{x})$ can be regarded as an "operator" in that it is a function of the dynamical variables $\mathbf{x}_\alpha$. The ensemble average (see Chapter 3 for a review of ensembles and thermodynamic averages) of the density operator is the average density $\langle n(\mathbf{x})\rangle$ at $\mathbf{x}$. In homogeneous, isotropic fluids, $\langle n(\mathbf{x})\rangle$ is independent of $\mathbf{x}$ and is simply the average density $n = N/V$. (The sum over delta functions in Eq. (2.3.1) has units of density since its volume integral is just the number of particles.) The independence of $\langle n(\mathbf{x})\rangle$ on either the magnitude or direction of $\mathbf{x}$ is a reflection of the rotational and translational invariance of the fluid state. All directions and positions in space are equivalent, and there can be no $\mathbf{x}$ dependence of $\langle n(\mathbf{x})\rangle$. In crystals, $\langle n(\mathbf{x})\rangle$ becomes a periodic function of $\mathbf{x}$, and, as we shall discuss in more detail shortly, both rotational and translational invariance are broken.

Correlation functions of the density are ensemble averages of products of the density operator at different points in space. The most important of these functions is the two-point density-density correlation function,

$$\begin{aligned} C_{nn}(\mathbf{x}_1,\mathbf{x}_2) &= \langle n(\mathbf{x}_1)n(\mathbf{x}_2)\rangle \\ &= \langle \sum_{\alpha,\alpha'} \delta(\mathbf{x}_1-\mathbf{x}_\alpha)\delta(\mathbf{x}_2-\mathbf{x}_{\alpha'})\rangle. \end{aligned} \tag{2.3.2}$$

The structure function, Eq. (2.1.9), is simply a Fourier transform of this function:

$$\begin{aligned} I(\mathbf{q}) &= \int e^{-i\mathbf{q}\cdot(\mathbf{x}_1-\mathbf{x}_2)}\langle n(\mathbf{x}_1)n(\mathbf{x}_2)\rangle d^dx_1d^dx_2 \\ &= \langle n(\mathbf{q})n(-\mathbf{q})\rangle, \end{aligned} \tag{2.3.3}$$

where

$$n(\mathbf{q}) = \int d^dx e^{-i\mathbf{q}\cdot\mathbf{x}}n(\mathbf{x}) = \sum_\alpha e^{-i\mathbf{q}\cdot\mathbf{x}_\alpha} \tag{2.3.4}$$

is the Fourier transform of the density. Thus, scattering measures the density-density correlation function. Note, however, that $C_{nn}(\mathbf{x}_1,\mathbf{x}_2)$ can be reconstructed from $I(\mathbf{q})$ only if $C_{nn}(\mathbf{x}_1,\mathbf{x}_2)$ depends only on $\mathbf{x}_1-\mathbf{x}_2$ as is the case in homogeneous fluids but not in periodic solids.

There are several more functions related to the density-density correlation function that are conventionally used. In the limit of large separation, $|\mathbf{x}_1-\mathbf{x}_2| \to \infty$, and $C_{nn}(\mathbf{x}_1,\mathbf{x}_2)$ tends to the product, $\langle n(\mathbf{x}_1)\rangle\langle n(\mathbf{x}_2)\rangle$, of the average densities. The *Ursell function*,†

$$\begin{aligned} S_{nn}(\mathbf{x}_1,\mathbf{x}_2) &= C_{nn}(\mathbf{x}_1,\mathbf{x}_2) - \langle n(\mathbf{x}_1)\rangle\langle n(\mathbf{x}_2)\rangle \\ &= \langle [n(\mathbf{x}_1)-\langle n(\mathbf{x}_1)\rangle][n(\mathbf{x}_2)-\langle n(\mathbf{x}_2)\rangle]\rangle \\ &\equiv \langle \delta n(\mathbf{x}_1)\delta n(\mathbf{x}_2)\rangle, \end{aligned} \tag{2.3.5}$$

decays to zero for distances $|\mathbf{x}_1 - \mathbf{x}_2|$ larger than some characteristic length, usually of the order of interparticle separations except near phase transitions.

† In the classical literature, the Ursell function is usually defined as $\langle \delta n(\mathbf{x}_1)\delta n(\mathbf{x}_2)\rangle - \langle n(\mathbf{x}_1)\rangle\delta(\mathbf{x}_1-\mathbf{x}_2)$, which differs from Eq. (2.3.5) by $-\langle n(\mathbf{x}_1)\rangle\delta(\mathbf{x}_1-\mathbf{x}_2)$.

The second form of Eq. (2.3.5) shows that $S_{nn}(\mathbf{x}_1, \mathbf{x}_1)$ is a measure of fluctuations $\delta n(\mathbf{x}_1)$ of the local density from the average density, while $S_{nn}(\mathbf{x}_1, \mathbf{x}_2)$ is a measure of the spatial correlations of these fluctuations.

Since the spatial range of $S_{nn}(\mathbf{x}_1, \mathbf{x}_2)$ is short, its Fourier transform,

$$S_{nn}(\mathbf{q}) = \frac{1}{V}\int d^d x_1 d^d x_2 e^{-i\mathbf{q}\cdot(\mathbf{x}_1-\mathbf{x}_2)} S_{nn}(\mathbf{x}_1, \mathbf{x}_2), \tag{2.3.6}$$

defined as a double integral over position times an inverse power of the volume $V \equiv \int d^d x$, is an intensive quantity that becomes independent of V in the large volume limit. The $\langle n(\mathbf{x}_1)\rangle\langle n(\mathbf{x}_2)\rangle$ part of $I(\mathbf{q})$, on the other hand, has two powers of the volume:

$$I(\mathbf{q}) = \left|\int e^{-i\mathbf{q}\cdot\mathbf{x}}\langle n(\mathbf{x})\rangle d^d x\right|^2 + V S_{nn}(\mathbf{q}). \tag{2.3.7}$$

If $\langle n(\mathbf{q})\rangle = \int d^d x e^{-i\mathbf{q}\cdot\mathbf{x}}\langle n(\mathbf{x})\rangle$ is nonzero, it will be proportional to the volume, so that the first term in Eq. (2.3.7) is proportional to V^2. In liquids, $\langle n(\mathbf{q})\rangle = V\langle n\rangle\delta_{\mathbf{q},0}$ is nonzero only for $\mathbf{q} = 0$, and there is a V^2 contribution to $I(\mathbf{q})$ only for forward scattering. In this case, the structure factor [Eq. (2.1.10)] becomes

$$S(\mathbf{q}) = S_{nn}(\mathbf{q}) + \langle n\rangle^2(2\pi)^3\delta(\mathbf{q}), \tag{2.3.8}$$

where we have used $V\delta_{\mathbf{q},0} = (2\pi)^3\delta(\mathbf{q})$. (See Appendix A2 for a review of Fourier transforms.) Thus for isotropic, homogeneous fluids, the structure function and the Fourier transform of the Ursell function are identical except at $\mathbf{q} = 0$. In an ideal gas, $S_{nn}(\mathbf{q}) = \langle n\rangle$ independent of $\mathbf{q}$. In periodic solids, as we shall see shortly, $\langle n(\mathbf{q})\rangle$ is nonzero on a lattice of vectors $\mathbf{G}$ giving rise to V^2 contributions to $I(\mathbf{q})$ (or equivalently δ-function contributions to $S(\mathbf{q})$) at many scattering angles. The general form for the structure factor from Eqs. (2.3.7) and (2.1.10) is:

$$S(\mathbf{q}) = \frac{1}{V}\left|\int e^{-i\mathbf{q}\cdot\mathbf{x}}\langle n(\mathbf{x})\rangle d^d x\right|^2 + S_{nn}(\mathbf{q}). \tag{2.3.9}$$

Again $S_{nn}(\mathbf{q})$ is intensive, but the first term has contributions proportional to the volume.

One of the most convenient functions for visualizing how structure and correlations are related to the interparticle forces and the scattering is the *pair distribution function*, $g(\mathbf{x}_1, \mathbf{x}_2)$, defined via

$$\begin{aligned}\langle n(\mathbf{x}_1)\rangle g(\mathbf{x}_1,\mathbf{x}_2)\langle n(\mathbf{x}_2)\rangle &\equiv \langle\sum_{\alpha\neq\alpha'}\delta(\mathbf{x}_1-\mathbf{x}_\alpha)\delta(\mathbf{x}_2-\mathbf{x}_{\alpha'})\rangle \\ &= \langle\sum_{\alpha,\alpha'}\delta(\mathbf{x}_1-\mathbf{x}_\alpha)\delta(\mathbf{x}_2-\mathbf{x}_{\alpha'})\rangle \\ &\quad -\langle\sum_{\alpha}\delta(\mathbf{x}_1-\mathbf{x}_\alpha)\delta(\mathbf{x}_2-\mathbf{x}_\alpha)\rangle \\ &= \langle n(\mathbf{x}_1)n(\mathbf{x}_2)\rangle - \langle n(\mathbf{x}_1)\rangle\delta(\mathbf{x}_1-\mathbf{x}_2).\end{aligned} \tag{2.3.10}$$

Note that $\mathbf{x}_\alpha$ cannot be $\mathbf{x}_{\alpha'}$ since the term $\alpha = \alpha'$ is excluded from the sum. Given a particle at $\mathbf{x}_1$, $g(\mathbf{x}_1, \mathbf{x}_2)$ is the probability of finding a different particle at $\mathbf{x}_2$ (actually in a volume $d^d x$ about $\mathbf{x}_2$ and normalized by the density). The pair distribution function is particularly useful when the system is homogeneous and translationally invariant. In that case $g(\mathbf{x}_1, \mathbf{x}_2) \to g(\mathbf{x}_1 - \mathbf{x}_2)$, and

$$\begin{aligned} \langle n \rangle^2 g(\mathbf{x}_1 - \mathbf{x}_2) &= \frac{1}{V} \int d^d x_2 \langle \sum_{\alpha \neq \alpha'} \delta(\mathbf{x}_1 - \mathbf{x}_\alpha) \delta(\mathbf{x}_2 - \mathbf{x}_{\alpha'}) \rangle \\ &= \left\langle \sum_{\alpha \neq \alpha'} \frac{1}{V} \int d^d x_2 \delta(\mathbf{x} + \mathbf{x}_2 - \mathbf{x}_\alpha) \delta(\mathbf{x}_2 - \mathbf{x}_{\alpha'}) \right\rangle \\ &= \frac{1}{V} \langle \sum_{\alpha \neq \alpha'} \delta(\mathbf{x} - \mathbf{x}_\alpha + \mathbf{x}_{\alpha'}) \rangle, \end{aligned} \tag{2.3.11}$$

where we introduced $\mathbf{x} = \mathbf{x}_1 - \mathbf{x}_2$. Since the sum over α' runs over all possible values of the difference $\mathbf{x}_\alpha - \mathbf{x}_{\alpha'}$ for each value of $\mathbf{x}_\alpha$, each term in the sum over α' is identical, and

$$g(\mathbf{x}) = \frac{1}{\langle n \rangle} \langle \sum_{\alpha \neq 0} \delta(\mathbf{x} - \mathbf{x}_\alpha + \mathbf{x}_0) \rangle. \tag{2.3.12}$$

A direct and intuitive method of determining $g(\mathbf{x})$ follows from this equation. Choose a configuration of particle positions (such as depicted in Fig. 2.3.1) in the ensemble of permitted configurations, and choose a coordinate system so that a particle, which we label with a 0, is at the origin. Then the integral of $\langle n \rangle g(\mathbf{x})$ over a volume element of size $d^d x$ at $\mathbf{x}$ is simply the number of particles in that volume element. Thus $g(\mathbf{x})$ can be determined by counting the number of particles in a small volume $d^d x$ at separation $\mathbf{x}$ from a particle at the origin. The average of this number over all (many) particles placed at the origin divided by $\langle n \rangle d^d x$ is $g(\mathbf{x})$. In an uncorrelated system, such as an ideal gas, the probability of finding a particle at any position is uniform and is independent of the positions of other particles. In this case, $g(\mathbf{x})$ does not depend on $\mathbf{x}$, and $g(\mathbf{x}) = (1 - N^{-1}) \to 1$ because $\int \langle n \rangle g(\mathbf{x}) d^d x = (N-1)$. As interparticle interactions are increased, spatial correlations build up and lead to nontrivial structure in $g(\mathbf{x})$. (Sometimes the correlated part is explicitly written as $g(\mathbf{x}) = 1 + h(\mathbf{x})$, where $h(\mathbf{x})$ is the *pair correlation function.)* From Eqs. (2.3.11), (2.3.2), and (2.3.5), the scattering intensity for homogeneous, isotropic fluids is directly written in terms of $g(\mathbf{x})$ as

$$S(\mathbf{q}) = \langle n \rangle [1 + \langle n \rangle \int g(\mathbf{x}) e^{-i\mathbf{q}\cdot\mathbf{x}} d^d x]. \tag{2.3.13}$$

We will use $g(\mathbf{x})$ extensively in the next section to describe qualitatively some interesting structures. Note that $g(\mathbf{x}) = 1$ in an ideal gas. This leads to $S(\mathbf{q}) = \langle n \rangle [1 + \langle n \rangle (2\pi)^3 \delta(\mathbf{q})]$ in agreement with Eq. (2.3.8) when $S_{nn} = \langle n \rangle$. When the system is, in addition, isotropic, $g(\mathbf{x}) \to g(r)$, where $r = |\mathbf{x}|$. In this case, $g(r)$ is known as the *radial distribution function.*

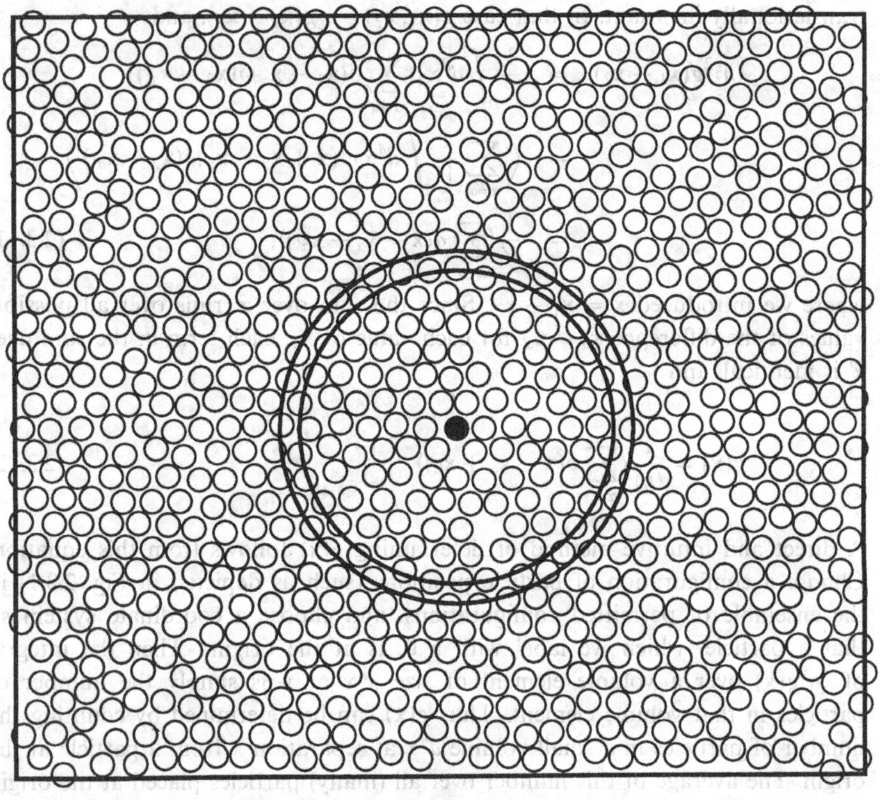

Fig. 2.3.1. Typical atomic configuration in a hard-sphere fluid. The pair distribution function can be obtained by choosing an arbitrary particle as the origin and counting the number of atoms whose centers lie within a distance dr of a circle of radius r of the origin.

2.4 Liquids and gases

Both liquids and gases are fluids. Fluids are *spatially homogeneous* and *rotationally isotropic*. This means that the *average* environment of any point in a fluid is identical to that of any other point and independent of direction. Thus the average properties of a fluid are *invariant* with respect to spatially uniform translations through any vector $\mathbf{R}$ and with respect to arbitrary rotations about any axis. It is instructive to see how these two invariance properties necessarily imply that densities are spatially uniform and that two-point correlation functions depend only on the magnitude of the difference between two spatial coordinates. Translational invariance implies

$$\langle n(\mathbf{x})\rangle = \langle n(\mathbf{x}+\mathbf{R})\rangle. \tag{2.4.1}$$

The displacement vector $\mathbf{R}$ is arbitrary. In particular, we can choose it to be equal to $-\mathbf{x}$ so that $\langle n(\mathbf{x})\rangle$ is equal to the density $\langle n(0)\rangle$ at the origin for every $\mathbf{x}$. Thus $\langle n(\mathbf{x})\rangle$ does not depend on $\mathbf{x}$. Similarly,

$$C_{nn}(\mathbf{x}_1, \mathbf{x}_2) = C_{nn}(\mathbf{x}_1 + \mathbf{R}, \mathbf{x}_2 + \mathbf{R}). \tag{2.4.2}$$

Here, the choice $\mathbf{R} = -\mathbf{x}_2$ implies $C_{nn}(\mathbf{x}_1, \mathbf{x}_2) = C_{nn}(\mathbf{x}_1 - \mathbf{x}_2, 0) \equiv C_{nn}(\mathbf{x}_1 - \mathbf{x}_2)$ depends only on $\mathbf{x}_1 - \mathbf{x}_2$. Rotational invariance implies

$$C_{nn}(\mathbf{x}_1 - \mathbf{x}_2) = C_{nn}(\mathcal{R}(\mathbf{x}_1 - \mathbf{x}_2)) \tag{2.4.3}$$

for any rotation matrix $\mathcal{R}$ and thus that $C_{nn}(\mathbf{x}_1 - \mathbf{x}_2) = C_{nn}(|\mathbf{x}_1 - \mathbf{x}_2|)$ is a function of the magnitude $|\mathbf{x}_1 - \mathbf{x}_2|$. These symmetries then imply

$$\begin{aligned} \langle n(\mathbf{q}_1)n(\mathbf{q}_2)\rangle &= \int d^d x_1 \int d^d x_2 e^{-i\mathbf{q}_1\cdot\mathbf{x}_1} e^{-i\mathbf{q}_2\cdot\mathbf{x}_2} C_{nn}(\mathbf{x}_1, \mathbf{x}_2) \\ &= C_{nn}(q_1)(2\pi)^d \delta^d(\mathbf{q}_1 + \mathbf{q}_2), \end{aligned} \tag{2.4.4}$$

where $q_1 = |\mathbf{q}_1|$ and

$$C_{nn}(q) = \int d^d x e^{-i\mathbf{q}\cdot\mathbf{x}} C_{nn}(|\mathbf{x}|). \tag{2.4.5}$$

Thus for a fluid, $I(\mathbf{q}) = VC_{nn}(q)$.

The set of operations that leave a system unchanged form a group called the *symmetry group*. The group of arbitrary translations, rotations, and reflections is the *Euclidean group*. Since a fluid is invariant under all of these operations, its symmetry group is the Euclidean group. Fluids have the highest possible symmetry (i.e., they have the largest number of symmetry operations). All other equilibrium phases of matter are invariant only under some subgroup of the Euclidean group and have lower symmetry than fluid phases. The description and consequences of reduced symmetry will be a major focus of this book.

What is conventionally meant by a liquid is a fluid phase with a high density. Unlike many of the other condensed phases that we will study, it is *not* distinguished from its higher temperature gas phase by a symmetry change. In fact, it is possible to go continuously from the liquid to the gas phase by going around a critical point (see Fig. 3.1.4 for a typical phase diagram). Except for ideal gases (which do not exist), there are always interactions and correlations between particles and consequently a pair distribution function that differs from unity and reflects these correlations. The usual distinction between gas and liquid comes only when the two coexist and a meniscus can be observed. In this case the phase separation between the dense and less dense phases is caused by the presence of an attractive interaction. However, the correlations in the liquid (and the correlations which eventually break the symmetries of the liquid state) are primarily due to the repulsive interactions. A typical form for the interparticle potential is the 6-12 or *Lennard-Jones* potential of Eq. (1.3.28) and Fig. 1.3.4. It has a steep short-range repulsive part and a longer-range attractive part.

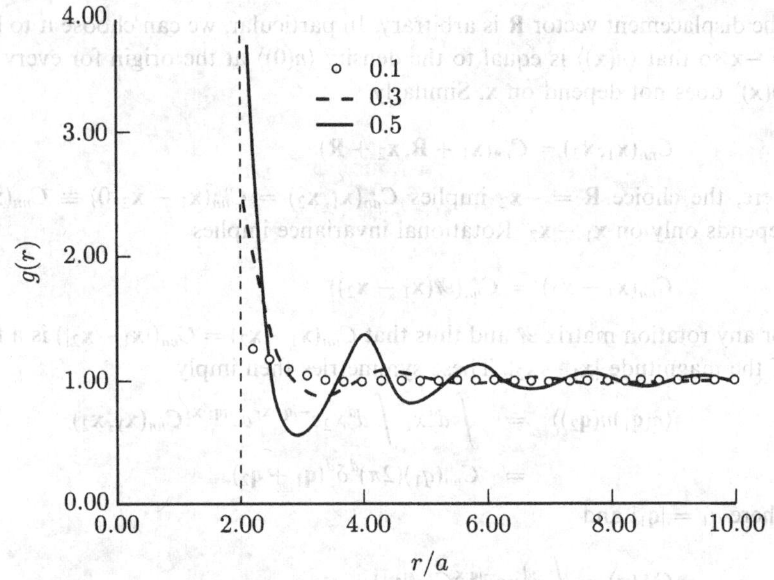

Fig. 2.4.1. The radial distribution function for a hard-sphere fluid for three different volume fractions (i.e., ratio of volume occupied by hard spheres to total volume). These curves were generated numerically using the Percus-Yevick equation.

1 Hard-sphere liquids

If in fact the repulsive part dominates the correlations, then the simplest physical model of a liquid we might take would involve an exclusion of the interpenetration of the particles. This is most easily represented by a *hard-sphere* interaction. Although this seems like an immense trivialization of the problem, there is a good deal of unusual and unexpected physics to be found in hard-sphere models ($U(r) = \infty$, $r < r_0$, $U(r) = 0$, $r > r_0$). What will the radial distribution function look like for such a potential? It is clear that $g(r)$ will have a hole up to $r = r_0$, and that $g(r)$ must rise above one at further distances to conserve the total density, but will there be further correlations?

The *Bernal model* of random-close-packing of hard spheres is a useful model for a liquid (or equally for an amorphous solid or a glass) for which the above questions can easily be addressed. Random packings of hard spheres can be generated by experiment (e.g. by packing ball-bearings or perhaps peas) or by computer simulation. A two-dimensional simulation for the packing of hard discs is shown in Fig. 2.3.1. From this picture, it is clear that there are strong short-range correlations in the positions of particles. There is a near-neighbor shell consisting of approximately six particles (twelve in three dimensions), then a dip in density caused by exclusion from this shell, then a next shell, etc. However, the

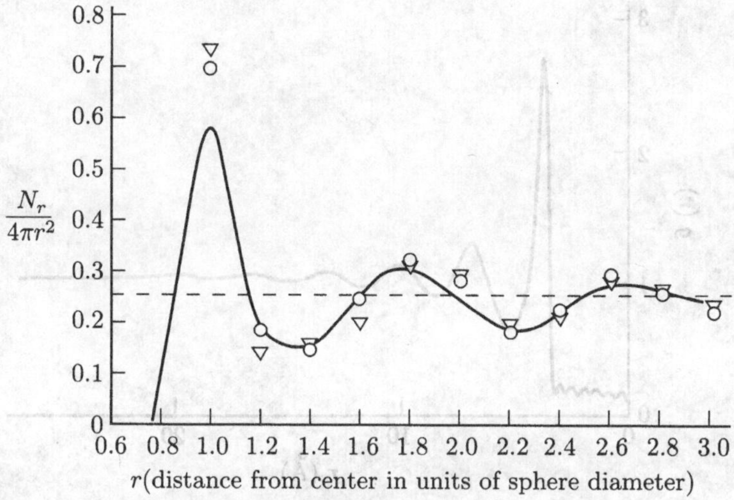

Fig. 2.4.2. The radial distribution function for the Bernal model and the experimentally observed radial distribution function for liquid argon. [J.M. Ziman, *Models of Disorder* (Cambridge University Press, 1979), p. 79.]

correlations rapidly die out on the scale of several particle diameters. The radial distribution function from Bernal's studies of ball-bearings in a bag are compared to $g(r)$ from early neutron scattering experiments in Fig. 2.4.2. More recent scattering experiments produce the static structure factor and radial distribution function shown Figs. 2.4.3 and 2.4.4. The hard-sphere model is not at all bad in describing the liquid correlations beyond the hard-sphere radius, and considerably more work is required to do better or to obtain an analytic theory.

Note that there is a strong peak in $g(r)$ at the average nearest neighbor spacing r_{nn} as would be expected from the fact that $g(r)$ is proportional to the number of particles in a spherical shell a distance r from a given particle. There are other less pronounced peaks at average next nearest and further neighbor separations. These peaks are reflected in the structure factor as peaks at wave vectors equal to 2π over a separation. Thus, the largest peak in $S(k)$ is at $k \approx 2\pi/r_{nn}$.

Some other remarks about hard-sphere systems.

- The density (volume fraction) of random close packed spheres (uniform size) in three dimensions is 0.638. This compares to a density of 0.7405 for the periodic close packing of spheres in the FCC (face-centered cubic) or HCP (hexagonal close packed) crystalline phases.
- A hard-sphere system is athermal (i.e., there is no temperature dependence to its phase diagram) because all of the energies are either zero or infinite, and hence all terms involving exponentials with temperature in the partition function are either zero or unity.

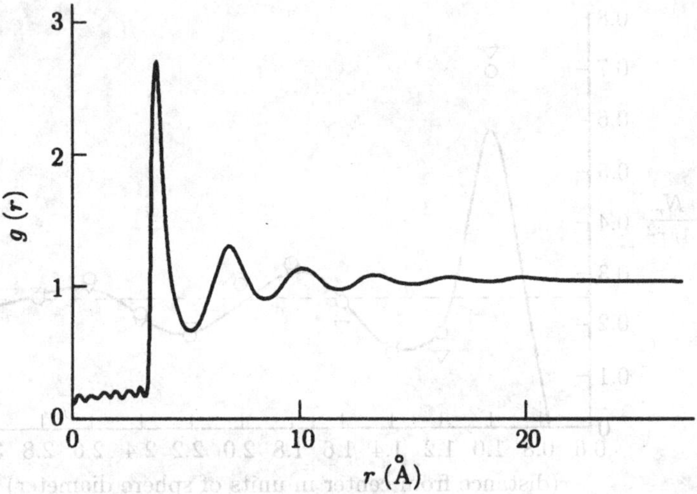

Fig. 2.4.3. The radial distribution function $g(r)$ for liquid argon at 85 K (density $2.13 \times 10^{22}\ \mathrm{cm}^{-3}$) as determined by neutron diffraction. [J.L. Yarnell, M.J. Katz, R.G. Wenzel, and S.H. Koenig, *Phys. Rev. A* 7, 2130 (1973).]

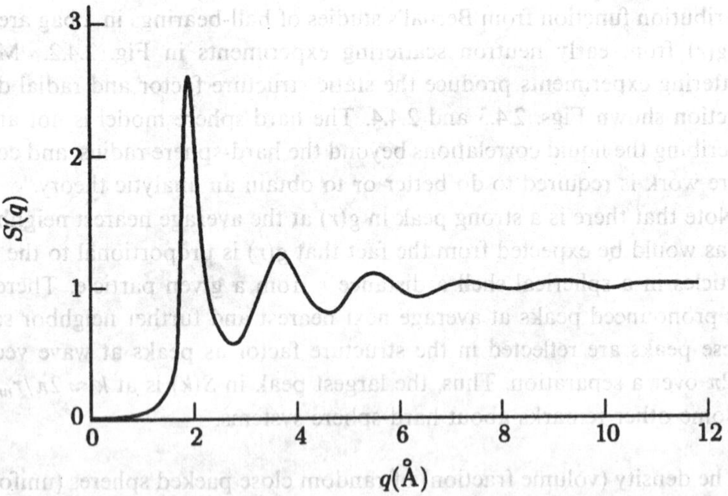

Fig. 2.4.4. The static structure factor $S(q)$ for liquid argon for the same conditions as Fig. 2.4.3. [J.L. Yarnell, M.J. Katz, R.G. Wenzel, and S.H. Koenig, *Phys. Rev. A* 7, 2130 (1973).]

- Nonetheless *there is a solid-liquid phase transition* as a function of particle density. On the phase boundary a liquid with 0.495 volume fraction is in equilibrium with an FCC solid with 0.545 volume fraction.

These are the results of many computer simulations. They appear to be relevant for many of the studied solid-liquid transitions observed, for example, in noble gases.

2.5 Crystalline solids

1 *Unit cells and the direct lattice*

Many condensed systems, for example all of the elements except helium, form crystalline solid phases at atmospheric pressure at sufficiently low temperatures. A perfect crystal consists of a space-filling array of periodically repeated identical copies of a single structural unit containing some distribution of mass and charge. An example of a two-dimensional crystal is shown in Fig. 2.5.1. In the simplest case, the structural unit contains a single atom; more generally, it may contain many different atoms or a continuous variation in the mass density about some mean. The repeated structural unit is called a *unit cell.* The unit cell with the smallest possible volume is called a *primitive* unit cell. If the unit cell contains more than one atom, the positions of the atoms relative to the center of the cell are called the *basis.* Equivalent points in unit cells in a d-dimensional perfect crystal lie on a *periodic lattice,* called a *Bravais lattice,* consisting of a mathematical array of points. Any lattice point can be specified by an integral linear combination of independent primitive translation vectors, $\mathbf{a}_1, ..., \mathbf{a}_d$, (for a d-dimensional lattice):

$$\mathbf{R_l} = l_1\mathbf{a}_1 + l_2\mathbf{a}_2 + ... + l_d\mathbf{a}_d, \tag{2.5.1}$$

where $\mathbf{l} = (l_1, ..., l_d)$ is a d-dimensional vector with components l_i. ($\mathbf{l}$ indexes a particular unit cell, $\mathbf{R_l}$ specifies its position in real space.) The set of vectors $\mathbf{a}_1, ..., \mathbf{a}_d$ completely define the mathematical lattice. A translation vector, or lattice vector, connects equivalent points in the lattice,

$$\mathbf{T} = \mathbf{R_l} - \mathbf{R_{l'}} \tag{2.5.2}$$

for any $\mathbf{l}$ and $\mathbf{l}'$.

The lattice of points in coordinate space is often called the *direct lattice.* All vectors in the set defined by Eq. (2.5.1) have a magnitude greater than or equal to that of the shortest length vector connecting vertices in a primitive unit cell. As is shown in Fig. 2.5.1, the set of primitive translation vectors for a lattice is not unique. It is always possible, however, to choose the set so that it contains the shortest vector in the lattice. The set of vectors $\mathbf{T}$ are closed under the operations of addition and multiplication by a minus sign; i.e., if vectors $\mathbf{T}_1$ and $\mathbf{T}_2$ are vectors in the periodic lattice, then the vectors $\pm\mathbf{T}_1$, $\pm\mathbf{T}_2$, $\mathbf{T}_1 \pm \mathbf{T}_2$, and $\pm\mathbf{T}_1 + \mathbf{T}_2$

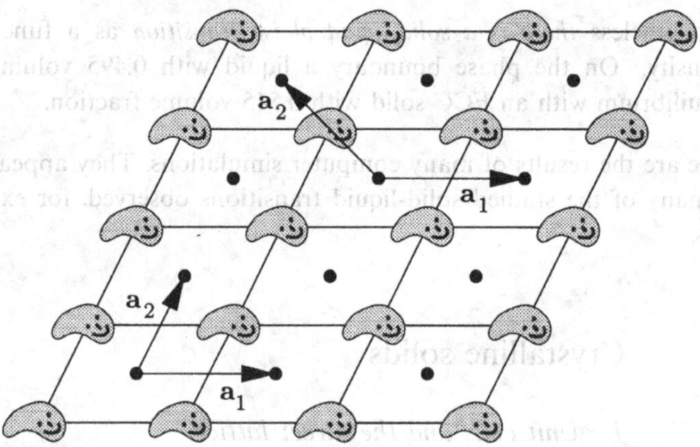

Fig. 2.5.1. A two-dimensional crystal consisting of identical unit cells periodically repeated to fill space. Two sets of primitive translation vectors are shown.

are also. This *closure* property generalizes to nonperiodic lattices, as we shall see in Secs. 2.10 and 2.11.

There is no unique choice for a primitive unit cell of a Bravais lattice. One choice is the parallelepiped whose edges are the primitive translation vectors. In three dimensions, the volume of the unit cell is thus $v_0 = \mathbf{a}_1 \cdot (\mathbf{a}_2 \times \mathbf{a}_3)$. An alternative commonly used unit cell is the *Wigner-Seitz* unit cell obtained by constructing perpendicular bisectors to all lattice vectors emerging from a given lattice point. The smallest volume enclosed by planes constructed in this way defines the Wigner-Seitz cell. The construction of the Wigner-Seitz cell for a two-dimensional lattice is shown in Fig. 2.5.2.

A complete description of a perfect crystal requires the specification of a periodic lattice and the distribution of mass in the unit cell surrounding each lattice point. In an ideal crystal consisting of a single type of (pointlike) atom located at each lattice site, the number density is simply

$$n(\mathbf{x}) = \sum_{\mathbf{l}} \delta(\mathbf{x} - \mathbf{R}_{\mathbf{l}}) . \tag{2.5.3}$$

If the lattice has a basis with atoms of mass m_α located at sites $\mathbf{c}_\alpha$ in the unit cell, the mass density is

$$\rho(\mathbf{x}) = \sum_{\mathbf{l},\alpha} m_\alpha \delta(\mathbf{x} - \mathbf{R}_{\mathbf{l}} - \mathbf{c}_\alpha). \tag{2.5.4}$$

The mass density of a perfect crystal is invariant with respect to translations through a lattice vector: $\rho(\mathbf{x}) = \rho(\mathbf{x} + \mathbf{T})$. There are no perfect crystals in nature with a mass density given precisely by Eq. (2.5.4). Thermal or quantum mechanical fluctuations cause instantaneous deviations from this ideal form. In addition any

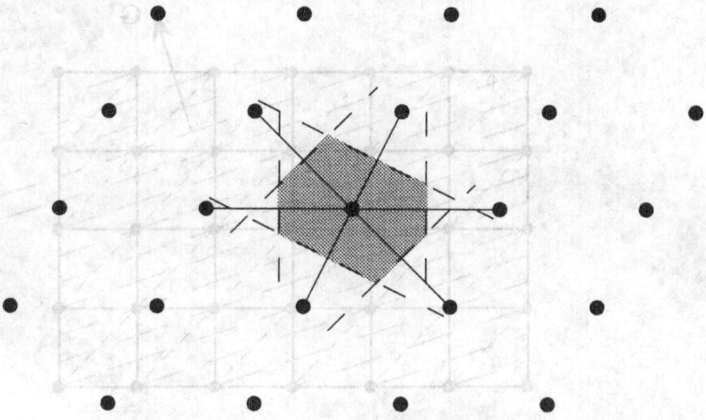

Fig. 2.5.2. Construction of the Wigner-Seitz unit cell for a low symmetry lattice in two dimensions. Lattice sites are indicated by black dots. Full lines, whose perpendicular bisectors are dashed lines, connect the central lattice site to other sites. The Wigner-Seitz cell is shaded. All Wigner-Seitz cells, except those of the square and rectangular lattices, are hexagonal.

real crystal will have imperfections of various sorts (vacancies, dislocations, etc.). Nevertheless, the average density,

$$\langle \rho(\mathbf{x}) \rangle = \langle \rho(\mathbf{x} + \mathbf{T}) \rangle, \tag{2.5.5}$$

has the periodicity of a perfect crystal. A physical crystal is a material whose average mass density is a periodic function of space.

2 *The reciprocal lattice*

Associated with any periodic lattice is a set of equispaced parallel planes containing all lattice points as shown in Fig. 2.5.3. Each set of these planes can be defined by its normal vector **G**. Lattice vectors in a given plane perpendicular to **G** satisfy $\mathbf{G} \cdot \mathbf{T} =$ const. For this set of parallel planes all lattice vectors lie in some plane which satisfies:

$$\mathbf{G} \cdot \mathbf{T} = 2\pi n \tag{2.5.6}$$

for some integer n. The coefficient 2π is chosen by convention so that $\exp(i\mathbf{G}\cdot\mathbf{T}) = 1$. Any point $\mathbf{x}_n$ (not just a lattice point **T**) in the nth plane associated with **G** satisfies $\mathbf{G} \cdot \mathbf{x}_n = 2\pi n$. The difference $\mathbf{x}_n - \mathbf{x}_{n-1}$ between points in adjacent planes satisfies $\mathbf{G} \cdot (\mathbf{x}_n - \mathbf{x}_{n-1}) = 2\pi$. The distance l between adjacent planes is the component of $\mathbf{x}_n - \mathbf{x}_{n-1}$ parallel to **G**. Thus $l = 2\pi/|\mathbf{G}|$. For any set of primitive translation vectors, $\mathbf{a}_1, ..., \mathbf{a}_d$, it is always possible to construct a set of *reciprocal* vectors, $\mathbf{b}_1, ..., \mathbf{b}_d$ satisfying

$$\mathbf{a}_i \cdot \mathbf{b}_j = 2\pi\delta_{ij}\,, \quad i, j = 1, ..., d\,. \tag{2.5.7}$$

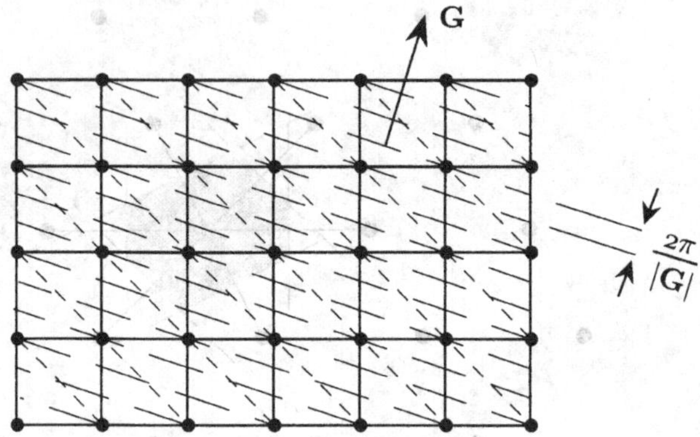

Fig. 2.5.3. Some sets of equispaced parallel planes containing all sites in a square lattice. The set of long dashed planes is perpendicular to the vector **G**.

In three dimensions, $\mathbf{b}_1 = 2\pi(\mathbf{a}_2 \times \mathbf{a}_3)/[\mathbf{a}_1 \cdot (\mathbf{a}_2 \times \mathbf{a}_3)]$, and $\mathbf{b}_2$ and $\mathbf{b}_3$ are obtained from $\mathbf{b}_1$ by cyclically permuting $\mathbf{a}_1$, $\mathbf{a}_2$, and $\mathbf{a}_3$. Any vector satisfying Eq. (2.5.6) can be written as

$$\mathbf{G} = n_1\mathbf{b}_1 + \cdots + n_d\mathbf{b}_d, \tag{2.5.8}$$

where $n_1, ..., n_d$ are positive or negative integers or zero. The vectors **G**, therefore, form a periodic lattice, called a *reciprocal lattice,* with primitive translation vectors $\mathbf{b}_1, ..., \mathbf{b}_d$. The Wigner-Seitz unit cell for the reciprocal lattice is called the first *Brillouin zone.*

3 Periodic functions

Any periodic function of position $f(\mathbf{x}) = f(\mathbf{x}+\mathbf{T})$ can be decomposed into Fourier components with wave vectors in the reciprocal lattice (see Appendix 2A):

$$f(\mathbf{x}) = \sum_{\mathbf{G}} f_{\mathbf{G}} e^{i\mathbf{G}\cdot\mathbf{x}}. \tag{2.5.9}$$

This can be derived by taking the general Fourier transform of $f(\mathbf{x})$:

$$\begin{aligned} f(\mathbf{q}) &= \int d^d x e^{-i\mathbf{q}\cdot\mathbf{x}} f(\mathbf{x}) = \sum_{\mathbf{T}} \int_0 d^d x e^{-i\mathbf{q}\cdot(\mathbf{x}+\mathbf{T})} f(\mathbf{x}+\mathbf{T}) \\ &= \left(\sum_{\mathbf{T}} e^{-i\mathbf{q}\cdot\mathbf{T}}\right) \int_0 d^d x f(\mathbf{x}) e^{-i\mathbf{q}\cdot\mathbf{x}}, \end{aligned} \tag{2.5.10}$$

where $\int_0$ means an integral over a unit cell. The sum over **T** in the last expression is equal to the number N_c of cells in the lattice if **q** is a reciprocal lattice vector and is zero otherwise. Therefore,

$$f(\mathbf{q}) = N_c v_0 \sum_{\mathbf{G}} \delta_{\mathbf{q},\mathbf{G}} f_{\mathbf{G}} = \sum_{\mathbf{G}} (2\pi)^d \delta(\mathbf{q}-\mathbf{G}) f_{\mathbf{G}}, \tag{2.5.11}$$

where

$$f_{\mathbf{G}} = \frac{1}{v_0} \int d^d x f(\mathbf{x}) e^{-i\mathbf{G}\cdot\mathbf{x}}, \tag{2.5.12}$$

where v_0 is the volume of a unit cell. The Fourier representation of Eq. (2.5.9) applies, in particular, if $f(\mathbf{x})$ is the scattering potential $U(\mathbf{x})$ [Eq. (2.1.5)], the average mass density, $\langle \rho(\mathbf{x}) \rangle$, or number density, $\langle n(\mathbf{x}) \rangle$:

$$\langle n(\mathbf{x}) \rangle = \sum_{\mathbf{G}} \langle n_{\mathbf{G}} \rangle e^{i\mathbf{G}\cdot\mathbf{x}}. \tag{2.5.13}$$

Thus, the average number density in a periodic solid is fully specified by its Fourier components $\langle n_{\mathbf{G}} \rangle$ at reciprocal lattice vectors $\mathbf{G}$.

4 *Bragg scattering*

If scatterers are rigidly fixed at sites on a periodic lattice, the scattering matrix element [Eq. (2.1.2)] becomes

$$M_{\mathbf{k},\mathbf{k}'} = V \sum_{\mathbf{G}} U_{\mathbf{G}} \delta_{\mathbf{q},\mathbf{G}}, \tag{2.5.14}$$

where $\mathbf{q} = \mathbf{k} - \mathbf{k}'$ is the difference between incident and outgoing wave vectors. The scattering cross-section then becomes

$$\frac{d^2\sigma}{d\Omega} = V^2 \sum_{\mathbf{G}} |U_{\mathbf{G}}|^2 \delta_{\mathbf{q},\mathbf{G}}. \tag{2.5.15}$$

Thus, there will be peaks in the scattering pattern at every reciprocal lattice vector with intensity proportional to the square of the volume of the sample and to the square of the Fourier component of the scattering potential at wave vector $\mathbf{G}$. These are the *Bragg* scattering peaks of the solid. As we shall see in Chapter 7, the scattering into Bragg peaks is elastic so that the magnitude of the incident and scattered wave vectors is the same, i.e., $|\mathbf{k}| = |\mathbf{k}'|$. This leads to a variation of Bragg's law, known as the *Laue condition,* which states that *incoming wave vectors* which lie on the perpendicular bisectors of reciprocal lattice vectors (i.e., on the Brillouin zone faces) will be scattered. This can be seen by setting $\mathbf{q} = \mathbf{G}$. Then

$$\begin{aligned} \mathbf{k}' &= \mathbf{k} - \mathbf{G}, \\ |\mathbf{k}'|^2 &= |\mathbf{k}|^2 + |\mathbf{G}|^2 - 2\mathbf{k}\cdot\mathbf{G}\,, \\ \mathbf{k}\cdot(\mathbf{G}/2) &= |\mathbf{G}/2|^2. \end{aligned} \tag{2.5.16}$$

These relations are equivalent to the simple Bragg condition (2.1.1) as can be seen by substituting $|\mathbf{G}| = 2\pi/d$ and $|\mathbf{k}| = 2\pi/\lambda$ into the above and recalling that θ in Eq. (2.1.1) is the angle between $\mathbf{k}$ and the scattering planes rather than the normal to the planes. In Fig. 2.5.4 we see that when an incident wave lies on a Brillouin zone face it is Bragg scattered across the zone to the opposite face.

We will now study in more detail the scattering from crystals. The structure factor [Eq. (2.1.10)] is

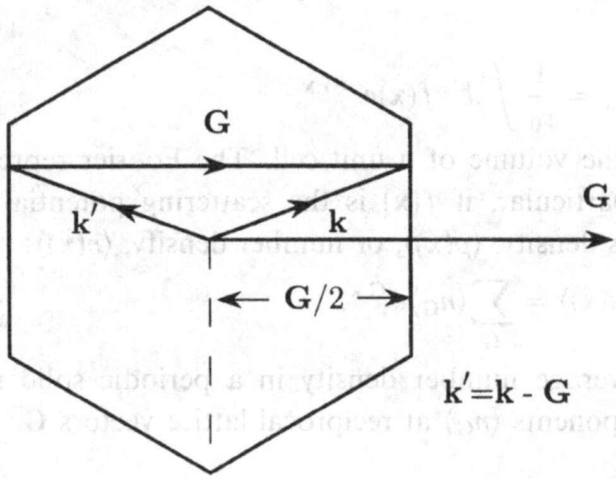

Fig. 2.5.4. Illustration of the Laue condition that Bragg scattering occurs when the incident wave vector **k** lies on the perpendicular bisector of a reciprocal lattice vector **G**.

$$S(\mathbf{q}) = \sum_{\mathbf{G}} |\langle n_{\mathbf{G}}\rangle|^2 (2\pi)^d \delta(\mathbf{q} - \mathbf{G}) + S_{nn}(\mathbf{q}), \tag{2.5.17}$$

and the intensity of the Bragg peak at **G** is proportional to the square $|\langle n_{\mathbf{G}}\rangle|^2$ of the Fourier component of the density at **G**. Along with the dominant scattering into Bragg peaks, there is also *diffuse* scattering at non-Bragg angles given by $S_{nn}(\mathbf{q})$. The simplest crystal is one in which there is a single atom rigidly fixed at each lattice site. In this case, $n_{\mathbf{G}} = v_0^{-1}$ for every **G**, and the Ursell function $S_{nn}(\mathbf{q})$ is identically zero. Thus if the atomic form factor were unity, the X-ray scattering pattern would consist of equally intense peaks at every reciprocal lattice vector. The atomic form factor generally dies off at large **q** so that the intensities of Bragg peaks would die off at large **G** even for this idealized model of rigidly fixed atoms. In a somewhat more complex model crystal, each unit cell has a basis consisting of two or more rigidly fixed identical atoms, located at positions $\mathbf{c}_\alpha$ relative to the origin of the cell [Eq. (2.5.4)]. An example of such a crystal is the two-dimensional honeycomb crystal shown in Fig. 2.5.5. In this case,

$$\langle n_{\mathbf{G}}\rangle = \frac{1}{v_0} \sum_{\alpha} e^{-i\mathbf{G}\cdot\mathbf{c}_\alpha} \tag{2.5.18}$$

with $\mathbf{c}_1 = 0$ and $\mathbf{c}_2 = (\mathbf{a}_1 + 2\mathbf{a}_2)/3$. For the honeycomb lattice, the intensities of the Bragg peaks will vary. For other lattices, there may actually be extinctions at particular reciprocal lattice vectors. Physically this can be understood as resulting whenever the atoms in the basis form planes that lie half way between lattice planes and, therefore, cause destructive interference where previously there was constructive interference. Thus, even though one can in general expect Bragg peaks at every reciprocal lattice vector, the distribution of matter in a unit cell may actually lead to zero intensity at some peaks.

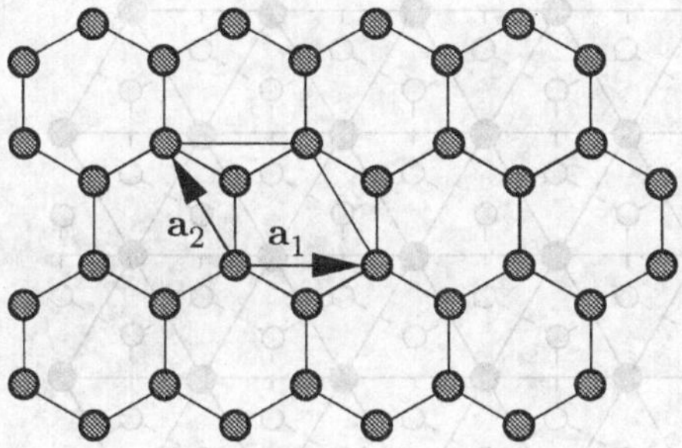

Fig. 2.5.5. The two-dimensional honeycomb lattice showing the primitive translation vectors and the two-atom unit cell.

The two examples just given assumed the atoms in the crystal were rigidly fixed. In real systems, this is never the case. Quantum mechanics and finite temperature always lead to fluctuations about the ideal rigid state that in turn cause the intensity of Bragg peaks at large **G** to decrease exponentially with the Debye-Waller factors that will be discussed in Chapters 6 and 7. In addition, when atomic positions are not fixed, the Ursell function is nonzero, and there will be a diffuse scattering background arising from $S_{nn}(\mathbf{q})$ in Eq. (2.5.17). Crystal imperfections such as dislocation and vacancies also reduce the high-**G** scattering intensity. The generalization of the above discussion to crystals with many types of atoms is straightforward and does not lead to qualitatively different behavior.

2.6 Symmetry and crystal structure

Consider a crystal fixed in some laboratory frame of reference. An observer in the same laboratory frame will identify the crystal by its spatially periodic density. If the crystal is now translated through a lattice vector **T**, it will be absolutely indistinguishable from the untranslated crystal to the fixed observer. Translations by lattice vectors are symmetry operations that leave the density of the crystal invariant. The set of all lattice translations form a group, and the crystal is said to be invariant under operations in this group. Crystals are also invariant under *point group* operations consisting of rotations, reflections, and inversions about special symmetry points. For example, the triangular lattice shown in Fig. 2.6.1a is invariant under rotations of $2\pi p/6$ for $p = 0, ..., 5$ about any lattice point. A six-fold symmetry axis passes through each lattice point. The lattice in Fig. 2.6.1a is also invariant under reflections through the six lines passing through nearest and next nearest neighbor lattice sites as shown in Fig. 2.6.1b. It has six mirror planes. The triangular lattice with a basis in Fig. 2.6.1c, on the other hand, is

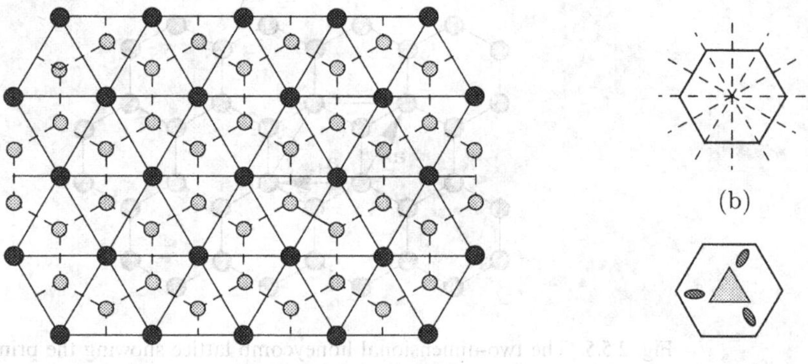

Fig. 2.6.1. (a) A triangular lattice showing lattice sites as dark circles. Bonds connecting nearest neighbor sites are shown as full lines. Also shown is the honeycomb lattice of the centers of triangles (indicated by lighter circles connected by dashed lines) formed by connecting nearest neighbor sites of the triangular lattice. Note that the bonds connecting nearest neighbor sites of the honeycomb lattice are perpendicular bisectors of the bonds of the triangular lattice and vice versa. Lattices in two dimensions whose bonds have this property are said to be *dual.* Every two-dimensional lattice has a dual lattice obtained by constructing perpendicular bisectors to its bonds. (b) The undecorated Wigner-Seitz unit cell of the triangular lattice showing six-fold rotational symmetry and six mirror planes indicated by dashed lines. (c) A decorated Wigner-Seitz unit cell with only three-fold symmetry and no mirror planes.

invariant only under rotations through $2\pi p/3$ and not under reflections; it has three-fold rotational symmetry but no mirror planes.

1 Two-dimensional Bravais lattices

Molecules and finite size objects can have symmetry axes of arbitrary order. The requirement that a crystal be invariant under translations through any vector in its direct lattice, which, as we have seen, contains no vector shorter than some minimum length vector, places severe restrictions on possible rotational symmetries. To illustrate how this comes about, we will show that it is impossible for a periodic crystal to have five-fold symmetry, i.e., to be invariant with respect to rotations through $2\pi/5$. Assume that a crystal does have five-fold symmetry, and let $\mathbf{a}_0 = (1, 0)$ be the shortest vector in the lattice. Since the crystal is assumed to have five-fold symmetry, the vectors $\mathbf{a}_n = [\cos(2\pi n/5),\ \sin(2\pi n/5)]$, with n an integer, must also be in its direct lattice. But by the closure property of any lattice, the vector

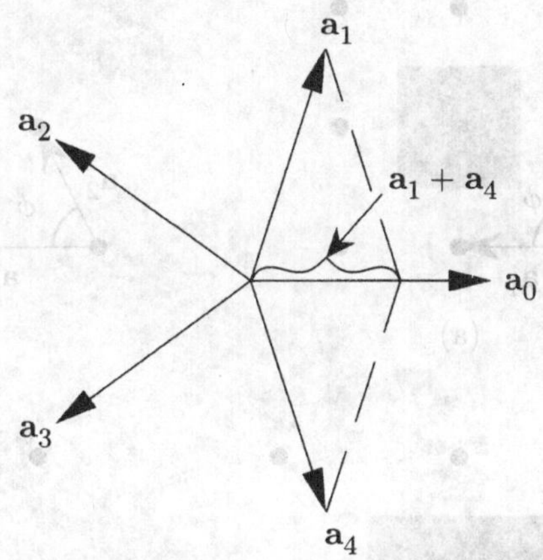

Fig. 2.6.2. The five vectors $\mathbf{a}_n = [\cos(2\pi n/5), \sin(2\pi n/5)]$ generated by applying five-fold symmetry operations to the vector $\mathbf{a}_0$. The vector $\mathbf{a}_1 + \mathbf{a}_4$ is parallel to and shorter than the vector $\mathbf{a}_0$.

$$\begin{aligned}\mathbf{T} &= \mathbf{a}_4 + \mathbf{a}_1 \\ &= [\cos(8\pi/5) + \cos(2\pi/5),\ \sin(8\pi/5) + \sin(2\pi/5)] \\ &= \tau^{-1}(1,0) = \tau^{-1}\mathbf{a}_0, \end{aligned} \tag{2.6.1}$$

where $\tau = (1 + \sqrt{5})/2 = 2\cos(2\pi/10)$, must also be in the direct lattice. The number τ is a special irrational number called the *golden mean* that satisfies $\tau^2 = \tau + 1$. The vector $\mathbf{T}$, as illustrated in Fig. 2.6.2, is shorter than $\mathbf{a}_0$, contradicting the assumption that $\mathbf{a}_0$ was the shortest vector in the lattice. Thus, it is impossible for a periodic lattice in two dimensions to have five-fold symmetry. Similar arguments rule out all periodic lattices in two dimensions with other than two-, three-, four-, or six-fold symmetry. These restrictions lead to only five distinct types of Bravais lattices in two dimensions, as illustrated in Fig. 2.6.3. Point groups that are compatible with periodic translational symmetry are called *crystallographic* point groups. Note that the centered rectangular lattice is formed by placing a lattice site at the center of rectangles (with sides $\mathbf{a}_1$ and $\mathbf{a}_2$) in a rectangular lattice. The primitive unit cell of the rectangular lattice with edges equal to the primitive translation vectors is a non-primitive unit cell for the centered rectangular lattice. This rectangular non-primitive unit cell is called the *conventional unit cell* of the centered rectangular lattice.

The triangular lattice with six-fold symmetry deserves special attention. It is the two-dimensional lattice with the highest rotational symmetry. In addition, the densest possible packing of hard circles of radius R is obtained by placing their centers on a triangular lattice with lattice parameter $a = 2R$ as shown in Fig. 2.6.4. Note that each circle is tangent to six other circles. A lattice in d-dimensions that provides the densest possible packing of hard spheres is said to be *close*

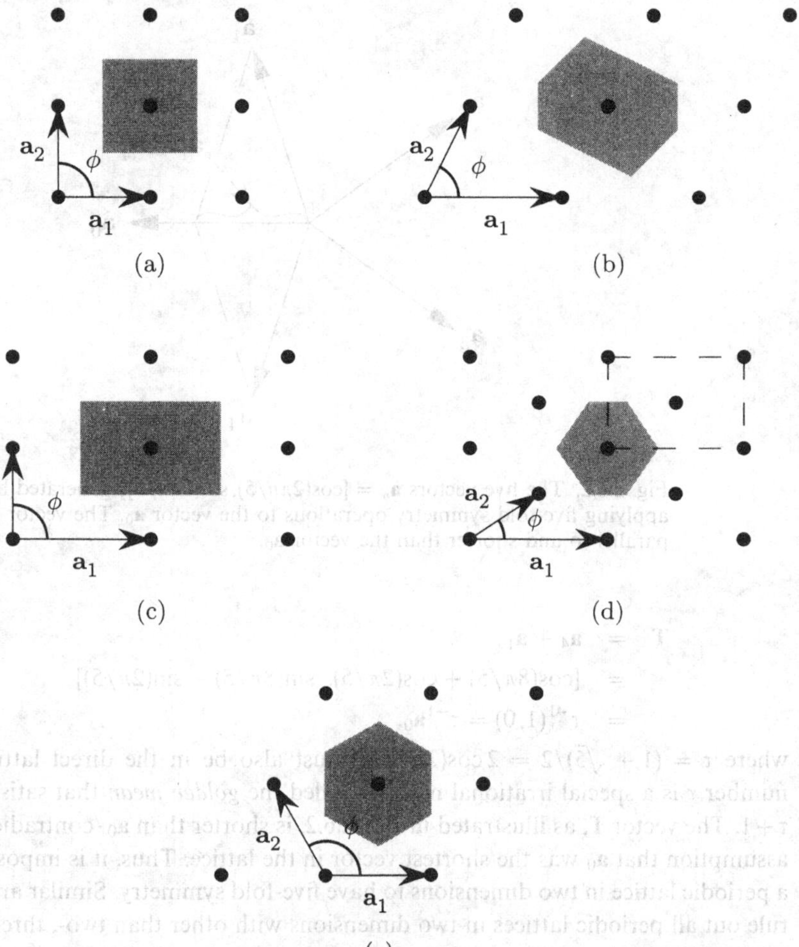

Fig. 2.6.3. The five two-dimensional Bravais lattices: (a) square, (b) oblique, (c) rectangular, (d) centered rectangular, and (e) hexagonal lattice. $\mathbf{a}_1$ and $\mathbf{a}_2$ are primitive translation vectors and ϕ is the angle between $\mathbf{a}_1$ and $\mathbf{a}_2$. Wigner-Seitz unit cells are shaded. The conventional unit cell of the centered rectangular lattice is enclosed with dashed lines.

packed. The triangular lattice is the only close packed lattice in two dimensions. The fractional area occupied by the hard circles is the area of a circle divided by the area of the hexagonal Wigner-Seitz cell with side of length $s = 2R/\sqrt{3}$: $\pi R^2/(3sR) = \pi\sqrt{3}/6 = 0.907$. (Note that the vacant spaces between circles lie on a honeycomb lattice (Fig. 2.5.5) with two sites per unit cell.) The primitive translation vectors of the direct triangular lattice can be chosen as

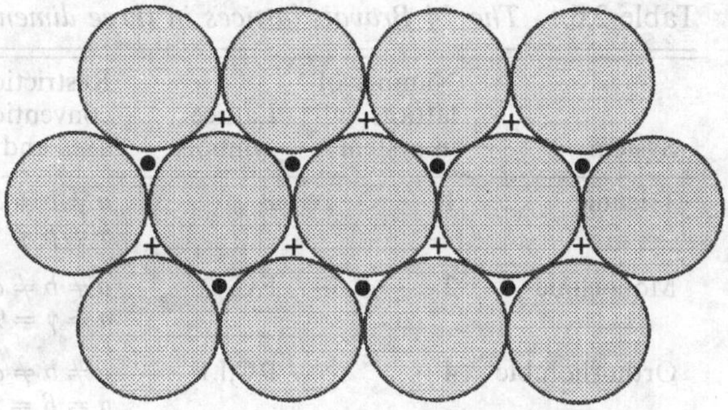

Fig. 2.6.4. Close packed hard circles on a triangular lattice. The centers of the circles are at lattice sites of the triangular lattice shown in Fig. 2.6.1a. The empty spaces between circles lie on a honeycomb lattice. The two inequivalent sites in the lattice are marked with + and •, respectively.

$$\mathbf{a}_1 = a(1,0),$$
$$\mathbf{a}_2 = a(\cos 2\pi/3, \sin 2\pi/3) = a(-1/2, \sqrt{3}/2), \tag{2.6.2}$$

and those for the reciprocal triangular lattice as

$$\mathbf{b}_1 = \frac{2\pi}{(a\sqrt{3}/2)}(\cos\pi/6, \sin\pi/6) = \frac{2\pi}{(a\sqrt{3}/2)}(\sqrt{3}/2, 1/2),$$
$$\mathbf{b}_2 = \frac{2\pi}{(a\sqrt{3}/2)}(0,1). \tag{2.6.3}$$

These vectors satisfy the orthogonality relations, Eq. (2.5.7).

2 *Three-dimensional Bravais lattices*

In three dimensions, there are 14 distinct Bravais lattices, as shown in Fig. 2.6.5. They range from the highest symmetry cubic lattices with four three-fold and three four-fold axes and three mirror planes to the triclinic lattice whose point group consists only of the inversion operation. There are three types of cubic lattices: the simple cubic (SC), the body-centered cubic (BCC), and face-centered cubic (FCC). The primitive translation vectors of the SC lattice are the edges of a cube. Its primitive unit cell is a conventional cube. Those for the BCC lattice point to the centers of non-primitive conventional cubic unit cells of side a:

$$\mathbf{a}_1 = \frac{1}{2}a(1,1,1); \quad \mathbf{a}_2 = \frac{1}{2}a(-1,1,1); \quad \mathbf{a}_3 = \frac{1}{2}a(1,-1,1), \tag{2.6.4}$$

whereas those for the FCC lattice point to face centers of the non-primitive cubic cell:

$$\mathbf{a}_1 = \frac{1}{2}a(1,1,0); \quad \mathbf{a}_2 = \frac{1}{2}a(0,1,1); \quad \mathbf{a}_3 = \frac{1}{2}a(1,0,1). \tag{2.6.5}$$

Table 2.6.1. *The 14 Bravais lattices in three dimensions.*

System	Number of lattices in system	Lattice symbols	Restrictions on conventional cell axes and angles
Triclinic	1	P	$a \neq b \neq c$ $\alpha \neq \beta \neq \gamma$
Monoclinic	2	P,C	$a \neq b \neq c$ $\alpha = \gamma = 90^\circ \neq \beta$
Orthorhombic	4	P,C,I,F	$a \neq b \neq c$ $\alpha = \beta = \gamma = 90^\circ$
Tetragonal	2	P,I	$a = b \neq c$ $\alpha = \beta = \gamma = 90^\circ$
Cubic	3	P or SC I or BCC F of FCC	$a = b = c$ $\alpha = \beta = \gamma = 90^\circ$
Trigonal	1	R	$a = b = c$ $\alpha = \beta = \gamma < 120^\circ, \neq 90^\circ$
Hexagonal	1	P	$a = b \neq c$ $\alpha = \beta = 90^\circ$ $\gamma = 120^\circ$

The primitive unit cells for the BCC and FCC lattices are not cubes. From the definition of the reciprocal lattice, it can be shown that the point group symmetry of the direct and reciprocal lattices is the same. The reciprocal lattice of the SC lattice is clearly an SC lattice. It is also clear that the reciprocal lattice of a BCC lattice is neither SC nor BCC so it must be FCC. Similarly, the reciprocal lattice of an FCC lattice is a BCC lattice. The Bragg scattering patterns from SC, BCC, and FCC lattices are, therefore, different. The FCC structure is of particular interest because it has the largest number of nearest neighbors (12) as well as the highest symmetry. The conventional unit cell of all of the cubic lattices is the unit cube with all sides equal and 90° angles between all edges. This is, however, not the primitive unit cell of either the BCC or FCC lattice. The Wigner-Seitz unit cells for the BCC and FCC lattices are shown in Fig. 2.6.6.

Lattices with lower than cubic symmetry have conventional unit cells that are anisotropic and can have angles between edges that differ from 90° as shown in Fig. 2.6.5 and outlined in Table 2.6.1. The lengths of the unit cell edges are commonly denoted by the symbols *a*, *b* and *c*. In tetragonal, trigonal, and hexagonal lattices, $b = c$. The ratio c/a provides a measure of the anisotropy of these lattices.

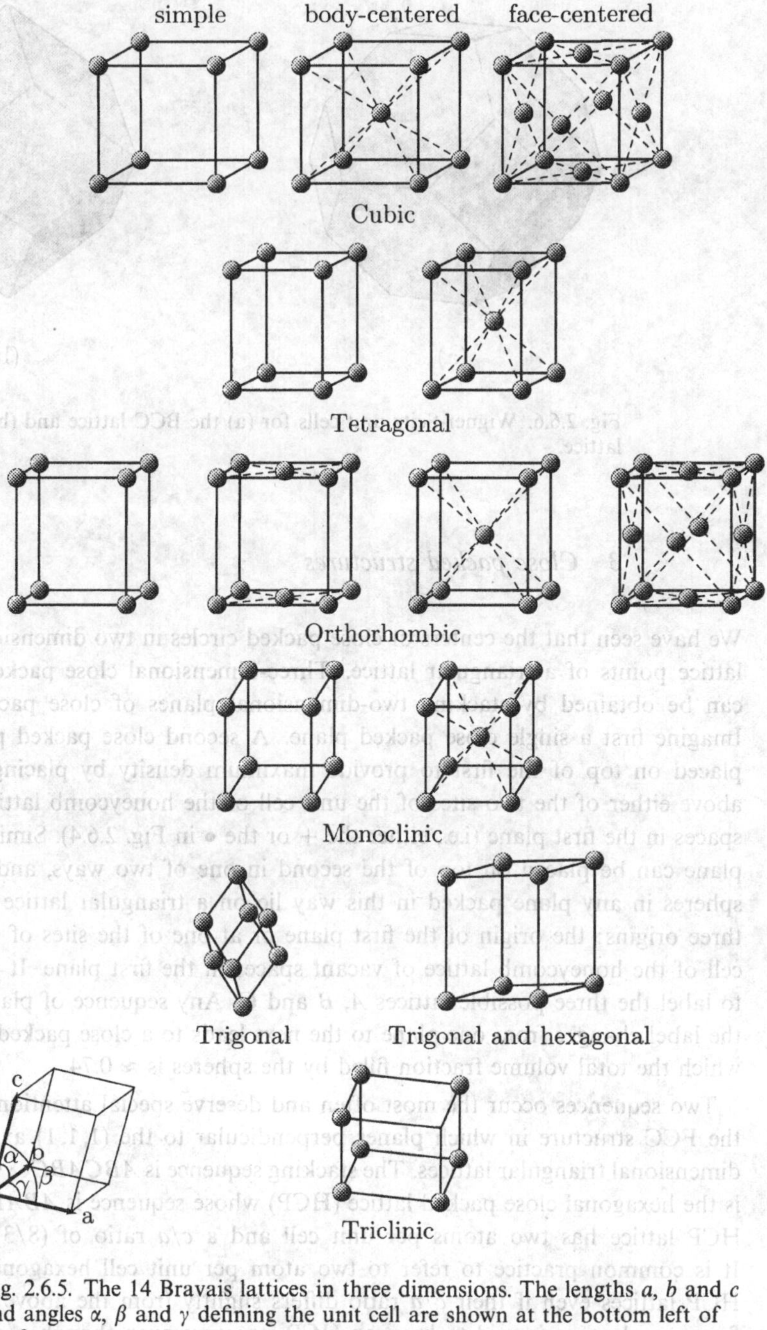

Fig. 2.6.5. The 14 Bravais lattices in three dimensions. The lengths a, b and c and angles α, β and γ defining the unit cell are shown at the bottom left of the figure.

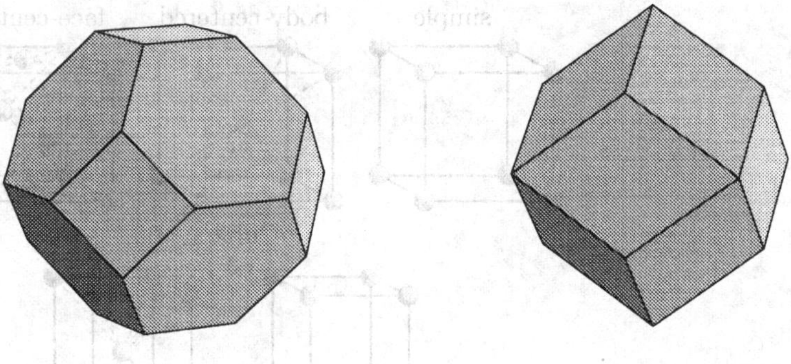

(a) (b)

Fig. 2.6.6. Wigner-Seitz unit cells for (a) the BCC lattice and (b) the FCC lattice.

3 Close packed structures

We have seen that the centers of close packed circles in two dimensions lie at the lattice points of a triangular lattice. Three-dimensional close packed structures can be obtained by stacking two-dimensional planes of close packed spheres. Imagine first a single close packed plane. A second close packed plane can be placed on top of the first to provide maximum density by placing its spheres above either of the two sites of the unit cell of the honeycomb lattice of vacant spaces in the first plane (i.e., either the + or the • in Fig. 2.6.4). Similarly, a third plane can be placed on top of the second in one of two ways, and so on. The spheres in any plane packed in this way lie on a triangular lattice with one of three origins: the origin of the first plane or at one of the sites of the first unit cell of the honeycomb lattice of vacant spaces in the first plane. It is customary to label the three possible lattices A, B and C. Any sequence of planes in which the label changes from one plane to the next leads to a close packed structure in which the total volume fraction filled by the spheres is ≈ 0.74.

Two sequences occur the most often and deserve special attention. The first is the FCC structure in which planes perpendicular to the $(1,1,1)$ axis form two-dimensional triangular lattices. The stacking sequence is $ABCABC\cdots$. The second is the hexagonal close packed lattice (HCP) whose sequence is $ABABAB\cdots$. The HCP lattice has two atoms per unit cell and a c/a ratio of $(8/3)^{1/2} = 1.633$. It is common practice to refer to two atom per unit cell hexagonal lattices as HCP lattices even if their c/a ratio differs slightly from the above ideal. Thus, for example, zinc is said to have an HCP structure even though $c/a = 1.86$. (Of the elements, 23 crystallize as FCC, 21 as HCP, and 14 as BCC, indicating the tendency of crystals to be both close packed and highly symmetric.)

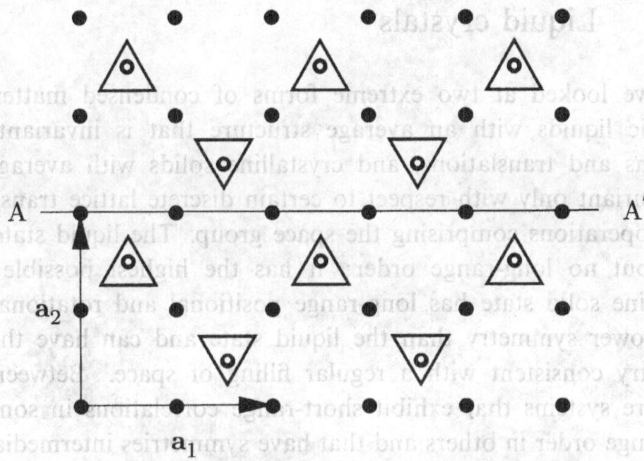

Fig. 2.6.7. A two-dimensional lattice with a glide plane AA'.

Though the sequence nomenclature for stacked hexagonal structures is most commonly used for FCC and HCP structures, it can be applied to other systems, such as molecules intercalated between graphite layers. In these systems, random or large unit cell stacking structures can occur.

4 Space groups

The group of all translations and rotations that leave a crystal invariant is called the *space group*. Often the space group consists only of point group operations about symmetry points and translations by vectors in the direct lattice. In this, the *symmorphic* case, the space group is a direct product of the point group and the translation group. In some cases, however, there may be operations in the space group consisting of a combination of point group and translation operations that individually are not in the space group. An example of such an operation is illustrated by the lattice shown in Fig. 2.6.7. This is a lattice with a multi-atom basis and primitive translation vectors $\mathbf{a}_1 = (2, 0)$ and $\mathbf{a}_2 = (0, 2)$. The lattice is not invariant under reflection about the line AA'. It is, however, invariant under first a reflection through the line AA' followed by a translation through the vector $\mathbf{a}_1/2$ not in the direct lattice. The line AA' is a two-dimensional version of a *glide plane.* A symmetry operation involving a rotation about some symmetry axis followed by a translation through a vector not in the direct lattice along the symmetry axis gives rise to a *screw* axis. Space groups with glide planes or screw axes are called *nonsymmorphic*.

2.7 Liquid crystals

We have looked at two extreme forms of condensed matter: homogeneous, isotropic liquids with an average structure that is invariant under arbitrary rotations and translations, and crystalline solids with average structures that are invariant only with respect to certain discrete lattice translations and point group operations comprising the space group. The liquid state has short-range order but no long-range order: it has the highest possible symmetry. The crystalline solid state has long-range positional and rotational order; it has a much lower symmetry than the liquid state and can have the lowest possible symmetry consistent with a regular filling of space. Between these extremes, there are systems that exhibit short-range correlations in some directions and long-range order in others and that have symmetries intermediate between those of liquids and the crystals. One form of this intermediate order is orientational order. In a periodic crystal, there is only a discrete set of directions defined by vectors between nearest neighbor particles, which occupy sites on a lattice. These directions are the same throughout the lattice and define a long-range orientational order often called *bond-angle* order. Remarkably, it is possible to have long-range orientational order in the absence of translational order.

Among the materials that show intermediate order, the most widely studied are liquid crystals. Some examples of liquid crystal forming molecules are shown in Fig. 2.7.1. The molecules are highly anisotropic, and to a good approximation can be modeled as rigid rods or ellipsoids of revolution with lengths l greater than their widths a as shown in Fig. 2.7.2. As is the case with solids, the orientational order is caused mostly by the repulsive interactions.†

1 *Isotropic, nematic and cholesteric phases*

At high temperatures the axes of the anisotropic molecules are randomly oriented and their centers of mass are randomly distributed as depicted in Fig. 2.7.3a. Globally the system is an isotropic liquid. The structure factor (Fig. 2.7.4a) is isotropic but shows liquid-like rings at wave numbers corresponding to the two characteristic lengths of the individual molecules – their length l and diameter a.

† The names of the different liquid crystalline phases are due to G. Friedel who studied many of their properties in the first part of the twentieth century. Nematic is from the Greek $\nu\eta\mu\sigma$ for thread. When observed between crossed polarizers, the defects in nematics produce a threadlike structure. Smectics are from $\sigma\mu\epsilon\gamma\mu\alpha$ for soap from which many layered mesophases are made. Cholesteryl nonanoate was the first liquid crystal discovered by R. Reinitzer and O. Lehman (in the late nineteenth century) and gives its name to cholesterics. Lyotropic refers to liquid crystals which undergo phase changes as a function of solvent concentration while thermotropics change with temperature as a control parameter. The names "smectic-*A*", "smectic-*C*", etc. merely indicate the historical order in which the phases were discovered. There is not the slightest clue about the arrangement of the molecules, the symmetries or the physical properties in the letter designation. [G. Friedel, *Ann. Physique (Paris)* **18**, 273 (1922); F. Reinitzer, *Monatsch Chem.* **9**, 421 (1988); O. Lehman, *Z. Physikal Chem.* **4**, 462 (1989).]

4,4' - dimethyloxyazoxybenzene (p-azoxyanisole)

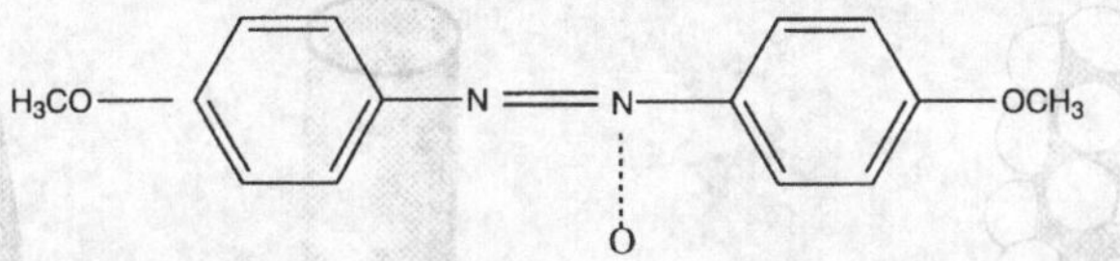

$$\text{solid} \underset{29.57\text{kJ}}{\overset{118.2\text{C}}{\rightleftarrows}} N \underset{0.57\text{kJ}}{\overset{135.3\text{C}}{\rightleftarrows}} I$$

2-(4-n-pentylphenyl)-5-(4-n-pentyloxyphenyl)-pyrimidine

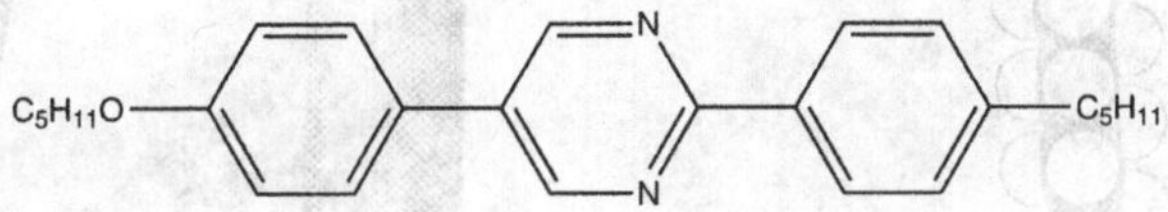

$$\text{solid} \underset{34.98\text{kJ}}{\overset{79\text{C}}{\rightleftarrows}} \text{Sm-}G \underset{0.58\text{kJ}}{\overset{102.7\text{C}}{\rightleftarrows}} \text{Sm-}F \underset{0.5\text{kJ}}{\overset{113.8\text{C}}{\rightleftarrows}} \text{Sm-}C \underset{11.45\text{kJ}}{\overset{144\text{C}}{\rightleftarrows}} \text{Sm-}A \underset{0.2\text{kJ}}{\overset{210\text{C}}{\rightleftarrows}} I$$

4-n-pentylbenzenethio-4'-n-heptyloxybenzoate ($\bar{7}$S5)

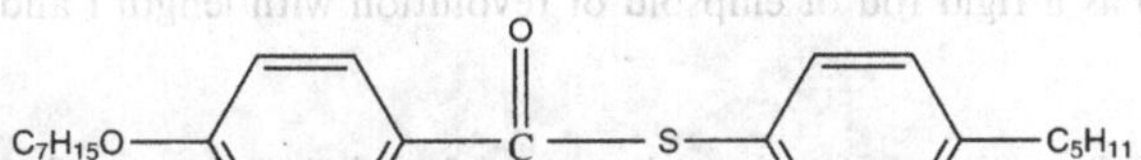

$$\text{Sm-}C \overset{37\text{C}}{\rightleftarrows} N \overset{81\text{C}}{\rightleftarrows} I$$

Fig. 2.7.1. Some molecules forming liquid crystal phases and their phase sequences as a function of temperature. The benzene rings give rigidity to the molecules.

When the isotropic liquid is cooled, the first phase that condenses is the *nematic* (N) phase in which long molecules align so that they are on average parallel to a particular direction specified by a unit vector $\mathbf{n}$ called the *director*. The positions of the molecular centers of mass remain randomly distributed as they are in an isotropic fluid. The nematic phase breaks rotational isotropy but not translational invariance. Rotations about an axis parallel to $\mathbf{n}$ leave the nematic phase unchanged, whereas rotations about axes perpendicular to $\mathbf{n}$ do not. The nematic phase still has an axial rotational symmetry. If each molecule is regarded as a rigid rod whose long axis makes an angle θ with respect to $\mathbf{n}$, then a measure of the degree of order in the nematic phase is provided by $S = \langle \cos^2 \theta - \frac{1}{3} \rangle$. Onsager (1949) has shown that the transition from the isotropic

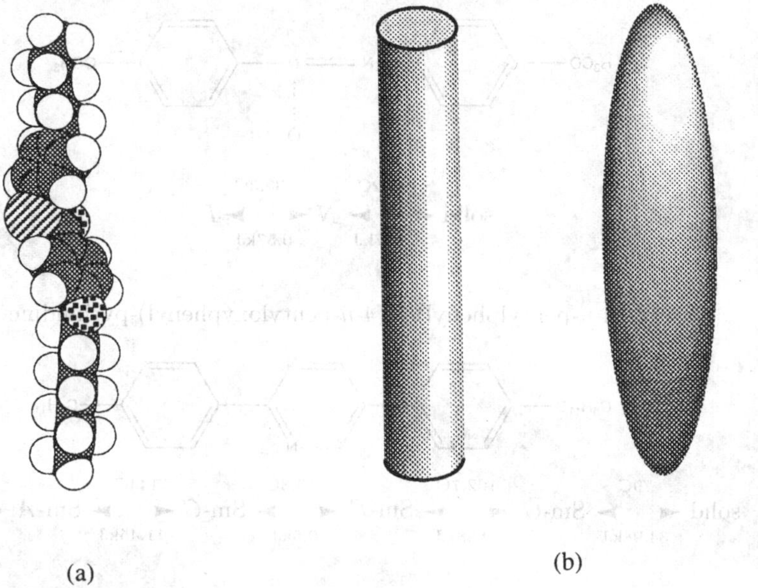

Fig. 2.7.2. (a) Space filling model for the molecule $\overline{7}S5$ (Fig. 2.7.1). (b) Model of (a) as a rigid rod or ellipsoid of revolution with length l and diameter a.

to the nematic phase is predominantly a result of the shape of the molecules and excluded volume (anisotropic, hard core repulsion). His calculation preceded the simulations of the transition of hard spheres from liquid to solid. Basically it is easier to achieve a high packing density of matchsticks by aligning their axes along a common direction rather than by allowing them to be randomly oriented. (This consideration enters the thermodynamics by way of an increase in entropy for the aligned state.) The structure factor of the nematic phase reflects the breaking of rotational symmetry. It is axially symmetric in any plane perpendicular to **n** but has only two-fold symmetry in any plane containing **n**. The intensity of the large wave number (short distance) sphere of the isotropic phase is compressed toward the plane perpendicular to **n** whereas the intensity of the small wave number sphere is compressed toward the **n** axis. Two possible forms for the X-ray intensity from the nematic phase are shown in Figs. 2.7.4b and c. In the first there are diffuse spots at $\mathbf{q} = q_0\mathbf{n}$ with $q_0 = 2\pi/l$, and in the second there are diffuse rings centered at the same values of $\mathbf{q}$.

Chiral molecules (such as cholesterol nonanoate shown in Fig. 2.7.5a) are molecules with no mirror plane. When such molecules are added to a nematic liquid crystal, a *twisted* or *chiral nematic* (N^*) state results. This state is often referred to as the *cholesteric* state. In this state, the direction of average molecular alignment rotates in a helical pattern as shown in Fig. 2.7.5b. The pitch (distance between equivalent planes) depends on the concentration and degree of chirality of the chiral molecules but is typically of order several thousand angstroms. This means that cholesterics Bragg scatter visible light. Their X-ray scattering intensity is, however, generally similar to that of a nematic.

2 Smectics-*A* and -*C*

As temperature is further reduced, molecules begin to segregate into planes giving rise to a *smectic-A* (Sm-*A*) liquid crystal. The usual picture of this smectic phase is one with molecules situated in well-defined layers with a spacing that is essentially the rod length as shown in Fig. 2.7.3c. There is liquid-like motion of the rods in each layer and no correlation of the positions of the molecules from one layer to the next. In smectic-*A* liquid crystals, molecules are aligned perpendicular to the layers. The introduction of the layering indicates the presence of a mass density wave perpendicular to the layers. There is, therefore, positional correlation in the system which can be described as a sinusoidal modulation of the average molecular number-density,

$$\langle n(\mathbf{x})\rangle = n_0 + 2n_{q_0}\cos(q_0 z), \tag{2.7.1}$$

where $q_0 = 2\pi/l$ and the z-axis is along the layer normals and parallel to $\mathbf{n}$. The Fourier transform of this equation leads to two Bragg peaks away from $\mathbf{q} = 0$ in the structure function:

$$S(\mathbf{q}) = |\langle n_{q_0}\rangle|^2(2\pi)^3[\delta(\mathbf{q}_z - q_0\mathbf{e}_z) + \delta(\mathbf{q}_z + q_0\mathbf{e}_z)]. \tag{2.7.2}$$

We shall see in Chapter 6 that thermal fluctuations destroy the ideal long-range periodic order of the smectic phase and that there are power-law singularities rather than delta-function spikes in $S(\mathbf{q})$ at $\mathbf{q} = \pm q_0\mathbf{e}_z$. Thus, the peaks in the smectic structure function are called quasi-Bragg rather than Bragg peaks, and smectics are thus said to be characterized by *quasi-long-range order* (QLRO) rather than true *long-range order* (LRO). As the smectic phase is approached on cooling from the nematic phase, the diffuse spots (Fig. 2.7.4b) in the nematic structure factor sharpen and eventually become the quasi-Bragg peaks of the smectic phase. An experimental intensity profile of these peaks is shown in Fig. 2.7.6. The transition from the nematic to the smectic-*A* phase can be second order with the mass density amplitude growing continuously from zero as we will see in more detail in Chapter 4. The experimental structure factor of most thermotropic smectics has only two quasi-Bragg peaks (at $\pm q_0$) indicating that the single sine-wave in Eq. (2.7.1) is a good representation of the actual density in the smectic in contrast to the square wave with many Fourier components implied by the usual schematic representation (Fig. 2.7.3c).

In some systems, molecules align along an axis tilted relative to the smectic planes as shown in Fig. 2.7.3d. This is the smectic-*C* phase. It has a lower symmetry than the smectic-*A* phase because the tilted molecules pick out a special direction in the smectic plane, i.e., their projections in the xy-plane align,

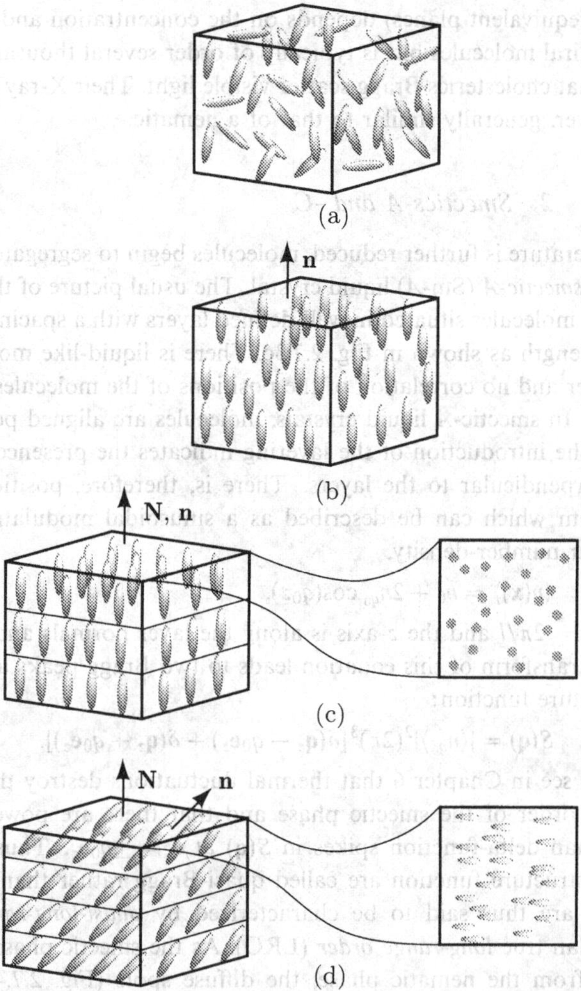

Fig. 2.7.3. Schematic representation of the position and orientation of anisotropic molecules in (a) the isotropic, (b) the nematic, (c) the smectic-*A*, and (d) the smectic-*C* phases. The direction of average molecular alignment in all but the isotropic phase is specified by a unit vector **n**. The layer normal in the smectic phases is indicated by the unit vector **N**. In the smectic-*A* phase, **n** is parallel to **N**, whereas in the smectic-*C* phase, it is not. In the text $\mathbf{N} = \mathbf{e}_z$ is parallel to the *z*-axis. (c) and (d) also show the arrangement of molecules in the smectic planes in the smectic-*A* and -*C* phases. In the smectic-*C* phase, the projections of molecular axes onto the plane perpendicular to **N** align on average along the **c**-director.

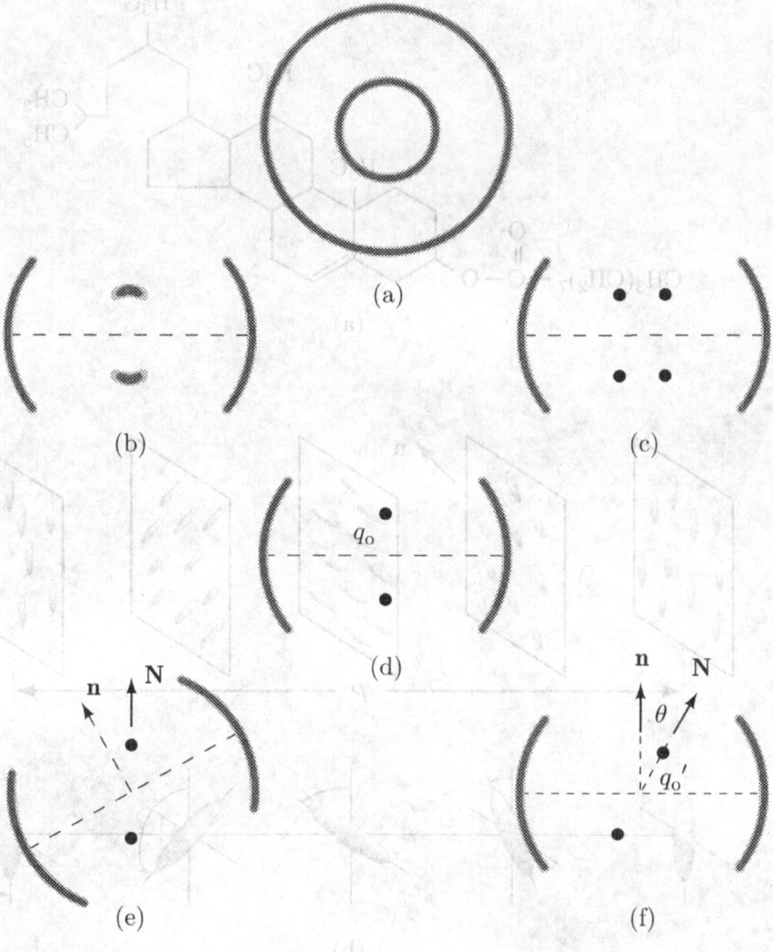

Fig. 2.7.4. Schematic representation of X-ray scattering intensities profiles from (a) the isotropic, (b) and (c) the nematic, (d) the smectic-A, and (e) and (f) the smectic-C phases. All intensities profiles are cross-sections in a plane containing the director **n**. The full three-dimensional profiles for (a) through (d) are obtained by rotating the cross-sectional profiles about **n**. (b) and (c) apply, respectively, to nematic phases near transitions to the smectic-A and smectic-C phases. The former has diffuse spots at $\mathbf{q} = \pm q_0\mathbf{n}$ whereas the latter has diffuse rings in the plane perpendicular to **n** very nearly centered at $\mathbf{q} = \pm(2\pi/l)\mathbf{n}$. The smectic-$A$ phase has quasi-Bragg peaks at $\mathbf{q} = \pm q_0\mathbf{n}$, and the smectic-$C$ phase has peaks at $\mathbf{q} = \pm q_0'\mathbf{N}$ where q_0' is very nearly $2\pi/(l\cos\theta)$ where θ is the angle between **N** and **n**. (e) and (f) are identical except for alignment of axes. The full three-dimensional scattering profile for the smectic-C phase is not invariant with respect to rotations about **n**.

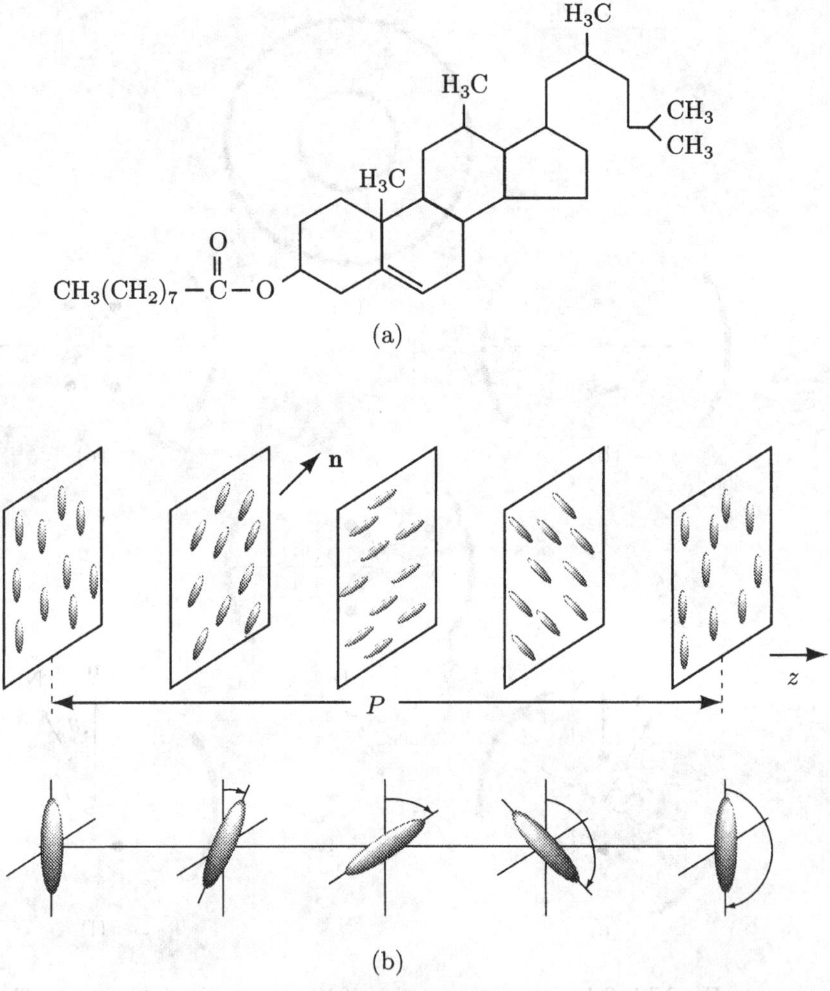

Fig. 2.7.5. (a) Cholesterol nonanoate molecule. (b) Schematic representation of the molecules in the cholesteric phase. The director $\mathbf{n} = (\cos k_0 z, \sin k_0 z, 0)$ rotates in a helical fashion. Because no physical quantities depend on the sign of $\mathbf{n}$, the physical pitch of the cholesteric phase is $P = \pi/k_0$ rather than $2\pi/k_0$.

like the molecules in a nematic, along a common direction denoted by a unit vector $\mathbf{c}$, called the $\mathbf{c}$-director. There are in fact transitions from the smectic-*A* to the smectic-*C* phase in which the tilt angle grows continuously from zero. The structure factor of the nematic phase just above a smectic-*C* phase has diffuse rings rather than diffuse spots as shown in Fig. 2.7.4c.

The nematic-to-smectic-*A* transition provides insight into how to describe the transition from the isotropic liquid to a conventional crystalline phase. Instead

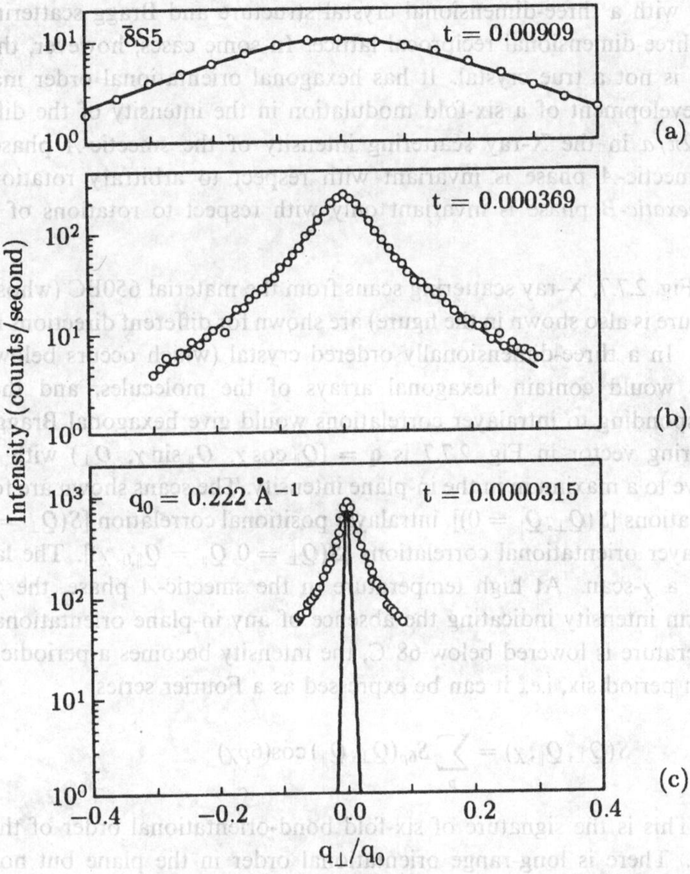

Fig. 2.7.6. $S(\mathbf{q})$ for $\mathbf{q} \approx q_0\mathbf{n}$ (a) in the nematic phase far from the nematic-to-smectic-A (NA) transition, (b) in the nematic phase near the NA transition, and (c) in the smectic-A phase. Here $t = (T - T_{NA})/T_{NA}$ is the reduced temperature where T_{NA} is the NA transition temperature. [Courtesy of Cyrus Safinya, MIT thesis (1981).]

of introducing a single density wave, we could imagine the development of many density waves (corresponding to the reciprocal lattice vectors as in Eq. (2.5.8)) with several directions and a complete set of harmonics. The usual liquid-solid transition, however, is first order (discontinuous), so that the density wave amplitudes do not grow smoothly from zero.

3 Hexatic phases

When smectic-A phases are cooled, they condense into what was historically called a smectic-B phase. It is now known that a smectic-B phase can be a crystalline-B

phase with a three-dimensional crystal structure and Bragg scattering at points in a three-dimensional reciprocal lattice. In some cases, however, the smectic-B phase is not a true crystal. It has hexagonal orientational order manifested by the development of a six-fold modulation in the intensity of the diffuse ring at $q = 2\pi/a$ in the X-ray scattering intensity of the smectic-A phase. Whereas the smectic-A phase is invariant with respect to arbitrary rotations about $\mathbf{n}$, this *hexatic-B* phase is invariant only with respect to rotations of $2\pi/6$ about $\mathbf{n}$.

In Fig. 2.7.7, X-ray scattering scans from the material 650BC (whose molecular structure is also shown in the figure) are shown for different directions in reciprocal space. In a three-dimensionally ordered crystal (which occurs below 60°C), the layers would contain hexagonal arrays of the molecules, and the scattering corresponding to intralayer correlations would give hexagonal Bragg spots. The scattering vector in Fig. 2.7.7 is $\mathbf{q} = (Q_\parallel \cos\chi,\ Q_\parallel \sin\chi,\ Q_\perp)$ with χ measured relative to a maximum in the in-plane intensity. The scans shown are for interlayer correlations $[S(Q_\perp, Q_\parallel = 0)]$, intralayer positional correlation $[S(Q_\perp = 0, Q_\parallel)]$, and intralayer orientational correlations $[S(Q_\perp = 0, Q_\parallel = Q_{\parallel,0}, \chi)]$. The latter scan is called a χ-scan. At high temperature in the smectic-A phase, the χ-scan gives uniform intensity indicating the absence of any in-plane orientational order. As temperature is lowered below 68°C, the intensity becomes a periodic function of χ with period six, i.e., it can be expressed as a Fourier series

$$S(Q_\perp, Q_\parallel, \chi) = \sum_p S_{6p}(Q_\perp, Q_\parallel)\cos(6p\chi) \tag{2.7.3}$$

in χ. This is the signature of six-fold bond-orientational order of the smectic-B phase. There is long-range orientational order in the plane but no long-range positional order. Just below the transition to the hexatic-B phase, only the first harmonic $S_6(Q_\perp, Q_\parallel)$ is measurable. As temperature is further lowered, more and more harmonics appear in the χ-scan so that eventually a Bragg peak characteristic of long-range crystalline order of the crystalline-B phase appears for temperature below 64°C.

The existence of long-range orientational order in hexatics is in a sense quite remarkable. The breaking of rotational symmetry in the nematic phase is easy to accept. The molecules comprising the nematic phase have a rigid core produced by strong chemical bonds. Orientational order is produced by the collective alignment of rigid bar-like molecules. In hexatics, on the other hand, there is no chemical bond between neighboring molecules, and orientational order is a reflection of the long-range alignment of the position vectors connecting nearest neighbor molecules as depicted in Fig. 2.7.8. In fact the hexatic-bond-angle order can be viewed as resulting from the loss of long-range positional but not orientational order of a hexagonal crystal. Indeed, the original theoretical prediction of the hexatic phase was based on this kind of reasoning. A measure

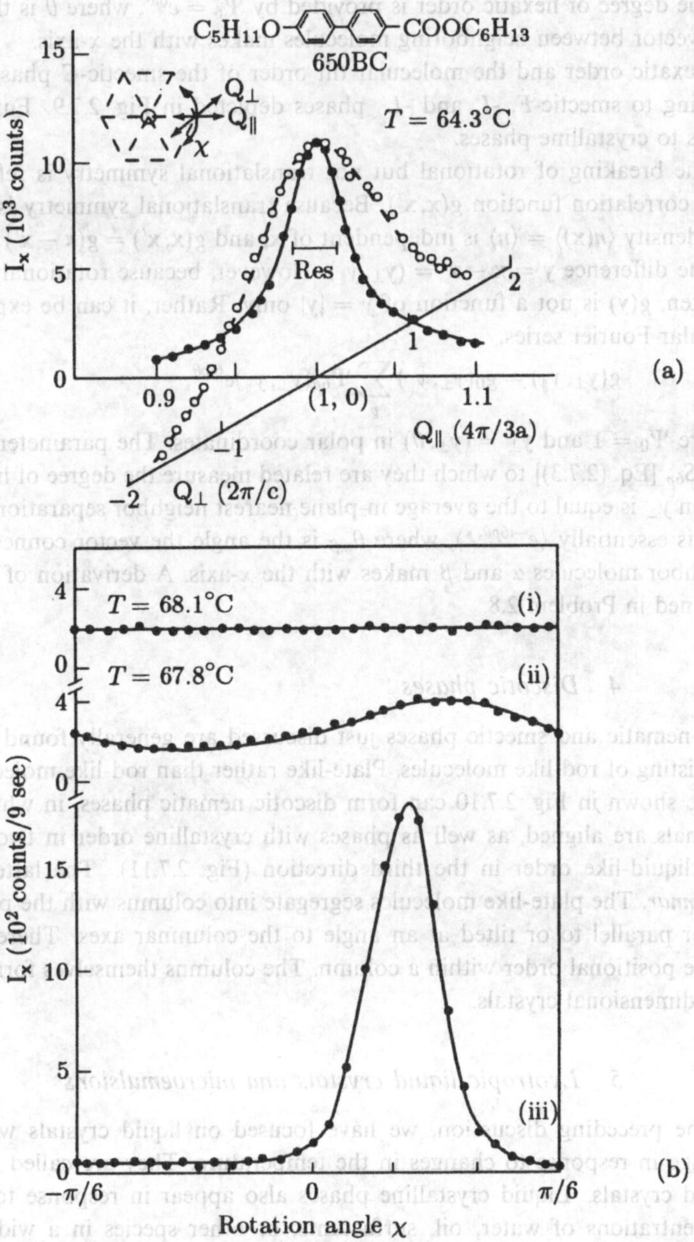

Fig. 2.7.7. (a) χ-averaged X-ray scattering intensity for a $Q_{\parallel}$ scan (closed circles) and a $Q_{\perp}$ scan in the *B* phase of 650BC. The $Q_{\parallel}$ scale has been expanded relative to the $Q_{\perp}$ scale. The scattering along $Q_{\perp}$ is a diffuse rod. The resolution width for the $Q_{\perp}$ scan ($\Delta Q_{\perp} = 0.006\text{Å}^{-1}$) was too small to illustrate. The inset describes the three scan directions. (b) χ-scans at three different temperatures: (i) in the smectic-*A* phase; (ii) just below the smectic-*A*-to-*B* transition; and (iii) well into the *B* phase. [R. Pindak, D.E. Moncton, S.C. Davey, and J.W. Goodby, *Phys. Rev. Lett.* **46**, 1135 (1981).]

of the degree of hexatic order is provided by $\Psi_6 = e^{6i\theta}$, where θ is the angle that the vector between neighboring molecules makes with the x-axis.

Hexatic order and the molecular tilt order of the smectic-C phase can coexist leading to smectic-F, -I, and -L phases depicted in Fig. 2.7.9. Further cooling leads to crystalline phases.

The breaking of rotational but not translational symmetry is reflected in the pair correlation function $g(\mathbf{x}, \mathbf{x}')$. Because translational symmetry is not broken, the density $\langle n(\mathbf{x})\rangle = \langle n\rangle$ is independent of $\mathbf{x}$, and $g(\mathbf{x}, \mathbf{x}') = g(\mathbf{x} - \mathbf{x}')$ is a function of the difference $\mathbf{y} = \mathbf{x} - \mathbf{x}' = (\mathbf{y}_\perp, y_\parallel)$. However, because rotational symmetry is broken, $g(\mathbf{y})$ is not a function of $y = |\mathbf{y}|$ only. Rather, it can be expanded in an angular Fourier series,

$$g(\mathbf{y}_\perp, y_\parallel) = g_0(y_\perp, y_\parallel)\sum_p \Psi_{6p}(y_\perp, y_\parallel)e^{6ip\theta}, \tag{2.7.4}$$

where $\Psi_0 = 1$ and $\mathbf{y}_\perp = (y_\perp, \theta)$ in polar coordinates. The parameters $\Psi_{6p}(y_\perp, y_\parallel)$ like S_{6p} [Eq. (2.7.3)] to which they are related measure the degree of hexatic order. When $y_\perp$ is equal to the average in-plane nearest neighbor separation and $y_\parallel = 0$, Ψ_{6p} is essentially $\langle e^{-i6p\theta_{\alpha,\beta}}\rangle$, where $\theta_{\alpha,\beta}$ is the angle the vector connecting nearest neighbor molecules α and β makes with the x-axis. A derivation of this result is outlined in Problem 2.8.

4 *Discotic phases*

The nematic and smectic phases just discussed are generally found in materials consisting of rod-like molecules. Plate-like rather than rod-like molecules such as those shown in Fig. 2.7.10 can form discotic nematic phases, in which the plate normals are aligned, as well as phases with crystalline order in two dimensions and liquid-like order in the third direction (Fig. 2.7.11). The latter are called *columnar*. The plate-like molecules segregate into columns with the plate normals either parallel to or tilted at an angle to the columnar axes. There is no long-range positional order within a column. The columns themselves form any of the two-dimensional crystals.

5 *Lyotropic liquid crystals and microemulsions*

In the preceding discussion, we have focused on liquid crystals whose phases change in response to changes in the temperature. They are called *thermotropic* liquid crystals. Liquid crystalline phases also appear in response to changes in concentrations of water, oil, surfactants, or other species in a wide variety of molecular mixtures. These are called *lyotropic* liquid crystals and are generally formed by amphiphilic molecules consisting of two parts that repel each other and/or are soluble in different solvents. The most widely studied of such systems are those composed of molecules called lipids with hydrophilic (water "liking") and hydrophobic (water "fearing") parts. The hydrophobic part consists of one

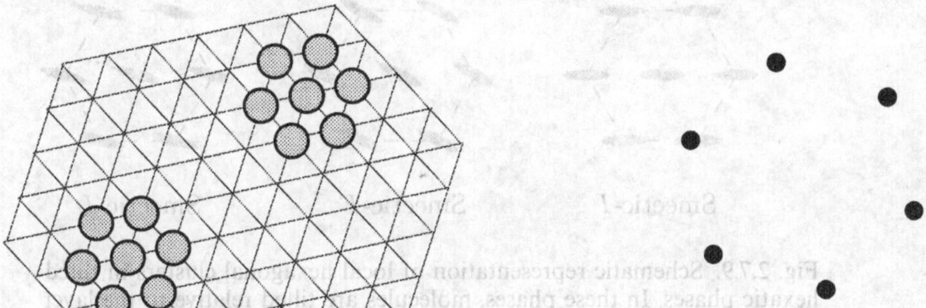

(a)

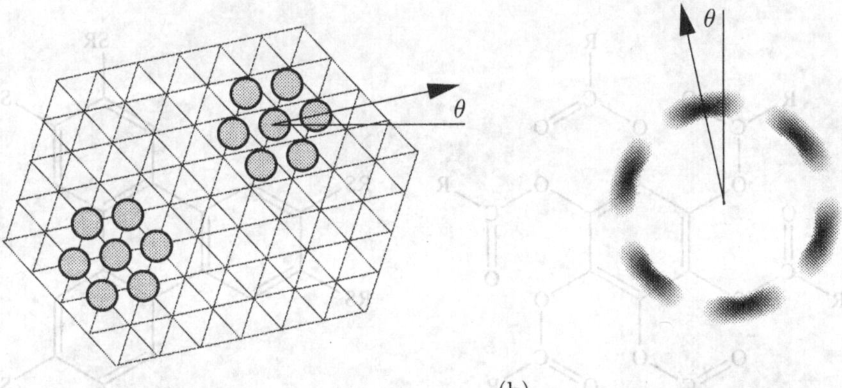

(b)

Fig. 2.7.8. (a) Separated groups of nearest neighbor atoms in a hexagonal crystalline phase. Atoms occupy lattice sites on a triangular lattice, indicated by grid lines. Each atom has six nearest neighbors forming a hexagon in orientational registry with the lattice. There is both long-range translational and orientational order. The figure at the side shows the hexagonal pattern of Bragg peaks in the X-ray scattering intensity at the shortest reciprocal lattice vectors. (b) Separated groups of atoms in a hexatic phase. Each atom has six nearest neighbors forming a local hexagon. Distant hexagons have the same orientation relative to some fixed axis. Atoms do not, however, occupy sites of a triangular lattice. There is long-range orientational but no long-range positional order. (While the orientational order would be easy to understand if the particles were on an oriented substrate or coupled with translational order, orientational order of spherical particles is remarkable in cases where neither a substrate nor crystalline order is present as in the case schematized above.) The figure at the side shows the X-ray scattering intensity from a hexatic phase. The intensity has a six-fold symmetry but no Bragg peaks. The scattering intensities are for three-dimensional systems. In two dimensions, fluctuations (as we shall see in Chapter 6) modify these intensities.

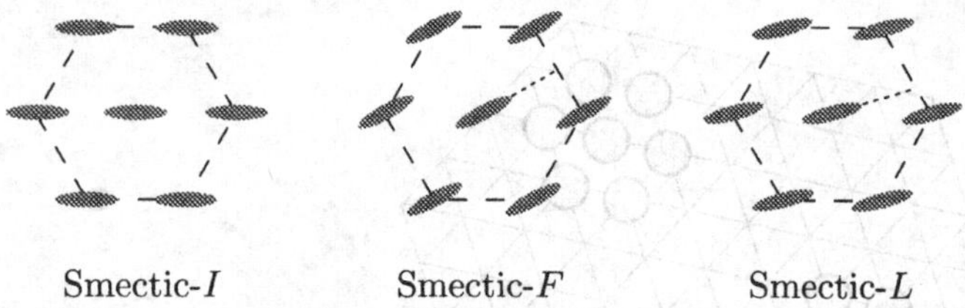

Fig. 2.7.9. Schematic representation of local hexagonal clusters in tilted hexatic phases. In these phases, molecules are tilted relative to the layer normals as in the smectic-*C* phase. Each molecule has six nearest neighbors (which in general occupy positions on a distorted rather than a regular hexagon). In the smectic-*F* phase, the **c**-director points to the midpoint of the bond connecting adjacent nearest neighbors (i.e., to next-nearest neighbors). In the smectic-*I* phase, it points to nearest neighbors, and in the smectic-*L* phase it points in a direction between next-nearest and nearest neighbors.

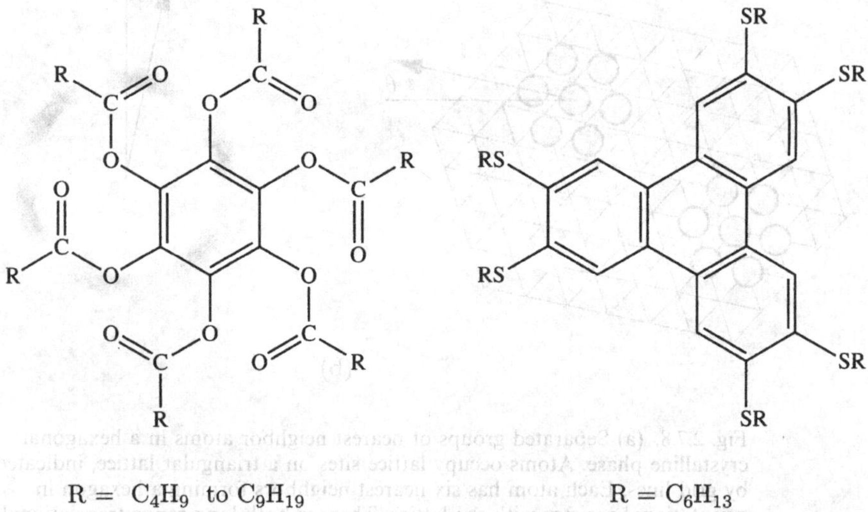

Fig. 2.7.10. Some plate-like molecules forming discotic liquid crystalline phases.

or two hydrocarbon chains containing 8 to 20 carbon atoms. The hydrophilic group generally has a charge or a dipole moment. Examples of lipids are shown in Fig. 2.7.12 (p. 72).

When in contact with water, lipids will self-organize into structures in which hydrophobic tails are shielded from contact with water. Common structures include spherical and cylindrical micelles, inverted micelles, bilayer sheets, and vesicles, as depicted in Fig. 2.7.13 (p. 73). The origin of these geometrical structures can often be understood in terms of the packing of lipids of different shapes. These structures can arrange into a bewildering array of equilibrium

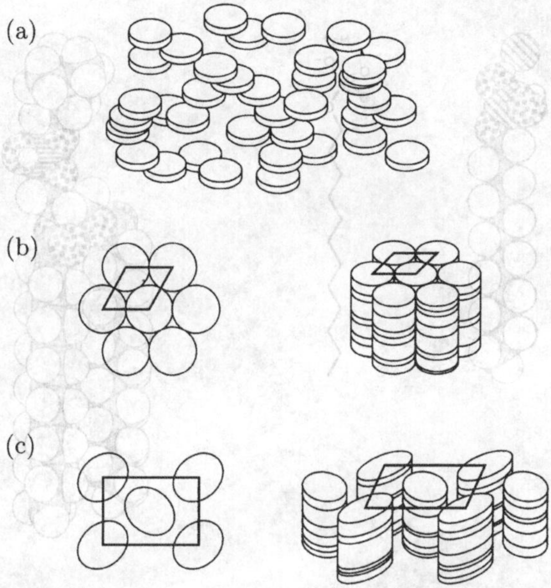

Fig. 2.7.11. Plate-like molecules in (a) a discotic nematic, (b) a hexagonal columnar discotic, and (c) a rectangular columnar discotic.

phases, including nematic, lamellar (smectic), and columnar phases. A phase diagram showing different equilibrium structures is shown in Fig. 2.7.14 (p. 74). An amusing example of a phase with three-dimensional crystalline symmetry (cubic) is the "plumber's nightmare" phase shown in Fig. 2.7.15 in which there is a single connected bilayer surface separating two *identical* water regions.

The hydrophobic hydrocarbon tails of amphiphiles are soluble in oil, whereas their hydrophilic charged heads dissolve in water. Amphiphiles can, therefore, be used to create equilibrium bulk mixtures of water and oil by providing monolayer interfaces between water and oil. These equilibrium mixtures of water, oil, and surfactant (the amphiphiles) are called *microemulsions.* They exhibit phases similar to those of water-lipid mixtures. A lamellar phase and a random bicontinuous phase are depicted schematically in Fig. 2.7.16 (p. 75).

2.8 One- and two-dimensional order in three-dimensional materials

The smectic-*A* and -*C* phases provide examples of order that is not complete – a type of one-dimensional ordering in a three-dimensional system. Another type of one-dimensional order, which is reciprocal to smectic order, is that of one-

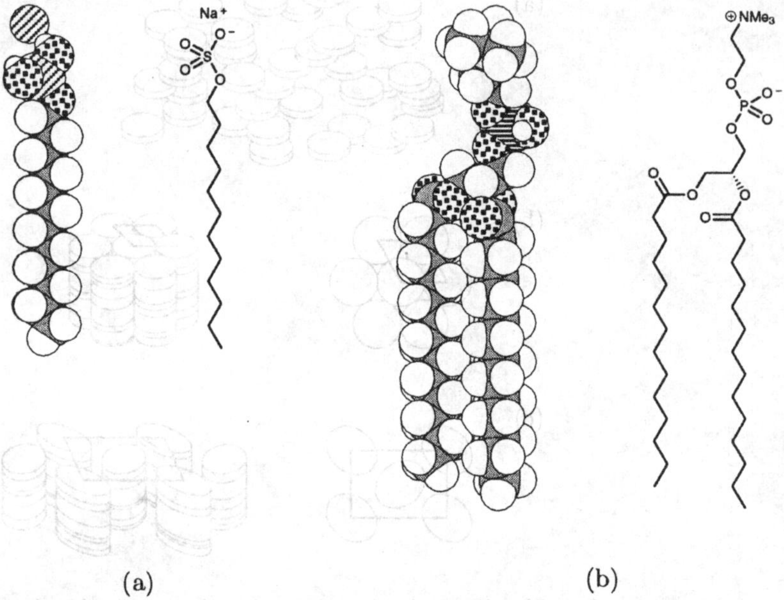

Fig. 2.7.12. Examples of lipids: (a) SDS, a soap with a single tail, and (b) DMPC, a phospholipid with two tails. Phospholipids are the primary constituents of cell walls.

dimensional chains. The scattering from a single chain consisting of periodically spaced mass points can be calculated from the density function

$$\langle n(\mathbf{x})\rangle = \rho_0\delta(x)\delta(y)\sum_p \delta(z-pc), \tag{2.8.1}$$

where p is an integer and c is the distance between plates. The Fourier transform of the density,

$$\begin{aligned}\langle n_{\mathbf{q}}\rangle &= \int \rho_0\delta(x)e^{-iq_x x}\delta(y)e^{-iq_y y}\sum_p \delta(z-pc)e^{-iq_z z}dxdydz \\ &= \rho_0\sum_p e^{-iq_z pc} = 2\pi(\rho_0/c)\sum_n \delta(q_z - 2\pi n/c),\end{aligned} \tag{2.8.2}$$

consists of a series of equally spaced delta-functions along the z-axis and is independent of q_x and q_y. The scattering intensity implied by Eq. (2.8.2) is distributed evenly on a set of parallel sheets in $\mathbf{q}$-space as shown in Fig. 2.8.1.

Now consider an array of chains each with $N_\parallel$ atoms arranged on a two-dimensional lattice with $N_\perp$ lattice vectors $\mathbf{R}_\mathbf{l}$. The density of atoms is then

$$n(\mathbf{x}) = \sum_{\mathbf{l},p} \delta^{(2)}(\mathbf{x}_\perp - \mathbf{R}_\mathbf{l})\delta(z - pc - \phi_\mathbf{l}), \tag{2.8.3}$$

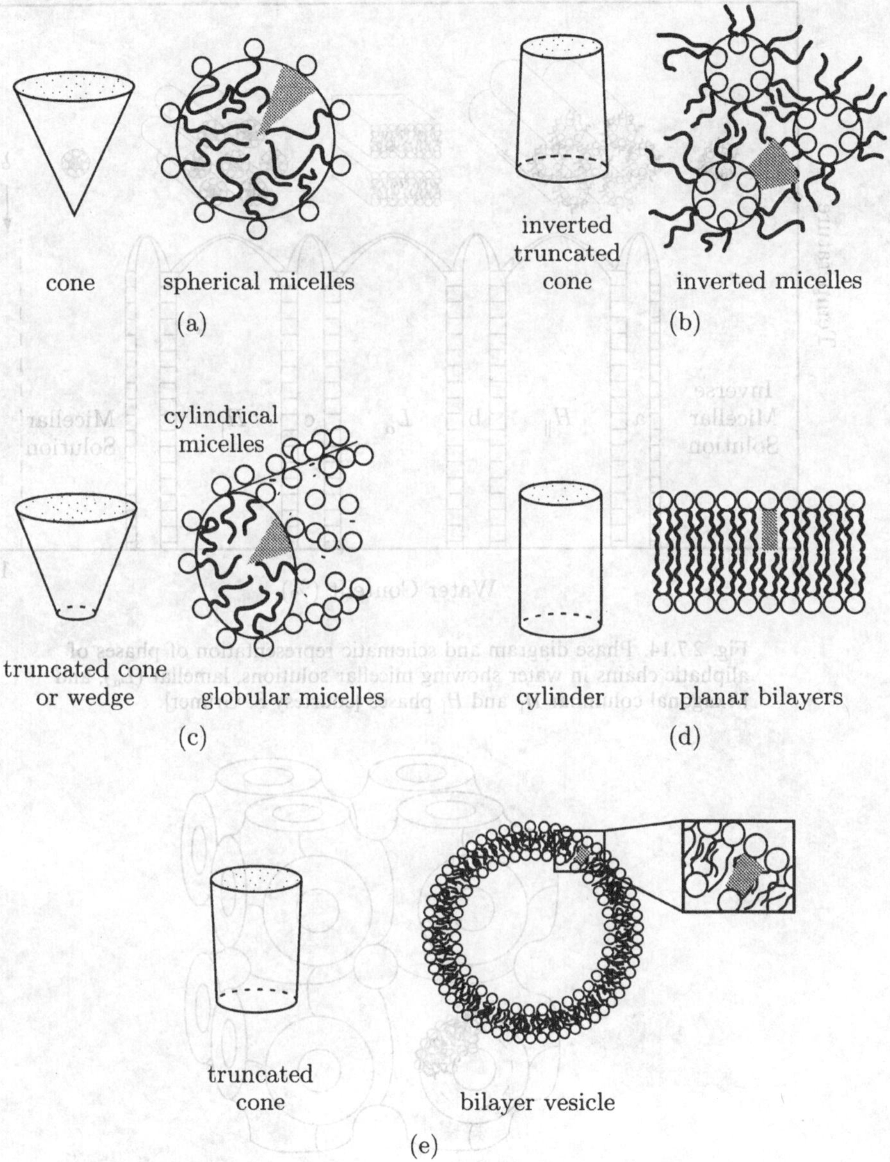

Fig. 2.7.13. Different local structures formed by lipids: (a) micelle, (b) inverted micelle, (c) cylindrical micelle, (d) flat bilayer, and (e) closed vesicle. Average shapes of the lipid molecules favoring the various structures are also shown. Note that asymmetric shapes favor nonzero curvature. [Adapted from J.N. Israellachvilli, *Physics of Amphiphiles: Micelles, Vesicles and Microemulsions* (North Holland, New York, 1985).]

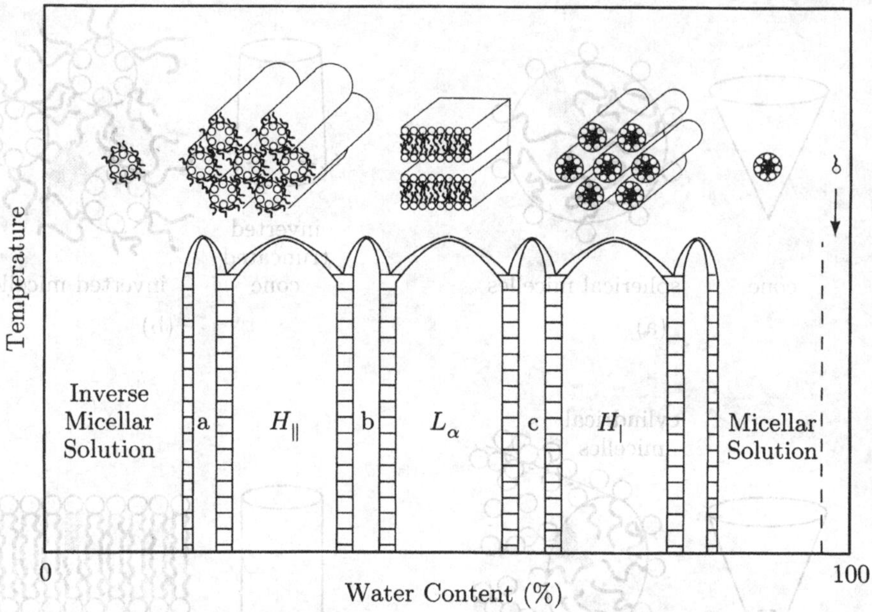

Fig. 2.7.14. Phase diagram and schematic representation of phases of aliphatic chains in water showing micellar solutions, lamellar (L_α), and hexagonal columnar $H_{\|}$ and $H_|$ phases [courtesy S. Gruner].

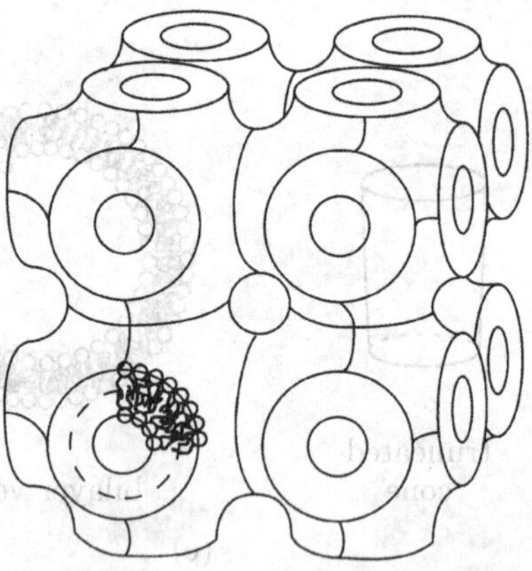

Fig. 2.7.15. Schematic representation of a surfactant surface in the triply periodic "plumber's nightmare" phase [D. M. Anderson, S. M. Gruner, and S. Leibler, *Proc. Acad. Sci. USA* **85**, 5364 (1988)]. This phase has the symmetry of a periodic crystal.

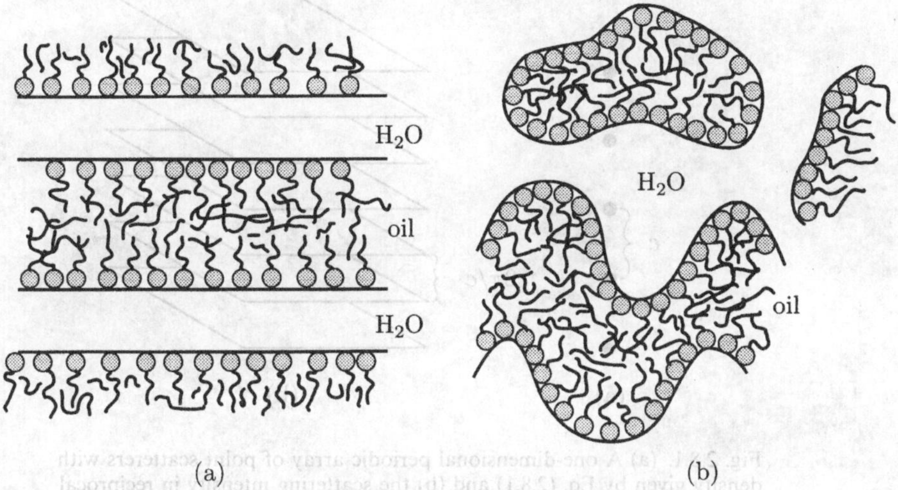

Fig. 2.7.16. (a) A lamellar microemulsion phase with water, oil, and surfactant layers. (b) Schematic representation of a random bicontinuous phase in which there is a random surfactant surface separating oil and water regions.

where $\mathbf{x} = (\mathbf{x}_\perp, z)$ and the phase $\phi_\mathbf{l}$ specifies the shift of the origin of the periodically spaced masses on chain $\mathbf{l}$. The scattering intensity is

$$I(\mathbf{q}) = \sum_{\mathbf{l},\mathbf{l}',p,p'} e^{-iq_z c(p-p')} e^{-i\mathbf{q}_\perp\cdot(\mathbf{R}_\mathbf{l}-\mathbf{R}_{\mathbf{l}'})} [e^{-iq_z(\phi_\mathbf{l}-\phi_{\mathbf{l}'})}], \tag{2.8.4}$$

where [ ] signifies an average over the variables $\phi_\mathbf{l}$. If neighboring chains are uncorrelated, then $\phi_\mathbf{l}$ will be an independent random variable at each $\mathbf{l}$, and

$$[e^{-iq_z(\phi_\mathbf{l}-\phi_{\mathbf{l}'})}] = \begin{cases} \delta_{\mathbf{l}\mathbf{l}'} & \text{if } q_z \neq 0; \\ 1 & \text{if } q_z = 0. \end{cases} \tag{2.8.5}$$

The sum in Eq. (2.8.4) can now be evaluated,

$$I(\mathbf{q}) = \begin{cases} N_\perp^2 N_\parallel^2 \sum_\mathbf{G} \delta_{\mathbf{q},\mathbf{G}} & \text{if } q_z = 0; \\ N_\perp N_\parallel^2 \sum_n \delta_{q_z, 2\pi n/c} & \text{if } q_z \neq 0. \end{cases} \tag{2.8.6}$$

Thus, there are real Bragg peaks in $I(\mathbf{q})$ with intensity $N_\perp^2 N_\parallel^2$ in the plane $q_z = 0$ that reflect the two-dimensional lattice of chains. In addition, there are Bragg sheets with intensity $N_\perp N_\parallel^2$ independent of $q_\perp$ at $q_z = 2\pi n/c$ (n is an integer) reflecting the periodic order along each chain. Away from $q_z = 0$ the uncorrelated chains look just like the single one-dimensional chain but with a larger intensity. As correlations between chains build up, the phases $\phi_\mathbf{l}$ cannot be treated as independent random variables on each chain, and the average in Eq. (2.8.5) develops a nonzero value for $\mathbf{l} \neq \mathbf{l}'$ even when $q_z \neq 0$. These correlations lead

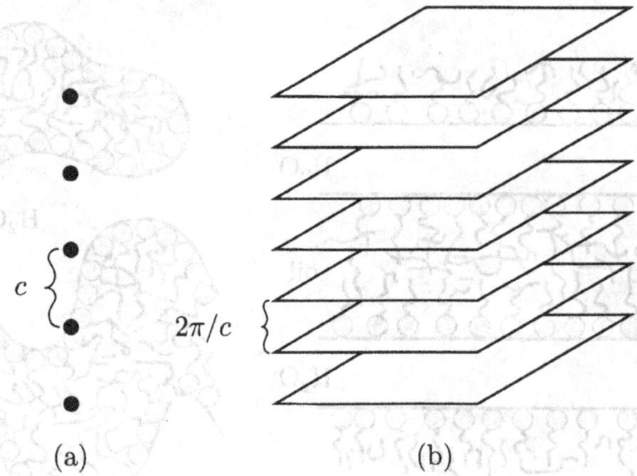

Fig. 2.8.1. (a) A one-dimensional periodic array of point scatterers with density given by Eq. (2.8.1) and (b) the scattering intensity in reciprocal space with delta-function intensity on sheets perpendicular to the *c*-axis.

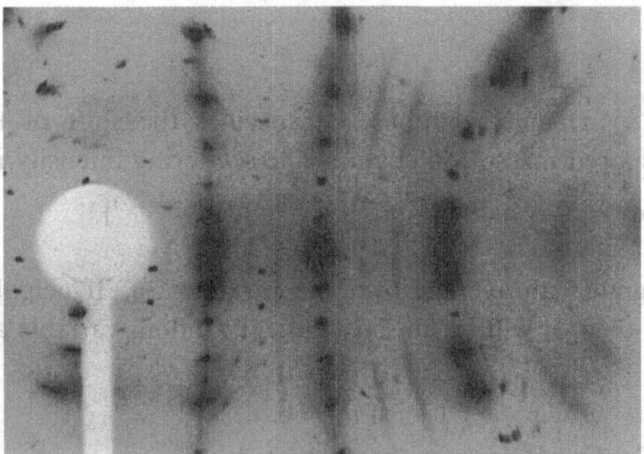

Fig. 2.8.2. X-ray diffuse scattering pattern from the organic chain salt MNTSF-TCNQ showing one-dimensional order. The diffuse lines between the lines of Bragg peaks arise from one-dimensional order. The modulation in the intensity of these lines arises from correlations between one-dimensional chains. [Courtesy of J. Pouget.]

to diffuse spots of intensity in the planes of the Bragg sheets that eventually develop into true Bragg peaks when long-range phase coherence between chains is established.

The above discussion assumes that each chain does indeed have ideal periodic order. When the chains are completely decoupled, however, they are effectively

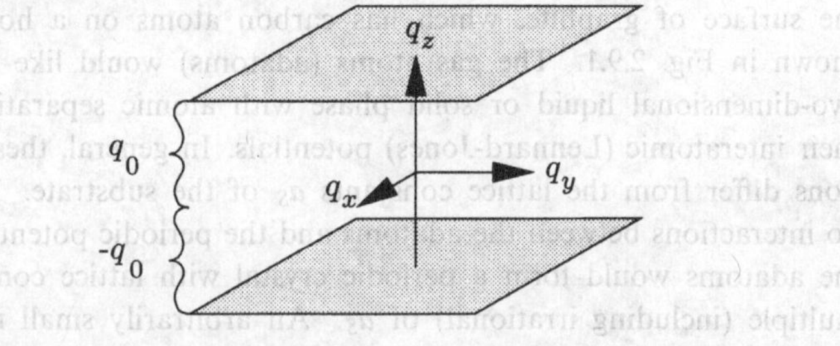

Fig. 2.8.3. The scattering intensity from a one-dimensional sine-wave density confined to the z-axis [Eq. (2.8.7)]. There are two Bragg planes at $q_z = \pm q_0$ but no Bragg plane at $q_z = 0$.

one-dimensional solids. We will see in Chapter 6 that there can be no long-range order in a one-dimensional system (with finite range forces). Fluctuations will convert the ideal Bragg sheets into diffuse sheets in reciprocal space. Since many scattering techniques look at a plane in **q**-space, the intersection of this plane with the diffuse sheets above gives diffuse lines. Correlations between chains will cause variations in the intensity of the diffuse lines, which, as discussed above, will become Bragg spots when long-range correlation is established. The development from diffuse streaks to Bragg spots is illustrated for an organic conductor MNTSF-TCNQ in Fig. 2.8.2. The one-dimensional ordering is a density wave along individual chains in the crystal. The coupling between the density waves eventually leads to their three-dimensional order. For the case of a simple one-dimensional sine wave there is only a single pair of diffuse sheets, as shown in Fig. 2.8.3:

$$\begin{aligned} \langle n(\mathbf{x}) \rangle &= \rho_0 \delta(x) \delta(y) \sin(q_0 z), \\ \langle n_{\mathbf{q}} \rangle &= (\rho_0 / c)[\delta(q_z + q_0) - \delta(q_z - q_0)]/2. \end{aligned} \tag{2.8.7}$$

Note that a smectic with its real space layers produces a periodic chain in reciprocal space in the same way that a real space chain produces layers in reciprocal space. Also, just as for a real smectic, where only the first harmonic is present and only two spots appear in the scattering function, a single harmonic wave in one dimension produces only two diffuse sheets.

2.9 Incommensurate structures

There are structures in nature that are neither random nor periodic but that exhibit spatial modulations with two or more relatively irrational periods. Such structures are called *quasi-periodic*. They usually result from competition between two different length scales.

An example of a system with two competing length scales producing incommensurate structures is that of noble gas (Xe, Kr, etc.) atoms adsorbed on the surface of graphite, which has carbon atoms on a honeycomb lattice as shown in Fig. 2.9.1. The gas atoms (adatoms) would like to condense into a two-dimensional liquid or solid phase with atomic separations determined by their interatomic (Lennard-Jones) potentials. In general, these preferred separations differ from the lattice constants a_S of the substrate. Thus, if there were no interactions between the adatoms and the periodic potential of the substrate, the adatoms would form a periodic crystal with lattice constants an arbitrary multiple (including irrational) of a_S. An arbitrarily small interaction between atoms and the periodic substrate potential will introduce six-fold anisotropy into adatom correlations, converting the adatom liquid phase to a hexatic-like phase and causing the adatom crystal phase to align with the substrate.

If the periodic substrate potential is strong, adatoms will sit at particular sites in the graphite lattice and form lattices containing an integral number of graphite unit cells. These are called *commensurate* lattices. In the case of krypton, the hard-sphere radius is such that one atom can occupy every third graphite hexagon as shown in Fig. 2.9.1. Since the adatom lattice is rotated by 30° relative to the substrate and has a lattice constant $\sqrt{3}a_S$, it is called a $\sqrt{3} \times \sqrt{3}$R30° structure.

In the opposite limit, when the periodic substrate potential is not too large, the adatom lattice will differ only slightly from its ideal form, with modulations in atomic positions determined by the period of the graphite lattice. In this case, which can be interpreted as a limit of commensurate lattices in which the number Q of graphite unit cells per adatom unit cell tends to infinity, the adatoms are said to form an *incommensurate* structure. Fig. 2.9.2 is a pressure-temperature phase diagram for Kr on graphite, showing fluid (F), commensurate solid (C), and incommensurate solid (IC) phases.

Incommensurate structures are most easily visualized in one dimension. Suppose we have a one-dimensional metal with atoms spaced periodically with separation a. The one-dimensional metal has an instability toward forming a charge-density insulator at low temperatures. Instead of having a homogeneous charge density, the system develops a spatial modulation in the electron density which, because of coupling to the lattice, induces a slight modulation of atomic positions. The deviation of the electron charge density from its average spatially uniform value is a periodic function of position that is well-approximated by a single cosine,

$$\delta\rho(\mathbf{x}) = \rho_1 \cos(2\pi x/\lambda), \tag{2.9.1}$$

with a periodicity λ. If a/λ is a rational number P/Q, where P and Q are relatively prime integers, a new unit cell can be formed from Q atoms of the original linear chain, and the structure is commensurate. This case is illustrated in Fig. 2.9.3 for $\lambda = 3.5a$, $P/Q = 2/7$, and $Q = 7$. If a/λ is irrational, the modulation is incommensurate. Any irrational number can be approached as a sequence of rational numbers of the form P/Q in the limit that $Q \to \infty$.

One of the first materials (Tanisaki 1961) in which an incommensurate phase

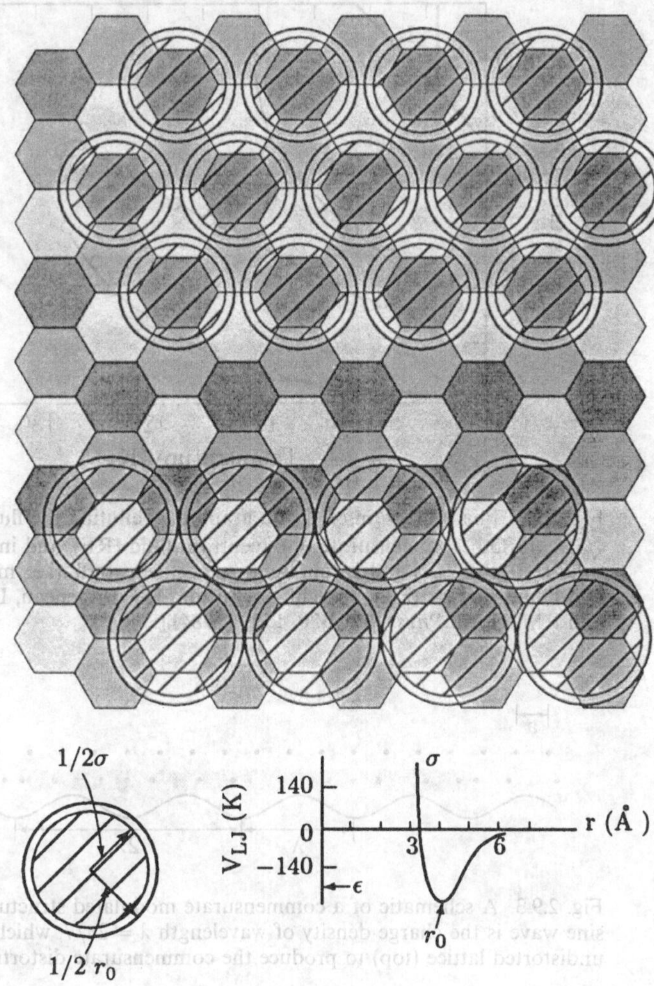

Fig. 2.9.1. Krypton (top) and xenon (bottom) on a graphite lattice. The size of krypton is such that it can occupy every third hexagon of the graphite lattice to produce a $\sqrt{3} \times \sqrt{3}R30°$ commensurate lattice. Xe is larger than Kr, and the Kr commensurate structure is prohibited. An incommensurate phase is often preferred. [R.J. Birgeneau and P. M. Horn, *Science* **232**, 329 (1986).]

was observed was $NaNO_2$. This crystal has a low temperature ($T < T_f = 162.5°C$) ferroelectric phase with a body-centered orthorhombic lattice. The V-shaped NO_2^- groups lie in the two-dimensional (b, c) plane with their axes aligned along the b-axis as shown in Fig. 2.9.4. At high temperatures ($T > T_i = 164°C$), the directions of the NO_2^- groups are randomly oriented. At intermediate temperatures the component of the NO_2^- dipole along the b-axis is a periodic function of position

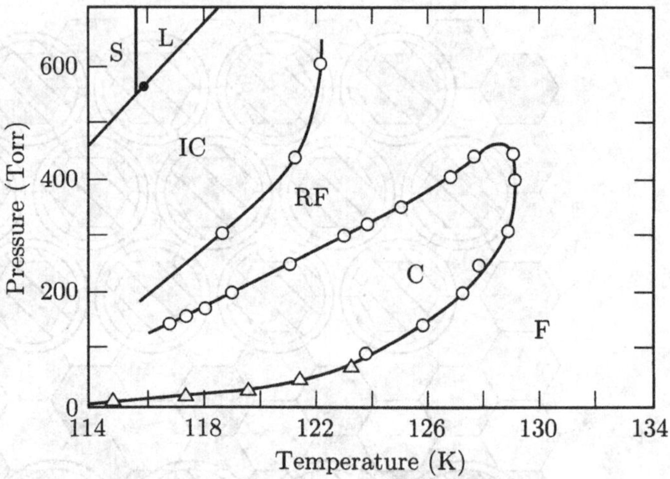

Fig. 2.9.2. Phase diagram of Kr on graphite exhibiting the fluid (F), $\sqrt{3} \times \sqrt{3}R30°$ commensurate (C), reentrant fluid (RF), and incommensurate solid (IC) phases. S and L represent, respectively, bulk (i.e., multilayer) solid and liquid phases. [E.D. Sprecht, M. Sutton, R.J. Birgeneau, D.E. Moncton, and P.M. Horn, *Phys. Rev.* **B30**, 1589 (1984).]

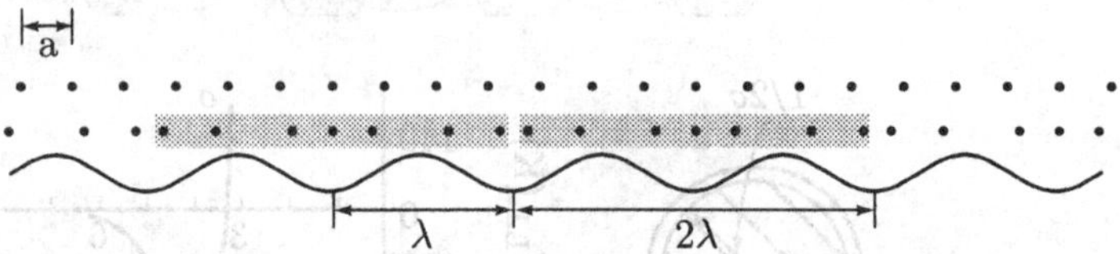

Fig. 2.9.3. A schematic of a commensurate modulated structure. The solid sine wave is the charge density of wavelength $\lambda = 2a/7$, which couples to the undistorted lattice (top) to produce the commensurate distortion (middle).

such as that of Eq. (2.9.1) with λ a continuous function of position for $T_f < T < T_i$. This periodic modulation gives rise to satellite peaks in the X-ray scattering at wave vectors $\mathbf{q} = \mathbf{G} \pm (2\pi/\lambda)\mathbf{b}$, where $\mathbf{G}$ is a vector of the undistorted orthorhombic reciprocal lattice. Another example of an incommensurate phase and subsequent commensurate phase at lower temperature occurs in the two-dimensional charge-density wave (CDW) system 1T-$TaSe_2$ whose high and low temperature diffraction patterns are shown in Fig. 2.9.5.

As we have seen, the signature of an incommensurate crystal is the appearance of satellite peaks in the X-ray scattering intensity at irrational multiples of reciprocal lattice vectors of some underlying crystal. In general, if the incommensurate modulation is strong, there will be many satellite peaks at positions [e.g. at

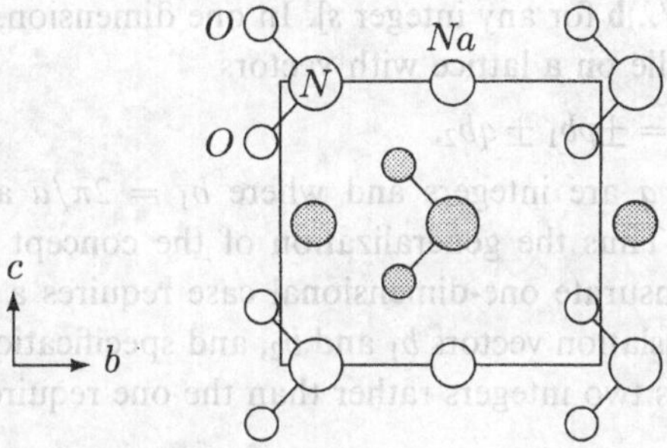

Fig. 2.9.4. The (b, c) plane of the low-temperature ferroelectric phase of $NaNO_2$.

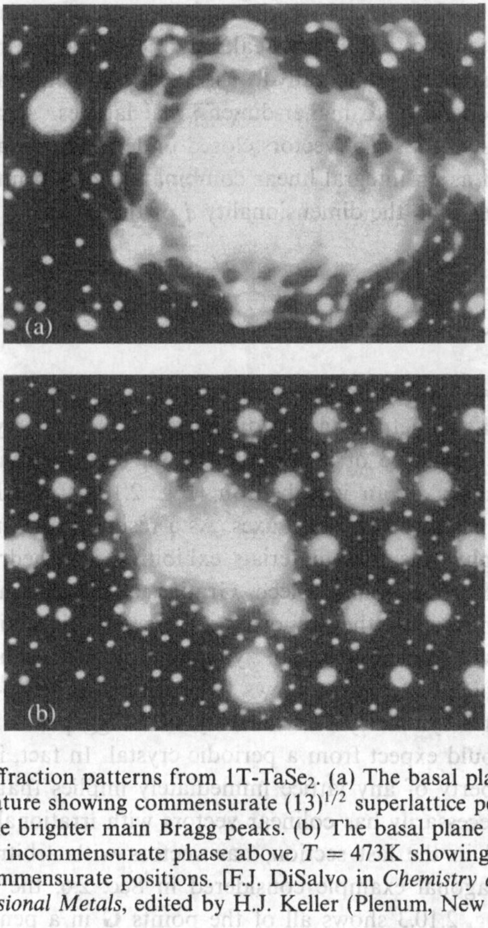

Fig. 2.9.5. Diffraction patterns from 1T-$TaSe_2$. (a) The basal plane pattern at room temperature showing commensurate $(13)^{1/2}$ superlattice peaks in addition to the brighter main Bragg peaks. (b) The basal plane diffraction pattern in the incommensurate phase above $T = 473$K showing satellite peaks at incommensurate positions. [F.J. DiSalvo in *Chemistry and Physics of One-Dimensional Metals*, edited by H.J. Keller (Plenum, New York, 1977).]

$\mathbf{q} = \mathbf{G} \pm s(2\pi/\lambda)\mathbf{b}$ for any integer s]. In one dimension, the peaks in the scattering intensity will lie on a lattice with vectors

$$G = \pm p b_1 \pm q b_2, \tag{2.9.2}$$

where p and q are integers and where $b_1 = 2\pi/a$ and $b_2 = 2\pi/\lambda$, where a/λ is irrational. Thus the generalization of the concept of the reciprocal lattice to the incommensurate one-dimensional case requires a reciprocal lattice with two primitive translation vectors b_1 and b_2, and specification of the positions of Bragg peaks requires two integers rather than the one required for a periodic reciprocal lattice.

An important property of the incommensurate lattice defined by Eq. (2.9.2) is that it contains vectors of arbitrarily small magnitude since for irrational b_1/b_2, it is always possible to find integers p and q such that $|pb_1 - qb_2|$ is less than any preassigned number. This means that the vectors of an incommensurate reciprocal lattice form a dense set in reciprocal space. There is large variation in the scattering intensities into Bragg peaks at different points in the reciprocal lattice so that it is possible experimentally to observe incommensurate structures. The above properties generalize to higher-dimensional lattices. A general incommensurate lattice consists of a set of vectors closed under addition and subtraction that can be expressed as an integral linear combination of r primitive translation vectors with r greater than the dimensionality d of the lattice.

2.10 Quasicrystals

In Sec. 2.6, we showed that five-fold rotational symmetry is incompatible with the translational symmetry of a two-dimensional periodic crystal. Similar arguments rule out the existence of a periodic crystal in three dimensions with the point group symmetry of an icosahedron (Fig. 2.10.1), which has six five-fold, ten three-fold, and fifteen two-fold axes. As a result, a fundamental tenet of classical crystallography was that materials exhibiting icosahedral symmetry could not exist. In 1984, Shechtman, Blech, Gratias, and Cahn shook the foundations of crystallography when they reported an electron diffraction pattern for an alloy of Al-Mn with the point-group symmetry of an icosahedron. This diffraction pattern (Fig. 2.10.2) clearly shows five-, three-, and two-fold axes characteristic of icosahedral symmetry. The density of Bragg peaks in each plane is higher than one would expect from a periodic crystal. In fact, it is easy to see that the closure property of any lattice immediately implies that a lattice with five-fold symmetry necessarily has colinear vectors with irrational magnitude ratios, and, as discussed in the last section, has vectors with arbitrarily small separations. In the pentagonal example considered in Sec. 2.6, the vector $\mathbf{a}_4 + \mathbf{a}_1$ is equal to $\tau^{-1}\mathbf{a}_0$. Fig. 2.10.3 shows all of the points $\mathbf{G}$ in a pentagonal lattice that can be expressed as $\sum A_n \mathbf{a}_n$, where $\mathbf{a}_n$ is one of the five vectors pointing to the

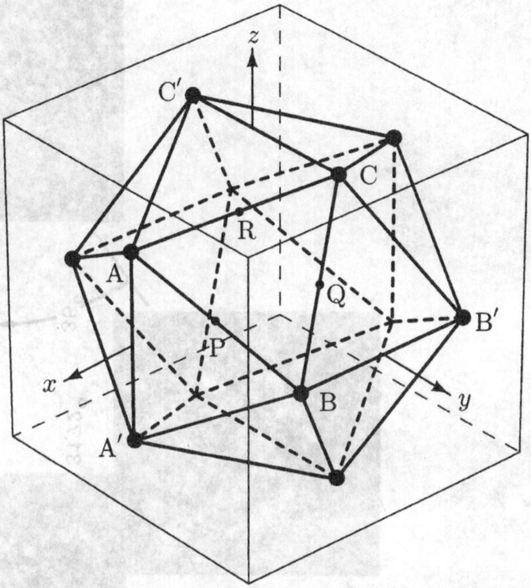

Fig. 2.10.1. An icosahedron showing its 12 vertices, 20 triangular faces, and 30 edges. The five-fold axes pass through the vertices, the three-fold axes through the centers of the faces, and the two-fold axes through the centers of the edges. The icosahedron with edges of unit length is inscribed in a cube. Its edges AA′, BB′, and CC′ are, respectively, parallel to the z-, x-, and, y-axes of the cube. The coordinates of A, A′, B, B′, C, and C′ are, respectively, $(\tau, 0, \pm 1)$, $(\pm 1, \tau, 0)$, and $(0, \pm 1, \tau)$, where $\tau = (1 + \sqrt{5})/2$ is the golden mean. The coordinates of P, Q and R are, respectively, $(1+\tau,\ \tau, 1)/2$, $(1, 1+\tau, \tau)/2$, and $(\tau, 1, 1+\tau)/2$.

vertices of a pentagon [Eq. (2.6.1)] and $A_n = 0, \pm 1, \pm 2$ and $|\mathbf{G}| \leq 4|\mathbf{a}_0|$. The area of each point is proportional to $|\mathbf{G}_\perp|^{-1}$, where $\mathbf{G}_\perp = \sum A_n \mathbf{a}_n^\perp$ and where $\mathbf{a}_n^\perp = [\cos 4\pi(n-1)/5, \sin 4\pi(n-1)/5]$. Note the similarity between this pattern and the experimental pattern for Al-Mn in the five-fold plane.

Pentagonal and icosahedral reciprocal lattices are incommensurate lattices. Unlike the examples considered in the last section, however, the irrational ratio of lengths is determined by the point group symmetry. Levine and Steinhardt (1984) introduced the term *quasicrystal* for this special type of incommensurate structure.

Though arbitrarily short vectors are permitted in a reciprocal lattice, atoms in real space cannot be arbitrarily close together. In a periodic solid, the existence of a shortest length in the primitive unit cell of the direct lattice ensures that distances between atoms are greater than some minimum distance. How can there be a minimum distance in a quasicrystal? The answer is provided by tilings of a two-dimensional plane with five-fold symmetry invented by R. Penrose (1974) and

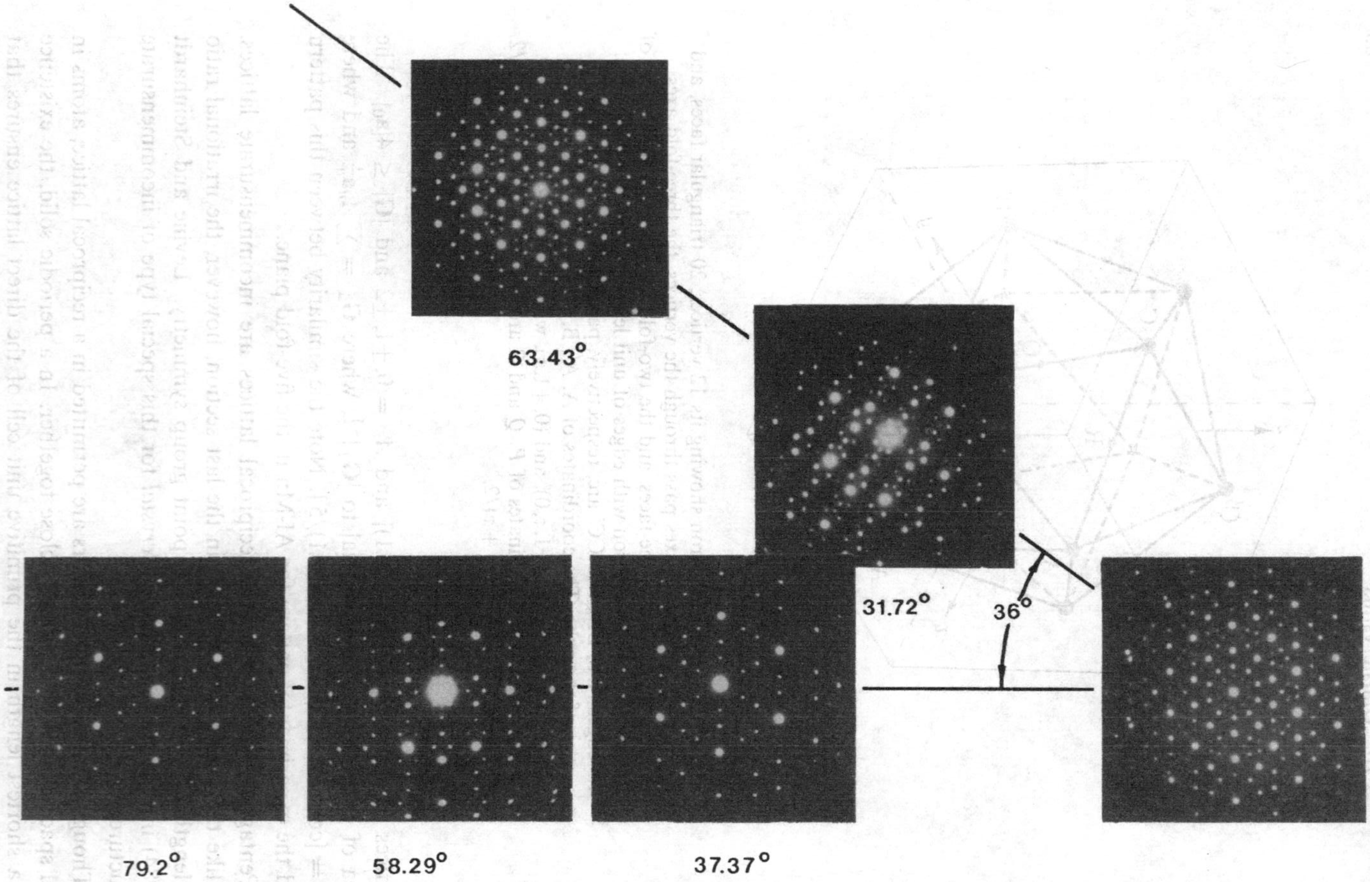

Fig. 2.10.2. Experimental diffraction from icosahedral $AlMn_6$ [D.S. Shechtman, I. Blech, D. Gratias, and J. W. Cahn, *Phys. Rev. Lett*. **53**, 1951 (1984)].

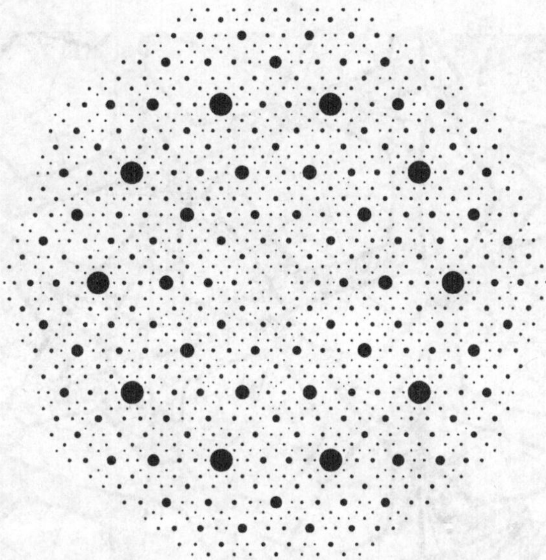

Fig. 2.10.3. Points **G** in a reciprocal lattice with ten-fold symmetry (the presence of the negative of vectors converts five-fold to ten-fold symmetry) generated by $\sum A_n \mathbf{a}_n$ with $A_n = 0, \pm 1, \pm 2$ with $|\mathbf{G}|/|\mathbf{a}_0| \leq 4$.

by their generalizations to icosahedral symmetry in three dimensions. Fig. 2.10.4 shows a Penrose tiling of a plane. The entire plane is filled with two types of tiles or unit cells rather than the single unit cell required to tile a plane periodically. The two types of tiles are called "fat" and "skinny". Each tile is decorated so that its different sides are distinguished, and adjacent tiles must be joined so that they obey certain matching rules. There is a shortest distance between tile vertices, and it is clearly possible to decorate the tiles with atoms in such a way that the atoms have a minimum separation. The diffraction pattern, first calculated by Levine and Steinhardt (1984), of the icosahedral generalization of Penrose tiles agrees well with the experimentally observed pattern.

2.11 Magnetic order

Spins in magnetic systems can exhibit ordered phases whose variety rivals that of the crystalline and liquid crystalline phases we have considered thus far. Magnetic systems have played a very important role in the development of our understanding of broken symmetry and how it arises, largely because they can be described with very simple models, which we will consider in some detail in succeeding chapters.

Often the spin degrees of freedom decouple from the electron charge density.

Fig. 2.10.4. Portion of Penrose tiling showing matching rules. Matching rules are enforced by the decoration shown on the tiles to the right. The rules are that the dark and dotted portions of one tile must join with the same portions of adjacent tiles. [Courtesy of Paul Steinhardt.]

Imagine that we have a crystalline solid where the atoms or molecules have unfilled shells and unpaired spins. In the case of transition metals or the lanthanide or actinide elements, these electrons may be *d* or *f* electrons, which are tightly bound to the nuclei, have small radii, and may be thought of as localized on atomic sites with little overlap with electrons on neighboring atoms. Because of intra-atomic electron Coulomb repulsion and spin-orbit coupling, the total spin, S, on a given atom may be $1/2$ or larger. Thus a good model for many magnetic systems is one in which spins localized at sites on a periodic lattice interact only among themselves via an interaction of the form of the exchange interaction discussed in Sec. 1.3.5. The only degrees of freedom in such a model are those associated with the spin. The energy scales for spin-interactions can be quite different from those for inter-atomic forces, and phase transitions in spin degrees of freedom can occur without significant changes in the crystal lattice.

At high temperature, spins are thermally disordered: they have no long-range orientational order. Because an external magnetic field **H** leads to a partial alignment of spins (via the $-\mu \mathbf{S} \cdot \mathbf{H}$ energy term, where μ is the gyromagnetic ratio) and a magnetization proportional to and parallel to **H**, this disordered phase is called the *paramagnetic phase.* As temperature is lowered, interactions among spins lead to ordered structures such as those shown in Fig. 2.11.1. If only nearest neighbor spins interact, and they prefer to be parallel (as they do if the exchange J (Eq. 1.3.29) is positive), then a *ferromagnetic phase* results. If neighboring spins prefer to be antiparallel, then *antiferromagnetic phases* with spins on different lattice sites alternating in sign results. The detailed form of antiferromagnetic order depends on the crystal lattice. On lattices, such as the square lattice in two dimensions and the BCC lattice in three dimensions, that can be decomposed into two sublattices (with all nearest neighbors of one lattice on the other – "alternate lattices"), the state will consist of up spins on one sublattice and down spins on the other as shown in Figs. 2.11.2a and c. On the FCC lattice, which cannot be decomposed into two sublattices, there is antiferromagnetic order with spins parallel in planes normal to the (111) axis, but alternating in sign from plane to plane as shown in Fig. 2.11.2d. Note that the development of antiferromagnetic order leads to an increase of the size of the magnetic unit cell and, therefore, to new magnetic Bragg superlattice peaks and a reduced size of the unit cell in reciprocal space as shown for the square lattice in Fig. 2.11.2b. The unit cell size is doubled on both the square and BCC lattices.

Various other ordered spin states are depicted in Fig. 2.11.1. These include ferrimagnetic states in which not only the sign but also the magnitude of the spin alternates from site to site, canted and fan states with both ferromagnetic and antiferromagnetic order, and helical states in which the spins precess in a helical fashion like the director in a cholesteric. Canted and fan phases usually result from a competition between local crystal fields and nearest neighbor interactions, whereas helical phases usually result from the competition between nearest neighbor and next nearest neighbor interactions.

Because neutrons have a spin and magnetic moment, they interact with the electron spin density (via the magnetic dipole interaction) as well as with the atomic nuclei. Thus neutron scattering provides information about magnetic order. The magnetic scattering cross-section is proportional to p^2, where the magnetic scattering length p is given by

$$p \propto \frac{\gamma e^2}{mc^2} \mathbf{S}_N \cdot [\mathbf{M}(\mathbf{q}) - \mathbf{q}(\mathbf{M}(\mathbf{q}) \cdot \mathbf{q})], \tag{2.11.1}$$

where $\mathbf{S}_N$ is the neutron spin, $\mathbf{M}(\mathbf{q})$ is the transform of the magnetization density, and γ is the neutron moment. In conventional neutron scattering the nucleus looks effectively like a point. In magnetic neutron scattering, there is a magnetic form factor that can give information about the spin density in a unit cell in much the same way that the X-ray scattering form factor gives us information about the charge density in a unit cell.

The magnetic scattering length p is usually much smaller than the nuclear scattering length b, and it is difficult to see the magnetic scattering from a crystal

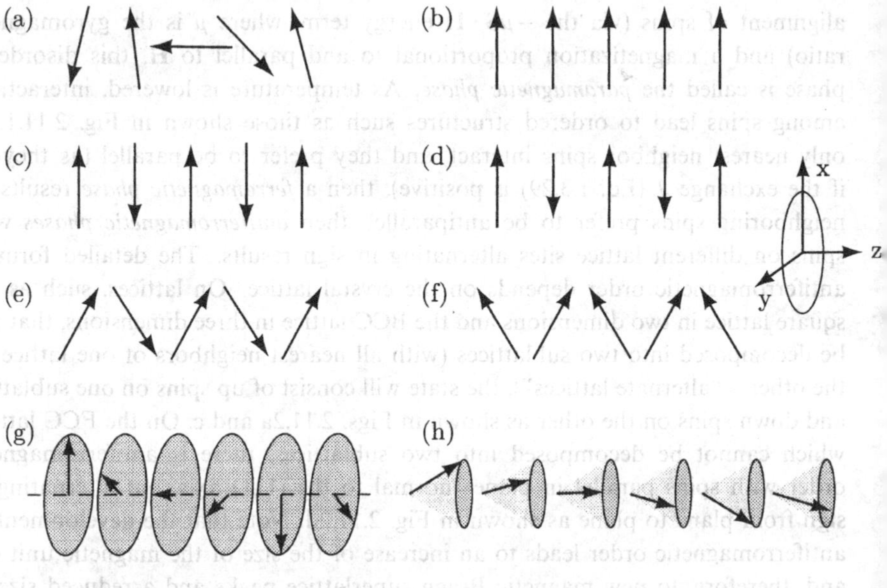

Fig. 2.11.1. Schematic representation of spins in (a) a paramagnetic, (b) a ferromagnetic, (c) an antiferromagnetic, (d) a ferrimagnetic, (e) a canted, (f) a fan, and (g) and (h) two helical phases. With the pitch axis parallel to the z-axis, the spins in the helical phases can be represented as $\mathbf{S}(\mathbf{x}) = (S_\perp \cos k_0 z, S_\perp \sin k_0 z, S_z)$ where k_0 is the twist wave vector. In (g) there is no spin component S_z along the pitch axis, whereas in (h) there is.

unless the magnetic unit cell (and hence magnetic reciprocal lattice vectors) differs from the chemical cell. However, in certain scattering geometries (when the polarization of the neutrons is perpendicular to the scattering plane and parallel or antiparallel to the sample magnetizations), the total scattering cross-section is proportional to $(b - p)^2$. Then by changing the polarization it is possible to separate the magnetic and nuclear scattering. An example of neutron scattering intensities from MnO showing the appearance of superlattice Bragg peaks indicative of antiferromagnetic order in shown in Fig. 2.11.3. These peaks are visible because the magnetic unit cell differs from the chemical unit cell, and there are magnetic Bragg peaks at positions where there are no nonmagnetic Bragg peaks.

Photons can also scatter from the magnetic moment of the electron. This is a relativistic effect which is smaller than conventional electron-photon (Compton) scattering by a factor of $(\hbar\omega/mc^2)^2$ where $\hbar\omega$ is the energy of the incident photon (X-ray) and mc^2 is the electron rest mass energy. Although the intensity of the magnetic scattering is down by approximately four orders of magnitude from charge scattering, the availability of intense photon sources from synchrotron radiation makes such experiments possible. In Fig. 2.11.4 we

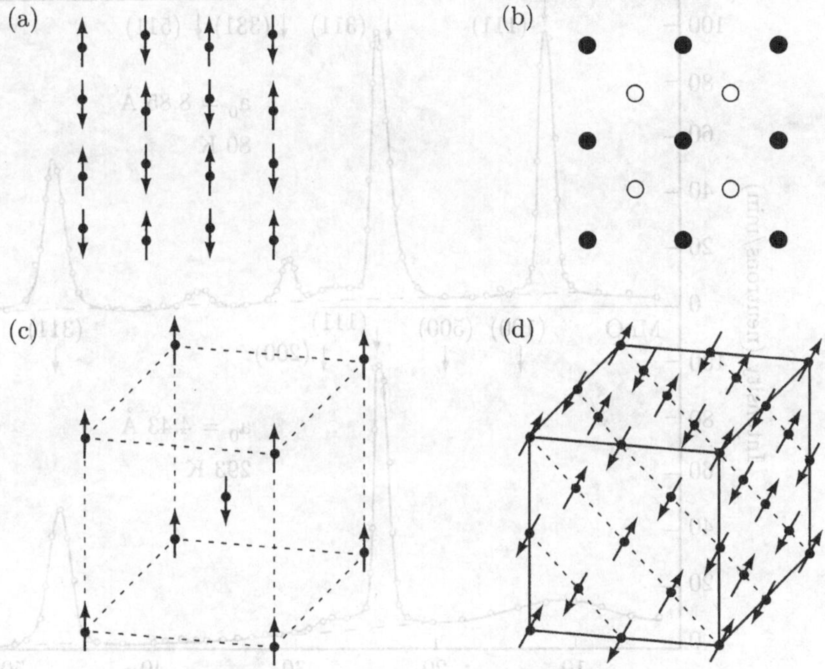

Fig. 2.11.2. (a) Antiferromagnetic order on a square lattice showing the doubling of the unit cell size. (b) Bragg peaks for antiferromagnetic order on a square lattice. The solid circles indicate Bragg peaks from the chemical unit cell. The open circles indicate magnetic superlattice peaks. (c) Antiferromagnetic order on a BCC lattice. (d) Antiferromagnetic order on an FCC lattice. The atoms shown correspond to the Mn^{2+} ions in MnO.

show recent experiments on the helical magnetic phase of the rare earth element holmium. The magnetic Bragg peaks from the twist wave vector are seen to vary with temperature as the sample is cooled below 130 K in Fig. 2.11.4a. This is a particularly interesting case of magnetic scattering since in general the twist wave vector may be incommensurate with the lattice wave vector. In Fig. 2.11.4b the temperature dependence of the twist wave vector as determined by magnetic neutron and magnetic X-ray scattering is shown. Here it is seen that although the twist wave vector varies quasi-continuously with temperature, the X-ray scattering can detect plateaus corresponding to a lock-in of the twist wave vector at various commensurate values. This results from magnetoelastic couplings – the spins are not completely independent of the lattice.

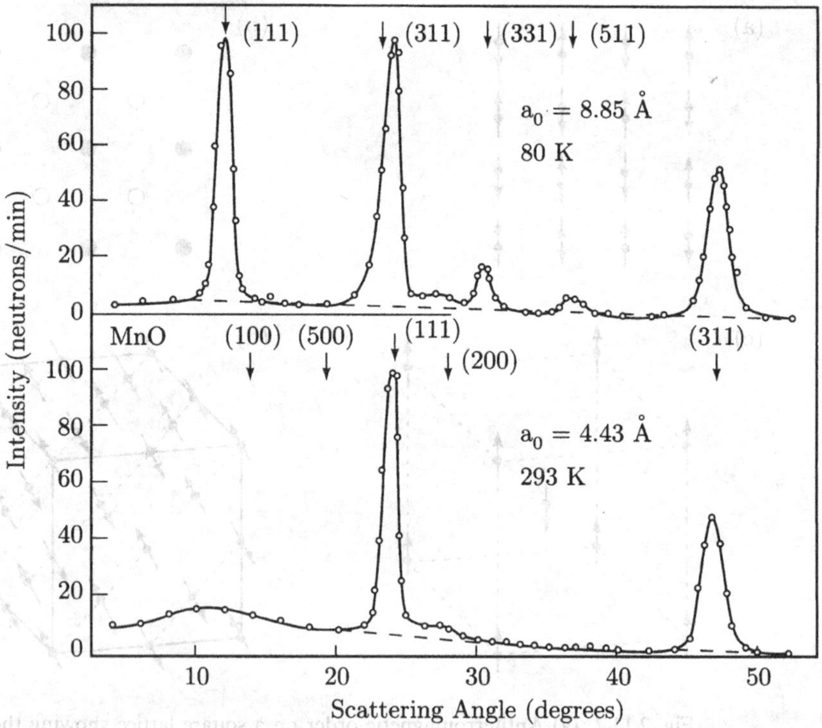

Fig. 2.11.3. Neutron diffraction patterns for MnO below and above the spin ordering temperature of 120K (see Fig. 2.11.2d). The reflection indices are based on an 8.85Åcell at 80K and a 4.43Åcell at 293K. [C.G. Shull, W.S. Strauser, and E.O. Wollan, *Phys. Rev.* **83**, 333 (1951). G.G. Low, Application of neutron scattering to magnetism, in *Magnetism, "Selected Topics"*, ed. S. Foner (Gordon and Breach, NY, 1976).]

2.12 Random isotropic fractals

One of the most useful illustrations of correlations and their effect on the pair correlation function and the static structure factor comes from dilation-symmetric or self-similar objects. The self-similarity reflects the property that the structure "looks the same" on any length scale, i.e., there is no characteristic size.

An example of such an object is an "ideal" polymer chain, composed of N monomers of length a. The polymer configuration can be described as a random walk of N steps of length a. We ignore the effects of monomer volume and repulsion for this treatment. For a random walk, the step directions are completely uncorrelated so that the average of the vector $\mathbf{R}$ from the position of the last monomer to the first is zero: $\langle \mathbf{R} \rangle = 0$. However, the average squared displacement for the random walk is proportional to the number of steps:

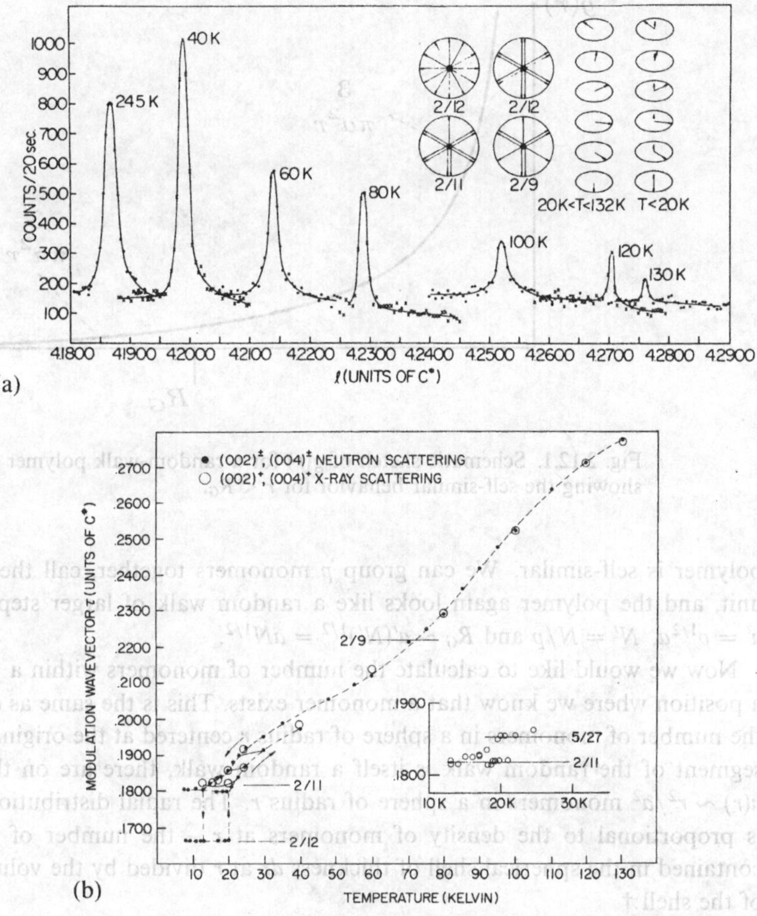

Fig. 2.11.4. X-ray and neutron magnetic scattering from the helical magnetic phase of Ho. (a) Magnetic Bragg peaks for several temperatures from the X-ray study. (b) Comparison of the temperature dependence of the twist wave vector from neutron and X-ray studies. Note the preference for commensurate values indicated by the plateaus. [D. Gibbs, D.E. Moncton, K.L. D'Amico, J. Bohr, and B.H. Grier, *Phys. Rev. Lett.* **55**, 234 (1985).]

$$\langle R^2 \rangle = Na^2. \tag{2.12.1}$$

The average radius of gyration, R_G ($R_G^2 \equiv \int \rho(\mathbf{r} - \mathbf{R}_{\rm cm})^2 d^3r / \int \rho(\mathbf{r}) d^3r$, where $\rho(\mathbf{r})$ is the monomer density at position $\mathbf{r}$ and $\mathbf{R}_{\rm cm}$ is the position of the center of mass of the polymer), determines the characteristic size of the polymer and is proportional to root mean square separation of end points, i.e.,

$$R_G \sim \langle R^2 \rangle^{1/2} = aN^{1/2}. \tag{2.12.2}$$

For distances large compared to a monomer size, but small compared to R_G, the

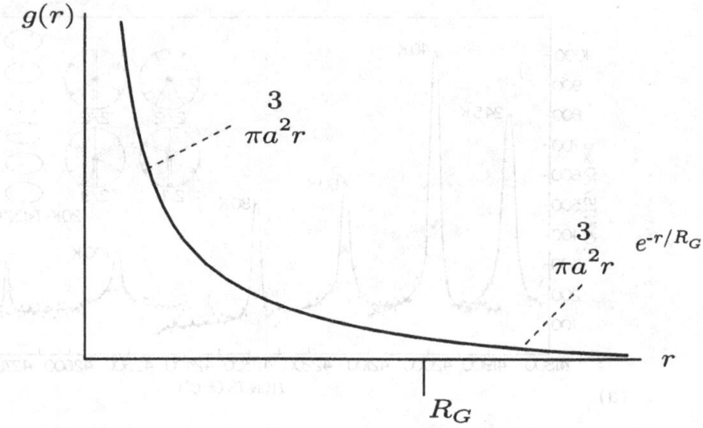

Fig. 2.12.1. Schematic sketch of $g(r)$ for a random-walk polymer chain showing the self-similar behavior for $r < R_G$.

polymer is self-similar. We can group p monomers together, call them a single unit, and the polymer again looks like a random walk of larger step size, with $a' = p^{1/2}a$, $N' = N/p$ and $R_G \sim a'(N')^{1/2} = aN^{1/2}$.

Now we would like to calculate the number of monomers within a radius r of a position where we know that a monomer exists. This is the same as calculating the number of monomers in a sphere of radius r centered at the origin. Since any segment of the random walk is itself a random walk, there are on the average $n(r) \sim r^2/a^2$ monomers in a sphere of radius r. The radial distribution function is proportional to the density of monomers at r – the number of monomers contained in the spherical shell of thickness dr at r divided by the volume $4\pi r^2 dr$ of the shell:†

$$g_F(r) \sim \frac{1}{4\pi r^2}\frac{dn(r)}{dr} \sim \frac{n(r)}{r^3} \sim \frac{1}{a^2 r}. \tag{2.12.3}$$

This result applies to a polymer in three dimensions. In d dimensions, the volume of a spherical shell is proportional to r^{d-1}, for a random walk $n(r) \sim r^2/a^2$ independent of dimension and $g(r) \sim r^{-(d-2)}$. For distances larger than R_G, $g(r)$ falls off exponentially. (One should worry about counting the possible contribution of parts of the polymer that extend outside the region r and then return. However, for a three-dimensional random walk, the return probability to the origin vanishes as the number of steps increases.) $g(r)$ is sketched in Fig. 2.12.1. We saw in Sec. 2.3 that the scattering intensity from a distribution of mass points is proportional to the Fourier transform of the pair distribution function, which

† Note that for a fractal object, $\langle n \rangle$ is length scale dependent. We therefore want to introduce a distribution function $g_F(\bar{x}) = \sum_{\alpha \neq 0} \delta(x - x_\alpha + x_0)$, which has units of volume and which, unlike $g(x)$ [Eq. (2.3.12)], is not normalized by $\langle n \rangle$.

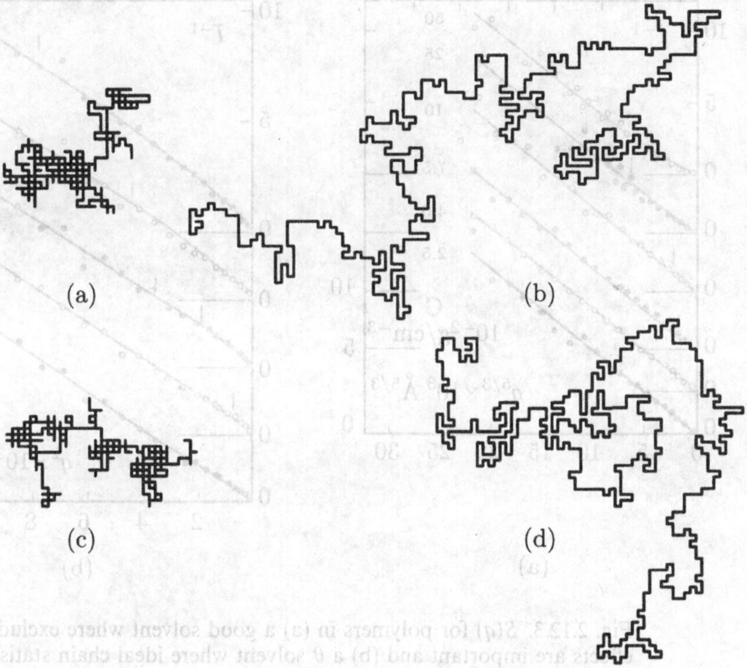

Fig. 2.12.2. (a) and (c) random walks and (b) and (d) self-avoiding random walks of $N = 500$ steps on a two-dimensional square lattice. Note that the random walk is much more compact than a self-avoiding random walk with the same number of steps.

for the isotropic case is the radial distribution function. The neutron scattering intensity from a polymer in d dimensions is, therefore, approximately proportional to

$$S(q) \propto \int g_F(r) e^{-i\mathbf{q}\cdot\mathbf{r}} d^d r. \tag{2.12.4}$$

If the pair distribution function has a general power-law singularity, $g(r) \sim r^{-\alpha}$, then

$$\begin{aligned} S(q) &\sim \int e^{-i\mathbf{q}\cdot\mathbf{r}}(d^d r/r^\alpha) = (q^\alpha/q^d)\int e^{i\mathbf{q}\cdot\mathbf{r}}[d^d(qr)/(qr)^\alpha] \\ &\sim q^{\alpha-d}\int e^{-ix}x^{-\alpha}d^d x. \end{aligned} \tag{2.12.5}$$

Thus, we then have

$$S(q) \sim \frac{1}{q^2 a^2}, \qquad qR_G \gg 1, \tag{2.12.6}$$

for the ideal chain in three dimensions ($\alpha = 1, d = 3$).

As an aside, we note that polymers in a "good" solvent are not a simple random walk but are "swollen" as a result of repulsive interactions among constituent

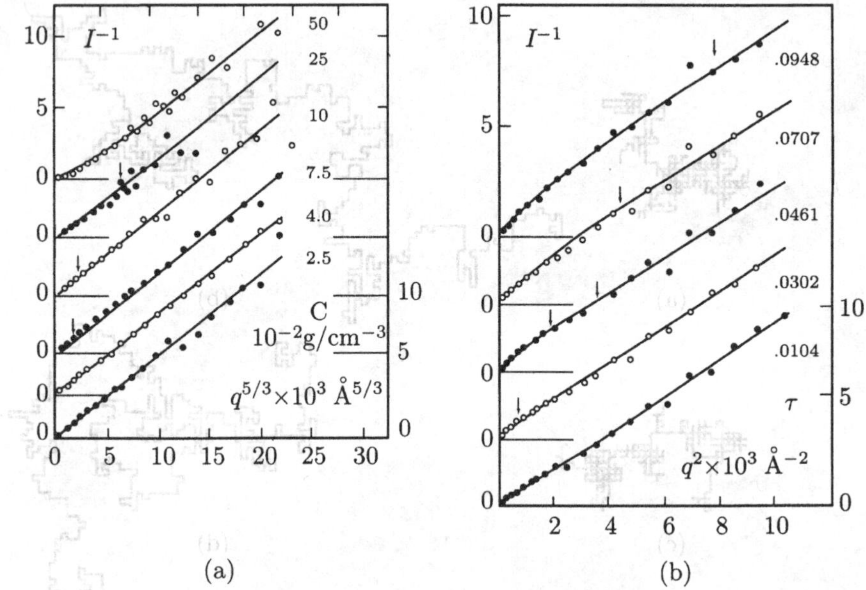

Fig. 2.12.3. $S(q)$ for polymers in (a) a good solvent where excluded volume effects are important and (b) a θ solvent where ideal chain statistics apply. [B. Farnoux, F. Bou'e, J. P. Cotton, M. Daoud, G. Jannink, M. Nierlich and P. G. de Gennes, *J. de Physique* **39**, 77 (1978).]

monomers. This "excluded volume" effect is often described in terms of a "self-avoiding random walk" (SAW) shown in Fig. 2.12.2. The radius of gyration of a self-avoiding walk obeys a relation similar to that of a random walk, [Eq. (2.12.2)] but with an exponent that differs from 1/2:

$$R_G = aN^{\nu} . \tag{2.12.7}$$

The value of the exponent ν is remarkably well approximated by the formula

$$\nu = \frac{3}{d+2}, \tag{2.12.8}$$

predicted by a mean-field theory due to Flory. The average number of monomers in a sphere of radius r is now

$$n(r) \sim (r/a)^{1/\nu} \quad \rightarrow \quad (r/a)^{5/3} \quad (d=3), \tag{2.12.9}$$

and the pair distribution function and structure factor are

$$g(r) \sim r^{-d}n(r) \sim r^{1/\nu-d} \rightarrow 1/(a^{5/3}r^{4/3}), \tag{2.12.10}$$

$$S(q) \sim (qa)^{-1/\nu} \rightarrow (qa)^{-5/3}. \tag{2.12.11}$$

$S(q)$ in polymers can be measured by low-angle neutron scattering. Both random walk and self-avoiding random walk statistics can be seen by changing the

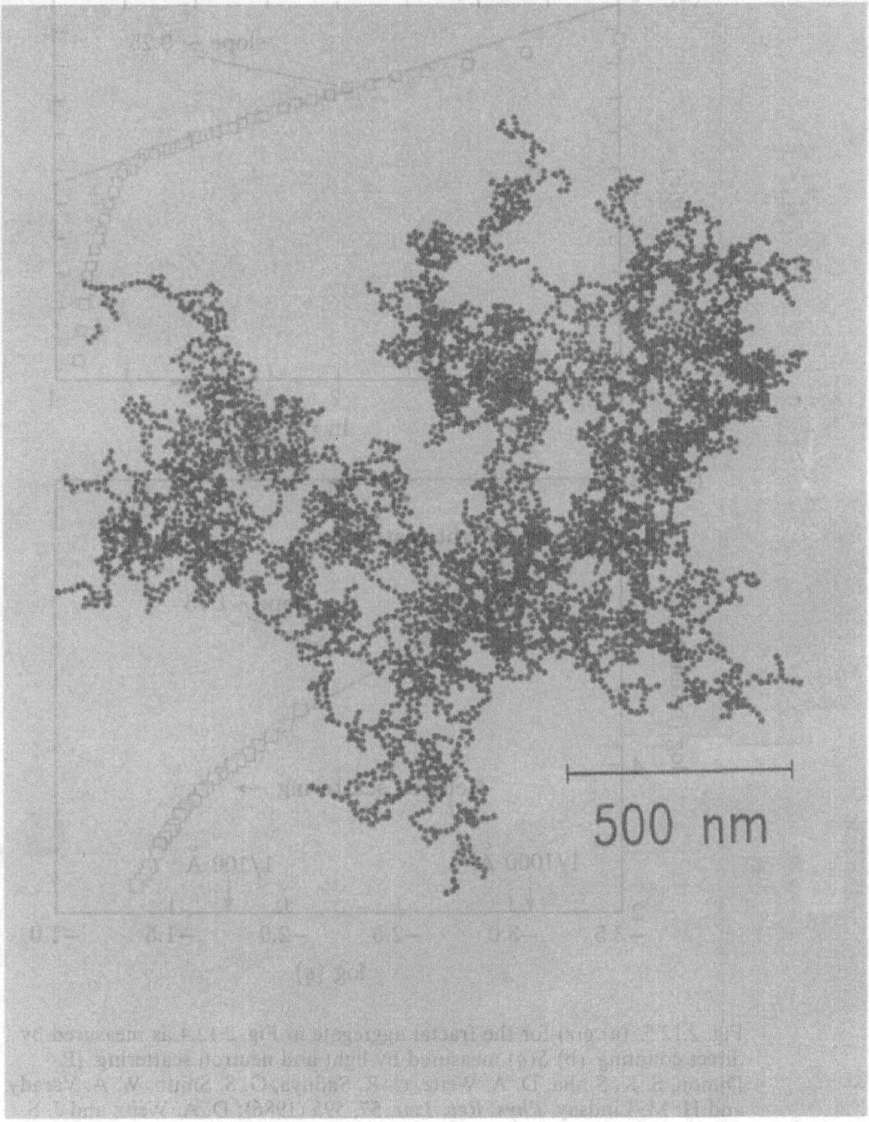

Fig. 2.12.4. A fractal aggregate of gold particles. [Courtesy of D. A. Weitz.]

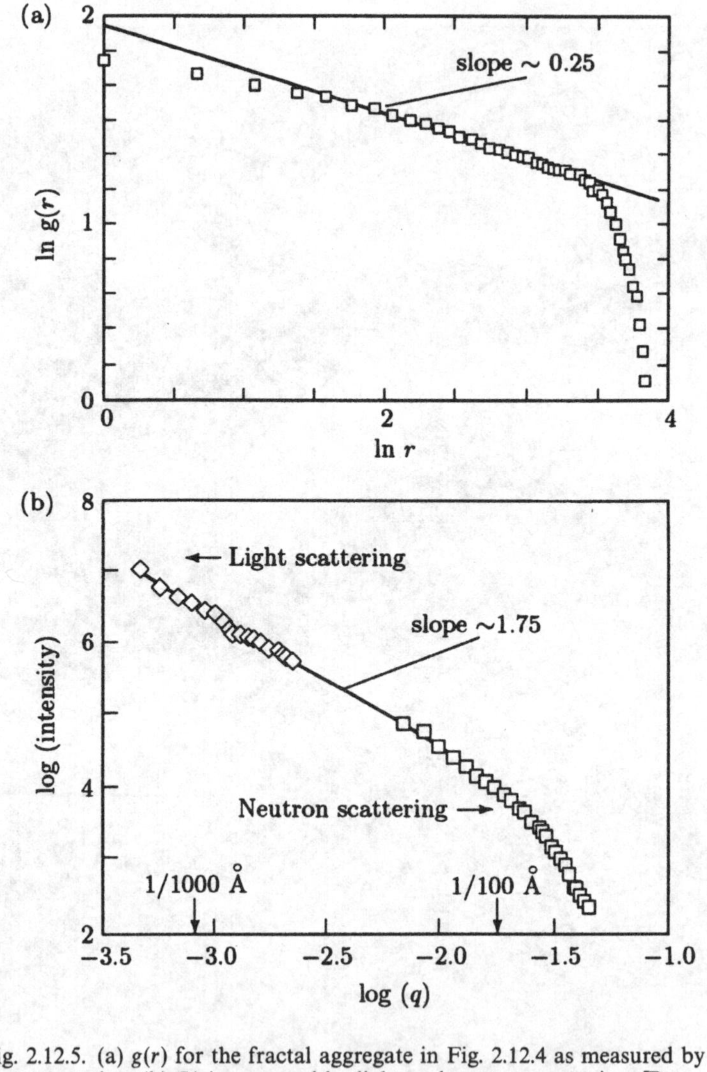

Fig. 2.12.5. (a) $g(r)$ for the fractal aggregate in Fig. 2.12.4 as measured by direct counting. (b) $S(q)$ measured by light and neutron scattering. [P. Dimon, S. K. Sinha, D. A. Weitz, C. R. Safinya, G. S. Smith, W. A. Varady, and H. M. Lindsay, *Phys. Rev. Lett.* **57**, 595 (1986); D. A. Weitz and J. S. Huang, in *Kinetics of Aggregation and Gelation*, eds. P. Family and D. P. Landau, p. 19 (Elsevier, New York, 1987).]

"quality" of the solvent. For a "θ" solvent the repulsion of the polymer from the solvent compensates on average for the hard-core excluded volume repulsion of the monomers with themselves (see Fig. 2.12.3). In this case, the effective interaction between monomers is zero, and the polymers obey random walk statistics with $R_G \sim N^{1/2}$.

In the general case of a random isotropic fractal, the *fractal* or *Hausdorff* dimension d_f is defined via

$$n(r) \sim (r/a)^{d_f}. \tag{2.12.12}$$

This is a generalization of the relation $n \sim r^d$ valid for normal compact objects. The fractal dimension of a polymer is thus $1/\nu$. The pair distribution function and static structure factor for a general fractal are given by

$$\begin{aligned} g_F(r) &\sim n/r^d \sim 1/(a_f^d r^{d-d_f}), \qquad \text{or } \alpha = d - d_f \\ S(q) &\sim (qa)^{-d_f}. \end{aligned} \tag{2.12.13}$$

It is very easy to get a physical feel for what the correlation and scattering functions mean by looking at Figs. 2.12.4 and 2.12.5. Fig. 2.12.4 is an electron micrograph of an aggregate of gold particles formed under highly nonequilibrium conditions. The uniform-size 50Å particles were in stable suspension because of repulsion between the like surface charges. The charges were then chemically removed. The particles diffused until they collided and then stuck wherever they hit. The process is known as *diffusion limited aggregation* (DLA). The mass correlation function was calculated from the electron micrograph by randomly picking a point in the cluster, drawing a circle of radius r and counting the number of particles intersecting the circle. The process was repeated for many origins and many radii. Since the picture is a projection of the structure in two dimensions, it is fractal with $g(r) \sim 1/r^{2-d_f}$, and the log-log plot indicates a slope of 0.25 ($d_f = 1.75 \pm 0.05$) until r approaches the size of the cluster at which point it rapidly decreases.

The same system was studied by both neutron scattering and light scattering. The resulting structure function is shown in Fig. 2.12.5. $S(q)$ should have the form q^{-d_f} until q reaches a crossover value. The $S(q)$ data on this sample give $d_f \sim 1.80$. For higher q the scattering probes distances smaller than a particle size, and the objects no longer look fractal. Fig. 2.12.5 also serves as a good example of the range covered by light ($q \geq 2\pi/\lambda \geq 2\pi/1500$Å) and neutron scattering ($q \geq 2\pi/1$Å).

Appendix 2A Fourier transforms

In this appendix, we will review Fourier transforms for functions of one- and d-dimensional continuous variables and for functions defined at lattice sites on one- and d-dimensional lattices.

1 One dimension

We begin with a function $f(x)$ of a single variable x in the interval $[-L/2, L/2]$ (i.e. $-L/2 \leq x \leq L/2$). If $f(x)$ satisfies reasonable continuity and boundedness conditions (e.g. it does not have an infinite number of zeros in some finite interval of x), it can be expanded in a uniformly convergent Fourier series:

$$f(x) = \sum_q \psi_q(x) f(q) , \tag{2A.1}$$

where $\psi_q(x)$ satisfies the same boundary conditions as $f(x)$. Common boundary conditions on $f(x)$ are $f(x = \pm L/2) = 0$ or $f'(x = \pm L/2) = 0$. In condensed matter physics, one is often interested in bulk systems in the thermodynamic limit, $L \to \infty$, for which most physical properties of interest do not depend on the boundary conditions. In this case, any physically reasonable boundary condition can be imposed. The periodic boundary condition requiring $f(x)$ to be a periodic function of period L,

$$f(x) = f(x + L) , \tag{2A.2}$$

is computationally the simplest and is almost universally used in situations where surface properties are not relevant. The condition (2A.2) is equivalent to wrapping the line of length L on a circle of circumference L and tying the two ends together. The functions $\psi_q(x)$ must satisfy the periodic boundary condition and can be chosen to be

$$\psi_q(x) = Ae^{iqx} , \tag{2A.3}$$

where

$$q = \frac{2\pi}{L} n, \qquad n = 0, \pm 1, \pm 2, ... \tag{2A.4}$$

and where A is an arbitrary normalization constant. The functions e^{iqx} satisfy the *orthogonality condition,*

$$\int_{-L/2}^{L/2} dx e^{i(q-q')x} = \frac{\sin[(q-q')L/2]}{[(q-q')/2]} = L\delta_{n,n'} = L\delta_{q-q',0} , \tag{2A.5}$$

where $\delta_{a,b}$ is the Kronecker delta ($\delta_{a,b} = 1$ if $a = b$ and $\delta_{a,b} = 0$ otherwise) and the *completeness condition,*

$$\sum_q e^{-iqx} = \lim_{N\to\infty} \sum_{N-1}^{N-1} e^{-i(2\pi n/L)x} = \lim_{N\to\infty} \frac{\sin[2\pi(N-1/2)x/L]}{\sin(\pi x/L)} = L\delta(x) , \tag{2A.6}$$

where $\delta(x)$ is the Dirac delta that is zero except at $x = 0$ but whose integral over x is unity. Thus, for periodic boundary conditions,

$$\left.\begin{aligned} f(x) &= A\sum_q e^{iqx} f(q), \\ f(q) &= \frac{1}{AL}\int_{-L/2}^{L/2} e^{-iqx} f(x) dx. \end{aligned}\right\} \tag{2A.7}$$

To treat systems in the limit $L \to \infty$, one takes the continuum limit in which $q = (2\pi/L)n$ is treated as a continuous variable and

$$\sum_q \equiv \sum_n \Delta n = \frac{L}{2\pi} \sum_q \Delta q \to L \int_{-\infty}^{\infty} \frac{dq}{2\pi} , \tag{2A.8}$$

where $\Delta n = 1$. Thus Eqs. (2A.7) can be rewritten as

$$f(x) = AL \int_{-\infty}^{\infty} \frac{dq}{2\pi} e^{iqx} f(q) \overset{LA=1}{\to} \int_{-\infty}^{\infty} \frac{dq}{2\pi} e^{iqx} f(q), \tag{2A.9}$$

$$f(q) = \frac{1}{AL} \int_{-\infty}^{\infty} dx e^{-iqx} f(q) \overset{LA=1}{\to} \int_{-\infty}^{\infty} dx e^{-iqx} f(q) . \tag{2A.10}$$

The normalization constant A is often chosen to be equal to L^{-1} so that the factors LA and $(LA)^{-1}$ become unity as shown in the final form on the right hand side of Eqs. (2A.9) and (2A.10). Other choices, such as $A = L^{-1/2}$ so that $L^{-1/2}$ appears as a factor in both Eqs. (2A.9) and (2A.10), are also used. In the continuum limit, the orthogonality and completeness relations (2A.5) and (2A.6) become

$$\int_{-\infty}^{\infty} dx e^{i(q-q')x} = \lim_{L\to\infty} L\delta_{q-q',0} \equiv 2\pi\delta(q-q') \tag{2A.11}$$

and

$$\int_{-\infty}^{\infty} \frac{dq}{2\pi} e^{-iq(x-x')} = \delta(x-x'). \tag{2A.12}$$

The identification of $L\delta_{q,q'}$ with $2\pi\delta(q-q')$ can be seen from

$$\sum_q \delta_{q,0} = 1 = \frac{L}{2\pi}\int dq\delta_{q,0} \to \int_{-\infty}^{\infty} dq\delta(q). \tag{2A.13}$$

2 d dimensions

The generalization of the above formulae to d dimensions is straightforward. Let $f(\mathbf{x})$ be a function of a d-component vector $\mathbf{x} = (x_1, x_2, ..., x_d)$ and impose periodic boundary conditions on each of the components of $\mathbf{x}$:

$$f(x_1, ..., x_i, ..., x_d) = f(x_1, ..., x_i + L_i, ..., x_d), \quad i = 1, 2, ..., d. \tag{2A.14}$$

Then $f(\mathbf{x})$ can be expanded in a Fourier series similar to Eqs. (2A.7):

$$f(\mathbf{x}) = A\sum_{\mathbf{q}} e^{i\mathbf{q}\cdot\mathbf{x}} f(\mathbf{q}) \tag{2A.15}$$

$$f(\mathbf{q}) = \frac{1}{AV}\int d^d x e^{-i\mathbf{q}\cdot\mathbf{x}} f(\mathbf{x}), \tag{2A.16}$$

where $V = L_1 L_2 ... L_d$ and

$$\mathbf{q} = \left(\frac{2\pi}{L_1}n_1, \frac{2\pi}{L_2}n_2, ..., \frac{2\pi}{L_d}n_d\right), \tag{2A.17}$$

where the coefficients n_i are integers. In the infinite volume limit, these relations become

$$f(\mathbf{x}) = A\sum_{\mathbf{q}} e^{i\mathbf{q}\cdot\mathbf{x}} f(\mathbf{q}) = AV\int \frac{d^d q}{(2\pi)^d} e^{i\mathbf{q}\cdot\mathbf{x}} f(\mathbf{q})$$
$$\overset{AV=1}{\to} \int \frac{d^d q}{(2\pi)^d} e^{i\mathbf{q}\cdot\mathbf{x}} f(\mathbf{q}) \tag{2A.18}$$

$$f(\mathbf{q}) = \frac{1}{AV}\int d^d x e^{-i\mathbf{q}\cdot\mathbf{x}} f(\mathbf{x}) \overset{AV=1}{\to} \int d^d x e^{-i\mathbf{q}\cdot\mathbf{x}} f(\mathbf{x}), \tag{2A.19}$$

where again the normalization factor A is often chosen to be equal to V^{-1} as indicated by the final form of these equations. It is understood that the **x**- and **q**-integrals in Eqs. (2A.18) and (2A.19) are over all space. Finally, in the infinite volume, continuum limit, the orthogonality and completeness conditions become

$$\int d^d x e^{i(\mathbf{q}-\mathbf{q}')\cdot\mathbf{x}} = V\delta_{\mathbf{q},\mathbf{q}'} = (2\pi)^d \delta^{(d)}(\mathbf{q}-\mathbf{q}') \tag{2A.20}$$

and

$$\int \frac{d^d q}{(2\pi)^d} e^{-i\mathbf{q}\cdot(\mathbf{x}-\mathbf{x}')} = \delta^{(d)}(\mathbf{x}-\mathbf{x}'), \tag{2A.21}$$

where $\delta^{(d)}(\mathbf{x})$ is a d-dimensional Dirac delta function.

3 Transforms on a lattice

Often one is interested in functions that are defined only at points on a regular periodic lattice rather than at all points in space. The Fourier transformation of these functions is the subject of this sub-section.

One-dimensional lattices Let f_l be a function of the integer l indexing the lattice site located at position $R_l = la$ of a one-dimensional lattice with lattice spacing a (see Sec. 2.5). The function f_l can be expanded in a discrete Fourier series

$$f_l = \sum_q \tilde{\psi}_q(l) f_q, \tag{2A.22}$$

where $\tilde{\psi}_q(l)$ satisfies the same boundary conditions as f_l. Again, we choose the periodic boundary condition,

$$f_l = f_{l+N}, \tag{2A.23}$$

where N is an integer. In this case, we can choose

$$\tilde{\psi}_q(l) = Ae^{iqR_l} = \tilde{\psi}_q(l+N), \tag{2A.24}$$

where

$$q = \frac{2\pi}{Na} n, \tag{2A.25}$$

where n is an integer. Because R_l is an integral multiple of the lattice spacing a, the function $\tilde{\psi}_q(x)$ in Eq. (2A.23) is periodic in q as well as in l:

$$\tilde{\psi}_q(l) = \tilde{\psi}_{q+(2\pi)/a}(l). \tag{2A.26}$$

Thus all the functions $\tilde{\psi}_q(l)$ and $f(q)$ are completely characterized by q in the interval $[-\pi/a, \pi/a]$, i.e., by q in the first *Brillouin zone* (BZ) of the one-dimensional lattice.

The number of points in the first BZ is equal to the number of sites N in the direct lattice. This follows because the number of points in some region of space is equal to its "volume" divided by the "volume" per point. The volume of the first BZ of a one-dimensional lattice is $2\pi/a$ and volume per point is simply $\Delta q = (2\pi)/Na$, so that

$$\text{number of points in first BZ} = \frac{2\pi}{a}\frac{1}{\Delta q} = \frac{2\pi/a}{(2\pi)/Na} = N. \tag{2A.27}$$

The functions e^{iqR_l} satisfy an orthogonality condition similar to that of the functions e^{iqx}:

$$\sum_{l=0}^{N} e^{i(q-q')R_l} = N\delta_{q,q'} \overset{N\to\infty}{\to} \frac{2\pi}{a}\delta(q-q'), \tag{2A.28}$$

where Eq. (2A.11) with $L = Na$ was used to relate the Kronecker delta to the Dirac delta. The completeness condition is

$$\sum_{q\in 1\text{st BZ}} e^{iqR_l} = \frac{1-e^{iNl}}{1-e^{il}} = N\delta_{l,0}. \tag{2A.29}$$

In the continuum limit, this equation becomes

$$a\int_{-\pi/a}^{\pi/a} \frac{dq}{2\pi} e^{iqR_l} = \delta_{l,0}. \tag{2A.30}$$

When the above results are combined, the lattice Fourier transforms can be written as

$$f_l \quad = \quad A \sum_{q\in 1\text{st BZ}} e^{iqR_l} f_q \quad \overset{N\to\infty}{\to} \quad A(Na)\int_{-\pi/a}^{\pi/a} \frac{dq}{2\pi} e^{iqR_l} f_q$$

$$\stackrel{ANa=1}{\rightarrow} \int_{-\pi/a}^{\pi/a} \frac{dq}{2\pi} e^{iqR_l} f_q \tag{2A.31}$$

$$f_q = \frac{1}{NA} \sum_l e^{-iqR_l} f_l \stackrel{ANa=1}{\rightarrow} a \sum_l e^{-iqR_l} f_l. \tag{2A.32}$$

Again, the choice of A is arbitrary. Often the choice $A = (1/Na)$ is made as shown on the far right hand side of these equations. In this case, $(NA)^{-1} = a$, and the sum over l in Eq. (2A.32) could be replaced by an integral over R_l in a spatial continuum limit.

***d*-dimensional lattices** The generalization of lattice Fourier transforms to d-dimensional lattices is again straightforward. If $f_{\mathbf{l}}$ is a function of the lattice index $\mathbf{l}$ satisfying periodic boundary conditions, $f_{\mathbf{l}} = f_{\mathbf{l+N}}$, where $\mathbf{N} = (N_1, N_2, ..., N_d)$, then

$$f_{\mathbf{l}} = \sum_{\mathbf{q}} \tilde{\psi}_{\mathbf{q}}(\mathbf{l}) f_{\mathbf{q}}, \tag{2A.33}$$

where, since $\mathbf{R}_{\mathbf{l+N}} = \mathbf{R}_{\mathbf{l}} + \mathbf{R}_{\mathbf{N}}$,

$$\tilde{\psi}_{\mathbf{q}}(\mathbf{l}) = Ae^{i\mathbf{q}\cdot\mathbf{R}_{\mathbf{l}}} \tag{2A.34}$$

with

$$\mathbf{q} = \left(\frac{2\pi}{N_1 a} n_1, \ \frac{2\pi}{N_2 a} n_2, ..., \ \frac{2\pi}{N_d a} n_d \right) . \tag{2A.35}$$

The restriction of $\mathbf{R}_{\mathbf{l}}$ to lattice points leads to

$$\tilde{\psi}_{\mathbf{q+G}}(\mathbf{l}) = \tilde{\psi}_{\mathbf{q}}(\mathbf{l}) , \tag{2A.36}$$

where $\mathbf{G}$ is a reciprocal lattice vector. Thus, as in the one-dimensional case, only wave vectors $\mathbf{q}$ in the first Brillouin zone need be considered. The number of points in the first Brillouin zone is again equal to the number of points, $N = N_1 N_2 ... N_d$ in the lattice. The orthogonality and completeness conditions are now

$$v_0 \sum_{\mathbf{l}} e^{i(\mathbf{q}-\mathbf{q}')\cdot\mathbf{R}_{\mathbf{l}}} = V\delta_{\mathbf{q},\mathbf{q}'} = (2\pi)^d \delta^{(d)}(\mathbf{q}-\mathbf{q}'), \tag{2A.37}$$

and

$$\frac{1}{N} \sum_{\mathbf{q}} e^{-i\mathbf{q}\cdot(\mathbf{R}_{\mathbf{l}}-\mathbf{R}_{\mathbf{l}'})} \rightarrow v_0 \int \frac{d^d q}{(2\pi)^d} e^{-i\mathbf{q}\cdot(\mathbf{R}_{\mathbf{l}}-\mathbf{R}_{\mathbf{l}'})} = \delta_{\mathbf{l},\mathbf{l}'}, \tag{2A.38}$$

where $v_0 = V/N$ is the volume of a unit cell and the q-integral is over the first BZ. The Fourier transform equations are

$$\begin{aligned} f_{\mathbf{l}} &= A \sum_{\mathbf{q}} e^{i\mathbf{q}\cdot\mathbf{R}_{\mathbf{l}}} f_{\mathbf{q}} \rightarrow AV \int \frac{d^d q}{(2\pi)^d} e^{i\mathbf{q}\cdot\mathbf{R}_{\mathbf{l}}} f_{\mathbf{q}} \\ &\stackrel{AV=1}{\rightarrow} \int \frac{d^d q}{(2\pi)^d} e^{i\mathbf{q}\cdot\mathbf{R}_{\mathbf{l}}} \end{aligned} \tag{2A.39}$$

and

$$f_{\mathbf{q}} = \frac{1}{NA} \sum_{\mathbf{l}} e^{-i\mathbf{q}\cdot\mathbf{R}_{\mathbf{l}}} f_{\mathbf{l}} \stackrel{AV=1}{\rightarrow} v_0 \sum_{\mathbf{l}} e^{-i\mathbf{q}\cdot\mathbf{R}_{\mathbf{l}}} . \tag{2A.40}$$

Bibliography

FLUIDS

J.P. Boon and S. Yip, *Molecular Hydrodynamics* (McGraw-Hill, New York, 1980).
Jean Pierre Hansen, *Theory of Simple Liquids* (Academic Press, New York, 1986).

Liquid crystals and complex fluids

S. Chandrasekhar, *Liquid Crystals* (Cambridge University Press, 1992).

Jean Charvolin and Annete Tardieu, Lytropic liquid crystals: structures and molecular motion, in *Solid State Physics,* supplement 14, ed. L. Liebert (Academic Press, New York, 1978).

P.G. de Gennes and J. Prost, *The Physics of Liquid Crystals,* 2nd edn (Clarendon Press, Oxford, 1993) .

G. Gompper and M. Schick, *Phase Transitions and Critical Phenomena,* Vol. 16 *Self-Assembling Amphiphilic Systems,* eds. C. Domb and J. Lebowitz (Academic Press, London, 1994).

J.N. Israellachvili, in *Physics of Amphiphiles: Micelles, Vesicles and Microemulsions* (North Holland, New York, 1985).

P.S. Pershan, *Structure of Liquid Crystal Phases* (World Scientific, Singapore, 1988).

Samuel A. Safran, *Statistical Thermodynamics of Surfaces, Interfaces, and Membranes,* (Addison Wesley, Reading, MA, 1994).

S.A. Safran and N.A. Clark, eds., *Physics of Complex and Supermolecular Fluids* (Wiley, New York, 1987).

Polymers and fractals

Pierre-Gilles de Gennes, *Scaling Concepts in Polymer Physics* (Cornell University Press, Ithaca, 1979).

M. Doi and S.F. Edwards, *The Theory of Polymer Dynamics* (Clarendon Press, Oxford, 1988).

Benoit Mandelbrot, *The Fractal Geometry of Nature* (Freeman, San Francisco, 1983).

Quasicrystals

D.P. DiVencenzo and P.J. Steinhardt, eds., *Quasicrystals, the State of the Art* (World Scientific, Singapore, 1991).

Martin Gardner, *Sci. Am.* **236**, 110 (1977).

Marko V. Jarić, ed., *Introduction to Quasicrystals* (Academic Press, New York, 1988).

Paul J. Steinhardt and Stellan Ostlund, eds., *The Physics of Quasicrystals* (World Scientific, Singapore, 1987).

Scattering

A. Guinier, *X-ray Diffraction* (Freeman, New York, 1963).

G.L. Squires, *Introduction to the Theory of Thermal Neutron Scattering* (Cambridge University Press, 1978).

Solids

P.W. Anderson, *Basic Notions of Condensed Matter Physics* (Benjamin, Menlo Park, CA, 1984).

N. Ashcroft and D. Mermin, *Solid State Physics* (Holt, Rinehart, Winston, 1976), introductory chapters.

R. Blinc and A.P. Levanyuk, eds., *Incommensurate Crystals,* Vols. I and II (North-Holland, New York, 1986).

C. Kittel, *Introduction to Solid State Physics* (John Wiley, New York, 1971), chaps. 1 and 2.

R. Zallen, *The Physics of Amorphous Solids* (Cambridge University Press, 1983).

J.M. Ziman, *Principles of the Theory of Solids* (Cambridge University Press, 1972).

J.M. Ziman, *Models of Disorder* (Cambridge University Press, 1979), chaps. 1, 2 and 4.

References

D.M. Anderson, S.M. Gruner, and S. Leibler, *Proc. Acad. Sci. USA* **85**, 5364 (1988).

R.J. Birgeneau and P. M. Horn, *Science* **232**, 329 (1986).

P. Dimon, S. K. Sinha, D. A. Weitz, C. R. Safinya, G. S. Smith, W. A. Varady, and H. M. Lindsay, *Phys. Rev. Lett.* **57**, 595 (1986).

F.J. DiSalvo, in *Chemistry and Physics of One-Dimensional Metals,* ed. H.J. Keller (Plenum, New York, 1977).

B. Farnoux, F. Bou'e, J. P. Cotton, M. Daoud, G. Jannink, M. Nierlich, and P. G. de Gennes, *J. de Phys.* **39**, 77 (1978).
G. Friedel, *Ann. Physique (Paris)* **18**, 273 (1922).
D. Gibbs, D. E. Moncton, K. L. D'Amico, J. Bohr, and B. H. Grier, *Phys. Rev. Lett.* **55**, 234 (1985).
Michael E. Huster, Paul Heiney, Victoria B. Cajipe, and John E. Fischer, *Phys. Rev.* **B35**, 3311 (1987).
O. Lehman, *Z. Physikal Chem.* **4**, 462 (1889).
D.I. Levine and P.J. Steinhardt, *Phys. Rev. Lett.* **53**, 2477 (1984).
G.G. Low, Application of neutron scattering to magnetism, in *Magnetism, "Selected Topics"*, ed. by S. Foner (Gordon and Breach, New York, 1976).
L.O. Onsager, *Ann. N.Y. Acad. Sci.* **51**, 627 (1949).
R. Penrose, *Bull. Inst. Math. Appl.* **10**, 266 (1974).
R. Pindak, D. E. Moncton, S. C. Davey, and J. W. Goodby, *Phys. Rev. Lett.* **46**, 1135 (1981).
P.N. Pusey, W. van Megen, P. Bartlett, B.J. Ackerson, J.G. Rarity, and S.M. Underwood, *Phys. Rev. Lett.* **25**, 2753 (1989).
F. Reinitzer, *Monatsch Chem.* **9**, 421 (1888).
D. Shechtman, I. Blech, D. Gratias, and J.W. Cahn, *Phys. Rev. Lett.* **53**, 1951 (1984).
C.G. Shull, W.S. Strauser, and E.O. Wollan, *Phys. Rev.* **83**, 333 (1951).
E.D. Sprecht, M. Sutton, R.J. Birgeneau, D.E. Moncton, and P.W. Horn, *Phys. Rev.* **B30**, 1589 (1984).
S. Tanisaki, *J. Phys. Japan* **16**, 579 (1961).
D. A. Weitz and J. S. Huang, in *Kinetics of Aggregation and Gelation*, eds. P. Family and D. P. Landau, p. 19 (Elsevier, New York, 1987).
J.L. Yarnell, M.J. Katz, R.G. Wenzel, and S.H. Koenig, *Phys. Rev. A* **7**, 2130 (1973).

Problems

2.1 (The Hendriks-Teller model) In a one-dimensional model of an alloy, mass points are separated by a unit distance with probability p and by a longer distance $1+\rho$ with probability $1-p$. Show that the structure factor for this model is

$$S(q) = \frac{p(1-p)[1-\cos(\rho q)]}{1-p(1-p)-p\cos q-(1-p)\cos[(1+\rho)q]+p(1-p)\cos q\rho}.$$

2.2 Calculate the coherent scattering intensity for the lattice shown in Fig. 2.6.7 assuming that the triangles are composed of three identical point particles. Let F_1 be the form factor for particles at the vertices of squares (solid circles) and F_2 be the form factor for the particles in the triangles. Compare your result to that of a square lattice with lattice vectors $\mathbf{a}_1/2$ and $\mathbf{a}_2/2$.

2.3 Calculate the structure factor in reciprocal space in three dimensions for

(a) a triangular lattice of points in the plane $z=0$;
(b) a triangular lattice of uniform density rods parallel to the z-axis as shown in Fig. 2P.1 (at left);
(c) a triangular lattice of uniform rods making an angle θ with respect to the z-axis as shown in Fig. 2P.1 (at right).

What happens to the scattering intensities in part (b) if the rods, rather than being uniform, consist of atoms with liquid-like correlations in a given rod but with no correlations in the positions of atoms in different rods? What happens if, in addition, correlations in the densities of adjacent rods develop?

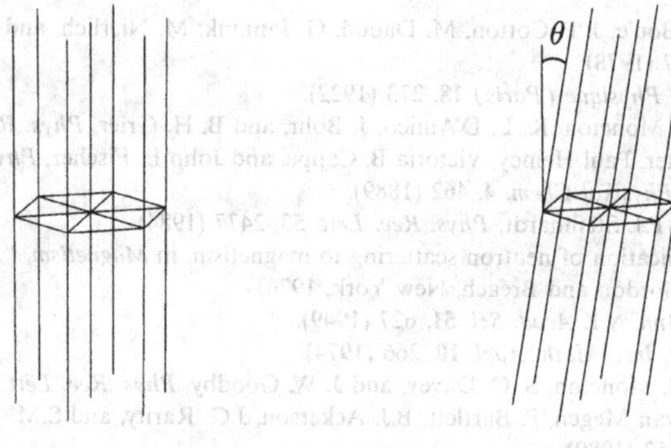

Fig. 2P.1. Regular two-dimensional lattices of rods. On the left, the rods are parallel to the z-axis, and on the right, they make an angle θ with the z-axis. In both cases, the rods intersect points forming a triangular lattice in the xy-plane.

[For an experimental application of these ideas to intercalated graphite, see Huster *et al.* (1987).]

2.4 The positions of atoms in an ideal one-dimensional incommensurate crystal satisfy

$$x_n = na + F(na),$$

where $F(y)$ is a periodic function of period b (i.e., $F(y+b) = F(y)$) with a/b irrational. Show that

$$n_k = N \sum_{p,q} A_q(k) \delta_{k,G_{pq}}$$

where p and q are integers, $G_{pq} = (2\pi p/a) + (2\pi q/b)$,

$$A_q(k) = \frac{1}{b} \int_0^b dy e^{2\pi i q y/b} e^{ikF(y)} .$$

Apply this formula to the sequence

$$x_n = n + \rho[n\sigma] = n(1+\rho\sigma) - \rho\{n\sigma\} \qquad \text{(2P.1)}$$

where $[x]$ is the greatest integer less than or equal to x and $\{x\} = x - [x]$ is the fractional part of x and show that

$$n_k = N \sum_{p,q} e^{iX_{pq}/2} \frac{\sin(X_{pq}/2)}{(X_{pq}/2)} \delta_{k,G_{pq}}$$

where

$$X_{pq} = \frac{2\pi\rho}{1+\rho\sigma} \left(\frac{q}{\rho} - p \right) ,$$

$a = (1+\rho\sigma)$, and $b = \sigma^{-1} + \rho$. The sequence defined by Eq. (2P.1) consists of points separated by short intervals of length 1 or long intervals of length $1+\rho$ with relative frequency $\sigma/(1-\sigma)$. When $\rho = \sigma = \tau^{-1} = \frac{1}{2}(\sqrt{5}-1)$, Eq. (2P.1) generates the so-called Fibonacci sequence.

Fig. 2P.2. Array of lines described by Eq. (2P.2)

2.5 (For the curious.) Choose $p = 1 - \tau^{-1}$ and $\rho = \tau^{-1}$ in Problem 2.1 and plot $S(q)$. (You may wish to use a computer.) Compare this function with n_k in Problem 2.4. for the Fibonacci sequence $[\sigma = \rho = \tau^{-1}]$.

2.6 Fig. 2P.2 shows a regular array of wiggly lines in which the x position of the pth line as a function of y is

$$Y_p(x) = pl + u_1 \cos[(2\pi x/l)\sin(2\pi/5)]\cos[2\pi p\cos(2\pi/5)]$$
$$+u_2\cos[(2\pi x/l)\sin(4\pi/5)]\cos[2\pi p\cos(4\pi/5)]. \qquad (2P.2)$$

Determine the set of points in reciprocal space for which there is Bragg scattering from this array.

2.7 Calculate the X-ray scattering intensity for the following close packed structures formed by stacks of hexagonal lattices.

(a) The sequence $ABAB\cdots$ (the HCP structure).

(b) The sequence $ABCABC\cdots$. Show that the scattering from this sequence is that of an FCC lattice.

(c) The random sequence $ABACBABCB\cdots$ in which there is an equal probability that B or C will follow A, etc.

Answer to (c):

$$I(\mathbf{q}) = N^2 \sum_{\mathbf{G}'} \delta_{\mathbf{q},\mathbf{G}'}$$
$$+N_{\|}N_{\perp}^2 \sum_{\mathbf{G}_\perp \neq \mathbf{G}'_\perp} \delta_{\mathbf{q},\mathbf{G}_\perp} \frac{\sin^2 \mathbf{G}_\perp \cdot \mathbf{c}}{1 - 2\cos q_z a_\| \cos \mathbf{G}_\perp \cdot \mathbf{c} + \cos^2 \mathbf{G}_\perp \cdot \mathbf{c}},$$

where $\mathbf{G}_\perp$ is a vector in the reciprocal lattice of a triangular lattice L with lattice spacing a, $\mathbf{G}'_\perp$ is a vector in the reciprocal lattice of the triangular lattice L' with lattice sites at the centers of the triangles in lattice L, $a_\| = \sqrt{3}a/2$, $\mathbf{G}' = (2\pi k/a_\|)\mathbf{e}_z + \mathbf{G}'_\perp$, and $\mathbf{c}$ is the vector from the vertex to the center of a triangle in lattice L. The vectors $\mathbf{G}'$ are the vectors common to both the FCC and HCP reciprocal lattices. Note that when $\mathbf{G}_\perp \neq \mathbf{G}'_\perp$, $I(\mathbf{q})$ depends

continuously on q_z with the ratio of intensities at $q_z a_\parallel$ an odd integral multiple of π to those with $q_z a_\parallel$ an even integral multiple of π equal to 9. (For an experimental system with this pattern, see Pusey *et al.* (1989).)
Solution hint: Vectors in B and C can, respectively, be obtained from those in A by the addition of the vectors $+\mathbf{c}$ and $-\mathbf{c}$. Thus vectors in the nth layer can be written as $\mathbf{R}_l + \sum_{k=1}^{n} \sigma_k \mathbf{c}$, where $\mathbf{R}_l$ is a vector in L and where σ_k takes on the values ± 1 with probability of 1/2.

2.8 Consider a two-dimensional fluid of circular particles with N-atic bond-angle order.

(a) Show that the pair correlation function can be written as

$$g(\mathbf{y}) = \sum_p g_p(r) e^{iNp\theta}$$

where $\mathbf{y} = (r, \theta)$ in polar coordinates, and

$$g_p(\mathbf{y}) = g_0(r)\Psi_{Np}(r),$$

where

$$g_0(\mathbf{y}) = \frac{1}{2\pi\langle n\rangle}\left\langle \sum_{\alpha\neq 0} \frac{\delta(r - R_{0,\alpha})}{|r - R_{0,\alpha}|}\right\rangle$$

and

$$\Psi_{Np}(r) = \frac{1}{2\pi\langle n\rangle g_0(r)}\left\langle \sum_{\alpha\neq 0} \frac{\delta(r - R_{0,\alpha})}{|r - R_{0,\alpha}|} e^{-iNp\theta_{0,\alpha}}\right\rangle$$
$$\equiv \langle e^{inp\theta_{0,\alpha}}\rangle_r,$$

where $R_{\alpha,\beta} = \mathbf{x}^\alpha - \mathbf{x}^\beta$, $R_{0,\alpha} = |\mathbf{R}_{0,\alpha}|$, and $\theta_{\alpha,\beta}$ is the angle $\mathbf{x}_\alpha - \mathbf{x}_\beta$ makes with the x-axis. This shows that when r is equal to the average separation between nearest neighbors, then only particles that are nearest neighbors to the particle 0 contribute to $\Psi_{Np}(r)$.

(b) Apply reasoning similar to the above to a three-dimensional liquid crystal to derive an expression for Ψ_{6p} in Eq. (2.7.4).

(c) Generalize the above results to general bond-angle order in three dimensions, and show that

$$g(\mathbf{y}) = g_0(r)\sum_{lm} \Psi_{lm}(r) Y_{lm}(\theta, \phi),$$

where

$$\Psi_{lm}(r) = \langle Y_{lm}(\theta_{0,\alpha}, \phi_{0,\alpha})\rangle_r.$$

(d) Explain how you might determine whether a configuration of circles in two dimensions has bond-angle order or not.

2.9 Consider a collection of N_p polymers each with $N+1$ identical monomer units connected by N identical flexible links. Let $\mathbf{R}_\alpha$ be the position of the initial (0th) monomer, and let $\mathbf{X}_{\alpha,i} = \mathbf{R}_\alpha + \mathbf{R}_{\alpha,i}$ be the position of the ith

(1, = 1, ..., N) monomer in polymer α. The structure factor for polymer α is

$$f_\alpha(\mathbf{q}) = 1 + \sum_i^N e^{-i\mathbf{q}\cdot\mathbf{R}_{\alpha,i}}.$$

(a) Show that the scattering intensity for a dilute solution of the above polymers is

$$I_{dp}(\mathbf{q}) = N_p \langle |f(\mathbf{q})|^2 \rangle,$$

where the average is over polymer configurations and $\langle |f(\mathbf{q})|^2\rangle \equiv \langle |f_\alpha(\mathbf{q})|^2\rangle$ independent of α.

(b) The unit cells of a simple cubic lattice are all decorated with identical polymers, each with identical configurations. Show that the scattering intensity is

$$I(\mathbf{q}) = N_p^2 \sum_{\mathbf{G}} \delta_{\mathbf{q},\mathbf{G}} \langle |f(\mathbf{q})|^2 \rangle,$$

where $\mathbf{G}$ is a reciprocal lattice vector of the cubic lattice.

(c) The same cubic crystal is decorated with the same polymers as above, but polymers in each cell perform independent random walks with their initial monomer at the centers of the cubic cells. Show that the scattering intensity in this case is

$$I(\mathbf{q}) = N_p^2 \sum_{\mathbf{G}} \delta_{\mathbf{q},\mathbf{G}} |\langle f(\mathbf{q})\rangle|^2 + N_p \left[\langle |f(\mathbf{q})|^2\rangle - |\langle f(\mathbf{q})\rangle|^2\right].$$

(d) Assume that the polymer steps $\mathbf{u}_{\alpha,i} = \mathbf{R}_{\alpha,i} - \mathbf{R}_{\alpha,i-1}$ are independent Gaussian random variables in d dimensions with zero mean and variance $\langle u_{\alpha,i} u_{\beta,j}\rangle = \delta_{\alpha,\beta}\delta_{ij}(a^2/d)$. Show that

$$\langle f(\mathbf{q})\rangle = \frac{1 - e^{-(N+1)\eta}}{1 - e^{-\eta}}$$

and

$$\langle |f(\mathbf{q})|^2\rangle = (N+1)\left[2\frac{1 - e^{-(N+1)\eta}}{1 - e^{-\eta}} - 1\right] + 2\left[\frac{(N+1)e^{-(N+1)\eta}}{1 - e^{-\eta}} - \frac{e^{-\eta}(1 - e^{-(N+1)\eta})}{(1 - e^{-\eta})^2}\right],$$

where $\eta = (qa)^2/(2d)$.

(e) Compare scattering intensities for parts (b) and (c) above for $\eta \ll 1/(N+1)$, $1/(N+1) \ll \eta \ll 1$, and $\eta \gg 1$. Be sure to discuss peak intensities and q-dependence in both cases. Also discuss behavior as a function of a/c, where c is the lattice constant of the cubic lattice.

3

Thermodynamics and statistical mechanics

Condensed matter physics by its very nature deals with systems with a large number of degrees of freedom, i.e. systems for which a statistical description is essential. In this chapter, we will present a rather comprehensive review of the fundamentals of thermodynamics and statistical mechanics. Much of this chapter, especially the parts dealing with homogeneous fluids (ideal and interacting gases and liquids), should be familiar to everyone. They are included here mostly to establish a basis for discussing more complicated ordered systems. As we saw in the preceding chapter, a great deal of useful and experimentally accessible information is contained in correlation functions such as the density-density correlation function $C_{nn}(\mathbf{x}, \mathbf{x}')$. This chapter will define these functions in a statistical mechanical context and develop a method of calculating them using functional differentiation with respect to spatially varying external fields. Functional differentiation will allow us to calculate position-dependent correlation and response functions by a simple generalization of the familiar technique used to calculate the magnetic susceptibility by differentiating the free energy with respect to the magnetic field. This is a very powerful tool that will be used throughout this book. It is presented here first in a familiar context that should make it easy to grasp.

After reviewing the properties of homogeneous fluids, we will introduce order parameters in Sec. 3.5 and show how they modify both thermodynamics and statistical mechanics. The formalism presented here will be used throughout this book, but especially in the next four chapters dealing with phase transitions and properties of broken-symmetry phases.

3.1 Thermodynamics of homogeneous fluids

An isotropic, homogeneous fluid is the simplest state of matter. It is composed of mobile atoms or molecules whose kinetic energy dominates the structure. A fluid at finite temperature with a finite number of particles cannot exist in equilibrium unless it is confined in some way to a finite volume V (otherwise particles will "evaporate" to less dense regions of space). We will, therefore, consider for the moment a single-phase fluid (either a liquid or a gas) in a volume V. If the fluid is thermally isolated so that it cannot exchange energy with the outside world, its

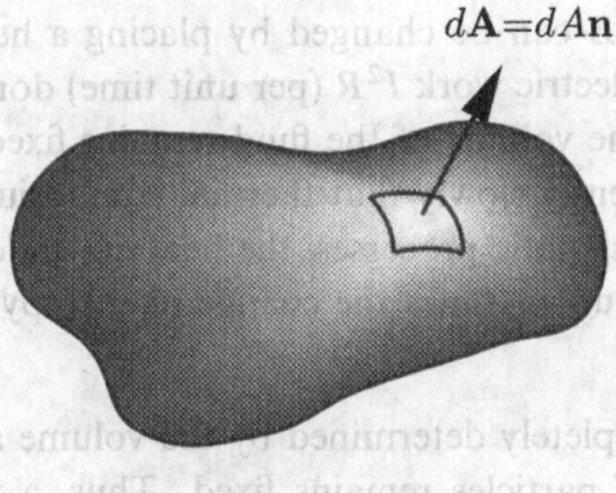

Fig. 3.1.1. A fluid contained in a volume element exerts an outward force normal to its surface. The force $d\mathbf{f}$ exerted on an element of area $d\mathbf{A} = dA\mathbf{n}$ with unit normal $\mathbf{n}$ is parallel to $\mathbf{n}$ and proportional to dA: $d\mathbf{f} = pd\mathbf{A}$. p is the pressure with units of force per unit area or energy per unit volume.

internal energy (kinetic plus potential energy) E is a constant. If external forces do work W_{ext} on the fluid, its internal energy changes by an amount

$$\Delta E = W_{\text{ext}}. \tag{3.1.1}$$

Work is done on a fluid by changing its volume V. Since the fluid in equilibrium always fills the volume to which it is confined, it must exert an outward force on the walls of its container. Conversely, the walls of the container exert an inward force on the fluid. A fluid contained in a volume surrounded by a surface A can only exert a force normal to A. Furthermore, the total force on a flat surface is proportional to its area. The outward force per unit area exerted by a fluid through its containing surface is called the *pressure* p. The force exerted on a surface element of area dA by the fluid is $pdA\mathbf{n}$, where $\mathbf{n}$ is the local outward unit normal to the surface, as shown in Fig. 3.1.1. The force exerted by container walls on a surface element is $-pdA\mathbf{n}$. The work done by external forces in displacing a particular surface element a distance dx along $\mathbf{n}$ is thus

$$\delta W_{\text{ext}} = -pdAdx \equiv -pdV. \tag{3.1.2}$$

The total work done in changing the volume is the integral of Eq. (3.1.2) between the initial and final volumes. Since the force exerted by the walls on the fluid is just the negative of the force exerted by the fluid on the walls, W_{ext} is minus the work W done by the fluid.

1 The first law of thermodynamics

If the fluid is thermally isolated, its internal energy can also be increased by the addition of heat. In this case, the general equation for the change in E becomes

$$\Delta E = Q - W \tag{3.1.3}$$

where Q is the added heat. More generally, Q is the amount of energy received by the system in forms which cannot be identified as the negative of work done by the system. For example, E can be changed by placing a heater in the fluid and applying a current. The electric work I^2R (per unit time) done by the current source adds heat, but, since the volume of the fluid remains fixed, W is zero.

Processes that occur sufficiently slowly that thermal equilibrium is maintained are called *quasi-static*. In quasi-static processes, the heat $\bar{d}Q$ added to a system is equal to its absolute temperature T times the change in entropy S:

$$\bar{d}Q = TdS. \tag{3.1.4}$$

S, like internal energy, is completely determined by the volume and temperature of the fluid if the number of particles remains fixed. Thus, a fluid, initially in equilibrium at temperature T_i and volume V_i with entropy S_i, subjected to a sequence of changes that terminate at final temperature T_f and volume V_f will have an entropy S_f that depends only on T_f and V_f. If the changes run through a cycle, $T_f = T_i$ and $V_f = V_i$, then the final entropy will equal the initial entropy, i.e., $S_f - S_i = \int_i^f dS = 0$. The total heat added, $\int_i^f \bar{d}Q$, need not, however, be zero in going around the cycle. dE and dS are perfect differentials, but $\bar{d}Q$ is not.

The *first law of thermodynamics* can be written as

$$dE = TdS - pdV \tag{3.1.5}$$

for quasi-static processes. This equation implies that temperature and pressure are derivatives of the internal energy with respect to entropy and volume,

$$T = \left.\frac{\partial E}{\partial S}\right)_V, \quad p = -\left.\frac{\partial E}{\partial V}\right)_S. \tag{3.1.6}$$

Similarly,

$$T^{-1} = \left.\frac{\partial S}{\partial E}\right)_V, \quad \frac{p}{T} = \left.\frac{\partial S}{\partial V}\right)_E. \tag{3.1.7}$$

Thus the functions $E(S, V)$ or $S(E, V)$ completely determine the thermodynamic state of a fluid with a fixed number of particles.

Changing the number N of particles in a fluid will lead to a change in the internal energy even if the entropy and volume are fixed. The chemical potential μ is the change in internal energy produced by the addition of one particle. The first law for quasi-static processes for a one-component fluid in which the number of particles can change is thus

$$dE = TdS - pdV + \mu dN. \tag{3.1.8}$$

A complete thermodynamic description of the fluid is contained in the function $E(S, V, N)$.

Temperature, pressure, and chemical potential are sometimes referred to as *generalized forces* (they are first derivatives of the energy). There is an important distinction between the generalized forces and the variables E, S, V, and N. Imagine dividing a fluid with energy E, entropy S, volume V, and particle number N into two equal parts. In each part E, V, S, and N are divided by two, but the generalized forces are the same as before. Variables such as E, S, V, and N whose magnitude increases upon increasing the size of the sample are called *extensive variables;* variables such as T, p, and μ whose values remain constant upon increasing the size of the sample are called *intensive variables.* Associated with each extensive variable, there is a generalized force such that the change in internal energy is the generalized force times the change in the extensive variable. The extensive variable and its associated generalized force are called *conjugate variables.* Thus, T and S, $-p$ and V, and μ and N are conjugate variable pairs.

2 The second law of thermodynamics

The first law of thermodynamics is merely a statement of the conservation of energy. The *second law* describes the nature of changes that can occur in statistical systems. It can be stated in various ways, the most physically intuitive of which is that heat cannot spontaneously flow "uphill" from a cold to a hot body. An equivalent statement is that any changes in a closed system arising from removal of constraints will lead to an increase in the entropy, i.e.,

$$\Delta S \geq 0. \tag{3.1.9}$$

An immediate consequence of this law is that the entropy of a closed system will be a maximum with respect to any changes of internal unconstrained variables. Consider, for example, a fluid isolated from the outside world in a container divided into two parts (labeled 1 and 2) by a movable partition that permits exchange of energy and particles, as depicted in Fig. 3.1.2. The total entropy S must be a maximum with respect to changes in volume, energy, and number of particles in the fluid on the two sides of the partition. The total energy, particle number, and volume cannot change since the system is isolated from the outside world. Thus, changes in particle number, volume, and internal energy in the fluid on one side of the partition must be accompanied by equal and opposite changes in the fluid on the other side: $\Delta E_1 = -\Delta E_2 = \Delta E$, $\Delta V_1 = -\Delta V_2 = \Delta V$, and $\Delta N_1 = -\Delta N_2 = \Delta N$. Since the entropy is a maximum, it must be stationary with respect to these changes, i.e.,

$$\Delta S = \left(\frac{1}{T_1} - \frac{1}{T_2}\right)\Delta E + \left(\frac{p_1}{T_1} - \frac{p_2}{T_2}\right)\Delta V - \left(\frac{\mu_1}{T_1} - \frac{\mu_2}{T_2}\right)\Delta N = 0, \tag{3.1.10}$$

implying $T_1 = T_2$, $p_1 = p_2$, and $\mu_1 = \mu_2$. The interface between two coexisting phases is essentially a movable partition that allows interchange of particles and energy. Thus, the above considerations imply that the temperatures, pressures, and chemical potentials of coexisting phases must be equal.

3 The third law of thermodynamics

The third law of thermodynamics (or Nernst theorem) relates to the behavior of systems as they approach the absolute zero of temperature. Nernst suggested that the entropy of physical systems would tend to zero as $T \to 0$. (Within quantum

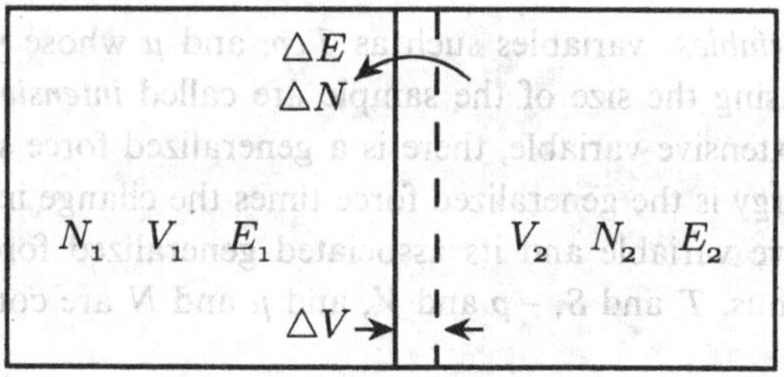

Fig. 3.1.2. A thermally insulated vessel of constant volume containing a fixed number of particles with a movable, permeable, heat conducting partition. The volumes, particles, and internal energies of the fluid on the two sides of the partition will adjust so as to maximize the entropy of the whole system.

statistical mechanics this is equivalent to having a non-infinite ground state degeneracy hence a nonextensive or zero entropy.) The consequences of this law are that most temperature derivatives of thermodynamic quantities go to zero at least as fast as T, and in particular the specific heat (see later in this section) should go to zero with T. There are presently several interesting physical and model systems which appear at first glance to violate this theorem, especially glasses and systems with "frustrated" ground states, retaining considerable entropy to quite low temperatures.

4 Thermodynamic potentials

As we saw above, all thermodynamic quantities can be obtained from the internal energy provided the latter is expressed as a function of its associated *natural variables*, S, V, and N. Functions that contain all thermodynamic information are called *thermodynamic potentials*. E is an extensive variable that is a natural function of extensive variables only. It is often desirable to consider other thermodynamic potentials that are natural functions of one or more of the extensive parameters that can be controlled by establishing contact with an appropriate reservoir as implied by Eq. (3.1.10). (Either side of Fig. 3.1.2 can be a reservoir if it is made sufficiently large so that its T, p, or μ do not change significantly when energy, volume or particles are removed.) These potentials are Legendre transforms of E. The Helmholtz free energy,

$$F = E - TS, \tag{3.1.11}$$

is a natural function of T, V, and N, as can be seen by the differential relation,

$$\begin{aligned} dF &= dE - TdS - SdT \\ &= -SdT - pdV + \mu dN, \end{aligned} \tag{3.1.12}$$

it satisfies. The Gibbs free energy is a natural function of T, p, and N:

$$G = E - TS + pV = F + pV,$$
$$dG = -SdT + Vdp + \mu dN. \tag{3.1.13}$$

The enthalpy $H = E + pV$ is a natural function of S, p, and N. Finally, the grand potential,

$$\mathscr{A} = E - TS - \mu N = F - \mu N \tag{3.1.14}$$

is a natural function of T, V, and μ satisfying

$$d\mathscr{A} = -SdT - pdV - Nd\mu. \tag{3.1.15}$$

Like E, all of the thermodynamic potentials just introduced contain all thermodynamic information, provided they are expressed as functions of their associated natural variables. For example, the entropy can be obtained by differentiation of F, G, or $\mathscr{A}$ with respect to T:

$$S = -\left.\frac{\partial F}{\partial T}\right)_{V,N} = -\left.\frac{\partial G}{\partial T}\right)_{p,N} = -\left.\frac{\partial \mathscr{A}}{\partial T}\right)_{V,\mu}. \tag{3.1.16}$$

5 Stability criteria

In equilibrium, the entropy is a maximum with respect to changes of unconstrained parameters as demonstrated below. In equilibrium, the free energies just introduced are minima with respect to similar changes. Consider a fluid in thermal and mechanical contact with a reservoir, whose variables will be indicated by a prime. The fluid and reservoir together form a closed system with constant total volume. The reservoir is at temperature T_0 and pressure p_0. The change, $\Delta S_T = \Delta S + \Delta S'$, in the total entropy occurring when any constraints are removed must be positive. The heat Q transferred from the reservoir to the fluid is, by the first law, equal to the change ΔE in the internal energy of the fluid plus the work done by the fluid. The work done by the fluid in changing its volume by ΔV is $p_0 \Delta V$. Thus, $Q = \Delta E + p_0 \Delta V$. The temperature of the reservoir does not change in transferring heat to the fluid. The change in entropy of the reservoir is, therefore, $\Delta S' = -Q/T_0$, and

$$\Delta S_T = \Delta S - \frac{\Delta E + p_0 \Delta V}{T_0} \geq 0, \tag{3.1.17}$$
$$\Delta(E - T_0 S + p_0 V) \leq 0. \tag{3.1.18}$$

Thus, in the absence of external forces, changes in the quantity $E - T_0 S + p_0 V$ resulting from the removal of constraints will be negative. If the volume and temperature of the fluid are fixed, then Eq. (3.1.18) implies

$$\Delta F \leq 0, \tag{3.1.19}$$

and if the pressure and temperature of the fluid are fixed, Eq. (3.1.18) implies

$$\Delta G \leq 0 \tag{3.1.20}$$

when there are no external forces. Thus, removal of constraints at constant volume and temperature lead to a decrease in the Helmholtz free energy, and removal of constraints at constant temperature and pressure lead to a decrease in the Gibbs free energy. In other words, to attain equilibrium nature adjusts any free variable to maximize S for fixed E and V, to minimize F for fixed T and V, and to minimize G for fixed T and p.

Since in equilibrium $E - T_0 S + p_0 V$ is a minimum, its deviations from thermodynamic equilibrium will be positive. Thus,

$$\delta E - T_0 \delta S + p_0 \delta V \geq 0, \tag{3.1.21}$$

where δE, δS, and δV are, respectively, the deviations of the internal energy, entropy, and volume from their equilibrium values in contact with the reservoir at temperature T_0 and pressure p_0. Eq. (3.1.21) is valid for arbitrary values of δE, δS, and δV. When applied to infinitesimal values of these parameters, Eq. (3.1.21) leads to positivity constraints on equilibrium derivatives. If $\delta E(\delta S, \delta V)$ is expanded in a power series, the first order terms in δS and δV on the left hand side of Eq. (3.1.21) will vanish because the equilibrium temperature and pressure of the fluid are equal to those of the reservoir. Thus, to second order in δS and δV, Eq. (3.1.21) becomes

$$\frac{1}{2}\frac{\partial^2 E}{\partial S^2}(\delta S)^2 + \frac{\partial^2 E}{\partial S \partial V}\delta S \delta V + \frac{1}{2}\frac{\partial^2 E}{\partial V^2}(\delta V)^2 \geq 0. \tag{3.1.22}$$

This equation is satisfied provided

$$\frac{\partial^2 E}{\partial S^2} = \left.\frac{\partial T}{\partial S}\right)_V \geq 0, \tag{3.1.23}$$

$$\frac{\partial^2 E}{\partial V^2} = -\left.\frac{\partial p}{\partial V}\right)_S \geq 0, \tag{3.1.24}$$

$$\frac{\partial^2 E}{\partial S^2}\frac{\partial^2 E}{\partial V^2} - \left(\frac{\partial^2 E}{\partial S \partial V}\right)^2 = -\left[\left.\frac{\partial T}{\partial S}\right)_V \left.\frac{\partial p}{\partial V}\right)_S - \left.\frac{\partial T}{\partial V}\right)_S \left.\frac{\partial p}{\partial S}\right)_V\right]$$
$$= -\frac{(\partial p/\partial V)_T}{(\partial S/\partial T)_V} \geq 0. \tag{3.1.25}$$

Thus, in thermodynamic equilibrium, the *heat capacity* at constant volume C_V, and the isothermal and isentropic compressibilities, κ_S and κ_T, must be non-negative:

$$C_V = T\left.\frac{\partial S}{\partial T}\right)_V \geq 0, \tag{3.1.26}$$

$$\kappa_S = -\frac{1}{V}\left.\frac{\partial V}{\partial p}\right)_S \geq 0, \tag{3.1.27}$$

$$\kappa_T = -\frac{1}{V}\left.\frac{\partial V}{\partial p}\right)_p \geq 0. \tag{3.1.28}$$

A straightforward exercise in partial derivatives yields the relation

$$C_p = T\frac{\partial S}{\partial T}\bigg)_p = C_V + T\frac{[(\partial V/\partial T)_p]^2}{V\kappa_T} \tag{3.1.29}$$

between C_V and the heat capacity at constant pressure C_p, implying that, in equilibrium, $C_p \geq C_V \geq 0$. Heat capacities are extensive. It is often useful to consider an intensive measure of heat capacity obtained by dividing C_V or C_p by an extensive measure of the quantity of matter such as the volume V, the particle number N, or the mass M, and one introduces the *specific heats*

$$c_V = C_V/V, \quad \tilde{c}_V = C_V/M = c_V/\rho, \tag{3.1.30}$$

where $\rho = M/V$ is the mass density. We will encounter the mass specific heat $\tilde{c}_V$ in our study of hydrodynamic modes of a fluid in Chapter 8.

6 Homogeneous functions

The fact that extensive parameters increase in proportion to the size of the system leads to simple but important constraints on the thermodynamic potentials. The internal energy is a function $Y(S, V, N)$ of the extensive variables S, V, and N. If the size of the system increases by a factor b, all of the extensive parameters increase by a factor b, but the function Y does not change. Therefore,

$$E = Y(S, V, N) \quad \text{and} \quad bE = Y(bS, bV, bN). \tag{3.1.31}$$

A function $f(x)$ is said to be *homogeneous of degree k* if it satisfies $f(x) = b^k f(bx)$. E is a homogeneous function of S, V, and N of degree -1. Eq. (3.1.31) is true for arbitrary b, and in particular for $b = V^{-1}$, leading to

$$E = V\varepsilon(s, n), \tag{3.1.32}$$

where ε, s, and n are, respectively, the energy density, the entropy density, and the particle density:

$$\varepsilon = \frac{E}{V}, \quad s = \frac{S}{V}, \quad n = \frac{N}{V}. \tag{3.1.33}$$

Similar arguments lead to

$$F(T, V, N) = Vf(n, T), \tag{3.1.34}$$

$$G(T, p, N) = Ng(p, T), \tag{3.1.35}$$

$$\mathscr{A}(T, V, \mu) = Va(T, \mu), \tag{3.1.36}$$

where $f = F/V$ is the Helmholtz free energy density, $a = \mathscr{A}/V$ is the grand potential density, and $g = G/N$ is the Gibbs free energy per particle.

Eqs. (3.1.31) to (3.1.36) show that the extensive thermodynamic potentials have a trivial dependence on one of the extensive parameters that lead to an identification of the free energy densities with intensive generalized forces. g is the chemical potential and a is minus pressure:

$$\frac{\partial G}{\partial N}\bigg)_{T,p} \equiv g(p, T) = \mu, \tag{3.1.37}$$

$$\left.\frac{\partial \mathscr{A}}{\partial V}\right)_{T,\mu} \equiv a(T,\mu) = -p = f - \mu n. \tag{3.1.38}$$

Note that the volume density of the extensive internal energy is a function only of the volume densities of the extensive quantities S and N, whereas the intensive pressure is a function only of the intensive parameters T and μ. Quasi-static changes in these quantities are related by the differential relations

$$d\varepsilon = T ds + \mu dn, \tag{3.1.39}$$

$$dp = n d\mu + s dT, \tag{3.1.40}$$

as can be seen by using $E = V\varepsilon$ in Eq. (3.1.8) and $\mathscr{A} = -Vp$ in Eq. (3.1.15). Eq. (3.1.39) says that the energy density changes in response to changes in the density and entropy density rather than to particle number, volume, and entropy separately. This allows us to discuss the properties of a system by considering some fixed sub-volume that is able to exchange particles and energy with the rest of the system through some imaginary wall. Eq. (3.1.40) says that changes in pressure at constant temperature are equivalent to the density times changes in the chemical potential. This implies, for example, that the isothermal compressibility can be expressed as a derivative of particle density with respect to chemical potential rather than of volume with respect to pressure:

$$\kappa_T = -\frac{1}{V}\left.\frac{\partial V}{\partial p}\right)_{T,N} = -\frac{1}{V}\left.\frac{\partial(N/n)}{n\partial\mu}\right)_{T,N} = n^{-2}\left.\frac{\partial n}{\partial \mu}\right)_T . \tag{3.1.41}$$

Thus, $n^2\kappa_T$ is the derivative of the density n with respect to its thermodynamically conjugate potential μ. This form of the compressibility is the one that makes the cleanest contact with more generalized susceptibilities to be discussed in Sec. 3.6.

7 Equations of state

Generalized forces and extensive parameters or their densities are related via *equations of state* obtained by the appropriate derivative of the thermodynamic potentials. For example, the pressure as a function of T, V, and N is obtained by differentiating F with respect to V

$$p(T,V,N) = -\left.\frac{\partial F}{\partial V}\right)_{T,N} . \tag{3.1.42}$$

Since p is an intensive function of the extensive variables V and N, it must be a function of the density only since

$$p(T,V,N) = p(T,bV,bN) \equiv p(T,n), \tag{3.1.43}$$

where the last expression was obtained by setting $b = V^{-1}$. An alternative equation of state expresses the chemical potential as a function of n and T:

$$\mu(T,n) = \left.\frac{\partial F}{\partial N}\right)_{T,V} \equiv \left.\frac{\partial f}{\partial n}\right)_T . \tag{3.1.44}$$

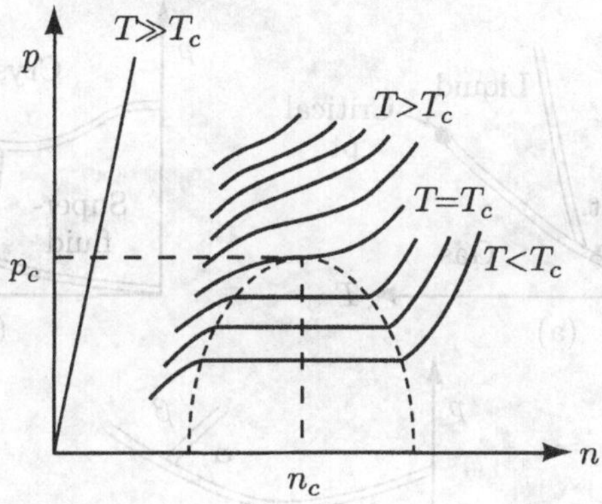

Fig. 3.1.3. Pressure as a function of density at different temperatures for a typical fluid.

Note that Eq. (3.1.42) implies (3.1.44) and vice versa since $-p = f - \mu n$.

Fig. 3.1.3 shows a typical equation of state which follows the experimental observations for a typical fluid as a function of density. Note that for each value of n, there is a unique value of p at fixed T. At low temperatures, however, there are two possible values of the density for a given pressure. This corresponds to the coexistence of liquid and solid phases with the same pressure, temperature and chemical potential. The phase diagrams for a "typical" system (say argon) as well as the quantum systems He^3 and He^4 are shown in Fig. 3.1.4. There is a line along which the liquid and gas phases coexist terminating in a critical point beyond which only a single value of the density is possible for a given pressure. At high pressure, both He^3 and He^4 solidify. At lower pressure, quantum zero point motion prevents solidification. In He^4, there is a superfluid phase below about 4 K. There are also superfluid phases (in the millikelvin range) in He^3 not shown in the phase diagram.

The negative of the pressure as a function of μ and T can be expressed via Eqs. (3.1.38) and (3.1.44) as a minimum over n of the function $w(\mu, T, n) = f(T, n) - \mu n$:

$$-p(\mu, T) = [f(T, n) - \mu n]_{\min n} \equiv w(\mu, T, n)|_{\min n}. \qquad (3.1.45)$$

This formula will be particularly useful in phenomenological treatments of the liquid-gas and liquid-solid transitions to be discussed in the next chapter.

3.2 Statistical mechanics: phase space and ensembles

Thermodynamics is an empirically based science that requires no knowledge of microscopic interactions. Statistical mechanics establishes a connection between the microscopic and macroscopic thermodynamic descriptions of a system. This

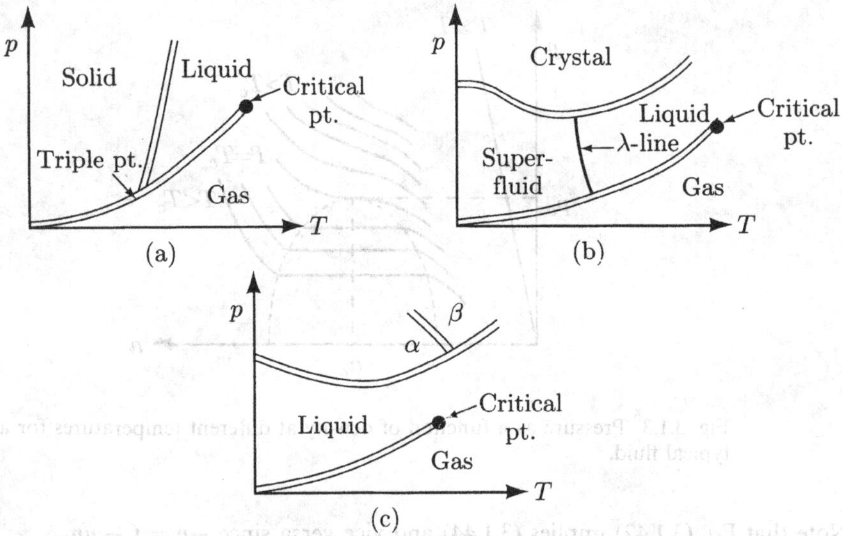

Fig. 3.1.4. Phase diagrams in the pressure-temperature plane for (a) a classical fluid such as argon, (b) He^4, and (c) He^3. Double lines represent discontinuous or first-order transitions and single lines represent continuous or second-order transitions

section will be devoted to the statistical mechanics of homogeneous fluids. In addition to reviewing ensembles and partition functions, it will establish notation, which will be used throughout this book in applications to more complex systems.

The state of a closed system consisting of N classical point particles in a d-dimensional space at time t is specified by the coordinates $\mathbf{x}_\alpha(t)$ and momenta $\mathbf{p}_\alpha(t)$, $\alpha = 1, ..., N$, of the N particles, or equivalently by the vector $\mathbf{R} = [\mathbf{x}_\alpha(t), \mathbf{p}_\alpha(t)]$ in an Nd-dimensional phase space. $\mathbf{x}_\alpha(t)$ and $\mathbf{p}_\alpha(t)$ evolve in time according to Newton's laws along a trajectory in phase space. In quantum mechanical systems, $\mathbf{R}$ is an operator, and the state of the system is specified by an N-particle wave function. If the state of the system (classical or quantum mechanical) is precisely known at some time t_0, its state at future times is, in principle, determined by its Hamiltonian $\mathcal{H}$. It is impossible, however, to have complete knowledge of the state of a system consisting of 10^{23} particles. In practice, detailed knowledge of only a few macroscopic properties such as the volume or the energy is possible, and a statistical description is the only reasonable one.

In an isolated fluid, the internal energy E is fixed. The Hamiltonian $\mathcal{H}$ depends on $\mathbf{x}_\alpha$, $\mathbf{p}_\alpha$, on the volume V, and possibly on other externally controllable parameters, which will not be considered at present. In the absence of any other knowledge about the system, there is no reason to favor any state (i.e., point in phase space) with energy E over any other. It is, therefore, postulated that all points in phase space with a given energy are equally likely and that the macroscopic properties of the fluid can be obtained by averaging over the ensemble of states with fixed energy. This is the *microcanonical ensemble.* In this ensemble, the probability of occurrence of a given point in phase space is

$$P_{\text{micro}}(\mathbf{R}) = \frac{\delta[E - \mathcal{H}(\mathbf{R}, V)]}{\omega(E, V)}, \tag{3.2.1}$$

where $\omega(E, V)$ is the density of states chosen so that the sum of P_{micro} over all states $\mathbf{R}$ is unity:

$$\omega(E, V) = \text{Tr}\delta[E - \mathcal{H}(\mathbf{R}, V)]. \tag{3.2.2}$$

In quantum mechanical systems, $\delta(E - \mathcal{H})$ is an operator or matrix called the microcanonical density matrix. The trace in Eq. (3.2.2) is over all states of the system. In classical systems, it is an integral over phase space:

$$\text{Tr} \equiv \frac{1}{N!} \prod_{\alpha} \int \frac{d^d p^\alpha d^d x^\alpha}{h^{dN}}, \tag{3.2.3}$$

where h is Planck's constant. The factor of $(N!h^{dN})^{-1}$ arises when the classical limit of the quantum mechanical trace operation is taken. It is, strictly speaking, not necessary in classical statistical mechanics for a fixed number of particles, but is needed for a proper definition of the entropy and evaluation of the chemical potential. The absence of this term leads to the Gibbs paradox in which the entropy of a classical gas is not extensive.

The entropy of a system in equilibrium is simply the logarithm of the number of configurations available at energy E,

$$S = \ln[\omega(E, V)\Delta E], \tag{3.2.4}$$

where ΔE is some energy interval representing, for example, the precision with which the energy of the system can actually be determined. Eqs. (3.2.4) and (3.2.1) provide a statistical mechanical interpretation of the second law. There are more configurations available to a system whenever constraints are removed. Thus the entropy increases. The most likely state of the system is one which occurs with maximum probability. If constraints are removed, the new equilibrium will have a higher entropy and thus higher probability than that with the constraints in effect.

The microcanonical ensemble is not the most useful ensemble for doing calculations because it involves difficult sums over states with constrained energies. Two other ensembles are far more useful. In the *canonical ensemble,* the density matrix is $e^{-\beta\mathcal{H}}$, where $\beta = T^{-1}$ (we use units in which the Boltzmann constant k_B is equal to one) and the probability matrix is

$$P_c = \frac{1}{Z_N(T, V)} e^{-\beta\mathcal{H}}, \tag{3.2.5}$$

where

$$\begin{aligned} Z_N(T, V) &= \text{Tr}e^{-\beta\mathcal{H}} \\ &= \frac{1}{N!h^{dN}} \int d\mathbf{R} e^{-\beta\mathcal{H}} \quad \text{(classical)} \end{aligned} \tag{3.2.6}$$

is the partition function. In the grand canonical ensemble, the number of particles is allowed to vary, and the probability of a given state is

$$P_{gc} = \frac{1}{\Xi(T,\mu,V)} e^{-\beta(\mathscr{H}-\mu N)}, \tag{3.2.7}$$

where

$$\Xi(T,\mu,V) = \mathrm{Tr} e^{-\beta(H-\mu N)} \tag{3.2.8}$$

is the grand partition function. The trace is over all states with all possible numbers of particles. For classical systems,

$$\Xi(T,\mu,V) = \sum_N e^{\beta\mu N} Z_N, \tag{3.2.9}$$

where it should be remembered that there is a factor of $(N!)^{-1}$ included in the definition of Z_N. In neither the canonical nor the grand canonical ensemble is the energy fixed.

The average energy is the average of the Hamiltonian over the ensemble:

$$\langle E \rangle \equiv \langle \mathscr{H} \rangle = \mathrm{Tr} P \mathscr{H}, \tag{3.2.10}$$

where P is either P_c or P_{gc}. $\langle E \rangle$ can be obtained in either ensemble by differentiating the partition function with respect to β,

$$\langle E \rangle = -\frac{1}{Z_N}\frac{\partial Z_N}{\partial \beta}\bigg|_{N,V} = -\frac{\partial \ln Z_N}{\partial \beta}\bigg|_{N,V}, \tag{3.2.11}$$

$$\langle E \rangle = -\frac{1}{\Xi}\frac{\partial \Xi}{\partial \beta}\bigg|_{\beta\mu,V} = -\frac{\partial \ln \Xi}{\partial \beta}\bigg|_{\beta\mu,V}. \tag{3.2.12}$$

Fluctuations in $\mathscr{H}$ can similarly be calculated:

$$\langle E^2 \rangle = \frac{1}{Z_N} \mathrm{Tr} e^{-\beta\mathscr{H}} \mathscr{H}^2 = \frac{1}{Z_N}\frac{\partial^2 Z_N}{\partial \beta^2} \tag{3.2.13}$$

and

$$\langle (\delta E)^2 \rangle \equiv \langle E^2 \rangle - \langle E \rangle^2 = \frac{\partial^2 \ln Z_N}{\partial \beta^2} = -\frac{\partial \langle E \rangle}{\partial \beta}. \tag{3.2.14}$$

Similar expressions apply for the grand canonical ensemble. In the grand canonical ensemble, the number of particles is not fixed. The average number of particles and the fluctuation in the number of particles are, respectively,

$$\langle N \rangle = \frac{1}{\Xi} \mathrm{Tr} e^{-\beta(\mathscr{H}-\mu N)} N, \tag{3.2.15}$$

and

$$\langle (\delta N)^2 \rangle \equiv \langle N^2 \rangle - \langle N \rangle^2 = \frac{1}{\Xi} \mathrm{Tr} e^{-\beta[\mathscr{H}-\mu N]} (N - \langle N \rangle)^2. \tag{3.2.16}$$

These quantities can be expressed as derivatives of the logarithm of the grand partition function with respect to $\beta\mu$. Because the particle number is conserved, the operator N commutes with the Hamiltonian $\mathscr{H}$: $[\mathscr{H}, N] = 0$. This implies

$$e^{-\beta(\mathscr{H}-\mu N)} = e^{-\beta\mathscr{H}} e^{\beta\mu N}, \tag{3.2.17}$$

$$\langle N \rangle = \frac{\partial \ln \Xi}{\partial \beta\mu}, \tag{3.2.18}$$

and

$$\langle(\delta N)^2\rangle = \frac{\partial^2 \ln \Xi}{\partial(\beta\mu)^2} = \frac{\partial\langle N\rangle}{\partial\beta\mu}. \tag{3.2.19}$$

The relation in Eq. (3.2.18) would be valid whether or not N commuted with $\mathscr{H}$. Eq. (3.2.19) on the other hand would not be correct if N did not commute with $\mathscr{H}$. We will return to the generalizations of Eq. (3.2.19) to variables that do not commute with $\mathscr{H}$ in Chapter 7.

Both $\langle E\rangle$ and $\langle N\rangle$ are extensive quantities that grow with the size of the sample. It is clear from Eqs. (3.2.14) and (3.2.16) that the fluctuations in E and N are also extensive quantities. Therefore, the relative fluctuations of the two quantities go to zero as the square root of the number of particles: $\langle(\delta E)^2\rangle^{1/2}/\langle E\rangle \sim N^{-1/2}$ and $\langle(\delta N)^2\rangle^{1/2}/\langle N\rangle \sim \langle N\rangle^{-1/2}$ so that, in large systems, the distinction between the fixed energy of the microcanonical ensemble and the average energy of the canonical ensemble is unimportant. The energy E of the microcanonical ensemble or the average energy $\langle E\rangle$ of the canonical and grand canonical ensembles is the thermodynamic internal energy. In the future, we will only distinguish between $\langle E\rangle$ and E when necessary.

The partition functions Z_N and Ξ are related to the thermodynamic potentials F and $\mathscr{A}$ introduced in the preceding sections:

$$F(T, N, V) = -T \ln Z_N(T, V), \tag{3.2.20}$$

$$\mathscr{A}(T, \mu, V) = -T \ln \Xi(T, \mu, V). \tag{3.2.21}$$

The first of these results is most easily obtained by recognizing that $\omega(E)e^{-\beta E}$ is a strongly peaked function in the vicinity of $E = \langle E\rangle$:

$$Z_N = \int dE\omega(E)e^{-\beta E} \sim e^{-\beta(\langle E\rangle - TS)}. \tag{3.2.22}$$

Similar expressions apply for Ξ.

The heat capacity and compressibility are related, respectively, to fluctuations in the energy and particle number,

$$C_V = -T\frac{\partial^2 F}{\partial T^2} = T^{-2}\frac{\partial^2 \ln Z_N}{\partial\beta^2} = \frac{\langle(\delta E)^2\rangle}{T^2} \geq 0, \tag{3.2.23}$$

and

$$\kappa_T = \langle n\rangle^{-2}\frac{\partial\langle n\rangle}{\partial\mu} = \frac{V}{T}\frac{\langle(\delta N)^2\rangle}{\langle N\rangle^2} \geq 0. \tag{3.2.24}$$

Thus, we see that the positivity of C_V and κ_T, which emerged as a condition for equilibrium in the thermodynamic analysis of the preceding section, emerges in a statistical mechanical analysis as a result of the positivity of the variances of random variables.

3.3 The ideal gas

One of the simplest statistical systems is the gas of non-interacting particles of mass m in a volume V. The Hamiltonian for this system consists only of the kinetic energy,

$$\mathcal{H} = \mathcal{H}_{\text{kin}} = \sum_{\alpha} \frac{\mathbf{p}_{\alpha}^2}{2m}, \tag{3.3.1}$$

plus the infinite potential barriers confining the particles to volume V. The partition function is easily evaluated as

$$Z_N = \frac{V^N}{N!}\left(\int \frac{dp}{h} e^{-\beta p^2/2m}\right)^{3N} = \frac{1}{N!}\frac{V^N}{\lambda^{3N}}, \tag{3.3.2}$$

where

$$\lambda = \frac{h}{(2\pi m T)^{1/2}} \tag{3.3.3}$$

is the thermal wavelength of a particle. The Helmholtz free energy density follows from Eq. (3.2.20)

$$f = Tn[\ln(n\lambda^3) - 1]. \tag{3.3.4}$$

The internal energy, chemical potential, pressure, and entropy follow from Eqs. (3.2.11), (3.1.44), (3.1.42), and (3.1.16):

$$E = \frac{3}{2}NT, \tag{3.3.5}$$

$$\mu = T\ln(n\lambda^3), \tag{3.3.6}$$

$$p = Tn, \tag{3.3.7}$$

$$S = N\left[\frac{5}{2} - \ln(n\lambda^3)\right]. \tag{3.3.8}$$

Eq. (3.3.7) is the familiar equation of state for an ideal gas, often written as $PV = Nk_BT$ (where we have taken the Boltzmann constant k_B as 1). Finally, the specific heat at constant volume and isothermal compressibility are

$$C_V = \frac{3}{2}N, \tag{3.3.9}$$

$$\kappa_T = \frac{1}{nT} = \frac{1}{p}. \tag{3.3.10}$$

There are a number of points to note about Eqs. (3.3.5) to (3.3.10), which are classical results that we expect to break down as quantum effects become important. The important unitless variable is $n\lambda^3$, the number of particles in a thermal volume λ^3. The density can be considered small as long as this number is small. When $n\lambda^3$ becomes of order *one* or more, quantum effects become important, and the classical approximation is no longer valid. The high density behavior of a non-interacting gas of particles depends critically on the statistics of the particles. The high density or low temperature behavior of a non-interacting Fermi gas is controlled by the localization kinetic energy imposed by the exclusion principle. A finite fraction of the particles in a non-interacting gas of Bose particles will condense into a single macroscopic quantum state at low temperature giving rise to the phenomenon of superfluidity. A superfluid phase exists in interacting systems as well as indicated in the phase diagram of Fig. 3.1.4.

The internal energy of a non-interacting gas is $T/2$ times the number of degrees of freedom (the "equipartition" theorem). It depends only on the temperature and not on the density of particles. Similarly, p/T depends only on the density and not on temperature. These simple behaviors are modified in the presence of interactions among the particles.

3.4 Spatial correlations in classical systems

In the preceding chapter, we saw that a great deal of information about condensed systems is contained in spatial correlation functions of the density and other observable parameters. In addition, these correlation functions can be measured in some detail by the appropriate experiment, often a scattering experiment. In this section, we discuss how spatial correlation functions in a classical fluid can be calculated in the context of statistical mechanics. We will delay until Chapter 7 a discussion of similar correlations in quantum systems where the noncommutativity of the local density operator with a Hamiltonian and with itself is important.

In the preceding section, we found that the average energy and average particle number in the grand canonical ensemble could be obtained by differentiating the partition function with respect to β or $\beta\mu$. It is extremely useful to be able to generate the spatially dependent density and its correlation functions by a similar process. In order to do so, it is necessary to introduce external potentials that depend on position. Let u be an externally controlled one-body potential. The potential energy must be augmented by a term

$$U_{\text{ext}} = -\sum_{\alpha} u(\mathbf{x}_\alpha) \equiv -\int d^d x u(\mathbf{x}) n(\mathbf{x}), \tag{3.4.1}$$

so that the total Hamiltonian becomes

$$\mathscr{H} = \mathscr{H}_{\text{kin}} + U + U_{\text{ext}}, \tag{3.4.2}$$

where $\mathscr{H}_{\text{kin}}$ is the kinetic energy and U is the potential energy associated with interactions among the constituent particles of the fluid. If u is changed by an amount $\delta u(\mathbf{x})$, U_{ext} changes according to

$$\delta U_{\text{ext}} = -\int d^d x \delta u(\mathbf{x}) n(\mathbf{x}). \tag{3.4.3}$$

The partition function can be expanded in powers of δu:

$$Z_N[T,V,u(\mathbf{x})+\delta u(\mathbf{x})] = \text{Tr}e^{-\beta H}[1+\int d^d x\beta\delta u(\mathbf{x})n(\mathbf{x})$$
$$+\frac{1}{2}\int d^d x\int d^d x'\beta\delta u(\mathbf{x})\beta\delta u(\mathbf{x}')n(\mathbf{x})n(\mathbf{x}')+...], \tag{3.4.4}$$

or

$$\frac{Z_N[T,V,u(\mathbf{x})+\delta u(\mathbf{x})]}{Z_N[T,V,u(\mathbf{x})]} = 1+\int d^d x\beta\delta u(\mathbf{x})\langle n(\mathbf{x})\rangle$$
$$+\frac{1}{2}\int d^d x\int d^d x\beta\delta u(\mathbf{x})\beta\delta u(\mathbf{x}')\langle n(\mathbf{x})n(\mathbf{x}')\rangle+..., \tag{3.4.5}$$

where it is understood that $\mathscr{H}$ is evaluated at $\delta u = 0$. The partition function $Z_N[T,V,u(\mathbf{x})]$ is a function of $u(\mathbf{x})$ at every point in space, i.e., it is a function of the function $u(\mathbf{x})$. A function of a function is called a *functional*, and $Z_N[T,V,u(\mathbf{x})]$ is a functional of $u(\mathbf{x})$.

A functional $R[t(\mathbf{x})+\eta(\mathbf{x})]$ of a function $t(\mathbf{x})$ can be expanded in a Taylor series in $\eta(\mathbf{x})$:

$$R[t(\mathbf{x})+\eta(\mathbf{x})] = R[t(\mathbf{x})]+\int d^d x\frac{\delta R}{\delta t(\mathbf{x})}\eta(\mathbf{x}) \tag{3.4.6}$$
$$+\frac{1}{2}\int d^d x\int d^d x'\frac{\delta^2 R}{\delta t(\mathbf{x})\delta t(\mathbf{x}')}\eta(\mathbf{x})\eta(\mathbf{x}')+....$$

$\delta R/\delta t(\mathbf{x})$ is the functional derivative of R with respect to the function $t(\mathbf{x})$. Note that each term in the expansion depends on $\delta t(\mathbf{x})$ at all points $\mathbf{x}$, indicating that functional derivatives are merely the continuum generalization of partial differentiation of a function of many variables. For further details about functional differentiation, see the appendix to this chapter.

The coefficient of $\delta u(\mathbf{x})$ in the integrand in the second term in the expansion of Eq. (3.4.5) is the average density $\langle n(\mathbf{x})\rangle$, and the coefficient of $\delta u(\mathbf{x})\delta u(\mathbf{x}')$ in the third term is the density-density correlation function $C_{nn}(\mathbf{x},\mathbf{x}') = \langle n(\mathbf{x})n(\mathbf{x}')\rangle$. Using the Eq. (3.4.5), we can now identify the density and density-density correlation function as functional derivatives of $Z_N[T,V,u(\mathbf{x})]$ with respect to $u(\mathbf{x})$:

$$\langle n(\mathbf{x})\rangle = \frac{1}{Z_N}\frac{\delta Z_N}{\beta\delta u(\mathbf{x})}\Bigg)_{T,V,N} = \frac{\delta\ln Z_N}{\delta\beta u(\mathbf{x})}\Bigg)_{T,N,V} = -\frac{\delta F}{\delta u(\mathbf{x})}\Bigg)_{T,N,V} \tag{3.4.7}$$

and

$$C_{nn}(\mathbf{x},\mathbf{x}') = \frac{1}{Z_N}\frac{\delta^2 Z_N}{\beta\delta u(\mathbf{x})\beta\delta u(\mathbf{x}')}\Bigg)_{T,V,N}. \tag{3.4.8}$$

The fluctuation in $n(\mathbf{x})$ is determined by the second derivative of $\ln Z_N[u(\mathbf{x})]$ with respect to $\beta u(\mathbf{x})$:

$$S_{nn}(\mathbf{x},\mathbf{x}') = \frac{\delta^2\ln Z_N}{\beta\delta u(\mathbf{x})\beta\delta u(\mathbf{x}')}\Bigg)_{T,V,N} = \frac{\delta\langle n(\mathbf{x})\rangle}{\delta\beta u(\mathbf{x}')}\Bigg)_{T,V,N}$$
$$= -T\frac{\delta^2 F}{\delta u(\mathbf{x})\delta u(\mathbf{x}')}\Bigg)_{T,V,N}, \tag{3.4.9}$$

where $S_{nn}(\mathbf{x},\mathbf{x}')$ is the Ursell function defined in Eq. (2.3.5).

To gain familiarity with the use of functional derivatives (whose properties are reviewed in the appendix), it is useful to derive the pair distribution function for the noninteracting classical gas from $Z_N[T,V,u(\mathbf{x})]$, which can easily be evaluated,

$$Z_N = \frac{1}{N!}(q\lambda^{-d})^N, \tag{3.4.10}$$

where

$$q = \int d^d x e^{\beta u(\mathbf{x})} \overset{u\to 0}{\to} V. \tag{3.4.11}$$

Thus we have

$$\langle n(\mathbf{x})\rangle = N\frac{\delta \ln q}{\beta \delta u(\mathbf{x})} = Nq^{-1}e^{\beta u(\mathbf{x})} \overset{u\to 0}{\to} \frac{N}{V} \tag{3.4.12}$$

and

$$\begin{aligned} S_{nn}(\mathbf{x},\mathbf{x}') &= Nq^{-1}e^{\beta u(\mathbf{x})}\delta(\mathbf{x}-\mathbf{x}') - Nq^{-2}e^{\beta u(\mathbf{x})}e^{\beta u(\mathbf{x}')} \\ &= \langle n(\mathbf{x})\rangle\delta(\mathbf{x}-\mathbf{x}') - \frac{1}{N}\langle n(\mathbf{x})\rangle\langle n(\mathbf{x}')\rangle \\ &\overset{N\to\infty}{\to} \langle n(\mathbf{x})\rangle\delta(\mathbf{x}-\mathbf{x}'). \end{aligned} \tag{3.4.13}$$

If $u(\mathbf{x})$ is nonzero and nonuniform, the equilibrium density is nonuniform. External nonuniform potentials provide a natural way to discuss inhomogeneous systems. It should be noted that Eq. (3.4.13) with the aid of Eqs. (2.3.5) and (2.3.11) yield the pair distribution function for a noninteracting gas:

$$g(\mathbf{x},\mathbf{x}') = (1-N^{-1}) \overset{N\to\infty}{\to} 1. \tag{3.4.14}$$

Thus, the formal manipulations presented here yield the result discussed in the preceding chapter that the pair distribution function is unity in the noninteracting limit.

The introduction of an external inhomogeneous potential is in many ways simpler and more instructive in the grand canonical ensemble. The external potential, like the chemical potential, couples linearly to the density, and one can interpret the external potential as a shift in the chemical potential:

$$\mu \to \mu(\mathbf{x}) = \mu + u(\mathbf{x}). \tag{3.4.15}$$

The grand partition function is now a functional of $\mu(\mathbf{x})$,

$$\Xi[T,V,\mu(\mathbf{x})] = \operatorname{Tr}\exp\left[-\beta\left(\mathscr{H} - \int d^d x \mu(\mathbf{x})n(\mathbf{x})\right)\right]. \tag{3.4.16}$$

Following the same steps as for the canonical ensemble, we obtain

$$\langle n(\mathbf{x})\rangle = \frac{\delta \ln \Xi[\mu(\mathbf{x})]}{\beta\delta\mu(\mathbf{x})} = -\frac{\delta\mathscr{A}[\mu(\mathbf{x})]}{\delta\mu(\mathbf{x})} \tag{3.4.17}$$

and

$$\beta S_{nn}(\mathbf{x},\mathbf{x}') = -\frac{\delta^2\mathscr{A}[\mu(\mathbf{x})]}{\delta\mu(\mathbf{x})\delta\mu(\mathbf{x}')} = \frac{\delta\langle n(\mathbf{x})\rangle}{\delta\mu(\mathbf{x}')}. \tag{3.4.18}$$

The last expression can be used to derive an important relation between the Ursell function (and hence the structure factor) and the compressibility. In Eq. (3.4.17),

$\langle n(\mathbf{x})\rangle$ is a functional of $\mu(\mathbf{x})$ and can be expanded in a Taylor series,

$$\begin{aligned}\langle n(\mathbf{x})\rangle - \langle n(\mathbf{x})\rangle_0 &= \int d^d x' \frac{\delta \langle n(\mathbf{x})\rangle}{\delta \mu(\mathbf{x}')} \delta\mu(\mathbf{x}') + ... \\ &= \int d^d x' \beta S_{nn}(\mathbf{x}, \mathbf{x}') \delta\mu(\mathbf{x}') + ... ,\end{aligned} \tag{3.4.19}$$

where $\langle n(\mathbf{x})\rangle_0$ is the density at $\delta\mu(\mathbf{x}) = 0$. The compressibility is related to the derivative of n with respect to a spatially constant chemical potential. If $\delta\mu$ is spatially constant, then $\delta\mu(\mathbf{x}) \equiv \delta\mu$ is independent of $\mathbf{x}$. Then, using Eq. (3.1.41) and Eq. (3.4.19), we obtain

$$\frac{\partial \langle n\rangle}{\partial \mu} = \langle n\rangle^2 \kappa_T = \beta \int d^d x S_{nn}(\mathbf{x}, \mathbf{x}') = \beta S_{nn}(\mathbf{q} = 0). \tag{3.4.20}$$

This is a very important and general equation relating the derivative of the density of an extensive parameter with respect to its conjugate field to the zero wave number fluctuations of the density of the same parameter.

We close this section with the observation that a Helmholtz free energy functional of the density $n(\mathbf{x})$ can be obtained via a generalized Legendre transformation on the grand potential $\mathscr{A}[\mu(\mathbf{x})]$ with a spatially varying chemical potential. Let

$$F[T, V, \langle n(\mathbf{x})\rangle] = \mathscr{A}[T, V, \mu(\mathbf{x})] + \int d^d x \langle n(\mathbf{x})\rangle \mu(\mathbf{x}), \tag{3.4.21}$$

then

$$\frac{\delta F(T, V, \langle n(\mathbf{x})\rangle)}{\delta \langle n(\mathbf{x})\rangle} = \mu(\mathbf{x}). \tag{3.4.22}$$

This is the generalization of the equation of state [Eq. (3.1.44)] to spatially inhomogeneous situations. If $\mu(\mathbf{x})$ is spatially uniform, F can be expressed as the volume integral of a local free energy functional of $\langle n(\mathbf{x})\rangle \equiv \langle n\rangle$, $F = \int d^d x f(T, \langle n\rangle)$, and Eq. (3.4.22) reduces to Eq. (3.1.44). More generally, as we shall see in subsequent chapters, F can depend on $\langle n(\mathbf{x})\rangle$ at different points in space. The concept that free energies can be functions of spatially varying densities of extensive variables is a very important one that has applications to many problems in condensed matter systems. Its generalizations to more complicated order will be used extensively in discussions that follow.

The expression [Eq. (3.1.45)] for the pressure as a minimum over n of the function $f(T, n) - \mu n$ has a straightforward generalization to situations in which $\langle n(\mathbf{x})\rangle$ is allowed to vary in space. Let

$$W[T, V, \mu(\mathbf{x}), \langle n(\mathbf{x})\rangle] = F[T, V, \langle n(\mathbf{x})\rangle] - \int d^d x \langle n(\mathbf{x})\rangle \mu(\mathbf{x}). \tag{3.4.23}$$

Then, when $\mu(\mathbf{x}) = \mu$ is independent of $\mathbf{x}$,

$$\begin{aligned}p(\mu, T) &= -\lim_{V\to\infty} \frac{1}{V} W[T, V, \mu, \langle n(\mathbf{x})\rangle]_{\min \langle n(\mathbf{x})\rangle} \\ &= -\frac{1}{V} \mathscr{A}(T, V, \mu).\end{aligned} \tag{3.4.24}$$

This relation will be used in future chapters to discuss the equations of state of fluids and solids.

3.5 Ordered systems

In the preceding sections, we have discussed homogeneous, isotropic fluids. The only macroscopic variables needed to characterize the thermodynamic state of fluids are their volume, particle number, and internal energy. As we saw in Chapter 2, however, condensed systems exhibit a wide variety of rotational and positional order not described by the above variables. Clearly, a complete description of ordered phases requires the introduction of new variables quantifying the degree of order and a modification of thermodynamics and statistical mechanics to describe the effects of these variables on energies and entropies.

The simplest order to describe is magnetic order. Consider an ideal gas in which each particle α carries a spin $\mathbf{s}_\alpha$ and an associated magnetic moment $\mu_s \mathbf{s}_\alpha$. In the presence of an external magnetic field $\mathbf{h}$, the gas will develop a magnetic moment $\mathbf{M} = \langle \sum_\alpha \mu_s \mathbf{s}_\alpha \rangle$ proportional to the total number of particles. It is an extensive variable whose density is the intensive magnetization,

$$\langle \mathbf{m}(\mathbf{x}) \rangle = \langle \sum_\alpha \mu_s \mathbf{s}_\alpha \delta(\mathbf{x} - \mathbf{x}_\alpha) \rangle. \tag{3.5.1}$$

This equation shows that the magnetization $\mathbf{m}(\mathbf{x})$ is an operator analogous to the density operator [Eq. (2.3.1)]: particle α contributes to $\mathbf{m}(\mathbf{x})$ only at its position $\mathbf{x}_\alpha$. The integral of $\mathbf{m}(\mathbf{x})$ over all $\mathbf{x}$ is the total magnetic moment, just as the integral of $n(\mathbf{x})$ over all $\mathbf{x}$ is the total number of particles. If the magnetization is spatially uniform, then $\langle \mathbf{m}(\mathbf{x}) \rangle = \mathbf{M}/V$. The internal energy and entropy clearly depend on the variable $\langle \mathbf{m} \rangle$, and it must be included along with the density for a complete thermodynamic description of the classical gas of magnetic particles.

The interaction between spins in the ideal gas just discussed is essentially zero (ignoring dipolar forces), and the magnetization is nonzero only in the presence of an external aligning magnetic field. If there are interactions among spins, it is possible to have a nonzero $\langle \mathbf{m} \rangle$ even when $\mathbf{h}$ is zero. To see how this can come about, it is useful to consider a model in which spins confined to sites on a regular periodic lattice (i.e., the spin positions $\mathbf{x}_\alpha$ are confined to sites $\mathbf{R}_\mathbf{l}$ on a periodic lattice) interact via a nearest neighbor exchange interaction (see Chapter 2). This model is called the Heisenberg model (Heisenberg 1928) of ferromagnetism, and its Hamiltonian can be expressed as

$$\mathscr{H}_{\text{Heis}} = -2J \sum_{<\mathbf{l},\mathbf{l}'>} \mathbf{s}_\mathbf{l} \cdot \mathbf{s}_{\mathbf{l}'}, \tag{3.5.2}$$

where the sum is over nearest neighbor sites on the lattice. When $\mathbf{h}$ is zero, the spins represent internal degrees of freedom, which will seek configurations that, at constant T, will minimize the Helmholtz free energy $F = E - TS$. At high T, F

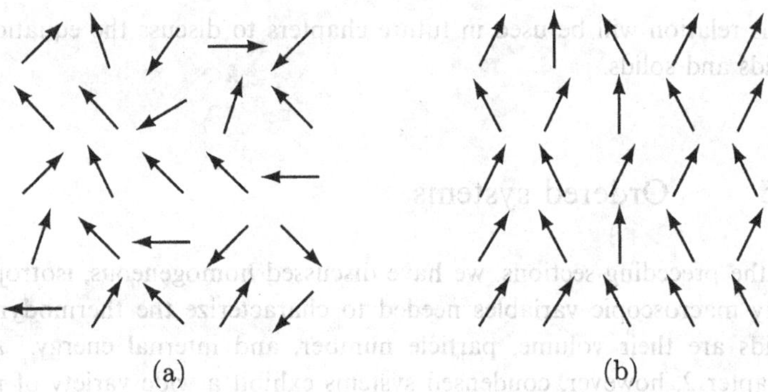

Fig. 3.5.1. (a) The paramagnetic state of the Heisenberg model with zero magnetization. (b) The ferromagnetic state of the Heisenberg model with nonzero magnetization. The direction of the magnetization relative to the lattice is not fixed in the Heisenberg model, whereas in the Ising model it must point either up or down.

is clearly minimized by maximizing the entropy. The maximally disordered state has the highest entropy, implying that the equilibrium state at high temperature is the *paramagnetic* state with no average alignment of spins, i.e., no magnetization (see Fig. 3.5.1). At low temperature, the internal energy dominates over TS, and the state that minimizes F is one that minimizes E. The ground states of $\mathcal{H}_{\text{Heis}}$ are clearly states in which all of the spins are aligned along a common axis. Thus, states which minimize E have a nonvanishing magnetization, and the low temperature equilibrium phase is the *ferromagnetic* phase with nonzero average spin $\langle \mathbf{s} \rangle = \langle \mathbf{s}_\mathbf{l} \rangle$ independent of site $\mathbf{l}$ or equivalently a magnetization $\langle \mathbf{m} \rangle = v_0{}^{-1}\mu_s \langle \mathbf{s} \rangle$, where v_0 is the volume of a unit cell. At some temperature T_c, there is a phase transition from the entropy dominated paramagnetic state to the energy dominated ferromagnetic state. The magnetization $\langle \mathbf{m} \rangle$ is called the *order parameter* of the ferromagnetic phase.

The Heisenberg Hamiltonian is invariant with respect to arbitrary rotations of every spin $\mathbf{s}_\mathbf{l}$. Thus there is no preferred direction for the magnetization in the ferromagnetic state. In order to obtain an unambiguous statistical mechanical characterization of the ferromagnetic state with $\langle \mathbf{m} \rangle$ pointing along a given direction, it is necessary to add an external magnetic field $\mathbf{h}$. The Hamiltonian describing the interaction of spins with a spatially varying external field $\mathbf{h}(\mathbf{x})$ is

$$\begin{aligned} \mathcal{H}_{\text{ext}} &= -\sum_\alpha \mu_s \mathbf{s}_\alpha \cdot \mathbf{h}(\mathbf{x}_\alpha) = -\sum_\mathbf{l} \mu_s \mathbf{s}_\mathbf{l} \cdot \mathbf{h}(\mathbf{R}_\mathbf{l}) \\ &= -\int d^d x \mathbf{h}(\mathbf{x}) \cdot \mathbf{m}(\mathbf{x}). \end{aligned} \tag{3.5.3}$$

The partition function for a lattice of N spins in the presence of $\mathbf{h}$ becomes

$$Z_N[T, \mathbf{h}(\mathbf{x})] = \text{Tr} e^{-\beta[\mathcal{H} - \int d^d x \mathbf{h}(\mathbf{x}) \cdot \mathbf{m}(\mathbf{x})]}, \tag{3.5.4}$$

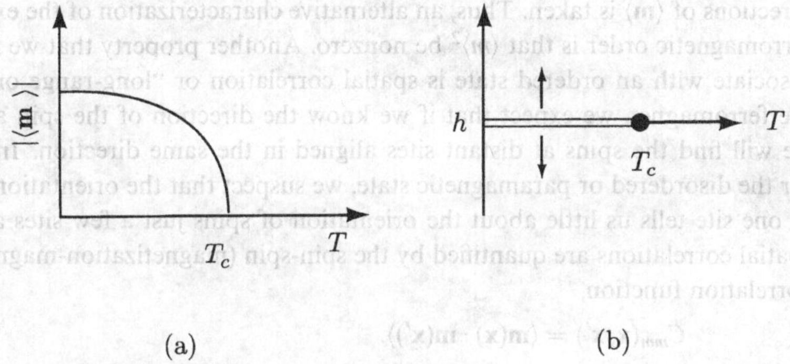

Fig. 3.5.2. (a) The magnetization as a function of temperature for the Heisenberg model. It grows continuously from zero below the ferromagnetic critical temperature T_c. (b) Phase diagram for the Heisenberg ferromagnet in the $h-T$ plane. The line $h=0$, $T \leq T_c$ is a coexistence line where all directions of $\langle \mathbf{m} \rangle$ are energetically equivalent. An approach to the coexistence from nonzero values of h will pick a particular direction of $\langle \mathbf{m} \rangle$.

where $\mathscr{H}$ is the total $\mathbf{h}$-independent part of the Hamiltonian (which for the Heisenberg model would simply be $\mathscr{H}_{\text{Heis}}$). The equilibrium magnetization in the thermodynamic limit is

$$\begin{aligned} \langle \mathbf{m}[T, \mathbf{h}(\mathbf{x}), \mathbf{x}] \rangle &= \lim_{N\to\infty} \frac{1}{Z_N[T, \mathbf{h}(\mathbf{x})]} \mathrm{Tr}[e^{-\beta[\mathscr{H} - \int d^d x \mathbf{h}(\mathbf{x}) \cdot \mathbf{m}(\mathbf{x})]} \mathbf{m}(\mathbf{x})] \\ &\equiv \lim_{N\to\infty} \langle \mathbf{m}_N[T, \mathbf{h}(\mathbf{x}), \mathbf{x}] \rangle. \end{aligned} \tag{3.5.5}$$

In the limit $N \to \infty$, $\langle \mathbf{m}(T, \mathbf{h}, \mathbf{x}) \rangle$ will align along the spatial uniform field $\mathbf{h}$ even for infinitesimally small fields. A zero-field ferromagnetic state with a given direction of $\langle \mathbf{m} \rangle$ can thus be obtained by taking the limit $\mathbf{h} \to 0$ after the thermodynamic limit:

$$\langle \mathbf{m}(T) \rangle = \lim_{\mathbf{h}\to 0} \lim_{N\to\infty} \langle \mathbf{m}_N(T, \mathbf{h}) \rangle. \tag{3.5.6}$$

(The spatial index $\mathbf{x}$ has been suppressed because $\mathbf{h}$ and thus $\langle \mathbf{m} \rangle$ are spatially uniform.)

The phase diagram for a Heisenberg magnet in the $h-T$ plane for a field $\mathbf{h} = h\mathbf{e}$ along a particular direction specified by the unit vector $\mathbf{e}$ is shown in Fig. 3.5.2b. For T less than the critical temperature T_c, there will be a spontaneous magnetization $\langle \mathbf{m}(T) \rangle$ defined by Eq. (3.5.6) aligned parallel to $\mathbf{e}$ for $h \to 0^+$ and antiparallel to $\mathbf{e}$ for $h \to 0^-$. At $h = 0$, there is a coexistence line (analogous to the liquid-gas coexistence curve) along which all directions of $\langle \mathbf{m} \rangle$ are energetically equivalent. The coexistence line terminates at the critical point, $T = T_c$, $h = 0$. Along the coexistence line, there is no reason to choose one direction of $\langle \mathbf{m} \rangle$ over another, and an average over all possible directions of $\langle \mathbf{m} \rangle$ will yield a zero magnetization. The magnitude of $\langle \mathbf{m} \rangle$, or equivalently $\langle m \rangle^2 = \langle \mathbf{m} \rangle \cdot \langle \mathbf{m} \rangle$, however, has the same value whether a particular direction or an average over all possible directions of $\langle \mathbf{m} \rangle$ is taken. Thus, an alternative characterization of the existence of ferromagnetic order is that $\langle m \rangle^2$ be nonzero. Another property that we intuitively associate with an ordered state is spatial correlation or "long-range order". For the ferromagnet, we expect that if we know the direction of the spin at one site we will find the spins at distant sites aligned in the same direction. In contrast, for the disordered or paramagnetic state, we suspect that the orientation of a spin at one site tells us little about the orientation of spins just a few sites away. The spatial correlations are quantified by the spin-spin (magnetization-magnetization) correlation function,

$$C_{mm}(\mathbf{x}, \mathbf{x}') = \langle \mathbf{m}(\mathbf{x}) \cdot \mathbf{m}(\mathbf{x}') \rangle. \tag{3.5.7}$$

In the paramagnetic phase, $C_{mm}(\mathbf{x}, \mathbf{x}')$ dies exponentially to zero in the large separation limit, $|\mathbf{x} - \mathbf{x}'| \to \infty$. In the ferromagnetic phase, it tends to $\langle m \rangle^2$ just as C_{nn} for a fluid [Eq. (2.3.5)] tends to $\langle n \rangle^2$. Thus,

$$C_{mm}(\mathbf{x}, \mathbf{x}') \xrightarrow{|\mathbf{x}-\mathbf{x}'| \to \infty} \langle m \rangle^2, \tag{3.5.8}$$

and we see that the large-separation behavior of a correlation function determines the existence of ferromagnetic order. A nonvanishing of the magnetization and long-range order go hand in hand.

The partition function defined in Eq. (3.5.4) has explicit dependence on the number of particles N but not on the volume of the sample, which might change, for example, if the lattice parameter changed. More generally, one should consider a generalization $Z_N[T, V, \mathbf{h}(\mathbf{x})]$ of the partition function of the canonical ensemble that is a function of T, N, V, and $\mathbf{h}(\mathbf{x})$ or of the partition function $\Xi[T, V, \mu(\mathbf{x}), \mathbf{h}(\mathbf{x})]$ of the grand canonical ensemble that is a function of T, V, $\mu(\mathbf{x})$, and $\mathbf{h}(\mathbf{x})$.

In what follows, we will assume that the number of particles and volume remain fixed and suppress any explicit reference to them in partition functions and free energies. $Z[T, \mathbf{h}(\mathbf{x})]$ in Eq. (3.5.4) has exactly the same form as the grand canonical partition function, Eq. (3.4.16), with $h(\mathbf{x})$ replacing the spatially varying chemical potential. Thermodynamic potentials as a function of $\mathbf{h}(\mathbf{x})$ or $\langle \mathbf{m}(\mathbf{x}) \rangle$ and correlation functions of $\mathbf{m}(\mathbf{x})$ follow by a straightforward generalization of the discussion of Sec. 3.4. The magnetic thermodynamic potential associated with $Z[T, \mathbf{h}(\mathbf{x})]$ is

$$\mathscr{A}[T, \mathbf{h}(\mathbf{x})] = -T \ln Z[T, \mathbf{h}(\mathbf{x})]. \tag{3.5.9}$$

The magnetization [Eq. (3.5.5)] is obtained by differentiating $\ln Z$ with respect to $\mathbf{h}$:

$$\langle m_i(\mathbf{x}) \rangle = \frac{1}{Z} \frac{\delta Z}{\delta \beta h_i(\mathbf{x})} = -\frac{\delta \mathscr{A}}{\delta h_i(\mathbf{x})}, \tag{3.5.10}$$

where $\langle m_i \rangle$ and h_i are respectively the ith Cartesian components of the vectors $\langle \mathbf{m} \rangle$ and $\mathbf{h}$. The potential $\mathscr{A}$ obeys the differential thermodynamic relation,

$$d\mathscr{A} = -S\,dT - \int d^d x \langle \mathbf{m}(\mathbf{x}) \rangle \cdot \delta \mathbf{h}(\mathbf{x}). \tag{3.5.11}$$

The generalized susceptibility measuring the change in $\langle m_i \rangle$ in response to an external field at $\mathbf{x}'$ is

$$\chi_{ij}(\mathbf{x}, \mathbf{x}') = \frac{\delta \langle m_i(\mathbf{x}) \rangle}{\delta h_j(\mathbf{x}')}. \tag{3.5.12}$$

In classical systems, the order parameter correlation function is the derivative of $\langle m_i(\mathbf{x}) \rangle$ with respect to $\beta h_j(\mathbf{x}')$ and is, therefore, T times $\chi_{ij}(\mathbf{x}, \mathbf{x}')$:

$$\begin{aligned} G_{ij}(\mathbf{x}, \mathbf{x}') &= \langle [m_i(\mathbf{x}) - \langle m_i(\mathbf{x}) \rangle][m_j(\mathbf{x}') - \langle m_j(\mathbf{x}') \rangle] \rangle, \\ &\equiv \langle \delta m_i(\mathbf{x}) \delta m_j(\mathbf{x}') \rangle, \\ &= \frac{\delta \langle m_i(\mathbf{x}) \rangle}{\delta \beta h_j(\mathbf{x}')} = T \chi_{ij}(\mathbf{x}, \mathbf{x}'). \end{aligned} \tag{3.5.13}$$

This equation is the analog of Eq. (3.4.18) for the Ursell function of a fluid. When we consider many fields, and when there is a possibility of ambiguity in the interpretation of the subscripts ij, we will use the notation $G_{m_i m_j}$ rather than G_{ij} for the $\delta m_i(\mathbf{x}) - \delta m_j(\mathbf{x}')$ correlation function. The uniform magnetic susceptibility is the derivative of the magnetization with respect to the external magnetic field. It is the analog of the compressibility in fluid systems and is proportional to the $\mathbf{q} = 0$ limit of the Fourier transform of $G_{ij}(\mathbf{x}, \mathbf{x}')$:

$$\chi_{ij} = \frac{\partial \langle m_i \rangle}{\partial h_j} = \lim_{\mathbf{q} \to 0} \beta G_{ij}(\mathbf{q}). \tag{3.5.14}$$

This equation should be compared with Eq. (3.4.20) relating $S_{nn}(\mathbf{q} = 0)$ to the compressibility. Because the magnetization is a conserved quantity (i.e., $\frac{\partial}{\partial t} \int \mathbf{m} d^d x = 0$), Eq. (3.5.14) is valid for both classical and quantum mechanical systems even though Eq. (3.5.13) is not valid for quantum mechanical systems. This point will be discussed in more detail in Chapter 7.

In many cases we will prefer to express thermodynamic quantities in terms of the order parameter $\langle \mathbf{m}(\mathbf{x}) \rangle$ rather than in terms of the field $\mathbf{h}(\mathbf{x})$. The thermodynamic potential that is a natural function of $\langle \mathbf{m}(\mathbf{x}) \rangle$ is obtained from $\mathscr{A}[T, \mathbf{h}(\mathbf{x})]$ by a Legendre transformation:

$$F[T, \langle \mathbf{m}(\mathbf{x}) \rangle] = \mathscr{A}[T, \mathbf{h}(\mathbf{x})] + \int d^d x \mathbf{h}(\mathbf{x}) \cdot \langle \mathbf{m}(\mathbf{x}) \rangle. \tag{3.5.15}$$

We will usually refer to F as the free energy. It satisfies the differential relation

$$dF = -S dT + \int d^d x \mathbf{h}(\mathbf{x}) \cdot \delta \langle \mathbf{m}(\mathbf{x}) \rangle. \tag{3.5.16}$$

The magnetic equation of state is, therefore,

$$\frac{\delta F}{\delta \langle m_i(\mathbf{x}) \rangle} = h_i(\mathbf{x}). \tag{3.5.17}$$

When $\langle \mathbf{m}(\mathbf{x}) \rangle = \langle \mathbf{m} \rangle$ is spatially uniform, F, like the Helmholtz free energy as a function of density, can be written as a volume integral of a free energy density that is a function of $\langle \mathbf{m} \rangle$,

$$F = \int d^d x f[T, \langle \mathbf{m} \rangle]. \tag{3.5.18}$$

In this case, the equation of state becomes

$$\frac{\partial f}{\partial \langle m_i \rangle} = h_i. \tag{3.5.19}$$

This is the analog of Eq. (3.1.44). It is clear now how the appearance of order is described mathematically. When $\mathbf{h}$ is zero, the equilibrium state is one which minimizes f. If solutions to Eq. (3.5.19) for $\mathbf{h} = 0$ with nonzero $\langle \mathbf{m} \rangle$ exist, there can be spontaneous order provided the free energy of the state with nonzero $\langle \mathbf{m} \rangle$ is lower than that with $\langle \mathbf{m} \rangle = 0$.

The magnetization correlation function can be obtained by differentiating $\mathscr{A}$ with respect to $\beta h_i(\mathbf{x})$ as indicated in Eqs. (3.5.12) and (3.5.13). Its inverse can be obtained directly by differentiating F with respect to $\langle m_i(\mathbf{x}) \rangle$. To see this, we differentiate $\langle m_i(\mathbf{x}) \rangle$ with respect to $\langle m_k(\mathbf{x}'') \rangle$ to obtain a delta function,

$$\begin{aligned} \frac{\delta \langle m_i(\mathbf{x}) \rangle}{\delta \langle m_k(\mathbf{x}'') \rangle} &= \delta_{ik}\delta(\mathbf{x} - \mathbf{x}'') \\ &= \int d^d x' \frac{\delta \langle m_i(\mathbf{x}) \rangle}{\delta h_j(\mathbf{x}')} \frac{\delta h_j(\mathbf{x}')}{\delta \langle m_k(\mathbf{x}'') \rangle}, \end{aligned} \tag{3.5.20}$$

where the last expression follows from the chain rule for functional differentiation and where we use the summation convention on repeated indices (i.e., the appearance of the index j twice in this expression implies that we sum over its values, therefore including terms from h_x, h_y and h_z). The inverse $\chi_{ij}^{-1}(\mathbf{x}, \mathbf{x}')$ of $\chi_{ij}(\mathbf{x}, \mathbf{x}')$ is defined via

$$\int d^d \mathbf{x}' \chi_{ij}(\mathbf{x}, \mathbf{x}') \chi_{jk}^{-1}(\mathbf{x}', \mathbf{x}'') = \delta_{ik}\delta(\mathbf{x} - \mathbf{x}''). \tag{3.5.21}$$

Comparing Eqs. (3.5.20) and (3.5.21) and using Eqs. (3.5.12) and (3.5.17), we find

$$\chi_{ij}^{-1}(\mathbf{x}, \mathbf{x}') = \frac{\delta h_i(\mathbf{x})}{\delta \langle m_j(\mathbf{x}') \rangle} = \frac{\delta^2 F}{\delta \langle m_i(\mathbf{x}) \rangle \delta \langle m_j(\mathbf{x}') \rangle}. \tag{3.5.22}$$

We will find in future chapters that Eq. (3.5.22) is the most efficient way of determining $\chi_{ij}(\mathbf{x}, \mathbf{x}')$. Relations between ordered systems and the fluids are reviewed in Table 3.5.1.

3.6 Symmetry, order parameters, and models

In Chapter 2, we explored some of the vast variety of ordered and disordered structures that are found in nature. In the preceding section, we discussed how thermodynamics and statistical mechanics can be modified to provide a description of a particular ordered phase: the ferromagnet. In this section, we will outline a general approach to the statistical and thermodynamic treatment of phases with arbitrary types of order. We will then introduce some simple model Hamiltonians that will provide a basis for much of our study of broken symmetry and order.

The dynamical properties of any system of particles and/or spins are deter-

Table 3.5.1. *Comparison of statistical quantities of fluids and magnetic systems.*

Grand partition	$\longleftrightarrow$	Magnetic partition
$\Xi[T,V,\mu(\mathbf{x})] = \mathrm{Tr}e^{-\beta[\mathscr{H}-\int d^dx\mu(\mathbf{x})n(\mathbf{x})]}$	$\longleftrightarrow$	$Z_N[T,\mathbf{h}(\mathbf{x})] = \mathrm{Tr}e^{-\beta[\mathscr{H}-\int d^dx\mathbf{h}(\mathbf{x})\cdot\mathbf{m}(\mathbf{x})]}$
μ	$\longleftrightarrow$	$\mathbf{h}$
n	$\longleftrightarrow$	$\mathbf{m}$
$\mathscr{A}[T,\mathbf{h}(\mathbf{x})] = -T\ln\Xi[T,V,\mu(\mathbf{x})]$	$\longleftrightarrow$	$\mathscr{A}[T,\mathbf{h}(\mathbf{x})] = -T\ln Z_N[T,\mathbf{h}(\mathbf{x})]$
$\langle n(\mathbf{x})\rangle = -\delta\mathscr{A}[\mu(\mathbf{x})]/\delta\mu(\mathbf{x})$	$\longleftrightarrow$	$\langle m_i(\mathbf{x})\rangle = -\delta\mathscr{A}/\delta h_i(\mathbf{x})$
$S_{nn}(\mathbf{x},\mathbf{x}') = \langle\delta n(\mathbf{x})\delta n(\mathbf{x}')\rangle$	$\longleftrightarrow$	$G_{ij}(\mathbf{x},\mathbf{x}') = \langle\delta m_i(\mathbf{x})\delta m_j(\mathbf{x}')\rangle$
$\beta S_{nn}(\mathbf{x},\mathbf{x}') = \delta\langle n(\mathbf{x})\rangle/\delta\mu(\mathbf{x}')$	$\longleftrightarrow$	$\beta G_{ij}(\mathbf{x},\mathbf{x}') = \delta\langle m_i(\mathbf{x})\rangle/\delta h_j(\mathbf{x}')$
$\partial\langle n\rangle/\partial\mu = \langle n\rangle^2\kappa_T = \beta S_{nn}(\mathbf{q}=0)$	$\longleftrightarrow$	$\chi_{ij} = \partial\langle m_i\rangle/\partial h_j = \lim_{\mathbf{q}\to 0}\beta G_{ij}(\mathbf{q})$
$F[T,V,\langle n(\mathbf{x})\rangle] = \mathscr{A}[T,V,\mu(\mathbf{x})] + \int d^dx\mu(\mathbf{x})\langle n(\mathbf{x})\rangle$	$\longleftrightarrow$	$F[T,\langle\mathbf{m}(\mathbf{x})\rangle] = \mathscr{A}[T,\mathbf{h}(\mathbf{x})] + \int d^dx\mathbf{h}(\mathbf{x})\cdot\langle\mathbf{m}(\mathbf{x})\rangle$
$\delta F/\delta\langle n(\mathbf{x})\rangle = \mu(\mathbf{x})$	$\longleftrightarrow$	$\delta F/\delta\langle m_i(\mathbf{x})\rangle = h_i(\mathbf{x})$

mined by a Hamiltonian $\mathscr{H}$, which is invariant under all transformations in some group $\mathscr{G}$. For example, the Hamiltonian of an ideal gas is invariant under time translations, time reversal, and the Euclidean group consisting of arbitrary translations, rotations and reflection. The Heisenberg Hamiltonian [Eq. (3.5.2)] is invariant under time translations, time reversal, and under the simultaneous rotations of all spins through an arbitrary angle about an arbitrary axis. The group $\mathscr{G}$ is the symmetry group of the Hamiltonian $\mathscr{H}$. In general, the high-temperature, entropy-dominated equilibrium phase is invariant under the same group as its Hamiltonian. This means that the only nonzero correlation functions and thermodynamic averages of operators are those that are unaffected by operations in $\mathscr{G}$. Thus, $\langle\mathbf{m}\rangle$ is zero, and $C_{m_im_j}(\mathbf{x},\mathbf{x}') = \frac{1}{3}\delta_{ij}C_{mm}(|\mathbf{x}-\mathbf{x}'|)$ in the paramagnetic phase. Ordered phases are distinguished from disordered phases by the appearance of thermodynamic averages $\langle\phi_a\rangle$ of operators ϕ_a, which are not invariant under $\mathscr{G}$. These expectation values are called *order parameters*. (Sometimes the operators ϕ_a rather than their average values are called order parameters.) The order parameter for the Heisenberg magnet is the magnetization $\langle\mathbf{m}\rangle$. It is invariant under rotations about an axis parallel to itself but changes under rotations about all axes perpendicular to itself. Thus, the ordered phase, i.e., its operator averages and correlations functions, is invariant under only a subgroup of $\mathscr{G}$ even though the Hamiltonian remains invariant under the full group $\mathscr{G}$. The ordered phase, therefore, breaks the symmetry of the Hamiltonian; it is a *broken symmetry* phase.

A complete description of an ordered phase requires a specification of how order parameters $\langle\phi_a\rangle$ transform under the group $\mathscr{G}$. In the case of the ferromagnet, the group $\mathscr{G}$ is the group of rotations. With each rotation $g \in \mathscr{G}$, there is an associated 3×3 matrix $U_{ij}(g)$ such that $\langle m_i\rangle \rightarrow U_{ij}(g)\langle m_j\rangle$ under g, i.e., the order parameter $\langle \mathbf{m}\rangle$ transforms under a three-dimensional representation of the rotation group. In the general case, the order parameters $\langle\phi_a\rangle$, $a = 1, ..., n$, will transform under an n-dimensional representation of the group $\mathscr{G}$: associated with each $g \in \mathscr{G}$, there is a matrix $T_{ab}(g)$ such that $\langle\phi_a\rangle \rightarrow T_{ab}(g)\langle\phi_b\rangle$. The theory of representations of groups in terms of matrices is beyond the scope of this book. We note, however, that any set of order parameters $\langle\phi_a\rangle$ that transforms under some representation of $\mathscr{G}$ can be decomposed into disjoint subsets that do not mix under the operations of $\mathscr{G}$. Each of these sets consists of k_I elements $\langle\phi_a^I\rangle$ that transform among themselves under a k_I-dimensional *irreducible* representation of $\mathscr{G}$. It is generally advantageous to describe ordered phases in terms of order parameters that transform under the irreducible representation with the lowest possible dimension.

The operations of the group $\mathscr{G}$ in the ordered phase can be viewed in a different way. If the ordered phase breaks a symmetry, then there are two or more equivalent minima in the free energy representing phases that can coexist in equilibrium. Each phase is characterized by a particular value of the order parameter. Thus, transformations of the order parameter induced by elements of $\mathscr{G}$ represent transformations between different equivalent equilibrium phases.

Once the order parameters ϕ_a have been identified, the thermodynamics and statistical mechanics of the ordered phase can be developed in a manner exactly analogous to that of the preceding section. First, introduce fields $h_a(\mathbf{x})$ conjugate to $\phi_a(\mathbf{x})$ by introducing the external Hamiltonian,

$$\mathscr{H}_{\text{ext}} = -\int d^d x h_a(\mathbf{x})\phi_a(\mathbf{x}), \tag{3.6.1}$$

and defining the partition function $Z[T, h_a(\mathbf{x})]$ and associated thermodynamic potential $\mathscr{A} = -T \ln Z[T, h_a(\mathbf{x})]$. Then, Legendre transform to obtain the free energy $F[T, \langle\phi_a(\mathbf{x})\rangle]$.

Broken symmetry plays a very important role in condensed matter physics. As we discussed in Chapter 2, high-temperature liquid and gas phases are invariant under the Euclidean group and under spin rotations if particles carry spin. As temperature is lowered, there are sequences of phase transitions, each of which generally leads to a lowering of symmetry. The symmetry group relevant to most phase transitions is that of some partially ordered state rather than the full Euclidean group. The number of possible symmetry changes is quite large.

There are, for example, 230 nonmagnetic space groups, and symmetry breaking transitions between almost any pair of these groups is possible. Though each of these phases and their associated transitions differ in detail, they share many properties in common, and it is useful to study phases and phase transitions associated with a few very simple groups.

Before introducing the groups that will be of greatest interest to us, it is useful to distinguish *discrete* groups from *continuous* groups. Discrete groups, such as the symmetry group of a cube, have a countable and, for our purposes, usually a finite number of elements. Continuous groups, such as the rotation group, have an uncountable continuum of elements. As we shall see again and again in the following chapters, static and dynamic properties and the nature of defects of ordered phases depend critically on whether they have a broken discrete or continuous symmetry. It is also useful to distinguish between *global* and *local* symmetries. A Hamiltonian has a global symmetry if it is invariant with respect to spatially uniform group operations, i.e., to operations which treat all constituents equally. A Hamiltonian has a local symmetry if it is, in addition, invariant under operations applied independently to different points in space. The Heisenberg Hamiltonian has global but not local symmetry. It is invariant with respect to simultaneous rotations of all spins, but not with respect to independent rotations of individual spins. Local symmetries are associated with gauge symmetries such as that of electromagnetism. Virtually all of the symmetries of interest in condensed matter physics are global symmetries, the only notable exception being the local gauge symmetry of a superconductor.

1 Discrete symmetries

The simplest discrete group is the group Z_2 consisting only of two elements: the identity and an element whose square is the identity. Realizations of this group include the group of reflections about a plane, time reversal, and the integers under addition modulo 2. In condensed matter physics, Hamiltonians that are invariant under Z_2 are said to have *Ising* symmetry. Ising symmetry is broken in any phase transition in which there are two and only two equivalent ordered states characterized by order parameters that can be chosen to differ only in sign. The simplest physical system in which Ising symmetry can be broken is the uniaxial ferromagnet in which the magnetization is constrained by crystal fields to lie along the z-axis. The order parameter is $\langle m_z \rangle$, the z-component of the magnetization. $\langle m_z \rangle$ can be either positive or negative, but its magnitude is fixed by the conditions of thermodynamic equilibrium determined by the equation of state. The conjugate field is simply the z-component of the external magnetic field.

Order-disorder and uniaxial antiferromagnetic transitions on bipartite lattices also break Ising symmetry. Bipartite lattices such as the BCC lattice can be

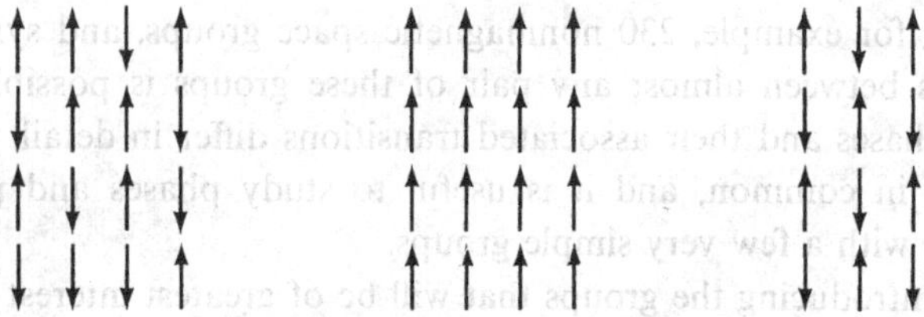

(a) (b) (c)

Fig. 3.6.1. Phases for an Ising magnet with Z_2 symmetry. Spins are constrained by crystal fields to point either up or down. (a) Disordered paramagnet, (b) ordered ferromagnet with spins aligned along a common direction, and (c) ordered antiferromagnet on a bipartite square lattice.

decomposed into two equivalent sublattices A and B. β-brass is a 50 – 50 mixture of Cu and Zn on a BCC lattice that undergoes an order-disorder transition. At high temperatures, the Cu and Zn atoms occupy the two sublattices with equal probability. In the ordered phase, there is a segregation of Cu atoms onto sublattice A and Zn atoms onto sublattice B or vice versa. The order parameter is thus $\langle \phi \rangle = \langle n_{\text{Cu},A} - n_{\text{Cu},B} \rangle$, where $n_{\text{Cu},A}$ is the number operator for Cu atoms on sublattice A. The conjugate field is the difference in chemical potential for Cu atoms on the two sublattices: $h = \mu_{\text{Cu},A} - \mu_{\text{Cu},B}$.

In the ordered phase of Ising antiferromagnets (Fig. 3.6.1), spins point up on one sublattice and down on the other or vice versa. The order parameter is thus the difference in magnetization between the two sublattices, $\langle \phi \rangle = \langle m_{z,A} - m_{z,B} \rangle$, and the conjugate field is an external magnetic field that points up on sublattice A and down on sublattice B: $h = h_{z,A} - h_{z,B}$. Note that the conjugate fields for both the β-brass and the antiferromagnetic transitions correspond to physical fields that oscillate in sign at the scale of a lattice spacing and cannot normally be produced in the laboratory. Nevertheless, these fields can be defined and used in theoretical treatments to generate order parameter correlation functions. There is another way to describe the development of order on bipartite lattices. Both the order-disorder transition and the antiferromagnetic transition lead to a doubling of the size of a unit cell and thus to the appearance of Bragg peaks at the Brillouin zone edge as depicted in Fig. 3.6.2. Thus, the respective order parameters for these transitions are equivalently mass- and spin-density wave amplitudes, $\langle \rho_{\mathbf{G}} \rangle$ and $\langle m_{z,\mathbf{G}} \rangle$, where $\mathbf{G}$ is a zone edge wave vector.

Other transitions that can have Ising symmetry are ferroelectric transitions, in which the order parameter is the electric polarization confined to lie along a particular crystal axis, and displacive transitions, in which some atom in a unit cell moves off of a symmetry position at low temperature as shown in Fig. 3.6.3.

There is no symmetry change in the transition from the liquid phase to the gas phase. Nevertheless, one can associate an order parameter with Ising symmetry

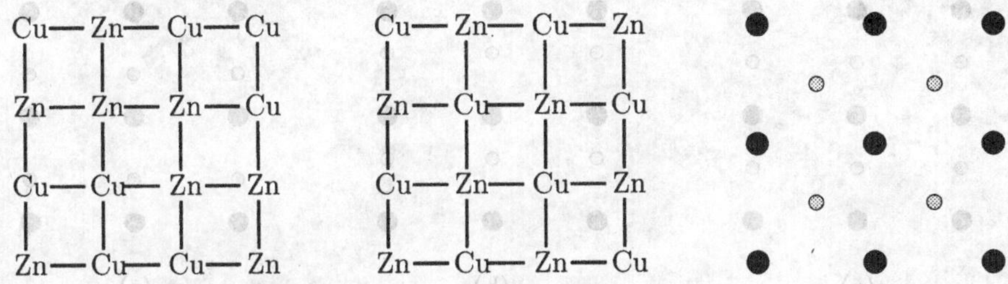

(a) (b) (c)

Fig. 3.6.2. (a) The disordered state of β-brass on a square lattice. Cu and Zn occupy lattice sites with equal probability. The unit cell is a square of side a. (b) The ordered state of β-brass. Cu atoms have a greater probability of occupying one sublattice while Zn has a greater probability of occupying the other. The unit cell is a square of side $2^{1/2}a$ containing two atoms. (c) Bragg scattering intensities. The dark spots on a square lattice with lattice parameter $2\pi/a$ are the Bragg scattering peaks from the disorder phases. The light spots are the superlattice peaks of the ordered phase.

with this transition because the density can take on one of two values along the coexistence curve. At the critical point, there is no difference between the liquid and gas phases. As one proceeds along any of the liquid-gas coexistence curves of Fig. 3.1.4, the density of the gas phase increases and that of the liquid phase decreases. An order parameter for the liquid-gas transition is, therefore, the difference in density between the liquid and gas phases: $\langle \phi \rangle = \langle n_L - n_G \rangle$. The conjugate field is just the chemical potential. The phase separation transition in two-component mixtures has a similar order parameter.

Though Ising, or Z_2, symmetry is the most important and most common discrete symmetry encountered in condensed matter physics, there are, of course, many others. A particularly useful hierarchy of discrete groups for studying properties of discrete symmetry is the set of groups Z_N of integers under addition modulo N. An example of a physical system with Z_3 symmetry is krypton adsorbed on graphite. As discussed in Sec. 2.9, there is a $\sqrt{3} \times \sqrt{3}$ commensurate ground state with one Kr atom for every three unit cells of the graphite lattice. There are, therefore, three equivalent ground states, and Z_3 symmetry is broken in the transition from the fluid to the commensurate state.

2 Continuous symmetries

The simplest continuous group is the two-dimensional orthogonal group O_2 of rotations in a two-dimensional plane. Since a two-dimensional vector is equivalent to a complex number, the group O_2 is isomorphic to the group $U(1)$ of transformations of the phase of a complex number. The symmetry associated with these groups is often called xy-symmetry because rotations are usually done

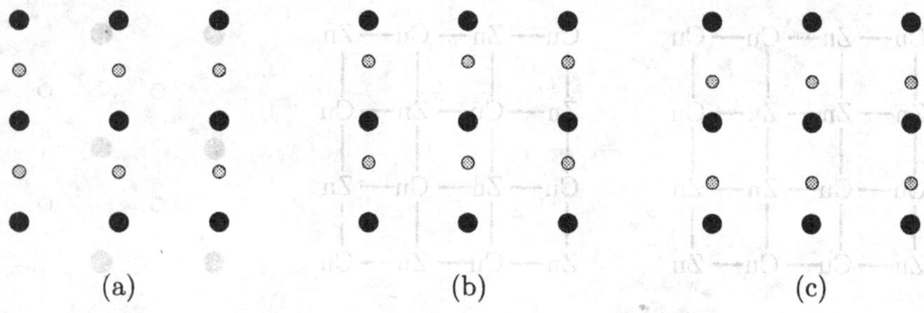

Fig. 3.6.3. A model displacive transition. At high temperature, the atom represented by a small circle sits at the center of a square unit cell, as shown in (a). At low temperature, it can move up or down, as shown in (b) and (c). (Such displacive transitions, where the distortion occurs in each unit cell, are often referred to as ferroelectric.)

in the *xy*-plane. The group O_2 is of enormous pedagogical importance and will be used extensively throughout this book. The simplest realization of a system with O_2 symmetry is an *easy-plane* ferromagnet in which spins are confined by crystal fields to lie in the *xy*-plane. We will repeatedly use this system and its companion, the easy-plane antiferromagnet, to study the properties of broken continuous symmetry, even though it has virtually no physical realizations. The most important phase with broken $U(1)$ symmetry is superfluid helium. The superfluid phase is characterized by a macroscopic condensate wave function with an amplitude and a phase. Its order parameter can be defined along the lines of the discussion of the preceding section as the average $\langle\psi(\mathbf{x})\rangle$ of the Bose annihilation operator $\psi(\mathbf{x})$. Its average $\langle\psi\rangle = |\langle\psi\rangle|e^{i\phi}$ has an amplitude and a phase. The field conjugate to $\psi(\mathbf{x})$ is a field that creates a particle at $\mathbf{x}$. It cannot be produced in a laboratory. Other examples of phases with broken $U(1)$ symmetry include the smectic-*C* and hexatic-*B* phases of liquid crystals. The order parameter in the smectic-*C* phase is the *c*-director, **c**, the projection of the director **n** onto the plane of the smectic layers. The order parameter of the hexatic-*B* phase is $\langle e^{6i\theta}\rangle$.

The group O_3 of rotations in three dimensions is another continuous group of considerable importance. Physical realizations of systems with this symmetry include Heisenberg ferromagnets and antiferromagnets on lattices where crystal fields aligning spins along crystal axes are unimportant. The nematic phase in liquid crystals also breaks O_3 symmetry. However, it is invariant, as we shall see, under a different subgroup of O_3 than is the ferromagnetic phase of the Heisenberg model.

We will often find it useful to study model systems invariant under the *n*-dimensional orthogonal group O_n of rotations in an *n*-dimensional space, even though physical realizations with $n > 3$ are difficult to find.

Table 3.6.1. *Symmetries and order parameters.*

Symmetry	System	Order parameter	Conjugate field	Physical example
Z_2/Ising	uniaxial FM	$\langle m_z \rangle$	h_z	
	uniaxial AF	$\langle m_{z,A} - m_{z,B} \rangle$	$h_{z,A} - h_{z,B}$	Rb_2NiF_4, K_2MnF_4
	order-disorder	$\langle n_A - n_B \rangle$	$\mu_A - \mu_B$	β-brass
	displacive	$\langle u_z \rangle$	f_z	$BaTiO_3$
	liquid-gas	$\langle n_L - n_G \rangle$	μ	many
$O_2/U(1)$	easy-plane FM	$\langle \mathbf{m} \rangle$	$\mathbf{h}$	
	easy-plane AF	$\langle \mathbf{m}_A - \mathbf{m}_B \rangle$	$\mathbf{h}_A - \mathbf{h}_B$	
	superfluid	$\langle \psi \rangle$	h_ψ	He_4
	smectic-*C*			Fig. 2.7.1
	hexatic-*B*	$\langle e^{6i\theta} \rangle$		Fig. 2.7.7
O_3	Heisenberg FM	$\langle \mathbf{m} \rangle$	$\mathbf{h}$	EuS, EuO, Fe, Ni
	Heisenberg AF	$\langle \mathbf{m}_A - \mathbf{m}_B \rangle$	$\mathbf{h}_A - \mathbf{h}_B$	$RbMnF_3$
O_0	SAW			polymer

3 Models

In our discussion of the development of order in the last section, we found it useful to consider the Heisenberg model of interacting spins on a lattice. Models such as this have had an enormous impact on our understanding of phase transitions and the properties of ordered phases. They are used throughout the literature to simplify the discussion of particular properties of real physical systems, and we will find it useful to refer to them in subsequent chapters. We will, therefore, introduce here some of the most commonly used models.

The simplest of all models is the Ising model (Ising 1925) with a global Z_2 symmetry. At each site $\mathbf{R}_\mathbf{l}$ on a lattice, there is a spin variable $\sigma_\mathbf{l}$ that can take on only the values +1 or −1. The spins interact via a nearest neighbor exchange interaction so that the Ising Hamiltonian is

$$\mathscr{H}_{\text{Ising}} = -J \sum_{<\mathbf{l},\mathbf{l}'>} \sigma_\mathbf{l} \sigma_{\mathbf{l}'} \tag{3.6.2}$$

where the sum is over nearest neighbor bonds $< \mathbf{l}, \mathbf{l}' >$ on the lattice. As in the Heisenberg model, there is a high-temperature paramagnetic phase with $\langle \sigma \rangle = 0$ and a low-temperature ferromagnetic phase with $\langle \sigma \rangle \neq 0$. The low-temperature phase breaks Z_2 symmetry. The Ising model differs from the Heisenberg model in that the spins are purely classical. They do not obey quantum commutation relations as do the spins in the Heisenberg model. The Ising model in two dimensions was solved exactly by Lars Onsager in 1944.

A class of models with Z_N symmetry, called *clock models*, can be defined by associating with each site on a lattice a spin variable $\mathbf{s}_\mathbf{l}$ of unit length that is constrained to point in one of N equally spaced directions on the unit circle, i.e., $\mathbf{s_l} = (\cos 2\pi n_l/N, \sin 2\pi n_l/N)$, where $n_l = 0, ..., N-1$. Similarly, models with O_n symmetry (often called n-vector models) can be defined by associating with each site on a lattice an n-dimensional unit length spin vector $\mathbf{s_l}$. The Hamiltonian for both the Z_N and O_n models has the same form as that of the Heisenberg model:

$$\mathscr{H} = -J \sum_{<\mathbf{l},\mathbf{l}'>} \mathbf{s_l} \cdot \mathbf{s_{l'}}. \tag{3.6.3}$$

This Hamiltonian for clock models can also be reexpressed in terms of the integers n_l as

$$\mathscr{H}_{\text{clock}} = -J \sum_{<\mathbf{l},\mathbf{l}'>} \cos[2\pi(n_l - n_{l'})/N]. \tag{3.6.4}$$

Similarly, the O_2 or xy-model can be reexpressed in terms of a local angle variable by setting $\mathbf{s_l} = (\cos \vartheta_l, \ \sin \vartheta_l)$. Then

$$\mathscr{H}_{xy} = -J \sum_{<\mathbf{l},\mathbf{l}'>} \cos[\vartheta_l - \vartheta_{l'}]. \tag{3.6.5}$$

The Ising model is identical to the Z_2 clock model. It can also be viewed as the O_n model with $n = 1$, even though, strictly speaking, rotations in a one-dimensional space are not defined. The O_3 model is the classical Heisenberg model. We shall see in the following chapters that the $n = \infty$ O_n model can be solved exactly. The O_n model with $n = 0$ describes the statistics of self-avoiding random walks or polymers in solution.

Another useful model that is invariant under Z_N (as a subgroup of the permutation group) is the N-state Potts model (Potts 1952, Wu 1982) in which there is a variable $\sigma(\mathbf{l})$ that can take on any of N discrete values. The Hamiltonian for the N-state Potts model associates one energy with nearest neighbor bonds in the same state and a second energy if they are in different states. The Hamiltonian is

$$\mathscr{H}_{\text{Potts}} = -J \sum_{<\mathbf{l},\mathbf{l}'>} [N\delta_{\sigma_l,\sigma_{l'}} - 1]. \tag{3.6.6}$$

The two-state Potts model is the Ising model. The one-state Potts model describes percolation. The three-state Potts model can be used to describe the transition from the fluid to the $\sqrt{3} \times \sqrt{3}$ structure of Kr adsorbed on graphite discussed in Sec. 2.9.

Appendix 3A Functional derivatives

The functional derivative introduced in Sec. 3.5 can be defined in various ways. Probably the most intuitive is to proceed by analogy with partial derivatives. Consider a functional Φ of the function $h(\mathbf{x})$. Φ can be obtained as a limit of a function $\tilde{\Phi}$ of a countable set of variables as follows: Divide space up into cells of volume ΔV whose centers are at positions $\mathbf{x}_\alpha$, $\alpha = 1, ...,$ and define $\tilde{\Phi}(h_\alpha) = \Phi[h(\mathbf{x}_\alpha)]$, where $h_\alpha = h(\mathbf{x}_\alpha)$. $\tilde{\Phi}$ is a *function* not a *functional* of h_α. It has a Taylor series expansion

$$\tilde{\Phi}(h_\alpha + \delta h_\alpha) = \tilde{\Phi}(h_\alpha) + \sum_\alpha \frac{\partial \tilde{\Phi}}{\partial h_\alpha}\delta h_\alpha + \frac{1}{2}\sum_{\alpha,\beta} \frac{\partial^2 \tilde{\Phi}}{\partial h_\alpha \partial h_\beta}\delta h_\alpha \delta h_\beta + ...,$$

where

$$\frac{\partial \tilde{\Phi}}{\partial h_\alpha} = \lim_{\delta h_\alpha \to 0} \frac{\tilde{\Phi}(h_1, ..., h_\alpha + \delta h_\alpha, ...) - \tilde{\Phi}(h_1, ..., h_\alpha, ...)}{\delta h_\alpha}.$$

To regain the functional $\Phi[h(\mathbf{x})]$ from $\tilde{\Phi}(h_\alpha)$, we take the limit $\Delta V \to 0$ and replace $\sum_\alpha$ by $\sum_\alpha \Delta V \to \int d\mathbf{x}$:

$$\begin{aligned}\Phi[h(\mathbf{x}) + \delta h_\alpha] &= \lim_{\Delta V \to 0} \tilde{\Phi}(h_\alpha + \delta h_\alpha) \\ &= \Phi[h(\mathbf{x})] + \sum_\alpha \Delta V \frac{1}{\Delta V}\frac{\partial \tilde{\Phi}}{\partial h_\alpha}\delta h_\alpha \\ &\quad + \frac{1}{2}\sum_{\alpha,\beta} (\Delta V)^2 \frac{1}{(\Delta V)^2}\frac{\partial^2 \tilde{\Phi}}{\partial h_\alpha \partial h_\beta}\delta h_\alpha \delta h_\beta \\ &= \Phi[h(\mathbf{x})] + \int d^d x \frac{\delta \Phi}{\delta h(\mathbf{x})} \\ &\quad + \frac{1}{2}\int d^d x d^d x' \frac{\delta^2 \Phi}{\delta h(\mathbf{x}) \delta h(\mathbf{x}')}\delta h(\mathbf{x})\delta h(\mathbf{x}') + ...\end{aligned}$$

so that

$$\frac{\delta \Phi}{\delta h(\mathbf{x})} = \lim_{\Delta V \to 0} \lim_{\delta h_\alpha \to 0} \frac{\tilde{\Phi}(h_1, ..., h_\alpha + \delta h_\alpha, ...) - \tilde{\Phi}(h_1, ..., h_\alpha, ...)}{\Delta V \delta h_\alpha}$$

where it is understood that $\mathbf{x}$ remains in cell α at all steps in the limiting procedure. Note that the functional derivative has a factor of one over volume relative to a partial derivative.

An alternative, more formal, definition of a functional derivative is

$$\frac{\delta \Phi}{\delta h(\mathbf{y})} = \lim_{\epsilon \to 0} \frac{\Phi[h(\mathbf{x}) + \epsilon \delta(\mathbf{x} - \mathbf{y})] - \Phi[h(\mathbf{x})]}{\epsilon},$$

i.e., $\delta \Phi / \delta h(\mathbf{y})$ is the change induced in Φ in response to a change in $h(\mathbf{x})$ at the point $\mathbf{x} = \mathbf{y}$ only.

It is clear from the above definitions that functional derivatives obey all of the usual rules of differentiation. For example,

$$\frac{\delta h(\mathbf{x})}{\delta h(\mathbf{y})} = \delta(\mathbf{x} - \mathbf{y}) \qquad \left(\frac{\partial h_\alpha}{\partial h_\beta} = \delta_{\alpha\beta} \ \text{discrete analog} \right).$$

If f is a *function* of $h(\mathbf{x})$, then

$$\frac{\delta f(h(\mathbf{x}))}{\delta h(\mathbf{y})} = f' \frac{\delta h(\mathbf{x})}{\delta h(\mathbf{y})} = f' \delta(\mathbf{x} - \mathbf{y}),$$

and

$$\frac{\delta f(g(h(\mathbf{x})))}{\delta h(\mathbf{y})} = f'g' \frac{\delta h(\mathbf{x})}{\delta h(\mathbf{y})} = f'g' \delta(\mathbf{x} - \mathbf{y}),$$

where $f'(z) = df/dz$. Finally, if $F[\phi(\mathbf{x})]$ can be expressed as an integral of a function of $\phi(\mathbf{x})$ and $\nabla\phi(\mathbf{x})$:

$$F[\phi(\mathbf{x})] = \int d\mathbf{x} f(\phi(\mathbf{x}), \nabla\phi(\mathbf{x})),$$

then

$$\begin{aligned}\frac{\delta F}{\delta\phi(\mathbf{y})} &= \int d^d x \frac{\delta f}{\delta\phi(\mathbf{y})} \\ &= \int d^d x \left[\frac{\partial f}{\partial\phi(\mathbf{x})}\frac{\delta\phi(\mathbf{x})}{\delta\phi(\mathbf{y})} + \frac{\partial f}{\partial\nabla\phi(\mathbf{x})}\cdot\frac{\delta\nabla\phi(\mathbf{x})}{\delta\phi(\mathbf{y})}\right] \\ &= \int d^d x \left[\frac{\partial f}{\partial\phi(\mathbf{x})}\delta(\mathbf{x}-\mathbf{y}) + \frac{\partial f}{\partial\nabla\phi(\mathbf{x})}\cdot\nabla\delta(\mathbf{x}-\mathbf{y})\right].\end{aligned}$$

Integrating the last term by parts and performing the integral over $\mathbf{x}$, we obtain

$$\frac{\delta F}{\delta\phi(\mathbf{y})} = \frac{\partial f}{\partial\phi(\mathbf{y})} - \nabla\cdot\frac{\partial f}{\partial\nabla\phi(\mathbf{y})}.$$

This equation is frequently used in Lagrangian mechanics.

Bibliography

E. Brézin, J.C. le Guillou, and J. Zinn-Justin, in *Phase Transitions and Critical Phenomena*, eds. by C. Domb and M.S. Green (Academic Press, New York, 1976).

L.D. Landau and E.M. Lifshitz, *Statistical Physics* (Addison-Wesley, Reading, Mass., 1969).

References

W. Heisenberg, *Z. Physik* **49**, 619 (1928).

E. Ising, *Z. Physik* **31**, 253 (1925).

L. Onsager, *Phys. Rev.* **65**, 117 (1944).

R. B. Potts, *Proc. Camb. Phil. Soc.* **48**, 106 (1952).

F.Y. Wu *Rev. Mod. Phys.* **54**, 235 (1982).

Problems

3.1 Calculate $A[T, \mu(\mathbf{x})]$ and $F[T, n(\mathbf{x})]$ for a non-interacting classical gas.

3.2 Calculate $A[T, \mu(\mathbf{x}), \mathbf{h}(\mathbf{x})]$ for a non-interacting classical gas of spin 1/2 particles. Then calculate the density $n(\mathbf{x})$, magnetization $m_i(\mathbf{x})$, density correlation function $G_{nn}(\mathbf{x}, \mathbf{x}')$, and magnetization correlation function $G_{m_i m_j}$.

3.3 A nearly flat fluctuating two-dimensional membrane can be characterized by two extensive quantities: its total area A and its area A_B projected onto the plane defined by its average normal. There is, therefore, a thermodynamic potential of the form $W(A, A_B, T) = A_B w(A/A_B, T)$ which obeys the differential thermodynamic relation

$$dW = \sigma dA - h dA_B - S dT,$$

where σ is the surface tension and h is a field conjugate to the projected area.

Define the thermodynamic potentials $F(\sigma, A_B, T)$ and $G(A, h, T)$ in terms of W, A, A_B, σ, and h, and show that they can, respectively, be expressed as $A_B f(\sigma, T)$ and $A g(h, T)$. Determine f and g in terms of σ, h, and T, and determine the differential thermodynamic relations satisfied by f and g.

4

Mean-field theory

In the preceding two chapters, we have discussed various types of order that can occur in nature and how the ordering process can be quantified by the introduction of order parameters. We also developed a formalism for dealing with the thermodynamics of ordered states. In this chapter, we will use mean-field theory to study phase transitions and the properties of various ordered phases. Mean-field theory is an approximation for the thermodynamic properties of a system based on treating the order parameter as spatially constant. It is a useful description if spatial fluctuations are not important. It becomes an exact theory only when the range of interactions becomes infinite. It, nevertheless, makes quantitatively correct predictions about some aspects of phase transitions (e.g. critical exponents) in high spatial dimensions where each particle or spin has many nearest neighbors, and it makes qualitatively correct predictions in physical dimensions. Mean-field theory has the enormous advantage of being mathematically simple, and it is almost invariably the first approach taken to predict phase diagrams and properties of new experimental systems.

Before proceeding, let us review some simple facts about phase transitions. At high temperatures, there is no order, and the order parameter $\langle\phi\rangle$ is zero. At a critical temperature, T_c, order sets in so that, for temperatures below T_c, $\langle\phi\rangle$ is nonzero. If $\langle\phi\rangle$ rises continuously from zero, as shown in Fig. 4.0.1a, the transition is *second order*. If $\langle\phi\rangle$ jumps discontinuously to a nonzero value just below T_c, as shown in Fig. 4.0.1b, the transition is *first order*. The entropy is also continuous at a second-order and discontinuous at a first-order transition. In a first-order transition, heat is adsorbed by the system in going from the low-temperature to the high-temperature phase. This heat is the latent heat $Q_L = T_c\Delta S$ of the transition, where ΔS is the entropy change and T_c is the transition temperature.

There are many formulations of mean-field theory, beginning with the van der Waals equation of state (van der Waals 1873) for the liquid-gas transition and the Weiss (1906) molecular field theory for ferromagnetism. In this chapter, we will introduce mean-field theory with the Bragg-Williams theory (W.L. Bragg and E.J. Williams 1934) for an Ising ferromagnet. This theory is equivalent to the Weiss molecular field theory and provides a qualitatively correct description of the properties of the Ising model at all temperatures, including the vicinity of zero temperature. This theory will set the stage for the Landau phenomenological

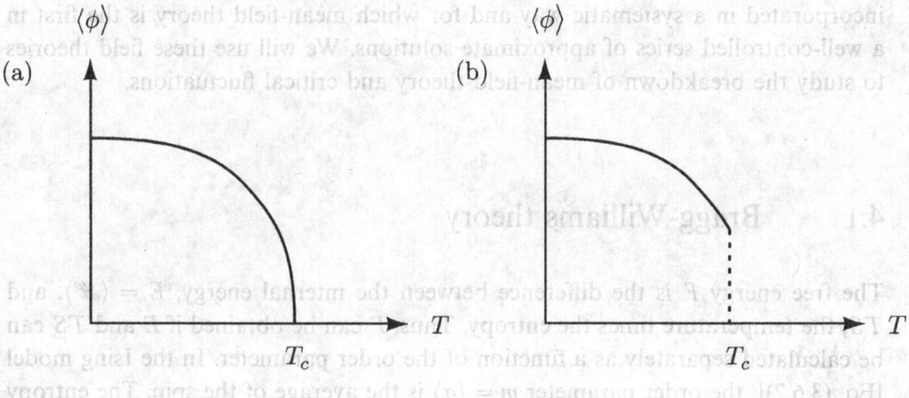

Fig. 4.0.1. Order parameter as a function of temperature for (a) a second-order and (b) for a first-order transition.

mean-field theory (see Landau 1937), to which most of this chapter will be devoted.

Landau theory is based on a power series expansion of the free energy in terms of the order parameter (or order parameters) for the transition of interest. It assumes that the order parameter is "small" so that only the lowest order terms required by symmetry (and to keep the energy from diverging) are kept. The form of a Landau phenomenological free energy is determined entirely by the nature of the broken symmetry of the ordered phase, i.e., by combinations of the order parameters that are left invariant under symmetry operations of the interaction Hamiltonian. The undisputed usefulness of Landau theory rests in its simplicity – most of its predictions can be determined by solving simple algebraic equations. It is, of course, most useful in the vicinity of second-order phase transitions, where the order parameter is guaranteed to be small. It, however, can be used with care to treat first-order transitions, where there are discontinuous changes in order parameters, or to determine properties of ordered phases rather than of phase transitions themselves. When more quantitative information about situations in which order parameters are not small (i.e., when the low-order power series of simple Landau theory is inapplicable) is desired, more complete mean-field theories, such as Bragg-Williams theory, are available. The most versatile technique for developing mean-field theories is based upon a variational principle and a single-site approximation for the many-particle density matrix. This technique will be reviewed and applied to the Potts model [Eq. (3.6.6)] and charged systems in Sec. 4.8.

None of the methods of deriving mean-field theory presented in this chapter is easily generalized to arrive at an understanding of how mean-field theory breaks down and how to correct it. In the next chapter, we will develop field theories for which fluctuations involving spatially nonuniform local order parameters can be incorporated in a systematic way and for which mean-field theory is the first in a well-controlled series of approximate solutions. We will use these field theories to study the breakdown of mean-field theory and critical fluctuations.

4.1 Bragg-Williams theory

The free energy F is the difference between the internal energy, $E = \langle \mathcal{H} \rangle$, and TS, the temperature times the entropy. Thus, F can be obtained if E and TS can be calculated separately as a function of the order parameter. In the Ising model [Eq. (3.6.2)], the order parameter $m = \langle \sigma \rangle$ is the average of the spin. The entropy for a given spatially uniform m can be calculated exactly. The total magnetic moment is simply the number $N_\uparrow$ of up spins minus the number $N_\downarrow$ of down spins, and

$$m = (N_\uparrow - N_\downarrow)/N , \tag{4.1.1}$$

where $N = N_\uparrow + N_\downarrow$ is the total number of sites in the lattice. The entropy for a given m is the logarithm of the number of configurations with a given $N_\uparrow$ and $N_\downarrow$:

$$\begin{aligned} S &= \ln \binom{N}{N_\uparrow} = \ln \binom{N}{N(1+m)/2} \\ &= \ln \left(\frac{N!}{(N(1+m)/2)!(N(1-m)/2)!} \right) \\ \frac{S}{N} &\equiv s(m) = \ln 2 - \frac{1}{2}(1+m)\ln(1+m) - \frac{1}{2}(1-m)\ln(1-m). \end{aligned} \tag{4.1.2}$$

This entropy is often called the *entropy of mixing*.

To evaluate E, one should calculate $\langle \mathcal{H} \rangle = Z_m^{-1} \mathrm{Tr}_m e^{-\beta \mathcal{H}} \mathcal{H}$, where Tr_m is a trace over all configurations with fixed m and $Z_m = \mathrm{Tr}_m e^{-\beta \mathcal{H}}$. An exact evaluation of this average would constitute an exact solution of the Ising model and is quite complicated. In Bragg-Williams theory, $\langle \mathcal{H} \rangle$ is approximated by replacing $\sigma_{\mathbf{l}}$ in $\mathcal{H}$ by its position independent average m:

$$E = -J \sum_{<\mathbf{l},\mathbf{l}'>} m^2 = -\frac{1}{2} J N z m^2 , \tag{4.1.3}$$

where z is the number of nearest neighbor sites in the lattice ($z = 2d$ for a d-dimensional hypercubic lattice). (Recall that $< \mathbf{l}\,\mathbf{l}' >$ signifies the bond between nearest neighbors $\mathbf{l}$ and $\mathbf{l}'$.)

The complete Bragg-Williams free energy is thus

$$\begin{aligned} f(T,m) &= (E - TS)/N \\ &= -\frac{1}{2} J z m^2 + \frac{1}{2} T [(1+m)\ln(1+m) + (1-m)\ln(1-m)] \\ &\qquad - T \ln 2. \end{aligned} \tag{4.1.4}$$

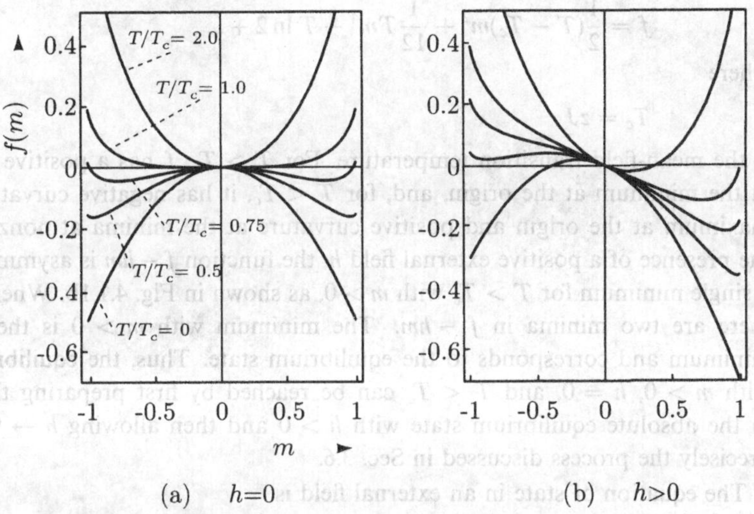

Fig. 4.1.1. (a) The Bragg-Williams free energy as a function of order parameter for various values of T/T_c. For $T > T_c$, there is a single minimum at $m = 0$. At $T = T_c$, the single minimum at $m = 0$ is very broad. For $T < T_c$, there are two minima with the same free energy at $\pm m$. At $T \equiv 0$, the minimum of the free energy occurs at $m = \pm 1$. (b) The Bragg-Williams free energy $f - hm$ in an external magnetic field of $h = 0.2T_c > 0$. Note that the minimum at $T > T_c$ occurs at positive m. The minimum at positive m for $T > T_c$ has a lower free energy than that at negative m. The order parameter m is restricted to lie between 0 and 1. At $T = 0$, the absolute minimum of the free energy occurs at $m = \pm 1$ when $h = 0$. The derivative $\partial f/\partial m$ is, however, not zero at these minima. For $T > 0$, $\partial f/\partial m$ is zero at the absolute minimum of f, which occurs at $|m| < 1$.

This function is plotted for various values of T in Fig. 4.1.1. Note that it is even under the operation $m \to -m$; it could, therefore, be expressed as a function of m^2 rather than m. For large T, f has a single minimum at $m = 0$. Below a critical temperature T_c (to be calculated below), it has two minima at $\pm m$. As $T \to 0$, the two minima occur at values of $|m|$ closer and closer to unity. In the absence of an external aligning magnetic field h, the equilibrium value of m is that which minimizes f. Thus, the disappearance of the minimum at $m = 0$ and the emergence of lower free energy minima at nonzero m corresponds to a phase transition. As we shall see shortly, the value of m at the new minima grows continuously from zero, so the transition is second order.

In the vicinity of T_c where m is small, we can expand $s(m)$ and $f(m)$ in powers of m:

$$s(m) = \ln 2 - \frac{1}{2}m^2 - \frac{1}{12}m^4 + \dots, \tag{4.1.5}$$

so that

$$f = \frac{1}{2}(T - T_c)m^2 + \frac{1}{12}Tm^4 - T\ln 2 + \ldots , \tag{4.1.6}$$

where

$$T_c = zJ \tag{4.1.7}$$

is the mean-field transition temperature. For $T > T_c$, f has a positive curvature at the minimum at the origin, and, for $T < T_c$, it has negative curvature at the maximum at the origin and positive curvature at the minima at nonzero m. In the presence of a positive external field h, the function $f - hm$ is asymmetric with a single minimum for $T > T_c$ with $m > 0$, as shown in Fig. 4.1.1b. When $T < T_c$, there are two minima in $f - hm$. The minimum with $m > 0$ is the absolute minimum and corresponds to the equilibrium state. Thus, the equilibrium state with $m > 0$, $h = 0$, and $T < T_c$ can be reached by first preparing the system in the absolute equilibrium state with $h > 0$ and then allowing $h \to 0$. This is precisely the process discussed in Sec. 3.6.

The equation of state in an external field is

$$\begin{aligned} \frac{\partial f}{\partial m} &= -zJm + \frac{1}{2}T\ln[(1+m)/(1-m)] = h \\ &= -zJm + T\tanh^{-1} m = h. \end{aligned} \tag{4.1.8}$$

Thus,

$$m = \tanh[(h + T_c m)/T]\,. \tag{4.1.9}$$

The quantity

$$h_m = h + T_c m \tag{4.1.10}$$

is the average local or molecular field at a given site. It arises both from the external field and from the exchange field produced by the neighboring spins with average spin m. The actual local field at a given site depends on the configurations of spins on neighboring sites as shown in Fig. 4.1.2. The quantity zJm is the average or mean field arising from neighboring sites. Weiss molecular field theory arrives at Eq. (4.1.9) by a slightly different route. The average magnetization for an Ising spin in a field h is: $m = (\langle N_\uparrow\rangle - \langle N_\downarrow\rangle)/N = (e^{h/T} - e^{-h/T})/(e^{h/T} + e^{-h/T}) = \tanh h/T$. If we take h to be the sum of the external field plus an internal field (due to interactions with neighboring spins) proportional to the magnetization ($h \to h + \alpha m$), then we have the self-consistency condition $m = \tanh(h + \alpha m)/T$ relating m to itself. Associating α with zJ or T_c we have Eq. (4.1.9).

The solutions to Eq. (4.1.9) are best visualized graphically, as shown in Fig. 4.1.3. When $h = 0$, the slope T_c/T of $\tanh(T_c m/T)$ at $m = 0$ is less than one for $T > T_c$ and greater than one for $T < T_c$. Thus for $T > T_c$, the only solution to Eq. (4.1.9) with $h = 0$ is $m = 0$. For $T < T_c$, there are three solutions, $\pm m(T)$ and zero. The tanh function lies between -1 and $+1$ so that $|m| \leq 1$. As $T \to 0$, $\tanh(T_c m/T)$ saturates at its maximum values of one at smaller and smaller m, and $|m(T)| \to 1$ as $T \to 0$ indicating full ferromagnetic order as expected in the fully aligned ground state of the Ising model. Near $T = 0$,

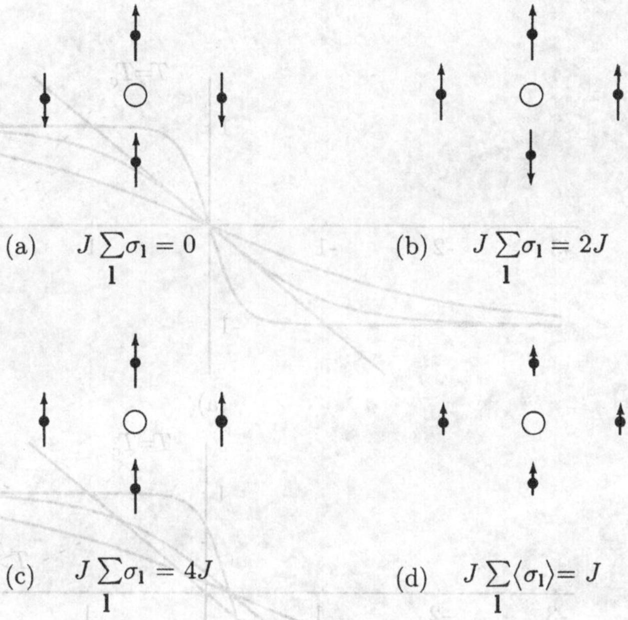

Fig. 4.1.2. Different configurations of neighboring spins giving rise to different local fields. In mean-field theory, the local field that depends on local spin configurations is approximated by the mean or average local field obtained by replacing the values of the local spins σ_1 by their mean value m. In figures (a) - (c), $|\sigma_1| = 1$. In figure (d), each spin has been replaced by its average value, $|\langle\sigma_1\rangle| = 1/2$.

$$m = \tanh(T_c m/T) \approx 1 - 2e^{-2zJ/T} \tag{4.1.11}$$

(compared with the exact result $1 - m = e^{-zJ/T}$) so that m tends to unity exponentially with temperature.

As $T \to T_c^-$, the solution to Eq. (4.1.9) can be obtained by expanding $\tanh x$ in powers of x:

$$m \approx (T_c/T)m - \frac{1}{3}(T_c/T)^3 m^3 \approx (T_c/T)m - \frac{1}{3}m^3 \tag{4.1.12}$$

or

$$m = \pm[3(T_c - T)/T]^{1/2}. \tag{4.1.13}$$

Thus m tends continuously to zero as $(T_c - T)^{1/2}$. This behavior is a general feature of second order mean-field phase transitions.

The solutions to Eq. (4.1.9) for finite (positive) h are also of some interest. They can be obtained graphically, as shown in Fig. 4.1.3b, by shifting the origin of the tanh function by $-h/T_c$. For $T > T_c$, there is a single solution with $m > 0$ corresponding to the single minimum of the free energy $f - hm$ in Fig. 4.1.1b.

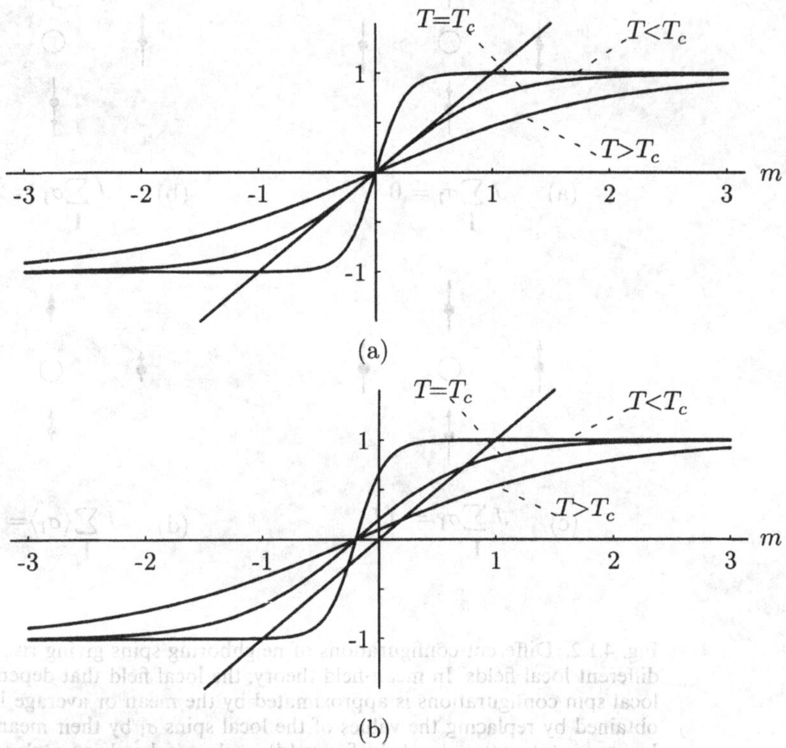

Fig. 4.1.3. (a) Graphical solution of the Ising mean-field equation of state [Eq. (4.1.9)] for $h = 0$. For $T > T_c$, the linear function m intersects the tanh function only at the origin. For $T < T_c$, the slope of $\tanh(T_c m/T)$ at the origin is greater than one, and there are three solutions to the equation of state corresponding to the two minima and one maximum (at the origin) of the free energy. (b) Graphical solution of the equation of state for $h > 0$.

For $T < T_c$, there are three solutions corresponding to the two minima and one maximum in Fig. 4.1.1b. The solution with $m > 0$ is the one with the lowest value (absolute minimum) of $f - hm$.

The Bragg-Williams free energy can be used to calculate thermodynamic quantities such as the internal energy and entropy and their derivatives. We will calculate most of these quantities using the more phenomenological Landau mean-field theory to be presented in the next section.

The Bragg-Williams approximation presented here assumes that the order parameter is spatially uniform. It can be generalized to spatially non-uniform order parameters $m_l \equiv \langle \sigma_l \rangle$ by using m_l rather than m in the internal energy and by assuming that the entropy is a sum of "local" entropies depending on m_l only:

$$F = -\frac{1}{2}\sum_{\mathbf{l},\mathbf{l}'} J_{\mathbf{l},\mathbf{l}'} m_{\mathbf{l}} m_{\mathbf{l}'} - T \sum_{\mathbf{l}} s[m_{\mathbf{l}}], \tag{4.1.14}$$

where $s(m)$ is defined in Eq. (4.1.2) and where, for nearest neighbor interactions, $J_{\mathbf{l},\mathbf{l}'}$ is equal to J if sites $\mathbf{l}$ and $\mathbf{l}'$ are nearest neighbors and is equal to zero otherwise. This generalized Bragg-Williams free energy is identical to the mean-field free energy arising from the local density matrix approximation discussed in Sec. 4.8. It is the basis for the Landau theory to be discussed in the next section.

4.2 Landau theory

Landau theory is remarkable in that, under the simple assumptions that the order parameter is small and uniform near T_c, it yields a wealth of information about phase transitions. Equilibrium thermodynamics is completely determined by the function $F[T, \langle\phi_i(\mathbf{x})\rangle]$, where $\langle\phi_i(\mathbf{x})\rangle$ is the local order parameter. F is a function that must be invariant under the symmetry group $\mathscr{G}$ of the disordered phase. This means that it can only be a function of those combinations of the order parameter $\langle\phi_i(\mathbf{x})\rangle$ that do not change (i.e. are scalars) under all of the operations in $\mathscr{G}$. For example, the Ising Hamiltonian [Eq. (3.6.2)] is invariant under spin inversion, $\sigma \to -\sigma$, and $F(T,m)$ is a function of $m^2 = \sigma^2$ as it is in the Bragg-Williams theory just discussed. F is, in general, a very complicated functional of $\langle\phi(\mathbf{x})\rangle$. However, since $\langle\phi(\mathbf{x})\rangle$ is zero for $T > T_c$, it is reasonable, following the example of Bragg-Williams theory, to expand F in a power series in $\langle\phi(\mathbf{x})\rangle$, at least in the vicinity of the critical point. Furthermore, it is possible in essentially all cases to define the order parameter so that it is spatially uniform in equilibrium in the ordered phase. This suggests that F be expressed in terms of a local free energy density $f(T, \langle\phi(\mathbf{x})\rangle)$ that is a *function* of the field $\langle\phi(\mathbf{x})\rangle$ at the point $\mathbf{x}$ only and a part that produces an energy cost for deviations from spatial uniformity. The simplest form that F can take is

$$F = \int d^d x f(T, \langle\phi(\mathbf{x})\rangle) + \int d^d x \frac{1}{2} c[\nabla\langle\phi(\mathbf{x})\rangle]^2, \tag{4.2.1}$$

where c is a phenomenological coefficient with units of energy×(length)$^{2-d}$ if $\phi(\mathbf{x})$ is unitless. f is then expanded in a power series in $\langle\phi(\mathbf{x})\rangle$. Above T_c, $\langle\phi\rangle$ must be zero when its conjugate field h is zero. Since $\langle\phi\rangle$ and h are related by an equation of state, Eq. (3.5.19), this requires that the linear term in the expansion of f be absent, so that

$$f(T,\phi) = \frac{1}{2} r\phi^2 - w\phi^3 + u\phi^4 + \dots . \tag{4.2.2}$$

Each of the terms in the above expansion must be invariant under the operations of $\mathscr{G}$. Thus, as we shall see shortly, odd order terms may often be absent, and on occasion there will be several terms of different symmetry of the same order in ϕ. All of the coefficients r, w, and u can, in principle, depend on temperature.

If f is truncated at a given power of $\langle\phi\rangle$ (say at order four as above), then the highest order term should be even with a positive coefficient to ensure that the equilibrium state will have a bounded value of $\langle\phi(\mathbf{x})\rangle$.

If the odd order terms in Eq. (4.2.2) are removed, f has exactly the same form as the free energy per site [Eq. (4.1.6)] of Bragg-Williams theory with $r \sim (T-T_c)$. This is one of the motivations for expanding $f(T,\phi)$ in an analytic power series in ϕ. The spatially-varying Bragg-Williams free energy [Eq. (4.1.14)] is a function of spatial derivatives of the order parameter via the spatial dependence of $J_{l,l'}$. In the continuum limit, when spatial variations of the order parameter are slow on a scale of the lattice spacing, the dominant contribution to F arising from the nonuniformity of the order parameter is proportional to the square of the gradient of the order parameter, as in Eq. (4.2.1). It should be emphasized that the neglect of higher gradients of $\phi(\mathbf{x})$ is only valid when spatial variations are slow on a microscopic length scale a determined by the range of interactions (the lattice spacing in the Ising model), i.e., when the wave number q of spatial variations in $\phi(\mathbf{x})$ is less than a *cutoff*

$$\Lambda \approx \frac{2\pi}{a}. \tag{4.2.3}$$

Thus, it is understood that any phenomenological model such as Eq. (4.2.1) carries with it an upper wave number cutoff Λ. The predictions of mean-field theory generally do not depend on Λ. Corrections to mean-field theory do, however.

Eqs. (4.2.1) and (4.2.2) are the essence of Landau mean-field theory. These forms for mean-field theory can easily be derived for models such as the n-vector models discussed in the preceding chapter. In the next few sections, we will determine what form f must take for a number of simple transitions and examine the predictions made by Landau mean-field theory.

4.3 The Ising and n-vector models

The order parameter for the Ising ferromagnet is the scalar magnetization $\langle\phi\rangle = m_z$. The free energy must be invariant under time reversal. Since $\langle\phi\rangle$ changes sign under time reversal, f must be invariant under $\langle\phi\rangle \to -\langle\phi\rangle$, i.e., only even powers of $\langle\phi\rangle$ are permitted in the expansion of f, Eq. (4.2.2). Thus we have

$$f(T,\langle\phi\rangle) = \frac{1}{2}r\langle\phi\rangle^2 + u\langle\phi\rangle^4, \tag{4.3.1}$$

in agreement with Eq. (4.1.6). In order for the partition function to be well defined, f must be positive definite at large values of $\langle\phi\rangle$, implying that u must be positive. $\langle\phi\rangle$ is determined by the equation of state, Eq. (3.5.19). At high temperatures, $\langle\phi\rangle$ must be zero when the external magnetic field h is zero. This means that f must have a minimum at $\langle\phi\rangle = 0$ at high T. At low temperatures, we expect a ferromagnetic state and f to have at least one minimum with nonzero $\langle\phi\rangle$. This

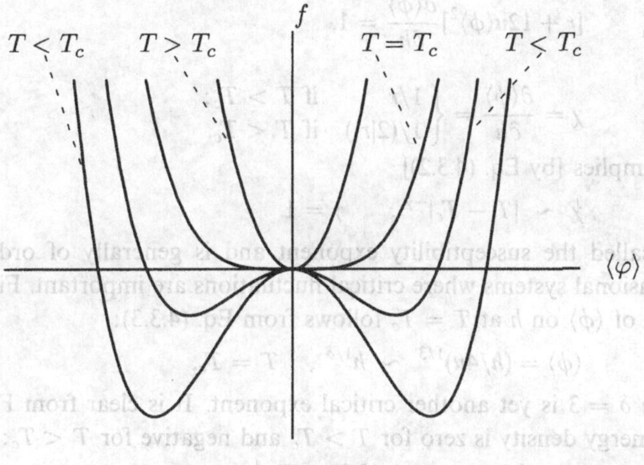

Fig. 4.3.1. $f = \frac{1}{2}r\langle\phi\rangle^2 + u\langle\phi\rangle^4$ as a function of $\langle\phi\rangle$ for different temperatures.

is most easily accomplished by allowing r to change sign at some temperature T_c:

$$r = a(T - T_c). \tag{4.3.2}$$

We will assume that u is independent of temperature. In Fig. 4.3.1, we sketch the function f for various values of T. For $T > T_c$, there is a single minimum at $\langle\phi\rangle = 0$. For $T = T_c$, there are two minima symmetrically placed about $\langle\phi\rangle = 0$. The equation of state for uniform h is, thus,

$$r\langle\phi\rangle + 4u\langle\phi\rangle^3 = h. \tag{4.3.3}$$

When $h = 0$, the solutions are

$$\langle\phi\rangle = \begin{cases} 0 & \text{if } T > T_c; \\ \pm(-r/4u)^{1/2} & \text{if } T < T_c. \end{cases} \tag{4.3.4}$$

Thus, mean-field theory predicts a second-order phase transition with

$$\langle\phi\rangle \sim (T_c - T)^\beta , \qquad \beta = 1/2 . \tag{4.3.5}$$

β is called a *critical exponent*, and it controls the temperature dependence of the order parameter in the vicinity of T_c. In mean-field theory, it has a value of 1/2. When critical fluctuations are important, β is generally less than its mean-field value, typically of order 1/3 in three-dimensional systems. Eq. (4.3.4) says that there are two possible values for $\langle\phi\rangle$ for $T < T_c$ and $h = 0$ corresponding to the two possible directions for the bulk magnetization. The sign can be fixed by allowing h to go to zero from positive or negative values. Both solutions for $\langle\phi\rangle$ have the same free energy so that the up and down phases coexist along the line $h = 0$ and $T < T_c$, as shown in the phase diagram of Fig. 3.5.2.

The susceptibility can be obtained by differentiating Eq. (4.3.3) with respect to h:

$$[r + 12u\langle\phi\rangle^2]\frac{\partial\langle\phi\rangle}{\partial h} = 1, \tag{4.3.6}$$

or

$$\chi = \frac{\partial\langle\phi\rangle}{\partial h} = \begin{cases} 1/r & \text{if } T > T_c; \\ 1/(2|r|) & \text{if } T < T_c. \end{cases} \tag{4.3.7}$$

This implies [by Eq. (4.3.2)]

$$\chi \sim |T - T_c|^{-\gamma}, \qquad \gamma = 1. \tag{4.3.8}$$

γ is called the susceptibility exponent and is generally of order 4/3 in three-dimensional systems where critical fluctuations are important. Finally, the dependence of $\langle\phi\rangle$ on h at $T = T_c$ follows from Eq. (4.3.3):

$$\langle\phi\rangle = (h/4u)^{1/3} \sim h^{1/\delta}, \quad T = T_c, \tag{4.3.9}$$

where $\delta = 3$ is yet another critical exponent. It is clear from Fig. 4.3.1 that the free energy density is zero for $T > T_c$ and negative for $T < T_c$:

$$f = \begin{cases} 0 & \text{if } T > T_c; \\ -r^2/(16u) & \text{if } T < T_c. \end{cases} \tag{4.3.10}$$

Thus the mean-field specific heat is zero for $T > T_c$ and positive for $T < T_c$:

$$c_V = -T\frac{\partial^2 f}{\partial T^2} = \begin{cases} 0 & \text{if } T > T_c; \\ Ta^2/(8u) & T < T_c. \end{cases} \tag{4.3.11}$$

This equation gives the specific heat associated with the establishment of order. The total specific heat includes a part analytic in temperature arising from degrees of freedom not associated with ordering. Thus a smoothly varying background must be added to Eq. (4.3.11) to obtain the total specific heat. This total specific heat will have a jump discontinuity at the transition, as shown in Fig. 4.3.2b.

1 The nonlocal susceptibility and the correlation length

We now turn to the calculation of the correlation function $\chi(\mathbf{x}, \mathbf{x}')$. This is most easily done using Eq. (3.5.22) and the mean-field form for F [Eq. (4.3.1)], including the gradient term [Eq. (4.2.1)]:

$$\begin{aligned} \chi^{-1}(\mathbf{x}, \mathbf{x}') &= \frac{\delta^2 F}{\delta\langle\phi(\mathbf{x})\rangle\delta\langle\phi(\mathbf{x}')\rangle} \\ &= (r + 12u\langle\phi\rangle^2 - c\nabla^2)\delta(\mathbf{x} - \mathbf{x}'), \end{aligned} \tag{4.3.12}$$

or

$$\chi(\mathbf{q}) = \frac{1}{r + 12u\langle\phi\rangle^2 + cq^2}. \tag{4.3.13}$$

Eq. (4.3.13) can be rewritten as

$$\chi(\mathbf{q}) = \frac{\chi}{[1 + (q\xi)^2]} \equiv \frac{1}{c}\frac{\xi^2}{1 + (q\xi)^2}, \tag{4.3.14}$$

where

$$\xi = c^{1/2}[r + 12u\langle\phi\rangle^2]^{-1/2}$$

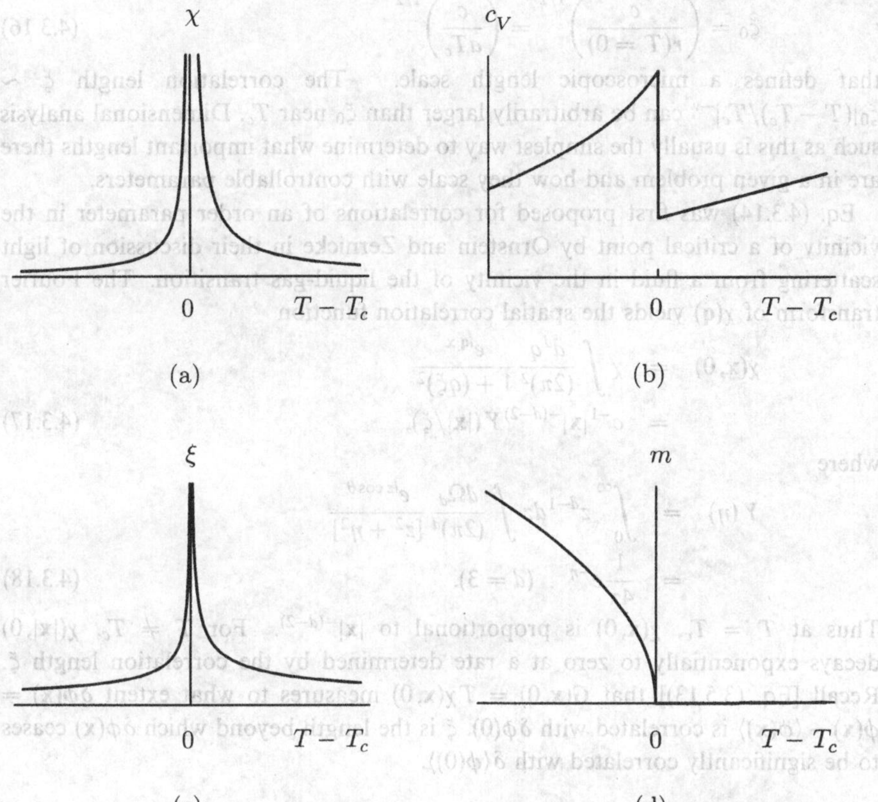

Fig. 4.3.2. (a) Mean-field susceptibility, (b) specific heat, (c) correlation length, and (d) order parameter as a function of temperature.

$$
\begin{aligned}
&= \begin{cases} (c/r)^{1/2} & \text{if } T > T_c; \\ [c/(-2r)]^{1/2} & \text{if } T < T_c \end{cases} \\
&\sim \;|T - T_c|^{-\nu}
\end{aligned} \tag{4.3.15}
$$

is the *correlation length* and ν is the correlation length exponent that is $1/2$ in mean-field theory and of order $2/3$ in most three-dimensional critical systems. The order-parameter correlation function, as discussed in the preceding chapter, is just T times the above nonlocal susceptibility.

The existence of the correlation length ξ follows from a simple dimensional analysis of the model free energy Eq. (4.2.1) with f given by Eq. (4.3.1). If $\langle\phi(\mathbf{x})\rangle$ is unitless, then r has units of (energy)×(length)$^{-d}$ $[EL^{-d}]$, and c has units of (energy)×(length)$^{d-2}$ $[EL^{d-2}]$. Therefore, $(c/r)^{1/2} \sim \xi$ must have units of length. These arguments allow us to introduce a temperature-independent bare correlation length

$$\xi_0 = \left(\frac{c}{r(T=0)}\right)^{1/2} = \left(\frac{c}{aT_c}\right)^{1/2} \tag{4.3.16}$$

that defines a microscopic length scale. The correlation length $\xi \sim \xi_0|(T-T_c)/T_c|^{-\nu}$ can be arbitrarily larger than ξ_0 near T_c. Dimensional analysis such as this is usually the simplest way to determine what important lengths there are in a given problem and how they scale with controllable parameters.

Eq. (4.3.14) was first proposed for correlations of an order parameter in the vicinity of a critical point by Ornstein and Zernicke in their discussion of light scattering from a fluid in the vicinity of the liquid-gas transition. The Fourier transform of $\chi(\mathbf{q})$ yields the spatial correlation function

$$\begin{aligned}\chi(\mathbf{x},0) &= \chi\int\frac{d^dq}{(2\pi)^2}\frac{e^{i\mathbf{q}\cdot\mathbf{x}}}{1+(q\xi)^2}\\ &= c^{-1}|\mathbf{x}|^{-(d-2)}Y(|\mathbf{x}|/\xi),\end{aligned} \tag{4.3.17}$$

where

$$\begin{aligned}Y(\eta) &= \int_0^\infty z^{d-1}dz\int\frac{d\Omega_d}{(2\pi)^d}\frac{e^{iz\cos\theta}}{[z^2+\eta^2]}\\ &= \frac{1}{4\pi}e^{-\eta}\qquad (d=3).\end{aligned} \tag{4.3.18}$$

Thus at $T = T_c$, $\chi(\mathbf{x},0)$ is proportional to $|\mathbf{x}|^{-(d-2)}$. For $T \neq T_c$, $\chi(|\mathbf{x}|,0)$ decays exponentially to zero at a rate determined by the correlation length ξ. Recall [Eq. (3.5.13)] that $G(\mathbf{x},0) = T\chi(\mathbf{x},0)$ measures to what extent $\delta\phi(\mathbf{x}) = \phi(\mathbf{x}) - \langle\phi(\mathbf{x})\rangle$ is correlated with $\delta\phi(0)$. ξ is the length beyond which $\delta\phi(\mathbf{x})$ ceases to be significantly correlated with $\delta\langle\phi(0)\rangle$.

2 O_n symmetry

The time reversal symmetry of the paramagnetic state of the Ising model does not allow any odd order terms in $\langle\phi\rangle$ in the expansion of f. Similarly the rotational invariance of the paramagnetic state of the n-vector model requires that f be a function only of the scalar,

$$\langle\phi\rangle^2 \equiv \sum_{i=1}^{n}\langle\phi_i\rangle^2. \tag{4.3.19}$$

The Landau free energy density for the n-vector model is identical to that of the Ising model with the above interpretation of $\langle\phi\rangle^2$. In the presence of an external field with components h_i, the equation of state for ϕ_i is

$$\frac{\partial f}{\partial\phi_i} = (r+4u\langle\phi\rangle^2)\langle\phi_i\rangle = h_i. \tag{4.3.20}$$

When $h_i = 0$, the solutions to this equation are $\langle\phi_i\rangle = 0$ for $T > T_c$ and $\langle\phi_i\rangle = (-r/4u)^{1/2}e_i$ for $T < T_c$, where $\mathbf{e}$ is any unit vector in the n-dimensional order parameter space. Thus, again there is a second-order transition with the same critical exponents β, γ, δ and ν as in the Ising model.

Correlation functions for $T < T_c$ require some further comment. Since a direction has been explicitly picked out (i.e., the symmetry of the high temperature phase has been broken), the correlation function,

$$G_{ij}(\mathbf{x},\mathbf{x}') = \langle\phi_i(\mathbf{x})\phi_j(\mathbf{x}')\rangle - \langle\phi_i(\mathbf{x})\rangle\langle\phi_j(\mathbf{x}')\rangle, \tag{4.3.21}$$

can be broken up into two distinct parts describing correlations parallel and perpendicular to the direction of order:

$$G_{ij}(\mathbf{x},\mathbf{x}') = G_{\parallel}(\mathbf{x},\mathbf{x}')e_ie_j + G_{\perp}(\mathbf{x},\mathbf{x}')(\delta_{ij} - e_ie_j). \tag{4.3.22}$$

With $\mathbf{e}$ parallel to the 1-axis, this means

$$\begin{aligned} G_{\parallel}(\mathbf{x},\mathbf{x}') &= \langle\phi_1(\mathbf{x})\phi_1(\mathbf{x}')\rangle - \langle\phi_1(\mathbf{x})\rangle\langle\phi_1(\mathbf{x}')\rangle, \\ G_{\perp}(\mathbf{x},\mathbf{x}') &= \langle\phi_i(\mathbf{x})\phi_i(\mathbf{x}')\rangle - \langle\phi_i(\mathbf{x})\rangle\langle\phi_i(\mathbf{x}')\rangle\ , \quad (i \neq 1). \end{aligned} \tag{4.3.23}$$

Differentiating the free energy with respect to $\phi_i(\mathbf{x})$ and $\phi_j(\mathbf{x}')$, we obtain

$$\chi_{ij}^{-1}(\mathbf{q}) = TG_{ij}^{-1}(\mathbf{q}) = (r + 4u\langle\phi\rangle^2 + cq^2)\delta_{ij} + 8u\langle\phi_i\rangle\langle\phi_j\rangle, \tag{4.3.24}$$

or

$$\chi_{\parallel}^{-1} = r + 12u\langle\phi\rangle^2 + cq^2 \tag{4.3.25}$$

and

$$\chi_{\perp}^{-1}(\mathbf{q}) = r + 4u\langle\phi\rangle^2 + cq^2 = \begin{cases} r + cq^2 & \text{if } T > T_c; \\ cq^2 & \text{if } T < T_c. \end{cases} \tag{4.3.26}$$

The $T < T_c$ form of $\chi_{\perp}$ follows from the temperature dependence of the order parameter [Eq. (4.3.4)]. Thus $G_{\parallel}(\mathbf{q}) = T\chi_{\parallel}(\mathbf{q})$ has the same form as $G(\mathbf{q})$ for the Ising model. $G_{\perp}(\mathbf{q}) = T\chi_{\perp}(\mathbf{q})$, on the other hand, has pure power law form for $T < T_c$:

$$G_{\perp}(\mathbf{q}) = \frac{T}{cq^2}. \tag{4.3.27}$$

This implies a power-law rather than exponential decay in real space with $G_{\perp}(\mathbf{x},0) \sim |\mathbf{x}|^{-(d-2)}$. This behavior, as we shall see in Chapter 6, is a direct consequence of the breaking of a continuous symmetry (rotational in this case). Physically, the difference between the parallel and perpendicular susceptibilities is that the former relates to changes in the magnitude of the order parameter while the latter relates to changes in its direction. The softness of the perpendicular response results from the lack of any restoring force for a tilt of the entire spin system.

3 Some mean-field transitions

There are a number of systems whose phase transitions are quantitatively described by the mean-field theory presented in this section. Here we will consider a few of these. In the next chapter, we will see why fluctuations that in general modify mean-field results are not important in these systems.

The transition from a normal metal to a superconductor is indisputably one of the transitions that is best described by mean-field theory. In Fig. 4.3.3, we show

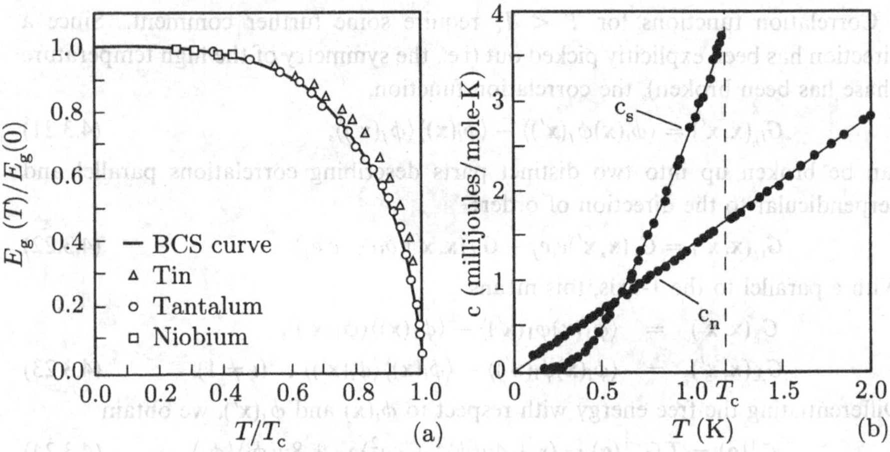

Fig. 4.3.3. (a) Superconducting order parameter as a function of temperature from electron tunneling measurements. (b) Specific heat of a superconductor [N.E. Phillips, *Phys. Rev.* **114**, 676 (1959)]. The transition to the superconducting phase is accompanied by a jump discontinuity in the specific heat, denoted c_s in the figure. When a small magnetic field is applied, there is no superconducting transition, and the specific heat c_n is linearly proportional to temperature.

the measured specific heat and order parameter as functions of temperature for some representative superconductors. Note that the specific heat has a mean-field jump at T_c identical to that predicted by Eq. (4.3.11). At low-temperature, the specific heat goes exponentially to zero indicating the existence of a gap in the low-temperature excitation spectrum (although this feature is not explained by our simple mean-field theory). The order parameter goes to zero as $(T - T_c)^{1/2}$ as predicted by mean-field theory and saturates at low temperature. The usual scattering probes do not couple to the superconducting order parameter so there are no direct spatial measurements of the order parameter susceptibility or correlation length. However, single-particle and Josephson tunneling directly measure the local energy gap and order parameter, and critical fields measure the correlation length indirectly.

Another mean-field transition is the smectic-*A* to smectic-*C* transition in liquid crystals. Here, the order parameter is the tilt angle of the director relative to the normal to the smectic planes. Fig. 4.3.4 shows the experimental measurements of the tilt angle, inverse order parameter susceptibility, specific heat, and inverse light scattering intensity for a single value of the scattering wave vector for butyloxybenzylidene heptylaniline. The inverse susceptibility (measured by magnetic birefringence) in the disordered phase goes linearly to zero (rather than with some power law) as $T \to T_c$, in agreement with mean-field theory. Similarly, the scattering intensity as a function of **q** follows Eq. (4.3.13). The wave number dependent susceptibility has correlation lengths parallel and perpendicular to layers that diverge as $|T - T_c|^{-1/2}$ but with different prefactors. The order parameter and specific heat, however, do not follow Eqs. (4.3.4) and (4.3.11) except very near T_c. Rather, they follow the predictions of a slightly generalized Landau free energy with a $\langle\phi\rangle^6$ term added to the simple free energy of Eq. (4.3.1):

$$f = \frac{1}{2}r\langle\phi\rangle^2 + u_4\langle\phi\rangle^4 + u_6\langle\phi\rangle^6. \tag{4.3.28}$$

It is straightforward to show that the susceptibility and correlation length in the disordered phase for this free energy are identical to those of the simple $\langle\phi\rangle^4$ free energy. The order parameter obtained by minimizing f with respect to $\langle\phi\rangle$ is

$$\langle\phi\rangle = (u_4/3u_6)^{1/2}[(1 - 3r/r_0)^{1/2} - 1]^{1/2} \tag{4.3.29}$$

and the specific heat is

$$c_V = \begin{cases} 0 & \text{if } T > T_c; \\ (Ta^2/8u_4)(1 - 3r/r_0)^{-1/2} & \text{if } T < T_c, \end{cases} \tag{4.3.30}$$

where $r_0 = 2u_4^2/u_6$. The solid curves in Figs. 4.3.4a and 4.3.4b follow, respectively, from Eqs. (4.3.29) and (4.3.30).

4.4 The liquid-gas transition

As we discussed in Chapter 2, there is no symmetry difference between the liquid and gas phases. The two phases can only be distinguished when they coexist in the same container and are separated by a meniscus. In a closed container with a fixed number of particles and a fixed volume, the meniscus is like a partition, on one side of which is the denser liquid phase and on the other side of which is the less dense gas phase. Particles and energy pass freely through the partition. In addition, the position of the meniscus is free to move to minimize the free energy. This implies that the temperature, chemical potential, and pressure are the same in the coexisting liquid and gas phases even though their respective densities, n_l and n_g, are different. Phase diagrams for classical and quantum fluids in the pressure-temperature plane are shown in Fig. 3.1.4. The liquid and gas (or vapor) phases coexist along the liquid-gas *coexistence curve* (or vapor pressure curve), terminating at a critical point with pressure $p = p_c$ and temperature $T = T_c$. The critical density n_c is determined from p_c and T_c by the equation of state. The phase diagram can also be represented in the chemical potential-temperature plane, as shown in the vicinity of the critical point (μ_c, T_c) in Fig. 4.4.1. The critical point can be approached along the coexistence curve, where there are two stable phases, or along various paths, along which there is only a single stable phase. A path of particular interest is the *critical isochore*, along which the density is equal to the critical density. Approach to the critical point along the critical isochore is easily implemented experimentally by preparing the fluid in a closed

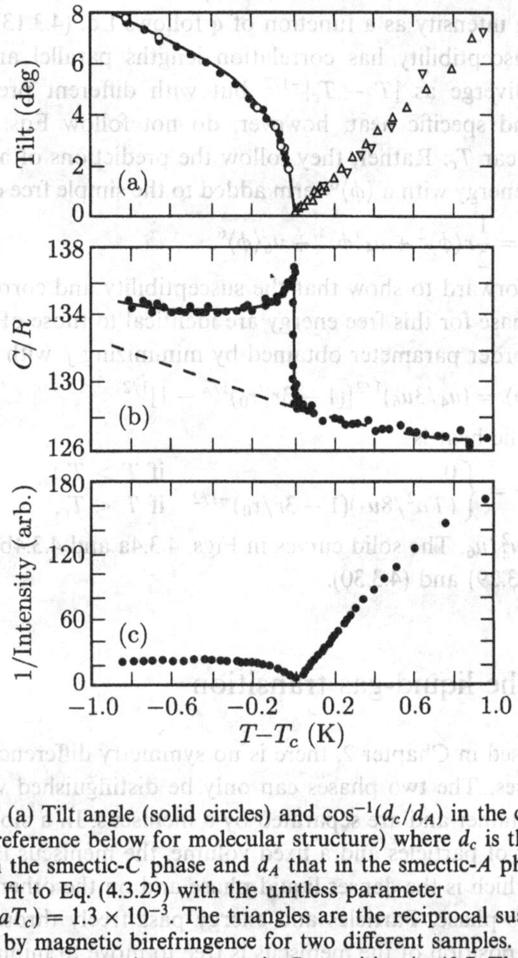

Fig. 4.3.4. (a) Tilt angle (solid circles) and $\cos^{-1}(d_c/d_A)$ in the compound 40.7 (see reference below for molecular structure) where d_c is the layer spacing in the smectic-C phase and d_A that in the smectic-A phase. The solid curve is a fit to Eq. (4.3.29) with the unitless parameter $t_0 = r_0/(2aT_c) = 1.3 \times 10^{-3}$. The triangles are the reciprocal susceptibility measured by magnetic birefringence for two different samples. (b) Heat capacity near the smectic-A to smectic-C transition in 40.7. The dashed curve is the background scaled from another compound (40.8) and the solid line is a fit to Eq. (4.3.30). (c) Inverse light scattering intensity from tilt fluctuations in 40.7. For this sample, $q = q_z = 7.0 \times 10^4$ cm^{-1}. [R.J. Birgeneau, G.W. Garland, A.R. Kortan, J.D. Litster, M. Meichle, B.M. Ocko, C. Rosenblatt, L.J. Yu, and J. Goodby, *Phys. Rev. A* **27**, 1251 (1983).]

volume at the critical density. The average density is then fixed. Other paths to the critical point include the *critical isobar*, along which $p = p_c$, and the *critical isotherm*, along which $T = T_c$.

The liquid-gas transition has much in common with the magnetic transition in an Ising model. In Fig. 4.4.1, we compare the phase diagram in the $h-T$ plane for the Ising model to the phase diagrams in the $p-T$ and $\mu-T$ planes for a fluid near its critical point. In both the Ising model and the fluid, there is a coexistence curve, terminating at a critical point, along which two distinct but

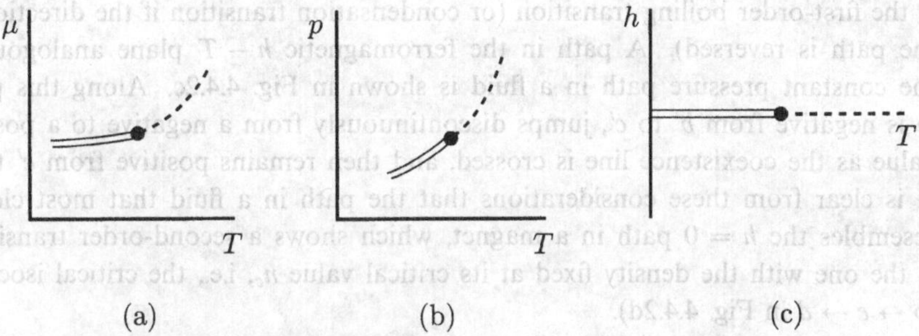

Fig. 4.4.1. (a) Liquid-gas phase diagram in the $\mu - T$ plane. The critical isochore is shown as a dashed line, the coexistence curve as a double line. Note that the two lines have a common slope but different curvatures at the critical point. (b) The critical isochore and coexistence curve in the $p - T$ plane. (c) Phase diagram from an Ising ferromagnet in the $h - T$ plane. The coexistence curve is shown as a solid line and the line $h = 0$, $T > T_c$ as a dashed line.

equal free energy phases coexist, and in both it is possible to go continuously around the critical point from one coexisting phase to the other by appropriately varying h and T or μ and T. In the magnet, the two coexisting phases are those with respective magnetization $m_+ = m$ and $m_- = -m$. The value of m at the critical point is $m_c = 0$, and one could express the magnetic order parameter as $m - m_c$ rather than as m. By analogy, one can introduce an order parameter $\phi = n - n_c$ for the liquid-gas transition. (We will use n rather than $\langle n \rangle$ throughout this section.) The magnetic and fluid phase diagrams shown in Fig. 4.4.1 differ in one obvious way: the coexistence curve for the magnet is a straight line $h = 0$, $T < T_c$, whereas that for the fluid is in general curved. The inversion ($m \to -m$) symmetry of a magnet forces the coexistence line to be the line $h = 0$ and the critical point value of h and m both to be zero. There is no such symmetry in a fluid and no special values of the critical point parameters p_c, T_c, or n_c.

In spite of the similarities between the magnetic and liquid-gas transitions, our intuition would suggest that they are quite different. Magnetic transitions are usually observed to be second order, in agreement with our analysis in Sec. 4.3. Our everyday experience with the change from the liquid to the gas state is with the boiling of water, which is clearly a first-order transition with an absorption of latent heat of vaporization. This difference of behavior reflects different paths through the transition in the two cases. In a ferromagnet, the natural experimental path ($b \to c \to d$ in Fig. 4.4.2a) is one with the external magnetic field h equal to zero. Because of reflection symmetry, the order parameter along this path is zero and equal to its critical point value for all $T > T_c$. For $T < T_c$, the order parameter grows continuously from zero. In a fluid, the natural experimental path is one in which temperature is varied at constant pressure ($b' \to c' \to d'$ in Fig. 4.4.2b). Along this path, there is a discontinuous change in the density. This is the first-order boiling transition (or condensation transition if the direction of the path is reversed). A path in the ferromagnetic $h-T$ plane analogous to the constant pressure path in a fluid is shown in Fig. 4.4.2c. Along this path, m is negative from b' to c', jumps discontinuously from a negative to a positive value as the coexistence line is crossed, and then remains positive from c' to d'. It is clear from these considerations that the path in a fluid that most closely resembles the $h=0$ path in a magnet, which shows a second-order transition, is the one with the density fixed at its critical value n_c, i.e., the critical isochore ($b \rightarrow c \rightarrow d$ in Fig. 4.4.2d).

A Landau theory for the liquid-gas transition can be developed most directly from the mixed thermodynamic function $w(T, \mu, n) = f - \mu n$ [Eq. (3.1.45)] whose minimum over n gives $-p(\mu, T)$. Using this function, we will first identify the conditions determining the critical parameters μ_c, T_c, and n_c. We will then proceed to determine the equations for the critical isochore and the coexistence curve and then to calculate the liquid and gas densities along the coexistence curve.

1 The critical point and the critical isochore

Three conditions are needed to specify the three parameters μ_c, T_c, and n_c. The first is the equation of state for n:

$$\frac{\partial w}{\partial n} = \frac{\partial f}{\partial n} - \mu = 0. \tag{4.4.1}$$

This equation is always satisfied in equilibrium for all μ and T. It provides the relation between μ and T along the critical isochore when n is set to its critical value n_c. The second condition determining the critical point is that the compressibility must be infinite or the inverse compressibility zero, as is clear from the pressure-density phase diagram shown in Fig. 3.1.3. Therefore,

$$\left.\frac{\partial^2 w(T_c, \mu_c, n)}{\partial n^2}\right|_{n=n_c} = \left.\frac{\partial^2 f(T_c, n)}{\partial n^2}\right|_{n=n_c} = 0. \tag{4.4.2}$$

The third condition is more subtle. The critical point terminates a line of coexistence. It must, therefore, be a point at which two solutions to Eq. (4.4.1) corresponding to the liquid and gas densities n_l and n_g merge into a single solution with density $n = n_c$. In a theory in which w is treated as an analytic function of n, this condition is reached by requiring the third derivative of w to be zero at the critical point:

$$\left.\frac{\partial^3 w(T_c, \mu_c, n)}{\partial n^3}\right|_{n=n_c} = \left.\frac{\partial^3 f(T_c, n)}{\partial n^3}\right|_{n=n_c} = 0. \tag{4.4.3}$$

Eqs. (4.4.1) to (4.4.3) determine T_c, μ_c, and n_c. The critical pressure is then $p_c = -w(T_c, \mu_c, n_c)$.

To study properties in the vicinity of the critical point, we expand $w(T, \mu, n)$ in powers of the order parameter $\phi \equiv n - n_c$:

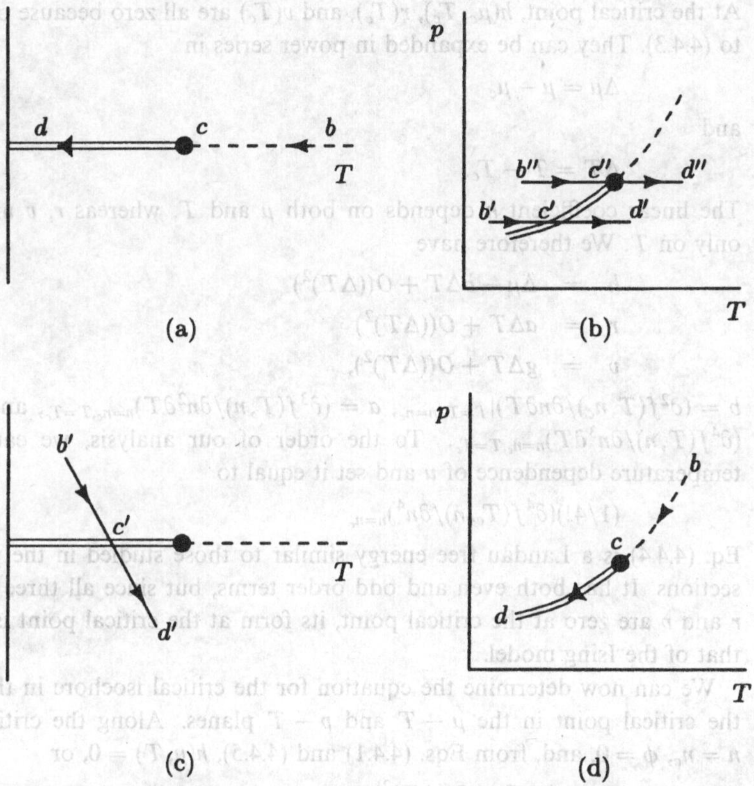

Fig. 4.4.2. (a) Path in the Ising $h - T$ plane obtained by reducing temperature at $h = 0$. The magnetization along this path is zero for $T > T_c$. (b) Two paths in the liquid-gas $p - T$ plane obtained by increasing temperature at constant pressure. The first ($b' \to c' \to d'$) crosses the coexistence line where there is a first-order change from the liquid to the gas state. The second ($b'' \to c'' \to d''$) is the critical isobar passing through the critical point. (c) A path in the Ising $h - T$ plane analogous to the constant pressure path in the liquid-gas system. (d) A path along the critical isochore in the $p - T$ plane of the liquid-gas phase diagram analogous to the $h = 0$, $m = 0$ path in the Ising model. The constant pressure paths will be curves rather than straight lines in the $\mu - T$ plane.

$$w = w(T, \mu, n_c) + \frac{1}{2}r\phi^2 - v\phi^3 + u\phi^4 - h\phi, \tag{4.4.4}$$

where

$$h = \mu - \left.\frac{\partial f(T,n)}{\partial n}\right)_{n=n_c}, \qquad r(T) = \left.\frac{\partial^2 f(T,n)}{\partial n^2}\right)_{n=n_c},$$

$$v(T) = -\left.\frac{1}{3!}\frac{\partial^3 f(T,n)}{\partial n^3}\right)_{n=n_c}, \qquad u(T) = \left.\frac{1}{4!}\frac{\partial^4 f(T,n)}{\partial n^4}\right)_{n=n_c}. \tag{4.4.5}$$

At the critical point, $h(\mu_c, T_c)$, $r(T_c)$, and $v(T_c)$ are all zero because of Eqs. (4.4.1) to (4.4.3). They can be expanded in power series in

$$\Delta\mu = \mu - \mu_c \tag{4.4.6}$$

and

$$\Delta T = T - T_c. \tag{4.4.7}$$

The linear coefficient h depends on both μ and T, whereas r, v and u depend only on T. We therefore have

$$\begin{aligned} h &= \Delta\mu - b\Delta T + O((\Delta T)^2) \\ r &= a\Delta T + O((\Delta T)^2) \\ v &= g\Delta T + O((\Delta T)^2), \end{aligned} \tag{4.4.8}$$

$b = (\partial^2 f(T, n_c)/\partial n \partial T)|_{T=T_c, n=n_c}$, $a = (\partial^3 f(T, n)/\partial n^2 \partial T)_{n=n_c, T=T_c}$, and $g = (1/3!)(\partial^4 f(T, n)/\partial n^3 \partial T)_{n=n_c, T=T_c}$. To the order of our analysis, we can ignore the temperature dependence of u and set it equal to

$$(1/4!)(\partial^4 f(T_c, n)/\partial n^4)_{n=n_c}.$$

Eq. (4.4.4) is a Landau free energy similar to those studied in the previous two sections. It has both even and odd order terms, but since all three potentials h, r and v are zero at the critical point, its form at the critical point is identical to that of the Ising model.

We can now determine the equation for the critical isochore in the vicinity of the critical point in the $\mu - T$ and $p - T$ planes. Along the critical isochore, $n = n_c$, $\phi = 0$, and, from Eqs. (4.4.1) and (4.4.5), $h(\mu, T) = 0$, or

$$\Delta\mu = b\Delta T + O((\Delta T)^2). \tag{4.4.9}$$

To determine the critical isochore in the $p - T$ plane, we recall that $w(T, \mu, n) = -p$ and $w(T_c, \mu_c, n_c) = p_c$. Expanding $w(T, \mu, n_c)$ in ΔT and $\Delta\mu$ we obtain

$$w(T, \mu, n_c) = -p_c - \Delta\mu n_c + e\Delta T + O((\Delta T)^2), \tag{4.4.10}$$

where $e = [\partial f(T, n_c)/\partial T]_{T=T_c}$. Then using Eq. (4.4.9) we obtain

$$p - p_c = (bn_c - e)\Delta T + O((\Delta T)^2). \tag{4.4.11}$$

Eqs. (4.4.9) and (4.4.11) show that the critical isochore in general has both finite slope and curvature at the critical point.

The equation of state,

$$\frac{\partial w}{\partial \phi} = r\phi - 3w\phi^2 + 4u\phi^3 - h = 0, \tag{4.4.12}$$

determines $\phi(\mu, T)$ away from the critical isochore. The inverse compressibility κ_T^{-1} goes to zero as the critical point is approached along the critical isochore:

$$\frac{\partial \mu}{\partial n} = \frac{\partial^2 w}{\partial n^2} = n_c^{-2}\kappa_T^{-1} = r. \tag{4.4.13}$$

Thus, like the susceptibility in the Ising model in zero external field, κ_T diverges along the critical isochore as $(\Delta T)^{-\gamma}$ with $\gamma = 1$. Spatial correlations can be calculated by adding a gradient term to the free energy as was done in the treatment of the Ising model of the preceding section. This leads to the expression

$$S_{nn}(\mathbf{q}) = \frac{T}{r + cq^2} = n^2\kappa_T \frac{T}{1 + (q\xi)^2} \tag{4.4.14}$$

for the density correlation function, implying that there is a correlation length $\xi = (c/r)^{-1/2}$ that diverges with a critical exponent $\nu = 1/2$ as in the Ising model. Eq. (4.4.14) is the form for the density correlation function first proposed by Ornstein and Zernicke. As we saw in Chapter 2, the intensity of light scattered at wave vector $\mathbf{q}$ is proportional to S_{nn}. Eq. (4.4.14) then says that the intensity of scattered light increases dramatically as the critical point is approached along the critical isochore. This is the phenomenon of *critical opalescence.*

2 The coexistence curve

The coexistence curve is determined by the condition that the pressure, chemical potential, and temperature of the liquid and gas phases be equal. The densities n_l and n_g of the liquid and gas phases will differ from the critical density according to $n_l = n_c + \phi_l$ and $n_g = n_c + \phi_g$. The pressures of the two phases are then $p_l = -w(T, \mu, n_l)$ and $p_g = -w(T, \mu, n_g)$. For a general function $w(T, \mu, n)$, the determination of the coexistence curve could be quite tedious. For the simple phenomenological form we are using, however, there is a fairly straightforward strategy for its determination. The idea is to break ϕ up into two parts,

$$\phi = \phi_0 + \Delta\phi, \tag{4.4.15}$$

and to choose ϕ_0 and μ as a function of T so that there are no odd order terms in an expansion of $w(T, \mu, n)$ in powers of $\Delta\phi$. In this case, there will be two values of $\Delta\phi$ which minimize the free energy, and $n_g = n_c + \phi_0 - \Delta\phi$ and $n_l = n_c + \phi_0 + \Delta\phi$. The expansion of w in powers of $\Delta\phi$ is

$$w = -p_c - h(\phi_0)\Delta\phi + \frac{1}{2}r(\phi_0)(\Delta\phi)^2 - v(\phi_0)(\Delta\phi)^3 + u(\Delta\phi)^4, \tag{4.4.16}$$

where

$$\begin{aligned} h(\phi_0) &= h - r\phi_0 + 3v\phi_0^2 - 4u\phi_0^3, \\ r(\phi_0) &= r - 6v\phi_0 + 12u\phi_0^2, \\ v(\phi_0) &= v - 4u\phi_0. \end{aligned} \tag{4.4.17}$$

The third-order term in the expansion of w is eliminated by choosing

$$\phi_0 = \frac{v}{4u} = \frac{g\Delta T}{4u}. \tag{4.4.18}$$

The first-order term is eliminated by choosing $\Delta\mu$ as a function of ΔT so that $h(\phi_0) = 0$. This yields

$$h = \frac{ag}{4u}(\Delta T)^2 + O((\Delta T)^3). \tag{4.4.19}$$

This says that the critical isochore and the coexistence curve have the same slopes in the $\mu - T$ plane at the critical point but that their curvatures are different, as a calculation of the $(\Delta T)^2$ term in Eq. (4.4.9) will show. Thus, along the line determined by Eqs. (4.4.17) and (4.4.19),

$$w(T,\mu,n) = w(T,\mu,n_c) + \frac{1}{2}r(\phi_0)(\Delta\phi)^2 + u(\Delta\phi)^4 \tag{4.4.20}$$

with

$$r(\phi_0) = a\Delta T + O((\Delta T)^2). \tag{4.4.21}$$

Eq. (4.4.20) has exactly the same form as the Ising free energy and is minimized by

$$\Delta\phi = \pm(-r(\phi_0)/4u)^{1/2} = \pm(-a\Delta T/4u)^{1/2} + O((\Delta T)^{3/2}). \tag{4.4.22}$$

Inserting Eq. (4.4.22) into Eq. (4.4.20), one finds that, as in the $\mu - T$ plane, the critical isochore and coexistence curve in the $p - T$ plane have the same slope but different curvatures at the critical point. The above analysis says that the difference in density of the liquid and gas phases, $2\Delta\phi$, tends to zero as $|\Delta T|^\beta$, with $\beta = 1/2$ in mean-field theory. The average density of the liquid and gas phases $(n_l + n_g)/2 = n_c + \phi_0$ tends to the critical density linearly in ΔT. This is the law of *rectilinear diameters*. Deviations from this law of the form $(\Delta T)^{1-\alpha}$, where α is a critical exponent associated with the specific heat, which will be discussed in the next chapter, are expected in real systems where mean-field theory does not apply.

For Ising magnetic systems, it is conventional to plot the order parameter versus temperature, whereas historically the phase separation and liquid-gas transitions are plotted with axes flipped, as shown schematically in Fig. 4.4.3. The actual liquid-gas phase boundary as measured for a number of different systems is shown in Fig. 4.4.4. This is one of the best examples of universality in phase transitions, which will be discussed in more detail in the next chapter: all eight materials in the figure fall on essentially the same curve. Note, however, that the boundary is better described by an exponent $\beta = 1/3$ rather than by the mean-field value $\beta = 1/2$.

In this section, we have seen that the liquid-gas transition is very similar to the Ising transition even though there is no symmetry difference between the liquid and gas phases. The important point is that there are two choices for the density in the coexistence region just as there are two choices for the magnetization along the coexistence line of an Ising ferromagnet. The presence of odd-order terms in the expansion of the free energy in powers of ϕ does not lead to quantitative differences between the liquid-gas and the Ising transitions. At the level of mean-field theory, they only lead to curvature in the coexistence curve and the average of the liquid and gas densities differing from the critical density $[n_l + n_g - 2n_c \sim \Delta T]$.

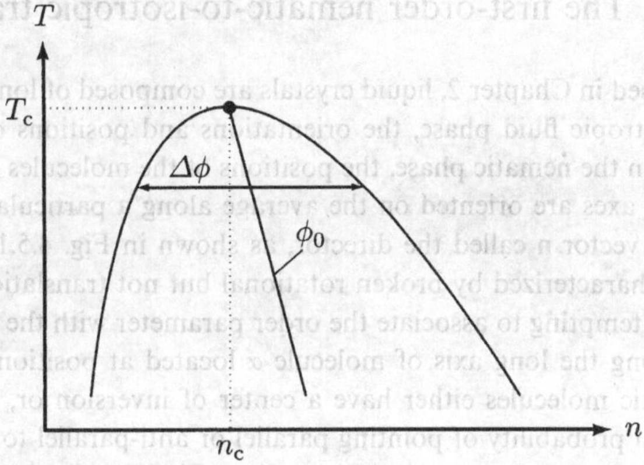

Fig. 4.4.3. Schematic phase boundary for the liquid-gas transition showing the asymmetry of the order parameter about the critical density.

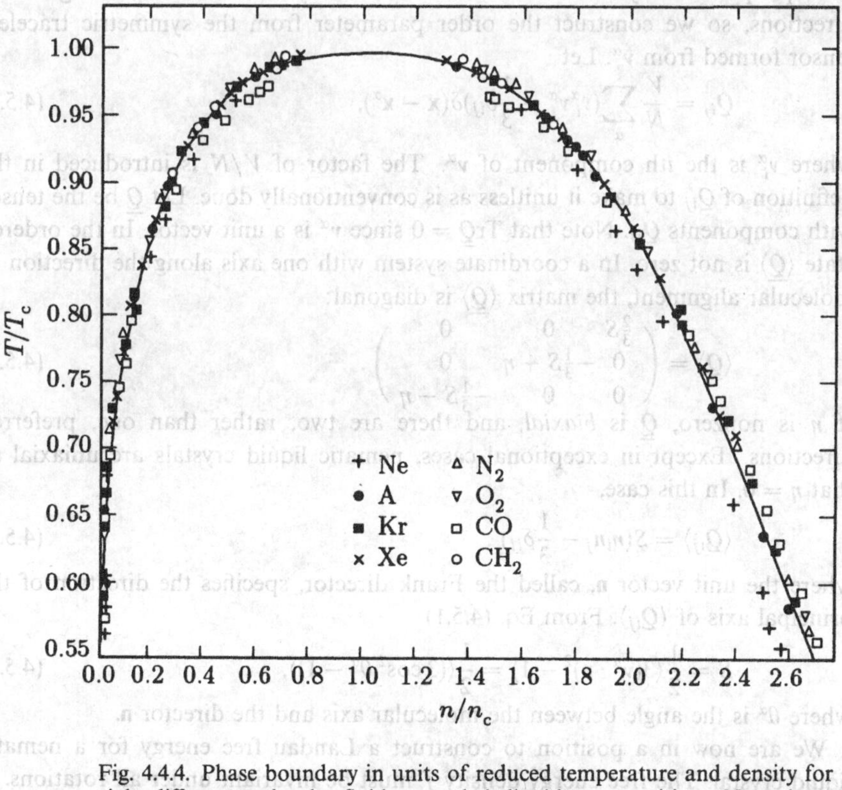

Fig. 4.4.4. Phase boundary in units of reduced temperature and density for eight different molecular fluids near their liquid-gas transitions. Note the universal behavior and the fact that the solid line is $\Delta\phi \propto (T_c - T)^{\beta}$ with $\beta = 1/3$ rather than the mean-field result $\beta = 1/2$. [E.A. Guggenheim, *J. Chem. Phys.* **13**, 253 (1945).]

4.5 The first-order nematic-to-isotropic transition

As discussed in Chapter 2, liquid crystals are composed of long, barlike molecules. In the isotropic fluid phase, the orientations and positions of the molecules are random. In the nematic phase, the positions of the molecules are still random, but their long axes are oriented on the average along a particular direction specified by a unit vector $\mathbf{n}$ called the director, as shown in Fig. 4.5.1. Thus, the nematic phase is characterized by broken rotational but not translational symmetry. It is, therefore, tempting to associate the order parameter with the unit vector $\boldsymbol{\nu}^\alpha$ which points along the long axis of molecule α located at position $\mathbf{x}^\alpha$. However, since the nematic molecules either have a center of inversion or, if they do not, they have equal probability of pointing parallel or anti-parallel to any given direction, both $\boldsymbol{\nu}^\alpha$ and $-\boldsymbol{\nu}^\alpha$ contribute to the order. Thus any order parameter must be even in $\boldsymbol{\nu}^\alpha$. Since a vector order parameter is insufficient, we can try a second rank tensor. We require the order parameter to be zero in the high-temperature isotropic phase. A symmetric traceless tensor will yield zero when averaged over directions, so we construct the order parameter from the symmetric traceless tensor formed from $\boldsymbol{\nu}^\alpha$. Let

$$Q_{ij} = \frac{V}{N}\sum_\alpha (\nu_i^\alpha \nu_j^\alpha - \frac{1}{3}\delta_{ij})\delta(\mathbf{x} - \mathbf{x}^\alpha), \tag{4.5.1}$$

where ν_i^α is the ith component of $\boldsymbol{\nu}^\alpha$. The factor of V/N is introduced in the definition of Q_{ij} to make it unitless as is conventionally done. Let $\underline{Q}$ be the tensor with components Q_{ij}. Note that $\text{Tr}\underline{Q} = 0$ since $\boldsymbol{\nu}^\alpha$ is a unit vector. In the ordered state $\langle \underline{Q} \rangle$ is not zero. In a coordinate system with one axis along the direction of molecular alignment, the matrix $\langle \underline{Q} \rangle$ is diagonal:

$$\langle \underline{Q} \rangle = \begin{pmatrix} \frac{2}{3}S & 0 & 0 \\ 0 & -\frac{1}{3}S + \eta & 0 \\ 0 & 0 & -\frac{1}{3}S - \eta \end{pmatrix}. \tag{4.5.2}$$

If η is nonzero, $\underline{Q}$ is *biaxial*, and there are two, rather than one, preferred directions. Except in exceptional cases, nematic liquid crystals are uniaxial so that $\eta = 0$. In this case,

$$\langle Q_{ij} \rangle = S(n_i n_j - \frac{1}{3}\delta_{ij}), \tag{4.5.3}$$

where the unit vector $\mathbf{n}$, called the Frank director, specifies the direction of the principal axis of $\langle Q_{ij} \rangle$. From Eq. (4.5.1)

$$S = \frac{1}{2}\langle 3(\boldsymbol{\nu}^\alpha \cdot \mathbf{n})^2 - 1 \rangle = \frac{1}{2}\langle (3\cos^2\theta^\alpha - 1) \rangle, \tag{4.5.4}$$

where θ^α is the angle between the molecular axis and the director $\mathbf{n}$.

We are now in a position to construct a Landau free energy for a nematic liquid crystal. The free energy density f must be invariant under all rotations. $\underline{Q}$ transforms like a tensor under the rotation group. f must, therefore, only be a function of the scalar combinations $\text{Tr}\langle \underline{Q} \rangle^p$, $p = 2, 3, \ldots$ that are invariant under rotations. The term with $p = 1$ is just the trace of $\langle \underline{Q} \rangle$ and is by definition zero.

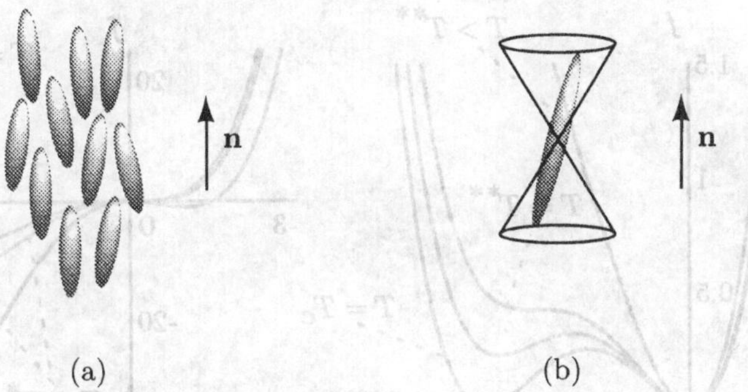

Fig. 4.5.1. (a) Schematic representation of barlike molecules in the nematic phase. They are oriented on the average along the director **n**. The direction of each molecule is effectively confined to lie within a cone, as shown in (b), rather than being free to choose any solid angle.

Thus, there is no term linear in $\langle \underline{Q} \rangle$ in the free energy. To fourth order in $\underline{Q}$, we therefore have

$$\begin{aligned} f &= \frac{1}{2}r(\frac{3}{2}\mathrm{Tr}\langle \underline{Q} \rangle^2) - w(\frac{9}{2}\mathrm{Tr}\langle \underline{Q} \rangle^3) + u(\frac{3}{2}\mathrm{Tr}\langle \underline{Q} \rangle^2)^2, \\ &= \frac{1}{2}rS^2 - wS^3 + uS^4. \end{aligned} \tag{4.5.5}$$

In general, there should be two fourth-order terms proportional, respectively, to $(\mathrm{Tr}\langle \mathrm{Q} \rangle^2)^2$ and $\mathrm{Tr}\langle \mathrm{Q} \rangle^4$. However, for 3×3 symmetric traceless tensors, they are strictly proportional, and we need only include the $(\mathrm{Tr}\langle \underline{\mathrm{Q}} \rangle^2)^2$ term. As before, r is positive at high T and negative at low T. We choose

$$r = a(T - T^*). \tag{4.5.6}$$

u and w are independent of temperature.

The free energy of Eq. (4.5.5) differs from that of the Ising model by the presence of the third-order term $-wS^3$. If the order parameter for the nematic phase were a vector (as might be imagined if the constituent molecules lacked inversion symmetry) rather than a tensor, then odd order terms would be prohibited in the free energy by rotational symmetry. However, the rodlike molecules have a quadrupolar rather than a dipolar symmetry, and the order parameter is a tensor for which rotational invariance does not rule out the odd terms. Note that the quadrupole symmetry is also reflected in the form of the order parameter in Eqs. (4.5.3) and (4.5.4). f is sketched as a function of S for various values of T in Fig. 4.5.2. Note that the cubic term leads to an asymmetry in f as a function of S and the emergence of a secondary minimum at finite S. The value of f at this minimum is greater than zero at high temperature but becomes equal to zero at a critical temperature T_c that is greater than the temperature T^* at which the extremum at the origin develops negative curvature. Since f is less than zero at the secondary minimum for all $T < T_c$, there is a phase transition with a

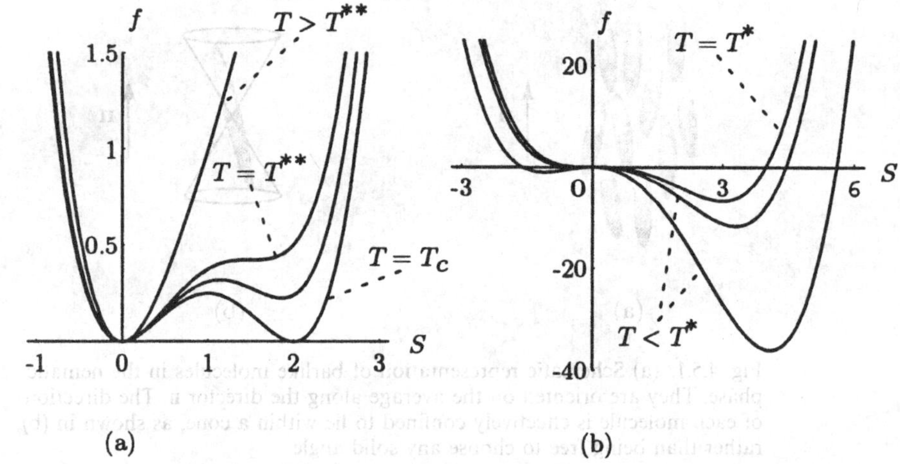

Fig. 4.5.2. Free energy density f as a function of order parameter S for different T for the isotropic-nematic transition. The transition is first order. Note the limits of metastability for supercooling (T^*) and superheating (T^{**}).

discontinuous change in S at T_c, i.e. there is a first-order transition at T_c. T^* is the *limit of metastability* of the isotropic phase since, for $T^* < T < T_c$, the origin is still a local minimum even though it is not a global minimum. The limit of metastability of the nematic phase occurs at the temperature T^{**} at which the secondary minimum disappears on heating.

The first-order transition temperature T_c and the value S_c of S at T_c are calculated by requiring that f be an extremum with respect to S in equilibrium and that the free energies of the disordered and ordered phases be equal at the transition. The latter condition implies that the isotropic and nematic phases can coexist at the transition temperature. If other variables, such as pressure or density, were included in our treatment, the two phases would coexist along a line rather than at a single point. The equations determining T_c and S_c are, therefore,

$$\frac{\partial f}{\partial S} = (r - 3wS + 4uS^2)S = 0 \tag{4.5.7}$$

$$f = (\frac{1}{2}r - wS + uS^2)S^2 = 0. \tag{4.5.8}$$

Thus,

$$S_c = \frac{w}{2u}, \quad r_c = a(T_c - T^*) = \frac{w^2}{2u}. \tag{4.5.9}$$

Since the transition is first order, there is an associated latent heat. The entropy per unit volume of the disordered phase is zero in mean-field theory, whereas that of the nematic phase is negative. This result can be obtained from the free energy of the nematic phase, which to lowest order in $r - r_c$ is

$$f = \frac{1}{2}(r - r_c)S_c^2 = \frac{1}{2}(r - r_c)(w/2u)^2. \tag{4.5.10}$$

The entropy density in the nematic phase (relative to that of the isotropic phase) is then

$$s = -\frac{\partial f}{\partial T} = -\frac{1}{2}aS_c^2 = -\frac{1}{2}a(w/2u)^2. \tag{4.5.11}$$

There are, of course, other contributions to the entropy not included in the present model that ensure that the total entropy is always positive. The latent heat absorbed in going from the nematic to the isotropic phase is thus

$$q = -T_c s = \frac{1}{2}aT_c(w/2u)^2. \tag{4.5.12}$$

The molecules comprising nematic liquid crystals (called *nematogens)* are anisotropic diamagnets. Typically they have lower energy when they are aligned with their long axis parallel to the magnetic field. The interaction Hamiltonian is

$$\begin{aligned} H_{\text{ext}} &= -\int d^d x \chi_a Q_{ij} H_i H_j, \\ &= -\frac{3}{2}\int d^d x \chi_a H^2 S, \end{aligned} \tag{4.5.13}$$

where χ_a is the difference in the magnetic susceptibility of a nematic molecule for directions parallel and perpendicular to its long axis and **H** is the external magnetic field along **n**. Thus $h = (3/2)\chi_a H^2$ is the field conjugate to S. The susceptibility associated with S can be calculated as before. It satisfies

$$\chi = \frac{\partial S}{\partial h} = (r - 6wS + 12uS^2)^{-1}. \tag{4.5.14}$$

χ appears to diverge at $T = T^*$ as temperature is lowered in the isotropic phase. The first-order phase transition at $T_c > T^*$ cuts off this divergence as shown in Fig. 4.5.3. T^* is thus the limit of metastability of the isotropic phase. To find the limit of metastability (T^{**}) of the nematic phase, we calculate the temperature at which $\chi^{-1} = 0$ in the free energy minimum with $S > 0$. S satisfies the equation of state [Eqs. (4.5.7)] as before, and

$$r^{**} = a(T^{**} - T^*) = \frac{9w^2}{16u}. \tag{4.5.15}$$

Note that $T^{**} - T_c = w^2/(16au) > 0$ as necessary.

A general phenomenon associated with first-order transitions is the presence of hysteresis in cycling through the transition and the related effects of superheating and supercooling. Thermodynamically the transition should occur at T_c. However, the high temperature phase with $S = 0$ is stable against small fluctuations until the temperature T^* is reached on cooling. Between T_c and T^*, the ordered phase will only occur if there is a sufficiently large fluctuation. Below T^* the system is unstable against infinitesimal fluctuations into the ordered state. Thus, on cooling, the actual transition to the low-temperature ordered phase will occur at some temperature between T_c and T^* depending on the particular sample and the experimental conditions. Similar arguments show that, on heating, the

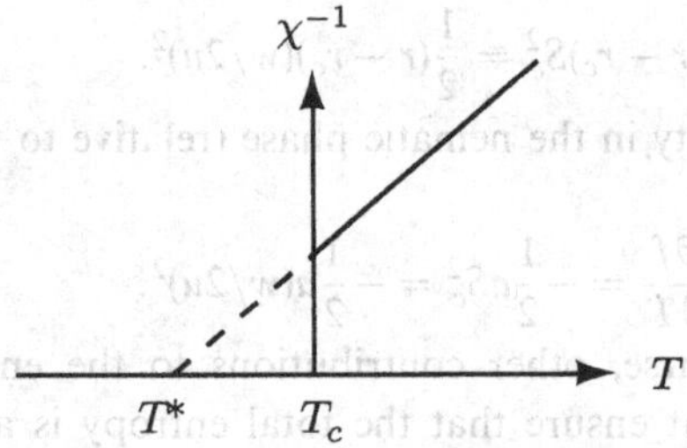

Fig. 4.5.3. χ^{-1} for a first-order transition. In mean-field theory, this function extrapolates to zero at the limit of metastability T^*.

transition from the ordered to the high temperature phase will occur between T_c and T^{**}.

The example of the isotropic-to-nematic transition is representative of phase transitions in which the order parameter possesses a third-order invariant. One expects in general that such transitions will be first order. Though the above Landau theory correctly predicts qualitative properties of first-order transitions, it certainly cannot make detailed quantitative predictions. This is because the order parameter is not zero at the transition. One is not justified, therefore, in truncating the power series expansion of f at fourth order. Even in mean-field theory, higher order terms in this expansion will lead to corrections both to T_c and S_c. If, however, the transition is nearly second order, as would be the case if the predicted value of $T_c - T^*$ is small, the truncated model is a reasonable approximation.

4.6 Multicritical points

The phase transitions discussed in Secs. 4.4 and 4.5 occurred in response to changes in a single variable, the temperature T, when the external ordering field (i.e., field conjugate to the order parameter) is zero. Thus, the phase diagram for these systems can be drawn on the temperature axis alone with a critical point separating the high-temperature disordered phase from the low-temperature ordered phase. In systems where there is more than one non-ordering field (such as pressure and temperature), phase diagrams become multidimensional, and there can occur critical points that can be reached only by fixing two or more non-ordering fields. In this section, we will consider a few of these *multicritical points*.

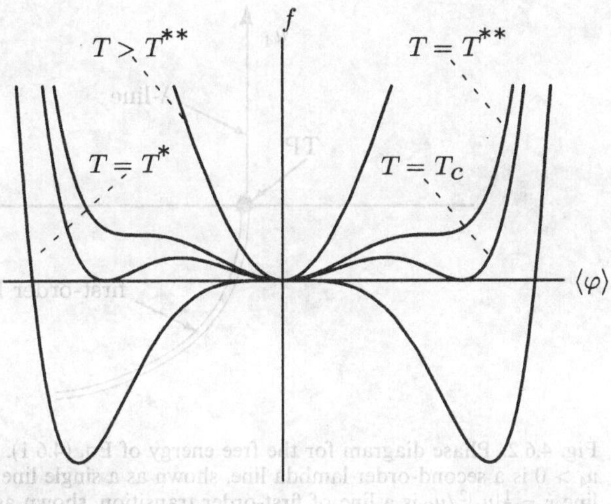

Fig. 4.6.1. f for a ϕ^6 potential [Eq. (4.6.1)] with u_4 negative. There is a first-order transition at $T = T_c$. T^{**} and T^* are, respectively, the limits of metastability on heating and cooling.

1 Tricritical points

In the preceding section, we found that third-order invariants lead to first-order transitions. First-order transitions can also occur if symmetry prohibits odd-order terms. Consider the following Landau free energy:

$$f = \frac{1}{2}r\phi^2 + u_4\phi^4 + u_6\phi^6, \tag{4.6.1}$$

where $r = a(T - T^*)$. If u_4 is positive, the sixth-order term can be neglected in the vicinity of the predicted second-order transition. If, on the other hand, u_4 is negative, the sixth-order term is required to maintain stability. In this case, secondary minima symmetrically placed about $\phi = 0$ develop as T is lowered, as shown in Fig. 4.6.1. When the free energies of the secondary minima with $\phi \neq 0$ pass through zero, there is a first-order transition as in the isotropic-to-nematic example.

When $u_4 > 0$, the mean-field transition temperature is the same as for the ϕ^4-model of Sec. 4.4, i.e., $r_c = 0$ and $T_c = T^*$. When $u_4 < 0$, however, the first-order transition temperature is determined by the conditions $f(r_c, \phi) = 0$ and $\partial f(r_c, \phi)/\partial\phi = 0$ just as for the nematic liquid crystal. This leads to

$$r_c = a(T_c - T^*) = \begin{cases} 0 & \text{if } u_4 > 0; \\ \frac{1}{2}|u_4|^2/u_6 & \text{if } u_4 < 0. \end{cases} \tag{4.6.2}$$

The phase diagram described by this equation in the $r - u_4$ plane is shown in Fig. 4.6.2. The line of second-order transitions for $u_4 > 0$ is called a *lambda* line. (It was first observed at the normal-to-superfluid transition in liquid helium mixtures; see Fig. 4.6.7. The superfluid transition is often referred to as a λ

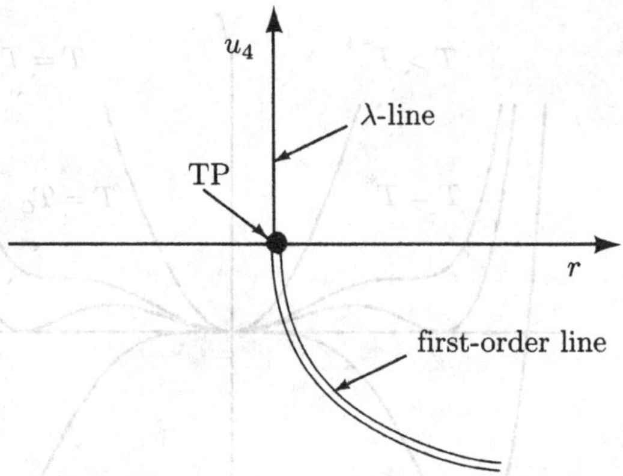

Fig. 4.6.2. Phase diagram for the free energy of Eq. (4.6.1). The line $r = 0$, $u_4 > 0$ is a second-order lambda line, shown as a single line in the figure. The line $r = \frac{1}{2}|u_4|^2/u_6$ is a line of first-order transition, shown as a double line in the figure. The point TP, $r = 0$, $u_4 = 0$, is a tricritical point.

transition since the specific heat curve resembles the Greek letter λ.) It meets the line of first-order transitions for $u_4 < 0$ at a *tricritical point*, $(r, u_4) = (0, 0)$.

The value of the order parameter, the latent heat along the first-order line, and the limit of metastability on heating can be calculated as in the previous section:

$$\phi_c = \pm[|u_4|/(2u_6)]^{1/2}, \tag{4.6.3}$$

$$q = -T_c s = \frac{a}{4}|u_4|/u_6, \tag{4.6.4}$$

$$r^{**} = a(T^{**} - T^*) = 2|u_4|^2/(3u_6).$$

Notice that both ϕ_c and q go to zero at the tricritical point where there is no longer a first-order transition. Note also that along the first-order line there is coexistence of three phases: the disordered phase with $\phi = 0$ and two ordered phases with $\phi = \pm|\phi_c|$. When $u_4 = 0$, there is a second-order transition but with an order parameter critical exponent β of 1/4 rather than 1/2:

$$\phi = \pm[-r/6u_6]^{1/4}. \tag{4.6.5}$$

Similarly, when an external ordering field h is applied at the tricritical point,

$$\phi = (h/6u_6)^{1/5}, \tag{4.6.6}$$

implying that the exponent δ is 5 rather than 3. The other critical exponents, γ and ν, for the tricritical point are the same in mean-field theory for $u_4 = 0$ and for $u_4 > 0$.

Fig. 4.6.2 depicts the phase diagram in the vicinity of a tricritical point in the most natural variables for the model free energy of Eq. (4.6.1). In real systems, all of the potentials are functions of the experimentally controllable parameters such as temperature, pressure, chemical potential, concentration of species, or external

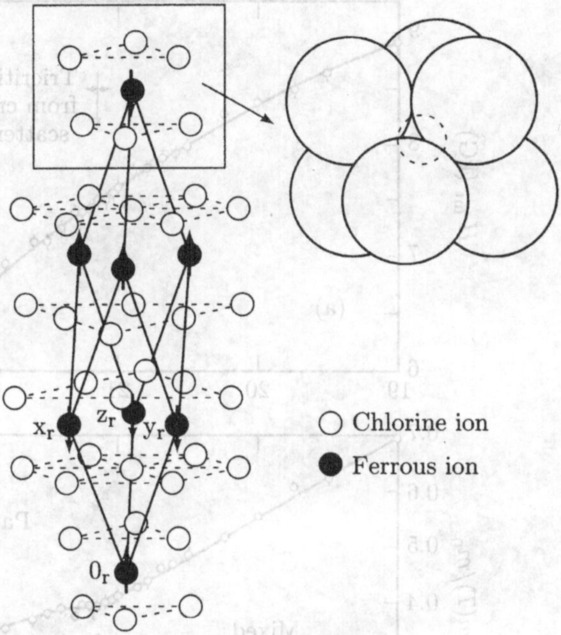

Fig. 4.6.3. Crystal structure of $FeCl_2$. Note the alternation in spin direction from layer to layer. [R.J. Birgeneau, W.B. Yelon, E. Cohen, and J. Markovski, *Phys. Rev. B* **5**, 2607 (1972).]

magnetic field. Physical phase diagrams with tricritical points will thus be rotated and stretched versions of Fig. 4.6.2. We will now consider some physical systems exhibiting tricritical points, the microscopic models used to describe them, and how their mean-field free energy can be cast into a form similar to Eq. (4.6.1).

2 Metamagnets and $FeCl_2$

Magnets which undergo first-order phase transitions in an increasing magnetic field are called *metamagnets*. There is a wide variety of metamagnets that exhibit tricritical points. One that has been studied in detail is $FeCl_2$. In this material, magnetic Fe^{2+} ions occupy sites on stacks of parallel triangular lattices separated by chlorine ions, as shown in Fig. 4.6.3. Each Fe^{2+} ion is surrounded by a cage of six chlorine ions and carries an effective spin of one. Lattice anisotropy favors alignment of spins perpendicular to the layers. There is positive intralayer and negative interlayer exchange. The ordered state is one in which spins in a given plane are parallel but alternate in direction from layer to layer, creating a two-sublattice antiferromagnetic structure. If an external magnetic field is applied perpendicular to the layers, the average spin in one sublattice will increase and that in the other will decrease such that there will be both staggered and uniform

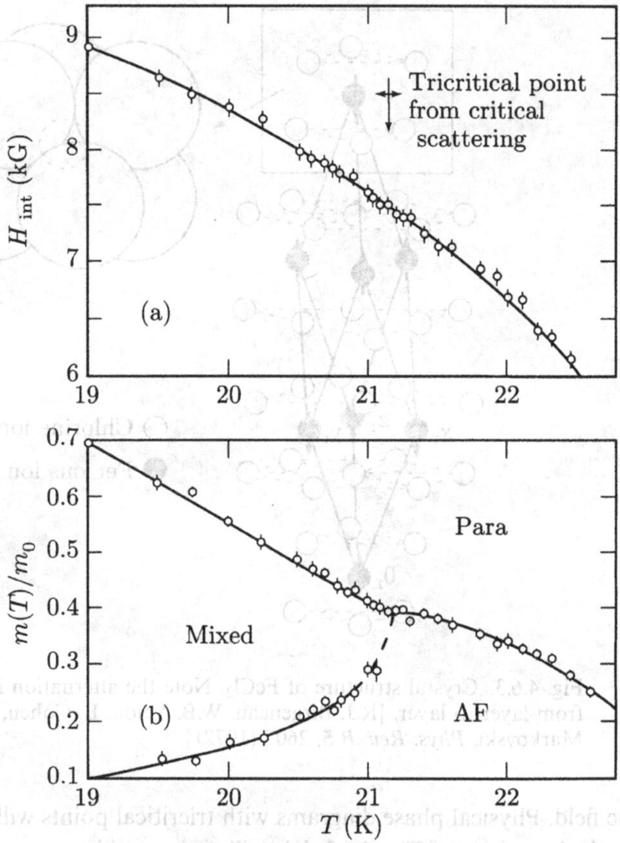

Fig. 4.6.4. Phase diagrams for the metamagnet $FeCl_2$ in (a) the internal field-temperature plane and (b) the magnetization-temperature plane. The lines in these curves are guides to the eye. [R.J. Birgeneau, G. Shirane, M. Blume, and W.C. Koehler, *Phys. Rev. Lett.* **33**, 1100 (1974).]

components to the magnetization. A sufficiently large field will lead to alignment of all spins and the destruction of long-range antiferromagnetic order.

The experimental phase diagram for $FeCl_2$ in the temperature-internal magnetic field plane is shown in Fig. 4.6.4a. (The internal magnetic field is the applied external magnetic intensity minus the demagnetizing field.) Note that the phase boundary separating the paramagnetic (P) and antiferromagnetic (AF) phases is perfectly smooth with no evidence of any multicritical point. The situation in the magnetization (m)-temperature plane is quite different, as can be seen in Fig. 4.6.4b. At high temperatures and small values of m, there is a single second-order phase boundary. Below a critical temperature, two phases with different values of m coexist, indicating the existence of a line of first-order transitions.

It is instructive to consider a microscopic model for a metamagnet and to see how it predicts a tricritical point and two-phase coexistence in qualitative agreement with the experiments discussed above. The Hamiltonian for a metamagnet is a generalization of the simple Heisenberg model that includes the possibility of anisotropic and further neighbor exchange and of lattice anisotropy favoring spin alignment along particular lattice directions. The following Hamiltonian is sufficient for our present discussion:

$$\mathscr{H} = -\frac{1}{2}\sum_{\mathbf{l},\mathbf{l}'} J_{\mathbf{l},\mathbf{l}'}\mathbf{S}_{\mathbf{l}}\cdot\mathbf{S}_{\mathbf{l}'} - D\sum_{\mathbf{l}}(S_{\mathbf{l}',\parallel})^2 - \sum_{\mathbf{l}}(H_{\parallel}S_{\mathbf{l},\parallel} + \mathbf{H}_{\perp}\cdot\mathbf{S}_{\mathbf{l},\perp}). \tag{4.6.7}$$

Here D is an anisotropy field resulting from spin-orbit interactions, $S_{\mathbf{l},\parallel}$ and $H_{\parallel}$ are, respectively, the spin and external magnetic field parallel to the preferred crystalline axis, and $\mathbf{H}_{\perp}$ is the component of the magnetic field perpendicular to that axis. The particular form for the exchange function $J_{\mathbf{l},\mathbf{l}'}$ will vary from system to system. In $FeCl_2$, $D > 0$ and

$$J_{\mathbf{l},\mathbf{l}'} = \begin{cases} J_1 & \text{if } \mathbf{l},\mathbf{l}' \text{ are n.n. in same plane;} \\ J_1' & \text{if } \mathbf{l},\mathbf{l}' \text{ are n.n.n. in same plane;} \\ -J_2 & \text{if } \mathbf{l},\mathbf{l}' \text{ are n.n. in adjacent planes,} \end{cases} \tag{4.6.8}$$

where n.n. and n.n.n. mean, respectively, nearest neighbor and next nearest neighbor. The antiferromagnetic coupling J_2 between planes favors an alternation in spin direction from one plane to the next.

If the anisotropy energy is sufficiently large, we may assume that the spins are effectively Ising spins that can point only up or down and, therefore, we can treat $\mathbf{S}_{\mathbf{l}}$ as an Ising spin variable. In this case, the mean-field free energy associated with the metamagnetic Hamiltonian can be calculated using Bragg-Williams theory. Let $m_A = \langle S^A_{\mathbf{l},\parallel}\rangle$ be the average spin (magnetization) in sublattice-A and m_B the average spin in sublattice-B. The entropy calculated according to the methods discussed in Sec. 4.2 reduces to the sum of the entropies of the two sublattices. If we ignore J_1', our mean-field free energy becomes

$$f = \frac{1}{2}z_2J_2m_Am_B - \frac{1}{4}z_1J_1(m_A^2 + m_B^2) - \frac{1}{2}T[s(m_A) + s(m_B)], \tag{4.6.9}$$

where $s(m)$ is the entropy of mixing [Eq. (4.1.3)] and where $z_2 = 2$ and $z_1 = 6$, the number of nearest neighbors in a triangular lattice. Now define the magnetization m and staggered magnetization m_s via

$$m = \frac{1}{2}(m_A + m_B)$$

$$m_s = \frac{1}{2}(m_A - m_B), \tag{4.6.10}$$

and expand f to sixth order in m_s with coefficients depending on m:

$$f = f_0(m) + \frac{1}{2}r_s(m)m_s^2 + u_4(m)m_s^4 + u_6(m)m_s^6. \tag{4.6.11}$$

Here, $f_0(m) = \frac{1}{2}T_mm^2 - Ts(m)$ with $T_m = z_2J_2 - z_1J_1$ is identical to the Bragg-Williams free energy of Eq. (4.1.4) with T_m replacing T_c. In addition

$$r_s(m) = T(1-m^2)^{-1} - T_N^0, \tag{4.6.12}$$

where $T_N = z_1J_1 + z_2J_2$ is the mean-field transition temperature to the antiferromagnetic state when $m = 0$ (i.e., $H = H_{\|} = 0$) and

$$u_4(m) = \frac{T}{12}\frac{1+3m^2}{(1-m^2)^3}. \tag{4.6.13}$$

We will not need an explicit form for $u_6(m)$ in what follows.

An effective free energy for the staggered magnetization can be obtained by using the equation of state $\partial f/\partial m = H$ to determine m as a function of T, H, and m_s. When m_s is small, this can be done in two steps. First solve for $m = m_0(T, H) = m_0$ when m_s is zero. The result is Eq. (4.1.9) with T_c replaced by $-T_m$. Then set $m = m_0 + \delta m$ and expand f for small δm:

$$\begin{aligned} g(m_s, \delta m) &\equiv f(m_0 + \delta m, m_s) - H\delta m - f(m_0, 0) \\ &= \frac{1}{2}\chi^{-1}(m_0)(\delta m)^2 + \frac{1}{2}r_s(m_0)m_s^2 \\ &\quad + u_4(m_0)m_s^4 + u_6(m_0)m_s^6 + \lambda(m_0)m_s^2\delta m, \end{aligned} \tag{4.6.14}$$

where $\chi^{-1}(m_0) = T_m + T(1-m_0^2)^{-2}$ and $\lambda(m_0) = m_0T(1-m_0^2)^{-1}$. Note that there is a linear coupling between m_s^2 and δm. Thus if m_s is nonzero,

$$\delta m = -\chi\lambda m_s^2 \tag{4.6.15}$$

is necessarily nonzero in equilibrium. Using Eq. (4.6.15) in Eq. (4.6.14), we obtain

$$g(m_0) = \frac{1}{2}r(m_s)m_s^2 + [u_4(m_0) - \frac{1}{2}\chi(m_0)\lambda^2(m_0)]m_s^4 + u_6'(m_0)m_s^6. \tag{4.6.16}$$

In this equation, we have explicitly displayed only the corrections to u_4 arising from couplings of δm to m_s. In general, there will be corrections to sixth and higher order potentials. For this reason, we have used u_6' to distinguish it from the unrenormalized potential u_6. Eq. (4.6.16) now has precisely the same form as the model free energy introduced at the beginning of this section. If u_6' is positive, then there is a tricritical point determined by the equations $r(m_0) = 0$, $u_4(m_0) - \frac{1}{2}\chi(m_0)\lambda^2(m_0) = 0$, and the equation of state relating m_0 to H. We will leave the calculation of H and T at the tricritical point to a homework problem (Problem 4.4).

On the first-order side of the tricritical point, a phase with $m_s = 0$ coexists with two antiferromagnetic phases with positive and negative staggered magnetization. Eq. (4.6.15) then implies the coexistence of two different values of the uniform magnetization: $m = m_0$ and $m = m_0 - \chi\lambda m_s^2$, in agreement with the results of experiments (Fig. 4.6.4).

Detailed calculations of the phase diagram associated with Eq. (4.6.16) show that there is only a lambda line with no tricritical point for $-1 < (z_1J_1/z_2J_2) < 0$. When $(z_1J_1/z_2J_2) > 3/5$, the sixth-order term in the expansion of $g(m_s)$ is positive and there is a tricritical point. However, when $0 < (z_1J_1/z_2J_2) < 3/5$, this coefficient is negative, and, rather than a tricritical point, there is a *critical endpoint* in which the second-order P-AF line terminates on a first-order line, which terminates at a liquid-gas like critical point within the region of antiferromagnet

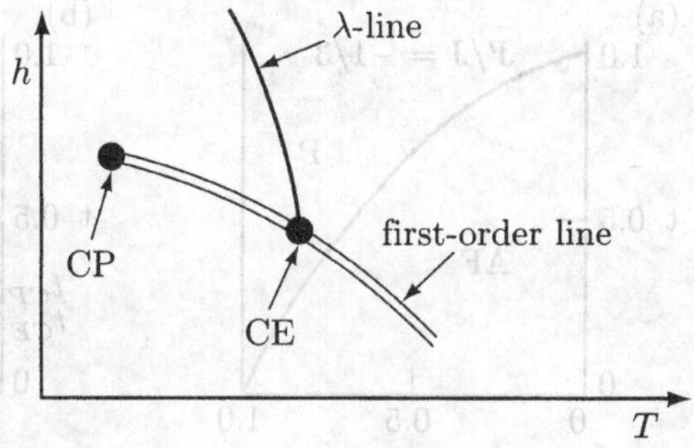

Fig. 4.6.5. Schematic representation of a critical endpoint (CE) in which a second-order line terminates at a first-order line that continues into an ordered region and itself terminates at a liquid-gas like critical point (CP).

order, as shown schematically in Fig. 4.6.5. Calculated phase diagrams for various values of the ratio z_1J_1/z_2J_2 are shown in Fig. 4.6.6.

3 $He^3 - He^4$ mixtures and the Blume-Emery-Griffiths model

Another system exhibiting an experimental tricritical point is that of mixtures of He^3 and He^4. In pure He^4, there is a transition from a normal fluid to a superfluid phase characterized by a complex order parameter. When He^4 is diluted with He^3, the superfluid transition temperature is depressed. At the same time, the tendency toward phase separation increases, and the second-order lambda line terminates at exactly the point where phase separation into He^3-rich and He^3-poor phases first takes place. An experimental phase diagram is shown in Fig. 4.6.7. Note the close resemblance of this phase diagram to the metamagnetic magnetization-temperature diagram (once the axes are flipped). The relative concentration of He^3 is the analog of the magnetization, and the superfluid order parameter is the analog of the staggered magnetization.

The Blume-Emery-Griffith (BEG) model is a spin-lattice model that successfully describes the essential features of the He^3-He^4 phase diagram. At each site **l** on a lattice, there is a spin variable $S_{\mathbf{l}}$ that can take on values $-1, 0, +1$. The order parameter for this model is the uniform average spin $\langle S_{\mathbf{l}}\rangle$. This is an Ising-like order parameter that fails to describe in detail the complex superfluid order parameter. At the level of mean-field theory, however, the distinction between the two types of order is unimportant. In the simple Ising model with the constraint $S_{\mathbf{l}} = \pm 1$, $S_{\mathbf{l}}^2 = 1$ at every site. In the present case, $S_{\mathbf{l}}^2$ can be either zero or unity, and one can interpret $\langle S_{\mathbf{l}}^2\rangle$ as the density of He^4 atoms and $1 - \langle S_{\mathbf{l}}^2\rangle$ as the density of He^3 atoms. The BEG Hamiltonian contains an exchange term favoring the development of nonzero $\langle S_{\mathbf{l}}\rangle$ and other terms describing the attraction or repulsion between atoms on different sites. The latter terms will be of the form

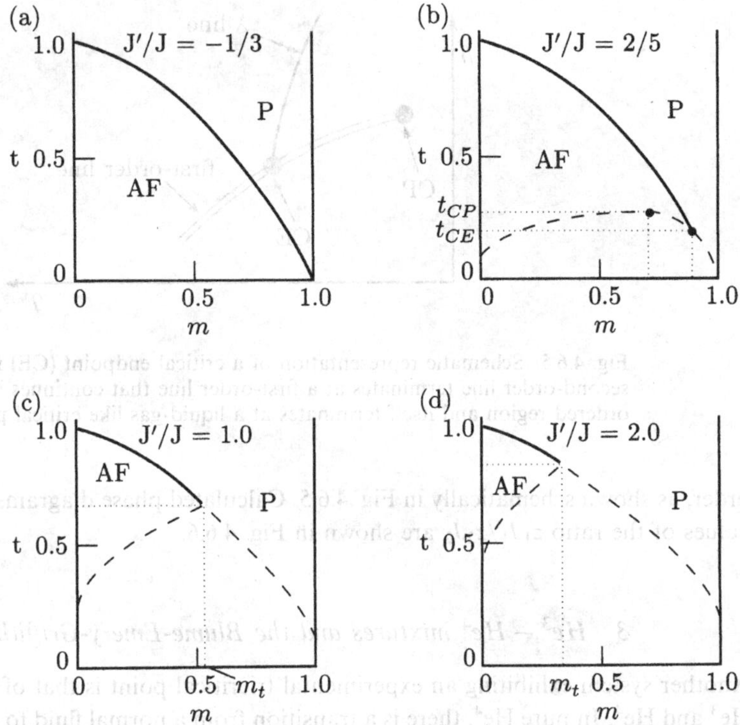

Fig. 4.6.6. Mean-field phase diagrams for the metamagnet. Here, $J = z_1 J_1$ and $J' = z_2 J_2$. (a) shows only a second-order P-AF transition. (b) shows a critical endpoint at t_{CE} where the second-order P-AF line meets the first-order coexistence line and a critical point at t_{CP} where two-phase coexistence ends. (c) and (d) show second-order P-AF lines terminating at a tricritical point. [J.M. Kincaid and E.G.D. Cohen, *Phys. Rep. C* **22**, 57 (1975).]

$S_l^2 S_{l'}^2$, $(1 - S_l^2)S_{l'}^2 + (1 - S_{l'}^2)S_l^2$ and $(1 - S_l^2)(1 - S_{l'}^2)$. In compact form the BEG Hamiltonian is

$$\mathscr{H} = -J \sum_{<l,l'>} S_l S_{l'} - K \sum_{<l,l'>} S_l^2 S_{l'}^2 + \Delta \sum_l S_l^2. \tag{4.6.17}$$

The mean-field phase diagram for this model is shown in Fig. 4.6.8.

As we briefly mentioned above, the first-order line near a tricritical point actually corresponds to coexistence of three phases: the paramagnetic and two antiferromagnetic phases. The degeneracy between the two AF phases can be lifted by the application of a field h (not physically realizable) conjugate to the order parameter (m_s for the metamagnets and S_l for the BEG model). The phase diagram for the BEG model in the $T - \Delta - h$ plane is shown in Fig. 4.6.9. Note

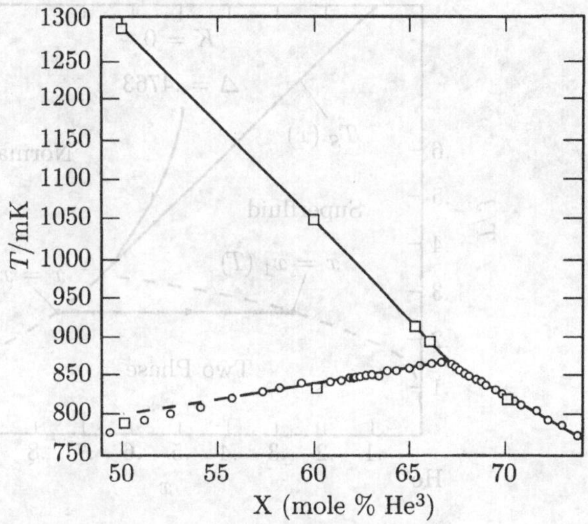

Fig. 4.6.7. Temperature-composition phase diagram for the system He^3 + He^4. [E.H. Graf, D.M. Lee, and John D. Reppy, *Phys. Rev. Lett.* **19**, 417 (1967).]

that in this three-dimensional space, three second-order lines meet at the tricritical point. This is in fact why it is called a tricritical point.

4 Bicritical and tetracritical points

Anisotropies favoring spin alignment along particular lattice directions can break the O_n symmetry of ideal Heisenberg models and give rise to a class of multicritical points that we will now discuss. Consider a system with two Ising order parameters ϕ_1 and ϕ_2 with the following Landau free energy:

$$f = \frac{1}{2}r(\phi_1^2 + \phi_2^2) - \frac{1}{2}g(\phi_1^2 - \phi_2^2) + u_1\phi_1^4 + u_2\phi_2^4 + 2u_{12}\phi_1^2\phi_2^2. \quad (4.6.18)$$

If $g = 0$ and $u_1 = u_2 = u_{12}$, this model reduces to the xy-model with a two-component vector order parameter $\phi = (\phi_1, \phi_2)$ and isotropic interactions. When $g > 0$, however, the field ϕ_1 will order before ϕ_2, and we expect an ordered phase with $\phi_1 \neq 0$ and $\phi_2 = 0$. When $g < 0$, the converse will occur. The details of the phase diagram for Eq. (4.6.18) depend on the relative magnitudes of the fourth-order potentials. When $u_1u_2 < u_{12}^2$, there is a first-order line along $g = 0$, $r < 0$ separating the phase with $\phi_1 \neq 0$ and $\phi_2 = 0$ from the phase with $\phi_1 = 0$ and $\phi_2 \neq 0$, as shown in Fig. 4.6.10a. Two distinct second-order lines meet at the point $r = 0$, $g = 0$, and this point is called a *bicritical point*. When $u_1u_2 > u_{12}^2$, there is an intermediate phase, with both ϕ_1 and ϕ_2 nonzero, separated by a second-order line from the phases with $\phi_2 = 0$ and $\phi_1 = 0$, as

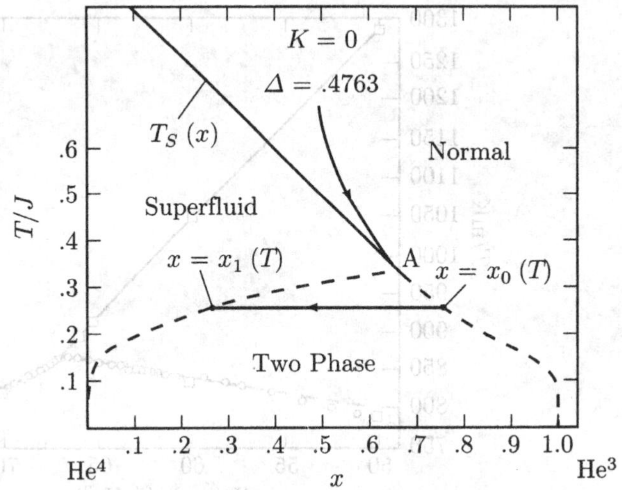

Fig. 4.6.8. Mean-field phase diagram in the temperature-concentration (x) plane for the Blume-Emery-Griffiths model [M. Blume, V.J. Emery, and Robert B. Griffiths, *Phys. Rev. A* **4**, 1071 (1971)].

shown in Fig. 4.6.10b. In this case, four second-order lines meet at the point $r = 0$, $g = 0$, which is now a *tetracritical point*.

Bicritical and tetracritical points can be found in a number of antiferromagnets in which the lattice anisotropy is not strong enough to enforce complete spin alignment. The zero-field state is antiferromagnetic with the staggered magnetization $\mathbf{m}_s$ aligned along the anisotropy axis. The application of a field parallel to the anisotropy axis initially increases the magnetization in one sublattice over the other but preserves the direction of $\mathbf{m}_s$. At a critical value of the field, the direction of $\mathbf{m}_s$ spontaneously flips from parallel to perpendicular to the field, as depicted schematically in Fig. 4.6.11a. This is called a *spin-flop* transition. It is a first-order transition that terminates at a bicritical point. At the bicritical point $(H, T) = (H_{\text{BP}}, T_{\text{BP}})$, both the parallel and perpendicular components of $\mathbf{m}_s$ are critical (as are the components of ϕ in our toy model). For $H > H_{\text{BP}}$, only the perpendicular components order, while for $H < H_{\text{BP}}$, only the parallel components order. Along the first-order line, two phases with different uniform magnetization coexist, as shown in Fig. 4.6.11b. If there is a tetracritical rather than a bicritical point, there will be an intermediate phase between the antiferromagnetic and spin-flop phases, as shown in Fig. 4.6.11c. Phase diagrams such as the ones shown in Fig. 4.6.11 can be obtained from a mean-field analysis of the Hamiltonian of Eq. (4.6.7) with appropriate values of anisotropy D and exchange energy $J_{\mathbf{l}\mathbf{l}'}$. The phase boundaries near either the bicritical or the tetracritical point predicted by mean-field theory will be straight lines. The effective free energy in the vicinity of the bicritical point will have the form of Eq. (4.6.18). The

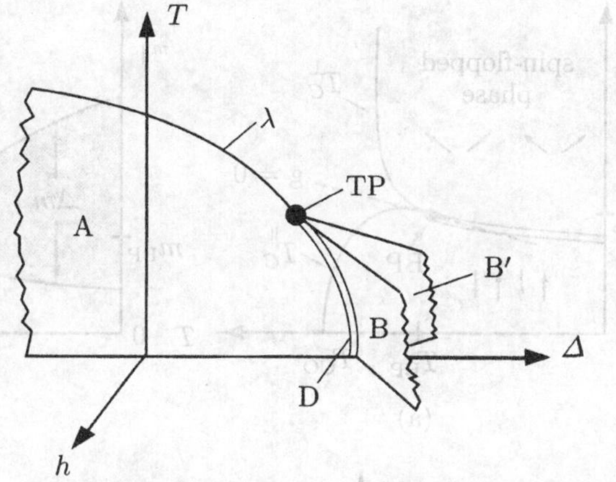

Fig. 4.6.9. Phase diagram for the BEG model in the $T-\Delta-h$ plane. The three solid lines are lambda lines terminating planes of two phase coexistence. They meet at a tricritical point (TP). The double line is a first-order line. Phases with positive and negative $\langle S \rangle$ coexist along the plane A, and phases with different values of $\langle S^2 \rangle$ coexist along the planes B and B'. Three phases coexist along the double line D. [R.B. Griffiths, *Phys. Rev. Lett.* **24**, 715 (1970).]

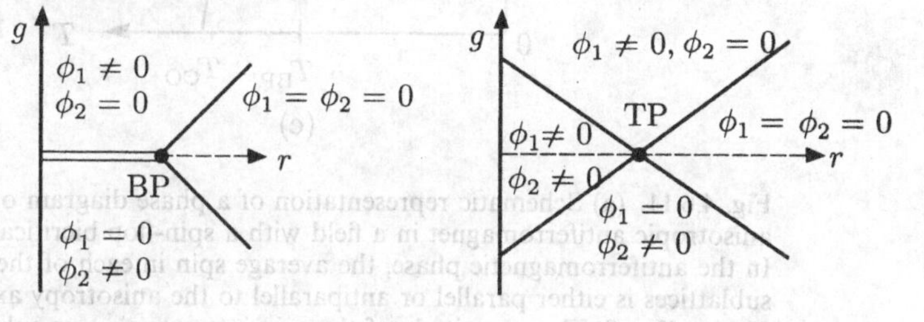

Fig. 4.6.10. Phase diagram from the model described by Eq. (4.6.18) showing (a) a bicritical point (BP) when $u_1u_2 < u_{12}^2$ and (b) a tetracritical point (TP) when $u_1u_2 > u_{12}^2$.

phase boundaries in the schematic phase diagrams of Fig. 4.6.11 have nonzero curvature near the critical points. We will see in the next chapter that critical fluctuations not included in mean-field theory predict such curvature. Fig. 4.6.12a shows a bicritical point in the experimental phase diagram of MnF_2. Fig. 4.6.12b shows a bicritical point in the experimental phase diagram for $GdAlO_3$ in an external field parallel to the anisotropy axis, and Fig. 4.6.12c shows a tetracritical point in the same material in an external field perpendicular to the anisotropy axis.

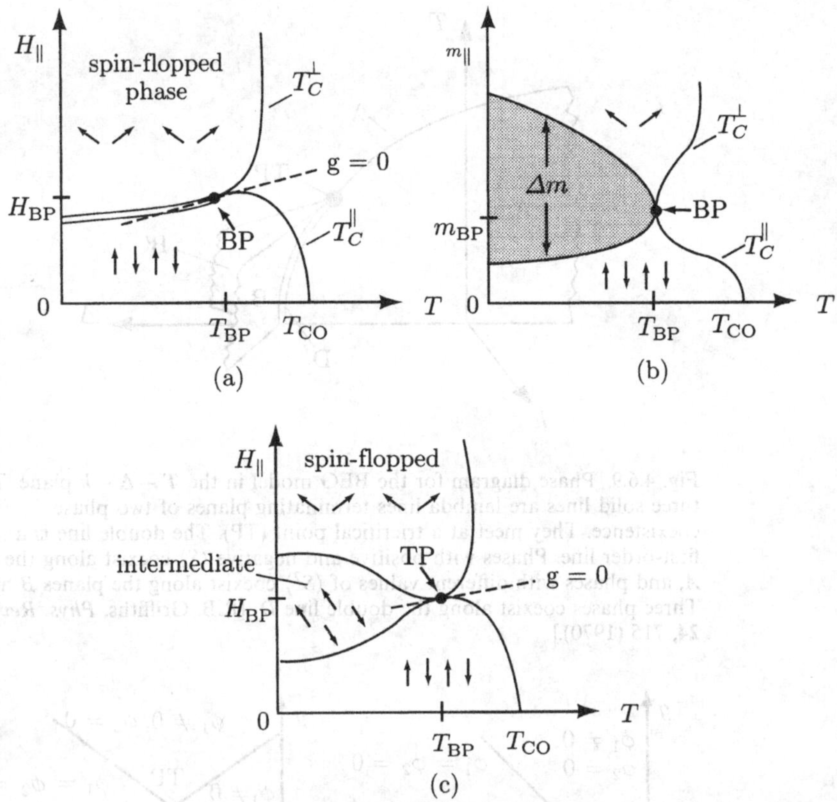

Fig. 4.6.11. (a) Schematic representation of a phase diagram of an anisotropic antiferromagnet in a field with a spin-flop bicritical point (BP). In the antiferromagnetic phase, the average spin in each of the two sublattices is either parallel or antiparallel to the anisotropy axis for $H_{BP} > H > 0$. The magnitude of the spin is greater in one sublattice than in the other. In the spin-flop phase, the component of the spin parallel to the field is the same in both sublattices, and the component perpendicular to the field alternates in sign. (b) Same as (a) but in the *m*-*T* plane. (c) Schematic representation of a phase diagram for an anisotropic antiferromagnet with a tetracritical point (TP). [Michael E. Fisher and David R. Nelson, *Phys. Rev. Lett.* **32**, 1350 (1974).]

5 Lifshitz points

There are systems in which the low-temperature ordered phase changes from one of spatially uniform order to one of spatially modulated order as a function of some potential or external field. As shown in Fig. 4.6.13, there are typically three phase boundaries in these systems separating, respectively, the high-temperature disordered phase from the spatially uniform ordered phase, the disordered phase from the spatially modulated phase, and the spatially uniform ordered phase from

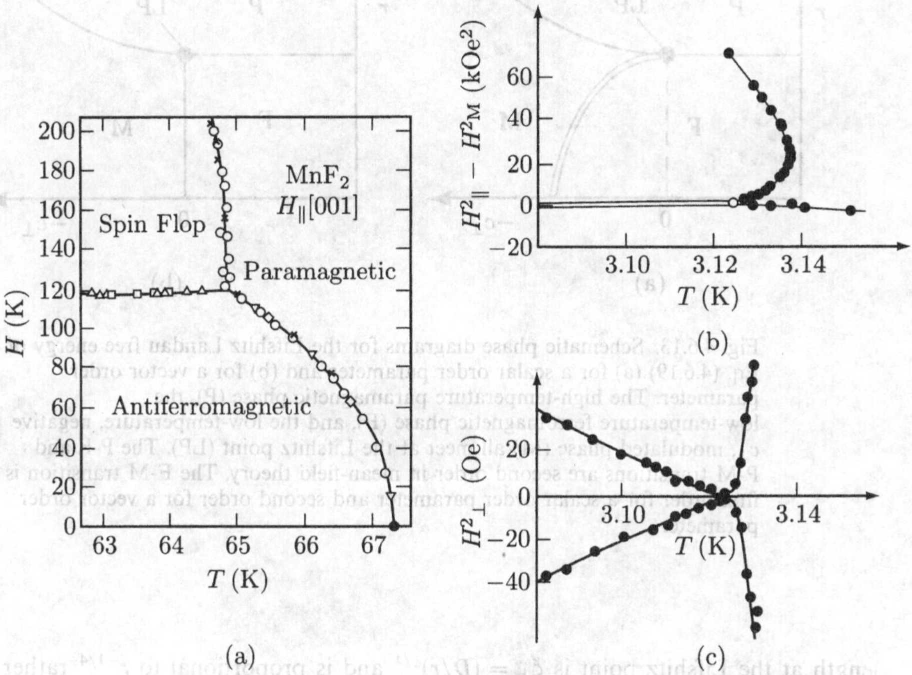

Fig. 4.6.12. (a) Phase diagram for MnF_2 in a field parallel to the anisotropy axis showing a bicritical point [Y. Shapira and S. Foner, *Phys. Rev.* **1**, 3083 (1970)]. (b) Phase diagram for $GdAlO_3$ in a field, $H_\parallel$, parallel to the anisotropy axis showing a bicritical point. (c) The same as (b) in a field, $H_\perp$, perpendicular to the anisotropy axis showing a tetracritical point [H. Rohrer and Ch. Gerber, *Phys. Rev. Lett.* **38**, 909 (1977)].

the spatially modulated ordered phase. These three phase boundaries meet at a *Lifshitz point* (Lifshitz 1941). A Landau free energy for a generic (d, m)-Lifshitz point is

$$F = \frac{1}{2}\int d^d x[r\phi^2 + c_\parallel(\nabla_\parallel\phi)^2 + c_\perp(\nabla_\perp\phi)^2 + D(\nabla^2\phi)^2] + u\int d^d x\phi^4, \tag{4.6.19}$$

where $\mathbf{x} = (\mathbf{x}_\parallel, \mathbf{x}_\perp)$ is divided into m perpendicular components $\mathbf{x}_\perp$ and $d - m$ parallel components $\mathbf{x}_\parallel$. The field ϕ can be a scalar or an n-component vector. When both $c_\parallel$ and $c_\perp$ are positive, the ordered phase will be spatially uniform. If, however, $c_\perp < 0$, then the system can lower its energy by creating spatially modulated structures with wave vectors of magnitude $|c_\perp|/2D$, as can be seen by seeking the minimum of the quadratic part of F over the spatially varying field $\phi(\mathbf{x})$. The point $r = 0$, $c_\perp = 0$ is a Lifshitz point. Because there is no term in the free energy proportional to $\nabla_\perp^2$, the mean-field correlation perpendicular

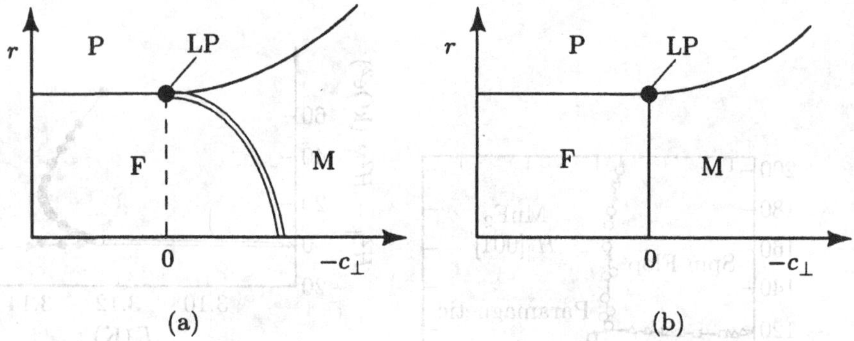

Fig. 4.6.13. Schematic phase diagrams for the Lifshitz Landau free energy of Eq. (4.6.19) (a) for a scalar order parameter and (b) for a vector order parameter. The high-temperature paramagnetic phase (P), the low-temperature ferromagnetic phase (F), and the low-temperature, negative $c_\perp$, modulated phase (M) all meet at the Lifshitz point (LP). The P-F and P-M transitions are second order in mean-field theory. The F-M transition is first order for a scalar order parameter and second order for a vector order parameter.

length at the Lifshitz point is $\xi_\perp = (D/r)^{1/4}$ and is proportional to $r^{-1/4}$ rather than to $r^{-1/2}$, i.e., the mean-field correlation length exponent ν is 1/4 rather than 1/2. Schematic phase diagrams for scalar and vector Lifshitz points are shown in Fig. 4.6.13. In both, there is a disordered paramagnetic phase (P), a spatially uniform ordered ferromagnetic phase (F), and a periodically modulated phase (M) meeting at the Lifshitz point. In mean-field theory, the P-F and P-M transitions are second order. In the modulated state with a vector order parameter, the vector ϕ_i rotates in helical fashion perpendicular to the axis of modulation. The F-M transition is first order for a scalar order parameter and second order for a vector order parameter (see Problem 4.7).

Lifshitz points can arise in metamagnets with the appropriate choice of exchange. A much studied example is the anisotropic-next-nearest-neighbor Ising model (ANNNI model). In this model, there is ferromagnetic exchange between Ising spins within a given layer and ferromagnetic exchanges between spins in nearest neighbor layers. In addition, however, there is an antiferromagnetic exchange between spins in next-to-nearest neighbor layers, as shown in Fig. 4.6.14. Let $\mathbf{l} = (\mathbf{r}, i)$ specify the position of a site in the lattice where $\mathbf{r}$ is a two-dimensional vector. The ANNNI model Hamiltonian is

$$\mathscr{H} = -J \sum_{i,<\mathbf{r},\mathbf{r}'>} S_{i,\mathbf{r}} S_{i,\mathbf{r}'} - J_1 \sum_{i,\mathbf{r}} S_{i,\mathbf{r}} S_{i+1,\mathbf{r}} + J_2 \sum_{i,\mathbf{r}} S_{i,\mathbf{r}} S_{i+2,\mathbf{r}}. \qquad (4.6.20)$$

The mean-field inverse nonlocal susceptibility for this model is easily calculated:

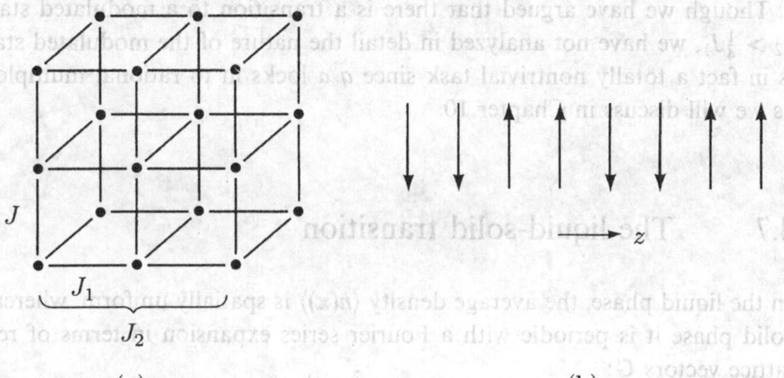

Fig. 4.6.14. (a) Nearest and next-nearest-neighbor interactions in the ANNNI model. (b) Ground state of the ANNNI model for $J_2 > J_1/2$.

$$\begin{aligned}\chi^{-1}_{\mathbf{l},\mathbf{l}'} &= \frac{\partial^2 F}{\partial\langle S_{\mathbf{l}}\rangle\partial\langle S_{\mathbf{l}'}\rangle} \\ &= T\delta_{\mathbf{l},\mathbf{l}'} - J\delta_{i,i'}\gamma_{\mathbf{r},\mathbf{r}'} - \delta_{\mathbf{r},\mathbf{r}'}[J_1(\delta_{i,i'+1}+\delta_{i,i'-1}) - J_2(\delta_{i,i'+2}+\delta_{i,i'-2})],\end{aligned} \tag{4.6.21}$$

where $\gamma_{\mathbf{r},\mathbf{r}'}$ is one if $\mathbf{r}$ and $\mathbf{r}'$ are nearest neighbors in the plane and zero otherwise. The Fourier transform of this function is

$$\chi^{-1}(q_\parallel, \mathbf{q}_\perp = 0) = T - zJ - 2J_1\cos q_\parallel a + 2J_2\cos 2q_\parallel a, \tag{4.6.22}$$

where a is the lattice spacing between layers and $q_\parallel$ and $\mathbf{q}_\perp$ are, respectively, the components of the wave vector $\mathbf{q}$ along the normal to the layers and within the layers. This function can now be minimized with respect to $q_\parallel$. When $J_2 < J_1/4$, the minimum occurs at $q_\parallel = 0$, and there is a transition to a uniform ferromagnetic state when

$$T = T_{FM} = zJ + 2J_1 - 2J_2. \tag{4.6.23}$$

When $J_2 > J_1/4$, the value of $q_\parallel$ minimizing χ^{-1} is

$$q_0 a = \cos^{-1}(J_1/4J_2). \tag{4.6.24}$$

As the temperature is lowered, the paramagnetic phase first becomes unstable with respect to fluctuations in $S_{\mathbf{l}}$ at this wave number. Thus, there will be a phase transition when $\chi^{-1}(q_0, 0)$ passes through zero at

$$T = T_M = zJ - \frac{1}{4}(J_1^2/J_2) - 2J_2 \tag{4.6.25}$$

to a spatially modulated state with $\langle S_{i,\mathbf{r}}\rangle = S_{q0}\cos(q_0 a + \alpha)$, where α is an arbitrary phase factor. Note that the maximum value of $q_0 a$ is $\pi/2$, corresponding to a unit cell size of $4a$. The zero temperature ground state of this model for $J_2 > J_1/2$ can easily be shown to consist of alternating pairs of parallel layers, as shown in Fig. 4.6.14b.

Though we have argued that there is a transition to a modulated state when $J_2 > \frac{1}{4}J_1$, we have not analyzed in detail the nature of the modulated state. This is in fact a totally nontrivial task since $q_{\|}a$ locks in to rational multiples of 2π, as we will discuss in Chapter 10.

4.7 The liquid-solid transition

In the liquid phase, the average density $\langle n(\mathbf{x})\rangle$ is spatially uniform, whereas in the solid phase it is periodic with a Fourier series expansion in terms of reciprocal lattice vectors $\mathbf{G}$:

$$\langle \delta n(\mathbf{x})\rangle =: \langle n(\mathbf{x})\rangle - n_0 = \sum_{\mathbf{G}} n_{\mathbf{G}} e^{i\mathbf{G}\cdot\mathbf{x}}, \tag{4.7.1}$$

where n_0 is the average uniform density and the vectors $\mathbf{G}$ are in some reciprocal lattice. Since $\langle \delta n(\mathbf{x})\rangle$ is real, $n_{\mathbf{G}}^* = n_{-\mathbf{G}}$. The order distinguishing the liquid from the solid is one of spatial modulations. Fluctuations in the liquid phase indicating an instability toward the solid phase have a maximum at nonzero wave number. Indeed, we found in Chapter 2 that the static structure function $S_{nn}(k)$ (see Fig. 2.4.4) for a liquid has a maximum on a sphere of radius $k_0 = 2\pi/l$, where l is the average interatomic separation. Although there are subsidiary peaks of lesser intensity at larger k, the dominant feature is the peak of S_{nn} at k_0. As temperature is lowered and the solid phase is approached, the magnitude of $S_{nn}(k_0)$ increases. Therefore (Alexander and McTague 1978), to describe the liquid-solid transition, it is reasonable to consider only the maximum peak and approximate $S_{nn}(k)$ in the vicinity of $k = k_0$ by

$$S_{nn}(k) = \frac{T}{[r + c(k^2 - k_0^2)^2]}, \tag{4.7.2}$$

where $r = a(T - T^*)$ decreases, leading to an increase in $S_{nn}(k)$ as T is lowered. The temperature T^*, which is in general a function of n_0, will turn out to be the mean-field limit of stability of the liquid phase. $S_{nn}(k)$ is the Fourier transform of the density-density correlation function,

$$S_{nn}(\mathbf{x}, \mathbf{x}') = \langle \delta n(\mathbf{x}) \delta n(\mathbf{x}')\rangle. \tag{4.7.3}$$

Since $\chi(\mathbf{x}, \mathbf{x}') = T S_{nn}(\mathbf{x}, \mathbf{x}')$ is the derivative of the free energy with respect to $\langle \delta n(\mathbf{x})\rangle$ and $\langle \delta n(\mathbf{x}')\rangle$, a phenomenological free energy which predicts Eq. (4.7.2) for $S_{nn}(k)$ in mean-field theory is simply

$$\begin{aligned} F_{SL} &= \int d^dx d^dx' \langle \delta n(\mathbf{x})\rangle \chi_0^{-1}(\mathbf{x}, \mathbf{x}') \langle \delta n(\mathbf{x}')\rangle \\ &\quad - w \int d^dx \langle \delta n(\mathbf{x})\rangle^3 + u \int d^dx \langle \delta n(\mathbf{x})\rangle^4, \end{aligned} \tag{4.7.4}$$

where

$$\chi_0^{-1}(\mathbf{x}, \mathbf{x}') = [r + c(\nabla^2 + k_0^2)^2]\delta(\mathbf{x} - \mathbf{x}'). \tag{4.7.5}$$

It is understood for the present purposes that $\langle\delta n(\mathbf{x})\rangle$ in Eq. (4.7.4) has Fourier components in the vicinity of $|\mathbf{k}| = k_0$, i.e., wave vectors near the origin are explicitly excluded. It should also be emphasized that k_0 appears as a parameter in the theory. It, and the other parameters, w, u, c, and T^*, can, and in general will, depend on other parameters, such as density n_0 or pressure, not explicitly being considered at the moment. At the end of this section, we will consider how variations in the uniform density n_0 can be treated.

Using the Fourier decomposition of Eq. (4.7.1), we immediately obtain

$$f_{SL} = \frac{F_{SL}}{V} = \sum_{\mathbf{G}} \frac{1}{2} r_{\mathbf{G}} |n_{\mathbf{G}}|^2 - w \sum_{\mathbf{G}_1,\mathbf{G}_2,\mathbf{G}_3} n_{\mathbf{G}_1} n_{\mathbf{G}_2} n_{\mathbf{G}_3} \delta_{\mathbf{G}_1+\mathbf{G}_2+\mathbf{G}_3,0}$$
$$+u \sum_{\mathbf{G}_1,\mathbf{G}_2,\mathbf{G}_3,\mathbf{G}_4} n_{\mathbf{G}_1} n_{\mathbf{G}_2} n_{\mathbf{G}_3} n_{\mathbf{G}_4} \delta_{\mathbf{G}_1+\mathbf{G}_2+\mathbf{G}_2+\mathbf{G}_3+\mathbf{G}_4,0}, \tag{4.7.6}$$

where

$$r_{\mathbf{G}} = r + c(G^2 - k_0^2)^2. \tag{4.7.7}$$

Note that there is a third-order term in this free energy. As in the case of a nematic liquid crystal, it will lead to a first-order transition. The solid phase is far more complicated than any of the ordered phases we have considered so far. To specify it completely, the order parameters of all of the infinite number of vectors $\mathbf{G}$ in the reciprocal lattice are needed. In addition, there are many different choices for the reciprocal lattice. A complete discussion of the liquid-solid transition, even in the simple approximation we are considering, requires a minimization of f with respect to $n_{\mathbf{G}}$ for all possible lattice candidates. The set of $n_{\mathbf{G}}$'s that give rise to the lowest value of f determines the equilibrium configuration. This collection may change with temperature, and transitions between different lattices (BCC to FCC for example) are expected. It therefore seems unlikely that anything very general about the liquid-solid transition can be said. However, the fact that only certain lattices have reciprocal lattice vectors that add to zero in triangles allows one to make some fairly general statements about this transition, at least when it is only weakly first order.

1 Are all crystals BCC?

Let us first simplify the problem by allowing c to go to infinity. This constrains all wave vectors $\mathbf{G}$ to have magnitude $G = k_0$, i.e., we need only consider vectors that lie on a sphere in reciprocal space. It is clear that the third-order term favors lattices with triads of vectors that add to zero. Since all $\mathbf{G}$'s have the same magnitude, three of them can add to zero only if they form an equilateral triangle. There are only three distinct sets L_G of equal length vectors $\mathbf{G}$ containing closed triangles and both $\mathbf{G}$ and $-\mathbf{G}$ that form symmetric structures. These are (1) the set of six edge vectors of an equilateral triangle and its inverted image (Fig. 4.7.1a), (2) the set of 12 edge vectors of an octahedron (Fig. 4.7.1b), and (3) the set of 30 edge vectors of an icosahedron (Fig. 4.7.1c). A solid with a reciprocal lattice generated by the edge vectors of an icosahedron is an icosahedral quasicrystal. These vectors can be translated so that their tails are at the origin and their heads are on the surface of a sphere of radius k_0. In this form, it is clear that the three cases correspond, respectively, to planar hexagonal, FCC, and icosahedral reciprocal lattices. The FCC lattice with lattice vectors $\mathbf{G} = 2^{-1/2}G\ (\pm1,\pm1,0),\ (\pm1,0,\pm1),\ (0,\pm1,\pm1)$ corresponds to a BCC direct lattice. The 30 vectors in the icosahedral case point to the vertices of what is called a triacontahedron. The astute reader will notice that the set of 12 vectors forming the edges of a tetrahedron and its inverted image (Fig. 4.7.1d) was not included in our list of possible vector sets. This is because this set is identical to that of the octahedral edges.

In order to calculate f, we need to evaluate f_2, f_3, and f_4, the terms in f of second, third, and fourth order in $n_{\mathbf{G}}$. The second-order term is proportional to the number of vectors m in L_G. We therefore choose $n_{\mathbf{G}} = m^{-1/2}n_G$ for every $\mathbf{G}$. This ensures that $f_2 = \frac{1}{2}rn_G^2$ has the same form for every lattice. f_3 depends on the number of triangles, p, to which each vector belongs. p is one for the hexagonal and two for the FCC and icosahedral lattices. The evaluation of f_3 proceeds as follows: In $n_{\mathbf{G}_1}n_{\mathbf{G}_2}n_{\mathbf{G}_3}$, there are m choices for $\mathbf{G}_1$. Once $\mathbf{G}_1$ is chosen, there are $2p$ choices for $\mathbf{G}_2$ since there are two free edges in each of the p triangles to which $\mathbf{G}_1$ belongs. Once $\mathbf{G}_1$ and $\mathbf{G}_2$ are fixed, there is a single choice for $\mathbf{G}_3$. Thus, including the factor of $m^{-1/2}$ in the definition of n_G, we have $f_3 = -2wpm^{-1/2}n_G^3$. The evaluation of f_4 is more complicated. There are two ways four vectors can add to zero: in two sets of equal and opposite vectors or in non-planar diamonds (Fig. 4.7.1b). Each vector belongs to q non-planar diamonds, where q is zero for the hexagonal and four for the FCC and icosahedral lattices. (Note that in the octahedron, there are sets of four edges that form a planar square. Since a square contains two sets of equal and opposite vectors, it is included in the first contribution to f_4.) There are six ways of choosing two sets of two $n_{\mathbf{G}}$'s from a set of four $n_{\mathbf{G}}$'s. In each set, $\mathbf{G}$ can run over all m vectors in L_G. Thus the contribution to f_4 from paired vectors is $6u(\sum|n_{\mathbf{G}}|^2)^2 = 6un_G^4$. The calculation of the contribution from non-planar diamonds is similar to the calculation of f_3. In $n_{\mathbf{G}_1}n_{\mathbf{G}_2}n_{\mathbf{G}_3}n_{\mathbf{G}_4}$, there are m ways to choose $\mathbf{G}_1$, $3q$ ways to choose $\mathbf{G}_2$, two ways to choose $\mathbf{G}_3$, and one way to choose $\mathbf{G}_4$, yielding a contribution of $6qm(m^{-1/2}n_G)^4 = (6q/m)n_G^4$ to f_4. Our final result for f is, therefore,

$$f = \frac{1}{2}rn_G^2 - 2wpm^{-1/2}n_G^3 + 6u(1 + \frac{q}{m})n_G^4. \tag{4.7.8}$$

It is clear that the favored phase will be the one with the largest value of f_3 compared to f_4. The largest values of f_3 are produced by the smallest m and the largest p. p^2m^{-1} is, respectively, 1/6, 1/3 and 2/15 for the hexagonal, FCC, and icosahedral reciprocal lattices. Thus, even though the hexagonal lattice has the smallest m, it should lose to the FCC lattice be-

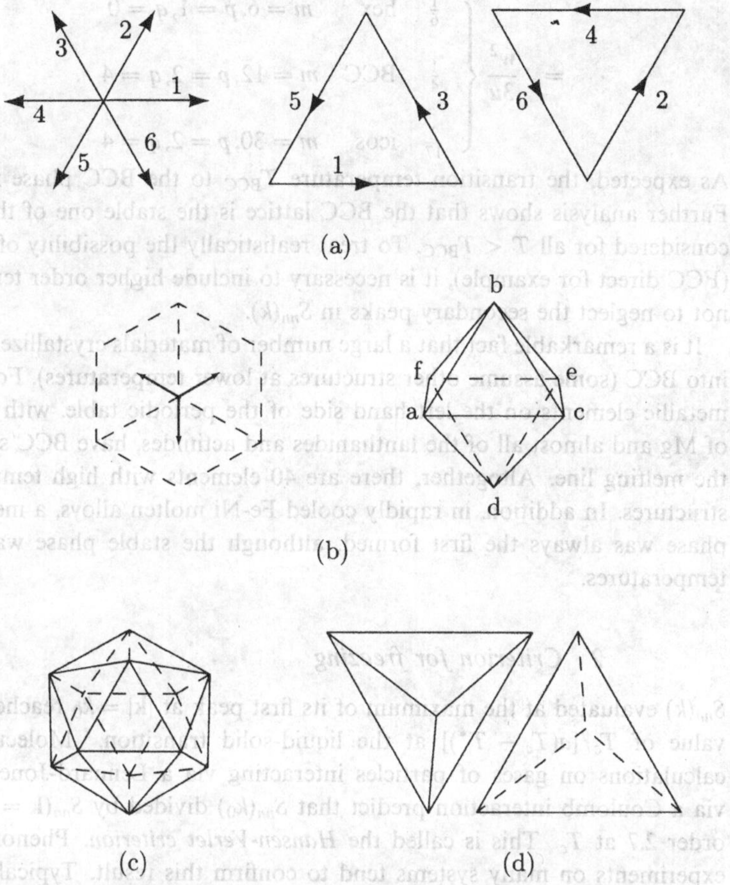

Fig. 4.7.1. (a) Six shortest vectors in a hexagonal reciprocal lattice and the two independent triangles that can be formed from these vectors. (b) Shortest vectors of the FCC reciprocal lattice: the figure on the left shows one octant of the FCC reciprocal lattice with three of the 12 shortest vectors, and the figure on the right shows the 12 shortest vectors arranged on the edges of an octahedron. The square acef in the octahedron is planar, and the diamond abcd is non-planar. (c) An icosahedron. (d) A tetrahedron and its inverted image.

cause each of its vectors is a member of only one triangle. The fourth-order term is important in determining the actual transition temperatures. It does not, however, change the preferred lattice. Using our analysis of the first-order isotropic-to-nematic transition to calculate the transition temperature, we obtain

$$r_c = a(T_c - T^*) = \frac{1}{2}\frac{(2wp)^2}{6u(m+q)}$$

$$= \frac{w^2}{3u} \begin{cases} \frac{1}{6} & \text{hex} \quad m = 6, p = 1, q = 0 \\ \frac{1}{4} & \text{BCC} \quad m = 12, p = 2, q = 4 \\ \frac{2}{17} & \text{icos} \quad m = 30, p = 2, q = 4 \end{cases} . \tag{4.7.9}$$

As expected, the transition temperature T_{BCC} to the BCC phase is the highest. Further analysis shows that the BCC lattice is the stable one of the set of three considered for all $T < T_{\text{BCC}}$. To treat realistically the possibility of other lattices (FCC direct for example), it is necessary to include higher order terms in $n_{\mathbf{G}}$ and not to neglect the secondary peaks in $S_{nn}(k)$.

It is a remarkable fact that a large number of materials crystallize from the melt into BCC (some assume other structures at lower temperatures). For example, all metallic elements on the left hand side of the periodic table, with the exception of Mg and almost all of the lanthanides and actinides, have BCC structures near the melting line. Altogether, there are 40 elements with high temperature BCC structures. In addition, in rapidly cooled Fe-Ni molten alloys, a metastable BCC phase was always the first formed, although the stable phase was FCC at all temperatures.

2 Criterion for freezing

$S_{nn}(k)$ evaluated at the maximum of its first peak at $|\mathbf{k}| = k_0$ reaches a maximum value of $T_c/[a(T_c - T^*)]$ at the liquid-solid transition. Molecular dynamics calculations on gases of particles interacting via a Lennard-Jones potential or via a Coulomb interaction predict that $S_{nn}(k_0)$ divided by $S_{nn}(\mathbf{k} = \infty) = n_0$ is of order 2.7 at T_c. This is called the *Hansen-Verlet criterion.* Phenomenologically, experiments on many systems tend to confirm this result. Typically, as a liquid is cooled, the first peak grows, and, as $S_{nn}(k)/n_0$ surpasses ~ 2.7, solidification occurs. Quantitative predictions of this ratio from the crude mean-field theory presented here require a specification of the values of the phenomenological parameters a, w and u. The parameter w has units of energy times volume squared and u has units of energy times volume cubed. Thus, w^2/u has units of energy times volume. The natural unit of energy in this problem is the transition temperature T_c, and the natural unit of volume is the inverse density n_0^{-1}. We can, therefore, write $w^2/u = bT_c n_0^{-1}$, where b is a numerical factor of order unity. We then have $S_{nn}(k_0)/n_0 = T_c/(2w^2 n_0/21u) = 21/2b$. Thus, with b of order four, the phenomenological Verlet criterion can be satisfied.

3 Improvements of the theory

How can we have structures other than BCC? The major approximation of the theory just presented, other than neglecting terms of order $n_{\mathbf{G}}^5$ and higher, was the restriction $|\mathbf{G}| = k_0$ obtained by setting $c = \infty$ and ignoring higher order peaks in the liquid structure factor. If this restriction is lifted, then the third-order term will couple $n_{\mathbf{G}}$ with $|\mathbf{G}| = k_0$ to $n_{\mathbf{G}_2}$ with $|\mathbf{G}_2| \neq k_0$. Schematically, one has contributions f of the form

$$f_2' = \frac{1}{2} m_2 r_{\mathbf{G}_2} |n_{\mathbf{G}_2}|^2 - vgn_{\mathbf{G}}^2 n_{\mathbf{G}_2}, \tag{4.7.10}$$

where m_2 is the number of wave vectors $\mathbf{G}_2$ equal to the sum of two vectors $\mathbf{G}$ of magnitude k_0, and g is a combinatorial factor. Minimizing f_2' with respect to $n_{\mathbf{G}_2}$, we find

$$n_{\mathbf{G}_2} = \frac{vg}{m_2 r_{\mathbf{G}_2}} |n_{\mathbf{G}}|^2 \tag{4.7.11}$$

and

$$f_2' = -\frac{1}{2} \frac{v^2 g^2}{m_2 r_{\mathbf{G}_2}} |n_{\mathbf{G}}|^4, \tag{4.7.12}$$

leading to a negative correction to the coefficient u in the total free energy. If only the first ring is $S_{nn}(k)$ is kept, but the restriction $c = \infty$ is relaxed, then $r_{\mathbf{G}_2} \neq \infty$. However, both $n_{\mathbf{G}_2}$ and the corrections to f_2' are small if $r_{\mathbf{G}_2} \gg 1$. For hexagonal and FCC reciprocal lattices, $G_2^2 - k_0^2$ is, respectively, $2k_0^2$ and k_0^2, and $r_{\mathbf{G}_2} \gg 1$ even for $c \neq \infty$. For BCC reciprocal (FCC direct) lattices, $G_2^2 - k_0^2 = k_0^2/3$ or $(G_2 - k_0)/k_0 = 0.15$. The magnitudes of these two vectors differ only by 25%, and both could be included under the first peak in the liquid structure factor. This undoubtedly plays a role in stabilizing FCC crystals, even though there are no closed triangles in its reciprocal lattice. The difference in length between vectors $\mathbf{G}_2$ from the origin to the vertices of an icosahedron is of order 5%, and $G_2^2 - G^2 = 0.1G^2$. In this case, $r_{\mathbf{G}_2}$ may be small even for fairly large values of c. This effect tends to favor icosahedral order within the model. If

$$\langle \delta n(\mathbf{x}) \rangle = \sum_{\mathbf{G}} n_{\mathbf{G}} e^{i\mathbf{G}\cdot\mathbf{x}} + \sum_{\mathbf{G}_2} n_{\mathbf{G}_2} e^{i\mathbf{G}_2\cdot\mathbf{x}}, \tag{4.7.13}$$

where the sum over $\mathbf{G}$ is over the 30 icosahedral edge vectors and the sum over $\mathbf{G}_2$ is over the 12 icosahedral vertex vectors, then f_3 is larger for icosahedral symmetry than it is for a BCC lattice, provided $ck_0^4 > 70r$. Unfortunately, when the fourth-order term is properly treated, the BCC phase is still the favored phase. In order to produce a stable icosahedral phase from a Landau theory, it is necessary to allow a second peak in $S_{nn}(k)$ to become large as temperature is lowered.

We have argued that the third-order term in the Landau free energy for the liquid-solid transition favors BCC structures for small latent heat transitions. Can the third-order potential w be adjusted to zero? If so, what is the nature of the liquid-solid transition? Is it second or first order, and what is the favored solid phase? The answer to the first question is probably affirmative. By choosing the appropriate atom or mixture of atoms, it is likely that w can be adjusted to zero. The liquid-solid transition remains first order in this case, however, even though the simple Landau theory would predict a second-order transition. Fluctuations depress the limit of metastability of the fluid to zero temperature (see Problem 5.2). Since the transition to the solid phase occurs necessarily before the limit of metastability is reached, it must be first order.

4 Changes in density

In our model treatment of the liquid-solid transition, we assumed that the average density does not change, i.e., we assumed that both the liquid and solid phases are incompressible. Though this is often a good approximation, it is not rigorously correct. We will now outline how density changes can be incorporated into a general theory of the liquid-solid transition.

We begin, as we did with our simple model, with the assumption that the properties of the liquid phase at temperature T and chemical potential μ_l are known (either from experiment or from some detailed microscopic theory). This will be used as a reference state to which the solid phase will be compared. We then consider the difference between the density-dependent grand potential [Eq. (3.4.23)] with arbitrary spatially dependent average density $\langle n(\mathbf{x})\rangle$ and that of the equilibrium liquid phase with uniform density n_l:

$$\begin{aligned}\Delta W[T,\Delta\mu,\langle n(\mathbf{x})\rangle] &= W[T,\mu,\langle n(\mathbf{x})\rangle] - W[T,\mu_l,\langle n(\mathbf{x})\rangle = n_l], \\ &= \Delta F[T,\langle n(\mathbf{x})\rangle, n_l] - \int d^3x[\mu\langle n(\mathbf{x})\rangle - \mu_l n_l],\end{aligned} \tag{4.7.14}$$

where $\Delta\mu = \mu - \mu_l$ is the chemical potential relative to that of the reference fluid and

$$\Delta F[T,\langle n(\mathbf{x})\rangle, n_l] = F[T,\langle n(\mathbf{x})\rangle] - F[T,n_l] \tag{4.7.15}$$

is the difference in Helmholtz free energies. Note that it is assumed that n_l is the density that minimizes $W[T,\mu_l,\langle n\rangle]$ so that $-V^{-1}W[T,\mu_l,n_l]$ is the pressure $p_l(\mu_l,T)$ of the liquid phase, and $(\delta F/\delta\langle n(\mathbf{x})\rangle)_{\langle n(\mathbf{x})\rangle=n_l} = \mu_l$. We now expand $\langle n(\mathbf{x})\rangle$ in a Fourier series about the average density n_s of the solid phase:

$$\langle n(\mathbf{x})\rangle = n_s + \sum_{\mathbf{G}} n_{\mathbf{G}} e^{i\mathbf{G}\cdot\mathbf{x}}. \tag{4.7.16}$$

The free energy density difference of Eq. (4.7.15) then becomes a function of n_s, n_l, and $n_{\mathbf{G}}$. For small $\delta n/n_l = (n_s - n_l)/n_l$, $\Delta F/V$ can be expressed as F_{SL} [Eq. (4.7.7)] plus terms arising from the change in the density:

$$\frac{\Delta F}{V} = f_{SL}(T,n_{\mathbf{G}},n_s) + \mu_l\delta n + \frac{1}{2}\kappa_l^{-1}(\delta n/n_l)^2, \tag{4.7.17}$$

where κ_l is the isothermal compressibility of the liquid. Finally, expanding f_{SL} in powers of δn and retaining only the lowest order term, we obtain

$$\Delta W/V = f_{SL}(T,n_{\mathbf{G}},n_l) - b\delta n\sum_{\mathbf{G}}|n_{\mathbf{G}}|^2 + \frac{1}{2}\kappa_l(\delta n/n_l)^2 - \Delta\mu(n_l+\delta n), \tag{4.7.18}$$

where $b = a\partial T^*(n_l)/\partial n_l$. Thus, there is a change in density,

$$\delta n = n_l^2 \kappa_l (\Delta\mu - b \sum_{\mathbf{G}} |n_{\mathbf{G}}|^2), \tag{4.7.19}$$

associated with the liquid-solid transition. Note that along the coexistence curve when $\Delta\mu = 0$, δn, like $\delta n_{\mathbf{G}_2}$, is proportional to $|n_{\mathbf{G}}|^2$ for small $n_{\mathbf{G}}$. When Eq. (4.7.19) is substituted into Eq. (4.7.17), a new effective free energy with essentially the same form as Eq. (4.7.7) is generated. It can be minimized over $n_{\mathbf{G}}$ to produce the pressure difference $p_s(T, \mu) - p_l(T, \mu_l)$ between the solid and liquid phases. The coexistence curve can be calculated by setting $p_s = p_l$ and $\Delta\mu = 0$.

5 *Density functional theory*

As we have seen, the liquid-solid transition is first order. It is in fact normally strongly first order. Thus, to obtain a mean-field theory that has any hope of making quantitative predictions about real liquid-solid transitions, it is necessary to have a free energy that provides a valid description for large order parameter changes, such as, for example, the Bragg-Williams theory of ferromagnetism, which gives reasonable results even near $T = 0$ where m is of order unity. The free energy should also incorporate as much information as is possible about the nature of the reference liquid state to which the solid phase is to be compared. In this section, we will outline a successful phenomenological theory incorporating these two goals that was introduced by Ramakrishnan and Yussouff (1979).

We begin with the free energy functional for a classical non-interacting gas with density $\langle n(\mathbf{x})\rangle$ compared to the free energy of a similar system with uniform density n_l. Using the generalization of the free gas free energy [Eq. (3.3.4)] for spatially varying $\langle n(\mathbf{x})\rangle$, we obtain

$$\begin{aligned}\Delta F_{\rm cl}[T, \langle n(\mathbf{x})\rangle, n_l] &= \int d^3x\ T[\langle n(\mathbf{x})\rangle \ln(\langle n(\mathbf{x})\rangle/n_l) \\ &\quad - (\langle n(\mathbf{x})\rangle - n_l)] + \mu_l(\langle n(\mathbf{x})\rangle - n_l),\end{aligned} \tag{4.7.20}$$

where we used $\mu_l = T\ln(n_l\lambda^3)$ [Eq. (3.3.6)]. The second derivative of this free energy with respect to $\langle n(\mathbf{x})\rangle$ evaluated at $\langle n(\mathbf{x})\rangle = n_l$ is the inverse density susceptibility $\chi_{nn}^{-1}(\mathbf{x}, \mathbf{x}') = TS_{nn}(\mathbf{x}, \mathbf{x}') = Tn_l\delta(\mathbf{x} - \mathbf{x}')$ of the classical gas. As expected, there are no correlations in the classical gas. To obtain a free energy that correctly describes the pair correlation function of the reference liquid, a term second order in $\delta n(\mathbf{x}) = \langle n(\mathbf{x})\rangle - n_l$ should be added to $\Delta F_{\rm cl}$ with coefficient

$$-C_l(\mathbf{x}, \mathbf{x}') = S_{nn}^{-1}(\mathbf{x}, \mathbf{x}') - n_l^{-1}\delta(\mathbf{x} - \mathbf{x}'), \tag{4.7.21}$$

where $S_{nn}(\mathbf{x}, \mathbf{x}')$ is the Ursell function of the liquid phase at density n_l. The function $C_l(\mathbf{x}, \mathbf{x}')$ is called the *direct pair correlation function* of the fluid. The Fourier transform of C_l is usually defined with an explicit power of the inverse density:

$$C_l(\mathbf{x}, \mathbf{x}') = n_l^{-1} \int \frac{d^3q}{(2\pi)^2} C_l(\mathbf{q}) e^{i\mathbf{q}\cdot(\mathbf{x}-\mathbf{x}')}. \tag{4.7.22}$$

The Fourier transform of the Ursell function has a simple relation to $C_l(\mathbf{x})$:

$$S_{nn}(\mathbf{q}) = \frac{n_l}{1 - C_l(\mathbf{q})}. \tag{4.7.23}$$

The exact free energy describing density deviations from a reference fluid has third and higher order derivatives with respect to $\langle \delta n(\mathbf{x})\rangle$ reflecting liquid correlations. Ramakrishnan and Yussouff ignore corrections to the classical gas results for these functions arising from interactions between particles in the liquid phase, and use the free energy of the non-interacting gas to describe large deviations $\langle \delta n(\mathbf{x})\rangle$.

Their expression for the grand potential relative to that of the reference fluid is, therefore,

$$\begin{aligned}\Delta W &= \int d^3x T[\langle n(\mathbf{x})\rangle \ln(\langle n(\mathbf{x})\rangle / n_l) - (\langle n(\mathbf{x})\rangle - n_l)] \\ &\quad - \frac{1}{2}T\int d^3x C_l(\mathbf{x}, \mathbf{x}')\delta\langle n(\mathbf{x})\rangle \delta\langle n(\mathbf{x}')\rangle - \int d^3x \Delta\mu \langle n(\mathbf{x})\rangle.\end{aligned} \tag{4.7.24}$$

The equation of state for $\langle n(\mathbf{x})\rangle$ follows from minimization of this equation:

$$\frac{\delta \Delta W/T}{\delta\langle n(\mathbf{x})\rangle} = \ln(\langle n(\mathbf{x})\rangle / n_l) - \int d^3x' C_l(\mathbf{x}-\mathbf{x}')\delta\langle n(\mathbf{x}')\rangle - \beta\Delta\mu = 0, \tag{4.7.25}$$

where as usual $\beta = 1/T$. Alternatively, this equation can be expressed as

$$\frac{\langle n(\mathbf{x})\rangle}{n_l} = e^{\beta U_e(\mathbf{x})}, \tag{4.7.26}$$

where

$$\beta U_e(\mathbf{x}) = \beta\Delta\mu + \int d^3x' C_l(\mathbf{x}-\mathbf{x}')\delta\langle n(\mathbf{x}')\rangle \tag{4.7.27}$$

is an effective or mean potential (analogous to the mean field of the Bragg-Williams theory), which is a functional of the average density.

We now seek solutions to these equations in which $\langle n(\mathbf{x})\rangle$ has periodic order of some crystal lattice, i.e., we set

$$\langle n(\mathbf{x})\rangle = n_s(1 + \sum_{\mathbf{G}\neq 0} \eta_{\mathbf{G}} e^{i\mathbf{G}\cdot\mathbf{x}}). \tag{4.7.28}$$

The order parameters $\eta_{\mathbf{G}} \equiv n_{\mathbf{G}}/n_s$ are the normalized mass-density wave amplitudes. The vectors are in some periodic reciprocal lattice with primitive translation vectors $\mathbf{b}_1$, $\mathbf{b}_2$, and $\mathbf{b}_3$. The volume of a unit cell in the associated direct lattice is

$$v_0 = \frac{(2\pi)^2}{\mathbf{b}_1 \cdot (\mathbf{b}_2 \times \mathbf{b}_3)}. \tag{4.7.29}$$

The order parameters $\eta_{\mathbf{G}}$ and the ratio n_s/n_l of the solid to liquid density can be calculated from Eqs. (4.7.26) and (4.7.27):

$$\frac{n_s}{n_l} = e^{\beta\Delta\mu + C_l(0)[(n_s/n_l)-1]} \left(\frac{1}{v_0}\int_0 d^3x e^{\beta\overline{U}_e(\mathbf{x},\eta_{\mathbf{G}})}\right), \tag{4.7.30}$$

and

$$\eta_{\mathbf{G}} = \frac{\int d^3x e^{-i\mathbf{G}\cdot\mathbf{x}} e^{\beta\overline{U}_e(\mathbf{x},\eta_{\mathbf{G}})}}{\int d^3x e^{\beta\overline{U}_e(\mathbf{x},\eta_{\mathbf{G}})}}, \tag{4.7.31}$$

where the integrals are over a unit cell of the direct lattice and

$$\beta\overline{U}_e(\mathbf{x},\eta_{\mathbf{G}}) = \sum_{\mathbf{G}\neq 0} C_l(\mathbf{G})(n_s/n_l)\eta_{\mathbf{G}} e^{i\mathbf{G}\cdot\mathbf{x}}. \tag{4.7.32}$$

Rather than expanding $\langle n(\mathbf{x})\rangle$ in a periodic Fourier series, it is often more convenient to expand the logarithm of $\langle n(\mathbf{x})\rangle$ in a similar series:

$$\frac{\langle n(\mathbf{x})\rangle}{n_l} = A\exp[\sum_{\mathbf{G}\neq 0} \zeta_{\mathbf{G}}(n_s/n_l)C_l(\mathbf{G})e^{i\mathbf{G}\cdot\mathbf{x}}]. \tag{4.7.33}$$

In equilibrium, the coefficients $\zeta_{\mathbf{G}}$ satisfy exactly the same self-consistency equations as the parameters $\eta_{\mathbf{G}}$:

$$\zeta_{\mathbf{G}} = \frac{\int d^3x e^{-i\mathbf{G}\cdot\mathbf{x}} e^{\beta\overline{U}_e(\mathbf{x},\zeta_{\mathbf{G}})}}{\int d^3x e^{\beta\overline{U}_e(\mathbf{x},\zeta_{\mathbf{G}})}}. \tag{4.7.34}$$

The amplitude A satisfies

$$A = e^{\beta\Delta\mu + C_l(\mathbf{q}=0)[(n_s/n_l)-1]}. \tag{4.7.35}$$

In practice, convergence to realistic solid densities is obtained more rapidly using the expansion of Eq. (4.7.34) rather than that of Eq. (4.7.31).

The atoms in a solid are usually very well localized at lattice sites even near the melting temperature. This means that many Fourier components of the density are needed to obtain a realistic picture of the density. This encourages one to look for a description of the density with a limited number of variational parameters that localize atoms in the vicinity of sites on some lattice. A form that agrees well with our intuition and with calculations using $\eta_{\mathbf{G}}$ evaluated at many different $\mathbf{G}$'s is a superposition of Gaussian densities on the lattice:

$$\langle n(\mathbf{x})\rangle = n_s \frac{v_0}{a^3}\frac{\alpha^{3/2}}{\pi^{3/2}} \sum_{\mathbf{l}} e^{-\alpha(\mathbf{x}-\mathbf{R}_\mathbf{l})^2/a^2} \equiv n_s(1+\sum \eta_{\mathbf{G}} e^{i\mathbf{G}\cdot\mathbf{x}}). \tag{4.7.36}$$

This density is normalized so that its integral over all space gives the total number of particles. The vectors $\mathbf{R}_\mathbf{l}$ are sites on a cubic (generalizations to non-cubic lattices are straightforward) direct lattice with primitive translation vectors of magnitude a, and v_0 is the volume of a direct lattice unit cell. Note that there is no constraint in this relation requiring the solid density n_s to be equal to the number of unit cells per unit volume. The equilibrium state will in general have a finite density of vacancies so that on average there will be less than one atom per unit cell. The order parameters $\eta_{\mathbf{G}}$ satisfy

$$\eta_{\mathbf{G}} = e^{-G^2a^2/4\alpha}. \tag{4.7.37}$$

If α is large, atoms are well localized around a given site, and overlaps between neighboring Gaussians can be neglected in the evaluation of ΔW. In this case, we have

$$\frac{\Delta W}{T n_l V} = (n_s/n_l)\left\{\ln[(n_s/n_l)(v_0/a^3)(\alpha/\pi)^{3/2}] - \frac{3}{2}\right\}$$
$$- [(n_s/n_l) - 1] - (n_s/n_l)\Delta\mu \tag{4.7.38}$$
$$- \frac{1}{2}C_l(\mathbf{q}=0)[1-(n_s/n_l)]^2 - \frac{1}{2}\sum_{\mathbf{G}\neq 0}(n_s/n_l)^2 C_l(\mathbf{G})e^{-G^2a^2/2\alpha}.$$

This expression is to be minimized for given liquid density n_l, temperature, T, and chemical potential difference $\Delta\mu$ with respect to the three variational parameters α, a, and n_s/n_l. If the system is incompressible, $n_s/n_l = 1$, and there are only two variational parameters: a specifying the size of a unit cell and α specifying the degree of localization of the average density in a unit cell.

4.8 Variational mean-field theory

In this chapter, we have presented two formulations of mean-field theory: the Bragg-Williams theory and the Landau phenomenological theory. The first, though it can be applied at all temperatures including the vicinity of zero temperature, is not easily extended to models having other than the simple up-down Ising symmetry. The second, though easily constructed once the symmetry of the order parameter is established, is not particularly useful far from the critical point. In this section, we outline a general approach to deriving a mean-field theory valid for all ranges of temperatures for systems with order parameters of essentially arbitrary complexity. This variational method is based upon approximating the total equilibrium density matrix by a product of local site or particle density matrices and is often referred to as $\mathrm{Tr}\rho\ln\rho$ mean-field theory.

1 *Two inequalities*

We begin with the derivation of two inequalities that will establish the variational theorem for our mean-field approximation. Let ϕ be a random variable, which can be either continuous or discrete, and let $P(\phi) \geq 0$ be its associated probability distribution. Then the expectation value of any function $f(\phi)$ is

$$\langle f(\phi)\rangle \equiv \mathrm{Tr}P(\phi)f(\phi), \tag{4.8.1}$$

where Tr signifies a sum or integral over all possible values of ϕ. The inequality

$$\langle e^{-\lambda\phi}\rangle \geq e^{-\lambda\langle\phi\rangle}, \tag{4.8.2}$$

valid for any probability distribution, may be proved as follows. The inequality

$$e^{\phi} \geq 1+\phi \tag{4.8.3}$$

applies for any real number ϕ regardless of its sign. Thus,

$$e^{-\lambda\phi} = e^{-\lambda\langle\phi\rangle}e^{-\lambda(\phi-\langle\phi\rangle)} \geq e^{-\lambda\langle\phi\rangle}[1-\lambda(\phi-\langle\phi\rangle)], \tag{4.8.4}$$

which, when averaged over $P(\phi)$, implies

$$\langle e^{-\lambda\phi}\rangle \geq \langle [1-\lambda(\phi-\langle\phi\rangle)]e^{-\lambda\langle\phi\rangle}\rangle = e^{-\lambda\langle\phi\rangle}, \tag{4.8.5}$$

which establishes Eq. (4.8.2).

The inequality in Eq. (4.8.2) will be used to establish variational theorems for classical partition functions. In quantum systems, we need an additional inequality. Let $\hat{A}$ be any quantum operator, and let $|\,n\rangle$ be any normalized state in the Hilbert space in which $\hat{A}$ operates, then

$$\langle n\mid e^{-\lambda\hat{A}}\mid n\rangle \geq e^{-\lambda\langle n|\hat{A}|n\rangle}. \tag{4.8.6}$$

This inequality can be proven using Eq. (4.8.3) and a complete set of states $\{|\,p\rangle\}$ in which $\hat{A}$ is diagonal:

$$\begin{aligned}\langle n|e^{-\lambda\hat{A}}\mid n\rangle &= e^{-\lambda\langle n|\hat{A}|n\rangle}\sum_p \langle n\mid p\rangle e^{-\lambda(\langle p|\hat{A}|p\rangle-\langle n|\hat{A}|n\rangle)}\langle p\mid n\rangle \qquad (4.8.7)\\ &\geq e^{-\lambda\langle n|\hat{A}|n\rangle}\sum_p \langle n\mid p\rangle[1-\lambda(\langle p\mid\hat{A}\mid p\rangle-\langle n\mid\hat{A}\mid n\rangle)]\langle p\mid n\rangle.\end{aligned}$$

Eq. (4.8.6) follows directly from this equation because $\{|\,p\rangle\}$ is a complete set of states.

Now consider a classical Hamiltonian $\mathscr{H}$ that is a function of a discrete or continuous classical field ϕ (which in general depends on a spatial coordinate **x** or a lattice site **l**). Let $\rho(\phi)$ be any classical probability distribution satisfying $\mathrm{Tr}\rho=1$ and $\rho(\phi)\geq 0$. $\rho(\phi)$ is thus a probability distribution for ϕ. The canonical partition function can be written as

$$\begin{aligned}Z &= \mathrm{Tr}e^{-\beta\mathscr{H}[\phi]} \equiv \mathrm{Tr}\rho e^{-\beta\mathscr{H}-\ln\rho}\\ &= \langle e^{-\beta\mathscr{H}-\ln\rho}\rangle_\rho = e^{-\beta F},\end{aligned} \tag{4.8.8}$$

where $\langle\,\rangle_\rho$ signifies an average with respect to the density matrix ρ and where F is the free energy (or more precisely the thermodynamic potential) associated with $\mathscr{H}$. Thus, using Eq. (4.8.8) and the inequality in Eq. (4.8.3), we obtain

$$e^{-\beta F} \geq \exp(-\beta\langle\mathscr{H}\rangle_\rho - \langle\ln\rho\rangle_\rho) \tag{4.8.9}$$

or

$$\begin{aligned}F\leq F_\rho &= \langle\mathscr{H}\rangle_\rho + T\langle\ln\rho\rangle_\rho\\ &= \mathrm{Tr}\rho\mathscr{H} + T\mathrm{Tr}\rho\ln\rho,\end{aligned} \tag{4.8.10}$$

where F_ρ is an approximate free energy associated with the density matrix ρ. This inequality is valid for any density matrix. F_ρ is a minimum with respect to variations in ρ subject to the constraint $\mathrm{Tr}\rho=1$ when

$$\rho = \frac{e^{-\beta\mathscr{H}}}{Z} \tag{4.8.11}$$

is the actual equilibrium density matrix. This can be seen from the equation

$$\frac{\delta F_\rho}{\delta\rho} = \mathscr{H} + T(\ln\rho + 1) = \zeta, \tag{4.8.12}$$

where ζ is a Lagrange multiplier whose value is chosen to impose the constraint $\mathrm{Tr}\rho = 1$. Thus, F_ρ at its minimum with respect to ρ is the actual free energy F.

In the quantum case, the inequality of Eq. (4.8.9) can be derived from

$$\begin{aligned} Z &= \mathrm{Tr}e^{-\beta\mathscr{H}} = \sum_n \langle n \mid e^{-\beta\mathscr{H}} \mid n\rangle \\ &\geq \sum_n e^{-\beta\langle n|\mathscr{H}|n\rangle} = \sum_n \rho_n e^{-\beta\langle n|\mathscr{H}|n\rangle - \ln\rho_n}, \end{aligned} \tag{4.8.13}$$

where ρ is a density matrix, the states $\mid n\rangle$ form a complete set with respect to which ρ is diagonal, and $\rho_n = \langle n \mid \rho \mid n\rangle \geq 0$. The last expression can be regarded as the expectation value with respect to the probability weight ρ_n of $\exp(-\beta\langle n \mid \mathscr{H} \mid n\rangle - \ln\rho_n)$ regarded as a function of n. Thus, using the inequality Eq. (4.8.2), we have

$$\begin{aligned} Z &\geq \exp[-\sum_n \rho_n(\beta\langle n \mid \mathscr{H} \mid n\rangle + \ln\rho_n)] \\ &\geq \exp[-\beta\langle\mathscr{H}\rangle_\rho - \langle\ln\rho\rangle_\rho], \end{aligned} \tag{4.8.14}$$

where $\langle\ \rangle_\rho$ signifies an average with respect to ρ. This equation is identical to Eq. (4.8.9) for the classical systems.

2 The mean-field approximation

The inequality, Eq. (4.8.9), provides the basis for variational approximations to the free energy that can be implemented as follows: a functional form with free unspecified parameters is chosen for a trial density matrix ρ to approximate the actual density matrix. The trial density matrix with the chosen functional form that best approximates the actual density matrix is obtained by minimizing the approximate free energy F_ρ with respect to the free parameters in ρ. Mean-field theory is obtained by a trial density matrix that is a product of independent single particle matrices. If ρ_α is the single particle density matrix depending only on the degree of freedom of particle α, the mean-field density matrix is

$$\rho = \prod_\alpha \rho_\alpha, \tag{4.8.15}$$

and the variational mean-field free energy is

$$F_\rho = \langle\mathscr{H}\rangle_\rho + T\sum_\alpha \mathrm{Tr}\rho_\alpha \ln\rho_\alpha. \tag{4.8.16}$$

The precise form of $\langle\mathscr{H}\rangle_\rho$ will, of course, depend on $\mathscr{H}$.

There are now two approaches one can use to determine variational minima to Eq. (4.8.16). (1) A parametrization of ρ_α in terms of the order parameter $\langle\phi_\alpha\rangle$ of a phase transition can be chosen. This parametrization must satisfy the constraints $\mathrm{Tr}\rho_\alpha = 1$ and $\mathrm{Tr}\rho_\alpha\phi_\alpha = \langle\phi_\alpha\rangle$. The variational parameter is simply the order parameter $\langle\phi_\alpha\rangle$. If there is no external field in $\mathscr{H}$ coupling linearly to ϕ, then F_ρ is simply the Helmholtz free energy $F(\langle\phi\rangle)$. If there is an external field h, then $F_\rho = F(\langle\phi\rangle) - Vh\langle\phi\rangle$ is a function analogous to

$W(T, \mu, \langle n(\mathbf{x})\rangle)$ introduced in Eq. (3.4.23). We will apply this procedure below to the s-state Potts model. (2) Alternatively, the single particle density matrix ρ_α itself may be regarded as a variational function, and the best functional form in terms of the dynamical variables of particle α is obtained by minimizing F_ρ with respect to ρ_α. This procedure is more general than the preceding one, but the connection between F_ρ and $F(\langle\phi\rangle)$ is not entirely straightforward. We will apply this procedure below to the O_n Heisenberg model and to a classical plasma.

3 *The s-state Potts model*

In the s-state Potts model, there is a variable $\sigma_\mathbf{l}$ at each site on a lattice that can take on the values $1, ..., s$. The Potts-model Hamiltonian was introduced in Eq. (3.6.6). If we add an external field $h_\mathbf{l}$ favoring occupancy of state 1 at site $\mathbf{l}$, $\mathcal{H}$ can be written as

$$\begin{aligned}\mathcal{H} &= -\frac{1}{2}\sum_{\mathbf{l},\mathbf{l}'} J_{\mathbf{l},\mathbf{l}'}(s\delta_{\sigma_\mathbf{l},\sigma_{\mathbf{l}'}}-1) - \sum_\mathbf{l} h_\mathbf{l}(s\delta_{\sigma_\mathbf{l},1}-1) \\ &\equiv \mathcal{H}_0 + \mathcal{H}_{\text{ext}},\end{aligned} \tag{4.8.17}$$

where $J_{\mathbf{l},\mathbf{l}'}$ is J for nearest neighbor sites and zero otherwise. To identify an order parameter distinguishing low- and high-temperature phases, we note that at high temperature, all states are occupied with equal probability and $\langle\delta_{\sigma_\mathbf{l},\sigma}\rangle = 1/s$ for every value of $\sigma = 1, ..., s$. At low temperature, the energy defined by $\mathcal{H}_0$ is clearly minimized if $\sigma_\mathbf{l}$ has the same value for all sites. There are s equivalent ground states characterized by the value of $\sigma_\mathbf{l}$. We can, therefore, choose

$$\phi_\mathbf{l}^\sigma = \frac{1}{s-1}(s\delta_{\sigma_\mathbf{l},\sigma}-1) \tag{4.8.18}$$

as the order parameter field. At high temperature, $\langle\phi_\mathbf{l}^\sigma\rangle = 0$, whereas at zero temperature, $\langle\phi_\mathbf{l}^\sigma\rangle = 1$ in the ground state with $\sigma_\mathbf{l} = \sigma$. Since all ground states are equivalent, we choose the order parameter $m_\mathbf{l} = \langle\phi_\mathbf{l}^1\rangle$, characterizing condensation into state 1. The single particle density matrix $\rho_\mathbf{l}$ is a function of $\sigma_\mathbf{l}$ and must satisfy

$$\text{Tr}\rho_\mathbf{l} = \sum_{\sigma_\mathbf{l}} \rho_\mathbf{l} = 1 \quad\text{and}\quad \text{Tr}\rho_\mathbf{l}\phi_\mathbf{l}^1 = m_\mathbf{l}. \tag{4.8.19}$$

Since these are the only two constraints on the density matrix, it must have the form $\rho_\mathbf{l} = a + b\delta_{\sigma_\mathbf{l},1}$. It is a straightforward exercise to calculate the parameters a and b and to obtain

$$\rho_\mathbf{l} = \frac{1}{s}\left[1 + m_\mathbf{l}(s\delta_{\sigma_\mathbf{l},1}-1)\right]. \tag{4.8.20}$$

This density matrix can now be used to calculate F_ρ. The result is

$$\frac{F_\rho}{s-1} = -\frac{1}{2}\sum_{\mathbf{l},\mathbf{l}'} J_{\mathbf{l},\mathbf{l}'} m_{\mathbf{l}} m_{\mathbf{l}}' - \sum_{\mathbf{l}} h_{\mathbf{l}} m_{\mathbf{l}} - \frac{NT}{s-1}\ln s$$
$$+\frac{T}{s}\sum_{\mathbf{l}}\left\{\frac{1}{s-1}[1+(s-1)m_{\mathbf{l}}]\ln[1+(s-1)m_{\mathbf{l}}] + (1-m_{\mathbf{l}})\ln(1-m_{\mathbf{l}})\right\}. \tag{4.8.21}$$

Note that this reduces to the Bragg-Williams free energy for the Ising model [Eqs. (4.1.4) and (4.1.14)] when $s = 2$ and $h_{\mathbf{l}} = 0$. If the external field $h_{\mathbf{l}}$ is uniform, $m_{\mathbf{l}} = m$ is independent of $\mathbf{l}$ and satisfies the equation of state

$$\frac{1}{N}\frac{\partial}{\partial m}\frac{F_\rho}{s-1} = -zJm + \frac{T}{s}\ln\frac{1+(s-1)m}{1-m} - h = 0, \tag{4.8.22}$$

where z is the number of nearest neighbors. Thus,

$$m = \frac{e^{sh^e}-1}{e^{sh^e}+(s-1)}, \tag{4.8.23}$$

where

$$h^e = zJm + h \tag{4.8.24}$$

is the effective field. This equation reduces to the mean-field equation of state for the Ising model [Eq. (4.1.9)] when $s = 2$. When $s > 2$, it predicts a first-order transition as can be seen via a direct analysis. Alternatively, one can expand the free energy of Eq. (4.8.21) for small m and $h = 0$ to find that there is a third-order term when $s \neq 2$ and a mean-field first-order transition:

$$\frac{F}{N(s-1)} = \frac{1}{2}(T-T_c)m^2 - \frac{T}{6}(s-2)m^3 + \frac{T}{12}(s^2-3s+3)m^4 + \cdots. \tag{4.8.25}$$

The Potts model in the limit that the number of states goes to one ($s \to 1$) describes percolation. In this case, there is a second-order transition even though there is a third-order term in F.

4 The O_n classical Heisenberg model

Though the simple parametrization of the trial density matrix in terms of the order parameter provides a good description of the phase transition and low-temperature properties of the Potts model, we will now show how allowing the functional form of the trial density matrix to vary provides a much more complete and more correct description of the classical Heisenberg model. As we have discussed in Secs. 3.6 and 4.3, the dynamical variables of the classical Heisenberg model are unit length spins $\mathbf{S}_{\mathbf{l}}$. Here we will parametrize these spins by the solid angle coordinate $\Omega_{\mathbf{l}}$, depending on $n-1$ angular coordinates, on a unit n-dimensional sphere: $\mathbf{S}_{\mathbf{l}} \equiv \mathbf{S}(\Omega_{\mathbf{l}})$. (For $n = 3$, $\Omega = (\vartheta, \varphi)$ and $d\Omega = \sin\vartheta d\vartheta d\varphi$.)

The single particle density matrix is $\rho_{\mathbf{l}} = \rho_{\mathbf{l}}(\Omega_{\mathbf{l}})$, and the free energy is

$$F_\rho = -\frac{1}{2}\sum_{\mathbf{l},\mathbf{l}'} J_{\mathbf{l},\mathbf{l}'}\langle \mathbf{S}_{\mathbf{l}}\rangle\langle \mathbf{S}_{\mathbf{l}'}\rangle - \sum_{\mathbf{l}} \mathbf{h}_{\mathbf{l}}\cdot\langle \mathbf{S}_{\mathbf{l}}\rangle + T\sum_{\mathbf{l}}\int d\Omega_{\mathbf{l}}\rho_{\mathbf{l}}(\Omega_{\mathbf{l}})\ln\rho_{\mathbf{l}}(\Omega_{\mathbf{l}}), \tag{4.8.26}$$

where

$$\langle \mathbf{S}_{\mathbf{l}}\rangle = \int d\Omega_{\mathbf{l}}\mathbf{S}(\Omega_{\mathbf{l}})\rho_{\mathbf{l}}(\Omega_{\mathbf{l}}). \tag{4.8.27}$$

Minimization of F_ρ with respect to the function $\rho_{\mathbf{l}}(\Omega_{\mathbf{l}})$ subject to the constraint $\mathrm{Tr}\rho(\Omega_{\mathbf{l}}) = \int d\Omega_{\mathbf{l}}\rho_{\mathbf{l}}(\Omega_{\mathbf{l}}) = 1$ leads to

$$\frac{\delta F_\rho}{\delta\rho_{\mathbf{l}}(\Omega_{\mathbf{l}})} = \mathbf{h}_{\mathbf{l}}^e\cdot\mathbf{S}_{\mathbf{l}} + T[\ln\rho_{\mathbf{l}}(\Omega_{\mathbf{l}}) + 1] = \zeta_{\mathbf{l}}, \tag{4.8.28}$$

where $\zeta_{\mathbf{l}}$ is a Lagrange multiplier and

$$\mathbf{h}_{\mathbf{l}}^e = \sum_{\mathbf{l}'} J_{\mathbf{l},\mathbf{l}'}\langle \mathbf{S}_{\mathbf{l}'}\rangle + \mathbf{h}_{\mathbf{l}} \tag{4.8.29}$$

is the effective field at site $\mathbf{l}$. The solution to Eq. (4.8.28) with $\zeta_{\mathbf{l}}$ chosen to satisfy $\mathrm{Tr}\rho = 1$ is

$$\rho_{\mathbf{l}}(\Omega_{\mathbf{l}}) = \frac{1}{Z_{\mathbf{l}}}e^{\beta\mathbf{h}_{\mathbf{l}}^e\cdot\mathbf{S}_{\mathbf{l}}}, \tag{4.8.30}$$

where

$$Z_{\mathbf{l}} = \int d\Omega_{\mathbf{l}}e^{\beta\mathbf{h}_{\mathbf{l}}^e\cdot\mathbf{S}_{\mathbf{l}}}. \tag{4.8.31}$$

The effective field at site $\mathbf{l}$ depends on the average spins $\langle \mathbf{S}_{\mathbf{l}'}\rangle$ at sites coupled to $\mathbf{l}$ via the exchange $J_{\mathbf{l}\mathbf{l}'}$. The average spin at site $\mathbf{l}$ must thus be determined self-consistently via the equation

$$\langle \mathbf{S}_{\mathbf{l}}\rangle = \frac{1}{Z_{\mathbf{l}}}\int d\Omega_{\mathbf{l}}e^{\beta\mathbf{h}_{\mathbf{l}}^e\cdot\mathbf{S}_{\mathbf{l}}}\mathbf{S}_{\mathbf{l}}. \tag{4.8.32}$$

When the external field $\mathbf{h}_{\mathbf{l}} = h\mathbf{e}_z$ is independent of $\mathbf{l}$, $\langle \mathbf{S}_{\mathbf{l}}\rangle = \langle S_z\rangle\mathbf{e}_z$ will also be independent of $\mathbf{l}$. It is always possible to parameterize the solid angle $\Omega_{\mathbf{l}}$ so that $S_z = \cos\vartheta$. The equation of state for $\langle S_z\rangle$ is thus

$$\langle S_z\rangle = \frac{\int d\Omega e^{\beta h^e\cos\vartheta}\cos\vartheta}{\int d\Omega e^{\beta h^e\cos\vartheta}} = \coth\beta h^e - \frac{1}{\beta h^e} \quad \text{when } n = 3, \tag{4.8.33}$$

where $h^e = \mathbf{h}^e\cdot\mathbf{e}_z$. This equation is very similar to the equation of state for the Ising model obtained from Bragg-Williams theory and can be solved using the graphical procedures discussed in Sec. 4.2. Expansion of its right hand side, which is an odd function of βh^e, for small βh^e at $h = 0$ shows that it predicts a second-order phase transition with a transition temperature $T_c = zJ\overline{\cos^2\vartheta}$, where the bar signifies an average over the unit sphere.

The trial density matrix determines the order parameter $\langle \mathbf{S}_{\mathrm{l}} \rangle$ as a function of temperature via the equation of state. In addition, the density matrix determines all moments of the order parameter. For example, it determines the average of the symmetric-traceless tensor $\langle S_{\mathrm{l},i} S_{\mathrm{l},j} - \delta_{ij}/n \rangle$, which is zero in the high temperature phase but nonzero in the ordered phase. That all moments are contained in ρ_{l} is particularly evident in the low-temperature limit where it becomes a Dirac delta function. If we choose $\mathbf{S} = (\boldsymbol{\pi}, \sigma)$, where $\sigma = (1 - \pi^2)^{1/2}$, we obtain

$$\rho = (\beta h^e / 2\pi)^{(n-1)/2} e^{-\beta h^e \pi^2/2} \to \delta^{(n-1)}(\boldsymbol{\pi}). \tag{4.8.34}$$

This is a mathematical statement of the fact that, at zero temperature, the components of $\mathbf{S}_{\mathrm{l}}$ perpendicular to the direction of order are zero.

5 Debye-Hückel theory

Our final application of variational mean-field theory will be to classical plasmas in equilibrium in which there are mobile charges of opposite signs. This theory is often applied to electrolytes in which there are various charged ions in solution and is usually referred to as Debye-Hückel theory (Debye and Hückel 1923). It is also used, as we shall see in Sec. 9.4, to describe the disorder phase of two-dimensional xy-models where topological point defects called vortices interact among themselves via a Coulomb potential. For simplicity, we will restrict our attention to plasmas with only two types of charge carriers, one with positive charge Q_+ and one with negative charge $Q_- = -|Q_+|$. Generalization to more types is straightforward. Overall charge neutrality requires that $Q_+ N_+ + Q_- N_- = 0$, where N_+ and N_- are, respectively, the number of positive and negative charge carriers. Let $\mathbf{x}_+^\alpha$ denote the positions of positive charges and $\mathbf{x}_-^\alpha$ the positions of negative charges. The Hamiltonian in the presence of an external electric potential ϕ^{ext} is then

$$\mathscr{H} = \frac{1}{2} \sum_{\alpha,\beta,\sigma,\sigma'} Q_\sigma Q_{\sigma'} U(\mathbf{x}_\sigma^\alpha - \mathbf{x}_{\sigma'}^\beta) + \sum_{\alpha,\sigma} Q_\sigma \phi^{\mathrm{ext}}(\mathbf{x}_\sigma^\alpha), \tag{4.8.35}$$

where $U(\mathbf{x}) = |\mathbf{x}|^{-1}$ is the Coulomb potential (in three dimensions).

Our goal is to determine charge densities and charge-density response functions in mean-field theory. We therefore choose a representation of the total density matrix ρ in terms of single particle density matrices $\rho_+(\mathbf{x}_+^\alpha)$ and $\rho_-(\mathbf{x}_-^\beta)$ for the positive and negative charges:

$$\rho = \prod_\alpha \rho_+(\mathbf{x}_+^\alpha) \prod_\beta \rho(\mathbf{x}_-^\beta). \tag{4.8.36}$$

The average number density of positive and negative charges are proportional to the single particle density matrices,

$$\langle n_\pm(\mathbf{x}) \rangle = \langle \sum_\alpha \delta(\mathbf{x} - \mathbf{x}_\pm^\alpha) \rangle = N_\pm \rho_\pm(\mathbf{x}). \tag{4.8.37}$$

Thus, the densities $\langle n_\pm(\mathbf{x})\rangle$, rather than the density matrices $\rho_\pm(\mathbf{x})$, can be used as variational functions. The total charge density is

$$\langle \rho_Q \rangle = Q_+\langle n_+(\mathbf{x})\rangle + Q_-\langle n_-(\mathbf{x})\rangle, \tag{4.8.38}$$

and the variational free energy is

$$\begin{aligned} F_\rho &= \frac{1}{2}\int d^3x d^3x' \langle \rho_Q(\mathbf{x})\rangle U(\mathbf{x}-\mathbf{x}')\langle \rho_Q(\mathbf{x}')\rangle \\ &+ \int d^3x \langle \rho_Q(\mathbf{x})\rangle \phi^{\text{ext}}(\mathbf{x}) \\ &+T\int d^3x \langle n_+(\mathbf{x})\rangle \ln\frac{\langle n_+(\mathbf{x})\rangle}{N_+} + T\int d^3x \langle n_-(\mathbf{x})\rangle \ln\frac{\langle n_-(\mathbf{x})\rangle}{N_-}. \end{aligned} \tag{4.8.39}$$

The mean-field equation of state for the average densities $\langle n_\pm(\mathbf{x})\rangle$ is obtained by minimizing F_ρ subject to the constraints that the total number of particles of each species be fixed:

$$\frac{\delta F_\rho}{\delta\langle n_\pm(\mathbf{x})\rangle} = T\left(\ln\frac{\langle n_\pm(\mathbf{x})\rangle}{N_\pm}+1\right) + Q_\pm\phi(\mathbf{x}) = \mu_\pm, \tag{4.8.40}$$

where μ_+ and μ_- are Lagrange multipliers (chemical potentials) and where

$$\begin{aligned} \phi(\mathbf{x}) &= \int d^3x' U(\mathbf{x}-\mathbf{x}')\langle \rho_Q(\mathbf{x}')\rangle + \phi^{\text{ext}}(\mathbf{x}) \\ &\equiv \phi^{\text{ind}}(\mathbf{x}) + \phi^{\text{ext}}(\mathbf{x}) \end{aligned} \tag{4.8.41}$$

is the electric potential consisting of an external part $\phi^{\text{ext}}(\mathbf{x})$ and an induced part ϕ^{ind} arising from the induced charge $\langle \rho_Q(\mathbf{x})\rangle$. Thus, when the Lagrange multipliers are chosen so that $\int d^3x\langle n_\pm(\mathbf{x})\rangle = N_\pm$, densities of positive and negative charges satisfy

$$\langle n_\pm(\mathbf{x})\rangle = \frac{N_\pm}{Z_\pm}e^{-Q_\pm\phi(\mathbf{x})/T}, \tag{4.8.42}$$

where

$$Z_\pm = \int d^3x e^{-Q_\pm\phi(\mathbf{x})/T}. \tag{4.8.43}$$

Since the total potential $\phi(\mathbf{x})$ depends on the induced charge density, Eq. (4.8.42) must be solved self-consistently for the densities $\langle n_\pm(\mathbf{x})\rangle$. When ϕ^{ext} is zero, the self-consistent solution is trivial: the average charge density $\langle \rho_Q\rangle$ and potential $\phi(\mathbf{x})$ are zero, $Z_\pm = V$, and $\langle n_\pm(\mathbf{x})\rangle = \langle n_\pm\rangle = N_\pm/V$, where V is the volume of the system.

The charge-density response function We will now use the mean-field equations to calculate the charge-density response function and the dielectric constant of a classical plasma. The charge-density response function is

$$\chi_{\rho\rho}(\mathbf{x},\mathbf{x}') = -\frac{\delta\langle \rho_Q(\mathbf{x})\rangle}{\delta\phi^{\text{ext}}(\mathbf{x}')}. \tag{4.8.44}$$

A negative sign appears on the right hand side of this equation because the external Hamiltonian is $+\int d^dx\phi^{\text{ext}}(\mathbf{x})\rho_Q(\mathbf{x})$ and has a positive rather than the negative sign we have usually associated with coupling to external fields. The linear response of $\langle n_\pm(\mathbf{x})\rangle$ to $\phi^{\text{ext}}(\mathbf{x}')$ can be calculated from Eq. (4.8.42):

$$\frac{\delta\langle n_\pm(\mathbf{x})\rangle}{\delta\phi^{\text{ext}}(\mathbf{x}')} = -\frac{Q_\pm}{T}\langle n_\pm\rangle\frac{\delta\phi(\mathbf{x})}{\delta\phi^{\text{ext}}(\mathbf{x}')} + \frac{\langle n_\pm\rangle^2}{N_\pm}\int d^d y\frac{\delta\phi(\mathbf{y})}{\delta\phi^{\text{ext}}(\mathbf{x}')}, \qquad (4.8.45)$$

where we used the fact that $\langle n_\pm(\mathbf{x})\rangle = \langle n_\pm\rangle$ are spatially uniform in equilibrium when ϕ^{ext} is zero. The charge-density response function is

$$\begin{aligned}\chi_{\rho\rho}(\mathbf{x},\mathbf{x}') &= Q_+\frac{\delta\langle n_+(\mathbf{x})\rangle}{\delta\phi^{\text{ext}}(\mathbf{x}')} + Q_-\frac{\delta\langle n_-(\mathbf{x})\rangle}{\delta\phi^{\text{ext}}(\mathbf{x}')} \\ &= \frac{\kappa^2}{4\pi}\frac{\delta\phi(\mathbf{x})}{\delta\phi^{\text{ext}}(\mathbf{x}')},\end{aligned} \qquad (4.8.46)$$

where we used $Q_+\langle n_+\rangle^2/N_+ + Q_-\langle n_-\rangle^2/N_- = V^{-2}(Q_+N_+ + Q_-N_-) = 0$ and where

$$\kappa^2 = \frac{4\pi}{T}(Q_+^2\langle n_+\rangle + Q_-^2\langle n_-\rangle) \qquad (4.8.47)$$

is the square of an inverse Debye-Hückel screening length. Thus, we see that the charge-density response function is proportional to the response of the total potential to an external potential, which can be calculated from Eq. (4.8.41):

$$\frac{\delta\phi(\mathbf{x})}{\delta\phi^{\text{ext}}(\mathbf{x}')} = \delta(\mathbf{x}-\mathbf{x}') - \int d^3y U(\mathbf{x}-\mathbf{y})\chi_{\rho\rho}(\mathbf{y},\mathbf{x}'). \qquad (4.8.48)$$

This is a self-consistent equation for $\chi_{\rho\rho}(\mathbf{x},\mathbf{x}')$, which can be solved using Eq. (4.8.46) and the expression $U(\mathbf{q}) = 4\pi/q^2$ for the Fourier transform of the Coulomb potential. We find

$$\chi_{\rho\rho}(\mathbf{q}) = -\frac{\kappa^2}{q^2}\chi_{\rho\rho}(\mathbf{q}) + \frac{\kappa^2}{4\pi} \qquad (4.8.49)$$

or

$$\chi_{\rho\rho}(\mathbf{q}) = \frac{\kappa^2}{4\pi}\frac{q^2}{q^2+\kappa^2} \qquad (4.8.50)$$

for the Fourier transform of the charge-density response function. Note that, unlike $\chi_{nn}(\mathbf{q})$ in neutral systems, $\chi_{\rho\rho}(\mathbf{q})$ tends to zero as q tends to zero, indicating that the charge density is incompressible at $\mathbf{q} = 0$. This is a result of the long-range nature of the Coulomb potential.

The dielectric constant and screening The dielectric constant is often of greater importance than density response functions in charged systems. In insulators, the dielectric tensor ϵ_{ij} relates the Maxwell electric displacement vector $\mathbf{D}$ to the electric field $\mathbf{E}$ via $D_i = \epsilon_{ij}E_j$. In conductors with mobile charges, the dielectric constant at zero wave number is infinite. A dielectric constant at nonzero wave number and zero frequency can, however, be defined through the response of the spatially varying displacement $\mathbf{D}(\mathbf{x})$ to the spatially varying electric field $\mathbf{E}(\mathbf{x}')$:

$$D_i(\mathbf{x}) = \int d^3x'\epsilon_{ij}(\mathbf{x}-\mathbf{x}')E_j(\mathbf{x}'). \qquad (4.8.51)$$

In isotropic systems, the dielectric tensor can be divided into a longitudinal part and a transverse part, i.e., its Fourier transform can be decomposed as

$$\epsilon_{ij}(\mathbf{q}) = \epsilon_L(\mathbf{q})\hat{q}_i\hat{q}_j + \epsilon_T(\mathbf{q})(\delta_{ij} - \hat{q}_i\hat{q}_j), \tag{4.8.52}$$

where $\hat{q}_i = q_i/|\mathbf{q}|$. In static situations, both $\mathbf{E}$ and $\mathbf{D}$ are irrotational and can be expressed as gradients of a potential. The electric field is determined by the total potential $\phi(\mathbf{x})$, whereas the displacement is determined by the external potential ϕ^{ext}:

$$\mathbf{E} = -\nabla\phi, \quad \mathbf{D} = -\nabla\phi^{\text{ext}}. \tag{4.8.53}$$

Thus, the response of $\mathbf{D}$ to $\mathbf{E}$ is determined by the longitudinal part of the dielectric tensor

$$\epsilon_L^{-1}(\mathbf{x} - \mathbf{x}') = \frac{\delta\phi(\mathbf{x})}{\delta\phi^{\text{ext}}(\mathbf{x}')}. \tag{4.8.54}$$

Then,

$$\epsilon_L^{-1}(\mathbf{q}) = 1 - U(\mathbf{q})\chi_{\rho\rho}(\mathbf{q}), \tag{4.8.55}$$

or

$$\epsilon_L(\mathbf{q}) = 1 + (\kappa^2/q^2), \tag{4.8.56}$$

and, as advertised, the dielectric constant diverges as $\mathbf{q} \to 0$. The dielectric constant gives the response of the total potential to the external potential, and allows us to calculate $\phi(\mathbf{x})$ if we know $\phi^{\text{ext}}(\mathbf{x})$. For example, if a point charge of charge Q is placed at the origin, it will give rise to a potential $\phi^{\text{ext}}(\mathbf{x}) = Q/|\mathbf{x}|$ or $\phi^{\text{ext}}(\mathbf{q}) = 4\pi Q/q^2$. The total potential in the presence of the external charge is thus

$$\phi(\mathbf{q}) = \frac{\phi^{\text{ext}}(\mathbf{q})}{\epsilon_L(\mathbf{q})} = \frac{4\pi Q}{q^2 + \kappa^2} \tag{4.8.57}$$

or

$$\phi(\mathbf{x}) = \frac{Q}{|\mathbf{x}|}e^{-\kappa|\mathbf{x}|}. \tag{4.8.58}$$

The total potential dies exponentially to zero with a length κ^{-1}, whereas the external potential dies algebraically to zero. This is the phenomenon of *screening*. The mobile charges in the plasma collect around the external charge to reduce the total charge in the vicinity of the external charge.

Though we have considered here explicitly only three-dimensional systems, it is clear that the expressions for the dielectric constant and charge density response functions depend only on the fact that $U(\mathbf{q}) \sim q^{-2}$. Thus they also apply to Coulomb interactions with $U(\mathbf{x}) \sim |\mathbf{x}|^{-(d-2)}$ in d dimensions. They apply, in particular, in two dimensions where $U(\mathbf{x}) \sim \ln|\mathbf{x}|$.

When Debye-Hückel theory is valid Mean-field theory provides a powerful yet relatively simple description of Coulomb gases, and it is of some interest to know when it breaks down. The Hamiltonian is a function of the positions $\mathbf{x}_{\pm}^{\alpha}$ of the positive and negative charge carriers. We know that the Coulomb potential favors the formation of charge-neutral bound states of positive and negative charges (i.e., atoms or molecules). The mean-field theory replaces individual position coordinates by average densities that are spatially uniform in the absence of external potentials. It thus ignores altogether the possibility of bound states. It is a good approximation when kinetic energy dominates over potential energy, i.e., at high temperatures in classical systems (or at high densities in Fermi systems). In classical systems, the kinetic energy per particle is $3T/2$, while the average potential energy is of order $Q^2/\langle n\rangle^{1/3}$ when $Q_+ = -Q_- = Q$ and $\langle n_+\rangle = \langle n_-\rangle = \langle n\rangle$. Mean-field theory then applies for $T\langle n\rangle^{-1/3} > Q^2$. Another way to arrive at this result is to use the criterion that fluctuations in charge density should be small compared to the square of the average density of either charged species, i.e., $\langle[\delta\rho_Q(\mathbf{x})]^2\rangle < Q^2\langle n\rangle^2$. Using Eq. (3.4.18), which relates density correlations to the density response function, we find

$$\langle[\delta\rho_Q(\mathbf{x})]^2\rangle = T\int\frac{d^3q}{(2\pi)^3}\chi_{\rho,\rho}(\mathbf{q}) \sim T\kappa^5 \tag{4.8.59}$$

for a classical plasma. This implies that the condition for the validity of mean-field theory is $T(Q^2\langle n\rangle/T)^{5/2} < Q^2\langle n\rangle^2$ or $T > Q^2\langle n\rangle^{1/3}$ in agreement with the preceding argument.

Again, the important feature of the mean-field theory is the replacement of the full Hamiltonian of Eq. (4.8.35) expressed in terms of the coordinates of individual charged particles by the free energy of Eq. (4.8.39) expressed in terms of the coarse-grained charge and number densities. The charge-density response function and dielectric constant can be obtained directly by expanding the free energy F to second order in deviations $\delta n_\pm = \langle n_\pm(\mathbf{x})\rangle - \langle n_\pm\rangle_0$ of the charge densities from their charge-neutral equilibrium values $\langle n_\pm\rangle_0$ with $Q_+\langle n_+\rangle + Q_-\langle n_-\rangle = 0$. Defining $\psi = -Q_-(\langle n_-\rangle_0/\langle n_+\rangle_0)^{1/2}\delta n_+ + Q_+(\langle n_+\rangle_0/\langle n_-\rangle_0)^{1/2}\delta n_-$ and expanding Eq. (4.8.39), we obtain

$$\begin{aligned}\delta F &= \frac{1}{2}\int d^3x d^3x'\langle\rho_Q(\mathbf{x})\rangle[U(\mathbf{x}-\mathbf{x}') + 4\pi\kappa^{-2}\delta(\mathbf{x}-\mathbf{x}')]\langle\rho_Q(\mathbf{x}')\rangle \\ &\quad +2\pi\kappa^{-2}\int d^3x\psi^2(\mathbf{x}).\end{aligned} \tag{4.8.60}$$

Because ψ and $\langle\delta\rho_Q(\mathbf{x})\rangle$ are decoupled, this equation immediately gives

$$\chi_{\rho\rho}^{-1} = 4\pi\kappa^{-2} + \frac{4\pi}{q^2} \tag{4.8.61}$$

in agreement with Eq. (4.8.50).

Bibliography

R.P. Feynman, *Statistical Mechanics, A Set of Lectures* (W.A. Benjamin, Reading, Mass., 1972).

L.D. Landau and E.M. Lifshitz, *Electrodynamics of Continuous Media* (Pergamon Press, New York, 1960).

L.D. Landau and E.M. Lifshitz, *Statistical Physics* (Addison-Wesley, Reading, Mass., 1969).

H. Eugene Stanley, *Introduction to Phase Transitions and Critical Phenomena* (Oxford University Press, New York, 1987).

References

S. Alexander and J.P. McTague, *Phys. Rev. Lett.* **41**, 702 (1978).
R.J. Birgeneau, W.B. Yelon, E. Cohen, and J. Markovski, *Phys. Rev. B* **5**, 2607 (1972).
R.J. Birgeneau, G. Shirane, M. Blume, and W.C. Koehler, *Phys. Rev. Lett.* **33**, 1100 (1974).
R.J. Birgeneau, G.W. Garland, A.R. Kortan, J.D. Litster, M. Meichle, B.M. Ocko, C. Rosenblatt, L.J. Yu, and J. Goodby, *Phys. Rev. A* **27**, 1251 (1983).
M. Blume, V.J. Emery, and Robert B. Griffiths, *Phys. Rev. A* **4**, 1071 (1971).
W.L. Bragg and E.J. Williams, *Proc. Roy. Soc. London A* **145**, 699 (1934); **151**, 540 (1935); **152**, 231 (1935).
J.-H. Chen and T.C. Lubensky, *Phys. Rev. A* **14**, 1202 (1976).
P. Debye and E. Hückel, *Z. Physik* **24**, 185 (1923).
Michael E. Fisher and David R. Nelson, *Phys. Rev. Lett.* **32**, 1350 (1974).
E.H. Graf, D.M. Lee, and John D. Reppy, *Phys. Rev. Lett.* **19**, 417 (1967).
R.B. Griffiths, *Phys. Rev. Lett.* **24**, 715 (1970).
E.A. Guggenheim, *J. Chem. Phys.* **13**, 253 (1945).
J.P. Hansen and L. Verlet, *Phys. Rev.* **184**, 151 (1969).
J.M. Kincaid and E.G.D. Cohen, *Phys. Rep. C* **22**, 57 (1975).
L.D. Landau, *Phys. Z. Sowejetunion* **11**, 26 (1937); reprinted in *Collected Papers of L.D. Landau*, ed. D. ter Haar (Pergamon, New York, 1965).
E.M. Lifshitz, *Zh. Eksp. Teor. Fiz.* **11**, 253 & 269 (1941).
W. Maier and W. Saupe, *Z. Naturf.* A**13**, 564 (1958); A**14**, 882 (1959); A**15**, 287 (1960).
N.E. Phillips, *Phys. Rev.* **114**, 676 (1959).
T.V. Ramakrishnan and M. Yussouff, *Phys. Rev. B* **19**, 2775 (1979).
H. Rohrer and Ch. Gerber, *Phys. Rev. Lett.* **38**, 909 (1977).
Y. Shapira and S. Foner, *Phys. Rev.* **1**, 3083 (1970).
J.D. van der Waals, *Over de Continuïteit van dpe Gas- en Vloeistoftoestand*, Thesis, Leiden (1873).
L. Verlet, *Phys. Rev.* **165**, 201 (1965).
P. Weiss, *Comptes Rendus* **143**, 1136 (1906); *J. Phys.* (France) **6**, 661 (1907).

Problems

4.1 (Tetracritical and bicritical points) Consider the model Landau free energy density

$$f = \frac{1}{2}r_1\phi_1^2 + \frac{1}{2}r_2\phi_2^2 + u_1\phi_1^4 + u_2\phi_2^4 + 2u_{12}\phi_1^2\phi_2^2$$

with $r_1 = a(T - T_1) = r - g$, $r_2 = a(T - T_2) = r + g$ and u_1, u_2, and u_{12} positive. Show that (i) when $u_1u_2 > u_{12}^2$, there are four phases in the mean-field phase diagram separated by second-order transition lines meeting at a tetracritical point as shown in Fig. 4.6.10b, and (ii) when $u_1u_2 < u_{12}^2$, there are only three phases with a first-order transition line meeting two second-order lines at a bicritical point as shown in Fig. 4.6.10a. A complete solution to this problem should verify that each phase is the lowest free energy locally stable state. If you are ambitious, you may wish to consider what happens when $u_{12} < 0$.

4.2 (Tricritical point in an antiferromagnet in a field) An external magnetic field h in an antiferromagnet couples to the magnetization m rather than to the order parameter (the staggered magnetization) m_s. Assume that the coupling between m_s and m is described phenomenologically via the free energy

$$f = \frac{1}{2}rm_s^2 + um_s^4 + \frac{1}{2}r_m m^2 - hm + \frac{1}{2}wm_s^2m^2,$$

where $r = a(T - T^*)$, $w > 0$, and r_m is independent of temperature. Show that this model has a tricritical point at temperature T_t and field h_t, where

$$T_t = T^* - \frac{(2ur_m)}{aw},$$

$$h_t^2 = \frac{2ur_m^3}{w^2}.$$

Show that the mean-field phase diagram is similar to that shown in Fig. 4.6.4, where the second-order transition temperature for $h < h_t$ is

$$T_c = T_t - \frac{wh_t^2}{ar_m^2}\eta$$

and the first-order transition for $h > h_t$ for small $|h - h_t|$ is

$$T_c = T_t - \frac{wh_t^2}{ar_m^2}[\eta - \frac{1}{4}\eta^2],$$

where $\eta = [1 - (h_t^2/h^2)]$.

4.3 As in the Blume-Emery-Griffiths model, there are three second-order λ lines meeting at the tricritical point ($r = 0$, $u_4 = 0$, $h = 0$) of the model tricritical Landau free energy of Eq. (4.6.1) in an external field h coupling linearly to the order parameter. Calculate the equations for these lines. Hint: the λ lines terminate a plane of two-phase coexistence and can be treated in mean-field theory in much the same way as the critical point of the liquid-gas transition.

4.4 Calculate the values of H and T at the tricritical point of the metamagnetic model free energy of Eq. (4.6.9), assuming $z_1J_1/z_2J_2 > 3/5$.

4.5 A phase diagram with a critical endpoint such as that depicted in Fig. 4.6.5 can be described by a Landau free energy expanded to eighth order in the potential with a positive eighth-order and a negative fourth-order coefficient. Using the fact that an eighth-order polynomial may have up to four minima, sketch the form of the Landau free energy (i) at the critical point of Fig. 4.6.5, where there is two-phase coexistence, (ii) between the critical point and the critical endpoint, where there is four-phase coexistence, (iii) at the critical endpoint, where there is three-phase coexistence, and (iv) to the right of the critical endpoint, where there is two-phase coexistence.

4.6 In Chapter 2, we saw that the X-ray scattering intensity in the nematic phase has strong diffuse spots along the axis parallel to the director above a smectic-A phase but two strong diffuse annuli centered along the same axis above a smectic-C phase. Develop a Landau theory along the lines of those used in Sec. 4.6 to describe Lifshitz points and in Sec. 4.7 to describe the liquid-solid transition. This theory should reproduce the X-ray scattering intensity in the nematic phase and predict nematic-to-smectic-A, nematic-to-smectic-C, and smectic-A-to-smectic-C transitions and a Lifshitz point. Are any of these transitions first order in mean-field theory? (See Chen and Lubensky 1976.)

4.7 Calculate the phase diagram for the model free energy of Eq. (4.6.19) in the vicinity of the Lifshitz point $(r, c_\perp) = (0,0)$ when $d = 3$ and $m = 1$ (i.e., when $\mathbf{x}_\perp = (0,0,z)$) for

(a) a scalar order parameter ϕ and
(b) an xy-order parameter $\boldsymbol{\phi} = (\phi_1, \phi_2)$.

For the scalar order parameter, you may assume $\phi = \phi_0 \cos q_0 z$ in the modulated phase. You should show that all phases are stable with respect to fluctuations (including fluctuations in q_0 in the modulated phases). For the scalar case, you should calculate the limit of metastability of the ferromagnetic (F) phase and that of the modulated phase if you are more ambitious.

4.8 (Maier-Saupe theory, 1958) Consider the following model for interparticle interactions in a nematic liquid crystal. Associated with each particle α is a symmetric traceless tensor $Q_{ij}^\alpha = Q_{ij}(\Omega_\alpha) = v_i^\alpha v_j^\alpha - \frac{1}{3}\delta_{ij}$, where $\mathbf{v}^\alpha$ is a unit vector pointing along the long axis of the molecule specified by its solid angle Ω_α on the unit sphere. The interaction Hamiltonian is

$$\mathscr{H} = \frac{1}{2}\sum_{\alpha,\alpha'} Q_{ij}^\alpha Q_{ij}^{\alpha'} U(\mathbf{x}^\alpha - \mathbf{x}^{\alpha'}).$$

Show that the variational free energy F_ρ associated with this Hamiltonian is

$$F_\rho = \frac{1}{2}\int d^3x d^3x' \langle R_{ij}(\mathbf{x})\rangle \langle R_{ij}(\mathbf{x}')\rangle U(\mathbf{x}-\mathbf{x}') + NT \int d^3x d\Omega \rho(\mathbf{x},\Omega) \ln \rho(\mathbf{x},\Omega),$$

where $\rho(\mathbf{x},\Omega)$ is a single particle density matrix,

$$\langle R_{ij}(\mathbf{x})\rangle = \langle \sum_\alpha Q_{ij}^\alpha \delta(\mathbf{x}-\mathbf{x}^\alpha)\rangle = N\int d\Omega \rho(\mathbf{x},\Omega) Q_{ij}(\Omega)$$

and N is the number of molecules. Then show that the best single particle density matrix is

$$\rho(\mathbf{x},\Omega) = \frac{1}{Z} e^{-U^e(\mathbf{x},\Omega)/T},$$

where

$$U^e(\mathbf{x},\Omega) = \int d^3x' U(\mathbf{x}-\mathbf{x}') \langle R_{ij}(\mathbf{x}')\rangle Q_{ij}(\Omega)$$

and

$$Z = \int d^3x d\Omega e^{-U^e(\mathbf{x},\Omega)/T}.$$

Finally, show that in spatially uniform nematics when there are no external forces that the nematic order parameter S defined via $\langle Q_{ij}\rangle = S(n_i n_j - \delta_{ij}/3)$, where $\mathbf{n}$ is the director, satisfies

$$S = \frac{1}{Z'}\int d(\cos\vartheta)e^{-SU_0(\cos^2\vartheta-(1/3))}\left(\cos^2\vartheta - \frac{1}{3}\right)$$

where $U_0 = (N/V)U(\mathbf{q} = 0)$ and

$$Z' = \int d(\cos\vartheta)e^{-SU_0(\cos^2\vartheta-(1/3))}.$$

Expand the right hand side of this equation in powers of S and show that it has an even number of terms so that it predicts a first-order transition just as the simple Landau theory of Sec. 4.6 does. What happens as $T \to 0$.

4.9 (a) Use $\mathrm{Tr}\rho\ln\rho$ mean-field theory to calculate the probability $P(\theta)$ that a spin makes an angle θ with the x-axis. Use this to calculate the order parameters $\psi_n = \langle e^{in\theta}\rangle$ near the critical point.

(b) Use symmetry arguments to develop a Landau theory for the xy-transition in terms of the parameters ψ_n including couplings between ψ_n and ψ_m for arbitrary n and m. Show that this theory predicts the same results as (a).

4.10 Consider the isotropic-to-nematic transition in the vicinity of the special point (the Landau point) where the coefficient of the third-order term in the free energy [Eq. (4.5.5)] is zero. Allow for both biaxial as well as unixial order using Eq. (4.5.2) for the nematic order parameter Q_{ij}. Show that two first-order lines and two second-order lines meet at the Landau point. On one side of the Landau point, there is a first-order transition from the isotropic fluid to a state with $S > 0$ and on the other a transition to a state with $S < 0$. The two second-order lines enclose a biaxial region in which both S and η are nonzero.

5

Field theories, critical phenomena, and the renormalization group

Mean-field theory, presented in the preceding chapter, correctly describes the qualitative features of most phase transitions and, in some cases, the quantitative features. Since mean-field theory replaces the actual configurations of the local variables (e.g. spins) by their average value, it neglects the effects of fluctuations about this mean. These fluctuations may or may not be important. The more spins that interact with a particular test spin, the more the test spin sees an effective average or mean field. If the test spin interacts with two neighbors, the averaging is minimal and the fluctuations are large and important. The number of spins producing the effective field increases with the range of the interaction and with the dimension. Thus we will find that mean-field theory is a good approximation in high dimensions but fails to provide a quantitatively correct description of second-order critical points in low dimensions. This chapter will be devoted to the study of second-order phase transitions when mean-field theory is not a good approximation.

We will begin by considering fluctuations of the order parameter about its spatially uniform mean-field value. We will find that for spatial dimensions below an *upper critical dimension* d_c (typically $d_c = 4$), fluctuations always become important and invalidate mean-field theory for temperatures sufficiently close to T_c. This will motivate a generalization of Landau theory that incorporates spatial fluctuations about a mean-field free energy minimum. This generalization is a *field theory* that emphasizes the special role of fluctuations of the order parameter. The partition function trace in this theory is an integral over all possible values of the order parameter at all points in space. We will discuss a simple approximate treatment of fluctuations that shows how critical exponents different from those predicted by mean-field theory can arise.

After the above theoretical introduction to the breakdown of mean-field theory, we will discuss the experimentally observed phenomena of *scaling* and *universality*. In the vicinity of critical points, the free energy, susceptibility, and other thermodynamic functions are generalized homogeneous functions of temperature and conjugate field, and, as a result, not all of the critical exponents introduced in the preceding chapter are independent. For example, near the critical point the only relevant length is the correlation length, which serves as a scale for all distances. Since we know how the correlation length varies with temperature, we know something about how the thermodynamic functions and correlations vary with temperature. This is scaling. Critical exponents show remarkably little variation among systems of a given dimensionality even when there is wide variation in the magnitude of the critical temperature or in the nature of microscopic interactions. Any variation in exponents, in a given spatial dimension, that does exist is associated with changes in symmetry of the order parameter. This constancy of exponents for different transitions with the same symmetry order parameter and dimension of space is called *universality*.

Universality and scaling were the essential clues that led to the development of the *renormalization group* by Kenneth Wilson in the early 1970s. The renormalization group is a general calculational scheme for treating problems where fluctuations at many length or time scales are important. It has been applied with great success to the study of second-order phase transitions and critical phenomena and to a number of other problems as well. It has had such a significant impact on how we think about fluctuation-dominated physical phenomena that every physicist should have some understanding of how it works. At the end of this chapter, we will present a simplified introduction to the basic concepts of the Wilson renormalization group.

5.1 Breakdown of mean-field theory

Mean-field theory is an approximation that replaces a fluctuating local order parameter by a spatially uniform average order parameter. This is a good approximation if fluctuations of the order parameter about its mean value are small. A quantitative measure (first introduced by V.L. Ginzburg, 1960) of the importance of fluctuations can be obtained by considering the average over a coherence volume, $V_\xi \sim \xi^d$ of the deviation $\delta\phi(\mathbf{x}) = \phi(\mathbf{x}) - \langle\phi\rangle$, of the local order parameter from its equilibrium value:

$$\delta\phi_{\text{coh}} \equiv V_\xi^{-1} \int_{V_\xi} d^d x \delta\phi(\mathbf{x}). \tag{5.1.1}$$

Fluctuations are negligible if, in the ordered phase, $\langle(\delta\phi_{\text{coh}})^2\rangle$ is much less than $\langle\phi\rangle^2$, i.e., if

$$V_\xi^{-2} \int_{V_\xi} d^d x d^d x' \langle\delta\phi(\mathbf{x})\delta\phi(\mathbf{x}')\rangle = V_\xi^{-1} \int_{V_\xi} d^d x G(\mathbf{x}, 0) < \langle\phi\rangle^2, \tag{5.1.2}$$

where $G(\mathbf{x}, \mathbf{x}') = G(\mathbf{x} - \mathbf{x}')$ is the order parameter correlation function introduced in Eqs. (3.5.13) and (4.3.21).

Remarkably, as discussed in Sec. 4.3, mean-field theory itself provides a prediction for $G(\mathbf{x}, \mathbf{x}') = T\chi(\mathbf{x}, \mathbf{x}')$, where $\chi(\mathbf{x}, \mathbf{x}')$ is the susceptibility relating $\delta\langle\phi(\mathbf{x})\rangle$ to a conjugate external field at $\mathbf{x}'$. Thus, mean-field theory gives rise to its own internal consistency criterion. If Eq. (5.1.2) is satisfied by the mean-field values for $G(\mathbf{x},0)$ and $\langle\phi\rangle$, then mean-field theory is internally consistent and a good approximation. If not, fluctuations are important, and an alternative theory must be found.

The left hand side of Eq. (5.1.2) can be obtained for ϕ^4 theories in mean-field theory using Eq. (4.3.17) for $\chi(\mathbf{x},0)$ and Eq. (4.3.4) for $\langle\phi\rangle$. We obtain

$$\begin{aligned}\langle(\delta\phi_{\rm coh})^2\rangle &= TV_\xi^{-1}c^{-1}\int_{V_\xi} d^dx|\mathbf{x}|^{-(d-2)}Y(|\mathbf{x}|/\xi)\\ &= A_dT\xi^{-(d-2)}/c < |r|/4u,\end{aligned}\tag{5.1.3}$$

where $\xi = (c/|r|)^{1/2}$ is the correlation length and A_d is a constant for fixed dimension d. This equation can be reexpressed in a unitless form,

$$\left(\frac{\xi}{\xi_0}\right)^{d-4} = \left(\frac{T-T_c}{T_c}\right)^{(4-d)/2} > \frac{A_d}{2\Delta c_V\xi_0^d},\tag{5.1.4}$$

where $\xi_0 = (c/aT_c)^{1/2}$ is the bare coherence length [Eq. (4.3.16)] and $\Delta c_V = Ta^2/8u$ is the mean-field specific heat jump per unit volume [Eq. (4.3.11)]. Though this equation was derived by consideration of fluctuations about the ordered phase (i.e., $T < T_c$), it also applies in the disordered phase (i.e., $T > T_c$).

For $d > 4$, ξ^{d-4} diverges as $T \to T_c$, and Eq. (5.1.4) is always satisfied near the critical point. For $d < 4$, ξ^{d-4} tends to zero as $T \to T_c$, and Eq. (5.1.4) is never satisfied near the critical point. Thus, mean-field theory provides an internally consistent description of the second-order phase transition of ϕ^4 and related models for all $d > 4$. It does not provide an adequate description for $d < 4$. The dimension d_c below which mean-field theory breaks down is called the *upper critical dimension*. For phase transitions described by ϕ^4 and related theories, $d_c = 4$. Other transitions can have other values of d_c (see Problem 5.1).

For $d < d_c$, mean-field theory is valid for temperatures sufficiently far from T_c that Eq. (5.1.4) is satisfied. As T approaches T_c (either from above or below), fluctuations become more important, and the inequality in Eq. (5.1.4) is eventually violated. The temperature T_G at which fluctuations become important is called the *Ginzburg temperature* and is determined by equality in Eq. (5.1.4):

$$t_G = \frac{|T_G - T_c|}{T_c} = \left(\frac{A_d}{2\Delta c_V\xi_0^d}\right)^{2/(4-d)}.\tag{5.1.5}$$

Here and in what follows, we will use upper case T to denote temperature and lower case $t = (T-T_c)/T_c$ to denote reduced temperature. Thus, t_G is the reduced Ginzburg temperature. (Later, when we discuss dynamics, t will denote time.) Alternatively, a *Ginzburg length*, ξ_G, can be defined via Eq. (5.1.4):

$$\xi_G^{4-d} \sim \Delta c_V\xi_0^4 = c^2/(8uT_c),\tag{5.1.6}$$

or

$$\xi_G \sim \xi_0(\Delta c_V\xi_0^d)^{1/(4-d)}.\tag{5.1.7}$$

Mean-field theory is valid when $\xi < \xi_G$ and invalid when $\xi > \xi_G$. Note that $|T_G - T_c| \to 0$ as $\xi_0 \to \infty$ for $d < 4$. Thus, mean-field theory will be valid

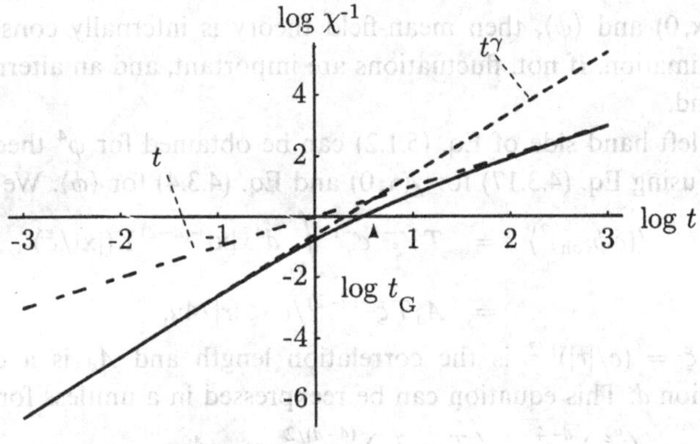

Fig. 5.1.1. Schematic of the inverse susceptibility as a function of reduced temperature $t = (T - T_c)/T_c$ showing crossover from linear mean-field dependence on t at large t to power-law behavior for $t < t_G$.

even very close to a critical point for $d < d_c$ if the bare coherence length ξ_0 is large. This is the case for systems with long-range forces, or, as we shall see, for superconductors. When $|T_G - T_c|$ is not small, one can expect a crossover from mean-field behavior to critical behavior when the reduced temperature $t = (T - T_c)/T_c$ becomes of order t_G. The inverse susceptibility showing crossover is sketched in Fig. 5.1.1. In three dimensions, a more careful evaluation of A_d yields

$$t_G = \frac{k_B^2}{32\pi^2(\Delta c_V)^2\xi_0^6}, \tag{5.1.8}$$

where Δc_V is measured in units of erg cm^{-3} K^{-1} and $k_B = 1.38 \times 10^{-16}$ erg K^{-1} is Boltzmann's constant. This is the form for t_G that is usually used in three dimensions.

1 Mean-field transitions revisited

In Sec. 4.4, we gave two examples of phase transitions that exhibited mean-field behavior. We can now ask why critical fluctuations are not observed in these transitions. In the case of the smectic-A to smectic-C transition (data in Fig. 4.3.4), the specific heat jump is of order 10^6 erg cm^{-3} K^{-1}, and the coherence length ξ_0 is of order 20Å. Thus, $t_G \approx 10^{-5}$. The transition temperature is of order 300 K so that $T_G - T_c \approx 3 \times 10^{-3}$ K. The experiments reported in Fig. 4.3.4 do not probe reduced temperatures of order 10^{-5}, and mean-field behavior is expected. The case of superconductors is even more dramatic. The specific heat jump in aluminum is of order 2×10^4 erg mole^{-1}K. The lattice spacing in Al is 4Å, whereas the coherence length is $\xi_0 \approx 1.6 \times 10^4$Å. Thus, $\Delta c_V \approx (2 \times 10^5/4^2)$ erg cm^{-3} K^{-1} and $t_G \approx 10^{-16}$. The transition temperature in Al is 1.19K so that access to its critical region is virtually impossible. The reason for the very small value of T_G is the enormous coherence length relative to the lattice spacing. This feature is common to most superconductors, with the exception of the the so-called high T_c superconductors. This means that mean-field theory should provide a very good description of superconductors, as indeed it does.

5.2 Construction of a field theory

In the preceding chapter, we studied the predictions of simple mean-field theory based on the phenomenological expansion of the free energy F in powers of the order parameter $\langle\phi(\mathbf{x})\rangle$. In order to treat fluctuations that become important for $T < T_G$ and $d < d_c$, it is necessary to have a more microscopic description of how partition functions and $F[\langle\phi(\mathbf{x})\rangle]$ are actually evaluated. One could, of course, begin with the microscopic Hamiltonian expressed either in terms of quantum mechanical operators or in terms of classical dynamical variables and attempt to evaluate the partition function directly. This is often a workable procedure. It is, however, usually very difficult to carry out in the vicinity of second-order phase transitions where, as we have seen, correlation lengths diverge. To study critical properties of phase transitions, it is often more useful to introduce semi-phenomenological field theories, where the trace operation is an integral over all possible values at all points in space or all points on a lattice of the local order parameter treated as a continuous classical field. Field theories of this sort are also of considerable use in studying systems far from any phase transition. Their derivation will be described below.

1 *Coarse graining*

The partition function discussed in the preceding chapter involves a trace over all possible states of the system. To facilitate discussion of the appearance of order and properties of the ordered phases, we would like a formulation of the partition function that emphasizes the role of the order parameter. One way to do this is to divide the system up into many cells with dimensions large compared to any microscopic length such as the interparticle spacing or range of the interparticle potential. Each cell contains a large number of particles. The order parameter field, $\phi(\mathbf{x})$, like the Hamiltonian, is a quantum mechanical operator or a function of the classical dynamical variables (e.g. the magnetization of Eq. (3.5.1)). Its average over the particles in a cell centered at position $\mathbf{x}$ is $\tilde{\phi}(\mathbf{x})$. Since there are many particles in each cell, $\tilde{\phi}(\mathbf{x})$ can be regarded as a continuous classical variable, which can vary from cell to cell. This process of averaging over many particles in some volume of space is called *coarse graining*. In essence, we take the Landau form for the free energy and treat it as the energy for a particular local configuration of the order parameter. Then instead of considering the minimum

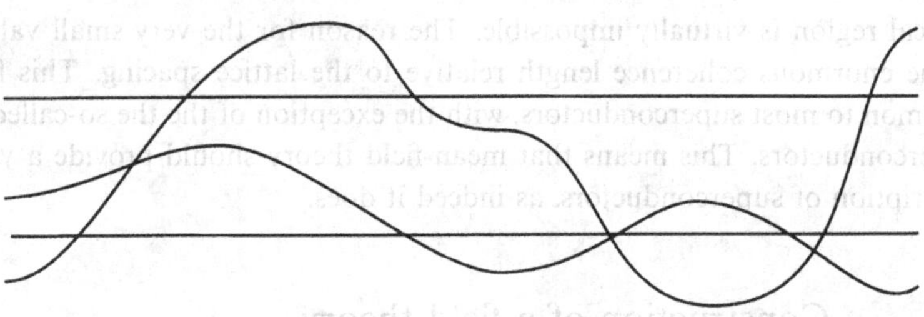

Fig. 5.2.1. Paths contributing to the functional integral of a one-dimensional field. (a) and (b) are paths with a spatially uniform order parameter, and (c) and (d) are paths with a spatially varying order parameter.

energy we allow for excitations according to their statistical weight. The states of the system can now be specified by the field $\tilde{\phi}(\mathbf{x})$ and have an effective energy $\tilde{\mathscr{H}}[\tilde{\phi}(\mathbf{x})]$. The partition function is thus an integral, or more properly a *functional integral,* over all possible values of $\tilde{\phi}(\mathbf{x})$ at all positions $\mathbf{x}$:

$$Z = \int \mathscr{D}\tilde{\phi}(\mathbf{x}) e^{-(\tilde{\mathscr{H}} - \int d^d x h(\mathbf{x})\tilde{\phi}(\mathbf{x}))/T}. \tag{5.2.1}$$

The integral over $\tilde{\phi}(\mathbf{x})$ is often called a path integral because in one-dimension it is an integral over all possible paths of $\tilde{\phi}(\mathbf{x})$ in space. Some paths for a one-dimensional field are shown in Fig. 5.2.1.

In purely classical systems, one can arrive at Eq. (5.2.1) by introducing a delta function and setting the field $\phi(\mathbf{x})$ equal to $\tilde{\phi}(\mathbf{x})$ into the expression for the partition function:

$$Z = \int \mathscr{D}\tilde{\phi}(\mathbf{x}) \mathrm{Tr} \prod_{\mathbf{x}} \delta[\phi(\mathbf{x}) - \tilde{\phi}(\mathbf{x})] e^{-[\mathscr{H} - \int d^d x h(\mathbf{x})\phi(\mathbf{x})]/T}. \tag{5.2.2}$$

This yields Eq. (5.2.1) with

$$e^{-\tilde{\mathscr{H}}/T} = \mathrm{Tr} \prod_{\mathbf{x}} \delta[\phi(\mathbf{x}) - \tilde{\phi}(\mathbf{x})] e^{-\mathscr{H}/T}. \tag{5.2.3}$$

For our present purposes, the distinction between the operator field $\phi(\mathbf{x})$ and its coarse-grained local average $\tilde{\phi}(\mathbf{x})$ is irrelevant, and we will denote both fields by $\phi(\mathbf{x})$. We will discuss in Appendix 5A how field theories, such as that of Eq. (5.2.1), can be derived directly for models, such as the Ising and Heisenberg models discussed in Sec. 3.6, for which there is a strong constraint on the value of the local order parameter field (e.g. $S_l^2 = 1$).

The effective energy $\tilde{\mathscr{H}}$ is usually called a Hamiltonian in the statistical mechanical and critical phenomena literature, even though it is not a function of variables for which commutation or Poisson bracket relations have been defined. It is closely related to the action of quantum field theories and is, therefore, referred to as the action in some literature. Since we will usually not consider in detail the original microscopic Hamiltonian, we will usually use the symbol

$\mathscr{H}$ rather than $\tilde{\mathscr{H}}$ for the effective energy determining the weight assigned to configurations in a functional integral representation of the partition function.

A phenomenological form for $\mathscr{H}$ can be derived in much the same way as the Landau form for the mean-field free energy F [Eq. (4.2.1)]. There should be a local part of $\mathscr{H}$ depending only on the order parameter at a single position in space and a part favoring a spatially uniform order parameter. We therefore write

$$\mathscr{H} = \int d^d x \tilde{f}[\phi(\mathbf{x})] + \frac{1}{2} \int d^d x c[\nabla\phi(\mathbf{x})]^2, \tag{5.2.4}$$

where $\tilde{f}$ is a local function of $\phi(\mathbf{x})$, which can be expanded in a power series in combinations which are invariant under the symmetry group of the disordered phase. For example, $\tilde{f}$ could be identical to the ϕ^4-Landau free energy density of Eq. (4.3.1). The Hamiltonian of Eq. (5.2.4) provides a description of the energy associated with long-wavelength, slow spatial variations of $\phi(\mathbf{x})$. It does not provide a realistic description of short-wavelength distortions. However, most phenomena in the vicinity of critical points are controlled by long-wavelength fluctuations, and f_{SL} in Eq. (5.2.4) is adequate provided short-wavelength fluctuations are suppressed. This can be accomplished by the so-called *hard cutoff* procedure whereby excitations with wave number greater than a cutoff $\Lambda \sim 2\pi/a$, where a is a length of order the range of interparticle interactions, are simply not permitted in the partition trace. Other methods of suppressing short-wavelength fluctuations, such as the addition of a term proportional to $[\nabla^2\phi(\mathbf{x})]^2$, are also possible. In what follows, we will employ the hard cutoff procedure almost exclusively.

2 Lattice field theories and their continuum limit

The definition of the functional integral in the partition trace of Eq. (5.2.1) is a little vague. We will now refine this definition by considering the continuum limit of a theory with fields $\phi_{\mathbf{l}}$ defined on N sites $\mathbf{l}$ of a d-dimensional lattice. The partition function for such a lattice model can be written as

$$Z = \prod_{\mathbf{l}} \int d\phi_{\mathbf{l}} e^{-\beta \mathscr{H}_L[\phi_{\mathbf{l}}]}, \tag{5.2.5}$$

where $\beta = 1/T$ and

$$\mathscr{H}_L[\phi_{\mathbf{l}}] = \sum_{\mathbf{l}} f_L(\phi_{\mathbf{l}}) + \frac{1}{2} \sum_{\mathbf{l},\mathbf{l}'} C_{\mathbf{l},\mathbf{l}'}(\phi_{\mathbf{l}} - \phi_{\mathbf{l}'})^2, \tag{5.2.6}$$

where $f_L(\phi_{\mathbf{l}})$ has a power series expansion about $\phi_{\mathbf{l}} = 0$ and $C_{\mathbf{l}\mathbf{l}'}$ has a finite range, typically a lattice spacing. Each $\phi_{\mathbf{l}}$ can take on any value between $+\infty$ and $-\infty$. A configuration entering the partition sum in Eq. (5.2.5) is specified by the value of $\phi_{\mathbf{l}}$ at each site $\mathbf{l}$. Typical configurations on a one-dimensional lattice are shown in Fig. 5.2.2. The local term $f_L(\phi_{\mathbf{l}})$ favors particular values of $\phi_{\mathbf{l}}$. For example, if f_L has the form of Eq. (4.3.1), it can have a single minimum

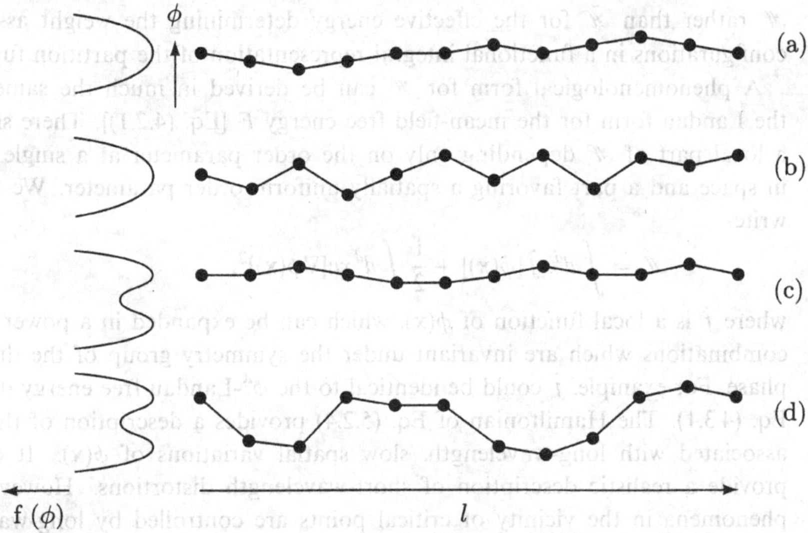

Fig. 5.2.2. Configurations of a one-dimensional lattice field theory. (a) Low-energy configuration when f_L has a single minimum at $\phi_l = 0$. (b) High-energy spatially varying configuration when f_L has a single minimum. (c) The same as (a) when f_L has two equivalent minima. (d) The same as (b) when f_L has two minima. When the lattice spacing goes to zero, these paths become those of a continuum theory depicted schematically in Fig. 5.2.1.

at $\phi_l = 0$ or two minima at $\phi_l = \pm(|r|/4u)^{1/2}$ if r is negative. The interaction term favors equal values of ϕ_l at all lattice sites. Thus, configurations such as those shown in Figs. 5.2.2b and 5.2.2d, where ϕ_l changes rapidly from site to site, have a larger value of $\mathscr{H}_L$ and thus a smaller weight in the partition trace than do configurations such as those in Figs. 5.2.2a and 5.2.2c, in which ϕ_l is spatially uniform with a value near a minimum of f_L.

The continuum limit of the lattice model defined in Eqs. (5.2.5) and (5.2.6) is obtained by allowing the volume per lattice site v_0 to tend to zero, $\mathbf{R}_l$ to become a continuous variable $\mathbf{x}$, and ϕ_l to become $\phi(\mathbf{x})$ while keeping the total volume $V = Nv_0$ fixed. In this limit,

$$v_0 \sum_{l} \to \int d^d x, \tag{5.2.7}$$

$$\sum_{l} f_L(\phi_l) \to \int d^d x \tilde{f}[\phi(\mathbf{x})], \tag{5.2.8}$$

$$\frac{1}{2} \sum_{l,l'} C_{l,l'}(\phi_l - \phi_{l'})^2 \to \frac{1}{2} \int d^d x c(\nabla \phi)^2, \tag{5.2.9}$$

where $\tilde{f} = v_0^{-1} f_L$ and

$$c = \frac{1}{dv_0}\sum_{\mathbf{l}} \mathbf{R}_{\mathbf{l}}^2 C_{\mathbf{l},0}. \tag{5.2.10}$$

Only the lowest order term in the gradient expansion of Eq. (5.2.9) was retained. As discussed previously, implicit in the right hand side of this equation is an upper wave number cutoff Λ. The functional integral entering the continuum partition function has the formal definition

$$\int \mathscr{D}\phi(\mathbf{x}) = \lim_{v_0\to 0}\prod_{\mathbf{l}}\int d\phi_{\mathbf{l}}. \tag{5.2.11}$$

Finally, the lattice paths of Fig. 5.2.2 become continuous paths such as those in Fig. 5.2.1.

Both lattice and continuum field theories are used extensively in condensed matter physics. In the study of critical phenomena, where interesting properties are independent of the way short-wavelength excitations are suppressed, the choice of which to use is largely a matter of personal taste. People with experience with microscopic models defined on a lattice tend to use lattice field theories, whereas those with more experience in quantum field theories tend to use continuum theories. We will usually use continuum theories partly because they are physically more appropriate for many of the systems we will study such as liquids and liquid crystals.

3 Gaussian integrals

When the Hamiltonian $\mathscr{H}$ is harmonic in $\phi(\mathbf{x})$, then the weight function $e^{-\beta\mathscr{H}}$ becomes Gaussian, and the partition trace in Eq. (5.2.1) can be evaluated exactly. This exact evaluation, which we will present in this section, is the first step in the development of perturbative calculations of the partition function when $\mathscr{H}$ has anharmonic terms.

The starting point for the evaluation of Gaussian functional integrals is the identity

$$\int_{-\infty}^{\infty} dy e^{-\frac{1}{2}Cy^2+\lambda y} = \left(\frac{2\pi}{C}\right)^{1/2} e^{\lambda^2/(2C)}. \tag{5.2.12}$$

We will now derive the generalization of this result to multidimensional integrals. Let $\underline{C}$ be an $n\times n$ matrix with components $C_{ij} = \langle i|\underline{C}|j\rangle$. If $\underline{C}$ is real and symmetric, then it can be diagonalized with real orthonormal eigenfunctions $\langle i|p\rangle$ such that $\langle p|\underline{C}|p'\rangle = \delta_{p,p'}C_p$, where C_p is an eigenvalue that is necessarily real. We can therefore write

$$\begin{aligned}\int\left(\prod_{i=1}^{n} dy_i\right) e^{-\frac{1}{2}y_iC_{ij}y_j+\lambda_iy_j} &= \prod_{p=1}^{n}\int dy_p e^{-C_py_p^2/2}e^{\lambda_py_p} \\ &= \prod_{p=1}^{n}\left(\frac{2\pi}{C_p}\right)^{1/2} e^{\lambda_p^2/(2C_p)} \end{aligned}\tag{5.2.13}$$

$$= (2\pi)^{n/2}(\det \underline{C})^{-1/2}e^{\frac{1}{2}\lambda_i C_{ij}^{-1}\lambda_j}$$
$$= \exp[-\frac{1}{2}\text{Tr}\ln(\underline{C}/2\pi) + \frac{1}{2}\lambda_i C_{ij}^{-1}\lambda_j],$$

where the summation convention on repeated indices is understood and where $y_p = \sum_i \langle p|i\rangle y_i$ and $\lambda_p = \sum_i \langle p|i\rangle \lambda_i$. The Jacobian of the transformation from the variable y_i to y_p is the determinant of the matrix $\underline{M}$ with components $\langle p|i\rangle$. It is one because $|\det \underline{M}| = |\det \underline{M}^2|^{1/2}$ and $\langle p|\underline{M}^2|p'\rangle = \sum_i \langle p|i\rangle\langle i|p'\rangle = \delta_{p,p'}$. The eigenfunctions $\langle p|i\rangle$ can always be chosen to be real for a real symmetric matrix. It is often convenient, however, to choose them to be complex (e.g. plane waves $e^{i\mathbf{q}\cdot\mathbf{R}_\mathbf{l}}$ with periodic boundary conditions). Since the complex functions can always be expressed as linear combinations of real eigenfunctions, the final result of Eq. (5.2.13) remains unchanged.

The identity, Eq. (5.2.13), for multiple Gaussian integrals can be applied directly to the evaluation of the partition function of an harmonic lattice model with

$$\mathscr{H}_L^0 = \frac{1}{2}\sum_{\mathbf{l},\mathbf{l}'} r_{\mathbf{l},\mathbf{l}'}\phi_\mathbf{l}\phi_{\mathbf{l}'} = \frac{1}{2V}\sum_\mathbf{q} r(\mathbf{q})|\phi_\mathbf{q}|^2, \tag{5.2.14}$$

where

$$\phi_\mathbf{l} = \frac{1}{N v_0}\sum_\mathbf{q} e^{i\mathbf{q}\cdot\mathbf{R}_\mathbf{l}}\phi_\mathbf{q} \tag{5.2.15}$$

and

$$r(\mathbf{q}) = \frac{1}{v_0}\sum_\mathbf{l} e^{-i\mathbf{q}\cdot\mathbf{R}_\mathbf{l}} r_{\mathbf{l},0}. \tag{5.2.16}$$

The convention for Fourier transformations used here is the same as that discussed in the appendix to Chapter 2 and is the one most suited for taking the continuum limit $v_0 \to 0$. An alternative convention in which the factors of v_0 do not appear is often more convenient for lattice models. In Eq. (5.2.14), $r_{\mathbf{l},\mathbf{l}'}$ are the components of a matrix $\underline{r}$. The orthonormal eigenfunctions diagonalizing $\underline{r}$ are the functions $\langle \mathbf{l}|\mathbf{q}\rangle = N^{-1/2}e^{i\mathbf{q}\cdot\mathbf{R}_\mathbf{l}}$. The eigenvalues of $\underline{r}$ are

$$\langle \mathbf{q}|\underline{r}|\mathbf{q}\rangle = \frac{1}{N}\sum_{\mathbf{l},\mathbf{l}'} e^{i\mathbf{q}\cdot(\mathbf{R}_\mathbf{l}-\mathbf{R}_{\mathbf{l}'})} r_{\mathbf{l},\mathbf{l}'} = v_0 r(\mathbf{q}). \tag{5.2.17}$$

Thus,

$$\begin{aligned} \mathscr{A}[T,h_\mathbf{l}] &= -T\ln Z = -T\ln\left(\prod_\mathbf{l}\int d\phi_\mathbf{l} e^{\beta(\mathscr{H}_L^0+\mathscr{H}_{\text{ext}})}\right) \\ &= \frac{1}{2}T\sum_\mathbf{q}\ln[\beta r(\mathbf{q})v_0/(2\pi)] - \frac{1}{2}\sum_{\mathbf{l},\mathbf{l}'} h_\mathbf{l}\beta G_{\mathbf{l},\mathbf{l}'}^0 h_{\mathbf{l}'}, \end{aligned} \tag{5.2.18}$$

where $\mathscr{H}_{\text{ext}} = -\sum h_\mathbf{l}\phi_\mathbf{l}$ and

$$G_{\mathbf{l},\mathbf{l}'}^0 = \int \frac{d^dq}{(2\pi)^d} e^{i\mathbf{q}\cdot(\mathbf{R}_\mathbf{l}-\mathbf{R}_{\mathbf{l}'})}\frac{T}{r(\mathbf{q})} = \langle\phi_\mathbf{l}\phi_{\mathbf{l}'}\rangle. \tag{5.2.19}$$

The free energy of a continuum model with

$$\mathscr{H}_0 = \frac{1}{2}\int d^dx d^dx' r(\mathbf{x},\mathbf{x}')\phi(\mathbf{x})\phi(\mathbf{x}') = \frac{1}{2V}\sum_{\mathbf{q}} r(\mathbf{q})|\phi_{\mathbf{q}}|^2 \qquad (5.2.20)$$

is merely the continuum limit of Eq. (5.2.18),

$$\begin{aligned}\mathscr{A}[T,h(\mathbf{x})] &= \frac{1}{2}TV\int \frac{d^dq}{(2\pi)^d}\ln[\beta r(\mathbf{q})v_0/(2\pi)] \\ &\quad -\frac{1}{2}\int d^dx d^dx' h(\mathbf{x})\beta G_0(\mathbf{x},\mathbf{x}')h(\mathbf{x}')\end{aligned} \qquad (5.2.21)$$

with $\beta r(\mathbf{q}) = G_0^{-1}(\mathbf{q})$ and

$$G_0(\mathbf{x},\mathbf{x}') = \int \frac{d^dq}{(2\pi)^d} e^{i\mathbf{q}\cdot(\mathbf{x}-\mathbf{x}')}\frac{T}{r(\mathbf{q})} = \langle\phi(\mathbf{x})\phi(\mathbf{x}')\rangle. \qquad (5.2.22)$$

With the Fourier transform convention of Eq. (5.2.16), $r(\mathbf{q})$ is the same in both the lattice and continuum models. Note that the free energy in the above equation is formally infinite in the limit $v_0 \to 0$ (assuming $r(\mathbf{q})$ remains finite in this limit as it must). This infinity has no physical significance and can be removed merely by redefining the continuum trace operation. The cell volume v_0 can be related to the cutoff Λ of the continuum theory. As discussed in Appendix 2A, the number of wave vectors $\mathbf{q}$ must be equal to the number of lattice points in a discrete theory: $\sum_{\mathbf{q}} 1 = N = V\int d^dq/(2\pi)^d$. Thus, if there is a hard spherical cutoff restricting $|\mathbf{q}|$ to be less than Λ, then $v_0 = V/N = d\Lambda^{-d}/K_d$ where $K_d = \Omega_d/(2\pi)^d$ with Ω_d the solid angle subtended by a d-dimensional sphere. In the continuum model, N has no meaning, but Λ does, and by the above relation, we can assign meaning to v_0. In two and three dimensions, v_0 is, respectively, $4\pi\Lambda^{-2}$ and $6\pi^2\Lambda^{-3}$.

It is worth noting that Eq. (5.2.22) is a restatement of the equipartition theorem for classical harmonic Hamiltonians:

$$\begin{aligned}\langle\phi(\mathbf{q}_1)\phi(\mathbf{q}_2)\rangle &= \frac{1}{Z}\int \mathscr{D}\phi(\mathbf{x})\exp[-\frac{1}{2VT}\sum_{\mathbf{q}} r(\mathbf{q})|\phi(\mathbf{q})|^2]\phi(\mathbf{q}_1)\phi(\mathbf{q}_2) \\ &= \delta_{\mathbf{q}_1,-\mathbf{q}_2}\frac{VT}{r(\mathbf{q}_1)}\end{aligned} \qquad (5.2.23)$$

Inverse Fourier transforming this equation leads to Eq. (5.2.22).

4 *Mean-field theory from functional integrals*

Eqs. (5.2.1) and (5.2.4) define a field theory from which the thermodynamic potential $\mathscr{A}[T,h(\mathbf{x})]$ and its conjugate potential $F[\langle\phi(\mathbf{x})\rangle]$ can be calculated. All possible configurations in functional space contribute to Z. As discussed above, some configurations contribute more to Z than others. The configuration contributing the most to Z is the one minimizing $\beta[\mathscr{H} - \int d^dx h(\mathbf{x})\phi(\mathbf{x})]$, i.e., the functional integral is dominated by the saddle point path along which $\phi(\mathbf{x}) = \phi_{\text{sad}}(\mathbf{x})$ determined by

$$\left.\frac{\delta\mathscr{H}}{\delta\phi(\mathbf{x})}\right|_{\phi(\mathbf{x})=\phi_{\text{sad}}(\mathbf{x})} = h(\mathbf{x}). \qquad (5.2.24)$$

Mean-field theory consists of approximating Z by its contribution from the saddle point path only. Thus the mean-field approximation for the partition function is

$$\begin{aligned} Z_{MF} &= \exp\{-\beta(\mathscr{H}[\phi_{\text{sad}}(\mathbf{x})] - \int d^d x h(\mathbf{x})\phi_{\text{sad}}(\mathbf{x}))\} \\ &= \exp\{-\beta(F_{MF}[\langle\phi(\mathbf{x})\rangle] - \int d^d x h(\mathbf{x})\langle\phi(\mathbf{x})\rangle)\}, \end{aligned} \tag{5.2.25}$$

where F_{MF} is the mean-field free energy. It is clear from Eqs. (5.2.24) and (5.2.25) that in mean-field theory $\langle\phi(\mathbf{x})\rangle = \phi_{\text{sad}}(\mathbf{x})$ and that $F_{MF} = \mathscr{H}[\phi_{\text{sad}}(\mathbf{x})]$.

Corrections to mean-field theory can be studied by expanding $\mathscr{H}$ in powers of

$$\delta\phi(\mathbf{x}) = \phi(\mathbf{x}) - \langle\phi(\mathbf{x})\rangle. \tag{5.2.26}$$

By definition, $\langle\delta\phi(\mathbf{x})\rangle = 0$. As we have just seen, $\langle\phi(\mathbf{x})\rangle = \phi_{\text{sad}}(\mathbf{x})$ in mean-field theory. In general, fluctuations will cause $\langle\phi(\mathbf{x})\rangle$ to differ from $\phi_{\text{sad}}(\mathbf{x})$ when $\langle\phi(\mathbf{x})\rangle$ is nonzero in ordered phases or in the presence of external fields. There are well controlled ways of calculating $\langle\phi(\mathbf{x})\rangle$ which are beyond the scope of this book. Gaussian fluctuations about the saddle solution can, however, be calculated using the material presented in this section. Expanding to second order in $\delta\phi(\mathbf{x})$, we find

$$\mathscr{H} - \int d^d x h(\mathbf{x})\phi(\mathbf{x}) = \mathscr{H}(\langle\phi(\mathbf{x})\rangle) - \int d^d x h(\mathbf{x})\langle\phi(\mathbf{x})\rangle + \mathscr{H}', \tag{5.2.27}$$

where

$$\beta\mathscr{H}' = \frac{1}{2}\int d^d x d^d x' \delta\phi(\mathbf{x}) G_0^{-1}(\mathbf{x},\mathbf{x}')\delta\phi(\mathbf{x}') \tag{5.2.28}$$

is the harmonic correction to the saddle point Hamiltonian and

$$\begin{aligned} G_0^{-1}(\mathbf{x},\mathbf{x}') &= \left.\frac{\delta\beta\mathscr{H}}{\delta\phi(\mathbf{x})\delta\phi(\mathbf{x}')}\right|_{\phi(\mathbf{x})=\langle\phi(\mathbf{x})\rangle} \\ &= \beta[r + 12u\langle\phi(\mathbf{x})\rangle^2 - c\nabla^2]\delta(\mathbf{x}-\mathbf{x}'), \end{aligned} \tag{5.2.29}$$

is the inverse of the mean-field correlation function, where the final form applies only to a ϕ^4-theory. Thus, harmonic fluctuations about the mean-field or saddle point solution are controlled by the mean-field order parameter correlation function. The lowest order fluctuation corrections to the mean-field free energy are obtained by evaluating the Gaussian integral over $\delta\phi(\mathbf{x})$, leading to

$$\begin{aligned} F - F_{MF} &= \frac{1}{2}TV\int\frac{d^d q}{(2\pi)^d}\ln[G_0^{-1}(\mathbf{q})v_0/(2\pi)] \\ &= \frac{1}{2}T\,\mathrm{Tr}\ln[G_0^{-1}v_0/(2\pi)]. \end{aligned} \tag{5.2.30}$$

The trace in this expression is the logical extension to continuous systems of the trace operation in discrete systems. The trace of a matrix $A(\mathbf{x},\mathbf{x}')$ is simply $\mathrm{Tr}A = \int d^d x A(\mathbf{x},\mathbf{x})$. If A diagonalizes under Fourier transformation, then $\mathrm{Tr}A = \sum_{\mathbf{q}} A(\mathbf{q})$. Eq. (5.2.30) is often called the one-loop approximation for the free energy. It depends on $\langle\phi(\mathbf{x})\rangle$ via the dependence of G_0 on $\langle\phi(\mathbf{x})\rangle$, and can be expanded in a power series in $\langle\phi(\mathbf{x})\rangle$. The second-order term in this expansion gives the one-loop correction to the inverse correlation function, $G^{-1}(\mathbf{x},\mathbf{x}')$, which in the disordered phase with $\langle\phi(\mathbf{x})\rangle = 0$ is

$$\begin{aligned} G^{-1}(\mathbf{x},\mathbf{x}') &= \frac{\delta\beta F}{\delta\langle\phi(\mathbf{x})\rangle\delta\langle\phi(\mathbf{x}')\rangle} \\ &= G_0^{-1}(\mathbf{x},\mathbf{x}') + \frac{1}{2}\mathrm{Tr}G_0\frac{\delta G_0^{-1}}{\delta\langle\phi(\mathbf{x})\rangle\delta\langle\phi(\mathbf{x}')\rangle} \\ &= G_0^{-1}(\mathbf{x},\mathbf{x}') + \frac{1}{2}\int d^dx_1 d^dx_2 G_0(\mathbf{x}_1,\mathbf{x}_2)\frac{\delta G_0^{-1}(\mathbf{x}_2,\mathbf{x}_1)}{\delta\langle\phi(\mathbf{x})\rangle\delta\langle\phi(\mathbf{x}')\rangle}, \end{aligned} \tag{5.2.31}$$

where the second two forms apply only to the disordered phase.

For a ϕ^4-theory, this reduces to

$$G^{-1}(\mathbf{x},\mathbf{x}') = G_0^{-1}(\mathbf{x},\mathbf{x}') + 12u\delta(\mathbf{x}-\mathbf{x}')G_0(\mathbf{x},\mathbf{x}), \tag{5.2.32}$$

which in the disordered phase when $\langle\phi(\mathbf{x})\rangle = 0$ is

$$TG^{-1}(\mathbf{q},r) = r + cq^2 + 12u\int\frac{d^dq}{(2\pi)^d}\frac{T}{r+cq^2}. \tag{5.2.33}$$

One-loop corrections to fourth- and higher order terms of the expansion of F in powers of $\langle\phi(\mathbf{x})\rangle$ can be calculated in a similar way.

5 Breakdown of mean-field theory revisited

In the preceding section, we showed how fluctuations can lead to a breakdown of mean-field theory. We have just seen that the replacement of Z by its contribution from the most probable path with a spatially uniform order parameter is equivalent to mean-field theory. We will now show that the criterion for the breakdown of this approximation is exactly the same as the Ginzburg criterion discussed in the previous section.

When $h(\mathbf{x})$ is spatially uniform, the saddle point path yielding mean-field theory is one with a spatially uniform $\langle\phi(\mathbf{x})\rangle$. This is a good approximation so long as the contribution of spatially non-uniform paths to Z is unimportant. This is the case when the energy, $\mathcal{H}_{n.u.} = \frac{1}{2}c\int d^dx(\nabla\phi)^2$, associated with non-uniform paths is large compared to T. We can estimate the magnitude of $\mathcal{H}_{n.u.}$ by noting that $\langle\phi\rangle^2 = |r|/4u$ at the saddle point and that spatial correlations only extend up to a distance of order $\xi = (c/|r|)^{-1/2}$. We can, therefore, estimate

$$\mathcal{H}_{n.u.} = \frac{1}{2}c\int d^dx(\nabla\phi)^2 \sim \frac{1}{2}c\xi^{d-2}\langle\phi\rangle^2 = \frac{c^2}{8u}\xi^{d-4}. \tag{5.2.34}$$

In order for mean-field theory to be valid, $F_{n.u.}$ must be greater than T, or

$$\xi^{d-4} > \frac{8u}{c^2}T = \frac{1}{2\xi_0^4\Delta c_V}, \text{ or } \left(\frac{\xi}{\xi_0}\right)^{d-4} > \frac{1}{2\xi_0^d\Delta c_V}. \tag{5.2.35}$$

This is identical to Eq. (5.1.4).

The Ginzburg temperature also indicates when perturbation theory will break down. Consider the one-loop expression for

$$\frac{dTG^{-1}(\mathbf{q},r)}{dr} = 1 - 12uT \int \frac{d^dq}{(2\pi)^d} \frac{1}{(r+cq^2)^2} \tag{5.2.36}$$

obtained from Eq. (5.2.33). The perturbation term in this expression proportional to u diverges as $(uT/c^2)\xi^{4-d}$ for $d < 4$ and becomes comparable to unity at the Ginzburg temperature T_G. Thus, for $T > T_G$, a perturbation expansion in u (or the number of loops) presents no problems. For $T < T_G$, low-order terms in perturbation theory become divergent, and simple perturbation theory will not work. The renormalization group, which we will discuss later in this chapter, maps a critical problem with $T < T_G$ to a non-critical problem with $T > T_G$, where perturbation theory can be used.

5.3 The self-consistent field approximation

We have just seen that mean-field theory breaks down below an upper critical dimension $d_c = 4$. One can see explicitly how this happens in a simple treatment of the ϕ^4-field theory. This approximation is variously called the self-consistent field, Hartree or random phase approximation (RPA). It becomes exact (as will be shown at the end of this section) for n-component fields in the limit $n \to \infty$. It consists of replacing one factor of ϕ^2 in the ϕ^4 term in f_{SL} by its average, $\langle\phi^2\rangle$, to be determined self-consistently. There are six ways of choosing the two factors of ϕ to be paired in $\langle\phi^2\rangle$ from the four factors of ϕ in ϕ^4, and the harmonic free energy of the RPA becomes $\frac{1}{2}r\phi^2 + 6u\langle\phi^2\rangle\phi^2$. In an n-component theory, the factor of 6 becomes $2(n+2)$. The correlation function $G(\mathbf{q})$ can be evaluated using the equipartition theorem as before [Eq. (5.2.22)]:

$$TG^{-1}(\mathbf{q}) = r + q^2 + 12u\langle\phi^2\rangle, \tag{5.3.1}$$

where as before $r = a(T - T^*)$. For T greater than the as yet to be determined transition temperature T_c,

$$\begin{aligned} \langle\phi^2\rangle \equiv G(\mathbf{x},\mathbf{x}) &= \int \frac{d^dq}{(2\pi)^d} G(\mathbf{q}) \\ &= \int \frac{d^dq}{(2\pi)^d} \frac{T}{r + cq^2 + 12u\langle\phi^2\rangle}. \end{aligned} \tag{5.3.2}$$

Eq. (5.3.2) provides a self-consistent equation for $\langle\phi^2\rangle$.

Before evaluating the integral in Eq. (5.3.2), we recall that the expansion of $G(\mathbf{q})$ to first order in q^2 is only valid for small q (long wavelength). For larger q, higher powers in the expansion are needed. Alternatively (see Sec. 4.3), one can say that the q^2 form is valid up to some cutoff Λ of order the inverse bare correlation length, $\xi_0 = (c/aT_c)^{1/2}$. In this case, the integral in Eq. (5.3.2) is restricted to a sphere of radius Λ. Using Eqs. (5.3.1) and (5.3.2), we can now calculate the temperature dependence of $\chi^{-1} = TG^{-1}(\mathbf{q} = 0)$:

$$\begin{aligned}\chi^{-1} \equiv \tau &= r + 12uT \int_0^\Lambda \frac{d^d q}{(2\pi)^d} \frac{1}{\tau + cq^2} \\ &= r + 12uTK_d \int_0^\Lambda q^{d-1} dq(\tau + cq^2)^{-1},\end{aligned} \tag{5.3.3}$$

where

$$K_d = \Omega_d/(2\pi)^d, \tag{5.3.4}$$

and Ω_d is the solid angle subtended by a sphere in d dimensions. At the actual transition temperature T_c, the susceptibility diverges: $\chi^{-1} = \tau = 0$. This allows us to determine the critical temperature for $d \geq 2$ via

$$r_c \equiv a(T_c - T^*) = -\frac{12uT_c}{c} \int_0^\Lambda \frac{d^d q}{q^2} = -\frac{12uT_c}{c} \frac{K_d \Lambda^{d-2}}{d-2}. \tag{5.3.5}$$

This gives

$$T_c = \left(1 + \frac{12uK_d \Lambda^{d-2}}{(d-2)ca}\right)^{-1} T^*. \tag{5.3.6}$$

The transition temperature is depressed below the mean-field limit of metastability. Note also that $T_c \to 0$ as $d \to 2$. This is the *lower critical dimension* d_L at which critical fluctuations become so violent that no phase transition at nonzero temperature is possible. For all $d < d_L$, there is no transition at all. We will encounter d_L again in the next chapter.

We now express χ^{-1} as a function of $r - r_c = a(T - T_c)$:

$$\begin{aligned}\chi^{-1} \equiv \tau &= r - r_c + 12uK_d \int q^{d-1} dq \left(\frac{T}{\tau + cq^2} - \frac{T_c}{cq^2}\right) \\ &= (a'/a)(r - r_c) - 12uT_c(K_d/c)\tau I_d(\tau),\end{aligned} \tag{5.3.7}$$

where we dropped a term of order $(T - T_c)\tau I_d(\tau)$, $(a'/a) = 1 + 12uT_c(K_d/ac)(\Lambda^{d-2}/(d-2))$, and

$$I_d(\tau) = \int_0^\Lambda q^{d-1} dq \frac{1}{q^2(\tau + cq^2)} = \int_0^\Lambda \frac{q^{d-3} dq}{\tau + cq^2}. \tag{5.3.8}$$

For $d > 4$, $I_d(\tau)$ is analytic and tends to $I_d(0) = \Lambda^{d-4}/[c(d-4)]$ as $\tau \to 0$, and τ is a linear function of small $r - r_c$. Solving for τ in terms of $r - r_c$ for small $T - T_c$, we obtain

$$\chi = \frac{1 + (12uT_c K_d/c) I_d(0)}{a'(T - T_c)} \sim (T - T_c)^{-1}. \tag{5.3.9}$$

The critical exponent $\gamma = 1$ is thus the same as in mean-field theory. The only modification in χ for $d > d_c = 4$ is in the value of T_c and in the prefactor of $(T - T_c)^{-1}$.

For $d < 4$, $I_d(\tau)$ is divergent at $\tau = 0$:

$$\begin{aligned}I_d &= c^{-(d-2)/2} \tau^{(d-4)/2} \int_0^{\Lambda(c/\tau)^{1/2}} \frac{y^{d-3} dy}{1 + y^2} \\ &\overset{\tau \to 0}{\to} c^{-(d-2)/2} \tau^{-\epsilon/2} B_d,\end{aligned} \tag{5.3.10}$$

where $\epsilon \equiv (4-d)$ and $B_d = \Gamma[(d-2)/2]\Gamma[(4-d)/2]/2$, $\Gamma(x)$ being the gamma function. Thus, if $\tau^{-\epsilon/2}$ is small, the term involving $I_d(\tau)$ in Eq. (5.3.7) can be neglected and $\tau \sim (r - r_c)$. This implies a mean-field form for χ for $T > T_G$, where

$$T_G = T_c + a^{-1}[12B_d K_d u T_c c^{-d/2}]^{2/(4-d)}, \tag{5.3.11}$$

in agreement with the Ginzburg temperature [Eq. (5.1.5)] calculated previously. For $T < T_G$, $\tau I_d(\tau)$ dominates in Eq. (5.3.7), and we have

$$r - r_c = 12u(K_d/c)c^{-(d-2)/2}B_d\tau^{(d-2)/2}, \tag{5.3.12}$$

$$\chi = \tau^{-1} \sim (r-r_c)^{-\gamma}; \quad \gamma = \frac{2}{(d-2)}. \tag{5.3.13}$$

Note that the susceptibility exponent $\gamma \to 1$ as $d \to 4$ as required.

In this approximation, there are no **q**-dependent corrections to the mean-field $\chi(\mathbf{q})$. Therefore,

$$\chi(\mathbf{q}) = (\tau + cq^2)^{-1} = \chi[1 + (q\xi)^2]^{-1}, \tag{5.3.14}$$

where

$$\xi^2 = (c/\tau) \sim |T - T_c|^{-2\nu}. \tag{5.3.15}$$

Thus, the correlation length exponent is

$$\nu = \frac{1}{2}\gamma = \frac{1}{(d-2)}, \tag{5.3.16}$$

which again reduces to its mean-field value of 1/2 when $d = 4$.

The crossover of the correlation length from its mean-field to its critical form is of some interest. From Eqs. (5.3.14), (5.3.7), (5.3.10), and (5.1.7), we find

$$\xi^2 = \xi_{MF}^2[1 + g(\xi/\xi_G)^\epsilon], \tag{5.3.17}$$

where $g = 12K_d B_d$, ξ_G is the Ginzburg length, and

$$\xi_{MF}^2 = \frac{c}{a|T - T_c|}. \tag{5.3.18}$$

These equations imply

$$\xi \sim \begin{cases} \xi_{MF}, & \text{if } \xi \ll \xi_G; \\ \xi_{MF}^{2/(d-2)}\xi_G^{-(4-d)/(d-2)} & \text{if } \xi \gg \xi_G. \end{cases} \tag{5.3.19}$$

Note that when $\xi_{MF} \sim \xi_G$, $\xi \sim \xi_G$.

This simple approximation shows the major effects of fluctuation corrections to mean-field theory. First, fluctuations in general reduce the transition temperature below its mean-field value. Secondly, they lead to critical exponents that differ from those of mean-field theory for physical dimension less than the upper critical dimension d_c. For $d < d_c$, there is a crossover from mean-field to critical behavior for reduced temperature t less than the Ginzburg temperature t_G.

1 The n-vector model in the limit $n \to \infty$

The self-consistent field solution just presented becomes exact in the $n \to \infty$ limit of the n-vector model. To see this, consider the n-vector Hamiltonian

$$\mathscr{H} = \frac{1}{2}\int d^d x r\boldsymbol{\phi}\cdot\boldsymbol{\phi} + c[\nabla\boldsymbol{\phi}]^2 + n^{-1}u\int d^d x(\boldsymbol{\phi}\cdot\boldsymbol{\phi})^2, \tag{5.3.20}$$

where $\boldsymbol{\phi} = (\phi_1, ..., \phi_n)$ is an n-component vector field and where the quartic term is explicitly proportional to n^{-1} to produce a meaningful $n \to \infty$ limit. The exponential of the quartic term in $\mathscr{H}$ can be expressed as the exponential of a quadratic term with a fluctuating coefficient with the aid of the Gaussian identity in Eq. (5.2.13):

$$\exp[-\beta u(\boldsymbol{\phi}\cdot\boldsymbol{\phi})^2/n] = C\int_{-i\infty}^{i\infty}\mathscr{D}\psi(\mathbf{x})\exp\left[\frac{1}{16u}n\beta\psi^2(\mathbf{x}) - \frac{1}{2}\beta\psi(\boldsymbol{\phi}\cdot\boldsymbol{\phi})\right], \tag{5.3.21}$$

where C is an unimportant constant. In this expression, the integration contour is along the imaginary rather than the real axis because of the negative sign in the argument of the exponential on the right hand side. The partition function can now be written as

$$\begin{aligned} Z &= \int\prod_i^n \mathscr{D}\phi_i(\mathbf{x})\mathscr{D}\psi(\mathbf{x})\exp\left[n\frac{\beta}{16u}\int d^d x\psi^2\right] \\ &\quad\times\exp\left[-\frac{\beta}{2}\int d^d x[(r+\psi(\mathbf{x}))\boldsymbol{\phi}\cdot\boldsymbol{\phi} + c(\nabla\boldsymbol{\phi})^2]\right] \qquad (5.3.22) \\ &= \int\mathscr{D}\psi(\mathbf{x})\exp\left[n\frac{\beta}{16u}\int d^d x\psi^2 - n\Phi[\psi(\mathbf{x})]\right], \qquad (5.3.23) \end{aligned}$$

where

$$\exp(-n\Phi[\psi(\mathbf{x})]) = \left(\int d\phi(\mathbf{x})\exp\left[-\frac{\beta}{2}\int d^d x[(r+\psi(\mathbf{x}))\phi^2(\mathbf{x}) + c(\nabla\phi)^2]\right]\right)^n. \tag{5.3.24}$$

The integral on the right hand side of this equation is n identical replicas of the same integral, and is, therefore, the nth power of an integral over a scalar variable. Thus, the argument of the exponential on the left hand side of Eq. (5.3.23) is proportional to n as indicated. The integral in Eq. (5.3.24) can be evaluated formally using Eq. (5.2.13) leading to

$$\Phi[\psi(\mathbf{x})] = \frac{1}{2}\mathrm{Tr}\ln[\beta G^{-1}(\mathbf{x},\mathbf{x}')/(2\pi)], \tag{5.3.25}$$

where

$$G^{-1}(\mathbf{x},\mathbf{x}') = [r+\psi(\mathbf{x})]\delta(\mathbf{x}-\mathbf{x}') - c\nabla^2\delta(\mathbf{x}-\mathbf{x}'). \tag{5.3.26}$$

The saddle point approximation to the partition function in Eq. (5.3.23) is *exact* in the limit $n \to \infty$ and yields

$$\psi = 4u\int\frac{d^d q}{(2\pi)^d}\frac{T}{r+\psi+cq^2} \tag{5.3.27}$$

and

$$TG^{-1}(\mathbf{q}) = r + \psi + cq^2. \tag{5.3.28}$$

Apart from the difference in prefactors of u, these two equations are identical to (5.3.1) and (5.3.2).

5.4 Critical exponents, universality, and scaling

Critical exponents for most experimental second-order transitions differ from those predicted by mean-field theory. Table 5.4.1 reviews the definitions of critical exponents and critical amplitudes defining the strength of leading singularities. Note that the definition of the critical amplitudes A and A' for the specific heat are multiplied by a factor of $1/\alpha$. This is to permit a description of situations in which the exponent α is either positive, negative, or zero. A representative sample of transitions and associated exponents are listed in Table 5.4.2. It is remarkable that there is very little variation in the critical exponents between systems for fixed spatial dimension d. In three dimensions, β is of order 1/3, γ of order 4/3, ν of order 2/3, and the specific heat exponent α of order zero. There is, however, a substantial difference between exponents in two- and three-dimensional systems. The renormalization group introduced by Kenneth Wilson in the early 1970s provided a method for calculating exponents and established that they should depend on the spatial dimension, the symmetry of the order parameter, and the symmetry and range of interactions, but not on the detailed form and magnitude of interactions. Thus, there are *universality classes,* and all transitions in the same universality class have the same critical exponents. For example, all transitions in which the order parameter has up-down symmetry ($n = 1$, Ising) should have the same exponents. Indeed, transitions such as the ferromagnetic transition in uniaxial magnets, the liquid-gas transition, and order-disorder transitions all have very nearly the same exponents as can be seen in Table 5.4.2. Similarly, experimentally determined critical exponents for different three-dimensional systems with O_3 symmetry have nearly the same exponents and again for different two-dimensional systems with $n = 1$ symmetry as can also be seen in Table 5.4.2.

1 Exponents and scaling relations

Another remarkable feature of second-order phase transitions is that not all of the critical exponents are independent. For example, γ is always of order 2ν and $\alpha + 2\beta + \gamma$ is of order 2. These relations are a result of the homogeneity or scaling properties of correlation functions and thermodynamic quantities near $T = T_c$ that can be derived using the renormalization group. This homogeneity is already present in the mean-field correlation function [see Eq. (4.3.17)],

Table 5.4.1. *Definition of critical exponents and amplitudes.*

Susceptibility	$\chi = \Gamma t^{-\gamma}$	$t > 0$
	$\chi = \Gamma'(-t)^{-\gamma'}$	$t < 0$
Specific heat	$C = \frac{A}{\alpha} t^{-\alpha}$	$t > 0$
	$C = \frac{A'}{\alpha'}(-t)^{-\alpha'}$	$t < 0$
Correlation length	$\xi = \xi_0 t^{-\nu}$	$t > 0$
	$\xi = \xi_0'(-t)^{-\nu'}$	$t < 0$
Order parameter	$\langle\phi\rangle = B(-t)^{\beta}$	$t < 0$
	$\langle\phi\rangle = D_c^{-1} h\lvert h\rvert^{(1-\delta)/\delta}$	$t = 0$
Correlation function	$G(q) = D_\infty q^{-(2-\eta)}$	$t = 0$

Table 5.4.2. *Some critical exponents from theory and experiment.*

Exponent	α	β	γ	ν	η
Property	specific heat	order parameter	susceptibility	coherence length	correlation function
Definition	$C \sim t^{-\alpha}$	$\langle\phi\rangle \sim t^{\beta}$	$\chi \sim t^{-\gamma}$	$\xi \sim t^{-\nu}$	$G(q) \sim q^{-2+\eta}$
Mean-field	0	0.5	1	0.5	0
3D theory					
$n = 0$ (SAW)	0.24	0.30	1.16	0.59	
$n = 1$ (Ising)	0.11	0.32	1.24	0.63	0.04
$n = 2$ (xy)	−0.01	0.35	1.32	0.67	0.04
$n = 3$ (Heisenberg)	−0.12	0.36	1.39	0.71	0.04
Experiment					
3D $n = 1$	$0.11^{+.01}_{-.03}$	$0.32^{+.16}_{-.04}$	$1.24^{.16}_{-.04}$	$0.63^{+.04}_{-.04}$	0.03 - 0.06
3D $n = 3$	$0.1^{+.05}_{-.04}$	$0.34^{+.04}_{-.04}$	$1.4^{+.07}_{-.07}$	$0.7^{+.03}_{-.03}$	
2D $n = 1$	$0.0^{+.01}_{-.003}$	$0.3^{+.04}_{-.04}$	$1.82^{+0.7}_{-.07}$	$1.02^{+.07}_{-.07}$	

Experiments on 3D $n = 1$ compiled from liquid-gas, binary fluid, ferromagnetic, and antiferromagnetic transitions.
Experiments on 3D $n = 3$ transitions compiled from some ferromagnetic and antiferromagnetic transitions.
Experiments on 2D $n = 1$ complied from some antiferromagnetic transitions.

$$G(\mathbf{x},0) \equiv G(|\mathbf{x}|) = |\mathbf{x}|^{-(d-2)} Y(|\mathbf{x}|/\xi). \tag{5.4.1}$$

The only length that appears in this function is ξ. The microscopic length is not important. At $T = T_c$, $\xi = \infty$, and $G \sim |\mathbf{x}|^{-(d-2)}$ is a homogeneous function (see Eq. (3.1.31) for the definition of a homogeneous function) of $|\mathbf{x}|$ in that

$$G(|\mathbf{x}|) = b^{-(d-2)} G(b^{-1}|\mathbf{x}|). \tag{5.4.2}$$

The susceptibility χ is the integral of $T^{-1}G(|\mathbf{x}|)$ over $\mathbf{x}$:

$$\begin{aligned} \chi &= T^{-1} \int d^d x G(|\mathbf{x}|) = T^{-1}\xi^2 \int d^d y y^{-(d-2)} Y(y) \\ &\sim \xi^2 \sim (T - T_c)^{-\gamma}. \end{aligned} \tag{5.4.3}$$

From this we see that the divergence of χ is due to the increasing spatial range of correlations in the order parameter field. Eq. (5.4.3) also says that $\gamma = 2\nu = 1$ since $\xi \sim |T - T_c|^{-\nu}$ with $\nu = 1/2$ in mean-field theory. This is, of course, the same value for γ obtained by direct calculation in Sec. 4.4. In this context, we see that γ is completely determined by ν and the way G falls off with $|\mathbf{x}|$ at $T = T_c$.

In critical systems, the behavior of $G(|\mathbf{x}|)$ at $T = T_c$ is characterized by an exponent η defined via

$$G(|\mathbf{x}|) \sim |\mathbf{x}|^{-(d-2+\eta)}. \tag{5.4.4}$$

This suggests a generalization of Eq. (5.4.1) to

$$G(|\mathbf{x}|) = |\mathbf{x}|^{-(d-2+\eta)} Y(|\mathbf{x}|/\xi). \tag{5.4.5}$$

In this form, $G(|\mathbf{x}|, t)$ $(t = (T - T_c)/T)$ satisfies the generalized homogeneity relation

$$G(|\mathbf{x}|, t) = b^{-(d-2+\eta)} G(b^{-1}|\mathbf{x}|, b^{1/\nu} t), \tag{5.4.6}$$

and $\chi(\mathbf{q}, t) = T^{-1}G(\mathbf{q}, t)$ satisfies the relation

$$\chi(\mathbf{q}, t) = b^{2-\eta} \chi(b\mathbf{q}, b^{1/\nu} t). \tag{5.4.7}$$

This equation is valid for all b. When $t = 0$, we can choose $b = |\mathbf{q}|^{-1} \equiv q^{-1}$ to obtain

$$\chi(\mathbf{q}, t = 0) \sim q^{-(2-\eta)} \tag{5.4.8}$$

since rotational invariance implies that χ should not depend on the direction of $\mathbf{q}$. Alternatively, we can choose $b = |t|^{-\nu}$. Then

$$\chi(\mathbf{q}, t) = |t|^{-\gamma} \chi(\mathbf{q}|t|^{-\nu}, t/|t|), \tag{5.4.9}$$

where

$$\gamma = (2 - \eta)\nu \tag{5.4.10}$$

is the susceptibility exponent that describes the divergence of $\chi(t) = \chi(\mathbf{q} = 0, t)$. This result could also have been obtained by integrating Eq. (5.4.5) for $G(\mathbf{x})$ over $\mathbf{x}$ to yield $\chi \sim \xi^{(2-\eta)} \sim |t|^{-\gamma}$.

Note that $\chi(\mathbf{q}, t)$ in Eq. (5.4.9) depends on $\mathbf{q}|t|^{-\nu}$ and on $t/|t| = \pm 1$. Thus, the scaling function for $\chi(\mathbf{q}, t)$ can be different for $t > 0$ and $t < 0$:

$$\chi(\mathbf{q}, t) = |t|^{-\gamma} X_2(q|t|^{-\nu}), \tag{5.4.11}$$

where $X_2(u) = \chi(\mathbf{q}u/q, \pm 1)$. When $q\xi = 0$, X_2 depends only on the sign of t. In commonly accepted notation, coefficients for $t < 0$ are denoted with primes and those for $t > 0$ without so that $X_2(0) = \Gamma$ for $t > 0$ and $X_2(0) = \Gamma'$ for $t < 0$, as indicated in Table 5.4.1. More generally, as already noted, $X_2(u) = X_2^+(u)$ for $t > 0$ and $X_2(u) = X_2^-(u)$ for $t < 0$. At $t = 0$, $\chi(\mathbf{q}, t)$ is independent of t. This is compatible with Eq. (5.4.11) only if $X_2(u) \sim u^{-\gamma/\nu}$ as $u \to \infty$. Thus,

$$X_2(u) \to \begin{cases} \Gamma, & \text{as } u \to 0\ , t > 0; \\ D_\infty u^{-\gamma/\nu}, & \text{as } u \to \infty; \\ \Gamma', & \text{as } u \to 0\ , t < 0 \end{cases} \tag{5.4.12}$$

summarizes the limiting forms of $X_2(u)$.

The scaling equation for $G(\mathbf{x}, t)$ [Eq. (5.4.6)] can be used to determine the order parameter exponent β in terms of ν and η. Since $\bar{G}(|\mathbf{x}|) = \langle\phi(\mathbf{x})\phi(0)\rangle$ and $\bar{G}(|\mathbf{x}|) = G(|\mathbf{x}|) + \langle\phi(\mathbf{x})\rangle\langle\phi(0)\rangle$ are identical for $T > T_c$, we expect $\bar{G}$ to obey the same scaling law as G. For $T < T_c$, $\langle\phi(\mathbf{x})\rangle$ is nonzero, and from Eq. (3.5.8)

$$\lim_{|\mathbf{x}|\to\infty} \bar{G}(|\mathbf{x}|) \to \langle\phi\rangle^2 \sim |T - T_c|^{\beta}. \tag{5.4.13}$$

The choice $b = |t|^{-\nu}$ in Eq. (5.4.6) leads to

$$\lim_{|\mathbf{x}|\to\infty} G(|\mathbf{x}|, t) \to |t|^{(d-2+\eta)\nu}, \tag{5.4.14}$$

which is independent of $\mathbf{x}$. Thus, comparing Eqs. (5.4.14) and (5.4.13), we obtain

$$\beta = \frac{1}{2}(d - 2 + \eta)\nu. \tag{5.4.15}$$

This relation involves the spatial dimension d. Relations involving d are called *hyperscaling relations*. Note that the mean-field exponents satisfy this relation only at the upper critical dimension $d_c = 4$. Combining Eqs. (5.4.10) and (5.4.15), we find

$$\gamma + 2\beta = d\nu. \tag{5.4.16}$$

Again this equation is satisfied by the mean-field exponents at the upper critical dimension.†

The existence of a single length scale allows the specific heat exponent α to be determined. The specific heat c_V is the second derivative of the free energy density f with respect to temperature and diverges as $|t|^{-\alpha}$ as indicated in Table 5.4.1. The free energy density has units of energy divided by volume. An energy scale is provided by the transition temperature T_c, and a volume is provided by ξ^d. Thus we estimate

$$f \sim T_c\xi^{-d} \sim T_c\xi_0^{-d}t^{d\nu}, \tag{5.4.17}$$

and

$$c_V = -T\frac{\partial^2 f}{\partial T^2} \sim |T - T_c|^{d\nu-2}, \tag{5.4.18}$$

from which we conclude

† Above the critical dimension, hyperscaling relations such as Eqs. (5.4.13) and (5.4.15) break down because of the existence of a *dangerous irrelevant variable*. See Sec. 5.8 for details.

$$\alpha = 2 - d\nu. \tag{5.4.19}$$

Again this is a hyperscaling relation satisfied by the mean-field exponents only at d_c. Combining Eqs. (5.4.16) and (5.4.19), we obtain the relation

$$\gamma + 2\beta + \alpha = 2. \tag{5.4.20}$$

Hyperscaling applies quite generally to transitions that are fluctuation dominated (i.e., non-mean-field). When it applies, there are only two independent critical exponents, say η and ν, and the critical point is said to obey *two scale factor universality*.

2 Scaled equation of state

The correlation function $G(|\mathbf{x}|, t)$ obeys a generalized homogeneity relation in which both $|\mathbf{x}|$ and t are rescaled. It is predicted by the renormalization group and found experimentally that any thermodynamic function obeys a homogeneity relation in which t and external fields (such as h) rescale by different factors. The free energy density as a function of external field satisfies

$$f(t, h) = b^{-d} f(b^{1/\nu} t, b^{\lambda} h). \tag{5.4.21}$$

(The free energy density f is really the field-dependent free energy density $a = \mathscr{A}/V$ of Eq. (3.5.9). Here, we adopt the common practice of using the symbol f for free energy regardless of what its natural variables are.) When $h = 0$, this implies $f \sim \xi^{-d}$ as previously predicted. This equation allows us to calculate both the order parameter $\langle\phi\rangle$ and the susceptibility:

$$\langle\phi\rangle = \frac{\partial f}{\partial h} = b^{-d+\lambda} f'(b^{1/\nu} t, b^{\lambda} h), \tag{5.4.22}$$

$$\chi = \frac{\partial^2 f}{\partial h^2} = b^{-d+2\lambda} f''(b^{1/\nu} t, b^{\lambda} h). \tag{5.4.23}$$

With $b = t^{-\nu}$, this leads to

$$\beta = d\nu - \Delta, \quad \gamma = -d\nu + 2\Delta, \tag{5.4.24}$$

with

$$\Delta \equiv \lambda\nu = \beta + \gamma. \tag{5.4.25}$$

The exponent Δ is often called the *gap exponent*.

With the choice $b = |t|^{-\nu}$, Eqs. (5.4.21) to (5.4.23) can be reexpressed as

$$\begin{aligned} f(t,h) &= |t|^{2-\alpha} X_0(h/t^{\Delta}), \\ \langle\phi\rangle &= |t|^{\beta} X_1(h/|t|^{\Delta}), \\ \chi &= |t|^{-\gamma} X_2(h/|t|^{\Delta}), \end{aligned} \tag{5.4.26}$$

where each X_i is a scaling function whose form depends in general on the sign of t. This scaling relation for the equation of state has been verified for some systems in great detail. Fig. 5.4.1 shows data for the Heisenberg ferromagnet EuO. The magnetization divided by $|t|^{\beta}$ is a function of $h/|t|^{\Delta}$ and not of h and t separately. The scaling function $X_1(u)$ is different for $t < 0$ and $t > 0$.

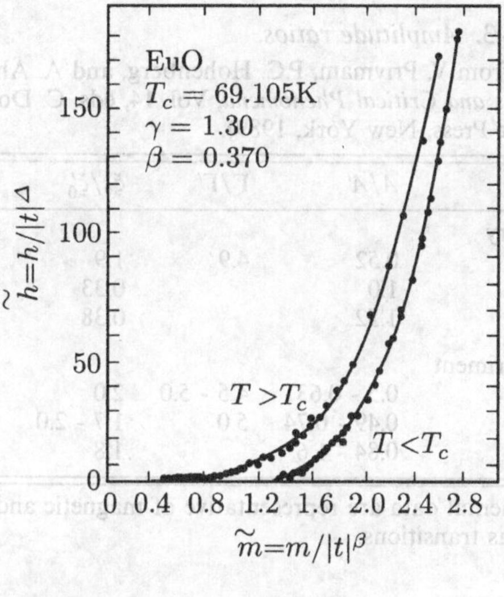

Fig. 5.4.1. Experimental equation of state near the ferromagnetic critical point of EuO [Cheng-Cher Huang and John T. Ho, *Phys. Rev. B* **12**, 5255 (1975)].

3 Multicritical points

As we saw in Sec. 4.6, multicritical points are reached by fixing at least one non-ordering field in addition to the temperature. This implies that there must be at least one additional field to describe scaling in the vicinity of a multicritical point. For example, the scaling relations of Eq. (5.4.26) must be modified in the vicinity of the bicritical and tetracritical points discussed in Sec. 4.6 to include the field g. The scaling relation for the susceptibility becomes

$$\begin{aligned} \chi(t,h,g) &= b^{-d+2\lambda} f''(b^{1/\nu} t, b^{\lambda} h, b^{\lambda_g} g) \\ &= |t|^{-\gamma} X_2(h/|t|^{\Delta}, g/|t|^{\phi}), \end{aligned} \tag{5.4.27}$$

where

$$\phi = \lambda_g \nu. \tag{5.4.28}$$

In the simple model of Eq. (4.6.18), two fields, ϕ_1 and ϕ_2, are critical at the bicritical or tetracritical point P, $t = h = g = 0$, in Fig. 4.6.10. Thus, the transition at P when $g = 0$ has $n = 2$ symmetry, whereas the transitions along the lambda lines with $g \neq 0$ have Ising symmetry. More generally, the point P could have O_n symmetry and the lambda lines some symmetry lower than O_n. The exponents appearing in Eq. (5.4.27) are thus those of the O_n critical point P. When g is nonzero, there is a second-order transition along the lambda lines with equations $t = t_c(g)$ meeting P. The susceptibility diverges along these lines, and we can use

Table 5.4.3. *Amplitude ratios.*
Adapted from V. Privmam, P.C. Hohenberg, and A. Aharony, in *Phase Transitions and Critical Phenomena*, Vol. 14, eds. C. Domb and J.L. Lebowitz (Academic Press, New York, 1989).

	A/A'	Γ/Γ'	ξ_0/ξ_0'
3*D* theory			
$n=1$	0.52	4.9	1.9
$n=2$	1.0		0.33
$n=3$	1.52		0.38
3*D* experiment			
$n=1$	0.5 - 0.63	4.5 - 5.0	2.0
$n=2$	0.49 - 0.74	5.0	1.7 - 2.0
$n=3$	0.84 - 1.6		1.8

Experimental data are representative of magnetic and liquid-gas transitions.

Eq. (5.4.27) to determine their equations. The susceptibility will diverge at some value of the argument of $X_2(h=0, g/|t|^{\phi})$. Thus

$$t_c(g) = g^{1/\phi}. \tag{5.4.29}$$

If $\phi > 1$, the two lambda lines will be tangent to the line $g = 0$ at P, as depicted in Fig. 4.6.11a. If $\phi < 1$, the lambda lines will approach P linearly because the scaling field t is really a linear combination of temperature and anisotropy field, and the linear term will dominate $g^{1/\phi}$ near $g = 0$. As we shall see in Sec. 5.8 [Eq. (5.8.75)], ϕ is expected to be greater than one so that the two lambda lines are tangent at P as shown in Fig. 4.6.11. Generalizations of Eq. (5.4.27) to tricritical, Lifshitz, and other multicritical points is straightforward.

4 Amplitude ratios

Not only are critical exponents universal within a given universality class, but unitless ratios of amplitudes governing the strengths of leading singularities and scaling functions expressed in appropriately scaled units are also. For example, the ratio Γ/Γ' of the amplitudes Γ and Γ' determining, respectively, the strength of the divergence of the order parameter susceptibility above and below T_c is universal. In mean-field theory, this ratio is 2 as can be seen from Eq. (4.3.7). Similarly, the amplitude ratios A/A' and ξ_0/ξ_0' for the specific heat and the correlation length are universal. As can be seen in Table 5.4.2, the variation of critical exponents among universality classes in a fixed spatial dimension is small. The variation in amplitude ratios is more pronounced, as can be seen from Table 5.4.3. Thus, amplitude ratios provide a more stringent test of universality classes than do critical exponents.

Table 5.4.4. *Critical exponents from three-dimensional renormalization group and series.*
From J.C. le Guillou and J. Zinn-Justin, *Phys. Rev. Lett.* **39**, 95 (1977); *Phys. Rev. B* **21**, 3976 (1980).

n	Exponent	RG	ϵ-expansion	Series
$n=0$	α	0.236 ± 0.0045		0.25
	β	0.302 ± 0.0015	0.305	
	γ	1.1615 ± 0.0020	1.163	$1.1615 - 1.167$
	ν	0.588 ± 0.0015	0.589	0.60
$n=1$	α	0.110 ± 0.0045		$0.11 - 0.13$
	β	0.325 ± 0.0015	0.330	$0.303 - 0.318$
	γ	1.241 ± 0.00020	1.242	$1.241 - 1.250$
	ν	0.630 ± 0.0020	0.632	0.638
$n=2$	α	-0.007 ± 0.006		
	β	0.3455 ± 0.0020	0.357	
	γ	1.316 ± 0.0025	1.324	
	ν	0.669 ± 0.0020	0.676	
$n=3$	α	-0.115 ± 0.009		
	β	0.3645 ± 0.0025	0.379	
	γ	1.386 ± 0.0040	1.395	$1.315 - 1.333$
	ν	0.705 ± 0.0030	0.713	$0.670 - 0.678$

5 *Theoretical calculations of critical exponents and amplitude ratios*

Critical exponents can now be calculated with considerable accuracy with a variety of techniques including high temperature series, renormalized field theories at fixed spatial dimension, ϵ-expansions about the upper critical dimension coupled with exact results in two dimensions, and Monte-Carlo. We will discuss calculations of critical exponents to first order in $\epsilon = 4 - d$ at the end of this chapter. Happily, the results of all of these techniques agree among themselves and reasonably well with experiments. In Table 5.4.4, we review theoretical predictions for critical exponents in three dimensions.

5.5 The Kadanoff construction

In the last section, we saw that thermodynamic functions near second-order critical points exhibit scaling with universal exponents determined by the symmetry of the order parameter. There remains, however, to develop a theoretical understanding of the origin of scaling and universality. In this section, we will present a theory due to Kadanoff (1966) that provides a heuristic explanation for the origin of scaling. It also provides a starting point from which to build the Wilson renormalization group, which, as we shall see, permits the actual calculation of critical exponents and scaling functions.

To be concrete, let us consider an Ising model with spins $s(\mathbf{y})$ at sites $\mathbf{y}$ on a d-dimensional lattice of N sites with lattice constant a. At the critical point, the correlation length ξ is infinite, and the spins at different spatial positions are strongly correlated. Thus, the average spin,

$$s_{\text{av}}(\mathbf{x}) = b^{-d} \sum_{\mathbf{y}\in c(\mathbf{x})} s(\mathbf{y}), \tag{5.5.1}$$

in a cell $c(\mathbf{x})$ containing b^d sites centered at $\mathbf{x}$ behaves at the critical point like the spin $s(\mathbf{y})$. In particular, the correlation function of the average spin dies off with distance with the same power law as the correlation function of the original spin:

$$b^{-2d} \sum_{\mathbf{y}_1\in c(\mathbf{x}_1)} \sum_{\mathbf{y}_2\in c(\mathbf{x}_2)} \langle s(\mathbf{y}_1)s(\mathbf{y}_2)\rangle = \langle s_{\text{av}}(\mathbf{x}_1)s_{\text{av}}(\mathbf{x}_2)\rangle \sim |\mathbf{x}_1 - \mathbf{x}_2|^{-2\omega}, \tag{5.5.2}$$

where $\omega = \frac{1}{2}(d-2+\eta) = \beta/\nu$ [Eqs. (5.4.4) and (5.4.15)].

In the Kadanoff construction, the original lattice is divided into $N' = b^{-d}N$ cells centered on a new lattice (Fig. 5.5.1) with lattice constant $a' = ba$. Each cell centered at $\mathbf{x}$ in the original lattice contains b^d sites of the original lattice and corresponds to a site in the new lattice. Distances in the new lattice are measured in terms of the new lattice constant a' so that the position of a site in the new lattice corresponding to a cell at $\mathbf{x}$ in the original lattice is $\mathbf{x}' = \mathbf{x}/b$. At each site $\mathbf{x}/b$ of the new lattice is a *block spin* variable,

$$s'(\mathbf{x}') = s'(\mathbf{x}/b) = b^{\omega_s} s_{\text{av}}(\mathbf{x}), \tag{5.5.3}$$

proportional to the average spin in the cell centered at that site. Since correlation functions of $s_{\text{av}}(\mathbf{x})$ and $s(\mathbf{y})$ for $\mathbf{y}$ in a cell centered at $\mathbf{x}$ scale in the same way at the critical point, it is not really necessary to distinguish between $s_{\text{av}}(\mathbf{x})$ and $s(\mathbf{x})$, and we can express Eq. (5.5.3) as

$$s'(\mathbf{x}/b) = b^{\omega_s} s(\mathbf{x}). \tag{5.5.4}$$

The exponent ω_s in the b-dependent proportionality constant in the above two equations will be determined shortly. Kadanoff argued that there is a Hamiltonian that is a function of the block spin variables on the new lattice. At the critical point, this new Hamiltonian is in some sense identical to the original Hamiltonian expressed in terms of $s(\mathbf{y})$ on the original lattice. In particular, the correlation function of the block spin variable as a function of distances measured in the new lattice should be identical to the correlation function of the original spin as a function of distance in the original lattice:

$$\begin{aligned} \langle s'(\mathbf{x}_1/b)s'(\mathbf{x}_2/b)\rangle &= b^{2\omega_s}\langle s_{\text{av}}(\mathbf{x}_1)s_{\text{av}}(\mathbf{x}_2)\rangle \\ &\sim \left|\frac{\mathbf{x}_1}{b} - \frac{\mathbf{x}_2}{b}\right|^{-2\omega}. \end{aligned} \tag{5.5.5}$$

Comparing Eqs. (5.5.5) and (5.5.2), we conclude that

$$\omega_s = \omega. \tag{5.5.6}$$

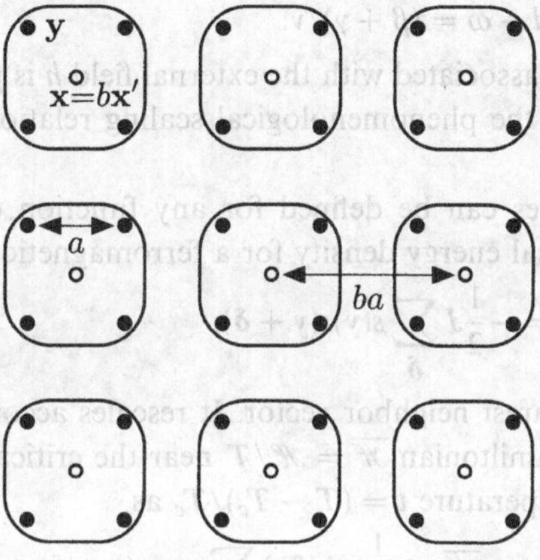

Fig. 5.5.1. The coarse-graining procedure of the Kadanoff construction. The original spins on sites **y** are grouped into cells and associated with a block spin at its central sites **x**. At the critical point, the block spins on the superlattice should exhibit the same correlations as the original spins on the original lattice. Here the scaling parameter b is 2.

The exponent ω, therefore, determines the relation between original and block spin variables.

The scaling of the external field $h(\mathbf{y})$ conjugate to the order parameter $s(\mathbf{y})$ under the above block spin transformation can be determined by requiring that the free energy associated with the external field does not change. If $h(\mathbf{x})$ varies so slowly in space that it can be assumed constant within any cell, then the Hamiltonian describing the interaction of spins with the external field can be expressed either in terms of the original spins on the original lattice or in terms of the block spins on the new lattice:

$$\mathscr{H}_{\text{ext}} = -\sum_{\mathbf{y}} h(\mathbf{y})s(\mathbf{y}) \equiv -\sum_{\mathbf{x}'} h'(\mathbf{x}')s'(\mathbf{x}'). \tag{5.5.7}$$

Since $h(\mathbf{y})$ is slowly varying and can be assumed to have the same value $h(\mathbf{x})$ for each $\mathbf{y}$ in a given cell $c(\mathbf{x})$, we can reexpress the Hamiltonian of the original lattice as

$$\begin{aligned} \mathscr{H}_{\text{ext}} &= -\sum_{\mathbf{x}} h(\mathbf{x})b^d \left(b^{-d} \sum_{\mathbf{y}\in c(\mathbf{x})} s(\mathbf{y}) \right) \\ &= -\sum_{\mathbf{x}} h(\mathbf{x})b^{d-\omega}s'(\mathbf{x}'), \end{aligned} \tag{5.5.8}$$

where we used Eq. (5.5.3) to express $s_{\text{av}}(\mathbf{x})$ in terms of $s'(\mathbf{x}')$. Comparing the first and last lines of this equation, we obtain

$$h'(\mathbf{x}') = h'(\mathbf{x}/b) = b^{d-\omega}h(\mathbf{x}) . \tag{5.5.9}$$

It is easy to verify from the scaling relations presented in the last section that

$$\lambda = d - \omega = (\beta + \gamma)/\nu. \tag{5.5.10}$$

The exponent λ associated with the external field h is precisely the same exponent that appears in the phenomenological scaling relations of the preceding section [Eq. (5.4.21)].

Block variables can be defined for any function of the spin operators. For example, the local energy density for a ferromagnetic Ising model is

$$\varepsilon(\mathbf{y}) = -\frac{1}{2}J\sum_{\boldsymbol{\delta}} s(\mathbf{y})s(\mathbf{y}+\boldsymbol{\delta}), \tag{5.5.11}$$

where δ is a nearest neighbor vector. It rescales according to $\varepsilon'(\mathbf{x}/b) = b^{\omega_\varepsilon}\varepsilon_{\text{av}}(\mathbf{x})$. The reduced Hamiltonian $\overline{\mathscr{H}} = \mathscr{H}/T$ near the critical point can be expanded in the reduced temperature $t = (T - T_c)/T_c$ as

$$\begin{aligned} \overline{\mathscr{H}} &= \overline{\mathscr{H}}_c + \frac{1}{2}(tJ/T_c)\sum_{\mathbf{y},\delta} s(\mathbf{y})s(\mathbf{y}+\delta) + O(t^2) \\ &= \overline{\mathscr{H}}_c - (t/T_c)\sum_{\mathbf{y}} \varepsilon(\mathbf{y}), \end{aligned} \tag{5.5.12}$$

where $\overline{\mathscr{H}}_c$ is the reduced Hamiltonian at the critical point. Thus the field conjugate to the energy density is proportional to the reduced temperature, and we can conclude, following the arguments leading to Eq. (5.5.9), that

$$t' = b^{d-\omega_\varepsilon}t \equiv b^{\lambda_t}t. \tag{5.5.13}$$

In the preceding section, we saw that the scaling exponent associated with t is the inverse of the correlation length exponent. Thus

$$\lambda_t = \frac{1}{\nu} = d - \omega_\varepsilon. \tag{5.5.14}$$

More generally, there are a wide variety of additional local fields, $\phi_\alpha(\mathbf{y})$, involving for example multi-spin interactions or anisotropies, that can be constructed from the original spin variables $s(\mathbf{y})$. External fields h_α coupling linearly to $\phi_\alpha(\mathbf{y})$ can be introduced into $\overline{\mathscr{H}}$ by the addition of term

$$\overline{\mathscr{H}}_\alpha = -h_\alpha \sum_{\mathbf{y}} \phi_\alpha(\mathbf{y}) \tag{5.5.15}$$

to the reduced Hamiltonian. Under the Kadanoff transformation, the field $\phi_\alpha(\mathbf{x})$ (or equivalently its average in a cell centered at $\mathbf{x}$) will transform with a dominant exponent ω_α according to

$$\phi'_\alpha(\mathbf{x}/b) = b^{\omega_\alpha}\phi_\alpha(\mathbf{x})\,. \tag{5.5.16}$$

Its conjugate field h_α will then transform as

$$h'_\alpha = b^{\lambda_\alpha}h_\alpha \tag{5.5.17}$$

with

$$\lambda_\alpha = d - \omega_\alpha\,. \tag{5.5.18}$$

Note that λ_α can be positive, as it is if ϕ_α is the order parameter or the energy density, or it can be negative. If λ_α is positive, h_α grows with successive rescalings associated with the transformations from original to block spin variables and is a *relevant* field; if λ_α is negative, h_α approaches zero with successive rescalings and is an *irrelevant* field. The field ϕ_α is also classified as relevant or irrelevant if its associated exponent ω_α is, respectively, less than or greater than d.

The Hamiltonian can be expressed either in terms of the original variables on a lattice of N sites or in terms of the block variables on a lattice of $N' = b^{-d}N$ sites. The free energy F calculated with either set of variables is the same. This implies a simple relation between the original free energy density $f(t, h, h_\alpha) = F/N$ and the block free energy density $f(t', h', h'_\alpha) = F/N'$:

$$Nf(t, h, h_\alpha) = b^{-d}Nf(t', h', h'_\alpha) \tag{5.5.19}$$

or

$$f(t, h, h_\alpha) = b^{-d}f(b^{1/\nu}t, b^{\lambda}h, b^{\lambda_\alpha}h_\alpha) \, . \tag{5.5.20}$$

This is precisely the scaling form of the free energy introduced in the last section. The arbitrary rescaling factor b can be eliminated, as discussed in the last section, by choosing $b = t^{-\nu}$. In this case,

$$f(t, h, h_\alpha) = |t|^{d\nu}X(h|t|^{-\Delta}, h_\alpha|t|^{-\Delta_\alpha}) \tag{5.5.21}$$

depends only on the fields h_α and h in the combinations $h_\alpha/|t|^{\Delta_\alpha}$ and $h/|t|^{\Delta}$, where

$$\Delta_\alpha = \lambda_\alpha \nu \tag{5.5.22}$$

and

$$\Delta = \lambda\nu = \gamma + \beta. \tag{5.5.23}$$

If $\lambda_\alpha < 0$, $h_\alpha|t|^{-\Delta_\alpha} = h_\alpha|t|^{|\Delta_\alpha|}$ tends to zero as $t \to 0$, and the field h_α does not affect the leading order singularities in f at the critical point. This is why such fields are called irrelevant. Note that if one is interested in the leading singularities of f, one could have set all irrelevant fields equal to zero initially. This fact, as we shall see, is in essence what is responsible for universality. If two Hamiltonians differ only by irrelevant fields, they will have the same leading critical exponents and thus be in the same universality class.

Though the irrelevant variables do not affect the leading singularities near the critical point, they do give rise to nonzero corrections to leading singularities that are called *corrections to scaling*. Consider, for example, a system with a single irrelevant variable h_1 with $\lambda_1\nu = -\Delta_1 < 0$. The susceptibility will satisfy a scaling relation $\chi = |t|^{-\gamma}X_2(h_1|t|^{\Delta_1})$. As $t \to 0$, $h_1|t|^{\Delta_1} \to 0$ for any h_1, and we can expand X_2 as a power series to obtain

$$\chi = \Gamma|t|^{-\gamma} + Eh_1|t|^{-\gamma+\Delta_1} + \cdots, \tag{5.5.24}$$

where $E = X'_2(0)$. Thus, the irrelevant term leads to subdominant singularities in χ, which may even diverge if $\Delta_1 < \gamma$. The existence of such corrections to scaling considerably complicates the determination of critical exponents from experimental data.

The introduction of block spins that interact via the same Hamiltonian as do the original spins leads naturally to scaling of the free energy. The Kadanoff construction, however, assumes but does not demonstrate that the original and block spins interact via the same Hamiltonian, and it does not provide an algorithm for actually calculating any critical exponents. The Wilson renormalization group to be discussed in the next few sections quantifies the ideas of the Kadanoff construction. It provides a precise meaning to identical original and block spin Hamiltonians and provides algorithms for calculating critical exponents.

5.6 The one-dimensional Ising model

In this section, we will introduce the renormalization group for the simplest possible example – the one-dimensional Ising model. We will begin by reviewing exact calculations of correlation functions and the free energy of the one-dimensional Ising model, showing how $T = 0$ can be treated as a second-order critical point. We will then show how decimation of spins leads to block spins analogous to those of the Kadanoff construction.

1 Exact solution

In the one-dimensional Ising model, there is a spin variable $\sigma_i = \pm 1$ at each site $i = 1, 2, ..., N$ on a one-dimensional lattice. The reduced Hamiltonian is

$$\begin{aligned} -\overline{\mathscr{H}} &= -\mathscr{H}/T = K\sum_i \sigma_i\sigma_{i+1} + L\sum_i \sigma_i + \sum_i C \\ &= K\sum_i \sigma_i\sigma_{i+1} + \frac{1}{2}L\sum_i(\sigma_i + \sigma_{i+1}) + \sum_i C \\ &= \sum_i \overline{K}(\sigma_i, \sigma_{i+1}), \end{aligned} \tag{5.6.1}$$

where $K = J/T$ and $L = h/T$ with J the exchange integral and h the external magnetic field. C is a constant that defines the zero of the free energy. It is introduced to facilitate discussion of the renormalization group. The partition function for this model can be calculated exactly using transfer matrices. The exponential of $\overline{K}(\sigma, \sigma')$ is a two-by-two matrix in the variable σ and σ' with entries

$$e^{\overline{K}(\sigma,\sigma')} = e^C \begin{bmatrix} e^{K+L} & e^{-K} \\ e^{-K} & e^{K-L} \end{bmatrix} \equiv \underline{M}(K, L, C)\,. \tag{5.6.2}$$

When $h = 0$ and $C = 0$, $\underline{M}(K, L, C)$ takes on a simple form,

$$\underline{M}(K, 0, 0) = \cosh K(1 + \sigma\sigma' \tanh K)\,, \tag{5.6.3}$$

which will be useful in what follows.

Bulk properties in the thermodynamic limit are insensitive to boundary conditions. Calculations are most easily carried out in a model with periodic boundary conditions in which σ_{N+1} is identical to σ_1 or equivalently in a model in which lattice sites lie on a circle with sites 1 and N connected. In this case, there are N

bonds connecting N spins, and the partition function can be expressed as a trace of a product of transfer matrices:

$$\begin{aligned} Z_N &= \sum_{\sigma_1,\sigma_2,\ldots,\sigma_N} e^{-\overline{\mathscr{H}}} \\ &= \sum_{\sigma_1,\sigma_2,\ldots,\sigma_N} e^{\overline{K}(\sigma_1,\sigma_2)} e^{\overline{K}(\sigma_2,\sigma_3)} \ldots e^{\overline{K}(\sigma_{N-1},\sigma_N)} e^{\overline{K}(\sigma_N,\sigma_1)} \\ &= \mathrm{Tr}\underline{M}^N = e^{NC}(\lambda_+^N + \lambda_-^N), \end{aligned} \tag{5.6.4}$$

where λ_+ and λ_- are the eigenvalues,

$$\lambda_\pm = e^K \cosh L \pm (e^{2K} \sinh^2 L + e^{-2K})^{1/2}, \tag{5.6.5}$$

of $\underline{M}(K, L, 0)$. When $h = 0$, the larger eigenvalue is $\lambda_+ = 2\cosh K$. In the limit of large N, λ_-^N can be ignored relative to λ_+^N, and the free energy per spin becomes

$$\begin{aligned} \frac{f}{T} &= \lim_{N\to\infty} \frac{1}{N}[-\ln Z_N] \\ &= -C - \ln[e^K \cosh L + (e^{2K} \sinh^2 L + e^{-2K})^{1/2}]. \end{aligned} \tag{5.6.6}$$

In the low T ($K \to \infty$) low h ($L \to 0$, $Le^{2K} \ll 1$) limit.

$$f - f_0 \to -Te^{-2K} - \frac{1}{2}e^{2K}(h^2/T), \tag{5.6.7}$$

where $f_0 = -J - TC$. This shows, as expected, that the ground state energy per spin is J (when $C = 0$) and that there is a gap in the excitation spectrum leading to an exponential approach with temperature to the ground state. The susceptibility at low temperature is

$$\chi = -\frac{\partial^2 f}{\partial h^2} = T^{-1}e^{2K}. \tag{5.6.8}$$

Thus, χ diverges as $T \to 0$, indicating that there is a critical point at $T = 0$ in the 1D Ising model. In the high-temperature limit, $e^K \to 1 + K + \frac{1}{2}K^2$. Then

$$f \to -T\ln 2 - \frac{1}{2}h^2/T \tag{5.6.9}$$

when $C = 0$. This implies that

$$\chi = T^{-1} \tag{5.6.10}$$

is the familiar Curie spin susceptibility.

Correlation functions are also easily calculated in the 1D Ising model. In particular

$$\begin{aligned} G(n) &= \frac{1}{Z_N} \sum_{\sigma_1,\ldots,\sigma_N} e^{-\overline{\mathscr{H}}} \sigma_1 \sigma_{n+1} \\ &= \frac{1}{Z_N} \sum_{\sigma_1,\ldots,\sigma_N} \sigma_1 e^{\overline{K}(\sigma_1,\sigma_2)} \cdots e^{\overline{K}(\sigma_n,\sigma_{n+1})} \sigma_{n+1} e^{\overline{K}(\sigma_{n+1},\sigma_{n+2})} \cdots e^{\overline{K}(\sigma_N,\sigma_1)} \\ &= \frac{1}{Z_N} \sum_{\sigma_1,\sigma_{n+1}} \sigma_1 e^{n\overline{K}(\sigma_1,\sigma_{n+1})} \sigma_{n+1} e^{(N-n)\overline{K}(\sigma_{n+1},\sigma_1)}. \end{aligned} \tag{5.6.11}$$

When h and C are zero, we can calculate

$$e^{n\overline{K}(\sigma,\sigma')} = 2^{n-1}(\cosh K)^n(1+\sigma\sigma'\tanh^n K) \tag{5.6.12}$$

from Eq. (5.6.3). Using this result in Eq. (5.6.11), we obtain

$$\begin{aligned} G(n) &= \frac{1}{Z_N}2^{N-2}(\cosh K)^N \sum_{\sigma_1,\sigma_{n+1}} \sigma_1(1+\sigma_1\sigma_{n+1}\tanh^n K)\sigma_{n+1} \\ &\quad \times(1+\sigma_{n+1}\sigma_1\tanh^{N-n} K) \\ &= \frac{1}{Z_N}(2\cosh K)^N \tanh^n K(1+\tanh^{N-2n} K). \end{aligned} \tag{5.6.13}$$

When $N \to \infty$ at fixed n, this reduces to

$$G(n) = \tanh^n K \equiv e^{-n/\xi}, \tag{5.6.14}$$

where

$$\xi = -[\ln(\tanh K)]^{-1} \overset{T\to 0}{\to} \frac{1}{2}e^{2K} \tag{5.6.15}$$

is the correlation length that diverges as the $T = 0$ critical point is approached. We can now see that there is scaling in the vicinity of $T = 0$ if functions are expressed in terms of the correlation length. The susceptibility is the sum over n of $T^{-1}G(n)$:

$$\begin{aligned} T\chi &= \sum_n G(n,0) = \sum_n (\tanh K)^n \\ &= (1-\tanh K)^{-1} \to \frac{1}{2}e^{2K} = \xi. \end{aligned} \tag{5.6.16}$$

The scaling relations of Sec. 5.5 predict $\chi \sim \xi^{\gamma/\nu}$, and we conclude from the above equation that

$$\gamma/\nu = 1 \tag{5.6.17}$$

for the 1D Ising model. Near $T = 0$, the free energy [Eq. (5.6.7)] can be expressed as a scaling function of L and ξ:

$$T^{-1}(f+J) \sim -e^{-2K}(1+\frac{1}{2}L^2e^{4K}) = \xi^{-1}f_a(L\xi). \tag{5.6.18}$$

This is to be identified with the scaling form for the free energy [Eq. (5.4.26)] expressed in terms of $\xi = |t|^{-\nu}$ rather than reduced temperature t:

$$f \sim \xi^{-(2-\alpha)/\nu}X(L\xi^{\Delta/\nu}). \tag{5.6.19}$$

Comparing Eqs. (5.6.18) and (5.6.17), we obtain

$$\frac{2-\alpha}{\nu} = 1, \qquad \frac{\Delta}{\nu} = \frac{\gamma+\beta}{\nu} = 1. \tag{5.6.20}$$

Since $\gamma/\nu = 1$, this leads to $\beta = 0$. Alternatively, we can use $\gamma/\nu = 1 = 2-\eta$ to obtain $\eta = 1$ and $\beta = \frac{1}{2}(d-2+\eta)\nu = 0$. To see that this is the correct identification of β, we note that as $T \to 0$,

$$\frac{m}{T} = \frac{\partial f}{\partial L} = e^{2K}L = \xi^0 m_a(L\xi). \tag{5.6.21}$$

In summary, the exact solution of the one-dimensional Ising model shows that there is a critical point at $T = 0$ at which the correlation length and susceptibility diverge. The free energy and related functions satisfy standard scaling relations with exponents $\gamma/\nu = (2-\alpha)/\nu = 1$, $\eta = 1$ and $\beta = 0$.

2 Decimation and renormalization

A new lattice with block spin variables can be obtained in the 1D Ising model merely by tracing (decimating) over blocks of $b-1$ spins leaving a spin at every bth site, as shown in Fig. 5.6.1. The partition function of the new lattice is identical to that of the original lattice and can be expressed as

$$Z_N(K,L,C) = \mathrm{Tr}\underline{M}^N = \mathrm{Tr}[\underline{M}^b]^{N'} = Z_{N'}(K',L',C'), \tag{5.6.22}$$

where $N' = N/b$ is the number of sites in the new lattice. The potentials of the decimated lattice are determined by

$$\underline{M}(K',L',C') = \underline{M}^b(K,L,C). \tag{5.6.23}$$

When $L = 0$, this leads, with the help of Eq. (5.6.3), to a simple relation

$$\tanh K' = (\tanh K)^b, \tag{5.6.24}$$

or equivalently

$$K' = \tanh^{-1}[(\tanh K)^b], \tag{5.6.25}$$

between K' and K that is independent of C. We will consider the equation for C later in this section. Eq. (5.6.24) is a renormalization group recursion relation that can be iterated an arbitrary number of times. After an infinite number of iterations, K reaches a *fixed point* value K^* determined by $K' = K = K^*$ in Eq. (5.6.23). There are only two fixed point solutions to this equation:

$$\begin{aligned} \tanh K &= 0 \quad (T = \infty), \\ \tanh K &= 1 \quad (T = 0). \end{aligned} \tag{5.6.26}$$

The function $\tanh K$ lies between zero and one. Thus, unless $K = \infty$, $\tanh K$ will diminish upon iteration of the recursion relation approaching the fixed point at $\tanh K = 0$ as the number of iterations tends to infinity. If $K = \infty$, $\tanh K$ will remain equal to unity after any number of iterations of the recursion relation. Since all values of K other than $K = \infty$ flow towards $K = 0$, the fixed point $\tanh K = 0$ ($T = \infty$) is said to be *stable.* The set of values of K flowing to $K = 0$ is called the *basin of attraction* of the $K = 0$ fixed point. The fixed point at $K = \infty$ is *unstable* and its basin of attraction consists of the single point $K = \infty$. The flow of $\tanh K$ and T under Eq. (5.6.24) is depicted in Fig. 5.6.2.

The stable fixed point describes all finite temperature behavior. It is associated with the paramagnetic phase. The unstable fixed point describes the Ising critical point at $T = 0$. The recursion relation, Eq. (5.6.23), can be expressed in terms of the correlation length [Eq. (5.6.15)]:

$$\xi' = \xi/b. \tag{5.6.27}$$

This is merely the statement that the correlation length measured in terms of the lattice constant of the new lattice is b^{-1} times that measured in terms of

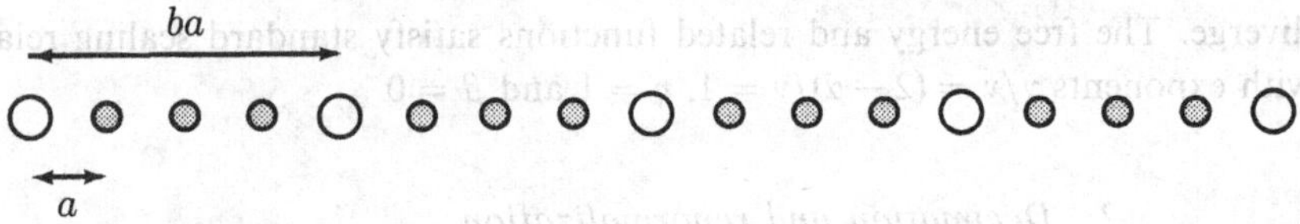

Fig. 5.6.1. Schematic representation of the one-dimensional Ising model showing sites of the original lattice with lattice constant a and sites on the new lattice with lattice constant ba. The $b-1$ spins between sites of the new lattice are removed by a trace or decimation operation.

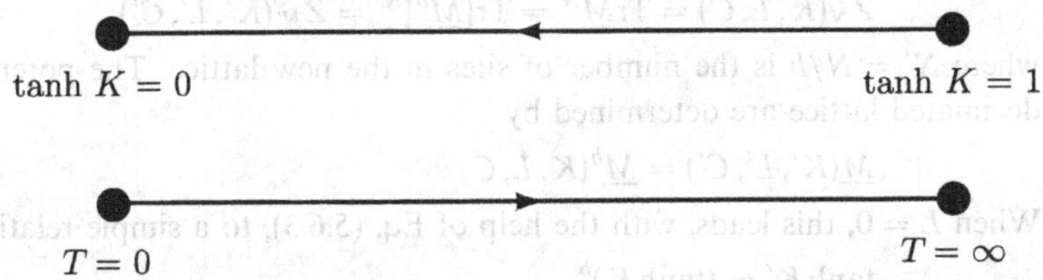

Fig. 5.6.2. Renormalization group flows for $\tanh K$ and for T showing the stable fixed points at $\tanh K = 0$ ($T = \infty$) and the unstable fixed points at $\tanh K = 1$ ($T = 0$).

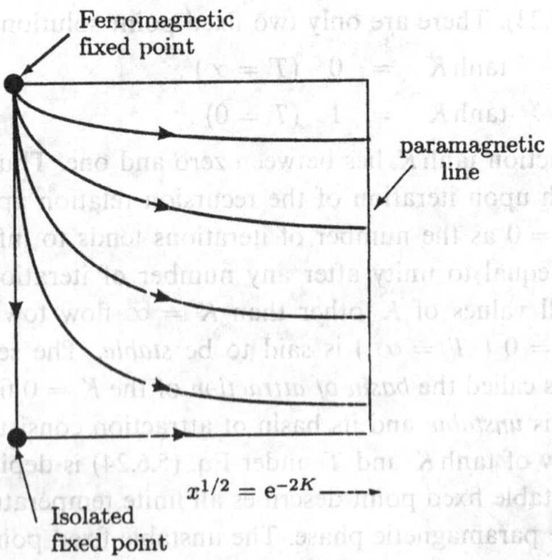

Fig. 5.6.3. Renormalization group flows for the one-dimensional Ising model in a field [Eq. (5.6.29)] showing the three fixed points of Eq. (5.6.30). [D. R. Nelson and M. E. Fisher, *Ann. Phys.* **91**, 226 (1975).]

the lattice constant of the original lattice. This is a general property of all renormalization group transformations. The correlation length decreases when the lattice is rescaled. There are only two fixed points to this recursion relation: $\xi = 0$ and $\xi = \infty$. The first corresponds to the non-critical high temperature fixed point, and the second corresponds to the critical point.

Recursion relations for L and C also follow from Eq. (5.6.23). When $b = 2$, we find

$$\begin{aligned} x' &= \frac{x(1+y)^2}{(x+y)(1+xy)} \\ y' &= \frac{y(x+y)}{(1+xy)} \\ w' &= \frac{w^2xy^2}{(1+y)^2(x+y)(1+xy)}, \end{aligned} \tag{5.6.28}$$

where

$$x = e^{-4K}, \quad y = e^{-2L}, \quad w = e^{-4C}. \tag{5.6.29}$$

The equations for x and y do not depend on w and have three fixed points. They are

$$\begin{aligned} &(1) \quad x^* = y^* = 0 \\ &(2) \quad x^* = 1, \quad y \text{ arbitrary}, \\ &(3) \quad x^* = 0, \quad y^* = 1. \end{aligned} \tag{5.6.30}$$

The first fixed point is an isolated fixed point corresponding to a frozen spin configuration (aligned spins at all temperatures as a result of an infinite external field). The second fixed point describes the high temperature paramagnetic phase. The third fixed point describes the $T = 0$ critical point that is our primary concern. The recursion relations can be linearized in the vicinity of the third fixed point to yield

$$\begin{aligned} \Delta y' &= (-2L') = 2\Delta y = 2(-2L), \\ \Delta x' &= 4\Delta x, \end{aligned} \tag{5.6.31}$$

where $\Delta y = y - y^*$ and $\Delta x = x - x^*$. These equations imply that the singular part of the free energy satisfies the scaling law derived from the exact solution of the Ising model. The free energy per spin is

$$\begin{aligned} \frac{f(K,L,C)}{T} \equiv \bar{f}(K,L,C) &= -\frac{1}{N}\ln Z_N(K,L,C) \\ &= -\frac{1}{bN'}\ln Z_{N'}(K',L',C'), \end{aligned} \tag{5.6.32}$$

where the second line follows from Eq. (5.6.22). The partition function depends on C only via the prefactor e^{NC}, and the free energy per spin can be written as $\bar{f}(K,L,C) = C + \bar{f}_{\text{sing}}(K,L)$. Thus, apart from terms involving the nonsingular C, Eqs. (5.6.32) and (5.6.31) imply

$$\bar{f}_{\text{sing}}(\Delta x, \Delta y) = b^{-1}\bar{f}_{\text{sing}}(b^2\Delta x, b\Delta y) \tag{5.6.33}$$

for rescaling parameter $b = 2$. With $\Delta y = -2L$ and $\Delta x = e^{-4K}$, we then have

$$\bar{f}_{\text{sing}}(e^{-K}, L) = e^{-2K} \bar{f}_{\text{sing}}(Le^{2K}) \tag{5.6.34}$$

in agreement with the exact result Eq. (5.6.19).

This simple example shows that critical points can be described by renormalization group recursion relations. The exponents obtained by linearizing the recursion relations in the vicinity of unstable critical fixed points are the exponents describing the scaling of the free energy. Thus, the renormalization group recursion relations yield critical exponents. These properties are generally valid. What is not apparent from this example is that the fixed point describing a critical point must be unstable in one or more directions, but it can be, and in general is, stable in other directions. All potentials or Hamiltonians in the subspace perpendicular to the unstable directions of the fixed point have the same critical exponents. This is the origin of universality.

5.7 The Migdal-Kadanoff procedure

In the preceding section, we saw that renormalization group recursion relations for the 1D Ising model can be obtained by decimating a regular sequence of spins. The $T = 0$ critical point of the $1D$ Ising model is clearly special. The challenge is to develop a renormalization group procedure applicable to finite-temperature critical points in higher dimensions. In this section, we will study a simple approximate renormalization procedure due to Migdal (1975) and Kadanoff (1976) that yields nontrivial critical points in dimensions greater than one and that gives reasonable predictions for phase diagrams of real physical systems such as xenon adsorbed on graphite. We will first apply this procedure to the Ising model just above one dimension. From this, we will be able to make some statements about the general structure of renormalizations and how they predict critical exponents and universality.

1 *The Ising model on a hypercubic lattice*

The 1D Ising model suggests an obvious approach to models in two or more dimensions: decimate spins on a subset of lattice points leaving spins on a new larger lattice-constant lattice interacting via a new Hamiltonian determined by the decimation procedure. Consider, for example, spins σ on a square lattice with lattice constant a and N sites interacting via a reduced Hamiltonian $\overline{\mathscr{H}}_N(\sigma, a)$. The square lattice can be decomposed into two sublattices (marked with squares and circles in Fig 5.7.1) with lattice constant $ba = \sqrt{2}a$. The trace over spins σ_1 on the sublattice marked with circles yields a new reduced Hamiltonian $\overline{\mathscr{H}}_{N'}(\sigma', ba)$ ($N' = b^{-2}N$) describing interactions among spins σ' on the "square" sublattice defined via

$$e^{-\overline{\mathscr{H}}_{N'}(\sigma',ba)} = \mathrm{Tr}_{\sigma_1} e^{-\mathscr{H}_N(\sigma,a)}. \tag{5.7.1}$$

Clearly, the partition functions associated with the two Hamiltonians $\overline{\mathscr{H}}_N$ and $\overline{\mathscr{H}}_{N'}$ are equal since

$$Z_{N'} = \mathrm{Tr}_{\sigma'} e^{-\overline{\mathscr{H}}_{N'}} \equiv \mathrm{Tr}_{\sigma'}\left(\mathrm{Tr}_{\sigma_1} e^{-\overline{\mathscr{H}}_N}\right) = Z_N. \tag{5.7.2}$$

Unfortunately, this procedure leads after several iterations to hopelessly complicated Hamiltonians. For example, if $\mathscr{H}_N$ is a nearest neighbor Ising Hamiltonian, then $\mathscr{H}_{N'}$ will have not only nearest neighbor but also next nearest neighbor and four-spin interactions of the form $\mathscr{H}_4 = \sum \sigma_1\sigma_2\sigma_3\sigma_4$, where the sum is over all four-spin plaquettes illustrated in Fig. 5.7.1. A second decimation applied to $\mathscr{H}_{N'}$ produces a new Hamiltonian $\mathscr{H}_{b^{-2}N'}$ with further neighbor and more multi-spin interactions. Clearly, some method of truncating the range and complexity of interactions is needed. Various truncation methods, none of which is fully controlled, have been proposed. The simplest, and in many ways the most successful, is the bond-moving scheme introduced by Migdal and elaborated by Kadanoff. The Midgal-Kadanoff procedure is a two step process illustrated for the square lattice in Fig. 5.7.2. First, bonds of nearest neighbor interaction strength K_y on $b-1$ lines parallel to the y-axis are broken and moved to create bonds of strength bK_y at every bth line. The sites originally connected by the broken bonds are no longer connected to their neighbors along the y-axis. The spins at these sites can be removed by the one-dimensional decimation procedure discussed in the previous section. The recursion relations of the Migdal-Kadanoff procedure are, therefore,

$$K_y' = bK_y\ , \quad K_x' = R^b(K_x)\ , \tag{5.7.3}$$

where R^b is the operator defined by the 1D decimation. In the Ising model, $R^b(K) = \tanh^{-1}[(\tanh K)^b]$, as can be seen from Eq. (5.6.24). The recursion relations in Eq. (5.7.3) have the undesirable property that K_y and K_x are treated unsymmetrically. To restore symmetry at least partially, the same process can be repeated with the lattice rotated by 90° so that bonds parallel to the x-axis are moved and decimation is along the y-axis.

The extension of the Migdal-Kadanoff approach to arbitrary dimension is straightforward. There are nearest neighbor interactions K_p ($p = 1, ..., d$) for bonds parallel to the d axes of a d-dimensional hypercubic lattice. Bond moving and decimation are alternated (as illustrated in Fig. 5.7.3 for the three-dimensional case) to produce

$$K_p' = b^{d-p} R^b(b^{p-1}K_p). \tag{5.7.4}$$

These equations again do not preserve the equality of coupling constants in different directions. At least partial symmetry can be restored by performing the same series of operations for all equivalent orientations of the lattice. For complicated problems, this is often the best procedure to follow. To illustrate how these equations describe a non-trivial critical point at finite temperature, it is more instructive, however, to consider them in the limit where $b \to 1$. Though it

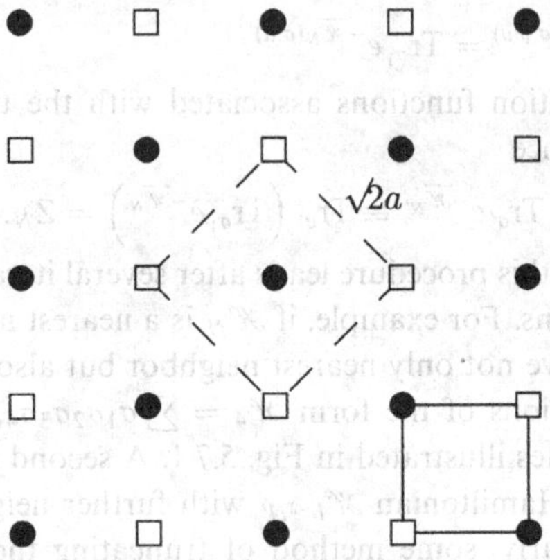

Fig. 5.7.1. An Ising model on a square lattice showing the original lattice with N sites (marked • and □) and lattice constant a and the new lattice with $N' = b^{-2}N$ sites and lattice constant $ba = \sqrt{2}a$ after spins at the sites marked by • have been removed by decimation. Also indicated are a plaquette of the new (dashed lines) and one of the original (solid lines) lattices. Four-spin interactions involve products of spins on a plaquette.

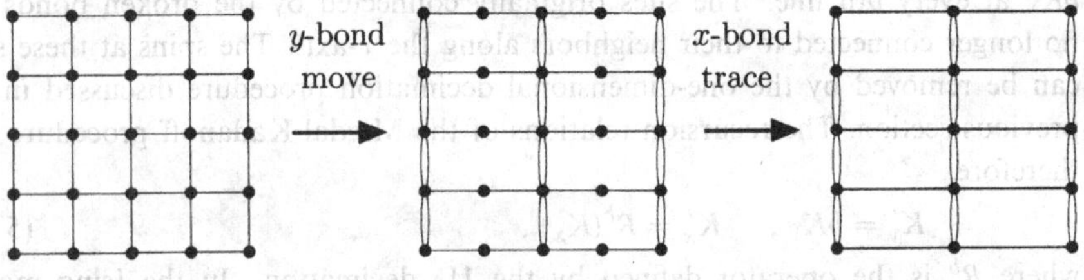

Fig. 5.7.2. The Migdal-Kadanoff procedure for a square lattice. First, $b - 1$ bonds parallel to the y-axis are broken and moved to the right to create new bonds of strength bK_y. Then, spins on the $(b - 1)$ sites per cell not connected with y-bonds are removed by decimation. This process can be repeated in the opposite order to restore the symmetry of the x- and y-directions.

is impossible to remove a non-integral number of bonds in the procedure used to arrive at the Migdal-Kadanoff recursion relations, the final equation [Eq. (5.7.4)] has a well defined analytic continuation to all b. Setting $b = e^{\delta l} \sim (1 + \delta l)$, we have

$$\begin{aligned} R^{1+\delta l}[(1+\delta l)^{(p-1)}K] &= \tanh^{-1}\left([\tanh((1+\delta l)^{(p-1)}K)]^{1+\delta l}\right) \\ &= \tanh^{-1}(\tanh[(1+(p-1)\delta l)K] \\ &\qquad \times[1+\delta l\ln(\tanh K)]) + O[(\delta l)^2] \quad (5.7.5) \\ &= (1+(p-1)\delta l)K + \delta l[\frac{1}{2}\sinh 2K\ln(\tanh K)]. \end{aligned}$$

This expansion and Eq. (5.7.4) allow us to calculate the rate of change of K resulting from an infinitesimal change $b = e^{\delta l}$ in the lattice constant,

$$\frac{dK}{dl} = (d-1)K + \frac{1}{2}[\sinh 2K \ln(\tanh K)] \,. \tag{5.7.6}$$

When $d = 1$, this equation is identical to Eq. (5.6.24) in the limit $b \to 1 + \delta l$. For small $d - 1$, new structure appears at small $T = K^{-1}$. To see this, we expand Eq. (5.7.6) to second order in T:

$$\frac{dT}{dl} = -\epsilon T + \frac{1}{2}T^2, \tag{5.7.7}$$

where $\epsilon = (d-1)$. The fixed points of this equation occur when $dT/dl = 0$, and are trivially

$$T^* = 0\,, \quad \text{and} \quad T^* = 2\epsilon. \tag{5.7.8}$$

The fixed point at $T = 0$ is stable, whereas the one at $T = 2\epsilon$ is not, as can be seen by the linear stability equation in the vicinity of $T = 2\epsilon$:

$$\frac{d\delta T}{dl} = (-\epsilon + T^*)\delta T = \epsilon\delta T \equiv \lambda_t \delta T\,, \tag{5.7.9}$$

where $\delta T = T - T^*$. Thus for $T < 2\epsilon$, the flow is towards the fixed point at $T = 0$ describing the ordered phase of the Ising model. For $T > 2\epsilon$, the flow is toward $T = \infty$ or the paramagnetic phase. These flows are illustrated in Fig. 5.7.4.

The Migdal-Kadanoff recursion relations yield a non-trivial fixed point that describes the Ising critical point. To see how the correlation length exponent follows from the above recursion relations, we note that, in the vicinity of $T = T^*$, Eq. (5.7.9) can be integrated to yield $\delta T(l) = e^{\epsilon l}\delta T(0)$. But, under the rescaling transformations, the lattice spacing goes from a to $e^l a$, and the correlation length measured in terms of the new lattice constant is $\xi(l) = e^{-l}\xi(0)$. On the other hand, $\xi(l) \sim (\delta T(l))^{-\nu}$. Therefore,

$$\frac{\xi(l)}{\xi(0)} = e^{-l} = \left(\frac{\delta T(l)}{\delta T(0)}\right)^{-\nu} = \left(e^{\lambda_t l}\right)^{-\nu}\,, \tag{5.7.10}$$

from which we conclude

$$\nu = \frac{1}{\lambda_t} = \frac{1}{\epsilon} \tag{5.7.11}$$

Thus stability exponents at fixed points of renormalization group recursion relations are equivalent to critical exponents of the critical point associated with the fixed point.

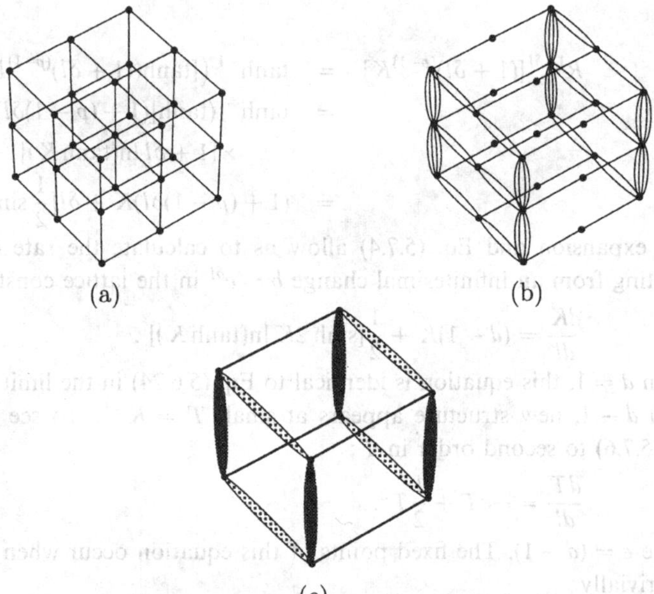

Fig. 5.7.3. The Migdal-Kadanoff procedure in three dimensions for $b = 2$. The original lattice is shown in (a). In step (1), bonds parallel to the z-axis are broken and moved to create bonds of strength $4K_z$. Then bonds parallel to the x-axis are broken and moved parallel to the y-axis to create new bonds of strength $2K_x$. The result is the lattice shown in (b). In step (2), sites, connected only by bonds parallel to the y-axis are removed by decimation, producing new bonds that are translated parallel to the x- and z-axes to produce bonds of strength $4R^2(K_y)$. Finally, in (3), sites connected only by bonds parallel to the x-axis are removed, producing new bonds that are translated to produce final bonds with strength $2R^2(2K_x)$, and sites connected only by z-bonds are removed to produce bonds of strength $R^2(4K_z)$. The result of steps (1) to (3) is depicted in (c).

2 General properties of recursion relations

The Migdal-Kadanoff treatment of the Ising model in $1+\epsilon$ dimensions provides a simple illustration of how the renormalization group describes critical points with divergent correlation lengths and how critical exponents can be calculated from the stability exponents of fixed points of recursion relations. This example is, however, artificially simple, and there are a number of features of more complete calculations that are of some interest. In a general renormalization group procedure, there is a Hamiltonian characterized by a set of potentials K_α (e.g. the nearest neighbor, next nearest neighbor, etc. interactions in an Ising model). Under rescaling and removal of degrees of freedom, the Hamiltonian changes and is characterized by a new set of potentials K'_α that are functions of the original set (i.e., $K'_\alpha = \Phi_\alpha(K_\alpha)$). The renormalization group equations,

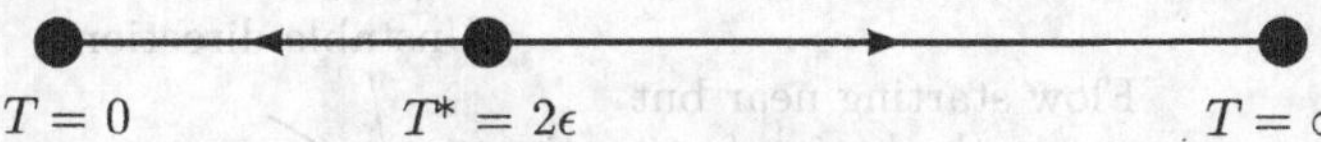

Fig. 5.7.4. Renormalization group flows for the Ising model in $1+\epsilon$ dimensions from the Migdal-Kadanoff procedure. The fixed point with $T^* = 2\epsilon$ describes the Ising critical point. It is unstable with respect to temperature. For $T < T^*$, the flow is toward the zero temperature fixed point of the ordered phase. For $T > T^*$, the flow is toward the the high temperature paramagnetic phase.

therefore, generate flows in a multi-dimensional space, as illustrated schematically in Fig. 5.7.5. Fixed points are typically unstable in a few directions in this space and stable in many others. In other words, associated with each fixed point, there is a local basin of attraction of dimensionality less than or equal to the total dimensionality of the space of potentials. All points in the basin of attraction of a fixed point flow toward the fixed point. Points outside of the basin of attraction flow away. Flow toward the fixed point is described by negative stability exponents, and flow away from the fixed point is described by positive exponents. Negative exponents are associated with *irrelevant potentials* (or fields), and positive exponents are associated with *relevant* potentials. There are two relevant fields: the temperature and the field conjugate to the order parameter, for fixed points describing standard second-order critical points. Higher order critical points, such as tricritical and tetracritical points, have more relevant fields. The positive exponents determine the dominant singularities of the free energy near a critical point. All points near the basin of attraction of a given fixed point flow away from the basin of attraction with rates determined by the positive exponents of the fixed points. Thus, all initial Hamiltonians near the basin of attraction of a fixed point have the same critical properties. This is universality. It follows directly from the renormalization group description of critical points.

3 *The Potts lattice gas and krypton on graphite*

As we have seen, the Migdal-Kadanoff scheme on a hypercubic lattice treats bonds along different directions differently. This is not the case for a triangular lattice. The bonds on a triangular lattice are on one of three grids of equally spaced parallel lines at relative angles of 120°. Divide the lines in each grid into two alternating sets as shown in Fig. 5.7.6a by full and dashed lines, respectively. Now translate every dashed bond one line over to create doubled bonds on each full line, as shown in Fig. 5.7.6b. The lattice parameter of the new lattice is twice that of the original lattice. Three of the four sites in each new unit cell are connected along only a single direction to neighboring sites by doubled bonds. A trace over these sites can now be performed as in a one-dimensional model.

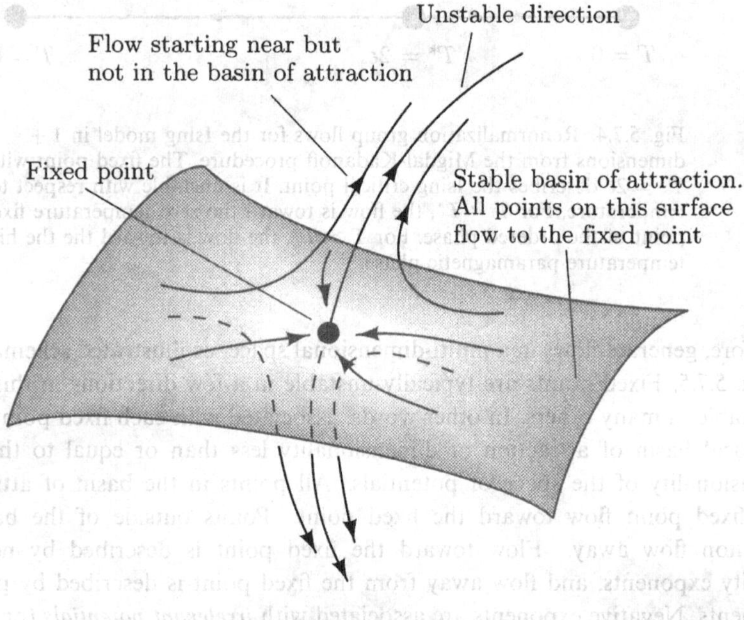

Fig. 5.7.5. Schematic representation of renormalization group flows in a high dimensional space. Fixed points describing ordered or disordered phases generally have a basin of attraction with dimension equal to that of the space of potentials. Fixed points describing critical points have a basin of attraction of lower dimensionality. Standard second-order critical points have fixed points unstable in two directions, and stable in the remaining dimensions.

After bond moving and the trace operation, the lattice has the same symmetry and same number of sites per cell as the original lattice.

Noble gas atoms adsorbed on a graphite substrate provide an example of a physical system with interesting phase transitions on a two-dimensional triangular lattice. We saw in Sec. 2.9 that krypton atoms are larger than the hexagons of graphite and that a $\sqrt{3} \times \sqrt{3}$ commensurate overlayer structure is favored. This suggests grouping the adsorption sites into a superlattice of cells each containing three sites, as shown in Fig. 5.7.7. Each cell can be in one of four states: it can be empty or it can have a krypton atom at any one of the three adsorption sites in the cell. Let $n_{\mathbf{l}}$ be the occupation number variable that is one if cell $\mathbf{l}$ is occupied and zero otherwise, and let $\sigma_{\mathbf{l}} = 1, 2, 3$ indicate which of the sites in the cell is occupied. The repulsive interaction between krypton atoms favors occupancy of equivalent adsorption sites in neighboring cells, i.e., it favors $\sigma_{\mathbf{l}} = \sigma_{\mathbf{l}'}$ if $\mathbf{l}$ and $\mathbf{l}'$ are nearest neighbors. A model which describes this repulsion and allows for partial occupancy of each cell is the *Potts lattice gas* with reduced Hamiltonian

$$\overline{\mathscr{H}} = -\sum_{<\mathbf{l},\mathbf{l}'>} [J(3\delta_{\sigma_{\mathbf{l}},\sigma_{\mathbf{l}'}} - 1) + K] n_{\mathbf{l}} n_{\mathbf{l}'} - \frac{1}{6}\Delta \sum_{<\mathbf{l},\mathbf{l}'>} (n_{\mathbf{l}} + n_{\mathbf{l}'}), \tag{5.7.12}$$

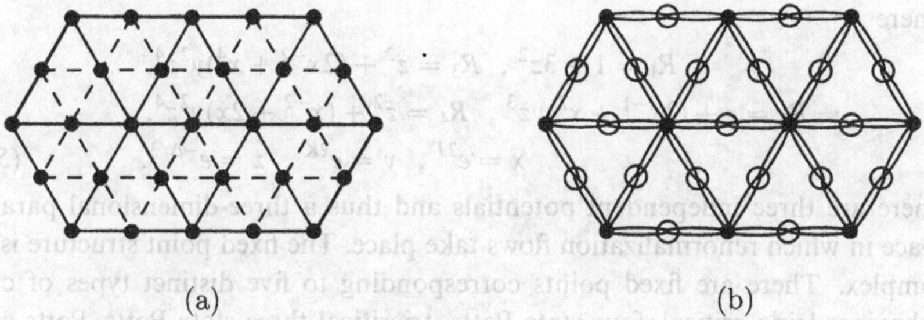

Fig. 5.7.6. (a) Triangular lattice showing three grids of parallel lines divided into two subsets shown, respectively, as full and dashed lines. (b) The same lattice after dashed bonds have been shifted over one line to create doubled bonds on the full lines. This lattice now has four sites per unit cell. Three of these (indicated by open circles) are connected to neighbors only along a single line.

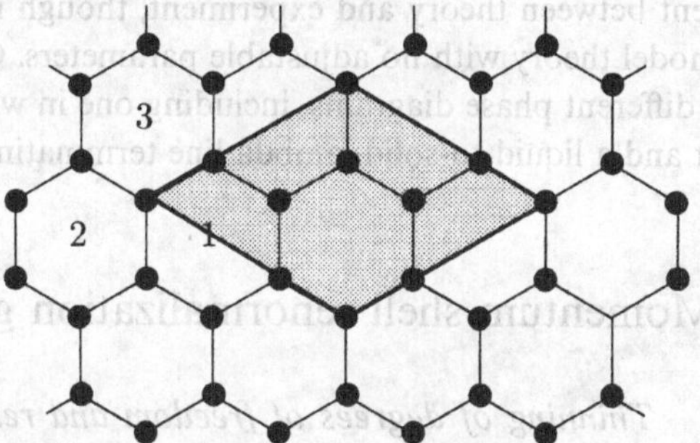

Fig. 5.7.7. Division of the triangular lattice of hexagonal adsorption sites into a superlattice of cells each containing three sites. A unit cell of the superlattice is shaded.

where the sum is over the nearest neighbor bonds $< \mathbf{l}, \mathbf{l}' >$ of the triangular lattice of cells.

Migdal-Kadanoff recursion relations can be derived for this Hamiltonian [Berker *et al.* (1978)]. They are

$$\begin{aligned} J' &= \frac{1}{3}\ln(R_3/R_4) \\ K' &= \frac{1}{3}\ln(R_1^3 R_3 R_4^2/R_2^6) \\ \Delta' &= 6\ln(R_1/R_2), \end{aligned} \tag{5.7.13}$$

where

$$R_1 = 1 + 3z^2 \ , \ R_3 = z^2 + (2x^{-2} + x^4)y^2z^4,$$
$$R_2 = z + (2x^{-1} + x^2)yz^3 \ , \ R_4 = z^2 + (x^{-2} + 2x)y^2z^4,$$
$$x = e^{2J} \ , \ y = e^{2K} \ , \ z = e^{-\Delta/3}. \tag{5.7.14}$$

There are three independent potentials and thus a three-dimensional parameter space in which renormalization flows take place. The fixed point structure is quite complex. There are fixed points corresponding to five distinct types of critical behavior: Ising critical, four-state Potts, tricritical three-state Potts, Potts critical endpoint, and a fourth-order critical point. In addition, there are fixed points describing first-order transitions at which the correlation length is zero rather than infinite. The connectivity of these fixed points is shown in Fig. 5.7.8.

From the Lennard-Jones potential between krypton atoms, the values of K/J and J are estimated to be 2.10 and $32.8/T$, where T is the temperature. With these values, the predicted phase diagram for krypton on graphite is shown in Fig. 5.7.9. Note that there is a liquid-to-solid lambda line described by the Potts-tricritical fixed point. Experimental points on this phase diagram are also plotted. The agreement between theory and experiment, though not quantitative, is quite good for a model theory with no adjustable parameters. Other values of the ratio K/J lead to different phase diagrams, including one in which there is a liquid-gas critical point and a liquid-to-solid lambda line terminating at a critical endpoint.

5.8 Momentum shell renormalization group

1 *Thinning of degrees of freedom and rescaling*

The renormalization group consists of a thinning of degrees of freedom followed by a rescaling of lengths. We have seen in the preceding sections how the thinning procedure can be implemented by tracing over variables on a subset of sites on a lattice. The renormalization group transformations can also be carried out by thinning degrees of freedom in momentum or wave number space rather than in real space. Recall that fields $\phi(\mathbf{x})$ defined on lattice sites $\mathbf{x}$ can be expressed in terms of their Fourier transformed fields $\phi(\mathbf{q})$, where $\mathbf{q}$ is restricted to the first Brillouin zone (BZ) of the reciprocal lattice. There are exactly N vectors $\mathbf{q}$ in the first BZ if the direct lattice contains N sites (see Appendix 2A). The volume of the first BZ in a reciprocal lattice associated with a hypercubic direct lattice with lattice constant a is $(2\pi/a)^d$, and the volume per point is $(2\pi/a)^d/N$. In the field theories introduced in Sec. 5.3, the lattice Brillouin zone is replaced by a spherical zone of the same volume with radius Λ defined via $\Omega_d\Lambda^d/d = (2\pi/a)^d$, where Ω_d is the solid angle of a d-dimensional sphere. The cutoff Λ is proportional to $(2\pi/a)$. The thinning of degrees of freedom now consists of tracing over all fields with wave vector $\mathbf{q}$ lying in the spherical shell defined by $\Lambda/b < q < \Lambda$. This

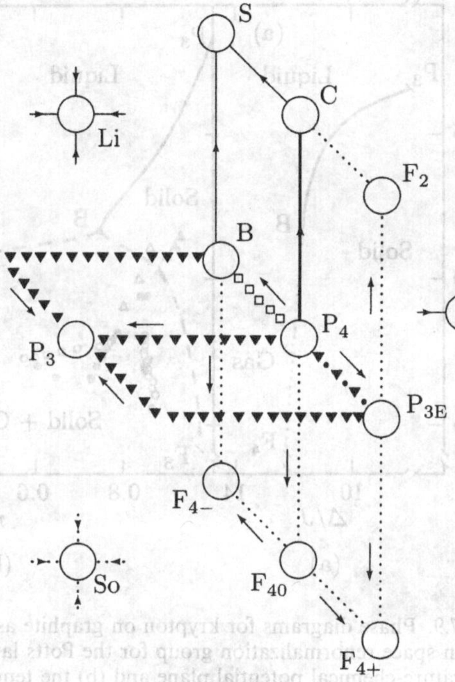

Fig. 5.7.8. Fixed-point connectivity for the recursion relations of Eq. (5.7.13). Renormalization group trajectories flowing through the various types of fixed points are indicated as follows: first-order (dotted line); Ising critical (dark solid line); Potts tricritical (filled triangles); Potts tricritical endpoint (triangle-dot); and fourth-order (open squares). The light solid lines (from C to S and from S to B) do not correspond to any phase transition. The fixed points in this diagram describing continuous transitions are as follows: C – Ising critical; P_4 – four-state Potts; P_3 – three-state Potts; P_{3E} – Potts tricritical endpoint; and B – fourth-order critical point. The fixed points describing discontinuous transitions are as follows: So – three-solid coexistence; $F_{4\pm}$ and F_{40} – three-solid-gas coexistence; and F_2 – liquid-gas coexistence. Finally, there are the fixed points Li, Ga, and S, describing, respectively, the liquid state, the gas state, and smooth continuation between the liquid and gas phases. [A.N. Berker, S. Ostlund, and F.A. Putman, *Phys. Rev. B* **17**, 3650 (1978).]

trace operation defines a new Hamiltonian with $N' = N/b^d$ degrees of freedom and an upper wave number cutoff Λ/b. The reduced cutoff is equivalent to an increased lattice spacing $a' = ba$. The next step is to rescale lengths so that distances are measured in the units natural to the new Hamiltonian. This is done by introducing a rescaled wave vector $\mathbf{q}' = b\mathbf{q}$. The cutoff for $\mathbf{q}'$ is Λ, as it was for $\mathbf{q}$ in the original system prior to thinning of degrees of freedom. This transformation is equivalent to the transformation $\mathbf{x}' = \mathbf{x}/b$ of the real space renormalization procedures. A final step in momentum-space renormalization

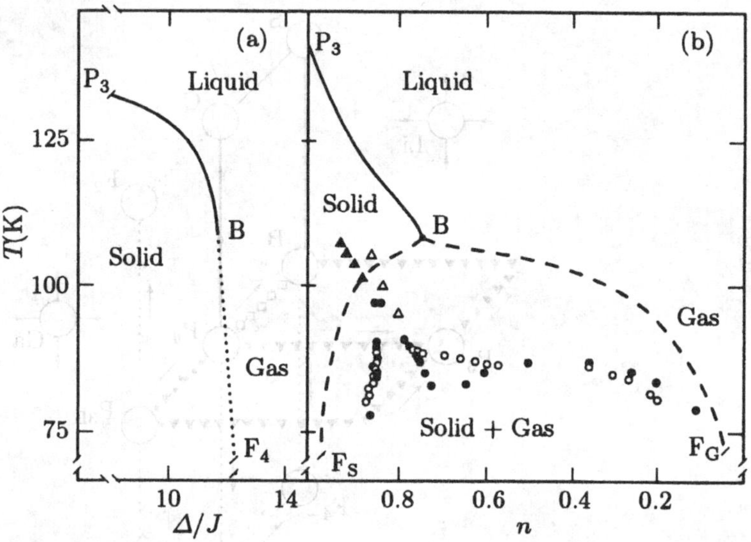

(a) (b)

Fig. 5.7.9. Phase diagrams for krypton on graphite as predicted by the position space renormalization group for the Potts lattice gas in (a) the temperature-chemical potential plane and (b) the temperature-density plane. The dotted line F_4-B in (a) is a line of first-order transitions corresponding to the solid-gas coexistence line F_S-B-F_G in (b). In both (a) and (b), P_3-B is a line of Potts tricritical transitions, and B is a fourth-order point. The experimental points in (b) were taken from various references. [A.N. Berker, S. Ostlund, and F.A. Putman, *Phys. Rev. B* **17**, 3650 (1978).]

group procedures is a rescaling of the fields $\phi(\mathbf{q})$ via $\phi(\mathbf{q}'/b) = \zeta\phi'(\mathbf{q}')$. The choice of the rescaling parameter ζ will be discussed below.

The counting of degrees of freedom is most straightforward if $\phi(\mathbf{x})$ is defined on a lattice. However, the low wave number Fourier transforms for fields defined on a lattice and those defined at all points in space are the same. Since the operations of the momentum shell renormalization group involve the Fourier transform fields $\phi(\mathbf{q})$, they can be used for either lattice or continuum models. In what follows, we will adopt the language of the latter approach.

To summarize, there are three steps in momentum-space renormalization group procedures.

(1) Thinning of degrees of freedom by tracing over fields $\phi(\mathbf{q})$ with $\Lambda/b < q < \Lambda$. This leads to a new Brillouin zone with radius (cutoff) Λ/b.

(2) Rescaling of lengths via $\mathbf{q}' = b\mathbf{q}$. The cutoff associated with $\mathbf{q}'$ is Λ.

(3) Rescaling of fields via $\phi(\mathbf{q}'/b) = \zeta\phi'(\mathbf{q}')$.

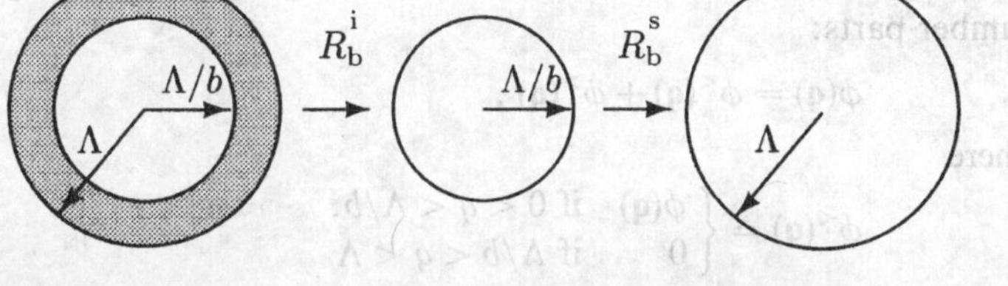

Fig. 5.8.1. Effect of thinning of degrees of freedom on the Brillouin zone. The original spherical Brillouin zone has radius Λ. All fields with wave vectors $\mathbf{q}$ satisfying $\Lambda/b < q < \Lambda$ are removed, leaving a new Brillouin zone with radius Λ/b. Next, wave vectors are rescaled via $\mathbf{q}' = b\mathbf{q}$. The Brillouin zone associated with $\mathbf{q}'$ is now Λ, the same as that for $\mathbf{q}$ before thinning. Note, however, that the total number of wave vectors $\mathbf{q}'$ is $N' = N/b$ rather than N.

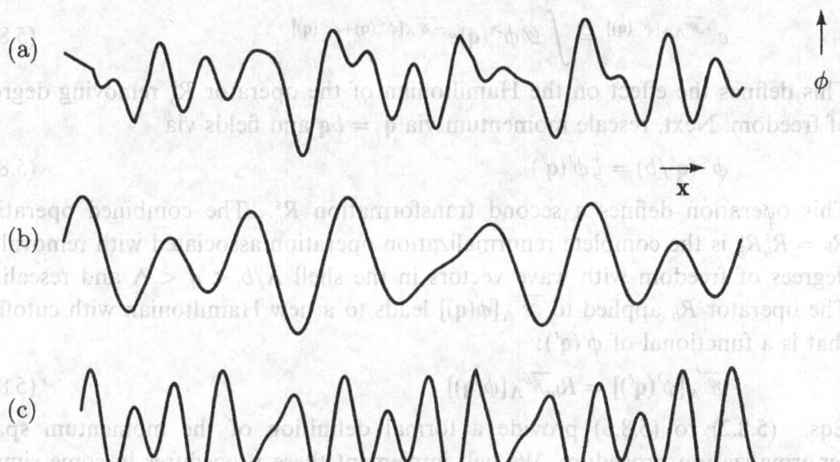

Fig. 5.8.2. (a) Schematic representation of the fields $\phi(\mathbf{x})$ with spatial variation in one dimension with maximum wave number in its Fourier transform equal to Λ. (b) The fields $\phi(\mathbf{x})$ after the high wave number degrees of freedom have been removed. The maximum wave number of oscillations is now Λ/b rather than Λ. (c) The rescaled field $\phi'(\mathbf{x}')$ as a function of the rescaled length $\mathbf{x}' = \mathbf{x}/b$. The maximum wave number in the rescaled units is now Λ so that the oscillations in $\phi'(\mathbf{x}')$ are similar to those of $\phi(\mathbf{x})$ in (a).

Steps (1) and (2) are depicted schematically in Fig. 5.8.1. The effects of these transformations on the real space field $\phi(\mathbf{x})$ are depicted in Fig. 5.8.2.

We now turn to how these renormalization procedures are actually implemented for field theories of the type discussed in Sec. 5.3. For simplicity, we will restrict our attention for the moment to systems with Ising symmetry and a single scalar field $\phi(\mathbf{x})$. First, the reduced Hamiltonian $\overline{\mathscr{H}}_\Lambda$ with cutoff Λ is expressed as a functional of the fields $\phi(\mathbf{q})$, which are decomposed into high and low wave number parts:

$$\phi(\mathbf{q}) = \phi^{<}(\mathbf{q}) + \phi^{>}(\mathbf{q}) , \tag{5.8.1}$$

where

$$\phi^{<}(\mathbf{q}) = \begin{cases} \phi(\mathbf{q}) & \text{if } 0 < q < \Lambda/b; \\ 0 & \text{if } \Lambda/b < q < \Lambda \end{cases} \tag{5.8.2}$$

$$\phi^{>}(\mathbf{q}) = \begin{cases} 0 & \text{if } 0 < q < \Lambda/b; \\ \phi(\mathbf{q}) & \text{if } \Lambda/b < q < \Lambda. \end{cases} \tag{5.8.3}$$

Fields $\phi^{<}(\mathbf{x})$ and $\phi^{>}(\mathbf{x})$ are defined as the Fourier transforms of $\phi^{<}(\mathbf{q})$ and $\phi^{>}(\mathbf{q})$. Clearly, $\phi(\mathbf{x}) = \phi^{<}(\mathbf{x}) + \phi^{>}(\mathbf{x})$. Thinning of degrees of freedom is accomplished by integrating out the fields $\phi^{>}(\mathbf{q})$. This leads to a new Hamiltonian $\overline{\mathscr{H}}_{\Lambda/b}[\phi^{<}(\mathbf{q})] \equiv R_b^i \overline{\mathscr{H}}_{\Lambda}[\phi(\mathbf{q})]$ with cutoff Λ/b defined via

$$e^{-\overline{\mathscr{H}}_{\Lambda/b}[\phi^{<}(\mathbf{q})]} = \int \mathscr{D}\phi^{>}(\mathbf{q}) e^{-\overline{\mathscr{H}}_{\Lambda}[\phi^{<}(\mathbf{q})+\phi^{>}(\mathbf{q})]} . \tag{5.8.4}$$

This defines the effect on the Hamiltonian of the operator R_b^i removing degrees of freedom. Next, rescale momentum via $\mathbf{q}' = b\mathbf{q}$ and fields via

$$\phi^{<}(\mathbf{q}'/b) = \zeta\phi'(\mathbf{q}') . \tag{5.8.5}$$

This operation defines a second transformation R_b^s. The combined operation $R_b = R_b^s R_b^i$ is the complete renormalization operation associated with removal of degrees of freedom with wave vectors in the shell $\Lambda/b < q < \Lambda$ and rescaling. The operator R_b applied to $\overline{\mathscr{H}}_{\Lambda}[\phi(\mathbf{q})]$ leads to a new Hamiltonian with cutoff Λ that is a functional of $\phi'(\mathbf{q}')$:

$$\overline{\mathscr{H}}'_{\Lambda}[\phi'(\mathbf{q}')] = R_b \overline{\mathscr{H}}_{\Lambda}[\phi(\mathbf{q})] . \tag{5.8.6}$$

Eqs. (5.8.2) to (5.8.6) provide a formal definition of the momentum space renormalization procedure. We will implement these procedures in some simple situations below.

2 Correlation functions

Before carrying out explicit calculations, however, it is useful to consider how correlation functions can be calculated using either the original Hamiltonian $\overline{\mathscr{H}}_{\Lambda}$ or the transformed Hamiltonian $\overline{\mathscr{H}}'_{\Lambda}$. This will provide us with some guidance as to how the field rescaling factor ζ should be chosen. If $q < \Lambda/b$, then $\phi(\mathbf{q}) = \phi^{<}(\mathbf{q}) = \zeta\phi'(b\mathbf{q})$, and we can use either $\overline{\mathscr{H}}_{\Lambda}$ or $\overline{\mathscr{H}}'_{\Lambda}$ to calculate its correlation function. In particular,

$$\begin{aligned} \langle\phi(\mathbf{q}_1)\phi(\mathbf{q}_2)\rangle &= \frac{1}{Z}\int \mathscr{D}\phi(\mathbf{q}) e^{-\overline{\mathscr{H}}_{\Lambda}[\phi(\mathbf{q})]} \phi(\mathbf{q}_1)\phi(\mathbf{q}_2), \\ &= \frac{1}{Z'}\int \mathscr{D}\phi'(\mathbf{q}') e^{-\overline{\mathscr{H}}'_{\Lambda}[\phi'(\mathbf{q}')]} \zeta^2 \phi'(b\mathbf{q}_1)\phi'(b\mathbf{q}_2) \end{aligned} \tag{5.8.7}$$

where $\mathbf{q}' = b\mathbf{q}$ and $Z' = \int \mathscr{D}\phi'(\mathbf{q}')e^{-\overline{\mathscr{H}}'_\Lambda[\phi'(\mathbf{q}')]}$. The correlation functions are non-zero in translationally invariant systems only if $\mathbf{q}_1 = -\mathbf{q}_2$, and they are proportional to the Fourier transform $G(\mathbf{q})$ of the correlation function $\langle\phi(\mathbf{x})\phi(\mathbf{x}')\rangle$. The first line of Eq. (5.8.7) is, therefore,

$$\langle\phi(\mathbf{q}_1)\phi(\mathbf{q}_2)\rangle = G(\mathbf{q}_1)V\delta_{\mathbf{q}_1,-\mathbf{q}_2} = G(\mathbf{q}_1)(2\pi)^d\delta(\mathbf{q}_1+\mathbf{q}_2). \tag{5.8.8}$$

Here it is understood that $G(\mathbf{q})$ depends on the potentials in the original Hamiltonian $\overline{\mathscr{H}}_\Lambda$. The second line of Eq. (5.8.7) yields

$$\langle\phi'(b\mathbf{q}_1)\phi'(b\mathbf{q}_2)\rangle = G'(b\mathbf{q}_1)(2\pi)^d\delta(b\mathbf{q}_2+b\mathbf{q}_2), \tag{5.8.9}$$

where $G'(\mathbf{q})$ depends on the potentials in the renormalized Hamiltonian $\overline{\mathscr{H}}'_\Lambda$. Because $\delta(b\mathbf{q}) = b^{-d}\delta(\mathbf{q})$, these equations imply that

$$G(\mathbf{q}) = \zeta^2 b^{-d}G'(b\mathbf{q})\,. \tag{5.8.10}$$

This equation, when applied at the critical point where both $G(\mathbf{q})$ and $G'(\mathbf{q})$ are proportional to $q^{-(2-\eta)}$, implies

$$\frac{A}{q^{2-\eta}} = \zeta^2 b^{-d}\frac{A'}{(bq)^{2-\eta}} = \zeta^2 b^{-(d+2-\eta)}\frac{A'}{q^{2-\eta}}\,, \tag{5.8.11}$$

where A is the critical point amplitude arising from $\overline{\mathscr{H}}_\Lambda$, and A' is that arising from the rescaled Hamiltonian $\overline{\mathscr{H}}'_\Lambda$. Thus

$$\zeta^2 = b^{(d+2-\eta)}(A/A')\,. \tag{5.8.12}$$

The ratio A/A' can depend on b. We can, however, use our freedom to choose ζ to fix $A = A'$. In this case

$$\zeta = b^{(d+2-\eta)/2}\,, \tag{5.8.13}$$

and $G(\mathbf{q}) = Aq^{-(2-\eta)}$ when calculated using either $\overline{\mathscr{H}}_\Lambda$ or $\overline{\mathscr{H}}'_\Lambda$. Though other choices for ζ are sometimes useful, this is the choice that is generally used.

3 *The Gaussian model*

The momentum space renormalization procedure can be carried out exactly for the Gaussian model with reduced Hamiltonian,

$$\begin{aligned}\overline{\mathscr{H}}_{0,\Lambda} &= \frac{1}{2}\int_{\mathbf{q}}^{\Lambda}(r+cq^2)|\phi(\mathbf{q})|^2 + V_0 \\ &= \int_{\mathbf{q}}^{<}(r+cq^2)|\phi^<(\mathbf{q})|^2 + \int_{\mathbf{q}}^{>}(r+cq^2)|\phi^>(\mathbf{q})|^2 + V_0 \\ &= \overline{\mathscr{H}}^<_{0,\Lambda} + \overline{\mathscr{H}}^>_{0,\Lambda} + V_0,\end{aligned} \tag{5.8.14}$$

where V_0 is a constant potential independent of $\phi(\mathbf{q})$ analogous to the constant potential C introduced in our discussion of the Ising model in Sec. 5.7, and where

$$\int_{\mathbf{q}}^{\Lambda} \equiv \int_0^{\Lambda}\frac{d^dq}{(2\pi)^d}\,, \quad \int_{\mathbf{q}}^{<} \equiv \int_0^{\Lambda/b}\frac{d^dq}{(2\pi)^2}\,, \quad \int_{\mathbf{q}}^{>} \equiv \int_{\Lambda/b}^{\Lambda}\frac{d^dq}{(2\pi)^d}\,. \tag{5.8.15}$$

(Throughout this section, we will use r, c, and u as parameters in the reduced Hamiltonian. They differ from our usual definitions by a factor of $1/T$.) The decomposition of $\overline{\mathscr{H}}_\Lambda$ into a part depending only on $\phi^<$ and a part depending only on $\phi^>$ makes the evaluation of $\overline{\mathscr{H}}_{\Lambda/b}$ trivial:

$$\begin{aligned} e^{-\overline{\mathscr{H}}_{\Lambda/b}[\phi^<(\mathbf{q})]} &= e^{-\overline{\mathscr{H}}^<_{0,\Lambda}} \int \mathscr{D}\phi^>(\mathbf{q}) e^{-\overline{\mathscr{H}}^>_{0,\Lambda}[\phi^>(\mathbf{q})]} e^{-V_0} \\ &\equiv e^{-\overline{H}^<_{0,\Lambda} - V_0} Z_0^<, \end{aligned} \tag{5.8.16}$$

or

$$\overline{\mathscr{H}}_{\Lambda/b} = \overline{\mathscr{H}}^<_{0,\Lambda} + V_0 - \ln Z_0^>. \tag{5.8.17}$$

Thus, after rescaling $\phi^<(\mathbf{q})$ according to Eq. (5.8.5),

$$\overline{\mathscr{H}}'_\Lambda = V_0' + \frac{1}{2} \int_{\mathbf{q}'}^{\Lambda} (r' + c' q'^2) |\phi'(\mathbf{q}')|^2, \tag{5.8.18}$$

where

$$r' = \zeta^2 b^{-d} r = b^2 r, \tag{5.8.19}$$

$$c' = \zeta^2 b^{-(d+2)} c = c, \tag{5.8.20}$$

where we used $\zeta^2 = b^{d+2}$ since η is trivially zero for this model. The constant part V_0' of $\overline{\mathscr{H}}'_\Lambda$ is $V_0 - \ln Z_0^>$ plus an additional part arising from the Jacobian associated with the change of variables $\phi^<(\mathbf{q}'/b) = \zeta\phi'(\mathbf{q}')$. We will not consider this term further. We note, however, that it is needed in the calculation of the free energy. Note that choosing ζ to satisfy Eq. (5.8.13) is equivalent to keeping the coefficient c constant. In fact, for the Gaussian model the amplitude of A of Eq. (5.8.11) is $1/c$. Usually, c is fixed at unity initially, and it remains unity under successive applications of R_b. The correlation length,

$$\xi' = (c'/r')^{1/2} = b^{-1}(c/r)^{1/2} = b^{-1}\xi, \tag{5.8.21}$$

of the rescaled Hamiltonian is b^{-1} times the correlation length ξ of the original Hamiltonian, regardless of the choice of ζ.

The recursion relation for r [Eq. (5.8.19)] has two fixed points: $r^* = 0$ and $r^* = \infty$. The latter is the high-temperature fixed point describing the disordered phase; the former is the Gaussian fixed point describing the critical point at $T = T_c$. The exponent $\lambda_t = 1/\nu$ controlling the growth of r in the vicinity of the critical fixed point is 2. Thus, the renormalization group predicts

$$\nu = \frac{1}{2} \tag{5.8.22}$$

for the Gaussian critical point. This is in agreement with the direct calculations presented in the preceding chapter. We have already indicated that the exponent η is zero. Thus, the values of the exponents ν and η for the Gaussian fixed point are the same as those of the mean-field theory.

As for the Migdal-Kadanoff equations, it is often more elegant to rescale lengths by an infinitesimal amount at each iteration of R_b. This is accomplished by removing an infinitesimal shell of width δl of wave vectors by setting $b = 1+\delta l$. A rescaling of lengths by a factor e^l results from repeated rescalings by $1 + \delta l$ via $e^l = \lim_{\delta l \to 0}(1 + \delta l)^{l/\delta l}$. Potentials and the correlation length $\xi(l)$ can thus be regarded as continuous functions of l. Recursion relations become differential equations. In particular, from Eq. (5.8.19) with $b = (1 + \delta l)$

$$r(l + \delta l) = (1 + \delta l)^2 r(l) \tag{5.8.23}$$

or

$$\frac{dr(l)}{dl} = 2r(l) \,. \tag{5.8.24}$$

Integration of this equation leads to

$$r(l) = e^{2l} r(0), \tag{5.8.25}$$

and $e^l = [r(l)/r(0)]^{1/2}$.

4 The ϵ-expansion

Exact momentum shell recursion relations can be obtained for the Gaussian model because there is no cross coupling between the high- and low-momentum fields $\phi^>$ and $\phi^<$. The addition of terms to the Gaussian Hamiltonian cubic or higher order in ϕ will lead to such couplings and make the evaluation of $\overline{\mathscr{H}}_{\Lambda/b}(\phi^<)$ non-trivial. Before attempting to deal with non-linear terms in $\mathscr{H}_\Lambda$, let us investigate how potentials scale in the vicinity of the Gaussian fixed point. Consider the Hamiltonian

$$\begin{aligned} \overline{\mathscr{H}}^p_\Lambda &= u_p \int d^d x \phi^p(\mathbf{x}) \\ &= u_p \int_{\mathbf{q}_1, \mathbf{q}_2, \ldots, \mathbf{q}_{p-1}} \phi(\mathbf{q}_1)\phi(\mathbf{q}_2)\ldots\phi(-\mathbf{q}_1 - \cdots - \mathbf{q}_{p-1}) \end{aligned} \tag{5.8.26}$$

of order ϕ^p. We can calculate how u_p scales near the Gaussian fixed point by replacing $\phi(\mathbf{q})$ by $\phi^<(\mathbf{q})$ and rescaling $\mathbf{q}$ according to $\mathbf{q}' = b\mathbf{q}$ and $\phi^<(\mathbf{q})$ according to Eq. (5.8.5) with $\zeta = b^{(d+2)/2}$. This leads to

$$u_p' = \zeta^p b^{-(p-1)d} u_p = b^{[-(p-2)d+2p]/2} u_p \equiv b^{\lambda_p} u_p \tag{5.8.27}$$

because there are $p-1$ $\mathbf{q}$-integrals in Eq. (5.8.26). The exponent $\lambda_p = [2p - (p-2)d]/2$ is $(d+2)/2$ for $p = 1$ and 2 for $p = 2$. Thus, the external field h $(p = 1)$ and the temperature r $(p = 2)$ are relevant at the Gaussian fixed point in all dimensions. For $p > 2$, λ_p is negative, and u_p will flow to zero and is irrelevant at the Gaussian fixed point for all dimensions

$$d > d_c(p) = \frac{2p}{p-2} \,. \tag{5.8.28}$$

Thus, $d_c(3) = 6$, $d_c(4) = 4$, and $d_c(p > 4) < 4$.

The Hamiltonian $\overline{\mathscr{H}}^p_\Lambda$ in Eq. (5.8.26) has a potential of zero range in that it couples fields at a single point in space. Potentials with nonzero range are also permitted by symmetry. For example, $\overline{\mathscr{H}}^4_\Lambda$ in Eq. (5.8.26) could be replaced by

$$\begin{aligned} \overline{\mathscr{H}}^4_\Lambda &= \int d^d x_1 d^d x_2 d^d x_3 d^d x_4 u(\mathbf{x}_1, \mathbf{x}_2, \mathbf{x}_3, \mathbf{x}_4) \\ &\quad \times \phi(\mathbf{x}_1)\phi(\mathbf{x}_2)\phi(\mathbf{x}_3)\phi(\mathbf{x}_4) \end{aligned} \tag{5.8.29}$$

$$= \int_{\mathbf{q}_1,\mathbf{q}_2,\mathbf{q}_3,\mathbf{q}_4} u(\mathbf{q}_1,\mathbf{q}_2,\mathbf{q}_3,\mathbf{q}_4)\phi(\mathbf{q}_1)\phi(\mathbf{q}_2)\phi(\mathbf{q}_3)\phi(\mathbf{q}_4) \quad (5.8.30)$$

where

$$u(\mathbf{q}_1,\mathbf{q}_2,\mathbf{q}_3,\mathbf{q}_4) = \int d_1^x d^d x_2 d^d x_3 d^d x_4 e^{-i(\mathbf{q}_1\cdot\mathbf{x}_1+\mathbf{q}_2\cdot\mathbf{x}_2+\mathbf{q}_3\cdot\mathbf{x}_3+\mathbf{q}_4\cdot\mathbf{x}_4)} \times u(\mathbf{x}_1,\mathbf{x}_2,\mathbf{x}_3,\mathbf{x}_4) \quad (5.8.31)$$

$$= u(\mathbf{q}_1,\mathbf{q}_2,\mathbf{q}_3)(2\pi)^d\delta(\mathbf{q}_1+\mathbf{q}_2+\mathbf{q}_3+\mathbf{q}_4). \quad (5.8.32)$$

The local interaction of Eq. (5.8.26) is retrieved by setting

$$u(\mathbf{x}_1,\mathbf{x}_2,\mathbf{x}_3,\mathbf{x}_4) = u\delta(\mathbf{x}_1-\mathbf{x}_2)\delta(\mathbf{x}_2-\mathbf{x}_3)\delta(\mathbf{x}_3-\mathbf{x}_4).$$

Then $u(\mathbf{q}_1,\mathbf{q}_2,\mathbf{q}_3) = u$ is independent of wave number. More generally, $u(\mathbf{q}_1,\mathbf{q}_2,\mathbf{q}_3)$ depends on wave number and, like $r(\mathbf{q}) = r + cq^2$, can be expanded in powers of wave number:

$$u(\mathbf{q}_1,\mathbf{q}_2,\mathbf{q}_3) = u + O(q^2), \quad (5.8.33)$$

where q^2 represents any quadratic combination of $\mathbf{q}_1$, $\mathbf{q}_2$, and $\mathbf{q}_3$ permitted by symmetry. Using the scaling procedures of the preceding paragraph, it is straightforward to show that the coefficient of q^2 in $u(\mathbf{q}_1,\mathbf{q}_2,\mathbf{q}_3)$ scales as $\zeta^4 b^{-3d} b^{-2} = b^{2-d}$ and is more irrelevant than $u = u_4$. In general, any potential of any order can be expanded in powers of $\mathbf{q}$. The lowest power of $\mathbf{q}$ is always the most relevant.

To simplify our discussion, we will now restrict our attention to Ising and O_n models with no external magnetic fields so that no odd order potentials are permitted. In this case, the Gaussian fixed point above four dimensions is stable with respect to all potentials except the temperature variable r. This means that the Gaussian fixed point with its mean-field exponents determines the critical properties of all O_n models above four dimensions. Just below four dimensions, the potential $u_4 = u$ in addition to r is relevant at the Gaussian fixed point, and we expect flow away from the Gaussian fixed point with $u = 0$ to a new fixed point with $u = u^* > 0$, with u^* approaching zero as $d \to 4^-$. Critical exponents at the new fixed point will depend on u^* and will approach mean-field values as $d \to 4^-$. Thus, the momentum shell renormalization group should predict a change from mean-field to non-mean-field critical exponents when d falls below the upper critical dimension $d_c = 4$, in agreement with the heuristic arguments of preceding sections.

Since u^* is expected to be small near four dimensions, and all other potentials except r are irrelevant, we can restrict our attention to a Hamiltonian parametrized only by r and u with u small. Other Hamiltonians with more potentials will flow toward this Hamiltonian under successive iterations of the renormalization group. Since u is small, we can calculate the renormalized Hamiltonian $\overline{\mathscr{H}}'_\Lambda$ and recursion relations perturbatively in u. Our Hamiltonian is thus

$$\overline{\mathscr{H}}_\Lambda = \overline{\mathscr{H}}_{0,\Lambda} + \overline{\mathscr{H}}^4_\Lambda, \quad (5.8.34)$$

where $u_4 = u$. The Hamiltonian $\overline{\mathscr{H}}_{\Lambda/b}(\phi^<)$ has the same form as $\overline{\mathscr{H}}_\Lambda$:

$$\overline{\mathcal{H}}_{\Lambda/b} = \int d^d x \left[\frac{1}{2} r^< (\phi^<)^2 + \frac{1}{2} c^< (\nabla \phi^<)^2 + u^< (\phi^<)^4 \right]. \tag{5.8.35}$$

It is obtained as before by integrating over $\phi^>$:

$$\begin{aligned} e^{-[\overline{\mathcal{H}}_{\Lambda/b}(\phi^<) - \overline{\mathcal{H}}^<_{0,\Lambda}(\phi^<)]} &= \int \mathcal{D}\phi^> e^{-\overline{\mathcal{H}}^>_{0,\Lambda}(\phi^>)} e^{-\overline{\mathcal{H}}^4_\Lambda(\phi^< + \phi^>)} \\ &= Z_0^> \langle e^{-\overline{\mathcal{H}}^4_\Lambda} \rangle_>, \end{aligned} \tag{5.8.36}$$

where

$$Z_0^> = \int \mathcal{D}\phi^> e^{-\overline{\mathcal{H}}^>_{0,\Lambda}} \tag{5.8.37}$$

and

$$\langle A \rangle_> = \frac{1}{Z_0^>} \int \mathcal{D}\phi^> e^{-\overline{\mathcal{H}}^>_{0,\Lambda}} A. \tag{5.8.38}$$

The exponential in Eq. (5.8.36) can be expanded in powers of $\mathcal{H}^4_\Lambda$:

$$\begin{aligned} \langle e^{-\overline{\mathcal{H}}^4_\Lambda} \rangle_> &= 1 - \langle \overline{\mathcal{H}}^4_\Lambda \rangle_> + \frac{1}{2} \langle (\overline{\mathcal{H}}^4_\Lambda)^2 \rangle_> + \cdots \\ &= \exp\{ -\langle \overline{\mathcal{H}}^4_\Lambda \rangle_> + \frac{1}{2} [\langle (\overline{\mathcal{H}}^4_\Lambda)^2 \rangle_> - \langle \overline{\mathcal{H}}^4_\Lambda \rangle_>^2] + \cdots \}. \end{aligned} \tag{5.8.39}$$

The evaluation of $\overline{\mathcal{H}}_{\Lambda/b}(\phi^<)$ is thus reduced to the evaluation of averages of powers of $\overline{\mathcal{H}}^4_\Lambda$ with respect to $\overline{\mathcal{H}}_{0,\Lambda}$.

Consider first $\langle \overline{\mathcal{H}}^4_\Lambda \rangle_>$. The binomial expansion of $(\phi^<(\mathbf{x}) + \phi^>(\mathbf{x}))^4$ leads to five distinct terms in this quantity, each with a different power of $\phi^>(\mathbf{x})$. The average with respect to $\overline{\mathcal{H}}^>_{0,\Lambda}$ of any odd power of $\phi^>(\mathbf{x})$ is zero, and only the three terms even in $\phi^>(\mathbf{x})$ are nonzero:

$$\begin{aligned} \langle \overline{\mathcal{H}}^4_\Lambda \rangle_> &= u \int d^d x (\phi^<(\mathbf{x}))^4 + 6u \int d^d x (\phi^<(\mathbf{x}))^2 \langle (\phi^>(\mathbf{x}))^2 \rangle_> \\ &\quad + u \int d^d x \langle (\phi^>(\mathbf{x}))^4 \rangle_>. \end{aligned} \tag{5.8.40}$$

This equation and Eq. (5.8.36) for $\overline{\mathcal{H}}_{\Lambda/b}$ imply $u^< = u$ and

$$r^< = r + 12u \int_{\mathbf{q}}^{>} G_0(\mathbf{q}), \tag{5.8.41}$$

where $G_0(\mathbf{q}) = (r + q^2)^{-1}$ when $c = 1$ and where $\langle (\phi^>(\mathbf{x}))^2 \rangle_>$ is expressed as an integral over $\mathbf{q}$ of $G_0(\mathbf{q})$. The contributions to $\overline{\mathcal{H}}_{\Lambda/b}$ to second order in u can be calculated in much the same way as the first-order contributions after expressing the average of products of $\phi^>(\mathbf{x})$ in terms of $G_0(\mathbf{q})$. In Appendix 5B, we will calculate both first-order and second-order contributions for O_n rather than Ising models, after outlining the rudiments of diagrammatic perturbation theory.

To obtain $\overline{\mathcal{H}}'_\Lambda$, we use the results of Appendix 5B for $r^<$, $c^<$ and $u^<$ and rescale $\phi^<(\mathbf{q})$ according to Eq. (5.8.5). The latter rescaling leads to $r' = b^{-d}\zeta^2 r^<$ and $u' = b^{-3d}\zeta^4 u^<$. The recursion relations for O_n models are then

$$r' = b^{-d}\zeta^2[r + 4(n+2)u \int_{\mathbf{q}}^{>} G_0(\mathbf{q})] \tag{5.8.42}$$

$$c' = b^{-(d+2)}\zeta^2 c = 1 \tag{5.8.43}$$

$$u' = b^{-3d}\zeta^4[u - 4(n+8)u^2 \int_{\mathbf{q}}^{>} G_0^2(\mathbf{q})]. \tag{5.8.44}$$

Note that we have retained contributions to the recursion relation for u' to second order in u, whereas we have retained only terms linear in u in the equations for r' and c'. This is because, as we shall see below, the second-order term in u in the equation for u' is essential if u is to have a nontrivial fixed point. The fixed point value for u is then of order $\epsilon = 4 - d$. Thus u^2 terms in the r' and c' equations are of order ϵ^2 and do not contribute at order ϵ. There are no terms in the equation for c' that are linear in u. Thus, to keep $c = 1$, we choose $\zeta^2 = b^{d+2}$ as in the Gaussian case. Taking the continuum limit $b \to (1 + \delta l)$, we obtain the differential momentum shell recursion relations:

$$\frac{dr(l)}{dl} = 2r(l) + 4K_d(n+2)\frac{u(l)}{1+r(l)} \tag{5.8.45}$$

$$\frac{du(l)}{dl} = \epsilon u(l) - 4K_d(n+8)\frac{u^2(l)}{(1+r(l))^2}, \tag{5.8.46}$$

where $\epsilon = 4 - d$. There are two fixed points to these equations as shown in Fig. 5.8.3: the Gaussian fixed point with $r = u = 0$ and a new Heisenberg fixed point with

$$u^* = \frac{\epsilon}{4(n+8)K_d} + O(\epsilon^2)$$
$$r^* = -\frac{1}{2}\frac{n+2}{n+8}\epsilon + O(\epsilon^2). \tag{5.8.47}$$

The Gaussian fixed point is unstable with respect to u and r. The Heisenberg fixed point, on the other hand, is stable with respect to u and unstable with respect to r. Linearizing the recursion relations Eqs. (5.8.45) and (5.8.46) about the Heisenberg fixed point, we obtain

$$\begin{pmatrix} d\delta r/dl \\ d\delta u/dl \end{pmatrix} = \begin{pmatrix} 2 - \frac{n+2}{n+8}\epsilon & \frac{4(n+2)K_d}{1+r^*} \\ 0 & -\epsilon \end{pmatrix} \begin{pmatrix} \delta r \\ \delta u \end{pmatrix}, \tag{5.8.48}$$

where $\delta r = r - r^*$ and $\delta u = u - u^*$. The eigenvalues of the matrix in the above equation are

$$\lambda_t = \frac{1}{\nu} = 2 - \frac{n+2}{n+8}\epsilon \tag{5.8.49}$$

$$\lambda_u = -\epsilon\,. \tag{5.8.50}$$

The flows implied by these equations are shown in Fig. 5.8.3. Any point along the line connecting the Heisenberg fixed point to the Gaussian fixed point will flow toward the Heisenberg fixed point. Thus, all initial Hamiltonians along this line are in the same universality class and have the same positive exponents.

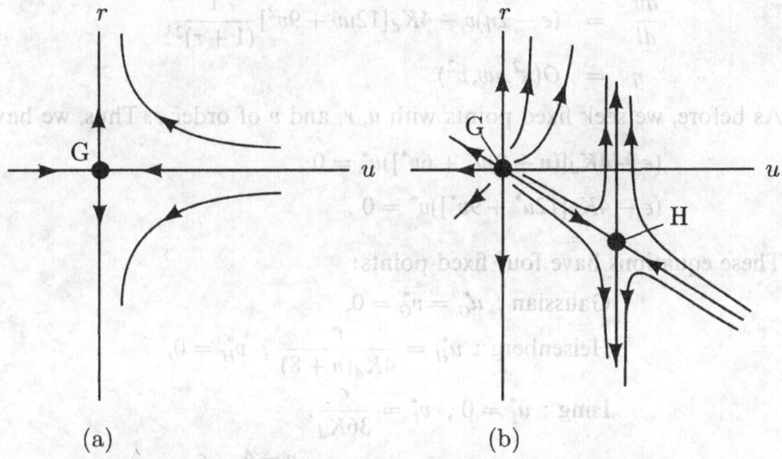

Fig. 5.8.3. Fixed points for the O_n model (a) for $d > 4$ and (b) for $d < 4$. When $d > 4$, the Gaussian fixed point G is stable with respect to u. When $d < 4$, the Gaussian fixed point is unstable with respect to u, and a new Heisenberg fixed point H with $u^* \sim \epsilon$ appears.

5 *n-vector model with cubic anisotropy*

In the n-vector model we have been studying, there are only two fixed points within the ϵ-expansion: the Gaussian fixed point and the Heisenberg fixed point. This structure remains valid to at least third order in ϵ and is almost certainly valid for all $2 < d < 4$. Far more complicated fixed point structures are possible. In this section, we will study the n-vector model with cubic anisotropy. This is the paradigm for systems with more complicated fixed point structure.

The reduced Hamiltonian for a Heisenberg model with cubic anisotropy is

$$\overline{\mathscr{H}} = \int d^d x \left\{ \sum_i \left[\frac{1}{2} r \phi_i^2 + \frac{1}{2} (\nabla \phi_i)^2 \right] + u (\sum_i \phi_i^2)^2 + v \sum_i \phi_i^4 \right\}. \quad (5.8.51)$$

This model is equivalent to that of Eq. (4.6.18) with $g = 0$ and $u_1 = u_2$ introduced in our discussion of mean-field bicritical and tetracritical points. It differs from the usual n-vector model Hamiltonian by the term containing ϕ_i^4. This term represents the effects of lattice anisotropy on spin order and should always be present in $n = 3$ Heisenberg systems on a cubic lattice. For $d > 4$, only r is relevant, and the Gaussian fixed point is the only fixed point. For $d < 4$, both u and v are relevant, and recursion relations for both must be treated together. Proceeding as before, we obtain to first order in ϵ,

$$\frac{dr}{dl} = (2-\eta) r + 4K_d [(n+2)u + 3v] \frac{1}{(1+r)}, \quad (5.8.52)$$

$$\frac{du}{dl} = (\epsilon - 2\eta) u - 4K_d [(n+8)u^2 + 6uv] \frac{1}{(1+r)^2}, \quad (5.8.53)$$

$$\frac{dv}{dl} = (\epsilon - 2\eta)v - 4K_d[12uv + 9v^2]\frac{1}{(1+r)^2}, \tag{5.8.54}$$

$$\eta = O(u^2, uv, v^2). \tag{5.8.55}$$

As before, we seek fixed points with u, r, and v of order ϵ. Thus, we have

$$(\epsilon - 4K_d[(n+8)u^* + 6v^*])u^* = 0\,, \tag{5.8.56}$$

$$(\epsilon - 4K_d[12u^* + 9v^*])v^* = 0\,. \tag{5.8.57}$$

These equations have four fixed points:

$$\text{Gaussian}: \; u_G^* = v_G^* = 0, \tag{5.8.58}$$

$$\text{Heisenberg}: \; u_H^* = \frac{\epsilon}{4K_d(n+8)}\,, \quad v_H^* = 0, \tag{5.8.59}$$

$$\text{Ising}: \; u_I^* = 0\,, \quad v_I^* = \frac{\epsilon}{36K_d}, \tag{5.8.60}$$

$$\text{cubic}: \; u_C^* = \frac{\epsilon}{12K_d n}\,, \quad v_C^* = \frac{n-4}{3}\frac{\epsilon}{12K_d n}. \tag{5.8.61}$$

The third fixed point is called the Ising fixed point because it corresponds to n non-interacting replicas of the Ising model. Note that $v_I^* = u_H^*$ $(n = 1)$. To study the stability, we linearize Eqs. (5.8.53) and (5.8.54) in the vicinity of each fixed point:

$$\frac{d\delta u}{dl} = (\epsilon - 4K_d[2(n+8)u^* + 6v^*])\delta u - 24K_d u^* \delta v, \tag{5.8.62}$$

$$\frac{d\delta v}{dl} = (\epsilon - 4K_d[12u^* + 18v^*])\delta v - 48K_d v^* \delta u. \tag{5.8.63}$$

The stability exponents arising from this equation are

$$\text{Gaussian}: \lambda_G^u = \lambda_G^v = \epsilon, \tag{5.8.64}$$

$$\text{Heisenberg}: \; \lambda_H^u = -\epsilon, \; \lambda_H^v = \frac{n-4}{n} 8\epsilon, \tag{5.8.65}$$

$$\text{Ising}: \; \lambda_I^u = \frac{1}{3}\epsilon, \; \lambda_I^v = -\epsilon, \tag{5.8.66}$$

$$\text{cubic}: \; \lambda_C^1 = -\epsilon, \; \lambda_C^2 = \frac{4-n}{3n}. \tag{5.8.67}$$

Note that the Heisenberg fixed point is stable and the cubic fixed point is unstable for $n < 4$, whereas the converse is true for $n > 4$. Eqs. (5.8.53), (5.8.54), (5.8.62), and (5.8.63) imply flow diagrams as shown in Fig. 5.8.4. For all values of n, there is a fixed point that is locally stable. Note, however, that there are regions of the $u - v$ plane that do not flow to a stable fixed point within the ϵ-expansion (e.g. $v < 0$ for all $n > n_c$). This is usually called a *runaway*. It is not always clear where runaways lead. In this (and indeed in many similar cases), one can show that this runaway implies a first-order transition.

Finally, we obtain the thermal exponent $\lambda_t = \nu^{-1}$ by linearizing Eq. (5.8.52):

$$\lambda_t = 2 - 4K_d[(n+2)u^* + 3v^*], \tag{5.8.68}$$

from which we obtain

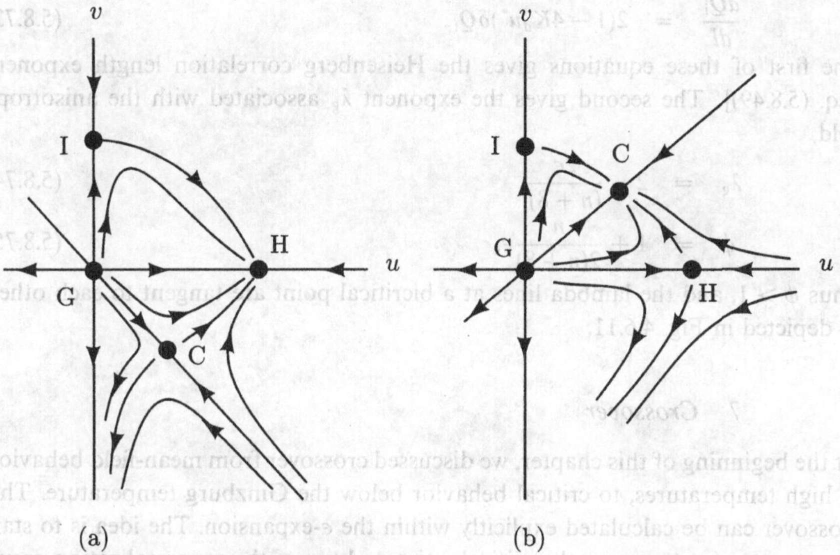

Fig. 5.8.4. Renormalization group flows in the $u-v$ plane showing the Gaussian (G), Heisenberg (H), Ising (I), and cubic (C) fixed points (a) for $n < n_c = 4 + O(\epsilon)$ and (b) for $n > n_c$.

$$\nu_C = \frac{1}{2} + \frac{n-1}{6n}\epsilon \tag{5.8.69}$$

for the cubic fixed point. The thermal exponents for the other fixed points have already been given.

6 *Quadratic anisotropy*

In Sec. 5.5, we argued that the shape of the lambda lines meeting at a bicritical point depended on the magnitude of the crossover exponent $\phi = \lambda_g \nu$ for the anisotropy field g. This exponent can easily be calculated to first order in ϵ. To do this, replace the quadratic term in the Heisenberg reduced Hamiltonian by

$$\overline{\mathscr{H}}_0 = \frac{1}{2}\int d^dx \sum_i r_i\phi_i^2. \tag{5.8.70}$$

The recursion relations for each r_i for an isotropic quartic term are

$$\frac{dr_i}{dl} = 2r_i + 4K_d u\left(\frac{2}{(1+r_i)} + \sum_{j=1}^{n}\frac{1}{(1+r_j)}\right). \tag{5.8.71}$$

Linearizing about the Heisenberg fixed point and setting $\delta r = \sum_i \delta r_i$ and $Q_i = \delta r_i - n^{-1}\delta r$, we obtain

$$\frac{d\delta r}{dl} = [2 - 4K_d u^*(n+2)]\delta r, \tag{5.8.72}$$

$$\frac{dQ_i}{dl} = 2(1 - 4K_d u^*)\delta Q_i. \tag{5.8.73}$$

The first of these equations gives the Heisenberg correlation length exponent [Eq. (5.8.49)]. The second gives the exponent λ_g associated with the anisotropy field,

$$\lambda_g = 2 - \frac{2\epsilon}{(n+8)}, \tag{5.8.74}$$

$$\phi = 1 + \frac{n}{2(n+8)}\epsilon. \tag{5.8.75}$$

Thus $\phi > 1$, and the lambda lines at a bicritical point are tangent to each other, as depicted in Fig. 4.6.11.

7 Crossover

At the beginning of this chapter, we discussed crossover from mean-field behavior, at high temperatures, to critical behavior below the Ginzburg temperature. This crossover can be calculated explicitly within the ϵ-expansion. The idea is to start at some temperature near the critical point and to use the renormalization group recursion relations to map the original critical Hamiltonian onto a non-critical Hamiltonian for which perturbation theory converges.

We will consider here crossover for the susceptibility of the isotropic O_n model. Our approach follows closely that of Rudnick and Nelson (1976). We begin by rewriting Eq. (5.8.10), relating G evaluated with the original Hamiltonian $\overline{\mathscr{H}}$ to G evaluated with the rescaled Hamiltonian $\overline{\mathscr{H}}'$ as

$$G(\mathbf{q}, r, u) = e^{(2-\eta)l} G(e^l\mathbf{q}, r(l), u(l)). \tag{5.8.76}$$

In general, the exponent η can be l-dependent, and ηl in the above should be replaced by $\int_0^l dl'\eta(l')$. In ϕ^4 theories, $\eta = O(\epsilon^2)$ and we can set $\eta = 0$ in our present treatment to first order in ϵ. Near the critical point, r (or at least $r - r_c$) and q are small, and the perturbation expansion for $G(\mathbf{q}, r, u)$ in powers of u is divergent. The scaling parameter l can be chosen so that $r(l)$ and/or $e^l q$ is of order unity so that the perturbation expansion of $G(e^l\mathbf{q}, r(l), u(l))$ presents no problems. The recursion relations Eqs. (5.8.45) and (5.8.46) for the potentials r and u for the O_n model are

$$\frac{dr}{dl} = 2r + A\frac{u}{1+r} \tag{5.8.77}$$

$$\frac{du}{dl} = \epsilon u - B\frac{u^2}{(1+r)^2}, \tag{5.8.78}$$

where $A = 4K_d(n+2)$ and $B = 4K_d(n+8)$. We wish to integrate these equations retaining only contributions to lowest order in ϵ. As we will show shortly, to linear order in ϵ, we can set $r = 0$ in the equation for u, which then becomes

$$\frac{du}{dl} = \epsilon u - Bu^2. \tag{5.8.79}$$

This equation can be integrated exactly. The result is

$$u(l) = u_0 \frac{e^{\epsilon l}}{Q(l)}, \tag{5.8.80}$$

where u_0 is the value of $u(l)$ at $l = 0$ (assumed to be of order ϵ or smaller), and

$$Q(l) = 1 + \frac{Bu_0}{\epsilon}(e^{\epsilon l} - 1). \tag{5.8.81}$$

Note that

$$Q(l) \to \begin{cases} 1, & \text{as } l \to 0; \\ (Bu_0/\epsilon)e^{\epsilon l}, & \text{as } l \to \infty \end{cases} \tag{5.8.82}$$

so that

$$u(l) \to \begin{cases} u_0, & \text{as } l \to 0; \\ \epsilon/B = u^*, & \text{as } l \to \infty, \end{cases} \tag{5.8.83}$$

where u^* is the value of u at the Heisenberg fixed point. To show that corrections to Eq. (5.8.80) for $u(l)$ coming from r are of order ϵ^2, we set $u(l) = f(l)u_1(l)$, where $u_1(l)$ satisfies Eq. (5.8.79). Then

$$\frac{df}{dl} = -Bu_1^2(f^2 - f) + f^2 u_1^2 \left[\frac{1}{(1+r)^2} - 1\right]. \tag{5.8.84}$$

u_1 is of order ϵ, and $f \to 1$ as $u_1 \to 1$, implying that $f = 1 + O(\epsilon^2)$. Thus, to order ϵ, we can, as advertised, ignore the r in the equation for u and use Eq. (5.8.80) for $u(l)$.

Next, we must integrate the equation for $r(l)$, remembering that $r(l)$ may become of order unity. To carry out the integration to order ϵ we write

$$\begin{aligned} \frac{dr}{dl} &= (2 - Au(l))r(l) + Au(l) \\ &\quad + Au(l)\left[\frac{1}{1+r(l)} - 1 + r(l)\right]. \end{aligned} \tag{5.8.85}$$

Then, setting $r(l) = e^{S(l)}\tilde{r}(l)$, where

$$S(l) = 2l - \int_0^l Au(l')dl' = 2l + O(\epsilon), \tag{5.8.86}$$

we find

$$\tilde{r}(l) = r_0 + \int_0^l e^{-S(l')}\left[Au(l') + Au(l')\left(\frac{1}{1+r(l')} - 1 + r(l')\right)\right] dl', \tag{5.8.87}$$

where $r_0 = r(l = 0)$. To evaluate $\tilde{r}(l)$ to order ϵ, we recall that $du/dl = O(\epsilon^2)$, $e^{-S(l)} = -(1/2)(de^{2l}/dl) + O(\epsilon)$, and $r(l) = e^{2l}r_0 + O(\epsilon)$. Therefore,

$$\int_0^l e^{-S(l')}Au(l')dl' = -\frac{A}{2}\left[e^{-S(l)}u(l) - u(0)\right] + O(\epsilon^2), \tag{5.8.88}$$

and

$$\begin{aligned} &\int_0^l dl' e^{-S(l')}Au(l')\left[\frac{1}{1+r(l')} - 1 + r(l')\right] \\ &= \int_0^l dl' Au(l')\left(\frac{r_0 r(l')}{1+r(l')} + O(\epsilon)\right) \end{aligned}$$

$$= \int_0^l dl' \frac{1}{2} Au(l') \left[\frac{d}{dl} r_0 \ln(1 + r(l')) + O(\epsilon) \right] \tag{5.8.89}$$

$$= \frac{1}{2} e^{-S(l)} Au(l) r(l) \ln[1 + r(l)] - \frac{1}{2} Au(0) r(0) \ln[1 + r(0)] + O(\epsilon^2).$$

Using these two equations, we can now write the equation for $r(l)$ as

$$t(l) = e^{S(l)} t(0), \tag{5.8.90}$$

where

$$t(l) = r(l) + \frac{1}{2} Au(l) - \frac{1}{2} Au(l) r(l) \ln(1 + r(l)) + O(\epsilon^2). \tag{5.8.91}$$

The variable $t(0)$ is actually the reduced temperature $(T - T_c)/T_c$. The phase transition occurs at $t(0) = 0$ or at $r(0) = -\frac{1}{2} Au(0) + O(\epsilon^2)$.

The quantity $S(l)$ can be evaluated analytically:

$$\begin{aligned} S(l) &= 2l - A \int_0^l dl' u(l') \\ &= 2l - Au_0 \int_0^l dl' \frac{e^{\epsilon l}}{1 + (Bu_0/\epsilon)(e^{\epsilon l} - 1)} \\ &= 2l - \frac{A}{B} \ln Q(l), \end{aligned} \tag{5.8.92}$$

from which we obtain

$$t(l) = e^{2l} [Q(l)]^{-A/B} t(0) \tag{5.8.93}$$

$$\rightarrow \begin{cases} e^{2l} t(0), & \text{as } l \to 0; \\ e^{(2 - A\epsilon/B) l} (Bu_0/\epsilon) t(0), & \text{as } l \to \infty. \end{cases} \tag{5.8.94}$$

This shows the crossover behavior we are after. For $(Bu_0/\epsilon)e^{\epsilon l} \ll 1$, we have mean-field behavior. For $(Bu_0/\epsilon)e^{\epsilon l} \gg 1$, we have Heisenberg critical behavior with $t(l) \sim e^{\nu^{-1} l} t(0)$ with $\nu^{-1} = 2 - [(n+2)/(n+8)]\epsilon$. Crossover between the two behaviors occurs at the Ginzburg reduced temperature $t = (Bu_0/\epsilon)^{2/\epsilon}$.

To complete the calculations we must express physical quantities such as the susceptibility in terms of $t(l)$. To do this, we choose l^* such that

$$t(l^*) = 1 \tag{5.8.95}$$

and perturbation theory presents no problem. To evaluate $G(r(l)) = G(\mathbf{q} = 0, r(l), u(l))$, we use the result (Eq. (5.2.33)) from one-loop perturbation theory, which in four dimensions is

$$\begin{aligned} G^{-1}(r(l)) &= r(l) + Au(l) \int \frac{d^4 q}{(2\pi)^4} \frac{1}{r(l) + q^2} \\ &= r(l) + \frac{1}{A} u(l) - \frac{1}{2} Ar(l) [\ln(1 + r(l) - \ln r(l)] & (5.8.96) \\ &= t(l) + \frac{1}{2} u(l) \ln t(l). & (5.8.97) \end{aligned}$$

Thus, at $l = l^*$,

$$G(r(l^*)) = \frac{1}{t(l^*)} = 1. \tag{5.8.98}$$

Finally, using Eq. (5.8.76) with $\eta = 0$, we have

$$G(r) = e^{2l^*} G(r(l^*)) = e^{2l^*}, \tag{5.8.99}$$

where l^* is determined by Eqs. (5.8.93) and (5.8.95):

$$e^{2l^*} \left[1 + \frac{Bu_0}{\epsilon}\left(e^{\epsilon l^*} - 1\right)\right]^{-A/B} t = 1, \tag{5.8.100}$$

where we set $t(0) = t$. An approximate solution to this equation, which reproduces correctly both the large and small l^* limits to lowest order in ϵ, is

$$e^{l^*} = t^{-1/2}\left[1 + B\frac{u_0}{\epsilon t^{\epsilon/2}}\right]^{A/2B}, \tag{5.8.101}$$

so that

$$\begin{aligned} G &= t^{-1}\left[1 + B\frac{u_o}{\epsilon t^{\epsilon/2}}\right]^{A/B} \\ &\sim \begin{cases} t^{-1} & \text{if } (Bu_0/\epsilon)\lambda t^{\epsilon/2}; \\ t^{-1-(\epsilon/2)(A/B)} & \text{if } (Bu_0/\epsilon) \gg t^{\epsilon/2}. \end{cases} \end{aligned} \tag{5.8.102}$$

The susceptibility exponent γ is 1 for $t \gg t_G$ and $1 + [(n+2)/(n+8)]\epsilon$ for $T \lambda t_G$.

A useful way to visualize crossover behavior in a function like $G(t)$ is via an effective temperature-dependent critical exponent defined via

$$\gamma_{\text{eff}}(t) = -\frac{d \log G(t)}{d \log t}. \tag{5.8.103}$$

This exponent can be calculated directly from Eqs. (5.8.99) and (5.8.100). Fig. 5.8.5 shows $\gamma_{\text{eff}}(t)$ evaluated from these equations with $n = 1$, $\epsilon = 1$, and $Bu_0 = 10^{-2}$. The crossover from Gaussian behavior with $\gamma = 1$ to Heisenberg critical behavior with $\gamma = 1.2$ occurs at $t = (Bu_0)^2 = 10^{-4}$.

8 *Dangerous irrelevant variables*

As we saw in Sec. 5.5, irrelevant variables are fields with a negative scaling exponent that scale to zero at the controlling fixed point. Normally, irrelevant variables do not contribute to the leading scaling behavior, though they do determine corrections to scaling. Sometimes, however, quantities of physical interest diverge as some inverse power of an irrelevant variable. In this case, the irrelevant variable cannot be ignored. It is called a *dangerous irrelevant variable*, and participates in determining leading singularities near a critical point.

Let v be an irrelevant variable with exponent $\lambda_v = -|\lambda_v|$ such that $v(l) = e^{\lambda_v l}$ in the vicinity of the fixed point of interest. Then let $A(t, v)$ be some observable that scales as

$$A(t, v) = e^{\lambda_A l} A(t(l), v(l)), \tag{5.8.104}$$

where $t(l) = e^{v^{-1} l} t$. Then, if $A(t(l), v(l)) \sim [v(l)]^{-\sigma}$,

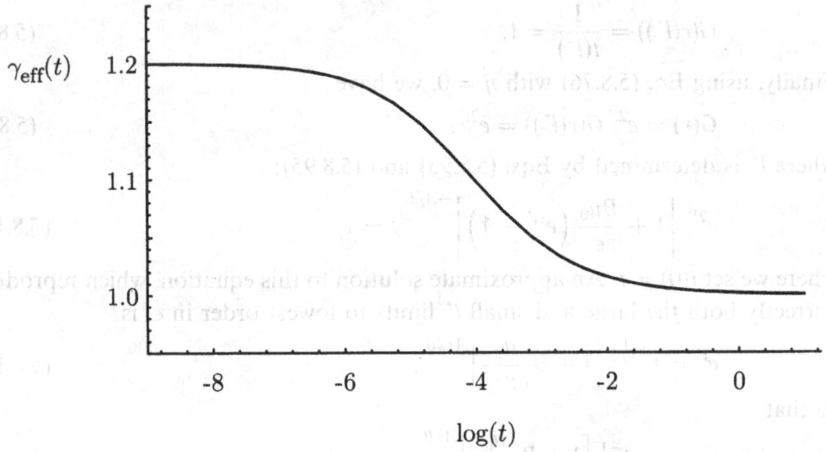

Fig. 5.8.5. The effective critical exponent $\gamma_{\text{eff}}(t)$, showing crossover from Gaussian to Heisenberg critical behavior.

$$A(t,v) \sim \frac{e^{\lambda_A l}}{[v(l)]^\sigma} \sim \frac{e^{\lambda_A l+\sigma|\lambda_v|l}}{v^\sigma} \sim \frac{1}{v^\sigma} t^{-(\lambda_A+\sigma|\lambda_v|)\nu}. \tag{5.8.105}$$

The leading divergence of A is determined both by λ_A and the exponent λ_v of the irrelevant variable v.

The simplest example of an observable, with a dangerous irrelevant variable, is the free energy above the upper critical dimensions. The free energy density f has dimensions of inverse volume. For a ϕ^4 model, it must therefore satisfy the renormalization equation

$$f(r,u) = e^{-dl} f(r(l), u(l)). \tag{5.8.106}$$

Below four dimensions, $u \to u^*$ at the critical point, and we can choose l^* such that $|t(l^*)| = e^{\nu^{-1}l^*}|t| = 1$ (see preceding subsection). In this case, $f(r(l^*), u(l^*)) = f^*$, and

$$f(r,u) = e^{-dl^*} f^* \sim |t|^{d\nu} \sim |t|^{2-\alpha}, \tag{5.8.107}$$

in agreement with the scaling predictions for Sec. 5.5.

Above $d = 4$, we know mean-field theory applies and predicts $f \sim -|t|^2$ for $t < 0$. The above argument would incorrectly predict $f \sim |t|^{d\nu} = |t|^{d/2}$. The problem is that, above four dimensions, $u(l) = e^{-\epsilon l} u_0$ is an irrelevant variable and f [diverges as $u \to 0$ Eq. (4.3.10)]:

$$f(r(l^*), u(l^*)) = -\frac{r^2(l^*)}{16u(l^*)}. \tag{5.8.108}$$

Thus, using Eq. (5.8.106) and setting $r(l^*) = e^{2l^*} r = 1$, we obtain

$$
\begin{aligned}
f(r,u) &= -e^{-dl^*}\frac{r^2(l^*)}{16e^{(4-d)l^*}u_0} \\
&= e^{-4l^*}\frac{1}{16u_0} = -\frac{r^2}{16u_0}
\end{aligned}
\tag{5.8.109}
$$

as required.

Another example where there is a dangerous irrelevant variable is the Heisenberg model when there is cubic anisotropy [Eq. (5.8.51)] and when the Heisenberg rather than the cubic fixed point is stable. When there is no cubic anisotropy, the direction of order is arbitrary, and the correlation function for directions perpendicular to the order diverges as q^{-2}. When there is cubic anisotropy, the order parameter aligns along one of the cubic axes, and the correlation function for directions perpendicular to the order is finite at $\mathbf{q} = 0$ and proportional to the inverse of the cubic potential v. When the Heisenberg fixed point is stable, v scales to zero and is irrelevant. Thus, v is a dangerous irrelevant variable for the perpendicular correlation function and determines the exponent of its power-law divergence as the critical point is approached from below. We will leave the detailed derivation of this result to Problem 5.7.

9 *The utility of the ϵ-expansion*

The first order in the ϵ-expansions that we have presented do not make numerically accurate predictions of the critical exponents in physical dimensions, and one may ask why one should bother at all with such calculations. Their great virtue is that they provide an analytically not-too-complicated way of determining what types of universality classes one can expect. Though the numerical values of exponents change considerably as one moves away from the upper critical dimension, the topology of flow diagrams does not. Thus, one can investigate in a well-controlled and relatively simple way which interactions will lead to new universality classes and which will not. We have seen how the addition of a cubic anisotropy can lead to a new universality class with exponents that differ from those of the isotropic models. Other anisotropies and potentials can also do this. For example, dipolar forces, which are always present in magnetic systems, can change the universality class of ferromagnets but not of antiferromagnets. Another example is the effect of quenched random impurities on critical points. The ϵ-expansion has been one of our most powerful tools to assess how randomness can change universality classes. The ϵ-expansion can also be used to calculate equations of state, crossover functions, and amplitude ratios. It is the one calculational tool that allows for straightforward calculations of virtually all functions of interest in the vicinity of a second-order critical point, and is in a real sense responsible for the rather detailed understanding we now have of critical phenomena.

The ϵ-expansion is really an asymptotic expansion. With considerable effort, reasonably high order terms in this series have been calculated. With sophisticated resummation techniques, the series can be resummed to yield exponents in three dimensions that are in very good agreement with those obtained by other techniques, as can be seen in Table 5.4.4.

Appendix 5A The Hubbard-Stratonovich transformation

Though the phenomenological continuum coarse grained field theories we introduced in Sec. 5.2 are usually quite adequate for the study of properties near second-order phase transitions, it is sometimes valuable to have an exact functional integral representation for the partition function of strongly constrained models such as the Ising and *n*-vector models. Here we will present a method for obtaining such a representation (Hubbard 1954, Stratonovich 1957). For simplicity, we will consider only the Ising model, though generalizations to any microscopic Hamiltonian that can be expressed as a quadratic form are straightforward.

We begin by expressing the Ising Hamiltonian as a sum over all sites $\mathbf{l}$ and $\mathbf{l}'$ rather than over bonds $<\mathbf{l},\mathbf{l}'>$:

$$\mathscr{H} = -\frac{1}{2}\sum_{\mathbf{l},\mathbf{l}'} J_{\mathbf{l},\mathbf{l}'} S_{\mathbf{l}} S_{\mathbf{l}'}, \tag{5A.1}$$

where $S_{\mathbf{l}} = \pm 1$ and $J_{\mathbf{l},\mathbf{l}'}$ is the exchange integral which in the simplest model is zero unless $\mathbf{l}$ and $\mathbf{l}'$ are nearest neighbor sites. We now use Eq. (5.2.13) to write

$$\begin{aligned} e^{-\beta\mathscr{H}} &= \exp\left[\frac{1}{2}\sum_{\mathbf{l},\mathbf{l}'} K_{\mathbf{l},\mathbf{l}'} S_{\mathbf{l}} S_{\mathbf{l}'}\right] \\ &= A\prod_{\mathbf{l}}\int d\phi_{\mathbf{l}} \exp\left[-\frac{1}{2}\sum_{\mathbf{l},\mathbf{l}'} \phi_{\mathbf{l}} K^{-1}_{\mathbf{l},\mathbf{l}'} \phi_{\mathbf{l}'} + \sum_{\mathbf{l}} \phi_{\mathbf{l}} S_{\mathbf{l}}\right] \end{aligned} \tag{5A.2}$$

where $K_{\mathbf{l},\mathbf{l}'} = \beta J_{\mathbf{l},\mathbf{l}'}$ and $A = (2\pi)^{-N/2}(\det K)^{-1/2}$. The Ising model partition function in an external aligning field is then

$$Z = A\prod_{\mathbf{l}}\int d\phi_{\mathbf{l}} \exp\left[-\frac{1}{2}\sum_{\mathbf{l},\mathbf{l}'} \phi_{\mathbf{l}} K^{-1}_{\mathbf{l},\mathbf{l}'} \phi_{\mathbf{l}'}\right] \mathrm{Tr}\exp\left[\sum_{\mathbf{l}} (\phi_{\mathbf{l}} + \beta h_{\mathbf{l}}) S_{\mathbf{l}}\right]. \tag{5A.3}$$

The trace (Tr) in this expression is over the Ising spin variables. Note the spin variable $S_{\mathbf{l}}$ appears linearly rather than quadratically in Eq. (5A.3). This simplification allows for an exact evaluation of the trace over $S_{\mathbf{l}}$ and is the great virtue of the Hubbard-Stratonovich transformation of Eq. (5A.2). The integrals over $\phi_{\mathbf{l}}$ are, of course, nontrivial. It is easy to show that $\langle S_{\mathbf{l}}\rangle = \partial \ln Z/\partial\beta h_{\mathbf{l}} = \sum_{\mathbf{l}'} K^{-1}_{\mathbf{l},\mathbf{l}'}\langle\phi_{\mathbf{l}}\rangle$ by integrating by parts after shifting the derivative with respect to $\beta h_{\mathbf{l}}$ to a derivative with respect to $\phi_{\mathbf{l}'}$. Thus, averages of the unconstrained field $\phi_{\mathbf{l}}$ are proportional to averages of the constrained field $S_{\mathbf{l}}$. The dependence of Z on $h_{\mathbf{l}}$ can be cast into a more convenient form by changing variables: if $\psi_{\mathbf{l}} = \phi_{\mathbf{l}} + \beta h_{\mathbf{l}}$, then

$$Z = A\prod_{\mathbf{l}}\int d\psi_{\mathbf{l}} \exp\left[-\frac{1}{2}\sum_{\mathbf{l},\mathbf{l}'}\psi_{\mathbf{l}}K^{-1}_{\mathbf{l},\mathbf{l}'}\psi_{\mathbf{l}'} - \frac{1}{2}\sum_{\mathbf{l},\mathbf{l}'}\beta h_{\mathbf{l}}K_{\mathbf{l},\mathbf{l}'}\beta h_{\mathbf{l}'} - \sum_{\mathbf{l}}\beta f_L(\psi_{\mathbf{l}}) + \sum_{\mathbf{l}} K^{-1}_{\mathbf{l},\mathbf{l}'}h_{\mathbf{l}}\psi_{\mathbf{l}'}\right] \tag{5A.4}$$

where

$$\beta f_L(\psi) = -\ln \mathrm{Tr}e^{S\psi} = -\ln(2\cosh\psi). \tag{5A.5}$$

Both Eqs. (5A.3) and (5A.4) have the form of a lattice field theory like that of Eqs. (5.2.5) and (5.2.6). Their continuum limit can be carried out as described Sec. 5.2.

Appendix 5B Diagrammatic perturbation theory

In this appendix, we present an outline of low-order diagrammatic perturbation theory and calculate the second-order contributions to the recursion relation for u in Eq. (5.8.46). Diagrammatic perturbation theory is used extensively in both relativistic and nonrelativistic field theories, and the reader is encouraged to look at some of the many excellent texts on the subject for further details.

The essential ingredient of perturbation theory for any field theory is that the average of a product of an even number of fields with respect to a Gaussian weight function produced by a harmonic Hamiltonian can be expressed as the sum of products of the Gaussian correlation function or *propagator* $G_0(\mathbf{x},\mathbf{x}')$. To see this, we use Eq. (5.2.21) for $\mathscr{A} = -T\ln Z$ and

$$\begin{aligned}\langle\phi(\mathbf{x}_1)\cdots\phi(\mathbf{x}_n)\rangle &= \frac{1}{Z}\frac{\delta^n Z}{\delta\beta h(\mathbf{x}_1)\cdots\delta\beta h(\mathbf{x}_n)} \\ &= \frac{\delta^n}{\delta\beta h(\mathbf{x}_1)\cdots\beta h(\mathbf{x}_n)}\exp\left(\frac{1}{2}\int d^dx d^dx' h(\mathbf{x})G_0(\mathbf{x},\mathbf{x}')h(\mathbf{x}')\right).\end{aligned} \tag{5B.1}$$

Direct application of this formula for $n=1$ and $n=2$ yields

$$\langle\phi(\mathbf{x}_1)\phi(\mathbf{x}_2)\rangle = G_0(\mathbf{x}_1,\mathbf{x}_2) \tag{5B.2}$$

and

$$\begin{aligned}\langle\phi(\mathbf{x}_1)\phi(\mathbf{x}_2)\phi(\mathbf{x}_3)\phi(\mathbf{x}_4)\rangle \;=\;& G_0(\mathbf{x}_1,\mathbf{x}_2)G_0(\mathbf{x}_3,\mathbf{x}_4) + G_0(\mathbf{x}_1,\mathbf{x}_3)G_0(\mathbf{x}_2,\mathbf{x}_4) \\ &+ G_0(\mathbf{x}_1,\mathbf{x}_4)G_0(\mathbf{x}_2,\mathbf{x}_3).\end{aligned} \tag{5B.3}$$

In general, if there are $2n$ fields in the product, then there will be a sum of $(2n-1)\cdot(2n-3)\cdots 3\cdot 1$ distinct products of propagators. This follows from the fact that each field will be paired or *contracted* with every other field to produce a propagator G_0. Thus, with a product of $2n$ fields to be averaged, the first field can be contracted with any of $(2n-1)$ others. There then remain $(2n-2)$ unpaired fields, the first of which can be contracted with $(2n-3)$ others, and so on. Thus, for $n=2$, there will be $3\times 1 = 3$ terms, as shown in Fig. 5B.1.

The above expansion also applies to averages of products of fields $\phi^>(\mathbf{x})$ with respect to the Hamiltonian $\overline{\mathscr{H}^>_{0,\Lambda}}$ in the presence of a field $h(\mathbf{x})$ coupling to $\phi^>(\mathbf{x})$. In this case, the factors of $G_0(\mathbf{x},\mathbf{x}')$ in Eq. (5B.1) are replaced by

$$G_0^>(\mathbf{x},\mathbf{x}') = \int_{\mathbf{q}} G_0^>(\mathbf{q})e^{i\mathbf{q}\cdot(\mathbf{x}-\mathbf{x}')} \tag{5B.4}$$

where

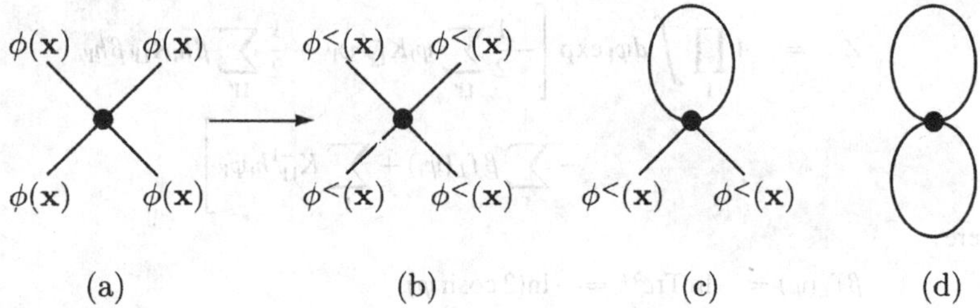

Fig. 5B.1. Diagrams in coordinate space to first order in u. The vertex u is represented by a dot. (a) Representation of the perturbation Hamiltonian $\overline{\mathscr{H}}^4_\Lambda$ before averaging over $\phi^>$. The four lines represent fields $\phi(\mathbf{x})$. Integrating over the fields $\phi^>$ leads to contraction of internal lines and diagrams (b) - (d).

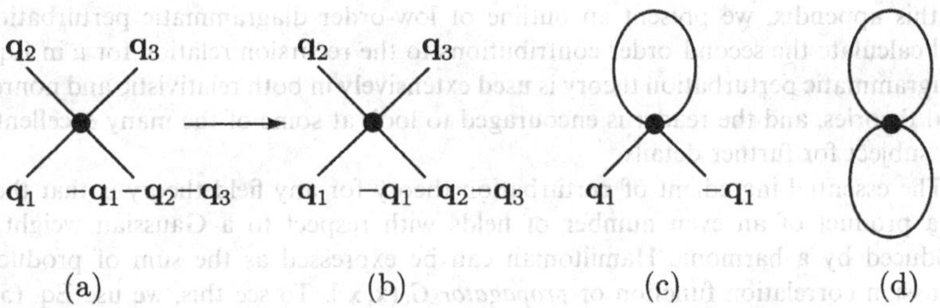

Fig. 5B.2. Same as Fig. 5B.1 in wave number space.

$$G_0^>(\mathbf{q}) = \begin{cases} G_0(\mathbf{q}) = (r+q^2)^{-1} & \text{if } \Lambda/b < a < \Lambda; \\ 0 & \text{otherwise} . \end{cases} \tag{5B.5}$$

We now turn to the representation of the perturbation expansion [Eq. (5.8.36)] of $\overline{\mathscr{H}}_{\Lambda/b}$ in terms of diagrams. The potential u in $\overline{\mathscr{H}}^4_\Lambda$ is represented by a dot and factors of ϕ by lines. The perturbation Hamiltonian $\overline{\mathscr{H}}^4_\Lambda$ is then depicted as a dot from which four lines emerge, as depicted in Fig. 5B.1. Each of the four lines has a free end. An integration over $\mathbf{x}$ to obtain $\overline{\mathscr{H}}^4_\Lambda$ is understood. Alternatively, in the wave number representation (Fig. 5B.2), the four lines represent $\phi(\mathbf{q}_1)$, $\phi(\mathbf{q}_2)$, $\phi(\mathbf{q}_3)$, and $\phi(-\mathbf{q}_1-\mathbf{q}_2-\mathbf{q}_3)$ because the integral over $\mathbf{x}$ constrains the sum over wave numbers of the four ϕ fields to be zero. An integral over $\mathbf{q}_1$, $\mathbf{q}_2$, and $\mathbf{q}_3$ is understood. Each line can be decomposed into a $\phi^<$ and a $\phi^>$ part. The average over $\phi^>$ leads to the contraction of pairs of $\phi^>$ fields to form propagators $G_0^>$. This process can be represented by connecting pairs of ϕ lines together. Uncontracted lines with free ends are called *external legs* and represent $\phi^<$ fields. Contracted lines, which always begin and end at a vertex, are called *internal lines* and represent a factor of $G_0^>$. If diagrams are represented in coordinate space, as in Fig. 5B.1, connected lines represent $G_0^>(\mathbf{x},\mathbf{x}')$ and connect vertices as positions $\mathbf{x}$ and $\mathbf{x}'$. Since the internal line in Fig. 5B.1(c) originates and terminates at the same point $\mathbf{x}$, it represents a factor of $G_0^>(\mathbf{x},\mathbf{x})$, and the contribution of this diagram is proportional to

$$u\int d^dx(\phi^<(\mathbf{x}))^2 G_0^>(\mathbf{x},\mathbf{x}). \tag{5B.6}$$

If diagrams are represented in wave number space, as in Fig. 5B.2, a connected line represents a factor of $G_0^>(\mathbf{q}_2)$. In this case, it should be remembered that $G_0^>(\mathbf{q}_2)$ results from the contraction of two fields $\phi^>(\mathbf{q}_2)$ and $\phi(\mathbf{q}_3)$ and implies a delta function setting $\mathbf{q}_3 = -\mathbf{q}_2$. The contribution of Fig. 5B.2(c) is thus proportional to

$$u \int_{\mathbf{q}_1} |\phi^<(\mathbf{q})|^2 \int_{\mathbf{q}_2} G_0^>(\mathbf{q}_2). \tag{5B.7}$$

Similarly the contribution from Figs. 5B.1(d) or 5B.2(d) is proportional to

$$u \int d^d x (G_0^>(\mathbf{x}, \mathbf{x}))^2 = u \left(\int_{\mathbf{q}} G_0^>(\mathbf{q}) \right)^2. \tag{5B.8}$$

The three terms in Eq. (5.8.40) are represented either by diagrams 5A.1b–d or by diagrams 5A.2b–d. Diagrams (b) have four external legs and contribute to $u^<$, diagrams (c) have two external legs and contribute to $r^<$, and diagrams (d) have no external legs and contribute to the constant terms V_0. To obtain Eqs. (5.8.40), weights must be assigned to each of these diagrams. This can be done by counting the number of ways of contracting lines to obtain a given diagram. The weight assigned to diagrams (b) is trivially 1 since there is only one way for all lines to be external $\phi^<$ legs. This agrees with the coefficient of $(\phi^<)^4$ in Eq. (5.8.40). There are six ways of picking two lines out of four to form the singly contracted line in diagrams (c). Thus diagrams (c) contribute to $r^<$ with a factor of 12 when the $1/2$ in the definition of r is included. Finally, the weight of diagrams (d) is 3 because a given line can be contracted with any of three others, and there is only one way to contract the remaining two lines. These weights agree with Eq. (5.8.40).

We now turn to the evaluation of $\langle(\overline{\mathscr{H}}_\Lambda^4)^2\rangle_>$ using diagrams. The factor of $(\overline{\mathscr{H}}_\Lambda^4)^2$ prior to the average over $\phi^>$ can be represented by two vertices each with four ϕ lines as shown in Fig. 5B.3a. The average over $\phi^>$ will contract pairs of ϕ lines in all possible ways. The diagrams resulting from this average fall into three categories, as shown in Figs. 5B.3b, c, and d. In Fig. 5B.3b, there are no contractions coupling the vertex at $\mathbf{x}$ to that at $\mathbf{x}'$. These are called *disconnected diagrams.* They are identical to the diagrams representing $\langle\overline{\mathscr{H}}_\Lambda^4\rangle_>^2$ and thus do not contribute to the cumulant, $\langle(\overline{\mathscr{H}}_\Lambda^4)^2\rangle_> - \langle\overline{\mathscr{H}}_\Lambda^4\rangle_>^2$, appearing in the expression Eq. (5.8.39) for $\overline{\mathscr{H}}_{\Lambda/b}$. The diagrams in Fig. 5B.3d are called *one-particle reducible* diagrams because they can be divided into two disconnected parts by cutting a single $G_0^>$ line. Diagram d(1) is zero because the internal line connecting the two vertices must have the same wave number as the external legs. The external legs represent $\phi^<$ fields whereas internal lines represent a contraction of $\phi^>$ fields and from the definition of $\phi^>$ and $\phi^<$, it is impossible for both internal lines and external legs to represent fields with the same wave number. A similar argument applies to diagram 5B.3d(2) with six external legs. Thus, the diagrams in Fig. 5B.3c are the only ones which contribute to $\overline{\mathscr{H}}_{\Lambda/b}$. Fig. 5B.3c(2) is proportional to

$$u^2 \int_{\mathbf{q}_1, \mathbf{q}_2} G_0^>(\mathbf{q}_1) G_0^>(\mathbf{q}_2) G_0^>(-\mathbf{q}_1 - \mathbf{q}_2 - \mathbf{q}). \tag{5B.9}$$

It is a second-order contribution to $r^>(\mathbf{q})$. Unlike Fig. 5B.1c, it depends on $\mathbf{q}$ and contributes a term of order u^2 to the equation for c'. It, therefore, leads to an order ϵ^2 correction to η. Its $\mathbf{q} = 0$ part contributes a ϵ^2 correction to ν. To first order in ϵ, however, it can be ignored. Fig. 5B.3c(1) is responsible for the u^2 contribution to the equation for u'. As can be seen from its representation in Fig. 5B.4, it is proportional to

$$u^2 \int_{\mathbf{q}} G_0^>(\mathbf{q}) G_0^>(-\mathbf{q} - \mathbf{q}_1 - \mathbf{q}_2) \overset{\mathbf{q}_1, \mathbf{q}_2 \to 0}{\to} u^2 \int_{\mathbf{q}} (G_0^>(\mathbf{q}))^2. \tag{5B.10}$$

It is the $\mathbf{q}_1 = \mathbf{q}_2 = 0$ limit of this diagram that contributes to the equation for u'. To

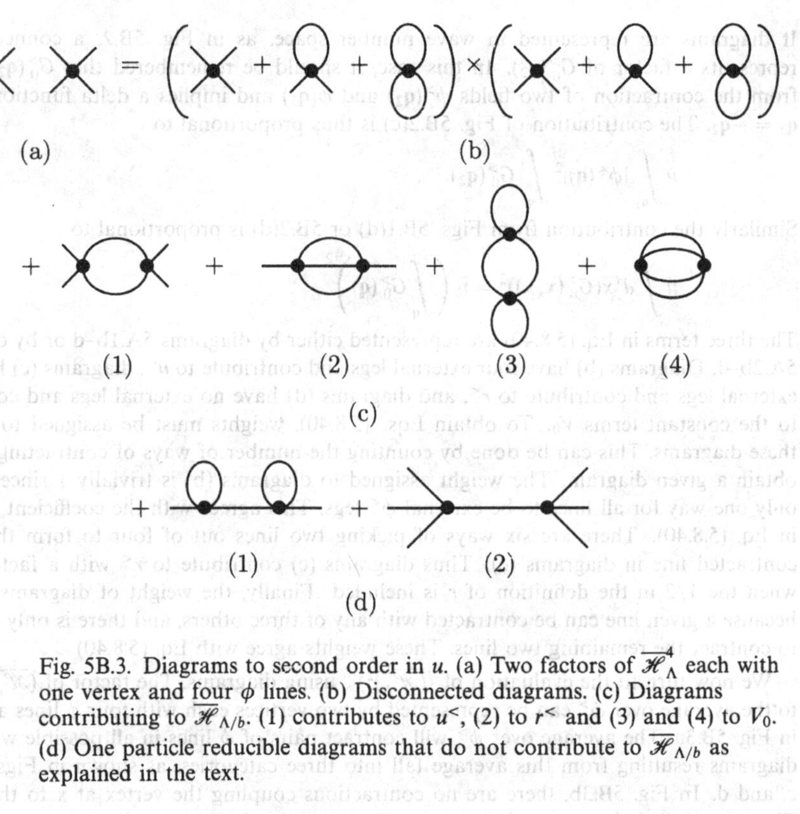

Fig. 5B.3. Diagrams to second order in u. (a) Two factors of $\overline{\mathscr{H}}^4_\Lambda$ each with one vertex and four ϕ lines. (b) Disconnected diagrams. (c) Diagrams contributing to $\overline{\mathscr{H}}_{\Lambda/b}$. (1) contributes to $u^<$, (2) to $r^<$ and (3) and (4) to V_0. (d) One particle reducible diagrams that do not contribute to $\overline{\mathscr{H}}_{\Lambda/b}$ as explained in the text.

determine its weight, we note that the two external legs at each vertex can be chosen from the four original legs in any of six ways for a total factor of $6 \times 6 = 36$. In addition, the internal lines connecting the two vertices can be contracted in two ways for an extra factor of 2. Finally, there is a factor of 1/2 in the coefficient of the cumulant in the expression for $\overline{\mathscr{H}}_{\Lambda/b}$ in terms of $\overline{\mathscr{H}}^4_\Lambda$ in Eq. (5.8.39). The total weight of this diagram in $u^<$ is $2 \times 36/2 = 36$ in agreement with Eq. (5.8.44) when $n = 1$.

A similar diagrammatic analysis of perturbation theory can be carried out for O_n models in which

$$\overline{\mathscr{H}}^4_\Lambda = u \int d^d x \phi_i(\mathbf{x})\phi_i(\mathbf{x})\phi_j(\mathbf{x})\phi_j(\mathbf{x}) \tag{5B.11}$$

where the summation convention on repeated indices is understood. In this case, it is useful to represent the vertex u by a dotted line as shown in Fig. 5B.5. The two ϕ-legs emanating from either end of the vertex must have the same Cartesian index. The average

$$\langle \phi_i^>(\mathbf{x})\phi_j^>(\mathbf{x}')\rangle_> = \delta_{ij} G_0^>(\mathbf{x}, \mathbf{x}') \tag{5B.12}$$

is zero unless $i = j$. The leading order diagrams contributing to r are shown in Fig. 5B.6 and those contributing to u in Fig. 5B.7. First consider the contributions to r in which there are two external legs. Each of these diagrams, like the contribution to r from Fig. 5B.2c in the Ising case is proportional to $\int_\mathbf{q} G_0^>(\mathbf{q})$. In Fig. 5B.6a, there are two choices for the end to be contracted. There is a factor of $G_0^>$ for each of the n indices j in the contracted loop. Finally, there is the factor of 2 coming from the factor of 1/2 in the definition of r. Thus

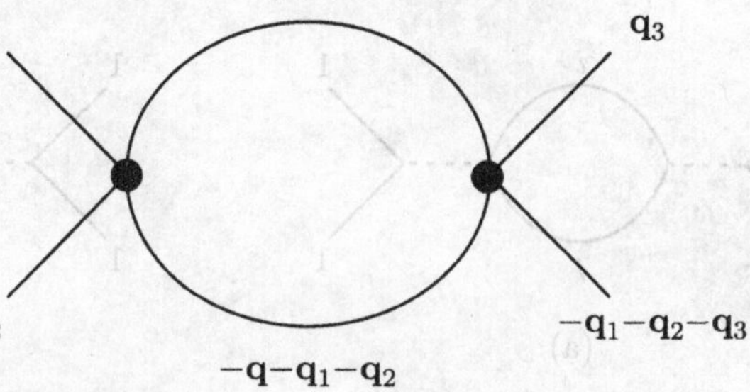

Fig. 5B.4. Diagram 5B.3c(1) showing wave number dependence of internal and external lines.

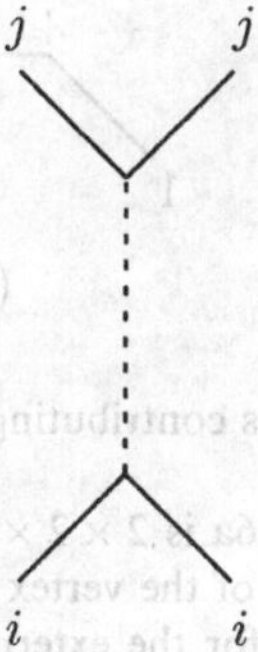

Fig. 5B.5. Representation of $\overline{\mathscr{H}_{\Lambda}^{4}}$ for O_n models. The dashed line represents the vertex u. At each end of this vertex are a pair of ϕ lines with the same Cartesian index i or $j = 1, \ldots, n$.

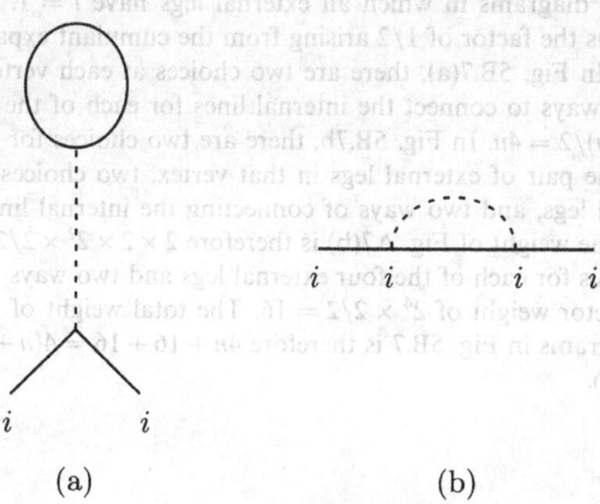

Fig. 5B.6. Diagrams contributing to r to linear order in u.

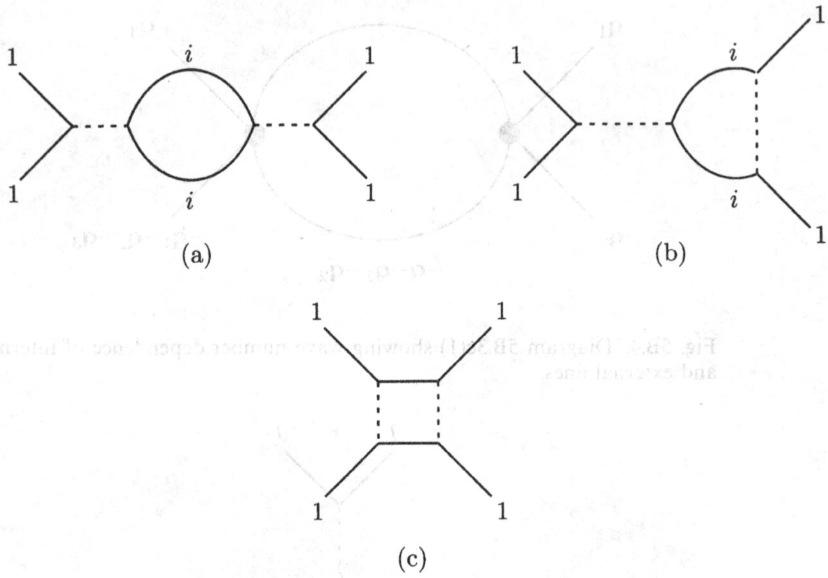

Fig. 5B.7. Diagrams contributing to u to second order in u.

the total weight of diagram 5B.6a is $2 \times 2 \times n = 4n$. In Fig. 5B.6b, there are two choices for the external leg at each end of the vertex and only one way of connecting the internal lines. (The index i is the same for the external legs and for the internal line.) Therefore, the weight of this diagram is $2 \times 2 \times 2 = 8$, and the total weight of the two diagrams in Fig. 5B.6 is $4(n+2)$. This is the weight in Eq. (5.8.42). Next consider contributions to u from the diagrams in Fig. 5B.7. Again the contribution of each of these diagrams to $u^{>}$ is proportional to $\int_{\mathbf{q}}[G_0^{>}(\mathbf{q})]^2$. Since interactions will always have O_n symmetry, we need only consider diagrams in which all external legs have $i = 1$. To obtain the weights of these diagrams the factor of $1/2$ arising from the cumulant expansion in Eq. (5.8.39) must be included. In Fig. 5B.7(a), there are two choices at each vertex for the pair of external legs and two ways to connect the internal lines for each of the n values of i for a weight of $(2 \times 2 \times 2n)/2 = 4n$. In Fig. 5B.7b, there are two choices for the right hand vertex, two choices for the pair of external legs in that vertex, two choices for each of the two right hand external legs, and two ways of connecting the internal lines. The internal legs must have $i = 1$. The weight of Fig. A7(b) is therefore $2 \times 2 \times 2^2 \times 2/2 = 16$. In Fig. 5B.7c, there are two choices for each of the four external legs and two ways of contracting the internal lines for a factor weight of $2^4 \times 2/2 = 16$. The total weight of the u^2 contributions to $u^{<}$ from the diagrams in Fig. 5B.7 is therefore $4n + 16 + 16 = 4(n+8)$. This is the result used in Eq. (5.8.44).

Bibliography

Daniel J. Amit, *Field Theory, the Renormalization Group, and Critical Phenomena*, revised 2nd edn (World Scientific, Singapore, 1984).

C. Domb and M.S. Green (eds.), *Phase Transitions and Critical Phenomena*, Vols. 5 and 6 (Academic Press, New York, 1976). See articles by A. Aharony; J. Als-Nielsen; E. Breézin, J.C. le Guillou, and J. Zinn-Justin; C. Di Castro and G. Jona-Lasino; L. Kadanoff; S.K. Ma; T.H. Niemeijer and J.M. van Leeuwen; D.J. Wallace; and K.G. Wilson.

Nigel Goldenfeld, *Lectures on Phase Transitions and the Renormalization Group* (Addison-Wesley, Reading, Mass., 1992).

S.K. Ma, *Modern Theory of Critical Phenomena* (Benjamin, Reading, Mass., 1976).

G. Parisi, *Statistical Field Theory* (Addison-Wesley, New York, 1988).

H. Eugene Stanley, *Introduction to Phase Transitions and Critical Phenomena* (Oxford University Press, New York, 1987).

K.G. Wilson and J.B. Kogut, *Phys. Reports* **12C**, 75 (1974).

J. Zinn-Justin, *Quantum Field Theory and Critical Phenomena* (Clarendon Press, Oxford, 1989).

References

A.N. Berker, S. Ostlund, and F.A. Putman, *Phys. Rev. B* **17**, 3650 (1978).

Michael Fisher, *Phys. Rev. Lett.* **40**, 1610 (1978).

C.M. Fortuin and P.W. Kasteleyn, *Physica* **57**, 536 (1972).

V.L. Ginzburg, *Fiz. Tverd. Tela.* **2**, 2031 (1960) [*Sov. Phys. Solid State* **2**, 1824 (1961)].

Cheng-Cher Huang and John T. Ho, *Phys. Rev. B* **12**, 5255 (1975).

J. Hubbard, *Phys. Rev. Lett.* **3**, 77 (1954).

L.P. Kadanoff, *Physics (NY)* **2**, 263 (1966).

L.P. Kadanoff, *Ann. Phys.* **100**, 359 (1976).

J.C. le Guillou and J. Zinn-Justin, *Phys. Rev. Lett.* **39**, 95 (1977); *Phys. Rev. B* **21**, 3976 (1980).

T.C. Lubensky, *Phys. Lett.* **67A**, 169 (1978).

T.C. Lubensky, in *Ill-Condensed Matter*, eds. Roger Balian, Roger Maynard, and Gerard Toulouse (North Holland, New York, 1979).

A.A. Migdal, *Z. Eksper. Teroet. Fiz.* **69**, 810, 1457 (1975) [*Sov. Phys., JETP* **42**, 413, 743 (1975)].

D.R. Nelson and M.E. Fisher, *Ann. Phys.* **91**, 226 (1975).

V. Privmam, P.C. Hohenberg, and A. Aharony, in *Phase Transitions and Critical Phenomena*, Vol. 14, eds. C. Domb and J.L. Lebowitz (Academic Press, New York, 1989).

J. Rudnick and D.R. Nelson, *Phys. Rev. B* **13**, 2208 (1976).

M.J. Stephen, *Phys. Lett.* **56A**, 149 (1976).

R.L. Stratonovich, *Doklady Akad. Nauk SSSR* **115**, 1097 (1957) [*Sov. Phys. Solid State* **2**, 1824 (1958)].

C.N. Yang and T.D. Lee, *Phys. Rev.* **87**, 404 (1952).

Problems

5.1 Calculate the upper critical dimension d_c for the following critical points:

(a) The tricritical point with free energy density

$$f = \frac{1}{2} r\phi^2 + u_6 \phi^6 .$$

(b) The (d, m) Lifshitz point described by the Hamiltonian

$$\mathscr{H} = \int d^d x \left[\frac{1}{2} r\phi^2 + \frac{1}{2} \sum_{i=1}^{m} (\nabla_i \phi)^2 + \frac{1}{2} \sum_{i=m+1}^{d} (\nabla_i^2 \phi)^2 + u\phi^4 \right].$$

Hint: It is useful to introduce correlation lengths $\xi_\parallel \sim r^{-1/2}$ and $\xi_\perp \sim r^{-1/4}$ to describe correlations in the two directions defined in the free energy. The correlation volume is then $\xi_\parallel^m \xi_\perp^{d-m}$.

(c) The strongly anisotropic dipolar magnet with Hamiltonian

$$\mathscr{H} = \int d^d x \left[\frac{1}{2} r\phi^2 + u\phi^4\right] + \int \frac{d^d q}{(2\pi)^d} [cq^2 + v(q_\parallel^2/q^2)]|\phi(\mathbf{q})|^2,$$

where the $\parallel$ specifies the single direction defined by anisotropy fields. Again, it is useful to introduce anisotropic correlation lengths via

$$\begin{aligned} G^{-1}(\mathbf{q}) &= [r + c(q_\parallel^2 + q_\perp^2) + vq_\parallel^2/(q_\parallel^2 + q_\perp^2)] \\ &\sim r[1 + (q_\perp \xi_\perp)^2 + v(q_\parallel \xi_\parallel / q_\perp \xi_\perp)^2] \\ &\quad + O[rq_\parallel^2 \xi_\parallel^2, r(q_\parallel \xi_\parallel / q_\perp \xi_\perp)^4], \end{aligned}$$

where $\xi_\parallel = r^{-\nu_\parallel}$ and $\xi_\perp = r^{-\nu_\perp}$ with $\nu_\parallel = 1$ and $\nu_\perp = 1/2$.

5.2 Show that the RPA predicts that the liquid-solid transition temperature in three dimensions for the model of the liquid-solid transition considered in Sec. 4.7 is zero when the coefficient of the third-order potential is zero. The model Hamiltonian is

$$F = \frac{1}{2}\int \frac{d^3 q}{(2\pi)^3} [r + c(q^2 - k_0^2)^2] n(\mathbf{q}) n(-\mathbf{q}) + \int d^3 x n^4(\mathbf{x}).$$

5.3 Assume the correlation function in an anisotropic system obeys the homogeneity relation

$$G(x_\parallel, x_\perp, t) = b^{-(d-2+\eta)} G(b^{-(1+\mu_\parallel)} x_\parallel, b^{-1} x_\perp, b^{1/\nu} t)$$

where $t = (T - T_c)/T_c$. Determine the critical exponents governing the behavior of $G(x_\parallel, x_\perp = 0, t = 0)$, $G(x_\parallel = 0, x_\perp, t = 0)$, the correlation lengths $\xi_\parallel$ and $\xi_\perp$, the susceptibility, and the order parameter for $t < 0$ in terms of the exponents η, $\mu_\parallel$ and ν.

5.4 (a) Calculate the transfer matrix and free energy for the one-dimensional Blume-Emery-Griffiths model defined in Eq. (4.6.17).

(b) Use the results of (a) to determine the Migdal-Kadanoff recursion relations on a triangular lattice for J, K, and Δ.

(c) (For the truly ambitious) Determine the fixed points and flows for the recursion relations derived in (b).

5.5 (a) Calculate the transfer matrix and free energy for the one-dimensional s-state Potts model with reduced Hamiltonian

$$H = -K \sum_{\langle l,l' \rangle} [s\delta_{\sigma_l,\sigma_{l'}} - 1],$$

where $\sigma_l = 1, ..., s$.

(b) Determine the continuous recursion relation for K. This equation should reduce to Eq. (5.7.6) in the Ising limit when $s = 2$. Then calculate the critical exponent ν in $1 + \epsilon$ dimensions for $s \geq 0$.

(c) Analytically continue the recursion relations derived in (b) to $s = 0$. Show that they lead to a lower critical dimension of 2 rather than 1, and calculate

ν in $2+\epsilon$ dimensions. The $s=0$ limit of the Potts model counts spanning trees on a lattice, and its correlation functions give the resistance between two points in a resistor network. (See Fortuin and Kasteleyn 1972; Lubensky 1978; Stephen 1976.)

5.6 Calculate the scaling exponent for u_p for $p=3$ and $p>4$ to first order in ϵ.

5.7 Consider the O_3 Heisenberg model with cubic anisotropy [Eq. (5.8.51)].

(a) Show that the order parameter aligns along one of the cubic directions when $v>0$.

(b) Assuming that ϕ aligns along the z-axis, show that in mean-field theory

$$G_{xx} = -\frac{4u}{vr}$$

for $r<0$ (i.e., that G_{xx} diverges in mean-field theory as $r^{-\gamma_\perp}$ with $\gamma_\perp = 1$).

(c) Use this result to show that, near four dimensions,

$$G_{xx} \sim |t|^{-\gamma_\perp},$$

where

$$\gamma_\perp = \gamma + |\lambda_H^v|\nu$$

with γ and ν the susceptibility and correlation length exponents of the Heisenberg fixed point and λ_H^v the exponent for v at the Heisenberg fixed point [Eq. (5.8.60)].

5.8 Bond percolation is defined mathematically as the formation of an infinite (or sample spanning) cluster of sites connected by occupied bonds on a lattice whose bonds are occupied with probability p and unoccupied with probability $1-p$. The formation of this infinite cluster is analogous to the passage of a fluid (coffee) through a porous medium (coffee grounds). An infinite cluster exists for p greater than a critical value p_c. The probability $\mathscr{P}(p) \sim (p-p_c)^\beta$ that a site is in an infinite cluster grows continuously from zero for $p>p_c$. Thus percolation is a geometrical phase transition with $\mathscr{P}(p)$ playing the role of the order parameter in a thermodynamic phase transition. The probability $S(\mathbf{x},\mathbf{x}')$ that two sites are in the same cluster is the analog of the nonlocal susceptibility $\chi(\mathbf{x},\mathbf{x}')$, and $S(p) = \int d^dx S(\mathbf{x},\mathbf{x}')$ is the analog of the susceptibility χ.

(a) The percolation problem can be described by the s-state Potts model in the limit $s\to 1$ (Fortuin and Kasteleyn 1972; Lubensky 1979). Use the mean-field theory for the Potts model developed in Sec. 4.8 to show that there is a continuous phase transition in which the order parameter is positive (as a probability must be) with $\gamma=\beta=1$ and $\nu=1/2$. Then show that the upper critical dimension d_c for percolation is 6.

(b) Let $\mathbf{e}_l$, $l=0,1,\ldots,s-1$ be a complete set of orthonormal s-dimensional vectors with $\mathbf{e}_0 = s^{-1/2}(1,1,\ldots,1)$. Show that

$$\sum_{l=1}^{s-1} e_l^\sigma e_l^{\sigma'} = s\delta^{\sigma,\sigma'} - 1, \qquad \sum_{\sigma=1}^{s} e_l^\sigma = 0 \text{ if } l \neq 0,$$

where e_l^σ are the components of $\mathbf{e}_l$. Use the Hubbard-Stratonovich transformation to derive the field theory

$$\mathscr{H} = \int d^d x \left[\frac{1}{2} \sum_{l=1}^{s} (\psi_l^2 + (\nabla \psi_l)^2) - w\lambda_{l_1 l_2 l_3} \psi_{l_1} \psi_{l_2} \psi_{l_3} + O(\psi_l^4) \right],$$

where $\lambda_{l_1 l_2 l_3} = s^{-1} \sum_{\sigma=1}^{s} e_{l_1}^\sigma e_{l_2}^\sigma e_{l_3}^\sigma$. From this, derive the recursion relations

$$\frac{dr}{dl} = (2-\eta)r - 18K_6(s-2)\frac{w^2}{(1+r)^2},$$

$$\frac{dw}{dl} = \frac{1}{2}(\epsilon - 3\eta)w + 36K_6(s-3)w^3,$$

$$\eta = 6K_6 w^2 (s-2),$$

where $\epsilon = 6 - d$, and show that the exponents for percolation are $\eta = -\epsilon/21$ and $\nu^{-1} = 2 - (5\epsilon/21)$. What are α, β, and γ? The integral

$$I = \int_S \frac{d^6 k}{(2\pi)^6} \frac{1}{r+k^2} \frac{1}{r+(\mathbf{k}+\mathbf{q})^2} = \frac{\delta l}{(1+r)^2} - \frac{\delta l}{3}(1+O(r)), \quad (5P.1)$$

where S is the region defined by $b^{-1} < k < 1$ and $b^{-1} < |\mathbf{k}+\mathbf{q}| < 1$ with $b = 1 + \delta l$, may prove useful. The exponent η is set by the requirement that the coefficient of $(\nabla\psi)^2$ be equal to unity under renormalization. Thus, you will need the recursion relation for this coefficient.

5.9 Yang and Lee (1952) showed that the partition function for the Ising model has zeros in the complex $H = H' + iH''$ plane that become densely distributed in the thermodynamic limit. For $T < T_c$, these zeros give rise to branch cuts in the free energy and magnetization along the imaginary H axis originating at the Yang-Lee edges $H = \pm iH_c''(T)$.

(a) Assume that the free energy has a singularity of the form

$$f(H, T) \sim |H'' - H_c''(T)|^\sigma.$$

Deduce from this that the gap exponent $\Delta = \frac{1}{2}(d + 2 - \eta)\nu$ is unity so that the exponents η and ν satisfy

$$\nu = \frac{2}{d+2-\eta}$$

and are thus not independent (Fisher 1978).

(b) Use the Landau free energy $f = \frac{1}{2}r\phi^2 + u\phi^4 - ih\phi$ to show that $\sigma = 3/2$, $\beta = \gamma = 1/2$, and $\nu = 1/4$ in mean-field theory. From this, show that the upper critical dimension d_c for the Yang-Lee edge is 6. At a given r (temperature), the values, h_c and ϕ_c, of h and ϕ at the Yang-Lee edge are determined by the conditions $\partial f/\partial\phi = 0$ and $\partial^2 f/\partial\phi^2 = 0$. The latter condition can be satisfied because ϕ_c is imaginary.

(c) Let $\phi = \phi_c + \psi$. Show that fluctuations near the Yang-Lee edge can be described by a field theory with

$$\mathcal{H} = \int d^d x \left[\frac{1}{2} c(\nabla\psi)^2 + iw\psi^3 - i\Delta h\psi \right],$$

where $\Delta h - h - h_c$. Derive recursion relations for w and c, and show that $\eta = -\epsilon/9$ and $\nu = \frac{1}{4} + \frac{1}{36}\epsilon$ to first order in ϵ. You may wish to use the integral in Eq. (5P.1).

6

Generalized elasticity

In the preceding several chapters, we have seen that the order established below a phase transition breaks the symmetry of the disordered phase. In many cases, the broken symmetry is continuous. For example, the vector order parameter $\mathbf{m}$ of the ferromagnetic phase breaks the continuous rotational symmetry of the paramagnetic phase, the tensor order parameter Q_{ij} of the nematic phase breaks the rotational symmetry of the isotropic fluid phase, and the set of complex order parameters ρ_G of the solid phase breaks the translational symmetry of the isotropic liquid. In these cases, there are an infinite number of equivalent ordered phases that can be transformed one into the other by changing a continuous variable θ. If rotational symmetry is broken, θ specifies the angle (or angles) giving the direction of the order parameter; if translational symmetry is broken, θ specifies the origin of a coordinate system. Uniform changes in θ do not change the free energy. Spatially non-uniform changes in θ, however, do. In the absence of evidence to the contrary, one expects the free energy density f to have an analytic expansion in gradients of θ. Thus we expect a term in f that is proportional to $(\nabla\theta)^2$ for θ varying slowly in space. We refer to this as the elastic free energy, f_{el}, since it produces a restoring force against distortion, and we will refer to θ as an elastic or hydrodynamic variable. Probably the most familiar example of such a free energy is elastic energy of solids where θ is the displacement of the lattice. It is, however, one of the most complicated of what we will call generalized elastic energies arising from broken continuous symmetry. In this chapter, we will study the generalized elasticity and its consequences for a number of systems exhibiting broken continuous symmetry.

The free energy of a system with a broken continuous symmetry is invariant with respect to spatially uniform displacements that take the system from one point in a ground state manifold to another. In general, there is one elastic variable (or degree of freedom) associated with each generator of translations within this manifold. Thus, there is a single elastic variable for an xy-model whose ground state manifold is simply the unit circle, and there are two elastic variables in an O_3 model whose ground state manifold is the unit sphere. Notable exceptions to this rule are systems, such as smectic liquid crystals and crystalline solids, with periodically broken translational symmetry. These systems are invariant with respect to both translation and rotations, and one might expect both translational and rotational elastic variables. Low-energy rotational distortions are, however,

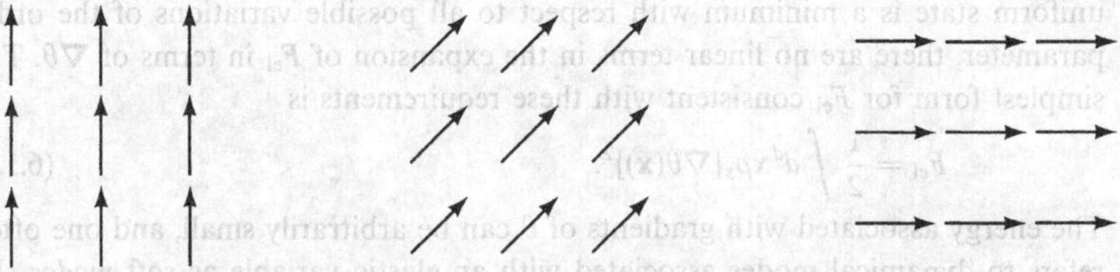

Fig. 6.1.1. Different ground states for the xy-Hamiltonian. The vector order parameter can point anywhere in a two-dimensional plane.

completely determined by spatially varying translational distortions, and there are only translational elastic variables.

6.1 The xy-model

The simplest continuous symmetry ($U(1)$ or O_2) is that of rotations in a two-dimensional plane. The order parameter that breaks this symmetry can be either a two-dimensional vector $\langle \mathbf{s} \rangle = s(\cos\theta, \sin\theta)$ or a complex number $\langle \psi \rangle = |\langle \psi \rangle| e^{i\theta}$, whose respective direction or phase is specified by the angle θ. A useful representation of systems with this symmetry is the xy-model on a lattice introduced in Sec. 3.6. In this model, there is a spin $\mathbf{s}(\mathbf{x}) = s(\cos\theta(\mathbf{x}), \sin\theta(\mathbf{x}))$ of magnitude s at each site $\mathbf{x}$ on a lattice. These spins interact with their nearest neighbors via an exchange interaction in the Hamiltonian of Eqs. (3.6.3) and (3.6.5). Different equivalent ground state configurations of this model are shown in Fig. 6.1.1. We will use the term xy-model to include all systems such as superfluid helium or hexatic liquid crystals with a complex or two-dimensional vector order parameter. It is useful to bear in mind, however, that the hexatic order parameter is usually written as $\langle \psi_6 \rangle = |\langle \psi_6 \rangle| e^{6i\theta}$ to emphasize its invariance with respect to rotations through $2\pi/6$.

1 The elastic free energy

The free energy in the ordered phase of the xy-model has the form of the base of a wine bottle, as shown in Fig. 6.1.2. The minimum value of the free energy occurs on a circle at the base of the wine bottle. The radius of the circle specifies the magnitude of the order parameter ($s = |\langle \mathbf{s} \rangle|$ or $|\langle \psi \rangle|$). The position on the circle is determined by the angle θ. Spatially uniform changes in θ lead to a rotation around the base of the wine bottle but do not change the energy. Spatially non-uniform changes in θ such as those depicted in Fig. 6.1.3 will, however, increase the free energy F. In the absence of any indications to the contrary, we expect $F_{\text{el}} = F[\theta(\mathbf{x})] - F[\theta = \text{const.}]$ to be analytic in $\nabla\theta$. Furthermore, since the uniform state is a minimum with respect to all possible variations of the order parameter, there are no linear terms in the expansion of F_{el} in terms of $\nabla\theta$. The simplest form for F_{el} consistent with these requirements is

$$F_{el} = \frac{1}{2}\int d^d x \rho_s [\nabla\theta(\mathbf{x})]^2. \tag{6.1.1}$$

The energy associated with gradients of θ can be arbitrarily small, and one often refers to dynamical modes associated with an elastic variable as *soft modes.* F_{el} is invariant both with respect to uniform displacements of θ and with respect to uniform displacements and rotations of space. This is because in the model we are considering, rotations of the spin variable are completely decoupled from rotations in space: the total symmetry group is a direct product of the spin and space symmetry groups. This is not always the case, as we shall see in the next section. The coefficient ρ_s is called the *spin-wave stiffness* or *helicity modulus* in magnetic systems, the *superfluid density* in superfluids, and is often referred to simply as a *rigidity.* In d dimensions, ρ_s has units of energy/(length)$^{d-2}$ or force/(length)$^{d-3}$. An energy scale is provided by the transition temperature T_c (which for the lattice model is of the order of the exchange J). At low temperatures, the length scale is that of the interparticle spacing a (the lattice parameter in the lattice model). Thus we expect ρ_s to be of order $T_c a^{-(d-2)}$ at low temperatures. For the lattice model of Eq. (3.6.3), $\rho_s = s^2 J z a^{-(d-2)}/4d$ at $T = 0$, where z is the coordination number of the lattice. Near T_c, the length scale is set by the correlation length, and one could argue that ρ_s should be of order $T_c\xi^{-(d-2)}$, i.e., ρ_s should go to zero at a second-order critical point as $|T - T_c|^{(d-2)\nu}$. Scaling in critical systems predicts this result, as we shall see shortly. In mean-field theory (see Sec. 4.3), $\rho_s = c|\langle \mathbf{s}\rangle|^2 \sim |T - T_c|$, which agrees with the scaling result only at the upper critical dimension $d_c = 4$, as expected of hyperscaling relations (Sec. 5.4) that involve the spatial dimension.

2 Boundary conditions and external fields

In the absence of external fields,

$$\frac{\delta F_{el}}{\delta\theta(\mathbf{x})} = -\rho_s \nabla^2\theta(\mathbf{x}) = 0. \tag{6.1.2}$$

This predicts that the lowest energy state is one of spatially uniform θ. If, however, boundary conditions are imposed, non-uniform solutions for θ may be required. For example if θ is constrained to be zero along the plane $z = 0$ and θ_0 along the plane $z = L$, we have

$$\begin{aligned} \theta &= \theta_0 \frac{z}{L}, \\ F_{el} &= \frac{1}{2}\rho_s L^{d-2}\theta_0^2. \end{aligned} \tag{6.1.3}$$

Note that the energy is independent of L when $d = 2$. We shall see in several different ways that two dimensions is quite special for this model. Eq. (6.1.3) can

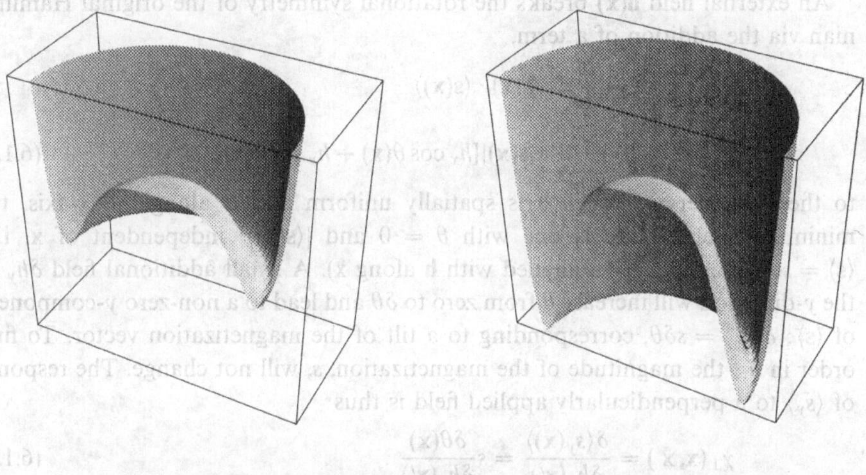

Fig. 6.1.2. The free energy as a function of a two-component order parameter in the ordered phase of the *xy*-model for temperatures near the transition temperature T_c (right) and for temperatures much less than T_c (left). It has the shape of a wine bottle, with all states on the circular minimum having the same energy. At low temperatures, the deep well fixes the magnitude of the spin at the value s_0 at the minimum of F_a. The steep well implies that fluctuations in the magnitude of the spin about s_0 are negligible.

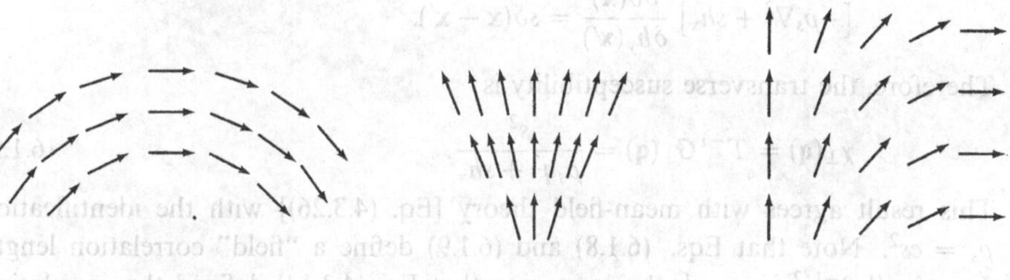

Fig. 6.1.3. (a)-(c) Spin configurations with a spatially non-uniform $\theta(\mathbf{x})$. These configurations have a higher free energy than do those with a spatially uniform $\theta(\mathbf{x})$. (c) Shows spin configurations corresponding to the boundary conditions of Eq. (6.1.3).

be rewritten as an expression for ρ_s in terms of the difference between the free energy $F[\theta_0]$, with boundary conditions $\theta = 0$ at $z = 0$ and $\theta = \theta_0$ at $z = L$, and the free energy $F[0]$, with boundary conditions $\theta = 0$ at $z = 0$ and $z = L$:

$$\rho_s = \lim_{L\to\infty} 2L^{2-d}(F[\theta_0] - F[0])/\theta_0^2. \tag{6.1.4}$$

This equation can be regarded as a definition of the helicity modulus ρ_s.

An external field $\mathbf{h}(\mathbf{x})$ breaks the rotational symmetry of the original Hamiltonian via the addition of a term,

$$F_{\text{ext}} = -\int d^d x \mathbf{h}(\mathbf{x}) \cdot \langle \mathbf{s}(\mathbf{x}) \rangle$$
$$= -\int d^d x |\mathbf{s}(\mathbf{x})| [h_x \cos\theta(\mathbf{x}) + h_y \sin\theta(\mathbf{x})], \tag{6.1.5}$$

to the free energy. When $\mathbf{h}$ is spatially uniform and is along the x-axis, the minimum energy state is one with $\theta = 0$ and $|\langle \mathbf{s}(\mathbf{x}) \rangle|$ independent of $\mathbf{x}$, i.e., $\langle \mathbf{s} \rangle = s\mathbf{e}_x$ (magnetization aligned with $\mathbf{h}$ along x). A small additional field δh_y in the y-direction will increase θ from zero to $\delta\theta$ and lead to a non-zero y-component of $\langle \mathbf{s} \rangle$: $\delta\langle s_y \rangle = s\delta\theta$, corresponding to a tilt of the magnetization vector. To first order in h_y, the magnitude of the magnetization, s, will not change. The response of $\langle s_y \rangle$ to a perpendicularly applied field is thus

$$\chi_\perp(\mathbf{x}, \mathbf{x}') = \frac{\delta\langle s_y(\mathbf{x}) \rangle}{\delta h_y(\mathbf{x}')} = s\frac{\delta\theta(\mathbf{x})}{\delta h_y(\mathbf{x}')} . \tag{6.1.6}$$

To calculate $\delta\theta/\delta h_y$, we minimize the total elastic free energy $F_T = F_{\text{el}} + F_{\text{ext}}$ expanded about $\theta = 0$ for small h_y:

$$F_T = \int d^d x \left[\frac{1}{2}\rho_s(\nabla\theta)^2 + \frac{1}{2}sh_x\theta^2 - sh_y\theta - sh_x + \ldots \right], \tag{6.1.7}$$

$$\frac{\delta F_T}{\delta\theta} = \left[-\rho_s\nabla^2 + sh_x \right] \theta - sh_y = 0, \tag{6.1.8}$$

$$\left[-\rho_s\nabla^2 + sh_x \right] \frac{\delta\theta(\mathbf{x})}{\delta h_y(\mathbf{x}')} = s\delta(\mathbf{x} - \mathbf{x}').$$

Therefore, the transverse susceptibility is

$$\chi_\perp(\mathbf{q}) = T^{-1}G_\perp(\mathbf{q}) = \frac{s^2}{\rho_s q^2 + sh_x}. \tag{6.1.9}$$

This result agrees with mean-field theory [Eq. (4.3.26)] with the identification $\rho_s = cs^2$. Note that Eqs. (6.1.8) and (6.1.9) define a "field" correlation length $\xi_h = (\rho_s/h_x)^{-1/2}$ in much the same way that Eq. (4.3.14) defined the correlation length near second-order phase transitions.

3 The Josephson scaling relation

In the critical region, we expect $G_\perp(\mathbf{q})$ to obey the generalized scaling relation discussed in Chapter 5:

$$G_\perp(\mathbf{q}, h_x) = q^{-(2-\eta)} Y(q\xi,\ h_x t^{-\Delta}) \quad (t \equiv (T - T_c)/T). \tag{6.1.10}$$

This relation predicts $G_\perp \sim t^\beta/h_x$ when $\mathbf{q} = 0$ because $\Delta = (\beta + \gamma)\nu$ and $\gamma = (2-\eta)\nu$, in agreement with the prediction required by symmetry that $G_\perp = Ts/h_x$ ($s \sim t^\beta$). When $h_x = 0$, it can be used to predict the temperature dependence of ρ_s:

$$G_{\perp}(\mathbf{q}) = q^{-(2-\eta)}Y(q\xi) = \frac{s^2}{\rho_s q^2}$$

$$\rho_s = \lim_{t\to 0} s^2 q^{-\eta} Y^{-1}(q\xi) \sim t^{2\beta}\xi^{\eta} \sim t^{(d-2)\nu}, \qquad (6.1.11)$$

where we used $s \sim t^{\beta}$ and the hyperscaling relation [Eq. (5.4.15)] for β. This result agrees with that obtained earlier by dimensional analysis. It was first derived by Brian Josephson (1966) and is called the *Josephson relation.* When both $\mathbf{q}$ and $\mathbf{h}$ are nonzero, a more complicated form than that derived above is expected for $G_{\perp}(\mathbf{q})$ in the critical region.

4 Fluctuations

The energies associated with elastic variables like θ can be arbitrarily small. The existence of such low energy distortions can have profound effects on other properties of the system – in particular on the value of the order parameter $\langle \mathbf{s} \rangle$ or $\langle \psi \rangle$. $\langle \mathbf{s} \rangle$ is the average of the field $\mathbf{s}$ with respect to some thermodynamic weight function. Assume $\theta = 0$, i.e., assume $\langle s \rangle = s\mathbf{e}_x$, and parameterize $\mathbf{s}(\mathbf{x})$ with its magnitude $s(\mathbf{x})$ and angle $\vartheta(\mathbf{x})$ with respect to the x-axis

$$\mathbf{s}(\mathbf{x}) = s(\mathbf{x})[\cos\vartheta(\mathbf{x}), \sin\vartheta(\mathbf{x})] . \qquad (6.1.12)$$

The lattice xy-model is a special case of the above in which $s(\mathbf{x})$ is constrained to be a constant. Note again the distinction between ϑ and θ. θ specifies the direction of the average spin at $\mathbf{x}$, whereas ϑ specifies the direction of the unaveraged spin field. If $\langle \mathbf{s} \rangle$ is nonzero, ϑ will not be uniformly distributed between 0 and 2π but will have a greater probability to have a value near θ.

With $\theta = 0$ we have

$$\langle s_x \rangle = \langle s(\mathbf{x}) \cos\vartheta(\mathbf{x}) \rangle = s,$$
$$\langle s_y \rangle = \langle s(\mathbf{x}) \sin\vartheta(\mathbf{x}) \rangle = 0. \qquad (6.1.13)$$

The free energy functional is invariant with respect to uniform rotations of $\mathbf{s}$. Thus, we expect the same kind of expansion of the Hamiltonian $\mathscr{H}$ in terms of ϑ as we developed for F in terms of θ:

$$\mathscr{H} = \mathscr{H}_1[s(\mathbf{x})] + \mathscr{H}_{\text{el}}[\vartheta(\mathbf{x})],$$
$$\mathscr{H}_{\text{el}} = \frac{1}{2}\int d^d x \rho'_s[s(\mathbf{x})][\nabla\vartheta(\mathbf{x})]^2. \qquad (6.1.14)$$

In general, $\mathscr{H}_{\text{el}}$ should include higher order terms in $\nabla\vartheta$. The value of ρ'_s depends on $s(\mathbf{x})$. In the mean-field theory, as we discussed at the end of the preceding chapter,

$$F[\langle s(\mathbf{x}) \rangle] = \mathscr{H}[\langle s(\mathbf{x}) \rangle] , \qquad (6.1.15)$$

so that $\rho'_s(\langle \mathbf{s} \rangle)$ is equal to ρ_s in this case. Another interesting limit is that of low temperatures, where the fluctuations in the value of $s(\mathbf{x})$ are small. As the temperature is lowered, the wells in $\mathscr{H}$ become steeper, as shown in Fig. 6.1.2. These steep wells fix the value of $s(\mathbf{x})$ at a preferred value s_0. In this case, we can set $\rho'_s(s(\mathbf{x})) = \rho'_s(s_0) = \rho_s$. This low-temperature limit can be obtained directly from an expansion to harmonic order of the lattice xy-model:

$$\begin{aligned}\mathcal{H} &= -J\sum_{<\mathbf{l},\mathbf{l}'>}\cos(\vartheta_{\mathbf{l}} - \vartheta_{\mathbf{l}'}) \\ &\approx -zNJ + \frac{1}{4}J\sum_{\mathbf{l},\mathbf{l}'}\gamma_{\mathbf{l},\mathbf{l}'}(\vartheta_{\mathbf{l}} - \vartheta_{\mathbf{l}'})^2 + O[(\vartheta_{\mathbf{l}} - \vartheta_{\mathbf{l}'})^4],\end{aligned} \tag{6.1.16}$$

where $\gamma_{\mathbf{l},\mathbf{l}'}$ is the nearest-neighbor matrix equal to one if $\mathbf{l}$ and $\mathbf{l}'$ are nearest neighbors on the lattice and zero otherwise. The continuum limit of this Hamiltonian produces Eq. (6.1.14) with $\rho'_s = zJa^{2-d}/(2d)$. There is, in general, a real distinction between the thermodynamic stiffness ρ_s and the local stiffness ρ'_s. ρ'_s is simply an interaction parameter of the system. ρ_s, on the other hand, necessarily includes the effects of fluctuations that in general reduce rigidity. In this chapter, we will not be too careful about the distinction between the two rigidities and will in fact use the same symbol for both. We will, however, return to this question in Chapters 9 and 10.

The above considerations imply that fluctuations in ϑ in the low temperature limit are controlled by $\mathcal{H}_{\text{el}}$ with $\rho'_s = \rho_s$ a constant. This allows us to calculate the average order parameter,

$$\begin{aligned}\frac{\langle s_x\rangle}{s_0} &= \langle\cos\vartheta\rangle \equiv e^{-W} \\ &= \text{Re}\left(\frac{1}{Z_{\text{el}}}\int\mathcal{D}\vartheta e^{-\beta\mathcal{H}_{\text{el}}}e^{i\vartheta(\mathbf{x})}\right),\end{aligned} \tag{6.1.17}$$

where

$$Z_{\text{el}} = \int\mathcal{D}\vartheta e^{-\beta\mathcal{H}_{\text{el}}}. \tag{6.1.18}$$

(It is understood that $\mathcal{D}\vartheta$ does not include the spatially uniform $\vartheta(q=0)$.) $\mathcal{H}_{\text{el}}$ is harmonic in ϑ, and we can use the properties of Gaussian functional integrals derived in Sec. 5.3 to evaluate

$$\begin{aligned}W &= \frac{1}{2}\langle\vartheta^2(\mathbf{x})\rangle = \frac{1}{2}T\int\frac{d^dq}{(2\pi)^d}\frac{1}{\rho_s q^2} \\ &= K_d\frac{T\Lambda^{d-2}}{2\rho_s(d-2)},\end{aligned} \tag{6.1.19}$$

where Λ is the wave number cutoff. e^{-2W} is called the *Debye-Waller factor*. In the classical systems we are considering here, it measures the degree to which the order parameter is depressed by thermal fluctuations from its zero temperature maximum value.

Note that $W \to \infty$ as $d \to 2$. Thus $\langle s_x\rangle$ is zero and there is no long-range order in two dimensions, even at low temperatures when its magnitude s_0 has achieved a well defined value. This is the phenomenon of *fluctuation destruction of long-range order*. The absence of long-range order in two-dimensional systems with a continuous symmetry is often referred to as the Mermin-Wagner-Berezinskii theorem

(Mermin and Wagner 1966; Mermin 1968; Berezinskii 1970, 1971). The lower critical dimension for this system is $d_L = 2$. Physically, a very long wavelength rotation of the direction of $\langle \mathbf{s} \rangle$ costs very little energy. Thermal excitation of these long wavelength modes changes the direction of the magnetization in space and time. Thus, even though the magnitude of the magnetization is locally saturated, spatial and time averages can greatly reduce the total magnetization from the value when all spins are aligned. The importance of these long wavelength, low q fluctuations is reduced in higher dimensions because of the phase space weighting by $q^{d-1}dq$.

5 Long-range order, quasi-long-range order, and disorder

The spin correlation function is easily calculated again using the properties of Gaussian integrals discussed in Sec. 5.3:

$$\begin{aligned} G'(\mathbf{x},0) &= \langle \mathbf{s}(\mathbf{x}) \cdot \mathbf{s}(0) \rangle = s_0^2 \langle \cos[\vartheta(\mathbf{x}) - \vartheta(0)] \rangle \\ &= s_0^2 \mathrm{Re} \langle e^{i(\vartheta(\mathbf{x}) - \vartheta(0))} \rangle = s_0^2 e^{-g(\mathbf{x})} , \end{aligned} \tag{6.1.20}$$

where

$$\begin{aligned} g(\mathbf{x}) &= T \int \frac{d^d q}{(2\pi)^d} \frac{1 - e^{i\mathbf{q}\cdot\mathbf{x}}}{\rho_s q^2} \\ &= \begin{cases} \frac{T K_d \Lambda^{(d-2)}}{(d-2)\rho_s} = 2W & |\mathbf{x}| \to \infty \quad (d > 2) \\ \frac{T}{2\pi\rho_s} \ln(\tilde{\Lambda}|\mathbf{x}|) & |\mathbf{x}| \to \infty \quad (d = 2) \\ \frac{T}{2\rho_s}|x| & |x| \to \infty \quad (d = 1) \end{cases} . \end{aligned} \tag{6.1.21}$$

Here, $\tilde{\Lambda} = \Lambda e^{\tilde{\gamma}}$, where Λ is the high wave number cutoff (see Sec. 5.2) and $\tilde{\gamma}$ is a constant to be evaluated below. It is clear from Eqs. (6.1.19) and (6.1.21) that $d = 2$ is a special dimension. For $d > 2$, $g(\mathbf{x}) \to$ const. and $G'(\mathbf{x}, 0) \to s_0^2 e^{-2W} =$ const. as $\mathbf{x} \to \infty$ in agreement with the condition for the existence of long-range order (LRO) discussed in Chapter 3 [see Eq. (3.5.8)]. For $d = 1$, $g(x)$ grows linearly with $|x|$, and

$$G'(x) = s_0^2 e^{-(T/2\rho_s)|x|} \tag{6.1.22}$$

decays exponentially with distance. Thus there is no long-range order for any $T > 0$ in one dimension. The disordered paramagnetic phase persists all of the way down to $T = 0$. The Fourier transform of Eq. (6.1.22),

$$G(q) = s_0^2 \frac{T/2\rho_s}{q^2 + (T/2\rho_s)^2}, \tag{6.1.23}$$

is a Lorentzian with a width that goes to zero with T.

In exactly two dimensions, $g(\mathbf{x})$ can be written as $(T/2\pi\rho_s)I(|\mathbf{x}|)$, where

$$I(|\mathbf{x}|) = \int_0^\Lambda \frac{dq}{q} \left[\frac{1}{2\pi} \int_0^{2\pi} d\theta \left(1 - e^{iq|\mathbf{x}|\cos\theta} \right) \right]$$

$$= \int_0^\Lambda \frac{1 - J_0(q|\mathbf{x}|)}{q} dq, \tag{6.1.24}$$

where $J_0(u)$ is the Bessel function of order 0. The large $|\mathbf{x}|$ behavior of this integral can be obtained by letting $u = q|\mathbf{x}|$ and breaking the integral into parts:

$$\begin{aligned} I(\mathbf{x}) &= \int_0^1 \frac{1 - J_0(u)}{u} du + \int_1^{\Lambda|\mathbf{x}|} \frac{du}{u} - \int_1^{\Lambda|\mathbf{x}|} \frac{J_0(u)}{u} du \\ &= \ln \Lambda|\mathbf{x}| + \tilde{\gamma} + O\left((\Lambda|\mathbf{x}|)^{-3/2}\right), \end{aligned} \tag{6.1.25}$$

where

$$\tilde{\gamma} = \int_0^1 \frac{1 - J_0(u)}{u} du - \int_1^\infty \frac{J_0(u)}{u} du \approx -0.116. \tag{6.1.26}$$

(This constant depends on cutoff. For a square lattice, it is $\gamma + \frac{1}{2}\ln 8$, where $\gamma = 0.5772\ldots$ is the Euler-Mascheroni constant.) $g(\mathbf{x})$ diverges logarithmically with $\mathbf{x}$, and

$$G'(\mathbf{x}, 0) = s_0^2(\tilde{\Lambda}|\mathbf{x}|)^{-\eta} \tag{6.1.27}$$

decays algebraically to zero with an exponent

$$\eta = \frac{T}{(2\pi\rho_s)}, \tag{6.1.28}$$

depending on the temperature. Systems with power-law decay of order-parameter correlation functions are said to have quasi-long-range order (QLRO). Thus, provided T/ρ_s is neither zero nor infinite, the spin correlation function $G'(\mathbf{x})$ of the xy-model in two dimensions exhibits the same power-law dependence on $\mathbf{x}$ as it would at a critical point in higher dimensions. In Fourier space,

$$G(\mathbf{q}) \sim |\mathbf{q}|^{-(2-\eta)} \tag{6.1.29}$$

again has the same form as a correlation function at a critical point. Note that $G(\mathbf{q} = 0)$ and thus the uniform susceptibility is infinite for all $\eta < 2$. When $T = 0$, $g(\mathbf{x}) = 0$ and $G'(\mathbf{x})$ is a constant independent of $\mathbf{x}$ so that, as expected, there is long-range order in the two-dimensional classical xy-model at zero temperature. If $\rho_s \to 0$, $T/\rho_s \to \infty$, and $G'(\mathbf{x})$ will tend to zero at large $|\mathbf{x}|$ more rapidly than algebraically. Thus, there is a transition from QLRO to disorder when $\rho_s \to 0$. In a pure harmonic model, there is QLRO at all temperatures. Vortex excitations, which we will study in Chapter 9, reduce ρ_s and lead to a QLRO-disorder transition. This rather unique transition is called the Kosterlitz-Thouless transition (1973).

On occasion, the precise form of fluctuations in the lattice model is of interest. In this case, W and $g(\mathbf{R}_l)$ are identical to Eqs. (6.1.19) and (6.1.21) with $\rho_s q^2$ replaced by $v_0^{-1}\epsilon(\mathbf{q})$, where

$$\epsilon(\mathbf{q}) = 2J\sum_{i=1}^{d}[1 - \cos(\mathbf{q} \cdot \mathbf{a}_i)] \tag{6.1.30}$$

on a hypercubic lattice, where $\mathbf{a}_i$ are the nearest-neighbor vectors of the lattice. More generally, if the exchange $J_{\mathbf{l},\mathbf{l}'}$ extends to further than nearest neighbors,

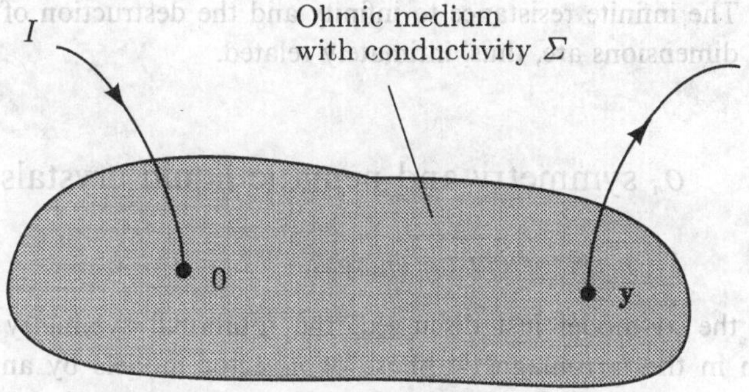

Fig. 6.1.4. If a current I is injected into an infinite ohmic medium at point 0 and extracted at point y, there will be a voltage difference $V = R_{\mathbf{y}} I$ between 0 and y. The resistance $R_{\mathbf{y}}$ is proportional to the angle correlation function $g(\mathbf{y})$ [Eq. (6.2.22)] of an xy-model. The relation $V \sim I$ is analogous to $\theta \sim h$ in a spin system.

$\epsilon(\mathbf{q}) = [J(0) - J(\mathbf{q})]$. Long wavelength fluctuations with $qa \ll 1$ are identical in lattice and continuum models, and the dominant singularities in $g(\mathbf{x})$ at large $|\mathbf{x}|$ are the same in both models. The precise value of W for $d > 2$ will, however, differ in the two models.

6 *Resistance of a conducting medium*

We close this section with the observation that there is a close analogy between the variable θ and the voltage V in an infinite conducting medium. In the absence of external current sources or fields, the voltage V satisfies Laplace's equation, as does θ in the absence of perturbing fields. The electric current density $\mathbf{J}$ in such a medium is equal to $\Sigma\mathbf{E}$, where Σ is the macroscopic electrical conductivity and $\mathbf{E} = -\nabla V$ is the electric field. When there are external current sources, we have

$$-\Sigma\nabla^2 V = \nabla \cdot \mathbf{J}. \tag{6.1.31}$$

If a current I is inserted at the origin and extracted at $\mathbf{y}$ (Fig. 6.1.4),

$$\nabla \cdot \mathbf{J} = I[\delta(\mathbf{x}) - \delta(\mathbf{x} - \mathbf{y})]. \tag{6.1.32}$$

The resistance between the origin and $\mathbf{y}$ is, therefore,

$$\begin{aligned} R_{\mathbf{y}} &= \frac{V(0) - V(\mathbf{y})}{I} \\ &= 2\int \frac{d^d q}{(2\pi)^d} \frac{1 - e^{i\mathbf{q}\cdot\mathbf{y}}}{\Sigma q^2} = \frac{2\rho_s}{\Sigma T} g(\mathbf{y}). \end{aligned} \tag{6.1.33}$$

$g(\mathbf{y})$ is the same function that appears in the spin-spin correlation [Eq. (6.1.21)] function. In two dimensions, it is logarithmically divergent at large $\mathbf{y}$. In dimensions greater than 2, it tends to a constant as $\mathbf{y} \to \infty$. This is the well known result that the resistance to infinity is infinite for $d = 2$ and finite for

$d > 2$. The infinite resistance to infinity and the destruction of long-range order in two dimensions are, thus, intimately related.

6.2 O_n symmetry and nematic liquid crystals

1 n-vector elastic energy

As in the *xy*-model just discussed, the rotational symmetry of spin space is broken in the ferromagnetic phase of *n*-vector models by an order parameter that picks out a particular average spin direction specified by $\mathbf{s} = |\mathbf{s}|\mathbf{n}$, where $\mathbf{n}$ is a unit *n*-component vector. Similarly, rotational symmetry is broken in the antiferromagnetic state of an *n*-vector model with staggered magnetization $\mathbf{N} = |\mathbf{N}|\mathbf{n}$. Since all directions of $\mathbf{n}$ are energetically equivalent, the free energy should be proportional to $(\nabla\mathbf{n})^2$. In addition, because spin and spatial coordinates rotate under separate groups, the free energy must be invariant with respect to independent rotations of $\mathbf{n}$ and position $\mathbf{x}$ (or equivalently ∇). Thus, there is an elastic free energy,

$$F_{\text{el}} = \frac{1}{2}\int d^d x \rho_s (\nabla_i n_j)(\nabla_i n_j)\,, \tag{6.2.1}$$

with a single elastic constant for the *n*-vector model. The constant ρ_s (often denoted by A in the magnetism literature) is again the spin wave stiffness. The unit vector $\mathbf{n}$, unlike ϕ in the *xy*-model, is a constrained variable, and F_{el} is necessarily anharmonic when expressed in terms of the $n-1$ independent variables in $\mathbf{n}$. For example, for $n = 3$, one can choose n_x and n_y to be independent variables; then, $n_z = \pm(1 - n_x^2 - n_y^2)^{1/2}$, and the $(\nabla n_z)^2$ term in Eq. (6.2.1) is anharmonic in n_x and n_y. For small n_x and n_y, F_{el} is effectively harmonic, and the correlation functions $|n_x(\mathbf{q})|^2$ and $|n_y(\mathbf{q})|^2$ are identical to $|\phi(\mathbf{q})|^2$ in the *xy*-model. The existence of these non-linearities leads to a crucial difference between the *xy*- and *n*-vector models in two dimensions. The coupling between long-wavelength fluctuations actually depresses the effective spin wave stiffness to zero for all $T > 0$ in two-dimensional *n*-vector models with $n > 2$, and there is no phase in these models with QLRO. We will see how this comes about at the end of this chapter (Sec. 6.7).

2 The Frank free energy of nematic liquid crystals

Rotational symmetry is also broken in nematic liquid crystals. In this case, $\mathbf{n}$ specifies the principal axis of a symmetric traceless tensor order parameter (Sec. 4.5). Nematic liquid crystals differ from the ferromagnetic phase of *n*-vector models in that rotations of spatial coordinates and the order parameter are produced by the same rotation operator. Thus, there is no longer a requirement that the elastic free energy as a function of $\mathbf{n}$ be invariant under independent rotations of $\mathbf{n}$ and ∇. The difference between this and the magnetic system is

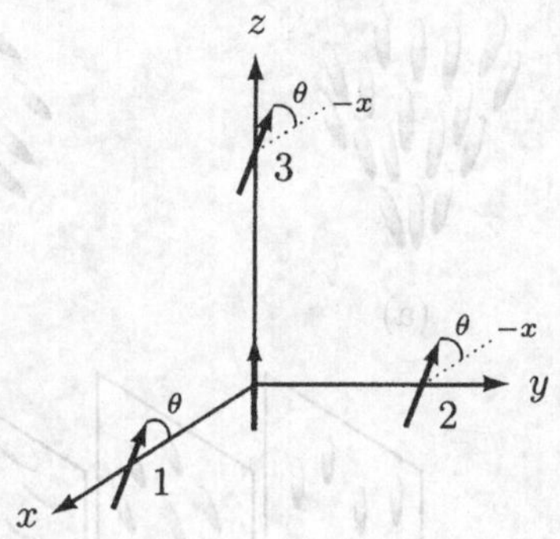

Fig. 6.2.1. Whereas the arrow at the origin is directed along the z-axis, the three neighboring arrows are identically tilted with a component along $-x$ and make an angle θ with respect to the x-axis. In a magnetic system with an $\mathbf{s}\cdot\mathbf{s}$ interaction between neighboring spins, the interaction energy of the spin at the origin with each of the spins along the x-, y- and z-axes is the same. However, the more complex interactions of rod-like molecules can give very different energies for (1) tilting a rod toward another (splay), (2) rotating so that the rod ends separate top and bottom (twist) or (3) taking colinear rods and changing their angle (bend).

illustrated in Fig. 6.2.1. The stiffness for the nematic becomes a fourth rank tensor:

$$F_{\rm el} = \frac{1}{2}\int d^d x K_{ijkl}\nabla_i n_j \nabla_k n_l \ . \qquad (6.2.2)$$

K_{ijkl} is a tensor that will in general depend on the local director $\mathbf{n}(\mathbf{x})$. This free energy must be invariant under uniform rotations of the whole sample and under the symmetry operations $\mathbf{n}\to-\mathbf{n}$ and $\mathbf{x}\to-\mathbf{x}$. In addition, $\mathbf{n}$ is a unit vector so that $n_i\nabla_j n_i$ is zero. These considerations imply that K_{ijkl} has three independent components, K_1, K_2 and K_3. The distortions whose energy is measured by the three elastic constants are shown in Fig. 6.2.2. They are (1) *splay*, with nonzero $\nabla\cdot\mathbf{n}$, (2) *twist*, with nonzero $\mathbf{n}\cdot(\nabla\times\mathbf{n})$, and (3) *bend*, with nonzero $\mathbf{n}\times(\nabla\times\mathbf{n})$. The elastic energy for the nematic phase, called the Frank free energy (Frank 1958), is

$$F_{\rm n} = \frac{1}{2}\int d^d x\{K_1(\nabla\cdot\mathbf{n})^2 + K_2[\mathbf{n}\cdot(\nabla\times\mathbf{n})]^2 + K_3[\mathbf{n}\times(\nabla\times\mathbf{n})]^2\} \qquad (6.2.3)$$

in three dimensions.

Since $\mathbf{n}$ is unitless, the elastic constants K_1, K_2 and K_3 (like ρ_s) have units of energy/length and are of order $k_B T_{NI}/a$, where T_{NI} is the isotropic-to-nematic transition temperature and a is a typical molecular length. T_{NI} is of order 400 K and a is of order 20Å. Thus we estimate $K_i \sim 1.4\times10^{-16}\times400/(20\times10^{-8}) \sim 3\times10^{-7}$ dynes.

If molecules are chiral, as they are in a cholesteric liquid crystal, the Frank free energy must be supplemented by an additional term,

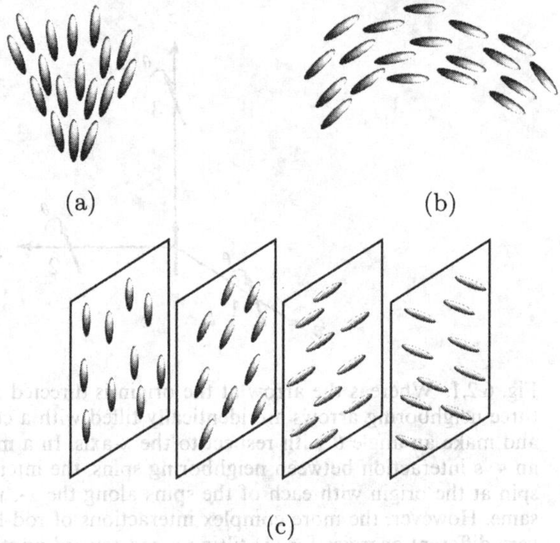

Fig. 6.2.2. Schematic representations of the (a) splay, (b) bend, and (c) twist distortions of a nematic liquid crystal.

$$F_{\mathrm{Ch}} = -h \int d^3x\mathbf{n} \cdot (\nabla \times \mathbf{n}), \tag{6.2.4}$$

favoring molecular twist. The magnitude of potential h depends on the degree of molecular chirality. It is easy to see that $F_n + F_{\mathrm{Ch}}$ is minimized with a director,

$$\mathbf{n}_c(\mathbf{x}) = (0, \sin k_0 x, \cos k_0 x), \tag{6.2.5}$$

where $k_0 = h/K_2$ because $\mathbf{n}_c \cdot (\nabla \times \mathbf{n}_c) = k_0$.

3 Cells with non-uniform n

The minimum energy state of the Frank free energy is clearly one with a uniform $\mathbf{n}$. Boundary conditions or external fields can, however, lead (as in the case of the xy-model) to non-uniform $\mathbf{n}$. Consider the simple geometry shown in Fig. 6.2.3a. The surface at $z = 0$ is prepared so that $\mathbf{n}$ is rigidly fixed to lie along the x-axis, and the surface at $z = L$ is fixed so that $\mathbf{n}$ makes an angle θ_0 relative to the x-axis. Then $\mathbf{n} = [\cos\theta(z), \sin\theta(z), 0]$, where θ depends only on the coordinate z. The distortion induced by the boundary conditions is one of twist. Thus

$$\begin{aligned} \frac{F_n}{A} &= \frac{1}{2}K_2 \int_0^L dz \left(\frac{d\theta}{dz}\right)^2, \\ K_2 \frac{d^2\theta}{dz^2} &= 0, \end{aligned} \tag{6.2.6}$$

where A is the area. This equation is easily solved for θ for the boundary conditions of Fig. 6.2.3a, yielding

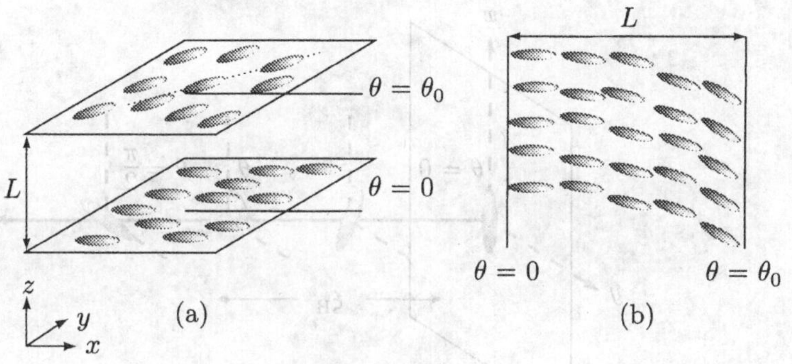

Fig. 6.2.3. Cell with boundary conditions leading (a) to twist and (b) to bend.

$$\theta = \theta_0(z/L)\,, \quad F_{\text{n}}/A = \frac{1}{2}K_2 L^{-1}\theta_0^2\,. \tag{6.2.7}$$

Other boundary conditions can lead to distortions such as shown in Fig. 6.2.3b. (Most of the useful properties of liquid crystals come from the ability to fix the orientation of the molecules at a boundary by surface treatment of the container walls.)

As discussed in Sec. 4.5, an external magnetic field **H** (or an electric field **E**) can align molecules if the susceptibility (or dielectric constant) is anisotropic ($\chi_a \neq 0$ or $\epsilon_a \neq 0$). In the presence of **H**, the total free energy is

$$\begin{aligned} F_T &= F_{\text{n}} + F_{\text{ext}}, \\ F_{\text{ext}} &= \begin{cases} -\frac{1}{2}\chi_a \int d^3x(\mathbf{H}\cdot\mathbf{n})^2 & \text{(magnetic)} \\ -\frac{1}{8\pi}\epsilon_a \int d^3x(\mathbf{E}\cdot\mathbf{n})^2 & \text{(electric)} \end{cases} \end{aligned} \tag{6.2.8}$$

This equation implies the existence of a magnetic coherence length ξ_H measuring the distance over which **n** can vary in the presence of an aligning field. It is obtained by comparing the Frank free energy with the aligning energy

$$\frac{1}{2}K\xi_H^{-2} = \frac{1}{2}\chi_a H^2 \;\Rightarrow\; \xi_H = [K/(\chi_a H^2)]^{1/2}\,, \tag{6.2.9}$$

where K is one of the elastic constants. χ_a is of order 10^{-7} c.g.s. so that $\xi_H \sim [3\times 10^{-7}/(10^{-7}\times 10^8)]^{1/2} \sim 2\ \mu\text{m}$ in a field of 10 kG. The physical significance of ξ_H becomes apparent in the simple example shown in Fig. 6.2.4. A surface at $z = 0$ rigidly fixes **n** to lie along the x-axis. A field $\mathbf{H} = H\mathbf{e}_y$ tends to align **n** along the y-axis far from the surface. Thus at $z = 0$, $\mathbf{n} = \mathbf{e}_x$, and for z large, $\mathbf{n} = \mathbf{e}_y$. This can be made more precise by minimizing F_T. Since the distortion induced by the external field is one of twist, we have $\mathbf{n} = [\cos\theta(z), \sin\theta(z), 0]$ and

$$F_{\text{n}}/A = \frac{1}{2}\int dz\left[K_2\left(\frac{d\theta}{dz}\right)^2 - \chi_a H^2\sin^2\theta\right]. \tag{6.2.10}$$

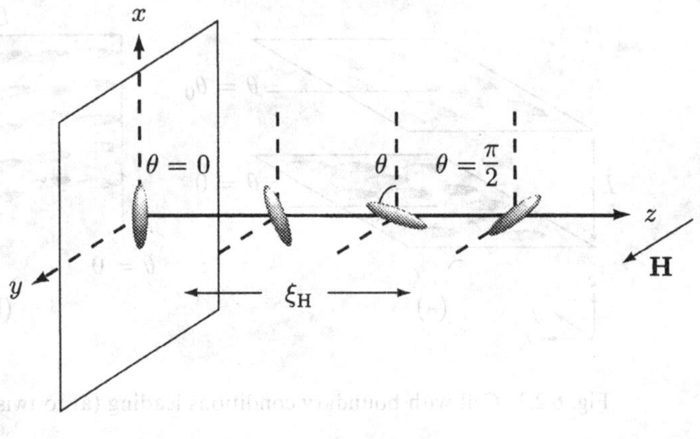

Fig. 6.2.4. The Frank director in the vicinity of an anchoring wall in the presence of an external magnetic field. The wall aligns the director along the z-direction, and the magnetic field is in the y-direction. The director reaches alignment with the field at a distance of order the coherence length ξ_H from the wall.

Minimization leads to the Euler-Lagrange equation

$$\xi_H^2 \frac{d^2\theta}{dz^2} = -\frac{1}{2}\frac{d\sin^2\theta}{d\theta}, \quad \xi_H = [K_2/(\chi_a H^2)]^{1/2}. \tag{6.2.11}$$

This equation can easily be reduced to quadratures with

$$\begin{aligned} \xi_H^2\left(\frac{d\theta}{dz}\right)^2 &= -\sin^2\theta + C' \\ &= \cos^2\theta + C, \end{aligned} \tag{6.2.12}$$

where $C = C' - 1$. The boundary conditions on θ are $\theta = \pi/2$ and $d\theta/dz = 0$ at $z \to \infty$. Thus the constant C is zero. Since $\theta \to \pi/2$ as $z \to \infty$, we introduce $\beta = \pi/2 - \theta$, which tends to zero as $z \to \infty$. Then, since β decreases with increasing z, we choose the negative sign in the solution to Eq. (6.2.12):

$$\begin{aligned} &\frac{d\beta}{\sin\beta} = \frac{-dz}{\xi_H}, \quad \ln(\tan\beta/2) = -\frac{z}{\xi_H}, \\ &\beta(z) = 2\tan^{-1} e^{-z/\xi_H}. \end{aligned} \tag{6.2.13}$$

This shows clearly that β is of the order of its large z value of zero when z is a few times ξ_H.

4 The Freedericksz transition

The competition between the elastic and field terms of the free energy provides a method of measuring the elastic constants. Consider a cell of height L in which $\mathbf{n}$ is anchored at the top and bottom surfaces so that the equilibrium configuration is one with a spatially uniform director $\mathbf{n}$. If $\xi_H \gg L$ (small field), the field

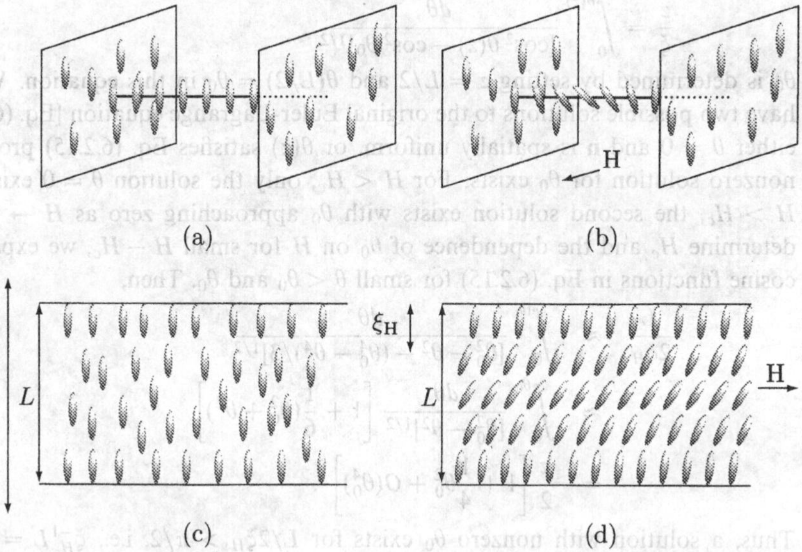

Fig. 6.2.5. (a) Twist cell with $H < H_c$ with the director aligned parallel to the cell boundary. (b) Same cell as in (a) but with $H > H_c$ so that there is alignment of the director along **H** in the center of the cell. (c) A cell with molecules aligned perpendicular to the cell walls and with $H < H_c$. (d) Same as (c) but with $H > H_c$

will be unable to overcome the effect of surface pinning and the equilibrium configuration will continue to be uniform. On the other hand, if $\xi_H \ll L$, it is clear that **n** will align along **H** in the center of the cell, as shown in Fig. 6.2.5. We therefore expect that, at a critical field determined by

$$\begin{aligned} \xi_{H_c}^{-1} L &= c = \text{const.}, \\ H_c L &= (K/\chi_a)^{1/2} c = \text{const.}, \end{aligned} \tag{6.2.14}$$

there will be a transition, called the Freedericksz transition, from a state with **n** spatially uniform to one in which **n** begins to align along the field, as shown in Fig. 6.2.5. Note that Eq. (6.2.14) says that, for a given material, the product $H_c L$ will be a constant. If an electric rather than a magnetic field were applied, there would be a critical voltage $V_c = E_c L$ independent of L.

We can easily evaluate the constant c appearing in Eq. (6.2.14) when the external field induces a twist as shown in Fig. 6.2.5b. The equation determining θ is still Eq. (6.2.12). The boundary conditions are, however, $\theta = 0$ at $z = 0$ and $z = L$, and $d\theta/dz = 0$ at $z = L/2$. By symmetry, $\theta(L/2 - z) = \theta(L/2 + z)$. At $z = L/2$, θ reaches a value, θ_0, to be determined. Because $d\theta/dz = 0$ when $\theta = \theta_0$ at $z = L/2$, the constant C in Eq. (6.2.12) is $-\cos^2\theta_0$. The equation for $\theta(z)$ is thus

$$\frac{z}{\xi} = \int_0^{\theta(z)} \frac{d\theta}{[\cos^2\theta(z) - \cos^2\theta_0]^{1/2}} . \tag{6.2.15}$$

θ_0 is determined by setting $z = L/2$ and $\theta(L/2) = \theta_0$ in this equation. We now have two possible solutions to the original Euler-Lagrange equation [Eq. (6.2.11)]: either $\theta = 0$ and $\mathbf{n}$ is spatially uniform, or $\theta(z)$ satisfies Eq. (6.2.15) provided a nonzero solution for θ_0 exists. For $H < H_c$, only the solution $\theta = 0$ exists. For $H > H_c$, the second solution exists with θ_0 approaching zero as $H \to H_c$. To determine H_c and the dependence of θ_0 on H for small $H - H_c$, we expand the cosine functions in Eq. (6.2.15) for small $\theta < \theta_0$ and θ_0. Then,

$$\begin{aligned} \frac{L}{2\xi_H} &\approx \int_0^{\theta_0} \frac{d\theta}{[\theta_0^2 - \theta^2 - (\theta_0^4 - \theta^4)/3]^{1/2}} \\ &\approx \int_0^{\theta_0} \frac{d\theta}{[\theta_0^2 - \theta^2]^{1/2}} \left[1 + \frac{1}{6}(\theta_0^2 + \theta^2)\right] \\ &= \frac{\pi}{2}\left[1 + \frac{1}{4}\theta_0^2 + O(\theta_0^4)\right] . \end{aligned} \tag{6.2.16}$$

Thus, a solution with nonzero θ_0 exists for $L/2\xi_H > \pi/2$, i.e., $\xi_{H_c}^{-1} L = \pi$. For $H < H_c$, there is no response to the applied field. For $H > H_c$,

$$\theta_0^2 = \frac{4L}{\pi}\left(\frac{\chi_a}{K_2}\right)^{1/2} (H - H_c). \tag{6.2.17}$$

We leave it as an exercise to show that when the solution with nonzero θ_0 exists, it has lower free energy than the solution with $\theta = 0$. A similar but mathematically more challenging problem is the cell shown in Fig. 6.2.5c and d where the surface alignment is director perpendicular to the boundary and the applied field is parallel to the boundary.

5 The twisted nematic display

Liquid crystal display devices are a practical and very widespread (over 500 million display cells are sold every year) application of the ideas just discussed. The simplest and most commonly used device in watches, computer terminals, etc., is the twisted nematic cell shown in Fig. 6.2.6. The top and bottom plates of the cell in the xy-plane are located at $z = 0$ and $z = L$, with L of order 10 μm. The surfaces are treated so that molecules at the top plate are twisted by an angle $\pi/2$ relative to those at the bottom plate. This corresponds exactly to the pure twist configuration discussed in Eqs. (6.2.6) and (6.2.7) with $\theta_0 = \pi/2$. A normal electric field (parallel to the z-axis) is created when the two plates are connected to a voltage source. The molecules in the cell have positive dielectric anisotropy so that the electric field will tend to rotate the director out of the xy-plane. Just as in the example considered above, there will be a critical voltage above which the zero field twisted state becomes unstable to one with a z-component of the

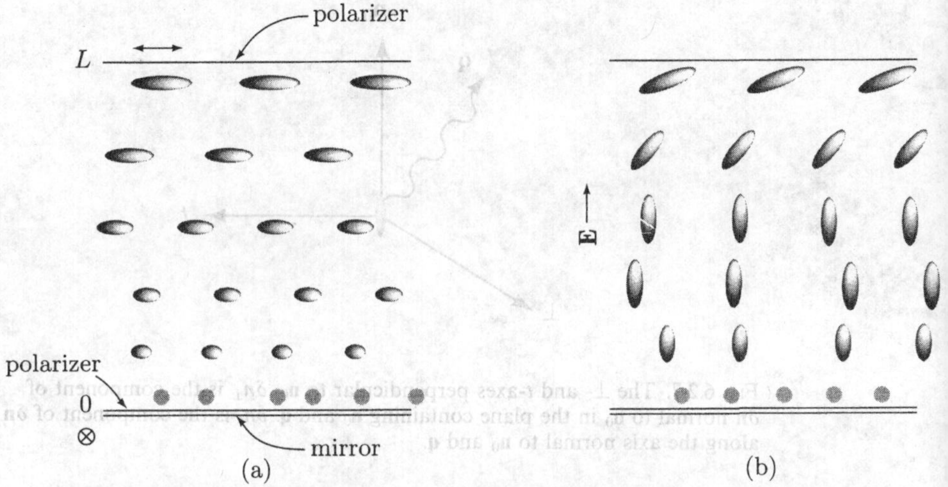

Fig. 6.2.6. The twisted nematic display device with top and bottom plates in the xy-plane and separated by a distance $L \sim 10\,\mu$m. The cell is placed between crossed polarizers allowing transmission of light polarized parallel to the plane of the paper at the top plate and perpendicular to the plane of the paper at the bottom plate. (a) The cell is below the threshold voltage. The surfaces are treated so that the director undergoes a uniform rotation of $\pi/2$ from the bottom plate to the top plate so that light is transmitted. (b) Molecular orientation above threshold. The director in the center is parallel to the z-axis so that the unrotated light is blocked by the polarizers.

director. This voltage can be calculated in a manner exactly analogous to that of the last example. It is

$$V_c = 2\pi^{3/2}\left[K_1 + \frac{1}{4}(K_3 - 2K_2)\right]^{1/2} \epsilon_a^{-1/2}. \tag{6.2.18}$$

The molecular arrangement for $V > V_c$ is depicted in Fig. 6.2.6b. The instability of the twisted state provides the switching mechanism for the display device. For typical materials, ϵ_a is of order 10, and V_c is of order one volt. This is a low voltage that can easily be produced by long-lived batteries.

The optical properties of this device are in fact quite complicated because the dielectric constant of the twisted nematic is spatially inhomogeneous. The qualitative explanation of what happens is, however, clear. In the low voltage state, the pitch ($4L$) of the twisted nematic configuration is much greater than the wavelength of visible light, and the direction of polarization of light in the cell will adiabatically follow the director. A linear polarizer is placed above the top plate such that only light with electric polarization parallel (or perpendicular) to the top plate enters the cell. The polarization follows the director (even after reflection) inside the cell so that light impinging on the top plate after a round trip in the cell will have a direction of polarization parallel to the axis of the polarizer and will be able to leave the cell. When the field is turned on, most of

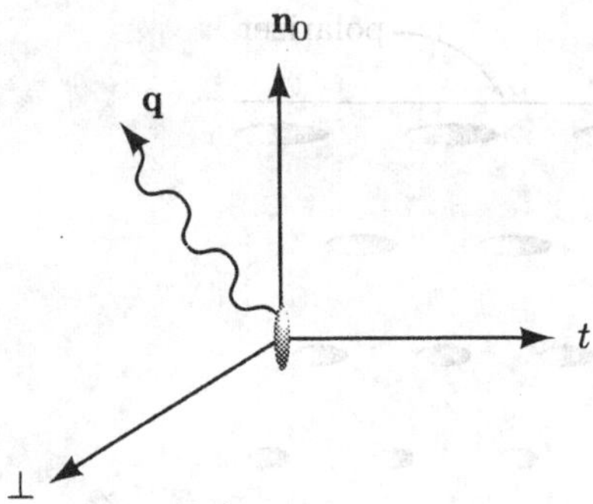

Fig. 6.2.7. The ⊥- and t-axes perpendicular to $\mathbf{n}_0$. $\delta n_\perp$ is the component of $\delta\mathbf{n}$ normal to $\mathbf{n}_0$ in the plane containing $\mathbf{n}_0$ and $\mathbf{q}$. δn_t is the component of $\delta\mathbf{n}$ along the axis normal to $\mathbf{n}_0$ and $\mathbf{q}$.

the molecules are normal to the plates and are unable to change the direction of polarization, so light is blocked from leaving the cell by the polarizer. Thus, when the voltage is off, the cell reflects ambient light and appears shiny; when the voltage is on, the cell "captures" light and appears dark.

6 Fluctuations and light scattering

In the absence of external fields or boundary conditions, $\mathbf{n}(\mathbf{x}) = \mathbf{n}_0$ is uniform in the equilibrium state. Fluctuations in which $\mathbf{n}(\mathbf{x}) = \mathbf{n}_0 + \delta\mathbf{n}(\mathbf{x})$ occur and have an energy determined by the Frank free energy. To order $(\delta\mathbf{n})^2$, the fluctuation energy is diagonalized by setting $\delta\mathbf{n}(\mathbf{q}) = [\delta n_\perp(\mathbf{q}), \delta n_t(\mathbf{q})]$, where $\delta\mathbf{n}$ lies in the plane perpendicular to $\mathbf{n}_0$, with $\delta n_\perp$ its component in the plane containing $\mathbf{q}$ and $\mathbf{n}_0$ and δn_t its component normal to that plane (Fig. 6.2.7).

In this coordinate system,

$$F = \frac{1}{2V}\sum_{\mathbf{q}}(K_3q_\parallel^2 + K_1q_\perp^2 + \chi_a H^2)|\delta n_\perp(\mathbf{q})|^2 + \frac{1}{2V}\sum_{\mathbf{q}}(K_3q_\parallel^2 + K_2q_\perp^2 + \chi_a H^2)|\delta n_t(\mathbf{q})|^2, \tag{6.2.19}$$

where $q_\parallel$ is the component of $\mathbf{q}$ along $\mathbf{n}_0$ and $q_\perp$ is the component perpendicular to $\mathbf{n}_0$. This implies

$$\langle|\delta n_\perp(\mathbf{q})|^2\rangle = \frac{TV}{K_3q_\parallel^2 + K_1q_\perp^2 + \chi_a H^2},$$
$$\langle|\delta n_t(\mathbf{q})|^2\rangle = \frac{TV}{K_3q_\parallel^2 + K_2q_\perp^2 + \chi_a H^2}. \tag{6.2.20}$$

These fluctuations scatter light strongly and lead to a high turbidity of the nematic phase. This can be seen by noting that the dielectric tensor has a part proportional to the nematic order parameter Q_{ij}: $\epsilon_{ij} = \epsilon_0\delta_{ij} + \epsilon_a Q_{ij}$. Light scatters from fluctuations in ϵ_{ij},

$$I(\mathbf{q}) \sim |e_i^I \delta\epsilon_{ij}(\mathbf{q}) e_j^T|^2 , \tag{6.2.21}$$

where e_i^I and e_i^T are the electric polarization directions of the incident and transmitted light, respectively. The dominant contributions to $\delta\epsilon_{ij}$ come from fluctuations in the director, $\delta\epsilon_{ij} \sim \epsilon_a S[n_{0i}\delta n_j + (i \leftrightarrow j)]$. Thus,

$$\begin{aligned} I(\mathbf{q}) &\sim (e^I \cdot \mathbf{n}_0)^2 \langle|\delta\mathbf{n}(\mathbf{q}) \cdot \mathbf{e}^T|^2\rangle + (\mathbf{e}^T \cdot \mathbf{n}_0)^2 \langle|\delta\mathbf{n}(\mathbf{q}) \cdot \mathbf{e}^I|^2\rangle \\ &\quad +2(\mathbf{e}^I \cdot \mathbf{n}_0)(\mathbf{e}^T \cdot \mathbf{n}_0)\mathbf{e}^I \cdot \langle\delta\mathbf{n}(\mathbf{q})\delta\mathbf{n}(-\mathbf{q})\rangle \cdot \mathbf{e}^T \\ &\sim \frac{T}{Kq^2}, \end{aligned} \tag{6.2.22}$$

where in the last formula we have suppressed tensor indices. Scattering from a nematic liquid crystal, as described by Eq. (6.2.22), should be compared with the scattering from an isotropic fluid. Light scattering results from fluctuations in the dielectric constant or the index of refraction on the length scale of the wavelength of light $\sim 1\,\mu$m. For isotropic fluids, these fluctuations arise only from density changes or local compressions. Compressional energy costs are similar in both isotropic and anisotropic systems because compression does not excite any soft mode at low q associated with a broken symmetry. In liquid crystals, fluctuations in the dielectric constant are dominated by the fluctuations in the orientation of anisotropic molecules. The restoring force for long wavelength orientation changes goes to zero as $q \to 0$. Thus an orientational fluctuation of wavelength $1\,\mu$m is thermally excited much more readily than a density fluctuation of the same wavelength. In an isotropic fluid, we can express I in terms of the unitless variable $\delta\rho/\rho_0$, where ρ_0 is the average density. (Here we use ρ rather than n for the number density to avoid confusion with the director $\mathbf{n}$.) We have from Eq. (4.4.14) $I(\mathbf{q}) \sim \rho^{-2}S_{\rho\rho}(\mathbf{q}) = \rho^{-2}T(\partial\rho/\partial\mu)[1 + (q\xi)^2]^{-1}$. Except near the critical point, $\partial\rho/\partial\mu \sim a^{-3}T^{-1}$, where a is the interparticle spacing. In addition, for optical wavelengths, $q\xi < 1$ except near the critical point. Thus the ratio of scattering intensity from a typical nematic to that of a typical fluid is of order

$$\frac{I_N}{I_{ISO}} \sim \frac{T_{NI}/Kq^2}{a^3} \sim \frac{1}{(qa)^2} . \tag{6.2.23}$$

With $a \sim 5-20$Å and $q = 2\pi/\lambda$, with λ the wavelength of light, this predicts that the scattering from a nematic is of order 10^3 to 10^6 times larger than from an isotropic fluid. Nematic liquid crystals are in fact very turbid, indicating a strong scattering of light.

Scattering of polarized light from single crystal nematics provides one of the most precise ways of measuring the elastic constants. For example, in the geometry shown in Fig. 6.2.8, the scattering intensity is proportional to $|\delta n_t(\mathbf{q})|^2$ and provides a measure of K_2 and K_3.

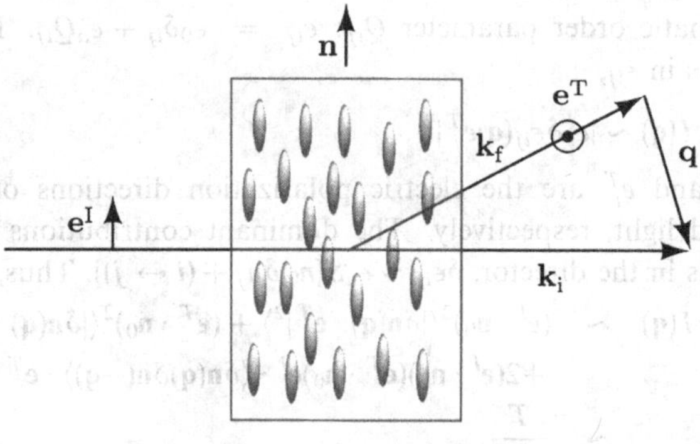

Fig. 6.2.8. A scattering geometry for measuring K_2 and K_3. The polarization of the incident light is parallel to the director **n**; the scattering vector **q** lies in the plane of the director and the incident beam; and the polarization of the scattered light is perpendicular to the $\mathbf{n}-\mathbf{q}$ plane.

6.3 Smectic liquid crystals

Ideal smectic-A liquid crystals have liquid-like correlations in two dimensions and a crystal-like periodic modulation of the density along the third direction. They can, therefore, be thought of as stacks of parallel planes separated by a distance d, as shown in Fig. 6.3.1a. In the idealized picture of the smectic phase presented in Chapter 2, these planes represent the boundaries between successive layers of molecules, as shown in Fig. 6.3.1b. In the lamellar phase of microemulsions, they could represent the center of bilayers separating layers of either water or oil, as shown in Figs. 2.7.14 and 2.7.16. More generally, the planes can be defined via the phase of the mass-density wave. The molecular density can be expanded in a Fourier series (see Sec. 2.7) with period d:

$$\begin{aligned}\rho(\mathbf{x}) &= \rho_0+\sum_n[\langle\psi_n\rangle e^{in\mathbf{q}_0\cdot\mathbf{x}}+\text{c.c.}]\\ &\sim \rho_0+[\langle\psi_1\rangle e^{i\mathbf{q}_0\cdot\mathbf{x}}+\text{c.c.}],\end{aligned} \tag{6.3.1}$$

where $\mathbf{q}_0=(2\pi/d)\mathbf{n}_0=(2\pi/d)\mathbf{e}_z$. As discussed in Chapter 2, the higher harmonics of the density modulation ($\langle\psi_n\rangle$) are not experimentally visible in a wide class of smectics. They are, however, visible in some lyotropic systems. $\langle\psi_1\rangle$ is the dominant order parameter that distinguishes the smectic-A phase from the nematic phase. It is a complex number with an amplitude and a phase

$$\langle\psi_1\rangle=|\langle\psi_1\rangle|e^{-iq_0u}. \tag{6.3.2}$$

The planes in Fig. 6.3.1a can be interpreted as the planes of constant phase of the mass-density wave at wave number $\mathbf{q}_0$:

$$\phi\equiv\mathbf{q}_0\cdot\mathbf{x}-q_0u=2\pi n\,,\quad n=0,\pm1,\pm2,\ldots\,. \tag{6.3.3}$$

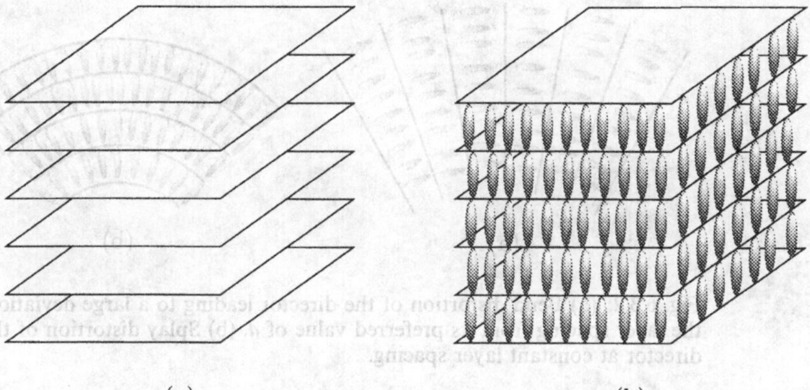

Fig. 6.3.1. (a) Parallel planes of a layered structure. (b) Planes interpreted as separating successive planes of molecules.

When u is independent of $\mathbf{x}$, Eq. (6.3.3) defines the set of uniformly spaced planes $z - u = nd$ perpendicular to the z-axis. A spatially uniform increment of u from zero to u corresponds to a uniform translation of the coordinate system through $-u$. Alternatively, it corresponds to a uniform translation, $\mathbf{x}^\alpha \to \mathbf{x}^\alpha + u\mathbf{e}_z$, of all the molecules of the smectic a distance u along the z-axis:

$$\begin{aligned} \langle \psi_n \rangle &= \frac{1}{V} \int d^3x e^{-in\mathbf{q}_0 \cdot \mathbf{x}} \langle \sum_\alpha \delta(\mathbf{x} - \mathbf{x}^\alpha) \rangle \\ &= \frac{1}{V} \sum_\alpha \langle e^{-in\mathbf{q}_0 \cdot \mathbf{x}^\alpha} \rangle \to \langle \psi_n \rangle e^{-inq_0 u}. \end{aligned} \tag{6.3.4}$$

Thus, a uniform translation by $u\mathbf{e}_z$ decreases the phase ϕ_n of the order parameter $\langle \psi_n \rangle$ by nq_0u:

$$\phi_n = \phi_n^0 - nq_0 u \, , \tag{6.3.5}$$

where ϕ_n^0 is the phase when $u = 0$. It is always possible to choose the origin of the coordinate system so that $\langle \psi_1 \rangle$ is real when $u = 0$. The phases ϕ_n^0 will in general not be zero. They will, however, be fixed at some energetically preferred value in any equilibrium state.

1 The elastic free energy

Since the free energy F of the system is invariant with respect to uniform translations, one could argue, in analogy with the xy-model of Sec. 6.1, that F should be proportional to $(\nabla u)^2$ for long wavelength distortions of u. The symmetry of smectics is, however, more complicated than that of the xy-model, and the above argument is not totally correct. The layering of the smectic breaks rotational as well as translational symmetry, and any spatial distortion of u that

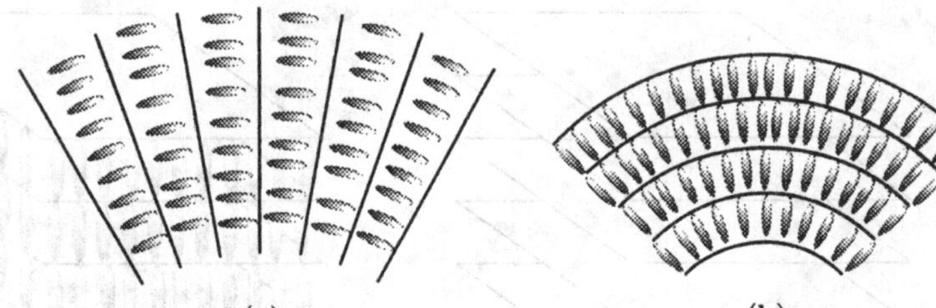

Fig. 6.3.2. (a) Bend distortion of the director leading to a large deviation of the layer spacing from its preferred value of d. (b) Splay distortion of the director at constant layer spacing.

corresponds to a rigid rotation of the layers must have zero energy. The simplest way to discuss the rotation of the layers is to introduce the unit normal to the layers defined in Eq. (6.3.3) via the relation

$$\mathbf{N} = \frac{\nabla\phi}{|\nabla\phi|} \sim (-\nabla_x u, -\nabla_y u, 1) + O[(\nabla u)^2]. \tag{6.3.6}$$

In the smectic-A phase, molecules align normal to the layers, and in equilibrium $\mathbf{N}$ is identical to the Frank director $\mathbf{n}$ introduced in the previous section. The constraint that $\mathbf{n}$ be normal to the layers implies that twist and bend distortions of $\mathbf{n}$ are energetically more costly compared to splay distortions because they cannot be produced at constant layer spacing, as shown in Fig. 6.3.2. For uniform layers normal to the z-axis, $\mathbf{n} = \mathbf{n}_0 = \mathbf{e}_z$. If the planes are rotated rigidly through an angle $\delta\boldsymbol{\Omega}$ about an axis in the xy-plane, $\mathbf{n} \to \mathbf{n}_0 + \delta\boldsymbol{\Omega} \times \mathbf{n}_0 = (\delta n_x, \delta n_y, 1)$. Thus from Eq. (6.3.6), a uniform rotation of $\mathbf{N}$ leads to a constant value for $\nabla_\perp u = (\nabla_x u, \nabla_y u, 0)$.

An alternative way to obtain this result is to determine the function $u(\mathbf{x})$ which will describe a uniform rotation of the planes through $\delta\boldsymbol{\Omega}$. Under such a rotation, $\mathbf{q}_0 \to \mathbf{q}_0' = \mathbf{q}_0 + \delta\boldsymbol{\Omega} \times \mathbf{q}_0$, and

$$\begin{aligned} \rho(\mathbf{x}) &= \rho_0 + \sum_n [\langle\psi_n\rangle e^{in\mathbf{q}'_0\cdot\mathbf{x}} + \text{c.c.}] \\ &= \rho_0 + \sum_n [\langle\psi_n\rangle e^{in\mathbf{q}_0\cdot\mathbf{x}} e^{-inq_0 u(\mathbf{x})} + \text{c.c.}] \, . \end{aligned} \tag{6.3.7}$$

Thus, if

$$q_0 u = (\mathbf{q}_0 - \mathbf{q}'_0)\cdot\mathbf{x} = -(\delta\boldsymbol{\Omega} \times \mathbf{q}_0)\cdot\mathbf{x}\,, \tag{6.3.8}$$

$$\nabla_\perp u = -q_0^{-1}(\delta\boldsymbol{\Omega} \times \mathbf{q}_0) = -\delta\mathbf{n}\,, \tag{6.3.9}$$

then the original density is transformed into that with rotated layers.

Since uniform rotations do not affect the free energy, there cannot be any $(\nabla_\perp u)^2$ term in F_{el}. The leading term in F_{el} involving only $\nabla_\perp$ and u is therefore $(\nabla_\perp^2 u)^2$. Since $\nabla_\perp^2 u = -\nabla_\perp \cdot \delta\mathbf{n} = \nabla \cdot \mathbf{n}$, $\nabla_\perp u$ is associated with splay distortions of the director. The elastic energy for a smectic is, therefore,

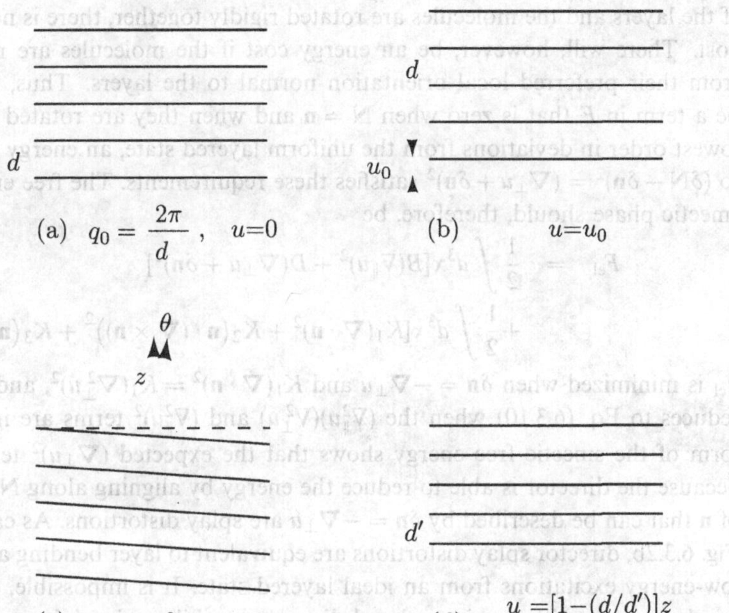

Fig. 6.3.3. (a) Smectic with equilibrium layer spacing. (b) A translated smectic with the same energy as that in (a). (c) A rotated smectic with the same energy as that in (a). (d) A smectic with compressed layers with an energy greater than that in (a).

$$F_{\rm el} = \frac{1}{2}\int d^3x[B(\nabla_\parallel u)^2 + K_1(\nabla_\perp^2 u)^2]. \qquad (6.3.10)$$

This energy was derived independently in the 1930s by L. Landau and Peierls before either knew that a smectic existed. In general, there could be terms in $F_{\rm el}$ proportional to $(\nabla_\parallel^2 u)(\nabla_\perp^2 u)$ or $(\nabla_\parallel^2 u)^2$. They are negligible compared to the $(\nabla_\parallel u)^2$ term and will be neglected in what follows. K_1 is the splay elastic constant of the Frank free energy. The radius of curvature of a bent layer (Fig. 6.3.2a) is $R = (\nabla^2 u)^{-1}$. Thus, the splay term $(\nabla_\perp^2 u)^2$ is, apart from terms involving $(\nabla_\parallel^2 u)^2$ that have been ignored, equal to R^{-2}. B measures the energy cost associated with compressing or stretching the layers such that their average spacing is no longer the preferred value d (see Fig. 6.3.3).

The free energy in Eq. (6.3.10) can be generalized to d dimensions by allowing $\nabla_\perp$ to be a derivative in the $(d-1)$-dimensional space perpendicular to $\mathbf{n}_0$. This generalization provides the correct description for a two-dimensional smectic.

Before examining the consequences of Eq. (6.3.10), we will derive it in yet another way that treats the phase variable u and the Frank director simultaneously.

If the layers and the molecules are rotated rigidly together, there is no free energy cost. There will, however, be an energy cost if the molecules are rotated away from their preferred local orientation normal to the layers. Thus, there should be a term in F that is zero when $\mathbf{N} \doteq \mathbf{n}$ and when they are rotated together. To lowest order in deviations from the uniform layered state, an energy proportional to $(\delta\mathbf{N} - \delta\mathbf{n})^2 = (\nabla_\perp u + \delta\mathbf{n})^2$ satisfies these requirements. The free energy for the smectic phase should, therefore, be

$$F_{\text{el}} = \frac{1}{2}\int d^3x[B(\nabla_\parallel u)^2 + D(\nabla_\perp u + \delta\mathbf{n})^2] \tag{6.3.11}$$
$$+\frac{1}{2}\int d^3x[K_1(\nabla\cdot\mathbf{n})^2 + K_2\big(\mathbf{n}\cdot(\nabla\times\mathbf{n})\big)^2 + K_3\big(\mathbf{n}\times(\nabla\times\mathbf{n})\big)^2].$$

F_{el} is minimized when $\delta\mathbf{n} = -\nabla_\perp u$ and $K_1(\nabla\cdot\mathbf{n})^2 = K_1(\nabla_\perp^2 u)^2$, and Eq. (6.3.11) reduces to Eq. (6.3.10) when the $(\nabla_\parallel^2 u)(\nabla_\perp^2 u)$ and $(\nabla_\parallel^2 u)^2$ terms are ignored. This form of the smectic free energy shows that the expected $(\nabla_\perp u)^2$ term is absent because the director is able to reduce the energy by aligning along $\mathbf{N}$. Distortions of $\mathbf{n}$ that can be described by $\delta\mathbf{n} = -\nabla_\perp u$ are splay distortions. As can be seen in Fig. 6.3.2b, director splay distortions are equivalent to layer bending and represent low-energy excitations from an ideal layered state. It is impossible, on the other hand, to create either twist or bend distortions while maintaining constant layer spacing. These represent high energy excitations from the ideal layered state and are said to be "expelled" by the smectic phase. The director in a twist or bend configuration is perpendicular to $\nabla_\perp$, and Eq. (6.3.11) predicts a volume energy cost of $D(\delta\mathbf{n})^2$ for such configurations, in agreement with the geometric argument. There are two new lengths, $\lambda_2 = (K_2/D)^{1/2}$ and $\lambda_3 = (K_3/D)^{1/2}$, describing respectively the penetration depths of twist and bend into the smectic phase.

Both translational and rotational degrees of freedom appear in Eq. (6.3.11). The low-energy rotational distortions $\delta\mathbf{n}$ are determined completely by $\nabla_\perp u$. Thus, even though both rotational and translational symmetry are broken in the smectic-A phase, long-wavelength elastic distortions are described entirely by the translational elastic variable u.

2 Fluctuations

We can now investigate the effects of fluctuations in the phase variable u on the magnitude of the smectic order parameters $\langle\psi_n\rangle = |\langle\psi_n\rangle|e^{i\phi_n^0 - iq_0 nu}$. As in the xy-model, we can decouple the phase and amplitude fluctuations so that†

$$\begin{aligned}\langle\psi_n\rangle &= |\langle\psi_n\rangle|\langle e^{-iq_0 u}\rangle e^{i\phi_n^0} \\ &= |\langle\psi_n\rangle|e^{i\phi_n^0}e^{-\frac{1}{2}n^2 q_0^2\langle u^2(\mathbf{x})\rangle}\end{aligned} \tag{6.3.12}$$

† Strictly speaking, we should distinguish here between the phase of $\langle\psi_n\rangle$ and the fluctuating phase of ψ_n, just as in the xy-model we distinguished between the angle θ describing $\langle\mathbf{s}\rangle$ and the angle ϑ describing $\mathbf{s}$.

where

$$\begin{aligned}
\langle u^2(\mathbf{x})\rangle &= \int \frac{dq_\parallel d^{d-1}q_\perp}{(2\pi)^d}\frac{T}{Bq_\parallel^2+K_1q_\perp^4} \\
&= K_{d-1}\int \frac{dq_\parallel q_\perp^{d-2}dq_\perp}{(2\pi)^d}\frac{T}{Bq_\parallel^2+K_1q_\perp^4} \\
&= \frac{K_{d-1}T}{4\pi B^{1/2}K_1^{(d-1)/4}}\int_{-B^{1/2}\Lambda}^{B^{1/2}\Lambda} dy_\parallel \int_0^{\Lambda K_1^{1/2}} \frac{y_\perp^{(d-3)/2}dy_\perp}{y_\parallel^2+y_\perp^2},
\end{aligned} \tag{6.3.13}$$

where K_d is the d-dimensional solid angle divided by $(2\pi)^d$ introduced in Eq. (5.3.4). $\langle u^2(\mathbf{x})\rangle$ is finite for $2+(d-3)/2>2$. At $d=3$, however, $\langle u^2(\mathbf{x})\rangle$, like $\langle\phi^2(\mathbf{x})\rangle$ for the xy-model in $d=2$, becomes infinite because of infrared singularities in the integral at small argument. The lower critical dimension, d_c, is three. This means that $\langle\psi_n\rangle$ is zero and there is no long-range order in a three-dimensional layered system at any finite temperature. As in the two-dimensional xy-model, this implies power-law decay of correlation functions (see Problem 6.9 for more details):

$$\langle\psi_n(\mathbf{x})\psi_n^*(0)\rangle = |\langle\psi_n\rangle|^2 e^{-n^2 g(\mathbf{x})} \equiv G_n(\mathbf{x}),$$

$$\begin{aligned}
g(\mathbf{x}) &= \frac{1}{2}q_0^2\langle[u(\mathbf{x})-u(0)]^2\rangle = q_0^2T\int\frac{d^3q}{(2\pi)^3}\frac{1-e^{i\mathbf{q}\cdot\mathbf{x}}}{Bq_\parallel^2+K_1q_\perp^4} \\
&\sim \frac{q_0^2T}{8\pi(K_1B)^{1/2}}\times\begin{cases}\ln x_\parallel, & \text{if } \mathbf{x}_\perp=0;\\ \ln x_\perp^2, & \text{if } x_\parallel=0.\end{cases}
\end{aligned} \tag{6.3.14}$$

Thus

$$G_n(\mathbf{x}) \sim \begin{cases} x_\parallel^{-n^2\eta_c}, & \text{if } \mathbf{x}_\perp=0;\\ |\mathbf{x}_\perp|^{-2n^2\eta_c}, & \text{if } x_\parallel=0,\end{cases} \tag{6.3.15}$$

where

$$\eta_c = \frac{q_0^2 T}{8\pi(K_1B)^{1/2}}. \tag{6.3.16}$$

The power law form of spatial correlations implies that the X-ray structure factor will have a power law rather than a delta function singularity at $\mathbf{q}=q_0\mathbf{n}$:

$$\begin{aligned}
I(\mathbf{q}) &\sim G_n(\mathbf{q}+n\mathbf{q}_0)+G_n(\mathbf{q}-n\mathbf{q}_0) \\
G_n(\mathbf{q}-n\mathbf{q}_0) &\sim \begin{cases}(q_\parallel-nq_0)^{-2+n^2\eta_c}, & \text{if } q_\perp=0;\\ q_\perp^{-4+2n^2\eta_c}, & \text{if } q_\parallel=0.\end{cases}
\end{aligned} \tag{6.3.17}$$

This behavior has been observed in both thermotropic smectics (Als-Nielsen *et al.* 1980) and in lyotropic lamellar phases (Safinya *et al.* 1986). The X-ray scattering data on lamellar smectics are shown in Fig. 6.3.4. The elastic constants in the lamellar phases are entropy dominated, and as a result $\eta_c(T)$ is independent of temperature (see Sec. 10.5). Note that $G_n(q_\parallel, 0)$ is divergent for $n^2\eta_c<2$ but only has a cusp singularity for $n^2\eta_c>2$. Thus the maximum intensity of quasi-Bragg peaks decreases rapidly with n, and, if η_c is large enough, not even the first peak will have a divergent maximum intensity. In most thermotropic smectics, only

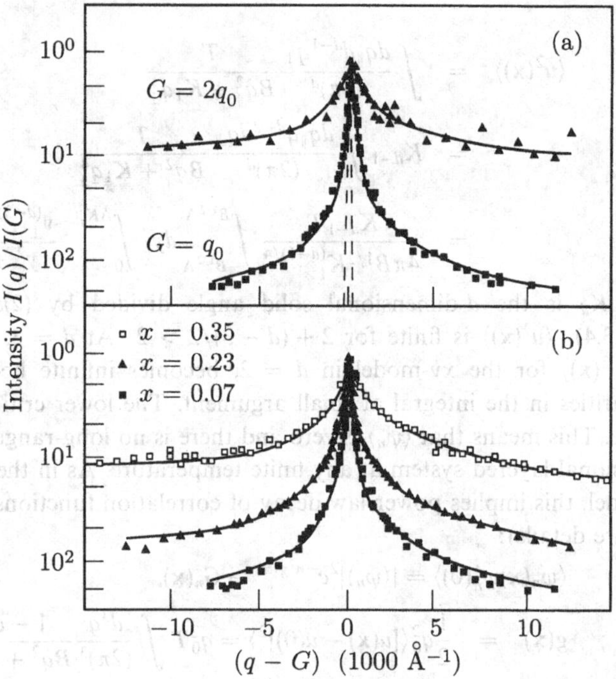

Fig. 6.3.4. X-ray structure factors for SDS-pentanol-water lamellar smectic. (a) $G(q_\parallel) = I(q_\parallel)/I(nq_0)$ for $n = 1$ and $n = 2$. (b) $G(q_\parallel)$ for $n = 1$ for different concentrations x of water. [C.R. Safinya, D. Roux, G.S. Smith, S.K. Sinha, P. Dimon, N.A. Clark, and A.M. Belloq, *Phys. Rev. Lett.* **57**, 2718 (1986); see also D. Roux and C.R. Safinya, *J. Phys. (Paris)* **49**, 307 (1988).]

the first peak is clearly visible, at least partially as a result of a small value for B and thus a large η. Lyotropic smectics are more incompressible, B is larger, and more peaks are visible. Fig. 6.3.4a shows $I(q)$ for the peaks at q_0 and $2q_0$ in a lyotropic smectic. The second peak has only a cusp singularity. In thermotropic systems, B, and thus η_c, can be strongly temperature dependent, especially near the nematic-to-smectic-A transition.

3 Nonlinearities

The preceding analysis treated F_{el} as a harmonic function of u in which there are no terms of order $(\delta\boldsymbol{\Omega})^2$ when the system is rotated through an angle $\delta\Omega$. F should, however, remain unchanged under rotations through an arbitrary angle. Under rotations through an angle θ about the y-axis, $\mathbf{q}_0$ goes to $\mathbf{q}_0 = q_0(\cos\theta\mathbf{e}_z - \sin\theta\mathbf{e}_x)$. Then, from Eq. (6.3.9),

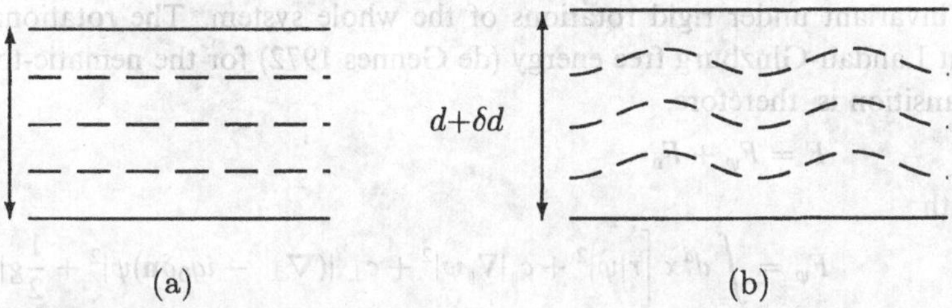

Fig. 6.3.5. Buckling instability of a smectic. (a) Equilibrium configuration and (b) stretched configuration showing undulations.

$$\begin{aligned} q_0 u &= q_0[(1-\cos\theta)z - \sin\theta x], \\ \nabla_\parallel u &\equiv \nabla_z u = 1 - \cos\theta, \\ \nabla_\perp u &\equiv \nabla_x u = -\sin\theta. \end{aligned} \qquad (6.3.18)$$

The combination $\nabla_z u - (\nabla u)^2/2$ is independent of θ since $1-\cos\theta - [(1-\cos\theta)^2 + \sin^2\theta]/2 = 0$. Thus, rotational invariance of F through arbitrary angles can be guaranteed by replacing $\nabla_z u$ by $\nabla_z u - (\nabla u)^2/2$:

$$F_{\text{el}} = \frac{1}{2}\int d^3x \left\{ B\left[\nabla_\parallel u - \frac{1}{2}(\nabla u)^2\right]^2 + K_1(\nabla_\perp^2 u)^2 \right\}. \qquad (6.3.19)$$

The non-linear terms in F have important consequences. They lead, for example, to a buckling instability in a stretched smectic and to fluctuations that renormalize (Grinstein and Pelcovits 1981) the elastic constants B and K_1 respectively to zero and infinity in a three-dimensional system. A discussion of the latter effect is beyond the scope of this course. The former is easily understood by the following example. A smectic is in equilibrium between two glass plates, as shown in Fig. 6.3.5a. The top plate is moved upward relative to the bottom plate, creating a positive $\nabla_\parallel u$. If $\nabla_\parallel u$ is large enough, the system can clearly reduce its energy by creating a non-zero $\nabla_\perp u$, i.e., by creating undulations of the layers, as shown in Fig. 6.3.5b.

4 The nematic-to-smectic-A transition

As we have discussed in this chapter, and previously in Chapter 2, the order parameter distinguishing the smectic-A phase from the nematic phase is the complex mass density wave amplitude $\psi_1 = \psi$. Following the rules for constructing Landau free energies discussed in Chapters 3 and 5, one would naively expect that the appropriate continuum free energy to describe this transition could be constructed from combinations such as $|\psi|^2$ or $|\nabla\psi|^2$, invariant under changes in the phase of ψ. It is important, however, to remember that rotations of the director cause the equilibrium phase of ψ to have a part, described by Eq. (6.3.9), equal to $-q_0\delta\mathbf{n}\cdot\mathbf{x}$. Thus, the covariant derivative, $(\nabla_\perp - iq_0\delta\mathbf{n})\psi$, rather than $\nabla_\perp\psi$, is invariant under rigid rotations of the whole system. The rotationally invariant Landau-Ginzburg free energy (de Gennes 1972) for the nematic-to-smectic-A transition is, therefore,

$$F = F_\psi + F_n \tag{6.3.20}$$

with

$$F_\psi = \int d^d x \left[r|\psi|^2 + c_\parallel |\nabla_\parallel \psi|^2 + c_\perp \, |(\nabla_\perp \; - iq_0 \delta \mathbf{n})\psi|^2 + \frac{1}{2} g|\psi|^4 \right] \tag{6.3.21}$$

where F_n is the Frank free energy for the director [Eq. (6.2.3)]. In mean-field theory, $|\psi| = (-r/u)^{1/2}$ when $r < 0$, and this free energy reduces to Eq. (6.3.11) with $B = c_\parallel q_0^2 |\psi|^2$ and $D = c_\perp \; q_0^2 |\psi|^2$. In the nematic phase, the director is an elastic variable with power-law correlations. Fluctuations in the director either cause the nematic-to-smectic-A transition to be first order or to be in a universality class different from that of the xy-model (Halperin *et al.* 1974).

6.4 Elasticity of solids: strain and elastic energy

As discussed in Sec. 2.5, atoms or molecules in a perfect three-dimensional solid occupy positions in the unit cells of a periodic lattice, with lattice vectors $\mathbf{R}_l = l_1 \mathbf{a}_1 + l_2 \mathbf{a}_2 + l_3 \mathbf{a}_3$, where l_1, l_2 and l_3 are integers and $\mathbf{a}_1$, $\mathbf{a}_2$ and $\mathbf{a}_3$ are primitive translation vectors. The density $\langle n(\mathbf{x}) \rangle$ is periodic in three directions with a Fourier expansion [Eq. (2.5.13)] with non-vanishing amplitudes at reciprocal lattice vectors $\mathbf{G} = m_1 \mathbf{b}_1 + m_2 \mathbf{b}_2 + m_3 \mathbf{b}_3$, where m_1, m_2 and m_3 are integers and $\mathbf{b}_1$, $\mathbf{b}_2$ and $\mathbf{b}_3$ are the primitive translation vectors of the reciprocal lattice. A crystalline solid can be described either as a periodic array of atoms or as a collection of planes of constant phase orthogonal to the vectors $\mathbf{G}$.

1 The strain tensor

The complex mass density amplitudes have an amplitude and a phase

$$\langle n_{\mathbf{G}} \rangle = |\langle n_{\mathbf{G}} \rangle| e^{i\phi_{\mathbf{G}}}. \tag{6.4.1}$$

Uniform translations of the sample through $\mathbf{u}$, i.e., $\mathbf{x}^\alpha \rightarrow \; \mathbf{x}^\alpha + \mathbf{u}$ for every particle α, are described, as in the case of smectics, by uniform changes of the phases:

$$\phi_{\mathbf{G}} = \phi_{\mathbf{G}}^0 - \mathbf{G} \cdot \mathbf{u}. \tag{6.4.2}$$

Since the energy is invariant with respect to uniform translations of the sample, we expect an elastic energy proportional to $(\nabla_i u_j)^2$. As in the case of smectic liquid crystals, arbitrary gradients of $\mathbf{u}$ cannot appear because the energy is also invariant with respect to uniform rigid rotations. The reciprocal lattice vectors transform according to

$$\mathbf{G} \rightarrow \mathbf{G}' = \mathbf{G} + \delta \mathbf{\Omega} \times \mathbf{G} \tag{6.4.3}$$

under an infinitesimal rotation through an angle $\delta\mathbf{\Omega}$. This means that **u**'s that satisfy

$$\begin{aligned}\mathbf{G}\cdot\mathbf{u} &= (\mathbf{G}-\mathbf{G}')\cdot\mathbf{x} \\ &\sim -(\delta\mathbf{\Omega}\times\mathbf{G})\cdot\mathbf{x}\end{aligned} \tag{6.4.4}$$

correspond to rigid rotations. Since Eq. (6.4.4) is valid for every **G**, it follows that, for rigid rotations,

$$u_i = (\delta\mathbf{\Omega}\times\mathbf{x})_i = \epsilon_{ijk}\delta\Omega_j x_k, \tag{6.4.5}$$

where ϵ_{ijk} is the Levi-Cevita symbol that is one if ijk is an even permutation of 123, minus one if ijk is an odd permutation of 123, and zero otherwise. Thus, the rotation angle $\delta\mathbf{\Omega}$ is proportional to the anti-symmetric part of $\nabla_i u_j$:

$$\delta\Omega_k = \frac{1}{4}\epsilon_{ijk}(\nabla_i u_j - \nabla_j u_i)\ . \tag{6.4.6}$$

Since the free energy $F_{\rm el}$ must be independent of $\delta\mathbf{\Omega}$, it must be independent of the antisymmetric part $u_{ij}^A = \frac{1}{2}(\nabla_i u_j - \nabla_j u_i)$ of the tensor $\nabla_i u_j$. In other words, $F_{\rm el}$ can depend only on the symmetric derivative

$$u_{ij} = \frac{1}{2}(\nabla_i u_j + \nabla_j u_i)\ . \tag{6.4.7}$$

u_{ij} is the linearized *strain tensor*. Non-linear contributions to the strain tensor will be discussed in Sec. 6.5.

Before considering the form of the elastic free energy, it is useful to investigate the types of distortions that can be described by u_{ij}. First consider the case of an isotropic infinitesimal volume change δV. Positive δV corresponds to dilation, and negative δV corresponds to compression. In either case, the relative change in the volume $\Omega_0 = \mathbf{a}_1\cdot(\mathbf{a}_2\times\mathbf{a}_3)$ of a primitive cell will be equal to the relative change in the volume of the crystal: $\delta\Omega_0/\Omega_0 = \delta V/V$. Since this distortion is isotropic, the relative change of the magnitude of each primitive lattice vector will be the same and equal to $\epsilon = \delta V/3V$. (We ignore for the moment vacancies and interstitials and assume that the number of atoms per unit cell remains constant under strain.) Because primitive translation vectors of the reciprocal lattice are defined via $\mathbf{b}_i\cdot\mathbf{a}_j = 2\pi\delta_{ij}$, the relative change in the magnitude of each $\mathbf{b}_i$ will be $-\epsilon$ for small ϵ. Thus, under uniform compression or dilation, $\mathbf{G}\rightarrow\mathbf{G}' = (1-\epsilon)\mathbf{G}$. Using the first line of Eq. (6.4.4), which is valid for every **G**, we can calculate the strain,

$$u_{ii} = 3\epsilon = \frac{\delta\Omega_0}{\Omega_0} = \frac{\delta V}{V}, \tag{6.4.8}$$

associated with isotropic compression or dilation. A similar analysis shows that

$$u_{xx} = \frac{\delta L_x}{L_x} \tag{6.4.9}$$

if there is a change δL_x in the length L_x of the sample along the x-direction. A *shear distortion* is produced by a displacement **u** with $\nabla\cdot\mathbf{u} = 0$ and $\nabla\times\mathbf{u} = 0$. For example, the distortion

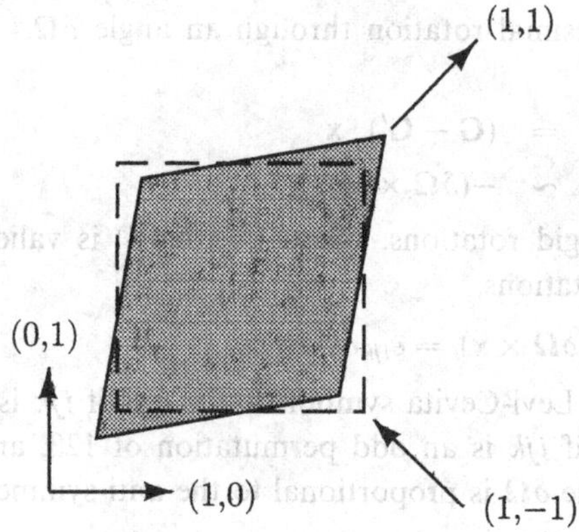

Fig. 6.4.1. The shear distortion of a square described by Eq. (6.4.10). This distortion is equivalent to a compression along the $(1,-1)$-axis and a dilation along the $(1,1)$-axis.

$$u_x = \epsilon y\,, \quad u_y = \epsilon x\,, \quad u_{xy} = u_{yx} = \epsilon \tag{6.4.10}$$

sketched in Fig. 6.4.1 corresponds to a shear in the chosen coordinate system. Note, however, that this same distortion is equally well described by a dilation along the $(1,1)$-axis and a compression along the $(1,-1)$-axis.

2 *The elastic free energy*

The elastic energy for a solid is quadratic in the symmetrized strains

$$F_{\text{el}} = \frac{1}{2}\int d^d x K_{ijkl} u_{ij} u_{kl} = \int d^d x f_{\text{el}}(u_{ij}), \tag{6.4.11}$$

where $f_{\text{el}}(u_{ij})$ is the elastic free energy density. Because u_{ij} is symmetric and f_{el} is invariant under interchange of the pairs of dummy indices ij and kl, the elastic constant tensor satisfies the symmetry relations

$$K_{ijkl} = K_{klij} = K_{jikl} = K_{ijlk} = K_{jilk}\,. \tag{6.4.12}$$

In addition, the entire free energy F_{el} must be invariant under the point group symmetries of the crystal (e.g. rotations through 90° about four-fold axes in cubic crystals, reflections about mirror planes, etc.). These symmetries restrict considerably the number of components of the tensor K_{ijkl}. In general, the number of independent elastic constants increases as the point group symmetry of the solid decreases: the most isotropic solids have the smallest number of elastic constants. The highest symmetry three-dimensional crystalline solid has cubic symmetry and three independent elastic constants. The highest symmetry two-dimensional crystalline solid has hexagonal symmetry and has only two independent elastic constants. In two- and three-dimensional crystals of lower symmetry, there are more elastic constants. For example, a two-dimensional

Table 6.4.1. *Number of elastic constants for crystal systems.*

Crystal system	No. of elastic constants
Triclinic	21
Monoclinic	13
Orthorhombic	9
Tetragonal	6 or 7
Rhombohedral	6 or 7
Hexagonal	5
Cubic	3

Table 6.4.2. *Elastic constants for some common solids in units of* 10^{12} *dynes/cm*^{2}.

Material	K_{11}	K_{12}	K_{44}
Li	0.148	0.125	0.108
Cu	1.68	1.21	0.75
Al	1.07	0.61	0.28
Fe	2.34	1.36	1.18
NaCl	0.487	0.124	0.126

From N. W. Ashcroft and N. D. Mermin, *Solid State Physics* (Holt, Rinehart, and Winston, New York, 1976), p. 447.

crystal with four- rather than six-fold symmetry, like a three-dimensional crystal with cubic symmetry, has three elastic constants. The number of independent elastic constants for three-dimensional crystals is listed in Table 6.4.1.

The order of magnitude of a typical elastic constant K appearing in K_{ijkl} can be estimated using dimensional analysis. Since strain is unitless, K has units of energy/volume. The energy scale is set by the binding energy per atom of the solid. As discussed in Chapter 1, this is typically of the order of a few electron volts (1.6×10^{-16} erg). The length scale is set by the interparticle spacing, which is typically of the order of a couple of angstroms. Thus we estimate K to be of order $(1.6 \times 10^{-12}/8 \times 10^{-24}) \sim 0.2 \times 10^{12}$ dynes/cm^2. Elastic constants for some common materials are listed in Tables 6.4.2 and 6.4.3.

3 *Isotropic and cubic solids*

Real materials are often composed of many randomly oriented microcrystals. On length scales much larger than the typical size of a microcrystal, the material is rotationally isotropic. Alternatively, some "solids", such as glasses, are microscopically isotropic. In these systems, as well as in two-dimensional hexagonal crystals, there are only two independent elastic constants in K_{ijkl}. Physically, we can compress or shear the solid, and the responses are independent of direction.

Table 6.4.3. *Elastic constants for some common isotropic materials in units of dynes/cm²*.

Material	Shear modulus	Bulk modulus
Tungsten carbide	2.2×10^{12}	3.2×10^{12}
Steel	0.83×10^{12}	1.5×10^{12}
Gold	0.28×10^{12}	1.7×10^{12}
Pyrex	0.25×10^{12}	0.4×10^{12}
Nylon	0.12×10^{12}	0.59×10^{12}
Rubber	$\sim 10^7$	0.03×10^{12}
Jello	$\sim 10^4$	0.02×10^{12}
Polystyrene foam	1.3×10^8	2×10^8
Shaving foam	$\sim 10^3$	$\sim 10^6$
Ice	0.025×10^{12}	0.073×10^{12}
Water	0	0.02×10^{12}
Air	0	10^6 (1 atm)

Most values are taken from the AIP Handbook. Values for conventionally crystalline materials are for polycrystalline samples with small grains.

Mathematically, the only tensor available to construct higher order tensors in such systems is the Kronecker delta δ_{ij}. The only two fourth rank tensors formed from the Kronecker delta satisfying the symmetry relations Eq. (6.4.12) are $\delta_{ij}\delta_{jk}$ and $\delta_{ik}\delta_{jl} + \delta_{il}\delta_{jk}$. Thus we have

$$K_{ijkl} = \lambda\delta_{ij}\delta_{kl} + \mu(\delta_{ik}\delta_{jl} + \delta_{il}\delta_{jk}), \tag{6.4.13}$$

where λ and μ are called the *Lamé* coefficients. From Eq. (6.4.11), we obtain

$$F_{\text{el}} = \frac{1}{2}\int d^d x[\lambda u_{ii}^2 + 2\mu u_{ij}u_{ij}] \tag{6.4.14}$$

for a d-dimensional isotropic solid or a two-dimensional hexagonal crystal. (The Einstein summation convention is understood in this equation so that $u_{ii}^2 = u_{ii}u_{jj}$.) An alternative representation of F_{el} is often more useful, namely u_{ij} can be decomposed into a scalar and a symmetric traceless tensor,

$$u_{ij} = \frac{1}{d}\delta_{ij}u_{kk} + \left(u_{ij} - \frac{1}{d}\delta_{ij}u_{kk}\right). \tag{6.4.15}$$

The first term measures volume changes, and the second measures distortions in which the volume does not change (shear). With this decomposition,

$$F = \frac{1}{2}\int d^d x\left[Bu_{kk}^2 + 2\mu\left(u_{ij} - \frac{1}{d}\delta_{ij}u_{kk}\right)^2\right], \tag{6.4.16}$$

where $B = \lambda + (2\mu/d)$ is the *bulk modulus* and μ is the *shear modulus*. Thermodynamic stability requires both B and μ to be positive.

In three-dimensional systems with cubic symmetry, the elastic free energy can be written as

$$\begin{aligned} F_{\text{cubic}} &= \frac{1}{2}\int d^3x[K_{11}(u_{xx}^2+u_{yy}^2+u_{zz}^2) \\ &\qquad +K_{12}(u_{xx}u_{yy}+u_{xx}u_{zz}+u_{yy}u_{zz}) \\ &\qquad +2K_{44}(u_{xy}^2+u_{xz}^2+u_{yz}^2)]. \end{aligned} \tag{6.4.17}$$

Because the elastic tensor K_{ijkl} is invariant under interchanges of i and j and of k and l, there are only six independent combinations of ij and kl. These are denoted by $1 = xx$, $2 = yy$, $3 = zz$, $4 = yz$, $5 = xz$, and $6 = xy$. In a cubic crystal, $K_{11} = K_{22} = K_{33}$, $K_{12} = K_{13} = K_{23}$, and $K_{44} = K_{55} = K_{66}$. All other components of K_{ijkl} are zero. The free energy of a cubic solid reduces to that of an isotropic solid when $K_{12} = K_{44}$. Thus, a measure of the extent to which a cubic solid differs from an isotropic solid is provided by the ratio $(K_{12} - K_{44})/K_{12}$, which varies from about 0.05 to 0.4 for the materials listed in Table 6.4.2. The elastic energy for a two-dimensional crystal with the point group symmetry of a square has the same form as Eq. (6.4.17) with $u_{zz} = u_{xz} = u_{yz} = 0$. Finally, the elastic free energy of a uniaxial solid is

$$\begin{aligned} F_{\text{uniaxial}} &= \frac{1}{2}\int d^3x[K_{11}(u_{xx}^2+u_{yy}^2)+2K_{12}u_{xx}u_{yy}+2(K_{11}-K_{12})u_{xy}^2 \\ &\qquad +K_{33}u_{zz}^2+K_{44}(u_{yz}^2+u_{xz}^2)+2K_{13}(u_{xx}+u_{yy})u_{zz}]. \end{aligned} \tag{6.4.18}$$

Note that there is only one elastic constant, K_{13}, that couples strains in the xy-plane to those along the z-axis.

4 *Fluctuations*

As in previous examples considered in this chapter, macroscopic elasticity is intimately related to the fluctuations in the local value of the phase vector $\mathbf{u}$. Using $u_{ij}(\mathbf{q}) = (iq_iu_j + iq_ju_i)/2$ and the symmetry properties of K_{ijkl}, we obtain

$$\mathscr{H}_{\text{el}} = \frac{1}{2}\int \frac{d^dq}{(2\pi)^d} C_{ik}(\mathbf{q})u_i(\mathbf{q})u_k(\mathbf{q}) \tag{6.4.19}$$

for the elastic Hamiltonian of a crystal, where

$$C_{ik}(\mathbf{q}) = K_{ijkl}q_jq_l \,. \tag{6.4.20}$$

For the d-dimensional isotropic and two-dimensional hexagonal solids, this becomes

$$\mathscr{H}_{\text{el}} = \frac{1}{2}\int \frac{d^dq}{(2\pi)^d}[(\lambda+2\mu)q^2u_l^2 + \mu q^2\mathbf{u}_T\cdot\mathbf{u}_T]\,, \tag{6.4.21}$$

where u_l and $\mathbf{u}_T$ are the *longitudinal* and *transverse* parts of $\mathbf{u}$ defined via

$$u_l(\mathbf{q}) = \hat{\mathbf{q}}\cdot\mathbf{u}(\mathbf{q})\,, \quad \mathbf{u}_T(\mathbf{q}) = \mathbf{u} - \hat{\mathbf{q}}u_l(\mathbf{q}) \tag{6.4.22}$$

with $\hat{\mathbf{q}} = \mathbf{q}/|\mathbf{q}|$. Thermodynamic stability with respect to spatially non-uniform distortions requires that $\lambda + 2\mu$ as well as μ be positive. Note, however, that $\lambda + 2\mu = B + 2(d-1)\mu/d \geq B$, with the strict inequality holding for all $d > 1$. Thus, if a system is stable against spatially uniform volume changes with $u_{ij} = (\delta V/dV)\delta_{ij}$, it will also be stable against long wavelength longitudinal distortions.

As usual, fluctuations in $u_i(\mathbf{q})$ are determined by the inverse elastic tensor:

$$\begin{aligned} G_{u_iu_j}(\mathbf{q}) &= \frac{1}{V}\langle u_i(\mathbf{q})u_j(-\mathbf{q})\rangle \\ &= TC_{ij}^{-1}(\mathbf{q}) \sim \frac{T}{Kq^2}, \end{aligned} \tag{6.4.23}$$

where the matrix inverse is in the indices i and j. In isotropic solids, the **u**-correlation function has a simple form:

$$G_{u_iu_j}(\mathbf{q}) = \frac{T}{(\lambda+2\mu)q^2}\hat{q}_i\hat{q}_j + \frac{T}{\mu q^2}(\delta_{ij} - \hat{q}_i\hat{q}_j). \tag{6.4.24}$$

These equations imply that fluctuations in **u** reduce the intensities of the Bragg peaks at a reciprocal lattice vector **G** according to

$$I(\mathbf{G}) \sim \langle n_{\mathbf{G}}\rangle\langle n_{-\mathbf{G}}\rangle \sim |\langle n_{\mathbf{G}}\rangle|^2 e^{-2W} \tag{6.4.25}$$

with

$$W = \frac{1}{2}G_i\langle u_i(\mathbf{x})u_j(\mathbf{x})\rangle G_j \sim \frac{1}{2}G^2\int\frac{d^dq}{(2\pi)^d}\frac{T}{Kq^2}. \tag{6.4.26}$$

As for the xy-model, the Debye-Waller factor e^{-2W} is nonzero for all $d > 2$ [see (Eq. 6.1.19)]. At $d = 2$, W becomes infinite, implying that long-range periodic crystalline order cannot exist in dimensions less than or equal to two. As in the xy-model, there is quasi-long-range periodic order in a two-dimensional "crystal" and only short-range order at any finite temperature in a one-dimensional crystal. The lower critical dimension for a crystal is 2.

5 *Mercury chain salts – one-dimensional crystals*

It is interesting to treat the case of an ideal classical one-dimensional crystal exactly, especially since there is a physical system that closely approximates this system. Mercury can be intercalated, as shown in Fig. 6.4.2, into linear channels in AsF_6, creating a compound $Hg_{3-\delta}AsF_6$, where $\delta \sim 0.18$ at 300 K. In each channel, there is a linear chain of Hg atoms that interact among themselves but interact only weakly with Hg atoms in other channels. Transverse motion in a channel is negligible so that only a single coordinate is needed to specify the position of a given Hg atom. Finally, the interaction between Hg atoms and AsF_6 molecules along a channel does not lead to a preferred position of the Hg atoms relative to the AsF_6 lattice. The interaction is especially weak because the finite value of δ means a non-integral number of Hg atoms per unit cell, a difference in periodicity between the Hg atoms and the AsF_6 lattice, and hence an incommensurate crystal (see Sec. 2.10). Thus, to a very good approximation, the Hamiltonian for the Hg atoms is an independent sum of channel Hamiltonians,

$$\mathscr{H} = \sum_l \frac{p_l^2}{2m} + U, \tag{6.4.27}$$

where U is the interaction potential among atoms in a given chain. If only nearest neighbor atoms interact,

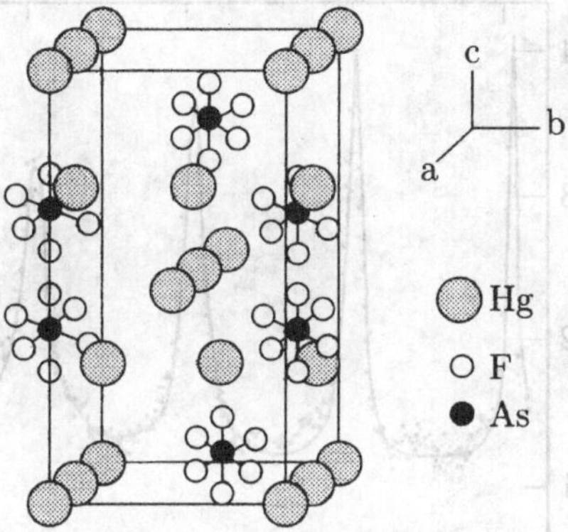

Fig. 6.4.2. Crystal structure of $Hg_{3-\delta}AsF_6$. The AsF_6 molecules form a regular lattice with channels, into which are intercalated chains of mercury atoms. The chains lie parallel to two orthogonal directions in the lattice.

$$U = \sum_l U_2(x_{l+1} - x_l), \tag{6.4.28}$$

where U_2 is a pair potential. At low temperatures, U_2 can be approximated by an harmonic potential,

$$U = \frac{1}{2}K \sum_l (x_{l+1} - x_l - l_0)^2, \tag{6.4.29}$$

where l_0 is the preferred separation between atoms. In the continuum limit, the harmonic approximation for U reduces to the elastic free energy of Eq. (6.4.11).

The structure factor for the classical chain with nearest neighbor interactions can be calculated exactly as

$$\begin{aligned} S(q) &= \frac{1 - Z^2}{1 + 2Z^2 - 2Z\cos ql_0} \\ &= \frac{\sinh[\sigma^2 q^2/2]}{\cosh[\sigma^2 q^2/2] - \cos ql_0} \quad \text{(harmonic)} \end{aligned} \tag{6.4.30}$$

where

$$Z(q) = \frac{\int dx e^{-iqx - \beta U_2(x)}}{\int dx e^{-\beta U_2(x)}} \tag{6.4.31}$$

and $\sigma = T/K$. These functions have strong Lorentzian-like (not delta-function) peaks in the vicinity of the reciprocal lattice peaks at $q = 2\pi n/l_0$ of the ideal lattice with long range order. As discussed in Sec. 2.8, $S(q)$ describes scattering on sheets in reciprocal space because the positions of atoms in parallel chains are uncorrelated. To provide a complete description of real Hg chains, the effects of both further neighbor interactions and quantum fluctuations must be included in $S(q)$. These lead to a q-dependent prefactor and the addition of a q-dependent background term in Eq. (6.4.30). These additional factors vary slowly in the vicinity of each peak in $S(q)$. Thus, a very good approximation for $S(q)$ in

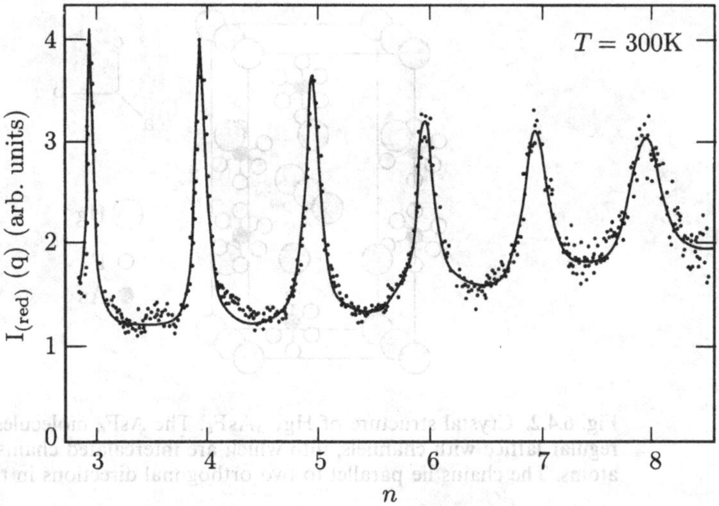

Fig. 6.4.3. Experimental structure factor at $T = 300$ K for mercury chains in $Hg_{3-\delta}AsF_4$ compared to the theoretical predictions of Eq. (6.4.32) (solid curve). [R. Spal, C.F. Chen, T. Egami, P.F. Nigrey, and A.J. Heeger, *Phys. Rev. B* **21**, 3110 (1980); R. Spal, thesis (University of Pennyslvania, 1980).]

the vicinity of its nth peak that includes both quantum fluctuations and further neighbor interactions is

$$S_{\text{fit}}(q) = A_n \frac{1 - Z_n^2}{1 + Z_n^2 - 2Z_n \cos[(q - Q_n)l_0]} + B_n q + C_n. \tag{6.4.32}$$

Fig. (6.4.3) shows the experimentally determined $S(q)$ (dots) compared to the theoretical prediction (smooth curve) of Eq. (6.4.32) with coefficients chosen for each peak.

6 Xenon on graphite – a two-dimensional crystal

As discussed in Sec. 2.8, rare earth atoms such as Xe adsorbed on graphite can form solid phases whose lattice parameter is incommensurate with that of the underlying graphite substrate. Because the two lattices are incommensurate, there is no preferred position (though there is a preferred orientation, see Problem 10.7) of the Xe lattice relative to the graphite lattice, and fluctuations of the graphite lattice can be described by an elastic Hamiltonian that is essentially independent of the substrate. (We will reconsider this point in more detail in Chapter 10).

Since the Xe lattice is hexagonal, its elasticity is described by the two elastic constant Hamiltonians of Eqs. (6.4.14) and (6.4.21). Using Eq. (6.4.24), we can calculate

$$\langle [u_i(\mathbf{x}) - u_i(0)][u_j(\mathbf{x}) - u_j(0)] \rangle = \delta_{ij} \frac{T}{4\pi} \left(\frac{1}{\lambda + 2\mu} + \frac{1}{\mu} \right) \ln(|\mathbf{x}|\Lambda) \tag{6.4.33}$$

and

$$\langle e^{-i\mathbf{G}\cdot[\mathbf{u}(\mathbf{x})-\mathbf{u}(0)]}\rangle \sim |\mathbf{x}|^{-\eta_{\mathbf{G}}}, \tag{6.4.34}$$

where

$$\eta_{\mathbf{G}} = \frac{T}{4\pi}\left(\frac{1}{\lambda+2\mu}+\frac{1}{\mu}\right)G^2. \tag{6.4.35}$$

The X-ray structure factor will have power-law peaks at the hexagonal lattice reciprocal lattice vectors **G**, which can be calculated using the Fourier expansion of the density,

$$\langle n(\mathbf{x})\rangle = \sum_{\mathbf{G}} |\langle n_{\mathbf{G}}\rangle| e^{i\phi^0_{\mathbf{G}}} e^{i\mathbf{G}\cdot[\mathbf{x}-\mathbf{u}(\mathbf{x})]}, \tag{6.4.36}$$

and treating the mass density amplitude $|\langle n_{\mathbf{G}}\rangle| \exp(i\phi^0_{\mathbf{G}})$ but not $\mathbf{u}(\mathbf{x})$ as constant. The result is that

$$S(\mathbf{q}) \sim |\mathbf{q}-\mathbf{G}|^{-(2-\eta_{\mathbf{G}})} \tag{6.4.37}$$

in the vicinity of $\mathbf{q} = \mathbf{G}$. Fig. 6.4.4 shows X-ray scattering data for Xe on graphite for a range of temperatures from 100 K to 160 K. The solid lines in these figures at low temperatures were obtained by the appropriate average of Eq. (6.4.37) over the different directions of graphite planes in a polycrystalline sample convoluted with an experimental resolution function. The asymmetry of the observed structure factor is a result (Warren 1941) of mosaic averaging in two-dimensional crystals whose Bragg or quasi-Bragg peaks are actually rods in reciprocal space. At higher temperatures (above about 152 K), the solid lines are powder averaged Lorentzians describing correlations in the fluid rather than the solid phase. These experiments permit $\eta_{\mathbf{G}}$ to be determined in the solid phase and the correlation length ξ in the liquid phase. These functions are plotted in Fig. 6.4.5.

7 *Vacancies and interstitials*

In the analysis just presented, it was tacitly assumed that the number of particles remained fixed and that the density changes are entirely fixed by the strain via Eq. (6.4.8):

$$\frac{\delta V}{V} = -\frac{\delta n}{n} = u_{ii}. \tag{6.4.38}$$

This relation may be violated if the average mass in a unit cell changes under strain, as it can if there are empty sites (vacancies) in the ideal lattice or atoms occupying sites other than ideal lattice sites (interstitials). Consider a periodic crystal whose ground state structure has one atom per unit cell. Let N_s be the number of lattice sites and let N_v be the number of vacancies. The total number of atoms in the solid is $N = N_s - N_v$, and the density $n = N/V$ of atoms is related to the volume $\Omega_0 = V/N_s$ of the unit cell of the lattice and the vacancy density $n_v = N_s/V$ via

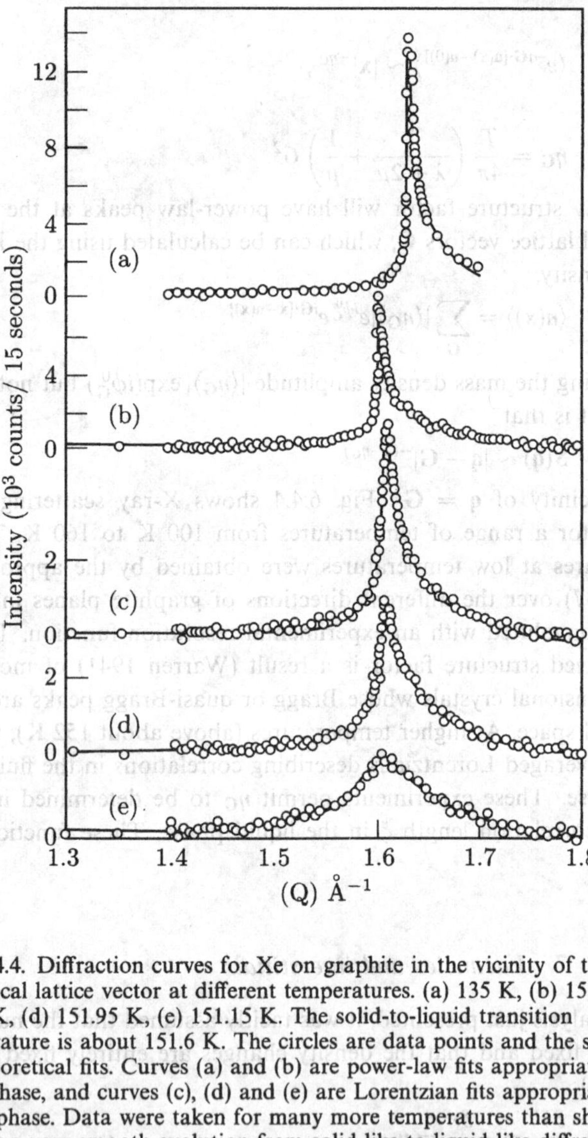

Fig. 6.4.4. Diffraction curves for Xe on graphite in the vicinity of the (1,0) reciprocal lattice vector at different temperatures. (a) 135 K, (b) 151.3 K, (c) 151.9 K, (d) 151.95 K, (e) 151.15 K. The solid-to-liquid transition temperature is about 151.6 K. The circles are data points and the solid lines are theoretical fits. Curves (a) and (b) are power-law fits appropriate to the solid phase, and curves (c), (d) and (e) are Lorentzian fits appropriate to the liquid phase. Data were taken for many more temperatures than shown, and there is a very smooth evolution from solid-like to liquid-like diffraction patterns. [Adapted from P.A. Heiney, P.W. Stephens, R.J. Birgeneau, P.M. Horn, and D.E. Moncton, *Phys. Rev. B* **28**, 6416 (1983).]

$$n = \Omega_0^{-1} - n_v. \tag{6.4.39}$$

Strain leads to changes in the volume of the unit cell via $u_{ii} = \delta\Omega_0/\Omega_0$ [Eq. (6.4.8)] without changing the number of vacancies. Thus, changes in density are brought about by changes in both strain and vacancy density:

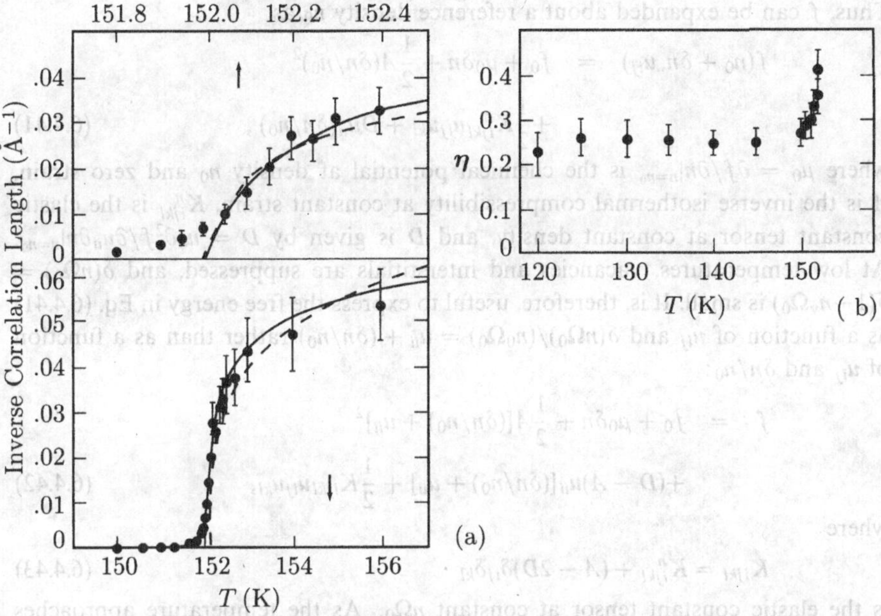

Fig. 6.4.5. (a) Inverse scattering lengths from Lorentzian fits to the structure function. Solid lines are fits to the dislocation mediated melting theory (see Chapter 9). Top and bottom panels are the same fits with different scales. (b) $\eta = \eta_{G=(1,0)}$ for the solid phase of Xe on graphite. [P. A. Heiney, P. W. Stephens, R. J. Birgeneau, P. M. Horn, and D. E. Moncton, *Phys. Rev. B* **28**, 6416 (1983).]

$$\Omega_0 \delta n = -u_{ii} - \Omega_0 \delta n_v. \tag{6.4.40}$$

If there are no vacancies, $n\Omega_0 = 1$, and this equation reduces to Eq. (6.4.38). In more complicated situations, in which there are many atoms per unit cell, such as, for example, the plumber's nightmare cubic phase of lyotropic liquid crystals (Fig. 2.7.15), it may be difficult to define a vacancy. Nevertheless, the average mass in a unit cell can change and lead to a breakdown of Eq. (6.4.38). In most crystalline solids at low temperature, the concentration of vacancies is very low (tending exponentially to zero with temperature in equilibrium) and does not change significantly in response to strain, and then Eq. (6.4.38) is a very good approximation. Furthermore, vacancy diffusion is a very slow process (see Chapter 7), and it may take geological times for the vacancy density to relax to its equilibrium value in the presence of strain. Thus, it is often the case that Eq. (6.4.38) is satisfied almost exactly over the lifetime of a laboratory experiment.

The Helmholtz free energy density f in a solid can be expressed as a function of the average density $n = n_0 + \delta n$ and the strain u_{ij}. This free energy must, of course, transform like a scalar under rotations. The lowest order scalar coupling density changes δn to the strain is $\delta n u_{ii}$, which for linearized strain is $\delta n \nabla \cdot \mathbf{u}$.

Thus, f can be expanded about a reference density n_0 as

$$f(n_0 + \delta n, u_{ij}) = f_0 + \mu_0 \delta n + \frac{1}{2} A(\delta n/n_0)^2 + \frac{1}{2} K^n_{ijkl} u_{ij} u_{kl} + D u_{ii} (\delta n/n_0) , \qquad (6.4.41)$$

where $\mu_0 = \partial f/\partial n|_{n=n_0}$ is the chemical potential at density n_0 and zero strain, A is the inverse isothermal compressibility at constant strain, K^n_{ijkl} is the elastic constant tensor at constant density, and D is given by $D = n_0 \partial^2 f/\partial u_{ii} \partial n|_{n=n_0}$. At low temperatures, vacancies and interstitials are suppressed, and $\delta(n\Omega_0) = \delta(1 - n_v \Omega_0)$ is small. It is, therefore, useful to express the free energy in Eq. (6.4.41) as a function of u_{ij} and $\delta(n\Omega_0)/(n_0\Omega_0) = u_{ii} + (\delta n/n_0)$ rather than as a function of u_{ij} and $\delta n/n_0$:

$$f = f_0 + \mu_0 \delta n + \frac{1}{2} A[(\delta n/n_0) + u_{ii}]^2 + (D - A) u_{ii} [(\delta n/n_0) + u_{ii}] + \frac{1}{2} K_{ijkl} u_{ij} u_{kl}, \qquad (6.4.42)$$

where

$$K_{ijkl} = K^n_{ijkl} + (A - 2D)\delta_{ij}\delta_{kl} \qquad (6.4.43)$$

is the elastic constant tensor at constant $n\Omega_0$. As the temperature approaches zero, the constant strain inverse compressibility A diverges as $e^{U/T}$, where U is the energy of an interstitial or vacancy and $D - A$ and K_{ijkl} are non-divergent (see Problem 6.8). The components of the elastic constant tensor K_{ijkl} will be of the order of the binding energy divided by (lattice spacing)^d.

8 Bond-angle order and rotational and translational elasticity

In our discussion of the elasticity of smectic liquid crystals in Sec. 6.3, we considered a free energy describing both layer translations and director rotations [Eq. (6.3.11)]. The elastic free energy for a crystalline solid depends only on the translational elastic variable **u** and then only on its symmetrized spatial derivative. A crystal also breaks rotational symmetry, and one should understand why rotational elastic variables do not appear in its elastic free energy. A free energy, analogous to Eq. (6.3.11), involving both rotational and translational variables, can be derived for a crystal. First imagine a bond-angle ordered state in which rotational symmetry is broken in two orthogonal directions. Such a state would be created, for example, by the development of hexatic bond-angle order in the plane perpendicular to the Frank director of a nematic liquid crystal. More generally, a bond-angle ordered state would be invariant under some set of point group operations such as the symmetry operations of a cube or of an icosahedron. Bond-angle phases invariant under these operations would be called, respectively, "cubatics" or "icosahedratics". Rotations of such bond-angle ordered states are described by an angle variable $\delta\boldsymbol{\Omega}$ (for example $\delta\mathbf{n} = \mathbf{n}_0 \times \delta\boldsymbol{\Omega}$ in a nematic) with an associated elastic free energy density

$$f_\Omega = \frac{1}{2} K_{ijkl}^\Omega \nabla_i \delta\Omega_j \nabla_k \delta\Omega_l . \tag{6.4.44}$$

A crystalline solid has both bond-angle and translational order. Translational invariance tells us that the free energy depends on $\nabla_i u_j$ and not on $\mathbf{u}$. Rotational invariance tells us that there is no energy change if both $\mathbf{u}$ and bond direction rotate simultaneously. Since, under rotation, $\delta u_i = \epsilon_{ijk}\delta\Omega_j x_k$, the combination $\nabla_i u_j - \epsilon_{ijk}\delta\Omega_k$ does not change under a uniform rotation. Thus, the free energy of a crystalline solid can be written as

$$f = \frac{1}{2} B_{ijkl}(\nabla_i u_j - \epsilon_{ijm}\delta\Omega_m)(\nabla_k u_l - \epsilon_{kln}\delta\Omega_n) + f_\Omega . \tag{6.4.45}$$

Minimization of this energy over $\delta\Omega$ yields $\delta\Omega_i = \epsilon_{ijk}\nabla_i u_k(1 + O(\nabla^2))$ and eliminates the antisymmetric part of $\nabla_i u_j$ from the free energy. Then B_{ijkl} properly symmetrized is the elastic constant tensor K_{ijkl}. The energy f_Ω is proportional to $(\nabla^2 u)^2$ and is subdominant compared with the leading elastic energy.

9 *Elastic constants from density functional theory*

The free energy density in Eqs. (6.4.41) and (6.4.42) can be calculated using the density functional approaches outlined in Sec. 4.7. As discussed there, the Helmholtz free energy F is a functional of the spatially varying density $n(\mathbf{x})$. In a solid, $n(\mathbf{x})$ can be expanded in the Fourier series of Eq. (4.7.1). The free energy density f then becomes a function of the average density n, the mass density amplitudes $n_{\mathbf{G}}$, and the size and shape of the periodically repeated unit cell as measured by the primitive translation vectors $\mathbf{b}_\alpha$ of the reciprocal lattice. In equilibrium, at chemical potential μ_0, $\partial f/\partial n = \mu_0$, $\partial f/\partial n_{\mathbf{G}} = 0$ and $\partial f/\partial b_{\alpha i} = 0$, and $n = n_0$, $n_{\mathbf{G}} = n_{\mathbf{G}}^0$ and $\mathbf{b}_\alpha = \mathbf{b}_\alpha^0$. Changes $\delta\mathbf{b}_\alpha = \mathbf{b}_\alpha - \mathbf{b}_\alpha^0$ are described by lattice strains as

$$\delta b_{\alpha i} = -b_{\alpha j}\nabla_i u_j , \tag{6.4.46}$$

as can be seen by differentiating the top part of Eq. (6.4.4) with respect to x_i. f can now be expanded about the equilibrium state described above in powers of $\nabla_i u_j$, δn and $\delta n_{\mathbf{G}} = n_{\mathbf{G}} - n_{\mathbf{G}}^0$. The invariance of f with respect to rigid rotations of the vectors $\mathbf{b}_\alpha$ ensures that this expansion will depend only on the symmetric strain u_{ij} and that the leading coupling of δn to the strain will be $\delta n u_{ii}$. Finally, there will be terms of quadratic and higher order in $\delta n_{\mathbf{G}}$ and couplings of $\delta n_{\mathbf{G}}$ to δn and u_{ij} of the form $n_{-\mathbf{G}}^0 \delta n_{\mathbf{G}} \delta n$, $n_{-\mathbf{G}}^0 \delta n_{\mathbf{G}} u_{ii}$ and $G_i G_j n_{-\mathbf{G}}^0 \delta n_{\mathbf{G}} u_{ij}$. The linear coupling between $\delta n_{\mathbf{G}}$ and δn and u_{ij} imply that changes in the latter two quantities will lead to changes in $\delta n_{\mathbf{G}}$. This means that, in general, the distribution of mass inside a unit cell will change in response to both changes in density and strain. The final free energy density of Eq. (6.4.41) is the free energy as a function of δn and u_{ij}, with $\delta n_{\mathbf{G}}$ evaluated at the equilibrium or relaxed value of $\delta n_{\mathbf{G}}$ in the presence of δn and u_{ij}.

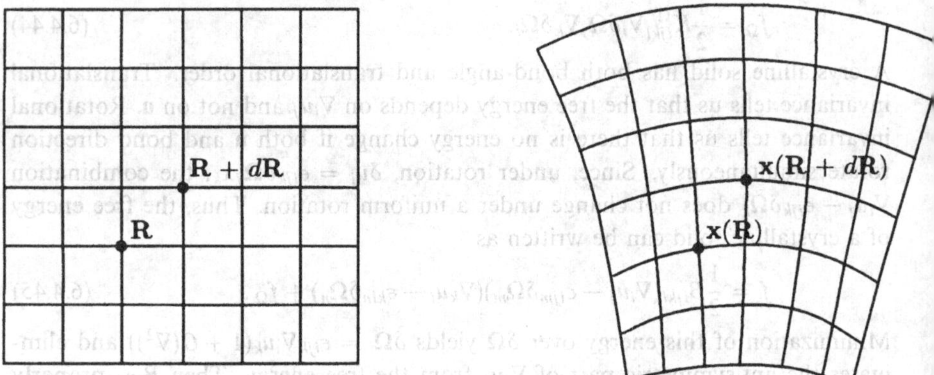

Fig. 6.5.1. An undistorted (at left) and a distorted (at right) elastic medium showing the initial points **R** and **R** $+ d$**R** and their images **x**(**R**) and **x**(**R** $+ d$**R**) after distortion.

6.5 Lagrangian elasticity

The theory of elasticity is quite old. It was essentially fully developed by the middle of the nineteenth century, before there was any definite concept of the periodic nature of ideal solids. The theory views a solid as a continuum of mass points that can be distorted in response to external stress but that will return to its initial at-rest configuration when stresses are removed. Each mass point in an elastic body can be indexed in its unstressed configuration by its position **R** with respect to a coordinate fixed in space, as shown in Fig. 6.5.1. Under stress, the body will distort like a piece of rubber, and the mass point that was initially at **R** will be at a new position,

$$\mathbf{x}(\mathbf{R}) = \mathbf{R} + \mathbf{u}(\mathbf{R}), \tag{6.5.1}$$

relative to the fixed coordinate system. Note here the double meaning assigned to **R**: it is both an initial position and a label for a mass point that does not lose its identity under distortion. To be more precise, we could have introduced a continuous parameter such as time t and specified the position $\mathbf{x}(\mathbf{R}, t)$ of the mass point **R** as a function of t with an initial condition that $\mathbf{x}(\mathbf{R}, 0) = \mathbf{R}$. The initial and final positions of two mass points **R** and $\mathbf{R} + d\mathbf{R}$ are shown in Fig. 6.5.1.

1 Classical theory of elasticity

An elastic energy is introduced by arguing that the ideal separation between nearby points is the distance $dR = (d\mathbf{R} \cdot d\mathbf{R})^{1/2}$ of the rest state. Their separation $dx = (d\mathbf{x} \cdot d\mathbf{x})^{1/2}$ in the distorted state with $d\mathbf{x} = \mathbf{x}(\mathbf{R} + d\mathbf{R}) - \mathbf{x}(\mathbf{R})$ will differ from dR and lead to an increase in energy just as stretching or compressing a

Hooke's law spring will lead to an increase in its energy. Therefore, the energy of the distorted body should depend on the parameters appearing in $(dx)^2 - (dR)^2$. We now have a choice as to which variables to choose as the fundamental ones. We can either choose to express everything in terms of the mass point index $\mathbf{R}$ or in terms of the position of $\mathbf{x}$ in the fixed coordinate system. Because the identity of the mass points does not change, the relation between $\mathbf{x}$ and $\mathbf{R}$ in Eq. (6.5.1) is unique and invertible, and either representation is acceptable. The first choice, called *Lagrangian* coordinates, yields

$$dx_i = dR_i + \frac{\partial u_i}{\partial R_j} dR_j \tag{6.5.2}$$

and

$$(dx)^2 - (dR)^2 = 2u_{ij}^L dR_i dR_j, \tag{6.5.3}$$

with

$$u_{ij}^L = \frac{1}{2}\left(\frac{\partial u_i}{\partial R_j} + \frac{\partial u_j}{\partial R_i} + \frac{\partial u_k}{\partial R_i}\frac{\partial u_k}{\partial R_j}\right) \tag{6.5.4}$$

the Lagrangian strain tensor. (This is the tensor with a positive sign for the non-linear term that appears in many text books on elasticity including Landau and Lifshitz (1970).) The second choice, called *Eulerian* coordinates, yields

$$\mathbf{R} = \mathbf{x} - \mathbf{u}[\mathbf{R}(\mathbf{x})], \quad \text{or} \quad dR_i = dx_i - \frac{\partial u_i}{\partial x_j} dx_j \tag{6.5.5}$$

and

$$(dx)^2 - (dR)^2 = 2u_{ij}^E dx_i dx_j, \tag{6.5.6}$$

with

$$u_{ij}^E = \frac{1}{2}\left(\frac{\partial u_i}{\partial x_j} + \frac{\partial u_j}{\partial x_i} - \frac{\partial u_k}{\partial x_i}\frac{\partial u_k}{\partial x_j}\right) \tag{6.5.7}$$

the Eulerian strain tensor.

The harmonic elastic energy [Eq. (6.4.11)] has the same form when expressed either as a function of the Lagrangian or as a function of the Eulerian strain. The integral appearing in Eq. (6.4.11) is over the mass point index $\mathbf{R}$ (or equivalently over the volume of the undistorted body) in Lagrangian coordinates and over the space occupied by the distorted body in Eulerian coordinates. In comparing the free energies in the two pictures in detail, it is important to remember that the relation between the volume elements d^dR and d^dx involves the strain via the Jacobian of the transformations in Eqs. (6.5.2) and (6.5.4).

Thus, the elastic energy of the classical Lagrangian theory and that presented in the preceding section are the same. It remains to establish the connection between the two definitions of strain. If the positions $\mathbf{R}$ of the sites in the unstrained body form an ideal periodic lattice, they satisfy

$$\mathbf{G}\cdot\mathbf{x} = \mathbf{G}\cdot\mathbf{R} = 2\pi m, \tag{6.5.8}$$

where m is an integer. In the strained lattice, when $\mathbf{u}$ is nonzero, the positions $\mathbf{x}$ of the lattice sites satisfy

$$\mathbf{G}\cdot(\mathbf{x}-\mathbf{u}(\mathbf{x}))=2\pi m\,. \tag{6.5.9}$$

This equation defines planes of constant phase in the distorted lattice. Thus,

$$\mathbf{R}=\mathbf{x}-\mathbf{u}(\mathbf{x}) \tag{6.5.10}$$

determines the positions of the atoms that were at **R** prior to distortion. This equation is identical to Eq. (6.5.5). Note, however, that the natural variables are the positions **x** of the atoms and not the indexes **R**. The coordinates of the previous sections are, therefore, the Eulerian coordinates. This explains the minus sign in the non-linear strain field for smectic liquid crystals introduced in Eq. (6.3.19). Similarly, the non-linear strain field that would guarantee total rotational invariance of the solid elastic energy is the Lagrangian strain of Eq. (6.5.4).

Having established that the classical theory of elasticity and that presented here are formally identical, it is necessary to emphasize the totally different philosophies used in their respective derivations. The elasticity discussed in Sec. 6.4.1 required the existence of mass density waves that break the translational symmetry of space. It applies whenever there is an equilibrium periodic mass density wave, even if there is substantial mass rearrangement under strain. The displacement variables **u** and associated strains u_{ij} are the generalizations of the angle variable θ and its gradient $\nabla\theta$ describing the broken rotational symmetry of the xy-model. The classical theory, on the other hand, applies to any medium for which there is a unique invertible map between mass points in stretched and initial unstretched configurations. It views the solid as a continuum limit of masses connected by fixed springs that do not break under stretching. It therefore provides a correct description for materials, like rubber, which are made up of randomly crosslinked polymers that do not break under stress. It also provides a good description for glasses over time scales short compared to the often very long times required for atoms to rearrange into new low-energy configurations that do not reduce to the initial configuration after removal of external forces.

2 Elasticity of classical harmonic lattices

As we have discussed many times, the equilibrium state of any system is one which minimizes the free energy. In classical systems at zero temperature, minimizing the free energy is equivalent to minimizing the potential energy. Let $U(\mathbf{X}_\mathbf{l})$ be the potential energy of a collection of atoms with positions $\mathbf{X}_\mathbf{l}$. This potential can be expanded in deviations $\mathbf{u}_\mathbf{l}=\mathbf{X}_\mathbf{l}-\mathbf{R}_\mathbf{l}$ from a set of reference sites $\mathbf{R}_\mathbf{l}$ according to

$$U=U(\mathbf{R}_\mathbf{l})+\sum_{\mathbf{l},i}\frac{\partial U}{\partial R_{\mathbf{l},i}}u_{\mathbf{l},i}+\frac{1}{2}\sum_{\mathbf{l},i,\mathbf{l}',k}\frac{\partial^2 U}{\partial R_{\mathbf{l},i}\partial R_{\mathbf{l}',k}}u_{\mathbf{l},i}u_{\mathbf{l}',j}+\cdots, \tag{6.5.11}$$

where $\partial U/\partial R_{\mathbf{l},i}\equiv\partial U/\partial u_{\mathbf{l},i}\,|_0$. If the positions $\mathbf{R}_\mathbf{l}$ correspond to equilibrium positions, the term linear in $\mathbf{u}_\mathbf{l}$ in this expression is zero, and to harmonic order we have

$$U_{\text{har}} = U - U(\mathbf{R}_\mathbf{l}) = \frac{1}{2} \sum_{\mathbf{l},i,\mathbf{l}',k} C_{ik}(\mathbf{R}_\mathbf{l} - \mathbf{R}_{\mathbf{l}'}) u_{\mathbf{l},i} u_{\mathbf{l}',k}, \tag{6.5.12}$$

where

$$C_{ik}(\mathbf{R}_\mathbf{l} - \mathbf{R}_{\mathbf{l}'}) = \frac{\partial^2 U}{\partial R_{\mathbf{l},i} \partial R_{\mathbf{l}',k}}. \tag{6.5.13}$$

$C_{ik}(\mathbf{R}_\mathbf{l} - \mathbf{R}_{\mathbf{l}'})$ obeys certain symmetry relations. Because it is defined as the second derivative of a potential,

$$C_{ik}(\mathbf{R}_\mathbf{l} - \mathbf{R}_{\mathbf{l}'}) = C_{ki}(\mathbf{R}_{\mathbf{l}'} - \mathbf{R}_\mathbf{l}). \tag{6.5.14}$$

If the equilibrium lattice is a Bravais lattice with one atom per unit cell, then

$$C_{ik}(\mathbf{R}_\mathbf{l}) = C_{ik}(-\mathbf{R}_\mathbf{l}) = C_{ki}(\mathbf{R}_\mathbf{l}) \tag{6.5.15}$$

because every Bravais lattice has a center of inversion. When there is more than one atom per unit cell, the position vectors $\mathbf{R}_\mathbf{l}$ and the above relation have to be modified to include the additional atoms. The invariance of the potential energy with respect to uniform translations and rotations of all particles leads to additional constraints on $C_{ik}(\mathbf{R}_\mathbf{l})$. U does not change if all particles are displaced by $\mathbf{u}$, and

$$U(\mathbf{R}_\mathbf{l} + \mathbf{u}) - U(\mathbf{R}_\mathbf{l}) = \frac{1}{2} N \sum_{\mathbf{l},i,k} C_{ik}(\mathbf{R}_\mathbf{l}) u_i u_k = 0, \tag{6.5.16}$$

where N is the number of particles. Since this relation applies for any $\mathbf{u}$,

$$\sum_{\mathbf{l}} C_{ik}(\mathbf{R}_\mathbf{l}) = 0 \tag{6.5.17}$$

for every ik, and $C_{ik}(\mathbf{q} = 0) = 0$. This relation allows the harmonic potential energy to be rewritten as

$$U_{\text{har}} = -\frac{1}{4} \sum_{\mathbf{l},i,\mathbf{l}',k} C_{ik}(\mathbf{R}_\mathbf{l} - \mathbf{R}_{\mathbf{l}'})(u_{\mathbf{l},i} - u_{\mathbf{l}',i})(u_{\mathbf{l},k} - u_{\mathbf{l}',k}). \tag{6.5.18}$$

In the continuum limit, this reduces to the elastic energy of Eq. (6.4.11), with

$$\begin{aligned} K_{ijkl} \quad = \quad & -\frac{1}{8\Omega_0} \sum_{\mathbf{l}} [R_{\mathbf{l},j} R_{\mathbf{l},l} C_{ik}(\mathbf{R}_\mathbf{l}) + R_{\mathbf{l},i} R_{\mathbf{l},l} C_{jk}(\mathbf{R}_\mathbf{l}) \\ & + R_{\mathbf{l},j} R_{\mathbf{l},k} C_{il}(\mathbf{R}_\mathbf{l}) + R_{\mathbf{l},i} R_{\mathbf{l},k} C_{jl}(\mathbf{R}_\mathbf{l})], \end{aligned} \tag{6.5.19}$$

where we used the fact that Eq. (6.5.18) must be invariant under uniform rotations to produce the correct symmetrized form for K_{ijkl}.

The harmonic free potential can also be expressed in Fourier transform variables:

$$U_{\text{har}} = \frac{1}{2} \int \frac{d^d q}{(2\pi)^d} C_{ik}(\mathbf{q}) u_i(\mathbf{q}) u_k(-\mathbf{q}) \,. \tag{6.5.20}$$

This is identical to Eq. (6.4.19), except that now the integral is over the first Brillouin zone, and deviations from q^2 behavior away from $\mathbf{q} = 0$ are permitted.

6.6 Elasticity of solids: the stress tensor

In this section, we will discuss forces and the stress tensor in elastically deformable solids. We will first consider the stress tensor using the classical Lagrangian description. We will then consider regular periodic crystals in the Eulerian picture when density as well as size and shape of the unit cell can change.

1 *The Lagrangian stress tensor*

As we have seen, the variable **u** describes displacements of mass points. Thus, the field thermodynamically conjugate to **u** is a force. To understand the change in energy resulting from such a force, we need to be able to describe forces in the interior of a solid medium. Recall that, in the Lagrangian picture, all mass points are indexed by their position **R** in an undistorted reference configuration. Mass points interior to some volume element Ω enclosed by a surface $\partial\Omega$ in the undistorted solid are mapped when the solid is distorted to new positions $\mathbf{x}(\mathbf{R})$ interior to a distorted volume element Ω' enclosed by a surface $\partial\Omega'$. The local connectivity of neighboring volume elements does not change under the distortion. Any force **F** exerted on mass points in a volume element Ω can be expressed as an integral over Ω of a force density **f**:

$$\mathbf{F} = \int_{\Omega} d^3R\mathbf{f}. \tag{6.6.1}$$

f is the force per unit volume of the undistorted solid and not per unit volume of the solid after distortion. It differs from the force per unit volume in space by the ratio $(1 + u_{ii})$ of the distorted to undistorted volume elements.

As we saw in Chapter 2, forces between atoms or molecules generally have a finite microscopic range. Even in metals with mobile charges, the Coulomb force is short range because of screening. Thus, in the absence of macroscopic electric or magnetic fields, it is safe to assume that interactions between mass points in a solid are short range. This means that the force **F** on the volume element Ω can only be transmitted to it by nearby mass points, i.e., **F** is an *internal* force transmitted to a volume element Ω through the surface $\partial\Omega$ by mass in surrounding volume elements. This implies that **f** can be expressed as a gradient ($\nabla_i = \partial/\partial R_i$) with respect to the undistorted positions of a stress tensor σ_{ij}:

$$f_i = \nabla_j \sigma_{ij} \, . \tag{6.6.2}$$

The volume integral in Eq. (6.6.1) can then be transformed into an integral over the surface $\partial\Omega$, and the ith component of **F** becomes

$$F_i = \int_{\partial\Omega} dS_j \sigma_{ij}. \tag{6.6.3}$$

σ_{ij} is the *stress tensor*. It is the force per unit area of the undistorted solid in direction i exerted by the surrounding medium on a volume element across its surface oriented in direction j. If the surrounding medium exerts a force along the outward (inward) surface normal, σ_{ij} is positive (negative).

The torque τ on a volume element, like the force, must be transmitted across its surfaces. This requires σ_{ij} to be symmetric under interchange of i and j. To see this, consider the torque on a volume element that has been only slightly distorted so that the actual position $\mathbf{x}(\mathbf{R})$ can be replaced by the position $\mathbf{R}$ in the reference state. Then,

$$\begin{aligned} \tau_i &= \int_\Omega d^3R(\mathbf{R}\times\mathbf{f})_i = \int_\Omega d^3R(\epsilon_{ijk}R_j\nabla_l\sigma_{kl}) \\ &= -\int d^3R[\epsilon_{ijk}\sigma_{kj}] + \int dS_l\epsilon_{ijk}R_j\sigma_{kl}. \end{aligned} \tag{6.6.4}$$

The volume term is zero only if σ_{ij} is symmetric. The surface term giving the torque transmitted across the boundary by neighboring volume elements can be nonzero for symmetric σ_{ij}.

In equilibrium, the internal force on each volume element must be zero. Thus, if there are no volume external forces such as gravity, $\nabla_j\sigma_{ij}$ must be zero, and the stress tensor must be constant throughout the body. Furthermore, the stress tensor must be continuous across any external surface of the solid. Thus, a solid in equilibrium surrounded by an isotropic fluid at pressure p will experience an isotropic inward force per unit area $-p$, and its stress tensor,

$$\sigma_{ij} = -p\delta_{ij}, \tag{6.6.5}$$

is negative and isotropic like that of a fluid. On the other hand, if a bar is subjected to a positive tension with a force per unit area T exerted across opposite surfaces normal to the z-axis, as shown in Fig. 6.6.1, then

$$\sigma_{zz} = T \tag{6.6.6}$$

is positive and all other components of σ_{ij} are zero. If the solid were subjected to a compression $-T$ along the z-axis, $\sigma_{zz} = -T$ would be negative.

When there are external volume forces such as gravity, the total force on a volume element must still be zero in mechanical equilibrium. In this case, the forces arising from internal stresses must balance the external forces. Near the surface of the earth, the force per unit volume exerted by gravity on a volume element with mass density ρ is $\rho\mathbf{g}$, where $\mathbf{g}$ is the acceleration of gravity. The equation of mechanical equilibrium is thus

$$\nabla_j\sigma_{ij} = \rho g_i. \tag{6.6.7}$$

This equation, of course, applies equally well to an isotropic fluid where the stress tensor is simply $-p\delta_{ij}$.

The force density $\mathbf{f}$ measures the internal force exerted on a volume element by neighboring volume elements. The work done by this internal force in displacing volume elements by $\delta\mathbf{u}(\mathbf{x})$ is

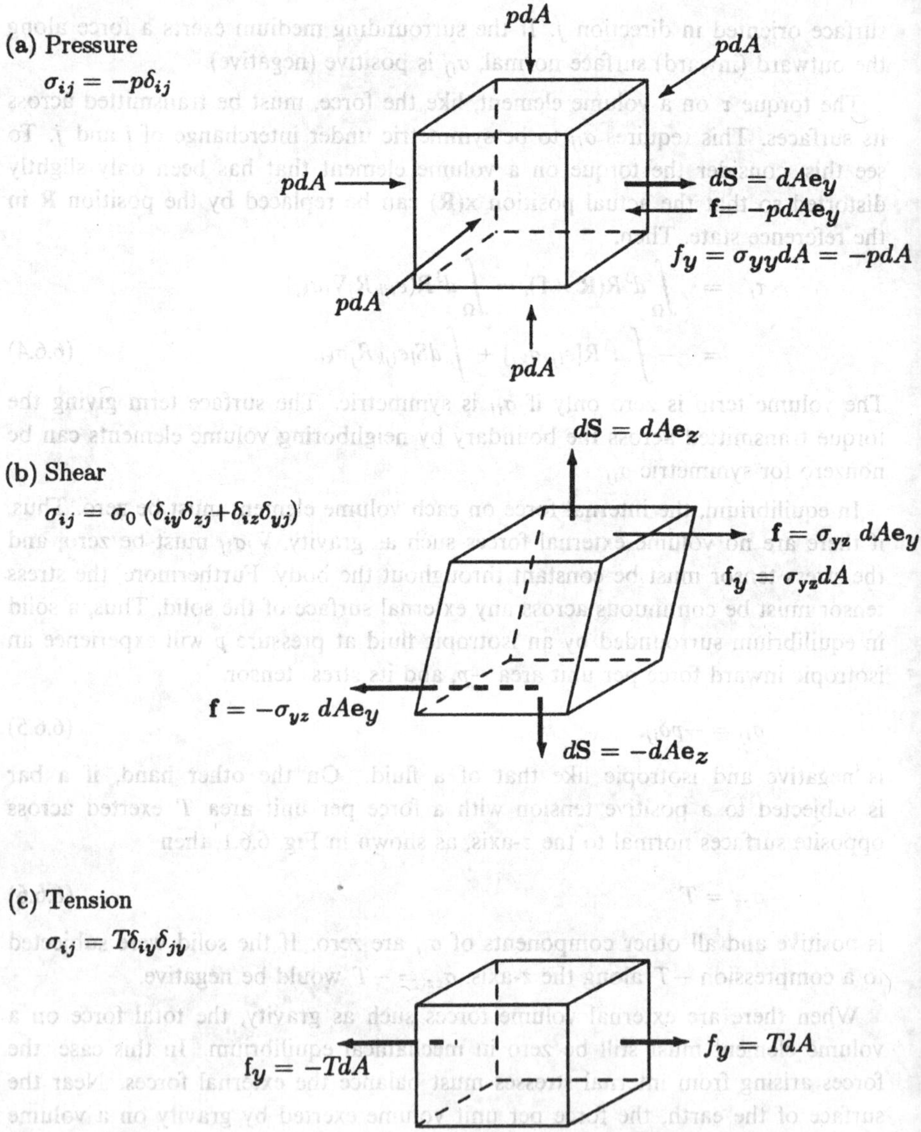

Fig. 6.6.1. The force exerted on a volume element by its surrounding medium is **F**. The force exerted across an infinitesimal surface element $d\mathbf{S}$ is $f_i = \sigma_{ij}dS_j$. f_i and σ_{ij} are positive for a force exerted along the outer normal to the surface. (a) A solid element under isotropic pressure. Forces on all faces are inward. (b) A solid element under shear. Tangential forces on the top and bottom faces are in opposite directions. (c) A solid element under tension. Normal forces on right and left faces are in opposite directions.

$$\begin{aligned}\delta W &= \int d^3R\mathbf{f}\cdot\delta\mathbf{u} = \int d^3R\nabla_j\sigma_{ij}\delta u_i \\ &= -\int d^3R\sigma_{ij}\delta u_{ij},\end{aligned} \tag{6.6.8}$$

where we used the fact that σ_{ij} is symmetric to replace $\nabla_i\delta u_j$ by δu_{ij}. Surface terms arising from the integration by parts have been eliminated by choosing the outer surface to be outside matter where $\mathbf{u}$ is not defined. The change in free energy brought about by changes in the strain field is the negative of the work done by internal forces:

$$dF = -SdT - \delta W = -SdT + \int d^3R\sigma_{ij}du_{ij}. \tag{6.6.9}$$

The stress tensor is, therefore, the derivative of F with respect to u_{ij} at constant temperature:

$$\sigma_{ij} = \left.\frac{\delta F}{\delta u_{ij}}\right)_T = \left.\frac{\partial f^L}{\partial u_{ij}}\right)_T , \tag{6.6.10}$$

where f^L is the Lagrangian elastic free energy density.

2 Stress-strain relations

Eq. (6.6.10) provides a relation between the stress and strain in a solid. In isotropic three-dimensional solids, F^L is given by Eq. (6.4.16) (with $d = 3$), and

$$\sigma_{ij} = B\delta_{ij}u_{kk} + 2\mu\left(u_{ij} - \frac{1}{3}\delta_{ij}u_{kk}\right) , \tag{6.6.11}$$

which is easily inverted to give the strain in terms of the stress. Trivially,

$$\sigma_{kk} = 3Bu_{kk}, \tag{6.6.12}$$

from which we obtain

$$u_{ij} = \frac{1}{9B}\sigma_{kk}\delta_{ij} + \frac{1}{2\mu}\left(\sigma_{ij} - \frac{1}{3}\delta_{ij}\sigma_{kk}\right) . \tag{6.6.13}$$

We can now calculate the strain produced by specified stresses in some simple geometries (see Fig. 6.6.1). If a solid is subjected to isotropic pressure, then

$$u_{ii} = \nabla\cdot\mathbf{u} = -\frac{p}{B} = \frac{\sigma_{kk}}{3B}. \tag{6.6.14}$$

If it is subjected to a uniaxial tension T, then

$$\begin{aligned} u_{zz} &= \frac{1}{3}\left(\frac{1}{3B} + \frac{1}{\mu}\right)T; \\ u_{xx} = u_{yy} &= -\frac{1}{3}\left(\frac{1}{2\mu} - \frac{1}{3B}\right)T. \end{aligned} \tag{6.6.15}$$

Note that the change in length along the direction of tension always has the same sign as T. The length in the perpendicular direction may, however, increase or decrease depending on the sign of $(2\mu)^{-1} - (3B)^{-1}$. In any case, the relative volume change is $\nabla\cdot\mathbf{u} = T/3B$. The dilation along z determines the *Young's modulus* Y:

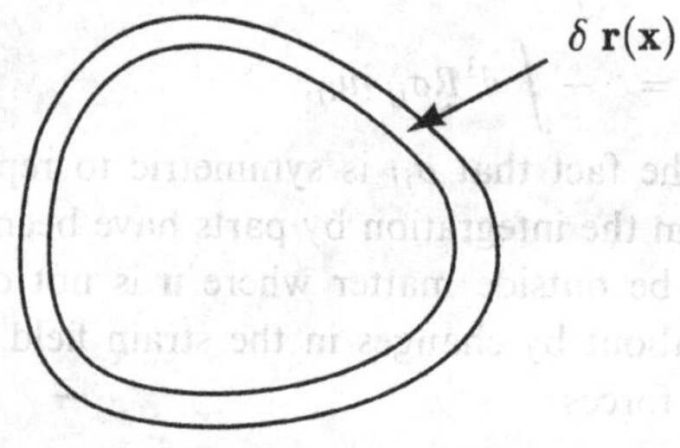

Fig. 6.6.2. The distortion of a small volume element by $\delta\mathbf{r}(\mathbf{x})$ used in the derivation of the Eulerian stress tensor.

$$u_{zz} = \frac{T}{Y}, \quad Y = \frac{9B\mu}{3B+\mu}, \tag{6.6.16}$$

and the strain along the normal direction determines Poisson's ratio σ:

$$u_{xx} = -\sigma u_{zz}\,, \quad \sigma = \frac{1}{2}\frac{(3B-2\mu)}{(3B+\mu)}\,. \tag{6.6.17}$$

If a rod of material is strained along the z-axis, one expects that there will be a contraction in the orthogonal direction so that the Poisson ratio should be positive. Normally this is the case. There is, however, no thermodynamic constraint on the sign of σ. Stability of the solid phase requires only $B > 0$ and $\mu > 0$, or $3B - 2\mu > -2\mu$. It is, in fact, possible to make materials with negative σ. In Sec. 10.4, we will investigate two-dimensional polymerized membranes that have a negative Poisson ratio. Three-dimensional materials with $\sigma < 0$ also exist.

Young's modulus and Poisson's ratio can easily be calculated for isotropic solids in d dimensions. Of particular interest are their values,

$$Y_2 = \frac{4B\mu}{B+\mu}, \quad \sigma_2 = \frac{B-\mu}{B+\mu}, \tag{6.6.18}$$

in two dimensions. We will have occasion to use these quantities in our discussion of dislocations and dislocation mediated melting in Chapter 9.

3 *The Eulerian stress tensor*

In the Eulerian picture, one considers volumes in real space occupied by matter as opposed to volumes in a reference material. The force $\mathbf{F}$ on a volume element Ω_0 in space is the integral of a force per unit volume $\mathbf{f}$:

$$\mathbf{F} = \int_{\Omega_0} d^3x\mathbf{f}. \tag{6.6.19}$$

Again, the interparticle forces are short range, and $\mathbf{F}$ can be expressed as the gradient of a stress tensor. Thus,

$$F_j = \int_{\partial\Omega_0} dS_j \sigma_{ij}, \tag{6.6.20}$$

where the integral is over the surface $\partial\Omega_0$ in the space of the volume element Ω_0. To calculate the work done by internal stresses in the Eulerian picture, we imagine displacing the boundary $\partial\Omega_0$ of a small volume element by $\delta\mathbf{r}(\mathbf{x})$ as shown, in Fig. 6.6.2. The work done in this displacement is the force exerted across each surface element times the displacement of that element:

$$\delta W = \int_{\partial\Omega_0} dS_j \delta r_i(\mathbf{x}) \sigma_{ij} , \tag{6.6.21}$$

where it is understood that $\mathbf{x}$ is on the surface $\partial\Omega_0$. We can also calculate the change in free energy brought about by the displacements $\delta\mathbf{r}(\mathbf{x})$. In a small volume element, we can assume that each point $\mathbf{x}$ in Ω_0 undergoes a displacement to a new position $\mathbf{x} + \delta\mathbf{r}(\mathbf{x})$, so that the surface displacement is $\delta\mathbf{r}(\mathbf{x})$ when $\mathbf{x}$ is on a surface. We require that the total number of particles in each volume element remains fixed under these deformations. The volume element and the density then change according to

$$\begin{aligned} d^3x &\to [1 + \nabla \cdot \delta\mathbf{r}(\mathbf{x})] d^3x \\ n &\to [1 + \nabla \cdot \delta\mathbf{r}(\mathbf{x})]^{-1} n \sim [1 - \nabla \cdot \delta\mathbf{r}(\mathbf{x})] n. \end{aligned} \tag{6.6.22}$$

In addition, the distance between particles and planes of constant phase changes by $\delta\mathbf{r}(\mathbf{x})$, implying that u_{ij} changes according to

$$u_{ij}(\mathbf{x}) \to u_{ij} + \frac{1}{2}(\nabla_i \delta r_j + \nabla_j \delta r_i). \tag{6.6.23}$$

Thus, the change in free energy of the volume element Ω_0 is

$$\begin{aligned} \delta F_{\Omega_0} &= \int_{\Omega_0} [\nabla \cdot \delta\mathbf{r}(\mathbf{x}) f + \delta f] d^3x \\ &= \int_{\Omega_0} \left[\nabla \cdot \delta\mathbf{r}(\mathbf{x}) f + \frac{\partial f}{\partial n}(-n \nabla \cdot \delta\mathbf{r}(\mathbf{x})) + \frac{\partial f}{\partial u_{ij}} \nabla_i \delta r_j \right] d^3x \\ &= \int_{\partial\Omega_0} dS_j \delta r_i(\mathbf{x}) \left[\delta_{ij} \left(f - n \left.\frac{\partial f}{\partial n}\right)_{u_{ij}} \right) + \left.\frac{\partial f}{\partial u_{ij}}\right)_n \right], \end{aligned} \tag{6.6.24}$$

where f is the free energy per unit volume of space. Comparing Eqs. (6.6.24) and (6.6.21), we obtain

$$\sigma_{ij} = \left[f - n \left.\frac{\partial f}{\partial n}\right)_{u_{ij}} \right] \delta_{ij} + \left.\frac{\partial f}{\partial u_{ij}}\right)_n . \tag{6.6.25}$$

Though it is not immediately apparent, this stress tensor reduces to the Lagrangian tensor of Eq. (6.6.10) when vacancies and interstitials are prohibited and $\delta n/n = -u_{ii}$. The Lagrangian energy density f^L and the Eulerian density f differ by a factor of the ratio of the stretched to unstretched volume elements. If non-linear contributions are ignored, we need not distinguish between Eulerian and Lagrangian strain and $f^L = (1 + u_{ii}) f$. Then, to lowest order in the strain, $\partial f^L / \partial u_{ij} = f \delta_{ij} + \partial f / \partial u_{ij}$. Furthermore, $\partial f / \partial u_{ij})_{\delta n = -n u_{ii}} =$

$\partial f/\partial u_{ij})_n - n\partial f/\partial n)_{u_{ij}}\delta_{ij}$. This establishes the connection between the Eulerian and Lagrangian stress tensors to lowest order in the strain. They can be shown to be equivalent in general.

The first term in Eq. (6.6.25) is identical to the formula [Eq. (3.1.38)] for the pressure of an isotropic fluid. The second term is the field h_{ij} thermodynamically conjugate to the strain and is not present in isotropic fluids. In equilibrium,

$$\left.\frac{\partial f}{\partial u_{ij}}\right)_n = h_{ij},$$
$$\left.\frac{\partial f}{\partial n}\right)_{u_{ij}} = \mu. \tag{6.6.26}$$

Thus, the stress tensor can change at constant temperature in response to changes in both μ and h_{ij}. The two fields are physically different. h_{ij} couples to the separation between layers in a periodic solid, whereas μ couples to changes in the density. In normal situations, the equilibrium state of a solid is established in the absence of forces favoring a particular layer spacing. In this case, h_{ij} is zero. The equilibrium stress tensor σ_{ij}^0 in the solid is then simply $-p\delta_{ij}$, where p is the pressure of its surrounding medium (e.g. a coexisting liquid or gas phase). Changes in σ_{ij} are brought about by changes in temperature, chemical potential, and h_{ij}:

$$\delta\sigma_{ij} = -(s dT + n d\mu)\delta_{ij} + dh_{ij}\ . \tag{6.6.27}$$

This is a generalization to solids of Eq. (3.1.40) for fluids. Note both Eqs. (6.6.27) and (6.6.25) imply that isotropic contributions to σ_{ij} arise from changes in T, μ and the isotropic part of h_{ij}, but that anisotropic contributions to σ_{ij} arise only from h_{ij}

Eqs. (6.6.26) and (6.4.41) imply that $\delta\mu$ and δh_{ij} induce changes in density and strain via

$$K^n_{ijkl}u_{kl} + D\frac{\delta n}{n_0}\delta_{ij}\lambda = \delta h_{ij},$$
$$A\frac{\delta n}{n_0^2} + \frac{D}{n_0}u_{ii} = \delta\mu\ . \tag{6.6.28}$$

The relation between stress and strain and density at constant temperature is thus

$$\begin{aligned}\delta\sigma_{ij} &= [(D-A)(\delta n/n_0) - Du_{ii}]\delta_{ij} + K^n_{ijkl}u_{kl}\\ &= (D-A)[(\delta n/n_0) + u_{ii}] + K_{ijkl}u_{kl},\end{aligned} \tag{6.6.29}$$

where K_{ijkl} is the elastic constant tensor at constant $n\Omega_0$ introduced in Eq. (6.4.43). In some situations, either h_{ij} or $\delta\mu$ is zero. For example, if a piece of iron is bent or stretched in air, there is essentially no change in chemical potential, and $A(\delta n/n_0) = -Du_{ii}$. The stress-strain relation then becomes

$$\begin{aligned}\delta\sigma_{ij} &= [K^n_{ijkl} - (D^2/A)\delta_{ij}\delta_{kl}]u_{kl}\\ &= K^\mu_{ijkl}u_{kl},\end{aligned} \tag{6.6.30}$$

where $K^{\mu}_{ijkl} = K^{n}_{ijkl} - (D^2/A)\delta_{ij}\delta_{kl}$ is the elastic constant tensor at constant chemical potential. Alternatively, one might increase the chemical potential without altering h_{ij} by increasing the pressure in a fluid surrounding and coexisting with the solid. In this case, in an isotropic solid with K^{n}_{ijkl} of the form of Eq. (6.4.13), $B^n u_{ii} = -D(\delta n/n_0)$, and

$$\delta\sigma_{ij} = [A - (D^2/B^n)](\delta n/n_0)\delta_{ij} \equiv B^h u_{kk}\delta_{ij}, \tag{6.6.31}$$

where B^n and B^h are, respectively, the bulk modulus at constant density and constant field h. This relation is somewhat more complicated for anisotropic solids. The inverse compressibility A is nonzero in the fluid as well as the solid phase, whereas both D and B^n are zero in the fluid. The constant field bulk modulus B^h approaches the liquid inverse compressibility as the solid-liquid transition is approached. Finally, we recall that, in crystalline solids, A and D diverge exponentially, whereas $D - A$ approaches a constant as $T \to 0$. In this case, $K^{\mu}_{ijkl} = K_{ijkl} - [(A - 2D) - D^2/A]\delta_{ij}\delta_{kl}$ [Eq. (6.6.30)] and K_{ijkl} [Eq. (6.4.43)] approaches well-defined values as $T \to 0$.

6.7 The nonlinear sigma model

In Sec. 6.3, we noted that the elastic free energies of the *xy*- and *n*-vector models differ in that the latter is necessarily anharmonic whereas the former is not. In this section, we will show how these anharmonicities lead to a renormalization of the spin wave stiffness and ultimately to its disappearance altogether at any nonzero temperature in two dimensions (Migdal 1975; Polyakov 1975; Brézin and Zinn-Justin 1976; Nelson and Pelcovits 1977). Thus, there is no phase with quasi-long-range order in a two-dimensional *n*-vector model with $n > 2$. There is, however, long-range order at exactly zero temperature, and, as we shall see, the correlation length and susceptibility, like those of the one-dimensional Ising model, diverge as $\exp(1/T)$ as $T \to 0$. Our approach will be to derive momentum shell renormalization group recursion relations along the lines used in our treatment of the ϵ-expansion in Chapter 5.

Before undertaking a formal derivation and analysis of recursion relations, it is useful to consider why anharmonicities renormalize the spin-wave stiffness. The spin-wave stiffness is determined by the change in free energy in response to boundary conditions imposing an average gradient in the spin direction. In the absence of such twist boundary conditions, there will be thermally excited spin excitations at any nonzero temperature that depress the average local spin from its zero temperature maximum. When the twist boundary conditions are imposed, the distribution of spin excitations will rearrange so as to minimize the free energy. There are more excitations the larger the distance between walls at which the boundary conditions are imposed. Thus one expects length dependent renormalizations such as we encountered in our study of critical phenomena. In the xy-model without vortices, the system is linear and there is no coupling to thermally excited degrees of freedom and no renormalization of the spin-wave stiffness.

To study fluctuations in the n-vector model, we employ a reduced Hamiltonian with a part whose form is identical to the free energy of Eq. (6.2.1) and a part arising from an external field $\mathbf{h}$,

$$\overline{\mathscr{H}} = \frac{1}{2}K \int d^d x (\nabla_i \mathbf{n})^2 - (\mathbf{h}/T) \cdot \int d^d x \mathbf{n}(\mathbf{x}), \tag{6.7.1}$$

where $K = \rho_s/T$ and $|\mathbf{n}(\mathbf{x})|^2 = 1$. In what follows, we will measure temperature in units set by ρ_s, and set $K = 1/T$, where T is now the reduced temperature. In the perfectly ordered ground state, $\mathbf{n}$ is spatially uniform and can be parameterized as $\mathbf{n} = (0, ..., 1)$. Thermal excitations will reduce the component of $\mathbf{n}$ along the direction of order. A useful parameterization of $\mathbf{n}$ at low temperature is thus $\mathbf{n} = (\boldsymbol{\pi}, \sigma)$, where $\boldsymbol{\pi}$ is an $n-1$ component vector describing excitations transverse to the assumed direction of order. If $\mathbf{h}$ is aligned along the σ-direction with magnitude h, then the partition function associated with Eq. (6.7.1) is

$$Z = \int \mathscr{D}\sigma(\mathbf{x}) \int \mathscr{D}\pi(\mathbf{x}) \prod_{\mathbf{x}} \delta[\sigma^2(\mathbf{x}) + |\boldsymbol{\pi}(\mathbf{x})|^2 - 1] e^{-\overline{\mathscr{H}}}. \tag{6.7.2}$$

The vectors $\mathbf{n}(\mathbf{x})$ are initially taken to be on a lattice, but we will eventually take a continuum limit in the manner described in Sec. 5.2. In the ordered phase, we may, to a good approximation, ignore configurations in which the sign of $\sigma(\mathbf{x})$ changes from site to site. If we assume $\sigma(\mathbf{x})$ has the same sign throughout the sample, we can perform the integral over $\sigma(\mathbf{x})$ in Eq. (6.7.2). The integral over the δ function leads to a Jacobian $(1 - |\boldsymbol{\pi}(\mathbf{x})|^2)^{-1/2}$ at each site of the lattice and

$$\begin{aligned} \int \mathscr{D}\sigma(\mathbf{x}) \prod_{\mathbf{x}} \delta(\sigma^2(\mathbf{x}) + \boldsymbol{\pi}^2(\mathbf{x}) - 1) &= \exp\left\{ -\frac{1}{2} \sum_{\mathbf{x}} \ln[1 - \boldsymbol{\pi}^2(\mathbf{x})] \right\} \\ &\to \exp\left(-\frac{1}{2}\rho \int d^d x \ln[1 - \boldsymbol{\pi}^2(\mathbf{x})] \right), \end{aligned} \tag{6.7.3}$$

where the final form corresponds to the continuum limit discussed in Sec. 5.2 [Eq. (5.2.7) in particular] with $\rho = N/V$ the number of degrees of freedom per unit volume. As in our treatment of the momentum shell renormalization group in Sec. 5.8, we replace the lattice Brillouin zone by a spherical zone with the same volume. In this case,

$$\rho = \frac{1}{V} \sum_{\mathbf{q} \in B.Z.} = \int_0^\Lambda \frac{d^d q}{(2\pi)^d} = \frac{K_d \Lambda^d}{d}. \tag{6.7.4}$$

The partition trace over all states with a single sign of $\sigma(\mathbf{x})$ is then

$$Z = \int \mathscr{D}\boldsymbol{\pi}(\mathbf{x}) e^{-\overline{\mathscr{H}}(\boldsymbol{\pi})}, \tag{6.7.5}$$

where

$$\begin{aligned}\overline{\mathscr{H}}(\pi) &= \frac{1}{2T}\int d^dx\left\{(\nabla\pi)^2+[\nabla(1-\pi^2)^{1/2}]^2\right\}-\frac{h}{T}\int d^dx(1-\pi^2)^{1/2}\\ &\quad+\frac{1}{2}\rho\int d^dx\ln(1-\pi^2) \qquad (6.7.6)\\ &\cong \frac{1}{2T}\int d^dx\Big[(\nabla\pi)^2+(\pi\cdot\nabla\pi)^2\\ &\quad -h\left(2-\pi^2-\frac{1}{4}(\pi\cdot\pi)^2\right)-T\rho\pi^2\Big].\end{aligned}$$

The final form of this equation is a power series expansion in which all terms up to order T have been retained. This can be seen by using simple dimensional analysis. Let $\pi=\sqrt{T}\pi'$. Then $\pi^2/T=O(1)$, $\pi^4/T=O(T)$ and $\rho\pi^2=O(T)$.

Eq. (6.7.6) defines a field theory with nonlinear interaction terms that can be analyzed using a momentum shell renormalization group analogous to that discussed in Sec. 5.8. The field $\pi(\mathbf{q})$ is decomposed into low and high wave number parts $\pi^>(\mathbf{q})$ and $\pi^<(\mathbf{q})$, and the integral in the partition trace over $\pi^>(\mathbf{q})$ is carried out. Finally, $\pi^<(\mathbf{q})$ is rescaled via $\pi^<(\mathbf{q})=\zeta\pi'(b\mathbf{q})$. The original Hamiltonian in Eq. (6.7.1) has O_n rotational symmetry broken by an external field $\mathbf{h}$ coupling linearly to $\mathbf{n}$. This rotational symmetry leads to specific ratios between terms in the power series expanded Hamiltonian of Eq. (6.7.6). For example, the ratio of the coefficient of the $(\pi\cdot\nabla\pi)^2$ to the $(\nabla\pi)^2$ is one, and the ratio of the coefficients of the π^4 and π^2 terms multiplying h is $1/4$. If these ratios were different, the Hamiltonian would not have O_n symmetry. Renormalized Hamiltonians must retain the O_n symmetry of the original Hamiltonian, i.e., maintain the same ratios of potentials as the original Hamiltonian. This means that we need only consider recursion relations for the two independent potentials $K=1/T$ and h/T of the original Hamiltonian; rotational invariance will then fix the values of other potentials in terms of these.

We are interested in behavior near zero temperature, and we will treat the $O(T)$ terms, $T\rho\pi^2$, $(\pi\cdot\nabla\pi)^2/T$ and $(h/T)\pi^4$, as perturbations relative to the low-temperature harmonic Hamiltonian:

$$\overline{\mathscr{H}}_0=\frac{1}{2T}\int_{\mathbf{q}}(q^2+h)|\pi(\mathbf{q})|^2 \qquad (6.7.7)$$

with propagator

$$G(\mathbf{q})=\frac{T}{q^2+h}. \qquad (6.7.8)$$

The treatment of the $-\rho\pi^2$ term under removal of $\pi^>$ requires special consideration. Because the coefficient $\rho=\int_{\mathbf{q}}$ is an integral over the entire Brillouin zone, it can be decomposed into a $>$ part and a $<$ part: $\rho=\rho^<+\rho^>\equiv\int_{\mathbf{q}}^<+\int_{\mathbf{q}}^>$. Under removal of $\pi^>$, the second term is treated perturbatively in the diagrams shown in Fig. 6.7.1, leaving the first terms to be treated in subsequent iterations of the renormalization procedure. Diagrams contributing to K and (h/T) are shown in

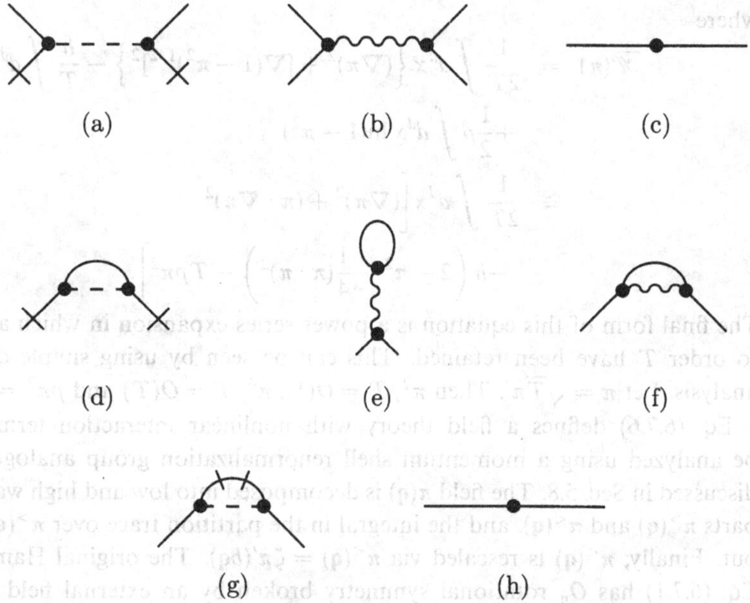

Fig. 6.7.1. (a) Diagrammatic representation of the $(1/2T)(\boldsymbol{\pi}\cdot\nabla\boldsymbol{\pi})^2$ term in $\mathscr{H}(\boldsymbol{\pi})$. The slashes, which appear on two of the four $\boldsymbol{\pi}$ legs, represent the spatial derivative ∇. (b) and (c) Diagrammatic representations of the $(h/8T)\boldsymbol{\pi}^4$ and $-(\rho/2)\boldsymbol{\pi}^2$ terms. (d) Diagram for $(1/T)$. The internal line represents the propagator $G(\mathbf{q})$ [Eq. (6.7.8)], and external legs represent $\boldsymbol{\pi}^<(\mathbf{q})$. (e)-(h) Diagrams for h/T. The weight of (d) is $1/T$, while those of (e) and (f) are, respectively, $(1/8)(h/T)\times 2\times 2(n-1)$ and $(1/8)(h/T)\times 2\times 2^2$. Diagram (g) has weight $(1/T)$ and contributes $(1/T)\int^> q^2G(\mathbf{q}) = \int^> q^2(q^2+h)^{-1} = \rho^> - (h/T)\int^> G(\mathbf{q})$ to (h/T). Diagram (h) contributes $-\rho^>$ to (h/T) and cancels the $\rho^>$ part of diagram (e). Diagrams (d) and diagrams (e)-(h) lead to Eqs. (6.7.9) and (6.7.10) for $(1/T')$ and $(h/T)'$.

Fig. 6.7.1. As explained in the figure caption, they lead to the momentum shell recursion relations

$$(1/T)' = \zeta^2 b^{-(d+2)}\left[\frac{1}{T} + \int^> \frac{1}{q^2+h}\right], \qquad (6.7.9)$$

$$(h/T)' = \zeta^2 b^{-d}\left[(h/T) + \frac{1}{2}(n-1)\int \frac{h}{q^2+h}\right], \qquad (6.7.10)$$

where b is the rescaling factor, the integrals are over the shell $b^{-1} < |\mathbf{q}| < 1$ for $\Lambda = 1$, and ζ is the field renormalization factor [Eq. (5.8.5)] defined via

$$\boldsymbol{\pi}^<(\mathbf{q}'/b) = \zeta\boldsymbol{\pi}'(\mathbf{q}'). \qquad (6.7.11)$$

This last equation implies $\boldsymbol{\pi}^<(\mathbf{x}) = b^{-d}\zeta\boldsymbol{\pi}(\mathbf{x}/b) \equiv b^{-\omega}\boldsymbol{\pi}(\mathbf{x}/b)$, in agreement with Eq. (5.5.4). The arguments presented in Sec. 5.5 show that the scaling of the field (h/T) is determined entirely by the scaling of $\mathbf{n}(\mathbf{x})$. Because of the O_n symmetry

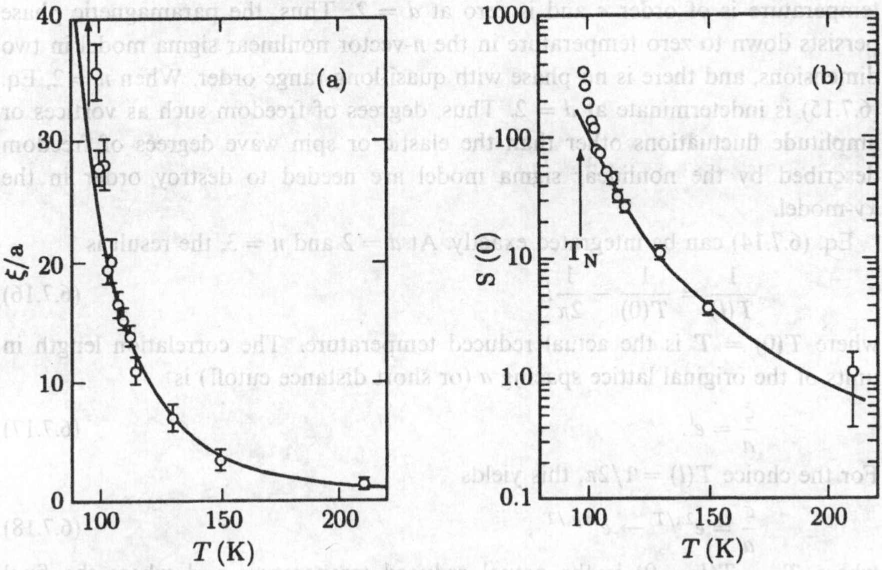

Fig. 6.7.2. (a) The correlation length in K_2NiF_4 showing experimental data points and the theoretical curve from Eq. (6.7.19) (solid line). (b) The $\mathbf{q} = 0$ structure factor with experimental data compared with the theoretical curve of Eq. (6.7.20). [R.J. Birgeneau, *Phys. Rev. B* **41**, 2514 (1990).]

of the Hamiltonian [Eq. (6.7.1)] with $\mathbf{h} = 0$, all components of $\mathbf{h}$ must scale in the same way. We can therefore determine the scaling of $\mathbf{h}$ by looking at its components along $\boldsymbol{\pi}$ rather than along σ. Invariance of the external Hamiltonian under rescaling requires $(h/T)\int d^dx\boldsymbol{\pi}(\mathbf{x}) = (h'/T')\int d^dx'\boldsymbol{\pi}'(\mathbf{x}')$, where $\mathbf{x}' = \mathbf{x}/b$, and hence

$$h'/T' = \zeta h/T. \tag{6.7.12}$$

Eqs. (6.7.12) and (6.7.10) then lead to

$$\zeta = b^d\left(1 - \frac{1}{2}(n-1)\int^{>}\frac{T}{q^2+h}\right). \tag{6.7.13}$$

Using Eq. (6.7.13) in Eq. (6.7.9) and setting $b = \exp(-\delta l)$, we obtain

$$\frac{dT(l)}{dl} = -\epsilon T(l) + \frac{n-2}{2\pi}T^2(l) \tag{6.7.14}$$

at $h = 0$ near two dimensions, where $\epsilon = (d-2)$.

This equation can be analyzed in exactly the same way as the Migdal- Kadanoff recursion relations for the Ising magnet. If $\epsilon > 0$ and $n - 2 > 0$, there is a fixed point at

$$T^* = [2\pi/(n-2)]\epsilon. \tag{6.7.15}$$

This is the transition temperature separating the low-temperature ferromagnetic state from the high-temperature paramagnetic state. For $n > 2$, this transition temperature is of order ϵ and is zero at $d = 2$. Thus, the paramagnetic phase persists down to zero temperature in the n-vector nonlinear sigma model in two dimensions, and there is no phase with quasi-long-range order. When $n = 2$, Eq. (6.7.15) is indeterminate at $d = 2$. Thus, degrees of freedom such as vortices or amplitude fluctuations other than the elastic or spin wave degrees of freedom described by the nonlinear sigma model are needed to destroy order in the xy-model.

Eq. (6.7.14) can be integrated exactly. At $d = 2$ and $n = 3$, the result is

$$\frac{1}{T(l)} = \frac{1}{T(0)} - \frac{1}{2\pi}l, \tag{6.7.16}$$

where $T(0) = T$ is the actual reduced temperature. The correlation length in units of the original lattice spacing a (or short distance cutoff) is

$$\frac{\xi}{a} = e^{l}. \tag{6.7.17}$$

For the choice $T(l) = 1/2\pi$, this yields

$$\frac{\xi}{a} = e^{2\pi/T} \rightarrow e^{2\pi\rho_s/T}, \tag{6.7.18}$$

where $T = T(l = 0)$ is the actual reduced temperature and where the final expression is in real rather than reduced units. Thus, the correlation length diverges faster than any power of T as the zero temperature critical point is approached.

Eq. (6.7.18) describes the dominant temperature dependence of the correlation length near zero temperature. It needs to be corrected by higher order terms in the classical renormalization group recursion relations and by the effects of quantum fluctuations. These effects have been calculated for antiferromagnets (Chakravarty *et al.* 1989). In addition, the spin susceptibility (or the magnetic structure function at $\mathbf{q} = 0$) can be calculated. The results are

$$\xi/a = C_\xi \frac{e^{2\pi\rho_s/T}}{1 + (T/2\pi\rho_s)} \tag{6.7.19}$$

$$S(0) = C_s \frac{(T/2\pi\rho_s)^2 e^{4\pi\rho_s/T}}{[1 + (T/2\pi\rho_s)]^4}, \tag{6.7.20}$$

where C_ξ and C_s are constants. The spin stiffness can be calculated as a series in $1/S$, where S is the spin:

$$2\pi\rho_s = 2\pi J S^2 (1 + 0.158/2S)^2 (1 - 0.552/2S), \tag{6.7.21}$$

where J is the nearest neighbor exchange. The constant C_ξ is estimated to be of order 0.5 for $S = 1/2$ and 0.17 for $S = 1$.

In the antiferromagnet K_2NiF_4, there are well-separated planes of spins that behave as two-dimensional magnets over a wide range of temperature. In addition, the lattice anisotropy in this material is very small. Fig. 6.7.2 shows the correlation length and $q = 0$ magnetic structure factor as determined by neutron scattering. The fit to the zero free parameter theoretical predictions of Eqs. (6.7.19) and (6.7.20) is remarkable.

Bibliography

P.G. de Gennes and J. Prost, *The Physics of Liquid Crystals*, 2nd edn, (Clarendon Press, Oxford, 1993).

L.P. Landau and I.M. Lifshitz, *Electrodynamics of Continuous Media* (Addison-Wesley, Reading, Mass., 1960).

L.P. Landau and I.M. Lifshitz, *Theory of Elasticity* (Pergamon Press, New York, 1970).

References

J. Als-Nielsen, J.D. Litster, R.J. Birgeneau, M. Kaplan, C.R. Safinya, A. Lindegaard-Andersen, and S. Mathiesen, *Phys. Rev. B* **22**, 312 (1980).

V.L. Berezinskii, *Ah. Eksp. Teor. Fiz.* **59**, 907 (1970) [*Sov. Phys. JETP* **32**, 493 (1971)]; *Ah. Eksp. Teor. Fiz* **61**, 1144 (1971) [*Sov. Phys. JETP* **34**, 610 (1972)].

R.J. Birgeneau, *Phys. Rev. B* **41**, 2514 (1990).

E. Brézin and J. Zinn-Justin, *Phys. Rev. Lett.* **36**, 691 (1976).

S. Charkravarty, B.I. Halperin, and David R. Nelson, *Phys. Rev. B* **39**, 2344 (1989).

Ming Cheng, John T. Ho, S.W. Hui, and R. Pindak, *Phys. Rev. Lett.* **61**, 550 (1988).

F.C. Frank, *Disc. Faraday Soc.* **25**, 19 (1958).

P.G. de Gennes, *Sol. St. Commun.* **10**, 753 (1972).

G. Grinstein and R. Pelcovits, *Phys. Rev. Lett.* **47**, 856 (1981).

B.I. Halperin, T.C. Lubensky, and Shang-keng Ma, *Phys. Rev. Lett.* **47**, 1469 (1974).

P.A. Heiney, P.W. Stephens, R.J. Birgeneau, P.M. Horn, and D.E. Moncton, *Phys. Rev. B* **28**, 6416 (1983).

B.D. Josephson, *Phys. Lett.* **21**, 608 (1966).

J.M. Kosterlitz and D.J. Thouless, *J. Phys. C* **6**, 1181 (1973).

N.D. Mermin, *Phys. Rev.* **176**, 250 (1968); *Phys. Rev. B* **20**, 4762(E) (1979).

N.D. Mermin, *J. Math Phys.* **8**, 1061 (1976).

N.D. Mermin and H. Wagner, *Phys. Rev. Lett.* **17**, 1133 (1966).

A.A. Migdal, *Ah. Eksp. Teor. Fiz.* **69**, 810 (1975).

David R. Nelson and Robert A. Pelcovits, *Phys. Rev. B* **16**, 2191 (1977).

A.M. Polyakov, *Phys. Lett.* **59B**, 79 (1975).

C.R. Safinya, D. Roux, G.S. Smith, S.K. Sinha, P. Dimon, N.A. Clark, and A.M. Belloq, *Phys. Rev. Lett.* **57**, 2718 (1986). See also D. Roux and C.R. Safinya, *J. Phys. (Paris)* **49**, 307 (1988).

R. Spal, Thesis (University of Pennsylvania, 1980).

R. Spal, C.F. Chen, T. Egami, P.F. Nigrey, and A.J. Heeger, *Phys. Rev. B* **21**, 3110 (1980).

B.E. Warren, *Phys. Rev.* **59**, 693 (1941).

Problems

6.1 Consider an anisotropic *xy*-magnet on a *d*-dimensional hypercubic lattice with Hamiltonian

$$\begin{aligned}\mathscr{H} &= -J\sum_{<\mathbf{l},\mathbf{l}'>}\mathbf{S}_{\mathbf{l}}\cdot\mathbf{S}_{\mathbf{l}'} - J_2\sum_{\mathbf{r},i}\hat{\mathbf{e}}_z\cdot(\mathbf{S}_{\mathbf{r},i}\times\mathbf{S}_{\mathbf{r},i+1})\\ &= -J\sum_{<\mathbf{l},\mathbf{l}'>}\cos(\phi_{\mathbf{l}}-\phi_{\mathbf{l}'}) - J_2\sum_{\mathbf{r},i}\sin(\phi_{\mathbf{r},i}-\phi_{\mathbf{r},i+1}),\end{aligned}$$

where $\mathbf{S}$ is a three-dimensional spin vector and where $\mathbf{l}=(\mathbf{r},i)$ is a d-dimensional lattice vector and $\mathbf{r}$ a $(d-1)$-dimensional lattice vector. Determine the classical ground state of this Hamiltonian. Identify the invariances of the ground state and find the elastic Hamiltonian describing low-energy, long-wavelength excitations from this ground state. Discuss the nature of order or lack thereof for $d = 1$, 2, and 3.

6.2 Fluctuations destroy $U(1)$ order in infinite two-dimensional systems. This means that the pair-correlation function $g(\mathbf{x}, \mathbf{x}')$ of an infinite two-dimensional hexatic should be a homogeneous and isotropic function of $\mathbf{y} = \mathbf{x} - \mathbf{x}'$ and that the scattering intensity $I(\mathbf{q})$ should depend only on $q = |\mathbf{q}|$ and not on the angle between $\mathbf{q}$ and some axis. In a finite sample of length L, however, the Debye Waller factor e^{-2W} is nonzero. Thus, the scattering intensity from a finite region of a hexatic illuminated for a finite period of time will show the same sort of angular modulation as seen in three-dimensional hexatics. Let $\psi_{6p} = \langle e^{6ip\vartheta} \rangle$ and show that

$$\psi_{6p} = (\psi_6)^{p^2}$$

for finite two-dimensional samples. Fig. 6P.1 shows a photographic image of electrons scattered from a free-standing hexatic film. The parameters ψ_{6p} for this intensity pattern obey the above scaling relation.

6.3 Calculate the energy associated with the Freedericksz transition to lowest order in θ_0 and show that the transition is continuous.

6.4 Calculate the critical magnetic field as a function of L for the Freedericksz transition in the geometry shown in Fig. 6.2.5b.

6.5 Derive Eq. (6.2.18) for the critical voltage of the twisted nematic display.

6.6 Determine the elastic energy for the hexagonal columnar liquid crystal phase shown in Fig. 2.7.11. Recall that this is a phase that has crystalline order in two dimensions and fluid "order" in one dimension.

6.7 (a) Develop an elastic theory for the low energy, long wavelength distortions of the cholesteric state in liquid crystals. (Hint - This state, like the smectic state, has a layered structure.) What are the effects of fluctuations on the three-dimensional cholesteric structure? You should provide estimates of any relevant parameters and lengths.

(b) (For the ambitious) Use the Frank free energy,

$$F = \frac{1}{2}\int d^3x\{K_1(\nabla\cdot\mathbf{n})^2 + K_2[\mathbf{n}\cdot(\nabla\times\mathbf{n})]^2 + K_3[\mathbf{n}\times(\nabla\times\mathbf{n})]^2\} - h\int d^3x\mathbf{n}\cdot(\nabla\times\mathbf{n}),$$

for a cholesteric to derive the results of part (a).

6.8 Determine the elastic energy for a negative dielectric anisotropy smectic-A liquid crystal in an external electric field normal to the equilibrium director $\mathbf{n}$, as shown in Fig. 6P.2. The additional term in the free energy due to the interaction with the electric field is

$$F_{\text{ext}} = -\frac{1}{2}\epsilon_a\int d^3x(\mathbf{n}\cdot\mathbf{E})^2$$

with $\epsilon_a < 0$. Find a suitable generalization of this elastic energy to arbitrary

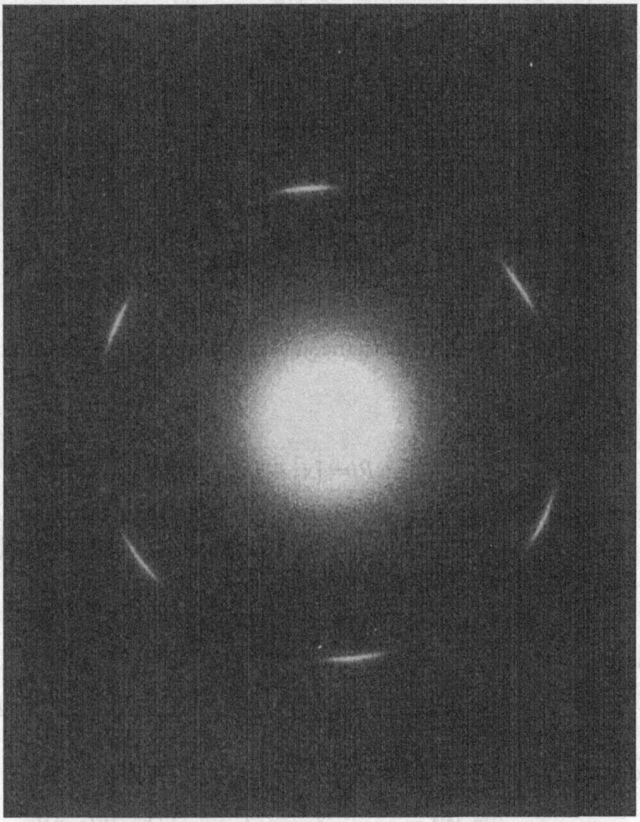

Fig. 6P.1. Intensity profile for electrons scattered from a finite region of a free standing hexatic film. [Ming Cheng, John T. Ho, S.W. Hui, and R. Pindak, *Phys. Rev. Lett.* **61**, 550 (1988).]

dimension and use it to calculate the lower critical dimension d_L for the negative anisotropy smectic in a field.

6.9 This problem will lead you through a derivation of Eqs. (6.3.14) to (6.3.17). Assume that the integral in Eq. (6.3.14) is over a cylindrical Brillouin zone $-\Lambda_\parallel \le q_\parallel \le \Lambda_\parallel$, $0 < q_\perp < \Lambda_\perp$. By introducing the variable $y = \lambda q_\perp^2$, show that

$$g(\mathbf{x}) = \frac{q_0^2 T}{B\lambda}\frac{1}{4\pi^2}\int_0^{\Lambda_\parallel} dq_\parallel \int_0^{\Lambda_\perp^2 \lambda} dy \frac{1 - J_0((yx_\perp^2/\lambda)^{1/2})\cos q_\parallel x_\parallel}{q_\parallel^2 + y^2},$$

where $\lambda = (K_1/B)^{1/2}$. Next, define R and an angle α via

$$x_\parallel = R\cos\alpha, \qquad x_\perp^2/\lambda = R\sin\alpha \tag{6P.1}$$

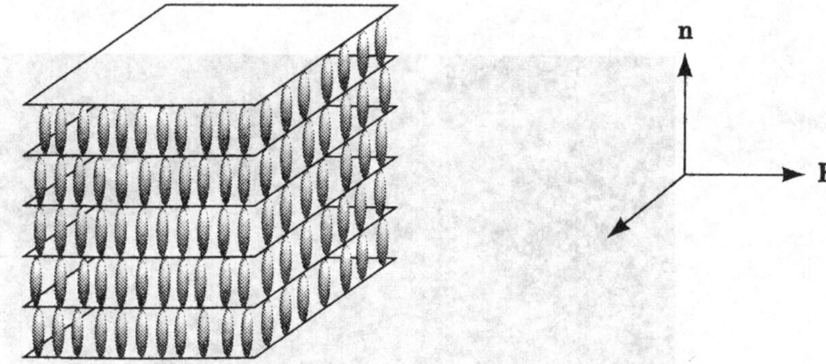

Fig. 6P.2. Negative dielectric anisotropy smectic in an external electric field.

so that

$$R = [x_{\parallel}^2 + (x_{\perp}^4/\lambda^2)]^{1/2}.$$

Then, by breaking the integral over the rectangular domain $\Gamma = [0 \le q_{\parallel} < \Lambda_{\parallel}, 0 \le y < \Lambda_{\perp}^2 \lambda]$ (assume $\Lambda_{\parallel} < \Lambda_{\perp}^2 \lambda$) into two parts $\Gamma = S \bigcup C$, as shown in Fig. 6P.3, show that

$$g(\mathbf{x}) = g_1(\mathbf{x}) + g_2(\mathbf{x}),$$

where

$$g_2(\mathbf{x}) = \frac{q_0^2 T}{(BK_1)^{1/2}} \frac{1}{4\pi^2} \int_C dq_{\parallel} dy \frac{1 - J_0((yR\sin\alpha)^{1/2})\cos(q_{\parallel} R\cos\alpha)}{q_{\parallel}^2 + y^2}$$

and

$$g_1(\mathbf{x}) = \frac{1}{8\pi} \frac{q_0^2 T}{(BK_1)^{1/2}} \int_0^{\Lambda_{\parallel} R} \frac{d\eta}{\eta} [1 - F(\alpha, \eta)],$$

where

$$F(\alpha, \eta) = \frac{2}{\pi} \int_0^{\pi/2} d\phi J_0(\eta \sin\phi \sin\alpha)^{1/2} \cos(\eta \cos\phi \cos\alpha).$$

Use these results to show that when $R \to \infty$,

$$g(\mathbf{x}) = \frac{1}{8\pi} \frac{q_0^2 T}{(BK_1)^{1/2}} \ln(\tilde{\Lambda} R),$$

where

$$\tilde{\Lambda} = \Lambda_{\parallel} e^{D_1 + D_2}.$$

Also

$$D_1 = \int_0^1 \frac{d\eta}{\eta} [1 - F(\alpha, \eta)] - \int_1^{\infty} \frac{d\eta}{\eta} F(\alpha, \eta),$$

and

$$D_2 = \frac{2}{\pi} \int_C \frac{dq_{\parallel} dy}{q_{\parallel}^2 + y^2}.$$

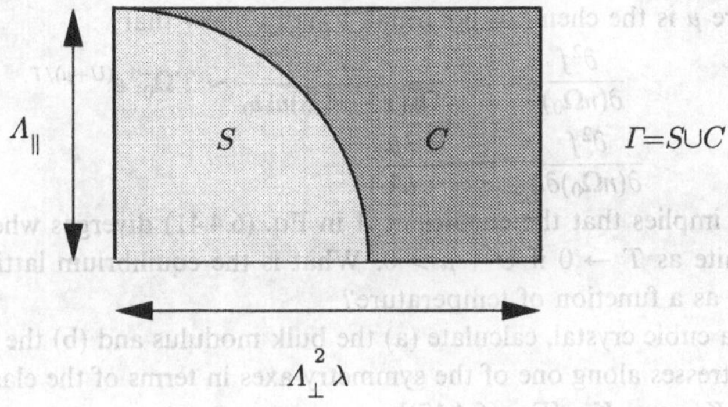

Fig. 6P.3. Domain of integration Γ and the subdomains S and C for Problem 6.9.

Finally, use the above asymptotic form for $g(\mathbf{x})$, which reproduces Eq. (6.3.14) to derive Eq. (6.3.17).

6.10 Calculate the elastic constant tensor K_{ijkl}, the compression modulus A, and the density-strain modulus D [Eq. (6.4.41)] for a solid at its melting point using the phenomenological model for the liquid-solid transition discussed in Sec. 4.7. Use only the octagonal vectors generating the reciprocal lattice of the BCC structure. Assume that $c > 0$ and that k_0 is a function of the density n that can be expanded around the reference density n_0 according to $k_0(n) = k_0 + g\delta n$.

6.11 Consider a model of a low-temperature cubic solid consisting of N_s unit cells with lattice parameter a. Each unit cell can either be unoccupied or occupied by a single atom. Let $n_v(\mathbf{x})$ equal one if lattice site $\mathbf{x}$ is unoccupied (i.e., has a vacancy) and equal zero if site $\mathbf{x}$ is occupied. The energy as a function of a and $n_v(\mathbf{x})$ is

$$E = N_s \epsilon(a) + \sum_{\mathbf{x}} n_v(\mathbf{x}) U(a),$$

where N_s is the number of lattice sites and $\epsilon(a) = (K/2)[(a - a_0)/a_0]^2$ is the energy per site of the lattice without vacancies. Show that the free energy per unit volume for this model is

$$\begin{aligned} f(T, a, n) &= \frac{1}{a^3}[\epsilon(a) + U(1 - n\Omega_0) \\ &\quad + T(1 - n\Omega_0)\ln(1 - n\Omega_0) + Tn\Omega_0 \ln n\Omega_0], \end{aligned}$$

where $\Omega_0 = a^3$ is the volume of a unit cell and n is the particle density. Then show that

$$n\Omega_0 = \left[1 + e^{-(U+\mu)/T}\right]^{-1},$$

where μ is the chemical potential. Finally, show that

$$\frac{\partial^2 f}{\partial(n\Omega_0)^2} = \frac{T}{\Omega_0(1-n\Omega_0)n\Omega_0} \sim T\Omega_0^{-1}e^{(U+\mu)/T}$$
$$\frac{\partial^2 f}{\partial(n\Omega_0)\partial a} = -\frac{3\mu}{a^4}.$$

This implies that the coefficient A in Eq. (6.4.41) diverges whereas $(D-A)$ is finite as $T \to 0$ if $U+\mu > 0$. What is the equilibrium lattice parameter $a(T)$ as a function of temperature?

6.12 For a cubic crystal, calculate (a) the bulk modulus and (b) the Poisson ratio for stresses along one of the symmetry axes in terms of the elastic constants K_{11}, K_{12} and K_{44} [Eq. (6.4.17)].

6.13 (a) Consider a square lattice of mass points connected by central force harmonic springs that act only between nearest neighbors. The potential energy is

$$U = \frac{1}{2}K\sum_{\mathbf{l},\boldsymbol{\delta}}[|\mathbf{X_l}-\mathbf{X_{l+\delta}}|-a]^2,$$

where $\boldsymbol{\delta} = (\pm a, 0)$, $(0, \pm a)$ are the nearest neighbor sites on a square lattice. Calculate the harmonic elasticity matrix $C_{ik}(\mathbf{q})$ and elastic constant tensor K_{ijkl} for this model. What is the response of this model to a shear stress?

(b) Now add central force springs connecting next nearest neighbors so that there is an additional term,

$$U' = \frac{1}{2}K'\sum_{\mathbf{l},\delta_2}[|\mathbf{X_l}-\mathbf{X}_{\mathbf{l}+\delta_2}|-\sqrt{2}a]^2,$$

where $\delta_2 = (\pm a, \pm a)$ is the potential energy. Calculate $C_{ik}(\mathbf{q})$ and K_{ijkl} for this model. Is the response to a shear stress different from that in part (a). Why?

7

Dynamics: correlation and response

Much of what we observe in nature is either time- or frequency-dependent. In this chapter, we will introduce language to describe time- and frequency-dependent phenomena in condensed matter systems near thermal equilibrium. We will focus on dynamic correlations and on linear response to time-dependent external fields that are described by time-dependent generalizations of correlation functions and susceptibilities introduced in Chapters 2 and 3. These functions, whose definitions are detailed in Sec. 7.1, contain information about the nature of dynamical modes. To understand how and why, we will consider linear response in damped harmonic oscillators in Secs. 7.2 and 7.3, and in systems whose dynamics are controlled by diffusion in Sec. 7.4. These examples show that complex poles in a complex, frequency-dependent response function determine the frequency and damping of system modes. Furthermore, the imaginary part of this response function is a measure of the rate of dissipation of energy of external forces.

A knowledge of phenomenological equations of motion in the presence of external forces is sufficient to determine dynamical response functions. The calculation of dynamical correlation functions in dissipative systems requires either a detailed treatment of many degrees of freedom or some phenomenological model for how thermal equilibrium is approached. In Sec. 7.5, we follow the latter approach and introduce Langevin theory, in which thermal equilibrium is maintained by interactions with random forces with well prescribed statistical properties. Frequency-dependent correlation functions for a diffusing particle and a damped harmonic oscillator are proportional to the imaginary part of a response function. This is the classical version of the very important *fluctuation-dissipation theorem*.

Having discussed correlation and response in simple, phenomenological models, we turn in Sec. 7.6 to a general formal treatment of response and correlation functions. This treatment is valid at all temperatures for both classical and quantum systems, and includes a discussion of symmetry and sum rules and a derivation of the general fluctuation-dissipation theorem. Finally, in Sec. 7.7, we will show how inelastic scattering of neutrons measures dynamic correlation functions.

7.1 Dynamic correlation and response functions

1 Correlation functions

The time dependence of both classical and quantum mechanical dynamical variables is governed by equations of motion determined by a Hamiltonian $\mathcal{H}$. A quantum mechanical operator (or field) $\phi_i(\mathbf{x}, t)$ evolves in time in the Heisenberg representation according to

$$\phi_i(\mathbf{x}, t) = e^{i\mathcal{H}t/\hbar}\phi_i(\mathbf{x}, 0)e^{-i\mathcal{H}t/\hbar}. \tag{7.1.1}$$

We will often be interested in the frequency rather than time dependence of operators, and it is useful to introduce the temporal Fourier transforms,

$$\begin{aligned} \phi_i(\mathbf{x}, t) &= \int_{-\infty}^{\infty} \frac{d\omega}{2\pi} e^{-i\omega t}\phi_i(\mathbf{x}, \omega), \\ \phi_i(\mathbf{x}, \omega) &= \int_{-\infty}^{\infty} dt e^{i\omega t}\phi_i(\mathbf{x}, t). \end{aligned} \tag{7.1.2}$$

We will frequently study time-dependent correlations of variables such as the position $\mathbf{x}^\alpha(t)$ of particle α or simple functions of such variables such as the density,

$$n(\mathbf{x}, t) = \sum_\alpha \delta(\mathbf{x} - \mathbf{x}^\alpha(t)). \tag{7.1.3}$$

Here, both $n(\mathbf{x}, t)$ and $\mathbf{x}^\alpha(t)$ evolve according to Eq. (7.1.1). Classically, operators such as $\mathbf{x}^\alpha(t)$ evolve according to Newton's equations.

Time-dependent correlation functions can be introduced in strict analogy with the static correlations introduced in Secs. 3.5 and 3.6. Thus, we define

$$C_{\phi_i\phi_j}(\mathbf{x}, \mathbf{x}', t, t') = \langle\phi_i(\mathbf{x}, t)\phi_j(\mathbf{x}', t')\rangle \tag{7.1.4}$$

and

$$\begin{aligned} S_{\phi_i\phi_j}(\mathbf{x}, \mathbf{x}', t, t') &= \langle(\phi_i(\mathbf{x}, t) - \langle\phi_i(\mathbf{x}, t)\rangle)(\phi_j(\mathbf{x}', t') - \langle\phi_j(\mathbf{x}', t')\rangle)\rangle \\ &\equiv C_{\phi_i\phi_j}(\mathbf{x}, \mathbf{x}', t, t') - \langle\phi_i(\mathbf{x}, t)\rangle\langle\phi_j(\mathbf{x}', t')\rangle, \end{aligned} \tag{7.1.5}$$

where, as for the static case, $\langle\,\rangle$ signifies an average with respect to an equilibrium ensemble. Because the time evolution of the fields $\phi_i(\mathbf{x}, t)$ is governed by the Hamiltonian according to Eq. (7.1.1), there is no ambiguity in the meaning of these averages: for each value of t and t', they are evaluated by tracing over all points in phase space or all quantum states weighted by the appropriate equilibrium weight function. When $t = t'$, these correlation functions reduce to the static correlation functions discussed in Chapter 3:

$$\begin{aligned} C_{\phi_i\phi_j}(\mathbf{x}, \mathbf{x}', t, t) &\equiv C_{\phi_i\phi_j}(\mathbf{x}, \mathbf{x}'), \\ S_{\phi_i\phi_j}(\mathbf{x}, \mathbf{x}', t, t) &\equiv S_{\phi_i\phi_j}(\mathbf{x}, \mathbf{x}'). \end{aligned} \tag{7.1.6}$$

Unless otherwise specified, we will consider only Hamiltonians that are independent of time so that all thermodynamic averages are invariant under time translations. This implies that $\langle\phi_i(\mathbf{x}, t)\rangle \equiv \langle\phi_i(\mathbf{x})\rangle$ is independent of time and that the correlation functions $C_{\phi_i\phi_j}(\mathbf{x},\mathbf{x}',t,t')$ and $S_{\phi_i\phi_j}(\mathbf{x},\mathbf{x}',t,t')$ depend only on the difference $t-t'$ rather than on t and t' individually. Thus, the correlation function of the temporal Fourier transform variables can be written as

$$\langle \phi_i(\mathbf{x},\omega)\phi_j(\mathbf{x}',\omega')\rangle = C_{\phi_i\phi_j}(\mathbf{x},\mathbf{x}',\omega)2\pi\delta(\omega+\omega'), \tag{7.1.7}$$

where

$$C_{\phi_i\phi_j}(\mathbf{x},\mathbf{x}',\omega) = \int_{-\infty}^{\infty} d(t-t')e^{i\omega(t-t')}C_{\phi_i\phi_j}(\mathbf{x},\mathbf{x}',t-t'). \tag{7.1.8}$$

The correlation function $C_{\phi_i\phi_i}(\mathbf{x},\mathbf{x},\omega)$ is often called the *power spectrum* of $\phi_i(\mathbf{x},t)$. Eqs. (7.1.7) and (7.1.8) are generalizations of the *Wiener-Khintchine* theorem relating a power spectrum to the Fourier transform of a time-dependent correlation function. Similarly, we define

$$S_{\phi_i\phi_j}(\mathbf{x},\mathbf{x}',\omega) = \int_{-\infty}^{\infty} d(t-t')e^{i\omega(t-t')}S_{\phi_i\phi_j}(\mathbf{x},\mathbf{x}',t-t'). \tag{7.1.9}$$

Eq. (7.1.5) then implies

$$C_{\phi_i\phi_j}(\mathbf{x},\mathbf{x}',\omega) = S_{\phi_i\phi_j}(\mathbf{x},\mathbf{x}',\omega) + \langle\phi_i(\mathbf{x})\rangle\langle\phi_j(\mathbf{x}')\rangle 2\pi\delta(\omega), \tag{7.1.10}$$

indicating that the $\langle\phi_i(\mathbf{x})\rangle\langle\phi_j(\mathbf{x}')\rangle$ contributes only to the zero-frequency or static part of $C_{\phi_i\phi_j}(\mathbf{x},\mathbf{x}',\omega)$.

2 *Response functions*

Just as the static susceptibilities $\chi_{\phi_i\phi_j}(\mathbf{x},\mathbf{x}')$ relate changes $\delta\langle\phi_i(\mathbf{x})\rangle$ in averages of fields to changes in external fields $\delta h_j(\mathbf{x}')$ conjugate to $\phi_j(\mathbf{x})$, the dynamic response function $\tilde{\chi}_{\phi_i\phi_j}(\mathbf{x},\mathbf{x}',t,t')$ relates changes $\delta\langle\phi_i(\mathbf{x},t)\rangle$ in averages of time-dependent fields to time-dependent changes $\delta h_j(\mathbf{x}',t')$ in external fields:

$$\delta\langle\phi_i(\mathbf{x},t)\rangle = \int d^dx'dt'\tilde{\chi}_{\phi_i\phi_j}(\mathbf{x},\mathbf{x}',t,t')\delta h_j(\mathbf{x}',t'). \tag{7.1.11}$$

It is important to recognize the difference between the temporal and spatial variables in this equation. Disturbances at $\mathbf{x}'$ can lead to changes in $\langle\phi_i(\mathbf{x},t)\rangle$ at all points $\mathbf{x}$. Disturbances at time t' can lead to changes in $\langle\phi_i(\mathbf{x},t)\rangle$ only for times t *later* than t', i.e., the response of $\langle\phi_i(\mathbf{x},t)\rangle$ to $h_j(\mathbf{x}',t')$ is *causal*. This means that the response function $\tilde{\chi}_{\phi_i\phi_j}(\mathbf{x},\mathbf{x}',t,t')$ can be nonzero only for $t>t'$. It is very useful to incorporate this step-function dependence on time into the definition of the response function by writing

$$\tilde{\chi}_{\phi_i\phi_j}(\mathbf{x},\mathbf{x}',t,t') = 2i\eta(t-t')\tilde{\chi}''_{\phi_i\phi_j}(\mathbf{x},\mathbf{x}',t,t'), \tag{7.1.12}$$

where

$$\eta(t-t') = \begin{cases} 1 & \text{if } t>t'; \\ 0 & \text{if } t<t' \end{cases} \tag{7.1.13}$$

is the Heaviside unit step function. The factor of $2i$ ($i=\sqrt{-1}$) is at this stage arbitrary, but it will make comparisons with our later more formal development more straightforward. Eq. (7.1.12) can be viewed as a definition of $\tilde{\chi}''_{\phi_i\phi_j}(\mathbf{x},\mathbf{x}',t,t')$, which is pure imaginary if ϕ_i and ϕ_j are both real. Time translational invariance again implies that $\tilde{\chi}(\mathbf{x},\mathbf{x}',t,t')$ and $\tilde{\chi}''_{\phi_i\phi_j}(\mathbf{x},\mathbf{x}',t,t')$ depend only on $t-t'$.

We will now discuss some of the analytic properties of the response function and its Fourier transform with respect to time. In order to keep notation compact, we will consider the response of a single position independent field $\phi(t)$ to its conjugate external field $h(t)$. In this case, we have

$$\langle\delta\phi(t)\rangle = \int_{-\infty}^{\infty} dt'\tilde{\chi}(t-t')\delta h(t'), \tag{7.1.14}$$

where $\tilde{\chi}(t) = 2i\eta(t)\tilde{\chi}''(t)$. Both $\langle\phi(t)\rangle$ and $h(t)$ are real so that $\tilde{\chi}''(t)$ is pure imaginary. We will be interested in response as a function of frequency rather than time. We therefore need to calculate the temporal Fourier transform of $\tilde{\chi}(t)$. Because of the causal step-function prefactor in $\tilde{\chi}(t)$, it is useful to introduce the Laplace transform as a function of complex frequency z:

$$\chi(z) = \int_{-\infty}^{\infty} e^{izt}\tilde{\chi}(t)dt = \int_0^{\infty} e^{izt}\tilde{\chi}(t)dt. \tag{7.1.15}$$

The function $\tilde{\chi}''(t)$ is bounded as $t\to\infty$ because a disturbance at time $t=0$ will only produce a finite change in $\phi(t)$ at later times. Thus, because t is positive in the above integral, $\chi(z)$ is *analytic in the upper half z-plane* (Imz > 0). The function $\tilde{\chi}''(t,t') = \tilde{\chi}''(t-t')$ is bounded, and we can define its Fourier transform with respect to a real frequency variable,

$$\begin{aligned}\tilde{\chi}''(t) &= \int_{-\infty}^{\infty}\frac{d\omega}{2\pi}e^{-i\omega t}\chi''(\omega)\\ \chi''(\omega) &= \int_{-\infty}^{\infty} dte^{i\omega t}\tilde{\chi}''(t).\end{aligned} \tag{7.1.16}$$

If $\tilde{\chi}''(t)$ approaches a constant as $t\to\infty$, then $\chi''(\omega)$ will have *delta*-function parts. Quite general arguments to be discussed in Sec. 7.6 show that $\tilde{\chi}''(t) = -\tilde{\chi}''(-t)$. This, along with the fact that $\tilde{\chi}''(t)$ is pure imaginary, implies that $\chi''(\omega)$ is *real* and *odd* in ω. Eqs. (7.1.12), (7.1.15) and (7.1.16) imply

$$\begin{aligned}\chi(z) &= \int_0^{\infty} dte^{izt}2i\int_{-\infty}^{\infty}\frac{d\omega}{2\pi}e^{-i\omega t}\chi''(\omega)\\ &= \int_{-\infty}^{\infty}\frac{d\omega}{\pi}\frac{\chi''(\omega)}{\omega-z}\end{aligned} \tag{7.1.17}$$

for z in the upper half plane. This representation of $\chi(z)$ shows clearly that it only has singularities on the real axis and is, therefore, analytic in the upper half plane. The time-dependent response function $\tilde{\chi}(t)$ is the inverse Laplace transform of $\chi(z)$, which in the present case is an integral along a contour in the upper half plane:

$$\tilde{\chi}(t) = \int_{-\infty+ic}^{\infty+ic}\frac{dz}{2\pi}e^{-izt}\chi(z), \tag{7.1.18}$$

where c is any real number. This result is most easily derived using Eq. (7.1.17). If $t>0$, the contour $[-\infty+ic,\infty+ic]$ can be closed in the lower half plane, and there is a contribution to the integral at $z=\omega$. If $t<0$, the contour can be closed in the upper half plane where $\chi(z)$ is zero. Thus $\tilde{\chi}(t)$ is zero for $t < 0$ and equal to $2i\tilde{\chi}''(t)$ for $t > 0$.

The response function $\chi(\omega)$ relating $\langle\delta\phi(\omega)\rangle$ to $\delta h(\omega)$ can be obtained by using the Fourier representation,

$$\eta(t) = \lim_{\epsilon\to 0}\int_{-\infty}^{\infty}\frac{d\omega}{2\pi i}e^{i\omega t}\frac{1}{\omega - i\epsilon}, \tag{7.1.19}$$

for the step function. From this and Eq. (7.1.14) we obtain

$$\begin{aligned}\langle\delta\phi(\omega)\rangle &= \int_{-\infty}^{\infty} dte^{i\omega t}\int_{-\infty}^{\infty} dt' 2i\eta(t-t')\tilde{\chi}''(t-t')\delta h(t') \\ &= \chi(\omega)\delta h(\omega),\end{aligned} \tag{7.1.20}$$

with

$$\chi(\omega) \equiv \lim_{\epsilon\to 0}\chi(\omega + i\epsilon), \tag{7.1.21}$$

where $\chi(\omega + i\epsilon)$ is given by Eq. (7.1.17) with $z = \omega + i\epsilon$. Thus the response function $\chi(\omega)$ is the limit as z approaches the real axis of the function $\chi(z)$, which is analytic in the upper half plane. When the frequency of the external perturbation tends to zero, $\chi(\omega)$ must reduce to the static susceptibility:

$$\lim_{\omega\to 0}\chi(\omega) = \int_{-\infty}^{\infty}\frac{d\omega'}{\pi}\frac{\chi''(\omega')}{\omega'} = \frac{\partial\langle\phi\rangle}{\partial h} = \chi. \tag{7.1.22}$$

This is a *sum rule* relating an integral over $\chi''(\omega)$ to a static quantity, the static susceptibility. Because the static quantity is a thermodynamic derivative, this is often called the *thermodynamic* sum rule. It is one of a hierarchy of sum rules which we will discuss in more detail in Sec. 7.6.

The function $\chi(\omega)$, unlike its static limit, has a real part and an imaginary part, as can be seen using the identity

$$\frac{1}{\omega' - \omega - i\epsilon} = \mathscr{P}\frac{1}{\omega' - \omega} + i\pi\delta(\omega - \omega') \tag{7.1.23}$$

($\mathscr{P}$ signifies the principal part) in Eqs. (7.1.17) and (7.1.20). The result is

$$\chi(\omega) = \chi'(\omega) + i\chi''(\omega), \tag{7.1.24}$$

where

$$\chi'(\omega) = \mathscr{P}\int_{-\infty}^{\infty}\frac{d\omega'}{\pi}\frac{\chi''(\omega')}{\omega' - \omega}. \tag{7.1.25}$$

Since $\chi''(\omega)$ is a real function, $\chi'(\omega)$ is also. Thus, $\chi'(\omega)$ and $\chi''(\omega)$ are, respectively, the real and imaginary parts of the complete response function $\chi(\omega)$. Eq. (7.1.25) is a *Kramers-Kronig* relation between the real and imaginary parts of $\chi(\omega)$. There is also a complementary expression relating $\chi''(\omega)$ to $-\chi'(\omega)$. This is most easily derived by using the Cauchy representation for $\chi(z)$:

$$\chi(z) = \oint_{\Gamma}\frac{d\zeta}{2\pi i}\frac{\chi(\zeta)}{\zeta - z}, \tag{7.1.26}$$

where the contour Γ is the semicircle shown in Fig. 7.1.1. This equation follows because $\chi(z)$ is analytic in the upper half plane. As we shall see in Sec. 7.7, $\chi(z)$ tends to zero faster than $1/z$, as $z \to \infty$ in most cases of interest. In this case, the

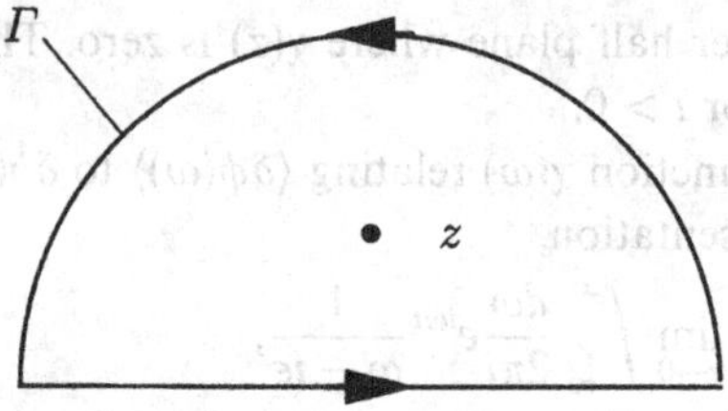

Fig. 7.1.1. Contour in the complex plane for the integral in Eq. (7.1.26).

integral in Eq. (7.1.26) reduces to an integral along a line just above the real axis, i.e., from $-\infty + i\epsilon'$ to $\infty + i\epsilon'$. Then, setting $z = \omega + i\epsilon$ with $\epsilon' < \epsilon$, we obtain

$$\chi(\omega + i\epsilon) = \mathscr{P}\int \frac{d\omega'}{2\pi i}\frac{\chi(\omega' + i\epsilon')}{\omega' - \omega} + \frac{1}{2}\chi(\omega + i\epsilon). \tag{7.1.27}$$

Using Eqs. (7.1.24) and (7.1.25), we obtain

$$\chi''(\omega) = -\mathscr{P}\int \frac{d\omega'}{\pi}\frac{\chi'(\omega')}{\omega' - \omega}. \tag{7.1.28}$$

Because $\chi'(\omega)$ is real and $\chi''(\omega)$ is imaginary, we could also have obtained this result simply by taking the imaginary part of Eq. (7.1.27). The real part of Eq. (7.1.27) yields Eq. (7.1.25). Eqs. (7.1.25) and (7.1.28) are the usual Kramers-Kronig relations. They require slight modification if $\chi(z)$ does not fall off more rapidly than $1/z$ at infinity. Often it is easier to measure $\chi''(\omega)$ (say by an absorption experiment) than $\chi'(\omega)$. If the measurements of $\chi''(\omega)$ are made over a sufficiently large frequency range, the real response can be obtained via Eq. (7.1.25).

The above analysis of the response of a single scalar field applies without change to more general response functions. Thus, the Laplace transform of $\tilde{\chi}_{\phi_i\phi_j}(\mathbf{x}, \mathbf{x}', t, t')$ satisfies

$$\chi_{\phi_i\phi_j}(\mathbf{x}, \mathbf{x}', z) = \int_{-\infty}^{\infty} \frac{d\omega}{\pi}\frac{\chi''_{\phi_i\phi_j}(\mathbf{x}, \mathbf{x}', \omega)}{\omega - z}. \tag{7.1.29}$$

Following Eqs. (7.1.21) and (7.1.24), we have

$$\chi_{\phi_i\phi_j}(\mathbf{x}, \mathbf{x}', \omega) = \chi'_{\phi_i\phi_j}(\mathbf{x}, \mathbf{x}', \omega) + i\chi''_{\phi_i\phi_j}(\mathbf{x}, \mathbf{x}', \omega), \tag{7.1.30}$$

where $\chi'_{\phi_i\phi_j}(\mathbf{x}, \mathbf{x}', \omega)$ is related to $\chi''_{\phi_i\phi_j}(\mathbf{x}, \mathbf{x}', \omega)$ by a Kramers-Kronig relation analogous to Eq. (7.1.25). In Sec. 7.6, we will show that $\chi''_{\phi_i\phi_j}(\mathbf{x}, \mathbf{x}', \omega)$ is real provided ϕ_i and ϕ_j have the same sign under time reversal and there are no external fields or order parameters that break time reversal symmetry. In this case, $\chi'_{\phi_i\phi_j}(\mathbf{x}, \mathbf{x}', \omega)$ is the real part of and $\chi''_{\phi_i\phi_j}(\mathbf{x}, \mathbf{x}', \omega)$ the imaginary part of the complex response function $\chi_{\phi_i\phi_j}(\mathbf{x}, \mathbf{x}', \omega)$. The zero-frequency limit of Eq. (7.1.30) leads to the thermodynamic sum rule,

$$\chi_{\phi_i\phi_j}(\mathbf{x}, \mathbf{x}') = \frac{\delta\langle\phi_i(\mathbf{x})\rangle}{\delta h_j(\mathbf{x}')} = \int_{-\infty}^{\infty} \frac{d\omega}{\pi}\frac{\chi''_{\phi_i\phi_j}(\mathbf{x}, \mathbf{x}', \omega)}{\omega}. \tag{7.1.31}$$

The spatial Fourier transform in the zero-wavenumber limit of this equation gives, as before, the usual static susceptibility.

7.2 The harmonic oscillator

1 The undamped oscillator

The dynamical properties of condensed matter systems are very often dominated by harmonic oscillator-like modes. These modes include sound waves in fluids, elastic waves and phonons in solids, and spin waves in magnets. Information about the frequency and damping of these modes is contained in both dynamical response and correlation functions. In this section, we will explore in detail the response function of a simple damped harmonic oscillator. Its properties generalize directly to any system with well defined modes at finite frequency.

The Hamiltonian for an undamped oscillator of mass m and spring constant k is

$$\mathscr{H} = \frac{p^2}{2m} + \frac{1}{2}kx^2. \tag{7.2.1}$$

The equations of motion for position $x(t)$ and momentum $p(t)$ are calculated by taking their Poisson brackets with the Hamiltonian:

$$\dot{x} \equiv v = \{\mathscr{H}, x\} = \left(\frac{\partial \mathscr{H}}{\partial p}\frac{\partial x}{\partial x} - \frac{\partial \mathscr{H}}{\partial x}\frac{\partial x}{\partial p}\right) = \frac{p}{m}, \tag{7.2.2}$$

$$\dot{p} = \{\mathscr{H}, p\} = -\frac{\partial \mathscr{H}}{\partial x} = -kx. \tag{7.2.3}$$

The mode structure implied by these equations is obtained by assuming that both $x(t)$ and $p(t)$ are proportional to $e^{-i\omega t}$ and solving the resulting characteristic equation

$$\det\begin{bmatrix} -i\omega & -1/m \\ k & -i\omega \end{bmatrix} = -\omega^2 + k/m = 0. \tag{7.2.4}$$

There are *two* solutions to this equation:

$$\omega = \pm\omega_0 \equiv \pm\sqrt{k/m}. \tag{7.2.5}$$

Each of these solutions corresponds to a *mode* of the harmonic oscillator. Note that there is one mode per degree of freedom (x and p). The time dependence of each degree of freedom is governed by a *first-order* differential equation in time. Thus, there is one mode per first-order differential equation in the equations of motion. This property is quite general and will be encountered again in our study of hydrodynamics of conserved and broken symmetry variables.

The variables $x(t)$ and $p(t)$ have opposite signs under the operation of time reversal (i.e., under change in the sign of time t): $x(-t) = +x(t)$, whereas $p(-t) = -p(t)$. The Hamiltonian [Eq. (7.2.1)] and its associated equations of motion [Eqs. (7.2.2) and (7.2.3)] are invariant under time reversal. The equations of motion relate the time derivative of a variable with one sign under time reversal to the variable with the opposite sign. These relations lead to the off-diagonal terms in the characteristic determinant and to *real* and *non-zero* solutions to the characteristic equation. This property is again quite general: modes at non-zero real frequency invariably arise from the coupling of variables with opposite sign under time reversal via first-order differential equations in time.

We have taken the trouble to discuss the undamped oscillator in terms of the first-order differential equations determined by the Poisson bracket relations with the Hamiltonian to point out features of such equations that will generalize to more complicated dynamical problems. The first-order Poisson bracket relations can of course be converted into the second-order differential equation of Newton's second law by substituting Eq. (7.2.2) into Eq. (7.2.3). The result is

$$\ddot{x} + \omega_0^2 x = 0. \tag{7.2.6}$$

This equation, like Eqs. (7.2.2) and (7.2.3), is invariant under time reversal and predicts modes with frequencies $\pm\omega_0$.

2 The damped oscillator

To introduce damping in an intuitive way, we place the particle of mass m into a viscous fluid. In constant motion, it experiences a friction force proportional to its velocity at small velocities. This force can be written as

$$f_{\text{vis}} = -\alpha v, \tag{7.2.7}$$

where α is a friction constant with units of [mass]/[time]. Alternatively, $v = -(1/\alpha)f_{\text{vis}}$, and α^{-1} is a *mobility*. For a sphere of radius a moving in a fluid with shear viscosity η, α is given by Stokes's law

$$\alpha = 6\pi\eta a. \tag{7.2.8}$$

We will discuss the meaning of the shear viscosity in the next chapter. The viscosity η has units of [energy×time]/[volume] (poise) and is of order $n_{\text{fl}}\tau_c T$ in a fluid with number density n_{fl} at temperature T in which the average time between molecular collisions is τ_c. For the moment, both α and η can be regarded as phenomenological parameters. The viscous force law, Eq. (7.2.7), is strictly speaking only valid for a time-independent (i.e., zero frequency) velocity. It must approach zero, as we shall see in Sec. 7.7, at frequencies greater than τ_c^{-1}. For low frequencies or for masses with densities much larger than that of the surrounding fluid (see Problem 7.6), however, it is a very good approximation to the exact force, and we will use it without further apology.

In the presence of a viscous force and an external force f, Newton's equation for a one-dimensional harmonic oscillator becomes

$$\ddot{x} + \omega_0^2 x + \gamma\dot{x} = f/m, \tag{7.2.9}$$

where

$$\gamma = \alpha/m. \tag{7.2.10}$$

The characteristic decay time $\gamma^{-1} = m/(6\pi\eta a)$ is of order $m/(an_{\rm fl}\tau_c T)$. If the average interparticle spacing $d = n_{\rm fl}^{-1/3}$ and the mean free path $v\tau_c = (2T/m_{\rm fl})^{1/2}\tau_c$, where $m_{\rm fl}$ is the mass of a fluid particle, are of the same order, then $\gamma^{-1} \sim (m/m_{\rm fl})(d/a)\tau_c$. Thus, for all but the most microscopic of particles, $m \gg m_{\rm fl}$ and $\gamma^{-1} \gg \tau_c$.

The viscous force breaks time-reversal invariance in Eq. (7.2.9). Any microscopic Hamiltonian and its associated equations of motion must be invariant under time reversal. In the present case, the microscopic Hamiltonian is that describing the harmonic oscillator *and* all of the degrees of freedom of the fluid in which it moves. The viscous force describes the average effect on the harmonic oscillator of interactions with the many incoherent degrees of freedom of the fluid. In general, any energy in the harmonic oscillator will tend to flow irreversibly into the many modes of the fluid. This is reflected in the sign of the viscous force which leads to the decay of $x(t)$ with time. The irreversible flow of energy into incoherent degrees of freedom is called *dissipation*, and $f_{\rm vis}$ is a dissipative force. We will return in Sec. 7.5 to a description of the harmonic oscillator when it is in thermal equilibrium with the fluid so that it receives energy from as well as transmits energy to the fluid.

The mode structure of the damped harmonic oscillator is determined by the equation

$$-\omega^2 + \omega_0^2 - i\gamma\omega = 0 \tag{7.2.11}$$

with solutions

$$\omega = \pm[\omega_0^2 - (\gamma^2/4)]^{1/2} - i\gamma/2 \equiv \pm\omega_1 - i\gamma/2. \tag{7.2.12}$$

If $\omega_0^2 > \gamma^2/4$, ω_1 is real, and solutions for $x(t)$ will oscillate with frequency ω_1 and decay in time with time constant $\tau = 2/\gamma$. If $\omega_0^2 < \gamma^2/4$, ω_1 is imaginary, and there will be no oscillatory component to $x(t)$. In this case, the oscillator is said to be *overdamped*, with inverse decay times

$$\begin{aligned} \tau_f^{-1} &= \frac{1}{2}\gamma[1 + (1 - 4\omega_0^2\gamma^{-2})^{1/2}] \overset{\omega_0 \ll \gamma/2}{\longrightarrow} \gamma, \\ \tau_s^{-1} &= \frac{1}{2}\gamma[1 - (1 - 4\omega_0^2\gamma^{-2})^{1/2}] \overset{\omega_0 \ll \gamma/2}{\longrightarrow} \omega_0^2/\gamma = k/\alpha. \end{aligned} \tag{7.2.13}$$

When $\omega_0^2 \ll \gamma^2/4$, the fast decay time τ_f is much shorter than the slow decay time τ_s. Thus for times long compared to τ_f, the first mode can be neglected. This corresponds in the original equations of motion to neglecting the *inertial* term $m\ddot{x}$. The resulting equation of motion is

$$\alpha\dot{x} = -kx + f. \tag{7.2.14}$$

This approximate equation of motion is often written as

$$\dot{x} = -\frac{k}{\alpha}x + \frac{1}{\alpha}f = -\Gamma\frac{\partial \mathcal{H}_T}{\partial x}, \tag{7.2.15}$$

where $\Gamma = \alpha^{-1}$ and $\mathcal{H}_T = \mathcal{H} - fx$ is the total Hamiltonian including $\mathcal{H}_{\text{ext}} = -fx$. It is very useful in describing the dynamics of systems, such as polymers in solution, dominated by viscous effects.

3 The response function

The frequency-dependent response of x to an external force is easily calculated using Eqs. (7.1.20) and (7.2.9):

$$\chi(\omega) = \frac{x(\omega)}{f(\omega)} = \frac{1}{m}\frac{1}{-\omega^2 + \omega_0^2 - i\omega\gamma}. \tag{7.2.16}$$

The denominator of this equation is precisely the characteristic equation [Eq (7.2.11)] determining the mode structure. Thus, *there are poles in $\chi(\omega)$ at complex mode frequencies of the oscillator.* This result is quite general. A static external force f will lead to an equilibrium displacement of $x = f/k$. This result is correctly described by the zero-frequency limit of Eq. (7.2.16):

$$\lim_{\omega\to 0}\chi(\omega) = \frac{1}{m\omega_0^2} = \frac{1}{k} = \frac{\partial x}{\partial f} = \chi. \tag{7.2.17}$$

At high frequency, $\chi(\omega)$ is negative and falls off as ω^{-2} with a coefficient that depends only on the mass:

$$\lim_{\omega\to\infty}\chi(\omega) = -\frac{1}{m\omega^2}. \tag{7.2.18}$$

We will reconsider this result in Sec. 7.6.

The imaginary part of the response function is

$$\begin{aligned}\chi''(\omega) &= \frac{1}{m}\frac{\omega\gamma}{(\omega^2-\omega_0^2)^2 + (\omega\gamma)^2} \\ &= \frac{1}{2m\omega_1}\left[\frac{\gamma/2}{(\omega-\omega_1)^2 + (\gamma/2)^2} - \frac{\gamma/2}{(\omega+\omega_1)^2 + (\gamma/2)^2}\right] \\ &\xrightarrow{\gamma\to 0} \frac{\pi\omega}{m\mid\omega\mid}\delta(\omega^2-\omega_0^2) = \frac{\pi}{2m\omega_0}[\delta(\omega-\omega_0) - \delta(\omega+\omega_0)].\end{aligned} \tag{7.2.19}$$

We see from this that $\chi''(\omega)$ is *real* and *odd* in ω, and it has peaks with Lorentzian line shapes centered at $\omega = \pm\omega_1$ (when ω_1 is real) with half-width at half-maximum equal to $\gamma/2$. Furthermore, when the viscous damping is set to zero, $\chi''(\omega)$ has delta-function spikes at the frequencies $\pm\omega_0$ of the undamped oscillator. The real part of the response function is

$$\chi'(\omega) = \frac{1}{m}\frac{\omega_0^2 - \omega^2}{(\omega^2-\omega_0^2)^2 + \omega^2\gamma^2}. \tag{7.2.20}$$

$\chi'(\omega)$ is positive for $\omega < \omega_0$, tending to $1/k$ as $\omega \to 0$; it is negative for $\omega > \omega_0$, tending to $-1/(m\omega^2)$ as $\omega \to \infty$; and it is zero at exactly $\omega = \omega_0$. $\chi''(\omega)$ and $\chi'(\omega)$ are plotted in Fig. 7.2.1.

The steady-state time dependence of $x(t)$ in the presence of a force $f(t) = f_0 \cos\omega t$ is obtained from the real part of $\chi(\omega) f_0 e^{-i\omega t} = \mid\chi(\omega)\mid f_0 e^{-i(\omega t - \phi)}$:

$$x(t) = f_0 \mid\chi(\omega)\mid \cos[\omega t - \phi(\omega)], \tag{7.2.21}$$

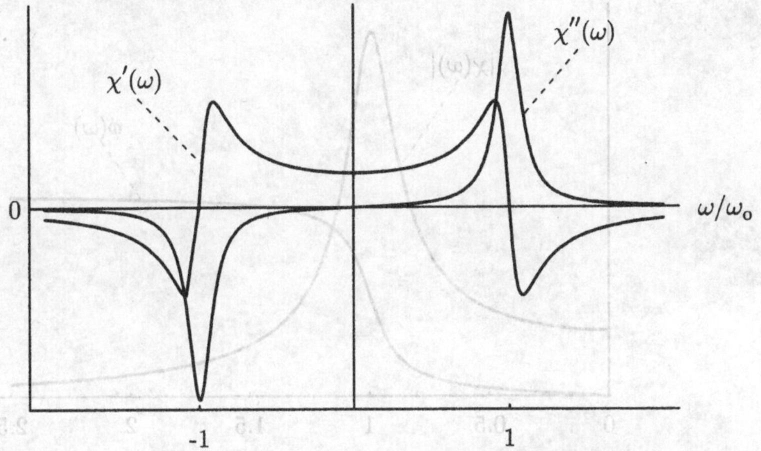

Fig. 7.2.1. $\chi''(\omega)$ and $\chi'(\omega)$ for a harmonic oscillator when ω_1 is real.

where

$$|\chi(\omega)| = \frac{1}{m}\frac{1}{[(\omega^2-\omega_0^2)^2+\omega^2\gamma^2]^{1/2}} \tag{7.2.22}$$

and

$$\tan\phi(\omega) = \frac{\chi''(\omega)}{\chi'(\omega)} = \frac{\omega\gamma}{\omega_0^2-\omega^2}. \tag{7.2.23}$$

Thus, the amplitude of $x(t)$ reaches a maximum for driving frequencies in the vicinity of the natural frequency ω_0 of the oscillator. Furthermore, the phase shift describing the degree to which $x(t)$ lags behind $f(t)$ passes through $\pi/2$ at precisely ω_0. $|\chi(\omega)|$ and $\phi(\omega)$ are plotted in Fig. 7.2.2.

In the overdamped case, the imaginary part of $\chi(\omega)$ is peaked at the origin rather than at nonzero frequencies. In the extreme overdamped limit at frequencies $\omega\tau_f \ll 1$ where inertial terms can be ignored,

$$\chi(\omega) = \frac{1}{m}\frac{1}{\omega_0^2-i\omega\gamma} = \chi\frac{1}{1-i\omega\tau_s} \tag{7.2.24}$$

and

$$\frac{\chi''(\omega)}{\omega} = \chi\frac{\tau_s^{-1}}{\omega^2+\tau_s^{-2}}. \tag{7.2.25}$$

Thus, $\chi''(\omega)/\omega$ is a Lorentzian centered at the origin with width $\tau_s^{-1} = \Gamma\chi$, as shown in Fig. 7.2.3. Its integral over ω trivially satisfies the thermodynamic sum rule, Eq. (7.1.22).

The high frequency behavior of $\chi(z)$ is determined by the frequency moments of $\chi''(\omega)$, as can be seen by expanding the integral representation [Eq. (7.1.17)] in

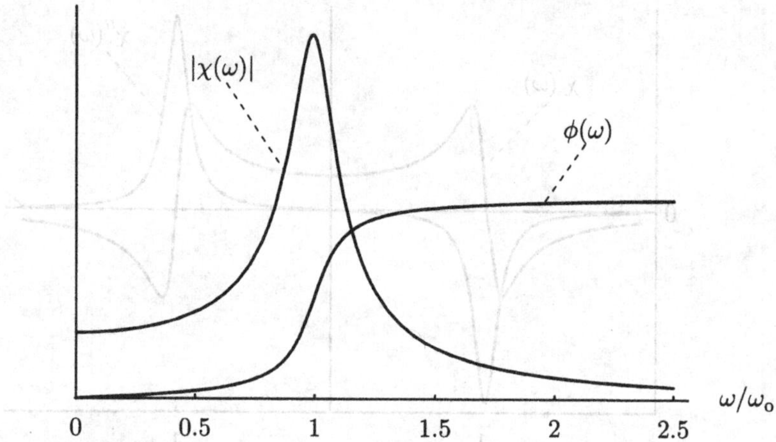

Fig. 7.2.2. The amplitude and phase functions $|\chi(\omega)|$ and $\phi(\omega)$ for a harmonic oscillator.

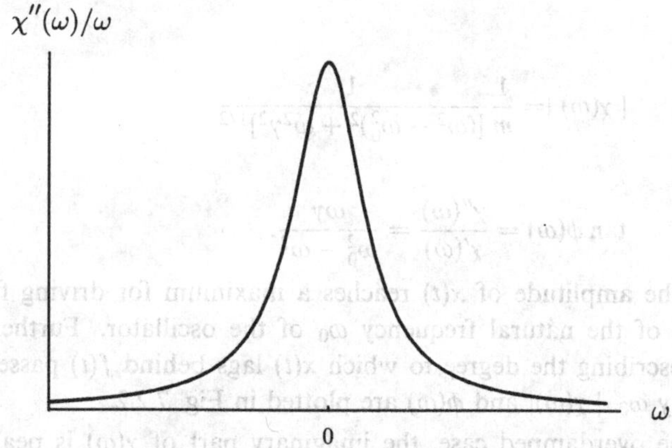

Fig. 7.2.3. $\chi''(\omega)/\omega$ in the overdamped limit when $\tau_f \ll \tau_s$.

powers of $1/z$:

$$\begin{aligned}\chi(z) &= -\frac{1}{z}\int \frac{d\omega}{\pi}\frac{\chi''(\omega)}{1-\omega/z} \\ &= -\frac{1}{z}\int \frac{d\omega}{\pi}\omega\frac{\chi''(\omega)}{\omega} - \frac{1}{z^2}\int \frac{d\omega}{\pi}\omega^2\frac{\chi''(\omega)}{\omega} + \cdots. \qquad (7.2.26)\end{aligned}$$

We shall see in Secs. 7.5 and 7.6 that $\chi''(\omega) \equiv \chi''_{xx}(\omega)$ is related via the fluctuation-dissipation theorem to $S(\omega) \equiv S_{xx}(\omega)$ measuring fluctuations in x via $\chi''(\omega)/\omega = S(\omega)/2T$. Frequency moments of $S(\omega)$ are simply equal-time correlation functions of $x(t)$ and its time derivatives, which are all finite:

$$\int \frac{d\omega}{2\pi}\omega^n S(\omega) = i^n \left\langle \left(\frac{d}{dt}\right)^n x(t)x(t') \right\rangle_{t'=t} = T\int \frac{d\omega}{2\pi}\omega^n \frac{\chi''(\omega)}{\omega}. \quad (7.2.27)$$

This equation says that all moments of $\chi''(\omega)/\omega$ exist and are finite. The odd n moments are all zero because $\chi''(\omega)$ is odd in ω. The first two nonzero moments of the phenomenological form for $\chi''(\omega)$ in Eq. (7.2.19) are finite. The zeroth moment is simply χ, as required by the thermodynamic sum rule. The second moment is $-\langle \ddot{x}(t)x(t)\rangle/T = \langle(\dot{x}(t))^2\rangle/T = 1/m$ because the average kinetic energy $m\langle v^2\rangle/2$ is $T/2$. This agrees with Eq. (7.2.18) and the high-frequency expansion Eq. (7.2.27). The higher moments of Eq. (7.2.19) are infinite. The problem is that the phenomenological damping parameter γ does not provide a correct description of high-frequency behavior. In order for all moments of $\chi''(\omega)/\omega$ to exist, γ must be replaced by a function $\gamma(z)$ of complex z that tends to zero more rapidly than any power of z. An often-used phenomenological form for γ is $\gamma(z) = \gamma/(1 - iz\tau)$, where τ is some microscopic collision time. This form leads to a finite fourth moment of $\chi''(\omega)/\omega$ but to infinite higher moments.

4 Dissipation

In steady state, the external force does work on the oscillator that is eventually dissipated as heat in the viscous fluid. The rate at which the external force does work is

$$\frac{dW}{dt} = f(t)\dot{x}(t). \quad (7.2.28)$$

Since in the steady state, both $f(t)$ and $\dot{x}(t)$ are periodic functions of t with period $T = 2\pi/\omega$, the average power dissipated is

$$P = \frac{1}{T}\int_0^T dt f(t)\dot{x}(t) = -\frac{1}{T}\int_0^T dt x(t)\dot{f}(t). \quad (7.2.29)$$

Using Eq. (7.2.21) for $x(t)$, we obtain

$$\begin{aligned} P &= -\frac{f_0^2}{T}\int_0^T dt\omega \mid \chi(\omega) \mid \cos\omega t \sin[\omega t - \phi(\omega)] \\ &= \frac{1}{2}\omega f_0^2 \mid \chi(\omega) \mid \sin\phi(\omega) = \frac{1}{2}f_0^2\omega\chi''(\omega). \end{aligned} \quad (7.2.30)$$

Thus, we arrive at the very important result that the rate of energy dissipation is proportional to $\omega\chi''(\omega)$. For this reason, $\chi''(\omega)$ is sometimes referred to as the dissipation. Note that $\chi''(\omega)$ is odd in ω so that $\omega\chi''(\omega)$ is even. In thermodynamic equilibrium, the power dissipation must be positive, implying that $\omega\chi''(\omega)$ must be positive. The positivity of $\omega\chi''(\omega)$ in the present case is associated with the positivity of the dissipative coefficient γ. Its sign was chosen so that the viscous force opposes motion of the oscillator mass. This sign is consistent with energy transfer to the incoherent degrees of freedom of the fluid and to positive power absorption.

7.3 Elastic waves and phonons

1 *Sound waves in an elastic continuum*

As we discussed in Chapter 6, an elastic medium in the Eulerian picture is the continuum limit of a collection of mass points connected by Hooke's law springs. The position of each point in the medium relative to its unstretched position is given by the displacement variable $\mathbf{u}(\mathbf{x})$. The velocity of each mass point is therefore $\mathbf{v}(\mathbf{x},t) = \dot{\mathbf{u}}(\mathbf{x},t)$, and the kinetic energy of mass motion is

$$K.E. = \frac{1}{2}\int d^d x \rho(\mathbf{x}) v^2(\mathbf{x}), \tag{7.3.1}$$

where $\rho(\mathbf{x})$ is the mass density at $\mathbf{x}$. In the absence of dissipation, Newton's equation determining the time dependence of the displacement of each mass point is

$$\rho \ddot{u}_i = -\frac{\delta \mathscr{H}_T}{\delta u_i(\mathbf{x})} = \nabla_j \sigma_{ij} + f_i^{\text{ext}}(\mathbf{x}), \tag{7.3.2}$$

where σ_{ij} is the elastic stress tensor of Eq. (6.6.10) and $\mathscr{H}_T = \mathscr{H}_{\text{el}} + \mathscr{H}_{\text{ext}}$, with $\mathscr{H}_{\text{el}}$ the elastic Hamiltonian of Eq. (6.4.11) and $\mathscr{H}_{\text{ext}} = -\int d^d x \mathbf{u}(\mathbf{x}) \cdot \mathbf{f}^{\text{ext}}(\mathbf{x})$ the Hamiltonian arising from an external force density $\mathbf{f}^{\text{ext}}$. Dissipation can be introduced by adding a phenomenological term to the stress tensor proportional to the velocity and thus odd under time reversal. A spatially uniform velocity is equivalent to a Galilean transformation to a moving coordinate system, which will not lead to any dissipation. The dissipative part of the stress tensor is, therefore, proportional to the gradient of the velocity rather than to the velocity itself:

$$\sigma_{ij}^{\text{dis}} = \eta_{ijkl} \nabla_k v_l, \tag{7.3.3}$$

where η_{ijkl} is the viscosity tensor of the solid (viscosity tensors in fluids and solids will be discussed in more detail in Chapter 8). In an isotropic elastic medium, the viscosity tensor, like the elastic tensor, has two independent components,

$$\eta_{ijkl} = \zeta \delta_{ij}\delta_{kl} + \eta\left(\delta_{ik}\delta_{jl} + \delta_{il}\delta_{jk} - \frac{2}{d}\delta_{ij}\delta_{kl}\right), \tag{7.3.4}$$

where ζ is the bulk viscosity and η is the shear viscosity.

The equations of motion for the longitudinal and transverse parts of $\mathbf{u}$ decouple in an isotropic medium. In Fourier space they are

$$[-\omega^2 \rho + q^2 \mu - i\omega\eta q^2]\mathbf{u}_T(\mathbf{q},\omega) = \mathbf{f}_T^{\text{ext}}(\mathbf{q},\omega), \tag{7.3.5}$$

$$[-\omega^2 \rho + q^2(\lambda + 2\mu) - i\omega(\zeta + 2(d-1)\eta/d)q^2]u_L = f_L^{\text{ext}}(\mathbf{q},\omega), \tag{7.3.6}$$

where $\mathbf{u}(\mathbf{q}) = (\mathbf{q}/q)u_L + \mathbf{u}_T$ and similarly for $\mathbf{f}^{\text{ext}}$. These equations yield sound modes whose frequencies go to zero linearly with wave number q and whose widths (imaginary parts) are of order q^2

$$\omega_T = \pm c_T q - i\frac{\eta}{2\rho}q^2, \tag{7.3.7}$$

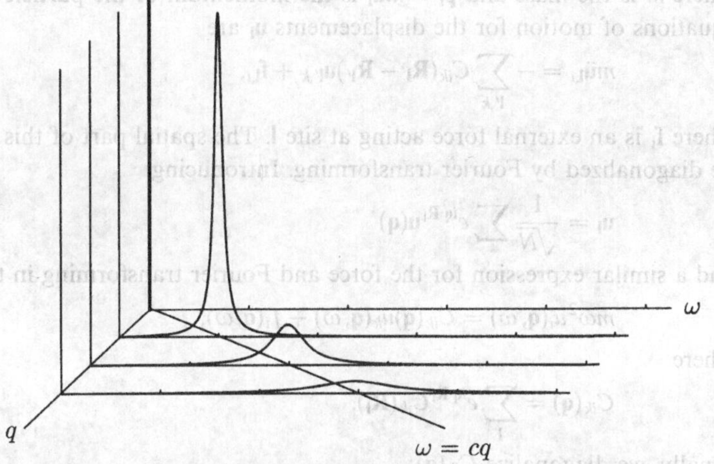

Fig. 7.3.1. The imaginary part of the transverse elastic response function showing Lorentzian peaks at frequencies proportional to q.

$$\omega_L = \pm c_L q + \frac{i}{2\rho}\left[\zeta + \frac{2(d-1)}{d}\eta\right] q^2, \tag{7.3.8}$$

where $c_T = (\mu/\rho)^{1/2}$ is the transverse sound velocity and $c_L = [(\lambda+2\mu)/\rho]^{1/2} > c_T$ is the longitudinal sound velocity. Note there are $2(d-1)$ transverse and two longitudinal modes for each $\mathbf{q}$. This corresponds to one mode per degree of freedom.

The transverse response function is the ratio of $\mathbf{u}_T$ to $\mathbf{f}_T^{\text{ext}}$:

$$\chi_T(\mathbf{q},\omega) = \frac{1}{\rho}\frac{1}{-\omega^2 + (\mu - i\omega\eta)q^2/\rho}. \tag{7.3.9}$$

This function reduces to the static susceptibility $1/(\mu q^2)$ [Eq. (6.4.24)] when $\omega = 0$. The imaginary part of the response function is

$$\chi_T''(\mathbf{q},\omega) = \frac{1}{\rho^2}\frac{\omega\eta q^2}{(\omega^2 - \mu q^2/\rho)^2 + (\eta\omega q^2/\rho)^2}. \tag{7.3.10}$$

This function is sketched in Fig. 7.3.1. A similar expression applies for longitudinal sound waves.

2 Acoustic phonons in a harmonic lattice

The dynamical properties of the classical harmonic lattice described in Sec. 6.6 are easily calculated. The complete Hamiltonian for this system, including the kinetic energy, is

$$\mathscr{H} = \sum_{\mathbf{l}} \frac{\mathbf{p}_{\mathbf{l}}^2}{2m} + \frac{1}{2}\sum_{\mathbf{l},\mathbf{l}',i,k} C_{ik}(\mathbf{R}_{\mathbf{l}} - \mathbf{R}_{\mathbf{l}'})u_{\mathbf{l},i}u_{\mathbf{l}',k}, \tag{7.3.11}$$

where m is the mass and $\mathbf{p_l} = m\dot{\mathbf{u}}_\mathbf{l}$ is the momentum of the particle at site $\mathbf{l}$. The equations of motion for the displacements $\mathbf{u_l}$ are

$$m\ddot{u}_{\mathbf{l},i} = -\sum_{\mathbf{l}',k} C_{ik}(\mathbf{R_l} - \mathbf{R_{l'}})u_{\mathbf{l}',k} + f_{\mathbf{l},i}, \tag{7.3.12}$$

where $\mathbf{f_l}$ is an external force acting at site $\mathbf{l}$. The spatial part of this equation can be diagonalized by Fourier transforming. Introducing

$$\mathbf{u_l} = \frac{1}{\sqrt{N}}\sum e^{i\mathbf{q}\cdot\mathbf{R_l}}\mathbf{u}(\mathbf{q}) \tag{7.3.13}$$

and a similar expression for the force and Fourier transforming in time, we have

$$m\omega^2 u_i(\mathbf{q},\omega) = C_{ik}(\mathbf{q})u_k(\mathbf{q},\omega) + f_i(\mathbf{q},\omega), \tag{7.3.14}$$

where

$$C_{ik}(\mathbf{q}) = \sum_{\mathbf{l}} e^{i\mathbf{q}\cdot\mathbf{R_l}} C_{ik}(\mathbf{R_l}). \tag{7.3.15}$$

Finally, we diagonalize $C_{ik}(\mathbf{q})$:

$$C_{ik}(\mathbf{q})e_k^\lambda(\mathbf{q}) = m\omega_\lambda^2(\mathbf{q})e_i^\lambda(\mathbf{q}). \tag{7.3.16}$$

$m\omega_\lambda^2(\mathbf{q})$ ($\lambda = 1, ..., d$) are the d eigenvalues of $C_{ik}(\mathbf{q})$ and $e_i^\lambda(\mathbf{q})$ are the associated orthonormalized eigenvectors satisfying

$$\begin{aligned} \sum_\lambda e_i^{\lambda*}(\mathbf{q})e_j^\lambda(\mathbf{q}) &= \delta_{ij}, \\ \sum_i e_i^{\lambda'*}(\mathbf{q})e_i^\lambda(\mathbf{q}) &= \delta^{\lambda\lambda'}. \end{aligned} \tag{7.3.17}$$

We can now write Eq. (7.3.14) in terms of independent normal modes

$$\omega^2 u_\lambda(\mathbf{q},\omega) = \omega_\lambda^2(\mathbf{q})u_\lambda(\mathbf{q},\omega) + f_\lambda(\mathbf{q},\omega)/m, \tag{7.3.18}$$

where

$$u_\lambda(\mathbf{q},\omega) = e_i^{\lambda*}(\mathbf{q})u_i(\mathbf{q},\omega) \tag{7.3.19}$$

and similarly for $f_\lambda(\mathbf{q},\omega)$. Eq. (7.3.18) is identical to that of the simple harmonic oscillator, Eq. (7.2.9). It implies, therefore, a response function

$$\chi_\lambda(\mathbf{q},\omega) = \frac{u_\lambda(\mathbf{q},\omega)}{f_\lambda(\mathbf{q},\omega)} = \frac{1}{m}\frac{1}{[-\omega^2 + \omega_\lambda^2(\mathbf{q})]}. \tag{7.3.20}$$

This, in turn, implies that the response function for a displacement at $\mathbf{l}$ in response to a force at $\mathbf{l}'$ is

$$\chi_{ij}(\mathbf{l},\mathbf{l}',\omega) = \frac{1}{N}\sum_\mathbf{q} e^{i\mathbf{q}\cdot(\mathbf{R_l}-\mathbf{R_{l'}})}e_i^{\lambda*}(\mathbf{q})\frac{1}{m}\frac{1}{[-\omega^2 + \omega_\lambda^2(\mathbf{q})]}e_j^\lambda(\mathbf{q}). \tag{7.3.21}$$

The spatial Fourier transform of the imaginary part of this response function is

$$\chi_{ij}''(\mathbf{q},\omega) = \sum_\lambda e_i^{\lambda*}(\mathbf{q})e_j^\lambda(\mathbf{q})\frac{\pi}{m}\frac{\omega}{|\,\omega\,|}\delta(\omega^2 - \omega_\lambda^2(\mathbf{q})). \tag{7.3.22}$$

Thus, $\chi_{ij}''(\mathbf{q},\omega)$ provides a direct measure of the phonon spectrum. We will see in Sec 7.7 how neutron scattering determines this function. In an anharmonic lattice where there are interactions among phonons, the plane wave phonon states will generally be damped, and the δ function in Eq. (7.3.22) should be replaced by a Lorentzian.

7.4 Diffusion

1 Fick's law

Consider particles dissolved or suspended in a fluid. The nature of these particles is, for the moment, arbitrary. They can simply be molecules of a species different from those composing the fluid, or they can be specks of dust or polystyrene spheres with diameters of order 0.1 microns. The number of these particles does not change with time. Thus, their number density $n(\mathbf{x}, t)$ (represented in terms of particle positions by Eq. (7.1.3)) obeys a conservation law:

$$\frac{\partial n(\mathbf{x}, t)}{\partial t} + \nabla \cdot \mathbf{j}(\mathbf{x}, t) = 0, \tag{7.4.1}$$

where

$$\mathbf{j}(\mathbf{x}, t) = \sum_{\alpha} \mathbf{v}_{\alpha}(t)\delta(\mathbf{x} - \mathbf{x}_{\alpha}(t)) \tag{7.4.2}$$

is the particle current, with $\mathbf{v}_{\alpha}(t) = \dot{\mathbf{x}}_{\alpha}(t)$ the particle velocity. In thermal equilibrium, the particles are distributed uniformly throughout the fluid, and the thermal average $\langle n(\mathbf{x}, t)\rangle$ of the density is independent of both $\mathbf{x}$ and t. What happens, however, if, as a result either of spontaneous fluctuations or of an external force, there exists at some time a spatially non-uniform density, as depicted in Fig. 7.4.1? If external forces are turned off, the density must eventually tend to the spatially uniform equilibrium state. This can only occur as a result of particle motion. Thus, we expect a spatially non-uniform density to give rise to a non-zero current $\mathbf{j}$. If the density varies very slowly in space, then the density is nearly in equilibrium at each point in space, and currents should be very small. These considerations lead one to expect the current to be proportional to gradients of the density. The current $\mathbf{j}$ must transform like a vector so that the simplest relation between $\mathbf{j}$ and ∇n is

$$\mathbf{j} = -D\nabla n. \tag{7.4.3}$$

This equation is known as *Fick's law*. It is a phenomenological relation analogous to that of Eq. (7.2.7) relating the viscous force to the velocity. It says that a spatially non-uniform density will lead to currents in directions opposite to the direction of changes in densities, i.e., to currents tending to reestablish spatial uniformity of $n(\mathbf{x}, t)$. The coefficient D is a *diffusion constant*. It has units of $[\text{length}]^2/[\text{time}]$. The current $\mathbf{j}$ is odd under time reversal, whereas $n(\mathbf{x}, t)$ and its gradient are even. Thus, the two sides of Eq. (7.4.2) have opposite signs under time reversal, and the diffusion constant is a type of dissipative coefficient.

When Fick's law for the current is substituted into the conservation law, the result is the diffusion equation,

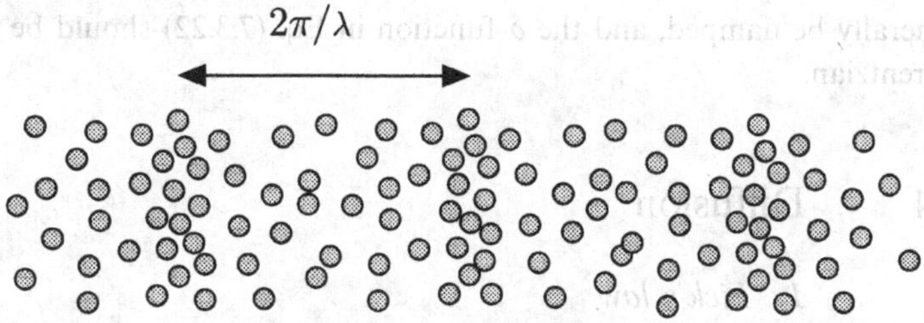

Fig. 7.4.1. Spatially modulated distribution of particles. Random motion of these particles will restore spatial homogeneity in times that diverge as the square of the wavelength of the spatial modulation.

$$\frac{\partial n}{\partial t} = D\nabla^2 n. \tag{7.4.4}$$

The modes predicted by this equation are again obtained by assuming $n(\mathbf{x}, t) \sim e^{-i\omega t}$. The resulting mode frequency is

$$\omega = -iDq^2, \tag{7.4.5}$$

where $q = 2\pi/\lambda$ is the wave number of the spatial modulation of the density. This frequency is purely imaginary, implying, as for the overdamped oscillator, that the response of $n(\mathbf{x}, t)$ to external forces or non-equilibrium boundary conditions will decay exponentially to zero in times of order $D^{-1}\lambda^2$. There will be no oscillatory part to this decay.

2 The Green function and dynamic response

The density at position $\mathbf{x}$ and time t is related to the density at position $\mathbf{x}'$ and time t' via

$$n(\mathbf{x}, t) = \int d^d x' G(\mathbf{x} - \mathbf{x}', t - t')n(\mathbf{x}', t'), \tag{7.4.6}$$

where $G(\mathbf{x}, t)$ is the diffusion Green function satisfying the boundary condition

$$G(\mathbf{x}, t = 0) = \delta(\mathbf{x} - \mathbf{x}'). \tag{7.4.7}$$

For times $t > 0$, $G(\mathbf{x}, t)$ satisfies the same equation as $n(\mathbf{x}, t)$:

$$\frac{\partial G(\mathbf{x}, t)}{\partial t} - D\nabla^2 G(\mathbf{x}, t) = 0. \tag{7.4.8}$$

The solution to this equation subject to the boundary condition Eq. (7.4.7) can be obtained via Laplace transformation in time and Fourier transformation in space. The results are

$$G(\mathbf{q}, z) = \frac{1}{-iz + Dq^2} \tag{7.4.9}$$

and

$$G(\mathbf{q}, t) = \int_{-\infty+i\epsilon}^{\infty+i\epsilon} \frac{dz}{2\pi} e^{-izt} G(\mathbf{q}, z) = e^{-Dq^2|t|}, \tag{7.4.10}$$

$$G(\mathbf{x},t) = \int \frac{d^d q}{(2\pi)^d} e^{i\mathbf{q}\cdot\mathbf{x}} G(\mathbf{q},t) = \frac{1}{(4\pi D \mid t\mid)^{d/2}} e^{-|\mathbf{x}|^2/(4D|t|)}.$$

This says that particle density initially localized at the origin will spread out with time, occupying a region with mean-square radius

$$\langle \mid \mathbf{x} \mid^2 \rangle = 2dD \mid t \mid, \tag{7.4.11}$$

where as usual d is the dimension of space. The diffusion constant, therefore, measures the mean-square displacement per unit time interval.

3 *The response function*

The Green function allows us to determine the density at time t if we know the density at some earlier time t'. It does not, however, give us directly the density response function. To obtain the response function, we create a spatially non-uniform density at time $t < 0$ that is in equilibrium with an external chemical potential with a small spatially varying part $\delta\mu(\mathbf{x})$. We then turn off the external chemical potential at time $t = 0$. For $t > 0$, the equilibrium state is again spatially uniform, and the decay to equilibrium is controlled by the Green function. The external Hamiltonian creating the spatially varying $\langle n(\mathbf{x}, t = 0)\rangle$ is

$$\mathcal{H}_{\text{ext}} = -\int d^d x n(\mathbf{x},t)\delta\mu(\mathbf{x})\eta(-t)e^{\epsilon t}, \tag{7.4.12}$$

where ϵ is an infinitesimal. The Fourier transform of the change in the density at $t = 0$ brought about by this external Hamiltonian is

$$\langle \delta n(\mathbf{q}, t = 0)\rangle = \chi(\mathbf{q})\delta\mu(\mathbf{q}), \tag{7.4.13}$$

where $\chi(\mathbf{q})$ is the static density response function discussed in Chapter 3. The density for $t > 0$ is then determined by its Laplace-Fourier transform satisfying

$$\langle \delta n(\mathbf{q}, z)\rangle = G(\mathbf{q}, z)\langle \delta n(\mathbf{q}, t = 0)\rangle = \frac{\chi(\mathbf{q})\delta\mu(\mathbf{q})}{-iz + Dq^2}. \tag{7.4.14}$$

Our next step is to determine how $\langle \delta n(\mathbf{q}, z)\rangle$ is related to the dynamic susceptibility $\chi_{nn}(\mathbf{q}, z) \equiv \chi(\mathbf{q}, z)$. From the definition of $\tilde{\chi}(\mathbf{x}, \mathbf{x}', t, t')$ [Eq. (7.1.11)], and the fact that $\delta\mu(\mathbf{x}, t) = \eta(-t)e^{\epsilon t}\delta\mu(\mathbf{x})$, we have

$$\delta\langle n(\mathbf{q}, t)\rangle = \int_{-\infty}^{0} dt' \tilde{\chi}(\mathbf{q}, t - t')e^{\epsilon t'}\delta\mu(\mathbf{q}), \tag{7.4.15}$$

from which we obtain

$$\begin{aligned}
\delta\langle n(\mathbf{q}, z)\rangle &= \int_0^\infty dt e^{izt} \int_{-\infty}^0 dt' \int \frac{d\omega}{2\pi} 2i\chi''(\mathbf{q}, \omega)e^{-i\omega(t-t')}e^{\epsilon t'}\delta\mu(\mathbf{q}) \\
&= \int \frac{d\omega}{\pi i} \frac{\chi''(\mathbf{q}, \omega)}{(\omega - z)(\omega - i\epsilon)}\delta\mu(\mathbf{q}) \\
&= \int \frac{d\omega}{\pi i}\chi''(\mathbf{q}, \omega)\frac{1}{z}\left(\frac{1}{\omega - z} - \frac{1}{\omega}\right)\delta\mu(\mathbf{q}) \\
&= \frac{1}{iz}[\chi(\mathbf{q}, z) - \chi(\mathbf{q})]\delta\mu(\mathbf{q}).
\end{aligned} \tag{7.4.16}$$

This equation and Eq. (7.4.14) then imply

$$G(\mathbf{q}, z) = \frac{1}{iz}\left[\frac{\chi(\mathbf{q}, z)}{\chi(\mathbf{q})} - 1\right] \tag{7.4.17}$$

and

$$\chi(\mathbf{q}, z) = \chi(\mathbf{q})\frac{Dq^2}{-iz + Dq^2}. \tag{7.4.18}$$

This response function has exactly the same form as that of the overdamped oscillator [Eq. (7.2.24)] except that the inverse decay time Dq^2 now depends on wave number. The imaginary part of $\chi(\mathbf{q}, \omega)$ is

$$\frac{\chi''(\mathbf{q}, \omega)}{\omega} = \chi(\mathbf{q})\frac{Dq^2}{\omega^2 + (Dq^2)^2}. \tag{7.4.19}$$

This is a Lorentzian with integrated intensity $\chi(\mathbf{q})$, height $\chi(\mathbf{q})/Dq^2$, and a width that goes to zero as $q \to 0$, as shown in Fig. 7.4.2.

As in the case of the harmonic oscillator discussed in the preceding section, the high-frequency moments of $\chi''(\omega)/\omega$ must all be finite. As for the harmonic oscillator, the dissipative function $\chi''(\mathbf{q}, \omega)$ is related to the correlation function $S_{nn}(\mathbf{q}, \omega)$ via the fluctuation-dissipation theorem, which in the classical limit is

$$\frac{\chi''(\mathbf{q}, \omega)}{\omega} = \frac{S_{nn}(\mathbf{q}, \omega)}{2T}. \tag{7.4.20}$$

The second moment of $\chi''(\mathbf{q}, \omega)/\omega$ is thus

$$\begin{aligned}\int \frac{d\omega}{\pi}\omega^2\frac{\chi''(\mathbf{q}, \omega)}{\omega} &= \frac{1}{T}\int\frac{d\omega}{2\pi}\omega^2 S_{nn}(\mathbf{q}, \omega) = \frac{1}{VT}\left\langle\frac{\partial}{\partial t}n(\mathbf{q}, t)\frac{\partial}{\partial t}n(-\mathbf{q}, t)\right\rangle \\ &= \frac{1}{VT}q^2\langle\mathbf{j}(\mathbf{q}, t)\cdot\mathbf{j}(-\mathbf{q}, t)\rangle = \frac{1}{VT}q^2\sum_\alpha\langle v_\alpha^2\rangle \\ &= \frac{nq^2}{m},\end{aligned} \tag{7.4.21}$$

where we used $\langle v_\alpha^2\rangle = T/m$, where m is the mass of the diffusing particles. This is the f-sum rule, which, as we shall see in Sec. 7.6, is always valid for both classical and quantum systems.

Only the zeroth moment of Eq. (7.4.19) is finite. In order to make all moments finite and to reproduce Eq. (7.4.18) at low frequency, we can introduce a frequency-dependent diffusion constant $D(z)$ that tends to zero faster than any power of $1/z$ at large z and reduces to D at $z = 0$. $D(z)$ will have an integral representation similar to Eq. (7.1.17) for $\chi(z)$ (see Problem 7.4). The phenomenological expression $D(z) = D/(1 - iz\tau)$ produces a finite second moment but infinite higher moments. Following Eq. (7.2.26), the second moment of $\chi''(\mathbf{q}, \omega)/\omega$ is the coefficient of $-1/z^2$ in the high-frequency expansion of $\chi(\mathbf{q}, z)$, which for the above form for $D(z)$ is $-\chi Dq^2/\tau$. Identifying this result with the f-sum rule, we find $D = n\tau/m\chi$. This provides a phenomenological connection between the diffusion constant, the susceptibility, and a "microscopic" collision time τ. We will obtain this result in a different way shortly.

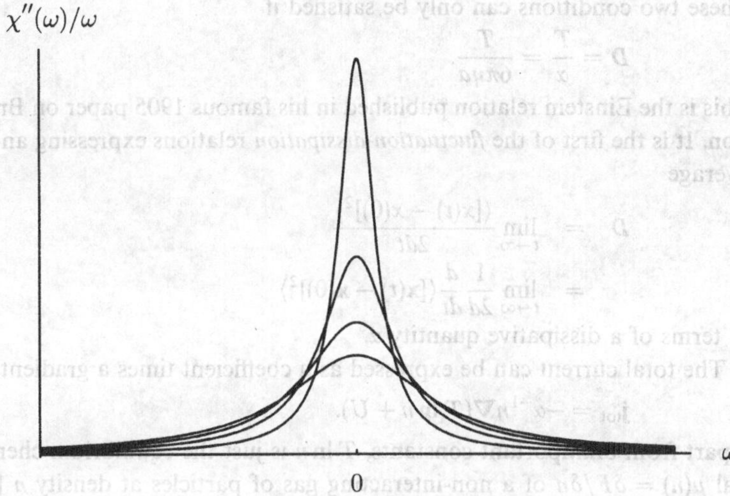

Fig. 7.4.2. The imaginary part of the diffusive response function over ω [Eq. (7.4.19)] at different values of q. The half width at half maximum is Dq^2, so that a measurement of this quantity as a function of q gives the diffusion constant D. The static susceptibility is the area under this curve, or alternatively Dq^2 times its height.

4 *External potentials and the Einstein relation*

Fick's law is appropriate to situations when there is no external potential, such as that of a gravitational field. When there are external potentials, it must be modified. To see why, consider, as Einstein did, densities of diffusing particles sufficiently dilute that interactions between them can be neglected. In this case, the only forces acting on a given particle are those arising from external potentials and from collisions with molecules comprising the fluid. We have already argued [Eq. (7.2.7)] that the effect of the latter is to introduce a friction force on a particular particle proportional to its velocity. In steady state, this force must equal any external forces, implying particles will drift with velocity

$$\mathbf{v}_D = \frac{1}{\alpha}\mathbf{f}^{\text{ext}} \equiv -\frac{1}{\alpha}\boldsymbol{\nabla} U \,, \tag{7.4.22}$$

where α is the friction constant introduced in Eq. (7.2.7) and U is the external potential determining the force $\mathbf{f}^{\text{ext}} = -\boldsymbol{\nabla} U$. This drift gives rise to a drift current $\mathbf{j}_D = n\mathbf{v}_D$, in addition to the diffusion current, about any average flow predicted by Fick's law. The total current is, therefore,

$$\mathbf{j}_{\text{tot}} = -D\boldsymbol{\nabla} n + \mathbf{j}_D = -D\boldsymbol{\nabla} n - \alpha^{-1} n \boldsymbol{\nabla} U. \tag{7.4.23}$$

In thermodynamic equilibrium, the total particle current must be zero, and the density must satisfy the Boltzmann relation

$$n_{\text{eq}} \sim e^{-U(\mathbf{x})/T}. \tag{7.4.24}$$

These two conditions can only be satisfied if

$$D = \frac{T}{\alpha} = \frac{T}{6\pi\eta a}. \tag{7.4.25}$$

This is the Einstein relation published in his famous 1905 paper on Brownian motion. It is the first of the *fluctuation-dissipation* relations expressing an equilibrium average

$$\begin{aligned} D &= \lim_{t\to\infty} \frac{\langle [x(t)-x(0)]^2\rangle}{2dt} \\ &= \lim_{t\to\infty} \frac{1}{2d}\frac{d}{dt}\langle [\mathbf{x}(t)-\mathbf{x}(0)]^2\rangle \end{aligned} \tag{7.4.26}$$

in terms of a dissipative quantity α.

The total current can be expressed as a coefficient times a gradient of a scalar:

$$\mathbf{j}_{\text{tot}} = -\alpha^{-1} n \nabla (T \ln n + U). \tag{7.4.27}$$

Apart from unimportant constants, $T \ln n$ is just the equilibrium chemical potential $\mu(n) = \delta F/\delta n$ of a non-interacting gas of particles at density n [Eq. (3.3.6)]. The external potential U is equivalent to minus the externally fixed chemical potential $\mu^{\text{ext}}(\mathbf{x}) \equiv \mu(\mathbf{x})$ appearing in $\mathscr{H}_{\text{ext}}$ [Eq. (7.4.12)]. The total current can, therefore, be written as

$$\mathbf{j}_{\text{tot}} = -\Gamma(n)\nabla[\mu(n) - \mu^{\text{ext}}(\mathbf{x})] = -\Gamma(n)\nabla[\delta F_T/\delta n(\mathbf{x})], \tag{7.4.28}$$

where $F_T = F[n(\mathbf{x})] - \int d^d x \mu^{\text{ext}}(\mathbf{x}) n(\mathbf{x})$ and

$$\Gamma(n) = \frac{n}{\alpha} = \frac{n}{m\gamma} \tag{7.4.29}$$

is a density-dependent dissipative coefficient. Note the appearance of the "total" free energy F_T in Eq. (7.4.28). This is similar to the appearance of the total Hamiltonian in Eqs. (7.2.15) and (7.3.2). F_T is identical to the function W introduced in Eq. (3.4.23). In equilibrium, the equation of state,

$$\frac{\delta F}{\delta n(\mathbf{x})} = \mu^{\text{ext}}(\mathbf{x}), \tag{7.4.30}$$

is satisfied, and there is no current. Current only flows when the function $\mu(n)$ differs from the externally imposed chemical potential μ^{ext}.

When $n(\mathbf{x}, t)$ differs from its equilibrium value n_{eq} determined by the equation of state, there will be a current, which for small and slowly varying $\delta n(\mathbf{x}, t) = n(\mathbf{x}, t) - n_{\text{eq}}$ is

$$\begin{aligned} \mathbf{j} &= -\Gamma(n)\nabla[(\partial\mu(n)/\partial n)\delta n] \\ &= -\Gamma(n)(\partial\mu/\partial n)\nabla n. \end{aligned} \tag{7.4.31}$$

This leads to an alternative expression for the diffusion constant:

$$D = \Gamma/\chi = \Gamma/(\partial n/\partial \mu). \tag{7.4.32}$$

This is exactly the result for D we obtained using the sum and a phenomenological frequency-dependent $D(z)$ with the identification γ with τ^{-1}. Though Eq. (7.4.27) was motivated by considerations of a dilute gas of diffusing particles, it and Eq. (7.4.32) for D are also applicable to denser systems when interactions between

Fig. 7.4.3. Schematic representation of two trajectories of a Brownian particle.

particles become important. In this case, the dissipative coefficient Γ is not simply a linear function of n, and the chemical potential $\mu(n)$ is that appropriate to the interacting system.

5 Brownian motion

As just discussed, particles do not interact with each other in the dilute limit. In this case, we can focus on an individual diffusing particle. It is constantly subjected to collisions with the molecules of the fluid, and it describes an erratic trajectory in space. Such erratic motion was first reported in 1828 by the botanist Robert Brown, who used a microscope to observe particles of pollen floating on the surface of water. He found that the pollen particles would appear to jump some distance in a random direction, then remain at rest for a period, then jump again in another direction. He subsequently observed similar motion of very fine particles of a number of substances including minerals and fragments of the Sphinx. He concluded that this motion occurred independently of the composition and origin of the particle. The explanation of the origin of this phenomenon of Brownian motion is due to Einstein, who apparently was unaware of Brown's observations. Random motion consisting of a sequence of apparently discrete steps is often referred to as a *random walk* or more colorfully as a drunkard's walk (see Sec. 2.12).

The position $\mathbf{x}(t)$ of the Brownian particle is a random function of time. Such a randomly fluctuating variable is called a *stochastic* variable. The series of values of a random variable as a function of time is generally called a *stochastic process*. The conditional probability $P(\mathbf{x}, t \mid \mathbf{x}_0, t_0)$ that the particle is at position $\mathbf{x}$ at time t, given that it was at position $\mathbf{x}_0$ at time t_0, can be expressed as

$$P(\mathbf{x}, t \mid \mathbf{x}_0, t_0) = \langle \delta(\mathbf{x} - \mathbf{x}(t)) \rangle_{\mathbf{x}_0, t_0}, \tag{7.4.33}$$

where $\mathbf{x}(t)$ is the instantaneous position of the particle moving under the influence of a random force originating from collisions with fluid particles. The brackets in the above equation signify an average over this random force, and the condition that the particle was at position $\mathbf{x}_0$ at t_0 is implemented by the boundary condition $\mathbf{x}(t_0) = \mathbf{x}_0$.

In the dilute limit, the probability $\langle \delta(\mathbf{x}-\mathbf{x}^\alpha(t))\rangle_{\mathbf{x}_0^\alpha,t_0}$ that particle α is at $\mathbf{x}$ at time t, given that it was at $\mathbf{x}_0^\alpha$ at time t_0, is simply $P(\mathbf{x},t \mid \mathbf{x}_0^\alpha t_0)$. The average density of particles at $\mathbf{x}$ at time t, given that the density was $n(\mathbf{x},t_0) = \sum_\alpha \delta(\mathbf{x}-\mathbf{x}_0^\alpha)$ at $t = t_0$, is, therefore,

$$\begin{aligned}\langle n(\mathbf{x},t)\rangle &= \sum_\alpha \langle \delta(\mathbf{x}-\mathbf{x}^\alpha(t))\rangle_{\mathbf{x}_0^\alpha,t_0} = \sum_\alpha P(\mathbf{x},t \mid \mathbf{x}_0^\alpha, t_0) \\ &= \int d^d x_0 P(\mathbf{x},t \mid \mathbf{x}_0,t_0) \sum_\alpha \delta(\mathbf{x}_0 - \mathbf{x}_0^\alpha) \\ &= \int d^d x_0 P(\mathbf{x},t \mid \mathbf{x}_0,t_0) n(\mathbf{x}_0,t_0). \end{aligned} \tag{7.4.34}$$

This equation is identical to Eq. (7.4.6) and allows us to identify $P(\mathbf{x},t \mid \mathbf{x}_0,t_0)$ with the diffusion Green function in the dilute limit:

$$P(\mathbf{x},t \mid \mathbf{x}_0,t_0) = G(\mathbf{x}-\mathbf{x}_0, t-t_0). \tag{7.4.35}$$

The mean-square displacement of a single Brownian particle, therefore, satisfies Eq. (7.4.11) with $d = 3$ and with the diffusion constant Eq. (7.4.25) appropriate to a non-interacting particle diffusing in a fluid with viscosity η:

$$\langle (\Delta\mathbf{x})^2\rangle = \langle [\mathbf{x}(t)-\mathbf{x}_0]^2\rangle = 6Dt = \frac{k_B T}{\pi\eta a} t, \tag{7.4.36}$$

where we have explicitly displayed Boltzmann's constant k_B. This equation was used in one of the early determinations of k_B. The fluid viscosity and radius of a diffusing particle can be measured with reasonable accuracy. Measurements of $\mathbf{x}(t)$ by observations under a microscope then yield $\langle (\Delta\mathbf{x})^2\rangle$ as a function of time. It is then straightforward to determine k_B from Eq. (7.4.36). A typical fluid such as water has a viscosity of order 0.01 poise. Eq. (7.4.36) then predicts that a particle with a radius of order 0.1 microns will diffuse a distance of order 1 micron in 1 second. Thus, diffusion of a particle of this size is observable in laboratory times.

6 Cooperative diffusion versus self-diffusion

We have considered diffusion of both the average density of particles and of an individual particle. These two processes are referred to, respectively, as *cooperative diffusion* and *self-diffusion*. They are different and are controlled by different diffusion constants D_c and D_s that become equal only when interactions between diffusing particles can be ignored (as they can in the dilute limit discussed above). Both constants can be measured experimentally. As just discussed, self-diffusion can be detected in dilute systems by observations under a microscope.

It can also be observed by more sophisticated techniques in which individual particles are tagged, either by rendering them radioactive or by treating them with a photochromic dye that changes from transparent to opaque when exposed to ultraviolet light. In the latter case, quantitative measurements can be made by "forced Rayleigh" scattering experiments in which a suspension is first irradiated with spatially modulated ultraviolet light. Light from a second source is then diffracted from the sample. Its diffracted intensity determines D_s. Self-diffusion can also be detected, as we shall see in Sec. 7.8, via incoherent neutron scattering, which measures the function

$$S_{\text{self}}(\mathbf{x}, \mathbf{x}', t - t') = \sum_{\alpha} \langle \delta(\mathbf{x} - \mathbf{x}^{\alpha}(t)) \delta(\mathbf{x}' - \mathbf{x}^{\alpha}(t')) \rangle. \tag{7.4.37}$$

This is a self-correlation function because it involves temporal and spatial correlations of a single particle rather than a collection of particles. It can be related to the conditional probability of Eq. (7.4.33) using

$$P_{\alpha}(\mathbf{x}t \mid \mathbf{x}'t') = \frac{\langle \delta(\mathbf{x} - \mathbf{x}^{\alpha}(t)) \delta(\mathbf{x}' - \mathbf{x}^{\alpha}(t')) \rangle}{\langle \delta(\mathbf{x}' - \mathbf{x}^{\alpha}(t')) \rangle}, \tag{7.4.38}$$

where we have indicated the possibly different behavior for different particles by the subscript α. For translationally invariant systems with volume V, $\langle \delta(\mathbf{x}' - \mathbf{x}^{\alpha}(t')) \rangle = V^{-1}$, P_{α} is independent of α, and from Eqs. (7.4.38) and (7.4.37)

$$S_{\text{self}}(\mathbf{x}, \mathbf{x}', t - t') = nP(\mathbf{x}t \mid \mathbf{x}'t'), \tag{7.4.39}$$

implying

$$\begin{aligned} S_{\text{self}}(\mathbf{q}, \omega) &= n \int_{-\infty}^{\infty} dt e^{i\omega t} e^{-D_s q^2 |t|} \\ &= 2n \frac{D_s q^2}{\omega^2 + (D_s q^2)^2} \end{aligned} \tag{7.4.40}$$

is a Lorentzian with a width determined by the self-diffusion constant D_s.

Cooperative diffusion gives rise to density changes and can be probed by inelastic light scattering that, as we shall see at the end of this chapter, measures the density correlation function $S_{nn}(\mathbf{q}, t)$ related to $\chi''_{nn}(\mathbf{q}, \omega)$ via the fluctuation-dissipation theorem,

$$S_{nn}(\mathbf{q}, t) = \int \frac{d\omega}{\pi} \frac{\chi''_{nn}(\mathbf{q}, \omega)}{\beta\omega} e^{-i\omega t} = T\chi(\mathbf{q}) e^{-D_c q^2 |t|}. \tag{7.4.41}$$

Thus both $S_{nn}(\mathbf{q}, t)$ and $S_{\text{self}}(\mathbf{q}, t)$ show diffusive behavior with characteristic inverse decay times that increase linearly with q^2.

Fig. 7.4.4. shows experimental data obtained from forced Rayleigh scattering from polystyrene spheres in water. The inverse decay time is linear in q^2. Fig. 7.4.5a shows the exponential decay in time of the correlation function for the intensity of scattered light (which is proportional to $S_{nn}^2(\mathbf{q}, t)$ as discussed in Problem 7.13) from polystyrene spheres in methanol. The inverse decay time is again proportional to q^2, as shown in Fig. 7.4.5b.

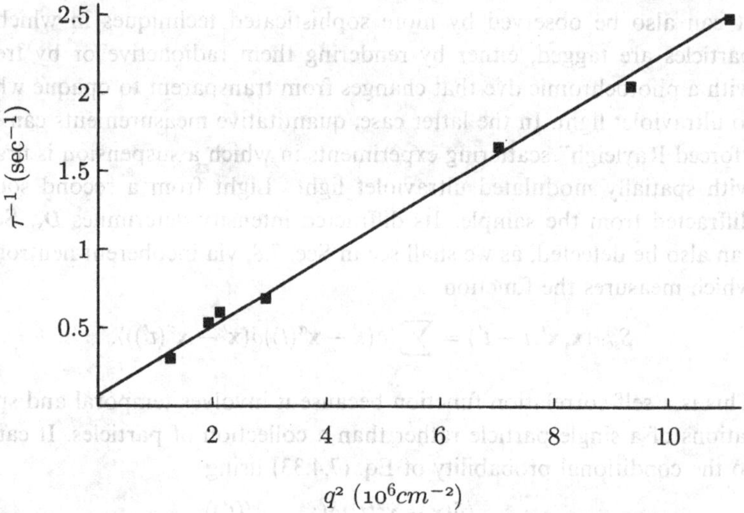

Fig. 7.4.4. Inverse decay time as a function of q^2 for self-diffusion measured by forced Rayleigh scattering. [Courtesy of W.D. Dozier.]

7 Master equation for diffusion on a lattice

Diffusion is not limited to particles in solution. It often occurs when there are processes whose time dependence is controlled by random processes. Here we will consider diffusion on a lattice. We imagine that sites $\mathbf{l}$ on a lattice can be occupied by an atom or some localized excitation. The probability that site $\mathbf{l}$ is occupied at time t is $P(\mathbf{l}, t)$. As time progresses, the atom can hop to other sites. Let $P(\mathbf{l}, t+\Delta t \mid \mathbf{l}', t) \equiv R(\mathbf{l}, \mathbf{l}', \Delta t)$ be the probability that the atom is at site $\mathbf{l}$ at time $t+\Delta t$, given that it was at site $\mathbf{l}'$ at time t. Then the probability that the atom is at site $\mathbf{l}$ at time $t+\Delta t$ is

$$P(\mathbf{l}, t+\Delta t) = \sum_{\mathbf{l}'} R(\mathbf{l}, \mathbf{l}', \Delta t) P(\mathbf{l}', t). \tag{7.4.42}$$

Because $R(\mathbf{l}, \mathbf{l}', \Delta t)$ is a probability, it must satisfy

$$\sum_{\mathbf{l}} R(\mathbf{l}, \mathbf{l}', \Delta t) = 1. \tag{7.4.43}$$

As the time difference Δt goes to zero, the probability that an atom initially at site $\mathbf{l}'$ is at a site different from $\mathbf{l}'$ must go to zero. Thus, for small Δt, we can write

$$R(\mathbf{l}, \mathbf{l}', \Delta t) = \begin{cases} 1 - \sum_{\mathbf{l}_1} w_{\mathbf{l}\to\mathbf{l}_1} \Delta t & \text{if } \mathbf{l} = \mathbf{l}'; \\ w_{\mathbf{l}'\to\mathbf{l}} \Delta t & \text{if } \mathbf{l} \neq \mathbf{l}', \end{cases} \tag{7.4.44}$$

where $w_{\mathbf{l}'\to\mathbf{l}}$ is a transition rate (with units of 1/[time]) from site $\mathbf{l}'$ to $\mathbf{l}$ defined to be zero for $\mathbf{l} = \mathbf{l}'$. Eqs. (7.4.42) and (7.4.44) lead to a differential equation for $P(\mathbf{l}, t)$:

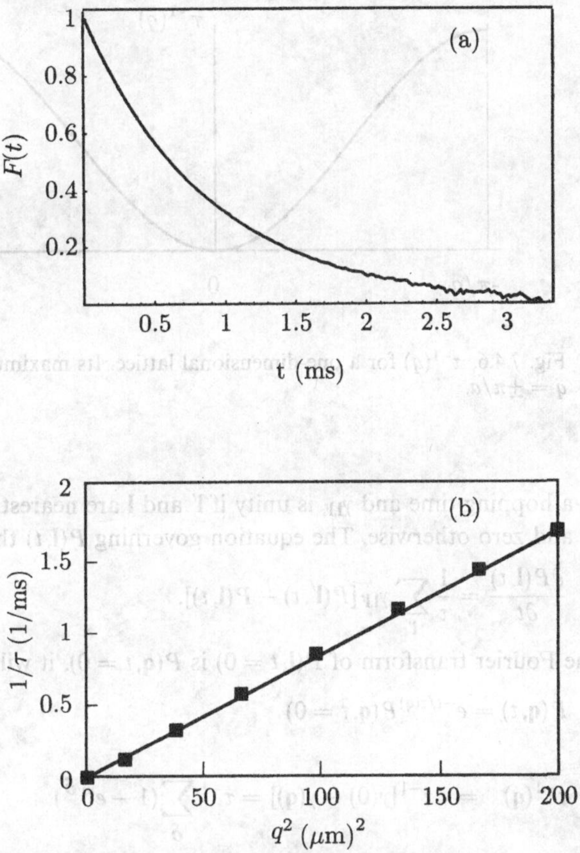

Fig. 7.4.5. (a) Intensity correlation function $F(t) = (\langle I(q,0)I(q,t)\rangle / \langle I(q)\rangle^2) - 1$ where $I(q,t)$ is the intensity of scattered light, and (b) characteristic decay time as a function of q^2 for polystyrene spheres in methanol. [Courtesy of J. Xue.]

$$\frac{\partial P(\mathbf{l},t)}{\partial t} = \sum_{\mathbf{l}'} w_{\mathbf{l}'\to\mathbf{l}} P(\mathbf{l}',t) - \left(\sum_{\mathbf{l}_1} w_{\mathbf{l}\to\mathbf{l}_1}\right) P(\mathbf{l},t). \tag{7.4.45}$$

This equation has a simple interpretation: $P(\mathbf{l},t)$ increases as a result of hops, which occur at rates $w_{\mathbf{l}'\to\mathbf{l}}$, from sites $\mathbf{l}' \neq \mathbf{l}$ to the site $\mathbf{l}$, and it decreases as a result of hops, which occur at rates $w_{\mathbf{l}\to\mathbf{l}_1}$, from $\mathbf{l}$ to sites $\mathbf{l}_1 \neq \mathbf{l}$.

There are no restrictions on the hopping rates $w_{\mathbf{l}'\to\mathbf{l}}$ in Eq. (7.4.45). They can connect any sites on the lattice, and the rate $w_{\mathbf{l}'\to\mathbf{l}}$ does not necessarily have to equal $w_{\mathbf{l}\to\mathbf{l}'}$. A simplified model is one in which there is hopping only between nearest neighbor sites on a lattice and in which the rate for hopping from $\mathbf{l}$ to $\mathbf{l}'$ is equal to that for hopping from $\mathbf{l}'$ to $\mathbf{l}$. In this case,

$$w_{\mathbf{l}'\to\mathbf{l}} = w_{\mathbf{l}\to\mathbf{l}'} = \tau^{-1}\gamma_{\mathbf{l},\mathbf{l}'}, \tag{7.4.46}$$

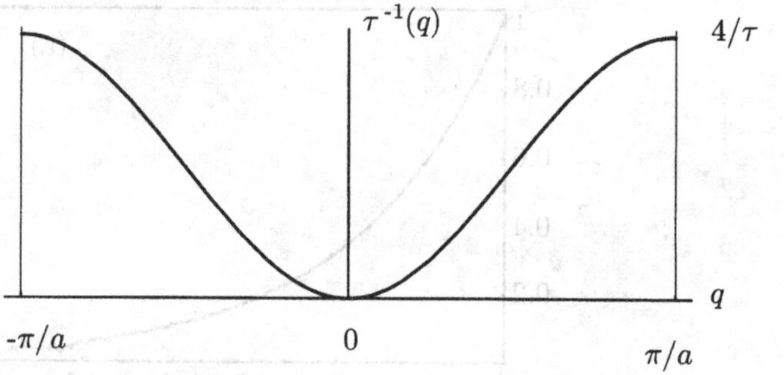

Fig. 7.4.6. $\tau^{-1}(q)$ for a one-dimensional lattice. Its maximum $4/\tau$ is at $q = \pm\pi/a$.

where τ is a hopping time and $\gamma_{\mathbf{l},\mathbf{l}'}$ is unity if $\mathbf{l}'$ and $\mathbf{l}$ are nearest neighbor sites on the lattice and zero otherwise. The equation governing $P(\mathbf{l}, t)$ then becomes

$$\frac{\partial P(\mathbf{l}, t)}{\partial t} = \frac{1}{\tau}\sum_{\mathbf{l}'} \gamma_{\mathbf{l},\mathbf{l}'}[P(\mathbf{l}', t) - P(\mathbf{l}, t)]. \tag{7.4.47}$$

Thus, if the Fourier transform of $P(\mathbf{l}, t = 0)$ is $P(\mathbf{q}, t = 0)$, it will decay to zero as

$$P(\mathbf{q}, t) = e^{-t/\tau(\mathbf{q})}P(\mathbf{q}, t = 0) \tag{7.4.48}$$

with

$$\begin{aligned}\tau^{-1}(\mathbf{q}) &= \tau^{-1}[\gamma(0) - \gamma(\mathbf{q})] = \tau^{-1}\sum_{\boldsymbol{\delta}}(1 - e^{i\mathbf{q}\cdot\boldsymbol{\delta}}) \\ &\sim \frac{a^2}{\tau}q^2 \ \text{ for } \mathbf{q} \to 0,\end{aligned} \tag{7.4.49}$$

where $\boldsymbol{\delta}$ is a nearest neighbor vector of magnitude a of the lattice and where the numerical coefficient of q^2 in the last equation is that appropriate to a hypercubic lattice. Thus, at long wavelengths, $P(\mathbf{q}, t)$ decays diffusively [see Eq. (7.4.10)] with diffusion constant

$$D = a^2/\tau. \tag{7.4.50}$$

This is the form of D that one might have predicted simply on the basis of dimensional analysis. The value of $\tau(\mathbf{q})$ at higher values of $\mathbf{q}$ depends on the lattice in question. In one dimension,

$$\tau^{-1}(\mathbf{q}) = 2\tau^{-1}(1 - \cos qa). \tag{7.4.51}$$

This function is plotted in Fig. 7.4.6.

The decay time τ can vary widely from system to system. It is often determined by processes involving thermal activation over some barrier. In this case,

$$\tau^{-1} \sim e^{-\Delta E/T}, \tag{7.4.52}$$

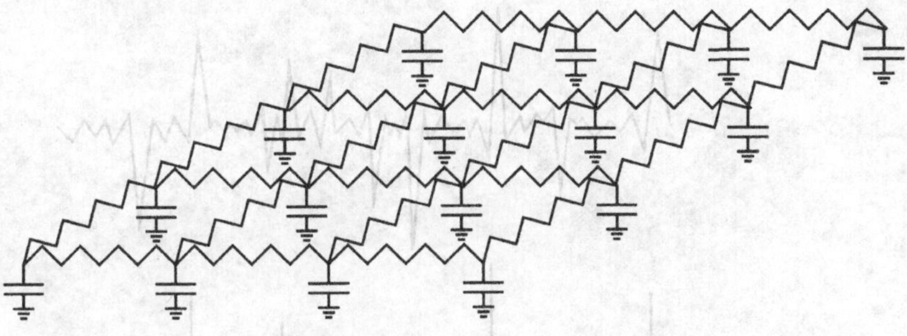

Fig. 7.4.7. Schematic representation of a resistor network with capacitances to ground.

where ΔE is the barrier energy. For temperatures much less than ΔE, τ becomes very long. In some situations, it can be so long that the probability that a hop occurs in the time scale of a laboratory experiment can be vanishingly small.

It is interesting to observe that Eq. (7.4.47) is precisely the equation governing the voltage in a resistor network consisting of sites connected by resistances of conductance $\sigma_{\mathbf{l},\mathbf{l}'}$ with capacitance to ground C. The equation for the voltage $V(\mathbf{l}, t)$ at site $\mathbf{l}$,

$$C\frac{\partial V(\mathbf{l},t)}{\partial t} = \sum_{\mathbf{l}'} \sigma_{\mathbf{l},\mathbf{l}'}[V(\mathbf{l}',t) - V(\mathbf{l},t)], \tag{7.4.53}$$

is determined by Kirchhoff's laws. When resistors connect only nearest neighbor sites, $\sigma_{\mathbf{l},\mathbf{l}'} = \sigma\gamma_{\mathbf{l},\mathbf{l}'}$, and Eq. (7.4.53) reduces to Eq. (7.4.47) with $\tau = (C/\sigma)$. Resistor networks are often used to model diffusive transport problems (see Fig. 7.4.7).

7.5 Langevin theory

1 *Random forces and thermal equilibrium*

The erratic motion of a Brownian particle is due to collisions with molecules in the fluid in which it moves. These collisions allow an exchange of energy between the fluid at temperature T and the Brownian particle and for the establishment of thermal equilibrium between the degrees of freedom of the particle and those of the fluid. This means that the mean-square of each component of the velocity of the Brownian particle averaged over a sufficiently long time must have the value T/m predicted by Boltzmann statistics. This average is maintained through constant collisions.

To understand how thermal equilibrium can be brought about by random forces, let us focus on a particle diffusing in one dimension. Individual molecules of the fluid collide with the diffusing particle in a random fashion and exert a force whose time average is simply the viscous force $-\alpha v$ introduced in Eq. (7.2.7).

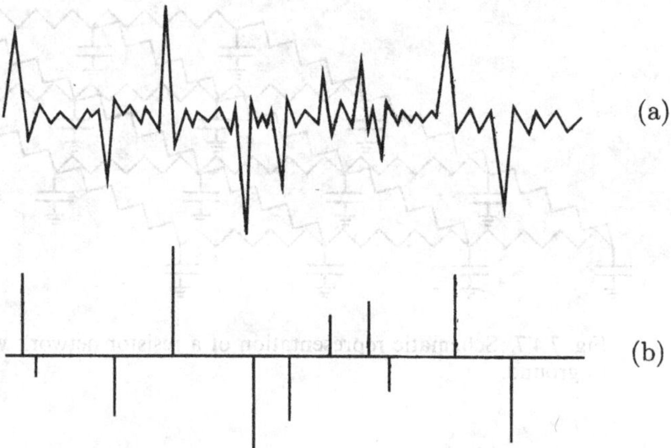

Fig. 7.5.1. (a) Schematic representation of the random force $\zeta(t)$ as a function of time. (b) $\zeta(t)$ approximated by a series of random impulses.

We can, therefore, break the force exerted on the particle by the fluid into two parts: the average viscous force $-\alpha v$ and a random force $\zeta(t)$ whose time average is zero.

This random force is well approximated by a sequence of independent impulses of random sign and magnitude as shown in Fig. 7.5.1; it is a stochastic process whose time average is zero. Rather than considering averages over time, we will consider averages over the ensemble of possible random forces and represent averages over this ensemble with brackets, $\langle\,\rangle$, in the same way that we represented averages over equilibrium thermodynamic ensembles. We will choose the ensemble of random forces so that averages over it are identical to averages over an equilibrium ensemble. Thus, we have

$$\langle \zeta(t) \rangle = 0. \tag{7.5.1}$$

If each impulse is considered an independent random event, then the probability distribution for $\zeta(t)$ is independent of $\zeta(t')$ for $t' \neq t$. This implies

$$\langle \zeta(t)\zeta(t') \rangle = A\delta(t-t') \tag{7.5.2}$$

is local in time. A is a constant that remains to be determined. Finally, $\zeta(t)$ in the independent impulse approximation is the sum of a large number of independent functions. The central limit theorem then implies that the probability distribution for $\zeta(t)$ is Gaussian with a width determined by its variance, Eq. (7.5.2):

$$P\left[\zeta(t)\right] = \frac{1}{\sqrt{2\pi A}} e^{-\frac{1}{2A}\int dt \zeta^2(t)}. \tag{7.5.3}$$

Random forces such as $\zeta(t)$ give rise to erratic or *noisy* behavior of observables and are often referred to as *noise sources*, especially in the context of electrical circuits.

Eqs. (7.5.1) to (7.5.3) provide a sufficiently precise characterization of the stochastic collision force $\zeta(t)$ to allow us to discuss the establishment of thermal equilibrium. The detailed form of $\zeta(t)$ is determined by the temporal statistics of the molecules of the fluid. Thus, one expects the approximation of independent random events to break down for time differences $t - t'$ less than a characteristic collision time τ_c of the fluid. As discussed in Sec. 7.3, however, the characteristic time γ^{-1} for motion of the Brownian particle is much larger than τ_c, and the independent collision approximation will be very good for times of interest.

The power spectrum of $\zeta(t)$, or the Fourier transform of $\langle\zeta(t)\zeta(t')\rangle$,

$$I(\omega) \equiv C_{\zeta\zeta}(\omega) = A \tag{7.5.4}$$

is independent of ω in the independent collision approximation. A noise source with a frequency-independent power spectrum is called a *white noise* source.

2 Correlation functions for diffusion

We will now show how knowledge of statistics of the stochastic force allows us to calculate correlation functions rather than response functions (Langevin 1908). In the absence of external forces, the equation of motion of a diffusing particle is

$$m\dot{v} + \alpha v = \zeta(t). \tag{7.5.5}$$

The solution to this equation for $v(t)$ has a homogeneous part determined by initial conditions and an inhomogeneous part proportional to $\zeta(t)$. Since the homogeneous part, which depends on initial conditions, will decay to zero in a time of order γ^{-1}, the long-time properties of $v(t)$ will be determined entirely by the inhomogeneous part and be independent of initial conditions. In Fourier space the inhomogeneous part of v is simply

$$v(\omega) = \frac{\zeta(\omega)}{-i\omega m + \alpha}. \tag{7.5.6}$$

Using Eqs. (7.1.7) and (7.5.4), we can calculate $C_{vv}(\omega)$ by averaging $v(\omega)v(-\omega)$ over the random forces:

$$C_{vv}(\omega) = \frac{I(\omega)}{|-i\omega m + \alpha|^2} = \frac{A}{m^2[\omega^2 + \gamma^2]}. \tag{7.5.7}$$

The constant A characterizing the variance of the random force is as yet unspecified. We can now use Eq. (7.5.7) to calculate the instantaneous mean-square velocity in terms of A and thereby determine the value of A needed to ensure thermal equilibrium:

$$\langle v^2\rangle = \int \frac{d\omega}{2\pi} C_{vv}(\omega) = \frac{A}{2m\alpha}. \tag{7.5.8}$$

In thermal equilibrium $\langle v^2\rangle = T/m$, and we conclude

$$A = 2\alpha T = 2m\gamma T \text{ and } \langle\zeta(t)\zeta(t')\rangle = 2\alpha T\delta(t - t'). \tag{7.5.9}$$

Thus, the amplitude of white noise fluctuations is fixed by the requirements of thermal equilibrium.

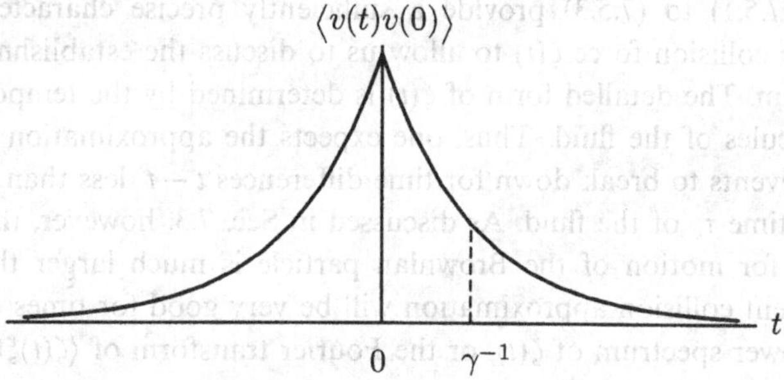

Fig. 7.5.2. The velocity correlation function $\langle v(t)v(0)\rangle$ showing exponential decay to zero.

The correlation function $C_{vv}(\omega)$ determines $C_{vv}(t,t') = \langle v(t)v(t')\rangle$ as well as the instantaneous average $\langle v^2\rangle$. Neglect of the homogeneous term in the solution for $v(\omega)$ is only valid for times long compared to γ^{-1}. The Fourier transform of $C_{vv}(\omega)$ in Eq. (7.5.7) therefore gives the function $C_{vv}(t) = \lim_{\tau\to\infty} C_{vv}(\tau+t,\tau)$:

$$C_{vv}(t-t') = \int \frac{d\omega}{2\pi} e^{-i\omega(t-t')} C_{vv}(\omega) = \frac{T}{m} e^{-\gamma|t-t'|}. \tag{7.5.10}$$

This equation shows that $v(t)$ and $v(0)$ become decorrelated for times greater than γ^{-1}, and that $\langle v(t)v(0)\rangle$ is of order the equal-time thermal average T/m for times less than γ^{-1}, as shown in Fig. 7.5.2.

The Fourier transform $x(\omega)$ of position is $v(\omega)/(-i\omega)$. Thus, we can determine the position correlation function,

$$C_{xx}(\omega) = \frac{2\gamma T}{m\omega^2(\omega^2+\gamma^2)}, \tag{7.5.11}$$

from Eq. (7.5.7) for $C_{vv}(\omega)$. The integral of $C_{xx}(\omega)$ over ω gives the mean-square displacement $\langle x^2(t)\rangle$. This integral is divergent because of the extra factor of ω^2 in the denominator of $C_{xx}(\omega)$, and we correctly conclude that $\langle x^2(t)\rangle$ is infinite. This result is analogous to the result, discussed in Chapter 6, that the mean-square of an elastic variable is infinite below its lower critical dimension. The average $\langle [x(t)-x(t')]^2\rangle \equiv \langle [\Delta x(t-t')]^2\rangle$ is, however, finite. Using Eq. (7.5.11), we obtain

$$\begin{aligned}\langle [\Delta x(t)]^2\rangle &= \int \frac{d\omega}{2\pi} 2C_{xx}(\omega)[1-e^{-i\omega t}] \\ &= \frac{4T}{m\gamma}\int \frac{d\omega}{2\pi}\left(\frac{1}{\omega^2} - \frac{1}{\omega^2+\gamma^2}\right)\left(1-e^{-i\omega t}\right). \end{aligned} \tag{7.5.12}$$

The second term in this expression is easily evaluated by contour integration; the first term, which is proportional to $|t|$, can be obtained from the second by taking the limits $\gamma \to 0$. The result is

$$\langle [\Delta x(t)]^2\rangle = 2D\left(|\,t\,| - \frac{1-e^{-\gamma|t|}}{\gamma}\right), \tag{7.5.13}$$

where we used the Einstein relation $D = T/m\gamma$. For times $t \gg \gamma^{-1}$, this equation reduces to the result, Eq. (7.4.11), predicted by the diffusion equation in one spatial dimension. At short times,

$$\langle[\Delta x(t)]^2\rangle \sim D\gamma t^2 = \langle v^2\rangle t^2, \tag{7.5.14}$$

indicating that the Brownian particle moves ballistically in this limit.

The Einstein relation [Eq. (7.4.26)] can be reexpressed in various ways in terms of the velocity correlation function. First we have

$$\begin{aligned}\langle[\Delta x(t)]^2\rangle &= \left\langle\left(\int_0^t dt' v(t')\right)^2\right\rangle = \int_0^t dt_1 \int_0^t dt_2 C_{vv}(t_1 - t_2)\\ &= 2\int_0^t dt_1 \int_0^{t_1} dt_2 C_{vv}(t_1 - t_2) = 2\int_0^t dt_1 \int_0^{t_1} d\tau C_{vv}(\tau)\\ &= 2\int_0^t (t-\tau) C_{vv}(\tau) d\tau,\end{aligned} \tag{7.5.15}$$

where the final expression in this equation can be obtained from the preceding line by integrating by parts. From this, we can easily calculate a time-dependent diffusion constant,

$$\begin{aligned}D(t) &= \frac{1}{2}\frac{d}{dt}\langle(\Delta x(t))^2\rangle,\\ &= \int_0^t d\tau C_{vv}(\tau),\end{aligned} \tag{7.5.16}$$

that approaches the diffusion constant,

$$D = \int_0^\infty d\tau C_{vv}(\tau) = \frac{1}{2}C_{vv}(\omega = 0), \tag{7.5.17}$$

in the infinite time limit. The last expression could have been obtained directly from Eqs. (7.5.7), (7.5.9), and the relation $D = T/(m\gamma)$.

3 Short-time behavior

In the above analysis, we argued we could neglect initial conditions if we are interested in long-time limits and thermal equilibrium. Initial conditions are, however, often of interest and can be treated almost as easily as the long-time limits. The solution to Eq. (7.5.5) for $v(t)$ subject to the boundary condition that $v(t = 0) = v_0$ is

$$v(t) = v_0 e^{-\gamma t} + \int_0^t dt_1 e^{-\gamma(t-t_1)}\zeta(t_1)/m. \tag{7.5.18}$$

The average velocity is then

$$\langle v(t)\rangle = v_0 e^{-\gamma t}. \tag{7.5.19}$$

The velocity correlation function is

$$\begin{aligned}\langle v(t)v(t')\rangle &= v_0^2 e^{-\gamma(t+t')} + \int_0^t dt_1 \int_0^{t'} dt_2 e^{-\gamma(t-t_1)-\gamma(t'-t_2)} \frac{2\gamma T}{m}\delta(t_1 - t_2) \\ &= \left(v_0^2 - \frac{T}{m}\right) e^{-\gamma(t+t')} + \frac{T}{m} e^{-\gamma|t-t'|}, \end{aligned} \tag{7.5.20}$$

and the variance of the velocity is

$$\Delta_v(t) = \langle [v(t) - \langle v(t)\rangle]^2\rangle = \frac{T}{m}(1 - e^{-2\gamma|t|}), \tag{7.5.21}$$

where we used Eq. (7.5.2) for $\langle \zeta(t)\zeta(t')\rangle$ with $A = 2m\gamma T$. These equations show that the velocity correlation function tends to the thermal equilibrium result of Eq. (7.5.10) for times t and/or t' much greater than the decay time γ^{-1}, regardless of the initial velocity v_0. Furthermore, if the initial velocity is averaged over an equilibrium Maxwell-Boltzmann distribution at temperature T, then $\langle v(t)v(t')\rangle$ has its thermal equilibrium form at all times.

The displacement variable $x(t)$ can be obtained from the velocity by simple integration:

$$x(t) = x_0 + \int_0^t v(t_1)dt_1, \tag{7.5.22}$$

where $x(t = 0) = x_0$. From this and Eq. (7.5.19), we can calculate the average displacement as a function of time:

$$\langle x(t)\rangle = x_0 + (v_0/\gamma)(1 - e^{-\gamma t}). \tag{7.5.23}$$

Similarly, we can calculate correlations in the displacement at different times $t > 0$ and $t' > 0$:

$$\langle [x(t) - x(t')]^2\rangle = \left\langle \left(\int_{t'}^t dt_1 v(t_1)\right)^2\right\rangle. \tag{7.5.24}$$

Using Eq. (7.5.20) for the velocity correlation function, we obtain

$$\begin{aligned}\langle [x(t) - x(t')]^2\rangle &= \left(v_0^2 - \frac{T}{m}\right)\frac{1}{\gamma^2}\left(e^{-\gamma t'} - e^{-\gamma t}\right)^2 \\ &\quad + \frac{2T}{\gamma m}\left[|t - t'| - \frac{1}{\gamma}\left(1 - e^{-\gamma|t-t'|}\right)\right]. \end{aligned} \tag{7.5.25}$$

If both t and t' are much greater than γ^{-1}, this reduces to Eq. (7.5.13) independent of v_0. If, on the other hand, both t and t' are much less than γ^{-1}, $\langle [x(t) - x(t')]\rangle = v_0^2(t - t')^2$, i.e., the Brownian particle moves ballistically with the specified initial velocity. The average of Eq. (7.5.25) over an equilibrium ensemble of initial velocities also reduces to Eq. (7.5.13). The average $\langle [x(t) - x_0]^2\rangle$ is obtained from Eq. (7.5.25) by setting $t' - 0$ so that $x(t') = x_0$. Finally, we can calculate the variance of the position,

$$\begin{aligned}\Delta_x(t) &= \langle [x(t) - \langle x(t)\rangle]^2\rangle \\ &= 2\frac{T}{\gamma m}\left[t - \frac{1}{\gamma}\left(1 - e^{-\gamma t}\right) - \frac{1}{2\gamma}\left(1 - e^{-\gamma t}\right)^2\right]. \end{aligned} \tag{7.5.26}$$

Note that $\Delta_x(t)$ is not equal to $\langle[\Delta x(t)]^2\rangle$ in Eq. (7.5.13) because it explicitly retains the memory that initial motion was ballistic rather than diffusive. The variance, $\langle[x(t) - x(t') - \langle x(t) - x(t')\rangle]^2\rangle$ does not, and is in fact identical to $\langle[\Delta x(t - t')]^2\rangle$.

The noise $\zeta(t)$ is a Gaussian random variable. Both $v(t)$ and $x(t)$ are linear functions of $\zeta(t)$. Since linear functions of Gaussian random variables are also Gaussian random variables, the probability distribution functions for $v(t)$ and $x(t)$ are Gaussian and are completely determined by the expectation values and variances of these variables. We leave a formal derivation of these results to the problems at the end of the chapter.

7.5.4 *Fluctuation-dissipation theorem for the harmonic oscillator*

A harmonic oscillator in a viscous fluid, like a free particle in the same fluid, will reach thermal equilibrium as a result of collisions with the fluid molecules. This means that the average energy per degree of freedom of the oscillator will be $T/2$, or that $\langle x^2(t)\rangle = T/(m\omega_0^2)$ and $\langle v^2(t)\rangle = T/m$. The equation of motion for an oscillator in a random force is Eq. (7.2.9), with f replaced by $\zeta(t)$. In the long-time limit, we need only concern ourselves with the inhomogeneous solution to this equation, which as a function of frequency is

$$x(\omega) = \chi(\omega)\zeta(\omega) = \frac{\zeta(\omega)}{m[-\omega^2 + \omega_0^2 - i\omega\gamma]}, \tag{7.5.27}$$

where $\chi(\omega)$ is the response function of Eq. (7.2.16). The nature of the random force $\zeta(t)$ does not depend on whether our particle is attached to a spring or not. The noise correlation function $C_{\zeta\zeta}(\omega)$ is thus independent of ω_0 and has the same form as for $\omega_0 = 0$. From this, and the correlation function for $\zeta(t)$, we obtain

$$C_{xx}(\omega) = 2m\gamma T|\chi(\omega)|^2 = \frac{2\gamma T}{m}\frac{1}{(\omega^2 - \omega_0^2)^2 + \omega^2\gamma^2}. \tag{7.5.28}$$

We leave it as an exercise to verify that $\langle x^2\rangle$ obtained by integrating this function over ω is, in fact, $T/(m\omega_0^2)$. Then, using Eq. (7.2.19) and the fact that $\langle x(t)\rangle = 0$ so that $C_{xx}(\omega) = S_{xx}(\omega)$, we obtain the very important result

$$\chi''_{xx}(\omega) = \frac{1}{2}\beta\omega S_{xx}(\omega), \tag{7.5.29}$$

where $\beta \equiv 1/T$. This is the classical *fluctuation-dissipation theorem*, the complete quantum mechanical version of which was originally derived by Callen and Welton (1952). It relates $\chi''_{xx}(\omega)$, which, as we saw in Sec. 7.3, is proportional to the rate at which work done by external forces is dissipated as heat, to the Fourier transform of the mean-square fluctuation $\langle[x(t) - \langle x(t)\rangle][x(0) - \langle x(0)\rangle]\rangle$. Thus, absorption or response experiments that probe $\chi''_{xx}(\omega)$ contain the same information as scattering or related measurements that probe $S_{xx}(\omega)$. Although we derived the fluctuation-dissipation theorem for a single classical oscillator in equilibrium in a viscous fluid, the theorem applies to all response and correlation functions of systems in equilibrium. Furthermore, it is applicable, as we shall see in the next section, to situations where the classical approximation is not applicable.

5 The Fokker-Planck and Smoluchowski equations

In the preceding discussion, we focused on the correlation functions of velocity and position. The Langevin equations can be used to derive not only these correlation functions but also the equations determining the entire probability distribution function for these variables. The equation for the velocity probability function for a diffusing particle is called the Fokker-Planck equation; its generalization to displacement and other variables is generally called the Smoluchowski equation. These equations show how probability distributions decay to Maxwell-Boltzmann distributions describing thermal equilibrium at long times. They are applicable not only for harmonic Hamiltonians but also for anharmonic Hamiltonians containing other than quadratic terms in the fundamental variables. This latter result is important because it implies that Langevin equations provide a correct phenomenological description of dynamics for all arbitrarily complicated interacting systems as well as for the simple free particles and harmonic oscillator we have considered so far.

We will begin our derivation of the Fokker-Planck equation by rewriting the equation of motion in terms of the momentum p to produce a form that will most easily generalize to other variables:

$$\frac{dp}{dt} = -\gamma p + \zeta = -\Gamma \frac{\partial \mathscr{H}}{\partial p} + \zeta, \tag{7.5.30}$$

where $\Gamma \equiv \alpha = \gamma m$ and

$$\langle \zeta(t)\zeta(t') \rangle = 2\Gamma T \delta(t-t'). \tag{7.5.31}$$

This equation is now in a form that could in general include anharmonicities in the Hamiltonian $\mathscr{H}$.

We now consider the probability

$$P(p,t \mid p_0,t_0) = \langle \delta(p - p(t)) \rangle_{p_0,t_0} \tag{7.5.32}$$

that a diffusing particle has momentum p at time t, given that it had momentum p_0 at time t_0. The probability that the particle has a momentum p at time $t + \Delta t$ is

$$P(p,t+\Delta t \mid p_0,t_0) = \int dp' P(p,t+\Delta t \mid p',t) P(p',t \mid p_0,t_0). \tag{7.5.33}$$

The conditional probability

$$P(p,t+\Delta t \mid p',t) = \langle \delta(p - p(t+\Delta t)) \rangle_{p',t} \tag{7.5.34}$$

can be calculated from the equation of motion for $p(t)$:

$$p(t+\Delta t) = p' - \Gamma \frac{\partial \mathscr{H}}{\partial p'} \Delta t + \int_t^{t+\Delta t} dt' \zeta(t'). \tag{7.5.35}$$

The average of the third term in this equation is zero; its square, however, is proportional to Δt:

$$\int_t^{t+\Delta t} dt_1 \int_t^{t+\Delta t} dt_2 \langle \zeta(t_1)\zeta(t_2)\rangle = 2\Gamma T \Delta t. \tag{7.5.36}$$

Terms higher order in $\int dt \zeta(t)$ are higher order in Δt because $\zeta(t)$ is a Gaussian random variable and averages of products of $\zeta(t)$ can be expressed as products of the variance $\langle \zeta(t_1)\zeta(t_2)\rangle$. Thus, for example $\langle (\int dt \zeta(t))^4\rangle \sim (\int dt_1 dt_2 \langle \zeta(t_1)\zeta(t_2)\rangle)^2 \sim (\Delta t)^2$. Using this result, we now expand the left hand side of Eq. (7.5.34) to first order in Δt:

$$\langle \delta(p - p(t+\Delta t))\rangle_{p',t} = \left[1 + \Delta t \Gamma \frac{\partial \mathscr{H}}{\partial p'}\frac{\partial}{\partial p} + \Delta t \Gamma T \frac{\partial^2}{\partial p^2}\right] \delta(p - p'). \tag{7.5.37}$$

This result and Eq. (7.5.32) then allow us to calculate

$$\frac{\partial P}{\partial t} = T\Gamma \frac{\partial}{\partial p}\left[\left(\frac{1}{T}\frac{\partial \mathscr{H}}{\partial p} + \frac{\partial}{\partial p}\right) P\right]. \tag{7.5.38}$$

The left hand side of this equation is zero when

$$P = P_{\text{eq}} \sim e^{-\mathscr{H}(p)/T}, \tag{7.5.39}$$

i.e., when P has the equilibrium form predicted by Maxwell-Boltzmann statistics. In fact, P decays in time to P_{eq}.

The probability distribution for *any* variable ϕ satisfying a linear differential equation in time of the form of Eq. (7.5.30) will satisfy Eq. (7.5.38) with p replaced by ϕ. For example, the equation for an overdamped oscillator Eq. (7.2.15) has exactly the same form of Eq. (7.5.30). The equation for $P(x,t)$, which is identical to Eq. (7.5.38) with p replaced by x, is the Smoluchowski equation.

The probability distribution [Eq. (7.5.32)] appearing in the Fokker-Planck equation [Eq. 7.5.38] is for the momentum subject to the boundary condition $p(t=0) = p_0$. We have calculated both the expectation value $\langle p(t)\rangle = m\langle v(t)\rangle$ and the variance $\Delta_p(t) = m^2 \Delta_v(t)$ of the momentum subject to this boundary condition. The force $\zeta(t)$ is a Gaussian random process governed by the probability distribution of Eq. (7.5.3). Since the velocity is linearly proportional to $\zeta(t)$, it should also be a Gaussian random process with a Gaussian probability distribution (i.e., characterized only by its mean and variance). One can easily verify that

$$P(p,t \mid p_0, 0) = \frac{1}{(2\pi\Delta_p(t))^{1/2}} e^{-(p-\langle p(t)\rangle)^2/2\Delta_p(t)} \tag{7.5.40}$$

satisfies the Fokker-Planck equation.

7.6 Formal properties of response functions

1 Response to external fields

In Sec. 7.1, we defined the dynamic response function $\tilde{\chi}_{\phi_i\phi_j}(\mathbf{x},\mathbf{x}',t,t')$ relating the deviation $\delta\langle\phi_i(\mathbf{x},t)\rangle$ of the average of the field $\phi_i(\mathbf{x},t)$ from its equilibrium value to first-order changes in the time-dependent external field $h_j(\mathbf{x}',t')$ thermodynamically conjugate to $\phi_j(\mathbf{x}',t')$. We then showed how this function could be calculated from phenomenological equations of motion. We also found that the imaginary part of the frequency-dependent response function for a classical harmonic oscillator was related in a simple way to an equilibrium correlation function [Eq. (7.5.29)]. In this section, we will develop a general formalism for describing dynamic response functions. The important result of this general treatment is that response functions can be expressed in terms of equilibrium expectation values of commutators of operators. From this follow a number of general symmetry properties of response functions and the general quantum mechanical fluctuation-dissipation theorem relating the dissipative part of the susceptibility to an equilibrium correlation function. Our development will be fully quantum-mechanical. Classical results follow simply from the classical limit of quantum mechanics.

The Hamiltonian of a system in the presence of an external field $h_j(\mathbf{x},t)$ can be expressed as

$$\mathscr{H}_T = \mathscr{H} + \mathscr{H}_{\text{ext}}, \tag{7.6.1}$$

where $\mathscr{H}$ is the Hamiltonian describing the system when h_j is zero, and

$$\mathscr{H}_{\text{ext}} = -\int d^d x \sum_j \phi_j(\mathbf{x}) h_j(\mathbf{x},t), \tag{7.6.2}$$

expressed in the Schrödinger representation, where the field $\phi_j(\mathbf{x})$ is independent of time. $\mathscr{H}_{\text{ext}}$ is a perturbation introduced to measure response. There can, of course, be terms in $\mathscr{H}$ linear in $\phi_j(\mathbf{x})$ that look like $\mathscr{H}_{\text{ext}}$. We will be interested in the limit $h_j(\mathbf{x},t) \to 0$, and we will assume that $h_j(\mathbf{x},t)$ is zero for times less than some time t_0, which we will eventually allow to go to $-\infty$. Expectation values of operators in the presence of the external field can be expressed as

$$\langle\phi_i(\mathbf{x},t)\rangle_h = \mathrm{Tr}\rho_h(t,t_0)\phi_i(\mathbf{x},t_0), \tag{7.6.3}$$

where $\rho_h(t,t_0)$ is the time-dependent density matrix for nonzero $h_j(\mathbf{x},t)$ that reduces to the statistical equilibrium density matrix $\rho_{\text{eq}}(h=0)$ for times less than t_0. The time evolution of $\rho_h(t,t_0)$ is governed by the Schrödinger equation,

$$i\hbar\frac{\partial\rho_h}{\partial t} = [\mathscr{H} + \mathscr{H}_{\text{ext}}, \rho_h], \tag{7.6.4}$$

with the boundary condition $\rho_h(t_0,t_0) = \rho_{\text{eq}} \equiv \rho$. Thus,

$$\rho_h(t,t_0) = U(t,t_0)\rho U^{-1}(t,t_0), \tag{7.6.5}$$

where

$$i\hbar\frac{dU(t,t_0)}{dt} = (\mathscr{H} + \mathscr{H}_{\text{ext}})U(t,t_0) \tag{7.6.6}$$

with $U(t_0, t_0) = 1$ and $U(t, t_0)U^{-1}(t, t_0) = 1$. Because of the cyclic invariance property of the trace, Eq. (7.6.3) can also be expressed as

$$\langle\phi_i(\mathbf{x}, t)\rangle_h = \text{Tr}\rho_h(t, t_0)\phi_i(\mathbf{x}, t_0) = \text{Tr}\rho U^{-1}(t, t_0)\phi_i(\mathbf{x}, t_0)U(t, t_0). \tag{7.6.7}$$

The final form puts time variation in the field operator, expressed in the Heisenberg representation, rather than in the density matrix.

In order to discuss the time dependence of $\phi_i(\mathbf{x}, t)$ when h_j is nonzero, it is convenient to introduce the interaction representation for $U(t, t_0)$ via

$$U(t, t_0) = U_0(t, t_0)U'(t, t_0), \tag{7.6.8}$$

where $U'(t_0, t_0) = 1$ and

$$i\hbar\frac{dU_0}{dt} = \mathscr{H}U_0. \tag{7.6.9}$$

From this and Eqs. (7.6.6) and (7.6.8), it follows that

$$i\hbar\frac{dU'}{dt} = [U_0^{-1}\mathscr{H}_{\text{ext}}U_0]U' \equiv \mathscr{H}^I_{\text{ext}}U', \tag{7.6.10}$$

where the superscript on $\mathscr{H}^I_{\text{ext}}$ indicates that it is expressed in the interaction representation where time evolution is determined by $\mathscr{H}$ rather than by the total Hamiltonian $\mathscr{H} + \mathscr{H}_{\text{ext}}$. Eq. (7.6.10) can be integrated perturbatively to yield $U'(t, t_0)$ as a power series in $\mathscr{H}^I_{\text{ext}}$. The result is

$$\begin{aligned} U'(t, t_0) &= 1 + \frac{1}{i\hbar}\int_{t_0}^{t} \mathscr{H}^I_{\text{ext}}(t')dt' \\ &\quad + \left(\frac{1}{i\hbar}\right)^2 \int_{t_0}^{t} \mathscr{H}^I_{\text{ext}}(t') \int_{t_0}^{t'} \mathscr{H}^I_{\text{ext}}(t'')dt'dt'' + \cdots \\ &\equiv \left[\exp\left(\frac{1}{i\hbar}\int_{t_0}^{t} \mathscr{H}^I_{\text{ext}}(t')dt'\right)\right]_+, \end{aligned} \tag{7.6.11}$$

where $[\;]_+$ indicates that all operators within the brackets are to be ordered from right to left according to increasing time t. Defining $\phi_i^I(\mathbf{x}, t)$ to be the operator $\phi_i^I(\mathbf{x}, t) = U_0^{-1}(t, t_0)\phi_i(\mathbf{x}, t_0)U_0(t, t_0)$ in the interaction representation, we can write

$$\begin{aligned} \langle\phi_i(\mathbf{x}, t)\rangle_h &= \text{Tr}\rho U'^{-1}(t, t_0)\phi_i^I(\mathbf{x}, t)U'(t, t_0) \\ &\approx \text{Tr}\rho\left[\left(1 - \frac{1}{i\hbar}\int_{t_0}^{t} \mathscr{H}^I_{\text{ext}}(t')dt'\right)\phi_i^I(\mathbf{x}, t)\right. \\ &\quad \left.\times\left(1 + \frac{1}{i\hbar}\int_{t_0}^{t} \mathscr{H}^I_{\text{ext}}(t')dt'\right)\right] \\ &\approx \text{Tr}\rho\phi_i^I(\mathbf{x}, t) + \text{Tr}\rho\left(\frac{1}{i\hbar}\int_{t_0}^{t} dt'\,[\phi_i^I(\mathbf{x}, t), \mathscr{H}^I_{\text{ext}}(t')]\right) \end{aligned} \tag{7.6.12}$$

where $[\phi_i^I, \mathscr{H}^I_{\text{ext}}]$ is the commutator of ϕ_i^I with $\mathscr{H}^I_{\text{ext}}$. The first term in this equation is merely the equilibrium expectation value $\langle\phi_i(\mathbf{x}, t)\rangle$ in the absence of the external field h_j, and the second term reflects the effects of the external potential to lowest (i.e., linear) order in $h_j(\mathbf{x}, t)$. Allowing the initial time t_0 to go to $-\infty$, we obtain

$$\begin{aligned}\delta\langle\phi_i(\mathbf{x},t)\rangle &= \langle\phi_i(\mathbf{x},t)\rangle_h - \langle\phi_i(\mathbf{x},t)\rangle \\ &= -\int_{-\infty}^{t} dt\frac{1}{i\hbar}\int d^d\mathbf{x}'\langle[\phi_i(\mathbf{x},t),\phi_j(\mathbf{x}',t')]\rangle h_j(\mathbf{x}',t') \\ &= \int_{-\infty}^{\infty} dt\int d^d\mathbf{x}'\frac{i}{\hbar}\eta(t-t')\langle[\phi_i(\mathbf{x},t),\phi_j(\mathbf{x}',t')]\rangle h_j(\mathbf{x}',t'),\end{aligned} \tag{7.6.13}$$

where we have dropped the now superfluous superscript I and where, as usual, $\langle\,\rangle$ signifies an average with respect to the equilibrium density matrix ρ. The field $\phi_i(\mathbf{x},t)$ is the Heisenberg operator, evolving in time according to the Hamiltonian $\mathscr{H}$. Comparing Eqs. (7.1.11), (7.1.12), and (7.6.13), we obtain

$$\begin{aligned}\tilde{\chi}_{\phi_i\phi_j}(\mathbf{x},\mathbf{x}',t-t') &= \eta(t-t')\frac{i}{\hbar}\langle[\phi_i(\mathbf{x},t),\phi_j(\mathbf{x}',t')]\rangle, \\ \tilde{\chi}''_{\phi_i\phi_j}(\mathbf{x},\mathbf{x}',t-t') &= \frac{1}{2\hbar}\langle[\phi_i(\mathbf{x},t),\phi_j(\mathbf{x}',t')]\rangle,\end{aligned} \tag{7.6.14}$$

where we have indicated explicit dependence on the time difference $t-t'$. This equation gives us the desired expressions for $\tilde{\chi}''_{\phi_i\phi_j}(\mathbf{x},\mathbf{x}',t-t')$ and $\tilde{\chi}_{\phi_i\phi_j}(\mathbf{x},\mathbf{x}',t-t')$ in terms of average values of products of operators evaluated in an equilibrium ensemble.

In classical systems, the commutators in Eq. (7.6.14) become Poisson brackets:

$$\begin{aligned}\tilde{\chi}''_{\phi_i\phi_j}(\mathbf{x},\mathbf{x}',t-t') &= -\frac{i}{2}\langle\{\phi_i(\mathbf{x},t),\phi_j(\mathbf{x}',t')\}\rangle \\ &\equiv \frac{i}{2}\sum_\alpha\left\langle\frac{\partial\phi_i(\mathbf{x},t)}{\partial q^\alpha(t)}\frac{\partial\phi_j(\mathbf{x}',t')}{\partial p^\alpha(t')} - \frac{\partial\phi_i(\mathbf{x},t)}{\partial p^\alpha(t)}\frac{\partial\phi_j(\mathbf{x}',t')}{\partial q^\alpha(t')}\right\rangle\end{aligned} \tag{7.6.15}$$

where (q^α, p^α) is a complete set of canonically conjugate coordinates.

2 Symmetry properties of response functions

There are a number of symmetry properties of $\tilde{\chi}''_{\phi_i\phi_j}(\mathbf{x},\mathbf{x}',t-t')$ that follow directly because it is an equilibrium average of a commutator. It must be antisymmetric under interchange of all indices:

$$\begin{aligned}\tilde{\chi}''_{\phi_i\phi_j}(\mathbf{x},\mathbf{x}',t-t') &= -\tilde{\chi}''_{\phi_j\phi_i}(\mathbf{x}',\mathbf{x},t'-t) \\ \chi''_{\phi_i\phi_j}(\mathbf{x},\mathbf{x}',\omega) &= -\chi''_{\phi_j\phi_i}(\mathbf{x}',\mathbf{x},-\omega).\end{aligned} \tag{7.6.16}$$

The fields $\phi_i(\mathbf{x},t)$ are observables and thus Hermitian operators. Therefore,

$$\begin{aligned}[\tilde{\chi}''_{\phi_i\phi_j}(\mathbf{x},\mathbf{x}',t-t')]^* &= \left(\frac{1}{2\hbar}\langle[\phi_i(\mathbf{x},t),\phi_j(\mathbf{x}',t')]\rangle\right)^* \\ &= \frac{1}{2\hbar}\langle[\phi_j(\mathbf{x}',t'),\phi_i(\mathbf{x},t)]\rangle,\end{aligned} \tag{7.6.17}$$

and

$$[\tilde{\chi}''_{\phi_i\phi_j}(\mathbf{x},\mathbf{x}',t-t')]^* = \tilde{\chi}''_{\phi_j\phi_i}(\mathbf{x}',\mathbf{x},t'-t). \tag{7.6.18}$$

Or, from Eq. (7.6.16),

$$[\tilde{\chi}''_{\phi_i\phi_j}(\mathbf{x},\mathbf{x}',t-t')]^* = -\tilde{\chi}''_{\phi_i\phi_j}(\mathbf{x},\mathbf{x}',t-t'). \tag{7.6.19}$$

Hence, $\tilde{\chi}''_{\phi_i\phi_j}(\mathbf{x},\mathbf{x}',t-t')$ is pure imaginary and $\tilde{\chi}_{\phi_i\phi_j}(\mathbf{x},\mathbf{x}',t-t')$ is pure real as they should be. Finally, using Eq. (7.6.18), we obtain

$$[\chi''_{\phi_i\phi_j}(\mathbf{x},\mathbf{x}',\omega)]^* = \chi''_{\phi_j\phi_i}(\mathbf{x}',\mathbf{x},\omega). \tag{7.6.20}$$

It follows from Eqs. (7.6.16) and (7.6.20) that if $\chi''_{\phi_i\phi_j}(\mathbf{x},\mathbf{x}',\omega)$ is *even* under interchange of $(\mathbf{x},i)$ and $(\mathbf{x}',j)$, it is *real* and *odd* in ω. If it is odd under the same interchange, it is *imaginary* and *even* in ω. The former case is the most common, but the latter can occur.

The behavior of the fields $\phi_i(\mathbf{x},t)$ and the density matrix ρ under time reversal determine further symmetry properties of $\tilde{\chi}''_{\phi_i\phi_j}(\mathbf{x},\mathbf{x}',t,t')$. The time-reversal operator θ applied to an operator ϕ_i leads to a new operator $\phi'_i = \theta\phi_i\theta^{-1}$. Fields $\phi_i(\mathbf{x},t)$ can be classified according to their signature $\epsilon_{\phi_i} = \pm 1$ under time reversal:

$$\phi'_i(\mathbf{x},t) = \theta\phi_i(\mathbf{x},t)\theta^{-1} = \epsilon_{\phi_i}\phi_i(\mathbf{x},-t). \tag{7.6.21}$$

Operators such as mass and energy density are even ($\epsilon = +1$), whereas those such as momentum density and magnetization are odd ($\epsilon = -1$) under time reversal. The density matrix depends on externally applied fields (such as magnetic fields) and the nature of order in the system. Thus, in the absence of external fields breaking time-reversal symmetry and in states with no order parameter breaking time reversal, ρ is invariant under θ. If, however, there are external fields or order parameters, which we represent by B, breaking time reversal, then ρ will change under θ: $\theta\rho(B)\theta^{-1} = \rho(-B)$. The time-reversal operator θ is antiunitary. Let $|m'\rangle = (\theta|m\rangle)$ and $\langle m'| = (\langle m|\theta)$ be, respectively, the images of $|m\rangle$ and $\langle m|$ under time reversal. Then for any operator A, $\langle m|A|n\rangle = \langle m|\theta\theta A|n\rangle = [\langle m'|(\theta A|n\rangle)]^* = \langle m'|A'|n'\rangle^* = \langle n'|A^\dagger|m'\rangle$ where $A' = \theta A\theta^{-1}$. Applying this relation to $A = \rho\phi_i\phi_j$ and noting that if $\{|m\rangle\}$ is a complete set of states, then $\{|m'\rangle\}$ is also, we obtain

$$\begin{aligned}
\tilde{\chi}''_{\phi_i\phi_j}(\mathbf{x},\mathbf{x}',t-t',B) &= -\epsilon_{\phi_i}\epsilon_{\phi_j}\tilde{\chi}''_{\phi_i\phi_j}(\mathbf{x},\mathbf{x}',t'-t,-B) \\
&= \epsilon_{\phi_i}\epsilon_{\phi_j}\tilde{\chi}_{\phi_j\phi_i}(\mathbf{x}',\mathbf{x},t-t',-B) \\
\chi''_{\phi_i\phi_j}(\mathbf{x},\mathbf{x}',\omega,B) &= -\epsilon_{\phi_i}\epsilon_{\phi_j}\chi''_{\phi_i\phi_j}(\mathbf{x},\mathbf{x}',-\omega,-B) \\
&= \epsilon_{\phi_i}\epsilon_{\phi_j}\chi''_{\phi_j\phi_i}(\mathbf{x}',\mathbf{x},\omega,-B).
\end{aligned} \tag{7.6.22}$$

In a classical system, these results follow from Eq. (7.6.15) and the fact that Poisson brackets change sign under time reversal because $\partial/\partial p^\alpha(t) \to -\partial/\partial p^\alpha(-t)$. When there is no external field or order parameter breaking time-reversal invariance, Eqs. (7.6.16), (7.6.18), and (7.6.22) imply that $\chi''_{\phi_i\phi_j}(\mathbf{x},\mathbf{x}',\omega)$ is real, odd in ω, and symmetric under interchange of $(\mathbf{x},i)$ and $(\mathbf{x}',j)$ when ϕ_i and ϕ_j have the same sign under time reversal and imaginary, even in ω, and antisymmetric under interchange of $(\mathbf{x},i)$ and $(\mathbf{x}',j)$ when they have the opposite sign. Symmetry properties of $\chi'_{\phi_i\phi_j}(\mathbf{x},\mathbf{x}',\omega)$ follow from

$$\chi'_{\phi_i\phi_j}(\mathbf{x},\mathbf{x}',\omega) = \mathscr{P}\int_{-\infty}^{\infty}\frac{d\omega'}{\pi}\frac{\chi''_{\phi_i\phi_j}(\mathbf{x},\mathbf{x}',\omega')(\omega'+\omega)}{\omega'^2-\omega^2}. \tag{7.6.23}$$

For example, when $B=0$, $\chi'_{\phi_i\phi_j}(\mathbf{x},\mathbf{x}',\omega)$ is real, even in ω, and symmetric under interchange of $(\mathbf{x},i)$ and $(\mathbf{x}',j)$ when $\epsilon_{\phi_i}=\epsilon_{\phi_j}$.

In addition to the symmetries involving time coordinates, there are also symmetries involving spatial coordinates. For example, in homogeneous, isotropic systems, $\chi''_{\phi_i\phi_j}(\mathbf{x},\mathbf{x}',t-t')$ must be a function only of $|\mathbf{x}-\mathbf{x}'|$ and its spatial Fourier transform must be a function only of $q=|\mathbf{q}|$. In more complicated crystalline systems, point and space group symmetries lead to other spatial symmetries for $\chi''_{\phi_i\phi_j}(\mathbf{x},\mathbf{x}',\omega)$.

3 Dissipation

In our discussion of the harmonic oscillator, we found that power dissipated was proportional to $\omega\chi''(\omega)$. We will now show that this result is more generally valid. The rate at which work is done on the system can be calculated using a generalization of Eq. (7.2.28). The rate dW/dt at which the external field $h_i(\mathbf{x},t)$ does work on the system is equal to the rate of change of the total energy of the system: $dE/dt = (d/dt)\mathrm{Tr}\rho(t)\mathscr{H}_T = \mathrm{Tr}\rho(t)d\mathscr{H}_T/dt + \mathrm{Tr}(d\rho(t)/dt)\mathscr{H}_T$. In the Schrödinger representation, the fields $\phi_i(\mathbf{x})$ are independent of time, and $d\mathscr{H}_T/dt = d\mathscr{H}_{\mathrm{ext}}/dt$. In addition, $\mathrm{Tr}(d\rho/dt)\mathscr{H}_T = (1/i\hbar)\mathrm{Tr}[\mathscr{H}_T,\rho]\mathscr{H}_T = (1/i\hbar)\mathrm{Tr}\rho[\mathscr{H}_T,\mathscr{H}_T]=0$. Therefore, the rate at which work is done on the system is

$$\begin{aligned}\frac{dW}{dt} &= -\sum_i\int d^dx\langle\phi_i(\mathbf{x},t)\rangle_h\dot{h}_i(\mathbf{x},t)\\ &= -\sum_i\int d^dx\langle\phi_i(\mathbf{x},t)\rangle_0\dot{h}_i(\mathbf{x},t)\\ &\quad -\sum_{ij}\int d^dx d^dx'\int_{-\infty}^{t}[dt'\dot{h}_i(\mathbf{x},t)2i\eta(t-t')\\ &\quad\times\tilde{\chi}''_{\phi_i\phi_j}(\mathbf{x},\mathbf{x}',t-t')h_j(\mathbf{x}',t')]+O(h_j^3).\end{aligned} \tag{7.6.24}$$

Now, consider an external field oscillating at a single frequency ω:

$$h_i(\mathbf{x},t) = \mathrm{Re}h_i(\mathbf{x})e^{-i\omega t} = \frac{1}{2}[h_i(\mathbf{x})e^{-i\omega t}+h_i^*(\mathbf{x})e^{i\omega t}]. \tag{7.6.25}$$

Then

$$\begin{aligned}\frac{dW}{dt} &= -\sum_{ij}\int d^dx d^dx'\frac{1}{4}\left\{\int_{-\infty}^{t}dt'i\omega[h_i^*(\mathbf{x})e^{i\omega t}-h_i(\mathbf{x})e^{-i\omega t}]\right.\\ &\quad\left.\times 2i\tilde{\chi}''_{\phi_i\phi_j}(\mathbf{x},\mathbf{x}',t-t')[h_j^*(\mathbf{x}')e^{i\omega t'}+h_j(\mathbf{x}')e^{-i\omega t'}]\right\}\end{aligned} \tag{7.6.26}$$

where the first term in Eq. (7.6.24) has been dropped because it will disappear on averaging over a cycle of the external potential. Upon performing this average, we obtain

$$\begin{aligned}\overline{\frac{dW}{dt}} &= -\frac{1}{4}i\omega\int_{-\infty}^{t}dt'\int d^dx d^dx'[2ih_i^*(\mathbf{x})\tilde{\chi}''_{\phi_i\phi_j}(\mathbf{x},\mathbf{x}',t-t')h_j(\mathbf{x}')e^{i\omega(t-t')}\\ &\quad -2ih_i(\mathbf{x})\tilde{\chi}''_{\phi_i\phi_j}(\mathbf{x},\mathbf{x}',t-t')h_j^*(\mathbf{x}')e^{-i\omega(t-t')}]\end{aligned} \tag{7.6.27}$$

since the terms proportional to $e^{\pm i\omega(t+t')}$ disappear on averaging. Changing variables to $t_1 = t - t'$, we obtain

$$\begin{aligned}\overline{\frac{dW}{dt}} &= \frac{1}{2}\omega\sum_{ij}\int d^dx d^dx'\int_0^{\infty}dt_1[h_i^*(\mathbf{x})\tilde{\chi}''_{\phi_i\phi_j}(\mathbf{x},\mathbf{x}',t_1)h_j(\mathbf{x}')e^{i\omega t_1}\\ &\quad -h_i(\mathbf{x})\tilde{\chi}''_{\phi_i\phi_j}(\mathbf{x},\mathbf{x}',t_1)h_j^*(\mathbf{x}')e^{-i\omega t_1}].\end{aligned} \tag{7.6.28}$$

Then interchanging $(\mathbf{x}, i)$ and $(\mathbf{x}, j)$, and letting $t_1 \to -t_1$ in the second term of this expression, and using the symmetry property of Eq. (7.6.16), we obtain

$$\overline{\frac{dW}{dt}} = \frac{1}{2}\omega\sum_{ij}\int d^dx d^dx' h_i^*(\mathbf{x})\chi''_{\phi_i\phi_j}(\mathbf{x},\mathbf{x}',\omega)h_j(\mathbf{x}'). \tag{7.6.29}$$

For systems in thermal equilibrium, power is absorbed from external sources, and $\overline{dW/dt}$ must be positive definite. Thus, $\omega\chi''_{\phi_i\phi_j}(\mathbf{x},\mathbf{x}',\omega)$ is a *positive definite* matrix. This in turn implies that the real part of the response function $\chi_{\phi_i\phi_j}(\mathbf{x},\mathbf{x}',\omega)$ is positive for small ω and negative for large ω when ρ is invariant under time reversal and $\epsilon_{\phi_i} = \epsilon_{\phi_j}$:

$$\begin{aligned}\chi'_{\phi_i\phi_j}(\mathbf{x},\mathbf{x}',\omega) &= \mathscr{P}\int_{-\infty}^{\infty}\frac{d\omega'}{\pi}\frac{\chi''_{\phi_i\phi_j}(\mathbf{x},\mathbf{x}',\omega')}{\omega'-\omega}\\ &= \mathscr{P}\int_{-\infty}^{\infty}\frac{d\omega'}{\pi}\frac{\omega'\chi''_{\phi_i\phi_j}(\mathbf{x},\mathbf{x}',\omega')}{\omega'^2-\omega^2}\\ &\to \mathscr{P}\int_{-\infty}^{\infty}\frac{d\omega'}{\pi}\frac{1}{\omega'^2}\omega'\chi''_{\phi_i\phi_j}(\mathbf{x},\mathbf{x}',\omega') > 0 \qquad \omega\to 0,\\ &\to -\frac{1}{\omega^2}\mathscr{P}\int_{-\infty}^{\infty}\frac{d\omega'}{\pi}\omega'\chi''_{\phi_i\phi_j}(\mathbf{x},\mathbf{x}',\omega') < 0 \qquad \omega\to\infty.\end{aligned} \tag{7.6.30}$$

The simple harmonic oscillator discussed in Sec. 7.2 demonstrated a simple example of this behavior.

4 Spectral representations of $\chi''_{\phi_i\phi_j}$

It is instructive to express $\chi''_{\phi_i\phi_j}(\mathbf{x},\mathbf{x}',\omega)$ in terms of the matrix elements of $\phi_i(\mathbf{x},t)$ with respect to energy eigenstates of the system. Let $|n\rangle$ be an eigenstate of $\mathscr{H}$ with energy $\hbar\omega_n$, and assume that ρ is diagonal in the energy basis with matrix elements $\langle n \mid \rho \mid m\rangle = \rho_n\delta_{nm}$. Then we can write

$$\begin{aligned}\langle[\phi_i(\mathbf{x},t),\phi_j(\mathbf{x}',0)]\rangle &= \mathrm{Tr}\rho[e^{i\mathscr{H}t/\hbar}\phi_i(\mathbf{x},0)e^{-i\mathscr{H}t/\hbar},\phi_j(\mathbf{x}',0)]\\ &= \sum_{nm}\rho_n[e^{i(\omega_n-\omega_m)t}\langle n \mid \phi_i(\mathbf{x},0)\mid m\rangle\langle m \mid \phi_j(\mathbf{x}',0)\mid n\rangle\\ &\quad -e^{-i(\omega_n-\omega_m)t}\langle n \mid \phi_j(\mathbf{x}',0)\mid m\rangle\langle m \mid \phi_i(\mathbf{x},0)\mid n\rangle].\end{aligned} \tag{7.6.31}$$

This then implies

$$\begin{aligned}\chi''_{\phi_i\phi_j}(\mathbf{x},\mathbf{x}',\omega) &= \frac{\pi}{\hbar}\sum_{nm}\rho_n\{\langle n\mid\phi_i(\mathbf{x},0)\mid m\rangle\langle m\mid\phi_j(\mathbf{x}',0)\mid n\rangle\delta(\omega+\omega_n-\omega_m)\\ &\quad -\langle n\mid\phi_j(\mathbf{x}',0)\mid m\rangle\langle m\mid\phi_i(\mathbf{x},0)\mid n\rangle\delta(\omega-\omega_n+\omega_m)\}.\end{aligned} \tag{7.6.32}$$

If $\phi_i(\mathbf{x},t)=\phi(t)$ is independent of $\mathbf{x}$ and i, Eq. (7.6.32) reduces in the canonical ensemble to

$$\begin{aligned}\chi''(\omega) &= \frac{\pi}{\hbar}\sum_{nm}\frac{e^{-\beta\hbar\omega_n}}{Z}\mid\langle n\mid\phi\mid m\rangle\mid^2[\delta(\omega+\omega_n-\omega_m)\\ &\quad -\delta(\omega-\omega_n+\omega_m)],\end{aligned} \tag{7.6.33}$$

where $Z=\sum_n e^{-\beta\hbar\omega_n}$ is the partition function and $\beta=1/T$.

Thus, $\chi''_{\phi_i\phi_j}$ consists of a series of delta function spikes with weights determined by the equilibrium density matrix and the matrix elements $\langle n\mid\phi_i(\mathbf{x},0)\mid m\rangle$. This in turn implies that the complex response function has poles along the real axis at frequencies $\pm(\omega_n-\omega_m)$, corresponding to the possible excitation frequencies of the system. Using Eqs. (7.1.29) and (7.6.32), we obtain

$$\chi_{\phi_i\phi_j}(\mathbf{x},\mathbf{x}',z) = \frac{1}{\hbar}\sum_{nm}\rho_n\left\{\frac{\langle n|\phi_i(\mathbf{x},0)\mid m\rangle\langle m\mid\phi_j(\mathbf{x}',0)\mid n\rangle}{\omega_m-\omega_n-z}\right. \tag{7.6.34}$$
$$\left.-\frac{\langle n|\phi_j(\mathbf{x}',0)|m\rangle\langle m|\phi_i(\mathbf{x},0)\mid n\rangle}{\omega_n-\omega_m-z}\right\}.$$

For a finite system, there is a minimum excitation energy and thus a minimum distance between poles of $\chi(z)$, as shown in Fig. 7.6.1. This means that $\chi(z)$ can be analytically continued to the negative half plane directly using Eq. (7.6.34). As the characteristic length L of the system tends to infinity, as it does in most systems of interest in condensed matter physics, the energy level spacing goes to zero as L^{-2}. In this case, the poles in $\chi(z)$ push closer and closer together until, finally, when $L\to\infty$, the discrete set of poles becomes a branch cut. $\chi(z)$ defined by the spectral representation of Eq. (7.1.29) is an analytic function for z in the upper half plane or in the lower half plane. However, it reaches different values on opposite sides of the cut:

$$\lim_{\eta\to 0}\chi(\omega+i\eta)\neq\lim_{\eta\to 0}\chi(\omega-i\eta). \tag{7.6.35}$$

Each pole in $\chi(z)$ corresponds to a delta function in $\chi''(\omega)$. Thus, in a finite system, $\chi''(\omega)$ will consist of separated spikes with intensities determined by the matrix elements in Eq. (7.6.34) as shown in Fig. 7.6.2a. When $L\to\infty$, the spikes merge into a continuous curve, as shown in Fig. 7.6.2b, that can, for example, have the Lorentzian shape discussed in Sec. 7.2. This illustrates how dissipation characterized by a finite width in $\chi''(\omega)$ results when $L\to\infty$.

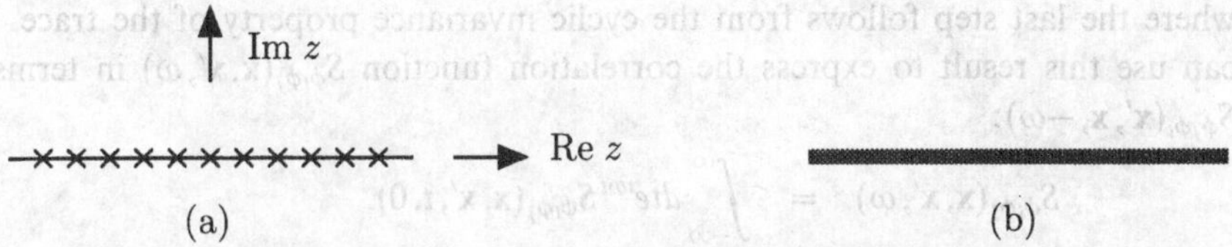

Fig. 7.6.1. (a) Pole structure for $\chi(z)$ for finite systems. There are poles along the real axis separated by a minimum distance. $\chi(z)$ is analytic everywhere in the complex z-plane, except at these poles, and is in particular analytic in the upper half plane as required by the considerations of Sec. 7.2. (b) Singularity structure for an infinite system. The poles merge together to form a branch cut separating the upper and lower half planes. $\chi(z)$ remains analytic in the upper half plane.

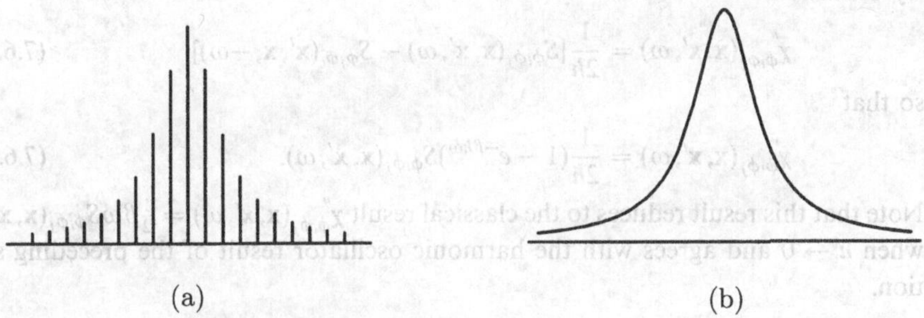

Fig. 7.6.2. (a) $\chi''(\omega)$ consisting of a set of discrete delta-function spikes corresponding to the pole structure of Fig. 7.6.1a. (b) $\chi''(\omega)$ when $L \to \infty$ (corresponding to the branch cut of Fig. 7.6.1b).

5 *The fluctuation-dissipation theorem*

In the preceding section, we saw that there is a simple relation between the dissipation function $\chi''_{xx}(\omega)$ and the correlation function $S_{xx}(\omega)$ for the classical harmonic oscillator. We will now use the formalism developed in this section to show that an obvious generalization of this result applies to all systems in thermal equilibrium. For simplicity, we will consider only systems described by the canonical density matrix,

$$\rho = \frac{1}{Z} e^{-\beta \mathscr{H}}, \qquad Z = \mathrm{Tr} e^{-\beta \mathscr{H}}. \tag{7.6.36}$$

A proof of the theorem rests on the observation that $e^{-\beta\mathscr{H}}$ is an imaginary time translation operator. Thus, we can write

$$\begin{aligned} \mathrm{Tr} e^{-\beta\mathscr{H}} \phi_i(\mathbf{x},t)\phi_j(\mathbf{x}',t') &= \mathrm{Tr} e^{-\beta\mathscr{H}} \phi_i(\mathbf{x},t) e^{\beta\mathscr{H}} e^{-\beta\mathscr{H}} \phi_j(\mathbf{x}',t') \\ &= \mathrm{Tr} \phi_i(\mathbf{x},t+i\beta\hbar) e^{-\beta\mathscr{H}} \phi_j(\mathbf{x}',t') \\ &= \mathrm{Tr} e^{-\beta\mathscr{H}} \phi_j(\mathbf{x}',t') \phi_i(\mathbf{x},t+i\beta\hbar), \end{aligned} \tag{7.6.37}$$

where the last step follows from the cyclic invariance property of the trace. We can use this result to express the correlation function $S_{\phi_i\phi_j}(\mathbf{x},\mathbf{x}',\omega)$ in terms of $S_{\phi_j\phi_i}(\mathbf{x}',\mathbf{x},-\omega)$:

$$\begin{aligned} S_{\phi_i\phi_j}(\mathbf{x},\mathbf{x}',\omega) &= \int_{-\infty}^{\infty} dt e^{i\omega t} S_{\phi_i\phi_j}(\mathbf{x},\mathbf{x}',t,0) \\ &= \int_{-\infty}^{\infty} dt S_{\phi_j\phi_i}(\mathbf{x}',\mathbf{x},-t-i\beta\hbar,0)e^{i\omega t} \\ &= e^{\beta\hbar\omega}\int_{-\infty}^{\infty} dt S_{\phi_j\phi_i}(\mathbf{x}',\mathbf{x},t,0)e^{-i\omega t}. \end{aligned} \tag{7.6.38}$$

Thus

$$S_{\phi_i\phi_j}(\mathbf{x},\mathbf{x}',\omega) = e^{\beta\hbar\omega} S_{\phi_j\phi_i}(\mathbf{x}',\mathbf{x},-\omega). \tag{7.6.39}$$

By definition,

$$\chi''_{\phi_i\phi_j}(\mathbf{x},\mathbf{x}',\omega) = \frac{1}{2\hbar}[S_{\phi_i\phi_j}(\mathbf{x},\mathbf{x}',\omega) - S_{\phi_j\phi_i}(\mathbf{x}',\mathbf{x},-\omega)] \tag{7.6.40}$$

so that

$$\chi''_{\phi_i\phi_j}(\mathbf{x},\mathbf{x}',\omega) = \frac{1}{2\hbar}(1-e^{-\beta\hbar\omega})S_{\phi_i\phi_j}(\mathbf{x},\mathbf{x}',\omega). \tag{7.6.41}$$

Note that this result reduces to the classical result $\chi''_{\phi_i\phi_j}(\mathbf{x},\mathbf{x}',\omega) = \frac{1}{2}\beta\omega S_{\phi_i\phi_j}(\mathbf{x},\mathbf{x}',\omega)$ when $\hbar \to 0$ and agrees with the harmonic oscillator result of the preceding section.

6 Sum rules and moment expansions

The representation of $\chi''_{\phi_i\phi_j}$ as the expectation value of a commutator allows us to express the high-frequency moments of $\chi''_{\phi_i\phi_j}$ in terms of expectation values of equilibrium commutators. At high frequency, we can expand in powers of $1/z$:

$$\begin{aligned} \chi_{\phi_i\phi_j}(\mathbf{x},\mathbf{x}',z) &= \int \frac{d\omega}{\pi}\frac{\chi''_{\phi_i\phi_j}(\mathbf{x},\mathbf{x}',\omega)}{\omega - z} \\ &= -\sum_{p=1}^{\infty}\frac{1}{z^p}\int_{-\infty}^{\infty}\frac{d\omega}{\pi}\omega^p\frac{\chi''_{\phi_i\phi_j}(\mathbf{x},\mathbf{x}',\omega)}{\omega} \\ &\equiv -\sum_{p=1}^{\infty}\frac{1}{z^p}[\omega^p_{ij}(\mathbf{x},\mathbf{x}')]\chi_{\phi_i\phi_j}(\mathbf{x},\mathbf{x}',0), \end{aligned} \tag{7.6.42}$$

where we introduced the pth frequency moment of $\chi''_{\phi_i\phi_j}(\mathbf{x},\mathbf{x}',\omega)$ defined as

$$[\omega^p_{ij}(\mathbf{x},\mathbf{x}')]\chi_{\phi_i\phi_j}(\mathbf{x},\mathbf{x}',0) \equiv \int_{-\infty}^{\infty}\frac{d\omega}{\pi}\omega^p\frac{\chi''_{\phi_i\phi_j}(\mathbf{x},\mathbf{x}',\omega)}{\omega}. \tag{7.6.43}$$

These frequency moments are defined so that the zeroth moment is one. They can be evaluated in terms of commutators of the time derivatives of the field

$\phi_i(\mathbf{x},t)$. It follows directly from Eq. (7.6.14) that

$$\frac{1}{\hbar}\langle[(id/dt)^n\phi_i(\mathbf{x},t),\phi_j(\mathbf{x}',0)]\rangle = \int_{-\infty}^{\infty}\frac{d\omega}{\pi}\omega^n\chi''_{\phi_i\phi_j}(\mathbf{x},\mathbf{x}',\omega)$$
$$= \frac{1}{\hbar}\langle[[[...[\phi_i(\mathbf{x},t),\mathscr{H}/\hbar],\mathscr{H}/\hbar]...],\phi_j(\mathbf{x}',t)]\rangle. \tag{7.6.44}$$

Thus, the high-frequency moments can be obtained from equal-time commutators. Note that since the left hand side of Eq. (7.6.44) is bounded, all moments of $\chi''_{\phi_i\phi_j}(\mathbf{x},\mathbf{x}',\omega)$ for $p \geq 0$ must exist, and as a result $\chi''_{\phi_i\phi_j}(\mathbf{x},\mathbf{x}',\omega)$ must die off at large ω faster than any power of ω.

One particularly important sum rule is the f-sum rule for the particle density. We first note that

$$\int\frac{d\omega}{\pi}\omega\chi''_{nn}(\mathbf{x},\mathbf{x}'\omega) = \frac{1}{\hbar}\langle[\partial n(\mathbf{x},t)/\partial t, n(\mathbf{x}',t)]\rangle$$
$$= \frac{1}{\hbar}\nabla_i\langle[j_i(\mathbf{x},t),n(\mathbf{x}',t)]\rangle, \tag{7.6.45}$$

where $n(\mathbf{x},t)$ is the density operator and $j_i(\mathbf{x},t) = \sum_\alpha p_i^\alpha\delta(\mathbf{x}-\mathbf{x}_\alpha(t))/m$ is the current operator. The current-density commutator is easily calculated to be

$$\int\frac{d\omega}{\pi}\omega\chi''_{nn}(\mathbf{x},\mathbf{x}',\omega) = \frac{1}{m}\nabla\cdot\nabla'\langle n(\mathbf{x})\rangle\delta(\mathbf{x}-\mathbf{x}'). \tag{7.6.46}$$

The Fourier transform of this is Eq. (7.4.21).

7.7 Inelastic scattering

1 Scattering geometry and partial cross-sections

In Chapter 2, we found that quasi-elastic scattering experiments, X-ray scattering experiments in particular, provide a direct measure of static correlations functions. In these experiments, the energy change of scattered particles is not monitored. In inelastic scattering experiments, the energy change of scattered particles is monitored, and dynamic rather than static correlations functions can be measured. In this section, we will derive the fundamental formulae relating experimentally measured scattering cross-sections to dynamic correlation functions and discuss their application in a number of particular cases.

The typical geometry of a scattering experiment is depicted in Fig. 7.7.1. An incident beam of probe particles with momentum $\mathbf{p} = \hbar\mathbf{k}$ and energy E scatters off a target and emerges with momentum $\mathbf{p}' = \hbar\mathbf{k}'$ and energy E'. The probe particles can be neutrons, electrons, photons, or even ions. The flux of incident probe particles is the particle current

$$\Phi = \frac{\#\text{particles}}{\text{area}\times\text{sec}} = \frac{\#\text{particles}}{\text{volume}}\times\text{velocity}. \tag{7.7.1}$$

The scattered particles are detected by placing a detector along the direction defined by the unit vector $\hat{\mathbf{\Omega}} = \mathbf{k}'/k'$ specified by the polar angles (θ,ϕ) relative to

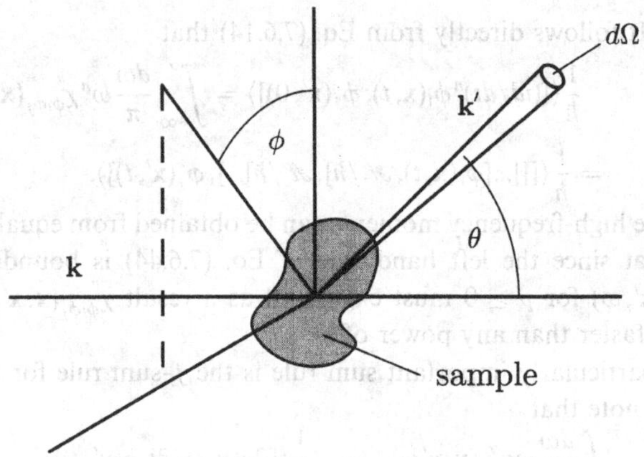

Fig. 7.7.1. Geometry for a typical scattering experiment. An incident beam of particles with momentum $\mathbf{p} = \hbar\mathbf{k}$ and energy E is scattered by a target into a state with momentum $\mathbf{p}' = \hbar\mathbf{k}'$ and energy E'. The scattered particles are counted in a detector placed along a radius from the target parallel to the unit vector $\hat{\mathbf{\Omega}} = \mathbf{k}'/k'$ with polar angles (θ, ϕ) relative to the incident direction of the incident beam.

the incident beam, as shown in Fig. 7.7.1. The detector collects all particles within a solid angle $d\Omega$ about $\hat{\mathbf{\Omega}}$ and can be arranged to count only particles within an energy dE' of E'. The partial differential cross-section is simply the counting rate at $\hat{\mathbf{\Omega}}$ per unit solid angle per energy interval normalized by the incident flux:

$$\frac{d^2\sigma}{d\Omega dE'} = \frac{\#\text{neutrons/sec in } d\Omega \text{ at } \hat{\mathbf{\Omega}} \text{ between } E' \text{ and } E' + dE'}{\Phi d\Omega dE'}. \quad (7.7.2)$$

Note that σ has units of area. The differential cross-section specifying the total count rate at $\hat{\mathbf{\Omega}}$,

$$\frac{d^2\sigma}{d\Omega} = \int dE' \frac{d^2\sigma}{d\Omega dE'}, \quad (7.7.3)$$

is proportional to the scattering intensity $I(\mathbf{q})$ discussed in Chapter 2. Finally, the total cross-section,

$$\sigma_{\text{tot}} = \int dE' d\Omega \frac{d^2\sigma}{d\Omega dE'}, \quad (7.7.4)$$

determines the total scattering rate out of the incident beam. If the probe particles have an internal degree of freedom such as spin or polarization, the partial cross-section for scattering from spin (or polarization) s to spin s' can be defined in an obvious way.

2 Fermi golden rule and neutron scattering

The cross-sections defined above are quite general and apply to any scattering process. They can be related to the properties of the target via Fermi's golden rule. For simplicity, we will now restrict our attention to neutron scattering in which the probe particle is a spin-1/2 neutron of mass m_n. Let $|\lambda\rangle$ be a state of the target with energy E_λ, and let $|\mathbf{k},s\rangle = |\mathbf{k}\rangle\,|s\rangle$ be a plane-wave state of a probe particle with momentum $\hbar\mathbf{k}$ and spin s. In the absence of external magnetic fields, the energy $E = \hbar^2 k^2/2m_n$ of the state $|\mathbf{k},s\rangle$ is independent of s. We may consider the neutron to be confined to a large volume V and to have normalized wavefunctions,

$$\langle \mathbf{x} \mid \mathbf{k}\rangle = \frac{1}{\sqrt{V}} e^{i\mathbf{k}\cdot\mathbf{x}}. \tag{7.7.5}$$

The transition rate from the combined neutron-target state $|\mathbf{k}s\lambda\rangle = |\mathbf{k}s\rangle\,|\lambda\rangle$ to the final state $|\mathbf{k}'s'\lambda'\rangle$ from the Fermi golden rule is

$$W_{\mathbf{k}s\lambda\rightarrow\mathbf{k}'s'\lambda'} = \frac{2\pi}{\hbar} \mid \langle \mathbf{k}s\lambda \mid U \mid \mathbf{k}'s'\lambda'\rangle \mid^2 \delta(E + E_\lambda - E' - E_{\lambda'}), \tag{7.7.6}$$

where U is the interaction potential between neutron and target. Eq. (7.7.6) specifies the transition rate between two particular states of the combined neutron-target system. We are really interested in the transition rate $W_{\mathbf{k}s\rightarrow\mathbf{k}'s'}$ between different neutron plane-wave states. This can be obtained by summing Eq. (7.7.6) over all final target states λ' and all initial target states λ weighted by their probability of occurrence p_λ:

$$W_{\mathbf{k}s\rightarrow\mathbf{k}'s'} = \frac{2\pi}{\hbar} \sum_{\lambda,\lambda'} p_\lambda W_{\mathbf{k}s\lambda\rightarrow\mathbf{k}'s'\lambda'}. \tag{7.7.7}$$

If the target is in thermal equilibrium, $p_\lambda = Z^{-1}\exp(-\beta E_\lambda)$ in the canonical ensemble and $\Xi^{-1}\exp[-\beta(E_\lambda - \mu N_\lambda)]$ in the grand canonical ensemble (N_λ is the number of particles in state λ).

If at a given time there are N neutrons in the state $|\mathbf{k}s\rangle$, then the number of neutrons per second scattered into the final neutron state $|\mathbf{k}'s'\rangle$ is $NW_{\mathbf{k}s\rightarrow\mathbf{k}'s'}$. The total number of neutrons scattered into all possible final states with spin s' is the incident flux Φ times the total cross-section for scattering from spin state s to spin state s', i.e.,

$$\Phi\sigma_{\text{tot}}\,|_{s\rightarrow s'} = N\sum_{\mathbf{k}'} W_{\mathbf{k}s\rightarrow\mathbf{k}'s'} = NV\int \frac{d^3k'}{(2\pi)^3} W_{\mathbf{k}s\rightarrow\mathbf{k}'s'}. \tag{7.7.8}$$

From Eq. (7.7.1), $\Phi V/N$ is the velocity of the incident beam, which for neutrons is $\hbar k/m_n$, and $NV/\Phi = V^2 m_n/\hbar k$. The integral over $\mathbf{k}'$ in Eq. (7.7.8) can be converted to an integral over E' and Ω using

$$d^3k' = k'^2 dk' d\Omega = \frac{m_n}{\hbar^2} k' dE' d\Omega. \tag{7.7.9}$$

Comparing Eqs. (7.7.9) and (7.7.3), we obtain the partial differential cross-section

$$\left.\frac{d^2\sigma}{d\Omega dE'}\right)_{s\rightarrow s'} = \frac{m_n^2 k'}{\hbar^3 k} \frac{V^2}{(2\pi)^3} W_{\mathbf{k}s\rightarrow\mathbf{k}'s'}. \tag{7.7.10}$$

The factor of V^2 in this expression is a result of the box-normalization of our wavefunction. Clearly, the measured cross-section should not depend on the size of the box used to define normalized plane-wave states. The transition rate defined in Eq. (7.7.6) contains the product of four plane-wave wavefunctions and is thus proportional to V^{-2}. Therefore, the factor of V^2 appearing in Eq. (7.7.10) can be absorbed into the transition rate by replacing box-normalized wavefunctions by the wavefunctions $\langle \mathbf{x} \mid \mathbf{k} \rangle = e^{i\mathbf{k}\cdot\mathbf{x}}$. With this definition of plane-wave wavefunctions, we obtain

$$\left.\frac{d^2\sigma}{d\Omega dE'}\right)_{s\to s'} = \frac{k'}{k}\frac{m_n{}^2}{(2\pi\hbar^2)^2}\sum_{\lambda,\lambda'} p_\lambda \mid \langle \mathbf{k}s\lambda \mid U \mid \mathbf{k}'s'\lambda' \rangle \mid^2 \delta(E_\lambda - E_{\lambda'} + \hbar\omega), \tag{7.7.11}$$

where $\hbar\omega = E - E'$ is the energy transferred from the neutron to the target. If the incident neutron beam has a probability p_s of being in spin state s, and no spin information is requested in the detection process, the spin-independent partial cross-section becomes

$$\frac{d^2\sigma}{d\Omega dE'} = \frac{k'}{k}\frac{m_n{}^2}{(2\pi\hbar^2)^2}\sum_{\lambda,\lambda',s,s'} p_\lambda p_s \mid \langle \mathbf{k}s\lambda \mid U \mid \mathbf{k}'s'\lambda' \rangle \mid^2 \delta(E_\lambda - E_{\lambda'} + \hbar\omega). \tag{7.7.12}$$

Eqs. (7.7.11) and (7.7.12) are the fundamental equations for the partial differential scattering cross-section when probe particle and target interact weakly enough that multiple scattering processes can be ignored.

3 The Fermi pseudopotential

To relate the cross-sections just derived to target correlation functions, we need an explicit form for the interaction potential U. Neutrons interact with the atomic nuclei in the target. The wavelength of thermal neutrons used to probe structures at the atomic scale is of order 10^{-8} cm. The range of nuclear forces on the other hand is of the order of a fermi (10^{-13} cm). Hence the scattering potential of an individual atom is pointlike, isotropic and well represented for an atom at the origin by

$$U(\mathbf{x}) = \frac{2\pi\hbar^2}{m_n} b\delta(\mathbf{x}) \equiv a\delta(\mathbf{x}). \tag{7.7.13}$$

The factor b is called the scattering length. It depends on the particular nucleus and, in particular, is different for different isotopes. If the nucleus has a spin, then b should be regarded as an operator depending on the combined spin states of the neutron and the nucleus. For simplicity, we will for the moment ignore spin degrees of freedom of the nucleus. For the energies of interest to condensed matter physics, the scattering from a single nucleus is elastic since nuclear transitions involve energies of order MeV whereas thermal neutrons have energies of order meV. The matrix elements entering the formulae for the scattering cross-section are

$$\langle \mathbf{k}\lambda_N \mid \delta(\mathbf{x}) \mid \mathbf{k}'\lambda_N' \rangle = \langle \lambda_N \mid \lambda_N' \rangle \langle \mathbf{k} \mid \delta(\mathbf{x}) \mid \mathbf{k}' \rangle \tag{7.7.14}$$

$$= \delta_{\lambda_N \lambda'_N} \int d^d x \langle \mathbf{k} \mid \mathbf{x} \rangle \delta(\mathbf{x}) \langle \mathbf{x} \mid \mathbf{k}' \rangle = \delta_{\lambda_N \lambda'_N},$$

where $\mid \lambda_N \rangle$ is a nuclear eigenstate. The differential cross-section for scattering from a single nucleus is thus

$$\frac{d^2\sigma}{d\Omega} = b^2, \tag{7.7.15}$$

which as expected is isotropic and has units of area.

In general, the target will contain many atoms consisting of a nucleus and electrons, which may be tightly bound to the nucleus or mobile, as in a metal. Each nucleus interacts with neutrons with a potential of the form of Eq. (7.7.13). The scattering potential for a many atom target is then

$$U = \sum_\alpha U_\alpha(\mathbf{x} - \mathbf{x}^\alpha) = \sum_\alpha a_\alpha \delta(\mathbf{x} - \mathbf{x}^\alpha), \tag{7.7.16}$$

where $\mathbf{x}^\alpha$ is the position of the nucleus α. Again the scattering length $b_\alpha = (m_n/2\pi\hbar^2)a_\alpha$ depends on the type of nucleus α. The matrix elements in the Fermi transition rates are thus

$$\begin{aligned}
\langle \mathbf{k} s \lambda \mid U \mid \mathbf{k}' s' \lambda' \rangle &= \langle \mathbf{k} s \lambda \mid \sum_\alpha a_\alpha \delta(\mathbf{x} - \mathbf{x}^\alpha) \mid \mathbf{k}' s' \lambda \rangle \\
&= \int d^3 x \delta_{ss'} \langle \mathbf{k} \mid \mathbf{x} \rangle \langle \mathbf{x} \mid \mathbf{k}' \rangle \langle \lambda \mid \sum_\alpha a_\alpha \delta(\mathbf{x} - \mathbf{x}^\alpha) \mid \lambda' \rangle \\
&= \delta_{ss'} \int d^3 x e^{-i\mathbf{q}\cdot\mathbf{x}} \langle \lambda \mid \sum_\alpha a_\alpha \delta(\mathbf{x} - \mathbf{x}^\alpha) \mid \lambda' \rangle \\
&= \delta_{ss'} \sum_\alpha \langle \lambda \mid a_\alpha e^{-i\mathbf{q}\cdot\mathbf{x}^\alpha} \mid \lambda' \rangle,
\end{aligned} \tag{7.7.17}$$

where $\hbar\mathbf{q} = \hbar(\mathbf{k} - \mathbf{k}')$ is the momentum transfer from the neutron to the target. The operators appearing in the differential cross-section can be converted from Schrödinger to Heisenberg time-dependent operators with the aid of the integral representation of the energy conserving δ-function,

$$\delta(E_\lambda - E_{\lambda'} + \hbar\omega) = \int_{-\infty}^{\infty} \frac{d(t-t')}{2\pi\hbar} e^{(i/\hbar)(E_\lambda - E_{\lambda'} + \hbar\omega)(t-t')} \tag{7.7.18}$$

and with the identity

$$\langle \lambda \mid f(\mathbf{x}^\alpha) \mid \lambda' \rangle e^{+(i/\hbar)(E_\lambda - E_{\lambda'})t} = \langle \lambda \mid f(\mathbf{x}^\alpha(t)) \mid \lambda' \rangle, \tag{7.7.19}$$

where $f(\mathbf{x}^\alpha)$ is any function of the particle coordinate operators $\mathbf{x}^\alpha$, including $\delta(\mathbf{x} - \mathbf{x}^\alpha)$, $e^{-i\mathbf{q}\cdot\mathbf{x}^\alpha}$ or the density operator $n(\mathbf{x})$. Eqs. (7.7.17) to (7.7.19) used in the expression, Eq. (7.7.12), for the partial differential cross-section yield

$$\begin{aligned}
\frac{d^2\sigma}{d\Omega dE'} &= \frac{k'}{2\pi\hbar k} \sum_{\lambda,\lambda'} \int dt e^{i\omega(t-t')} \\
&\times \left[p_\lambda \sum_{\alpha,\alpha'} \langle \lambda \mid b_\alpha e^{-i\mathbf{q}\cdot\mathbf{x}^\alpha(t)} \mid \lambda' \rangle \langle \lambda' \mid b_{\alpha'} e^{i\mathbf{q}\cdot\mathbf{x}^{\alpha'}(t')} \mid \lambda \rangle \right].
\end{aligned} \tag{7.7.20}$$

The states $| \lambda' \rangle$ form a complete set of target states, and

$$\sum_{\lambda'} | \lambda' \rangle \langle \lambda' | = 1. \tag{7.7.21}$$

In addition, the equilibrium expectation value of any target operator O is

$$\langle O \rangle = \sum_{\lambda} p_\lambda \langle \lambda | O | \lambda \rangle. \tag{7.7.22}$$

Thus, Eq. (7.7.20) can be rewritten as

$$\frac{d^2\sigma}{d\Omega dE'} = \frac{k'}{2\pi\hbar k} \int dt e^{i\omega(t-t')} \langle \sum_{\alpha,\alpha'} b_\alpha b_{\alpha'} e^{-i\mathbf{q}\cdot\mathbf{x}^\alpha(t)} e^{i\mathbf{q}\cdot\mathbf{x}^{\alpha'}(t')} \rangle. \tag{7.7.23}$$

This is the fundamental formula for neutron scattering.

4 Coherent and incoherent scattering

The states $| \lambda \rangle$ form a complete set of states for all degrees of freedom of the target. The degree of freedom specifying the type of isotope of a nucleus α is, however, dynamically decoupled from the other degrees of freedom such as the positions of the nuclei. The average over $b_\alpha b_{\alpha'}$ in Eq. (7.7.23) can be decoupled from the average over the positions $\mathbf{x}^\alpha(t)$. The thermodynamic average in Eq. (7.7.23) can thus be expressed in the form

$$S = \sum_{\alpha,\alpha'} \overline{b_\alpha b'_\alpha} \langle \alpha; \alpha' \rangle, \tag{7.7.24}$$

where $\langle \alpha; \alpha' \rangle$ is an average depending only on the coordinates $\mathbf{x}^\alpha(t)$ and $\mathbf{x}^{\alpha'}(t')$. The probability that a nucleus at position $\mathbf{x}^\alpha$ is a particular isotope is independent of α. In particular, if the target contains N_1 nuclei of isotope 1 and N_2 nuclei of isotope 2, the probability that a nucleus at $\mathbf{x}^\alpha$ is isotope 1 (2) is simply N_1/N (N_2/N), where $N = N_1 + N_2$. Thus, the scattering b_α is statistically independent of $b_{\alpha'}$ for $\alpha \neq \alpha'$, implying

$$\overline{b_\alpha b_{\alpha'}} = \overline{b^2} \delta_{\alpha\alpha'} + |\overline{b}|^2 (1 - \delta_{\alpha\alpha'}) \tag{7.7.25}$$

and

$$S = | \overline{b} |^2 \sum_{\alpha,\alpha'} \langle \alpha; \alpha' \rangle + \left[\overline{b^2} - | \overline{b} |^2 \right] \sum_{\alpha} \langle \alpha; \alpha \rangle, \tag{7.7.26}$$

where $\overline{b}$ is the mean and $\overline{b^2}$ the mean-square scattering length.

The partial differential cross-section in Eq. (7.7.23) can now be expressed as a sum of two parts. One, the *coherent cross-section,* is proportional to $| \overline{b} |^2$ and the other, the *incoherent cross-section,* is proportional to $[\overline{b^2} - | \overline{b} |^2]$:

$$\left. \frac{d^2\sigma}{d\Omega dE'} \right)_{\text{coh}} = \frac{\sigma_{\text{coh}}}{4\pi} \frac{k'}{k} \frac{1}{2\pi\hbar} \sum_{\alpha,\alpha'} \int dt e^{i\omega(t-t')} \langle e^{-i\mathbf{q}\cdot\mathbf{x}^\alpha(t)} e^{i\mathbf{q}\cdot\mathbf{x}^{\alpha'}(t')} \rangle \tag{7.7.27}$$

and

$$\left. \frac{d^2\sigma}{d\Omega dE'} \right)_{\text{inc}} = \frac{\sigma_{\text{inc}}}{4\pi} \frac{k'}{k} \frac{1}{2\pi\hbar} \sum_{\alpha} \int dt e^{i\omega(t-t')} \langle e^{-i\mathbf{q}\cdot\mathbf{x}^\alpha(t)} e^{i\mathbf{q}\cdot\mathbf{x}^{\alpha}(t')} \rangle, \tag{7.7.28}$$

Table 7.7.1. *Coherent and incoherent scattering amplitudes.*

Element	Z	$\sigma_{\mathrm{coh}}(10^{-22}\mathrm{cm}^2)$	$\sigma_{\mathrm{inc}}(10^{-22}\mathrm{cm}^2)$
^{1}H	1	1.8	80.2
^{2}H	1	5.6	2.0
C	6	5.6	0
Al	13	1.5	0
V	23	0.02	5.0

where

$$\sigma_{\mathrm{coh}} = 4\pi \mid \bar{b} \mid^2 \tag{7.7.29}$$

and

$$\sigma_{\mathrm{inc}} = 4\pi [\overline{b^2} - \mid \bar{b} \mid^2] \tag{7.7.30}$$

are, respectively, the total coherent and incoherent cross-sections from a single atom.

Our derivation of the coherent and incoherent scattering cross-sections followed from the existence in the target of different spinless nuclear isotopes with different scattering lengths. If the nuclei have spin, then the nuclear spin degrees of freedom, which, like the isotope degrees of freedom, are dynamically decoupled from the positional degrees of freedom, can contribute significantly to σ_{inc}. The coherent and incoherent cross-sections for a number of elements are listed in Table 7.7.1. The large incoherent scattering in some elements arises from the nuclear spin degrees of freedom.

5 *Cross-sections and correlation functions*

The coherent and incoherent partial differential cross-sections are proportional, respectively, to the density-density correlation function and the self-correlation function introduced in Secs. 7.1 and 7.4. The coherent cross-section [Eq. (7.7.27)] can be reexpressed as

$$\left.\frac{d^2\sigma}{d\Omega dE'}\right)_{\mathrm{coh}} = \frac{\sigma_{\mathrm{coh}}}{4\pi}\frac{k'}{k}\frac{V}{2\pi\hbar}C_{nn}(\mathbf{q},\omega), \tag{7.7.31}$$

where

$$\begin{aligned} C_{nn}(\mathbf{q},\omega) &= \int dt e^{i\omega(t-t')} C_{nn}(\mathbf{q}, t-t') \qquad (7.7.32) \\ &= \frac{1}{V}\int d^3x \int d^3x' \int e^{-i\mathbf{q}\cdot(\mathbf{x}-\mathbf{x}')} \int dt e^{i\omega(t-t')} C_{nn}(\mathbf{x},\mathbf{x}',t,t'). \end{aligned}$$

The function

$$\langle n\rangle^{-1} C_{nn}(\mathbf{q}, t-t') = \frac{1}{N}\sum_{\alpha,\alpha'}\langle e^{-i\mathbf{q}\cdot\mathbf{x}^{\alpha}(t)} e^{i\mathbf{q}\cdot\mathbf{x}^{\alpha'}(t')}\rangle \tag{7.7.33}$$

is often called the *intermediate function* in the neutron scattering literature. Alternatively, the coherent cross-section can be expressed with the aid of Eq. (7.1.5) as

$$\left.\frac{d^2\sigma}{d\Omega dE'}\right)_{\rm coh} = \frac{\sigma_{\rm coh}}{4\pi}\frac{k'}{k}\frac{1}{2\pi\hbar}[VS_{nn}(\mathbf{q},\omega)+ \mid \langle n_{\mathbf{q}}\rangle \mid^2 2\pi\delta(\omega)]. \tag{7.7.34}$$

The differential cross-section obtained by integrating this expression over ω is proportional to the quasi-static scattering intensity of Eq. (2.3.7). Note, however, that the term containing the factor $\mid \langle n_{\mathbf{q}}\rangle \mid^2$, which provides information about long-range spatially periodic order, is proportional to $\delta(\omega)$ and involves no energy change in the scattered neutron. It is totally elastic. Thus, the density wave amplitudes $\langle n_{\mathbf{G}}\rangle$ determining the degree of periodic order are measured by the elastic part of the dynamic neutron scattering cross-section. Information about the target dynamics measured by $S_{nn}(\mathbf{q},\omega)$ is contained in the inelastic scattering cross-section.

The incoherent scattering cross-section can be expressed as

$$\left.\frac{d^2\sigma}{d\Omega dE'}\right)_{\rm inc} = \frac{\sigma_{\rm inc}}{4\pi}\frac{k'}{k}\frac{1}{2\pi\hbar}S_{\rm self}(\mathbf{q},\omega), \tag{7.7.35}$$

where $S_{\rm self}(\mathbf{q},\omega)$ is the Fourier transform of the individual particle correlation function introduced in Sec. 7.4. As we saw in that section, $S_{\rm self}(\mathbf{q},t)$ is proportional to $e^{-D_s q^2|t|}$ in fluids. Thus, neutron scattering from fluids with an incoherent scattering cross-section can measure both the self-diffusion constant D_s and the density correlation function.

6 Neutron scattering from crystals

In crystals, we may assume to a very good approximation that atomic nuclei fluctuate about lattice sites with instantaneous positions that can be determined by the displacement variables $\mathbf{u}_{\mathbf{l}}(t)$. The density correlation function in this case is

$$C_{nn}(\mathbf{x},\mathbf{x}',t,t') = \langle\sum_{\mathbf{l},\mathbf{l}'}\delta(\mathbf{x}-\mathbf{R}_{\mathbf{l}}-\mathbf{u}_{\mathbf{l}}(t))\delta(\mathbf{x}'-\mathbf{R}_{\mathbf{l}'}-\mathbf{u}_{\mathbf{l}'}(t'))\rangle \tag{7.7.36}$$

and

$$\begin{aligned} VC_{nn}(\mathbf{q},t) &= \sum_{\mathbf{l},\mathbf{l}'} e^{-i\mathbf{q}\cdot(\mathbf{R}_{\mathbf{l}}-\mathbf{R}_{\mathbf{l}'})}\langle e^{-i\mathbf{q}\cdot\mathbf{u}_{\mathbf{l}}(t)}e^{i\mathbf{q}\cdot\mathbf{u}_{\mathbf{l}'}(t')}\rangle \\ &= \sum_{\mathbf{l},\mathbf{l}'} e^{-i\mathbf{q}\cdot(\mathbf{R}_{\mathbf{l}}-\mathbf{R}_{\mathbf{l}'})}[\langle e^{-i\mathbf{q}\cdot\mathbf{u}_{\mathbf{l}}(t)}e^{i\mathbf{q}\cdot\mathbf{u}_{\mathbf{l}'}(t')}\rangle - \langle e^{-i\mathbf{q}\cdot\mathbf{u}_{\mathbf{l}}(t)}\rangle\langle e^{i\mathbf{q}\cdot\mathbf{u}_{\mathbf{l}'}(t')}\rangle] \\ &+ \sum_{\mathbf{l},\mathbf{l}'} e^{-i\mathbf{q}\cdot(\mathbf{R}_{\mathbf{l}}-\mathbf{R}_{\mathbf{l}'})}\langle e^{-i\mathbf{q}\cdot\mathbf{u}_{\mathbf{l}'}(t)}\rangle\langle e^{i\mathbf{q}\cdot\mathbf{u}_{\mathbf{l}'}(t')}\rangle. \end{aligned} \tag{7.7.37}$$

The final term in this equation is independent of time and contributes only to the elastic scattering cross-section. If there is only one atom per unit cell,

$$\left(\frac{d^2\sigma}{d\Omega}\right)^{\text{el}}_{\text{coh}} = \frac{\sigma_{\text{coh}}}{4\pi} V^2 e^{-2W_{\mathbf{G}}} \sum_{\mathbf{G}} \delta_{\mathbf{q},\mathbf{G}}, \tag{7.7.38}$$

where, as usual, $\mathbf{G}$ is a reciprocal lattice vector and $e^{-2W_{\mathbf{G}}}$ is the crystalline Debye-Waller factor defined in Eqs. (6.4.25) and (6.4.26).

The inelastic part of Eq. (7.7.37) contains information about phonon excitations. In a harmonic classical crystal,

$$\begin{aligned} \langle e^{-i\mathbf{q}\cdot\mathbf{u}_{\mathbf{l}}(t)} e^{i\mathbf{q}\cdot\mathbf{u}_{\mathbf{l}'}(t')}\rangle &= e^{-\langle[\mathbf{q}\cdot(\mathbf{u}_{\mathbf{l}}(t)-\mathbf{u}_{\mathbf{l}'}(t'))]^2\rangle/2} \\ &= e^{-2W_{\mathbf{q}}} e^{q_i q_j G_{ij}(\mathbf{l},\mathbf{l}',t,t_1)}, \end{aligned} \tag{7.7.39}$$

where

$$\begin{aligned} G_{ij}(\mathbf{l},\mathbf{l}',t-t') &= \langle u_i(\mathbf{l},t)u_i(\mathbf{l}',t')\rangle \\ &= \int \frac{d\omega}{2\pi} \frac{2\hbar}{(1-e^{-\beta\hbar\omega})} \chi''_{ij}(\mathbf{l},\mathbf{l}',\omega) \end{aligned} \tag{7.7.40}$$

is the phonon correlation function expressed in terms of the response function introduced in Eq. (7.3.21). This result follows from the general properties of Gaussian integrals discussed in Sec. 5.3. It also applies to quantum harmonic crystals. In an anharmonic crystal, there will be corrections to the above equation appearing as extra terms, involving higher cumulants of the displacement operator, in the argument of the exponential.

The exponential in Eq. (7.7.39) can be expanded in powers of G_{ij}. The first non-trivial term in this expansion gives rise to the *one-phonon* differential cross-section,

$$\frac{d^2\sigma}{d\Omega dE'} = \frac{\sigma_{\text{coh}}}{4\pi} \frac{k'}{k} \frac{1}{2\pi\hbar} \sum_{\mathbf{G}} \delta_{\mathbf{q}',\mathbf{q}+\mathbf{G}}\, e^{-2W_{\mathbf{q}}} q_i q_j G_{ij}(\mathbf{q}',\omega). \tag{7.7.41}$$

Thus, the one-phonon part of the differential cross-section provides a direct measure of the displacement correlation function. As we saw in Sec. 7.3, this function consists of delta-function spikes at the phonon frequencies in an undamped system and Lorentzians in the presence of dissipation. Thus, neutron scattering provides a direct measure of the phonon spectrum of crystals. Higher order terms in the expansion of Eq. (7.7.39) in powers of G_{ij} contribute to the scattering cross-section. These terms, since they involve the excitation of many phonons, do not have pronounced peaks as a function of frequency and contribute only an incoherent background to the cross-section.

7 *Magnetic scattering*

Neutrons have a spin that couples to magnetic moments in the target via the magnetic dipole interaction. Consider for simplicity targets in which only electron spin and not orbital angular momentum contributes to local magnetic moments. In this case, the magnetic potential energy describing the interaction between probe neutrons and the target is

$$U_m = \gamma\mu_N \boldsymbol{\sigma}^N \cdot \mathbf{H}, \tag{7.7.42}$$

where $\gamma = 1.91$ is the neutron magnetic moment, μ_N is the nuclear magneton, and

$$\mathbf{H} = -2\mu_B \sum_\alpha \boldsymbol{\nabla} \times \left(\mathbf{S}^\alpha \times \boldsymbol{\nabla} \frac{1}{|\mathbf{x} - \mathbf{x}^\alpha|} \right), \tag{7.7.43}$$

where μ_B is the Bohr magneton, and $\mathbf{S}^\alpha$ is the spin of electron α. The scattering matrix element associated with U_m is

$$\begin{aligned}
\langle \mathbf{k}s \mid U_m \mid \mathbf{k}'s' \rangle &= -2\gamma\mu_N\mu_B \langle s \mid \boldsymbol{\sigma}^N \mid s' \rangle \\
&\quad \cdot \int d^3x e^{-i\mathbf{q}\cdot\mathbf{x}} \sum_\alpha \boldsymbol{\nabla} \times \left(\mathbf{S}^\alpha \times \boldsymbol{\nabla} \frac{1}{|\mathbf{x} - \mathbf{x}^\alpha|} \right) \\
&= -2\gamma\mu_N \langle s \mid \boldsymbol{\sigma}^N \mid s' \rangle \qquad\qquad (7.7.44) \\
&\quad \cdot \int d^3x d^3x' e^{-i\mathbf{q}\cdot\mathbf{x}} \boldsymbol{\nabla} \times \left(\mathbf{m}(\mathbf{x}') \times \boldsymbol{\nabla} \frac{1}{|\mathbf{x} - \mathbf{x}'|} \right),
\end{aligned}$$

where

$$\mathbf{m}(\mathbf{x}) = \mu_B \sum_\alpha \mathbf{S}^\alpha \delta(\mathbf{x} - \mathbf{x}^\alpha) \tag{7.7.45}$$

is the magnetization operator. With the aid of the identity

$$\frac{1}{|\mathbf{x} - \mathbf{x}'|} = \int \frac{d^dq}{(2\pi)^3} \frac{4\pi}{q^2} e^{i\mathbf{q}\cdot(\mathbf{x}-\mathbf{x}')}, \tag{7.7.46}$$

the magnetic matrix element of Eq. (7.7.44) can be expressed as

$$\langle \mathbf{k}s \mid U_m \mid \mathbf{k}'s' \rangle = 2\gamma\mu_N \langle s \mid \boldsymbol{\sigma}^N \mid s' \rangle \cdot \int d^3x e^{-i\mathbf{q}\cdot\mathbf{x}} \frac{4\pi}{q^2} [\mathbf{q} \times (\mathbf{m}(\mathbf{x}) \times \mathbf{q})] . \tag{7.7.47}$$

If the initial neutron beam is unpolarized,

$$\sum_{s,s'} p_s \langle s \mid \sigma_i^N \mid s' \rangle \langle s' \mid \sigma_j^N \mid s \rangle = \delta_{ij}. \tag{7.7.48}$$

Using the above results in the expression [Eq. (7.7.12)] for the partial differential cross-section, we find

$$\begin{aligned}
\frac{d^2\sigma}{d\Omega dE'} &= \frac{k'}{k} \left(\frac{m_n^2}{2\pi\hbar^2} \right)^2 \frac{(8\pi\gamma\mu_N)^2}{2\pi\hbar} (\delta_{ij} - \hat{q}_i\hat{q}_j)[V S_{m_i m_j}(\mathbf{q}, \omega) \\
&\quad + \langle m_i(\mathbf{q}) \rangle \langle m_j(-\mathbf{q}) \rangle 2\pi\delta(\omega)].
\end{aligned} \tag{7.7.49}$$

Thus, the elastic part of the magnetic neutron scattering cross-section provides information about long-range magnetic order, and the inelastic part provides information about spin excitations.

8 How neutron scattering experiments are actually done

Particularly for neutron scattering, it is important to have information about the energy as well as the direction of both the incident and the scattered beams. At present, there are two ways of performing these experiments. The tradi-

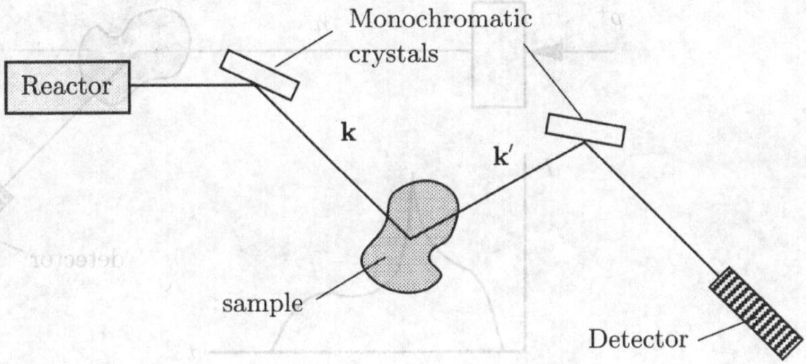

Fig. 7.7.2. Experimental set-up for dynamic neutron scattering. Monochromatic neutrons are selected by Bragg scattering. The energy and momentum of incident and scattered neutrons are thus well determined.

tional way is to use a reactor with different moderators to cool the neutrons to the thermal wavelengths required for the neutron source and then to use Bragg scattering from a crystal to select a narrow energy window from this source. The monochromatic beam is directed to the sample, and the scattered beam is then reanalyzed by Bragg scattering from another crystal before entering the neutron detector. The experiment is shown schematically in Fig. 7.7.2. Note that for a typical experiment in which there is energy resolution of the scattered beam, the sample, analyzer and detector must all be moved. In addition, each of the three scatterings before detection greatly attenuates the beam current from the reactor. Counting rates are often slow (several counts per second).

A more recent technique uses a pulsed neutron source and time-of-flight measurements to resolve the energy. Charged particles are accelerated in a particle accelerator and then collide with a material that produces neutrons from nuclear reactions in a short pulse. The pulse is directed onto the sample and from the sample to a neutron detector, where it is recorded as a function of time. As the distances from the source to the sample and from the sample to the detector are known, the neutron velocity and hence its energy can be measured. Thus, a single pulse measurement gives the energy spectrum of scattering at a particular angle or wave vector, and the incident beam is not attenuated by the analyzer. A schematic of a pulse apparatus is shown in Fig. 7.7.3.

Although it is preferable to use incident wavelengths comparable to the size of the structure of interest, it is possible to do low-angle scattering to probe structures considerably larger than λ. For example, there are now detectors with a spatial resolution better than 1 mm which can be located several meters from the scattering object. For X-rays or neutrons with wavelengths of 1Å, it is then possible to look at scattered wavelengths of the order of microns.

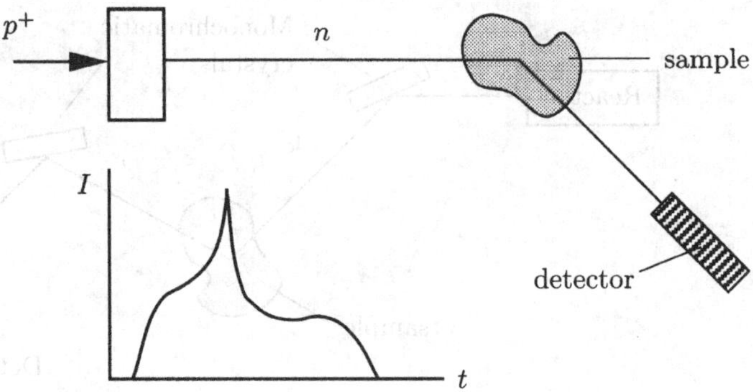

Fig. 7.7.3. Experimental set-up for pulsed neutron experiments. The time dependence of the neutron intensity I at the detector provides information about the energy distribution of neutrons scattered at a particular angle.

9 Scattering of charged particles and photons

The analysis just presented for neutron scattering applies with minor modification to scattering of any probe particles, including charged particles and photons, provided the single-scattering approximation is valid. Charged particles interact strongly with matter via the Coulomb interaction, and the single-scattering approximation applies only to fast particles scattered through thin samples. The single-scattering approximation for light scattering is, on the other hand, often valid.

To determine the differential cross-section for the scattering of particles of charge Q and mass M from particles in the sample of charge Q_α, we have only to replace the potential in Eq. (7.7.13) by the Coulomb potential whose Fourier transform is $4\pi QQ_\alpha/q^2$. Introducing the charge density $\rho(\mathbf{x}) = \sum_\alpha Q_\alpha \delta(\mathbf{x} - \mathbf{x}_\alpha(t))$, we obtain

$$\frac{d^2\sigma}{d\Omega dE'} = \left(\frac{M}{2\pi\hbar^2}\right)^2 \left(\frac{4\pi Q}{q^2}\right)^2 \frac{k'}{k}\frac{1}{2\pi\hbar} V C_{\rho\rho}(\mathbf{q},\omega). \tag{7.7.50}$$

If the scattered particles are electrons, $Q = e$ and $M = m_e$.

Light scatters from inhomogeneities in the dielectric constant ϵ, which depends on the density and, possibly, on other order parameters in the system. In isotropic fluids, where the dominant dependence of the dielectric constant is on the particle density n, a classical calculation gives

$$\frac{d^2\sigma}{d\Omega d\omega_f} = \left(\frac{1}{2}\frac{\partial\epsilon}{\partial n}\right)^2 \left(\frac{\omega_i}{c}\right)^4 \frac{(\mathbf{e}_i \times \hat{\mathbf{k}}')^2}{(2\pi)^3} \sqrt{\epsilon} S_{nn}(\mathbf{q},\omega), \tag{7.7.51}$$

where $\mathbf{e}_i$ is the polarization vector of the incident light, $\hat{\mathbf{k}}' = \mathbf{k}'/k'$, and ω_i and ω_f are, respectively, the frequency of the incident and scattered light. In liquid crystals, the dielectric constant is anisotropic and depends on the nematic order parameter Q_{ij} as well as on the particle density n. The scattering cross-section measures correlations in Q_{ij} and depends on the relative direction of the polarization vectors of the incident and scattered light and the director $\mathbf{n}$. Inelastic light scattering provides an ideal probe of sound waves and thermal diffusion in liquids and of director modes in liquid crystals.

Bibliography

Dynamical correlation and response functions

Dieter Forster, *Hydrodynamic Fluctuations, Broken Symmetry, and Correlation Functions* (Addison Wesley, Reading, Mass., 1983).

P.C. Martin, in *Many-Body Physics*, eds. C. De Witt and R. Balian (Gordon and Breach, New York, 1967).

Langevin theory and stochastic processes

R. Kubo, M. Toda, and N. Hashitsume, *Statistical Physics II*, 2nd ed. (Springer-Verlag, Berlin, 1978).

Nelson Wax, ed., *Selected Papers on Noise and Stochastic Processes* (Dover Publications, Inc., New York, 1954). The articles by S. Chandrasekhar, G.E. Uhlenbeck and L.S. Ornstein, and by Ming Chen and G.E. Uhlenbeck are particularly useful.

J. Zinn-Justin, *Quantum Field Theory and Critical Phenomena* (Clarendon Press, Oxford, 1990).

Neutron scattering

G. L. Squires, *Thermal Neutron Scattering* (Cambridge University Press, 1978).

References

H.B. Callen and R.F. Welton, *Phys. Rev.* **86**, 702 (1952).

A. Einstein, *Ann. d. Physik* **17**, 549 (1905).

P. Langevin, *Comptes Rendus Acad. Sci. Paris* **146**, 530 (1908).

Problems

7.1 Consider the coupled *LRC* electrical circuit shown in Fig. 7P.1.

(a) Calculate the frequency-dependent response function $\chi_{ij}(\omega)$ for the charge q_i on capacitor $i = 1,2$ produced by a voltage at source $j = 1,2$ at frequency ω.

(b) Calculate $\chi''_{11}(\omega)$ and plot it as a function of frequency. Discuss the limit $R \to 0$.

(c) Calculate $\lim_{\omega\to 0} \chi_{ij}(\omega)$ and $\lim_{\omega\to\infty} \chi_{ij}(\omega)$ and compare these results with sum rules that can be calculated independently (you may assume $R = 0$ here).

(d) Calculate the thermal equilibrium charge correlation function $S_{q_iq_j}(\omega)$.

7.2 Calculate the transverse position-velocity and velocity-velocity response functions for the harmonic continuum elastic model of Sec. 7.3. Determine, in particular, the low-frequency form of $\chi_{vv}(\mathbf{q}, z)$ and the long-wavelength form for $\chi''_{vv}(\mathbf{q}, \omega)$.

7.3 Calculate the density-density response function using Eq. (7.4.28) for the current and a time-dependent external chemical potential $\mu^{\text{ext}}(\mathbf{x},t)$.

7.4 If the diffusion constant depends on frequency, then

$$\chi(\mathbf{q},z)=\chi(\mathbf{q})\frac{D(\mathbf{q},z)q^2}{-iz+D(\mathbf{q},z)q^2}.$$

Express $D(\mathbf{q},z)$ in terms of $\chi(\mathbf{q},z)$ and then, using the analyticity properties of $\chi(\mathbf{q},z)$, show that $D(\mathbf{q},z)$ is analytic in the upper half plane and can be written as

$$D(\mathbf{q},z)=\int\frac{d\omega}{i\pi}\frac{D'(\mathbf{q},\omega)}{\omega-z},$$

where $D'(\mathbf{q},\omega)=D'(\mathbf{q},-\omega)$ is the real part of $D(\mathbf{q},\omega+i\epsilon)$.

7.5 Generalize the resistor network equation [Eq. (7.4.47)] to include a current $I(t)$ inserted at the origin and a current $-I(t)$ extracted at site $\mathbf{x}$. Show that the complex impedance for current flowing from the origin to $\mathbf{x}$ is

$$Z(\mathbf{x},\omega)=\int\frac{d^dq}{(2\pi)^d}\frac{2(1-\cos\mathbf{q}\cdot\mathbf{x})}{-i\omega C+\sigma[\gamma(0)-\gamma(\mathbf{q})]}.$$

From this, calculate the time-dependent impedance $Z(\mathbf{x},t)$ for $t>0$ and express your result in the continuum limit in terms of the diffusion Green function [Eq. (7.4.10)].

7.6 (a) Show that the equilibrium displacement correlation function for a damped harmonic oscillator is

$$C_{xx}(t)=\frac{T}{m\omega_0^2}\left(\cos\omega_1 t\frac{\gamma}{2\omega_1}+\sin\omega_1 t\right)e^{-\gamma t/2}.$$

Use this function to calculate the velocity-velocity correlation function $C_{vv}(t)$ and the velocity-displacement correlation functions $C_{xv}(t)$ and $C_{vx}(t)$.

(b) Calculate the frequency-dependent correlation functions $C_{vv}(\omega)$, $C_{xv}(\omega)$, and $C_{vx}(\omega)$. Then use the fluctuation-dissipation theorem to calculate $\chi''_{vv}(\omega)$, $\chi''_{xv}(\omega)$, and $\chi''_{vx}(\omega)$. Discuss the symmetry properties of these response functions.

7.7 (Memory effects) In our treatment of a Brownian particle in a viscous fluid, we assumed that the friction force was simply a constant friction coefficient times the velocity. In general, however, the dissipative force at time t should depend on the velocity at earlier times, and the dissipative coefficient γ should be a time-dependent memory function $\tilde{\gamma}(t)=2\eta(t)\tilde{\gamma}'(t)$. In the presence of an external force $F(t)$, the velocity equation then becomes

$$\frac{\partial v}{\partial t}=-\int_{-\infty}^{\infty}\tilde{\gamma}(t-t')v(t')dt'+\frac{1}{m}F(t)+\frac{1}{m}\zeta(t).$$

(a) Using arguments similar to those used to derive Eq. (7.1.17), show that

$$\gamma(z)=\int_0^{\infty}dte^{izt}\tilde{\gamma}(t)=\int\frac{d\omega}{\pi i}\frac{\gamma'(\omega)}{\omega-z}$$

and that $\gamma(\omega)=\lim_{\epsilon\to 0}\gamma(\omega+i\epsilon)=\gamma'(\omega)+i\gamma''(\omega)$, where the imaginary

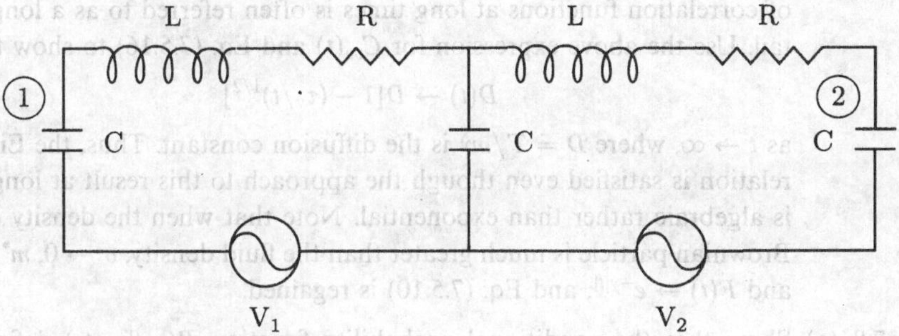

Fig. 7P.1. *LRC* circuit for Problem 7.1.

part $\gamma''(\omega)$ is related to the real part $\gamma'(\omega)$ by a Kramers-Kronig relation. Show also that $\gamma^*(\omega) = \gamma(-\omega)$.

(b) Calculate the mobility $\mu(\omega)$ relating the velocity to the external force via $v(\omega) = \mu(\omega)F(\omega)$ in terms of $\gamma(\omega)$. What is the complex electrical conductivity of a system with a density n, of non-interacting Brownian particles, each carrying a charge e?

(c) Show that noise correlations must satisfy

$$C_{\zeta\zeta}(\omega) = I(\omega) = 2Tm\gamma'(\omega)$$

to produce thermal equilibrium. Then show that the velocity correlation function is

$$C_{vv}(\omega) = \frac{2T}{m}\frac{\gamma'(\omega)}{|-i\omega + \gamma(\omega)|^2}.$$

You may wish to use the fact that $\gamma(\omega)$ is analytic in the upper half plane to obtain this result.

(d) (difficult) If a particle of radius a and density ρ_0 moves with a time-dependent velocity in a fluid of density ρ, it will excite viscous shear waves in the fluid with frequency $\omega_v = -i\eta q^2/\rho$ (see the next chapter). This leads to a singular memory function

$$\gamma(\omega) = -2i\omega\rho/\rho_0 + \gamma[(-i\omega\tau_v)^{1/2} + 1],$$

where $\tau_v = \rho a^2/\eta$ is the viscous relaxation time (the time for a shear wave to diffuse across a particle radius). Show that

$$C_{vv}(t) = \frac{T}{m^*}F(\tau),$$

where $m^* = m[1 + \rho/(2\rho_0)]$, $\tau = (m/m^*)\gamma t$, and

$$F[\tau] = \frac{\sigma}{\pi}\tau^{-3/2}\int_0^\infty \frac{e^{-u^2}u^2 du}{(\tau^{-1}u - 1)^2 + \sigma^2\tau^{-1}u^2},$$

where $\sigma^2 = (m/m^*)(\gamma\tau_v) = (9/2)\rho/[\rho_0 + (\rho/2)]$. This implies that $F(\tau) \to \sigma\tau^{-3/2}/(2\sqrt{\pi})$ as $\tau \to \infty$. Such algebraic rather than exponential fall-off of correlation functions at long times is often referred to as a long-time tail. Use the above expression for $C_{vv}(t)$ and Eq. (7.5.16) to show that

$$D(t) \to D[1 - (\tau_v/t)^{1/2}]$$

as $t \to \infty$, where $D = T/m\gamma$ is the diffusion constant. Thus, the Einstein relation is satisfied even though the approach to this result at long time is algebraic rather than exponential. Note that when the density of the Brownian particle is much greater than the fluid density, $\sigma \to 0$, $m^* \to m$, and $F(t) \to e^{-\gamma|t|}$, and Eq. (7.5.10) is regained.

7.8 (a) Show that the conditional probability function $P(v, t|v_0, t_0)$ defined in Eq. (7.5.32) satisfies

$$\int dv P(v, t|v_0, t_0) = 1, \qquad \int dv F(v) P(v, t|v_0, t_0) = \langle F(v(t)) \rangle.$$

(b) Show that

$$P(v, t|v_0, t_0) = \int_{-\infty}^{\infty} d\lambda \, \langle e^{i\lambda(v - v(t))} \rangle_{v_0, t_0} .$$

Then use Eq. (7.5.18) and the fact that

$$\left\langle \exp\left[i\lambda \int_0^t f(t')\zeta(t')dt' \right] \right\rangle = \exp\left[-m\gamma T\lambda^2 \int_0^t f^2(t')dt' \right]$$

for any function $f(t)$ to show that

$$P(v, t|v_0, t_0) = \frac{1}{\sqrt{2\pi\Delta_v(t)}} e^{-(v - \langle v(t) \rangle)^2 / 2\Delta_v(t)}.$$

(c) The function

$$P(x, t|x_0, v_0, t_0) = \langle \delta(x - x(t)) \rangle_{x_0, v_0, t_0}$$

is the probability that $x = x(t)$ at time t, given that $x = x_0$ and $v = v_0$ at time t_0. Using Eqs. (7.5.18) and (7.5.22), show that

$$x(t) - \langle x(t) \rangle = \int_0^t dt' \left(1 - e^{\gamma(t'-t)} \right) \zeta(t')/m,$$

where $\langle x(t) \rangle$ is given in Eq. (7.5.23). Then, following the same reasoning as in part (a), show that

$$P(x, t|x_0, v_0, t_0) = \frac{1}{\sqrt{2\pi\Delta_x(t)}} e^{-(x - \langle x(t) \rangle)^2 / 2\Delta_x(t)},$$

where $\Delta_x(t)$ is given in Eq. (7.5.26). Finally, average the probability $P(x, t|x_0, v_0, t_0)$ over a Maxwell-Boltzmann distribution for v_0 to obtain

$$P(x, t|x_0) = \frac{1}{\sqrt{2\pi\langle [\Delta x(t)]^2 \rangle}} e^{-(x - x_0)^2 / 2\langle [\Delta x(t)]^2 \rangle},$$

where $\langle [\Delta x(t)]^2 \rangle$ is given by Eq. (7.5.13).

7.9 (Diffusion in an external force field)

(a) Show that the stochastic equation for a three dimensional position $\mathbf{x}(t)$

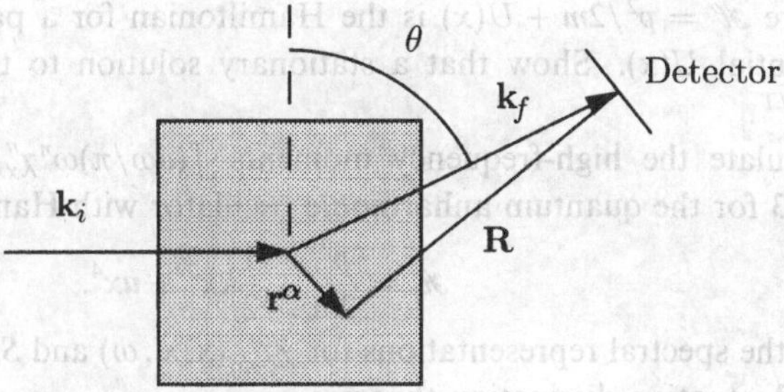

Fig. 7P.2. Scattering geometry for Problem 7.13.

for a Brownian particle in an external force $\mathbf{F}$ for times long compared to $1/\gamma$ where inertial effects can be ignored is

$$\frac{\partial \mathbf{x}}{\partial t} = \Gamma \mathbf{F} + \boldsymbol{\eta} = -\Gamma \nabla_{\mathbf{x}} \mathscr{H} + \boldsymbol{\eta},$$

where $\Gamma = 1/\alpha$ and $\boldsymbol{\eta} = \boldsymbol{\zeta}/\gamma$ is a vector noise.

(b) Now consider a particle of mass density ρ in a fluid of density ρ_0 in a gravitational field $\mathbf{g} = g\mathbf{e}_z$. Show that the Smoluchowski equation for the probability $P(\mathbf{x}, t)$ that the particle is at position $\mathbf{x}$ at time t satisfies the equation

$$\frac{\partial P}{\partial t} = D\nabla^2 P + \mathbf{c} \cdot \nabla P \equiv -\nabla \cdot \mathbf{j},$$

where $D = T\Gamma = T/(\gamma m)$, $\mathbf{c} = \mathbf{g}[1 - (\rho/\rho_0)]/\gamma$, and $\mathbf{j} = -(D\nabla P + \mathbf{c}P)$ is the current. This defines a directed diffusion problem in which there is an average drift along $\mathbf{e}_z$.

(c) Show that the solution to this equation in an infinite system subject to the boundary condition $P(\mathbf{x}, 0) = \delta(\mathbf{x} - \mathbf{x}_0)$ is

$$P(\mathbf{x}, t) = \frac{1}{(4\pi D|t|)^{3/2}} e^{-|\mathbf{x}-\mathbf{x}_0+\mathbf{c}t|^2/(4D|t|)}.$$

This shows that there is an average drift in the direction of the applied force. As a result, diffusion in an external field is often referred to as directed diffusion

(d) (Sedimentation: difficult) A physical container is not infinite, and no current can flow through the boundaries of the container. Assume $P(\mathbf{x}, t) = P(z, t)$ does not depend on x or y and solve the directed diffusion equation for $P(z, t)$ subject to the boundary condition $P(z, 0) = \delta(z - z_0)$ and $j_z = 0$ at $z = 0$.

7.10 Show that the probability $P(x, p, t|x_0, p_0, t_0)$ that a particle has position x and momentum p at time t, given that it had position x_0 and momentum p_0 at time t_0, satisfies Kramer's equation

$$\frac{\partial P}{\partial t} = \alpha T \frac{\partial}{\partial p}\left[\frac{1}{T}\frac{\partial \mathscr{H}}{\partial p} + \frac{\partial}{\partial p}\right] P + \frac{\partial}{\partial p}\left(\frac{\partial \mathscr{H}}{\partial x} P\right) - \frac{\partial}{\partial x}\left(\frac{\partial \mathscr{H}}{\partial p} P\right),$$

where $\mathscr{H} = p^2/2m + U(x)$ is the Hamiltonian for a particle in an external potential $U(x)$. Show that a stationary solution to this equation is $P = e^{-\mathscr{H}/T}$.

7.11 Calculate the high-frequency moments $\int (d\omega/\pi)\omega^n \chi''_{xx}(\omega)$ for $n = 1$ and $n = 3$ for the quantum anharmonic oscillator with Hamiltonian

$$\mathscr{H} = \frac{p^2}{2m} + \frac{1}{2}kx^2 + ux^4.$$

7.12 Use the spectral representations for $\chi''_{\phi_i\phi_j}(\mathbf{x},\mathbf{x}',\omega)$ and $S_{\phi_i\phi_j}(\mathbf{x},\mathbf{x}',\omega)$ to derive the fluctuation-dissipation theorem.

7.13 This problem is about using light scattering to measure the self-diffusion constant of non-interacting spherical particles of volume V_0 suspended in a fluid. The geometry of the scattering process is sketched in Fig. 7P.2. There are N particles at positions $\mathbf{r}^\alpha(t)$. The incident electric field is

$$\mathbf{E}_i = \mathbf{E}_0 e^{i(\mathbf{k}_i\cdot\mathbf{r}-\omega_0 t)}.$$

The detector is at position $\mathbf{R}$ relative to the average position of the scattering particles. The wave vector of the scattered field $\mathbf{k}_f$ is parallel to $\mathbf{R}$.

(a) Show that the total electric field at the detector is

$$\mathbf{E}(t) = \sum_\alpha \mathbf{E}_\alpha(t) = \sum_\alpha \mathbf{E}'_0 e^{i[\mathbf{q}\cdot\mathbf{r}^\alpha(t')-\omega_0 t']},$$

where $t' = t - R/c$, $\mathbf{q} = \mathbf{k}_i - \mathbf{k}_f$, and

$$\mathbf{E}'_0 = \mathbf{E}_0(\omega_0/c)^2(e^{i\mathbf{k}_f\cdot\mathbf{R}}/R)(\alpha-\alpha_0)V_0 \sin\Phi,$$

where $\alpha - \alpha_0$ is the difference between the dipolar polarizability of the particles and the fluid and Φ is the angle between $\mathbf{k}_f$ and the normal to $\mathbf{k}_i$, as shown in the figure.

(b) Assuming the particles do not interact, show that

$$\begin{aligned} S_{EE}(\omega) &= \int_{-\infty}^{\infty} d(t-t')\langle \mathbf{E}(t)\cdot\mathbf{E}^*(t')\rangle e^{i\omega(t-t')} \\ &= 2N\,|\,E_0\,|^2\frac{Dq^2}{(\omega-\omega_0)^2+(Dq^2)^2}. \end{aligned}$$

(c) Show that the intensity-intensity correlation function satisfies

$$\begin{aligned} S_{II}(\omega) &= \int_{-\infty}^{\infty} d(t-t')\langle |\,\mathbf{E}(t)\,|^2 |\,\mathbf{E}(t')\,|^2\rangle e^{i\omega(t-t')} \\ &= N^2\,|\,E_0\,|^2\left[2\pi\delta(\omega) + 2\frac{2Dq^2}{\omega^2+(2Dq^2)^2}\right]. \end{aligned}$$

The photocurrent in a phototube (the detector) is proportional to the electric field intensity $I(t) = |\,\mathbf{E}(t)\,|^2$ so that $S_{II}(\omega)$, or its Fourier transform $S_{II}(t)$, are directly measurable. Note the advantage of measuring S_{II} rather than S_{EE}. Dq^2 is of order 500 Hz. To measure a frequency shift of 500 Hz relative to the frequency 10^{14} Hz of visible laser light is difficult. To measure a 500 Hz shift in frequency relative to zero is, however, fairly straightforward.

8

Hydrodynamics

Thermodynamics provides a description of the equilibrium states of systems with many degrees of freedom. It focuses on a small number of macroscopic degrees of freedom, such as internal energy, temperature, number density, or magnetization, needed to characterize a homogeneous equilibrium state. In systems with a broken continuous symmetry, thermodynamics can be extended to include slowly varying elastic degrees of freedom and to provide descriptions of spatially nonuniform states produced by boundary conditions or external fields. Since the wavelengths of the elastic distortions are long compared to any microscopic length, the departure from ideal homogeneous equilibrium is small. In this chapter, we will develop equations governing dynamical disturbances in which the departure from ideal homogeneous equilibrium of each point in space is small at all times.

8.1 Conserved and broken-symmetry variables

Thermodynamic equilibrium is produced and maintained by collisions between particles or elementary excitations that occur at a characteristic time interval τ. In classical fluids, τ is of order 10^{-10} to 10^{-14} seconds. In low-temperature solids or in quantum liquids, τ can be quite large, diverging as some inverse power of the temperature T. The mean distance λ between collisions (mean free path) of particles or excitations is a characteristic velocity v times τ. In fluids, v is determined by the kinetic energy, $v \sim (T/m)^{1/2}$, where m is a mass. In solids, v is typically a sound velocity. Imagine now a disturbance from the ideal equilibrium state that varies periodically in time and space with frequency ω and wave number q. If $\omega\tau \ll 1$ and $q\lambda \ll 1$, the disturbance varies slowly on time and length scales set by τ and λ, and there will be many equilibrating collisions in each of its temporal and spatial cycles. Thus, each point in space is close to thermodynamic equilibrium at each instant of time, and one would expect to be able to treat such disturbances as perturbations from thermodynamic equilibrium even though they vary in time.

Most disturbances in many body systems have characteristic frequencies that are of order τ^{-1}. If excited, they decay rapidly to equilibrium. There are, however, certain classes of variables that are guaranteed to have slow temporal variations at long wavelengths. These are

(1) densities of conserved variables, and

(2) broken-symmetry elastic variables.

A conserved density such as the number density n obeys a conservation law of the form

$$\frac{\partial n}{\partial t} + \nabla \cdot \mathbf{j} = 0, \tag{8.1.1}$$

where $\mathbf{j}$ is the particle current. When Fourier transformed, such equations imply frequencies ω that go to zero with wave number q. Indeed, in the preceding chapter, we found that the characteristic frequency associated with the conserved density of particles suspended in a fluid was $-iDq^2$. In Chapter 6, we saw that the free energy is invariant with respect to the spatially uniform displacement of broken-symmetry elastic variables such as the angle θ in the xy-model. Thus, spatially uniform changes in elastic variables lead to new equilibrium states that are stationary in time. This implies that the frequency associated with zero wave number displacements of broken-symmetry elastic variables is zero. Spatially non-uniform displacements will, however, have nonzero characteristic frequencies.

Historically, the first system whose long-wavelength, low-frequency dynamics was given serious attention was water. The dynamics of water in motion is called *hydrodynamics*. Today, the term hydrodynamics is used for the long-wavelength, low-frequency dynamics of conserved and broken-symmetry variables in any system. Thus, for example, spin systems and crystalline solids as well as water have a well defined hydrodynamics. They will be the subject of this chapter.

The time dependence of each conserved variable is determined by a current as in Eq. (8.1.1). Currents for broken-symmetry variables can, as we shall see, also be introduced. For slowly varying disturbances, these currents are local functions of the fields thermodynamically conjugate to the hydrodynamical variables. Thus, as we saw in Chapter 7, the current for particles suspended in a fluid is proportional to the gradient of their chemical potential. The equations relating currents to thermodynamic fields are called *constitutive relations*. The hydrodynamics of a given system is determined by its hydrodynamical variables and their currents and associated constitutive relations. As we discussed in the preceding chapter, fields can be classified according to their sign under time reversal. Currents have the opposite sign under time reversal from their associated hydrodynamical variables. Coefficients relating a current and a field with the same sign under time reversal are *nondissipative* and are ultimately responsible for propagating modes such as sound waves. These nondissipative coefficients can usually be determined by straightforward invariance arguments. For example, the time derivative $\partial\mathbf{u}/\partial t$ of the displacement variable $\mathbf{u}$ in a crystal in motion at constant velocity $\mathbf{v}$ must be equal to $\mathbf{v}$. In addition, the momentum density $\mathbf{g}$ must be equal to the mass density times $\mathbf{v}$. The equation $\partial\mathbf{u}/\partial t = \mathbf{v}$ provides a coupling between the hydrodynamical variables $\mathbf{u}$ and $\mathbf{g}$ that does not involve any dissipation since it applies to a steady state situation. Relations between a current and a field with the opposite sign under time reversal are necessarily irreversible. They imply entropy production and are thus *dissipative.*

Our program for obtaining hydrodynamical equations is the following. First, since hydrodynamics is basically a perturbation about thermodynamic equilibrium, we have to generalize our treatment of thermodynamics and statistical mechanics to include all conserved and broken-symmetry variables, including for example the momentum in fluids. Secondly, we must identify the time dependence induced in hydrodynamical variables of one sign under time reversal by nonzero values of variables of the opposite sign, i.e., we must identify reactive couplings. Finally, we must derive irreversible dissipative couplings.

The hydrodynamics of water is quite complicated because there are five hydrodynamical variables arising from the five conservation laws (mass, energy and momentum) in a one-component fluid. The imposition of a broken symmetry, as in a liquid crystal, leads to even greater complexity. We will, therefore, study first the hydrodynamics of a simple model with only two conserved variables and a single broken-symmetry variable in the low-temperature, ordered phase. This model will introduce all of the ingredients essential to the understanding of hydrodynamics, including generalization of thermodynamics to include variables describing states of nonzero motion, derivation of reactive and dissipative constitutive relations, determination of the linearized mode structure, and calculation of response functions. We will then derive and discuss the hydrodynamics of spin systems, one- and two-component fluids, liquid crystals, crystalline solids and superfluids.

It is possible for variables other than conserved or broken-symmetry variables to have characteristic frequencies that are much slower than the inverse collision time. For example, decay times of a nonconserved order parameter diverge as a second-order phase transition is approached. In Sec. 8.6, we will discuss dynamic scaling of correlation functions and characteristic frequencies near second-order critical points. This is a natural generalization of the static scaling discussed in Chapter 5. We will then discuss stochastic dynamical equations for both hydrodynamical and slow critical variables. These are the natural generalizations to continuous fields of the Langevin equations discussed in Sec. 7.5.

8.2 A tutorial example – rigid rotors on a lattice

The hydrodynamics of real physical systems is either quite complicated because of the large number of hydrodynamic variables (e.g. fluids, liquid crystals and solids) or confusing because of possibly unfamiliar time evolution (e.g. spin systems). We will, therefore, study a simple model system that has no known physical realization but that will illustrate all of the essential features of hydrodynamics. The study of real systems will then be almost straightforward though sometimes tedious.

1 Description of the model

The model we will investigate is one with a symmetric rigid rotor or bar at each of N sites $\mathbf{l}$ on a d-dimensional lattice, as shown in Fig. 8.2.1. Each rotor can rotate frictionlessly in the two-dimensional xy-plane. The direction of the rotor at site $\mathbf{l}$ can be specified by the unit vector

$$\boldsymbol{\nu}_{\mathbf{l}} = (\cos \vartheta_{\mathbf{l}}, \sin \vartheta_{\mathbf{l}}), \tag{8.2.1}$$

where $\vartheta_{\mathbf{l}}$ is the angle of the rotor relative to the x-axis. Neighboring rotors do not touch, but there is an exchange-like interaction potential favoring parallel alignment. At high temperature, the rotors are randomly oriented like spins in the paramagnetic phase. At low temperature, they align along a common axis like the molecules of a nematic liquid crystal, as shown in Fig. 8.2.1b. The ordered phase, like the nematic phase of liquid crystals, is characterized by a symmetric, traceless tensor order parameter,

$$\begin{aligned}\langle Q_{ij}(\mathbf{x})\rangle &= \left\langle \sum_{\mathbf{l}} \left(\nu_{\mathbf{l}i}\nu_{\mathbf{l}j} - \frac{1}{2}\delta_{ij}\right)\delta(\mathbf{x}-\mathbf{R}_{\mathbf{l}})\right\rangle \\ &= (N/V)S\left(n_i n_j - \frac{1}{2}\delta_{ij}\right) \quad (i,j = x,y),\end{aligned} \tag{8.2.2}$$

where V is the volume and $\mathbf{n}(\mathbf{x}) = [\cos\theta(\mathbf{x}), \sin\theta(\mathbf{x})]$ is the director specifying the direction of average alignment at $\mathbf{x}$. The distinction between the microscopic angle $\vartheta_{\mathbf{l}}$ specifying the direction of the local rotor at site $\mathbf{l}$ and the coarse-grained angle $\theta(\mathbf{x})$ specifying the direction of the director $\mathbf{n}(\mathbf{x})$ is identical to that introduced in Sec. 6.1 in our discussion of fluctuations in the xy-model. A potential energy favoring parallel rotors which is invariant under the inversion operation $\boldsymbol{\nu}_{\mathbf{l}} \to -\boldsymbol{\nu}_{\mathbf{l}}$ $(\vartheta_{\mathbf{l}} \to \vartheta_{\mathbf{l}} + \pi)$ is

$$U[\vartheta_{\mathbf{l}}] = -J\sum_{<\mathbf{l},\mathbf{l}'>} \cos[2(\vartheta_{\mathbf{l}} - \vartheta_{\mathbf{l}'})]. \tag{8.2.3}$$

Except for the factor of two in the cosine assuring inversion symmetry, this is identical to the classical xy-Hamiltonian of Eq. (6.1.16).

Because the lattice is rigidly fixed and the rotors rotate without friction, the rotational angular momentum of the rotors about the z-axis is conserved. Thus, there are two, and only two, conserved variables in this system: the energy E and the angular momentum L. The energy density and angular momentum density operators (in the sense of Chapter 3), $\hat{\varepsilon}(\mathbf{x},t)$ and $\hat{l}(\mathbf{x},t)$, obey local conservation laws,

$$\frac{\partial \hat{\varepsilon}}{\partial t} + \nabla\cdot\hat{\mathbf{j}}^{\varepsilon} = 0, \qquad \frac{\partial \hat{l}}{\partial t} + \nabla\cdot\hat{\boldsymbol{\tau}} = 0, \tag{8.2.4}$$

where $\hat{\mathbf{j}}^{\varepsilon}$ and $\hat{\boldsymbol{\tau}}$ are, respectively, the energy and angular momentum current operators. Ensemble or coarse-grained averages of the above conservation equations lead to identical equations relating the averaged densities, $\varepsilon(\mathbf{x},t) = \langle\hat{\varepsilon}(\mathbf{x},t)\rangle$ and $l(\mathbf{x},t) = \langle\hat{l}(\mathbf{x},t)\rangle$, to the averaged currents, $\mathbf{j}^{\varepsilon}(\mathbf{x},t) = \langle\hat{\mathbf{j}}^{\varepsilon}(\mathbf{x},t)\rangle$ and $\boldsymbol{\tau}(\mathbf{x},t) = \langle\hat{\boldsymbol{\tau}}(\mathbf{x},t)\rangle$.

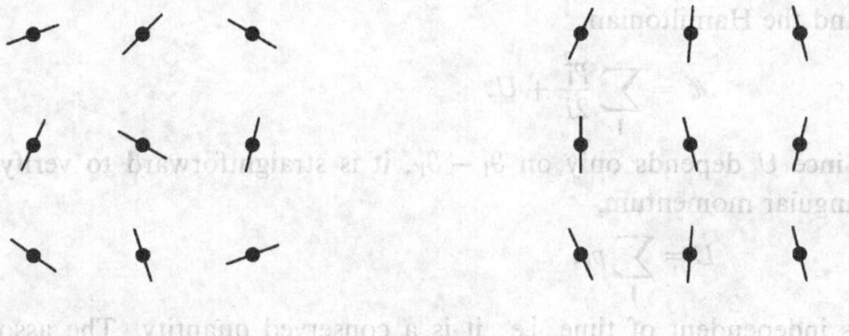

(a) (b)

Fig. 8.2.1. (a) Disordered phase of the rigid rotor model. Rigid rotors at lattice sites $\mathbf{l}$ rotate freely about a fixed axis. The direction of the rotors is random. (b) Ordered phase of the rigid rotor model. Rotors align along a common axis as in a nematic liquid crystal. The order parameter $\langle Q_{ij} \rangle$ is nonzero.

It is important to remember that ε and $\boldsymbol{\tau}$ are even under time reversal whereas l and $\mathbf{j}^{\varepsilon}$ are odd under the same operation.

In the ordered phase, the angle $\theta(\mathbf{x})$ is an elastic variable with associated free energy [Eq. (6.1.1)],

$$F_{\text{el}} = \frac{1}{2} \int d^d x \rho_s (\nabla \theta)^2 \equiv \frac{1}{2} \int d^d x \rho_s \mathbf{v}_\theta^2, \quad \mathbf{v}_\theta \equiv \nabla \theta. \tag{8.2.5}$$

If $\theta = 0$ in the equilibrium state, then $\mathbf{n} = (1, 0)$, $\langle Q_{xx} \rangle = -\langle Q_{yy} \rangle = (N/V)S/2$, and $\langle Q_{xy} \rangle = 0$. Small changes of θ produce a non-vanishing $\langle Q_{xy} \rangle = 2\langle Q_{xx} \rangle \delta\theta$. Thus, for small deviations from the ground state, $\theta = \langle Q_{xy} \rangle / (2\langle Q_{xx} \rangle)$. As in a nematic liquid crystal, the bars are inversion invariant, and all physical quantities must be invariant under the transformation $\mathbf{n} \to -\mathbf{n}$. θ is even under time reversal.

2 The disordered phase

In the disordered phase, the only hydrodynamic variables are ε and l. Before we can study the hydrodynamics of this system, we must first include the angular momentum in its statistical mechanics and thermodynamics. We begin by constructing the Lagrangian, from which we can obtain the Hamiltonian that controls statistical averages. If each rod has a moment of inertia I, then the Lagrangian is

$$\mathscr{L} = \frac{1}{2} \sum_{\mathbf{l}} I \dot{\vartheta}_{\mathbf{l}}^2 - U[\vartheta_{\mathbf{l}}], \tag{8.2.6}$$

where U is the potential energy like that of Eq. (8.2.3). From this, we can construct the angular momentum of each site,

$$p_{\mathbf{l}} = \frac{\partial \mathscr{L}}{\partial \dot{\vartheta}_{\mathbf{l}}} = I \dot{\vartheta}_{\mathbf{l}}, \tag{8.2.7}$$

and the Hamiltonian,

$$\mathscr{H} = \sum_{\mathbf{l}} \frac{p_{\mathbf{l}}^2}{2I} + U. \tag{8.2.8}$$

Since U depends only on $\vartheta_{\mathbf{l}} - \vartheta_{\mathbf{l}'}$, it is straightforward to verify that the total angular momentum,

$$L = \sum_{\mathbf{l}} p_{\mathbf{l}}, \tag{8.2.9}$$

is independent of time, i.e., it is a conserved quantity. The associated angular momentum density is

$$\hat{l}(\mathbf{x}, t) = \sum_{\mathbf{l}} p_{\mathbf{l}} \delta(\mathbf{x} - \mathbf{R}_{\mathbf{l}}). \tag{8.2.10}$$

Since L is conserved, it is possible to have stationary states (i.e., states that do not vary in time) in which $\langle L \rangle = V \langle \hat{l} \rangle$ is nonzero. Thus, stationary states are characterized by their angular momentum density $l = \langle \hat{l} \rangle$ as well as by their energy density ε. If $\langle \hat{l}(\mathbf{x}, t) \rangle$ is independent of $\mathbf{x}$, then the average angular velocity of each rotor is the same, and we can introduce an angular velocity Ω via

$$l = \langle \hat{l} \rangle = \tilde{I} \Omega, \tag{8.2.11}$$

where $\tilde{I} = NI/V$ is the moment of inertia per unit volume. The angular frequency $\Omega(l)$ is determined completely by the value of the stationary angular momentum density. More generally, $\tilde{I}$ could be a function of l and ε, and $\Omega(l, \varepsilon)$ would be a function of both l and ε.

Having established that stationary states with a nonzero l are possible, we next need to identify ensembles that lead to equilibrium nonzero values of l. Clearly, the canonical ensemble constructed from the Hamiltonian of Eq. (8.2.8) will lead to $l = 0$ because it is a minimum at $p_{\mathbf{l}} = 0$ for every $\mathbf{l}$ and it is an even function of $p_{\mathbf{l}}$. To create an ensemble favoring $l \neq 0$, we need only add a term $-\Omega_e L = -\Omega_e \sum_{\mathbf{l}} p_{\mathbf{l}}$ to $\mathscr{H}$. Rather than adding such a term directly, we will show how it arises naturally from a Lagrangian expressed in terms of angular velocities relative to a coordinate system rotating with angular velocity Ω_e. Define the relative angle $\vartheta'_{\mathbf{l}}$ and its angular velocity $\omega_{\mathbf{l}} = \dot{\vartheta}'_{\mathbf{l}}$ via

$$\begin{aligned} \vartheta_{\mathbf{l}} &= \Omega_e t + \vartheta'_{\mathbf{l}} \\ \dot{\vartheta}_{\mathbf{l}} &= \Omega_e + \omega_{\mathbf{l}}. \end{aligned} \tag{8.2.12}$$

Then, because $\vartheta_{\mathbf{l}} - \vartheta_{\mathbf{l}'} = \vartheta'_{\mathbf{l}} - \vartheta'_{\mathbf{l}'}$, the Lagrangian as a function of $\vartheta'_{\mathbf{l}}$ is

$$\mathscr{L} = \frac{1}{2} I \sum_{\mathbf{l}} (\Omega_e + \omega_{\mathbf{l}})^2 - U[\vartheta'_{\mathbf{l}}]. \tag{8.2.13}$$

This Lagrangian differs by a total time derivative from one in which the kinetic energy is $\frac{1}{2} I \sum_{\mathbf{l}} \omega_{\mathbf{l}}^2$ rather than $\frac{1}{2} I \sum_{\mathbf{l}} (\Omega_e + \omega_{\mathbf{l}})^2$. The equations of motion for $\vartheta'_{\mathbf{l}}$ predicted by these two Lagrangians are thus identical. Either can be used to construct momentum conjugate to $\omega_{\mathbf{l}}$ and a Hamiltonian. The Lagrangian of Eq. (8.2.13) will be more useful to us. The momentum conjugate to $\omega_{\mathbf{l}}$ is

$$p_l = \frac{\partial \mathscr{L}}{\partial \omega_l} = I(\Omega_e + \omega_l) = I \sum_l \dot{\vartheta}_l. \tag{8.2.14}$$

Thus, the value of the canonical momentum p_l is $I\dot{\vartheta}_l$, independent of the value of Ω_e. The Hamiltonian associated with the Lagrangian of Eq. (8.2.13) is then

$$\begin{aligned} \mathscr{H}_T &= \sum_l p_l \omega_l - \mathscr{L} \\ &= \frac{1}{2}\sum_l (p_l^2/I) + U[\theta_l] - \Omega_e L \\ &\equiv \mathscr{H} - \Omega_e L, \end{aligned} \tag{8.2.15}$$

where $\mathscr{H}$ is the Hamiltonian in the frame with $\Omega_e = 0$ and $L = \sum_l p_l = \sum_l I(\Omega_e + \omega_l)$ is the total angular momentum. It is a straightforward exercise to verify the Poisson bracket relation, $\{\mathscr{H}, L\} = 0$, and that L is conserved. Note that L is the rest frame angular momentum regardless of the value of Ω_e. Note also that the Ω_e is thermodynamically conjugate to the angular momentum (see Chapter 2).

Thermodynamic functions can now be determined via the partition function,

$$\begin{aligned} \Xi(T, \Omega_e) &= \text{Tr} e^{-(\mathscr{H} - \Omega_e L)/T} \equiv e^{S - (E - \Omega_e \langle L \rangle)/T} \\ &= \frac{1}{N!}(2\pi I T)^{dN/2} e^{NI\Omega_e^2/(2T)} \prod_l \int d\vartheta_l e^{-U[\vartheta_l]/T}, \end{aligned} \tag{8.2.16}$$

where S is the entropy and $E = \langle \mathscr{H} \rangle$ is the internal energy. From this, the average angular momentum in thermal equilibrium is easily found to be

$$\langle L \rangle = \frac{\partial \ln \Xi}{\partial \beta \Omega_e} = NI\Omega_e. \tag{8.2.17}$$

Thus, in thermal equilibrium, the frequency Ω introduced in Eq. (8.2.11) is equal to Ω_e. This is really just the statement that when there is an equilibrium angular momentum, the average angular velocity of each rotor will be $\langle \dot{\vartheta}_l \rangle = \Omega = l/\tilde{I}$, and the average angular velocity measured relative to the frame rotating with frequency $\Omega_e = \Omega$ will be zero, since by Eq. (8.2.12)

$$\langle \omega_l \rangle = \Omega - \Omega_e. \tag{8.2.18}$$

In what follows, we will generally measure all angular velocities in the rest frame so that Ω_e will be zero. It is useful, however, to remember that there is the same distinction between Ω and Ω_e that there is between the equilibrium and external chemical potential μ and μ^{ext} that we encountered in our discussion of diffusion in Sec. 7.4.

The thermodynamic potential associated with the partition function of Eq. (8.2.16) is

$$W(T, \Omega) = -T \ln \Xi(T, \Omega) = E - \Omega \langle L \rangle - TS. \tag{8.2.19}$$

It satisfies the differential thermodynamic relation

$$dW = -SdT - \langle L \rangle d\Omega. \tag{8.2.20}$$

Here, since we are considering only relations valid in thermodynamic equilibrium, we have dropped the distinction between Ω and Ω_e; they are equal in this case. This equation implies a relation (the volume V is constant),

$$T\,ds = d\varepsilon - \Omega dl, \tag{8.2.21}$$

between intensive densities ε, l and $s = S/V$ of extensive quantities. This equation is the analog of Eq. (3.1.39) for isotropic fluids. The entropy density s is a thermodynamic potential that is a function of the variables ε and l. The variables Ω and T are also functions of ε and l via the relations

$$\Omega(\varepsilon, l) = -T\frac{\partial s}{\partial l}\Big)_\varepsilon, \qquad T^{-1}(\varepsilon, l) = \frac{\partial s}{\partial \varepsilon}\Big)_l. \tag{8.2.22}$$

We will be principally interested in states near the equilibrium state with $l = 0$. In this case, the entropy density can be expanded in powers of l as

$$s(\varepsilon, l) = s_0(\varepsilon) - \frac{1}{2T\tilde{I}}l^2, \tag{8.2.23}$$

where $\tilde{I}^{-1} = \partial\Omega/\partial l)_\varepsilon$.

The fundamental thermodynamic identity, Eq. (8.2.21), and the continuity equations (8.2.4) can now be used to tell us how the entropy changes in response to changes in the conserved hydrodynamical variables:

$$\begin{aligned} T\frac{\partial s}{\partial t} &= \frac{\partial \varepsilon}{\partial t} - \Omega\frac{\partial l}{\partial t} = -\nabla\cdot\mathbf{j}^\varepsilon + \Omega\nabla\cdot\boldsymbol{\tau} \\ &= -\nabla\cdot(\mathbf{j}^\varepsilon - \Omega\boldsymbol{\tau}) - \boldsymbol{\tau}\cdot\nabla\Omega. \end{aligned} \tag{8.2.24}$$

Then, using the identity

$$\nabla\cdot\mathbf{Q} = T\nabla\cdot(\mathbf{Q}/T) + \mathbf{Q}\cdot(\nabla T/T), \tag{8.2.25}$$

we obtain

$$T\left(\frac{\partial s}{\partial t} + \nabla\cdot(\mathbf{Q}/T)\right) = -\mathbf{Q}\cdot(\nabla T/T) - \boldsymbol{\tau}\cdot\nabla\Omega, \tag{8.2.26}$$

where

$$\mathbf{Q} = \mathbf{j}^\varepsilon - \Omega\boldsymbol{\tau} \tag{8.2.27}$$

is the *heat current*. Integrating Eq. (8.2.26) over a large volume subject to the boundary condition that the heat current is zero at its outer surface, we obtain an expression for the total rate of entropy production,

$$T\frac{dS}{dt} = \int d^dx[-\mathbf{Q}\cdot(\nabla T/T) - \boldsymbol{\tau}\cdot\nabla\Omega], \tag{8.2.28}$$

which must be non-negative.

In reversible, non-dissipative processes, the entropy remains constant, i.e., dS/dt is zero. Thus, *in the absence of dissipation,* the currents $\boldsymbol{\tau}$ and $\mathbf{Q}$ must be zero:

$$\boldsymbol{\tau} = 0; \qquad \mathbf{Q} = \mathbf{j}^\varepsilon - \Omega\boldsymbol{\tau} = 0. \tag{8.2.29}$$

Since by the second law of thermodynamics, the entropy always increases when constraints are removed from the system, the rate of entropy production must be strictly positive when dissipation is allowed. When T and Ω are spatially uniform, the currents are zero. We therefore expect $\boldsymbol{\tau}$ and $\mathbf{Q}$ to be linearly proportional to ∇T and $\nabla\Omega$. The *constitutive relations* between these variables must be chosen so that dS/dt is positive. In addition, dissipative currents of one sign under time reversal must be proportional to variables of the opposite sign. These constraints imply

$$\mathbf{Q} = -\kappa\nabla T, \qquad \boldsymbol{\tau} = -\Gamma\nabla\Omega, \tag{8.2.30}$$

with $\kappa > 0$ and $\Gamma > 0$ because

$$T\frac{dS}{dt} = \int d^d x[\kappa(\nabla T)^2/T + \Gamma(\nabla\Omega)^2] > 0. \tag{8.2.31}$$

κ and Γ are *transport coefficients;* κ is the *thermal conductivity*. Eqs. (8.2.30) are constitutive relations expressing the currents in terms of a spatial derivative of the fields T and Ω conjugate to E and $\langle L\rangle$. There is no term relating $\boldsymbol{\tau}$ to ∇T because they have the same signs under time reversal (or l and T have opposite signs). Similarly, there is no term coupling $\mathbf{Q}$ to $\nabla\Omega$.

The equations for the energy and angular momentum densities linearized about the state with zero angular momentum and $\Omega = 0$ are, therefore,

$$\begin{aligned} \frac{\partial\varepsilon}{\partial t} &= -\nabla\cdot\mathbf{j}^{\varepsilon} = \kappa\nabla^2 T, \\ \frac{\partial l}{\partial t} &= -\nabla\cdot\boldsymbol{\tau} = \Gamma\nabla^2\Omega. \end{aligned} \tag{8.2.32}$$

These equations can be closed with the aid of the thermodynamic relations

$$d\Omega = \tilde{I}^{-1}dl, \qquad dT = C_l^{-1}d\varepsilon, \tag{8.2.33}$$

where C_l is the specific heat at constant l. C_l and $\tilde{I}$ are particular realizations of a susceptibility relating changes of conjugate variables. We now obtain

$$\frac{\partial\varepsilon}{\partial t} = D_\varepsilon\nabla^2\varepsilon, \qquad \frac{\partial l}{\partial t} = D_l\nabla^2 l, \tag{8.2.34}$$

where

$$D_\varepsilon = \kappa/C_l, \qquad D_l = \Gamma/\tilde{I} \tag{8.2.35}$$

are, respectively, the thermal and angular momentum diffusion constants. Thus, both ε and l relax diffusively. D_ε and D_l, like the diffusion constant [Eq. (7.4.32)] for particles in a fluid, are the ratio of a transport coefficient Γ to a susceptibility χ. This form is quite general. The dissipative parts of the response functions can be obtained directly using the methods introduced in Sec. 7.4:

$$\frac{\chi''_{\varepsilon\varepsilon}(\mathbf{q},\omega)}{\omega} = C_l\frac{D_\varepsilon q^2}{\omega^2 + (D_\varepsilon q^2)^2}, \tag{8.2.36}$$

$$\frac{\chi''_{ll}(\mathbf{q},\omega)}{\omega} = \tilde{I}\frac{D_l q^2}{\omega^2 + (D_l q^2)^2}. \tag{8.2.37}$$

From this we see that there is one mode (one peak in a response function) for each of the conserved variables ε and l. This result is quite general. There is always one mode associated with each conserved variable and, as we shall see shortly, with each broken-symmetry variable.

Eqs. (8.2.34) are the phenomenological diffusive dynamical equations for the conserved densities ε and l. Identical equations control the time development of the conjugate fields T and Ω as long as the linear thermodynamic relations of Eqs. (8.2.33) hold. Thus, for example, $\partial T/\partial t = D_\varepsilon \nabla^2 T$. This is the equation of thermal diffusion. We regard the equations in terms of ε and l as more fundamental than those for T and Ω because the former variables obey microscopic conservation laws.

3 The ordered phase

The free energy of the ordered phase is completely independent of the spatially uniform angle variable θ. Gradients of θ do, however, increase the free energy, and $\mathbf{v}_\theta = \nabla\theta$ must be included as an independent thermodynamic variable. $\mathbf{v}_\theta$ is also a broken-symmetry hydrodynamical variable whose characteristic excitation frequencies go to zero with wave number q. ε, l and $\mathbf{v}_\theta$ are the only hydrodynamic variables. All other variables relax to equilibrium values determined by local values of ε, l and $\mathbf{v}_\theta$ in microscopic times τ; they do not need to be considered at any stage in the derivation of hydrodynamical equations. It is possible, however, to include non-hydrodynamical variables in a non-rigorous way by a slight generalization of the methods presented here.

Thermodynamic functions describing the ordered phase must be a function either of $\mathbf{v}_\theta$ or its conjugate field $\mathbf{h}_\theta$ in addition to T and Ω. The function $\tilde{W}'(T,\Omega,\mathbf{h}_\theta) = E - \Omega\langle L\rangle - TS - \int d^dx\mathbf{h}_\theta\cdot\mathbf{v}_\theta$ is a natural function of $\mathbf{h}_\theta$, whereas

$$\tilde{W} = \tilde{W}'(T,\Omega,\mathbf{h}_\theta) + \int d^dx\mathbf{h}_\theta\cdot\mathbf{v}_\theta = E - \Omega\langle L\rangle - TS \tag{8.2.38}$$

is a natural function of T, Ω and $\mathbf{v}_\theta$ satisfying

$$d\tilde{W} = -SdT - Ld\Omega + \int d^dx\mathbf{h}_\theta\cdot d\mathbf{v}_\theta. \tag{8.2.39}$$

From this follows the fundamental relation

$$T\,ds = d\varepsilon - \Omega dl - \mathbf{h}_\theta\cdot d\mathbf{v}_\theta \tag{8.2.40}$$

among intensive quantities. The potential $\tilde{W}(T,\Omega,\mathbf{v}_\theta)$ can be expanded in a power series in $\mathbf{v}_\theta$ for small $\mathbf{v}_\theta$. It must reduce to $W(T,\Omega)$ when $\mathbf{v}_\theta$ is zero. There is no linear term in $\mathbf{v}_\theta$ because $\mathbf{v}_\theta$ is zero when $\mathbf{h}_\theta$ is zero. The coefficient of v_θ^2 is independent of Ω for small Ω. Thus, for small $\mathbf{v}_\theta$ and Ω, we have

$$\tilde{W}(T,\Omega,\mathbf{v}_\theta) = W(T,\Omega) + F_{\text{el}}(T,\mathbf{v}_\theta). \tag{8.2.41}$$

The conjugate field $\mathbf{h}_\theta$ satisfies

$$\begin{aligned} h_{\theta i} &= -T\left.\frac{\partial s}{\partial v_{\theta i}}\right)_{\varepsilon,l} = \frac{1}{V}\left.\frac{\partial\tilde{W}}{\partial v_{\theta i}}\right)_{T,\Omega} \\ &= \rho_s v_{\theta i}. \end{aligned} \tag{8.2.42}$$

The first line of Eq. (8.2.42) is generally valid. The second line is valid only to lowest order in $\mathbf{v}_\theta$. Since we will be most interested in modes associated with the equilibrium state with $\mathbf{v}_\theta = 0$, the latter form will suffice for our purposes. In this case, the entropy density can be expanded about the state with $l = 0$ and $\mathbf{v}_\theta = 0$ as

$$s(\varepsilon, l, \mathbf{v}_\theta) = s_0(\varepsilon) - \frac{1}{2T\tilde{I}} l^2 - \frac{1}{2T} \rho_s \mathbf{v}_\theta \cdot \mathbf{v}_\theta. \tag{8.2.43}$$

To treat the dynamical properties of θ and $\mathbf{v}_\theta$, we define the "current" X via

$$\frac{d\theta}{dt} + X = 0. \tag{8.2.44}$$

We first consider the reactive part of X. In the ordered phase with nonzero $\langle Q_{ij} \rangle$, the director $\mathbf{n}(\mathbf{x}, t)$ will rotate with the average angular frequency of the rotors. Thus, in an ordered stationary state with nonzero angular momentum,

$$\frac{d\theta}{dt} = \Omega = \tilde{I}^{-1} l. \tag{8.2.45}$$

Alternatively, the angle θ' measured with respect to a frame rotating with frequency Ω_e will satisfy $d\theta'/dt = \Omega - \Omega_e$. There is no dissipation in the above relation; it will be satisfied so long as there are no external perturbations. Ω is the *reactive* or *nondissipative* part of the current X. As we have seen, reactive parts of currents always couple the time derivative of one variable to another variable with opposite sign under time reversal. The dissipative parts of currents couple the time derivative of one variable to other variables with the same sign under time reversal. We define the dissipative "current" X' via

$$X = -\Omega + X'. \tag{8.2.46}$$

Since θ is a broken-symmetry hydrodynamic variable, we expect X' to tend to zero with wave number. The equation of motion for $\mathbf{v}_\theta$ is

$$\frac{\partial \mathbf{v}_\theta}{\partial t} = -\nabla(X' - \Omega). \tag{8.2.47}$$

Then the thermodynamic relation [Eq. (8.2.40)] and the conservation laws [Eqs. (8.2.4)] for ε and l imply

$$T\frac{\partial s}{\partial t} = -\nabla \cdot \mathbf{j}^\varepsilon + \Omega \nabla \cdot \boldsymbol{\tau} - \mathbf{h}_\theta \nabla \cdot (\Omega - X') \tag{8.2.48}$$

so that the entropy production equation becomes

$$T\left(\frac{\partial s}{\partial t} + \nabla \cdot (\mathbf{Q}/T)\right) = -\mathbf{Q}\cdot(\nabla T/T) - (\boldsymbol{\tau} + \mathbf{h}_\theta)\cdot \nabla\Omega - X' \nabla \cdot \mathbf{h}_\theta, \tag{8.2.49}$$

where

$$\mathbf{Q} = \mathbf{j}^\varepsilon - \Omega \boldsymbol{\tau} - \mathbf{h}_\theta X' \tag{8.2.50}$$

is the heat current.

In the absence of dissipation, entropy production is zero,

$$\boldsymbol{\tau} = -\mathbf{h}_\theta, \qquad \mathbf{Q} = 0, \qquad X' = 0, \tag{8.2.51}$$

and

$$\frac{\partial l}{\partial t} = \nabla \cdot \mathbf{h}_\theta = \rho_s \nabla^2 \theta, \tag{8.2.52}$$

$$\frac{\partial \theta}{\partial t} = \Omega = \tilde{I}^{-1} l. \tag{8.2.53}$$

These are the equations determining the modes of the system in the absence of dissipation. Note their similarity to Poisson bracket equations [Eq. (7.2.2)] for the harmonic oscillator.

We should pause at this point to assess what our formal manipulations have told us. We began with a statement that $d\theta/dt$ is equal to Ω in steady state situations when there is a nonzero angular momentum. The requirement that there be no entropy production for non-dissipative processes then told us that there must be a reactive term in the angular momentum current equal to $-\mathbf{h}_\theta$. This term could not have been predicted using arguments with spatially uniform fields since $\mathbf{h}_\theta$ is nonzero only when there is spatial variation of θ. The end result is that the equations for l and θ, which have opposite signs under time reversal, are coupled. This is analogous to the coupling between p and x in the simple harmonic oscillator discussed in Sec. 7.2. A linear relation between the time derivative of a variable u with one sign under time reversal and a variable v with the opposite sign will invariably lead to a reciprocal linear relation between the time derivatives of v and u. Usually, one relation can be obtained using invariance arguments (such as those used to obtain the relation between θ and Ω); the other then follows from requirement of zero entropy production. The derived relation usually involves more gradients than the fundamental relation following from invariance arguments. The time derivative of either of Eqs. (8.2.52) or (8.2.53) leads to second-order sound-like equations,

$$\frac{\partial^2 l}{\partial t^2} = \rho_s \tilde{I}^{-1} \nabla^2 l \quad \text{or} \quad \frac{\partial^2 \theta}{\partial t^2} = \rho_s \tilde{I}^{-1} \nabla^2 \theta, \tag{8.2.54}$$

and predict undamped propagating modes with a sound-like dispersion relation

$$\omega = \pm \left(\frac{\rho_s}{\tilde{I}} \right)^{1/2} q. \tag{8.2.55}$$

Reactive couplings between variables with opposite sign under time reversal usually lead to propagating modes with linear dispersion in q and a velocity proportional to the square root of a rigidity divided by some measure of inertia. There are cases, as we shall see, however, where these modes can become overdamped and be effectively diffusive.

Constitutive relations for the dissipative parts of the currents can be derived just as in the disordered state. They are

$$\begin{aligned} Q'_i &= -\kappa_{ij} \nabla_j T, \\ \boldsymbol{\tau} &= -\mathbf{h}_\theta + \boldsymbol{\tau}', \qquad \tau'_i = -\Gamma_{ij} \nabla_j \Omega, \\ X' &= -\gamma \nabla \cdot \mathbf{h}_\theta = -\gamma \rho_s \nabla^2 \theta. \end{aligned} \tag{8.2.56}$$

γ is a new dissipative coefficient not present in the disordered phase. It must be positive for positive entropy production. The ordered phase is anisotropic, and dissipative currents can depend on the direction of spatial variation (∇) relative to the local director $\mathbf{n}(\mathbf{x})$. Thus, the dissipative coefficients κ and Γ of the disordered phase become tensors,

$$\begin{aligned}\kappa_{ij} &= \kappa_\parallel n_i n_j + \kappa_\perp(\delta_{ij} - n_i n_j),\\ \Gamma_{ij} &= \Gamma_\parallel n_i n_j + \Gamma_\perp(\delta_{ij} - n_i n_j),\end{aligned} \tag{8.2.57}$$

in the ordered phase. X' and $\mathbf{Q}$ have the same sign under time reversal, and in general there could be dissipative cross-couplings of the form

$$X' \sim \nabla T, \qquad \mathbf{Q} \sim \nabla \cdot \mathbf{h}_\theta. \tag{8.2.58}$$

Other symmetries, however, prevent such couplings. $\mathbf{Q}$ is a vector. The only possible way to create a vector from $\nabla \cdot \mathbf{h}_\theta$ would be to use the director $\mathbf{n}$ in $\mathbf{n}\nabla \cdot \mathbf{h}_\theta$. This is not permitted because all physical quantities must be invariant under $\mathbf{n} \to -\mathbf{n}$.

The complete linearized hydrodynamic mode equations in the ordered phase are

$$\begin{aligned}\frac{\partial \varepsilon}{\partial t} &= C_l^{-1}\kappa_{ij}\nabla_i\nabla_j\varepsilon,\\ \frac{\partial \theta}{\partial t} &= \Omega + \gamma\rho_s\nabla^2\theta = \tilde{I}^{-1}l + \gamma\rho_s\nabla^2\theta,\\ \frac{\partial l}{\partial t} &= \rho_s\nabla\cdot\mathbf{v}_\theta + \Gamma_{ij}\nabla_j\Omega = \rho_s\nabla^2\theta + \tilde{I}^{-1}\Gamma_{ij}\nabla_i\nabla_j l.\end{aligned} \tag{8.2.59}$$

The energy mode decouples from the others and remains diffusive as it was in the disordered phase. Its frequency,

$$\omega = -i(\kappa_\parallel \cos^2\theta_0 + \kappa_\perp \sin^2\theta_0)C_l^{-1}q^2, \tag{8.2.60}$$

depends on the direction of $\mathbf{q}$ relative to $\mathbf{n}_0$, the uniform equilibrium direction of $\mathbf{n}$ ($\mathbf{q}\cdot\mathbf{n}_0 = q\cos\theta_0$). The "sound wave" arising from the coupling of θ and l is now damped:

$$\begin{aligned}\omega &= \frac{1}{2}\left[-i(D_\theta + D_l)q^2 \pm [4\rho_s\tilde{I}^{-1}q^2 - (D_\theta - D_l)^2q^4]^{1/2}\right]\\ &\approx \pm cq - i\frac{1}{2}Dq^2,\end{aligned} \tag{8.2.61}$$

where

$$D_\theta = \gamma\rho_s, \qquad D_l = \tilde{I}^{-1}(\Gamma_\parallel\cos^2\theta_0 + \Gamma_\perp\sin^2\theta_0) \tag{8.2.62}$$

and

$$D = (D_\theta + D_l) \tag{8.2.63}$$

and where

$$c = (\rho_s/\tilde{I})^{1/2} \tag{8.2.64}$$

is the "sound velocity". There are in reality two sound modes at positive and negative frequencies. Thus, the ordered phase, which has one more hydrodynamical variable than the disordered phase, has one more mode than the disordered phase, with a total of three.

The response functions for θ and l can be obtained from the equations of motion, Eqs. (8.2.59), via the Laplace transform technique we used to obtain the diffusive response function in Sec. 7.4. The Laplace transform of Eqs. (8.2.59) yields the matrix equation,

$$\begin{pmatrix} \theta(\mathbf{q},\omega) \\ l(\mathbf{q},\omega) \end{pmatrix} = \frac{1}{\Delta}\begin{pmatrix} -i\omega + D_l q^2 & \tilde{I}^{-1} \\ -\rho_s q^2 & -i\omega + D_\theta q^2 \end{pmatrix}\begin{pmatrix} \theta(\mathbf{q},t=0) \\ l(\mathbf{q},t=0) \end{pmatrix}, \quad (8.2.65)$$

for $\theta(\mathbf{q},\omega)$ and $l(\mathbf{q},\omega)$, where

$$\begin{aligned} \Delta &= (-i\omega + D_l q^2)(-i\omega + D_\theta q^2) + \rho_s \tilde{I}^{-1} q^2 \\ &\approx -\omega^2 + c^2 q^2 - i\omega D q^2. \end{aligned} \quad (8.2.66)$$

To calculate response functions from Eq. (8.2.65), we need the matrix generalization of Eqs. (7.4.14) and (7.4.17),

$$\theta_\alpha(\mathbf{q},\omega) = \frac{1}{i\omega}[\chi_{\alpha\beta}(\mathbf{q},\omega) - \chi_{\alpha\beta}(\mathbf{q})]\chi^{-1}_{\beta\gamma}(\mathbf{q})\theta_\gamma(\mathbf{q},t=0), \quad (8.2.67)$$

where $\theta_\alpha \equiv (\theta, l)$ and summation over repeated indices is understood. Both $\chi''_{\theta\theta}(\mathbf{q},\omega)$ and $\chi''_{ll}(\mathbf{q},\omega)$ are real and odd in ω. They can, therefore, be obtained by taking the real parts of the $\theta-\theta$ and $l-l$ components of Eq. (8.2.65):

$$\begin{aligned} \frac{\chi''_{\theta\theta}(\mathbf{q},\omega)}{\omega} &= \frac{1}{\rho_s q^2}\frac{\omega^2 D q^2 - (\omega^2 - c^2 q^2) D_l q^2}{(\omega^2 - c^2 q^2)^2 + (\omega D q^2)^2}, \\ \frac{\chi''_{ll}(\mathbf{q},\omega)}{\omega} &= \tilde{I}\,\frac{\omega^2 D q^2 - (\omega^2 - c^2 q^2) D_\theta q^2}{(\omega^2 - c^2 q^2)^2 + (D\omega q^2)^2}. \end{aligned} \quad (8.2.68)$$

Note that the same modes appear in both $\chi''_{\theta\theta}$ and χ''_{ll}. The intensity of $\chi''_{\theta\theta}$ is much larger, however, because the susceptibility $\chi_{\theta\theta} = (\rho_s q^2)^{-1}$ diverges at small q. The angular momentum density l and the angle θ have opposite signs under time reversal. Thus, the general symmetry arguments of Sec. 7.6 require that $\chi''_{\theta,l}(\mathbf{q},\omega)$ be imaginary and even in ω and $\chi'_{\theta,l}(\mathbf{q},\omega)$ be imaginary and odd in ω. This implies that $\chi''_{\theta,l}(\mathbf{q},\omega) = -i\mathrm{Re}\chi_{\theta,l}(\mathbf{q},\omega)$ or, from Eq. (8.2.65):

$$\chi''_{\theta l}(\mathbf{q},\omega) = i\frac{\omega^2 D q^2}{(\omega^2 - c^2 q^2)^2 + (\omega D q^2)^2}. \quad (8.2.69)$$

When the dissipation goes to zero, the peaks in $\chi''_{\alpha\beta}$ become delta functions at the "sound" wave frequencies

$$\frac{\chi''_{\theta\theta}}{\omega} = \frac{1}{\rho_s q^2 \tilde{I}}\frac{\chi''_{ll}}{\omega} = \frac{1}{\rho_s q^2}\frac{\chi''_{\theta l}}{i} = \frac{1}{\rho_s q^2}\frac{\pi}{2}[\delta(\omega - cq) + \delta(\omega + cq)]. \quad (8.2.70)$$

The correlation functions,

$$\begin{aligned} S_{\theta\theta}(\mathbf{q},\omega) &= \frac{2\hbar}{1 - e^{-\beta\hbar\omega}}\chi''_{\theta\theta}(\mathbf{q},\omega), \\ S_{\varepsilon\varepsilon}(\mathbf{q},\omega) &= \frac{2\hbar}{1 - e^{-\beta\hbar\omega}}\chi''_{\varepsilon\varepsilon}(\mathbf{q},\omega), \end{aligned} \quad (8.2.71)$$

are plotted in Fig. 8.2.2. Note that the intensity of the "sound" peak at negative frequency is less than that at positive frequency.

4 Excitations from the classical ground state

The hydrodynamical equations derived above determine the long wavelength, low frequency dynamics throughout the ordered phase. Near zero temperature, when

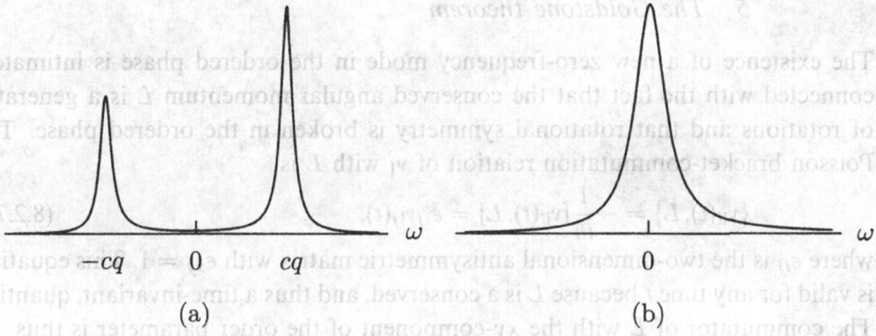

Fig. 8.2.2. (a) $S_{\theta\theta}(\mathbf{q},\omega)$ and (b) $S_{\varepsilon\varepsilon}(\mathbf{q},\omega)$ for fixed q. Note that the thermal factor in $S_{\theta\theta}$ causes the peak at $-cq$ to have a lower intensity for $T>0$ than the peak at $+cq$. Both the diffusive and propagating peaks have widths proportional to q^2.

there is nearly perfect alignment of all spins, the interaction energy between rotors [Eq. (8.2.3)] can be expanded in powers of $\vartheta_{\mathbf{l}}-\vartheta_{\mathbf{l}'}$. The leading term leads to the harmonic Hamiltonian,

$$H=\sum_{\mathbf{l}} p_{\mathbf{l}}^2/(2I)+J\sum_{\mathbf{l},\mathbf{l}'}\gamma_{\mathbf{l},\mathbf{l}'}(\vartheta_{\mathbf{l}}-\vartheta_{\mathbf{l}'})^2-zJN, \tag{8.2.72}$$

whose mode structure is easily calculated. ($\gamma_{\mathbf{l},\mathbf{l}'}$ is the nearest-neighbor matrix introduced in Eq. (6.1.16).) The equation of motion for $\vartheta_{\mathbf{l}}$ is

$$I\ddot{\vartheta}_{\mathbf{l}}=-4J\sum_{\mathbf{l}'}\gamma_{\mathbf{l},\mathbf{l}'}(\vartheta_{\mathbf{l}}-\vartheta_{\mathbf{l}'}), \tag{8.2.73}$$

or

$$\omega^2(\mathbf{q},\omega)=(4J/I)[\gamma(0)-\gamma(\mathbf{q})]\approx 4(2zJ/dI)a^2q^2. \tag{8.2.74}$$

Thus, the elementary excitations from the ground state are propagating waves with a linear dispersion. The velocity of these modes is $c=2(zJ/dIa^2)^{1/2}$. On the other hand, the low temperature rigidity [Eq. (6.1.16)] is $\rho_s(T\approx 0)=4(zJ/d)a^{2-d}$, and $\tilde{I}=I/a^d$. Thus $c=(\rho_s/\tilde{I})^{1/2}$, in agreement with the hydrodynamical result. The hydrodynamical result is, however, valid throughout the ordered phase, even when thermal (or quantum) fluctuations depress $\rho_s(T)$ considerably below its zero temperature classical value of $4zJa^{2-d}$. In addition, the hydrodynamical equations determine the form of the damping (imaginary part of ω) of modes. In a harmonic theory, each mode is independent and there is no damping. When anharmonic terms are added to the harmonic Hamiltonian, collisions between elementary excitations occur, and there can be damping. The dissipative coefficients κ and Γ can, therefore, be calculated at low temperatures by considering collisions between elementary excitations.

5 The Goldstone theorem

The existence of a new zero-frequency mode in the ordered phase is intimately connected with the fact that the conserved angular momentum L is a generator of rotations and that rotational symmetry is broken in the ordered phase. The Poisson bracket-commutation relation of $\mathbf{v}_\mathrm{l}$ with L is

$$\{v_{\mathrm{l}i}(t), L\} = -\frac{1}{i\hbar}[v_{\mathrm{l}i}(t), L] = \epsilon_{ij} v_{\mathrm{l}j}(t), \tag{8.2.75}$$

where ϵ_{ij} is the two-dimensional antisymmetric matrix with $\epsilon_{xy} = 1$. This equation is valid for any time t because L is a conserved, and thus a time-invariant, quantity. The commutator of L with the xy-component of the order parameter is thus

$$[Q_{xy}(\mathbf{x}, t), L] = i\hbar[Q_{xx}(\mathbf{x}, t) - Q_{yy}(\mathbf{x}, t)] = 2i\hbar Q_{xx}(\mathbf{x}, t). \tag{8.2.76}$$

When L is expressed as the integral over $\mathbf{x}'$ of the angular momentum density $l(\mathbf{x}', t')$, this equation implies

$$\int d^d x' \tilde{\chi}''_{Q_{xy},l}(\mathbf{x}, \mathbf{x}', t, t') = i\langle Q_{xx}(\mathbf{x}, t)\rangle \tag{8.2.77}$$

or

$$\chi''_{Q_{xy},l}(\mathbf{q} = 0, \omega) = 2\pi i \langle Q_{xx}\rangle \delta(\omega). \tag{8.2.78}$$

There is no factor of $2\hbar$ on the right hand side of Eqs. (8.2.77) and (8.2.78) because of the factor of $1/2\hbar$ in the definition [Eq. (7.6.14)] of $\chi''_{Q_{xy},l}(\mathbf{x}, \mathbf{x}', t, t')$ in terms of the commutator of $Q_{xy}(\mathbf{x}, t)$ and $l(\mathbf{x}', t')$. Thus, the existence of a broken continuous symmetry implies that there is a zero-frequency pole in the zero wave number order-parameter generator response function, $\chi_{Q_{xy},l}(\mathbf{q} = 0, z)$, or, equivalently, that there is a zero-frequency mode at zero wave number. This is the content of the *Goldstone theorem* (Nambu 1960; Goldstone 1961). The new mode is generally called the *Goldstone mode.*

The Goldstone theorem, Eq. (8.2.78), strictly speaking applies only at $\mathbf{q} = 0$. In the absence of long range forces, however, one can usually argue that the limit of response functions as $\mathbf{q} \to 0$ is equal to their value at $\mathbf{q} = 0$. In this case, the Goldstone theorem implies that there is a mode whose frequency goes continuously to zero as the wave number goes to zero. Indeed, our hydrodynamical analysis leads to precisely such a mode. If continuity is assumed, then, with the replacement $\delta\langle Q_{xy}\rangle = 2\langle Q_{xx}\rangle \delta\theta$, Eq. (8.2.78) can be rewritten as

$$\lim_{\mathbf{q}\to 0} \chi''_{\theta,l}(\mathbf{q}, \omega) = i\pi\delta(\omega), \tag{8.2.79}$$

in agreement with the hydrodynamical predictions of Eq. (8.2.70)

6 Kubo formulae

The fluctuation-dissipation theorem provides a correspondence between equilibrium correlation functions and response functions that lead, for example, to relations between the static thermodynamic susceptibility of conserved variables and equal time correlation functions [Eqs. (3.4.18) and (3.5.13)]. There are analogous relations, called *Kubo formulae,* between dissipative transport coefficients and current correlation functions. To see how such relations can be derived, consider the imaginary part of the angular momentum density response function [Eq. (8.2.37)]. It is straightforward to show that

$$\Gamma = D_l\tilde{I} = \lim_{\omega\to 0}\lim_{q\to 0}\frac{\omega}{q^2}\chi''_{ll}(\mathbf{q},\omega) = \lim_{\omega\to 0}\lim_{q\to 0}\frac{\beta}{2}\frac{\omega^2}{q^2}S_{ll}(\mathbf{q},\omega), \tag{8.2.80}$$

where the fluctuation-dissipation theorem [Eq. (7.6.41)] was used to obtain the final formula. The conservation law for angular momentum [Eq. (8.2.4)] implies

$$\begin{aligned}\omega^2 S_{ll}(\mathbf{q},\omega) &= q_i q_j S_{\tau_i\tau_j}(\mathbf{q},\omega) \qquad (8.2.81)\\ &= \frac{1}{dV}q^2\int d^dx d^dx'\int_{-\infty}^{\infty}dt e^{i\omega t}e^{-i\mathbf{q}\cdot\mathbf{x}}\langle\boldsymbol{\tau}(\mathbf{x},t)\cdot\boldsymbol{\tau}(\mathbf{x}',0)\rangle,\end{aligned}$$

where we used the fact that the disordered state is rotationally isotropic. Combining Eqs. (8.2.80) and (8.2.81), we obtain

$$\begin{aligned}\Gamma &= \frac{\beta}{2}\frac{1}{dV}\int d^dx\int d^dx'\int_{-\infty}^{\infty}dt\langle\boldsymbol{\tau}(\mathbf{x},t)\cdot\boldsymbol{\tau}(\mathbf{x}',0)\rangle\\ &= \frac{\beta}{2}\frac{1}{dV}\int d^dx\int d^dx'\int_{0}^{\infty}dt\left\langle\frac{1}{2}\{\tau_i(\mathbf{x},t),\tau_i(\mathbf{x}',0)\}_+\right\rangle, \qquad (8.2.82)\end{aligned}$$

where $\{A,B\}_+ = AB+BA$ is the anticommutator of A and B. Thus, the dissipative coefficient Γ is related to the integral over time of the current-current correlation function. Similar expressions apply for the thermal conductivity and, indeed, for any dissipative coefficient associated with a conserved variable (see Problem 8.3). There are also related generalized Kubo formulae for the ordered phase. Often, Kubo formulae provide the best way to calculate dissipative coefficients from microscopic models.

7 Summary

In this section, we have studied the hydrodynamics of a simple model system. Many of the concepts and results introduced here apply quite generally to the hydrodynamics of all systems. The most important of these are listed below.

- Long-wavelength, low-frequency excitations are associated with conservation laws and broken symmetry.
- There is exactly one mode associated with each conservation law and each broken symmetry.
- Currents of hydrodynamical variables contain reactive and dissipative parts. The reactive parts of currents couple variables of opposite sign under time reversal and lead to propagating modes.
- In the absence of reactive couplings, the hydrodynamical modes are diffusive.
- Diffusion constants are the ratio of a transport coefficient to a susceptibility.
- The velocities of propagating modes are square roots of the ratio of a reactive transport coefficient to a susceptibility.

- Dissipative coefficients are related to current correlation functions via Kubo formulae.
- Elementary excitations from the ground state can be described by a harmonic Hamiltonian. These excitations are sound-like propagating modes with a velocity that agrees with the predictions of the hydrodynamic theory in the low-temperature limit.
- The Goldstone theorem states that there must be a zero-frequency, zero wave number mode in systems with a continuous broken symmetry. The new hydrodynamic mode in the ordered phase is the continuation of the zero-frequency Goldstone mode to nonzero wave number.

8.3 Spin systems

In preceding chapters, we have studied a variety of ordered phases in spin systems, including ferromagnetic, antiferromagnetic and modulated phases. Many of these phases have broken continuous symmetries and will, therefore, have associated hydrodynamic Goldstone modes. In spin systems, these Goldstone modes are usually called *spin waves*. They will be the subject of this section.

1 Spin dynamics

Before we can discuss spin hydrodynamics, we need to review simple spin dynamics, which is fundamentally quantum mechanical. The commutation relations for distinct spins $\mathbf{S}_\alpha$ and $\mathbf{S}_\beta$ are

$$[S_\alpha^i, S_\beta^k] = i\hbar\delta_{\alpha\beta}\epsilon^{ijk}S_\alpha^k, \tag{8.3.1}$$

where i, j, k are the Cartesian indices x, y, z. These can alternatively be expressed in terms of the raising and lowering operators $S_\alpha^+ = S_\alpha^x + iS_\alpha^y$ and $S_\alpha^- = S_\alpha^x - iS_\alpha^y$:

$$\begin{aligned} [S_\alpha^+, S_\beta^-] &= 2\hbar\delta_{\alpha\beta}S_\alpha^z, \\ [S_\alpha^+, S_\beta^z] &= -\hbar\delta_{\alpha\beta}S_\alpha^+, \\ [S_\alpha^-, S_\beta^z] &= \hbar\delta_{\alpha\beta}S_\alpha^-. \end{aligned} \tag{8.3.2}$$

The time development of a Heisenberg spin operator is determined by

$$\frac{\partial \mathbf{S}_\alpha}{\partial t} = \frac{i}{\hbar}[\mathscr{H}, \mathbf{S}_\alpha], \tag{8.3.3}$$

where $\mathscr{H}$ is the spin-dependent Hamiltonian. If there is no interaction among spins, and there is an external magnetic field $\mathbf{h}_e$, then the Hamiltonian is

$$\mathscr{H} = \mathscr{H}_{\text{ext}} = -\mathbf{h}_e \cdot \sum_\alpha \mathbf{S}_\alpha. \tag{8.3.4}$$

In this case, the equation of motion for each spin reduces to

$$\frac{\partial \mathbf{S}_\alpha}{\partial t} = \mathbf{S}_\alpha \times \mathbf{h}_e. \tag{8.3.5}$$

If $\mathbf{h}_e = h_e \mathbf{e}_z$, then S_α^z is independent of time, and

$$\frac{\partial S_\alpha^+(t)}{\partial t/\hbar} = -ih_e S_\alpha^+(t),$$
$$S_\alpha^+(t) = S_\alpha^+(0)e^{-ih_e t/\hbar} = e^{i\mathscr{H}_e t/\hbar} S_\alpha^+(0) e^{-i\mathscr{H}_e t/\hbar}. \tag{8.3.6}$$

Thus, an external field simply causes the phase of $S_\alpha^+(t)$ to diminish linearly in time.

2 Generalized Heisenberg models

In order to discuss the dynamics of spin systems, it is useful to consider specific model Hamiltonians. The anisotropic Heisenberg Hamiltonian,

$$\begin{aligned} \mathscr{H} &= -2J_\parallel \sum_{<\mathbf{l},\mathbf{l}'>} S_\mathbf{l}^z S_{\mathbf{l}'}^z - 2J_\perp \sum_{<\mathbf{l},\mathbf{l}'>} [S_\mathbf{l}^x S_{\mathbf{l}'}^x + S_\mathbf{l}^y S_{\mathbf{l}'}^y] \\ &= -2J_\parallel \sum_{<\mathbf{l},\mathbf{l}'>} S_\mathbf{l}^z S_{\mathbf{l}'}^z - J_\perp \sum_{<\mathbf{l},\mathbf{l}'>} [S_\mathbf{l}^- S_{\mathbf{l}'}^+ + S_\mathbf{l}^+ S_{\mathbf{l}'}^-], \end{aligned} \tag{8.3.7}$$

is sufficient for our purposes. As usual, $\mathbf{S_l}$ is the three-component spin operator at site $\mathbf{R_l}$ on a d-dimensional lattice. The exchange integrals $J_\parallel$ and $J_\perp$ for spin directions parallel and perpendicular to the z-axis are, in general, not equal. The nature of the ordered phase depends on the signs and relative magnitudes of $J_\parallel$ and $J_\perp$. The following forms are of particular interest.

(1) $J_\parallel = J_\perp = J > 0$ – the ferromagnetic Heisenberg model. The low-temperature order parameter is the spatially uniform magnetization $\mathbf{m}$. The total spin $\mathbf{S}_T = \sum_\mathbf{l} \mathbf{S_l}$ commutes with the Hamiltonian so that all three components of the magnetization are conserved.

(2) $J_\parallel = J_\perp = J < 0$ – the antiferromagnetic Heisenberg model. The order parameter is the three-component staggered magnetization $\mathbf{N} = \langle \hat{\mathbf{N}} \rangle$, where

$$\hat{\mathbf{N}}(\mathbf{x}) = \sum_\mathbf{l} \eta_\mathbf{l} \mathbf{S_l} \delta(\mathbf{x} - \mathbf{R_l}), \tag{8.3.8}$$

where $\eta_\mathbf{l} = +1$ if $\mathbf{l}$ is a site in sublattice A and $\eta_\mathbf{l} = -1$ if $\mathbf{l}$ is in sublattice B. As in the ferromagnetic case, $\mathbf{S}_T$ commutes with $\mathscr{H}$, and all three components of the magnetization are conserved.

(3) $J_\perp > | J_\parallel | > 0$ – the planar ferromagnet. The classical energy is minimized for parallel alignment of spins in the xy-plane. The order parameter is, thus, the two-component magnetization vector $\mathbf{m} = (m_x, m_y)$ or, equivalently, the complex field $\psi = m_x + im_y$. The Hamiltonian is no longer rotationally invariant, and only the z-component of the total spin commutes with $\mathscr{H}$. Thus, m_z, but not m_x or m_y, is conserved.

(4) $J_\perp < -|J_\parallel| < 0$ – the planar antiferromagnet. The classical energy is minimized for antiparallel alignment of spins in the xy-plane. The order parameter is the two-component staggered magnetization $\mathbf{N} = (N_x, N_y)$, and m_z is a conserved density.

(5) $|J_\parallel| >| J_\perp |$ – the ferromagnetic ($J_\parallel > 0$) or antiferromagnetic ($J_\parallel < 0$) Ising model. In this case, the broken symmetry is discrete, and conservation laws alone determine the hydrodynamics. Again, m_z is conserved so that there are diffusive energy and m_z-modes in both the ordered and disordered phases.

The ordered phases were identified by treating $\mathbf{S}_l$ as a classical variable and by identifying the configurations that minimize $\mathscr{H}$. When the order parameter does not commute with the Hamiltonian (cases (2), (3), (4), and (5)), it is impossible simultaneously to diagonalize $\mathscr{H}$ and the order parameter. The result is that there will be quantum fluctuations that will lead to a reduction of the order parameter from its classical value even at zero temperature. However, as long as these fluctuations do not actually destroy long-range order, the symmetry of the quantum ordered state will be the same as that of the classical ordered state, and its hydrodynamics will be correctly determined by the approach described in this section.

3 The planar magnet

As discussed above, the z-component of the magnetization is conserved and satisfies the operator conservation law,

$$\frac{\partial \hat{m}_z}{\partial t} + \nabla \cdot \hat{\boldsymbol{\tau}} = 0, \tag{8.3.9}$$

where $\hat{\boldsymbol{\tau}}$ is a spin current operator. As in our preceding example, an identical conservation law applies to the averaged or coarse-grained magnetization $m_z(\mathbf{x}, t) = \langle \hat{m}_z \rangle$ with current $\boldsymbol{\tau}(\mathbf{x}, t) = \langle \hat{\boldsymbol{\tau}}(\mathbf{x}, t) \rangle$. In the disordered paramagnetic phase, there are two conserved densities, the energy density, ε, and m_z, and, as in our rigid rotor example, there are two associated diffusive modes. (m_z, like l, is odd under time reversal so that there is no cross-dissipative coefficient coupling $\boldsymbol{\tau}$ to ∇T.)

The ordered phase with $\psi = m_x + im_y =| \psi | e^{i\theta} \neq 0$ has xy-symmetry with θ a broken-symmetry hydrodynamic variable. The fundamental thermodynamic relation,

$$T ds = d\varepsilon - h dm_z - \mathbf{h}_\theta \cdot d\mathbf{v}_\theta, \tag{8.3.10}$$

where $\mathbf{v}_\theta = \nabla \theta$, is thus identical to that of the rigid rotor [Eq. (8.2.40)]. The entropy can be expanded in powers of m_z and $\mathbf{v}_\theta$ as

$$s = s_0(\varepsilon) - \frac{1}{2T}\chi^{-1} m_z^2 - \frac{1}{2T}\rho_s v_\theta^2. \tag{8.3.11}$$

As in our rigid rotor example, we need to determine any reactive couplings between $\mathbf{v}_\theta$ and the other hydrodynamic variables. To do this, we explore the possibility of having stationary or equilibrium states with a constant precession rate $d\theta/dt = -X$. Equilibrium states are determined completely by the conserved variables, ε and m_z, and $\mathbf{v}_\theta$ if it is nonzero. Assume now that a state is prepared with a given ε, m_z and $\mathbf{v}_\theta = 0$ for which X is nonzero when the external magnetic field h_e in the z-direction is zero. Now turn on an external field. The time evolution of ψ is determined by the total Hamiltonian $\mathscr{H} + \mathscr{H}_{\text{ext}} = \mathscr{H} - h_e S_T^z$ [Eq. (8.3.4)] rather than by $\mathscr{H}$. The operator $-h_e S_T^z$ commutes with $\mathscr{H}$, and

$$e^{i(\mathscr{H}-h_e S_T^z)t/\hbar} = e^{i\mathscr{H}t/\hbar}e^{-ih_e S_T^z t/\hbar}. \tag{8.3.12}$$

Therefore,

$$\begin{aligned}\psi(t,h_e) &= V^{-1}\left\langle \sum_{\mathbf{l}} e^{i\mathscr{H}t/\hbar}e^{-ih_e S_T^z t/\hbar}S_{\mathbf{l}}^{+}(t)e^{ih_e S_T^z t/\hbar}e^{-i\mathscr{H}t/\hbar}\right\rangle \\ &= e^{-ih_e t/\hbar}\psi(t,0),\end{aligned} \tag{8.3.13}$$

where V is the volume and where we have used Eq. (8.3.6). Thus, when h_e is not zero,

$$\frac{d\theta}{dt} = -(X + h_e). \tag{8.3.14}$$

In thermodynamic equilibrium, $d\theta/dt$ must be zero, and $X(\varepsilon, m_z) = -h_e$. The thermodynamic function that is equal to h_e in true equilibrium, i.e., in the stationary state with the maximum entropy, is $h(\varepsilon, m_z) = -T(\partial s/\partial m_z)_{\varepsilon,\mathbf{v}_\theta=0}$, and we conclude that

$$\frac{d\theta}{dt} = h(\varepsilon, m_z) - h_e. \tag{8.3.15}$$

This relation ensures that $d\theta/dt$ is zero in true thermodynamic equilibrium. It permits, however, stationary states for which $h(\varepsilon, m_z)$ is not equal to h_e. When $h_e = 0$, m_z is small, and a dissipative part X' of the "current" X is introduced, Eq. (8.3.15) becomes

$$\frac{d\theta}{dt} = \chi^{-1}m_z - X', \tag{8.3.16}$$

where χ is the magnetic susceptibility. This equation is often called a Josephson relation. Its analog for superfluids, which we will discuss in Sec. 8.5, was first derived by Brian Josephson in 1962. It is also the exact analog of Eqs. (8.2.44) and (8.2.46) for the rigid rotor. In fact, we have now established a one-to-one correspondence between the rigid rotor and the planar magnet: $\theta \leftrightarrow \theta$, $\chi \leftrightarrow \tilde{I}$, $m_z \leftrightarrow l$. The mode structure in the ordered phase of the planar magnet is thus identical to that of the rigid rotor. There is a diffusive energy mode and a propagating spin wave mode with frequency $\omega = \pm(\rho_s/\chi)^{1/2}q^2 + iDq^2$.

In the planar antiferromagnet, the two-dimensional staggered magnetization, $N_x + iN_y = \psi = |\psi| e^{i\theta}$, is a two-component broken-symmetry hydrodynamical variable. The z-component of the magnetization is a conserved variable. Arguments identical to those used for the planar ferromagnet imply that $d\theta/dt = h(\varepsilon, m_z) - h_e$ in the antiferromagnet. Thus, the hydrodynamical equations and mode structure for the planar ferro- and antiferromagnets are identical. Dynamics at $\omega\tau > 1$ will, however, differ in the two cases.

4 The isotropic antiferromagnet

In an isotropic antiferromagnet, the total vector magnetization is conserved and obeys a conservation law,

$$\frac{\partial m_i}{\partial t} + \nabla_j \tau_{ij} = 0. \tag{8.3.17}$$

Thus, there are four conserved densities, ε, m_x, m_y and m_z, and four diffusive modes in the disordered phase of the isotropic antiferromagnet.

In the ordered phase, the staggered magnetization $\mathbf{N} = N_0\mathbf{n}$, with $\mathbf{n}$ a unit vector, is nonzero. The magnitude, N_0, of the staggered magnetization is a non-hydrodynamical variable and decays to its equilibrium value in times of order the collision time τ. For small deviations from the uniform equilibrium state with $\mathbf{n} = \mathbf{n}_0 = \mathbf{e}_z$, we can set $\mathbf{n}(\mathbf{x}) = \mathbf{n}_0 + \delta\mathbf{n}$ with $\delta\mathbf{n} = \delta\boldsymbol{\theta}(\mathbf{x}) \times \mathbf{n}_0$ or

$$\delta\mathbf{n} \approx (\delta n_x, \delta n_y, 0) = (\delta\theta_y, -\delta\theta_x, 0). \tag{8.3.18}$$

The entropy density for small $\mathbf{m}$ and $\mathbf{v}_\alpha = \boldsymbol{\nabla}\theta_\alpha$ is

$$\begin{aligned} s(\varepsilon, \mathbf{m}, \mathbf{v}_\alpha) &= s_0(\varepsilon) - \frac{1}{2T}\left[\chi_\parallel^{-1} m_z^2 + \chi_\perp^{-1}(m_x^2 + m_y^2)\right] \\ &\quad - \frac{\rho_s}{2T}[|\,\mathbf{v}_x\,|^2 + |\,\mathbf{v}_y\,|^2], \end{aligned} \tag{8.3.19}$$

and the fundamental thermodynamic relation is

$$T ds = d\varepsilon - \mathbf{h}\cdot d\mathbf{m} - \mathbf{h}_{\theta x}\cdot d\mathbf{v}_{\theta x} - \mathbf{h}_{\theta y}\cdot d\mathbf{v}_{\theta y} \tag{8.3.20}$$

with

$$\begin{aligned} h_x &= \chi_\perp^{-1} m_x, \qquad h_y = \chi_\perp^{-1} m_y, \qquad h_z = \chi_\parallel^{-1} m_z, \\ \mathbf{h}_{\theta x} &= \rho_s \mathbf{v}_{\theta x}, \qquad \mathbf{h}_{\theta y} = \rho_s \mathbf{v}_{\theta y}. \end{aligned} \tag{8.3.21}$$

We should note that there is no term in Eq. (8.3.19) coupling the staggered magnetization $\mathbf{N}$ to $\mathbf{m}$. This is because $\mathbf{m}$ alternates in sign from site to site in the antiferromagnetic phase whereas $\mathbf{N}$ does not. Thus, $\mathbf{N}\cdot\mathbf{m}$ is a scalar that alternates in sign, and its integral over the volume is zero. In addition, terms coupling $\mathbf{v}_\alpha$ to $\mathbf{m}$ are prohibited by inversion symmetry.

The equation for $d\theta_\alpha/dt$ follows from exactly the same arguments used for the planar magnet. An external magnetic field leads to a uniform precession of each spin $\mathbf{S}_l$ and thus to a uniform precession of the staggered magnetization defined in Eq. (8.3.8). Thus, when $\mathbf{h}_e$ is zero,

$$\frac{d\theta_\alpha}{dt} = h_\alpha(\varepsilon, \mathbf{m}) - X_\alpha' = \chi_\perp^{-1} m_\alpha - X_\alpha', \qquad \alpha = x, y, \tag{8.3.22}$$

where $h_\alpha = -T\partial s/\partial m_\alpha$. These reactive couplings lead in the usual way to a reactive term $h_{\theta\alpha,j}$ in the current $\tau_{j\alpha}$. The complete linearized hydrodynamic equations for an isotropic antiferromagnet are, therefore,

$$\begin{aligned} \frac{\partial \varepsilon}{\partial t} &= \kappa\nabla^2 T = C^{-1}\kappa\nabla^2\varepsilon, \\ \frac{\partial m_z}{\partial t} &= \Gamma_\parallel\nabla^2 h_z = \Gamma_\parallel\chi_\parallel^{-1}\nabla^2 m_z, \end{aligned}$$

$$\frac{\partial m_\alpha}{\partial t} = \rho_s \nabla^2 \theta_\alpha + \Gamma_\perp \chi_\perp^{-1} \nabla^2 m_\alpha,$$

$$\frac{\partial \theta_\alpha}{\partial t} = \chi_\perp^{-1} m_\alpha + \Gamma \rho_s \nabla^2 \theta_\alpha. \tag{8.3.23}$$

These equations yield diffusive modes for ε and m_z and two sets of propagating modes with frequencies

$$\omega = \pm(\rho_s/\chi_\perp)^{1/2} q + i\frac{1}{2}(\Gamma_\perp \chi_\perp^{-1} + \Gamma \rho_s) q^2 \tag{8.3.24}$$

for a total of six hydrodynamical modes.

5 Isotropic ferromagnets

Ferromagnets differ from the systems we have so far considered in that the magnetization in the ordered phase is both a conserved and a broken-symmetry variable. Since a variable cannot be counted as a hydrodynamical variable twice, we expect four modes corresponding to ε, m_x, m_y and m_z in *both* the paramagnetic and ferromagnetic phases. In the paramagnetic phase, these modes are diffusive as they are in the antiferromagnet.

The fundamental thermodynamic relation for a ferromagnet is

$$T\,ds = d\varepsilon - \mathbf{h} \cdot d\mathbf{m} \tag{8.3.25}$$

in both the ordered and disordered phases. The equation of state relating $\mathbf{h}$ to $\mathbf{m}$ is, however, different in the two phases. In the disordered phase, $\mathbf{h} = \chi^{-1}\mathbf{m}$. In the ordered phase, $\mathbf{m} = M_0 \mathbf{e}_z$ in equilibrium. For small deviations from equilibrium,

$$\mathbf{m} = (M_0 + \delta m_z)\mathbf{e}_z + m_x \mathbf{e}_x + m_y \mathbf{e}_y, \tag{8.3.26}$$

and

$$s = s_0(\varepsilon, M_0) - \frac{1}{2T}\chi_\parallel^{-1}(\delta m_z)^2 - \frac{1}{2T}\frac{\rho_s}{M_0^2}[(\boldsymbol{\nabla} m_x)^2 + (\boldsymbol{\nabla} m_y)^2]. \tag{8.3.27}$$

The last term in this expression is the usual elastic term in a phase with a broken continuous symmetry. For small $\delta\mathbf{m}$, the angular deviation from the state with $\mathbf{m} = M_0\mathbf{e}_z$ is $\delta\boldsymbol{\theta} = \mathbf{e}_z \times \delta\mathbf{m}/M_0$. From this, we obtain

$$h_z = \chi_\parallel^{-1} m_z, \qquad h_\alpha = -\frac{\rho_s}{M_0^2}\nabla^2 m_\alpha, \qquad \alpha = x, y. \tag{8.3.28}$$

Again using arguments identical to those used to obtain the reactive equation for $d\theta/dt$ in the planar model, we obtain

$$\frac{\partial m_i}{\partial t} = -(\mathbf{m} \times \mathbf{h})_i - \nabla_j \tau'_{ij}, \tag{8.3.29}$$

where τ'_{ij} is the dissipative part of the spin current. Note the minus sign in the reactive part of this equation compared to the positive sign in Eq. (8.3.5) for a spin in an external field. If an external field is added, the reactive part of Eq. (8.3.29) becomes $\mathbf{m} \times (\mathbf{h}_e - \mathbf{h})$, which agrees with Eq. (8.3.5) when there is no interaction among spins ($\mathbf{h} = 0$) and reduces to zero in true thermodynamic equilibrium when $\mathbf{h} = \mathbf{h}_e$. The reactive part of this equation makes no contribution to the entropy production equation, and we have

$$T\left(\frac{\partial s}{\partial t} + \nabla \cdot \frac{\mathbf{Q}}{T}\right) = -\mathbf{Q} \cdot \frac{\nabla T}{T} - \tau'_{ij}\nabla_i h_j. \tag{8.3.30}$$

Since the spins are presumed not to be coupled to the lattice, spin and space can be rotated independently, and

$$\begin{aligned} \tau'_{zj} &= -\Gamma_{\parallel}\nabla_j h_z, \\ \tau'_{\alpha j} &= -\Gamma_{\perp}\nabla_j h_\alpha = \Gamma_{\perp}\frac{\rho_s}{M_0^2}\nabla_j \nabla^2 m_\alpha. \end{aligned} \tag{8.3.31}$$

The complete linearized hydrodynamical equations are, therefore,

$$\begin{aligned} \frac{\partial \varepsilon}{\partial t} &= \kappa C^{-1}\nabla^2 \varepsilon, \\ \frac{\partial m_z}{\partial t} &= \Gamma \chi_{\parallel}^{-1}\nabla^2 m_z, \\ \frac{\partial m_x}{\partial t} &= -\frac{\rho_s}{M_0^2}\nabla^2 m_y - \Gamma\frac{\rho_s}{M_0^2}\nabla^4 m_x, \\ \frac{\partial m_y}{\partial t} &= \frac{\rho_s}{M_0^2}\nabla^2 m_x - \Gamma\frac{\rho_s}{M_0^2}\nabla^4 m_y. \end{aligned} \tag{8.3.32}$$

These equations yield diffusive energy and m_z-modes and a pair of spin wave modes with

$$\omega = \pm(\rho_s/M_0^2)q^2 + i\tilde{D}q^4, \tag{8.3.33}$$

where $\tilde{D} = \Gamma\rho_s/M_0^2$. Here, we see that the effect of having $\mathbf{m}$ as both a conserved and a broken-symmetry hydrodynamic variable is that the spin-wave mode has a quadratic dispersion relation rather than the linear dispersion relation we have encountered in all previous examples.

8.4 Hydrodynamics of simple fluids

In this section, we will derive the equations governing hydrodynamics in one- and two-component fluids and discuss some of their implications. The principal feature distinguishing fluids from our previous examples is the existence of mass motion and the possibility of relative motion of different parts of the fluid. There are conservation laws for mass and momentum in addition to the conservation law for energy, leading to a total of five conserved hydrodynamic variables and five hydrodynamic modes in a one-component fluid. In multicomponent fluids, there is an additional mass conservation law for each molecular species. Our derivation of the equations of fluid hydrodynamics will follow closely our previous examples, but we will make contact with alternative derivations based more closely on the local application of Newton's laws.

1 Conservation laws

In a one-component fluid of molecules of mass m, energy, mass and momentum are conserved. Their respective density operators are

$$\begin{aligned}\hat{\varepsilon}(\mathbf{x},t) &= \sum_{\alpha}(\mathbf{p}^{\alpha})^2/(2m) + \frac{1}{2}\sum_{\alpha\neq\beta} U(\mathbf{x}^{\alpha}-\mathbf{x}^{\beta})\delta(\mathbf{x}-\mathbf{x}^{\alpha}),\\ \hat{\rho}(\mathbf{x},t) &= \sum_{\alpha} m\delta(\mathbf{x}-\mathbf{x}^{\alpha}),\\ \hat{\mathbf{g}}(\mathbf{x},t) &= \sum_{\alpha} \mathbf{p}^{\alpha}\delta(\mathbf{x}-\mathbf{x}^{\alpha}),\end{aligned} \tag{8.4.1}$$

where $U(\mathbf{x})$ is the two-body interaction potential. These densities obey local conservation laws

$$\frac{\partial\hat{\varepsilon}}{\partial t} + \boldsymbol{\nabla}\cdot\hat{\mathbf{j}}^{\varepsilon} = 0, \tag{8.4.2}$$

$$\frac{\partial\hat{\rho}}{\partial t} + \boldsymbol{\nabla}\cdot\hat{\mathbf{g}} = 0, \tag{8.4.3}$$

$$\frac{\partial\hat{g}_i}{\partial t} + \nabla_j\hat{\pi}_{ij} = 0, \tag{8.4.4}$$

where $\hat{\mathbf{j}}^{\varepsilon}$ is the energy current and $\hat{\pi}_{ij}$ is the momentum current tensor, which, as we shall see shortly, is closely related to the stress tensor. Note that the conserved momentum density is itself a current for another conserved density, the mass density. This is a feature that we did not encounter in our previous examples of hydrodynamics in which all modes in disordered phases were diffusive. It is responsible for many of the essential physical properties of isotropic fluids such as the existence of sound waves. As in our previous examples, the conservation laws apply to averaged or coarse-grained densities $\varepsilon(\mathbf{x},t) = \langle\hat{\varepsilon}(\mathbf{x},t)\rangle$, $\rho(\mathbf{x},t) = \langle\hat{\rho}(\mathbf{x},t)\rangle$ and $\mathbf{g}(\mathbf{x},t) = \langle\hat{\mathbf{g}}(\mathbf{x},t)\rangle$ as well as to the microscopic operators.

Angular momentum as well as momentum is conserved. The total angular momentum is obtained by integrating $\mathbf{x}\times\mathbf{g}$ over the volume of the fluid and is not independent of $\mathbf{g}$. Arguments identical to those used in Sec. 6.6 to show that the elastic stress tensor is symmetric imply that the momentum flux tensor π_{ij} must also be symmetric to guarantee conservation of angular momentum. No additional local variables other than those in Eqs. (8.4.2)–(8.4.4) are needed to describe angular momentum, and there are no hydrodynamical modes associated with its conservation law.

It is useful to consider two coordinate systems S and S' moving relative to each other with constant velocity, as shown in Fig. 8.4.1. We take S to be the frame at rest in the laboratory and S' to be moving relative to S with a velocity $\mathbf{v}_e$. Variables in S' are marked with primes and those in S are not. Thus,

$$\mathbf{x}^{\alpha}(t) = \mathbf{v}_e t + \mathbf{x}'^{\alpha}(t), \qquad \mathbf{p}^{\alpha}(t) = m\mathbf{v}_e + \mathbf{p}'^{\alpha}(t). \tag{8.4.5}$$

The transformation from S to S' is called a *Galilean transformation.* Newton's laws, which control the dynamical properties of a classical fluid, are invariant

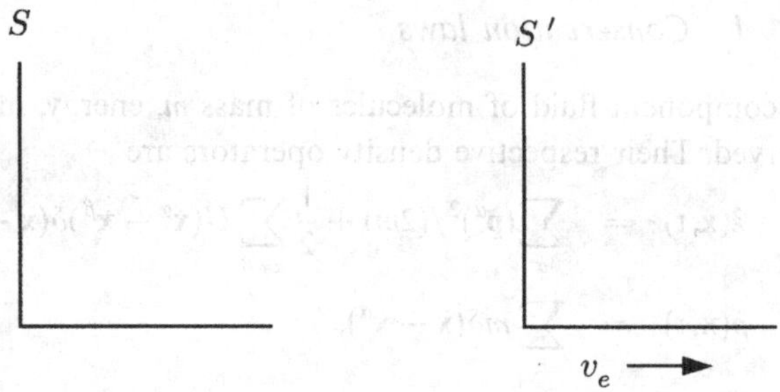

Fig. 8.4.1. Coordinate systems S and S'. S is the laboratory rest frame, and S' moves with velocity $\mathbf{v}_e$ relative to S. When $\mathbf{v}_e = \mathbf{v}$, S' is the rest frame of the fluid.

under such Galilean transformations for any value of $\mathbf{v}_e$. The mass density transforms under the Galilean transformation according to

$$\rho(\mathbf{x}, t) = \left\langle \sum_\alpha m\delta(\mathbf{x} - \mathbf{x}'^\alpha - \mathbf{v}_e t) \right\rangle \equiv \rho'(\mathbf{x}', t), \tag{8.4.6}$$

where $\mathbf{x}' = \mathbf{x} - \mathbf{v}_e t$ and $\rho'(\mathbf{x}, t) \equiv \langle \sum_\alpha \delta(\mathbf{x} - \mathbf{x}'^\alpha(t)) \rangle$. Similarly, the momentum and energy densities transform according to

$$\begin{aligned} \mathbf{g}(\mathbf{x}, t) &= \mathbf{g}'(\mathbf{x}', t) + \rho'(\mathbf{x}', t)\mathbf{v}_e, \\ \varepsilon(\mathbf{x}t) &= \varepsilon'(\mathbf{x}', t) + \mathbf{g}'(\mathbf{x}', t) \cdot \mathbf{v}_e + \frac{1}{2}\rho'(\mathbf{x}', t)v_e^2, \end{aligned} \tag{8.4.7}$$

where ε' and $\mathbf{g}'$ are obtained from averages [Eqs. (8.4.1)] with $\mathbf{x}^\alpha$ and $\mathbf{p}^\alpha$ replaced by $\mathbf{x}'^\alpha$ and $\mathbf{p}'^\alpha$.

Of all the moving coordinate systems, S', the most useful is the one that is locally at rest with respect to the fluid. In this, the fluid rest frame, the average momentum density $\mathbf{g}'$ is zero. We will denote the velocity of this frame by $\mathbf{v}$. The distinction between $\mathbf{v}_e$ and $\mathbf{v}$ is the same as that between Ω_e and Ω in our rigid rotor example. The microscopic definitions of the conserved variables [Eqs. (8.4.2)–(8.4.4)], Eq. (8.4.6) relating ρ to ρ', and the constraint $\mathbf{g}'(\mathbf{x}', t) = 0$ imply

$$\begin{aligned} \varepsilon(\mathbf{x}, t) &= \varepsilon_0(\mathbf{x}, t) + \frac{1}{2}\rho(\mathbf{x}, t)v^2, \\ \mathbf{g}(\mathbf{x}, t) &= \rho(\mathbf{x}, t)\mathbf{v}. \end{aligned} \tag{8.4.8}$$

Here, $\varepsilon_0(\mathbf{x}) = \varepsilon'(\mathbf{x} - \mathbf{v}t)$ is the rest-frame energy density at the laboratory point $\mathbf{x}$. It is the average of the energy density of Eq. (8.4.1) with $\mathbf{p}^\alpha$ replaced by $\mathbf{p}'^\alpha$ but $\mathbf{x}^\alpha$ not replaced by $\mathbf{x}'^\alpha$. These equations are valid for uniform translations of the entire fluid. They will, however, generalize to situations in which there are slow spatial variations.

2 *Thermodynamics with mass motion*

In order to derive the hydrodynamical equations for classical fluids, we have to include variations of all conserved variables in their thermodynamic potentials. This means that we must include momentum as well as energy and particle number. To do this, we can proceed exactly as we proceeded in the tutorial example of rotors on a lattice. The laboratory frame Lagrangian expressed in terms of coordinates with respect to the fluid rest frame is

$$\mathscr{L} = \frac{1}{2}\sum_{\alpha} m(\mathbf{v}+\mathbf{v}'^{\alpha})^2 - U[\mathbf{x}'^{\alpha}(t)], \tag{8.4.9}$$

and the momentum is

$$p_i^{\alpha} = \frac{\partial \mathscr{L}}{\partial v_i'^{\alpha}} = m(v_i + v_i'^{\alpha}). \tag{8.4.10}$$

Here we have chosen S' to be the fluid rest frame so that $\mathbf{v}_e = \mathbf{v}$, and $\langle \mathbf{v}'^{\alpha}\rangle = 0$. As in the rotor example, we could have chosen $\mathbf{v}_e$ to be arbitrary, but the conditions of thermal equilibrium will eventually require $\mathbf{v}_e$ to equal $\mathbf{v}$. The total Hamiltonian is then

$$\begin{aligned}\mathscr{H}_T &= \sum_{\alpha} \mathbf{p}^{\alpha}\cdot\mathbf{v}'^{\alpha} - \mathscr{L} \\ &\equiv \mathscr{H} - \hat{\mathbf{P}}\cdot\mathbf{v},\end{aligned} \tag{8.4.11}$$

where

$$\hat{\mathbf{P}} = \sum_{\alpha} \mathbf{p}^{\alpha} = \int d^d x \hat{\mathbf{g}}(\mathbf{x},t) \tag{8.4.12}$$

is the total momentum operator and

$$\mathscr{H} = \sum_{\alpha} \frac{(p^{\alpha})^2}{2m} + U \tag{8.4.13}$$

is the Hamiltonian, which has the same form as a function of $\mathbf{p}^{\alpha}$ regardless of the value of $\mathbf{v}$. The average of $\mathscr{H}$ is the internal energy E. The canonical partition function for a fluid with N particles in a volume V is then

$$Z_N(T,V,\mathbf{v}) = \mathrm{Tr}e^{-\beta(\mathscr{H}-\hat{\mathbf{P}}\cdot\mathbf{v})}. \tag{8.4.14}$$

The thermodynamic potential,

$$F(T,V,N,\mathbf{v}) = -T\ln Z_N = E - TS - \mathbf{P}\cdot\mathbf{v}, \tag{8.4.15}$$

satisfies

$$dF = -SdT - pdV + \mu dN - \mathbf{P}\cdot d\mathbf{v}, \tag{8.4.16}$$

where $\mathbf{P} = \langle\hat{\mathbf{P}}\rangle$. In classical systems, the trace over momentum variables is independent of the trace over coordinates, and

$$Z_N(T,V,\mathbf{v}) = e^{\beta N m v^2/2} Z_N(T,V,0), \tag{8.4.17}$$

or

$$F(T,V,N,\mathbf{v}) = F_0(T,V,N) - \frac{1}{2}Nmv^2, \tag{8.4.18}$$

where $F_0 = -T \ln Z_N(T, V, 0)$ is the Helmholtz free energy in the rest frame of the fluid. This equation and Eq. (8.4.14) imply as required that the total momentum is the total mass times the velocity,

$$P_i = -\left.\frac{\partial F}{\partial v_i}\right)_{T,V,N} = Nmv_i, \tag{8.4.19}$$

and that the internal energy is the rest-frame energy E_0 plus the kinetic energy $\frac{1}{2}Nmv^2 = P^2/2mN$ associated with center of mass motion,

$$E(S, V, N, \mathbf{P}) = F + TS + \mathbf{P} \cdot \mathbf{v} = E_0(S, V, N) + \frac{1}{2}\frac{P^2}{mN}, \tag{8.4.20}$$

where $E_0(S, V, N) = F_0 + TS$. Note that E is a natural function of extensive quantities only. The fundamental thermodynamic relation, Eq. (8.4.16), and Eq. (8.4.18) imply that pressure,

$$p = -\left.\frac{\partial F}{\partial V}\right)_{T,N,\mathbf{v}} = -\left.\frac{\partial E}{\partial V}\right)_{S,N,\mathbf{P}}, \tag{8.4.21}$$

and entropy, $S = -\partial F/\partial T$, expressed as functions of T, V and N are independent of $\mathbf{v}$. In addition, $\mathbf{g}$ and $\varepsilon = V^{-1}\langle \mathscr{H} \rangle$ are given by Eq. (8.4.8). The chemical potential in the lab frame is related to that of the fluid rest frame, $\mu_0 = \partial F_0/\partial N$, via

$$\mu = \frac{\partial F}{\partial N} = \mu_0 - \frac{1}{2}mv^2. \tag{8.4.22}$$

As in quiescent fluids, it is useful to introduce the grand potential,

$$\mathscr{A}(T, \mu, \mathbf{v}, V) = F - \mu N, \tag{8.4.23}$$

which is a function of only one extensive variable V and satisfies

$$d\mathscr{A} = -SdT - pdV - Nd\mu - \mathbf{P} \cdot d\mathbf{v}. \tag{8.4.24}$$

Arguments identical to those used in Sec. 3.2 imply that $\mathscr{A} = -Vp(\mu, T, \mathbf{v})$ and

$$p = -(\varepsilon - \alpha\rho - Ts - \mathbf{g} \cdot \mathbf{v}), \tag{8.4.25}$$

where we have introduced the chemical potential per unit mass

$$\alpha = \mu/m. \tag{8.4.26}$$

Note that p is indeed independent of $\mathbf{v}$ and equal to $-(\varepsilon_0 - \alpha_0\rho - Ts)$ if the density ρ (rather than α) is fixed, as can be seen from Eq. (8.4.8) for ε and $\mathbf{g}$, $\alpha = \mu/m = \alpha_0 - \frac{1}{2}v^2$, and the fact that T and s are independent of $\mathbf{v}$. Eqs. (8.4.24) and (8.4.25) lead to the entropy equation,

$$Tds = d\varepsilon - \alpha d\rho - \mathbf{v} \cdot d\mathbf{g}, \tag{8.4.27}$$

which, as in our previous examples, is essential to the derivation of hydrodynamical equations.

3 The entropy production equation

To determine the form of the constitutive equations relating the currents of conserved quantities to the fields T, α and $\mathbf{v}$, we proceed exactly as in our previous two examples. We use the entropy equation [Eq. (8.4.27)] and the conservation laws [Eqs. (8.4.2)–(8.4.4)] expressed in terms of the coarse-grained variables to arrive at an equation for the rate of change of entropy. As before, the right hand side of this equation will include the time rate of change of the entropy density and the divergence of an entropy current. Unlike previous cases, the entropy current must include a term $\mathbf{v}s$ describing the secular transport of entropy in addition to the heat current $\mathbf{Q}$. The time and spatial derivatives arising from the entropy equation are

$$\begin{aligned} T\frac{\partial s}{\partial t} &= \frac{\partial \varepsilon}{\partial t} - \alpha\frac{\partial \rho}{\partial t} - v_j\frac{\partial g_j}{\partial t} \\ &= -\nabla \cdot \mathbf{j}^{\varepsilon} + \alpha \nabla \cdot \mathbf{g} + v_j \nabla_i \pi_{ji}, \\ T\mathbf{v}\cdot\nabla s &= \mathbf{v}\cdot\nabla\varepsilon - \alpha\mathbf{v}\cdot\nabla\rho - v_i v_j \nabla_i g_j. \end{aligned} \tag{8.4.28}$$

Following the same procedures as before, we can use Eqs. (8.4.28) to obtain

$$\begin{aligned} T\left[\frac{\partial s}{\partial t} + \nabla\cdot\left(\mathbf{v}s + \frac{1}{T}\mathbf{Q}\right)\right] &= -\ \mathbf{Q}\cdot(\nabla T/T) - (\mathbf{g} - \rho\mathbf{v})\cdot\nabla\alpha \\ &\quad -\ [\pi_{ji} - p\delta_{ij} - v_i g_j]\nabla_i v_j, \end{aligned} \tag{8.4.29}$$

where p is the pressure determined by Eq. (8.4.25) and

$$Q_i = j_i^{\varepsilon} - \alpha(g_i - \rho v_i) - v_i\varepsilon + (\mathbf{v}\cdot\mathbf{g})v_i - v_j\pi_{ji} \tag{8.4.30}$$

is the heat current. Eq. (8.4.29), along with the requirement of non-negativity of entropy production, will provide us with constitutive relations for the currents.

4 Dissipationless hydrodynamics

In the absence of dissipation, dS/dt is zero, and the right hand side of Eq. (8.4.29) must be zero, implying

$$\begin{aligned} \mathbf{g} &= \rho\mathbf{v}, \\ \pi_{ij} &= p\delta_{ij} + v_j g_i = -\sigma_{ij} + \rho v_i v_j, \\ \mathbf{j}^{\varepsilon} &= (\varepsilon + p)\mathbf{v} = \left(\varepsilon_0 + p + \frac{1}{2}\rho v^2\right)\mathbf{v}, \end{aligned} \tag{8.4.31}$$

where $\sigma_{ij} = -p\delta_{ij}$ is the fluid stress tensor and where $\mathbf{g} = \rho\mathbf{v}$ was used to produce the final form of the equations for π_{ij} and $\mathbf{j}^{\varepsilon}$. Note that zero entropy production leads to the relation $\mathbf{g} = \rho\mathbf{v}$, which was previously obtained by Galilean invariance and the requirement that the rest-frame momentum density $\mathbf{g}'$ is zero. The momentum density and the mass density are observable averages of microscopically defined variables. Their ratio is the local velocity field $\mathbf{v}$.

Eqs. (8.4.31) and the conservation laws, Eqs. (8.4.2)–(8.4.4), yield the equations governing *inviscid* or *dissipationless* flow in a fluid. We will now express these equations in a more transparent and familiar form. First, the mass conservation equation is simply

$$\frac{\partial \rho}{\partial t} + \nabla\cdot(\rho\mathbf{v}) = 0. \tag{8.4.32}$$

The momentum conservation equation can be converted into an equation for the local velocity:

$$\begin{aligned}\frac{\partial g_i}{\partial t} &= \frac{\partial \rho v_i}{\partial t} = -\nabla_j \pi_{ij} \\ &= -\nabla_i p - \nabla_j(\rho v_i v_j).\end{aligned} \tag{8.4.33}$$

This equation is *Euler's equation.* It is one of the earliest (1755) and one of the most important equations of fluid dynamics. It is usually written, with the aid of the mass conservation equation Eq. (8.4.32), as

$$\frac{\partial \mathbf{v}}{\partial t} + (\mathbf{v} \cdot \nabla)\mathbf{v} = -\frac{1}{\rho}\nabla p. \tag{8.4.34}$$

Our treatment focuses on a fixed volume in space, i.e., it uses Eulerian coordinates. An alternative approach (as we discussed in Sec. 6.6) is that of Lagrange, in which one follows the motion of individual mass elements. A mass element will have an instantaneous velocity $\mathbf{v}(\mathbf{x}(t), t)$ determined by the Eulerian velocity at its instantaneous position $\mathbf{x}(t)$. The rate of change of the velocity of a mass point is, therefore,

$$\frac{dv_i}{dt} = \frac{\partial v_i}{\partial t} + \frac{\partial v_i}{\partial x_j}\frac{dx_j(t)}{dt}. \tag{8.4.35}$$

However, $d\mathbf{x}(t)/dt = \mathbf{v}$, and $d\mathbf{v}/dt = \partial \mathbf{v}/\partial t + (\mathbf{v} \cdot \nabla)\mathbf{v}$. Thus, Euler's equation expressed in Lagrangian coordinates is simply Newton's law, $\mathbf{f} = m\mathbf{a}$, for the mass point at $\mathbf{x}$,

$$\rho\frac{d\mathbf{v}}{dt} = -\nabla p + \mathbf{f}^{\text{ext}}, \tag{8.4.36}$$

where we have added an external force density $\mathbf{f}^{\text{ext}}$. Usually $\mathbf{f}^{\text{ext}}$ arises from the gravitational field and is equal to $\rho\mathbf{g}$, where $\mathbf{g}$ is the acceleration due to gravity. The external force could also be added to Euler's equation [Eq. (8.4.34)].

There are five independent conserved variables that determine the hydrodynamics of a one-component fluid, and five equations are needed to determine their time dependence. In an inviscid fluid, four of those equations are the mass conservation law and the three components of Euler's equation. The fifth equation could be the equation for energy conservation with the last of Eqs. (8.4.31) for the energy current. Alternatively, an entropy equation could be used. Since there is no dissipation, $\mathbf{Q} = 0$, and Eq. (8.4.29) is simply

$$\frac{\partial s}{\partial t} + \nabla \cdot (\mathbf{v}s) = 0. \tag{8.4.37}$$

This is an equation of entropy continuity. It implies that if ever the entropy is homogeneous in space, it will remain constant and homogeneous at all future times, i.e., dissipationless hydrodynamic processes are isentropic processes.

5 Dissipation

When there is dissipation, the right hand side of Eq. (8.4.29) must be positive. As in our previous examples, this requirement restricts the form of dissipative couplings between currents and fields with the opposite sign under time reversal. The dissipative heat current $\mathbf{Q}$ will be nonzero, and there will be a dissipative contribution, σ'_{ij}, to the stress tensor defined as

$$\pi_{ij} = p\delta_{ij} + \rho v_i v_j - \sigma'_{ij}. \tag{8.4.38}$$

Eq. (8.4.29) has a term proportional to $(\mathbf{g} - \rho\mathbf{v}) \cdot \nabla\alpha$, and, following our previous hydrodynamical examples, it would seem that there should be a part $\mathbf{g}'$ of the momentum density proportional to $\nabla\alpha$ and/or ∇T. $\mathbf{g}' = \mathbf{g} - \rho\mathbf{v}$ is, however, the momentum density in the fluid rest frame, and is, by definition, zero. There are, therefore, no dissipative contributions to $\mathbf{g}$, and $\mathbf{g} = \rho\mathbf{v}$ always. This result can also be proven using Kubo formulae (Forster 1983). The entropy production equation is therefore

$$T\frac{dS}{dt} = -\int d^3x[\mathbf{Q} \cdot \nabla T/T - \sigma'_{ij}\nabla_i v_j]. \tag{8.4.39}$$

$\mathbf{Q}$ and $\mathbf{v}$ are odd under time reversal whereas T and σ'_{ij} are even. Thus, there will be dissipative couplings between $\mathbf{Q}$ and T and between σ'_{ij} and $\mathbf{v}$. A fluid is isotropic, so any second rank tensor must be a scalar, implying there is only one dissipative coefficient, the thermal conductivity κ, coupling $\mathbf{Q}$ to ∇T:

$$\mathbf{Q} = -\kappa\nabla T. \tag{8.4.40}$$

The dissipative coefficient coupling the stress tensor to the velocity, like the elastic constant tensor, is a fourth rank tensor:

$$\sigma'_{ij} = \eta_{ijkl}\nabla_k v_l. \tag{8.4.41}$$

η_{ijkl} is the *viscosity* tensor. σ'_{ij} must be symmetric under interchange of i and j. Furthermore, the total entropy production is proportional to $\eta_{ijkl}\nabla_i v_j \nabla_k v_l$, so that only viscosities that are invariant under the interchange $ij \leftrightarrow kl$ contribute to dissipation. Kubo formulae for η_{ijkl} confirm this symmetry. Thus, the viscosity tensor has the same symmetry and will have the same number of independent components as the elastic constant tensor discussed in Chapter 6. In isotropic fluids, there are only two independent fourth rank tensors (see Sec. 6.4), and the dissipative part of the stress tensor can be written (in three dimensions) as

$$\sigma'_{ij} = \eta\left(\nabla_i v_j + \nabla_j v_i - \frac{2}{3}\delta_{ij}\nabla \cdot \mathbf{v}\right) + \zeta\delta_{ij}\nabla \cdot \mathbf{v}. \tag{8.4.42}$$

η is the *shear* viscosity and ζ is the *bulk* viscosity. Note that as required σ'_{ij} depends only on the symmetric combination $(\nabla_i v_j + \nabla_j v_i)/2$ of velocity gradients, called the *strain rate,* because it is the time derivative of the strain tensor introduced in Chapter 6. The decomposition of the viscosity tensor into a bulk and a shear part is identical to the decomposition of the elastic constant tensor in isotropic solids. The bulk viscosity measures the dissipative contribution to the stress arising from time-dependent volume changes.

Viscosity has units of [(energy/volume)×time] or poise = erg s/cm^3 in c.g.s. units. On dimensional grounds alone, one would expect viscosities to be of order the kinetic energy density times the collision time τ. Indeed, lowest order

Table 8.4.1. *Viscosities η for some common materials in units of centipoise (10^{-2} erg s/cm^3).*

Substance	Temperature	Viscosity (cp)
Air	18°C	0.018
Water	0°C	1.8
Water	20°C	1
Water	100°C	0.28
Glycerin	20°C	1500
Mercury	20°C	1.6
n-Pentane	20°C	0.23
Argon	85K	0.28
He^4	4.2K	0.033
Superfluid He^4	< 2.1K	0
Glass		> 10^{15}

Note that, by popular convention, the designation "glass" is applied to any disordered material once its viscosity exceeds 10^{15}cp.

kinetic theory calculations predict a shear viscosity of order $(\rho/m)T\tau$. The bulk viscosity is more subtle and is usually negligible in dilute gases. A useful way to think of the shear viscosity in a dense fluid is to imagine the fluid as being instantaneously a solid with a high frequency shear modulus $\mu(\infty)$. Then $\eta \approx \mu(\infty)\tau$. This relation is certainly dimensionally correct, and is often used to describe viscoelastic phenomena, very viscous fluids such as polymers in solution, or solids under flow. Viscosities of various fluids under various conditions are listed in Table 8.4.1.

The dissipative generalization of Euler's equation follows from Eqs. (8.4.33), (8.4.38) and (8.4.42):

$$\rho\left[\frac{\partial \mathbf{v}}{\partial t} + (\mathbf{v}\cdot\nabla)\mathbf{v}\right] = -\nabla p + \eta\nabla^2\mathbf{v} + \left(\zeta + \frac{1}{3}\eta\right)\nabla(\nabla\cdot\mathbf{v}). \qquad (8.4.43)$$

This, along with the equations of mass and energy conservation (the latter is now quite complicated when nonlinear velocity terms are important) and the thermodynamic equation of state, provide a complete description of the long wavelength, low-frequency dynamics of a one-component fluid, even if there are large differences in velocity between different points in the fluid.

6 *The Navier-Stokes equations*

A fluid like water is nearly incompressible, and the most useful equations for describing its flow are those obtained from the general hydrodynamic equations by imposing the constraint that the density be a constant in space and time. This constraint, along with the mass conservation equation, implies $\nabla\cdot\mathbf{v} = 0$ since

$$\frac{\partial \rho}{\partial t} = 0 = -\nabla \cdot (\rho \mathbf{v}) = -\rho \nabla \cdot \mathbf{v}. \tag{8.4.44}$$

The *Navier-Stokes* equation,

$$\rho \frac{\partial \mathbf{v}}{\partial t} + \rho(\mathbf{v} \cdot \nabla)\mathbf{v} = -\nabla p + \eta \nabla^2 \mathbf{v}, \tag{8.4.45}$$

along with the constraint $\nabla \cdot \mathbf{v} = 0$, describes the flow of incompressible fluids. Because of the nonlinear term $(\mathbf{v} \cdot \nabla)\mathbf{v}$, the solutions to these equations can be very complex.

7 Hydrodynamic modes

There are five conserved hydrodynamic variables, and we expect five low-frequency hydrodynamic modes. If the reactive coupling between ρ and $\mathbf{g}$ were absent, all of these modes would be diffusive; its presence leads to a pair of longitudinal sound modes. The linearized momentum equation expressed in terms of $\mathbf{g}$ rather than $\mathbf{v}$ is

$$\frac{\partial \mathbf{g}}{\partial t} + \nabla p - \frac{\eta}{\rho} \nabla^2 \mathbf{g} - \frac{1}{\rho}\left(\zeta + \frac{1}{3}\eta\right) \nabla(\nabla \cdot \mathbf{g}) = 0. \tag{8.4.46}$$

Because ρ is dynamically coupled only to the longitudinal part of $\mathbf{g}$ via the mass conservation equation, it is useful to introduce longitudinal and transverse parts of $\mathbf{g}$:

$$\mathbf{g} = \mathbf{g}_l + \mathbf{g}_t, \qquad \nabla \times \mathbf{g}_l = 0, \qquad \nabla \cdot \mathbf{g}_t = 0. \tag{8.4.47}$$

∇p is longitudinal, so that the equation for $\mathbf{g}_t$ obtained by taking the transverse part of Eq. (8.4.46) is independent of p:

$$\frac{\partial \mathbf{g}_t}{\partial t} = \frac{\eta}{\rho} \nabla^2 \mathbf{g}_t. \tag{8.4.48}$$

This is a canonical diffusion equation for each of the two independent components of $\mathbf{g}_t$. There are thus two diffusive transverse momentum modes with frequency

$$\omega = -i\frac{\eta}{\rho} q^2. \tag{8.4.49}$$

The ratio η/ρ is often called the *kinematic viscosity*. The transverse momentum correlation function will have the standard diffusive form we have encountered before.

Three modes remain to be determined. They involve the mass density, the longitudinal momentum and the energy density. We are interested in the linearized mode structure. We therefore write variables as equilibrium parts plus small deviations from equilibrium, e.g. $\rho(\mathbf{x}, t) = \rho + \delta\rho(\mathbf{x}, t)$, $p(\mathbf{x}, t) = p + \delta p(\mathbf{x}, t)$, etc. Taking the divergence of Eq. (8.4.46), using the mass conservation equation, we obtain

$$\left[-\frac{\partial^2}{\partial t^2} + \frac{1}{\rho}\left(\frac{4}{3}\eta + \zeta\right) \frac{\partial}{\partial t} \nabla^2\right] \delta\rho(\mathbf{x}, t) + \nabla^2 \delta p(\mathbf{x}, t) = 0. \tag{8.4.50}$$

This is a second-order differential equation in time and combines the information in first-order conservation laws. Finally, the linearized heat equation is

$$\frac{\partial}{\partial t}\left[\varepsilon(\mathbf{x},t)-\frac{\varepsilon+p}{\rho}\delta\rho(\mathbf{x},t)\right]-\kappa\nabla^2\delta T(\mathbf{x},t)=0, \tag{8.4.51}$$

where we used $\mathbf{j}^\varepsilon=(\varepsilon+p)\mathbf{v}-\kappa\boldsymbol{\nabla}T$ [Eqs. (8.4.31)] for the energy current (with $\varepsilon=\varepsilon_0$ since we are considering linear excitations from the quiescent equilibrium state) and the fact that $\boldsymbol{\nabla}\cdot\mathbf{v}=-\rho^{-1}\partial\rho(\mathbf{x},t)/\partial t$ for linear deviations from the spatially uniform state with density ρ. The temperature and pressure in Eqs. (8.4.50) and (8.4.51) are functions of ε and ρ via the equation of state of the quiescent fluid. It is more convenient, however, to use entropy and density as independent variables. For small deviations from equilibrium, the thermodynamic identity (Eq. (3.1.39) or Eq. (8.4.27) with $\mathbf{v}=0$) of the quiescent fluid can be recast as

$$T\rho d\tilde{s}\equiv d\tilde{q}=d\varepsilon-\frac{\varepsilon+p}{\rho}d\rho, \tag{8.4.52}$$

where $\tilde{s}=s/\rho=S/mN$ is the entropy per unit mass. Thus, Eq. (8.4.52) involves the time derivative of $\tilde{q}$. In addition, T and p are thermodynamic functions of the independent variables ρ and $\tilde{s}$.

Derivatives of T and p with respect to $\tilde{q}$ are equivalent to derivatives with respect to entropy at constant volume and particle number:

$$\left.\frac{\partial T(\rho,\tilde{s})}{\partial\tilde{q}}\right)_\rho=\left.\frac{1}{T\rho}\frac{\partial T}{\partial\tilde{s}}\right)_\rho=\left.\frac{V}{T}\frac{\partial T}{\partial S}\right)_{V,N}=\frac{1}{\rho\tilde{c}_V}, \tag{8.4.53}$$

$$\left.\frac{\partial p}{\partial\tilde{q}}\right)_\rho=\left.\frac{1}{T\rho}\frac{\partial p}{\partial\tilde{s}}\right)_\rho\equiv\left.\frac{V}{T}\frac{\partial p}{\partial S}\right)_\rho, \tag{8.4.54}$$

where $\tilde{c}_V=c_V/\rho$ is the mass specific heat (heat capacity per unit mass) often used in preference to the volume specific heat c_V in discussions of modes in fluids. In addition, derivatives with respect to ρ at constant $\tilde{q}$ are equivalent to derivatives at constant entropy and particle number. Thus, Eqs. (8.4.50) and (8.4.51) can be rewritten as

$$\left[-\frac{\partial^2}{\partial t^2}+\frac{1}{\rho}\left(\frac{4}{3}\eta+\zeta\right)\frac{\partial}{\partial t}\nabla^2\right]\delta\rho+\left.\frac{\partial p}{\partial\rho}\right)_S\nabla^2\delta\rho+\left.\frac{1}{T\rho}\frac{\partial p}{\partial\tilde{s}}\right)_\rho\nabla^2\tilde{q}=0 \tag{8.4.55}$$

and

$$\begin{aligned}\frac{\partial\tilde{q}}{\partial t}&=\kappa\nabla^2T(\mathbf{x},t)\\&=\left.\kappa\frac{\partial T}{\partial\rho}\right)_S\nabla^2\delta\rho+\left.\frac{\kappa}{T\rho}\frac{\partial T}{\partial\tilde{s}}\right)_\rho\nabla^2\tilde{q}.\end{aligned} \tag{8.4.56}$$

When there is no dissipation, κ, η and ζ are zero, $\tilde{q}=0$, and Eq. (8.4.55) predicts a pair of sound waves with velocity

$$c^2=\left.\frac{\partial p}{\partial\rho}\right)_S=\left.\frac{\tilde{c}_p}{\tilde{c}_V}\frac{\partial p}{\partial\rho}\right)_T, \tag{8.4.57}$$

where $\tilde{c}_p$ is the mass specific heat at constant pressure. Note that the speed of sound is determined by the isentropic compressibility. A purely mechanical theory predicts that the sound velocity is determined by the Newtonian or isothermal compressibility. It is the coupling between mass density and energy density that converts the isothermal sound velocity to the isentropic or adiabatic velocity.

When dissipation is included, there is a diffusive heat mode in addition to the two sound modes. This mode will have a frequency

$$\omega = -iD_T q^2, \tag{8.4.58}$$

where D_T is the thermal diffusion constant. In an incompressible system, $\nabla^2 p = 0$, and, from Eq. (8.4.56), $D_T = \kappa/\rho\tilde{c}_V = \kappa/c_V$. Thus, as in our previous examples, the thermal diffusion constant is the ratio of the thermal conductivity to a specific heat. In a compressible system, $\nabla^2 p$ is not zero. To find the thermal diffusion constant in this case, we can use the fact that the frequency of the thermal diffusion mode is proportional to q^2 and, at small q, is much less than the frequency, proportional to q, of sound modes. Setting $\partial/\partial t \sim \omega \sim q^2$ and using $\omega^2 \sim \omega q^2 \ll q^2$ in Eqs. (8.4.50) and (8.4.55), we obtain $\delta p(\mathbf{q},\omega) = O(q^4)$. Then, to order q^2,

$$\delta\rho(\mathbf{q},\omega) = -\frac{1}{T\rho}\frac{\partial p/\partial\tilde{s})_\rho}{\partial p/\partial\rho)_S}\tilde{q}(\mathbf{q},\omega). \tag{8.4.59}$$

When this density is used in Eq. (8.4.56), the resulting heat mode frequency is that given in Eq. (8.4.58) with

$$\begin{aligned} D_T &= \frac{\kappa}{\rho\tilde{c}_p} = \frac{\kappa}{c_p} \\ &= \frac{\kappa}{\rho}\left(\frac{1}{\tilde{c}_V} - \frac{\partial T/\partial\rho)_S \partial p/\partial\tilde{s})_\rho}{\partial p/\partial\rho)_S}\right). \end{aligned} \tag{8.4.60}$$

Thus, coupling to density fluctuations converts the constant-volume specific heat to the constant-pressure specific heat in the denominator of the thermal diffusion coefficient.

If $\tilde{q}$ were zero (i.e., if the thermal conductivity were zero), the width of the sound wave mode would be $(4\eta/3 + \zeta)/\rho$. When κ is nonzero, there is a contribution to the sound wave damping arising from entropy fluctuations. To determine this damping, we use the fact that the sound wave frequency is large compared to the thermal diffusion frequency. Eq. (8.4.56) can be used to solve for $\tilde{q}$ in terms of ρ. Since $\omega \gg q^2$, we can neglect the $\nabla^2\tilde{q}$ term in this equation, and we find

$$\tilde{q} = \kappa\frac{q^2}{i\omega}\frac{\partial T}{\partial\rho}\bigg)_S \delta\rho = -i\omega\kappa\frac{\partial T/\partial\rho)_S}{\partial p/\partial\rho)_S}\delta\rho, \tag{8.4.61}$$

where we used $\omega^2 = c^2q^2 + O(q^4)$ and Eq. (8.4.57) for the sound wave velocity. When this $\tilde{q}$ is used in Eq. (8.4.55), the resulting equation for the sound mode frequency is

$$\omega^2 - c^2q^2 + i\omega\Gamma q^2 = 0, \tag{8.4.62}$$

where

$$\Gamma = \frac{1}{\rho}\left(\frac{4}{3}\eta + \zeta\right) + \frac{\kappa}{\rho \tilde{c}_p}\left(\frac{\tilde{c}_p}{\tilde{c}_V} - 1\right). \tag{8.4.63}$$

The complex sound frequencies are thus

$$\omega = \pm cq - i\frac{1}{2}\Gamma q^2. \tag{8.4.64}$$

As in previous hydrodynamical examples, the imaginary part of the frequency tends to zero with wave number faster than the real part. The thermodynamic relation,

$$\tilde{c}_p - \tilde{c}_V = \frac{T}{\rho^2}\left(\frac{\partial p}{\partial T}\right)^2_\rho \left(\frac{\partial \rho}{\partial p}\right)_T, \tag{8.4.65}$$

implies that the constant-pressure and constant-volume specific heats become equal in the incompressible limit when $\partial\rho/\partial p \to 0$.

8 Light scattering

Light scattering measures fluctuations in the local dielectric constant, which to a good approximation in isotropic fluids is a function of the local mass (or particle) density. Thus, light scattering provides a direct measure of the density-density response function $S_{nn}(\mathbf{q},\omega) = m^{-2}S_{\rho\rho}(\mathbf{q},\omega)$. The susceptibility $\chi''_{nn}(\mathbf{q},\omega)$ can be calculated from the hydrodynamic equations using the techniques discussed in Secs. 7.4 and 8.2. It is

$$\begin{aligned}\frac{\chi''_{nn}(\mathbf{q},\omega)}{\omega} = n\left(\frac{\partial n}{\partial p}\right)_T & \left[\frac{\tilde{c}_V}{\tilde{c}_p}\frac{c^2\Gamma q^4}{(\omega^2 - c^2q^2)^2 + (\omega\Gamma q^2)^2}\right. \\ & + \left(1 - \frac{\tilde{c}_V}{\tilde{c}_p}\right)\frac{D_T q^2}{\omega^2 + (D_T q^2)^2} \\ & \left. - \left(1 - \frac{\tilde{c}_V}{\tilde{c}_p}\right)\frac{(\omega^2 - c^2q^2)q^2 D_T}{(\omega^2 - c^2q^2)^2 + (\omega\Gamma q^2)^2}\right].\end{aligned} \tag{8.4.66}$$

This result was first derived by Landau and Placzek in 1934. It shows that the density correlation function contains peaks arising from both thermal diffusion and from sound waves. $S_{nn}(\mathbf{q},\omega)$ is plotted in Fig. 8.4.2. The thermal peak of width $D_T q^2$ centered about $\omega = 0$ is the *Rayleigh peak*. The sound peaks at $\omega = \pm cq$ with width Γq^2 are the *Brillouin peaks*. The third term in Eq. (8.4.66) does not contribute significantly to light scattering intensities in the vicinity of the Rayleigh and Brillouin peaks and is often neglected. It is necessary, however, to guarantee that the first high-frequency sum rule of $\chi''_{nn}(\mathbf{q},\omega)$ is satisfied. The Landau-Placzek ratio $\tilde{c}_p/\tilde{c}_V$ can be obtained from the ratio $(\tilde{c}_p/\tilde{c}_V) - 1$ of the intensities (integrated area) of the Rayleigh and Brillouin peaks. Note that this ratio tends to infinity near the liquid-gas transition where $(\partial n/\partial p) \to \infty$ and, from Eq. (8.4.65), $\tilde{c}_p \to \infty$. Thus, near the liquid-gas transition, most of the light scattering intensity is in the Rayleigh peak, whose width also approaches zero because $\kappa/\tilde{c}_p \to 0$.

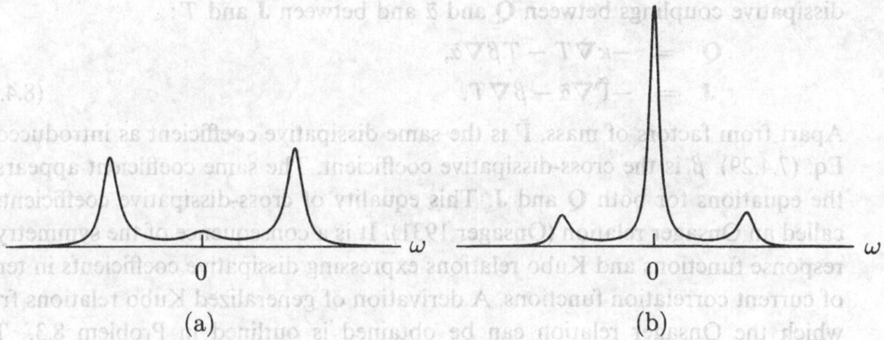

Fig. 8.4.2. Normalized light scattering intensity $nS_{nn}(\mathbf{q},\omega)/(\partial n/\partial p)_T = [2\hbar/(1-e^{-\beta\hbar\omega})]n\chi''_{nn}(\mathbf{q}\omega)/(\partial n/\partial p)_T$ as a function of ω at fixed $\mathbf{q}$. (a) Far from the critical point, where the Brillouin peaks dominate. (b) Closer to the critical point, where the Rayleigh peak dominates.

9 Two-component fluids

In two-component fluids containing molecules of mass m_a and m_b, there are two conserved mass densities, ρ_a and ρ_b, and, correspondingly, two mass currents, $\mathbf{g}_a$ and $\mathbf{g}_b$, related by the conservation laws

$$\frac{\partial \rho_a}{\partial t} + \nabla \cdot \mathbf{g}_a = 0,$$
$$\frac{\partial \rho_b}{\partial t} + \nabla \cdot \mathbf{g}_b = 0. \tag{8.4.67}$$

Only the total momentum,

$$\mathbf{g} = \mathbf{g}_a + \mathbf{g}_b, \tag{8.4.68}$$

rather than the momenta of the individual species, is conserved. The momenta for the individual species can then be expressed in terms of the total momentum $\mathbf{g}$ and a relative momentum $\mathbf{J}$:

$$\mathbf{g}_a = \frac{\rho_a}{\rho}\mathbf{g} + \mathbf{J}, \qquad \mathbf{g}_b = \frac{\rho_b}{\rho}\mathbf{g} - \mathbf{J},$$
$$\rho\mathbf{J} = \rho_b\mathbf{g}_a - \rho_a\mathbf{g}_b, \tag{8.4.69}$$

where $\rho = \rho_a + \rho_b$ is the total mass density. As in one-component fluids, the total momentum density and velocity are related by $\mathbf{g} = \rho\mathbf{v}$. Following the same procedures as for one-component fluids, we obtain the equation for entropy production,

$$T\left(\frac{\partial s}{\partial t} + \nabla \cdot (\mathbf{v}s + \mathbf{Q}/T)\right)$$
$$= -\mathbf{Q}\cdot\nabla T/T - \mathbf{J}\cdot\nabla\tilde{\alpha} - (\pi_{ji} - p\delta_{ij} - v_i g_j)\nabla_i v_j, \tag{8.4.70}$$

where $\tilde{\alpha} = \alpha_a - \alpha_b$ is the difference of the chemical potentials per unit mass of the two species. $\mathbf{Q}$ and $\mathbf{J}$ have the same sign under time reversal, and there will be dissipative couplings between **Q** and $\tilde{\alpha}$ and between **J** and T:

$$\begin{aligned} \mathbf{Q} &= -\kappa\nabla T - T\beta\nabla\tilde{\alpha}, \\ \mathbf{J} &= -\tilde{\Gamma}\nabla\tilde{\alpha} - \beta\nabla T. \end{aligned} \tag{8.4.71}$$

Apart from factors of mass, $\tilde{\Gamma}$ is the same dissipative coefficient as introduced in Eq. (7.4.29). β is the cross-dissipative coefficient. The same coefficient appears in the equations for both **Q** and **J**. This equality of cross-dissipative coefficients is called an Onsager relation (Onsager 1931). It is a consequence of the symmetry of response functions and Kubo relations expressing dissipative coefficients in terms of current correlation functions. A derivation of generalized Kubo relations from which the Onsager relation can be obtained is outlined in Problem 8.3. The dissipative part of the stress tensor is identical to that of a one-component fluid.

There are now six conservation laws, and we expect six hydrodynamical modes. They are two transverse momentum modes, two longitudinal sound modes, and two coupled thermal-diffusion–relative-mass-diffusion modes, whose frequencies can be calculated using the techniques we have developed in this section.

8.5 Liquid crystals, crystalline solids, and superfluid helium

In this chapter, we have studied the hydrodynamics of both the ordered and disordered phases of systems with no mass motion. We have also studied one- and two-component simple fluids in which conservation laws completely determine the set of hydrodynamical variables. In this section, we will study the hydrodynamics of liquid crystals, crystalline solids and superfluids in which there are one or more broken-symmetry hydrodynamic variables in addition to the five conserved variables of a one-component fluid. We will leave generalizations to multicomponent ordered fluids to the problems at the end of the chapter.

1 *Nematic liquid crystals*

As we discussed in Sec. 6.2 on generalized elasticity, the director **n** in a nematic liquid crystal is an elastic variable whose uniform changes do not change the free energy of the system. It is thus a hydrodynamical variable as well. In a theory linearized about a uniform state with $\mathbf{n} = \mathbf{n}_0 = \mathbf{e}_z$, we may replace **n** by its deviations δn_x and δn_y from $\mathbf{n}_0$. The linearized entropy equation for a nematic is, therefore,

$$T ds = d\varepsilon - \alpha d\rho - \mathbf{v}\cdot d\mathbf{g} - h_{ij} d(\nabla_j n_i), \tag{8.5.1}$$

where

$$h_{ij} = \left.\frac{\partial\varepsilon}{\partial\nabla_j n_i}\right)_{\rho,s,\mathbf{g}} = K_{jikl}\nabla_k n_l, \tag{8.5.2}$$

where K_{ijkl} is the Frank elastic constant tensor. As in previous examples, our next task is to find the reactive part of the current for $\mathbf{n}$. The fields $\mathbf{n}$ and $\mathbf{v}$ have different signs under time reversal, and there may be reactive couplings between them. The director $\mathbf{n}$ does not change under rigid translation at constant velocity. Therefore, the leading coupling of $\mathbf{n}$ to $\mathbf{v}$ must involve the gradient of $\mathbf{v}$. The dynamical equation for $\mathbf{n}$ including a dissipative current X'_i is then

$$\frac{\partial n_i}{\partial t} - \lambda_{ijk}\nabla_j v_k + X'_i = 0. \tag{8.5.3}$$

Since $\mathbf{n}\cdot\partial\mathbf{n}/\partial t = 0$, there are only two independent components of the tensor λ_{ijk}, which can be taken as the symmetric and antisymmetric with respect to interchange of the indices j and k:

$$\lambda_{ijk} = \frac{1}{2}\lambda(\delta^T_{ij}n_k + \delta^T_{ik}n_j) + \frac{1}{2}\lambda_2(\delta^T_{ij}n_k - \delta^T_{ik}n_j), \tag{8.5.4}$$

where $\delta^T_{ij} = \delta_{ij} - n_i n_j$. Under a uniform rigid rotation,

$$\frac{\partial \mathbf{n}}{\partial t} = \boldsymbol{\omega}\times\mathbf{n} = \frac{1}{2}(\boldsymbol{\nabla}\times\mathbf{v})\times\mathbf{n}. \tag{8.5.5}$$

This implies that the coefficient λ_2 of the antisymmetric part must be equal to -1. The magnitude of the coefficient λ is not determined by symmetry arguments. The coupling of $\mathbf{n}$ to $\mathbf{v}$ leads to an additional linear term in the reactive part of the stress tensor:

$$\sigma^R_{ij} = -p\delta_{ij} + \lambda_{kji}h_k, \tag{8.5.6}$$

where $p = -(\varepsilon - Ts - \alpha\rho - \mathbf{g}\cdot\mathbf{v}) = -\partial E/\partial V)_{S,N,\mathbf{P},\nabla_i n_j}$ is the normal fluid pressure and where

$$h_j = \nabla_j h_{ij} = K_{jikl}\nabla_j\nabla_k n_l = -\frac{\delta F}{\delta n_k}, \tag{8.5.7}$$

where F is the Frank free energy for the nematic. Note that σ^R_{ij} in Eq. (8.5.6) is not symmetric and is thus not the average of a microscopic stress tensor, which is symmetric. This is not really a problem for hydrodynamics where only the quantity $\nabla_i\sigma_{ij}$ enters. It is possible to construct a symmetric stress tensor with exactly the same value of $\nabla_i\sigma_{ij}$.

When dissipation is included, the entropy production equation becomes

$$T\frac{dS}{dt} = -\int d^3x\,[\mathbf{Q}\cdot(\boldsymbol{\nabla}T)/T - \sigma'_{ij}\nabla_i v_j - X'_i\nabla_j h_{ij}]. \tag{8.5.8}$$

Since the nematic liquid crystal is uniaxial, all dissipative coefficients are tensors. There are two thermal conductivity coefficients,

$$\kappa_{ij} = \kappa_\parallel n_i n_j + \kappa_\perp\delta^T_{ij}, \tag{8.5.9}$$

five viscosities,

$$\begin{aligned}\sigma'_{ij} \;=\;& 2\nu_2 A_{ij} + 2(\nu_3-\nu_2)[A_{ik}n_k n_l + A_{jk}n_i n_k] - (\nu_4-\nu_2)\delta_{ij}A_{kk}\\ &-2(\nu_1+\nu_2-2\nu_3)n_i n_j n_k n_l A_{kl}\\ &+(\nu_5-\nu_4+\nu_2)[\delta_{ij}n_k n_l A_{kl} + n_i n_j A_{kk}],\end{aligned} \tag{8.5.10}$$

where $A_{ij} = \frac{1}{2}(\nabla_i v_j + \nabla_j v_i)$ is the strain rate tensor, and one dissipative coefficient for the director,

$$X'_i = -\gamma^{-1}\nabla_j h_{ij}. \tag{8.5.11}$$

(We used ν rather than η for the viscosities to avoid confusion with an alternative convention for specifying viscosities using the symbol η.) The coefficient γ has units of viscosity (poise). There are no couplings between $\mathbf{n}$ and ∇T because of the invariance under $\mathbf{n} \to -\mathbf{n}$.

Since there are two additional broken-symmetry variables, δn_x and δn_y, there are two more modes in a nematic than there are in an isotropic liquid. The compressional sound mode retains an isotropic velocity. Its damping and the heat diffusion mode become anisotropic. The transverse velocity modes couple to the director and lead to composite diffusive modes with frequencies that can be expressed when $K\rho/\gamma\nu \ll 1$ as

$$\begin{aligned}
i\omega_{s1}(K_3 q_3^2 + K_1 q_1^2)^{-1} &= \gamma^{-1} + \frac{1}{4}\frac{[q_3^2 + q_1^2 + \lambda(q_3^2 - q_1^2)]^2}{\nu_3(q_1^2 - q_2^2)^2 + 2(\nu_1 + \nu_2)q_1^2 q_3^2} \\
i\omega_{s2}(K_2 q_1^2 + K_3 q_3^2)^{-1} &= \gamma^{-1} + \frac{1}{4}(1+\lambda)^2 \frac{q_3^2}{\nu_3 q_3^2 + \nu_2 q_1^2}, \\
i\omega_{f1} &= \frac{\nu_3(q_1^2 - q_3^2)^2 + 2(\nu_1 + \nu_2)q_1^2 q_3^2}{\rho(q_1^2 + q_3^2)}, \\
i\omega_{f2} &= \frac{\nu_3 q_3^2 + \nu_2 q_1^2}{\rho}
\end{aligned} \tag{8.5.12}$$

where the 3 direction is along $\mathbf{n}$.

2 Smectic-A liquid crystals

In a smectic-A liquid crystal, the layer phase u is the single broken-symmetry hydrodynamic variable. The director relaxes in microscopic times to its preferred orientation normal to the layers and is no longer a hydrodynamical variable. The entropy equation for a smectic is

$$T\,ds = d\varepsilon - \alpha\,d\rho - \mathbf{v}\cdot d\mathbf{g} - h_i d(\nabla_i u), \tag{8.5.13}$$

where

$$h_i = \left.\frac{\partial \varepsilon}{\partial \nabla_i u}\right)_{s,\rho,\mathbf{g}} = \left.\frac{\partial f}{\partial \nabla_i u}\right)_{T,\rho,\mathbf{g}}, \tag{8.5.14}$$

where $f = \varepsilon - Ts$ is the Helmholtz free energy density. In our study of the elastic properties of smectics in Chapter 6, we considered only processes at constant temperature and chemical potential. In this case, f is a quadratic function of u [Eq. (6.3.11)], and

$$h_i = B\delta_{iz}\nabla_z u - K_1 \nabla_{\perp i}\nabla_\perp^2 u. \tag{8.5.15}$$

As we saw in our treatment of isotropic fluids, hydrodynamical modes involve changes in entropy and density (and thus local temperature and chemical potential). In a smectic, $\nabla_z u$ is a scalar, and there are couplings between it and changes in the density $\delta\rho$ and the entropy $\delta\tilde{q}$. Thus, to obtain a complete description of the mode structure of a smectic, h_i should be treated as a function of u, ρ and $\tilde{q}$, and its expansion about the equilibrium state should include terms linear in $\delta\rho$ and $\delta\tilde{q}$. To keep our discussion simple, we will investigate only modes in the isothermal, incompressible limit where Eq. (8.5.15) applies. The isothermal limit results when the thermal conductivity κ is infinite.

If the smectic is translated with a constant velocity $\mathbf{v}$, $\partial u/\partial t$ will be equal to v_z, implying

$$\frac{\partial u}{\partial t} + X' = v_z, \tag{8.5.16}$$

where X' is the dissipative current associated with u. The v_z term in this equation leads to a reactive term in the stress tensor:

$$\sigma_{ij}^R = -p\delta_{ij} + \delta_{iz}h_j, \tag{8.5.17}$$

where the $p = -(\varepsilon - Ts - \alpha\rho - \mathbf{g}\cdot\mathbf{v}) = -\partial E/\partial V)_{S,N,\mathbf{P},\nabla_i u}$, as in an isotropic fluid. The longitudinal property of $\nabla_i u$ allowing $v_j h_i \nabla_j(\nabla_i u)$ to be replaced by $v_j h_i \nabla_i(\nabla_j u)$ was used to obtain this expression for the pressure. The dissipative currents are

$$\begin{aligned} Q_i &= -\kappa_{ij}\nabla_j T - \beta n_i^0 \boldsymbol{\nabla}\cdot\mathbf{h}, \\ X' &= -\zeta\boldsymbol{\nabla}\cdot\mathbf{h} - \beta\mathbf{n}^0\cdot\boldsymbol{\nabla}T \\ \sigma'_{ij} &= \eta_{ijkl}\nabla_k v_l, \end{aligned} \tag{8.5.18}$$

where $\mathbf{n}^0 = \mathbf{e}_z$ is the unit layer normal, β is a cross-dissipative coefficient, and where, as in a nematic, there are five independent components of the viscosity tensor η_{ijkl} and two independent components of the thermal conductivity tensor κ_{ij}.

The longitudinal sound mode of a smectic is essentially the same as that of a normal liquid except for anisotropies in the sound velocity and dissipative coefficients. The new Goldstone mode of the smectic involves u and the transverse momentum. To analyze the $u-\mathbf{g}$ modes, we will for simplicity assume that the temperature and density are constant (as would be the case if the thermal conductivity were infinite and the compressibility were zero). We will also assume an isotropic viscosity tensor with a single shear viscosity η. Because the smectic is uniaxial, we may, without loss of generality, choose $\mathbf{q} = (q_x, 0, q_z)$ to lie in the xz-plane. Then, the longitudinal and transverse components of $\mathbf{g}$ in the xz-plane are

$$g_l = (q_x g_x + q_z g_z)/q, \qquad g_\perp = (q_z g_x - q_x g_x)/q. \tag{8.5.19}$$

The equations for u and $g_\perp$ in the incompressible limit when $g_l = 0$ are

$$-i\omega u = \frac{q_x}{q}\frac{g_\perp}{\rho} - \zeta(Bq_z^2 + K_1 q_x^4)u \tag{8.5.20}$$

and

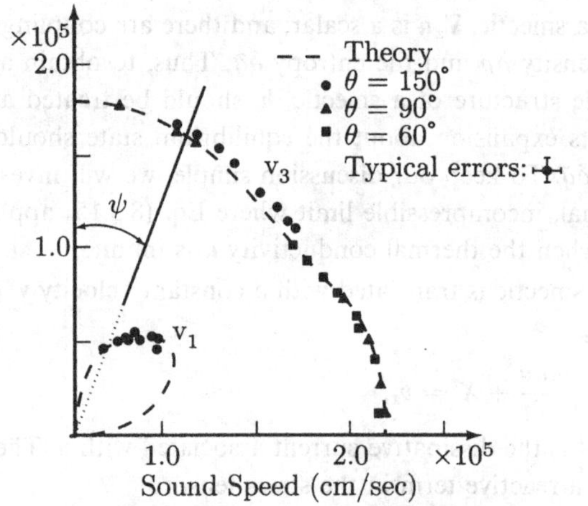

Fig. 8.5.1. Brillouin scattering measurements of the sound velocities in a smectic-*A* liquid crystal as a function of angle ψ between the wave vector $\mathbf{q}$ and the director $\mathbf{n}_0$. v_3 is the velocity of longitudinal sound and v_1 that of transverse sound. It is clear that v_1 is a maximum at $\psi = 45°$. The experiments are unable to resolve the transverse sound velocity near $\psi = 0°$ and $\psi = 90°$. v_1 and v_3, however, lie on the curves predicted by the hydrodynamical theory (see Problem 8.4). The depression in the longitudinal sound velocity at $\psi = 45°$ results from a coupling between longitudinal and transverse sound not present in isotropic fluids. [York Liao, Noel A. Clark, and Peter Pershan, *Phys. Rev. Lett.* **30**, 639 (1973).]

$$-i\omega g_\perp = -\frac{q_x}{q}(Bq_z^2 + K_1 q_x^4)u - \eta q^2 g_\perp. \tag{8.5.21}$$

In the dissipationless limit, these equations predict a shear sound mode with dispersion

$$\omega^2 = \frac{q_x^2}{q^2}\frac{Bq_z^2 + K_1 q_x^4}{\rho} \approx \frac{Bq_x^2 q_z^2}{\rho q^2}. \tag{8.5.22}$$

Thus the shear sound velocity is maximum when $\mathbf{q}$ is at an angle of 45° with respect to the z-axis, as shown in Fig. 8.5.1. In a compressible smectic, the longitudinal and shear sound modes are coupled. As a result, the angular dependence of the shear sound velocity is more complicated than that predicted by Eq. (8.5.22). We leave it as an exercise to verify this result.

When $q_x = 0$, u and $g_\perp$ decouple: there is a transverse momentum diffusion mode with $i\omega = \eta q^2/\rho$ and a diffusive u mode with frequency

$$i\omega = \zeta B q_z^2. \tag{8.5.23}$$

This is a mode in which there is relative motion of mass and the periodic structure of the smectic, i.e., one in which molecules diffuse from one layer to the next without changing the average periodic structure. It is called the *permeation* mode. When $q_x = 0$ there are slow and fast coupled $u - g_\perp$ diffusive modes. The component of $\mathbf{g}$ perpendicular to the xz-plane is not coupled to u and always has a diffusive mode.

A real smectic is not incompressible, and there is a longitudinal sound mode that couples to the transverse sound mode where it exists. We leave it as an exercise to show that the coupled sound velocities are identical to those of a uniaxial elastic solid described by Eq. (6.4.18) with $K_{12} = K_{11}$ and $K_{44} = 0$.

3 *Crystalline solids*

In solids, there are three broken-symmetry phase variables u_i, $i = x, y, z$, for each of the three directions in space. The entropy equation is

$$T\,ds = d\varepsilon - \alpha d\rho - \mathbf{v}\cdot d\mathbf{g} - h_{ij}d(\nabla_j u_i), \tag{8.5.24}$$

where

$$h_{ij} = \left.\frac{\partial \varepsilon}{\partial \nabla_i u_j}\right)_{s,\rho,\mathbf{g}} = \left.\frac{\partial \varepsilon}{\partial u_{ij}}\right)_{s,\rho,\mathbf{g}} = \left.\frac{\partial f}{\partial u_{ij}}\right)_{T,\rho,\mathbf{g}}, \tag{8.5.25}$$

where $f = \varepsilon - Ts$ is the Helmholtz free energy density. The free energy and internal energy depend only on the symmetric strain u_{ij} rather than on all components of $\nabla_i u_j$. Therefore, h_{ij} is symmetric, and $h_{ij}du_{ij} = h_{ij}d\nabla_i u_j$. The latter form used in Eq. (8.5.24) is, as we shall see, the most useful for the derivation of hydrodynamical equations. The free energy $f(T, \rho, u_{ij})$ including coupling between density and strain was derived in Eq. (6.4.14). In isothermal systems, it yields

$$h_{ij} = K^n_{ijkl}u_{kl} + (D/\rho_0)\delta_{ij}\delta\rho \tag{8.5.26}$$

where ρ_0 is the equilibrium density. (As in smectics, an additional term proportional to $\delta\tilde{q}$ is needed for a general treatment of hydrodynamical modes.) K^n_{ijkl} is the constant density elastic constant tensor, and D is the density-strain coupling introduced in Eq. (6.4.41). K^n_{ijkl} must be symmetric in ij and kl in order to ensure a symmetric stress tensor. Under uniform translations of velocity $\mathbf{v}$, the phases $\mathbf{u}$ grow uniformly:

$$\frac{\partial \mathbf{u}}{\partial t} = \mathbf{v}, \tag{8.5.27}$$

leading to an additional term in the reactive part of the stress tensor,

$$\sigma_{ij} = -p\delta_{ij} + h_{ij}. \tag{8.5.28}$$

The pressure p in this equation is $-(f - \rho\partial f/\partial\rho) = -(\varepsilon - Ts - \alpha\rho)$. Thus, Eq. (8.5.28), which is identical to Eq. (6.6.25), was derived using static arguments only.

There are five conserved and three broken-symmetry hydrodynamic variables in a solid. We therefore expect a total of eight modes. These are a heat diffusion mode and three pairs of propagating sound modes. In addition, there is a vacancy diffusion mode, analogous to the permeation mode of a smectic, which is present even in one-component solids. The diffusion constant for this mode is often very small because of the large potential barriers inhibiting mass diffusion. It is often ignored.

To illustrate how sound modes can be obtained from these equations, we will consider the dissipationless, isothermal limit (which can be obtained by letting the thermal conductivity go to infinity and all other dissipative coefficients go to zero). In this limit, $\delta p = \rho_0 \partial f/\partial \rho = A(\delta\rho/\rho_0) + D\nabla \cdot \mathbf{u}$ to linear order, where A is the inverse compressibility introduced in Eq. (6.4.41). Then,

$$\frac{\partial u_i}{\partial t} = \frac{g_i}{\rho}, \qquad \frac{\partial \rho}{\partial t} = -\nabla \cdot \mathbf{g}, \tag{8.5.29}$$

and

$$\frac{\partial g_i}{\partial t} = \nabla_j[(D-A)(\delta\rho/\rho_0)\delta_{ij} - D\nabla \cdot \mathbf{u}\delta_{ij} + K^n_{ijkl}u_{kl}], \tag{8.5.30}$$

where we used the expressions for h_{ij} and p discussed above. Equivalently, we used Eq. (6.6.29) for σ_{ij}. Eq. (8.5.29) implies that $\delta\rho/\rho_0 = -\nabla \cdot \mathbf{u}$ and

$$\rho \frac{\partial^2 u_i}{\partial t^2} = \nabla_j K_{ijkl} u_{kl}, \tag{8.5.31}$$

where $K_{ijkl} = K^n_{ijkl} + (A - 2D)\delta_{ij}\delta_{kl}$ is the elastic constant tensor at constant $n\Omega_0$ (i.e., constant vacancy density) introduced in Eq. (6.4.43). Eq. (8.5.31) is identical to Eq. (7.3.2) for elastic waves in an elastic continuum. In an isotropic solid, it predicts two pairs of transverse sound modes with velocity $c_T = \pm(\mu/\rho)^{1/2}$ and a pair of longitudinal modes with velocity $c_L = \pm[(\lambda + 2\mu)/\rho]^{1/2}$, where the Lamé coefficients λ and μ are those at constant $n\Omega_0$.

4 Superfluid helium

Superfluid helium was the first system with a broken continuous symmetry to have its hydrodynamics investigated. In his landmark 1941 paper, Landau not only discussed the microscopic energy spectrum of helium II, but also derived the hydrodynamic equations governing its normal and superfluid flow. Here we will show how these hydrodynamical equations can be obtained using the techniques developed in this chapter.

The complex order parameter for superfluid helium is the condensate wave-function $\psi = \langle\hat{\psi}\rangle = |\psi|e^{i\theta}$, where $\hat{\psi}$ is the annihilation operator for helium atoms. If the condensate is put into motion with velocity $\mathbf{v}_s$, then the phase θ of ψ will increase by $m\mathbf{v}_s \cdot \mathbf{x}/\hbar$. The velocity $\mathbf{v}_s = (\hbar/m)\nabla\theta$ is the superfluid velocity. If $\mathbf{v}_s$ is nonzero, the free energy will increase as $\rho_s \int d^d x v_s^2/2$, where ρ_s, the superfluid density, has units of mass density. As in our previous example, this is the energy associated with a spatially nonuniform order parameter. At this stage, our description of a superfluid is identical to that of all of the other systems with xy symmetry we have studied. Unlike our other examples, however, $\mathbf{v}_s$ really is a velocity, and we must inquire what happens under a Galilean transformation in which the whole system moves with velocity $\mathbf{v}_n$ relative to the laboratory rest frame. Under this transformation, the free energy acquires an extra term

$-Nmv_n^2/2$ [Eq. (8.4.18)]. On the other hand, there is no energy associated with the phase of the order parameter unless $\mathbf{v}_s$ differs from $\mathbf{v}_n$, i.e., unless the condensate is in motion with respect to the fluid. These considerations imply that the free energy as a function of volume V, particle number N, temperature T, normal velocity $\mathbf{v}_n$ and superfluid velocity $\mathbf{v}_s$ is

$$F(T, V, N, \mathbf{v}_n, \mathbf{v}_s) = F_0(T, V, N) - \frac{1}{2}mNv_n^2 + \frac{1}{2}\int d^d x \rho_s(\mathbf{v}_s - \mathbf{v}_n)^2 \equiv Vf. \tag{8.5.32}$$

As in a normal fluid, the momentum density is obtained via differentiation with respect to $\mathbf{v}_n$:

$$g_i = -\left.\frac{\partial f}{\partial v_{ni}}\right)_{T,V,N,\mathbf{v}_s} = \rho v_{ni} - \rho_s(v_{ni} - v_{si}). \tag{8.5.33}$$

If we define the normal density ρ_n via

$$\rho = \rho_n + \rho_s, \tag{8.5.34}$$

then

$$\mathbf{g} = \rho_n \mathbf{v}_n + \rho_s \mathbf{v}_s. \tag{8.5.35}$$

This equation suggests that superfluid helium can be viewed as consisting of normal and superfluid parts with different mass densities, each of which can carry momentum. Indeed, the hydrodynamic theory of superfluids is often referred to as the two-fluid model. The field conjugate to the superfluid velocity,

$$h_i = \frac{\partial f}{\partial v_{si}} = \rho_s(v_{si} - v_{ni}), \tag{8.5.36}$$

is zero when $\mathbf{v}_s = \mathbf{v}_n$.

Other thermodynamic functions follow using the procedure employed throughout this chapter. The free energy of Eq. (8.5.32), like that of Eq. (8.4.15), is $E - TS - \mathbf{P} \cdot \mathbf{v}_n$. (There is no term involving $\int \mathbf{h} \cdot \mathbf{v}_s d^d x$ because F is a function of $\mathbf{v}_s$ and not $\mathbf{h}$.) The pressure is thus $p = -(\varepsilon - \alpha\rho - Ts - \mathbf{g} \cdot \mathbf{v}_n)$ (i.e., Eq. (8.4.25) with $\mathbf{v}$ replaced by $\mathbf{v}_n$), and

$$T ds = d\varepsilon - \alpha d\rho - \mathbf{v}_n \cdot d\mathbf{g} - \mathbf{h} \cdot d\mathbf{v}_s. \tag{8.5.37}$$

The energy, mass and momentum densities obey the same conservation laws as in a normal fluid.

As in our previous examples, we need to determine any reactive couplings between $\mathbf{v}_s$, which is odd under time reversal, and fields that are even under time reversal. To do this, we proceed exactly as we did in deriving spin hydrodynamics in Sec. 8.3. The number operator $\hat{N}$ commutes with the Hamiltonian $\mathscr{H}$. Its commutation relation with ψ,

$$[\hat{\psi}, \hat{N}] = \hat{\psi}, \tag{8.5.38}$$

is analogous to the commutation relation between S_z and $S^+ = S_x + iS_y$ in spin systems. Consider now the effect on the time dependence of $\hat{\psi}$ of adding an external chemical potential μ_e:

$$\begin{aligned}\hat{\psi}(t,\mu_e) &= e^{i(\mathscr{H}-\mu_e\hat{N})t/\hbar}\hat{\psi}(0,0)e^{-i(\mathscr{H}-\mu_e\hat{N})t/\hbar} \\ &= e^{i\mu_e t/\hbar}\hat{\psi}(t,\mu_e=0).\end{aligned} \tag{8.5.39}$$

In thermal equilibrium, $\psi(t,\mu_e) = \langle\hat{\psi}(t,\mu_e)\rangle$ must be independent of time, and we conclude

$$\frac{\partial\theta}{\partial t} = \frac{1}{\hbar}(\mu_e - \mu), \tag{8.5.40}$$

where $\mu = -T(\partial s/\partial n)$ is the equilibrium chemical potential. As in the rigid rotor and our magnetic examples, we introduce a "current" $X = (\hbar/m)d\theta/dt$. Eq. (8.5.40) implies that X must have a part equal to the chemical potential per unit mass $\alpha = \mu/m$, i.e., $X = \alpha + X'$, when $\mu_e = 0$. The equation for $\mathbf{v}_s$ when $\mu_e = 0$ is then

$$\frac{\partial\mathbf{v}_s}{\partial t} + \boldsymbol{\nabla}(\alpha + X') = 0. \tag{8.5.41}$$

In our previous examples, we were able to identify the analog of X' with a dissipative current. In the present case, we will find that X' is purely dissipative in a linearized theory but that it has nonlinear non-dissipative parts.

We now have the information needed to derive the complete hydrodynamic equations for superfluid helium. The equations for entropy production are identical to those of a normal fluid, with additional contributions arising from $\mathbf{v}_s$. These are

$$T\frac{\partial s}{\partial t} = \cdots + \mathbf{h}\cdot\boldsymbol{\nabla}(\alpha + X'), \tag{8.5.42}$$

$$T\mathbf{v}_n\cdot\boldsymbol{\nabla}s = \cdots - v_{nj}h_i\nabla_j v_{si} = \cdots - v_{nj}h_i\nabla_i v_{sj}, \tag{8.5.43}$$

where $\cdots$ represents the contributions to these equations identical to those of a normal fluid. The last equation follows because $v_{si} = \nabla_i\theta$ is longitudinal. With these additions to the equations for a normal fluid, we find

$$\begin{aligned}T\left[\frac{\partial s}{\partial t} + \boldsymbol{\nabla}\cdot\left(\mathbf{v}_n s + \frac{\mathbf{Q}}{T}\right)\right] &= \mathbf{Q}\cdot\frac{\boldsymbol{\nabla}T}{T} - (\mathbf{g} - \rho\mathbf{v}_n - \mathbf{h})\cdot\boldsymbol{\nabla}\alpha \\ &\quad -[\pi_{ji} - p\delta_{ij} - v_{ni}g_j - h_i v_{sj}]\nabla_i v_{nj} \\ &\quad +(-X' + \mathbf{v}_s\cdot\mathbf{v}_n)\boldsymbol{\nabla}\cdot\mathbf{h}\end{aligned} \tag{8.5.44}$$

with

$$Q_i = j_i^\varepsilon - \alpha(g_i - \rho v_{ni}) - \varepsilon v_{ni} - (\mathbf{v}_n\cdot\mathbf{g})v_{ni} - v_{nj}\pi_{ji} + (\mathbf{v}_n\cdot\mathbf{v}_s - X')h_i. \tag{8.5.45}$$

In the non-dissipative limit, the right hand side of this equation is zero, implying $\mathbf{g} = \rho\mathbf{v}_n + \mathbf{h}$, in agreement with Eqs. (8.5.33) and (8.5.34), $X' = \mathbf{v}_s\cdot\mathbf{v}_n$, and

$$\begin{aligned}\pi_{ji} &= p\delta_{ij} + v_{ni}g_j + h_i v_{sj} \\ &= p\delta_{ij} + \rho_n v_{ni}v_{nj} + \rho_s v_{si}v_{sj}.\end{aligned} \tag{8.5.46}$$

The non-dissipative part of the energy current is

$$\mathbf{j}_R^\varepsilon = \mathbf{v}_n Ts + \alpha\mathbf{g} + \mathbf{v}_n\mathbf{g}\cdot\mathbf{v}_n - \rho_s(\mathbf{v}_n\cdot\mathbf{v}_s)(\mathbf{v}_n - \mathbf{v}_s). \tag{8.5.47}$$

Dissipative currents can be determined in the usual way. Here $\nabla \cdot \mathbf{h}$ and $\nabla \cdot \mathbf{v}_n$ have the same sign under time reversal, and there is a dissipative coefficient coupling X' to $\nabla \cdot \mathbf{v}_n$ and σ'_{ij} to $\nabla \cdot \mathbf{h}$:

$$\begin{aligned} \sigma'_{ij} &= \eta\left[\nabla_i v_{nj} + \nabla_j v_{ni} - \frac{2}{3}\delta_{ij}\nabla \cdot \mathbf{v}_n\right] \\ &\quad + \delta_{ij}[\zeta_2 \nabla \cdot \mathbf{v}_n + \zeta_1 \nabla \cdot \mathbf{h}], \\ \mathbf{Q} &= -\kappa \nabla T, \qquad (8.5.48) \\ X'' &= -\zeta_3 \nabla \cdot \mathbf{h} - \zeta_1 \nabla \cdot \mathbf{v}_n, \end{aligned}$$

where $X'' = X' - \mathbf{v}_s \cdot \mathbf{v}_n$. The dissipative coefficient ζ_1 appears in both X'' and σ'_{ij} because of Onsager reciprocity theorem (see Eq. (8.4.71) and Problem 8.3). We now have a complete set of thermodynamic and hydrodynamic equations for superfluid helium. These equations are usually expressed in a coordinate system in which $\mathbf{v}_s = 0$. We leave it as an exercise to derive thermodynamic parameters and currents in this coordinate system.

To determine the nature of the Goldstone mode of superfluid helium, we will determine the linearized modes predicted by the above hydrodynamical equations in the non-dissipative limit. In this limit, $\mathbf{Q} = 0$, the energy current $\mathbf{j}^\varepsilon$ can be expressed in terms of $\mathbf{g}$ and $\mathbf{v}_s$ as

$$\mathbf{j}^\varepsilon = \frac{1}{\rho_n}(\alpha\rho_n + Ts)\mathbf{g} - \frac{\rho_s}{\rho_n}Ts\mathbf{v}_s. \qquad (8.5.49)$$

Using this in the time derivative of the energy conservation equation [Eq. (8.4.2)], we obtain

$$\frac{\partial^2 \varepsilon}{\partial t^2} - \frac{\alpha\rho_n + Ts}{\rho_n}\frac{\partial^2 \rho}{\partial t^2} + \frac{\rho_s}{\rho_n}Ts\nabla^2\alpha = 0 \qquad (8.5.50)$$

with the aid of Eq. (8.5.41) with $X' = 0$. As in normal fluids, it is convenient to use the heat variable $\tilde{q}$ rather than the energy density. To linear order in $\mathbf{v}_s$, the relation $d\tilde{q} = d\varepsilon - (\varepsilon + p)(d\rho/\rho)$ [Eq. (8.4.52)] between $\tilde{q}$ and ε also applies to superfluids. This equation can be used to obtain an equation for the second time derivative of $\tilde{q}$, which, with Eq. (8.5.49) for the energy current, is

$$\frac{\partial^2 \tilde{q}}{\partial t^2} - \frac{\rho_s}{\rho_n}T\tilde{s}\frac{\partial^2 \rho}{\partial t^2} + \frac{\rho_s}{\rho_n}T\rho\tilde{s}\nabla^2\alpha = 0, \qquad (8.5.51)$$

where $\tilde{s} = s/m$ is the entropy per unit mass. In the normal fluid in the non-dissipative limit, $\tilde{q} = 0$. Here, there are reactive couplings to ρ and α that vanish when $\rho_s = 0$. The stress tensor to linear order in $\mathbf{v}_s$ and $\mathbf{v}_n$ is $-p\delta_{ij}$ so that the equation for the total density ρ is the same as in a normal fluid:

$$\frac{\partial^2 \rho}{\partial t^2} - \nabla^2 p = 0. \qquad (8.5.52)$$

When α and p are expanded to linear order in $\tilde{q}$ and $\delta\rho$ using the relation $dp = sdT + \rho d\alpha$, Eqs. (8.5.50) and (8.5.51) lead to the following characteristic polynomial for the frequency:

$$\omega^4 - \omega^2 q^2 \left[\frac{\partial p}{\partial \rho}\bigg)_S + \frac{\rho_s}{\rho_n}\tilde{s}^2 \frac{T}{\tilde{c}_V} \right] + q^4 \frac{\rho_s}{\rho_n} \frac{T\tilde{s}^2}{\tilde{c}_V} \frac{\partial p}{\partial \rho}\bigg)_T = 0. \tag{8.5.53}$$

This equation yields two pairs of sound-like modes with dispersion $\omega = \pm c_1 q$ and $\omega = \pm c_2 q$. The first sound velocity c_1 tends to the longitudinal sound velocity of a normal fluid when $\rho_s \to 0$. The second sound velocity tends to zero with ρ_s. At the low temperatures where superfluidity sets in, $\tilde{c}_p \approx \tilde{c}_V$ and $\partial p/\partial \rho)_S \approx \partial p/\partial \rho)_T$, and

$$\begin{aligned} c_1^2 &= \frac{\partial p}{\partial \rho} \\ c_2^2 &= \frac{\rho_s}{\rho_n} \frac{T\tilde{s}^2}{\tilde{c}_V}. \end{aligned} \tag{8.5.54}$$

The second sound mode results from the coupling of the heat current to the superfluid velocity.

8.6 Stochastic models and dynamic critical phenomena

1 *Critical slowing down and the conventional theory*

As we have seen throughout this chapter, conservation laws and broken symmetry determine all those variables whose characteristic frequencies ω are guaranteed to tend to zero with wave number q, i.e., all those variables for which $\omega\tau \ll 1$ whenever $q\lambda \ll 1$. Hydrodynamical equations determine the low-frequency, long-wavelength temporal and spatial behavior of these variables. The great virtue of hydrodynamics is that it provides a correct and closed description of the dynamics of a small set of macroscopic observables such as density or magnetization. It does not require an examination of all of the order 10^{23} microscopic degrees of freedom of a typical condensed matter system, though these degrees of freedom do ultimately determine the values of transport coefficients.

It is natural to extend the hydrodynamical approach to other "slow" variables whose characteristic frequencies are less than the inverse microscopic collision time τ^{-1}. Since hydrodynamics includes all variables for which $\omega \to 0$ as $q \to 0$, the new variables will have a nonzero characteristic frequency $\omega_{s0} \equiv \tau_{s0}^{-1}$ at $q = 0$. In order for the hydrodynamic approach to make sense for these new variables, it must be that $\omega_{s0}\tau = \tau/\tau_{s0} \ll 1$, i.e., there must be a clear separation between microscopic and collective time scales. Such a separation occurs quite naturally near a mechanical instability or a second-order phase transition. The overdamped oscillator discussed in Sec. 7.2 [Eq. (7.2.14)] provides the simplest example of a division into slow and fast time scales. The slow frequency, $\omega_{s0} = (k/\alpha)$, tends to zero as the spring constant k tends to zero, whereas the fast frequency, $\omega_f = \gamma$, is independent of k. Thus, by controlling k, it is always possible to reach $\omega_{s0}/\omega_f \ll 1$. In this limit, the inertial term can be neglected, and the equation of motion for x becomes a simple first-order differential equation. Near a second-order phase transition with a *non-conserved* order parameter ϕ, one can expect a similar separation of time scales with a dissipative mode for ϕ with a width $\tau_s^{-1} \sim \Gamma/\chi$, where χ is the susceptibility for ϕ and Γ is a phenomenological dissipative coefficient. In the conventional theory (often called the Van Hove theory) (Van Hove 1954, Landau and Khalatnikov 1954), it is argued that Γ approaches a constant as $T \to T_c$ and that τ_s diverges as $|T - T_c|^{-\gamma}$ and becomes arbitrarily large compared to any microscopic collision time. This is the phenomenon of *critical slowing down*. The approach to a critical point leads quite generally to critical slowing down even for hydrodynamical modes associated with a conserved order parameter. Thus, for example, in the Van Hove theory, the diffusion constant $D = \Gamma/\chi$ vanishes as $|T - T_c|^{\gamma}$ near the phase separation critical point of a binary mixture. The frequency $\omega = -iDq^2$ decreases as $T \to T_c$ at fixed q, i.e., there is critical slowing down.

In this section, we will develop stochastic equations that generalize hydrodynamics to include non-hydrodynamic variables with mode frequencies that are slow compared to inverse microscopic collision times. In addition, these equations will introduce noise sources, reflecting the presence of high-frequency modes, that will guarantee the approach to thermal equilibrium. Stochastic equations are the generalization to continuous fields (or to fields on a lattice with many degrees of freedom) of the Langevin equations for a single variable discussed in Sec. 7.5. They are also the natural dynamical generalization of the Landau-Ginzburg-Wilson (LGW) phenomenological field theories discussed in Chapter 5. They determine dynamical properties of variables whose static properties are determined by the LGW field theories. We will then discuss generalizations of scaling near a critical point to dynamical, i.e., time- or frequency-dependent, quantities, and show, in particular, how the form of ordered phase hydrodynamical modes determines the temperature dependence of dissipative coefficients Γ near a critical point.

Stochastic equations describing the low-frequency properties of a given system must include hydrodynamics. Thus, the stochastic equations for fields in any LGW field theory will depend on whether or not those fields are conserved or describe broken continuous symmetries. It is, therefore, possible to associate many different dynamics with a given static LGW theory. After discussing dynamic scaling, we will show how reactive couplings leading to propagating modes can be obtained from Poisson bracket relations. The resulting non-linear stochastic equations give rise to probability distributions that decay in time to the equilibrium distribution described by the associated LGW Hamiltonian. We will then consider a number of explicit models with Poisson bracket terms that have been used in the study of critical dynamics. In low-temperature ordered phases, these models reproduce broken-symmetry hydrodynamics discussed in this chapter. They also describe non-hydrodynamic slow modes in disordered phases.

Finally, we will show how nonlinear couplings between slow modes inherent in the above theories lead to wave number dependent renormalizations of transport coefficients.

2 Dissipative dynamics

Purely dissipative stochastic models are often referred to as kinetic Ising models. The simplest stochastic dynamical model is one in which there is a single non-conserved scalar field $\phi(\mathbf{x})$ in contact with a constant temperature heat bath. This model is variously called the Glauber (1963) model, the time-dependent-Ginzburg-Landau (TDGL) model, or model A. It provides a good description of Ising spins on a lattice coupled to phonons with a thermal conductivity very large compared to that of the spins alone. The energy of the spins in such a system is transferred quickly to the lattice and is thus not conserved. The only "slow" variable is $\phi(\mathbf{x}, t)$, whose equation of motion is

model A

$$\begin{aligned} \frac{\partial \phi}{\partial t} &= -\Gamma \frac{\delta \mathscr{H}_T}{\delta \phi} + \zeta(\mathbf{x}, t) \\ &= -\Gamma \frac{\partial \mathscr{H}}{\delta \phi} + \Gamma h(\mathbf{x}, t) + \zeta(\mathbf{x}, t), \end{aligned} \tag{8.6.1}$$

where $\mathscr{H}_T$ is the total Hamiltonian, $h(\mathbf{x}, t)$ is the external field conjugate to ϕ, and $\zeta(\mathbf{x}, t)$ is a noise source. This equation is simply the generalization to continuous fields of the Langevin equation [Eq. (7.5.30)] for a velocity. In the case of a diffusing particle, the noise described the effects of microscopic degrees of freedom of its host fluid. Here, it describes in a phenomenological way the effects of degrees of freedom with wave number q greater than the cutoff Λ in the phenomenological Hamiltonian $\mathscr{H}$ or of other fast degrees of freedom not described by ϕ. As in the case of Langevin theory for a single particle, we assume that $\zeta(\mathbf{x}, t)$ is an independent random variable at each point in space with zero mean and a white noise spectrum:

$$\langle \zeta(\mathbf{x}, t) \zeta(\mathbf{x}', t') \rangle = 2T\Gamma \delta(\mathbf{x} - \mathbf{x}') \delta(t - t'). \tag{8.6.2}$$

Note that Eq. (8.6.1) is non-linear if $\mathscr{H}$ is anharmonic. We will return to this point later.

Response and correlation functions for ϕ are easily calculated when $\mathscr{H}$ is a Gaussian Hamiltonian (see Sec. 5.8). In this case,

$$\frac{\partial \phi}{\partial t} = -\Gamma[r - c\nabla^2]\phi + \Gamma h + \zeta, \tag{8.6.3}$$

or, after Fourier transformation,

$$\phi(\mathbf{q}, \omega) = G(\mathbf{q}, \omega)[\Gamma h(\mathbf{q}, \omega) + \zeta(\mathbf{q}, \omega)], \tag{8.6.4}$$

where

$$G(\mathbf{q}, \omega) = \frac{1}{-i\omega + \Gamma(r + cq^2)}. \tag{8.6.5}$$

The last equation allows us to calculate the dynamic response function,

Table 8.6.1. *Properties of stochastic dynamical models.*

Model	Designation	System	Dimension of order parameter	Non-conserved fields	Conserved fields	Non-vanishing Poisson bracket
	A	Kinetic Ising anisotropic magnets	n	ϕ	none	none
Relaxational	B	Kinetic Ising uniaxial ferromagnet	n	none	ϕ	none
	C	Anisotropic magnets structural transitions	n	ϕ	m	none
Fluid	H	Gas-liquid binary fluid	1	none	$\phi\ \mathbf{j}$	$\{\phi, \mathbf{j}\}$
Symmetric planar magnet	E	Easy-plane magnet, $h_z = 0$	2	ψ	m	$\{\psi, m\}$
Asymmetric planar magnet	F	Easy-plane magnet, $h_z = 0$ superfluid helium	2	ψ	m	$\{\psi, m\}$
Isotropic antiferromagnet	G	Heisenberg antiferromagnet	3	ψ	$\mathbf{m}$	$\{\psi, \mathbf{m}\}$
Isotropic ferromagnet	J	Heisenberg ferromagnet	3	none	$\mathbf{m}$	$\{\mathbf{m}, \mathbf{m}\}$

$$\chi(\mathbf{q},\omega) = \Gamma G(\mathbf{q},\omega) = \left[\frac{-i\omega}{\Gamma} + \chi^{-1}(\mathbf{q})\right]^{-1}, \tag{8.6.6}$$

where $\chi(\mathbf{q}) = (r + cq^2)^{-1}$ is the static wave number dependent susceptibility and the correlation function,

$$\langle\phi(\mathbf{q},\omega)\phi(\mathbf{q}',\omega')\rangle = G(\mathbf{q}\omega)G(\mathbf{q}',\omega')\langle\zeta(\mathbf{q},\omega)\zeta(\mathbf{q}',\omega')\rangle \tag{8.6.7}$$

or

$$C_{\phi\phi}(\mathbf{q},\omega) = \frac{2T}{\Gamma}|\chi(\mathbf{q},\omega)|^2 = 2\frac{T}{\omega}\chi''(\mathbf{q},\omega). \tag{8.6.8}$$

This equation is the classical fluctuation-dissipation theorem. The noise fluctuation in Eq. (8.6.2) was chosen so that this relation would be satisfied. The dynamic response function can be rewritten in a form analogous to that [Eq. (7.2.24)] for an overdamped oscillator

$$\chi(\mathbf{q},\omega) = \frac{\chi(\mathbf{q})}{1 - i\omega/\omega(\mathbf{q})}, \tag{8.6.9}$$

where $\omega(\mathbf{q}) = \Gamma/\chi(\mathbf{q})$. Thus, $\omega(\mathbf{q} = 0) = \Gamma/\chi(0)$ goes to zero linearly with $r \sim T - T_c$.

Model A describes the slow dynamics of a non-conserved variable but not that of a conserved variable. A purely dissipative dynamics for a conserved variable can be obtained by replacing Γ by $-\lambda\nabla^2$:

model B

$$\frac{\partial\phi}{\partial t} = \lambda\nabla^2\frac{\delta\mathscr{H}_T}{\delta\phi} + \zeta. \tag{8.6.10}$$

This defines the Cahn-Hilliard model (Cahn and Hilliard 1958) or model B. The noise correlations must satisfy

$$\langle\zeta(\mathbf{x},t)\zeta(\mathbf{x}',t')\rangle = -2T\lambda\nabla^2\delta(\mathbf{x}-\mathbf{x}')\delta(t-t') \tag{8.6.11}$$

to ensure that the fluctuation-dissipation theorem is satisfied. If $\mathscr{H}$ is the Gaussian Hamiltonian, the response and correlation functions for model B are identical to those of model A with Γ replaced by $\Gamma(\mathbf{q}) = \lambda q^2$:

$$\chi^{-1}(\mathbf{q},\omega) = \frac{-i\omega}{\lambda q^2} + \chi^{-1}(\mathbf{q}). \tag{8.6.12}$$

This is identical to the diffusive response function of Eq. (7.4.18) with $D = \lambda/\chi$. Models A and B are easily generalized to n-component fields by replacing ϕ by ϕ_i.

A non-conserved order parameter may be coupled to a conserved but possibly non-critical density such as the energy density or the magnetization in an anisotropic antiferromagnet. The static LGW Hamiltonian for such a system will be a functional of both ϕ and the new conserved variable, which we will denote by m. In a ϕ^4 model, for example, the Hamiltonian would be

$$\mathscr{H} = \int d^dx\left[\frac{1}{2}\tilde{r}\phi^2 + \tilde{u}\phi^4 + \frac{1}{2}c(\nabla\phi)^2 + wm\phi^2 + \frac{1}{2}C_m^{-1}m^2\right]. \tag{8.6.13}$$

If m is non-critical, i.e., if C_m^{-1} does not approach zero, then it can be removed by integration to produce an effective Hamiltonian for ϕ alone. The dynamical equations ensuring conservation of m are

model C
$$\frac{\partial \phi}{\partial t} = -\Gamma \frac{\delta \mathscr{H}_T}{\delta \phi} + \zeta,$$
$$\frac{\partial m}{\partial t} = \lambda_m \nabla^2 \frac{\delta \mathscr{H}_T}{\delta m} + \zeta_m, \tag{8.6.14}$$

where $\mathscr{H}_T = \mathscr{H} - \int d^d x (h\phi + h_m m)$ and

$$\langle \zeta_m(\mathbf{x}, t)\zeta_m(\mathbf{x}', t')\rangle = -2T\lambda_m \nabla^2 \delta(\mathbf{x} - \mathbf{x}')\delta(t - t'). \tag{8.6.15}$$

Eqs. (8.6.14) and (8.6.15) define model C with a non-conserved critical field and a non-critical conserved density. In the harmonic approximation, these equations yield frequencies $\omega_\phi(\mathbf{q}) = -i\Gamma/\chi(\mathbf{q})$ and $\omega_m(\mathbf{q}) = -i(\lambda/C_m)q^2$.

One could also imagine a critical conserved density ϕ coupled to a non-critical density m. The dynamics of this model (model D) is described by Eq. (8.6.14) with $\Gamma = -\lambda_\phi \nabla^2$.

These simple models can be used to define characteristic frequencies in terms of response and correlation functions whose form is quite generally applicable. The dissipative coefficient Γ_ϕ for a field ϕ with dissipative dynamics is

$$\frac{1}{\Gamma_\phi(\mathbf{q})} = \left.\frac{\partial \chi^{-1}_{\phi\phi}(\mathbf{q}, \omega)}{\partial(-i\omega)}\right)_{\omega=0}, \tag{8.6.16}$$

and a characteristic frequency $\omega_\phi(\mathbf{q})$ can be defined as

$$\omega_\phi(\mathbf{q}) = \Gamma_\phi/\chi_\phi(\mathbf{q}). \tag{8.6.17}$$

When the field ϕ is conserved, then

$$\lambda_\phi = \lim_{q\to 0} q^{-2}\Gamma_\phi(\mathbf{q}). \tag{8.6.18}$$

This definition of a characteristic frequency is adequate when modes are dissipative and spectral weight is concentrated near $\omega = 0$. When there are propagating modes, however, an alternative characteristic frequency $\overline{\omega}_\phi(\mathbf{q})$ defined via

$$\int_{-\overline{\omega}_\phi(\mathbf{q})}^{\overline{\omega}_\phi(\mathbf{q})} \frac{d\omega}{2\pi} S_{\phi\phi}(\mathbf{q}, \omega) = \frac{1}{2} S_{\phi\phi}(\mathbf{q}) \tag{8.6.19}$$

is preferable.

3 Dynamic scaling

In Chapter 5 we learned that, near a second-order critical point, thermodynamic functions obey generalized homogeneity relations leading to scaling laws. In particular, we found that the order-parameter susceptibility satisfies $\chi(q, r) = b^{2-\eta}\chi(bq, b^{1/\nu}r)$, where $r \sim T - T_c$ is the temperature variable. The phenomenon of dynamic slowing down would lead one to expect that dynamic response and correlation functions would obey similar scaling relations. Indeed, the Gaussian response functions for models A and B obey scaling relations that are direct generalizations of the mean-field scaling relations discussed in Chapter 5:

$$\chi(\mathbf{q}, \omega, r) = b^{2-\eta}\chi(b\mathbf{q}, b^z\omega, b^{1/\nu}r), \tag{8.6.20}$$

where $\eta = 0$ and the *dynamic scaling exponent* z is equal to 2 for model A and to 4 for model B. Below the upper critical dimensions, where critical fluctuations become important, one can expect Eq. (8.6.20) to hold, with η the critical point exponent and z, in general, not equal to 2 or 4. Eq. (8.6.20) then implies the scaling relations

$$\chi(\mathbf{q}, \omega, r) = |r|^{-\gamma} X(q\xi, \omega\xi^z) \tag{8.6.21}$$

and

$$\omega(\mathbf{q}) = q^z \Omega(q\xi) \equiv \xi^{-z}\Omega_1(q\xi), \tag{8.6.22}$$

for the dynamic susceptibility and the characteristic frequency. Here, $\xi \sim |r|^{-\nu}$ is the correlation length. If the transport coefficients Γ and λ remain constant near the critical point, then $\chi^{-1}(\mathbf{q}, \omega, r) = [(-i\omega/\Gamma(q)) + \chi^{-1}(\mathbf{q}, r)]$ with $\Gamma(q) = \Gamma$ for model A and λq^2 for model B. In this case, $\chi(\mathbf{q}, \omega, r)$ and $\omega(q)$ obey the scaling laws of Eqs. (8.6.21) and (8.6.22) with $z = 2 - \eta$ for model A and $z = 4 - \eta$ for model B. These are the scaling exponents predicted by the classical Van Hove theory. In general, however, one can expect transport coefficients to become singular at the critical point. Renormalization group and $1/n$ calculations for model A confirm this expectation and yield

$$z = 2 + c\eta \qquad \text{(model A)}. \tag{8.6.23}$$

Near four dimensions, $c = 0.7621(1 - 1.687\epsilon) + O(\epsilon^4)$, where $\epsilon = 4 - d$. The quantity $c\eta$ has also been calculated to order $1/n$ and to first order in $\epsilon' = d - 2$ for $n > 2$. The latter calculation is done with a dynamic generalization of the nonlinear sigma model discussed in Sec. 6.7. Similar calculations for model B yield the Van Hove result

$$z = 4 - \eta \qquad \text{(model B)} \tag{8.6.24}$$

to all orders in ϵ. In model C, the order parameter is coupled to a conserved variable with diffusive dynamics. This coupling leads to

$$z = 2 + \frac{\alpha}{\nu} \qquad \text{(model C, } n = 1\text{)} \tag{8.6.25}$$

for systems with Ising symmetry, where α is the specific heat exponent. When $n > 1$ and $\alpha < 0$, the coupling to the non-critical conserved variable is unimportant, and model C is equivalent to model A.

In the purely dissipative models just discussed, the characteristic frequencies are proportional to ξ^{-z} for small q both above and below T_c. In models with a broken continuous symmetry in the ordered phase, there are Goldstone modes whose long-wavelength frequencies tend to zero with q, with coefficients determined entirely by thermodynamic rigidities and susceptibilities. In the disordered phase, on the other hand, the characteristic frequency of a non-conserved order parameter will tend to a finite constant as $q \to 0$. In this situation, the function $\Omega(q\xi)$ will have different small argument limits for $T > T_c$ and $T < T_c$. In addition, the exponent z is a function of static exponents only. Consider first the planar magnet (whose ordered phase dynamics is essentially the same as that of the rigid rotor, the isotropic antiferromagnet, and superfluid helium). In this model, $\omega(\mathbf{q}) = (\rho_s/\chi_m)^{1/2}q$ (see Secs. 8.2 and 8.3). The rigidity obeys the Josephson scaling relation $\rho_s \sim \xi^{-(d-2)}$. The susceptibility χ_m is associated with the non-critical field $m = m_z$. It is non-divergent (case 1) when the system is invariant under $m \to -m$, and terms in an effective Hamiltonian of the form $m\phi^2$ are prohibited. This symmetry applies to the planar magnet with no external magnetic fields but not to one with an external field, or to the superfluid where m is really the density. In the latter case (case 2), χ_m will diverge as $\xi^{\alpha/\nu}$ if $\alpha > 0$. Thus, in the ordered phase,

$$\omega(q) \sim \begin{cases} \xi^{-(d-2)/2}q, & \text{(case 1);} \\ \xi^{-(d-2)/2}\xi^{-\tilde{\alpha}/2\nu}q, & \text{(case 2),} \end{cases} \tag{8.6.26}$$

where $\tilde{\alpha} = \max(\alpha, 0)$ for case 2. From this and Eq. (8.6.22), we conclude

$$z = \begin{cases} d/2 & \text{(case 1);} \\ d/2 + \tilde{\alpha}/2 & \text{(case 2).} \end{cases} \tag{8.6.27}$$

We can now construct the behavior of $\omega(q)$ in the three regions shown in Fig. 8.6.1: (I) $q\xi \to 0$, $T < T_c$; (II) $q\xi \to \infty$; and (III) $q\xi \to 0$, $T > T_c$:

$$\omega(q) \sim \begin{cases} \xi^{-z+1}q & \text{(I)} \\ q^z & \text{(II)} \\ \xi^{-z} & \text{(III)}\,. \end{cases} \tag{8.6.28}$$

There is a continuous crossover between behaviors (I) and (II) and between behaviors (II) and (III) at $q\xi \sim 1$.

The hydrodynamical analysis at the beginning of the chapter also predicted a damping of the Goldstone modes proportional to q^2, i.e., $\text{Im}\,\omega(q) = Dq^2$. Near the critical point, scaling predicts $D \sim \xi^{-z+2}$.

In the isotropic ferromagnet, the Goldstone mode has frequency $\omega(q) = (\rho_s/M_0)q^2 \sim \xi^{-(d-2)+\beta/\nu}q^2$, implying

$$z = d - (\beta/\nu) = \frac{1}{2}(d + 2 - \eta). \tag{8.6.29}$$

In the disordered phase, $\omega(q) = -iDq^2$ because the magnetization $\mathbf{m}$ is a conserved variable. Scaling implies $D \sim \xi^{-z+2}$.

In high dimensions, where critical fluctuations are unimportant, the Van Hove theory with mean-field static critical exponents (i.e., $\eta = 0$) is valid. At the upper critical dimension d_c^{dyn}, where Van Hove theory breaks down and dynamical critical fluctuations become important, the exponent z predicted by scaling will be equal to the Van Hove result z_H. This permits us to determine d_c^{dyn} for the planar model and isotropic ferromagnet. The order parameter in the planar magnet is not conserved, and $z_H = 2$. Setting z in Eq. (8.6.27) equal to 2 (and using the fact that $\alpha = 0$ at $d = 4$) yields $d_c^{\text{dyn}} = 4$. The upper critical dimensions for static and dynamic properties are identical. In the isotropic ferromagnet, the order parameter is conserved and $z_H = 4$. Setting z in Eq. (8.6.29) equal to 4 with $\eta = 0$, we obtain $d_c^{\text{dyn}} = 6$ for the ferromagnet. Thus, $d_c^{\text{dyn}} > d_c = 4$, and the dynamic fluctuations become critical before the static thermodynamic

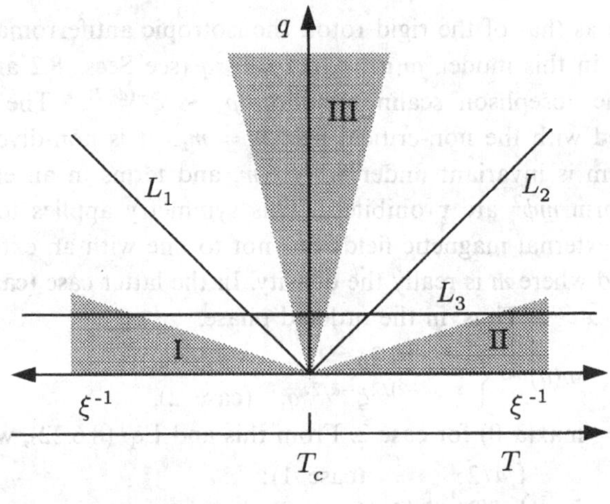

Fig. 8.6.1. The macroscopic domain of wave vector q and coherence length ξ. In the three shaded regions, the correlation functions have different characteristic behaviors. These regions are, respectively, defined by ($q\xi \ll 1$, $T < T_c$), ($q\xi \gg 1$, $T \approx T_c$) , and ($q\xi \ll 1$, $T > T_c$). The asymptotic forms for these regions merge when extrapolated to the lines L_1 or L_2 ($q\xi = 1$ for $T < T_c$ and $T > T_c$, respectively). An experiment done at constant q (line L_3) will pass through all three regions as temperature is varied. [B.I. Halperin and P.C. Hohenberg, *Phys. Rev.* **177**, 952 (1969).]

fluctuations as d is lowered. There is anomalous dynamical critical behavior, even though static properties are still described by mean-field theory. Calculations based on a stochastic model for ferromagnets to be discussed shortly confirm this result.

4 *Poisson bracket terms*

Models A–D are purely dissipative; they do not have reactive couplings between fields of opposite sign under time reversal and, as a consequence, they have no propagating modes. In this subsection, we will discuss how these couplings can be incorporated into stochastic models. The macroscopic hydrodynamic or quasi-hydrodynamic fields $\phi_\mu(\mathbf{x}, t)$ with low-frequency modes are coarse-grained averages of microscopic fields $\hat{\phi}_\mu(\mathbf{x}, t)$, which are functions of the microscopic canonical variables such as particle position and momentum. The time dependence of $\hat{\phi}_\mu(\mathbf{x}, t)$ is controlled by its Poisson bracket with the microscopic Hamiltonian $\mathscr{H}_m$:

$$\frac{\partial \hat{\phi}_\mu}{\partial t} = \{\mathscr{H}_m, \hat{\phi}_m\}. \tag{8.6.30}$$

Since averages of the microscopic field $\hat{\phi}_\mu$ and the coarse-grained field ϕ_μ are identical, we expect that the coarse-grained fields will have contributions to their equations of motion similar to Eq. (8.6.30) but with $\mathcal{H}_m$ replaced by a phenomenological coarse-grained Hamiltonian $\mathcal{H}$, which is a functional of the coarse-grained fields only. This leads to contributions to the equations of motion for ϕ_μ of the form

$$\frac{\partial \phi_\mu(\mathbf{x},t)}{\partial t} = \{\mathcal{H}, \phi_\mu(\mathbf{x},t)\} = -\int d^dx' Q_{\mu\nu}(\mathbf{x},\mathbf{x}') \frac{\delta \mathcal{H}}{\delta \phi_\nu(\mathbf{x})}, \tag{8.6.31}$$

where

$$Q_{\mu\nu}(\mathbf{x},\mathbf{x}') = -Q_{\nu\mu}(\mathbf{x}',\mathbf{x}) = \{\phi_\mu(\mathbf{x}), \phi_\nu(\mathbf{x}')\} \tag{8.6.32}$$

is the Poisson bracket of $\phi_\mu(\mathbf{x})$ and $\phi_\nu(\mathbf{x}')$ evaluated as though they were macroscopic fields. We will consider some explicit examples of $Q_{\mu\nu}(\mathbf{x},\mathbf{x}')$ shortly.

Equations of motion for stochastic variables (like the velocity in Sec. 7.5) must produce probability distributions that obey a Fokker-Planck-Smoluchowski equation and decay to equilibrium distributions at long times. This requirement leads to a slight modification of Eq. (8.6.31). Consider the phenomenological equations for a set of stochastic fields $\phi_\mu(\mathbf{x},t)$:

$$\frac{\partial \phi_\mu(\mathbf{x},t)}{\partial t} = V_\mu(\mathbf{x}) - \Gamma_{\mu\nu} \frac{\delta \mathcal{H}}{\delta \phi_\nu(\mathbf{x})} + \zeta_\mu(\mathbf{x},t), \tag{8.6.33}$$

where $V_\mu(\mathbf{x})$ is a non-dissipative "velocity" that contains the Poisson bracket of Eq. (8.6.31) and may depend on the fields ϕ_μ, $\Gamma_{\mu\nu}$ is a dissipative tensor that may have components proportional to $-\nabla^2$, and $\zeta_\mu(\mathbf{x},t)$ is a noise source with variance

$$\langle \zeta_\mu(\mathbf{x},t)\zeta_\nu(\mathbf{x}',t')\rangle = 2T\Gamma_{\mu\nu}\delta(\mathbf{x}-\mathbf{x}')\delta(t-t'). \tag{8.6.34}$$

A probability distribution for the fields $\{\phi_\mu(\mathbf{x})\}$ can be introduced in analogy with that for the momentum of diffusing particles discussed in Sec. 7.5. Let

$$P(\{\phi_\mu(\mathbf{x})\}, t|\{\phi_\mu^0(\mathbf{x})\}, t_0) = \left\langle \prod_{\mathbf{x},\mu} \delta(\phi_\mu(\mathbf{x}) - \phi_\mu(\mathbf{x},t)) \right\rangle_{\{\phi_\mu^0(\mathbf{x})\},t_0} \tag{8.6.35}$$

be the probability that the fields $\phi_\mu(\mathbf{x},t)$ take on values $\phi_\mu(\mathbf{x})$ at time t, given that they had values $\phi_\mu^0(\mathbf{x})$ at time $t = t_0$. An equation for the time evolution of this probability distribution can be derived using the procedures presented in Sec. 7.5. First, following Eq. (7.5.33), we write

$$\begin{aligned} &P(\{\phi_\mu(\mathbf{x})\}, t+\Delta t|\{\phi_\mu^0(\mathbf{x})\}, t_0) \\ &= \int \mathcal{D}\phi'_\mu(\mathbf{x}) P(\{\phi_\mu(\mathbf{x})\}, t+\Delta t|\{\phi'_\mu(\mathbf{x})\}, t) P(\{\phi'_\mu(\mathbf{x})\}, t|\{\phi_\mu^0(\mathbf{x})\}, t_0). \end{aligned} \tag{8.6.36}$$

As in our derivation of the Fokker-Planck equation in Sec. 7.6, we need to determine the equation for $P(\{\phi_\mu(\mathbf{x})\}, t+\Delta t|\{\phi'_\mu(\mathbf{x})\}, t)$ to first order in Δt. We do this by integrating Eq. (8.6.33) subject to the boundary condition that $\phi_\mu(\mathbf{x},t)$ be equal to $\phi'_\mu(\mathbf{x})$ at time t:

$$\begin{aligned} \phi_\mu(\mathbf{x}, t+\Delta t) &= \phi'_\mu(\mathbf{x}) \\ &+ \left[V'_\mu(\mathbf{x}) - \Gamma_{\mu\nu} \frac{\delta \mathcal{H}}{\delta \phi'_\nu(\mathbf{x})} \right] \Delta t + \int_t^{t+\Delta t} \zeta_\mu(\mathbf{x},t')dt', \end{aligned} \tag{8.6.37}$$

where V'_μ and $\delta\mathscr{H}/\delta\phi'_\nu$ are functionals of $\phi'_\mu(\mathbf{x})$. Using Eq. (8.6.35), we find

$$\begin{aligned} P(\{\phi_\mu(\mathbf{x})\}, t+\Delta t|\{\phi'_\mu(\mathbf{x})\}, t) \\ = \left[1 + \Delta t \int d^d x \left(V'_\mu(\mathbf{x}) - \Gamma_{\mu\nu}\frac{\delta\mathscr{H}}{\delta\phi'_\nu(\mathbf{x})}\right)\frac{\delta}{\delta\phi'_\mu(\mathbf{x})}\right. \\ \left. + \Delta t \Gamma_{\mu\nu}\int d^d x \frac{\delta}{\delta\phi'_\mu(\mathbf{x})}\frac{\delta}{\delta\phi'_\nu(\mathbf{x})}\right]\prod_{\mathbf{x},\mu}\delta(\phi_\mu(\mathbf{x}) - \phi'_\mu(\mathbf{x})) \end{aligned} \quad (8.6.38)$$

and

$$\begin{aligned} \frac{\partial P}{\partial t} &= T\int d^d x \Gamma_{\mu\nu}\frac{\delta}{\delta\phi_\nu(\mathbf{x})}\left(\frac{1}{T}\frac{\delta\mathscr{H}}{\delta\phi_\nu(\mathbf{x})} + \frac{\delta}{\delta\phi_\nu(\mathbf{x})}\right)P \\ &\quad - \int d^d x \frac{\delta}{\delta\phi_\mu(\mathbf{x})}[V_\mu(\mathbf{x})P]. \end{aligned} \quad (8.6.39)$$

The first term in this equation is clearly zero if P takes its equilibrium form $e^{-\mathscr{H}/T}$. The second term is zero if

$$V_\mu(\mathbf{x}) = -\int d^d x' \left(Q_{\mu\nu}(\mathbf{x},\mathbf{x}')\frac{\delta\mathscr{H}}{\delta\phi_\nu(\mathbf{x}')} - T\frac{\delta Q_{\mu\nu}(\mathbf{x},\mathbf{x}')}{\delta\phi_\nu(\mathbf{x}')}\right). \quad (8.6.40)$$

Thus, there is an extra term not predicted by the Poisson bracket relation for the microscopic fields. In most models that are studied, the second term in this expression is either zero or unimportant.

The Poisson brackets contain information about translational, rotational and other invariances of the system. They produce the same reactive terms in hydrodynamical equations as we derived in preceding sections of this chapter using invariance arguments alone. Consider, for example, the Poisson bracket of the mass density and momentum density:

$$\begin{aligned} \{\hat\rho(\mathbf{x}), \hat g_i(\mathbf{x}')\} &= \sum_{\alpha,j}\left(\frac{\partial\hat\rho(\mathbf{x})}{\partial p_j^\alpha}\frac{\partial\hat g_i(\mathbf{x}')}{\partial x_j^\alpha} - \frac{\partial\hat\rho(\mathbf{x})}{\partial x_j^\alpha}\frac{\partial\hat g_i(\mathbf{x}')}{\partial p_j^\alpha}\right) \\ &= w_0\nabla_i(\hat\rho(\mathbf{x})\delta(\mathbf{x}-\mathbf{x}')). \end{aligned} \quad (8.6.41)$$

The coefficient w_0 is unity with our standard choices for $\mathbf{p}^\alpha$ and $\mathbf{x}^\alpha$. It will be useful, however, to regard it as a parameter that can vary under renormalization in models we will shortly discuss. When integrated over $\mathbf{x}'$, this equation becomes $\{\hat\rho(\mathbf{x}), \hat P_i\} = \nabla_i\hat\rho(\mathbf{x})$, a relation that follows from the fact that the total momentum operator $\hat{\mathbf{P}}$ is the generator of infinitesimal translations. Eqs. (8.6.40) and (8.6.41) then imply the following contributions to the equations of motion from ρ and $\mathbf{g}$:

$$\frac{\partial\rho}{\partial t} = -\int d^d x'\{\rho(\mathbf{x}), g_i(\mathbf{x}')\}\frac{\delta\mathscr{H}}{\delta g_i(\mathbf{x}')} = -w_0\nabla_i\left(\rho(\mathbf{x})\frac{\delta\mathscr{H}}{\delta g_i(\mathbf{x})}\right). \quad (8.6.42)$$

The Hamiltonian $\mathscr{H}$ is a functional of the momentum density $\mathbf{g}(\mathbf{x})$ via the kinetic energy term

$$\mathscr{H}_{\text{kin}} = \int d^d x \frac{g^2(\mathbf{x})}{2\rho}, \quad (8.6.43)$$

implying that $\delta\mathscr{H}/\delta g_i(\mathbf{x}) = g_i(\mathbf{x})/\rho$ is the velocity $v_i(\mathbf{x})$ and that Eq. (8.6.42) is identical to the conservation law $\partial\rho/\partial t = -\boldsymbol{\nabla}\cdot(\rho\mathbf{v}) = -\boldsymbol{\nabla}\cdot\mathbf{g}$. The Poisson bracket relation between ρ and g_i also contributes to the equation for g_i:

$$\frac{\partial g_i}{\partial t} = -\int d^d x' \{g_i(\mathbf{x})\ \rho(\mathbf{x}')\} \frac{\delta\mathscr{H}}{\delta\rho(\mathbf{x}')} = -w_0\rho(\mathbf{x})\nabla_i \frac{\delta\mathscr{H}}{\delta\rho(\mathbf{x})}. \tag{8.6.44}$$

$\mathscr{H}$ is a functional of the density $\rho(\mathbf{x})$. For small deviations from equilibrium, $\rho(\mathbf{x}) = \rho_0 + \delta\rho(\mathbf{x})$, and the leading dependence of $\mathscr{H}$ on $\rho(\mathbf{x})$ is (1/2) $\int d^d x \rho_0^{-2}\kappa_s^{-1}(\delta\rho)^2$, where $\kappa_s = -\rho(\partial p/\partial\rho)_s$ is the isentropic compressibility. With this form for $\mathscr{H}$, Eq. (8.6.44) becomes

$$\frac{\partial g_i}{\partial t} = -\rho_0^{-1}\kappa_s\nabla_i\delta\rho = \nabla_i\left[\left(\frac{\partial p}{\partial\rho}\right)_S \delta\rho(\mathbf{x})\right], \tag{8.6.45}$$

in agreement with our analysis of the linearized hydrodynamics equations for a fluid in Sec. 8.4.

The complete nonlinear equations for a fluid can be derived from Poisson brackets. However, direct evaluation of some Poisson brackets, particularly those involving the energy density, are sometimes tedious.

In the simple example just given, the nonlinear character of $\delta\mathscr{H}/\delta\phi_\mu(\mathbf{x})$ was suppressed. In general, however, $\delta\mathscr{H}/\delta\phi_\mu(\mathbf{x})$ is a nonlinear function of the fields $\phi_\mu(\mathbf{x})$, and the Poisson bracket relations provide non-trivial nonlinear couplings among long-wavelength, low-frequency modes. These couplings lead to important renormalization of hydrodynamic reactive and dissipative coefficients, particularly near second-order phase transitions.

5 Models with Poisson brackets

We now turn to an enumeration of some specific stochastic models that have been used in the study of dynamical critical phenomena in systems with reactive couplings. In particular, we will consider model H dynamics of the liquid-gas transition critical point and the phase-separation critical point of binary fluids, models E and F for the planar ferromagnet and superfluid helium, model G for antiferromagnets, and model J for ferromagnets. These models reduce in the hydrodynamic limit, when only conserved and broken-symmetry variables are kept, to the hydrodynamic equations considered in this chapter.

We begin with model H for the liquid-gas transition. We saw in Sec. 8.4 that linearized hydrodynamics of a one-component fluid is best described in terms of the scalars $\rho(\mathbf{x})$ and $\tilde{q}(\mathbf{x}) = [\varepsilon(\mathbf{x}) - (\varepsilon + p)(\delta\rho/\rho)]$. Coupling between the pressure p and the entropy field $\tilde{q}$ occur only through sound waves in the dynamics and not through zero-frequency thermodynamics. The sound-wave frequency with $\omega \sim q$ is always much faster than that of diffusive modes with $\omega \sim q^2$. Near the critical point, the Rayleigh peak, rather than the Brillouin peak, dominates the density correlation function (see Eq. (8.4.66) and Fig. 8.4.2) and the sound mode is irrelevant. A good model for slow dynamics near the liquid-gas critical point is thus one that neglects both pressure and the longitudinal component of the momentum density. These considerations and the Poisson bracket of Eq. (8.6.41) lead to the following model in terms of the entropy density field $\phi = \tilde{q}$ and the transverse momentum density $\mathbf{g}$ (with $\nabla \cdot \mathbf{g} = 0$):

model H

$$\begin{aligned} \frac{\partial \phi}{\partial t} &= \lambda_0 \nabla^2 \frac{\delta \mathscr{H}_T}{\delta \phi} - w_0 \nabla \phi \cdot \frac{\delta \mathscr{H}_T}{\delta \mathbf{g}} + \zeta_\phi, \\ \frac{\partial g_i}{\partial t} &= \mathscr{P}_{ij}^{\perp} \left[\overline{\eta}_0 \nabla^2 \frac{\delta \mathscr{H}_T}{\delta g_j} + w_0 (\nabla_j \phi) \frac{\delta \mathscr{H}_T}{\delta \phi} + \zeta_{g_j} \right], \end{aligned} \tag{8.6.46}$$

where $\mathscr{P}_{ij}^{\perp} = (\delta_{ij} - \nabla_i \nabla_j / \nabla^2)$ is the transverse projection operator and $\mathscr{H}_T = \mathscr{H} + \mathscr{H}_{\text{ext}}$ with

$$\begin{aligned} \mathscr{H} &= \int d^d x \left[\frac{1}{2} r_0 \phi^2 + \frac{1}{2} (\nabla \phi)^2 + u_0 \phi^4 + \frac{1}{2} g^2 \right], \\ \mathscr{H}_{\text{ext}} &= -\int d^d x (h\phi + \mathbf{h}_g \cdot \mathbf{g}), \end{aligned} \tag{8.6.47}$$

where h and $\mathbf{h}_g$ are infinitesimal external applied fields and where we have chosen our units so that $\rho = 1$. If nonlinear terms are neglected, these equations predict mode frequencies $\omega_\phi = -i\lambda r_0 q^2$ and $\omega_g(\mathbf{q}) = -i\overline{\eta}_0 q^2$ that agree with the hydrodynamical results of Eqs. (8.4.60) and (8.4.49) if $r_0 = c_p^{-1}$, $\lambda_0 = \kappa$, and $\overline{\eta}_0 = \eta/\rho$.

Next we consider the planar ferromagnet and superfluid helium. Here the order parameter is the complex field $\psi = m_x + i m_y$. The z-component of the magnetization $m \equiv m_z$ is a conserved non-critical field. The Poisson bracket for ψ and m is

$$\{\psi(\mathbf{x}), m(\mathbf{x}')\} = i g_0 \psi(\mathbf{x}) \delta(\mathbf{x} - \mathbf{x}'). \tag{8.6.48}$$

Normally, g_0 is unity; it can, however, be regarded as a parameter that can change under renormalization group transformations. This Poisson bracket leads to the following dynamical models:

models E & F

$$\begin{aligned} \frac{\partial \psi}{\partial t} &= -2\Gamma_0 \frac{\delta \mathscr{H}_T}{\delta \psi^*} - i g_0 \psi \frac{\delta \mathscr{H}_T}{\delta m} + \zeta_\psi, \\ \frac{\partial m}{\partial t} &= \lambda_0^m \nabla^2 \frac{\delta \mathscr{H}_T}{\delta m} + 2 g_0 \text{Im} \left(\psi^* \frac{\delta \mathscr{H}_T}{\delta \psi^*} \right) + \zeta_m \end{aligned} \tag{8.6.49}$$

where

$$\begin{aligned} \mathscr{H} &= \int d^d x \left[r_0 |\psi|^2 + |\nabla \psi|^2 + \frac{1}{2} u_0 |\psi|^4 + \frac{1}{2} \chi_{m0}^{-1} m^2 + \gamma_0 m |\psi|^2 \right] \\ \mathscr{H}_{\text{ext}} &= -\int d^d x [h_m m + \text{Re}(h \psi^*)]. \end{aligned} \tag{8.6.50}$$

In the planar model, the Hamiltonian must be invariant under $m \to -m$. In this case, $\gamma_0 = 0$, and Γ_0 must be real. This is model E. When there is no symmetry constraint on m, as in the case for the planar model in a field or for superfluid helium, γ_0 is not zero, and Γ_0 may be complex; this is model F. In model E, the susceptibility χ_m is a constant at T_c. In model F, χ_m diverges as $\xi^{\alpha/\nu}$ for $\alpha > 0$. When linearized, Eqs. (8.6.49) predict $\omega_\psi = \pm c_\psi q$ with $c_\psi^2 = g_0^2 \rho_s / \chi_{m0}$, where

$\rho_s = |\langle\psi\rangle|^2$ below T_c and $\psi = \Gamma_0 r_0 = \Gamma_0/\chi_\psi$, and $\omega_m = (\lambda_0/\chi_{m0})q^2$ for $T > T_c$. Thus, they reproduce the modes predicted by hydrodynamics below T_c, and they yield an overdamped order parameter mode and a diffusive magnetization mode above T_c.

Finally, we turn to three-dimensional ferromagnets and antiferromagnets. In the ferromagnet, there is a single vector order parameter m_i with Poisson bracket relations

$$\{m_i, m_j\} = g_0 \epsilon_{ijk} m_k. \tag{8.6.51}$$

In the antiferromagnet, there is the staggered magnetization order parameter field $\mathbf{N} = \boldsymbol{\psi}$ in addition to the conserved magnetization $\mathbf{m}$ that satisfies

$$\{\psi_i, m_j\} = g_0 \epsilon_{ijk} \psi_k. \tag{8.6.52}$$

The dynamical equations for the antiferromagnet are then

model G

$$\begin{aligned} \frac{\partial \boldsymbol{\psi}}{\partial t} &= -\Gamma_0 \frac{\delta \mathscr{H}_T}{\delta \boldsymbol{\psi}} + g_0 \boldsymbol{\psi} \times \frac{\delta \mathscr{H}_T}{\delta \mathbf{m}} + \zeta_\psi \\ \frac{\partial \mathbf{m}}{\partial t} &= \lambda_0 \nabla^2 \frac{\delta \mathscr{H}}{\delta \mathbf{m}} + g_0 \boldsymbol{\psi} \times \frac{\delta \mathscr{H}_T}{\delta \boldsymbol{\psi}} + g_0 \mathbf{m} \times \frac{\delta \mathscr{H}_T}{\delta \mathbf{m}} + \boldsymbol{\zeta}_{\mathbf{m}}, \end{aligned} \tag{8.6.53}$$

where

$$\mathscr{H} = \int d^d x \left[\frac{1}{2} r_0 \psi^2 + \frac{1}{2} |\nabla \boldsymbol{\psi}|^2 + u_0 \psi^4 + \frac{1}{2} \chi_m^{-1} m^2 \right]. \tag{8.6.54}$$

Finally, the dynamical equations for an isotropic ferromagnet are

model J

$$\frac{\partial \mathbf{m}}{\partial t} = \lambda_0 \nabla^2 \frac{\delta \mathscr{H}_T}{\delta \mathbf{m}} + g_0 \mathbf{m} \times \frac{\delta \mathscr{H}_T}{\delta \mathbf{m}} + \boldsymbol{\zeta}_{\mathbf{m}}, \tag{8.6.55}$$

where $\mathscr{H}$ is the ϕ^4 Hamiltonian.

6 Mode-mode coupling

The stochastic equations we have just considered have nonlinear couplings among dynamical variables and, therefore, interactions among harmonic modes that lead to corrections to the harmonic correlation functions and transport coefficients. These mode-mode coupling corrections can be calculated using very elegant diagrammatic perturbation theories or dynamical field theories that are beyond the scope of this book. Here we will be content to give a non-rigorous calculation of the form that these corrections take for model H. The equations of motion for model H with only the lowest order nonlinear couplings are

$$\begin{aligned} \frac{\partial \phi}{\partial t} &= \lambda_0 \nabla^2 (r - \nabla^2)\phi - w_0 \nabla\phi \cdot \mathbf{g} + \zeta_\phi, \\ \frac{\partial g_i}{\partial t} &= \mathscr{P}_{ij}^{\perp} [\bar{\eta}_0 \nabla^2 g_j + w_0 (\nabla_j \phi)(r - \nabla^2)\phi + \zeta_{g_j}]. \end{aligned} \tag{8.6.56}$$

If $w_0 = 0$, the Green functions for ϕ and $\mathbf{g}$ are

$$\begin{aligned} G^{0\ -1}_{\phi\phi}(\mathbf{q}, \omega) &= -i\omega + \lambda_0 q^2 (r + q^2), \\ G^{0\ -1}_{gg}(\mathbf{q}, \omega) &= -i\omega + \bar{\eta}_0 q^2. \end{aligned} \tag{8.6.57}$$

When interactions are included, these functions can be written as

$$G_{\phi\phi}^{-1}(\mathbf{q},\omega) = G_{\phi\phi}^{0\;-1}(\mathbf{q},\omega) + \Sigma_{\phi\phi}(\mathbf{q},\omega),$$
$$G_{gg}^{-1}(\mathbf{q},\omega) = G_{gg}^{0\;-1}(\mathbf{q},\omega) + \Sigma_{gg}(\mathbf{q},\omega). \tag{8.6.58}$$

The non-interacting correlation functions for ϕ and g_i are, as usual, $C_{\phi\phi}^0(\mathbf{q},\omega) = 2T\lambda_0 q^2|G_{\phi\phi}^0(\mathbf{q},\omega)|^2$ and $C_{g_ig_j}^0(\mathbf{q},\omega) = \mathscr{P}_{ij}^\perp(\mathbf{q})2T\overline{\eta}_0 q^2|G_{gg}^0(\mathbf{q},\omega)|^2$. The interacting correlation functions are given by the fluctuation-dissipation theorem.

Using the Green functions of Eq. (8.6.57), one can solve for ϕ and $\mathbf{g}$ perturbatively: $g_i = G_{g0}\zeta_{g_i} + \delta g_i$ and $\phi = G_{\phi 0}\zeta_\phi + \delta\phi$, where

$$\delta\phi(\mathbf{q}\omega) = G_{\phi\phi}^0(\mathbf{q},\omega)\left[-w_0\int\frac{d^dk}{(2\pi)^d}\int\frac{d\omega'}{2\pi}i(\mathbf{q}-\mathbf{k})\cdot\mathbf{g}(\mathbf{k},\omega')\phi(\mathbf{q}-\mathbf{k},\omega-\omega')\right]$$
$$\delta g_i(\mathbf{q}\omega) = w_0 G_{gg}^0(\mathbf{q},\omega)\mathscr{P}_{ij}^\perp\int\frac{d^dk}{(2\pi)^d}\int\frac{d\omega'}{2\pi}i(q-k)_j(r+k^2)$$
$$\times\phi(\mathbf{q}-\mathbf{k},\omega-\omega')\phi(\mathbf{k},\omega'). \tag{8.6.59}$$

These results can be used to develop a controlled perturbation theory for G_ϕ and G_g. To obtain expressions for Σ_ϕ and Σ_g, one normally sums an infinite series. Rather than doing this explicitly, we will proceed with a decoupling procedure similar to the one we used in our discussion of self-consistent field theory in Sec. 5.3. If we expand Eq. (8.6.56) for ϕ to linear order in $\delta\phi$ and δg, the term in w_0 becomes

$$-w_0^2\int q_i\mathscr{P}_{ij}^\perp(\mathbf{k})k_j' G_{gg}^0(\mathbf{k}',\omega'')\phi(\mathbf{k}-\mathbf{k}',\omega'-\omega'')$$
$$\times\phi(\mathbf{k}',\omega'')\phi(\mathbf{q}-\mathbf{k},\omega-\omega'). \tag{8.6.60}$$
$$-w_0^2\int G_{\phi\phi}^0(\mathbf{k},\omega')k_i(k_j-k_j')g_i(\mathbf{q}-\mathbf{k},\omega-\omega')$$
$$\times g_j(\mathbf{k}',\omega'')\phi(\mathbf{k}-\mathbf{k}',\omega'-\omega''),$$

where the integral is over $\mathbf{k}$, $\mathbf{k}'$, ω' and ω''. Now decouple the product of three ϕ's into an average over two ϕ's and a free ϕ and the product over two g's and a ϕ into an average over two g's and a free ϕ. Using the Wiener-Khintchine relation [Eq. (7.1.7)] between the averages of fields and their correlation function, we then obtain

$$\Sigma_{\phi\phi} = (r+q^2)w_0^2\int\frac{d^dk}{(2\pi)^d}\int\frac{d\omega'}{2\pi}q_i\mathscr{P}_{ij}^\perp(\mathbf{k}_-)q_j G_{gg}^0(\mathbf{k}_-,\omega')$$
$$\times C_{\phi\phi}^0(\mathbf{k}_+,\omega'-\omega), \tag{8.6.61}$$
$$= w_0^2(r+q^2)\int\frac{d^dk}{(2\pi)^d}\frac{Tq_i\mathscr{P}_{ij}^\perp(\mathbf{k})q_j}{(r+k_+^2)[-i\omega+\overline{\eta}_0k_-^2+\lambda_0(r+k_-^2)]}$$

where $\mathbf{k}_\pm = \mathbf{k}\pm(\mathbf{q}/2)$. A similar analysis can be carried out to obtain the form of Σ_{gg}.

To lowest order in perturbation theory, the Green function and correlation function in this expression are those of the harmonic theory. One can, however, replace them by their actual values to obtain self-consistent equations for $G_{\phi\phi}$ and

G_{gg}. In general, these functions have all the complexity of dynamical correlation functions near a critical point. For the purposes of solving a self-consistent equation, however, one can assume a Lorentzian form for the dynamics but allow for nontrivial $\mathbf{q}$-dependent renormalization of the static susceptibilities and transport coefficients. If we set $G_{\phi\phi}^{-1} = -i\omega + \lambda(\mathbf{q}) q^2 \chi_{\phi\phi}^{-1}(\mathbf{q})$ and $G_{gg}^{-1} = -i\omega + \eta(\mathbf{q}) q^2$, then the self-consistent equations for $\lambda(\mathbf{q})$ and $\eta(\mathbf{q})$ are

$$\lambda(\mathbf{q}) = \lambda_0 + w_0^2 \int \frac{d^d k}{(2\pi)^d} \chi_{\phi\phi}(\mathbf{k}_+) \frac{\hat{q}_i \mathscr{P}_{ij}^{\perp}(\mathbf{k}_-) \hat{q}_j}{\overline{\eta}(\mathbf{k}_-) k_-^2 + \lambda(\mathbf{k}_+) \chi_{\phi\phi}^{-1}(\mathbf{k}_+) k_+^2} \tag{8.6.62}$$

$$\overline{\eta}(\mathbf{k}) = \overline{\eta}_0 + \frac{w_0^2 q^{-2}}{d-1} \int \frac{d^d k}{(2\pi)^d} \frac{\chi_{\phi\phi}^{-1}(\mathbf{k}_-)[\chi_{\phi\phi}^{-1}(\mathbf{k}_+) - \chi_{\phi\phi}^{-1}(\mathbf{k}_-)] k_i \mathscr{P}_{ij}^{\perp}(\mathbf{q}) k_j}{\lambda(\mathbf{k}_+) \chi_{\phi\phi}^{-1}(\mathbf{k}_+) k_+^2 + \lambda(\mathbf{k}_-) \chi_{\phi\phi}^{-1}(\mathbf{k}_-) k_-^2}. \tag{8.6.63}$$

Near the critical point, χ_ϕ is strongly divergent, and the $\lambda(\mathbf{k}_+)$ term can be neglected in Eq. (8.6.62). If the static scaling form for χ_ϕ [Eq. (5.4.9)] is used and λ_0 and $\overline{\eta}_0$ are neglected, then Eqs. (8.6.62) and (8.6.63) admit a scaling solution with

$$\begin{aligned} \lambda(\mathbf{q}) &= q^{-x_\lambda} L_\lambda(q\xi) \\ \overline{\eta}(\mathbf{q}) &= q^{-x_{\overline{\eta}}} L_{\overline{\eta}}(q\xi), \end{aligned} \tag{8.6.64}$$

where

$$x_\lambda + x_{\overline{\eta}} = 4 - d + \eta. \tag{8.6.65}$$

In the limit $q \to 0$, $\lambda \sim \xi^{x_\lambda}$ and $\overline{\eta} \sim \xi^{x_{\overline{\eta}}}$, with

$$\lambda \overline{\eta} = w_0^2 \chi_\phi(\mathbf{q} = 0) \xi^{2-d} R, \tag{8.6.66}$$

where R is a universal numerical constant. When expressed in terms of the thermal diffusivity $D = \lambda/\chi_\phi$ in physical units, this relation becomes $D = R k_B T / \overline{\eta} \xi^{d-2}$ (Kawasaki 1970). A numerical calculation is required to obtain the exponents x_λ and $x_{\overline{\eta}}$.

8.7 Nucleation and spinodal decomposition

The focus throughout this book has been on static and dynamic properties of condensed systems at or near thermodynamic equilibrium. If external conditions or fields, such as temperature or pressure, are suddenly changed, then a system initially in equilibrium will not be in equilibrium under the new conditions. The approach to a new equilibrium state involves very complex nonequilibrium processes. In this section, we will investigate some simple versions of these nonequilibrium processes.

1 Nucleation with a nonconserved order parameter

Consider first an Ising ferromagnet with a nonconserved order parameter obeying model A dynamics. Its phase diagram in the external field-temperature ($h - T$) plane is reproduced in Fig. 8.7.1. There is a first-order phase boundary at $h = 0$, $T < T_c$, along which the spin-up and spin-down phases coexist. Along this boundary, there are two equivalent minima in the local free energy density† $f(\phi)$, At points in the phase diagram (Fig. 8.7.1) on either side of the phase boundary, between the dashed curves terminating at the critical point, there are two inequivalent minima in f, the lower energy of which corresponds to the equilibrium state and the higher energy of which corresponds to a metastable state (see Sec. 4.5). Beyond the dashed curves, there is only a single well in the free energy. Thus, the dashed curves are the limits of metastability of the Ising magnet. The metastable region between these two curves is exactly analogous to the metastable region, between temperatures T^* and T^{**}, which we encountered in our study of the isotropic-to-nematic transition in Sec. 4.5.

Consider now the effect of a sudden reversal in the sign of the magnetic field, depicted schematically in Fig. 8.7.1, on a system initially in equilibrium with negative magnetic field and order parameter ϕ_0. If the field reversal is sufficiently rapid, then the order parameter immediately after reversal remains equal to ϕ_0. The free energy curve, on the other hand, has changed. It has local minima at $\phi_A < \phi_B$. ϕ_0 is close to, but in general not equal to, the metastable order parameter ϕ_A. However, the free energy has a finite slope at ϕ_0 ($\partial f/\partial\phi|_{\phi_0} \neq 0$), and in a short time the order parameter will relax to ϕ_A according to the rules of model A dynamics. At long times after the field reversal, the system should eventually reach the equilibrium state with order parameter $\phi = \phi_B$ corresponding to the absolute minimum of the free energy in the new field. In the absence of fluctuations or external disturbances, the system would remain indefinitely in the metastable phase. Thermal fluctuations, however, will create droplets of the lower energy equilibrium phase in the "sea" of the higher energy metastable phase. As we will show below, the energy of these droplets has a maximum at a critical radius R_c. For radius $R < R_c$, the system can lower its energy by decreasing the size of a droplet: but for $R > R_c$, it can lower its energy by increasing the size of the droplet. Thus, a critical droplet or *nucleus* with $R = R_c^+$ will spontaneously grow, creating larger and larger regions of the favored equilibrium phase. The formation of the critical droplet is called *nucleation.*

There are two competing contributions to the energy of a droplet, which we will assume to be a sphere of radius R. One is the gain in bulk energy arising from the creation of a region with the lower free energy of the equilibrium phase. The other is the energy of the spherical interface separating the metastable phase

† Recall that $f(\phi)$ is not the actual free energy density. It is the energy density entering the coarse-grained Hamiltonian $\mathscr{H}$ (Sec. 5.2). The actual free energy is the log of the partition function associated with $\mathscr{H}$.

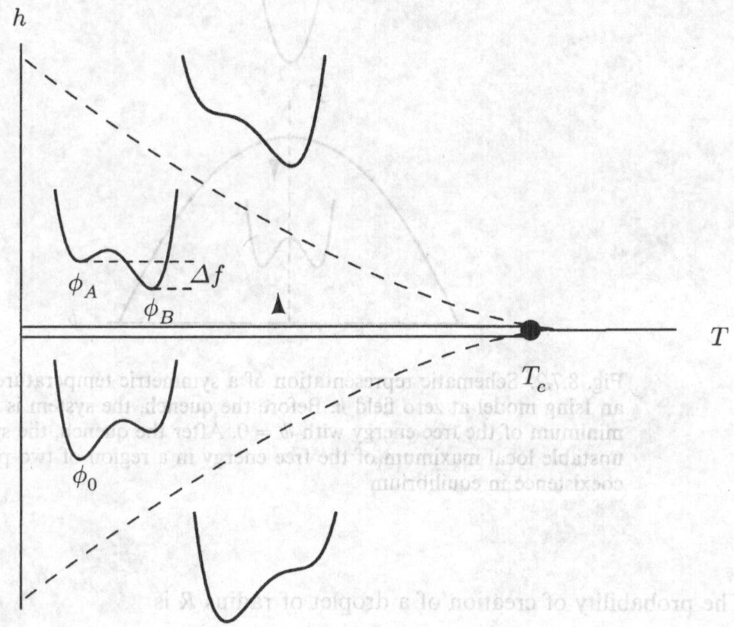

Fig. 8.7.1. Schematic representation of a field reversal across a first-order boundary (double lines in the figure) of an Ising ferromagnet. The dashed curves are the limits of metastability. Between these two curves, the free energy has a double-well structure. Outside these curves, there is a single minimum in the free energy. The inset curves are the function $f(\phi)$. The direction of the quench is indicated by the arrow. Before the reversal, the system is in equilibrium with an order parameter ϕ_0. Shortly after the field switch, the system relaxes to a metastable state with order parameter ϕ_A. The energy difference between the metastable state with order parameter ϕ_A and the equilibrium state with order parameter ϕ_B is Δf.

at infinity from the new equilibrium phase at the center of the droplet. Let $\Delta f = -(f_B - f_A) > 0$ be the magnitude of the difference in free energy density between the new equilibrium phase with order parameter ϕ_B and the metastable phase with order parameter ϕ_A, and let σ be the energy of a wall separating the two phases (see Secs. 10.2 and 10.3 for a discussion of walls). Then the energy of a droplet is

$$\mathscr{H}_{\text{drop}}(R) = -\frac{4\pi}{3}\Delta f R^3 + 4\pi\sigma R^2. \tag{8.7.1}$$

$\mathscr{H}_{\text{drop}}(R)$ attains its maximum value,

$$\mathscr{H}^c_{\text{drop}} = \frac{4\pi\sigma}{3}R_c^2 = \frac{16\pi\sigma^3}{3(\Delta f)^2}, \tag{8.7.2}$$

at the critical radius

$$R_c = \frac{2\sigma}{\Delta f}. \tag{8.7.3}$$

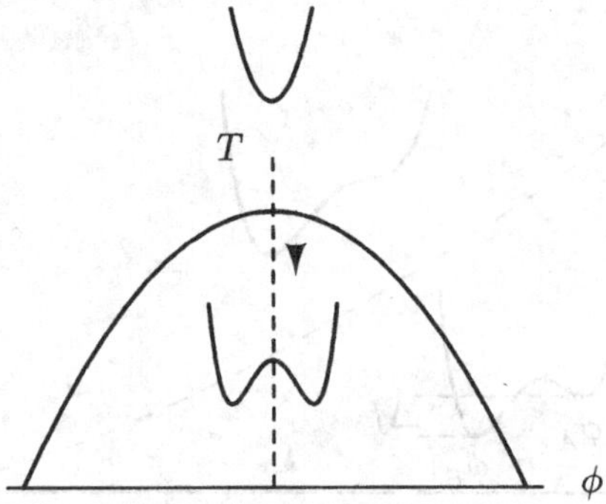

Fig. 8.7.2. Schematic representation of a symmetric temperature quench in an Ising model at zero field h. Before the quench, the system is in a global minimum of the free energy with $\phi = 0$. After the quench, the system is at an unstable local maximum of the free energy in a region of two-phase coexistence in equilibrium.

The probability of creation of a droplet of radius R is

$$P \sim \exp(-\mathscr{H}_{\text{drop}}(R)/T). \tag{8.7.4}$$

Thus, the critical nucleus, which maximizes $\mathscr{H}_{\text{drop}}$, is the least probable one. Once it forms, it will grow in the manner discussed above. The surface of a droplet is a domain wall (see Secs. 10.2 and 10.3) in which the order parameter differs by a finite amount from its value at spatial infinity. A droplet, therefore, is a localized fluctuation with a large deviation in order parameter.

If the field reversal takes the system outside the metastable region, then the order parameter will decay rapidly and homogeneously to the single minimum in the free energy. There will be no nucleation.

Nucleation occurs whenever a sudden change in external parameters (such as h or T) takes a system from one side of a first-order phase boundary into a metastable region on the other. If the order parameter is nonconserved and has discrete symmetry, then the energy and radius of the critical nucleus are identical in form to those of the Ising model just discussed. If the order parameter is nonconserved but has a continuous symmetry (as is the case, for example, in the isotropic-to-nematic transition), then the critical nucleus may trap a topological defect (see Chap. 9). Its energy will have an elastic contribution as well as surface and bulk contributions, and will, therefore, differ in detail from the Ising energy. If the order parameter is conserved, as it is in the liquid-gas transition or in a binary fluid, then a change in external parameters must conserve the order parameter, and it is not possible to access arbitrary points in a phase diagram. As we shall see below, however, it is still possible to reach metastable regions. Nucleation then proceeds in much the same way as described above.

2 *Symmetric unstable quench with model A dynamics*

A change in external parameters that is so rapid that the state of the system (in particular, its order parameter) immediately after the change is identical to its state immediately before is called a *quench*. Such a sudden change in temperature is a *temperature quench*, etc. One route to a nonequilibrium state is via a temperature quench from a one-phase region to a globally unstable point in the coexistence region. Consider for simplicity a temperature quench in an Ising model with external field $h = 0$ from $T > T_c$ to $T < T_c$ (Fig. 8.7.2). For $T > T_c$, the equilibrium value of ϕ is zero. For $T < T_c$, there are coexisting phases with $\phi = \pm\phi_A$. Immediately after the quench, $\phi = 0$. As time progresses, approximately equal size regions with $\phi = \pm\phi_A$ will form. These regions will be separated by domain walls. The characteristic size of domains with either sign of ϕ will grow with time t after the quench in a process called *coarsening*. Information about the time development of domains and correlations among them is contained in the time-dependent structure factor,

$$\tilde{S}(\mathbf{x} - \mathbf{x}', t) = \langle \phi(\mathbf{x}, t)\phi(\mathbf{x}', t)\rangle. \tag{8.7.5}$$

Note that $\tilde{S}(\mathbf{x}, t)$ is the average of the product of the order parameter at two points in space at the *same* time. In equilibrium situations, this function would be independent of time. Simulations show that this structure function obeys a scaling law of the form

$$\tilde{S}(\mathbf{x}, t) = X(|\mathbf{x}|/L(t)), \tag{8.7.6}$$

where the characteristic length $L(t)$ grows as a power law with time:

$$L(t) \sim t^{\omega}. \tag{8.7.7}$$

The Fourier transform of $\tilde{S}(\mathbf{x}, t)$ then obeys the scaling law

$$\tilde{S}(\mathbf{q}, t) = L^d(t) Y(qL(t)). \tag{8.7.8}$$

For the systems considered here, with a nonconserved order parameter obeying model A dynamics,

$$\omega = 1/2. \tag{8.7.9}$$

This can be inferred as follows. In model A dynamics [Eq. (8.6.1)], the time dependence of the order parameter is controlled by the equation

$$\frac{\partial \phi}{\partial t} = -\Gamma \frac{\delta \mathcal{H}}{\delta \phi} = -\Gamma[-c\nabla^2\phi + f'(\phi)]. \tag{8.7.10}$$

Near a domain wall described by a position function $\mathbf{R}(\mathbf{u}, t)$, the variation of ϕ is predominantly along the direction $\mathbf{N}$ normal to the wall, and $\boldsymbol{\nabla}\phi = \mathbf{N}\partial\phi/\partial N$. Thus, $\nabla^2\phi = \partial^2\phi/\partial N^2 + (\boldsymbol{\nabla}\cdot\mathbf{N})\partial\phi/\partial N$, and $\partial\phi/\partial t = v\partial\phi/\partial N$, where $v = \mathbf{N}\cdot\partial R/\partial t$ is the velocity of the wall along its normal direction. We therefore have

$$v\frac{\partial \phi}{\partial N} = \Gamma c \boldsymbol{\nabla}\cdot\mathbf{N}\frac{\partial \phi}{\partial N} - \Gamma\left[-c\frac{\partial^2\phi}{\partial N^2} + f'(\phi)\right]. \tag{8.7.11}$$

The term in square brackets on the right hand side of this equation is zero for an equilibrium wall [Eq. (10.2.5)]. In addition, $\nabla \cdot \mathbf{N}$ is just twice the mean curvature $2H_c \equiv 2/R$ of the wall [Eqs. (10.4.15) and (10.4.13)]. This gives us the Allen-Cahn (Allen and Cahn 1979) equation,

$$v \sim \frac{\partial R}{\partial t} = \frac{2\Gamma c}{R}. \tag{8.7.12}$$

This implies that the characteristic length scale $R(t)$ grows as $t^{1/2}$ after the quench. $R(t)$ is presumed to be the only length scale in the problem. Thus, it is a measure not only of the mean separation between domains but also of the average curvature of domain walls, and can be identified with $L(t)$ in Eq. (8.7.7). The power law dependence of $L(t)$ on time implies that $L(t)$ is finite for any finite time, and, therefore, that the infinite system does not reach true equilibrium in any finite time. It coarsens continuously.

3 Conserved order parameters and spinodal decomposition

Conserved order parameters and spinodal decomposition

Quenches in systems, such as one-component or two-component fluids, with a conserved order parameter are more complex than in those with a nonconserved order parameter because the integral, $\int d^3x\phi$, of the order parameter over the volume of the sample must remain constant in time. This constraint leads to interesting and nontrivial behavior in quenches into coexistence regions. Consider, for example, the temperature quench at constant order parameter ϕ_0 from a temperature $T = T_0 > T_c$ into the coexistence region with $T < T_c$, as depicted in Fig. 8.7.3b. At long times after the quench, the system will separate into regions in which the order parameter takes on one of its two coexisting values, ϕ_A or ϕ_B. The volume fractions x_A and $x_B = 1 - x_A$ are determined by the lever rule: $x_A = (\phi_B - \phi_0)/(\phi_B - \phi_A)$ and $x_B = (\phi_0 - \phi_A)/(\phi_B - \phi_A)$, where $\phi_B > \phi_0 > \phi_A$. In an external gravitational field, which favors the dense B phase at the bottom of the container, the eventual final state will be one with a single B region at the bottom of the sample separated by a single domain wall from the A phase at the top of the sample.

Fig. 8.7.3b depicts schematically a constant order parameter quench from $T > T_c$ into the coexistence region with $T < T_c$. There are two important curves in this figure: the coexistence curve, $T_{co}(\phi)$, and the *spinodal* curve, $T_{sp}(\phi)$, which lies below the coexistence curve.† For quench temperature T_1 lying between the spinodal and coexistence curves, the system is in metastable equilibrium, and decay towards the equilibrium state occurs initially via droplet nucleation. For quench temperatures T_2 below the spinodal curve, the system is globally unstable, and the system immediately develops a spatially modulated

† The spinodal curve is, strictly speaking, a mean-field concept. In real systems, where fluctuations are important, the boundary separating nucleation from spinodal decomposition is not perfectly sharp.

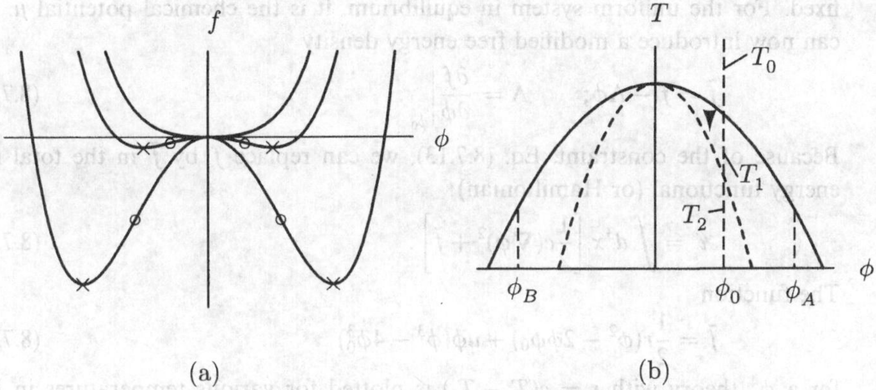

Fig. 8.7.3. (a) Free energy density as a function of the order parameter near the critical point for temperatures above and below the critical temperature. ×'s mark the equilibrium values of ϕ and o's indicate the inflection points. (b) The phase diagram predicted from free energies in (a). The solid curve is the curve $T_{co}(\phi)$ giving the equilibrium value for the order parameter as a function of temperature (from ×'s in (a)). The dashed curve is the spinodal line $T_{sp}(\phi)$ (from o's in (a)). Between the solid and dashed lines is a metastable region where finite energy and size fluctuations are needed to nucleate the phase separation. Inside the dashed spinodal line, the system is unstable and spontaneously decomposes into the two equilibrium values of ϕ at this temperature. The arrow indicates the direction of the temperature quench described in the text. The temperatures T_0, T_1, and T_2 indicate, respectively, equilibrium temperature with order parameter ϕ_0 in the one-phase region, a metastable quench at ϕ_0, and a spinodal quench at ϕ_0. ϕ_A and ϕ_B are the equilibrium values of the order parameter at coexistence at temperature T_2.

order parameter whose amplitude grows continuously from zero. Thus, in the spinodal region, the approach to equilibrium is initially via small amplitude changes in the order parameter throughout the sample rather than via the large amplitude but localized fluctuations encountered in nucleation.

We begin with a discussion of nucleation in the metastable region. To do this, we need to find states that extremize the free energy subject to the constraint

$$\int d^3x(\phi - \phi_0) = 0. \tag{8.7.13}$$

This constraint can be enforced via the introduction of a Lagrange multiplier Λ in the Euler-Lagrange equation for ϕ:

$$c\nabla^2\phi = \frac{\partial f}{\partial \phi} + \Lambda. \tag{8.7.14}$$

If we assume that the system is sufficiently large that the change in composition associated with droplet formation is small, then Λ can be evaluated by requiring the right hand side of Eq. (8.7.14) to be zero when $\phi = \phi_0$, i.e., $\Lambda = \partial f/\partial\phi|_{\phi_0}$. The coefficient Λ is a Lagrange multiplier chosen to keep the average order parameter fixed. For the uniform system in equilibrium, it is the chemical potential μ. We can now introduce a modified free energy density

$$\tilde{f} = f - \Lambda\phi; \qquad \Lambda = \left.\frac{\partial f}{\partial \phi}\right|_{\phi_0} . \tag{8.7.15}$$

Because of the constraint, Eq. (8.7.13), we can replace f by $\tilde{f}$ in the total free energy functional (or Hamiltonian):

$$\mathcal{H} = \int d^3x \left[\frac{1}{2}c(\nabla\phi)^2 + \tilde{f}\right] . \tag{8.7.16}$$

The function

$$\tilde{f} = \frac{1}{2}r(\phi^2 - 2\phi\phi_0) + u\phi(\phi^3 - 4\phi_0^3) \tag{8.7.17}$$

for a ϕ^4 theory with $r = a(T - T_c)$ is plotted for various temperatures in Fig. 8.7.4. This function always has an extremum at $\phi = \phi_0$. At a temperature T_0 above the coexistence curve, $\tilde{f}$ has an absolute minimum at $\phi = \phi_0$, and the system is in homogeneous equilibrium. When the temperature is lowered to T_1, the minimum at $\phi = \phi_0$ is not a global minimum, and the system is in metastable rather than stable equilibrium. When the temperature is lowered further to T_2, there is a local maximum in $\tilde{f}$ at $\phi = \phi_0$, and there is no restoring force to keep $\phi = \phi_0$. The system is globally unstable with respect to the formation of a spatially varying state. The system passes from being metastable to unstable at $\phi = \phi_0$ when the curvature $\partial^2\tilde{f}/\partial\phi^2|_{\phi_0}$ passes through zero. Thus, the spinodal line in Fig. 8.7.3 is determined by

$$\left.\frac{\partial^2\tilde{f}}{\partial\phi^2}\right|_{\phi_0} = \left.\frac{\partial^2 f}{\partial\phi^2}\right|_{\phi_0} = 0. \tag{8.7.18}$$

In the ϕ^4 theory, this condition yields a spinodal temperature $T_{sp} = T_c - (12u\phi/a)$. The coexistence temperature, on the other hand, is $T_{co}(\phi) = T_c - (4u\phi^2/a)$.

In the metastable region with $T_{sp} < T < T_{co}$, the initial approach to equilibrium takes place via the nucleation of droplets, as discussed at the beginning of this section. Thus, there is a surface tension σ determined by $\tilde{f}$ and a critical droplet radius $R_c = 2\sigma/\Delta\tilde{f}$. Note that $\tilde{f}$ is constructed so that one of its minima in the metastable region is always at ϕ_0. At neither minima of $\tilde{f}$ does the order parameter take on either of its values ϕ_A or ϕ_B on the coexistence curve. The constraint of Eq. (8.7.13) prevents either of these from being a local minimum of $\tilde{f}$. Thus, the droplets that are nucleated immediately after the quench will not have an order parameter equal to either of the equilibrium coexistence values. As time progresses, however, regions with order parameters equal to ϕ_A and ϕ_B will eventually form.

We have expressed this problem in general terms. When applied explicitly to a quench from the gas phase to the fluid phase, the order parameter is the particle density n, $\tilde{f}$ is the function $w(n, \mu, T)$ with $\mu = \Lambda$ [Eqs. (3.1.45) and (4.4.1)], and $\Delta f = \Delta p(\mu, T)$ (with $\mu = \lambda$) is the pressure difference between the interior and exterior of the droplet. In this case, the critical radius can be expressed as

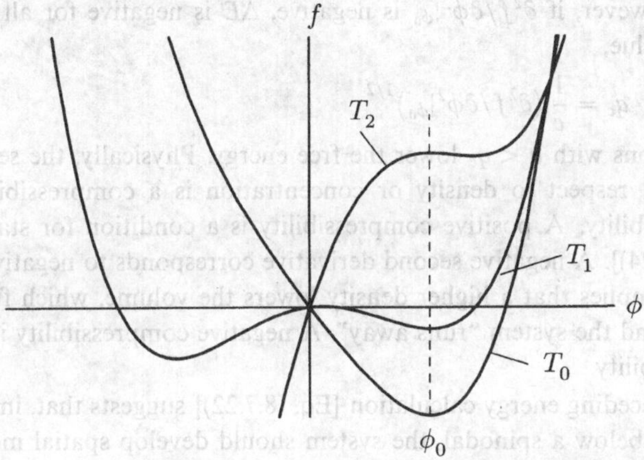

Fig. 8.7.4. Free energy density $\tilde{f}$ [Eq. (8.7.17)] including the Lagrange multiplier Λ. At T_0 the system is in equilibrium and homogeneous. At T_1 the system is metastable, and at T_2 it is unstable against small fluctuations in ϕ about ϕ_0. These three temperatures correspond to those indicated in Fig. 8.7.3b.

$$R_c = \frac{2\sigma}{\Delta p}. \tag{8.7.19}$$

When written as $\Delta p = 2\sigma/R_c$, this is the expression for the Laplace pressure (relating the internal to the external pressure to a droplet with equilibrium radius R_c). The energy of the critical-radius droplet is $16\pi\sigma^3/3(\Delta p)^2$.

We now turn to what happens inside the spinodal curve (Cahn 1961). For the earliest stages of growth, the order parameter remains close to its initial value:

$$\delta\phi = \phi - \phi_0 = A\cos\mathbf{q}\cdot(\mathbf{x}-\mathbf{x}_0). \tag{8.7.20}$$

The wave number $\mathbf{q}$ is nonzero, and this disturbance explicitly satisfies the constraint Eq. (8.7.13). Expanding the local free energy $\tilde{f}$ in a series about ϕ_0, we obtain

$$\tilde{f}(\phi) = \tilde{f}(\phi_0) + \frac{1}{2}\frac{\partial^2 f}{\partial\phi^2}\bigg|_{\phi_0}(\phi-\phi_0)^2 + \cdots, \tag{8.7.21}$$

where we have used the fact that the second derivatives of f and $\tilde{f}$ are equal. Note that there is no term linear in $\phi - \phi_0$ since Λ is defined so that $\tilde{f}$ is an extremum at ϕ_0. Putting the sinusoidal variation of the order parameter in the Hamiltonian and integrating over space, we find the energy difference (relative to the uniform state) to be

$$\Delta E = \frac{VA^2}{4}\left(2\frac{\partial^2 f}{\partial\phi^2} + cq^2\right), \tag{8.7.22}$$

where V is the volume. If the curvature of f is positive, then ΔE is positive for all q. However, if $\partial^2 f/\partial\phi^2|_{\phi_0}$ is negative, ΔE is negative for all q less than the critical value,

$$q_c = \frac{1}{c}\left(\partial^2 f/\partial\phi^2|_{\phi_0}\right)^{1/2}. \tag{8.7.23}$$

Fluctuations with $q < q_c$ lower the free energy. Physically, the second derivative of f with respect to density or concentration is a compressibility or osmotic compressibility. A positive compressibility is a condition for stable equilibrium [Eq. (3.2.24)]. A negative second derivative corresponds to negative compressibility, and implies that a higher density lowers the volume, which further increases density, and the system "runs away". A negative compressibility implies mechanical instability.

The preceding energy calculation [Eq. (8.7.22)] suggests that, immediately after a quench below a spinodal, the system should develop spatial modulations with all wave numbers $q \le q_c$. The actual development of modulations is governed by the dynamics of phase separation rather than by energy alone, and experiments show a somewhat different behavior. Since we are looking at long wavelength modulations of the order parameter, we can treat the time development of the modulation using phenomenological equations such as those discussed in Sec. 8.6. The order parameter is conserved, so we should use model B dynamics. (We neglect any coupling to the momentum, which may play a role in the case of a liquid-gas transition but would not be relevant for the case of a binary mixture of components with the same mass density.) From Eq. (8.6.10) and the form of the Hamiltonian in Eq. (8.7.16) we obtain the linearized equations of motion for ϕ:

$$\frac{\partial\phi}{\partial t} = \lambda\nabla^2\frac{\delta\mathscr{H}}{\delta\phi} = -\lambda\left[\left.\frac{\partial^2 f}{\partial\phi^2}\right|_{\phi_0}\frac{\nabla^2\phi}{2} - \frac{c}{2}\nabla^4\phi\right]. \tag{8.7.24}$$

Note that we do not need to distinguish between f and $\tilde{f}$ in this equation. The frequency of these modes is

$$\omega(\mathbf{q}) \equiv i\alpha(\mathbf{q}) = -i\lambda\left[\left.\frac{\partial^2 f}{\partial\phi^2}\right|_{\phi_0} q^2 + \frac{c}{2}q^4\right]. \tag{8.7.25}$$

If $\partial^2 f/\partial\phi^2|_{\phi_0}$ is positive, the mode is exponentially damped. This is the case outside the spinodal region. However, in the spinodal regime where $\partial^2 f/\partial\phi^2|_{\phi_0}$ is negative, there is a range of q values for which $\alpha(\mathbf{q})$ is positive. The amplitudes of order parameter fluctuation for this range of wave vectors grow as $\exp[\alpha(\mathbf{q})t]$. In Fig. 8.7.5, we schematically illustrate the growth rate as a function of wave vector. All modulations with $q < q_c$ grow, but the maximum growth rate is for $q_{\max} = q_c/\sqrt{2}$. Since the modulations start from infinitesimal amplitudes, the mode with the fastest exponential growth is the one that will dominate experimental observations. Note that the growth rate $\alpha(\mathbf{q})$ depends only on the magnitude of q and not on its direction (or on the phase $\mathbf{q}\cdot\mathbf{x}_0$ of the concentration modulation). We therefore expect that the modulation of ϕ will be an uncorrelated

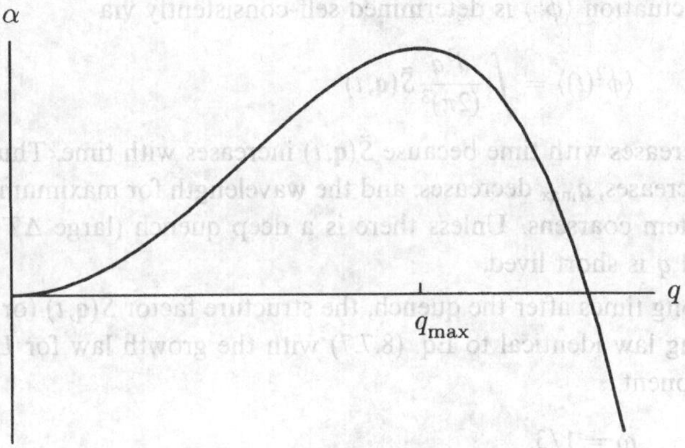

Fig. 8.7.5. The growth rate of a sinusoidal modulation of the order parameter as a function of wave vector for the spinodal regime.

superposition of sinusoidal modulations of wave vector magnitude $|\, q_c \,|$ but in random directions and displaced origins. The time-dependent structure function measured by scattering experiments is proportional to the square of $\phi(\mathbf{q}, t)$, determined by Eq. (8.7.24) and averaged over initial ensemble. Thus, the structure function [Eq. (8.7.5)],

$$\tilde{S}(\mathbf{q}, t) = \tilde{S}(\mathbf{q}, 0)e^{2\alpha(\mathbf{q})t}, \tag{8.7.26}$$

where $\tilde{S}(\mathbf{q}, 0) = S(\mathbf{q})$ is the static structure factor immediately before the quench. If the quench was from high temperatures, then the characteristic length scale is a microscopic length, and $S(\mathbf{q})$ is a constant for q up to many times q_c. Thus, $\tilde{S}(\mathbf{q}, t)$ has a single ring in q space with maximum intensity at $q = q_{\max}$.

Within the linearized theory just presented, the position of the peak remains constant in time, but its intensity increases. However, as the amplitude of modulations increases, the effective curvature of the free energy density will evolve to some average over the range of ϕ's around ϕ_0 represented by the amplitude of the modulation. Since at large values of $|\, \phi \,|$ the curvature of $f(\phi)$ is necessarily positive (as in Fig. 8.7.1), we expect $\partial^2 f/\partial\phi^2 \,|_{\text{eff}}$ to become less negative as time progresses and the value of $q_{\max}$ to decrease. To get an idea of how this comes about (Langer, Bar-on, and Miller 1975), we consider a self-consistent field theory (Sec. 5.3) of the ϕ^4 model around its symmetric point:

$$\begin{aligned} \frac{\partial\phi}{\partial t} &= -\lambda\left[a(T - T_c)\nabla^2\phi + 4u\nabla^2\phi^3 - \frac{c}{2}\nabla^4\phi\right] \\ \omega &= -i\lambda\left[(-a\Delta T + 12u\langle\phi^2\rangle)q^2 + \frac{c}{2}q^4\right] \end{aligned} \tag{8.7.27}$$

and

$$q_{\max} = \left(\frac{a\Delta T - 12u\langle\phi^2\rangle}{(c/2)}\right)^{1/2}. \tag{8.7.28}$$

The fluctuation $\langle \phi^2 \rangle$ is determined self-consistently via

$$\langle \phi^2(t) \rangle = \int \frac{d^3q}{(2\pi)^3} \tilde{S}(\mathbf{q}, t) \tag{8.7.29}$$

and increases with time because $\tilde{S}(\mathbf{q}, t)$ increases with time. Thus, as time evolves, $\langle \phi^2 \rangle$ increases, $q_{\max}$ decreases, and the wavelength for maximum growth increases: the system coarsens. Unless there is a deep quench (large ΔT), the initial stage of fixed q is short lived.

At long times after the quench, the structure factor $\tilde{S}(\mathbf{q}, t)$ (or $\tilde{S}(\mathbf{x}-\mathbf{x}', t)$) obeys a scaling law identical to Eq. (8.7.7) with the growth law for $L(t)$ determined by an exponent

$$\omega = 1/3. \tag{8.7.30}$$

This exponent can be derived heuristically as follows. Consider a quench in a binary fluid. The diffusion current is $\mathbf{J} = -\Gamma \boldsymbol{\nabla} \mu$, where μ is the chemical potential difference between species. The gradient of μ, and thus $\mathbf{J}$, is largest in the vicinity of domain walls. If we assume that there is only one length scale $L(t)$ in the system at large times, it will determine both the average distance between domain walls and the curvature of the walls. Thus we can estimate $\boldsymbol{\nabla} \mu \sim \Delta\mu / L$, where $\Delta\mu$ is the change in μ across the wall. Near coexistence, the difference in free energy between two phases is of order $\phi \Delta\mu$ (where ϕ is the magnitude of the order parameter in either of the phases at coexistence) and $L \sim 2\sigma/\phi\Delta\mu$ from Eq. (8.7.3). Thus, we have $J \sim (\Gamma\sigma/\phi)L^{-2}$. But J is also $\phi \dot{L}$. Thus, we have

$$\frac{\partial L}{\partial t} \sim \frac{\Gamma\sigma}{\phi^2} \frac{1}{L^2}. \tag{8.7.31}$$

This yields $L(t) \sim t^{1/3}$ at long times. Light scattering experiments on binary liquid mixtures tend to show this characteristic behavior. Directly following a pressure quench, a ring of constant q forms and grows in amplitude, while remaining at the same radius for the initial stages. Eventually, the nonlinear effects come in, the system coarsens, and the scattering ring collapses to smaller q. The $t^{1/3}$ behavior of $L(t)$ is often called the Lifshitz-Slyozov (1961) law.

In the case of a binary metal alloy, the quench is often from a molten phase above the phase separation boundary to a solid phase in the spinodal region. The initial decomposition takes place as outlined above with concentration modulations on the scale of $2\pi/q_{\max}$ as expected. However, before further processes can take over, the diffusion has decreased to such an extent that the modulations remain for geological times. Samples can then be split open and the modulation observed with a microscope. The period of the modulation is observed to depend inversely as the square of ΔT, the depth of the quench, as suggested above.

Model B dynamics describes diffusive behavior. The existence of a positive sign for the time exponent is sometimes referred to as a negative diffusion coefficient (diffusing up a concentration gradient) resulting from the metastability of the system, the negative value of the compressibility, and eventually the negative derivative of the free energy function with respect to concentration.

Bibliography

Hydrodynamics of broken-symmetry systems

Dieter Forster, *Hydrodynamic Fluctuations, Broken Symmetry, and Correlation Functions* (Addison Wesley, Reading, Mass., 1983).

P.C. Hohenberg and B.I. Halperin, *Phys. Rev.* **188**, 898 (1969).

P. Martin, O. Parodi, and P.S. Pershan, *Phys. Rev. A* **6**, 2401 (1972).

Fluid dynamics

L. Landau and I.M. Lifshitz, *Fluid Mechanics* (Pergamon Press, Oxford, 1959).

Superfluid helium

I.M. Khalatnikov, *An Introduction to the Theory of Superfluidity* (W.A. Benjamin, New York, 1965).

Dynamic scaling, dynamic critical phenomena, and stochastic models

R.A. Ferrel, N. Menyhárd, H. Schmidt, F. Schwabl, and P. Szépfalusy, *Phys. Rev. Lett.* **18**, 891 (1967); *Ann. Phys. (N.Y.)* **47**, 565 (1968).

P.C. Hohenberg and B.I. Halperin, *Rev. Mod. Phys.* **49**, 435 (1977).

K. Kawasaki, in *Phase Transitions and Critical Phenomena*, vol. 2, eds. C. Domb and M.S. Green (Academic, New York, 1976).

S. Ma, *Modern Theory of Critical Phenomena* (Benjamin, New York, 1976).

Nucleation and spinodal decomposition

A. J. Bray in NATO Advanced Study Institute on Phase Transitions and Relaxation in Systems with Competing Energy Scales, 1993.

John W. Cahn, *Trans. Met. Soc. AIME* **242**, 116 (1968).

J.D. Gunton, M. San Miguel, and Paramdeep S. Sahni, in *Phase Transitions and Critical Phenomena*, vol. 8, eds. C. Domb and J.L Lebowitz (Academic Press, New York, 1983).

J. S. Langer, in *Solids Far From Equilibrium*, ed. C. Godreche (Cambridge University Press, 1992).

References

S.M. Allen and J.W. Cahn, *Acta. Metall.* **27**, 1085 (1979).

J.W. Cahn, *Acta. Metall.* **9**, 795 (1961).

J.W. Cahn and J.E. Hilliard, *J. Chem. Phys.* **28**, 258 (1958).

Dieter Forster, *Hydrodynamic Fluctuations, Broken Symmetry and Correlation Functions* (Addison Wesley, Reading, Mass., 1983).

R. Glauber, *J. Math. Phys.* **4**, 294 (1963).

J. Goldstone, *Nuovo Cimento* **19**, 155 (1961).

B.I. Halperin and P.C. Hohenberg, *Phys. Rev.* **177**, 952 (1969).

B.D. Josephson, *Phys. Lett.* **1**, 251 (1962); see also *Adv. Phys.* **14**, 419 (1965), and *Rev. Mod. Phys.* **46**, 251 (1974).

K. Kawasaki, *Ann. Phys. (NY)* **61**, 1 (1970).

L.D. Landau, *J. Phys. USSR* **5**, 71 (1941). [A translation appears in *An Introduction to the Theory of Superfluidity* by I.M. Khalatnikov (1965).]

L. Landau and I.M. Khalatnikov, *Dokl. Akad. Nauk. SSSR* **96**, 469 (1954); reprinted in *Collected Papers of L.D. Landau*, ed. D. ter Haar (Pergamon, London, 1965).

L. Landau and G. Placzek, *Phys. Z. Sowjetunion* **5**, 172 (1934).

J.S. Langer, M. Bar-on, and H.D. Miller, *Phys. Rev.* **A11**, 1417 (1975).

York Liao, Noel A. Clark, and Peter Pershan, *Phys. Rev. Lett.* **30**, 639 (1973).
I.M. Lifshitz and V.V. Slyozov, *J. Phys. Chem. Solids* **19**, 35 (1961).
Y. Nambu, *Phys. Rev. Lett.* **4**, 380 (1960).
L. Onsager, *Phys. Rev.* **37**, 405 (1931); **38**, 2265 (1931).
L. Van Hove, *Phys. Rev.* **93**, 1374 (1954).

Problems

8.1 How many hydrodynamic modes do the following systems have?

(a) A hexagonal columnar discotic liquid crystal (Fig. 2.7.11).
(b) A lamellar smectic with surfactant layers separated by water (Fig. 2.7.14).
(c) A lamellar microemulsion with alternating water and oil layers separated by surfactant (Fig. 2.7.16).
(d) A cholesteric liquid crystal.
(e) A cubic plumber's nightmare (Fig. 2.7.15).
(f) Superfluid He^4-He^3 mixtures.
(g) Brass in both its ordered and disordered phases.

Make an educated guess as to the nature of the modes in each case.

8.2 Determine the hydrodynamical equations and linearized mode structure for a one-component hexagonal columnar liquid crystal.

8.3 This problem will lead you through the derivation of a general expression for transport coefficients in terms of current correlation functions and will establish the Onsager reciprocity theorem. Let $\{\rho_\alpha(\mathbf{x},t)\}$ be a set of conserved variables with associated conjugate fields $\{h_\alpha(\mathbf{x},t)\}$ and currents $\{\mathbf{J}_\alpha(\mathbf{x},t)\}$ satisfying $\partial\rho_\alpha + \nabla\cdot\mathbf{J}_\alpha = 0$. Using the fact that the conservation law implies

$$\frac{\partial}{\partial t}\tilde{\chi}''_{J_{\alpha i}\rho_\beta}(\mathbf{x},\mathbf{x}',t,t') = -\frac{\partial}{\partial t'}\tilde{\chi}''_{J_{\alpha i}\rho_\beta}(\mathbf{x},\mathbf{x}',t,t') = \nabla'_j\tilde{\chi}''_{J_{\alpha i}J_{\beta j}}(\mathbf{x},\mathbf{x}',t,t'),$$

show that

$$\begin{aligned}\frac{\partial}{\partial t}\langle\delta J_{\alpha i}(\mathbf{x},t)\rangle = \sum_\beta\Bigg\{&\int_{-\infty}^{t}dt'\int d^dx'\left[-2i\int d^dx'\tilde{\chi}''_{J_{\alpha i}J_{\beta j}}(\mathbf{x}-\mathbf{x}',t-t')\right.\\ &\left.\times\nabla'_j\delta h_\beta(\mathbf{x}',t')\right]\\ &+2i\int d^dx'\tilde{\chi}''_{J_{\alpha i}\rho_\beta}(\mathbf{x}-\mathbf{x}',t-t'=0)\delta h_\beta(\mathbf{x}',t)\Bigg\}.\end{aligned}$$

If the external fields vary slowly in space and time, then $\delta h_\beta(\mathbf{x}',t') = \delta h_\beta(\mathbf{x}+\mathbf{x}'-\mathbf{x},t+t'-t)\approx\delta h_\beta(\mathbf{x},t)$ and

$$\begin{aligned}\delta\langle J_{\alpha i}(\mathbf{x},\omega)\rangle &= -\frac{1}{i\omega}\sum_{\beta j}[\chi_{J_{\alpha i}J_{\beta j}}(\mathbf{q}=0,\omega)\\ &\quad-\chi_{J_{\alpha i}J_{\beta j}}(\mathbf{q}=0,\omega=0)]\nabla_j\delta h_\beta(\mathbf{x},\omega)\\ &= -\sum_{\beta j}\int\frac{d\omega'}{\pi i}\frac{\chi''_{J_{\alpha i}J_{\beta j}}(0,\omega')}{\omega'(\omega'-\omega)}\nabla_j\delta h_{\beta j}(\mathbf{x},\omega)\\ &= \mathscr{L}_{\alpha i,\beta j}(\omega)\nabla_j\delta h_\beta(\mathbf{x},\omega).\end{aligned}$$

If $\mathbf{J}_\alpha$ and $\mathbf{J}_\beta$ have the same sign under time reversal, show that

$$\lambda_{\alpha i,\beta j} = \mathrm{Re}\mathscr{L}_{\alpha i,\beta j} = \lim_{\omega\to 0}\lim_{\mathbf{q}\to 0}\frac{1}{\omega}\chi''_{J_{\alpha i}J_{\beta j}}(\mathbf{q},\omega).$$

Then, using the fluctuation-dissipation theorem, show that

$$\lambda_{\alpha i,\beta j} = \frac{1}{T}\int_0^\infty \int d^d x \left\langle \frac{1}{2}\{J_{\alpha i}(\mathbf{x},t), J_{\beta j}(0,0)\}_+ \right\rangle,$$

where $\{A,B\}_+ = AB + BA$. From this follows the Onsager reciprocity relation, $\lambda_{\alpha i,\beta j} = \lambda_{\beta j,\alpha i}$. In an isotropic system, $\lambda_{\alpha i,\beta j} = \delta_{ij}\lambda_{\alpha\beta}$ with $\lambda_{\alpha\beta} = \lambda_{\beta\alpha}$. It is possible in some systems for $\chi''_{J_{\alpha i}J_{\beta j}}/\omega$ to have a part proportional to $\delta(\omega)$. In this case, $\mathscr{L}_{\alpha i,\beta j}(\omega)$ will have a part proportional to $1/i\omega$. If $\mathbf{J}_\alpha$ and $\mathbf{J}_\beta$ have opposite signs under time reversal, the $\mathscr{L}_{\alpha i,\beta j}$ is imaginary in the low-frequency limit.

8.4 Determine the hydrodynamical equations and mode structure for a helical magnet described by the Hamiltonian in Problem 6.1.

8.5 Show that the sound velocities in a dissipationless smectic satisfy the equations for sound in a uniaxial solid with $K_{44} = 0$ and $K_{12} = K_{11}$:

$$c_1^2 + c_3^2 = \rho^{-1}[K_{11} + (K_{33} - K_{11})\cos^2\psi],$$
$$c_1^2 c_3^2 = \rho^{-1}[K_{11}K_{33} - K_{13}]\cos^2\psi\sin^2\psi,$$

where ψ is the angle that the wave vector $\mathbf{q}$ makes with the layer normal and where

$$K_{11} = \rho\left.\frac{\partial p}{\partial\rho}\right)_{S,N,\nabla_z u}, \quad K_{11} - K_{13} = \left.\frac{\partial p}{\partial\nabla_z u}\right)_{S,N,\rho} = \rho\left.\frac{\partial h_z}{\partial\rho}\right)_{S,N,\nabla_z u},$$

$$K_{33} - 2K_{13} + K_{11} = \left.\frac{\partial h_z}{\partial\nabla_z u}\right)_{S,N,\rho}.$$

8.6 Use the ordered phase hydrodynamical equations of Sec. 8.2 [Eqs. (8.2.59) and (8.2.68)] to calculate the order parameter correlation function

$$G(\mathbf{x},t) = S^2\langle\cos 2[\theta(\mathbf{x},t) - \theta(0,0)]\rangle$$

in a two-dimensional system. You may assume that $\theta(\mathbf{x},t)$ satisfies Gaussian statistics. Show that the $t\to 0$ limit of this function shows destruction of long-range order. Discuss the long-time behavior of $G(\mathbf{x},t)$.

8.7 Derive the hydrodynamical equations for a two-dimensional nematic liquid crystal and calculate the dynamical correlation function defined in the previous problem.

8.8 The hydrodynamical equations for superfluid helium can be expressed in many coordinate systems. One of particular interest is the one locally at rest with respect to the superfluid, i.e., the frame in which $\mathbf{v}_s = 0$. Landau and Khalatnikov (Landau 1941, Khalatnikov 1965) write the hydrodynamical equations in this frame. Show that, in this frame,

$$T\,ds = d\varepsilon_0 - \alpha_0 d\rho - (\mathbf{v}_n - \mathbf{v}_s)\cdot d\mathbf{g}_0,$$

where ε_0 and $\mathbf{g}_0$ are, respectively, the energy and momentum density in this frame and $\alpha_0 = \alpha - \frac{1}{2}(\mathbf{v}_n - \mathbf{v}_s)^2 - \frac{1}{2}v_n^2$. Note that there is no contribution to the entropy production from $\mathbf{h}$ in this frame. Show that the pressure is $p = -\varepsilon_0 + Ts + \alpha_0\rho + \rho_n(\mathbf{v}_n - \mathbf{v}_s)^2$. Finally, show that

$$\frac{\partial \mathbf{v}_s}{\partial t} + \nabla\left(\alpha_0 + \frac{1}{2}\mathbf{v}_s - X''\right) = 0$$

and

$$\mathbf{j}^\varepsilon = (\mathbf{Q} - X'')\mathbf{h} + \mathbf{v}_n Ts + \left(\alpha_0 + \frac{1}{2}v_s^2\right)\mathbf{g} + \rho_n\mathbf{v}_n[\mathbf{v}_n \cdot (\mathbf{v}_n - \mathbf{v}_s)].$$

8.9 Calculate to harmonic order the response and correlation functions for the dynamical variables in models H, E, G, and J.

8.10 Determine the stochastic equations governing the low-frequency dynamics of the rigid rotor model discussed in Sec. 8.2. Your dynamical variables should be $Q_{ij}(\mathbf{x}, t)$, $l(\mathbf{x}, t)$, and $\varepsilon(\mathbf{x}, t)$.
   (a) Show that these equations reproduce the hydrodynamics derived in Sec. 8.2 when $Q_{ij} = S(n_i n_j - \frac{1}{2}\delta_{ij})$ is non-zero.
   (b) Since Q_{ij} is a two-dimensional symmetric-traceless tensor, $\mathrm{Tr}Q^3$ is zero, and the transition from the disordered to the ordered phase can be second order. What does dynamical scaling predict about characteristic frequencies above and below this second order transition?

8.11 A thin liquid film on a solid substrate is a two-dimensional fluid in which momentum is not conserved (because it can be absorbed by the substrate). Identify all hydrodynamical variables of such a film and determine their hydrodynamical equations. Show, in particular, that the density mode is diffusive rather than propagating.

9

Topological defects

In Chapter 6, we studied the states of systems with broken continuous symmetry in which the slowly varying elastic variables described distortions from a spatially constant ground state configuration. These distortions arose from the imposition of boundary conditions, from external fields, or from thermal fluctuations. In this chapter, we will consider a class of defects, called *topological defects,* in systems with broken continuous symmetry. A topological defect is in general characterized by some core region (e.g., a point or a line) where order is destroyed and a far field region where an elastic variable changes slowly in space. Like an electric point charge, it has the property that its presence can be determined by measurements of an appropriate field on any surface enclosing its core. Topological defects have different names depending on the symmetry that is broken and the particular system in question. In superfluid helium and xy-models, they are called *vortices*; in periodic crystals, *dislocations*; and in nematic liquid crystals, *disclinations.*

Topological defects play an important role in determining properties of real materials. For example, they are responsible to a large degree for the mechanical properties of metals like steel. They are particularly important in two dimensions, where they play a pivotal role in the transition from low-temperature phases characterized by a non-vanishing rigidity to a high-temperature disordered phase.

This chapter begins (Sec. 9.1) with a discussion of how topological defects are characterized and a brief introduction to the concepts of homotopy theory. Examples of topological defects in a number of systems will be discussed in Sec. 9.2. Then, in Sec. 9.3, energies of individual and interacting defects will be calculated. The Kosterlitz-Thouless vortex unbinding and related transitions in two-dimensional systems will be discussed in Secs. 9.4 and 9.5. Some mathematical details associated with the renormalization group for the Kosterlitz-Thouless and lattice models relevant to topological defects will be discussed in Appendices 9A and 9B.

9.1 Characterization of topological defects

In our discussion of the xy-model in Chapter 6, we assumed that the angle variable θ was continuous everywhere and that the magnitude $\langle s(\mathbf{x})\rangle$ of the order parameter was everywhere nonzero. Because $\langle \mathbf{s}(\mathbf{x})\rangle = s(\cos\theta(\mathbf{x}),\ \sin\theta(\mathbf{x}))$ is a periodic

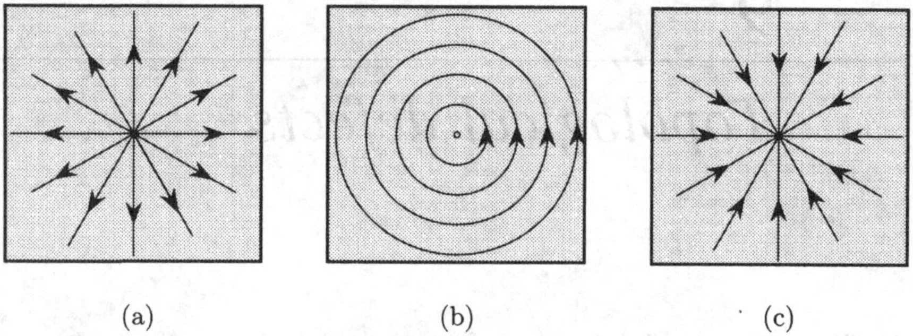

Fig. 9.1.1. Three $k = 1$ vortex configurations with (a) $\theta_0 = \phi$, (b) $\theta_0 = \phi + \pi/2$ and (c) $\theta_0 = \phi + \pi$.

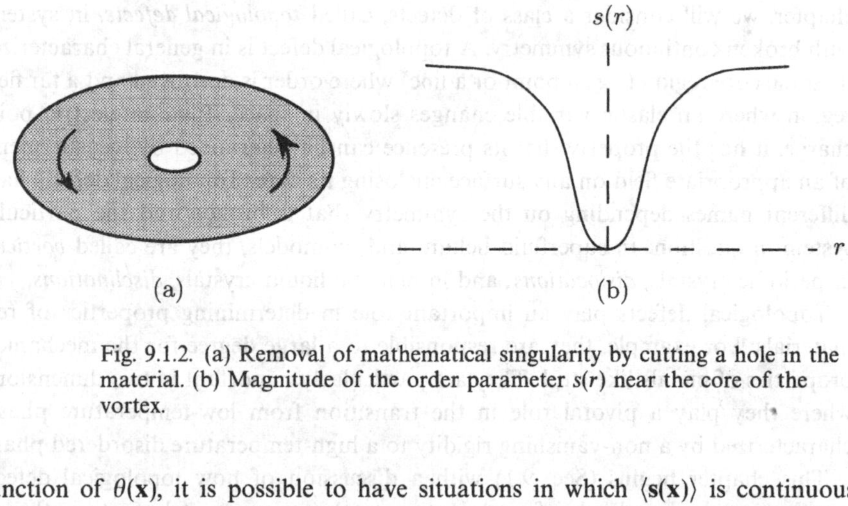

Fig. 9.1.2. (a) Removal of mathematical singularity by cutting a hole in the material. (b) Magnitude of the order parameter $s(r)$ near the core of the vortex.

function of $\theta(\mathbf{x})$, it is possible to have situations in which $\langle \mathbf{s}(\mathbf{x}) \rangle$ is continuous everywhere in d-dimensional space except in a subspace of dimensionality d_s less than d. For example, if $d = 2$ and

$$\theta(\mathbf{x}) = \phi + \theta_0, \tag{9.1.1}$$

where θ_0 is any constant and $\mathbf{x} = (r, \phi)$ in polar coordinates, then $\langle \mathbf{s}(\mathbf{x}) \rangle$ is continuous and $\nabla\theta = 1/r$ is finite everywhere except at the origin, as shown in Fig. 9.1.1. The mathematical singularity at the origin (a point of dimension $d_s = 0$) can be removed simply by cutting a hole of some radius ξ_0 out of the material, as shown in Fig. 9.1.2a. An alternative solution to cutting a hole in the material is to have the magnitude of the order parameter go to zero at the origin and to rise to its equilibrium value at a radius ξ_0, as shown in Fig. 9.1.2b. Since $\langle s(\mathbf{x}) \rangle = 0$ at the origin, the angle θ is no longer defined there, and the mathematical singularity has been removed.

The configuration of spins just defined is called a *vortex*. The angle θ, specifying the direction of the order parameter, changes by 2π in one circuit of any

(a) (b) (c)

Fig. 9.1.3. Point singularities of planar spins in two dimensions with winding numbers (a) −1, (b) 2, (c) −2.

closed contour enclosing the core of the vortex located at the origin. Since $\langle \mathbf{s}(\theta) \rangle = \langle \mathbf{s}(\theta + 2k\pi) \rangle$ for $k = 0, \pm 1, \pm 2, ...$, there are an infinite number of distinct singularities in which θ increases by $2k\pi$ in one complete circuit of the core. The integer k is called the *winding number,* or sometimes the *strength*, of the vortex. Spin configurations for low values of k for which

$$\theta(\mathbf{x}) = k\phi + \theta_0 \tag{9.1.2}$$

are shown in Fig. 9.1.3. A vortex is characterized by its winding number and not by Eq. (9.1.2). There are infinitely many configurations of $\langle \mathbf{s}(\mathbf{x}) \rangle$ for which $\theta(\mathbf{x})$ changes by $2k\pi$ in one circuit along any contour enclosing the core. For example, $\theta(\mathbf{x}) = k\phi + \sin\phi$ or $\theta(\mathbf{x}) = k\phi + \tanh(r\cos\phi)\cos 7\phi$ both describe spin configurations with winding number k. The existence of a nonzero integer winding number necessarily implies a singularity in $\nabla\theta$ at the core. This can be seen as follows: θ changes by $2k\pi$ in any circuit of the core. Thus, $|\nabla\theta|$ on a circuit a distance r from the core must be of order $2k\pi/r$ and diverge as $r \to 0$.

The physical variable $\langle \mathbf{s}(\mathbf{x}) \rangle$ is perfectly continuous throughout the plane. On the other hand, if $0 \le \phi < 2\pi$ and θ satisfies Eq. (9.1.2), then θ undergoes a discontinuous change of $-2k\pi$ as the positive x-axis is traversed in a counterclockwise direction. The x-axis can thus be viewed as a cut introducing two imaginary surfaces (lines) along which θ has respective values 0 and $2k\pi$, as shown in Fig. 9.1.4. Different parameterizations of θ in terms of ϕ and r and different choices of the ranges of ϕ (e.g. $-\pi \le \phi < \pi$) lead to cuts, which may be curved, extending in arbitrary directions away from the core to infinity. These cuts are purely mathematical and have no physical significance. They will, however, prove useful in evaluating the energies of vortices, as we shall see in Sec. 9.4.

Defects such as the vortices just described are important because they cannot be made to disappear by any continuous deformations of the order parameter. They are, therefore, called *topological defects.* Consider a singularity free distortion (Fig. 9.1.5a) of the ideal aligned state. It can be returned to the aligned state by

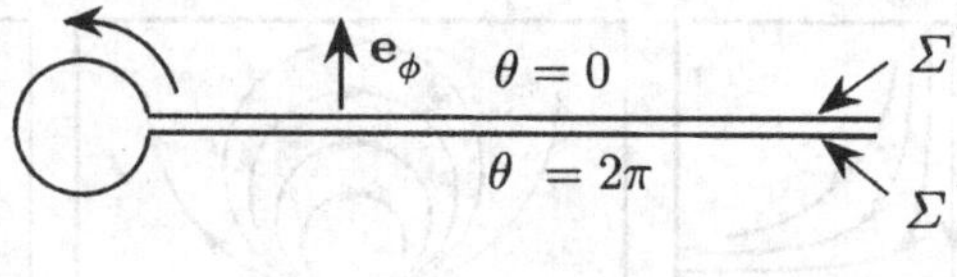

Fig. 9.1.4. Cut, across which θ is discontinuous.

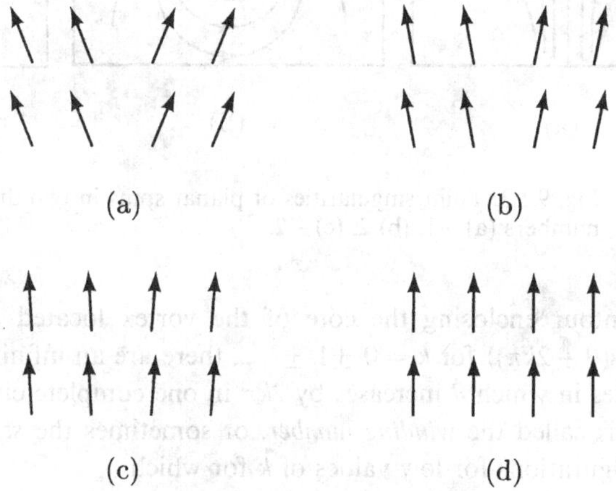

Fig. 9.1.5. Sequence of distortions leading continuously from singularity free state to state of total alignment.

a sequence of distortions in which each spin changes by an infinitesimal amount relative to the previous configuration, as shown in Figs. 9.1.5b–d. Now consider the $k = 1$ vortex. Spins rotate through 2π along *every* contour enclosing the core. It is, therefore, impossible to distort the spin configuration of a $k = 1$ vortex into that of a perfectly aligned state without tampering with spins at an arbitrary distance from the core. The $k = 1$ vortex is thus said to be *topologically stable.* Similar considerations apply to vortices with arbitrary winding number $k \neq 0$.

Topological stability is a mathematical concept distinct from physical stability, which depends on free energies of different configurations. In most cases, however, topological stability implies physical stability. Consider again the $k = 1$ vortex. Any attempt to carry out a sequence of distortions of this configuration to bring it to a state in which all of the spins are aligned along a single direction will invariably lead to configurations such as shown in Fig. 9.1.6b. The only way to achieve the aligned state is to flip lines of spins through an angle of π, i.e., discontinuous changes in the directions of the spins are necessary. The energy of the state shown in Fig. 9.1.6b is much higher (of order JL, where J is the exchange energy and L is the size of the sample) than that of the initial configuration shown in Fig. 9.1.6a (which is of order $J \ln L$, as we shall see shortly). Since it is necessary to go through an intermediate state, such as that of Fig. 9.1.6b, in order to reach

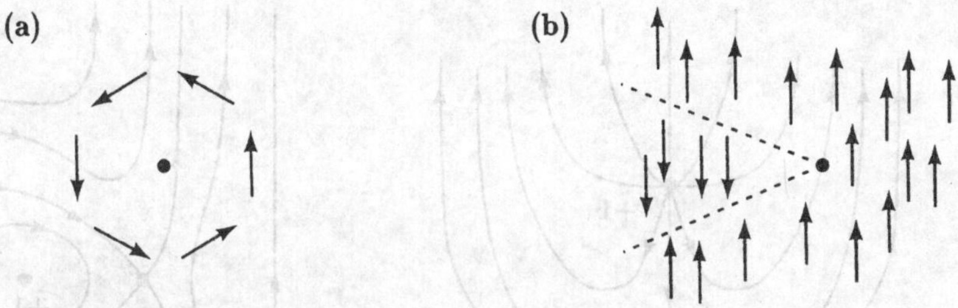

Fig. 9.1.6. (a) A $k = 1$ vortex. (b) An attempt to align spins by a continuous distortion. Note the appearance of boundaries (shown as dashed lines) where the spin changes rapidly from up to down.

the uniform state, it is extremely unlikely that a statistical fluctuation will destroy a single vortex. The vortex state could, of course, be destroyed by returning to the disordered state and reforming a state with uniform $\langle \mathbf{s}(\mathbf{x}) \rangle$. This process has an energy barrier of order JL^2 and is even more unlikely than the previous one. Thus, the vortex state, even though it has a much higher energy than the ground state, is stable because there is no path to the ground state that is not energetically costly. Similar arguments show that the $k = 1$ vortices shown in Fig. 9.1.1 can be converted into each other without passing over an energy barrier but that they cannot be converted to the $k \neq 1$ vortices shown in Fig. 9.1.3.

1 Vortex pairs

Pairs of vortices with opposite winding number lead to far-field configurations that can be distorted continuously to the uniform state. Two configurations of a $k = +1$, $k = -1$ pair of vortices are shown in Fig. 9.1.7. Note that far from the pair, the vectors $\langle \mathbf{s}(\mathbf{x}) \rangle$ are nearly parallel. As the two cores of the vortex pair approach each other, the region over which $\langle \mathbf{s}(\mathbf{x}) \rangle$ deviates significantly from spatial uniformity decreases. Continuous distortions of $\langle \mathbf{s}(\mathbf{x}) \rangle$ can bring the two vortices together and cause them to annihilate. Thus, the state with a vortex pair is topologically equivalent to the uniform state. Whereas a single vortex has energetically costly elastic distortions far from its core, a pair of vortices with opposite winding number does not. Vortex pairs represent important excitations from the ground state of two-dimensional systems with xy-symmetry, as we shall see in more detail in Sec. 9.4.

2 Order parameters with more than two components

Topological stability depends critically on the symmetry of the order parameter and the *codimension* $d' = d - d_s$, where d_s is the dimension of the core. In the example of the topologically stable vortex just considered, both d' and the

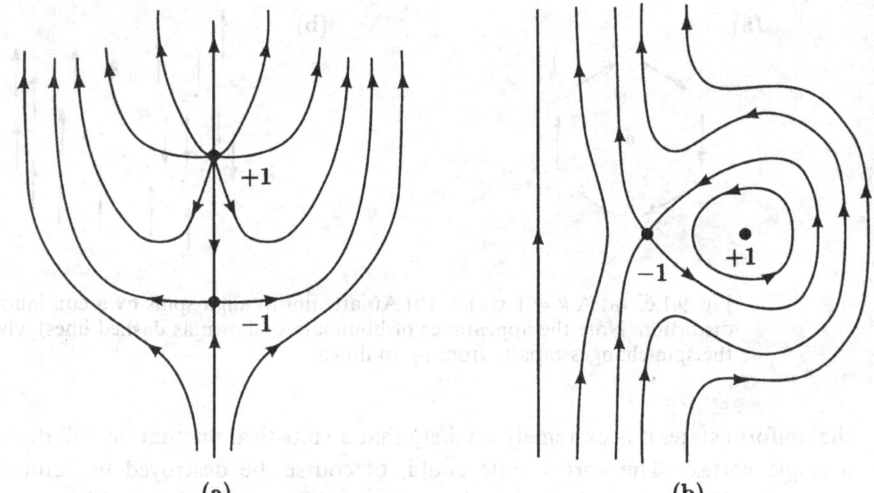

Fig. 9.1.7. Vortex pairs composed of vortices with winding numbers $k = +1$ and $k = -1$. (a) and (b) show two possible configurations of spins for such vortex pairs.

dimensionality n of the order parameter are 2. Vortices are also stable in three dimensions if the core is a line ($d_s = 1$). In this case, again $n = d' = 2$. In fact, the generalization of the vortex to n-dimensional vector order parameters is stable if $d' = n$. For example, the "hedgehog" configurations with $d_s = 0$ and $d' = 3$ (Fig. 9.1.8) of a three-dimensional spin in a three-dimensional space are stable. The analog of the loop enclosing the vortex core is the surface enclosing the center of the hedgehog. As for vortices, there is an integer-valued topological index k characterizing the strength of the hedgehog. The $k = +1$ hedgehog shown in Fig. 9.1.8a has spin vectors pointing outward from the core, like the electric field of a positive point charge. The $k = -1$ hedgehog in Fig. 9.1.8c has spin vectors pointing inward toward the core-like the electric field of a negative charge. Note, however, that alternative configurations can be obtained by rotating $\langle \mathbf{s}(\mathbf{x}) \rangle$ about any fixed axis. For example, the $k = +1$ configuration shown in Fig. 9.1.8b can be obtained from that of Fig. 9.1.8a by rotating all spins through π about the vertical axis.

If $n > d'$, there are no topologically stable defects. Consider, for example, three-dimensional ($n = 3$) spins in a two-dimensional space. A configuration of spins identical to that of a $k = 1$ vortex for two-component spins can be created as shown in Fig. 9.1.9a. It is possible, however, to reach the state in which all of the spins are aligned normal to the two-dimensional plane via a sequence of distortions involving only infinitesimal changes in the directions of the spins, as

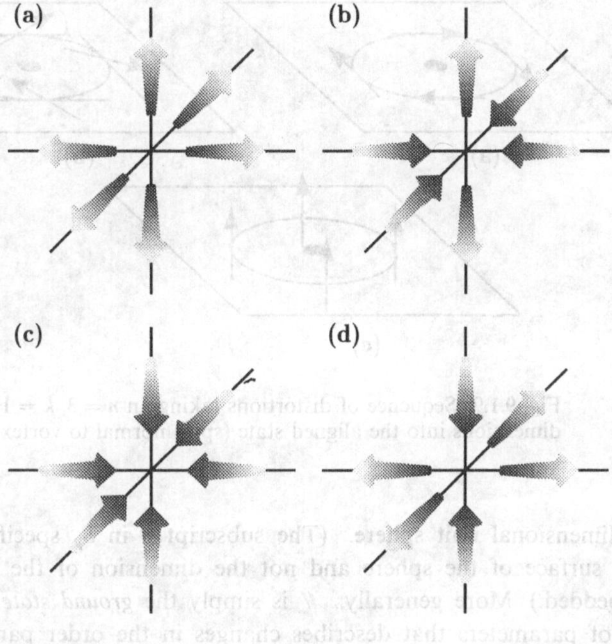

Fig. 9.1.8. "Hedgehog" defects of a 3-dimensional unit spin **s** (O_3 order parameter) in 3-space. (a) A $+1$ defect with **s** parallel to (x, y, z). (b) A $+1$ defect with **s** parallel to $(-x, -y, z)$. (c) A -1 defect with **s** parallel to $(-x, -y, -z)$. (d) A -1 defect with **s** parallel to $(x, y, -z)$. The defects in (c) and (d) are respectively obtained from those in (a) and (b) by rotating **s** through π about the z-axis.

shown in Figs. 9.1.9b,c. There is an "escape to the third dimension" that renders the vortex state of a three-component spin topologically unstable.

If $n < d'$, it is impossible to construct configurations in which the order parameter rotates continuously around the core. Consider, for example an Ising model ($n = 1$) in two dimensions. Any attempt to rotate $\mathbf{s} = s\mathbf{e}_x$ through 2π around the core will lead to a line rather than a point defect. For example, if $s(\phi) = \cos\phi/|\cos\phi|$, then $s = +1$ for $0 \leq \phi < \pi$ and $s = -1$ for $\pi \leq \phi < 2\pi$ and s changes discontinuously from -1 to $+1$ as the x-axis is crossed. There is therefore a line rather than a point defect. Line defects, or walls, will be discussed in more detail in the next chapter.

3 Order parameter spaces and homotopy

To discuss topological defects for more general order parameters, it is convenient to introduce the concept of an *order parameter space* $\mathcal{M}$. For two-component spins, $\mathcal{M}$ is the unit circle (S_1); for three-component spins, it is the surface (S_2) of the unit sphere; and for n-component spins, it is the surface (S_{n-1}) of

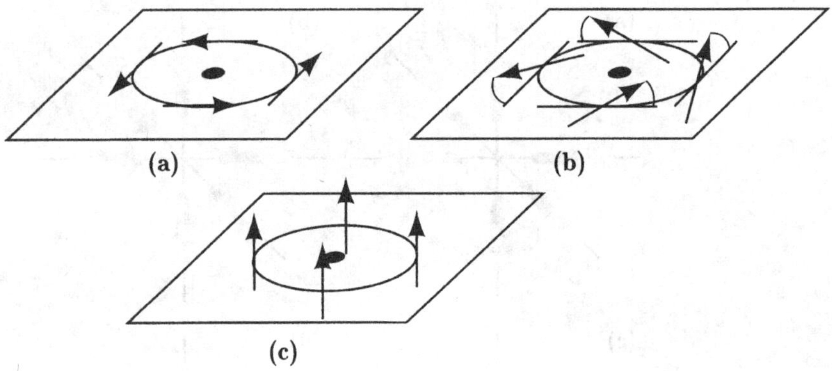

Fig. 9.1.9. Sequence of distortions taking an $n = 3$, $k = 1$ vortex state in two dimensions into the aligned state (spins normal to vortex plane).

an n-dimensional unit sphere. (The subscript p in S_p specifies the dimension of the surface of the sphere and not the dimension of the space in which it is embedded.) More generally, $\mathcal{M}$ is simply the *ground state manifold,* i.e., the space of parameters that describes changes in the order parameter that leave the equilibrium free energy unchanged. It is the space of parameters that enters into an elastic theory with points corresponding to the same physical value of the order parameter identified in some way. The characterization of topological defects and the way they combine can be related in a very elegant way to topological properties of the order parameter space $\mathcal{M}$ as we will now discuss.

Specification of the order parameter in some coordinate space domain $\mathcal{D}$ defines a mapping from $\mathcal{D}$ to $\mathcal{M}$. In the vortex example, specification of θ on some closed loop Γ in $\mathcal{D}$ (which is topologically equivalent to a circle S_1 in coordinate space) defines a mapping f from the loop Γ into $\mathcal{M} = S_1$, as shown in Fig. 9.1.10. Similarly specifying the spin direction on a closed surface $\mathcal{D}$ surrounding the core of a hedgehog defines a mapping from $\mathcal{D}$ in coordinate space to the order parameter space $\mathcal{M} = S_2$.

Consider again the xy-model in two dimensions and some loop Γ in the two-dimensional plane. Different spin configurations on Γ define different mappings into $\mathcal{M} = S_1$, i.e., different closed paths in $\mathcal{M}$, as shown in Fig. 9.1.11. When all spins in Γ are parallel, the image of Γ in $\mathcal{M}$ is simply a point, as shown in Fig. 9.1.11a. More generally, the image of Γ in $\mathcal{M}$ is a closed contour that wraps $\mathcal{M}$ one or more times if Γ encloses a vortex, as shown in Figs. 9.1.11c and d, or a path that closes in on itself, as shown in Fig. 9.1.11b, if Γ does not enclose a vortex. Distortions of spins on Γ lead to distortions of the closed image contours in $\mathcal{M}$. Continuous deformations of Γ itself also lead to continuous distortions of the image in $\mathcal{M}$ if Γ does not pass through the core. A sequence of distortions such as that shown in Fig. 9.1.5 causes the image of Γ in $\mathcal{M}$ to shrink to a point.

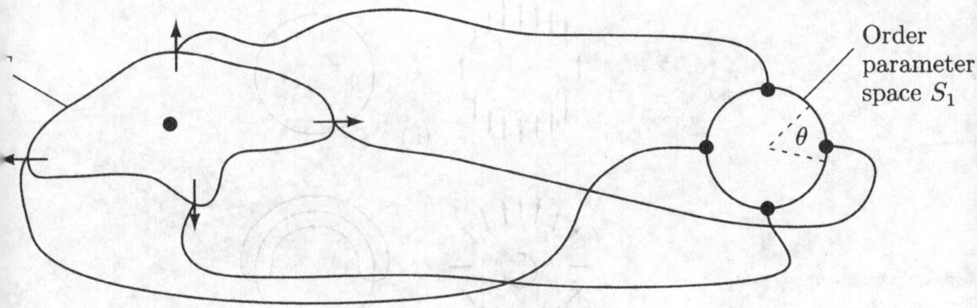

Fig. 9.1.10. Mapping of contour Γ surrounding a $k = 1$ vortex onto the order parameter space S_1.

Two mappings f_0 and f_1 are defined to be *homotopic* if they can be continuously deformed into each other, i.e., if there exists a one-parameter family of maps h_t such that $h_{t=0} = f_0$ and $h_{t=1} = f_1$. The explicit construction of the deformation of f_0 to f_1 is called a *homotopy*. Fig. 9.1.11 shows how closed paths on S_1 not encircling S_1 can be continuously distorted to points. This implies that all maps from a loop Γ not enclosing a vortex are homotopic to a map in which all spins are aligned. Similarly, all maps that wrap S_1 k times are homotopic. Finally, we note that two different paths Γ and Γ' enclosing identical defects define two homotopic maps f_0 and f_1 because it is possible to find a one-parameter family of maps h_t such that $h_{t=0} = f_0$ and $h_{t=1} = f_1$.

We can now classify defects in the two-dimensional *xy*-model into different homotopy classes. Defects are in the same homotopy class if the mappings of all paths enclosing them can be continuously deformed into each other. The homotopy classes of point defects of the two-dimensional *xy*-model are indexed by the number of times the paths wrap S_1, i.e., by the winding number k. The same is true of line defects in the three-dimensional *xy*-model. More generally, homotopy classes for any model can be defined by considering mappings from any closed surface or loop Γ into $\mathscr{M}$. In the O_3 model, all maps from closed loops to the order parameter space S_2 are homotopic because it is possible to slide any closed path off the surface of a sphere, as shown in Fig. 9.1.12. There is only one homotopy class. This is the escape to the third dimension depicted in Fig. 9.1.9. Maps from closed surfaces in three dimensions to S_2, however, are indexed by the number of times they wrap S_2. Point defects (hedgehogs) for the three-dimensional O_3 model can be divided into homotopy classes indexed by an integer "charge" k. Similar results apply to point defects of O_n models in $d = n$ dimensions.

The real value of homotopy theory is that it provides a natural group structure for combining defects. Since topological defects are associated with homotopy classes of the order parameter space, the rules for combining defects are equivalent

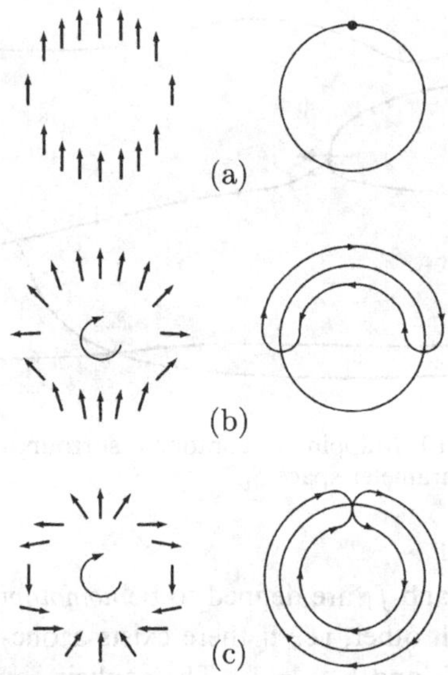

Fig. 9.1.11. Spin configurations on circular contours (left) and the maps they determine of the contours into order parameter space (right). (a) The spin is uniform over the entire contour. The contour is, therefore, mapped into a single point of order parameter space. (b) The spin is nonuniform with zero winding number. The resulting map of the contour into order parameter space can be shrunk to a point. (c) The spin is nonuniform with winding number 2. The resulting map wraps the contour twice around the circular order parameter space. [N.D. Mermin, *Rev. Mod. Phys.* **51**, 591 (1979).]

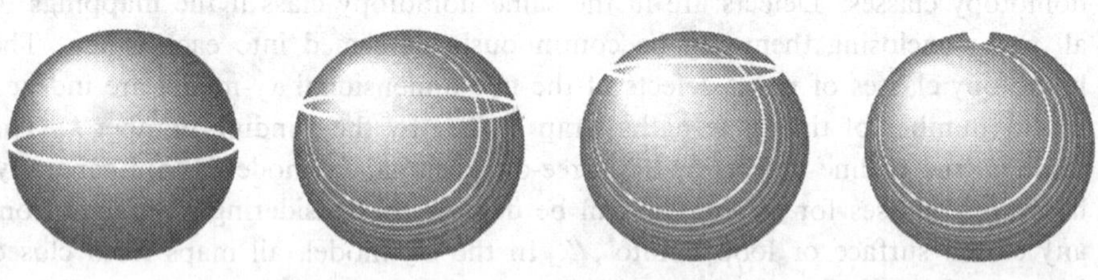

(a) (b) (c) (d)

Fig. 9.1.12. Continuous distortion of $n = 3$ vortex (line around equator of S_2) to the aligned state (point at a pole of S_2).

to rules for combining paths in different homotopy classes in the order parameter space. The group associated with closed loops in the order parameter space $\mathscr{M}$ is called the *fundamental group* or the *first homotopy group* of $\mathscr{M}$ and is denoted by the symbol $\pi_1(\mathscr{M})$. More generally, the group associated with an n-dimensional spherical closed surface in $\mathscr{M}$ is denoted by $\pi_n(\mathscr{M})$.

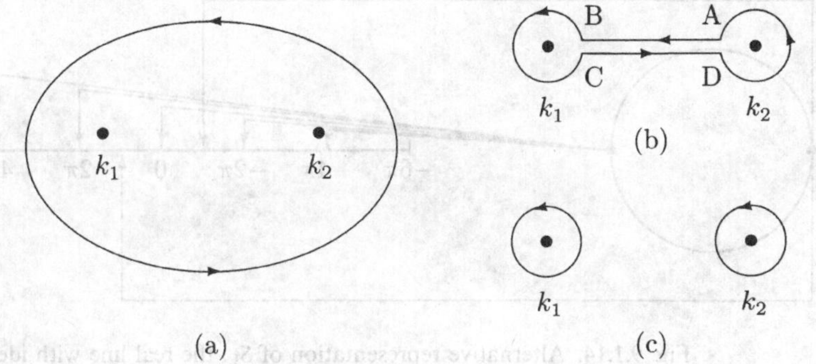

Fig. 9.1.13. (a)–(c) Sequence of distortions taking a loop around two vortices k_1 and k_2 to two loops around the individual vortices. This construction is possible because the change in θ going from A to B exactly cancels the change in going from C to D. For more complicated order parameters, this cancellation is not possible, and the addition law is more complicated.

It is easy to see that defects in the *xy*-model combine according to the sum of their winding numbers. Consider, for example, two vortices with winding numbers k_1 and k_2 located at different points in space, as shown in Fig. 9.1.13. The path enclosing both vortices is homotopic to the two paths enclosing the vortices individually, as can be seen by the sequence of continuous deformations of the original path depicted in Fig. 9.1.13. Since the rule for combining vortices is that their winding numbers add, the group in question is simply Z, the group of the integers under addition, and one writes $\pi_1(S_1) = Z$. Similarly, since all configurations of three-dimensional spins in $d = 2$ are homotopic, $\pi_1(S_2)$ is equivalent to the group consisting only of the identity element, and one writes $\pi_1(S_2) = 0$. In general,

$$\begin{aligned} \pi_n(S_n) &= Z \\ \pi_n(S_m) &= 0, \qquad m > n\,. \end{aligned} \tag{9.1.3}$$

This means that there are topologically stable configurations indexed by an integer for $n+1$-dimensional spins in an $n+1$-dimensional space for all $n \geq 1$. The groups π_n for order parameter spaces other than S_n are not, in general, equal to Z. The fundamental group π_1 may even be nonabelian. The group π_n for $n > 1$ is, however, always abelian.

Alternative representations of the order parameter space are sometimes useful. For example, S_1 can be represented by the real line with the rule that all points separated by $2k\pi$ are equivalent. A vortex of winding number k, therefore, defines a mapping from a loop in real space onto the real line beginning at the origin and terminating at the point $2k\pi$, as shown in Fig. 9.1.14.

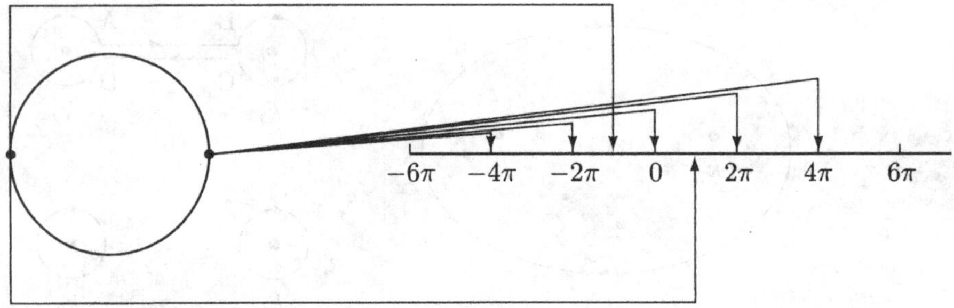

Fig. 9.1.14. Alternative representation of S_1: the real line with identification of points $2k\pi$.

9.2 Examples of topological defects

1 Vortices in xy-models

In the preceding section, we introduced the concept of a topological defect in somewhat general terms using the *xy*-model as an example. In this section, we will consider some examples of three-dimensional line and two-dimensional point defects in more detail. As we saw in Sec. 9.1, these defects are characterized by a mapping from some loop Γ in real space onto the order parameter space such that the physical order parameter remains single-valued in a complete circuit of Γ (Fig. 9.2.1). For the *xy*-model, this implies

$$\oint d\theta = \oint_\Gamma \frac{d\theta}{ds} ds = 2k\pi, \qquad k = 0, \pm 1, ..., s. \tag{9.2.1}$$

In two dimensions, the line integral about the core is taken in the counterclockwise direction. In three dimensions, the core is a line. Its unit tangent vector, chosen to be a continuous function of arclength, defines a direction **l** at each point along its length. The line integral in Eq. (9.2.1) is taken in the counterclockwise direction defined by the right hand rule and **l**.

Free standing smectic-*C* films provide striking visual proof of the existence of vortices. A schematic representation of such a film is shown in Fig. 9.2.2. The director **n** specifying the direction of average molecular alignment is tilted at an angle relative to the normal to the film. The projection of **n** onto the plane of the film is a two-component vector order parameter **c** with *xy*-symmetry. The intensity $I(\Phi)$ of depolarized light reflected from such a film is proportional to $\sin^2 \Phi$, where Φ is the angle between **s** and the axis of polarization of the incident light. In a typical experiment, incident light is polarized along an axis $\vec{P}$, and reflected light is passed through an analyzer that only accepts light that is polarized along an axis $\vec{A}$. If $\vec{A}$ is perpendicular to $\vec{P}$, the average intensity of reflected light passing through the analyzer will be zero if $\vec{P}$ is parallel to **s**. If there is a vortex (called in this case a disclination) in the sample, Φ will

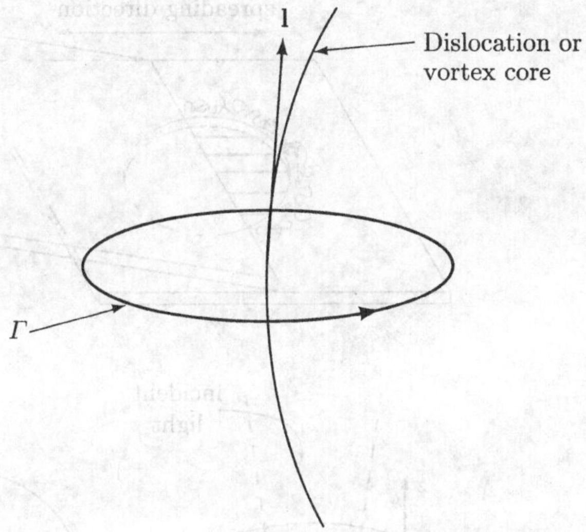

Fig. 9.2.1. Loop Γ enclosing a defect core with tangent vector **l**.

undergo a change of 2π about a core, and there will be four bright regions at $\Phi = \pi/4$, $3\pi/4$, $5\pi/4$ and $7\pi/4$ and four dark regions at $\Phi = 0$, $\pi/2$, π and $3\pi/2$. A photographic image of light reflected from a single disclination is shown in Fig. 9.2.3a. By rotating $\vec{A}$ relative to $\vec{P}$, it is possible to determine that the director rotates about the core of the disclination with the pattern shown in Fig. 9.1.1b. Fig. 9.2.3b shows a positive-negative disclination pair. The direction of molecular alignment of this pair is the same as that shown in Fig. 9.1.7b.

2 *Dislocations in smectic liquid crystals*

In a distorted smectic, planes are defined by the relation (see Sec. 6.3)

$$z - u(\mathbf{x}) = kd = 2\pi k/q_0, \tag{9.2.2}$$

where the displacement field $u(\mathbf{x})$ is a function of all of the components of the position vector $\mathbf{x}$. $u(\mathbf{x})$ is analogous to the angle $\theta(\mathbf{x})$ of the xy-model. The contribution from $u(\mathbf{x})$ to the phase of the mass density at wave number $\mathbf{G} = kq_0\mathbf{e}_z$ is $kq_0u(\mathbf{x})$. Thus, a change in $u(\mathbf{x})$ of $2k\pi/q_0 \equiv kd$ $(k = 0, \pm1, \pm2, ...)$ will leave the amplitude of every mass-density wave, and therefore the total density, unchanged. The order parameter space is, thus, the real line representing u with all points $u = kd$ identified (Fig. 9.2.4). This is identical to the alternative representation of the order parameter space for the xy-model discussed at the end of the preceding section.

Note that, because u represents displacements in the z-direction, the set of identified points for u is identical to the direct lattice of the smectic planes. To

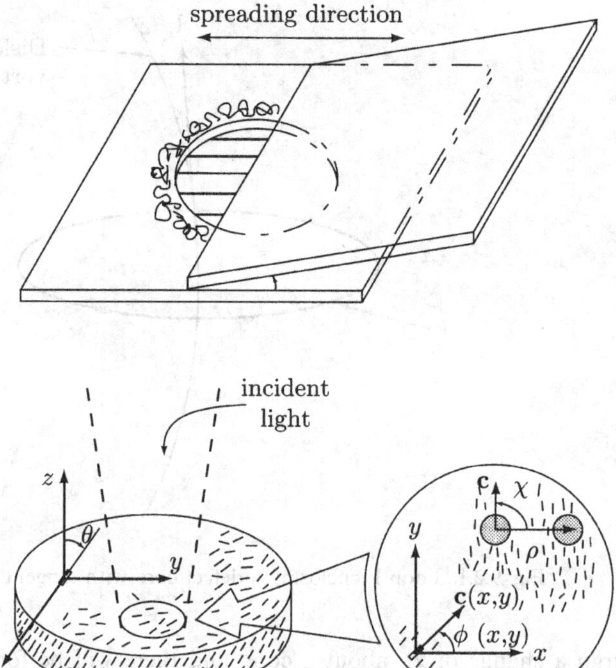

Fig. 9.2.2. Schematic representation of a free standing smectic-C film. [Adapted from David H. Van Winkle and Noel A. Clark, *Phys. Rev. A* **38**, 1573 (1988).]

emphasize this fact, and to facilitate contact with our discussion of solids, we define the vector

$$\mathbf{b}(\mathbf{x}) = kd\mathbf{e}_z. \tag{9.2.3}$$

The vector $\mathbf{b}$ is called a *Burgers vector*. The set of Burgers vectors is equivalent to the direct lattice specifying equilibrium positions of smectic planes. A line defect is thus characterized by the relation

$$\oint_\Gamma d\mathbf{u} = \mathbf{b} = kd\mathbf{e}_z. \tag{9.2.4}$$

Since $\mathbf{u}$ has a direction, the physical nature of the defect can depend on the direction, $\mathbf{l}$, of the core relative to $\mathbf{b}$ (the z-axis here). If $\mathbf{l}$ is parallel to the z-axis (i.e., parallel to $\mathbf{b}$), the spatial variation of $u(\mathbf{x})$ is in the xy-plane. If u changes by $+d$ in one circuit of the core (passing through the origin in the xy-plane), then the equation determining the position of the smectic planes is

$$z - \frac{d}{2\pi}\phi(x, y) = kd; \qquad \phi(x, y) = \tan^{-1}(y/x). \tag{9.2.5}$$

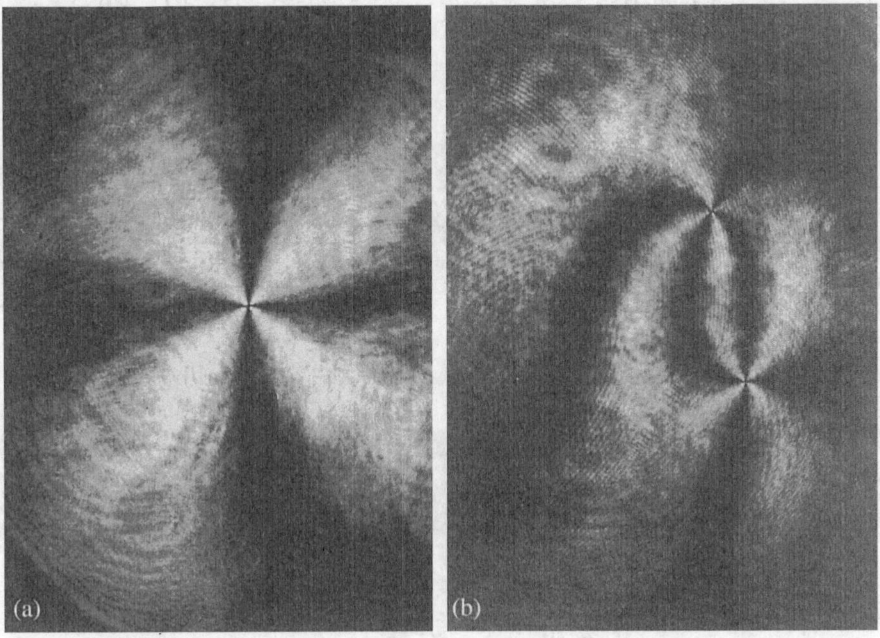

Fig. 9.2.3. (a) Photographic image of a single unit strength disclination in a free standing smectic-C film. (b) Photographic image of a positive-negative disclination pair. [David H. Van Winkle and Noel A. Clark, *Phys. Rev. A* **38**, 1573 (1988).]

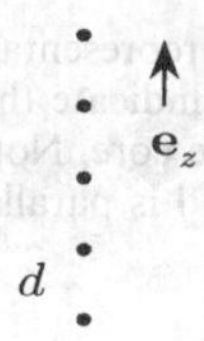

Fig. 9.2.4. Order parameter space for a smectic-A is a straight line with points separated by d identified. The vectors $\mathbf{b} = kd\mathbf{e}_z$, with k an integer, are Burgers vectors. They are equivalent to the vectors in the direct lattice specifying the equilibrium positions of smectic planes.

This equation says that a plane containing the positive x-axis at $z = 0$ will increase in height as the core is encircled in the counterclockwise direction. In one complete circuit of the core, the height will have changed by d so that the plane that was at $z = 0$ when $\phi = 0$ becomes the plane $z = d$ at $\phi = 2\pi$. Since the planes turn in a screw-like fashion, this defect is called a *screw dislocation*. Representations of simple configurations of screw dislocations are shown in Fig. 9.2.5.

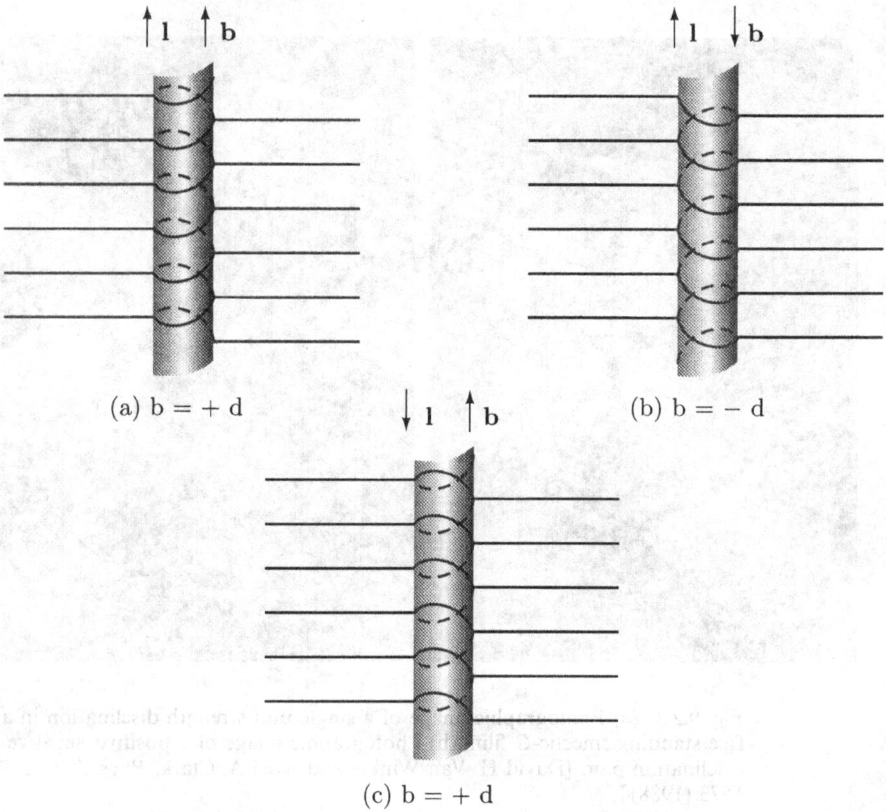

Fig. 9.2.5. Schematic representation of screw dislocations in a smectic-A. The cross-sections shown indicate that each layer changes height by half a layer in half a circuit of the core. Note that planes rise in going from left to right in front of the core if **l** is parallel to **b** and fall if **l** is antiparallel to **b**.

If **l** is perpendicular to **b**, the plane in which u varies contains the z-axis. These dislocations are called *edge dislocations*. If **l** is along the x-axis, the equation determining the position of the smectic planes becomes

$$z - \frac{d}{2\pi}\tan^{-1}(z/y) = kd + u_0, \tag{9.2.6}$$

where the constant u_0 defining the zero of the coordinate has been introduced to facilitate geometrical interpretation. (As we shall see in Sec. 9.3, the $\tan^{-1}$ form for u is not the one yielding the lowest energy for this dislocation. It suffices, however, for the present qualitative discussion.) Fig. 9.2.6 shows how to construct planes from Eq. (9.2.6) for $b = d$. Also shown is a dislocation for $b = -d$. For **l** pointing out of the paper as shown, a $b = d$ dislocation requires the insertion

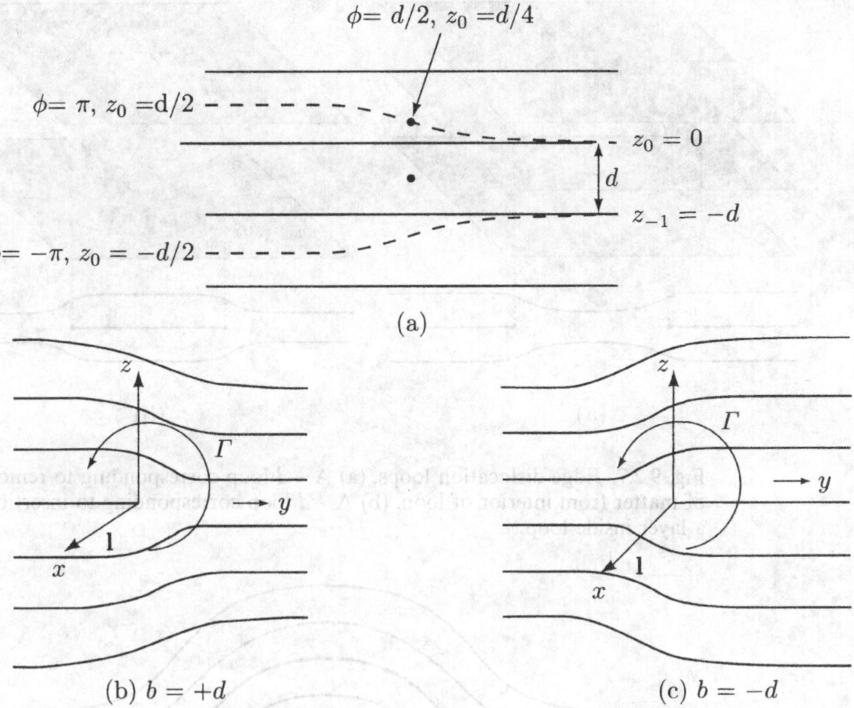

Fig. 9.2.6. (a) Representation of displacement of smectic planes as a function of $\phi = \tan^{-1} z/y$ for an edge dislocation with $u_0 = d/2$. (b) An edge dislocation with $b = +d$. (c) An edge dislocation with $b = -d$. An extra plane with an edge along $\mathbf{l}$ is inserted from the direction of $\mathbf{l} \times \mathbf{b}$ or alternatively removed from the direction of $-(\mathbf{l} \times \mathbf{b})$.

of an extra layer from the left (or the removal of a layer from the right) of the dislocation, whereas the dislocation with $b = -d$ requires the insertion of an extra layer from the left. These pictures show that $+d$ and $-d$ dislocations annihilate when they combine to reproduce a state with uniform layering. The direction of $\mathbf{l}$ relative to the z-axis can, of course, be arbitrary. In this general case, a dislocation can be decomposed into a screw part and an edge part.

Dislocation lines must either terminate at the boundary of the sample or form closed loops inside the sample. Closed loops lying entirely in the xy-plane are pure edge loops and describe the removal or addition of layers inside the loop, as shown in Fig. 9.2.7. The covering surface S of the loop marks the points where the field $\mathbf{u}$ undergoes a discontinuity of $\mathbf{b}$. S is the analog of the line of discontinuity for θ shown in Fig. 9.1.4. The surface S is not uniquely defined by its perimeter. It is merely the surface where one chooses to place the discontinuity in $\mathbf{u}$. It can be chosen to coincide exactly with the surface where matter was

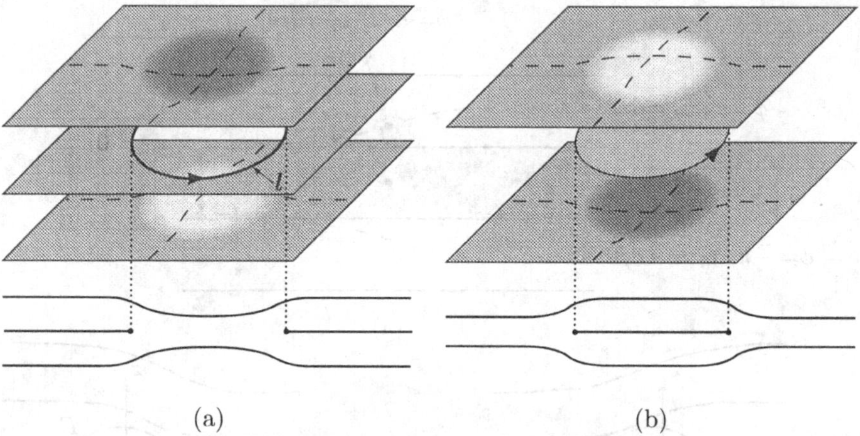

Fig. 9.2.7. Edge dislocation loops. (a) A $+d$ loop corresponding to removal of matter from interior of loop. (b) A $-d$ loop corresponding to insertion of a layer inside loop.

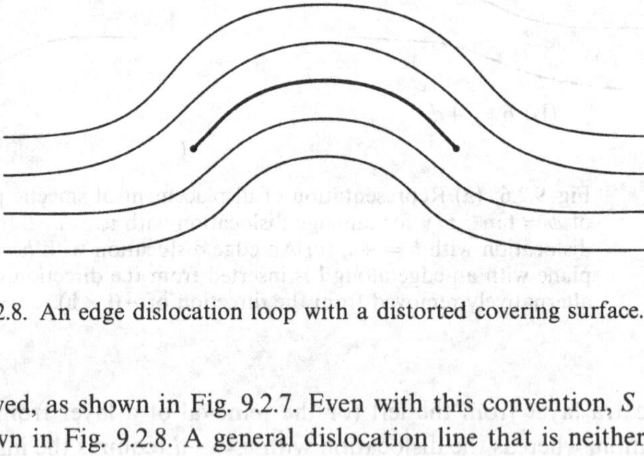

Fig. 9.2.8. An edge dislocation loop with a distorted covering surface.

inserted or removed, as shown in Fig. 9.2.7. Even with this convention, S can be distorted, as shown in Fig. 9.2.8. A general dislocation line that is neither in the xy-plane nor parallel to the z-axis yields both twist and edge distortions.

In smectic liquid crystals, edge dislocations with large Burgers vectors are often important. A large Burgers vector dislocation is shown in Fig. 9.2.9.

3 Periodic solids

The elastic variable analogous to θ in a periodic solid is the vector displacement field $\mathbf{u}(\mathbf{x})$. Changes of $\mathbf{u}$ by a direct lattice vector $\mathbf{R}$ leave the lattice unchanged. Thus the order parameter space for the solid is the three-dimensional space of displacements $\mathbf{u}$ with lattice points identified as shown in Fig. 9.2.10. An alternative representation in terms of a 3-torus (the generalization of the circle

Fig. 9.2.9. A large Burgers vector edge dislocation in a smectic liquid crystal.

for the *xy*-model) is also possible. A dislocation with a core along $\mathbf{l}$ is thus characterized by

$$\oint d\mathbf{u} = \oint_{\Gamma} \frac{d\mathbf{u}}{ds} ds = \mathbf{R} \equiv \mathbf{b}, \tag{9.2.7}$$

where Γ is a curve enclosing the core. This equation defines the mapping depicted in Fig. 9.2.10: a curve from the origin to $\mathbf{b}$ in the order parameter space is traced out as the core is encircled along Γ. The set of vectors $\mathbf{R}$ indexing the strength of the dislocations are the *Burgers vectors* $\mathbf{b}$ of the crystal. In periodic lattices, the Burgers vector lattice and the direct lattice of atomic positions are equivalent. As in the smectic, there are screw dislocations if $\mathbf{l}$ is parallel to $\mathbf{b}$ and edge dislocations if $\mathbf{l}$ is perpendicular to $\mathbf{b}$, as depicted in Fig. 9.2.11.

There is a simple way, depicted in Fig. 9.2.12, to determine the Burgers vector of a dislocation. Consider any closed path in an ideal crystal following nearest neighbor bonds in the lattice. Count the number of steps along nearest neighbor bonds that are made in each of the lattice directions in completing the circuit along this path. Now follow exactly the same sequence of steps in a path around a dislocation core starting at a point S and ending at a point E. This path is not closed. The vector from S to E, as shown in Fig. 9.2.12, is the Burgers vector characterizing the strength of the dislocation. Alternatively, the circuit around the dislocation may be traversed first. The same sequence of steps in the undislocated lattice will not close, and in this case the vector from the ending point to the starting point is the Burgers vector.

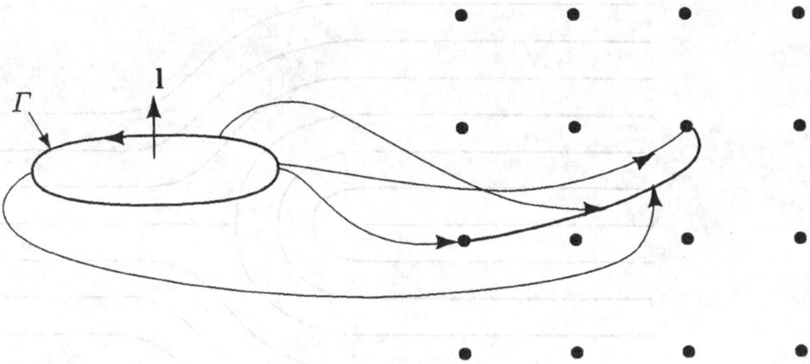

Fig. 9.2.10. Two-dimensional Burgers vector lattice and mapping of a loop Γ onto a path between Burgers vectors in the order parameter space.

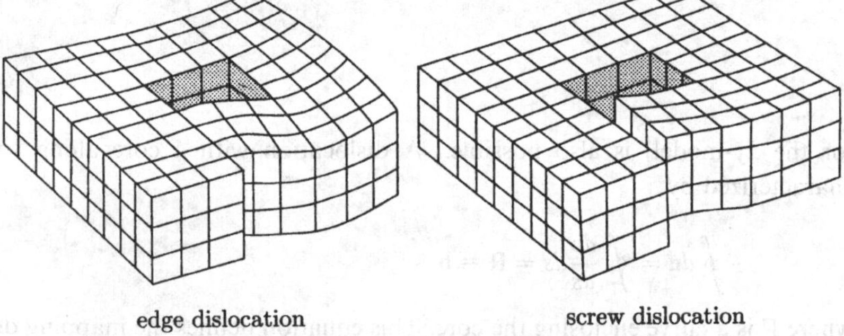

Fig. 9.2.11. Edge and screw dislocations in solids.

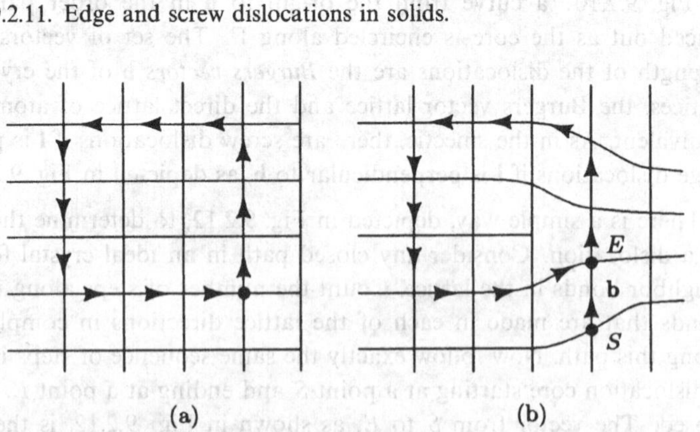

Fig. 9.2.12. Burgers circuit in an undislocated (a) and a dislocated lattice (b). The Burgers vector **b** is the vector SE.

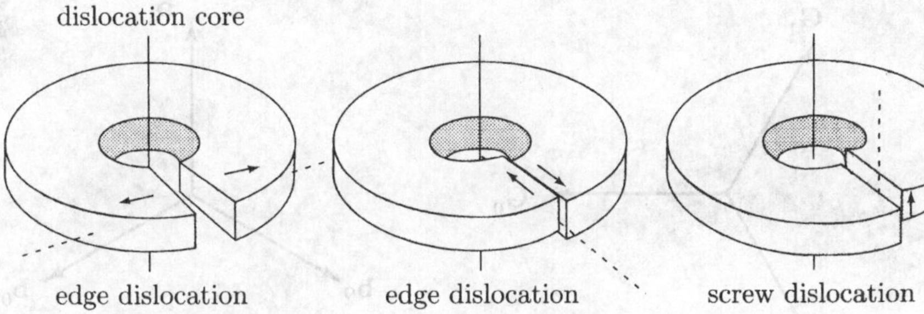

Fig. 9.2.13. The Volterra construction for edge and screw dislocations.

4 *Volterra construction*

A useful way of thinking of dislocations in solids is provided by the *Volterra* construction. A cylinder of material oriented along the z-axis is cut along the yz-plane. The material on the two sides of the plane are displaced by $\mathbf{b}$ relative to each other and then glued together (with additional material inserted if needed), as shown in Fig. 9.2.13. A screw dislocation is created if $\mathbf{b}$ is along the z-axis, and two types of edge dislocations are created when $\mathbf{b}$ is in the xy-plane. This construction places no constraint on the displacement b. If the material is periodic, however, the two sides will not glue together perfectly unless $\mathbf{b}$ is a vector of the direct lattice. The direct lattice provides a quantization of the possible values of $\mathbf{b}$.

5 *Hexagonal and close-packed lattices*

The hexagonal, BCC, FCC, and related lattices require special consideration. The sites on a hexagonal lattice are the points of intersection of the three sets of parallel grids defined by the equations

$$\mathbf{G}_0 \cdot \mathbf{x} = 2\pi k_0, \qquad \mathbf{G}_1 \cdot \mathbf{x} = 2\pi k_1, \qquad \mathbf{G}_2 \cdot \mathbf{x} = 2\pi k_2, \tag{9.2.8}$$

where $\mathbf{G}_0$, $\mathbf{G}_1$ and $\mathbf{G}_2$ are the three vectors in the hexagonal reciprocal lattice shown in Fig. 9.2.14. These equations may be understood in terms of the Fourier expansion of the density:

$$\rho(\mathbf{x}) = \rho_0 + \sum_{\mathbf{G}} \rho_{\mathbf{G}} e^{i\mathbf{G}\cdot\mathbf{x}} = \rho_0 + 2\,|\,\rho_{\mathbf{G}_0}\,|\sum_{n=0}^{2} \cos(\mathbf{G}_n \cdot \mathbf{x} + \phi_n) + \cdots, \tag{9.2.9}$$

where by symmetry $|\,\rho_{\mathbf{G}_0}\,| = |\,\rho_{\mathbf{G}_1}\,| = |\,\rho_{\mathbf{G}_2}\,|$. The phases ϕ_n of the mass-density waves at wave number $\mathbf{G}_n$ can be expressed as

$$\phi_n = \gamma/3 - \mathbf{G}_n \cdot \mathbf{u}. \tag{9.2.10}$$

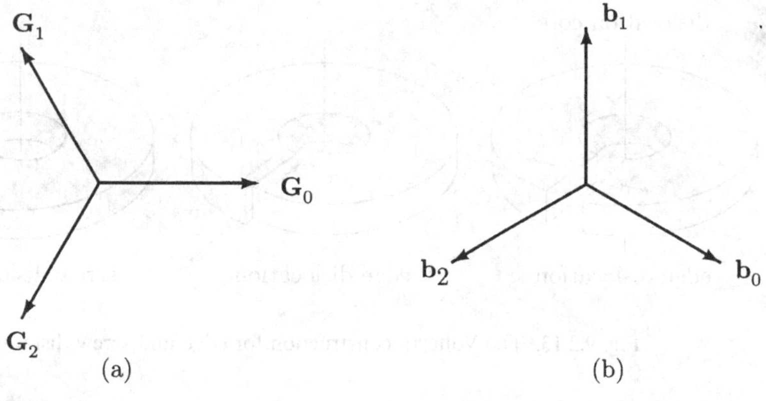

Fig. 9.2.14. (a) The vectors $\mathbf{G}_0$, $\mathbf{G}_1$ and $\mathbf{G}_2$ in a hexagonal reciprocal lattice. (b) Vectors $\mathbf{b}_0$, $\mathbf{b}_1$ and $\mathbf{b}_2$ in the direct lattice. The vectors $\mathbf{b}_0$ and $\mathbf{b}_1$ form a basis for the direct lattice and satisfy $\mathbf{b}_i \cdot \mathbf{G}_j = 2\pi\delta_{ij}$.

The constant γ is fixed in the equilibrium state, as can be seen by noting that the third order term in the liquid-solid free energy discussed in Sec. 4.6 is proportional to

$$-w\cos(\phi_1 + \phi_2 + \phi_3) = -w\cos\gamma. \tag{9.2.11}$$

Thus, if only third-order terms are present, $\gamma = 0$ is preferred for positive w. In general, other values of γ may be imposed by higher order terms in the expansion. Eqs. (9.2.8) define planes of constant phase of the mass-density waves.

Now consider a dislocation for which the displacement $\mathbf{u}(\mathbf{x})$ changes by a Burgers vector $\mathbf{b}$ in one circuit of the core. In this case, $\mathbf{u}(\mathbf{x})$ depends on $\mathbf{x}$, and Eqs. (9.2.8), defining planes of constant phase of the mass density, become

$$\mathbf{G}_n \cdot [\mathbf{x} - \mathbf{u}(\mathbf{x})] = 2\pi k_n, \quad k_n = \text{integer}. \tag{9.2.12}$$

For a dislocation with Burgers vector $-\mathbf{b}_1$ (Fig. 9.2.14), the phase of the mass-density wave with spatial modulation parallel to $\mathbf{G}_2$ changes by 2π and that with modulations parallel to $\mathbf{G}_1$ changes by -2π. Thus, a dislocation in a hexagonal lattice involves the insertion of extra lines of atoms in two directions (Fig. 9.2.15). In Fig. 9.2.15 both lines terminate at approximately the same point in space. This is not necessarily the case. It is possible, and even common, for dislocations in hexagonal lattices to be composed of two *partial dislocations* separated by several lattice spacings. An extra line of atoms ends at the core of each partial dislocation. The phase of only a single mass-density wave changes by 2π in a circuit around the core of a partial. Dislocations in FCC lattices also involve the insertion of more than one layer of atoms. Fig. 9.2.16 shows a dislocation pair in a magnetic bubble domain pattern.

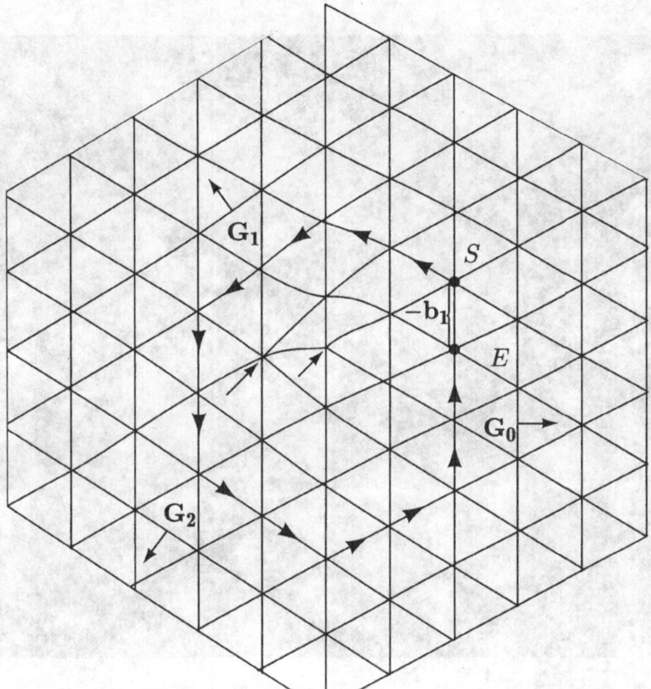

Fig. 9.2.15. A dislocation in a two-dimensional hexagonal crystal. The lines in this figure are lines of constant phase of the three mass-density waves of the crystal. The Burgers vector is $-\mathbf{b}_1$, and extra lines are inserted from the right in the grids perpendicular to $\mathbf{G}_1$ and $\mathbf{G}_2$. The Burgers circuit around the core starts at S and ends at E. The vector connecting S to E is the vector $-\mathbf{b}_1$ shown in Fig. 9.2.14. Note that, in the core region, there is a site with five neighbors and a site with seven neighbors, indicating that a dislocation can be viewed as a disclination pair.

6 *Disclinations in crystals*

There is another class of topological defects in solids analogous to twist and bend distortions that could occur in a smectic liquid crystal. They involve rotations of the lattice and are called *disclinations.* They are energetically very costly and occur only under special circumstances in the solid phase. They are, however, easily produced in the hexatic bond-angle-ordered phase that can intervene between the solid and isotropic liquid phases in two dimensions. The easiest way to visualize a disclination is via the Volterra construction. The two sides of the Volterra cut are twisted rather than translated relative to each other, as shown in Fig. 9.2.17. The point group of the lattice restricts the number of angles through which the two sides can be rotated and still glued together. In simple

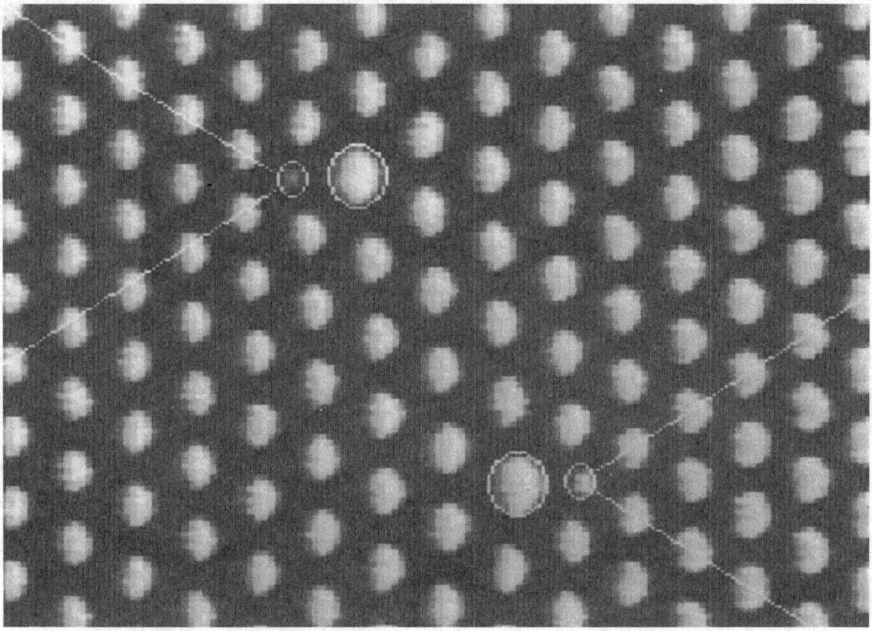

Fig. 9.2.16. Magnetic bubble domain pattern showing a dislocation pair in an otherwise ideal hexagonal lattice. The size of the magnetic bubbles can adjust to minimize energy. The bubbles at five-fold sites at dislocation cores contract whereas those at seven-fold sites expand. This pattern is on a magnetic garnet film of composition $(YGdTm)_3(FeGa)_6O_{12}$ grown to a thickness of approximately 13μm on a single crystal substrate of gadolinium gallium garnet in the (111) orientation. It was produced by cooling the film from the paramagnetic state in a small normal field ($H \sim 1$ oersted). [M. S. Seul and C.A. Murray, *Science* **262**, 558 (1993).]

cubic lattices, only rotations of multiples of $\pi/2$ are permitted. Wedge and twist disclinations of $\pm\pi/2$ in a cubic lattice are shown in Fig. 9.2.18. Disclinations of strength $\pm\pi/3$ for a hexagonal lattice are shown in Fig. 9.2.19. Note that the positive disclination has a site that is five- rather than six-fold coordinated, and the negative disclination has a site that is seven-fold coordinated. A single dislocation in a triangular lattice has one five- and one seven-fold coordinated site (Figs. 9.2.15 and 9.2.16). A dislocation is thus equivalent to a plus-minus disclination pair.

7 Strength of crystals

Dislocations play a very important role in determining the strength of materials. The *yield stress* σ_m of a solid is the maximum stress to which it can be subjected

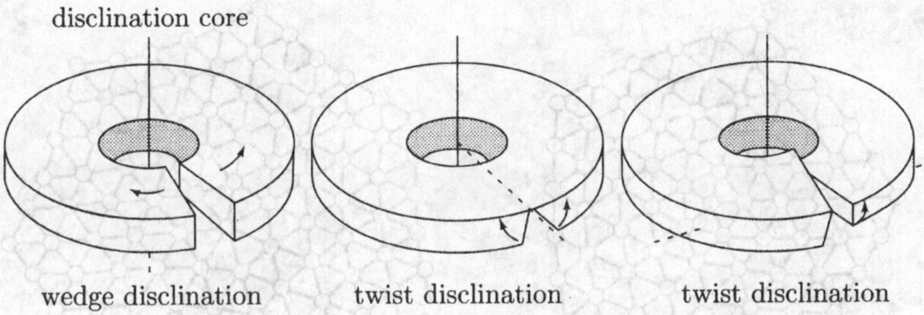

Fig. 9.2.17. Volterra construction for wedge and twist disclinations.

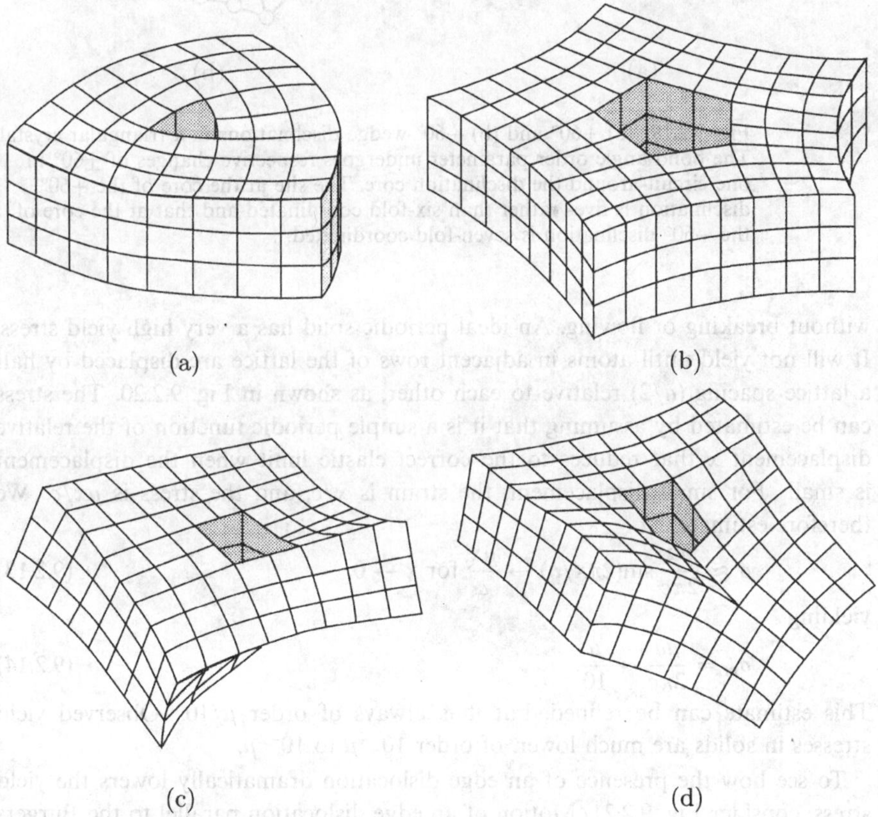

Fig. 9.2.18. Wedge and twist disclinations in a cubic solid: (a) +90° wedge, (b) −90° wedge, (c) +90° twist, (d) −90° twist.

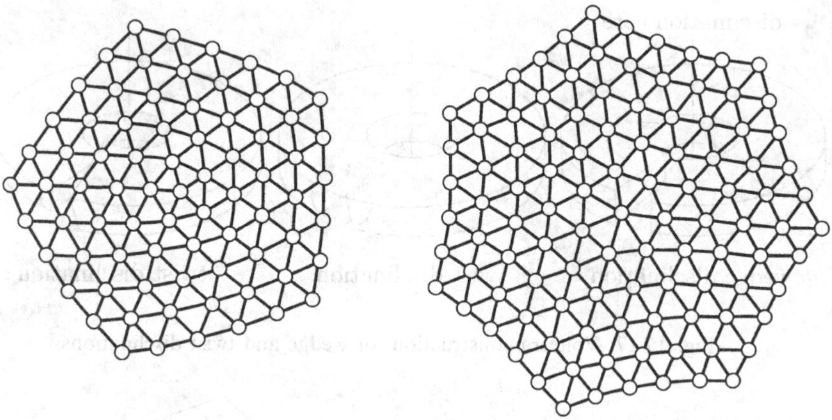

(a) (b)

Fig. 9.2.19. (a) $+60°$ and (b) $-60°$ wedge disclinations in a triangular crystal. The bond angle order parameter undergoes respective changes of $\pm 60°$ in one circuit around the disclination core. The site at the core of the $+60°$ disclination is five- rather than six-fold coordinated and that at the core of the $-60°$ disclination is seven-fold coordinated.

without breaking or flowing. An ideal periodic solid has a very high yield stress. It will not yield until atoms in adjacent rows of the lattice are displaced by half a lattice spacing ($a/2$) relative to each other, as shown in Fig. 9.2.20. The stress can be estimated by assuming that it is a simple periodic function of the relative displacement x that reduces to the correct elastic limit when the displacement is small. For small displacement, the strain is x/c, and the stress is $\mu x/c$. We therefore estimate

$$\sigma \approx \frac{\mu a}{2\pi c} \sin(2\pi x/a) \to \frac{\mu x}{c} \text{ for } x \to 0, \tag{9.2.13}$$

yielding

$$\sigma_m = \frac{\mu a}{2\pi c} \sim \frac{\mu}{10}. \tag{9.2.14}$$

This estimate can be refined, but it is always of order $\mu/10$. Observed yield stresses in solids are much lower, of order $10^{-4}\mu$ to $10^{-2}\mu$.

To see how the presence of an edge dislocation dramatically lowers the yield stress, consider Fig. 9.2.21. Motion of an edge dislocation parallel to the Burgers vector **b** requires essentially no mass motion. Only local displacements of atoms are needed to move the dislocation with an extra plane of atoms along AA' and core at site A to one with an extra plane along BB' and core at site B adjacent to A. For this reason, motion of a dislocation along **b** is called *conservative.* The plane along which conservative motion or *glide* takes place is called the glide or *slip* plane. Motion of a dislocation perpendicular to **b**, i.e., an upward

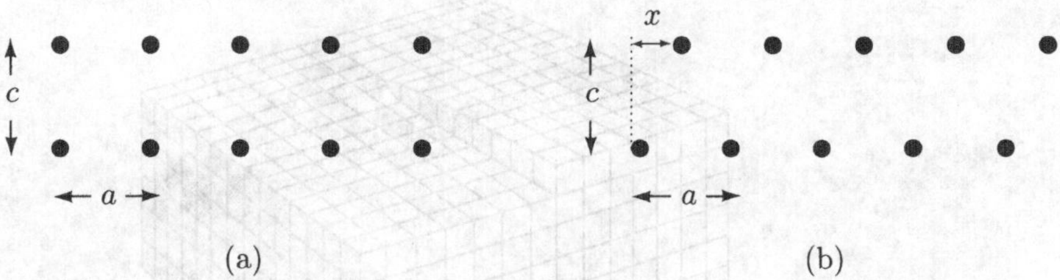

Fig. 9.2.20. (a) Unstrained and (b) strained rectangular lattice with lattice constants c and a.

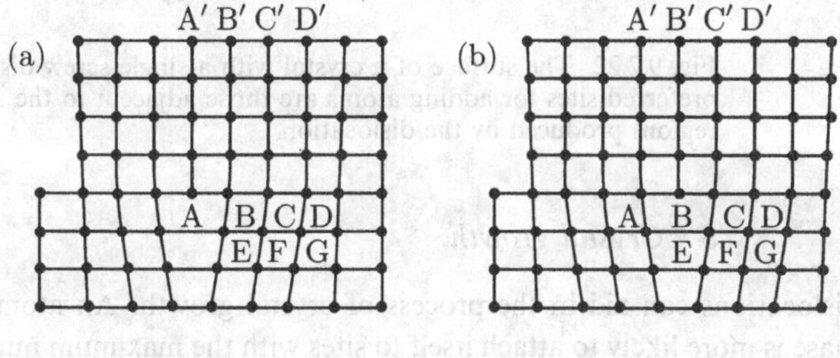

Fig. 9.2.21. A stressed crystal with an edge dislocation. In (a) the extra plane is AA'; in (b) it is BB'. The displacement of the top relative to the bottom is much less than half a lattice spacing. When the dislocation has moved off the right end of the crystal, there will be a perfect bulk crystal with a jog on the left surface.

displacement of the dislocation core at A in Fig. 9.2.21, requires motion of an entire plane of atoms, and it is much more difficult to produce. Such motion is called *climb*. Now consider the effect of a shear stress applied to the defected crystal shown in Fig. 9.2.21. The stress required to produce dislocation glide is far less than that needed to displace atoms by half a lattice spacing in an ideal crystal. Furthermore, moving the top half of the crystal by one lattice spacing relative to the bottom half corresponds to moving a single dislocation the entire length of the crystal. Thus crystal flow is produced with a far lower stress in a dislocated crystal than in an ideal crystal. Typical crystals that are not specially prepared have a dislocation density, interpreted either as the number of dislocation cores crossing a unit area of material or as the total length of dislocation core per unit volume of material, ranging from 10^2 to 10^{12} cm^{-2}. Freely mobile dislocations reduce the strength of a material. To make a material strong, one needs either to reduce the density of dislocations or inhibit dislocation motion. This can be accomplished by pinning dislocations to impurity sites or by putting in so many dislocations that they inhibit each other's motion. The latter process is called *work hardening*.

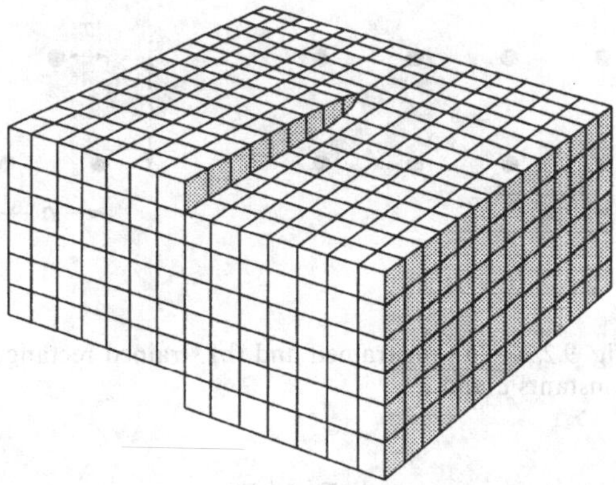

Fig. 9.2.22. The surface of a crystal with a single screw dislocation. The preferred sites for adding atoms are those adjacent to the step (darkened region) produced by the dislocation.

8 Crystal growth

Dislocations can aid in the process of crystal growth. An atom from the vapor phase is more likely to attach itself to sites with the maximum number of occupied neighbor sites. Thus, a corner or an edge site is preferable to an exposed site, as shown in Fig. 9.2.22. This leads to growth of crystalline whiskers with a single twist dislocation.

9 Grain boundaries

A low-angle grain boundary separating two crystals with a different orientation can be interpreted as a surface containing a sequence of dislocations. The angle of the grain boundary depends on the number of dislocations per unit length in the boundary. Fig. 9.2.23 shows an edge grain boundary and a twist grain boundary in a smectic-*A* liquid crystal. Each grain boundary consists of regularly spaced unit strength dislocations separated by a distance l_d. In both cases, the layer spacing far from the grain boundary is d. Simple geometry shows that the smectic layers rotate through an angle

$$\delta\theta = 2\sin^{-1}(d/2l_d) \approx d/l_d \tag{9.2.15}$$

from one side of the grain boundary to the other. This relation is equivalent to Eq. (9.2.4), $\oint_\Gamma du = N_d d$, where N_d is the number of unit strength dislocations enclosed in Γ. Consider first the edge grain boundary (Fig. 9.2.23a), which we take to be in the yz-plane. At large distances above and below the boundary, the normal to the smectic layers is, respectively, $\mathbf{N}_+$ and $\mathbf{N}_-$, with

$$\mathbf{N}_\pm = [0, \pm\sin(\delta\theta/2), \cos(\delta\theta/2)]. \tag{9.2.16}$$

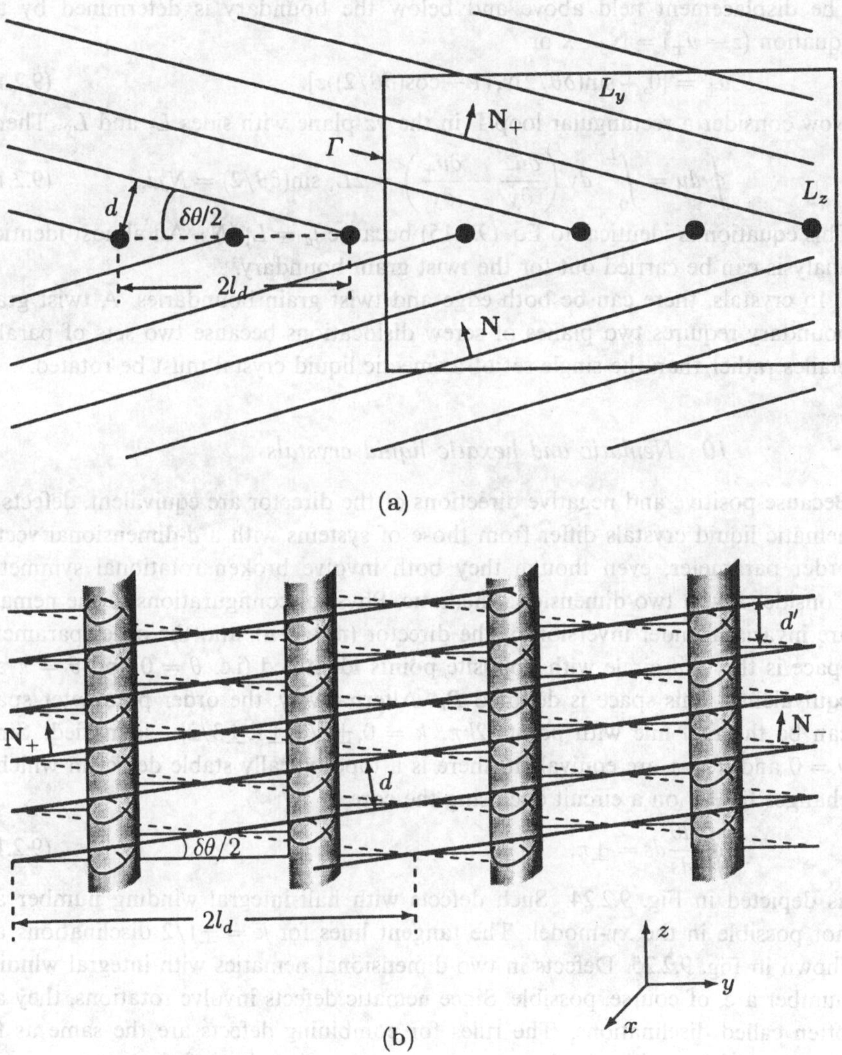

Fig. 9.2.23. (a) A low-angle edge grain boundary and (b) a low-angle twist grain boundary in a smectic-*A* liquid crystal. The normal **N** to the smectic layers rotates through an angle $\delta\theta$ from $\mathbf{N}_-$ to $\mathbf{N}_+$ from one side of the grain boundary to the other. Low-angle grain boundaries can be represented as a regular array of dislocations. The edge dislocations in (a) come out of the paper and pass through the terminated planes. Since the terminated planes enter from the left, the dislocations have positive Burgers vectors, as shown in Fig. 9.2.6. The layer spacing far from the grain boundary is the equilibrium spacing d. A rectangular contour Γ is shown in (a). The slanted dark lines in (b) show the smectic planes at large distances in front of the grain boundary, and the dashed slanted lines show the smectic planes behind the grain boundary. Also shown are the smectic layers in the plane of the grain boundary with separation $d' = d/\cos(\delta\theta/2)$ rather than d.

The displacement field above and below the boundary is determined by the equation $(z - u_\pm) = \mathbf{N}_\pm \cdot \mathbf{x}$ or

$$u_\pm = [0, \mp \sin(\delta\theta/2)y, (1 - \cos(\delta\theta/2))z]. \tag{9.2.17}$$

Now consider a rectangular loop Γ in the yz-plane with sides L_z and L_y. Then,

$$\oint du = \int_0^{L_y} dy \left(\frac{\partial u_-}{\partial y} - \frac{\partial u_+}{\partial y} \right) = 2L_y \sin(\partial\theta/2) = N_d d. \tag{9.2.18}$$

This equation is identical to Eq. (9.2.15) because $l_d = L_y/N_d$. An almost identical analysis can be carried out for the twist grain boundary.

In crystals, there can be both edge and twist grain boundaries. A twist grain boundary requires two planes of screw dislocations because two sets of parallel planes rather than the single set of a smectic liquid crystal must be rotated.

10 Nematic and hexatic liquid crystals

Because positive and negative directions of the director are equivalent, defects in nematic liquid crystals differ from those of systems with a d-dimensional vector order parameter, even though they both involve broken rotational symmetry. Consider first a two-dimensional nematic. Physical configurations in the nematic are invariant under inversion of the director ($\mathbf{n} \to -\mathbf{n}$), and the order parameter space is the unit circle with opposite points identified (i.e. $\theta = 0$ and $\theta = \pi$ are equivalent). This space is denoted P_1. Alternatively, the order parameter space can be the real line with points $2k\pi$, $k = 0, \pm 1/2, \pm 1, \pm 3/2, \ldots$ identified. Since $\theta = 0$ and $\theta = \pi$ are equivalent, there is a topologically stable defect in which θ changes by $\pm\pi$ on a circuit enclosing the core:

$$\oint_\Gamma \frac{d\theta}{ds} ds = \pm\pi, \tag{9.2.19}$$

as depicted in Fig. 9.2.24. Such defects with half-integral winding number are not possible in the xy-model. The tangent lines for $k = \pm 1/2$ disclinations are shown in Fig. 9.2.25. Defects in two-dimensional nematics with integral winding number are, of course, possible. Since nematic defects involve rotations, they are often called disclinations. The rules for combining defects are the same as for the xy-model so that the homotopy group remains that of the integers under addition, i.e. $\pi_1(P_1) = Z$.

The situation in three dimensions is quite different. We argued in Sec. 9.1 that there is no topologically stable line defect for a three-component vector order parameter in three dimensions because it is possible continuously to roll any loop off a sphere. The order parameter space P_2 for the three-dimensional nematic is the unit sphere S_2 with antipodal points identified. It is still possible to roll any loop closed in S_2 off of P_2. One therefore expects no integral winding number defects for the nematic. A defect with a winding number of exactly 1/2 is possible, however, since it is impossible to deform to a point a path starting at one pole and ending at the other. All other half-integer paths, starting at

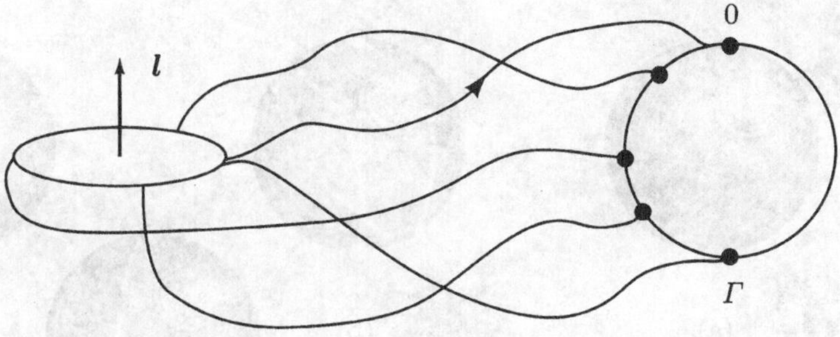

Fig. 9.2.24. Mapping between the circuit Γ and the order parameter space for a +1/2 disclination in a two-dimensional nematic.

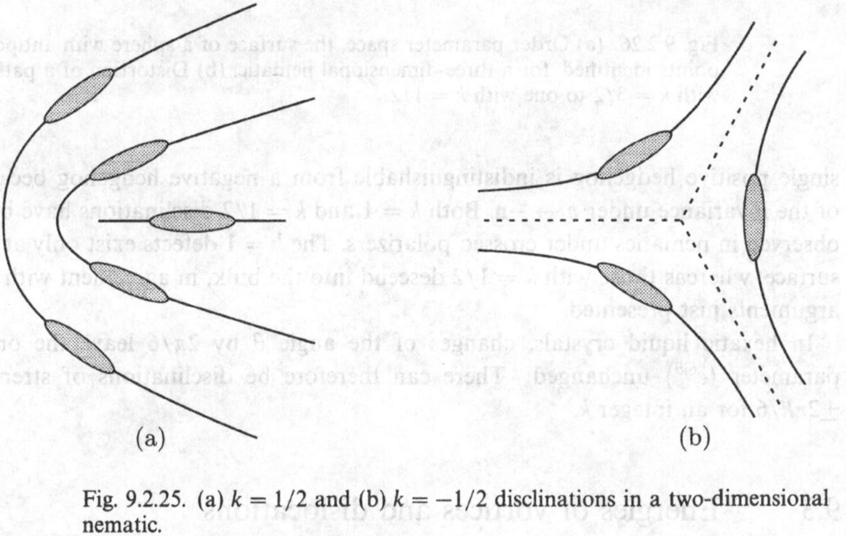

Fig. 9.2.25. (a) $k = 1/2$ and (b) $k = -1/2$ disclinations in a two-dimensional nematic.

one pole and winding several times around the sphere before terminating at the other pole, can be continuously deformed into the winding number 1/2 path starting and ending at a pole, as shown in Fig. 9.2.26. Thus, there is *one and only one* stable line defect in a nematic liquid crystal. The homotopy group is simply the integers modulo 2: $\pi_1(P_2) = Z_2$. This means, for example, that the disclination loops shown in Fig. 9.2.27 formed by rotating the +1/2 and −1/2 disclinations shown in Fig. 9.2.25 about a vertical axis displaced from the core are topologically equivalent. Inspection of the director configurations far from the core shows that this result is not surprising. The configurations far from the +1/2 disclination loop are equivalent to those of the +1 hedgehog of Fig. 9.1.8a with arrows removed, whereas those far from the −1/2 loop are equivalent to those of the +1 hedgehog of Fig. 9.1.8b, again with arrows removed. In a nematic, a

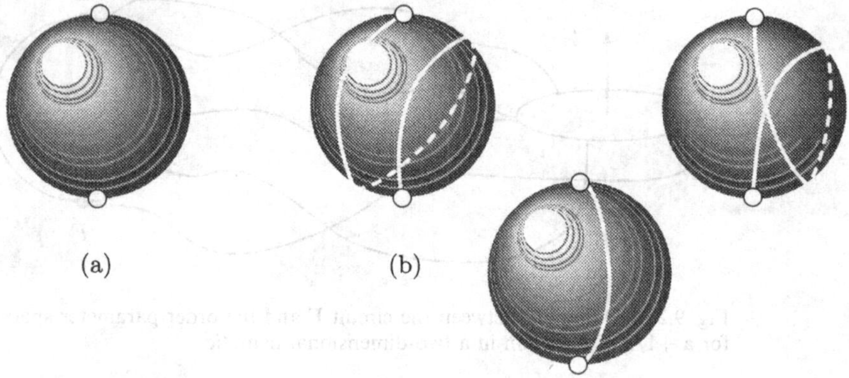

Fig. 9.2.26. (a) Order parameter space, the surface of a sphere with antipodal points identified, for a three-dimensional nematic. (b) Distortion of a path with $k = 3/2$ to one with $k = 1/2$.

single positive hedgehog is indistinguishable from a negative hedgehog because of the invariance under $\mathbf{n} \to -\mathbf{n}$. Both $k = 1$ and $k = 1/2$ disclinations have been observed in nematics under crossed polarizers. The $k = 1$ defects exist only at the surface, whereas those with $k = 1/2$ descend into the bulk, in agreement with the arguments just presented.

In hexatic liquid crystals, changes of the angle θ by $2\pi/6$ leave the order parameter $\langle e^{6i\theta}\rangle$ unchanged. There can therefore be disclinations of strength $\pm 2\pi k/6$ for an integer k.

9.3 Energies of vortices and dislocations

1 Simple calculation of xy-vortex energies

The energy E_d of topological defects can be divided into two parts: (1) the core energy E_c, and (2) the elastic or strain energy E_{el}. The core energy is that associated with the destruction of the order parameter at the core of the defect. Its detailed calculation requires some microscopic model for the order and is generally quite complicated. The order of magnitude of E_c can, however, be obtained by observing that the increase in free energy per unit volume due to the destruction of the order parameter is the condensation energy f_{cond} of the ordered state. In the ϕ^4 field theories of Chapter 3, $f_{\text{cond}} = r^2/16u$. E_c is of order the volume (or area) of the defect times f_{cond}. Thus, the core energy of a vortex in two dimensions is

$$E_c = Aa^2 f_{\text{cond}}, \tag{9.3.1}$$

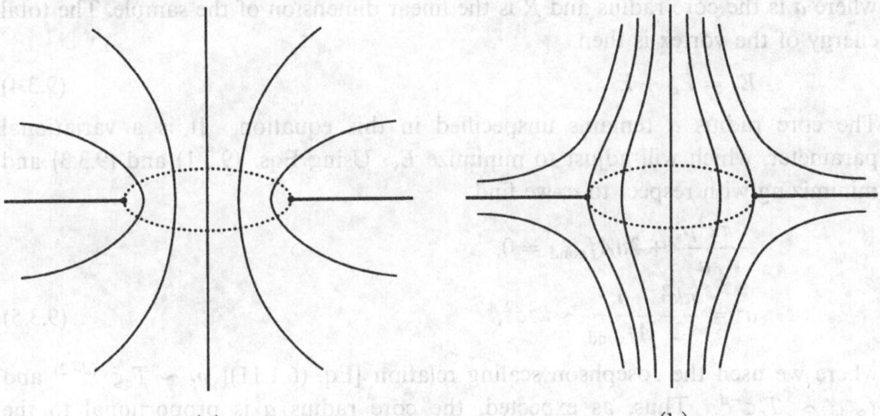

Fig. 9.2.27. Disclination loops in a nematic formed by rotating the (a) $k = 1/2$ and (b) $k = -1/2$ disclinations shown in Fig. 9.2.25 about a vertical axis. These two configurations are topologically equivalent and can be distorted one into the other via continuous transformations. The far-fields of both correspond to a $+1$ hedgehog with (a) equivalent to Fig. 9.1.8a and (b) to Fig. 9.1.8b.

where A is a numerical constant. This is also the core energy per unit length of a vortex line in three dimensions. The elastic energy is that associated with the slow spatial variation of the elastic variable (e.g., θ or $\mathbf{u}$) far from the core. It can be calculated using the theory of elasticity developed in Chapter 6 and the boundary conditions imposed by the defect.

Consider first the two-dimensional xy-vortex with winding number k. As we have seen, $\oint d\theta = 2k\pi$ for any loop enclosing the vortex. The actual configuration outside the vortex core is obtained by minimizing the elastic free energy subject to this topological constraint, i.e., θ a solution to Eq. (6.1.2) $[-\rho_s\nabla^2\theta = 0]$ with $\oint d\theta = 2k\pi$. The field

$$\begin{aligned} \theta &= k\phi, \\ \mathbf{v}_s &\equiv \nabla\theta = \frac{k}{r}\mathbf{e}_\phi, \qquad r = (x^2+y^2)^{1/2}, \end{aligned} \tag{9.3.2}$$

where $\phi = \tan(y/x)$, satisfies both the circuit constraint and the equilibrium condition everywhere except at the origin. We have introduced here the notation $\mathbf{v}_s$ for $\nabla\theta$. Using this expression for θ, the elastic energy of the vortex (strictly speaking the free energy) is easily calculated:

$$\begin{aligned} E_{\rm el} &= \frac{1}{2}\rho_s\int d^2x\mathbf{v}_s^2 = \frac{1}{2}\rho_s 2\pi k^2\int_a^R\frac{rdr}{r^2} \\ &= \pi k^2\rho_s\ln(R/a), \end{aligned} \tag{9.3.3}$$

where a is the core radius and R is the linear dimension of the sample. The total energy of the vortex is then

$$E_v = E_{el} + E_c. \tag{9.3.4}$$

The core radius a remains unspecified in this equation. It is a variational parameter, which will adjust to minimize E_v. Using Eqs. (9.3.1) and (9.3.3) and minimizing with respect to a, we find

$$-\frac{\pi k^2 \rho_s}{a} + 2aAf_{\text{cond}} = 0,$$

$$a^2 = \frac{\pi k^2}{2} \frac{\rho_s}{Af_{\text{cond}}} \sim k^2\xi^2, \tag{9.3.5}$$

where we used the Josephson scaling relation [Eq. (6.1.11)] $\rho_s \sim T_c\xi^{-(d-2)}$ and $f_{\text{cond}} \sim T_c\xi^{-d}$. Thus, as expected, the core radius a is proportional to the correlation length ξ. It also grows linearly with winding number k. The core energy at the optimal value of a grows quadratically with k:

$$E_c = \pi\rho_s k^2/2. \tag{9.3.6}$$

An identical calculation applies in three dimensions, in which case Eq. (9.3.6) is the energy per unit length of the vortex line.

An alternative calculation of E_{el} is instructive and is more easily generalized to the calculation of interaction energies. This calculation uses the fact that $\mathbf{v}_s$ is a continuous variable whereas θ is multivalued. A cut along which θ undergoes a discontinuity of $2k\pi$ is introduced as shown in Fig. 9.1.4. This cut has two surfaces labeled Σ^- and Σ^+. θ is zero on Σ^- and $2k\pi$ on Σ^+. The normals to Σ^- and Σ^+ are in the $-\mathbf{e}_\phi$ and $\mathbf{e}_\phi$ directions, respectively. The vortex energy is now obtained by an integration by parts using the fact that $\nabla^2\theta = 0$:

$$E_{el} = \frac{1}{2}\int d^2x \rho_s(\nabla\theta)^2 = \frac{1}{2}\int \theta\mathbf{h}_s \cdot d\mathbf{\Sigma} - \frac{1}{2}\rho_s \int d^2x\theta\nabla^2\theta, \tag{9.3.7}$$

where we introduced the field,

$$\mathbf{h}_s = \rho_s\mathbf{v}_s, \tag{9.3.8}$$

conjugate to $\mathbf{v}_s$. The surface integral is over the two sides of the cuts, and because $\nabla^2\theta = 0$, we have

$$\begin{aligned} E_{el} &= \frac{1}{2}\left(\int \theta^+\mathbf{h}_s \cdot d\mathbf{\Sigma}^+ + \int \theta^-\mathbf{h}_s \cdot d\mathbf{\Sigma}^-\right) \\ &= \frac{1}{2}(\theta^+ - \theta^-)\int_a^R dr\rho_s \mid \mathbf{v}_s \mid \\ &= \pi k^2\rho_s \ln(R/a), \end{aligned} \tag{9.3.9}$$

reproducing our first calculation.

The interaction energy between two vortices can be calculated in a similar manner by introducing a cut for each vortex, as shown in Fig. 9.3.1. The solution to Laplace's equation with two vortices at positions $\mathbf{x}_1$ and $\mathbf{x}_2$ is $\theta(\mathbf{x}) = \theta^{(1)} + \theta^{(2)}$, where $\theta^{(i)} = \tan^{-1}[(y - y_i)/(x - x_i)]$ (see Problem 9.1). Let $\mathbf{v}_s^{(1)} = \nabla\theta^{(1)}$ and

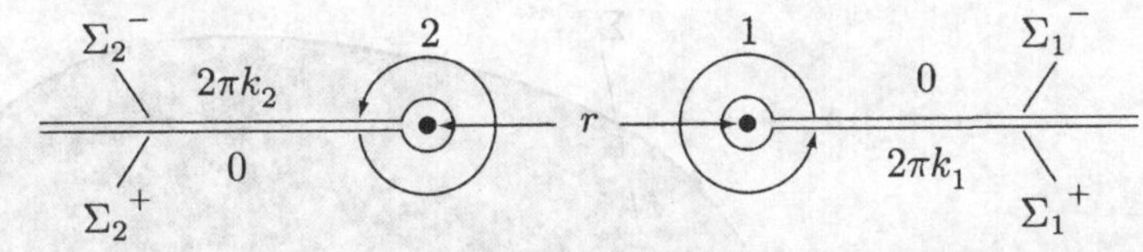

Fig. 9.3.1. Non-intersecting cuts for two vortices. The "+" surfaces always have their normals parallel to the vector $\mathbf{e}_\phi$ defined with respect to an origin at the core of the associated vortex.

$\mathbf{v}_s^{(2)} = \nabla\theta^{(2)}$. Then the energy of two vortices with respective winding numbers k_1 and k_2 is

$$\begin{aligned} E_{\text{el}} &= \frac{1}{2}\int d^2x\rho_s(\mathbf{v}_s^{(1)} + \mathbf{v}_s^{(2)})^2 \\ &= E_1 + E_2 + \frac{1}{2}\int d^2x(\mathbf{h}_s^{(1)}\cdot\mathbf{v}_s^{(2)} + \mathbf{h}_s^{(2)}\cdot\mathbf{v}_s^{(1)}) \\ &= E_1 + E_2 + \frac{1}{2}(\theta_2^+ - \theta_2^-)\int_r^R \rho_s\frac{k_1}{r}dr + \frac{1}{2}(\theta_1^+ - \theta_1^-)\int_r^R \rho_s\frac{k_2}{r}dr \\ &= E_1 + E_2 + 2\pi\rho_s k_1 k_2 \ln(R/r), \end{aligned} \tag{9.3.10}$$

where E_1 and E_2 are the energies of the isolated vortices, which are distance r apart. The R part of the interaction term can be combined with the R-dependent parts of E_1 and E_2 to yield

$$E_{\text{el}} = \rho_s\pi(k_1 + k_2)^2\ln(R/a) + 2\pi\rho_s k_1 k_2 \ln(a/r). \tag{9.3.11}$$

There are two things to note about this equation. First, the $\ln R$ divergence is eliminated if $k_1 = -k_2$. More generally, there will be no $\ln R$ term in a system with many vortices if the sum, $\sum_\alpha k_\alpha$, of the winding numbers of all of the vortices is zero. Thus, states containing vortices whose winding numbers satisfy $\sum_\alpha k_\alpha = 0$ have energies that do not diverge with the sample size. These states will be thermally excited for $T > 0$. Secondly, the sign of the interaction energy depends on the relative sign of k_1 and k_2. Since $\ln(a/r)$ is a monotonically decreasing function of increasing r, the interaction is clearly repulsive if k_1 and k_2 have the same sign and attractive if they have opposite signs. The force $\mathbf{F}_{21}$ exerted by vortex 1 on vortex 2 is

$$\mathbf{F}_{21} = -\nabla_2 E = 2\pi\rho_s k_1 k_2 \frac{(\mathbf{x}_2 - \mathbf{x}_1)}{|\ \mathbf{x}_1 - \mathbf{x}_1\ |^2}. \tag{9.3.12}$$

The above result also applies to the energy of two parallel vortex lines in three dimensions. As discussed in the preceding section, there can also be vortex loops in three dimensions, as shown in Fig. 9.3.2. The energy of such loops can be reduced to an integral over the covering surface Σ of the loop by the same technique we used to calculate the energy of two vortices. Let the discontinuity in θ take place across Σ. Then on the upper part Σ^- of Σ, $\theta = \theta^-$, and on the lower part Σ^+, $\theta = \theta^+ = 2k\pi$, and we have

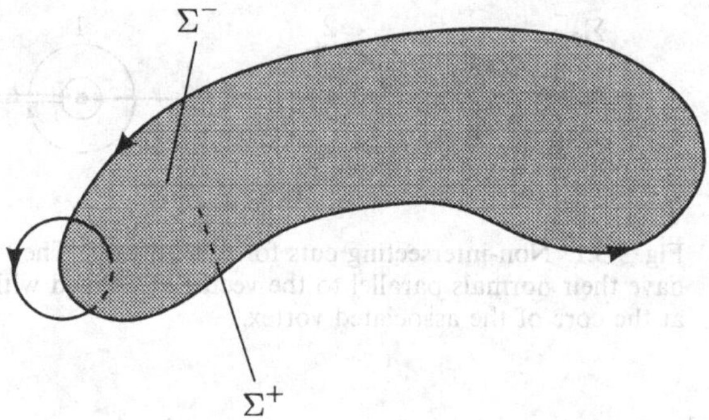

Fig. 9.3.2. Covering surface of a vortex loop. $\theta = \theta^-$ on the upper surface, and $\theta = \theta^+$ on the lower surface.

$$E = \frac{1}{2}\int d^3x \mathbf{h}_s \cdot \mathbf{v}_s = \frac{1}{2}(\theta^+ - \theta^-)\int \mathbf{h}_s \cdot d\boldsymbol{\Sigma} = \frac{1}{2}(2\pi k)\int \mathbf{h}_s \cdot d\boldsymbol{\Sigma}, \quad (9.3.13)$$

where k is the winding number of the loop.

2 Analogy with magnetism

There is an obvious analogy between vortex interactions and magnetic interactions among loops carrying constant currents. The line integral of $\mathbf{v}_s$ around a vortex line is 2π times the winding number. This is analogous to the line integral of the magnetic intensity $\mathbf{H}$ equaling the current I carried by the enclosed wire, i.e.,

$$\begin{array}{lcl} \int \mathbf{v}_s \cdot d\mathbf{l} = 2k\pi & & \mathbf{H} \leftrightarrow \mathbf{v}_s \\ & \Rightarrow & \\ \int \mathbf{H} \cdot d\mathbf{l} = I & & I \leftrightarrow 2k\pi \end{array} . \quad (9.3.14)$$

Similarly, the equation of elastic equilibrium implies that the divergence of $\mathbf{h}_s = \rho_s \mathbf{v}_s$ is zero. Thus, $\mathbf{h}_s$ is the analog of the magnetic induction $\mathbf{B}$ and ρ_s is the analog of the magnetic permeability μ:

$$\begin{array}{lcl} \rho_s \nabla^2 \theta = \boldsymbol{\nabla} \cdot \mathbf{h}_s = 0 & & \mathbf{B} \leftrightarrow \mathbf{h}_s \\ & \Rightarrow & \\ \boldsymbol{\nabla} \cdot \mathbf{B} = \mu \boldsymbol{\nabla} \cdot \mathbf{H} = 0 & & \mu \leftrightarrow \rho_s \end{array} . \quad (9.3.15)$$

Given this analogy, it is natural to introduce a field $\mathbf{m}$ in the vortex problem that is the analog of the current density $\mathbf{J}$ of the magnetic problem. If there are many vortex lines with winding numbers k_α passing through the surface S enclosed by a contour Γ, then

$$\oint_\Gamma \mathbf{v}_s \cdot d\mathbf{l} = \int \boldsymbol{\nabla} \times \mathbf{v}_s \cdot d\mathbf{S} = 2\pi \sum_\alpha k_\alpha = \int \mathbf{m} \cdot d\mathbf{S}, \quad (9.3.16)$$

or

$$\boldsymbol{\nabla} \times \mathbf{v}_s = \mathbf{m}. \quad (9.3.17)$$

The field $\mathbf{m}$ defined in this way is the vortex analog of the electric current density $\mathbf{J}$. The vortex density associated with a single vortex is clearly singular at the vortex core since any contour enclosing the core will yield $2\pi k$ for $\oint \mathbf{v}_s \cdot d\mathbf{l}$. Thus, for a line parallel to $\mathbf{e}_z$ at $\mathbf{x} = (\mathbf{x}_\perp, 0)$,

$$\mathbf{m}(\mathbf{x}) = 2\pi k \mathbf{e}_z \delta^{(2)}(\mathbf{x}_\perp - \mathbf{x}). \tag{9.3.18}$$

This is precisely analogous to the current density of a wire. If there are many vortex lines parallel to the z-axis with cores at positions $(\mathbf{x}_\perp^\alpha, 0)$, then

$$\mathbf{m}(\mathbf{x}) = \sum_\alpha 2\pi k_\alpha \mathbf{e}_z \delta^{(2)}(\mathbf{x}_\perp^\alpha - \mathbf{x}). \tag{9.3.19}$$

More generally, if there are vortex loops with winding numbers k_α with core positions as a function of arc length l specified by $\mathbf{x}^\alpha(l)$, then

$$\mathbf{m}(\mathbf{x}) = \sum_\alpha \int dl \frac{d\mathbf{x}^\alpha(l)}{dl} 2\pi k_\alpha \delta^{(3)}(\mathbf{x}^\alpha(l) - \mathbf{x}). \tag{9.3.20}$$

$\delta^{(3)}(\mathbf{x}^\alpha(l) - \mathbf{x})$ is a three-dimensional delta function. The vector $d\mathbf{x}^\alpha(l)/dl$ is the unit tangent to the vortex core at l. Because the lines of dislocation cores must either form closed loops or terminate at a surface, $\mathbf{m}$ must be divergenceless ($\nabla \cdot \mathbf{m} = 0$). Alternatively, $\nabla \cdot \mathbf{m} = 0$ directly from Eq. (9.3.17). This condition is satisfied by Eq. (9.3.20) because $d\delta^{(3)}(\mathbf{x}^\alpha(l) - \mathbf{x})/dl = -(d\mathbf{x}^\alpha/dl) \cdot \nabla \delta^{(3)}(\mathbf{x}^\alpha - \mathbf{x})$.

The energies associated with collections of vortices and vortex loops can be expressed in terms of the current $\mathbf{m}$ just as the magnetic energy can be expressed in terms of the current $\mathbf{J}$. Taking the curl of Eq. (9.3.17), we obtain

$$\nabla \times (\nabla \times \mathbf{v}_s) = -\nabla^2 \mathbf{v}_s = \nabla \times \mathbf{m}, \tag{9.3.21}$$

since $\nabla \cdot \mathbf{v}_s = 0$. Thus,

$$\mathbf{v}_s = \nabla \times \int d^d x G(\mathbf{x} - \mathbf{x}') \mathbf{m}(\mathbf{x}'), \tag{9.3.22}$$

where $G(\mathbf{x} - \mathbf{x}')$ is the Green function for minus the Laplacian in d dimensions. Using this expression in the equation for the energy, we obtain

$$\begin{aligned} E_{\text{el}} &= \frac{1}{2}\rho_s \int d^d x \int d^d x' \mathbf{m}(\mathbf{x}) \cdot G(\mathbf{x} - \mathbf{x}') \mathbf{m}(\mathbf{x}') \\ &= \frac{1}{2}\rho_s \int \frac{d^d q}{(2\pi)^d} \mathbf{m}(\mathbf{q}) \cdot \frac{1}{q^2} \mathbf{m}(-\mathbf{q}) \end{aligned} \tag{9.3.23}$$

for the elastic energy due to vortices. It is straightforward to verify that the interaction energy between two vortices derived from this equation is identical to Eq. (9.3.11). Because $\nabla \cdot \mathbf{h}_s = 0$, one could also introduce the analog of the vector potential $\mathbf{A}$ via $\mathbf{h}_s = \nabla \times \mathbf{A}$ to obtain the same result (see Problem 9.2).

3 *Energies of dislocations in crystals*

The energies for dislocation lines in solids are similar to those of vortex lines in the xy-system. The strain u_{ij} associated with a dislocation at the origin is of order $1/r$ so that the energy per unit length of a dislocation line is of order $\ln R$. The detailed calculation of dislocation energies is, however, much more complicated because of the greater complexity both of the elastic variables and the associated elastic energy.

The displacement field $\mathbf{u}$ is a solution to the equilibrium equations

$$\begin{aligned} \frac{\delta F}{\delta u_i} &= -\nabla_j \sigma_{ij} = 0, \\ &= -2\mu \nabla_j u_{ij} - \lambda \nabla_i u_{kk} = 0 \end{aligned} \tag{9.3.24}$$

everywhere but in the cores of dislocations and disclinations. If there is a single straight dislocation aligned along $\mathbf{l}$ with Burgers vector $\mathbf{b}$, then $\mathbf{u}$ is a solution to Eq. (9.3.24) subject to the Burgers circuit constraint $\oint d\mathbf{u} = \mathbf{b}$. In order to satisfy this constraint, $\mathbf{u}$ must have a singular part $\mathbf{w} = \mathbf{b}\phi/2\pi$, where ϕ is an angle in the plane perpendicular to $\mathbf{l}$. To satisfy the condition of equilibrium, $\mathbf{u}$ may have an additional nonsingular part $\tilde{\mathbf{u}}$, i.e., $\mathbf{u} = \tilde{\mathbf{u}} + \mathbf{w}$ with $\oint d\tilde{\mathbf{u}} = 0$.

Consider first a screw dislocation with $\mathbf{l}$ along the z-axis. In this case,

$$\mathbf{b} = b\mathbf{e}_z, \qquad \phi = \tan^{-1}(y/x), \tag{9.3.25}$$

and $\mathbf{u} = \mathbf{b}\phi/2\pi$ is a solution to Eq. (9.3.24), and no additional nonsingular part is needed. The components of the strain tensor produced by the above $\mathbf{u}$ are

$$\begin{aligned} u_{xz} &= u_{zx} = \frac{1}{2}\frac{b}{2\pi}\left(-\frac{y}{x^2+y^2}\right) = -\frac{b}{4\pi}\frac{\sin\phi}{r} \\ u_{yz} &= u_{zy} = \frac{1}{2}\frac{b}{2\pi}\frac{x}{x^2+y^2} = \frac{b}{4\pi}\frac{\cos\phi}{r}. \end{aligned} \tag{9.3.26}$$

All other strains (e.g. u_{xx}) are zero. Since $\nabla \cdot \mathbf{u} = 0$, the screw dislocation involves only shear, and

$$\sigma_{ij} = 2\mu u_{ij}, \tag{9.3.27}$$

from which we obtain

$$\begin{aligned} E_{\text{el}}/L &= \mu \int d^2x u_{ij} u_{ij} = \mu(b/4\pi)^2 2(2\pi) \int_a^R \frac{dr}{r} \\ &= \mu \frac{b^2}{4\pi} \ln(R/a), \end{aligned} \tag{9.3.28}$$

where L is the length along the z-axis. Thus, the energy of a screw dislocation in an isotropic solid is identical to the energy of a vortex in an xy-model.

For an edge dislocation with $\mathbf{l}$ along the z-axis and $\mathbf{b} = b\mathbf{e}_x$, $\mathbf{w} = b\mathbf{e}_x \tan^{-1}(y/x) = b\mathbf{e}_x\phi$. In this case,

$$w_{ji} = \nabla_j w_i = -\frac{b}{2\pi}\delta_{ix}\epsilon_{jk}\nabla_k \ln r, \tag{9.3.29}$$

where $\epsilon_{jk} = -\epsilon_{kj}$ is the antisymmetric tensor, leading to a *symmetrized* singular strain tensor $w_{ij}^s = (w_{ij} + w_{ji})/2$ that is not a solution of Eq. (9.3.24):

$$2\mu \nabla_j w_{ij}^s + \lambda \nabla_i w_{kk}^s = -\frac{b}{2\pi}(\mu\epsilon_{ik}\nabla_x + \lambda\epsilon_{xk}\nabla_i)\nabla_k \ln r \equiv f_i^w. \tag{9.3.30}$$

It should be noted that $\nabla_j w_{ij}$ is not equal to $\nabla_i(w_{jj})$ as it would be if $\mathbf{w}$ were not singular.

To satisfy Eq. (9.3.24), we add to $\mathbf{u}$ a nonsingular part $\tilde{\mathbf{u}}$ with associated strain $\tilde{u}_{ij}$ satisfying $\nabla_j \tilde{u}_{ij} = (\nabla^2 \tilde{u}_i + \nabla_i \boldsymbol{\nabla} \cdot \tilde{\mathbf{u}})/2$. For $\mathbf{u} = \tilde{\mathbf{u}} + \mathbf{w}$ to satisfy Eq. (9.3.24), $\tilde{\mathbf{u}}$ must satisfy

$$-2\mu \nabla^2 \tilde{u}_i - (\lambda + 2\mu)\nabla_i \boldsymbol{\nabla} \cdot \tilde{\mathbf{u}} = f_i^w. \tag{9.3.31}$$

This equation is most directly solved in Fourier space using the displacement susceptibility $\chi_{ij}(\mathbf{q}) = T^{-1} G_{ij}(\mathbf{q})$ of Eq. (6.4.24):

$$\tilde{u}_i(\mathbf{q}) = \chi_{ij}(\mathbf{q}) f_i^w(\mathbf{q}) = -\frac{b}{q^4}\left[\epsilon_{ik} q_k q_x + \frac{\lambda}{\lambda + 2\mu} q_y q_i\right]. \tag{9.3.32}$$

The coordinate space solution can be obtained from the function (Problem 9.9)

$$\begin{aligned} \mathscr{G}_{ij}(\mathbf{x}) &= -\int \frac{d^2 q}{(2\pi)^2} \frac{q_i q_j}{q^4} e^{i\mathbf{q}\cdot\mathbf{x}} \\ &= \frac{1}{8\pi}\left[(2\ln r - 1)\delta_{ij} + 2\frac{x_i x_j}{r^2}\right] + \text{const.} \end{aligned} \tag{9.3.33}$$

The constant term evaluated directly from the Fourier integral diverges as $\ln R$ for a sample of size R. When used to determine $\tilde{\mathbf{u}}$, however, it depends on boundary conditions imposed on $\tilde{\mathbf{u}}$, and we can choose it to be zero. The total displacement field, including both singular and nonsingular parts, is

$$\begin{aligned} u_x &= \frac{b}{2\pi}\left(\phi + \frac{1}{2}\frac{\lambda + \mu}{\lambda + 2\mu}\sin 2\phi\right), \\ u_y &= -\frac{b}{2\pi}\left(\frac{\mu}{\lambda + 2\mu}\ln r + \frac{1}{2}\frac{\lambda + \mu}{\lambda + 2\mu}\cos 2\phi\right). \end{aligned} \tag{9.3.34}$$

The associated stress tensor is

$$\begin{aligned} \sigma_{xx} &= -D\frac{\sin\phi(2 + \cos 2\phi)}{r}, \qquad \sigma_{yy} = D\frac{\sin\phi\cos 2\phi}{r} \\ \sigma_{xy} &= \sigma_{yx} = D\frac{\cos\phi\cos 2\phi}{r}, \end{aligned} \tag{9.3.35}$$

where

$$D = \frac{b}{2\pi}\frac{2\mu(\lambda + \mu)}{\lambda + 2\mu}. \tag{9.3.36}$$

These equations apply both to a line dislocation in three dimensions and to a point dislocation in two dimensions. In the former case, $D = \mu b/[2\pi(1 - \sigma)]$, where σ is the three-dimensional Poisson ratio [see Eq. (6.6.17)]; in the latter case, $D = bY_2/4\pi$, where Y_2 is the two-dimensional Young modulus [Eq. (6.6.18)]. The energy of an edge dislocation is then

$$\begin{aligned} E_{\text{el}}/L &= \frac{1}{2}\int d^2 x \sigma_{ij} u_{ij} = \frac{1}{2}\int d^2 x \sigma_{ij} \nabla_j u_i \\ &= \frac{1}{2}(u_i^+ - u_i^-)\int \sigma_{ij} d\Sigma_j = \frac{1}{2} b \int_a^R \sigma_{xy}(\phi = 0) dx \\ &= \frac{b^2}{2\pi}\frac{\mu(\lambda + \mu)}{\lambda + 2\mu}\ln(R/a), \end{aligned} \tag{9.3.37}$$

where, as in the case of vortices, Σ is the cut surface and $\mathbf{u}^+$ and $\mathbf{u}^-$ are the values of $\mathbf{u}$ on the upper and lower surfaces of the cut.

The displacement field produced by an edge dislocation gives rise to an antisymmetric part in $\partial_i u_j$, or, equivalently, to local rotations through an angle

$$\theta = \frac{1}{2}(\partial_x u_y - \partial_y u_x) = -\frac{b}{2\pi}\frac{x}{r^2} = -\frac{b}{2\pi r}\cos\phi. \tag{9.3.38}$$

The integral of θ around a closed loop is zero, indicating, as it should, that a dislocation is not a disclination.

In two dimensions, this result can be generalized to an arbitrary distribution of dislocations at positions $\mathbf{x}^\alpha$ with Burgers vectors $\mathbf{b}_\alpha$:

$$\begin{aligned}\theta(\mathbf{x}) &= -\frac{1}{2\pi}\sum_\alpha \frac{\mathbf{b}_\alpha \cdot (\mathbf{x}-\mathbf{x}^\alpha)}{|\,\mathbf{x}-\mathbf{x}^\alpha\,|^2} \\ &= -\frac{1}{2\pi}\int d^2x' \frac{\mathbf{b}(\mathbf{x}')\cdot(\mathbf{x}-\mathbf{x}')}{|\,\mathbf{x}-\mathbf{x}'\,|^2},\end{aligned} \tag{9.3.39}$$

where we have introduced the two-dimensional vector dislocation density

$$\mathbf{b}(\mathbf{x}) = \sum_\alpha \mathbf{b}_\alpha \delta(\mathbf{x}-\mathbf{x}^\alpha). \tag{9.3.40}$$

Equation (9.3.39) tells how dislocations influence bond-angle order. It will be used in Sec. 9.5 in our discussion of dislocation mediated melting.

An expression for the energy due to dislocations analogous to Eq. (9.3.23) for vortices is useful when there are many dislocations. In three dimensions, the expression is fairly complicated because of the tensor nature of the dislocation density. In two dimensions, however, it is quite elegant. To define a dislocation density, we proceed exactly as we did for vortices. Let $w_{ji} = \nabla_j u_i$ be the analog of $\mathbf{v}_s$ for the xy-model. Then,

$$\oint_\Gamma du_i = \oint_\Gamma w_{ji}dl_j = \int \epsilon_{kj}\nabla_k w_{ji} d^2x = \sum_\alpha b_{\alpha j}, \tag{9.3.41}$$

where the final sum is over the Burgers vectors $\mathbf{b}_\alpha$ of all dislocations enclosed by Γ. This equation can be reexpressed in terms of the dislocation density as

$$\epsilon_{kj}\nabla_k w_{ji} = b_i(\mathbf{x}). \tag{9.3.42}$$

If there are also disclinations of strength s_α, then

$$\oint_\Gamma d\theta = \oint_\Gamma \nabla_i\theta dl_i = \int d^2x \epsilon_{ij}\nabla_i\nabla_j\theta = \sum_\alpha s_\alpha, \tag{9.3.43}$$

where $\theta = \epsilon_{ij}w_{ij}/2$ is the bond angle. In a hexagonal lattice, $s_\alpha = \pm\pi/3$. Thus,

$$\epsilon_{ij}\nabla_i\nabla_j\theta = s(\mathbf{x}), \tag{9.3.44}$$

where

$$s(\mathbf{x}) = \sum_\alpha s_\alpha \delta(\mathbf{x}-\mathbf{x}^\alpha) \tag{9.3.45}$$

is the disclination density. If there are both dislocations and disclinations, then it is understood that Eq. (9.3.41) is still satisfied, i.e., that the line integral of $d\mathbf{u}$ around a closed loop does not pick up a contribution from disclinations.

We now want to find the energy associated with an arbitrary distribution of dislocations and disclinations. To do this, we could calculate the nonsingular parts of the strain field arising from the singular parts required by Eqs. (9.3.42) and (9.3.44). Alternatively, we can seek a stress tensor σ_{ij} satisfying Eq. (9.3.24). We choose the latter approach. Because σ_{ij} is symmetric and satisfies Eq. (9.3.24), it can be written as

$$\sigma_{ij} = \epsilon_{ik}\epsilon_{jl}\nabla_k\nabla_l\chi. \tag{9.3.46}$$

This equation is the analog of $\mathbf{B} = \mathbf{\nabla} \times \mathbf{A}$ in magnetism. The function χ is called the Airy stress function. The strain u_{ij} is related to the stress via the two-dimensional version of Eq. (6.6.13):

$$\begin{aligned} u_{ij} &= \frac{1}{2\mu}\sigma_{ij} - \frac{\lambda}{4\mu(\lambda+\mu)}\delta_{ij}\sigma_{kk} \\ &= \frac{1+\sigma_2}{Y_2}\epsilon_{ik}\epsilon_{kl}\nabla_k\nabla_l\chi - \frac{\sigma_2}{Y_2}\nabla^2\chi\delta_{ij}, \end{aligned} \tag{9.3.47}$$

where $Y_2 = 4B\mu/(B+\mu)$ is the two-dimensional Young's modulus and $\sigma_2 = (B-\mu)/(B+\mu)$ the two-dimensional Poisson ratio. Applying $\epsilon_{ik}\epsilon_{jl}\nabla_k\nabla_l$ to both sides of this equation, we find

$$\begin{aligned} \frac{1}{Y_2}\nabla^4\chi &= \frac{1}{2}\epsilon_{ik}\epsilon_{jl}\nabla_k\nabla_l(w_{ij}+w_{ji}) \\ &= \frac{1}{2}\epsilon_{ik}\epsilon_{jl}\nabla_k\nabla_l(w_{ij}-w_{ji}) + \epsilon_{ik}\epsilon_{jl}\nabla_k\nabla_l w_{ji} \\ &= s(\mathbf{x}) - \epsilon_{ik}\nabla_k b_i(\mathbf{x}) \equiv \tilde{s}(\mathbf{x}), \end{aligned} \tag{9.3.48}$$

where we used $(w_{ij}-w_{ji})/2 = \epsilon_{ij}\theta$, $\epsilon_{ik}\epsilon_{jl}\epsilon_{ij} = \epsilon_{kl}$ and Eqs. (9.3.42) and (9.3.44). In analogy with charge in a dielectric, $\tilde{s}(\mathbf{x})$ can be regarded as a total disclination density with a contribution $s(\mathbf{x})$ from "free" disclinations and a "polarization" contribution $-\epsilon_{ik}\nabla_k b_i$ from dislocations which, as indicated in Figs. 9.2.15 and 9.2.16, are bound disclination pairs.

The energy can be expressed in terms of σ_{ij} and, in turn, in terms of the Airy function χ:

$$E = \frac{1}{2Y_2}\int d^2x(\nabla^2\chi)^2 + \frac{1+\sigma_2}{Y_2}\int d^2x\epsilon_{ik}\epsilon_{jl}\nabla_k\nabla_l(\nabla_i\chi\nabla_j\chi). \tag{9.3.49}$$

The second term integrates to the boundary. It can be important, particularly when the total disclination number is nonzero. If, however, we restrict our attention to situations in which the total disclination number and total Burgers vector are zero, it can be neglected, and the energy expressed in Fourier space becomes

$$E = \frac{1}{2}Y_2\int\frac{d^2q}{(2\pi)^2}\frac{1}{q^4}\tilde{s}(\mathbf{q})\tilde{s}(-\mathbf{q}). \tag{9.3.50}$$

This equation predicts that the energy of a single disclination diverges as R^2 in a system of linear dimension R. When the free disclination density $s(\mathbf{x})$ is zero, Eq. (9.3.50) becomes an energy for dislocations alone:

$$E = \frac{1}{2}Y_2\int\frac{d^2q}{(2\pi)^2}\frac{1}{q^2}(\delta_{ij}-\hat{q}_i\hat{q}_j)b_i(\mathbf{q})b_j(-\mathbf{q}), \tag{9.3.51}$$

predicting a $\ln R$ energy for a single dislocation. Note that this energy depends only on the transverse part of the dislocation density $\mathbf{b}(\mathbf{q})$.

4 Dislocations in smectic liquid crystals

As we have seen, dislocations in smectic liquid crystals are topological excitations in which the phase variable $u(\mathbf{x})$ undergoes a change of kd in one circuit of the core. One might expect, in analogy with vortices in the xy-model, that the energy of a single dislocation in a smectic should grow logarithmically with the size of the sample. A smectic, however, has an additional variable, the director $\mathbf{n}$, which can adjust to lower the energy arising from the topological constraint on u. As a result, as we shall now see, the energy per unit length of a dislocation is finite rather than infinite, and the interaction potential between dislocations is, except for edge dislocations with separation vectors along the z-axis, an exponential rather than a logarithmic function of separation (Day, Lubensky, and McKane 1983). An exactly analogous phenomenon occurs in superconductors, where the superfluid velocity carries charge that couples to the vector potential (Abrikosov 1957; Fetter and Hohenberg 1969). The vector potential in a superconductor, like the director in a liquid crystal, can adjust to lower the strain energy associated with the existence of a vortex. As a result, vortices in superconductors have a finite rather than infinite energy per unit length.

Before deriving a general expression for the energy of interacting dislocations, it is instructive to consider a single screw dislocation for which the Burgers vector is parallel to the equilibrium director $\mathbf{n}_0$. In this case, all spatial variations in $\delta\mathbf{n} = \mathbf{n} - \mathbf{n}_0$ and u take place in the xy-plane perpendicular to $\mathbf{n}_0$. In this case, the smectic elastic energy [Eq. (6.3.1)] in the harmonic approximation, valid when $|\,\delta\mathbf{n}\,| \ll 1$, can be expressed as the energy per unit length:

$$F/L = \frac{1}{2}\int d^2x[D(\nabla u + \delta\mathbf{n})^2 + K_1(\nabla\cdot\delta\mathbf{n})^2 + K_2(\nabla\times\delta\mathbf{n})^2]. \quad (9.3.52)$$

The Euler-Lagrange equations,

$$\frac{\delta(F/L)}{\delta u} = -D\nabla\cdot(\nabla u + \delta\mathbf{n}) = 0, \quad (9.3.53)$$

$$\frac{\delta(F/L)}{\delta\mathbf{n}} = -K_1\nabla(\nabla\cdot\delta\mathbf{n}) + K_2\nabla\times(\nabla\times\delta\mathbf{n}) + D(\nabla u + \delta\mathbf{n}) = 0, \quad (9.3.54)$$

minimizing F/L must be satisfied at all points in space outside of the core region. To solve these equations in the presence of a single screw dislocation, we introduce cylindrical coordinates (ρ, ϕ) for the two-dimensional vector $\mathbf{x}$ and define $\mathbf{v} = \nabla u$. The boundary conditions involving $\mathbf{v}$ are then

$$\oint \mathbf{v}\cdot d\mathbf{l} = \int_0^{2\pi} v_\phi \rho\, d\phi = d \quad (9.3.55)$$

and

$$\lim_{\rho\to\infty}(\mathbf{v} + \delta\mathbf{n}) \to 0. \quad (9.3.56)$$

The first condition, valid for all ρ greater than the core radius a, is the topological statement that there is a dislocation at the origin, and the second is the requirement that the director $\mathbf{n}$ and the layer normal $\mathbf{N}$ [Eq. (6.3.6)] be parallel far from the dislocation core. In addition, because the director is well defined in both the smectic phase exterior to the core and in the nematic phase in the core interior, it must be nonsingular as $\rho \to 0$. To satisfy the condition in Eq. (9.3.55), we set

$$\mathbf{v} = \frac{d}{2\pi\rho}\mathbf{e}_\phi. \tag{9.3.57}$$

Then, $\nabla \cdot \mathbf{v} = 0$ for all $\rho > a$. Distortions produced by the dislocation should be azimuthally symmetric, implying that $|\,\delta\mathbf{n}\,|$ should depend on ρ only. From this, and the fact that Eq. (9.3.53) implies $\nabla \cdot \mathbf{n} = 0$ when $\mathbf{v}$ satisfies Eq. (9.3.57), we conclude that $\delta\mathbf{n}$ is parallel to $\mathbf{e}_\phi$. We can, therefore, define the vector,

$$\mathbf{Q}(\mathbf{x}) = \mathbf{v} + \delta\mathbf{n} = Q(\rho)\mathbf{e}_\phi, \tag{9.3.58}$$

measuring the difference between $\mathbf{n}$ and $\mathbf{N}$. Because $\nabla \cdot \delta\mathbf{n} = 0$ and $\nabla^2\mathbf{v} = 0$, we have $\nabla \times (\nabla \times \delta\mathbf{n}) = -\nabla^2\mathbf{Q}$. Thus, we can rewrite Eq. (9.3.54) as

$$\nabla^2\mathbf{Q} - \lambda_2^{-2}\mathbf{Q} = 0, \tag{9.3.59}$$

or

$$\rho^2 Q'' + \rho Q' - [(\rho/\lambda_2)^2 + 1]Q = 0, \tag{9.3.60}$$

where $\lambda_2 = (K_2/D)^{1/2}$ is the twist penetration depth introduced in Sec. 6.3. Two independent lengths, the penetration depth λ_2 and the core radius a, will enter into the dislocation solution. The core radius is of order the correlation length ξ. Both ξ and λ_2 diverge as the nematic to smectic-A transition is approached. Their ratio, however, should be constant.

The solution to Eq. (9.3.60) is an order one Bessel function of imaginary argument. Since $\delta\mathbf{n}$ is nonsingular at the origin, $Q \to v_\phi = d/(2\pi\rho)$ as $\rho \to 0$, and the appropriate solution is

$$Q(\rho) = \frac{d}{2\pi\lambda_2}\mathcal{K}_1(\rho/\lambda_2) \tag{9.3.61}$$

in the notation of Abramowitz and Stegun (1972, p. 376). The boundary condition as $\rho \to 0$ is satisfied because $\mathcal{K}_1(y) \to y^{-1}$ as $y \to 0$. At large y, $\mathcal{K}_1(y)$ dies off exponentially, and

$$Q(\rho) \to_{\rho\to\infty} \frac{d}{2(\pi\lambda_2\rho)^{1/2}}e^{-\rho/\lambda_2}. \tag{9.3.62}$$

Thus, although $\mathbf{v}$ dies only algebraically to zero as $\rho \to \infty$, $Q(\rho)$ dies exponentially to zero. Since the energy depends on Q rather than $\mathbf{v}$ alone, it will not diverge with sample size as it does in the xy-model. The dislocation energy, Eq. (9.3.52), expressed in terms of $\mathbf{Q}$ is

$$
\begin{aligned}
F/L &= \frac{1}{2}2\pi K_2 \int_a^\infty \rho d\rho [\lambda_2^{-2}Q^2 + (\nabla \times \mathbf{Q})^2] \\
&= \pi K_2 \int_a^\infty \frac{d\rho}{\rho}[(1+\rho^2\lambda_2^{-2})Q^2 + 2\rho Q Q' + \rho^2 (Q')^2] \\
&= \pi K_2 \int_a^\infty d\rho \frac{d}{d\rho}(Q^2 + \rho Q Q') \qquad (9.3.63)\\
&= -\pi K_2 \rho Q (Q' + \frac{1}{\rho} Q) \mid_{\rho=a} \\
&= -\pi K_2 [d^2/(2\pi\lambda_2)^2][(a/\lambda_2)\mathcal{K}_1(a/\lambda_2)][-\mathcal{K}_0(a/\lambda_2)],
\end{aligned}
$$

where we used Eq. (9.3.60) and the recursion $\mathcal{K}_1'(y) + y^{-1}\mathcal{K}_1(y) = -\mathcal{K}_0(y)$. This energy is perfectly finite as long as λ_2 is not infinite. In the limit $y \to 0$, $\mathcal{K}_0(y) \to -\ln y$ and $y\mathcal{K}_1(y) \to 1$ so that

$$
F/L \approx D\frac{d^2}{4\pi}\ln(\lambda_2/a) \qquad (9.3.64)
$$

when $\lambda_2 \gg a$. In this limit, the elastic energy associated with distortions of $\delta\mathbf{n}$ and u far from the core is greater than the core energy associated with the destruction of the smectic order parameter ψ. When λ_2 and a are comparable, however, the core and the far-field energies are comparable, and a complete calculation of the dislocation energy must include the spatial variation of $\mid \psi \mid$. Note that F/L diverges logarithmically as $\lambda_2 \to \infty$. When λ_2 becomes of order the sample size L, the argument of the logarithm should be replaced by L/a, and the xy-model result is regained.

We now turn to the more general problem of the edge and screw dislocations and their energy of interaction. The basic strategy will be to solve for $\delta\mathbf{n}$ and $\mathbf{v}$ as a function of a dislocation source $\mathbf{b}(\mathbf{x})$ analogous to the vortex source of Eq. (9.3.17). To simplify our analysis, we will set $D = B$ in Eq. (6.3.11). This is not an important constraint since we can always rescale lengths to satisfy this condition. As before, we will also assume that $\mathbf{n}$ never deviates significantly from $\mathbf{n}_0$ so that the harmonic form of the smectic free energy can be used. With these assumptions, the Euler-Lagrange equations arising from the minimization of Eq. (6.3.11) are

$$
\frac{\delta F}{\delta u} = -B\nabla \cdot (\nabla u + \delta \mathbf{n}) = 0 \qquad (9.3.65)
$$

$$
\frac{\delta F}{\delta n_i} = K_{ij}\delta n_j + B(\nabla_i u + \delta n_i) = 0, \qquad (9.3.66)
$$

where

$$
\begin{aligned}
K_{ij} &= -(K_1\nabla_\perp^2 + K_3\nabla_\parallel^2)e_{\perp i}e_{\perp j} - (K_2\nabla_\perp^2 + K_3\nabla_\parallel^2)e_{ti}e_{tj} \\
&\equiv -K_\perp(\nabla)e_{\perp i}e_{\perp j} - K_t(\nabla)e_{ti}e_{tj}, \qquad (9.3.67)
\end{aligned}
$$

where $\mathbf{e}_\perp$ and $\mathbf{e}_t$ are the unit vectors perpendicular to $\mathbf{n}_0$ depicted in Figs. 6.2.7 and 9.3.3. In the presence of dislocations, $\mathbf{v} = \nabla u$ will have a singular part satisfying

$$\nabla \times \mathbf{v} = \mathbf{b}(\mathbf{x}) = \sum_\alpha \int dl \frac{d\mathbf{x}_\alpha(l)}{dl} b_\alpha \delta^{(3)}[\mathbf{x}_\alpha(l) - \mathbf{x}], \tag{9.3.68}$$

where $\mathbf{b}(\mathbf{x})$ is the dislocation source whose form is identical to that of $\mathbf{m}(\mathbf{x})$ in Eq. (9.3.20) with $2\pi k_\alpha$ replaced by the Burgers vector component $b_\alpha = dk_\alpha$ along z. The direction of $\mathbf{b}(\mathbf{x})$ is along the tangent vectors $d\mathbf{x}_\alpha(l)/dl$ to the dislocation cores rather than along the Burger's vector, which point along z. In addition, $\mathbf{v}$ will have an analytic longitudinal part ∇h. Thus, in Fourier space, we have

$$\mathbf{v}(\mathbf{q}) = i\frac{\mathbf{q} \times \mathbf{b}}{q^2} + i\mathbf{q}h. \tag{9.3.69}$$

From Eq. (9.3.65), $q^2 h = -i\mathbf{q} \cdot \delta\mathbf{n}$ and

$$\mathbf{v} = i\frac{\mathbf{q} \times \mathbf{b}}{q^2} + \mathbf{q}\frac{\mathbf{q} \cdot \delta\mathbf{n}}{q^2}. \tag{9.3.70}$$

Using this expression for $\mathbf{v}$ in Eq. (9.3.66), we can solve for $\delta n_\perp$ and δn_t in terms of $\mathbf{b}$:

$$\begin{aligned} \delta n_t(\mathbf{q}) &= -iB\left(\frac{\mathbf{q} \times \mathbf{b}}{q^2}\right)_t \frac{1}{K_t(\mathbf{q}) + B} \\ \delta n_\perp(\mathbf{q}) &= -iB\left(\frac{\mathbf{q} \times \mathbf{b}}{q^2}\right)_\perp \frac{1}{K_\perp(\mathbf{q}) + B(q_\parallel/q)^2}. \end{aligned} \tag{9.3.71}$$

We now have expressions for $\mathbf{v}$ and $\delta\mathbf{n}$ in terms of $\mathbf{b}$. These can be used in Eq. (9.3.52) to obtain the dislocation energy as a function of $\mathbf{b}$. To cast our final result in its most useful form, we use the fact that $\mathbf{q} \cdot \mathbf{b} = 0$ and introduce the components of $\mathbf{b}$ perpendicular to $\mathbf{q}$. Let

$$\mathbf{e}_p = \frac{\mathbf{q} \times \mathbf{e}_t}{|\,\mathbf{q} \times \mathbf{e}_t\,|} = \frac{q_\perp}{q}\mathbf{n}_0 - \frac{q_\parallel}{q}\mathbf{e}_\perp. \tag{9.3.72}$$

The vectors $\mathbf{n}_0$, $\mathbf{e}_p$, $\mathbf{e}_\perp$ and $\mathbf{e}_t$ are shown in Fig. 9.3.3. The vectors $\mathbf{e}_t$ and $\mathbf{e}_p$ span the space perpendicular to $\mathbf{q}$, and we can decompose $\mathbf{b}$ as

$$\mathbf{b}(\mathbf{q}) = \mathbf{e}_t b_t(\mathbf{q}) + \mathbf{e}_p b_p(\mathbf{q}). \tag{9.3.73}$$

It then follows that

$$\frac{\mathbf{q} \times \mathbf{b}(\mathbf{q})}{q} = \mathbf{e}_p b_t(\mathbf{q}) - \mathbf{e}_t b_p(\mathbf{q}). \tag{9.3.74}$$

The components b_p and b_t measure, respectively, the screw and edge components of $\mathbf{b}$. This can be seen as follows: For a screw dislocation, $\mathbf{b}$ is parallel to $\mathbf{n}_0$, and all spatial variation is in the plane perpendicular to $\mathbf{n}_0$. In this case, $\mathbf{q} \cdot \mathbf{n}_0 = 0$, $\mathbf{e}_p \equiv \mathbf{n}_0$ and $\mathbf{b} = b_p\mathbf{e}_p = b_p\mathbf{n}_0$. For an edge dislocation, $\mathbf{b}$ is perpendicular to $\mathbf{n}_0$, and $\mathbf{q}$ (specifying the direction of spatial variation) is perpendicular to $\mathbf{b}$. Thus, $\mathbf{b}$ must be parallel to $\mathbf{e}_t$, or $\mathbf{b} = b_t\mathbf{e}_t$.

We can now use Eqs. (9.3.65), (9.3.66) and (9.3.70) to solve for $u(\mathbf{q})$ in the presence of dislocation sources. For example, if there is a single unit strength edge dislocation at the origin, $\mathbf{b}(\mathbf{x}) = d\mathbf{e}_y\delta(x)\delta(z)$ if $\mathbf{n}_0$ is along the z-axis. In this case, $q_\parallel = q_z$, $q_\perp = q_x$, and, from Eq. (9.3.65), $q^2 u = -iq_\perp \delta n_\perp$. Then, using

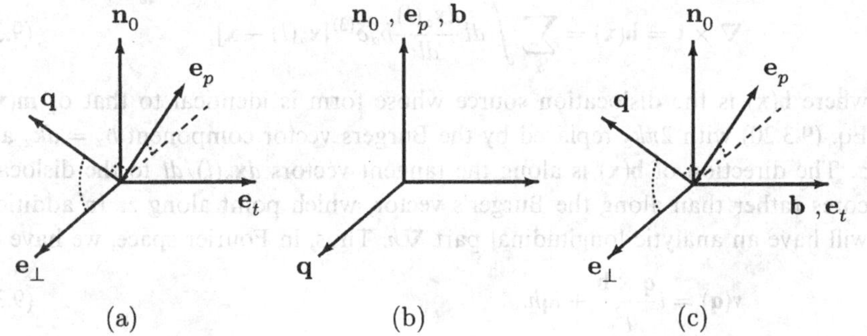

Fig. 9.3.3. The vectors $\mathbf{n}_0$, $\mathbf{e}_t$, $\mathbf{e}_\perp$ and $\mathbf{e}_p$ (a) for general $\mathbf{q}$, (b) for a screw dislocation with $\mathbf{b}$ parallel to $\mathbf{n}_0$ and $\mathbf{q}$ perpendicular to $\mathbf{n}_0$, and (c) for an edge dislocation with $\mathbf{b}$ parallel to $\mathbf{e}_t$ and perpendicular to $\mathbf{n}_0$ and with $\mathbf{q}$ perpendicular to $\mathbf{b}$.

Eq. (9.3.71) for $\delta n_\perp$, we obtain

$$\begin{aligned} u(\mathbf{q}) &= -\frac{q_\| q_\perp}{q^2}\frac{dB}{q^2(K_1 q_\perp^2 + K_3 q_\|^2) + B q_\|^2} \\ &\sim -d\frac{q_\|}{q_\perp}\frac{1}{q_\|^2 + \lambda_1^2 q_\perp^4}, \end{aligned} \tag{9.3.75}$$

where $\lambda_1 = (K_1/B)^{1/2}$. The final form of this equation provides the dominant large $|\,\mathbf{x}\,|$ ($|\,\mathbf{x}\,| \gg \lambda_3$) limit for $u(\mathbf{x})$. Using this form to calculate $\mathbf{u}(\mathbf{x})$, we find

$$\nabla_x u(\mathbf{x}) = \delta n_x(\mathbf{x}) = \frac{1}{4}\frac{d}{(\pi\lambda_1 \mid z \mid)^{1/2}}\operatorname{sgn}(z)\exp(-x^2/(4\lambda_1 \mid z \mid)) \tag{9.3.76}$$

for an edge dislocation at the origin. The strain is a maximum along the parabolas $z = \pm\lambda_1 x^2/4$, as shown in Fig. 9.3.4.

The general expressions for $\delta\mathbf{n}$ in terms of $\mathbf{b}$ permit us to calculate the dislocation energy. After some tedious but straightforward algebra, we find

$$F = \frac{1}{2}\int d^3x d^3x' b_i(\mathbf{x}) U_{ij}(\mathbf{x}-\mathbf{x}') b_j(\mathbf{x}') + E_c, \tag{9.3.77}$$

where E_c is the core energy proportional to the total length of dislocation lines and

$$U_{ij}(\mathbf{x}) = \int \frac{d^3q}{(2\pi)^3} U_{ij}(\mathbf{q}) e^{i\mathbf{q}\cdot\mathbf{x}} \tag{9.3.78}$$

with

$$\begin{aligned} U_{ij}(\mathbf{q}) &= B\left(\frac{K_1(q_\perp/q_\|)^2 + K_3}{[K_1(q_\perp/q_\|)^2 + K_3]q^2 + B} e_{ti} e_{tj} \right. \\ &\quad \left. + \frac{1}{q^2}\frac{K_2 q_\perp^2 + K_3 q_\|^2}{(K_2 q_\perp^2 + K_3 q_\|^2) + B} e_{\perp i} e_{\perp j}\right) \\ &\equiv U_e(\mathbf{x}) e_{ti} e_{tj} + U_s(\mathbf{x}) e_{\perp i} e_{\perp j}. \end{aligned} \tag{9.3.79}$$

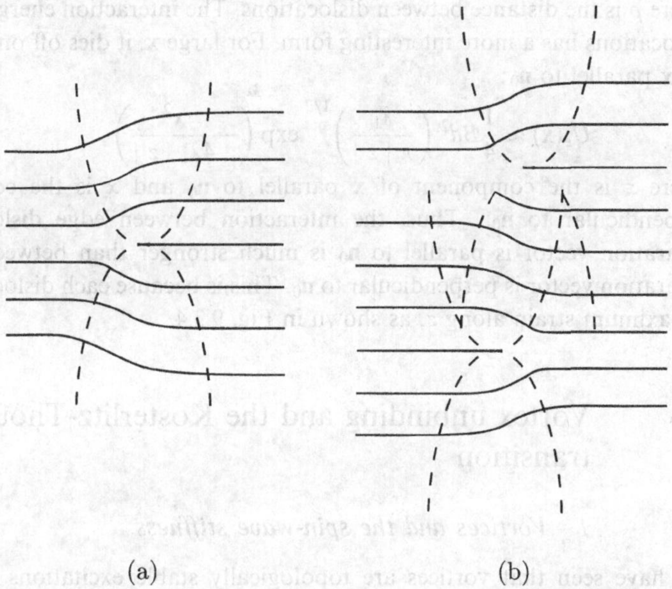

Fig. 9.3.4. (a) An edge dislocation in a smectic liquid crystal showing the parabolic regions of maximum strain. (b) A pair of edge dislocations in a smectic. When their cores are separated along the z-direction, their regions of maximum strain overlap and their interaction energy is the greatest.

The functions $U_e(\mathbf{x})$ and $U_s(\mathbf{x})$ are, respectively, the interaction potentials for edge and screw dislocations.

Eq. (9.3.79) gives both the elastic energy of a single dislocation and the interaction energy between dislocations. Consider, for example, the single screw dislocation discussed at the beginning of this section. In this case, $\mathbf{b}(\mathbf{x}) = \mathbf{n}_0\delta^2(\mathbf{x}_\perp)$, and

$$\begin{aligned} F_{\text{screw}}/L &= \frac{1}{2}d^2\int_0^{a^{-1}}\frac{d^2q_\perp}{(2\pi)^2}U_s(q_z=0,\mathbf{q}_\perp) \\ &= \frac{1}{2}d^2B\int_0^{a^{-1}}\frac{d^2q_\perp}{(2\pi)^2}\frac{1}{q_\perp^2+\lambda_2^{-2}} \\ &= \frac{d^2B}{4\pi}\ln(\lambda_2/a), \end{aligned} \tag{9.3.80}$$

where in the last equation we took $\lambda_2 \gg a$. This result agrees with our previous calculation [Eq. (9.3.64)] (recall that we set $B = D$). A similar analysis for the edge dislocation yields the same result provided $\lambda_3 = (K_3/B)^{1/2} \gg a$. The interaction potential for parallel screw dislocations dies off exponentially for separation $|\,\mathbf{x}\,| \gg \lambda_2$:

$$U_s(\rho) = d^2B\mathscr{K}_0(\rho/\lambda_2)/2\pi, \tag{9.3.81}$$

where ρ is the distance between dislocations. The interaction energy between edge dislocations has a more interesting form. For large $\mathbf{x}$, it dies off only algebraically for $\mathbf{x}$ parallel to $\mathbf{n}_0$:

$$U_e(\mathbf{x}) \approx \frac{1}{4} B d^2 \left(\frac{\lambda_1}{\pi \mid z \mid} \right)^{1/2} \exp \left(-\frac{x^2}{4\lambda_1 \mid z \mid} \right), \tag{9.3.82}$$

where z is the component of $\mathbf{x}$ parallel to $\mathbf{n}_0$ and x is the component of $\mathbf{x}$ perpendicular to $\mathbf{n}_0$. Thus, the interaction between edge dislocations whose separation vector is parallel to $\mathbf{n}_0$ is much stronger than between those whose separation vector is perpendicular to $\mathbf{n}_0$. This is because each dislocation produces a maximum strain along z, as shown in Fig. 9.3.4.

9.4 Vortex unbinding and the Kosterlitz-Thouless transition

1 *Vortices and the spin-wave stiffness*

We have seen that vortices are topologically stable excitations of the ordered phases of systems with a continuous $U(1)$ symmetry. These excitations are distinct from spin-wave excitations in which the phase of the superfluid velocity $\mathbf{v}_s$ remains longitudinal (i.e., proportional to the gradient of a scalar function). The magnitude $\mid \psi \mid$ of the order parameter goes to zero at the core of a vortex. Thus, vortices represent a type of amplitude fluctuation reducing the average magnitude of the order parameter. A sufficiently large number of vortices can destroy long range order altogether, and a thermally activated proliferation of vortices can lead to a phase transition to the high-temperature disordered phase. Because there is always a positive energy cost E_c associated with the creation of the core of the vortex, thermally excited vortices in thermal equilibrium always contribute terms proportional to $\exp(-E_c/T)$ to the partition function. These terms, in contrast to those arising from spin waves or amplitude fluctuations in the field theories of Chapter 3, have essential singularites in at $T = 0$.

Vortices reduce the spin-wave stiffness as well as the order parameter amplitude. They are mobile degrees of freedom that arrange themselves so as to minimize the free energy of any imposed macroscopic gradient of the phase of the order parameter: they screen and reduce the energy of elastic distortions. To see how this comes about, we will calculate the free energy of an xy-model with vortices in which the phase $\theta(\mathbf{x})$ of the average order parameter $\langle\psi\rangle$ has a uniform gradient. Let $\vartheta(\mathbf{x})$ be the phase of the fluctuating field ψ (this is the notation of Sec. 6.1). The partition function for a system in which $\theta(\mathbf{x})$ is spatially uniform can be calculated by imposing boundary conditions that $\vartheta(\mathbf{x})$ be zero at the edges of the sample. Thus, though $\vartheta(\mathbf{x})$ fluctuates, its volume averaged gradient is zero:

$$\Omega^{-1} \int d^d x \langle \nabla \vartheta(\mathbf{x}) \rangle = \Omega^{-1} \int d\mathbf{S} \langle \vartheta(\mathbf{x}) \rangle = 0, \tag{9.4.1}$$

where the final integral is over the surface of the sample and Ω is the volume of the sample. To calculate the energy associated with a spatially uniform gradient of θ, we let $\vartheta(\mathbf{x}) = \vartheta'(\mathbf{x}) + \mathbf{v}\cdot\mathbf{x}$, where ϑ' is constrained to be zero on the boundaries. Then,

$$\nabla\theta = \Omega^{-1}\int d^dx\langle\nabla\vartheta'(\mathbf{x}) + \mathbf{v}\rangle = \mathbf{v}. \tag{9.4.2}$$

The macroscopic spin-wave stiffness ρ_s^R renormalized by vortex and other fluctuations is the difference in free energy between the system with $\mathbf{v} \neq 0$ and that with $\mathbf{v} = 0$:

$$F(\mathbf{v}) - F(0) = \frac{1}{2}\Omega\rho_s^R \mathbf{v}^2. \tag{9.4.3}$$

At low temperatures, the *xy*-Hamiltonian $\mathscr{H}$ is a function of the "velocity" $\mathbf{v}_s = \nabla\vartheta(\mathbf{x})$. The microscopic angle variable ϑ can be decomposed into an analytic part ϑ_a and a singular part ϑ_{sing} arising from vortices. Then the velocity can be written as the sum of a longitudinal part $\mathbf{v}_s^{\parallel} = \nabla\vartheta_a$ and a transverse part $\mathbf{v}_s^{\perp} = \nabla\vartheta_{sing}$ determined by the density of vortices via [Eq. (9.3.17)]:

$$\mathbf{v}_s = \mathbf{v}_s^{\parallel} + \mathbf{v}_s^{\perp}, \qquad \nabla\times\mathbf{v}_s^{\parallel} = 0, \qquad \nabla\cdot\mathbf{v}_s^{\perp} = 0. \tag{9.4.4}$$

$\mathscr{H}$ consists of a "kinetic energy" part (Eq. (6.1.1) with $\mathbf{v}_s$ replacing $\nabla\theta(\mathbf{x})$) quadratic in $\mathbf{v}_s$ and a part independent of $\mathbf{v}_s$. When a uniform average gradient is imposed, $\mathbf{v}_s = \mathbf{v}_s^{\parallel} + \mathbf{v} + \mathbf{v}_s^{\perp}$, where it is understood that the angle $\vartheta_a(\mathbf{x})$ determining $\mathbf{v}_s^{\parallel}$ is zero on the boundaries so that the spatial average of $\mathbf{v}_s^{\parallel}$ is zero in every configuration of the system. The free energy of a system with a uniform gradient of the macroscopic phase is

$$F(\mathbf{v}) = -T\ln\mathrm{Tr}\exp[-\mathscr{H}(\mathbf{v})/T], \tag{9.4.5}$$

where

$$\mathscr{H}(\mathbf{v}) = \frac{1}{2}\rho_s\int d^dx(\mathbf{v}_s^{\parallel} + \mathbf{v} + \mathbf{v}_s^{\perp})^2 + \mathscr{H}'. \tag{9.4.6}$$

Here ρ_s is the bare spin-wave stiffness unrenormalized by vortices, and H' is independent of $\mathbf{v}_s$. Thus, ignoring $\mathscr{H}'$,

$$\begin{aligned} F(\mathbf{v}) &= \frac{1}{2}\Omega\rho_s\mathbf{v}^2 \\ &\quad -T\ln\mathrm{Tr}\exp[-\mathscr{H}(\mathbf{v}=0)/T]\exp\left\{-(\rho_s/T)\int d^dx[\mathbf{v}\cdot\mathbf{v}_s(\mathbf{x})]\right\} \\ &= \frac{1}{2}\Omega\rho_s\mathbf{v}^2 - \frac{1}{2}({\rho_s}^2/T)\int d^dx d^dx'\langle \mathrm{v}_{si}(\mathbf{x})\mathrm{v}_{sj}(\mathbf{x}')\rangle \mathrm{v}_i\mathrm{v}_j \\ &\quad +F(0) + O(\mathbf{v}^4). \end{aligned} \tag{9.4.7}$$

As discussed above, the boundary conditions we have imposed imply that the spatial average of $\mathbf{v}_s^{\parallel}$ is zero for every configuration appearing in the partition function trace. Thus, the longitudinal part of $\mathbf{v}_s$ does not contribute to the integral in Eq. (9.4.7). Then, using $\lim_{\mathbf{q}\to 0}\langle \mathrm{v}_{si}^{\perp}(\mathbf{q})\mathrm{v}_{sj}^{\perp}(-\mathbf{q})\rangle \sim \delta_{ij} - \hat{q}_i\hat{q}_j$, we obtain

$$\rho_s^R = \rho_s - \frac{{\rho_s}^2}{(d-1)T}\int d^dx\langle \mathbf{v}_s^\perp(\mathbf{x})\cdot\mathbf{v}_s^\perp(0)\rangle \tag{9.4.8}$$

(see Appendix 9A for further details). This equation shows that thermally excited vortices, which are responsible for a nonvanishing transverse part of $\mathbf{v}_s$, lead to a reduction in the macroscopic spin-wave stiffness ${\rho_s}^R$. It can be reexpressed in terms of the vortex source function $\mathbf{m(x)}$ using Eq. (9.3.22):

$$\rho_s^R = \rho_s - \left(\frac{1}{d-1}\right)\lim_{q\to 0} T^{-1}{\rho_s}^2\frac{1}{q^2}\langle \mathbf{m(q)}\cdot\mathbf{m(-q)}\rangle, \tag{9.4.9}$$

where we used the fact that $\mathbf{q}\cdot\mathbf{m(q)} = 0$. Finally, it should be remembered that $\mathbf{v}_s^\parallel\cdot\mathbf{v}_s^\perp = 0$ so that $\mathbf{v}_s^\parallel$ and $\mathbf{v}_s^\perp$ are completely decoupled in $\mathscr{H}$. Thus, $\langle \mathbf{v}_s^\parallel(\mathbf{q})\cdot\mathbf{v}_s^\parallel(-\mathbf{q})\rangle = T/\rho_s$ independent of the density of vortices. Thus, Eq. (9.4.8) is often written as

$$\rho_s^R = \frac{1}{T}\int d^dx\left(\langle \mathbf{g}^\parallel(\mathbf{x})\cdot\mathbf{g}^\parallel(0)\rangle - \frac{1}{d-1}\langle \mathbf{g}^\perp(\mathbf{x})\cdot\mathbf{g}^\perp(0)\rangle\right), \tag{9.4.10}$$

where $\mathbf{g(x)} \equiv \rho_s\mathbf{v}_s(\mathbf{x})$. The demonstration that the spin-correlation function is controlled by this renormalized ρ_s^R appears in Appendix 9A.

2 Vortex unbinding in two dimensions – the Kosterlitz-Thouless transition

As we saw in Chapter 6, systems with a broken continuous symmetry in two spatial dimensions do not have long-range order. Order parameter correlation functions at low temperature in systems with *xy*-symmetry do, however, die off algebraically rather than exponentially, as they would in completely disordered high-temperature phases. One can, therefore, ask whether there can be a transition from an algebraically ordered low-temperature phase to a disordered high-temperature phase in *xy*-models, i.e., a transition from quasi-long-range order to disorder. The renormalization group applied to the nonlinear sigma model (Sec. 6.7) predicts that the transition temperature of *n*-component Heisenberg models tends to zero as $d \to 2$ for all $n > 2$. Thus, these models are completely disordered (exponential decay of order parameter correlation) in two dimensions except at $T = 0$. The nonlinear sigma model gives an indeterminate result for the transition temperature of the two-component *xy*-model in two dimensions. In addition, there are topological point defects in two-dimensional *xy*-systems but not in Heisenberg systems with $n \geq 3$. It is thus natural to investigate whether thermally excited vortices might be responsible for a transition from algebraic order to disorder in two-dimensional *xy*-systems.

A simple heuristic argument due to Kosterlitz and Thouless (1973) indicates how vortices can lead to a second-order phase transition in the *xy*-model and to a remarkably accurate estimate of the transition temperature. We calculated in the preceding section that the energy of a single vortex of unit strength in a sample of linear dimension R is $\pi\rho_s\ln(R/a)$, where a is a short distance cutoff of order the core radius. The core can, of course, be anywhere in the sample. It therefore carries an entropy $\ln(R/a)^2$. The free energy of an xy-system with a single vortex is, thus,

$$F = E - TS = (\pi\rho_s - 2T)\ln(R/a). \tag{9.4.11}$$

When $T < \pi\rho_s/2$, F is clearly minimized if there is no vortex; if $T > \pi\rho_s/2$, F is minimized when there are vortices. Since vortices destroy phase rigidity, it is reasonable to identify the temperature,

$$T_c = \pi\rho_s/2, \tag{9.4.12}$$

at which it is first favorable to produce vortices with the transition temperature from the algebraically ordered to the disordered phase. In fact, as we shall see, this is the equation for T_c provided that ρ_s is the fully renormalized spin stiffness at T_c.

We will now turn to a more detailed analysis of the two-dimensional xy-model. We begin by recasting the vortex energy of Eq. (9.3.23) in a form more convenient for the special case of two dimensions. The contours enclosing vortices must be in the two-dimensional xy-plane so that the source function $\mathbf{m}$ must point in the direction $\mathbf{e}_z$ normal to the xy-plane. We can, therefore, define a scalar vortex density $n_v(\mathbf{x})$ via

$$\mathbf{m}(\mathbf{x}) = 2\pi\mathbf{e}_z n_v(\mathbf{x}), \tag{9.4.13}$$

where it is understood that $\mathbf{x}$ is a vector in the xy-plane. The Laplacian Green function in two dimensions is

$$G(\mathbf{x}) = \int \frac{d^2q}{(2\pi)^2}\frac{e^{i\mathbf{q}\cdot\mathbf{x}}}{q^2} = \frac{1}{2\pi}\ln(R/a) - \frac{1}{2\pi}\ln(|\,\mathbf{x}|/a) + C, \tag{9.4.14}$$

where C is a constant (see Eqs. (6.1.24)–(6.1.26)). The first term in this equation diverges with the sample size. It leads to a contribution,

$$\frac{\rho_s}{2}\frac{1}{2\pi}\ln(R/a)\left(\int d^2x n_v(\mathbf{x})\right)^2 \sim \ln(R/a)\left(\sum_\alpha k_\alpha\right)^2 \tag{9.4.15}$$

to the vortex energy of Eq. (9.3.23). Thus, in an infinite sample, there is an infinite energy cost associated with deviations of the total vorticity, $\sum k_\alpha$, from zero, and one need only consider states in which the total vorticity is zero. In this case, the Hamiltonian becomes

$$\mathscr{H} = \mathscr{H}_{SW} + \mathscr{H}_V, \tag{9.4.16}$$

where

$$\mathscr{H}_{SW}/T = \frac{1}{2}K\int d^2x(\nabla\vartheta_a)^2 \tag{9.4.17}$$

is the spin-wave part of the reduced Hamiltonian arising from the longitudinal part of $\mathbf{v}_s$ with $K = \rho_s/T$, and

$$\begin{aligned}\mathscr{H}_V/T &= -\pi K\int_{|\mathbf{x}-\mathbf{x}'|>a} d^2xd^2x' n_v(\mathbf{x})n_v(\mathbf{x}')\ln(|\,\mathbf{x}-\mathbf{x}'\,|/a)\\ &\quad +(E_c/T)\sum_\alpha k_\alpha^2\end{aligned} \tag{9.4.18}$$

is the vortex part of the reduced Hamiltonian. The last term in this expression is the core energy. The integral over positions in Eq. (9.4.18) contains a short distance cutoff preventing two vortices from occupying the same position in space. Since there is a minimum distance between vortices, it is often convenient to restrict the vortices to lie on a lattice with lattice parameter a. In this case, $\mathscr{H}_V/T$ becomes

$$\mathscr{H}_V/T = -\pi K \sum_{\mathbf{l},\mathbf{l}'} k_{\mathbf{l}} k_{\mathbf{l}'} \ln(|\mathbf{R}_{\mathbf{l}} - \mathbf{R}_{\mathbf{l}'}|/a) + (E_c/T)\sum_{\mathbf{l}} k_{\mathbf{l}}^2, \tag{9.4.19}$$

where $\mathbf{R}_{\mathbf{l}}$ are vectors in the lattice. Apart from the core contribution, $\mathscr{H}_V$ is identical to the Hamiltonian of a two-dimensional Coulomb gas with point charges of charge k_α and charge density $n_v(\mathbf{x})$. The constraint $\sum k_\alpha = 0$ is thus the constraint of charge neutrality.

The equation for the reduced renormalized spin rigidity $K_R = \rho_s{}^R/T$ can be expressed in terms of the vortex density using Eqs. (9.4.8), (9.4.9) and (9.4.13) as

$$\begin{aligned} K_R &= K - (2\pi)^2 K^2 \lim_{q\to 0} \langle n_v(\mathbf{q}) n_v(-\mathbf{q})\rangle / q^2 \\ &\equiv \lim_{\mathbf{q}\to 0} K_R(\mathbf{q}). \end{aligned} \tag{9.4.20}$$

The ratio of $K_R(\mathbf{q})$ to K is the inverse wave number dependent dielectric constant for the Coulomb gas described by Eq. (9.4.19):

$$\epsilon^{-1}(\mathbf{q}) = \frac{K_R(\mathbf{q})}{K}. \tag{9.4.21}$$

The last two equations will be of primary use in what follows.

At low temperatures, when the core energy is large compared to T, the number of vortices will be small, and physical properties can be calculated in a power series in the small fugacity

$$y = e^{-E_c/T}. \tag{9.4.22}$$

In particular, the vortex density correlation function appearing in Eq. (9.4.20) can be calculated in a power series in y. We first observe that $\lim_{q\to 0} n_v(\mathbf{q}) = 0$ because of the constraint of charge neutrality. Thus,

$$\langle n_v(\mathbf{q}) n_v(-\mathbf{q})\rangle = q^2 C_2 + O(q^4), \tag{9.4.23}$$

where

$$\begin{aligned} C_2 &= -\lim_{q\to 0}\frac{1}{4\Omega}\int d^2x d^2x' \langle n_v(\mathbf{x}) n_v(\mathbf{x}')\rangle (\mathbf{x}-\mathbf{x}')^2 \\ &= -\frac{1}{4\Omega}\sum_{\mathbf{l},\mathbf{l}'} (\mathbf{R}_{\mathbf{l}} - \mathbf{R}_{\mathbf{l}'})^2 \langle k_{\mathbf{l}} k_{\mathbf{l}'}\rangle; \end{aligned} \tag{9.4.24}$$

we have employed the notation of the lattice Hamiltonian in the last equation. The charge correlation function in this equation is easily evaluated as a power series in y using $\mathscr{H}_V$. To lowest nontrivial order in y, there is a single $\pm$ vortex pair. The only two-vortex terms that contribute to $\langle k_{\mathbf{l}} k_{\mathbf{l}'}\rangle$ are those with $k_{\mathbf{l}} = +1$, $k_{\mathbf{l}'} = -1$ or $k_{\mathbf{l}} = -1$, $k_{\mathbf{l}'} = +1$ and vorticity zero at all other sites in both cases. Thus, to lowest order in y,

$$\langle k_{\mathbf{l}} k_{\mathbf{l}'} \rangle = -2y^2 [| \mathbf{R}_{\mathbf{l}} - \mathbf{R}_{\mathbf{l}'} | /a]^{-2\pi K}. \tag{9.4.25}$$

This expression can now be used in Eq. (9.4.20) for K_R. Since we are interested in low temperatures where K is large, we will express this result in terms of $K^{-1} \sim T$ rather than K. Keeping only the lowest order term in K^{-1}, we obtain

$$K_R^{-1} = K^{-1} + 4\pi^3 y^2 \int_a^\infty \frac{dr}{a} \left(\frac{r}{a}\right)^{3-2\pi K}. \tag{9.4.26}$$

This expresses the renormalized spin rigidity in terms of the unrenormalized stiffness K, the vortex fugacity y and the short distance cutoff a. At low temperatures, when $K \geq 2/\pi$, the integral on the right hand side converges and there is a correction to K_R^{-1} of order $y^2 = e^{-2E_c/T}$. When, however, $K < 2/\pi$, the integral has a large r divergence, and perturbation theory breaks down.

The difficulty associated with the divergence at small K can be overcome by employing a renormalization procedure first used by José *et al.* (1977). The integral in Eq. (9.4.26) is broken up into two parts:

$$\int_a^\infty \to \int_a^{ae^{\delta l}} + \int_{ae^{\delta l}}^\infty . \tag{9.4.27}$$

The non-singular small-r part of this integral is evaluated and incorporated into the constant K^{-1}. This procedure can be carried out order by order in a perturbation series in y even though the coefficient of y^2 is formally divergent. This leads to a new equation for K_R,

$$K_R^{-1} = (K')^{-1} + 4\pi^3 y^2 \int_{ae^{\delta l}}^\infty \frac{dr}{a} \left(\frac{r}{a}\right)^{3-2\pi K}, \tag{9.4.28}$$

where

$$(K')^{-1} = K^{-1} + 4\pi^3 y^2 \int_a^{ae^{\delta l}} \frac{dr}{a} \left(\frac{dr}{a}\right)^{3-2\pi K}. \tag{9.4.29}$$

Finally, the cutoff in the remaining integral can be rescaled ($ae^{\delta l} \to a$) to yield an equation for K_R^{-1} identical to Eq. (9.4.26) but with shifted and rescaled parameters K and y:

$$K_R^{-1} = (K')^{-1} + 4\pi^3 (y')^2 \int_a^\infty \frac{dr}{a} \left(\frac{r}{a}\right)^{3-2\pi K'}, \tag{9.4.30}$$

where

$$y' = e^{(2-\pi K)\delta l} y. \tag{9.4.31}$$

The exponent in the integral in Eq. (9.4.30), rather than being proportional to K as simple algebraic manipulations would suggest, is proportional to K'. It can be shown that y^4 corrections to the equation for K_R lead to this replacement. The above equations are valid for arbitrary rescaling factors $e^{\delta l}$ and, in particular, for $\delta l \to 0$. The equations for K' and y' can thus be converted to differential renormalization equations,

$$\frac{dK^{-1}}{dl} = 4\pi^3 y^2(l) + O[y^4(l)], \tag{9.4.32}$$

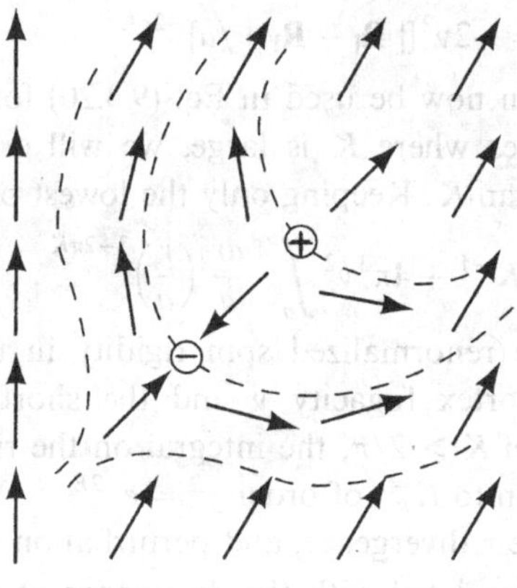

Fig. 9.4.1. A 5 × 5 array of spins containing a +−vortex pair. At the length scale a of the lattice spacing, the reduced rigidity is K. At a length scale of $5a$, the effective rigidity $K(5a)$ has been reduced by the vortices.

$$\frac{dy(l)}{dl} = [2 - \pi K(l)]y(l) + O[y^3(l)]. \tag{9.4.33}$$

The completely renormalized reduced stiffness can, by construction, be expressed in terms of $K(l)$ and $y(l)$:

$$K_R[K, y] = K_R[K(l), y(l)]. \tag{9.4.34}$$

This equation expresses the observable K_R in terms of rescaled parameters and is precisely a realization of the scaling relations discussed earlier in Chapter 5.

Before analyzing the recursion relations, Eqs. (9.4.32) and (9.4.33), let us consider what physics they represent. The cutoff a is the minimum distance between vortices and also the lattice parameter of our lattice model. By increasing a, we increase the minimum distance between vortices and in effect thin degrees of freedom in a manner analogous to the real space renormalization group discussed in Chapter 5. Consider the configuration of spins shown in Fig. 9.4.1. At the center of this 5 × 5 array of spins, there is a vortex-antivortex pair. The reduced stiffness $K = K(l = 0)$ is that measuring the energy of angle differences between nearest neighbor sites separated by a distance a. The stiffness at length scale $5a$ is reduced because of thermally excited vortex pairs with separations less than $5a$. This explains the increase in $K^{-1}(l)$ with increasing l. The vortex pair present at length scale a is no longer present when the system is viewed at length scale $5a$. The line integral of $\mathbf{v}_s$ around the perimeter of the 5 × 5 configuration is zero. Thus, if vortices exist in closely bound pairs, the density of vortices at the longer length scale is less than it is at the shorter length scale. This explains the decrease in $y(l)$ with increasing l when $K(l) > 2/\pi$. When $K(l) > 2/\pi$, $y(l)$ grows. This can be explained by the presence of unbound vortices distributed uniformly throughout the sample. When vortices are unbound, their presence continues to be felt when the length scale is changed. In the Coulomb analogy, unbound

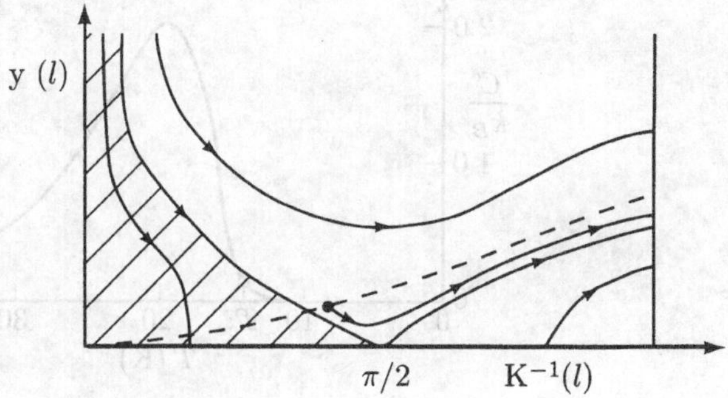

Fig. 9.4.2. Renormalization flows for the Kosterlitz recursion relations. The dashed line is a line of initial conditions as a function of T. The critical temperature is determined by the crossing of the dashed line and the separatrix terminating at $\pi/2$. The flow originating at the dot on the dashed line is for a temperature $T > T_c$.

vortices become free charges. Thus, the transition from decreasing vortex fugacity $y(l)$ to increasing $y(l)$ that occurs at $K(l) = 2/\pi$ is a transition from an insulating state with charges bound in "molecules" (vortex pairs) to a conducting plasma with mobile charges but charge neutrality.

The equations for $K(l)$ and $y(l)$ can be integrated analytically (see Appendix 9A) in the vicinity of the fixed point $K^* = 2/\pi$, $y^* = 0$ and numerically away from the fixed point. The resulting flows are shown in Fig. 9.4.2. For small y, and $K^{-1} < \pi/2$, all flows are towards $y(l) = 0$ and a finite value of K^{-1}, i.e., toward a state with a finite renormalized rigidity and no unbound vortices at the longest length scales:

$$K_R = \lim_{l\to\infty} K_R[K(l), y(l)] = \lim_{l\to\infty} K(l). \tag{9.4.35}$$

There is a separatrix passing through the critical point $y(l) = 0$, $K^{-1} = \pi/2$. Points above this separatrix flow towards large values of K^{-1} and large values of y, i.e., toward the phase with unbound vortices. Points exactly on the separatrix with $K^{-1} < \pi/2$ flow to the critical point. The starting point of flows is on the line $y = \exp(-E_c/T) = \exp(-E_c K/\rho_s)$. The transition temperature is then determined by the intersection of this line with the separatrix. Since the flow for $T < T_c$ is towards the line $y = 0$, spin correlations at the longest wavelengths are described by the simple spin-wave theory of Sec. 6.1, and we conclude that the spin correlation function dies off algebraically,

$$\bar{G}(\mathbf{x}) \sim |\,\mathbf{x}\,|^{-\eta(T)}, \tag{9.4.36}$$

with $\eta(T) = (2\pi K_R(T))^{-1}$, and that

$$\lim_{T\to T_c^-} K_R = 2/\pi. \tag{9.4.37}$$

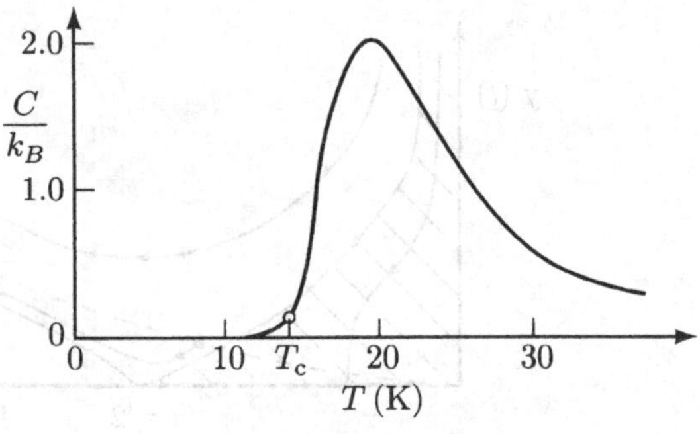

Fig. 9.4.3. Specific heat of an xy-model. There is an unobservable essential singularity at T_c and a nonuniversal maximum above T_c associated with the entropy liberated by the unbinding of vortex pairs.

Thus, the exponent $\eta(T)$ in the power law for $\bar{G}$ tends to a universal value as $T \to T_c^-$:

$$\lim_{T\to T_c^-} \eta(T) = \frac{1}{4}. \tag{9.4.38}$$

Exactly at $T = T_c$, there is a logarithmic correction to this result (see Appendix 9A):

$$\bar{G}(\mathbf{x}) \sim \frac{\ln^{1/8}(\mid \mathbf{x} \mid /a)}{\mid \mathbf{x} \mid^{1/4}}. \tag{9.4.39}$$

Experimental observation of such a logarithmic correction would be difficult.

The solution of the Kosterlitz recursion relations for T near T_c yield an exponentially divergent correlation length,

$$\xi(T)/a \sim e^{b'/|T-T_c|^{1/2}}, \tag{9.4.40}$$

and corrections to $\rho_s{}^R$,

$$\rho_s{}^R(T) \approx \rho_s{}^R(T_c^-)[1 + b(T - T_c)^{1/2}]. \tag{9.4.41}$$

The coefficients b and b' are nonuniversal; their product $bb' = 2\pi$ is, however, universal. The correlation length sets the scale for the density of unbound vortices of a given sign in the high-temperature phase and for the free energy density:

$$\langle n_{v+}(\mathbf{x})\rangle \sim \xi^{-2}, \qquad f \sim \xi^{-2}. \tag{9.4.42}$$

Note the essential, rather than power-law, singularity in both of these quantities. The specific heat obtained from the above free energy also has an essential singularity:

$$C_V^{\text{sing}} \sim \xi^{-2}. \tag{9.4.43}$$

This singularity is very weak and is essentially unobservable. There is however, a large nonuniversal peak in the specific heat associated with the entropy liberated by the unbinding of bound vortices above the Kosterlitz-Thouless temperature, as shown in Fig. 9.4.3.

Above T_c, there are free mobile charges interacting via a Coulomb potential. When the density of these charges is sufficiently high, the properties of the Coulomb gas of mobile charges can be calculated quite accurately using mean-field theory for the lattice Hamiltonian of Eq. (9.4.19) (which is a type of Debye-Hückel theory discussed in Sec. 4.8 (Problem 9.7)). The vortex density correlation function in this approximation is

$$G_{\rm vv} = \frac{1}{\Omega}\langle n_{\rm v}(\mathbf{q})n_{\rm v}(-\mathbf{q})\rangle = \frac{1}{B + (4\pi^2 K)/q^2}, \qquad (9.4.44)$$

where

$$\begin{aligned} B^{-1} &= \sum_k k^2 \langle n_{vk}\rangle = \langle k^2\rangle \\ &= \frac{\sum_k k^2 e^{-E_c k^2/T}}{\sum_k e^{-E_c k^2/T}} \sim \frac{T}{E_c}. \end{aligned} \qquad (9.4.45)$$

The simple high temperature final form for B is valid when there are unbound vortices with all vorticities. This equation implies that, in the high-temperature limit, vortex correlations can be described by treating $n_{\rm v}(\mathbf{x})$ as the independent fluctuating field in $\mathscr{H}_V$:

$$\mathscr{H}_V/T = \frac{1}{2}\int \frac{d^2q}{(2\pi)^2}[B + (4\pi^2 K/q^2)]n_{\rm v}(\mathbf{q})n_{\rm v}(-\mathbf{q}). \qquad (9.4.46)$$

Using Eqs. (9.4.44) and (9.4.20), we obtain a $\mathbf{q}$-dependent stiffness:

$$K_R(q) = \frac{B(l^*)K(l^*)}{B(l^*) + 4\pi^2 K(l^*)/(e^{2l^*}q^2)}, \qquad (9.4.47)$$

where l^* is an appropriate matching point well into the high temperature end of the renormalization group trajectory. This relation implies that the wave number dependent dielectric constant for the two-dimensional Coulomb gas is

$$\epsilon(q,T) = \frac{K}{K^R(q,T)} = \frac{1}{1 + 1/(q\lambda_s)^2}, \qquad (9.4.48)$$

where $\lambda_s \sim \xi$ is the screening length. This dielectric constant leads to the screened Coulomb potential $V(q) = 1/[q^2\epsilon(q)] = 1/(q^2 + \lambda_s^2)$, characteristic of a metal with mobile charges.

3 Superfluid helium films

Superfluid helium is one of the cleanest examples of a system with xy-symmetry, and helium films should undergo a Kosterlitz-Thouless transition from the superfluid to the normal fluid state. A slight modification of the results just derived is necessary before they can be applied to helium. The superfluid velocity $\mathbf{v}_s$ is equal to $(\hbar/m)\nabla\theta$ rather than simply $\nabla\theta$. With this definition, $\mathbf{v}_s$ has units of velocity, and the rigidity modulus ρ_s has units of mass per unit volume (or area in two dimensions) and is called the superfluid density. The elastic energy is thus

$$F_{\rm el} = \frac{1}{2}\int d^d x \rho_s \mathbf{v}_s^2 = \frac{1}{2}\rho_s(\hbar/m)^2 \int d^d x (\nabla\theta)^2, \qquad (9.4.49)$$

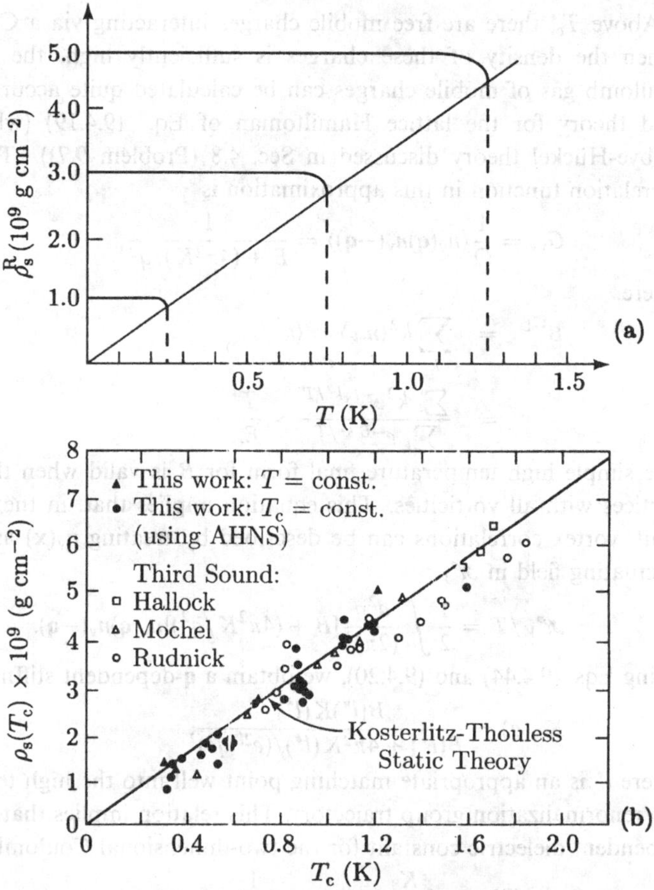

Fig. 9.4.4. (a) Schematic representation of the superfluid density $\rho_s(T)$ as a function of T for films of varying height, substrate, etc. All curves terminate on a universal line with slope given by Eq. (9.4.49). (b) $\rho_s(T_c^-)$ as a function of T_c^- as determined by third sound and torsion oscillator measurements. [D.J. Bishop and J.D. Reppy, *Phys. Rev. Lett.* **40**, 1727 (1978).]

and the predictions of the Kosterlitz-Thouless theory derived above can be applied to helium films provided ρ_s is replaced by $\rho_s(\hbar/m)^2$. The most striking prediction of the Kosterlitz-Thouless theory is the universal value of $K_R(T_c^-)$ [Eq. (9.4.37)]. When applied to helium films (Nelson and Kosterlitz 1977), this relation becomes

$$\frac{\rho_s(T_c^-)}{T_c} = \frac{2m^2}{\pi\hbar^2 k_B} = 3.491 \times 10^{-9} \,\text{g cm}^{-2} K^{-1}, \tag{9.4.50}$$

implying that, though the values of ρ_s at $T = 0$ might have widely different values in different films, the curves $\rho_s(T)$ should all terminate at T_C^- on a line with universal slope, as shown in Fig. 9.4.4a.

The best determinations of $\rho_s(T)$ come from measurements of velocity of third sound and from the response of torsion oscillators in which helium coats the surface of mylar films. Both of these measurements are done at nonzero frequency ω and measure a frequency dependent superfluid density $\rho_s(T,\omega)$, rather than the zero frequency density we have been discussing. Thus, these measurements must be extrapolated to $\omega = 0$ to yield the results shown in Fig. 9.4.4, which show a slope of $\rho_s(T_c^-)$ versus T_c^- equal to 0.96 of the value predicted by the Kosterlitz-Thouless theory. The theory of the dynamics of two-dimensional helium films is discussed in detail by Ambegaokar *et al.* (1980).

Third sound is a propagating mode (Atkins 1959; Atkins and Rudnick 1970) in superfluid helium films arising from the coupling between fluctuations in the two-dimensional mass density ρ_2 and the superfluid velocity $\mathbf{v}_s$. The mass density $\rho_2 = \rho d$ is the product of the three-dimensional mass density ρ and the thickness d of the film. The film is essentially incompressible, so that ρ is a constant and density changes are controlled by changes in thickness. The hydrodynamic equations governing the propagation of third sound are easily derived following the procedures discussed in Chapter 8. Because the substrate can absorb momentum, the momentum of the film is not conserved. Thus, only the energy and mass of the film are conserved variables. In the superfluid phase, the superfluid velocity is a broken-symmetry elastic variable, and the fundamental entropy relation is

$$T\,ds = d\epsilon - \alpha d\rho_2 - \mathbf{h}_s \cdot d\mathbf{v}_s, \tag{9.4.51}$$

where s is the entropy per unit area, ϵ is the energy per unit area, and $\mathbf{h} = \rho_s \mathbf{v}_s$. Both the chemical potential per unit mass α and the superfluid density ρ_s depend on film thickness. The dominant d dependence of α in thin films arises from the van der Waals attraction of the helium atoms to the substrate. As a result $\alpha = \alpha_0 - Hd^{-3}$, where α_0 is the chemical potential of the bulk and H is the Hameker constant. As in bulk helium, the reactive part of the "current" associated with $\mathbf{v}_s$ is $\nabla\alpha$. The thermal coupling between the substrate and the film is such that processes are essentially isothermal, implying that the energy conservation equation can be ignored. The remaining hydrodynamical equations are

$$\begin{aligned} \frac{\partial \rho_2}{\partial t} &= -\nabla\cdot\mathbf{g} = -\nabla\cdot\mathbf{h}_s + \Gamma\nabla^2\vartheta, \\ \frac{\partial \mathbf{v}_s}{\partial t} &= -\nabla\vartheta + \gamma\nabla\nabla\cdot\mathbf{h}_s, \end{aligned} \tag{9.4.52}$$

where Γ and γ are dissipative coefficients. Note that the momentum density $\mathbf{g}$ has a dissipative part $-\Gamma\nabla\vartheta$ not permitted when momentum is conserved. These equations lead to a propagating mode with velocity

$$c_3 = \left(\frac{\rho_s}{\chi}\right)^{1/2} = \left(\frac{\rho_s}{\rho}\frac{3H}{d^4}\right)^{1/2}, \tag{9.4.53}$$

where $\chi^{-1} = \partial\alpha/\partial\rho_2 = 3H/\rho d^4$. Thus, the velocity of third sound is determined by $\rho_s(T)$. Above T_c, thickness fluctuations are diffusive with frequency $\omega = -i\Gamma\chi^{-1}q^2 = (3\Gamma H d^{-4})q^2$.

The hydrodynamic equations [Eqs. (9.4.52)] are valid at the longest wavelengths and lowest frequencies. At wavelengths comparable with the average spacing between vortices in bound pairs, $\rho_s(\omega)$ will depart significantly from its static value. The detailed dynamical theory (Ambegaokar *et al.* 1980) shows that $\rho_s(T,\omega) = \rho_s(T)/\epsilon(T,\omega)$, where $\epsilon(T,\omega)$ is the frequency dependent dielectric constant of the Coulomb gas of vortices. The imaginary part of $\epsilon(T,\omega)$ grows as $T \to T_c^-$, making it difficult to measure the third sound velocity determined by the real part of $\rho_s(T,\omega)$ near T_c.

In torsion oscillator experiments (Bishop and Reppy 1978), a metal cylinder of radius R is wrapped with mylar sheets that form the substrate for the helium film. There are many layers so that the total film surface area A is quite large. The cylinder is attached to a torsion rod that is externally driven with a torque $\tau(\omega)$ at frequency ω. The combined mass of the cylinder and the helium film is M. The amplitude and phase of the angular displacement $\Theta(\omega)$ are detected. Only the normal component of the superfluid contributes to the moment of inertia,

$$I(T,\omega) = R^2[M - A\rho_s(T,\omega)], \tag{9.4.54}$$

of the cylinder. The dynamic response function of the cylinder-helium system is thus

$$\chi(\omega) = \frac{\Theta(\omega)}{\tau(\omega)} = \frac{1}{[-\omega^2 I(\omega) + i\omega\gamma + kR^2]}, \tag{9.4.55}$$

where kR^2 is the torsion constant of the torsion rod. The resonant frequency ω_R and damping $\gamma_R(\omega)$ are determined, respectively, by the real and imaginary parts of the complex frequency at the pole of this function (see Sec. 7.2). If we assume $M \gg \rho_s A$, the shift in period $P = 2\pi/\omega_R$ due to the presence of the superfluid is

$$\frac{\Delta P}{P_0} = \frac{\omega_{R0}^2 - \omega_R^2}{2\omega_{R0}^2} = \frac{A}{M}\rho_s(T)\mathrm{Re}[\epsilon^{-1}(T,\omega_{R0})], \tag{9.4.56}$$

where $\omega_{R0} = (k/M)^{1/2}$ is the frequency of the oscillator when all of the helium rotates. The change in the inverse quality factor, $Q^{-1} = \gamma_R(\omega)/\omega I(\omega)$, is

$$\Delta Q^{-1} = Q^{-1} - Q_0^{-1} = \frac{A}{M}\rho_s(T)\mathrm{Im}[\epsilon^{-1}(T,\omega)], \tag{9.4.57}$$

where $Q_0 = (\omega M R^2)/\gamma$ is the quality factor when all of the helium is normal. The period shift and quality factor for a helium film with a transition temperature of 1.215 K are shown in Fig. 9.4.5 along and compared with the dynamic theory of Ambegaokar *et al.* (1980). Note that ΔP is a smooth function near T_c and goes to zero at a higher temperature than the zero frequency superfluid density.

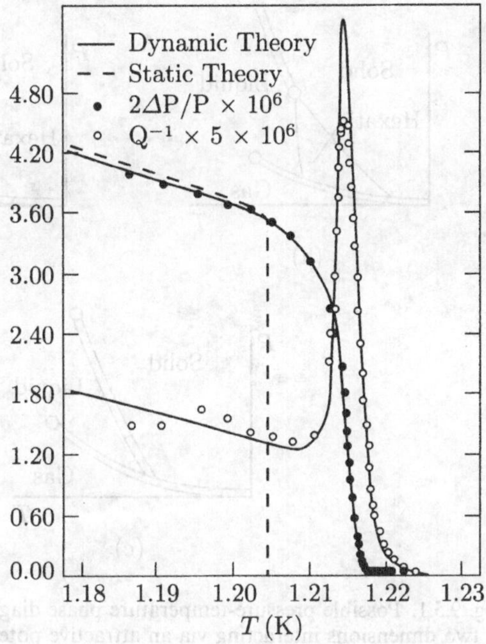

Fig. 9.4.5. The reduced period shifts and inverse quality factor for a superfluid film with transition temperature 1.215 K [D.J. Bishop and J.D. Reppy, *Phys. Rev. Lett.* **40**, 1727 (1978)]. The solid lines are fits using the dynamic theory of V. Ambegaokar, B.I. Halperin, D.R. Nelson, and E.D. Siggia [*Phys. Rev. Lett.* **40**, 783 (1980)], and the dashed curve is the result of a static theory.

9.5 Dislocation mediated melting

As we have seen in Chapter 6 and in this chapter, two-dimensional *xy*-models and solids have much in common. They both have fluctuation-induced quasi-long-range order (QLRO) rather than true long-range order, and they both have topological defects — vortices and dislocations — which can reduce local rigidity. It is, thus, natural to expect two-dimensional solids to exhibit a second-order dislocation unbinding transition, analogous to the vortex unbinding transition of the *xy*-model, at which the shear modulus vanishes. We should, however, expect differences between the two systems arising from the vector nature of the displacement field **u**. Our experience from three dimensions is that solids are quite different from *xy*-models. The former undergo first-order transitions to homogeneous and rotationally isotropic liquids whereas the latter undergo second-order transitions to the disordered phase. Indeed, mean-field theories and their fluctuation generalizations indicate that the direct solid-to-liquid transition is always first order. How then can there be a continuous transition out of the

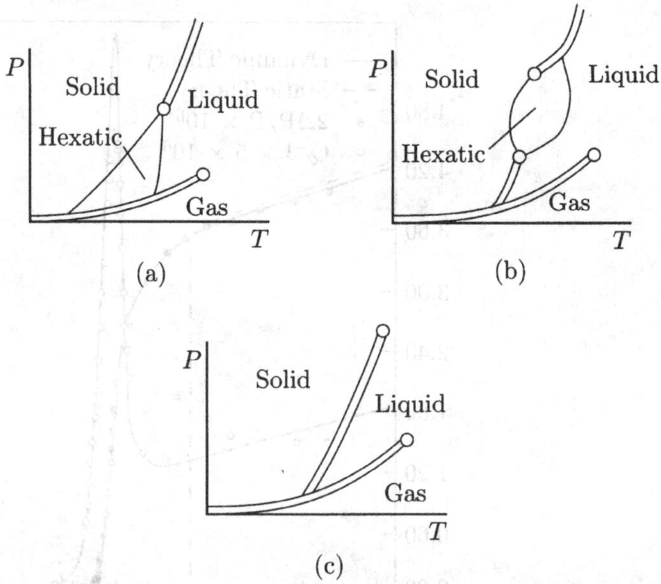

Fig. 9.5.1. Possible pressure-temperature phase diagrams for spherical matter in two dimensions interacting via an attractive potential with a repulsive core. Solid, liquid, gas and hexatic phases are shown. Second-order transitions are represented by lines and first-order transitions by double lines. In (a), the transition from the solid to the hexatic phase is always continuous, whereas in (b) it is discontinuous at high pressure and continuous at lower pressure. In (c), there is no hexatic phase at all, and, as in three-dimensional systems, the transition from the solid to the liquid phase is always first order. [Adapted from D.R. Nelson, in *Phase Transitions and Critical Phenomena*, eds. C. Domb and J.L. Lebowitz (Academic Press, New York, 1983).]

two-dimensional solid phase? The answer is that if the transition from the solid phase is continuous, it will be to a hexatic liquid-crystalline phase with local six-fold bond-angle order rather than to an isotropic fluid. The hexatic phase, like the the low-temperature phase of the *xy*-model, has QLRO and melts to the isotropic fluid phase via a disclination unbinding transition. Thus, the transition from the solid phase to the liquid phase can occur either directly via a first-order transition or via a two step process with an intervening hexatic liquid-crystalline phase. The transitions from the hexatic phase to the isotropic fluid and from the solid to the hexatic phase can both be continuous. Possible pressure-temperature phase diagrams are shown in Fig. 9.5.1. In this section, we will outline the theory of dislocation mediated melting of two-dimensional solids. This theory is a generalization of the Kosterlitz-Thouless theory due to Halperin and Nelson (1978, 1979) and to Young (1979), and is usually referred to as the KTHNY theory.

Using generalizations of the techniques discussed in the preceding section, we can derive an expression, analogous to Eq. (9.4.19) for the energy of interacting dislocations (see Problem 9.8):

$$\begin{aligned}\mathscr{H}_{\text{disc}}/T &= -\frac{1}{8\pi}\sum_{\mathbf{l},\mathbf{l}'}\Big[K_1\mathbf{b}_\mathbf{l}\cdot\mathbf{b}_{\mathbf{l}'}\ln(|\,\mathbf{R}_\mathbf{l}-\mathbf{R}_{\mathbf{l}'}\,|/a) \\ &\quad -K_2\frac{\mathbf{b}_\mathbf{l}\cdot(\mathbf{R}_\mathbf{l}-\mathbf{R}_{\mathbf{l}'})\mathbf{b}_{\mathbf{l}'}\cdot(\mathbf{R}_\mathbf{l}-\mathbf{R}_{\mathbf{l}'})}{|\,\mathbf{R}_\mathbf{l}-\mathbf{R}_{\mathbf{l}'}\,|^2}\Big]+\frac{E_c}{T}\sum_\mathbf{l}|\,\mathbf{b}_\mathbf{l}\,|^2,\end{aligned} \tag{9.5.1}$$

where a is the lattice spacing and $\mathbf{b}_\mathbf{l}$ is a dimensionless Burgers vector of the form $k_\mathbf{l}\mathbf{a}_1+m_\mathbf{l}\mathbf{a}_2$, where $k_\mathbf{l}$ and $m_\mathbf{l}$ are integers and $\mathbf{a}_1$ and $\mathbf{a}_2$ are primitive translation vectors of the lattice. The coupling constants K_1 and K_2 are equal to K, the unitless Young's modulus for a two-dimensional solid [Eqs. (6.6.18) and (9.3.51)]:

$$K_1=K_2\equiv K=\frac{4a^2}{T}\frac{\mu(\mu+\lambda)}{2\mu+\lambda}. \tag{9.5.2}$$

This energy differs from the vortex energy of Eq. (9.4.19) because $\mathbf{b}$ is a vector and because there is a second interaction term depending on the angle between $\mathbf{b}_\mathbf{l}$ and $\mathbf{R}_\mathbf{l}-\mathbf{R}_{\mathbf{l}'}$.

Recursion relations for $K(l)$ and the dislocation fugacity $y(l)$ can be derived by generalizations of the procedures outlined in the previous section. These equations lead to flows similar to those for the xy-model depicted in Fig. 9.4.2. The fugacity $y(l)$ flows to zero for all $K^{-1}(l)<1/(16\pi)$ and away from zero for $K^{-1}(l)>1/(16\pi)$. Thus, the melting temperature T_M is determined by

$$K(T_M)=\frac{16}{\pi}. \tag{9.5.3}$$

Throughout the solid phase, there will be a power-law decay of spatial correlations according to Eq. (6.4.34) and (6.4.35). The decay exponent $\eta_\mathbf{G}$ depends on a different combination of Lamé coefficients λ and μ than does K. Thus, recursion relations for λ and μ are needed to determine $\eta_\mathbf{G}$. These depend only on $K(l)$ and $y(l)$ and can be integrated. Eqs. (9.5.3) and (6.4.35) imply that

$$\eta_\mathbf{G}(T_M)=\frac{G^2a^2}{64\pi^2}\lim_{T\to T_M^-}[1+\sigma_2(T)][3-\sigma_2(T)], \tag{9.5.4}$$

where $\sigma_2=\lambda/(\lambda+2\mu)=(B-\mu)/(B+\mu)$ [Eq. (6.6.18)] is the two-dimensional Poisson ratio. Thus, $\eta_\mathbf{G}(T_M)$, unlike the exponent η for the xy-model, is not a universal quantity. The function $(1+\sigma_2)(3-\sigma_2)$ has a maximum value of four at the maximum value of $\sigma_2=1$ permitted by thermodynamic stability. The maximum value of $\eta_\mathbf{G}(T_M)$ at the first reciprocal lattice vector is, therefore, one-third. The temperature dependences of the shear modulus for $T<T_M$ and for the correlation length for $T>T_M$ are similar to those of the helicity modulus and correlation length in the xy-model:

$$\begin{aligned}\mu(T) &= \mu(T_M)(1+b\,|\,T-T_M\,|^{\bar{\nu}}) \quad \text{if } T<T_M, \\ \xi(T) &= \exp(b'\,|\,T-T_M\,|^{-\bar{\nu}}) \quad \text{if } T<T_M,\end{aligned} \tag{9.5.5}$$

where

$$\bar{\nu} = 0.36963477... \tag{9.5.6}$$

rather than 0.5 predicted for the *xy*-model.

Above the melting transition, there are unbound dislocations whose correlations can be described by a Debye-Hückel theory, like that discussed in the preceding section, in which the vector "charges" are treated as a continuous vector field. When $K_1 = K_2$, the dislocation energy of Eq. (9.5.1) in this approximation can be written as

$$\mathscr{H}_{\text{disc}}/T = \frac{1}{2\Omega}\sum_{\mathbf{q}}\left[\frac{K}{q^2}(\delta_{ij} - \hat{q}_i\hat{q}_j) + \frac{2E_c a^2}{T}\delta_{ij}\right] b_i(\mathbf{q})b_j(-\mathbf{q}), \tag{9.5.7}$$

where $\hat{q}_i = q_i/q$ and Ω is the sample area. The dislocation correlation function in the Debye-Hückel approximation is

$$\langle b_i(\mathbf{q})b_j(-\mathbf{q})\rangle = \frac{\Omega}{K/q^2 + 2E_c a^2/T}(\delta_{ij} - \hat{q}_i\hat{q}_j) + \frac{\Omega T}{2E_c a^2}\delta_{ij}. \tag{9.5.8}$$

Dislocations perturb the bond angle field according to Eq. (9.3.39), which in Fourier space is

$$\theta(\mathbf{q}) = -i\frac{a}{2\pi}\frac{\mathbf{q}}{q^2}\cdot \mathbf{b}(\mathbf{q}). \tag{9.5.9}$$

Note that only the longitudinal part of **b** contributes to θ. The contribution to $\mathscr{H}_{\text{disc}}$ from the longitudinal part of **b** is

$$\begin{aligned}\mathscr{H}_{\text{disc}} &= \frac{1}{2\Omega}\sum_{\mathbf{q}} K_A(T)q^2 \mid \theta(\mathbf{q}) \mid^2 \\ &= \frac{1}{2}\int d^2x K_A(T)(\nabla\theta)^2,\end{aligned} \tag{9.5.10}$$

where

$$K_A(T) \approx 2E_c a^2. \tag{9.5.11}$$

Thus, the phase formed by the unbinding of dislocations in a two-dimensional hexagonal solid has bond angle rigidity. It is a hexatic liquid crystal with QLRO in the bond angle order parameter $\psi = \exp(6i\theta)$:

$$\langle\psi^*(\mathbf{x})\psi(0)\rangle \sim \mid \mathbf{x} \mid^{-\eta_6(T)}, \tag{9.5.12}$$

where

$$\eta_6(T) = \frac{18T}{\pi K_A(T)}. \tag{9.5.13}$$

The hexatic phase can now melt via a disclination unbinding transition identical to the vortex unbinding transition discussed in the preceding section.

1 Effects of a substrate

As we saw in Sec. 6.4, rare gas atoms such as Kr or Xe adsorbed on graphite can form a two-dimensional solid. The coupling between the substrate and the adsorbed atoms can be sufficiently weak such that the adsorbate lattice has a lattice parameter incommensurate with that of the substrate. The adsorbate lattice will, however, always have a preferred orientation relative to that of the substrate. This gives rise to a term in the adsorbate elastic free energy proportional to the square of $\theta = (\partial_x u_y - \partial_y u_x)/2$:

$$\mathcal{H}_{\text{el}} = \frac{1}{2}\int d^2x[\lambda u_{ii}^2 + 2\mu u_{ij}^2 + \gamma(\partial_x u_y - \partial_y u_x)^2]. \tag{9.5.14}$$

The dislocation energy resulting from this modified elastic free energy is identical to Eq. (9.5.1) with

$$\begin{aligned} K_1 &= \frac{4a^2}{T}\left[\frac{\mu(\mu+\lambda)}{2\mu+\lambda} + \frac{\mu\gamma}{\mu+\lambda}\right] \\ K_2 &= \frac{4a^2}{T}\left[\frac{\mu(\mu+\lambda)}{2\mu+\lambda} - \frac{\mu\gamma}{\mu+\lambda}\right]. \end{aligned} \tag{9.5.15}$$

Again, recursion relations for $K_1(l)$, $K_2(l)$ and $y(l)$ can be developed. The result is that the exponent $\bar{\nu}$ is a continuous function of the ratio $\zeta = K_2(T_M)/K_1(T_M)$ equal to 2/5 at $\zeta = 0$ and to 0.369... at $\zeta = 1$. The disordered phase above T_M has bond angle oriented along the substrate determined preferred axis. The transition to the disordered fluid breaks a discrete symmetry and is Ising-like.

2 Experiments and numerical simulation

There are a number of two-dimensional systems that might be expected to exhibit the KTHNY melting transition. These include

- free standing liquid crystal films such as those discussed in Sec. 9.2,
- electrons on the surface of liquid helium,
- rare gases adsorbed on the surface of graphite, discussed in Secs. 2.9 and 6.4,
- colloids of micro-size polystyrene spheres or "polyballs".

It has proven quite difficult to establish unambiguously the existence of a continuous solid-to-hexatic KTHNY transition in any of these systems. Nevertheless, the observed transitions from the solid phase on heating are often not inconsistent with the KTHNY theory.

Thick liquid crystal films have both three-dimensional crystalline solid and hexatic phases (Sec. 2.7) and transitions between them. Two-dimensional crystalline phases with QLRO have been identified in X-ray scattering experiments (Moncton *et al.* 1982). Electron scattering also identifies a hexatic phase, as discussed in Chapter 6. The transition from the crystalline phase in two-layer films of the material $\overline{14}S5$ is, however, abrupt and strongly hysteretic, indicating a first-order transition.

A single layer of electrons can be confined about 100 Å above the surface of superfluid He^4 by the application of an electric field normal to the helium surface (Grimes and Adams 1979). The surface density n_s is low enough ($\approx 10^8 - 10^9 \text{cm}^{-2}$) that the electrons behave almost completely classically. They repel each other with a $1/R$ potential, and their properties can be characterized by a unitless parameter $\Gamma = (e^2\sqrt{\pi n_s}/T)$. For Γ above a critical value (i.e., at sufficiently low temperature or high density), the electrons form a hexagonal Wigner lattice. In this crystalline phase, there are dynamical modes which couple longitudinal phonons of the electron lattice and surface waves (ripplons) of the helium (Fisher, Halperin, and Platzman 1979). These modes give rise to resonances in experimentally measurable response functions. While these resonances appear rapidly as the system is cooled, there is no discontinuity in their frequency and amplitude. In addition, no hysteresis is observed. These observations are consistent with a continuous melting transition. The temperature where the intensity of the phonon-ripplon resonance grows strongly corresponds to a value of $\Gamma = \Gamma_M = 137 \pm 15$, which is consistent with that predicted by the KTHNY theory.

Xenon adsorbed on graphite was discussed in Sec. 2.9. X-ray scattering data from the solid phase show a power-law peak in $S(\mathbf{q})$ and an exponent η_G (Fig. 6.4.5) that is consistent with the KTHNY theory. The correlation length in the orientationally ordered fluid above the melting temperature also agrees with the predictions of the KTHNY theory. Thermodynamics measurements (Jin, Bjurstrom, and Chan 1989), however, show unambiguously that there is a first-order crystal-to-fluid transition with no intervening hexatic phase.

Polystyrene spheres ("polyballs") with a diameter of order 0.3 μm form stable colloidal suspensions in solutions with the proper pH. When confined between two glass plates with separations from 1 to 4 μm, these polyballs form a regular two-dimensional hexagonal lattice that can be seen under a microscope (Murray and Van Winkle 1987). Digitized images of the two-dimensional system can be obtained and Fourier transformed to produce a two-dimensional structure function $S(\mathbf{q})$. Different diffraction patterns are obtained at different positions along a wedge sample in which the spacing between the glass plates changes linearly from 1 μm to 4 μm. At the large spacing end of the wedge, where the polyballs are close together, the diffraction pattern is that of a two-dimensional solid with well-developed hexagonal peaks. At the opposite end, the diffraction pattern is that of a two-dimensional fluid with a ring of constant intensity. In the intermediate region, the diffraction pattern shows hexagonal modulation analogous to that observed in three-dimensional or finite-size two-dimensional hexatics. Though the visual data provided by these experiments would appear to support a continuous crystalline-to-hexatic transition as a function of chemical potential (varied by the separation of the glass plates), there are many unanswered questions regarding the establishment and the nature of thermal equilibrium in the inhomogeneous cell used in the experiments.

Numerical simulations of spherical particles interacting in two dimensions via a variety of potentials have been carried out (Strandburg 1988). They tend to yield strongly first-order solid-to-liquid transitions. They are, however, plagued by problems associated with finite-particle number and computer time insufficient to ensure thermal equilibrium of all degrees of freedom.

9.6 The twist-grain-boundary phase

Throughout this chapter, we have seen how topological defects disrupt order. Thermally excited vortices can destroy the rigidity of a two-dimensional superfluid, and thermally excited vortex loops can destroy long-range order in a three-dimensional superfluid. Similarly, thermally excited dislocations can destroy crystalline order. In this section, we will investigate a remarkable phase of matter in which topological defects, by arranging in a repeated spatial pattern, create a new symmetry rather than simply destroy an old one. This phase occurs in some chiral liquid crystals (Goodby *et al.* 1988) and is characterized by an array of equally spaced twist-grain boundaries (Fig. 9.2.23) in a smectic liquid crystal as shown in Fig. 9.6.1. For obvious reasons, it is usually called the *twist-grain-boundary (TGB) phase* (Renn and Lubensky 1988). However, because it is a chiral smectic, it is sometimes called the smectic-A^* phase. As is clear from Fig. 9.6.1, the TGB phase has an average director twist like a cholesteric and regions of regularly spaced molecular planes like a smectic. One could, therefore, reasonably expect it to appear between the cholesteric and smectic-A phases in phase diagrams.

In what follows, we will first describe the structure of the TGB phase in some detail. Next we will show that it can have a lower energy than either the smectic-A or cholesteric phases. Then, we will describe its unusual X-ray scattering profile. Finally, we will discuss briefly its relation to the Abrikosov vortex lattice phase in superconductors (Abrikosov 1957).

1 *Structure of the TGB phase*

As we saw in Sec. 6.2, molecular chirality converts the nematic phase into a twisted nematic or cholesteric phase, N^*, with an equilibrium director

$$\mathbf{n}_c = (0, \sin k_0 x, \cos k_0 x), \tag{9.6.1}$$

where we have chosen the pitch axis to lie along the x-direction. This director field has a spatially uniform twist

$$\mathbf{n}_c \cdot (\boldsymbol{\nabla} \times \mathbf{n}_c) = k_0, \tag{9.6.2}$$

and is incompatible with uniform smectic layering. Twist of smectic layers can, however, be produced by grain boundaries, as we saw in Sec. 9.2. If we set

$$\mathbf{n}(\mathbf{x}) = (0, \sin\theta(x), \cos\theta(x)), \tag{9.6.3}$$

for which

$$\mathbf{n} \cdot (\boldsymbol{\nabla} \times \mathbf{n}) = \frac{d\theta(x)}{dx}, \tag{9.6.4}$$

then [Eq. (9.2.15)] θ will change by an amount

$$\delta\theta = 2\sin^{-1}(d/2l_d) \sim d/l_d \tag{9.6.5}$$

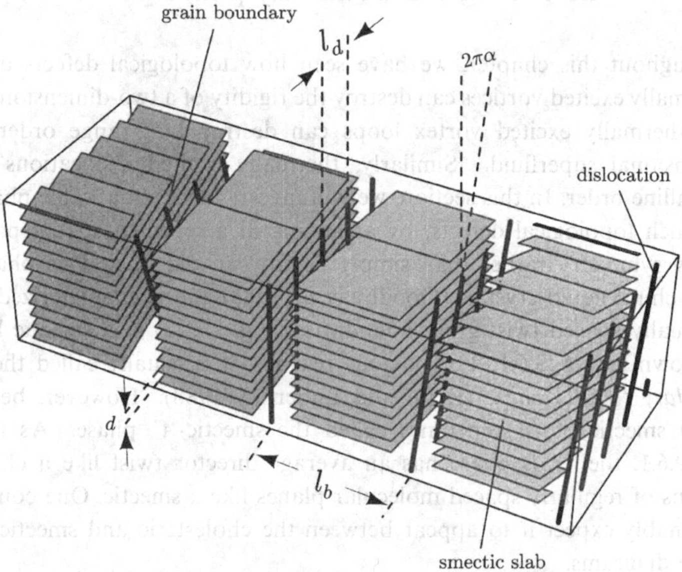

Fig. 9.6.1. Schematic representation of the TGB phase. There are smectic slabs with layer spacing d and length l_b. Between each pair of smectic slabs is a twist-grain boundary consisting of parallel screw dislocations separated by a distance l_d. The angle between dislocation cores in adjacent grain boundaries or, equivalently, between layer normals in adjacent smectic slabs is $2\pi\alpha = \delta\theta = 2\sin^{-1}(d/2l_d)$. If α is irrational, then no two slabs have identical layer normals, and there is an incommensurate TGB phase. If $\alpha = P/Q$ (with P and Q relatively prime integers) is a rational number, then there is a quasicrystalline TGB phase.

across each grain boundary composed of parallel screw dislocations separated by a distance l_d. The TGB phase consists of smectic slabs of length l_b separated by grain boundaries. In the smectic regions, $\theta(x)$ is approximately constant. Within a distance of order the twist penetration depth λ_2 (Secs. 6.2.1 and 9.3.4) of each grain boundary, $\theta(x)$ changes by an amount $\delta\theta$. Thus, $\theta(x)$ in the TGB phase has an average slope produced by a series of steps, as shown in Fig. 9.6.2, which contrasts this behavior with the uniform growth of $\theta(x)$ in the N^* phase. The spatial average of the twist in the TGB phase is nonzero:

$$\begin{aligned}\bar{k}_0 &= \frac{1}{V}\int d^3x\mathbf{n}\cdot(\nabla\times\mathbf{n}) \\ &= \frac{1}{L}\int_0^L dx\frac{d\theta}{dx} = \frac{1}{L}[\theta(L)-\theta(0)] \\ &= \frac{\delta\theta}{l_b} \approx \frac{d}{l_d l_b},\end{aligned} \tag{9.6.6}$$

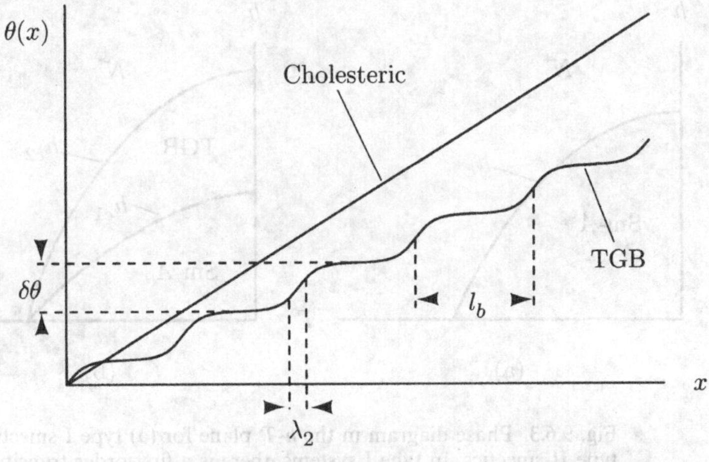

Fig. 9.6.2. Behavior of the director angle $\theta(x)$ in the cholesteric and TGB phases. In the cholesteric phase, $d\theta(x)/dx$ is a constant. In the TGB phase, $\theta(x)$ has an average linear growth produced by a series of steps across grain boundaries of height $\delta\theta$ and width λ_2. In general, the spatial average of $d\theta/dx$ diminishes in a transition from the cholesteric to the TGB phase.

where V is the sample volume and L is the length along the x-axis. In general, the average twist $\bar{k}_0$ in the TGB is less than the twist k_0 in a neighboring N^* phase.

Both the N^* and TGB phases have a nonvanishing director twist. The two phases, however, have different space group symmetries. The N^* phase has a continuous screw axis: it is invariant under an arbitrary translation x followed by a rotation through $\theta = k_0 x$. The TGB phase only has a discrete screw axis: it is invariant under translation by nl_b followed by a rotation through $\theta = n\delta\theta$ and possibly a translation in the yz-plane where n is an integer. Detailed symmetry properties of the TGB phase depend on whether

$$\alpha = \frac{\delta\theta}{2\pi} \approx \frac{d}{2\pi l_d} \tag{9.6.7}$$

is rational or irrational. If α is irrational, then all angles θ between 0 and 2π occur with equal probability, and no two smectic slabs will have exactly the same normal. This is the *incommensurate* TGB phase. If α is a rational number P/Q, with P and Q relatively prime integers, then there is a Q-fold screw axis. If Q is an integer other than one, two, three, four, or six, there is a quasicrystalline symmetry (see Sec. 2.10) with, as we shall see, a diffraction pattern with crystallographically disallowed Bragg (or quasi-Bragg) peaks.

The smectic slabs shown in Fig. 9.6.1 have a smectic-A structure with the director $\mathbf{n}$ parallel to the layer normal $\mathbf{N}$. There are materials in which these regions have a smectic-C structure with a nonzero angle between $\mathbf{n}$ and $\mathbf{N}$. When

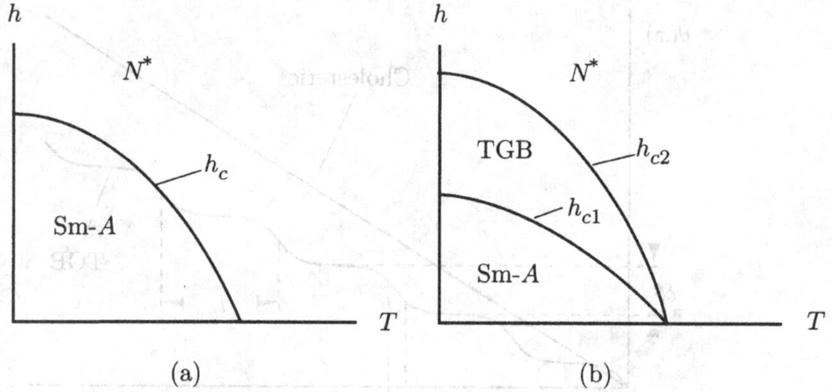

Fig. 9.6.3. Phase diagram in the *h*-*T* plane for (a) type I smectics and (b) type II smectics. In type I systems, there is a first-order transition from the N^* to the Sm-*A* phase at the thermodynamic critical field h_c. In type II systems, there is a second-order transition from the Sm-*A* to the TGB phase at the lower critical field h_{c1} and a second-order transition from the TGB to the N^* phase at the upper critical field h_{c2}. The TGB phase shown in this diagram may contain many distinct incommensurate phases and commensurate phases with quasicrystalline symmetry.

there is possible confusion, these two phases are denoted, respectively, by TGB_A and TGB_C.

2 *The thermodynamic critical field*

We saw in Sec. 6.2 that molecular chirality leads to an additional term,

$$F_{\mathrm{Ch}} = -h \int d^3x \mathbf{n} \cdot (\nabla \times \mathbf{n}) \tag{9.6.8}$$

in the Frank free energy and converts the nematic phase to a chiral nematic phase N^* with twist $k_0 = h/K_2$. Smectics can be characterized by the ratio $\kappa = \lambda_2/\xi$ of the twist penetration depth λ_2 to the smectic coherence length ξ. In type I smectics, $\kappa < 1/\sqrt{2}$, and in type II smectics, $\kappa > 1/\sqrt{2}$. As we shall verify below, in type I systems, there is (at least in the mean-field theory we consider here) a first-order transition between the Sm-*A* and the N^* phases at $h = h_c$, whereas in type II systems, the TGB phase intervenes between the Sm-*A* and N^* phases with a second-order Sm-*A*–TGB transition at h_{c1} and a second-order TGB–N^* transition at h_{c2}. Fig. 9.6.3 shows phase diagrams in the $h - T$ plane for type I and type II systems. In the language of superconductors, the field h_c is the *thermodynamic critical field*, h_{c1} is the *lower critical field*, and h_{c2} is the *upper critical field*.

To calculate the field h_c, we must compare the free energies of the N^* and Sm-A phases. The free energy density of the N^* phase relative to the nematic phase is

$$f_{N^*} = -\frac{h^2}{2K_2}. \tag{9.6.9}$$

In the smectic phase, the director $\mathbf{n}(\mathbf{x})$ is spatially uniform, and the Frank and chiral free energies are zero. The mean-field free energy of the Sm-A phase, therefore, comes entirely from the smectic order parameter ψ. Using Eq. (6.3.21) for the smectic free energy, we find

$$f_A = -\frac{r^2}{2g} \tag{9.6.10}$$

for the smectic free energy density. Equating f_{N^*} and f_A, we obtain the critical field h_c at which the transition from the Sm-A to the N^* phase takes place on increasing h:

$$h_c = (K_2 r^2/g)^{1/2}. \tag{9.6.11}$$

This is the thermodynamic critical field since it determines when the Sm-A phase becomes absolutely unstable to the N^* phase and vice versa. Alternatively, Eq. (9.6.11) gives the temperature at which a transition from the N^* phase takes place on lowering the temperature: $r_c = a(T - T_c) = (gh^2/K_2)^{1/2}$.

3 *The lower critical field*

If, upon increasing h, the Sm-A phase becomes unstable to the formation of the TGB phase before it becomes globally unstable to the N^* phase, then there will be a transition, which can be second order from the Sm-A to the TGB phase. The chiral energy, which for the director $\mathbf{n}(x)$ in Eq. (9.6.3) is

$$f_{\text{Ch}} \equiv \frac{F_{\text{Ch}}}{V} = -h k_0 \approx -\frac{hd}{l_d l_b}, \tag{9.6.12}$$

clearly favors twist and the formation of grain boundaries in the smectic phase. However, grain boundaries are composed of dislocations, which cost energy to create. In addition, dislocations repel each other, and there is a positive interaction energy associated with any array of dislocations. We saw in Sec. 9.3 that the interaction energy between screw dislocations in a smectic dies off exponentially with separation over λ_2. Thus, if l_d and l_b are much greater than λ_2, we can neglect interaction energy compared to creation energy. The energy associated with the creation of dislocations is $F_{\text{disc}} = L_{\text{disc}}\epsilon$, where ϵ is the energy per unit length of a dislocation [Eqs. (9.3.63), (9.3.64), and (9.3.81)] and L_{disc} is the total length of dislocation in the sample. In the model structure shown in Fig. 9.6.1, there are $N_d = L_\perp/l_d$ dislocations of length $L_\perp$ in each square section of grain boundary of side $L_\perp$ and $N_b = L/l_b$ grain boundaries in a length L. Thus $L_{\text{disc}} = N_d N_b L_\perp = V/(l_d l_b)$, where $V = L_\perp^2 L$ is the volume, and

$f_{\rm disc} = F_{\rm disc}/V = \epsilon/(l_d l_b)$. The free energy of the TGB structure (when dislocation interactions are ignored) compared to that of the untwisted Sm-A phase is then

$$\Delta f_{\rm TGB} = \frac{1}{l_d l_b}(\epsilon - hd). \tag{9.6.13}$$

The TGB phase has lower energy than the Sm-A phase when $h > h_{c1}$, where

$$\begin{aligned} h_{c1} &= \frac{\epsilon}{d} \\ &\approx \frac{dB}{2\pi}\ln(\lambda_2/\xi) = \frac{h_c}{\sqrt{2}\kappa}\ln\kappa, \end{aligned} \tag{9.6.14}$$

where, in the second line, we used Eq. (9.3.64) (with $D = B$) for the dislocation energy in extreme type II systems ($\kappa \gg 1$) with the core radius equal to the coherence length ξ. h_{c1} is the lower critical field. It is certainly less than h_c for sufficiently large κ. In this case, the TGB phase is favored over either the Sm-A or the N^* phases.

In the smectic phase, l_b and l_d are infinite. At $h = h_{c1}$, it becomes favorable for a single dislocation to penetrate the smectic. As h increases beyond h_{c1}, l_b^{-1} and l_d^{-1} increase continuously from zero in a way that depends in detail on the interaction energy between dislocations. We can get an idea of how these two inverse lengths grow for $h > h_{c1}$ by setting $l_d = l_b = l$ and assuming the interaction energy per unit volume is of the form we^{-l/λ_2}. Then $\Delta f_{\rm TGB} \approx l^{-2}d(h_{c1} - h) + we^{-l/\lambda_2}$. Minimization over l then yields $l \sim \lambda_2 \ln[w\lambda_2^3/2d(h - h_{c1})]$. From this we can see that l^{-1} tends to zero as $h \to h_{c1}^+$.

4 The upper critical field

At large values of h, the N^* phase is favored over the TGB phase because it can have a larger twist favored by h. As h is lowered in type II systems, the N^* phase becomes unstable with respect to the formation of the TGB phase at the upper critical field h_{c2}. To calculate h_{c2}, we need to determine when the cholesteric phase becomes unstable with respect to the establishment of a nonvanishing smectic order parameter $\psi(\mathbf{x})$, i.e., we need to determine when the inverse order parameter susceptibility $\delta F/\delta\psi(\mathbf{x})\delta\psi^*(\mathbf{x}')$ ceases to be positive definite. To describe the energy associated with the development of smectic order in an N^* phase with a nonuniform equilibrium director, we need to modify slightly the de Gennes free energy of Eq. (6.3.21), which was set up to describe the development of smectic order from a spatially uniform nematic. We express the mass density as

$$\rho(\mathbf{x}) = \rho_0 + \tilde{\psi}(\mathbf{x}) + \tilde{\psi}^*(\mathbf{x}), \tag{9.6.15}$$

where $\tilde{\psi}(\mathbf{x})$ does not have a Fourier component at zero wave number. $\tilde{\psi}(\mathbf{x})$ is related to the usual smectic order parameter $\psi(\mathbf{x})$ introduced in Eq. (6.3.1) via

$$\tilde{\psi}(\mathbf{x}) = e^{i\mathbf{q}_0\cdot\mathbf{x}}\psi(\mathbf{x}) \tag{9.6.16}$$

where $\mathbf{q}_0 = q_0\mathbf{n}_0$ with $\mathbf{n}_0$ the spatially uniform equilibirum director of the nematic phase. The de Gennes free energy for smectic order is now

$$F_{\tilde{\psi}} = \int d^3x[r|\tilde{\psi}|^2 + C|(\nabla - iq_0\mathbf{n})\tilde{\psi}|^2 + \frac{1}{2}g|\tilde{\psi}|^4], \tag{9.6.17}$$

where, for simplicity, we have chosen an isotropic energy by setting both $c_{\|}$ and $c_{\perp}$ of Eq. (6.3.21) equal to C. If $\mathbf{n}(\mathbf{x}) = \mathbf{n}_0$, this energy is minimized with $\tilde{\psi}(\mathbf{x}) = e^{iq_0\mathbf{n}_0\cdot\mathbf{x}}\psi$ with $\psi = (-r/g)^{1/2}$. The complete free energy for the development of smectic order from a cholesteric is

$$F = F_{\tilde{\psi}} + F_{\mathbf{n}} + F_{\mathrm{Ch}}, \tag{9.6.18}$$

where $F_{\mathbf{n}}$ is the Frank free energy of Eq. (6.2.3).

To determine the linearized stability of the N^* phase with respect to $\tilde{\psi}(\mathbf{x})$, we set $\mathbf{n}(\mathbf{x}) = \mathbf{n}_c(x)$ [Eq. (9.6.1)] in Eq. (9.6.17) and calculate

$$\begin{aligned} M(\mathbf{x}, \mathbf{x}') &= \frac{\delta F_{\tilde{\psi}}}{\delta\tilde{\psi}(\mathbf{x})\tilde{\psi}(\mathbf{x}')} \\ &= [-C\nabla^2 + 2iq_0\mathbf{n}_c(x)\cdot\nabla + Cq_0^2 + r]\delta(\mathbf{x}-\mathbf{x}'). \end{aligned} \tag{9.6.19}$$

The condition $\nabla \cdot \mathbf{n}_c(\mathbf{x}) = 0$ was used in the derivation of this equation. Unlike other inverse susceptibilities we have encountered, $M(\mathbf{x}, \mathbf{x}')$ is not a function of $\mathbf{x}-\mathbf{x}'$ only. It depends not only on $\mathbf{x}_{\perp} - \mathbf{x}'_{\perp}$, where $\mathbf{x}_{\perp} = (0, y, z)$, but explicitly on the coordinates x and $x - x'$. $M(\mathbf{x}, \mathbf{x}')$ cannot be diagonalized by plane waves and inverted by Fourier transformation. It is useful to think of $M(\mathbf{x}, \mathbf{x}')$ as the $\mathbf{x}-\mathbf{x}'$ matrix elements of an operator $\hat{M}$: $M(\mathbf{x}, \mathbf{x}') = \langle\mathbf{x}|\hat{M}|\mathbf{x}'\rangle$. The wavefunctions that diagonalize $\hat{M}$ (i.e., eigenfunctions of $\hat{M}$) are of the form

$$\Phi(\mathbf{x}) = A_{\mathbf{q}_{\perp}}(x)e^{i\mathbf{q}_{\perp}\cdot\mathbf{x}_{\perp}}, \tag{9.6.20}$$

where

$$\mathbf{q}_{\perp} = q_{\perp}(0, \sin\eta, \cos\eta) \equiv q_{\perp}\mathbf{n}_c(\eta/k_0). \tag{9.6.21}$$

When acting on states with this plane-wave dependence on transverse components of position, $\hat{M}$ reduces to the Mathieu operator:

$$\hat{M} = -c\partial_x^2 + c(q_{\perp} - q_0)^2 + r + 2cq_0q_{\perp}[1 - \cos(k_0x - \eta)]. \tag{9.6.22}$$

This is the Schrödinger operator for an electron in a cosine potential of period $2\pi/k_0$ with equivalent minima at positions $x_{\eta,l} = (2\pi l + \eta)/k_0$ for every integer l. In the vicinity of each of these minima, the cosine potential can be approximated by a harmonic oscillator potential, and $\hat{M}$ becomes

$$\hat{M} = -C\partial_x^2 + C(q_{\perp} - q_0)^2 + r + Cq_0q_{\perp}k_0^2(x - x_{\eta,l})^2. \tag{9.6.23}$$

The eigenfunctions of this equation are harmonic oscillator wavefunctions $A^n_{\mathbf{q}_{\perp}}(x)$ centered at $x_{\eta,l}$. The lowest energy normalized wavefunction is

$$A^0_{\mathbf{q}_{\perp}}(x) = \frac{1}{(2\pi\xi)^{1/2}}e^{-\frac{1}{2}\overline{\xi}^{-2}(x-x_{\eta,l})^2} \equiv A^0(x - x_{\eta,l}), \tag{9.6.24}$$

where

$$\overline{\xi}^{-2} = (q_0q_{\perp})^{1/2}k_0 \approx q_0k_0. \tag{9.6.25}$$

The distance between minima of the cosine potential is $2\pi/k_0$, so the amplitude of $A^0(x - x_{\eta,l})$ evaluated at a nearest neighbor site with $x = x_{\eta,l} + (2\pi/k_0)$ is $(2\pi\bar{\xi}^2)^{-1/2}e^{-2\pi^2(q_0/k_0)}$. In typical systems $q_0 \sim 2\pi/(30\text{Å})$ and $k_0 \sim 2\pi/(3000\text{Å})$ so that $(q_0/k_0) > 10^2$ and the overlap of wavefunctions centered on neighboring sites is vanishingly small. Thus, the matrix elements of $\hat{M}$ coupling $A^n(x - x_{\eta,l})$ with $A^n(x - x_{\eta,k})$ with $l \neq k$ can be neglected, and the eigenvalues of the operator $\hat{M}$ are simply harmonic oscillator energies:

$$\epsilon_n(k_0, q_\perp) = c[(2n+1)\bar{\xi}^{-2} + (q_\perp - q_0)^2 + r/c]. \tag{9.6.26}$$

Note that these eigenvalues depend only on the magnitude of $\mathbf{q}_\perp$ and not on its direction. We are particularly interested in these energies in the vicinity of their minimum. The value of $q_\perp$ that minimizes ϵ_n is

$$q_{\perp n} = q_0 \left[1 - \frac{1}{4}(2n+1)(k_0/q_0) + O(k_0^2/q_0^2)\right]. \tag{9.6.27}$$

The ground state energy $\epsilon_0(k_0, \mathbf{q}_\perp)$ passes through zero for $r < 0$ when

$$\bar{\xi}^{-2} = |r|/c = \xi^{-2} \tag{9.6.28}$$

or, equivalently, when $h = K_2 k_0 = h_{c2}$ with

$$h_{c2} = K_2 q_0^{-1}(-r/c)[1 + O(k_0/q_0)] = \sqrt{2}\kappa h_c[1 + O(k_0/q_0)]. \tag{9.6.29}$$

Thus, h_{c2} is greater than h_c, and the N^* phase becomes unstable to the TGB phase before it becomes globally unstable to the formation of the uniform smectic phase, provided $\kappa > 1/\sqrt{2}$ to lowest order in k_0/q_0.

The energies, Eq. (9.6.26), can now be expanded about $k_0 = k_{c2} = h_{c2}/K_2$ and $q_\perp = q_{\perp n}$. The lowest energy is

$$\begin{aligned}\epsilon_0(k_0, q_\perp) &= c\xi^{-2}\{[(k_0 - k_{c2})/k_{c2}] + \xi^2(q_\perp - q_0)^2\} \\ &= c\xi_2^{-2}[1 + \xi_2^2(q_\perp - q_0)^2],\end{aligned} \tag{9.6.30}$$

where

$$\xi_2 = \xi[(h - h_{c2})/h_{c2}]^{1/2} \tag{9.6.31}$$

is the twist lattice correlation length that diverges as $h \to h_{c2}$.

5 X-ray scattering

X-ray scattering profiles from TGB phases are quite unusual. They can be understood as a superposition of scattering intensities from an array of rotated smectic slabs consisting of N layers of length l_b along x and width $L \gg l_b$ along y, as shown in Fig. 9.6.4. The scattering intensity of one such slab with layer normal along the z-axis is proportional to $(\sin q_x l_b/q_x)\delta(q_y)[\delta(q_z - q_0) + \delta(q_z + q_0)]$, as shown in Fig. 9.6.4. Thus there are delta-function rods of length $2\pi/l_b$ parallel to the x-axis and centered at $q_x = q_y = 0$, $q_z = \pm q_0$. In an incommensurate TGB phase, slab normals point with equal probability in all directions, and the scattering amplitudes from different slabs add incoherently. The resulting pattern is a *Bragg cylinder* of radius q_0 and height $2\pi/l_b$. In a commensurate smectic,

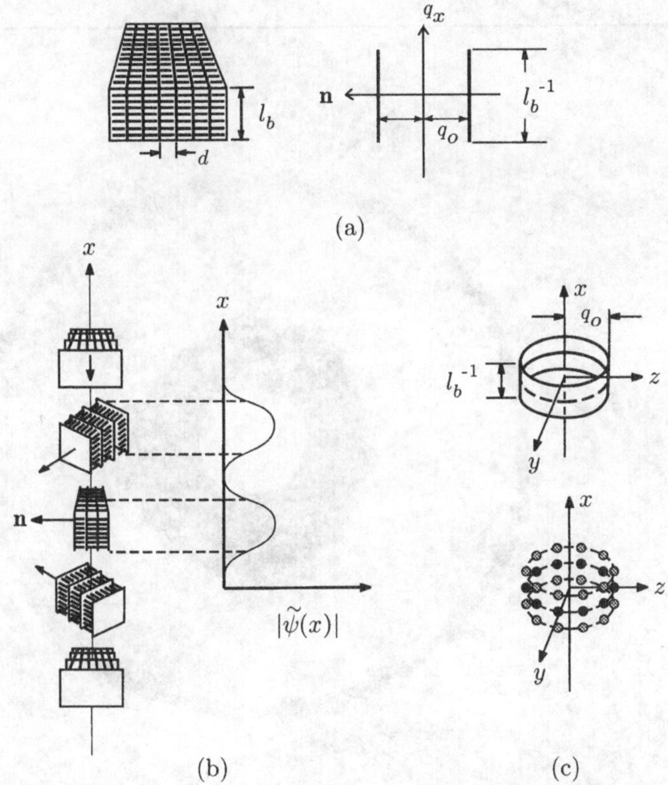

Fig. 9.6.4. (a) A smectic slab and its corresponding X-ray intensity. The latter is a pair of Bragg rods whose height is inversely proportional to the slab length l_b. These rods are perpendicular to the slab normal **n** (for a smectic-*A* slab) and are displaced from the pitch axis (the *x*-axis) by $\pm q_0\mathbf{n}$. (b) A stack of smectic slabs in the TGB phase with corresponding order parameter profile, $|\tilde{\psi}(\mathbf{x})|$. Near h_{c2}, this profile is the Gaussian eigenfunction [Eq. (9.6.24)] of $\hat{M}$. (c) X-ray scattering intensity from an incommensurate (top) and a commensurate or quasicrystalline (bottom) TGB phase. In the incommensurate TGB phase, the intensity is simply an in-plane powder average of the rods shown in (a). In commensurate TGB phases with $\alpha = P/Q$, intensities from every Qth slab add coherently, and there are Q Bragg spots for Q even and $2Q$ Bragg spots for Q odd on a series of rings at different values of q_x but with intensity decreasing with q_x^2. In the figure, $Q = 5$.

with $\alpha = P/Q$, every Qth slab is identical, and their scattering adds coherently. The result is that there will be Bragg spots with Q-fold symmetry rather than a Bragg cylinder, as shown in Figs. 9.6.4 and 9.6.5.

The X-ray scattering intensity in the cholesteric phase is simply a broadened cylinder of radius q_0, height $q\pi/l_b$, and radial width proportional to the inverse

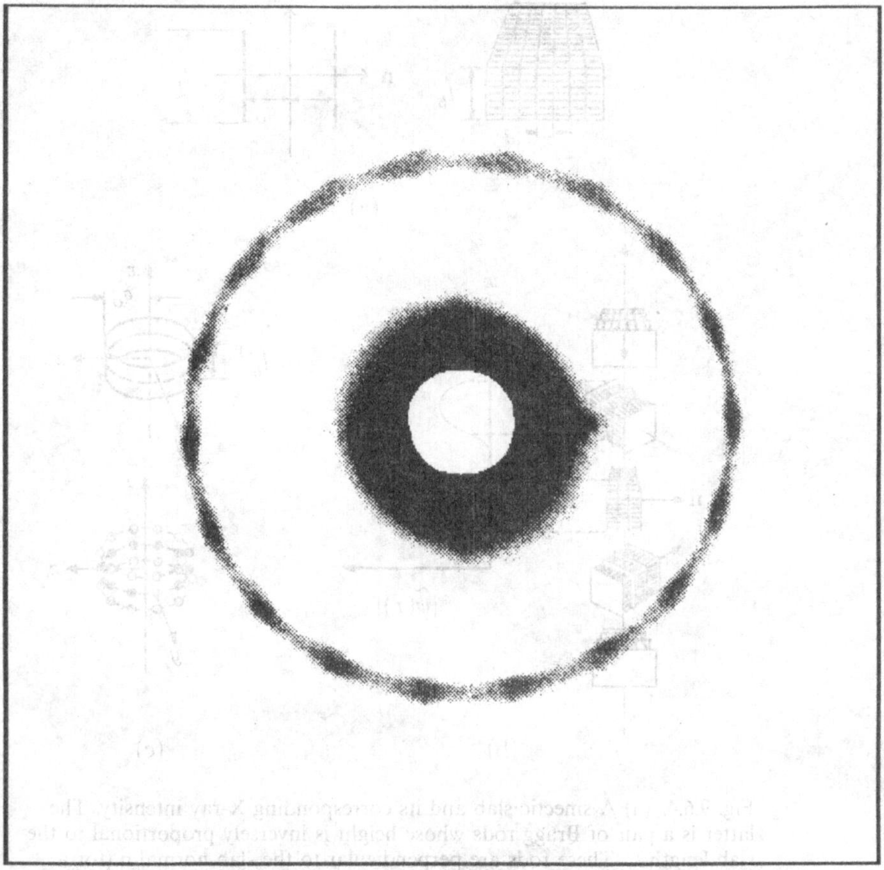

Fig. 9.6.5. Experimental scattering intensity from a quasicrystalline TGB phase showing 18 Bragg peaks around a ring. This material is a TGB$_C$ phase, in which the smectic slabs are smectic-C slabs in which the director is not parallel to the layer normal. It has either 9- or 18-fold quasicrystalline symmetry. [L. Navailles, P. Barois, and H. Nguyen, *Phys. Rev. Lett.* **71**, 545 (1993).]

coherence length ξ_2^{-1}. This intensity near h_{c2} in mean-field theory is the inverse of the operator $\hat{m}$, which we can represent in terms of the wavefunctions $A^n_{\mathbf{q}_\perp}(x)$ and energies $\epsilon_n(k_0, q_\perp)$ as

$$\begin{aligned} \chi(\mathbf{x}, \mathbf{x}') &= \langle \tilde{\psi}(\mathbf{x}) \tilde{\psi}^*(\mathbf{x}') \rangle \\ &= \sum_{n,l} \int \frac{d^2 q_\perp}{(2\pi)^2} \frac{A^n_{\mathbf{q}_\perp,l}(x) A^n_{\mathbf{q}_\perp,l}(x')}{\epsilon_n(k_0, q_\perp)} e^{i\mathbf{q}_\perp \cdot (\mathbf{x}-\mathbf{x}')}. \end{aligned} \tag{9.6.32}$$

Near h_{c2}, the $n = 0$ mode dominates, and we obtain

$$I(\mathbf{q}) = 2\frac{k_0\bar{\xi}}{\sqrt{\pi}}\frac{\xi_2^2}{C}\frac{e^{-q_x^2\bar{\xi}^2}}{[1+\xi_2^2(q_\perp - q_0)^2]}. \tag{9.6.33}$$

As a function of $q_\perp$ at constant q_x, the X-ray scattering profile is a Lorentzian centered at $q_\perp = q_0$ with halfwidth ξ_2^{-1} that tends to zero as $h \to h_{c2}$. As a function of q_x at constant $q_\perp$, the profile is a Gaussian with width $\bar{\xi}^{-1}$ that remains finite as $h \to h_{c2}$.

6 Analogy with superconductivity

We have alluded several times to the analogy between the de Gennes free energy for the nematic-to-smectic-*A* transition and the Landau-Ginzburg free energy for a superconductor. This analogy is so precise that there is a one-to-one correspondence between virtually all properties of superconductors and properties of the nematic or smectic-*A* phase of liquid crystals. The most striking correspondence is between the TGB phase in liquid crystals and the Abrikosov vortex lattice in superconductors, whose theoretical prediction and experimental discovery predated by many years those of the TGB phase.

Because of the important role the analogy with superconductivity plays in the development of our understanding of liquid crystals, we will very briefly review here the Landau-Ginzburg theory for superconductors. A superconductor, like a superfluid, has a complex order parameter ψ. In a superfluid, there is a mass current associated with the gradient of the phase of ψ, as we saw in Sec. 8.5. In a superconductor, gradients in the phase of ψ lead to electric currents, which couple to the electromagnetic field. These currents and the free energy must be invariant under gauge transformations. The Landau-Ginzburg free energy for a superconductor is the gauge invariant generalization of the superfluid energy with the addition of the magnetic field energy:

$$\begin{aligned} F_{\mathrm{LG}} &= \int d^3x \left[r|\psi|^2 + C\left|\left(\boldsymbol{\nabla} - \frac{2ie}{\hbar c}\mathbf{A}\right)\psi\right|^2 + \frac{1}{2}g|\psi|^4 \right] \\ &\quad + \frac{1}{8\pi\mu_0}\int d^3x(\boldsymbol{\nabla}\times\mathbf{A})^2 - \frac{1}{4\pi}\int d^3x\mathbf{H}\cdot(\boldsymbol{\nabla}\times\mathbf{A}), \end{aligned} \tag{9.6.34}$$

where $\mathbf{A}$ is the vector potential, e is the electron charge, c is the velocity of light, $\hbar$ is Planck's constant, μ_0 (= 1 in the cgs units we are using) is the magnetic permeability of free space, and $\mathbf{H}$ is the magnetic intensity whose source is external currents. The local magnetic field is $\mathbf{B} = \boldsymbol{\nabla}\times\mathbf{A}$. Many of the superconductor-liquid crystal analogies are immediately clear from Eqs. (9.6.34), (9.6.17), and (9.6.18). The vector $\mathbf{A}$, the magnetic intensity $\mathbf{H}$, and the magnetic field $\mathbf{B}$ are, respectively, the analog of the director $\mathbf{n}$, the chiral field h, and twist $k_0 = \mathbf{n}\cdot(\boldsymbol{\nabla}\times\mathbf{n})$ and bend $\mathbf{n}\times(\boldsymbol{\nabla}\times\mathbf{n})$ in liquid crystals. The inverse permeability $(4\pi\mu_0)^{-1}$ is the analog of the twist and bend elastic constant K_2 and K_3.

The important properties of superconducting and normal phases can be determined directly from F_{LG} and our already extensive investigations of liquid crystals. For $r > 0$ and $\mathbf{H} = 0$, there is a normal-metal phase with $\psi = 0$ and $\nabla \times \mathbf{A} = 0$. This is the analog of the nematic phase in liquid crystals. For $r > 0$ and $\mathbf{H} \neq 0$, $\psi = 0$, and $\mathbf{B} = \mu_0 \mathbf{H}$. This normal metal in a magnetic field with nonvanishing $\mathbf{H}$ and $\mathbf{B}$ is the analog of the cholesteric phase with nonzero h and twist $k_0 = h/K_2$. For $r < 0$, there is a superconducting phase with $\psi = (-r/g)^{1/2}$. This is the analog of the smectic phase obtained from a nematic. Magnetic fields are expelled from a superconductor with a penetration depth $\lambda = (\hbar^2 c^2/16\pi e^2 C|\psi|^2)^{1/2} = (\hbar^2 c^2 g/16\pi e^2 C|r|)^{1/2}$, just as twist and bend are expelled from a smectic with penetration depths λ_2 and λ_3. The ratio $\kappa = \lambda/\xi$ of the penetration depth to the correlation length in a superconductor is called the Ginzburg parameter. In type I systems, $\kappa < 1/\sqrt{2}$, and in type II systems, $\kappa > 1/\sqrt{2}$.

There are vortices in superconductors in which the phase of ψ changes by 2π in one circuit around a core. The vector potential, like the director in a smectic, can adjust to lower the energy associated with gradients in the phase of ψ imposed by topological constraints of the defect. The calculation of $\mathbf{A}$ and the energy of a superconducting vortex are *identical* to the analogous calculations for the screw dislocations presented in Sec. 9.3. One only has to exercise care to identify $-q_0 u$ with the phase ϕ of the superconducting order parameter and q_0 with $2e/\hbar c$. These calculations imply that a superconducting vortex carries *quantized* flux. This follows because $\nabla\phi - (2e/\hbar c)\mathbf{A}$ (the superconducting analog of $q_0 \mathbf{Q} = q_0(\nabla u + \delta \mathbf{n})$ in Eq. (9.3.58)) dies exponentially to zero with distance from the vortex core. Thus,

$$\Phi_0 = \oint \mathbf{A} \cdot ds = \frac{\hbar c}{2e} \oint \nabla\phi \cdot ds = \frac{hc}{2e} \tag{9.6.35}$$

is the quantum of flux carried by a single unit strength vortex in a superconductor.

We can now discuss the appearance of the Abrikosov phase in an external magnetic field. The external field superconducting order favors nonzero $\nabla \times \mathbf{A}$ and penetration of magnetic flux, just as molecular chirality favors nonzero $\mathbf{n} \cdot (\nabla \times \mathbf{n})$ and penetration of twist. Superconducting order expels magnetic flux, and smectic order expels twist. In type I superconductors, there is a first-order transition between the Meisner phase with expelled flux and the normal metal in a field at $H = H_c = (4\pi\mu_0 r^2/g)^{1/2}$ when the energies $-r^2/2g$ and $-H^2/8\pi\mu_0$ of the two phases are equal. In type II systems, we can calculate when it is first favorable for a vortex to penetrate the superconductor by neglecting interactions among vortices. In superconductors, vortices are all parallel to the magnetic field. If there are $N = nA$ vortices in an area A, then the energy due to vortices in a volume $V = AL$ is $F_{\text{vortex}} = LN\epsilon$. In addition, $\mathbf{H} \cdot \int \nabla \times \mathbf{A} = LHN\Phi_0 = VnH\Phi_0$. The free energy per unit volume of a state with vortices relative to the Meisner state is then

$$\Delta f = n\left(\epsilon - \frac{H\Phi_0}{4\pi}\right). \tag{9.6.36}$$

Thus,

$$H_{c1} = \frac{4\pi\epsilon}{\Phi_0} \approx \frac{\Phi_0}{4\pi\lambda^2} \ln\kappa = \frac{H_c}{\sqrt{2}\kappa} \ln\kappa. \tag{9.6.37}$$

For $H > H_{c1}$, the repulsive interaction between vortices causes them to form a triangular lattice. This is the Abrikosov vortex lattice.

The upper critical field H_{c2} is calculated in exactly the same way as h_{c2} was calculated in for liquid crystals. The limit of stability of the normal metal phase in a magnetic field producing a nonvanishing $\nabla \times \mathbf{A}$ determines H_{c2}. In the Landau Gauge in which $\mathbf{A} = (0, Bx, 0)$ the stability kernel for ψ is identical to the harmonic oscillator approximation for $\hat{M}$ for liquid crystals, and $H_{c2} = \sqrt{2}\kappa H_c$.

Appendix 9A Notes on the Kosterlitz-Thouless transition

1 Integration of the KT recursion relations

The Kosterlitz-Thouless recursion relations, Eqs. (9.4.32) and (9.4.33), can be integrated analytically in the vicinity of the critical point. The resulting solutions can be used to obtain the correlation length, Eq. (9.4.40), the rigidity modulus [Eq. (9.4.41)], and the spin-correlation function at T_c [Eq. (9.4.39)]. As discussed in Sec. 9.4, the fixed-point value of $K(l) = \rho_s/T$ is $K^* = 2/\pi$. To describe small deviations from the fixed point $K = K^*$, $y = y^* = 0$, we set

$$K(l) = K^*(1 - x(l)). \tag{9A.1}$$

The recursion relations for K^{-1} and y [Eqs. (9.4.32) and (9.4.33)] to lowest order in x are

$$\frac{dx}{dl} = 8\pi^2 y^2, \tag{9A.2}$$

$$\frac{dy}{dl} = 2xy. \tag{9A.3}$$

These equations imply

$$\frac{dx^2}{dl} = 16\pi^2 xy^2, \tag{9A.4}$$

$$\frac{dy^2}{dl} = 4xy^2, \tag{9A.5}$$

and

$$\frac{dx^2}{dy^2} = 4\pi^2. \tag{9A.6}$$

The last equation implies that all trajectories lie on the hyperbolae,

$$y^2 = \frac{1}{4\pi^2}(x^2 + C), \tag{9A.7}$$

where C is a constant. The lines $y = \pm x/(2\pi)$ are the asymptotes of the family of hyperbolae and are obtained by setting C to zero in Eq. (9A.7). All hyperbolae with $C > 0$ lie above the asymptotes $y = -x/(2\pi)$, $x < 0$, and $y = x/(2\pi)$, $x > 0$. Hyperbolae with $C < 0$ intersect the $y = 0$ axis at $x = -\sqrt{|C|}$. The only trajectory that passes through the critical fixed point ($x^* = 0$, $y^* = 0$) (Fig. 9.4.2) is that corresponding to the limiting hyperbola with $C = 0$. This is the critical trajectory, and the asymptote $y = -x/(2\pi)$ is the critical line: all Hamiltonians starting on this line flow to the critical fixed point.

Hyperbolae with nonzero values of C approach the critical trajectory as $C \to 0$. Thus, C measures distance from the critical point and is linearly proportional to $T - T_c$:

$$C = b^2(T - T_c), \tag{9A.8}$$

where b^2 is a constant with units of inverse temperature. Using Eq. (9A.7), we can convert Eq. (9A.2) to an equation for x alone:

$$\frac{dx}{dl} = 2(x^2 + C). \tag{9A.9}$$

This equation says that $x(l)$ increases with l for $x \neq 0$ regardless of its sign, i.e., that all flows are to the right, as shown in Fig. 9.4.2. If $C = 0$ and $x < 0$, flow is to the critical fixed point. If $C < 0$ and $x < 0$, flows are towards $y = 0$ and a state with no vortices. If $C > 0$, and x is initially negative, the flow will first be towards decreasing values of y and then, once $x(l)$ becomes greater than zero, towards the asymptote $y = x/(2\pi)$, $x > 0$. These trajectories are depicted in Fig. 9.4.2. The solution to Eq. (9A.9) thus depends on the sign of C, i.e., whether $T > T_c$ or $T < T_c$. If $C < 0$ ($T < T_c$), the equation for $u(l) = x(l)/\sqrt{|C|}$ is

$$\int_{u(0)}^{u(l)} \frac{du}{u^2 - 1} = \frac{1}{2}\ln\left(\frac{1 - u(l)}{1 + u(l)}\frac{1 + u(0)}{1 - u(0)}\right) = 2\sqrt{|C|}l \tag{9A.10}$$

or

$$u(l) = -\frac{1 - D_0 e^{-4\sqrt{|C|}l}}{1 + D_0 e^{-4\sqrt{|C|}l}}, \tag{9A.11}$$

where $D_0 = [1 + u(0)]/[1 - u(0)]$. The right-hand side of this equation tends to -1 as $l \to \infty$, and

$$\lim_{l\to\infty} x(l) = -\sqrt{|C|}. \tag{9A.12}$$

Note that the denominator of the right-hand side of Eq. (9A.11) is never singular for $l > 0$ because $-1 < D_0 < 0$ for $-\infty < u(0) < -1$. Because $K_R = \rho_s{}^R(T)/T = \lim_{l\to\infty} K(l)$ for $T < T_c$ [Eq. (9.4.35)], Eqs. (9A.1), (9A.8), and (9A.12) imply Eq. (9.4.41):

$$\rho_s{}^R(T) = \rho_s{}^R(T_c)[1 + b\sqrt{T - T_c}], \tag{9A.13}$$

where $\rho_s{}^R(T_c)/T_c = 2/\pi$.

When $C > 0$, the solution to Eq. (9A.9) is

$$\int_{x(0)}^{x(l)} \frac{dx}{x^2 + C} = \frac{1}{\sqrt{C}}\left(\tan^{-1}\frac{x(l)}{\sqrt{C}} - \tan^{-1}\frac{x(0)}{\sqrt{C}}\right) = 2l. \tag{9A.14}$$

Near the critical point for $T > T_c$, $x(0)$ is negative. As $T \to T_c$, $\sqrt{C} \to 0$ and $|x(0)| >> \sqrt{C}$ so that $\tan^{-1}(x(0)/\sqrt{C}) \approx -\pi/2$. To determine the correlation length, we choose $x(l^*)$ to be positive and of order unity and $\tan^{-1}(x(l^*)/\sqrt{C}) \approx \pi/2$. Thus, at the matching point,

$$2l^* = \frac{\pi}{\sqrt{|C|}} \tag{9A.15}$$

and

$$\frac{\xi}{a} = e^{l^*} = e^{b'/\sqrt{T - T_c}}; \tag{9A.16}$$

$bb' = \pi/2$ is universal.

Finally, to obtain the spin correlation function at the critical point, we observe that

$$g(\mathbf{x}) = \int \frac{d^2q}{(2\pi)^2}\frac{1 - e^{i\mathbf{q}\cdot\mathbf{x}}}{K_R(q)q^2}, \tag{9A.17}$$

where $K_R(q)$ is the renormalized reduced stiffness, which at $q = 0$ satisfies Eq. (9.4.26). Along the critical trajectory, $y^2 = x^2/(4\pi)$ and

$$\frac{dx}{dl} = 2x^2 \tag{9A.18}$$

or

$$x(l) = \frac{x(0)}{1 - 2lx(0)} \xrightarrow[l\to\infty]{} -\frac{1}{2l}. \tag{9A.19}$$

To leading order in y,

$$K_R^{-1} = \frac{\pi}{2}[1 + x(l)]. \tag{9A.20}$$

The rigidity ρ_s and thus K_R satisfy the Josephson scaling relation Eq. (6.1.11):

$$K_R(q) = e^{(d-2)l} K_R(e^l q). \tag{9A.21}$$

Thus, we can choose $e^{l^*} q = 1$ or $x(l^*) = -(\ln q^{-1})^{-1}/2$. Then, using Eqs. (9A.17) and (9A.20), we obtain

$$\begin{aligned} g(\mathbf{x}) \quad &\sim \quad \frac{1}{4}\int_{|\mathbf{x}|^{-1}}^{\Lambda} \frac{dq}{q}\left(1 - \frac{1}{2\ln q^{-1}}\right) \\ &\sim \quad \frac{1}{4}\ln|\mathbf{x}| - \frac{1}{8}\ln(\ln|\mathbf{x}|), \end{aligned} \tag{9A.22}$$

where Λ is the upper cutoff. Then,

$$G(\mathbf{x}) = e^{-g(\mathbf{x})} = \frac{\ln^{1/8}|\mathbf{x}|}{|\mathbf{x}|^{1/4}}, \tag{9A.23}$$

in agreement with Eq. (9.4.39).

2 Longitudinal and transverse response

Our derivation of Eq. (9.4.8) was somewhat sketchy. Here we will provide further details of this derivation and explore some of the subtleties of longitudinal and transverse momentum response in superfluids and xy-models. In normal fluids, the momentum density $\mathbf{g}$ is simply $\rho\mathbf{v}$, where ρ is the total mass density and $\mathbf{v}$ is the velocity. Thus, the momentum susceptibility is $\chi_{g_i g_j} = \partial g_i/\partial v_j = \rho\delta_{ij}$. In superfluids (see Sec. 8.5), there is a superfluid velocity $\mathbf{v}_s$ and a normal veclocity $\mathbf{v}_n$, and $\mathbf{g} = \rho_s\mathbf{v}_s + \rho_n\mathbf{v}_n$, where ρ_n and ρ_s are, respectively, the normal and superfluid densities with $\rho = \rho_n + \rho_s$. The superfluid velocity, being the gradient of the phase of the superfluid order parameter, is purely longitudinal. The normal velocity, on the other hand, has both longitudinal and transverse parts. As a result, the momentum susceptibility will have different longitudinal and transverse parts. We will show below that

$$\chi_{g_i g_j} = \rho\hat{q}_i\hat{q}_j + \rho_n(\delta_{ij} - \hat{q}_i\hat{q}_j) = \chi_l\hat{q}_i\hat{q}_j + \chi_t(\delta_{ij} - \hat{q}_i\hat{q}_j), \tag{9A.24}$$

where $\chi_l = \rho$ and $\chi_t = \rho_n$ are, respectively, the longitudinal and transverse momentum susceptibilities. As the normal fluid phase is approached (e.g. by raising the temperature), $\rho_s \to 0$, $\rho_n \to \rho$, and $\chi_{g_i g_j}$ approaches $\rho\delta_{ij}$. In the classical elastic version of the xy-model we study in Sec. 9.4, $\rho_n = 0$ at $T = 0$. It is the creation of vortex pairs that leads to transverse excitations and a nonzero ρ_n at finite temperature.

To derive Eq. (9A.24), we begin with the superfluid free energy density

$$f(\mathbf{v}_n, \mathbf{v}_s) = -\frac{1}{2}\rho v_n^2 + \frac{1}{2}\rho_s(\mathbf{v}_s - \mathbf{v}_n)^2 \tag{9A.25}$$

derived from Eq. (8.5.32). This leads to $g_i = -\partial f/\partial v_{ni} = \rho_n v_{ni} + \rho_s v_{si}$. The Legendre transformed free energy that is a natural function of $\mathbf{g}$ and $\mathbf{g}_s \equiv \rho_s \mathbf{v}_s$ is then

$$\begin{aligned}\tilde{f}(\mathbf{g}, \mathbf{g}_s) &= [f(\mathbf{v}_n, \mathbf{g}_s/\rho_s) + \mathbf{g}\cdot\mathbf{v}_n]_{\mathbf{v}_n=(\mathbf{g}-\mathbf{g}_s)/\rho_n} \\ &= \frac{1}{2\rho_n}g_t^2 + \frac{1}{2\rho_n}(g_l - g_s)^2 + \frac{1}{2\rho_s}g_s^2, \end{aligned} \tag{9A.26}$$

where $\mathbf{g}_l$ and $\mathbf{g}_t$ are, respectively, the longitudinal and transverse parts of $\mathbf{g}$. In deriving this equation, we used the fact that $\mathbf{g}_s$ is longitudinal so that $\mathbf{g}\cdot\mathbf{g}_s = g_l g_s$. The transverse momentum decouples from g_s and g_l, and Eq. (9A.26) implies that

$$\chi_t^{-1} = \partial^2\tilde{f}/\partial g_t^2 = \rho_n^{-1}. \tag{9A.27}$$

Since g_l and g_s are coupled, we introduce $\phi_\alpha = (g_l, g_s)$. Then

$$\chi_{\alpha\beta}^{-1} = \begin{pmatrix} \rho_n^{-1} & -\rho_n^{-1} \\ -\rho_n^{-1} & \rho_n^{-1} + \rho_s^{-1} \end{pmatrix}, \tag{9A.28}$$

and

$$\chi_{\alpha\beta} = \begin{pmatrix} \rho & \rho_s \\ \rho_s & \rho_s \end{pmatrix}. \tag{9A.29}$$

Thus

$$\chi_l = \partial g_s/\partial v_{nl} = \rho \tag{9A.30}$$

$$\chi_{g_l g_s} = \rho_s \partial g_l/\partial h = \rho_s \tag{9A.31}$$

$$\chi_{g_s g_s} = \rho_s \partial g_s/\partial h = \rho_s, \tag{9A.32}$$

where h is the field conjugate to v_s (see Sec. 8.5). Combining Eqs. (9A.27) and (9A.30), we obtain Eq. (9A.24).

To make contact with calculations in Sec. 9.4, we identify ρ_s in Eq. (9.4.6) with the zero temperature rigidity ρ and $\rho_s(\mathbf{v}_s^{\parallel} + \mathbf{v}_s^{\perp})$ with $\mathbf{g}$. In Eq. (9.4.6), we replaced $\mathbf{v}_s$ by $\nabla\vartheta' + \mathbf{v}_s^{\perp} + \mathbf{v}$ with the boundary condition $\vartheta' = 0$. If we lift the latter restriction for the moment, then the Hamiltonian in the presence of the external $\mathbf{v}$ is

$$\mathscr{H}(\mathbf{v}) = \mathscr{H}_0 + \int d^2x\mathbf{g}\cdot\mathbf{v} + \frac{1}{2}\int d^2x\rho v^2, \tag{9A.33}$$

where we have replaced, as discussed above, the bare rigidity ρ_s by ρ and where $\mathscr{H}_0 = \int d^2x g^2/2\rho$ with the identification of $\mathbf{g}$ discussed above. The free energy density associated with $\mathscr{H}(\mathbf{v}_n)$ is

$$f_1(\mathbf{v}_n) = -\frac{T}{\Omega}\ln \mathrm{Tr} e^{-\beta\mathscr{H}(\mathbf{v})} = f_2(\mathbf{v}_n) + \frac{1}{2}\rho v_n^2, \tag{9A.34}$$

where $f_2(\mathbf{v}_n) = -(T/\Omega)\mathrm{Tr}\ln\exp[-\beta(\mathscr{H}_0 + \int d^2x\mathbf{g}\cdot\mathbf{v}_n)]$ is the Legendre transform of $f(\mathbf{v}_s, \mathbf{v}_s)$ with respect to $\mathbf{v}_s$ with the field $\mathbf{h}$ conjugate to $\mathbf{v}_s$ equal to zero. The second derivative of $f_2(\mathbf{v}_n)$ with respect to $\mathbf{v}_n$ is $-\chi_{g_i g_j}$, so that

$$\frac{\partial^2 f_1}{\partial v_{ni} v_{nj}} = -\chi_{g_i g_j} + \rho\delta_{ij} = \rho_s(\delta_{ij} - \hat{q}_i\hat{q}_j). \tag{9A.35}$$

This implies that

$$f_1(\mathbf{v}_n) = f_1(0) + \frac{1}{2}\rho_s v_{nt}^2 \tag{9A.36}$$

depends only on the transverse component of $\mathbf{v}_n$ to quadratic order in $\mathbf{v}_n$. If we impose the constraint that ϑ' is zero at the boundaries, then $\int d^2x\mathbf{g}\cdot\mathbf{v}_n \to \int d^2x\mathbf{g}_t\cdot\mathbf{v}_n$, and $\mathbf{v}_n$ couples only to the transverse part of $\mathbf{g}$ in f_2. In this case, $\partial f_2/\partial v_{ni}\partial v_{nj} = -\rho_n(\delta_{ij} - \hat{q}_i\hat{q}_j)$, and

$$f_1(\mathbf{v}_n) \to f_1(0) + \frac{1}{2}\rho v_n^2 - \frac{1}{2}\rho_n v_{nt}^2, \tag{9A.37}$$

in agreement with Eqs. (9.4.7)–(9.4.9) with ρ_s identified with ρ and ρ_s^R identified with ρ_s.

3 The spin correlation function

In our treatment of the KT transition, we obtained Eqs. (9.4.9) and (9.4.26) for the renormalized stiffness by differentiating the free energy with respect to an externally imposed angle gradient. We did not, however, show explicitly that the same renormalized rigidity appears in the spin correlation function $\overline{G}(\mathbf{x}-\mathbf{x}') = \langle \cos(\vartheta(\mathbf{x}) - \vartheta(\mathbf{x}'))\rangle$. Here we will show that $\overline{G}(\mathbf{x}) = |\mathbf{x}|^{-1/(2\pi K_R(T))}$, where K_R is the reduced renormalized rigidity of Eq. (9.4.20). We begin, as in Sec. 9.4, by dividing ϑ into an analytic part and a singular part: $\vartheta = \vartheta_a + \vartheta_s$, with $\nabla\vartheta_a = \mathbf{v}_s^{\parallel}$ and $\nabla\vartheta_s = \mathbf{v}_s^{\perp}$. The transverse velocity is determined by the vortex density via $-\nabla^2\mathbf{v}_s = 2\pi(\nabla\times\mathbf{e}_z)n_v(\mathbf{x})$ or

$$v_{si}^{\perp}(\mathbf{x}) = 2\pi\int d^2x'\epsilon_{ij}\partial_j G(\mathbf{x}-\mathbf{x}')n_v(\mathbf{x}'), \tag{9A.38}$$

where ϵ_{ij} is the antisymmetric tensor, $G(\mathbf{x}) = (1/2\pi)\ln(|\mathbf{x}-\mathbf{x}'|/a)$, and $\nabla^2 G(\mathbf{x}) = \delta(\mathbf{x})$. The regular and singular parts of ϑ decouple, and the spin correlation function can be written as

$$\overline{G}(\mathbf{x}) = \overline{G}_{\rm SW}(\mathbf{x})\overline{G}_{\rm V}(\mathbf{x}), \tag{9A.39}$$

where

$$\overline{G}_{\rm SW}(\mathbf{x}) = \langle e^{i(\vartheta_a(\mathbf{x})-\vartheta_a(0))}\rangle = e^{-\ln(|\mathbf{x}|/a)/2K} \tag{9A.40}$$

is the spin-wave or longitudinal contribution of $\overline{G}$, and

$$\overline{G}_{\rm V}(\mathbf{x}) = \langle e^{i(\vartheta_s(\mathbf{x})-\vartheta_s(0))}\rangle = e^{-\langle(\vartheta_s(\mathbf{x})-\vartheta_s(0))^2\rangle/2} \equiv e^{-g_{\rm V}(\mathbf{x})} \tag{9A.41}$$

is the vortex or transverse contribution to $\overline{G}$. (We used a cumulant expansion to obtain the final form for $\overline{G}_{\rm V}$.)

To calculate $g_{\rm V}(\mathbf{x})$, we use Eq. (9A.38) to write

$$\vartheta_s(\mathbf{x}') - \vartheta_s(\mathbf{x}) = \int_{\mathbf{x}}^{\mathbf{x}'} d\mathbf{s}\cdot\mathbf{v}_s(\mathbf{s}) = 2\pi\int d^2y\mathbf{n}(\mathbf{y})u(\mathbf{y}), \tag{9A.42}$$

where the integral over $\mathbf{s}$ is along some path connecting $\mathbf{x}$ and $\mathbf{x}'$ and

$$u(\mathbf{y}) = \int_{\mathbf{x}}^{\mathbf{x}'} ds_i\epsilon_{ij}\partial_j^s G(\mathbf{s}-\mathbf{y}). \tag{9A.43}$$

(It is understood that $u(\mathbf{y})$ depends on $\mathbf{x}$ and $\mathbf{x}'$ as well.) The line integral in this equation can be done exactly using the Cauchy-Riemann relations

$$\epsilon_{ij}\partial_j G(\mathbf{x}) = -\partial_i G_{\perp}(\mathbf{x}), \qquad \epsilon_{ij}\partial_j G_{\perp}(\mathbf{x}) = \partial_i G(\mathbf{x}), \tag{9A.44}$$

where $G_{\perp}(\mathbf{x}) = (1/2\pi)\tan^{-1}(y/x)$:

$$u(\mathbf{y}) = -(G_{\perp}(\mathbf{x}'-\mathbf{y}) - G_{\perp}(\mathbf{x}-\mathbf{y})). \tag{9A.45}$$

We now use the fact that $\int d^2yn_v(\mathbf{y}) = 0$ to write

$$\begin{aligned} g_{\rm V}(\mathbf{x}-\mathbf{x}') &= \frac{1}{2}(2\pi)^2\int d^2y\int d^2y'\langle n_v(\mathbf{y})n_v(\mathbf{y}')\rangle u(\mathbf{y})u(\mathbf{y}') \\ &= -\frac{1}{4}(2\pi)^2\int d^2y\int d^2y'\langle n_v(\mathbf{y})n_v(\mathbf{y}')\rangle(u(\mathbf{y})-u(\mathbf{y}'))^2. \end{aligned} \tag{9A.46}$$

Setting $\mathbf{y} = \mathbf{R}+\mathbf{r}/2$ and $\mathbf{y}' = \mathbf{R}-\mathbf{r}/2$ and expanding in powers of $\mathbf{r}$, we obtain

$$\begin{aligned} g_{\rm V}(\mathbf{x}-\mathbf{x}') &= \frac{1}{2}(2\pi)^2\left(-\frac{1}{2}\int d^2r\langle n_v(\mathbf{r})n_v(0)\rangle r_ir_i\right)\int d^2R\partial_i u(\mathbf{R})\partial_j u(\mathbf{R}) \\ &= \frac{1}{2}(2\pi)^2C_2\int d^2R\partial_i u\partial_i u, \end{aligned} \tag{9A.47}$$

where $C_2 = -(1/4)\int d^2 r r^2 \langle n_v(\mathbf{r}) n_v(0)\rangle$ was introduced in Eq. (9.4.24). Next, using $\partial_i u \partial_i u = \epsilon_{ik}\partial_k u \epsilon_{il}\partial_l u$ and the relation

$$\epsilon_{ik}\partial_k u = -\frac{\partial}{\partial R_i}(G(\mathbf{x}' - \mathbf{R}) - G(\mathbf{y} - \mathbf{R})), \tag{9A.48}$$

which follows from Eqs. (9A.44) and (9A.45), we obtain, after integration by parts,

$$\begin{aligned} g_V(\mathbf{x} - \mathbf{x}') &= -\frac{1}{2}C_2(2\pi)^2 \int d^2R\{[G(\mathbf{x}' - \mathbf{R}) - G(\mathbf{x} - \mathbf{R})] \\ &\qquad \times \nabla_R^2 [G(\mathbf{x}' - \mathbf{R}) - G(\mathbf{x} - \mathbf{R})]\} \\ &= -\frac{1}{2}C_2(2\pi)^2 [G(\mathbf{x}' - \mathbf{x}') - G(\mathbf{x}' - \mathbf{x}) \\ &\qquad - G(\mathbf{x} - \mathbf{x}') + G(\mathbf{x} - \mathbf{x})]. \end{aligned} \tag{9A.49}$$

The function G evaluated at zero argument is formally infinite. However, we do not allow points to be closer together than a cutoff distance, and $G(0)$ is at worst a constant. We therefore have

$$g_V(\mathbf{x} - \mathbf{x}') = 2\pi C_2 \ln(|\mathbf{x} - \mathbf{x}'|/a), \tag{9A.50}$$

at $|\mathbf{x} - \mathbf{x}'| \gg a$, and

$$\overline{G}(\mathbf{x}) = |\mathbf{x}|^{-1/(2\pi K)} |\mathbf{x}|^{-2\pi C_2} = |\mathbf{x}|^{-1/(2\pi K_R)}, \tag{9A.51}$$

with $K_R^{-1} = K^{-1} + (2\pi)^2 C_2$, in agreement with Eqs. (9.4.20) (9.4.23), and (9.4.26).

Appendix 9B Duality and the Villain model

The Hamiltonians of a number of lattice models can be expressed as a sum over bonds of bond energies depending only on the differences of dynamical variables on the pairs of sites defining the bonds. When this is the case, the bond potential can be Fourier transformed, and partition functions can be expressed either as sums over the original dynamical variables defined on lattice sites or as sums over Fourier-transform variables defined on bonds but subject to constraints to be discussed below. The transformation to Fourier transform variables is called a *duality transformation*. Duality and related transformations provide valuable insight into phase transitions in two dimensions and have played a particularly important role in the theory of the Kosterlitz-Thouless transition. A duality transformation also provides a direct relation between the lattice Coulomb gas and the *discrete Gaussian model*, which, as we shall see in the next chapter, is used to describe the transition from rough to smooth crystal surfaces.

In this appendix, we will review some consequences of duality transformations for models on two-dimensional square lattices with bond energies only on nearest neighbor bonds. A square lattice $\mathscr{L}_\square$ with N sites $\mathbf{x}$ has $2N$ bonds $\langle \mathbf{x}, \mathbf{x}'\rangle$ connecting nearest neighbor sites $\mathbf{x}$ and $\mathbf{x}'$. Plaquettes on the lattice are squares with edges defined by bonds and vertices by sites, as shown in Fig. 9B.1. The reduced Hamiltonians for models of interest here can be expressed as

$$\overline{\mathscr{H}} \equiv \mathscr{H}/T = -\sum_{\langle \mathbf{x}, \mathbf{x}'\rangle} V(\sigma_{\mathbf{x}} - \sigma_{\mathbf{x}'}), \tag{9B.1}$$

where $\sigma_{\mathbf{x}}$ is the dynamical variable at the lattice site $\mathbf{x}$.

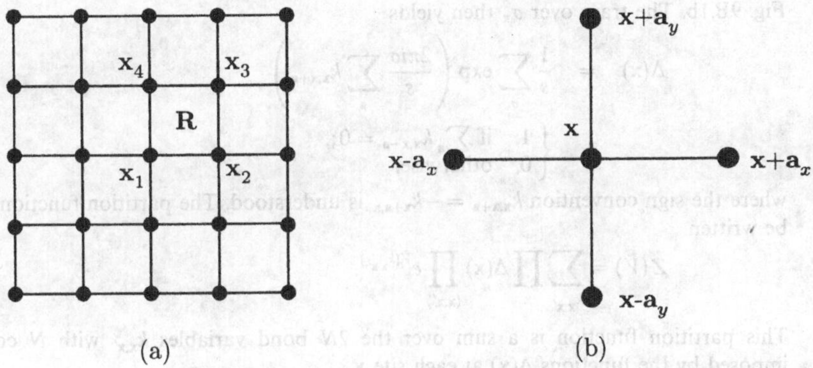

Fig. 9B.1. (a) Section of a square lattice $\mathscr{L}_\square$ with sites at positions **x** denoted by dots (•). Bonds, represented by lines, connect nearest neighbor sites. Plaquettes are squares whose four edges are bonds and four vertices are lattice sites. Shown is a plaquette centered at **R** with vertices $\mathbf{x}_1$, $\mathbf{x}_2$, $\mathbf{x}_3$ and $\mathbf{x}_4$ and bonds $\langle \mathbf{x}_1, \mathbf{x}_2\rangle$, $\langle \mathbf{x}_2, \mathbf{x}_3\rangle$, $\langle \mathbf{x}_3, \mathbf{x}_4\rangle$, and $\langle \mathbf{x}_4, \mathbf{x}_1\rangle$. (b) A site **x** and its nearest neighbors at $\mathbf{x} \pm \mathbf{a}_x$ and $\mathbf{x} \pm \mathbf{a}_y$, where $\mathbf{a}_{x,y} = a\mathbf{e}_{x,y}$. The site **x** is shared by the bonds $\langle \mathbf{x}, \mathbf{x}+\mathbf{a}_x\rangle$, $\langle \mathbf{x}, \mathbf{x}+\mathbf{a}_y\rangle$, $\langle \mathbf{x}, \mathbf{x}-\mathbf{a}_x\rangle$, and $\langle \mathbf{x}, \mathbf{x}-\mathbf{a}_y\rangle$.

1 Potts models

In an s-state Potts model (see Sec. 3.6), $\sigma_\mathbf{x}$ is an integer-valued variable taking on values $0, 1, ..., s-1$. Both $\sigma_\mathbf{x}$ and the difference $\sigma_\mathbf{x} - \sigma_{\mathbf{x}'}$ are defined mod s so that $\sigma_\mathbf{x} - \sigma_{\mathbf{x}'}$ only takes on the values $0, 1, ..., s-1$. Therefore, $e^{V(\sigma)}$ can be represented in a Fourier series (see Appendix 2A):

$$e^{V(\sigma)} = \sum_{k=0}^{s-1} e^{i2\pi k\sigma/s} e^{\tilde{V}(k)}, \tag{9B.2}$$

$$e^{\tilde{V}(k)} = \frac{1}{s}\sum_{\sigma=0}^{s-1} e^{-2\pi i k\sigma/s} e^{V(\sigma)}. \tag{9B.3}$$

The Potts model partition function can now be expressed as

$$\begin{aligned} Z(V) &= \sum_{\sigma(\mathbf{x})}\prod_{\langle \mathbf{x},\mathbf{x}'\rangle} e^{V(\sigma_\mathbf{x}-\sigma_{\mathbf{x}'})} \\ &= \left(\frac{1}{s}\right)^N \sum_{\sigma_\mathbf{x}}\sum_{k_{\mathbf{x},\mathbf{x}'}}\prod_{\langle \mathbf{x},\mathbf{x}'\rangle} e^{i(2\pi/s)k_{\mathbf{x},\mathbf{x}'}(\sigma_\mathbf{x}-\sigma_{\mathbf{x}'})} e^{\tilde{V}(k_{\mathbf{x},\mathbf{x}'})}, \end{aligned} \tag{9B.4}$$

where the sum in the last equation is over site variables $\sigma_\mathbf{x}$ and link variables $k_{\mathbf{x},\mathbf{x}'} = 1, ..., s-1$ defined on nearest neighbor bonds $\langle \mathbf{x}, \mathbf{x}'\rangle$. The nearest neighbor bond $\langle \mathbf{x}, \mathbf{x}'\rangle$ has not been assigned a direction and is, therefore, identical to the bond $\langle \mathbf{x}', \mathbf{x}\rangle$. Since the quantity $k_{\mathbf{x},\mathbf{x}'}(\sigma_\mathbf{x} - \sigma_{\mathbf{x}'})$ is defined on the undirected bond and is independent of the order of **x** and **x'**, it must be equal to $k_{\mathbf{x}',\mathbf{x}}(\sigma_{\mathbf{x}'} - \sigma_\mathbf{x})$, and we conclude that $k_{\mathbf{x},\mathbf{x}'} = -k_{\mathbf{x}',\mathbf{x}}$. The site variables appear linearly in the exponent in the partition function, and their partition trace can be carried out exactly.

On a square lattice, the variable $\sigma_\mathbf{x}$ appears four times, multiplying the link variables $k_{\mathbf{x},\mathbf{x}+\mathbf{a}}$ for $\mathbf{a} = \pm a\mathbf{e}_x, \pm a\mathbf{e}_y$ defined on the four bonds with an endpoint at **x** as shown in

Fig. 9B.1b. The trace over $\sigma_{\mathbf{x}}$ then yields

$$\begin{aligned}\Delta(\mathbf{x}) &= \frac{1}{s}\sum_{\sigma}\exp\left(\frac{2\pi i\sigma}{s}\sum_{\mathbf{a}}k_{\mathbf{x},\mathbf{x}+\mathbf{a}}\right) \\ &= \begin{cases}1, & \text{if } \sum_{\mathbf{a}}k_{\mathbf{x},\mathbf{x}+\mathbf{a}}=0;\\ 0, & \text{otherwise,}\end{cases}\end{aligned} \tag{9B.5}$$

where the sign convention $k_{\mathbf{x},\mathbf{x}+\mathbf{a}} = -k_{\mathbf{x}+\mathbf{a},\mathbf{x}}$ is understood. The partition function can now be written

$$Z(\tilde{V}) = \sum_{k_{\mathbf{x},\mathbf{x}'}}\prod_{\mathbf{x}}\Delta(\mathbf{x})\prod_{\langle\mathbf{x},\mathbf{x}'\rangle}e^{\tilde{V}(k_{\mathbf{x},\mathbf{x}'})}. \tag{9B.6}$$

This partition function is a sum over the $2N$ bond variables $k_{\mathbf{x},\mathbf{x}'}$ with N constraints imposed by the functions $\Delta(\mathbf{x})$ at each site $\mathbf{x}$.

A judicious parameterization of the variables $k_{\mathbf{x},\mathbf{x}'}$ guarantees that the constraint at each site is satisfied and converts $Z(\tilde{V})$ into a sum over N independent and unconstrained variables. To arrive at this parameterization, we show first that $\sum_{\mathbf{a}}k_{\mathbf{x},\mathbf{x}+\mathbf{a}} = 0$ is a discrete generalization of the curl operator on the link variables. Therefore, the constraint $\sum_{\mathbf{a}}k_{\mathbf{x},\mathbf{x}+\mathbf{a}} = 0$ implies that the lattice curl of the link variable $k_{\mathbf{x},\mathbf{x}'}$ is zero and can be expressed as a lattice divergence of a scalar field. We begin by assigning a direction to the bond $\langle\mathbf{x},\mathbf{x}'\rangle = \langle\mathbf{x},\mathbf{x}+\mathbf{a}\rangle$ pointing from $\mathbf{x}$ to $\mathbf{x}'$, as shown in Fig. 9B.2. We then associate with each bond $\langle\mathbf{x},\mathbf{x}'\rangle$ a perpendicular bond $\langle\mathbf{R},\mathbf{R}'\rangle$ passing through its center with a direction determined by the right hand rule, as shown in Fig. 9B.2. The bonds $\langle\mathbf{R},\mathbf{R}'\rangle$ terminate at sites $\mathbf{R}$ that form a regular square lattice $\mathscr{L}_{\Box}^{D}$, which is the *dual lattice*, of $\mathscr{L}_{\Box}$, as shown in Fig. 9B.2b. A given bond can be specified either by $\langle\mathbf{x},\mathbf{x}'\rangle$ or by its dual lattice version $\langle\mathbf{R},\mathbf{R}'\rangle$. We can now associate with each bond a vector $\mathbf{k}_{\mathbf{R},\mathbf{R}'} = k_{\mathbf{R},\mathbf{R}'}(\mathbf{R}'-\mathbf{R})$ pointing along the directed dual bond, where $k_{\mathbf{R},\mathbf{R}'} = k_{\mathbf{x},\mathbf{x}'}$. Then,

$$\sum_{\mathbf{a}}k_{\mathbf{x},\mathbf{x}+\mathbf{a}} = \frac{1}{a^2}\sum_{\text{plaquette}}\mathbf{k}_{\mathbf{R},\mathbf{R}'}\cdot(\mathbf{R}'-\mathbf{R}), \tag{9B.7}$$

where the final sum is over the bonds in the dual lattice plaquette centered at $\mathbf{x}$, as shown in Fig. 9B.2b. The right hand side of this equation is the lattice generalization of the integral $a^{-2}\oint\mathbf{k}\cdot d\mathbf{l} = a^{-2}\int(\nabla\times\mathbf{k})\cdot d\mathbf{S}$, i.e.,

$$\sum_{\mathbf{a}}k_{\mathbf{x},\mathbf{x}+\mathbf{a}} = (\nabla\times\mathbf{k})_L\cdot\mathbf{e}_z, \tag{9B.8}$$

where $(\nabla\times\mathbf{k})_L$ denotes the lattice curl of $\mathbf{k}_{\mathbf{R},\mathbf{R}'}$. Thus, the constraint imposed by $\Delta(\mathbf{x})$ is that the lattice curl of $\mathbf{k}_{\mathbf{R},\mathbf{R}'}$ on the plaquette centered at $\mathbf{x}$ is zero. This implies that $\mathbf{k}_{\mathbf{R},\mathbf{R}'}$ can be expressed as the lattice divergence of a scalar $m_{\mathbf{R}}$:

$$\mathbf{k}_{\mathbf{R},\mathbf{R}'} = (\mathbf{R}'-\mathbf{R})(m_{\mathbf{R}}-m_{\mathbf{R}'}) \tag{9B.9}$$

or

$$k_{\mathbf{R},\mathbf{R}'} = a^{-2}(\mathbf{R}'-\mathbf{R})\cdot\mathbf{k}_{\mathbf{R},\mathbf{R}'} = m_{\mathbf{R}}-m_{\mathbf{R}'}. \tag{9B.10}$$

Because $k_{\mathbf{R},\mathbf{R}'}$ is an integer, $m_{\mathbf{R}}$ and $m_{\mathbf{R}'}$ are also. The partition function can now be written as

$$Z(\tilde{V}) = \sum_{m_{\mathbf{R}}}\prod_{\langle\mathbf{R},\mathbf{R}'\rangle}e^{\tilde{V}(m_{\mathbf{R}}-m_{\mathbf{R}'})}. \tag{9B.11}$$

Eqs. (9B.4) and (9B.11) are equivalent expressions of the same partition function. The first, however, is a trace over integer valued variables $\sigma_{\mathbf{x}}$ on the sites $\mathbf{x}$ of $\mathscr{L}_{\Box}$, whereas the second is a trace over integer valued variables $m_{\mathbf{R}}$ on the sites $\mathbf{R}$ of the dual lattice, $\mathscr{L}_{\Box}^{D}$.

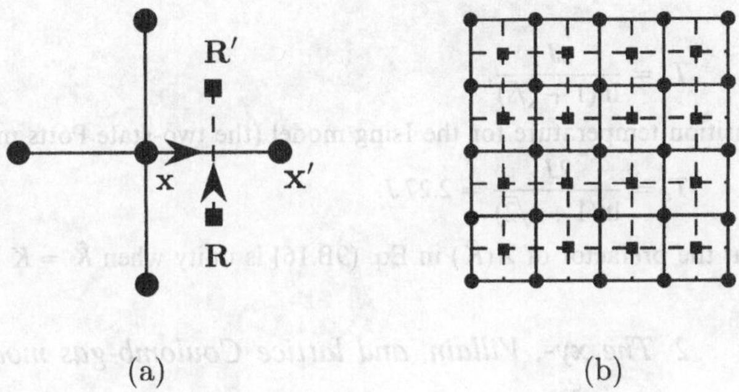

Fig. 9B.2. (a) This figure shows the construction of the nearest neighbor bond $\langle \mathbf{R}, \mathbf{R}' \rangle$ in the dual lattice $\mathscr{L}_\square^D$ from its associated bond $\langle \mathbf{x}, \mathbf{x}' \rangle$ in the lattice $\mathscr{L}_\square$. A direction pointing from the site $\mathbf{x}$ to the site $\mathbf{x}'$ is assigned to the bond $\langle \mathbf{x}, \mathbf{x}' \rangle$. The bond $\langle \mathbf{R}, \mathbf{R}' \rangle$ is the perpendicular bisector $\langle \mathbf{x}, \mathbf{x}' \rangle$ and points in a direction determined by the right hand rule relative to the normal pointing out of the page. (b) A section of the lattice $\mathscr{L}_\square$ and its associated dual lattice $\mathscr{L}_\square^D$. Sites and bonds in $\mathscr{L}_\square$ are, respectively, represented as full circles and solid lines; those in $\mathscr{L}_\square^D$ are shown as squares and dashed lines.

In the preceding analysis, we did not specify the precise form of $V(\sigma)$. In the usual definition of the s-state Potts model,

$$V(\sigma - \sigma') = K s \delta_{\sigma,\sigma'} = sK\delta_{\sigma-\sigma',0} \tag{9B.12}$$

and

$$e^{V(\sigma)} = 1 + (e^{sK} - 1)\delta_{\sigma,0}, \tag{9B.13}$$

where $K = J/T$ and where we have not included the constant $-K$ that appears in the definition of Eq. (3.6.6). With this form for $V(\sigma)$,

$$\begin{aligned} e^{\tilde{V}(k)} &= \frac{1}{s}\left[(e^{sK} - 1) + \sum_{\sigma=0}^{n-1} e^{-2\pi i k\sigma/s}\right] \\ &= \frac{1}{s}[e^{sK} - 1 + s\delta_{k,0}] \\ &= \frac{1}{s}(e^{sK} - 1)e^{s\tilde{K}\delta_{k,0}}, \end{aligned} \tag{9B.14}$$

where

$$e^{s\tilde{K}} = \frac{s}{e^{sK} - 1}. \tag{9B.15}$$

Thus, apart from a trivial prefactor, $e^{\tilde{V}(k)}$ and $e^{V(\sigma)}$ have identical forms. Eqs. (9B.4), (9B.12), and (9B.14) then imply

$$Z(K) = \left[\frac{1}{s}(e^{sK} - 1)^2\right]^N Z(\tilde{K}). \tag{9B.16}$$

If T increases, K decreases, and $\tilde{K}$ increases. The duality transformation maps the partition function at low temperatures onto the partition function at high temperatures. If the Potts model in question has a second order transition at some $K = K_c$, its free energy, and thus its partition function, must be singular at K_c. If $Z(K)$ has a singularity at K_c, then $Z(\tilde{K})$ must have the same singularity at $\tilde{K} = K_c$, and from Eq. (9B.15), K_c must satisfy

$$(e^{sK_c} - 1)^2 = s \tag{9B.17}$$

or

$$T_c = \frac{sJ}{\ln(1+\sqrt{s})}. \tag{9B.18}$$

The transition temperature for the Ising model (the two-state Potts model) is, therefore,

$$T_c = \frac{2J}{\ln(1+\sqrt{2})} = 2.27\,J. \tag{9B.19}$$

Note that the prefactor of $Z(\tilde{K})$ in Eq. (9B.16) is unity when $\tilde{K} = K = K_c$ as required.

2 The xy-, Villain, and lattice Coulomb-gas models

In the *xy*-model, the variable $\sigma_{\mathbf{x}}$ becomes the continuous angle variable $\vartheta_{\mathbf{x}}$, and the potential $V(\vartheta)$ becomes a periodic function of ϑ with period 2π. In the simplest version of the *xy*-model,

$$V(\vartheta) = K(\cos\vartheta - 1), \tag{9B.20}$$

where, for ease of presentation in this appendix, we have chosen $-V(\vartheta)$ to be equal to zero at its minimum at $\vartheta = 0$. The periodicity of $V(\vartheta)$ implies that it has a Fourier decomposition,

$$e^{V(\vartheta)} = \sum_{k=-\infty}^{\infty} e^{ik\vartheta} e^{\tilde{V}(k)}, \tag{9B.21}$$

where

$$e^{\tilde{V}(k)} = \frac{1}{2\pi}\int_0^{2\pi} d\vartheta e^{-ik\vartheta} e^{V(\vartheta)}. \tag{9B.22}$$

When $V(\vartheta)$ is the cosine function of Eq. (9B.20),

$$\begin{aligned} e^{\tilde{V}(k)} &= \frac{1}{2\pi}\int_0^{2\pi} d\vartheta e^{-ik\vartheta} e^{K(\cos\vartheta-1)} = e^{-K} I_k(K) \\ &\to \begin{cases} (K/2)^k/k!, & \text{for } K \to 0; \\ (2\pi K)^{-1/2} e^{-k^2/(2K)} & \text{for } K \to \infty, \end{cases} \end{aligned} \tag{9B.23}$$

where $I_k(K)$ is the modified Bessel function (Abramowitz and Stegun 1972, p. 374). The partition function for the *xy*-model can be transformed to dual variables following the same steps as outlined above for the Potts model:

$$\begin{aligned} Z(V) &= \prod_{\mathbf{x}} \int_0^{2\pi} d\vartheta_{\mathbf{x}} \prod_{\langle \mathbf{x},\mathbf{x}'\rangle} e^{V(\vartheta_{\mathbf{x}}-\vartheta_{\mathbf{x}'})} \\ &= \prod_{\mathbf{x}} \int_0^{2\pi} d\vartheta_{\mathbf{x}} \prod_{\langle \mathbf{x},\mathbf{x}'\rangle} \sum_{k_{\mathbf{x},\mathbf{x}'}} e^{ik_{\mathbf{x},\mathbf{x}'}(\vartheta_{\mathbf{x}}-\vartheta_{\mathbf{x}'})} e^{\tilde{V}(k_{\mathbf{x},\mathbf{x}'})} \\ &= \sum_{k_{\mathbf{x},\mathbf{x}'}} \prod_{\mathbf{x}} \Delta(\mathbf{x}) e^{\tilde{V}(k_{\mathbf{x},\mathbf{x}'})} = \sum_{k_{\mathbf{R}}} \prod_{\langle \mathbf{R},\mathbf{R}'\rangle} e^{\tilde{V}(k_{\mathbf{R}}-k_{\mathbf{R}'})} \end{aligned} \tag{9B.24}$$

where, as before, $\mathbf{R}$ is a site on the dual lattice, and $k_{\mathbf{R}} = 0, \pm 1, \pm 2, ...$ is an integer valued function. Eq. (9B.24) is the duality transformation for the *xy*-model. It is very similar to that for the Potts model except that $V(\vartheta)$ is a periodic function of the continuous variable ϑ and $\tilde{V}(k)$ is a function of the integer variable k, whereas, in the Potts case, both V and $\tilde{V}$ are functions of an integer mod s.

Further transformations employing the *Poisson summation formula* provide useful representations for the *xy*-model. The Poisson summation formula,

$$\sum_{k=-\infty}^{\infty} g(k) = \sum_{m=-\infty}^{\infty} \int_{-\infty}^{\infty} d\phi g(\phi) e^{-2\pi i \phi m}, \tag{9B.25}$$

where k is an integer, applies to any function $g(\phi)$ defined on the interval $(-\infty, \infty)$ and having a Fourier transform. When applied to the periodic potential of the xy-model, which can be written as a sum over integers k according to Eq. (9B.21), it implies

$$e^{V(\vartheta_{\mathbf{x}}-\vartheta_{\mathbf{x}'})} = \sum_{m_{\mathbf{x},\mathbf{x}'}} e^{V_0(\vartheta_{\mathbf{x}}-\vartheta_{\mathbf{x}'}-2\pi m_{\mathbf{x},\mathbf{x}'})}, \tag{9B.26}$$

where

$$e^{V_0(\vartheta)} = \int_{-\infty}^{\infty} d\phi e^{\tilde{V}(\phi)} e^{i\phi\vartheta}. \tag{9B.27}$$

With this expression, the partition function [Eq. (9B.24)] can be rewritten as

$$Z = \prod_{\mathbf{x}} \int_0^{2\pi} d\vartheta_{\mathbf{x}} \sum_{m_{\mathbf{x},\mathbf{x}'}} \prod_{\langle \mathbf{x},\mathbf{x}' \rangle} e^{V_0(\vartheta_{\mathbf{x}}-\vartheta_{\mathbf{x}'}-2\pi m_{\mathbf{x},\mathbf{x}'})}. \tag{9B.28}$$

At low temperature, $\tilde{V}(k) \approx -k^2/(2K) - \ln(2\pi K)/2$ [Eq. (9B.23)]. If this simple form is assumed for $\tilde{V}$, then

$$e^{V_0(\vartheta)} = e^{-K\vartheta^2/2} \tag{9B.29}$$

and

$$Z = \prod_{\mathbf{x}} \int_0^{2\pi} d\vartheta_{\mathbf{x}} \prod_{\langle \mathbf{x},\mathbf{x}' \rangle} \sum_{m_{\mathbf{x},\mathbf{x}'}} e^{-\frac{1}{2}K(\vartheta_{\mathbf{x}}-\vartheta_{\mathbf{x}'}-2\pi m_{\mathbf{x},\mathbf{x}'})^2}. \tag{9B.30}$$

This is the Villain model (Villain 1975). It is a version of the xy-model in which the periodic cosine potential of Eq. (9B.20) is replaced by the periodic function $\ln[\sum_m \exp(-K(\vartheta - 2\pi m)^2/2)]$. The link variable can be decomposed into a lattice version of longitudinal and transverse parts: $m_{\mathbf{x},\mathbf{x}'} = m_{\mathbf{x}} - m_{\mathbf{x}'} + m^{\perp}_{\mathbf{x},\mathbf{x}'}$, where $m^{\perp}_{\mathbf{x},\mathbf{x}'}$ is determined entirely by the vorticities $m_{\mathbf{R}'}$ on the dual lattice via $\sum_{\text{plaquette}} m^{\perp}_{\mathbf{x},\mathbf{x}'} = m_{\mathbf{R}'}$. The sums over the site integers $m_{\mathbf{x}}$ convert the integrals over $\vartheta_{\mathbf{x}}$ from 0 to 2π to integrals from $-\infty$ to $+\infty$. In this form, the Villain model is a lattice version of the low-temperature continuum model that we used to discuss vortices and the Kosterlitz-Thouless transition in the xy-model. The difference $\vartheta_{\mathbf{x}} - \vartheta_{\mathbf{x}'}$ is the lattice version of the longitudinal part of $\mathbf{v}_s = \nabla\theta$ [Eq. (9.4.4)] and $m^{\perp}_{\mathbf{x},\mathbf{x}'}$ is the lattice version of $\mathbf{v}_s^{\perp}$.

Finally, the Possion summation formula can be applied to the dual version of the xy-model [Eq. (9B.24)]:

$$\begin{aligned} Z &= \sum_{k_{\mathbf{R}}} \prod_{\langle \mathbf{R},\mathbf{R}' \rangle} e^{\tilde{V}(k_{\mathbf{R}}-k_{\mathbf{R}'})} \\ &= \sum_{m_{\mathbf{R}}} \prod_{\mathbf{R}} \int_0^{\infty} d\phi_{\mathbf{R}} e^{\tilde{V}(\phi_{\mathbf{R}}-\phi_{\mathbf{R}'})} e^{2\pi i \sum_{\mathbf{R}} m_{\mathbf{R}} \phi_{\mathbf{R}}}. \end{aligned} \tag{9B.31}$$

If the Villain form for $\tilde{V}$ is used, then

$$Z = \sum_{k_{\mathbf{R}}} \prod_{\mathbf{R},\mathbf{R}'} (2\pi K)^{-1/2} \exp\left[-\frac{1}{2K}(k_{\mathbf{R}} - k_{\mathbf{R}'})^2\right]. \tag{9B.32}$$

This is the *discrete Gaussian model*. It is dual to the Villain version of the xy-model, Eq. (9B.30). Then, using the Poisson summation formula [Eq. (9B.25), we find

$$Z = \sum_{m_{\mathbf{R}}} \int d\phi_{\mathbf{R}} \exp\left[-\frac{1}{2K}\sum_{\langle \mathbf{R},\mathbf{R}'\rangle}(\phi_{\mathbf{R}} - \phi_{\mathbf{R}'})^2\right]\exp\left[2\pi i \sum_{\mathbf{R}} m_{\mathbf{R}}\phi_{\mathbf{R}}\right]$$

$$= \sum_{m_{\mathbf{R}}} \exp\left[-2\pi^2 K \sum_{\mathbf{R},\mathbf{R}'} m_{\mathbf{R}} G(\mathbf{R}-\mathbf{R}') m_{\mathbf{R}'}\right], \qquad (9B.33)$$

where $G(\mathbf{R})$ is the lattice Green function

$$G(\mathbf{R}) = \int \frac{d^2q}{(2\pi)^2}\frac{e^{i\mathbf{q}\cdot\mathbf{R}}}{\epsilon(\mathbf{q})} \underset{(|\mathbf{R}|/a\gg 1)}{\longrightarrow} G(0) + 2\pi\ln(|\mathbf{R}|/a) + \text{const.}, \qquad (9B.34)$$

where $\epsilon(\mathbf{q}) = 2(2 - \cos q_x a - \cos q_y a)$. The final form is the lattice Coulomb gas [Eq. (9.4.19)] without the core energy.

Bibliography

General references on topological defects

Roger Balian, Maurice Kléman, and Jean-Paul Poirier (eds.), *Physics of Defects* (North Holland, NY, 1981).

G. Friedel, *Dislocations* (Pergamon Press, Oxford, 1964).

Maurice Kléman, *Points, Lines and Walls: in Liquid Crystals, Magnetic Systems, and Various Disordered Media* (J. Wiley, New York, 1983).

M.V. Kurik and O.D. Lavrentovich, *Usp. Fiz. Nauk.* **154**, 381 (1988) [*Sov. Phys. Usp.* **31**, 196 (1988)].

N.D. Mermin, *Rev. Mod. Phys.* **51**, 591 (1979).

F.R.N. Nabarro, *Theory of Crystal Dislocations* (Clarendon Press, Oxford, 1967).

G. I. Taylor, *Proc. Roy. Soc.* **145**, 362 (1934).

Kosterlitz-Thouless transition

J. José, L.P Kadanoff, S. Kirkpatrick, and D.R. Nelson, *Phys. Rev. B* **16**, 1217 (1977); E *Phys. Rev. B* **17**, 1477 (1978).

J.M. Kosterlitz, *J. Phys. C* **7**, 1046 (1974).

J.M. Kosterlitz and D.J. Thouless, *J. Phys. C* **5**, L124 (1972).

J.M. Kosterlitz and D.J. Thouless, *J. Phys. C* **6**, 1181 (1973).

Petter Minnhagen, *Rev. Mod. Phys.* **59**, 1001 (1987).

D.R. Nelson, "Defect-mediated phase transitions," in *Phase Transitions and Critical Phenomena*, eds. C. Domb and J.L. Lebowitz (Academic Press, New York, 1983).

Dislocation mediated melting

David R. Nelson and B.I Halperin, *Phys. Rev. B* **19**, 2457 (1979).

Katherine J. Strandburg, *Rev. Mod. Phys.* **60**, 161 (1988). [This is a complete review of numerical simulations and experiments.]

A.P. Young, *Phys. Rev. B* **19**, 1855 (1979).

References

Milton Abramowitz and Irene A. Stegun, eds., *Handbook of Mathematical Functions* (Dover, New York, 1972).

A.A. Abrikosov, *Zh. Eksp. Teor. Fiz.* **32**, 1442 (1957) [*Sov. Phys. JETP* **5**, 1174 (1957)].

V. Ambegaokar, B.I. Halperin, D.R. Nelson, and E.D. Siggia, *Phys. Rev. Lett.* **40**, 783 (1978).

V. Ambegaokar, B.I. Halperin, D.R. Nelson, and E. Siggia, *Phys. Rev. B* **21**, 1806 (1980).

K.R. Atkins, *Phys. Rev.* **113**, 962 (1959).
K.R. Atkins and I. Rudnick, *Prog. Low Temp. Phys.* **6**, 37 (1970).
D.J. Bishop and J.D. Reppy, *Phys. Rev. Lett.* **40**, 1727 (1978).
A.R. Day, T.C. Lubensky, and A.J. McKane, *Phys. Rev. A* **27**, 1461 (1983).
S. Fetter and P. Hohenberg, in *Superconductivity*, ed. R.D. Parks (Dekker, New York, 1969), vol. 2, p. 817.
D.S. Fisher, B.I. Halperin, and P.M. Platzman, *Phys. Rev. Lett.* **42**, 798 (1979).
J. Goodby, M.A. Waugh, S.M. Stein, E. Chin, R. Pindak, and J.S. Patel, *Nature* **337**, 449 (1988): *J. Am. Chem. Soc.* **111**, 8119 (1989).
C.C. Grimes and G. Adams, *Phys. Rev. Lett.* **42**, 795 (1979).
B.I. Halperin and D.R. Nelson, *Phys. Rev. Lett.* **41**, 121 (1978); E**41**, 519 (1978).
A.J. Jin, M.R. Bjurstrom, and M.H.W. Chan, *Phys. Rev. Lett.* **62**, 1372 (1989).
J. José, L.P. Kadanoff, S. Kirkpatrick, and D.R. Nelson, *Phys. Rev. B* **16**, 1217 (1977); E *Phys. Rev. B* **17**, 1477 (1978).
J.M. Kosterlitz and D.J. Thouless, *J. Phys. C* **6**, 1181 (1973).
O.D. Lavrentovich and E.M. Terent'ev, *Zh. Eksp. Fiz.* **91**, 2084 (1986) [*Sov. Phys. JETP* **64**, 1237 (1986)].
D.E. Moncton, R. Pindak, S.C. Davey, and G.S. Brown, *Phys. Rev. Lett.* **49**, 1861 (1982).
C.A. Murray and D.H. Van Winkle, *Phys. Rev. Lett.* **58**, 1200 (1987).
L. Navailles, P. Barois, and H. Nguyen, *Phys. Rev. Lett.* **71**, 545 (1993).
D.R. Nelson and J.M. Kosterlitz, *Phys. Rev. Lett.* **39**, 1201 (1977).
S.R. Renn and T.C. Lubensky, *Phys. Rev. A* **38**, 2132 (1988).
M.S. Seul and C.A. Murray, *Science* **262**, 558 (1993).
Katherine J. Strandburg, *Rev. Mod. Phys.* **60**, 161 (1988).
David H. Van Winkle and Noel A. Clark, *Phys. Rev. A* **38**, 1573 (1988).
J. Villain, *J. Phys. (Paris)* **36**, 581 (1975).
A.P. Young, *Phys. Rev. B* **19**, 1855 (1979).

Problems

9.1 Show that $\theta(\mathbf{x}) = \sum_i \theta_i$ with $\theta_i = k_i \tan^{-1}[(y-y_i)/(x-x_i)]$ is a solution in two dimensions to $\nabla^2\theta = 0$ with vortices of strength k_i at positions $\mathbf{x}_i = (x_i, y_i)$. One way to show this is to use the complex coordinates $z = x + iy$ and $\bar{z} = x - iy$. Then $\nabla^2 = 4(\partial/\partial\bar{z})(\partial/\partial z)$, and any function of z is a solution to Laplace's equation. Finally, $\operatorname{Im}\ln(z - z_i)$ is a solution with a unit strength vortex at z_i.

9.2 Derive Eq. (9.3.23) by introducing a vector potential $\mathbf{A}$ for $\mathbf{h}_s = \boldsymbol{\nabla} \times \mathbf{A}$.

9.3 A hedgehog in a three-dimensional O_3 model is a mapping from the sphere in three-dimensional coordinate space onto the unit sphere of the order parameter space. Let θ and ϕ be the spherical angles of the coordinate-space sphere and $\tilde{\theta}$, and let $\tilde{\phi}$ be the spherical angles of the order parameter sphere. Then $\tilde{\theta}$ and $\tilde{\phi}$ and the unit vector field $\mathbf{n}(\tilde{\theta}, \tilde{\phi})$ can be regarded as functions of θ and ϕ. The analog of the winding number of the xy-model is then

$$q = \frac{1}{4\pi}\int d\tilde{\theta}d\tilde{\phi}\sin\tilde{\theta} = \frac{1}{4\pi}\int d\theta d\phi \sin\tilde{\theta}\left|\frac{\partial(\tilde{\theta},\tilde{\phi})}{\partial(\theta,\phi)}\right| \tag{9P.1}$$

$$= \frac{1}{8\pi}\int dS_i \epsilon_{ijk}\mathbf{n}\cdot(\partial_j\mathbf{n}\times\partial_k\mathbf{n}). \tag{9P.2}$$

Use Eq. (9P.2) to show that $q = \pm 1$ for $\mathbf{n} = \pm \mathbf{x}/ \mid \mathbf{x} \mid$. Use Eq. (9P.1) to construct explicit forms for $\mathbf{n}$ with arbitrary "charge" q.

9.4 (Hedgehogs in a nematic) Calculate the elastic energy of a spherical nematic droplet of radius R with a radial hedgehog, in which $\mathbf{n} = (x, y, z)/|\mathbf{x}|$, and with a hyperbolic hedgehog, in which $\mathbf{n} = (-x, -y, z)/|\mathbf{x}|$ assuming, no surface anchoring. Show that the radial hedgehog has lower energy than the hyperbolic hedgehog for $K_3 > 6K_1$. In a real nematic droplet, surface anchoring cannot be neglected. Nevertheless, a transition from a hyperbolic to a radial hedgehog can be observed as the nematic-to-smectic-A transition where K_3 diverges is approached. (Lavrentovich and Terent'ev, 1986.)

9.5 Energies of dislocations in type I smectic liquid crystals, in which the twist and bend penetration depths are smaller than the correlation length, can be calculated using the Landau-Peierls elastic energy of Eq. (6.3.10) rather than the free energy of Eq. (6.3.11). In this problem, you are to use the Landau-Peierls free energy to calculate dislocation energies in terms of the dislocation source function $\mathbf{b}(\mathbf{x})$ introduced in Eq. (9.3.68). The field $\mathbf{v}(\mathbf{x}) = \nabla u(\mathbf{x})$ can be decomposed into a transverse part and a longitudinal part according to Eqs. (9.3.66), with the transverse part determined entirely by $\mathbf{b}(\mathbf{q})$. The longitudinal part $h(\mathbf{q})$ is determined by the Euler-Lagrange equation for $u(\mathbf{x})$, which, when expressed in terms of $\mathbf{v}$, is

$$\frac{\partial v_z}{\partial z} = \lambda_1^2 \nabla_\perp^2 \nabla_\perp \cdot \mathbf{v}_\perp,$$

where $\lambda_1 = (K_1/B)^{1/2}$. Show that

$$h(\mathbf{q}) = -\frac{q_z}{q^2(q_z^2 + \lambda_1^2 q_\perp^4)}(1 - \lambda_2^2 q_\perp^2)\mathbf{e}_z \cdot [\mathbf{q} \times \mathbf{b}(\mathbf{q})].$$

Use this result and the Landau-Peierls free energy to show that

$$F_{\text{disc}} = \frac{1}{2}\int \frac{d^3q}{(2\pi)^3}\left\{\frac{K_1\{|d\mathbf{e}_z \cdot [\mathbf{q} \times \mathbf{b}(\mathbf{q})]|^2\}}{\{q_z^2 + \lambda_1^2 q_\perp^4\}} + 2E_s|b_z(\mathbf{q})|^2 + 2E_e|\mathbf{b}_\perp(\mathbf{q})|^2\right\}$$

where E_s and E_e are, respectively, the core energy per unit length of screw and edge dislocations. Use this result to show that the interaction potential between screw dislocations in this approximation has zero range and that the interaction potential between edge dislocations is given by Eq. (9.3.82).

9.6 In this problem, you will calculate the contribution to the Frank elastic constant in the nematic phase arising from a distribution of unbound dislocation loops. Let

$$E_{\text{core}} = \frac{1}{2}\int \frac{d^3q}{(2\pi)^3} 2E_c|\mathbf{b}(\mathbf{q})|^2$$

in Eq. (9.3.77). Then, using $\delta\mathbf{n} = -\nabla_\perp u + \delta\mathbf{n}^s$, where $\delta\mathbf{n}^s$ is the contribution to $\delta\mathbf{n}$ arising from dislocations with components δn_t and $\delta n_\perp$ given by Eqs. (9.3.71). Calculate $\langle|\delta n_t(\mathbf{q})|^2\rangle$ and $\langle|\delta n_\perp(\mathbf{q})|^2\rangle$ using Debye-Hückel theory for

b, and show that the renormalized Frank elastic constants are $K_1^R = K_1$, $K_2^R = K_2 + 2E_c$, and $K_3^R = K_3 + 2E_c$.

9.7 (Mean-field theory for the lattice Coulomb gas) In the lattice Coulomb gas with Hamiltonian Eq. (9.4.19), each site **l** on the lattice can be occupied by a charge of strength $k_{\mathbf{l}} = 0, \pm 1, \pm 2, \cdots$. In mean-field theory, the density matrix is expressed as a product of site density matrices $\rho = \prod_{\mathbf{l}} \rho_{\mathbf{l}}$, where $\mathrm{Tr}\rho_{\mathbf{l}} = 1$ (see Sec. 4.8). For the Coulomb gas, $\rho_{\mathbf{l}}$ can be written as $\sum_k a_{k,\mathbf{l}} \delta_{k,k_{\mathbf{l}}}$ where $\sum_k a_{k,\mathbf{l}} = 1$, and where $a_{k,\mathbf{l}} = \langle \delta_{k,k_{\mathbf{l}}} \rangle = \langle n_{k,\mathbf{l}} \rangle$ is the average number of particles of charge **k** at site **l**.

(a) Show that the mean-field free energy can be written as

$$\begin{aligned} F &= \frac{1}{2} \sum_{\mathbf{l},\mathbf{l}'} \langle n_{\mathbf{l}} \rangle V_{\mathbf{l},\mathbf{l}'} \langle n_{\mathbf{l}'} \rangle + E_c \sum_{k,\mathbf{l}} k^2 \langle n_{k,\mathbf{l}} \rangle \\ &\quad + T \sum_{k,\mathbf{l}} \langle n_{k,\mathbf{l}} \rangle \ln \langle n_{k,\mathbf{l}} \rangle , \end{aligned}$$

where $\langle n_{\mathbf{l}} \rangle = \sum_k \langle n_{k,\mathbf{l}} \rangle$ is the average charge at site **l** and $V_{\mathbf{l},\mathbf{l}'}$ is the Coulomb potential.

(b) Derive the mean-field equation for $\langle n_{k,\mathbf{l}} \rangle$, and show that it has an **l**-independent solution,

$$\langle n_k \rangle \equiv \langle n_{k,\mathbf{l}} \rangle = \frac{e^{-k^2 E_c / T}}{\sum_k e^{-k^2 E_c / T}},$$

with average number of vortices $\langle n \rangle = \sum_k \langle n_k \rangle$ at each site equal to zero.

(c) Show that the charge density correlation function

$$G_{vv}(\mathbf{q}) = \sum e^{-i\mathbf{q}\cdot(\mathbf{R}_{\mathbf{l}} - \mathbf{R}_{\mathbf{l}'})} \langle n_{\mathbf{l}} n_{\mathbf{l}'} \rangle ,$$

where $n_{\mathbf{l}} = \sum_{k,\mathbf{l}} n_{k,\mathbf{l}}$ satisfies Eq. (9.4.44) with $B^{-1} = \sum_k k^2 \langle n_k \rangle$.

You may wish to introduce a chemical potential $\mu_{k,\mathbf{l}}$ conjugate to $n_{k,\mathbf{l}}$ to solve parts (b) and (c).

9.8 The Hamiltonian for vortices and dislocations in two dimensions can be written as [Eq. (9.3.50)]

$$\mathcal{H} = \frac{1}{2} Y_2 \int d^2x d^2x' U(\mathbf{x}, \mathbf{x}') \tilde{s}(\mathbf{x}) \tilde{s}(\mathbf{x}') + E_c \sum_\alpha b_\alpha^2 + E_s \sum_\alpha s_\alpha^2 ,$$

where $\tilde{s}(\mathbf{x}) = s(\mathbf{x}) - \epsilon_{ij} \nabla_j b_i(\mathbf{x})$,

$$U(\mathbf{x}, \mathbf{x}') = \int \frac{d^2q}{(2\pi)^2} \frac{1}{q^4} e^{-i\mathbf{q}\cdot(\mathbf{x}-\mathbf{x}')},$$

and E_c and E_s are, respectively, the dislocation and disclination core energies.

(a) Using this expression, show that the energy of a single isolated disclination at the center of a disc of radius R is proportional to $R^2 \ln R$.

(b) Use this energy to derive Eq. (9.5.1) when $s(\mathbf{x}) = 0$.

(c) Above the melting transition, the dislocation density $\mathbf{b}(\mathbf{x})$ can be treated as an independent continuum variable (as in the Debye-Hückel approximation). Calculate the equilibrium value of $\mathbf{b}(\mathbf{x})$ in the presence of a non-vanishing disclination density $s(\mathbf{x})$. Show that $\mathbf{b}(\mathbf{x})$ screens the interaction between disclinations, leading to an effective disclination Hamiltonian

$$\mathcal{H}_{\text{disc}} = \frac{1}{2} E_c a^2 \int \frac{d^2q}{(2\pi)^2} \frac{1}{q^2} s(\mathbf{q}) s(-\mathbf{q}) + \sum_\alpha E_s s_\alpha^2$$

with a logarithmic interaction between disclinations, where a is the dislocation core radius.

(d) Calculate the effective interaction between disclinations in the isotropic fluid phase where $s(\mathbf{x})$ is a free field.

9.9 This problem will lead you through the derivation of Eq. (9.3.33) for $\mathscr{G}_{ij}(\mathbf{x})$.

(a) Let $\mathscr{G}^{(2)}(\mathbf{x}) = -\ln(r/a)/(2\pi)$, where $r = (x^2 + y^2)^{1/2}$. Show by direct evaluation that

$$-\nabla^2 \mathscr{G}^{(2)}(\mathbf{x}) = \delta^{(2)}(\mathbf{x}).$$

(b) Show by direct evaluation of the integral that

$$\mathscr{G}^{(4)} = \int \frac{d^2q}{(2\pi)^2} \frac{e^{i\mathbf{q}\cdot\mathbf{x}}}{q^4} = \frac{1}{8\pi} r^2(\ln r + A) + B,$$

where A and B are constants (possibly depending on the sample size R. [Hint: write $e^{i\mathbf{q}\cdot\mathbf{x}} = 1 - (\mathbf{q}\cdot\mathbf{x})^2/2 - [1 - (\mathbf{q}\cdot\mathbf{x})^2/2) + e^{i\mathbf{q}\cdot\mathbf{x}}]$ and proceed along the lines developed in Eqs. (6.1.24) to (6.1.27).]

(c) Use the fact that $-\nabla^2 \mathscr{G}^{(4)} = \mathscr{G}^{(2)}$ to show that $A = -(3/2) - \ln a$.

(d) Derive Eq. (9.3.33) using $a = 1$.

9.10 This problem will lead you through a derivation of the connection between the partion functions of the two-dimensional Coulomb gas and the sine-Gordon model (see Secs. 10.2 and 10.6). Assume that there are only charges $q = \pm 1$ at respective positions $\mathbf{x}_\alpha^+$ and $\mathbf{x}_\beta^-$. The Coulomb gas Hamiltonian (without core energies) is

$$\beta\mathcal{H}_c = \frac{1}{2}(2\pi)^2 K \int d^2x d^2x' n_v(\mathbf{x}) \mathscr{G}^{(2)}(\mathbf{x} - \mathbf{x}') n_v(\mathbf{x}'),$$

where $\mathscr{G}^{(2)}(\mathbf{x})$ is defined in Problem 9.9. Using the Hubbard-Stratonovich (HS) transformation discussed in Appendix 5A, show that

$$e^{-\beta\mathcal{H}_c} = C \int \mathscr{D}\phi(\mathbf{x}) \exp\left[-\frac{1}{2}\frac{1}{(2\pi)^2 K} \int d^2x (\nabla\phi)^2\right] \exp\left[i \int d^3x \phi(\mathbf{x}) n_v(\mathbf{x})\right],$$

where C is a constant. The Coulomb gas partition function is

$$Z(y) = \sum_{N_+!N_-!} \frac{1}{N_+!} \frac{1}{N_-!} y^{N_+ + N_-} Z_c(N_+, N_-),$$

where y is the charge fugacity, and

$$Z_c(N_+, N_-) = \prod_{\alpha=1}^{N_+} \prod_{\beta=1}^{N_-} \int d^2x_\alpha^+ d^2x_\beta^- e^{-\beta\mathcal{H}_c}$$

is the partition function for N_+ positive charges and N_- negative charges. Use the HS form for $e^{-\beta\mathcal{H}_c}$ and the expression for $n_v(\mathbf{x})$ as a sum over δ-functions to show that $Z(y) = CZ_{SG} = C\int \mathcal{D}\phi(\mathbf{x})e^{-\beta\mathcal{H}_{SG}}$, where

$$\beta\mathcal{H}_{SG} = \frac{1}{2}\frac{1}{4\pi^2 K}\int d^2x(\nabla\phi)^2 - \frac{2y}{a^2}\int d^2x\cos\phi(\mathbf{x})$$

is the reduced sine-Gordon Hamiltonian. Thus, the sine-Gordon Hamiltonian is dual to the two-dimensional Coulomb gas Hamiltonian. This relation implies, as we shall see in more detail in Sec. 10.6, that the sine-Gordon model has a transition with the same properties as the Kosterlitz-Thouless transition.

9.11 Consider a low-angle grain boundary in a two-dimensional isotropic crystal consisting of edge dislocations at positions nl along the y-axis. The Burgers vector density for this grain boundary is

$$\mathbf{b}(\mathbf{x}) = \mathbf{e}_x b\sum_n \delta(y - nl)\delta(x).$$

(a) Show that the shear stress generated by this boundary is

$$\sigma_{xy}(\mathbf{x}) = 4\pi^2 D\frac{bx}{l^2}\sum_{k=1}^{\infty} k\cos(2\pi ky/l)e^{-2\pi k|x|/l},$$

where $D = bY_2/4\pi$. This shows that stress decays exponentially away from the grain boundary with a length scale set by the distance between dislocations. [Solution hints: this problem can be solved in several ways. One way is to sum the expression Eq. (9.3.35) for the stress generated by a single dislocation over all dislocations. The Poisson summation formula [Eq. (9B.25)] should be used to evaluate this sum. Another way is to solve for the Airy stress function [Eq. (9.3.48)] in Fourier space using $\mathbf{b}(\mathbf{q}) = \mathbf{e}_x(b/l)\sum_n \delta(q_y - (2\pi n/l))$ and to use Eq. (9.3.46) to evaluate $\sigma_{xy}(\mathbf{q})$ by Fourier transformation $\sigma_{xy}(\mathbf{x})$.]

(b) Show that the energy per unit length of a low-angle grain boundary with $l \gg a$, where a is the core size, is

$$\frac{E}{L} = \frac{1}{2}\frac{Db}{l}[1 - \ln(2\pi a/l)] = \frac{1}{2}D\theta[1 - \ln(\theta/2\pi)],$$

where, in the final expression, we used $a = b$ and $\theta \approx b/l$. [Solution hint: introduce cuts for each dislocation extending along the x-axis from a to ∞. Then use

$$E = \frac{1}{2}\int d^2x u_{ij}\sigma_{ij} = \frac{1}{2}\sum_{k=1}^{N}\int dS_j^k u_i^k\sigma_{ij} = \frac{1}{2}Nb\int_a^\infty dx\sigma_{xy},$$

where dS_j^k is the surface element associated with the cut of the kth dislocation and N is the number of dislocations in the boundary. You may also use Eq. (9.3.51), but then you will have to introduce a cutoff in the sum $\sum(1/k)$ at $k \approx l/(2\pi a)$.]

10

Walls, kinks and solitons

In Chapter 9, we studied topological defects in ordered systems with a broken continuous symmetry. In this chapter, we will study fundamental defects in systems with discrete symmetry such as the Ising model. These defects are surfaces, of one dimension less than the dimension of space, that separate regions of equal free energy but with different values of the order parameter. They are variously called *domain walls, kinks, solitons, discommensurations,* or simply *walls,* depending on the particular system and context. They can also be regarded as interfaces, such as, for example, the interface separating coexisting liquid and gas phases. They play an important, if not dominant, role in determining the physical properties of systems with discrete symmetry.

We begin this chapter (Sec. 10.1) with a number of examples of walls. In Sec. 10.2, we study the continuum mean-field theory for kinks and solitons. Then, in Sec. 10.3, we discuss in some detail the Frenkel-Kontorowa model for atoms adsorbed on a periodic substrate. This will introduce in a natural way a lattice of interacting kinks (called discommensurations in this case) to describe the incommensurate phase of adsorbed monolayers. After investigating the properties of interacting kinks at zero temperature, we will in Sec. 10.4 study thermal fluctuations of walls in dimensions greater than one and show that structureless walls in three dimensions or less have divergent height fluctuations that render them macroscopically rough. In Sec. 10.5, we will consider arrays of walls, beginning with the repulsive interaction between walls arising from the reduction in entropy produced by the steric confinement of a wall between two others. This will provide a quantitatively correct theory of the commensurate-to-incommensurate transition observed in monolayers adsorbed on graphite and discussed in Sec. 2.9. In Sec. 10.6, we will discuss the interface between coexisting solids and liquids or gases. Such interfaces are flat and smooth at low temperatures, where their shape is fixed by the periodic solid substrate. At high temperatures, they develop divergent height fluctuations and become rough. We will study the roughening transition between smooth and rough interfaces. Finally, we will study faceting and equilibrium shapes of crystals.

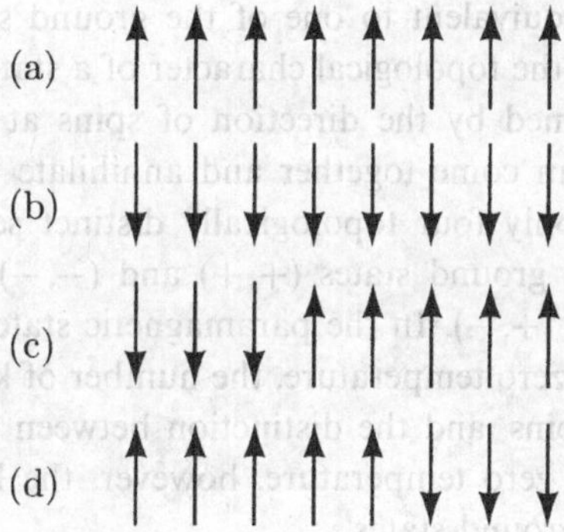

Fig. 10.1.1. Topologically inequivalent states of the one-dimensional Ising model: (a) the spin-up ground state, (b) the spin-down ground state, (c) a kink state with a positive domain wall separating a spin-down ground state from a spin-up ground state, and (d) a kink state with a negative domain wall. These four states can be labeled according to the sign of the spin at $-\infty$ and $+\infty$. Thus, the states depicted in (a) to (d) are, respectively, $(+,+)$, $(-,-)$, $(-,+)$ and $(+,-)$.

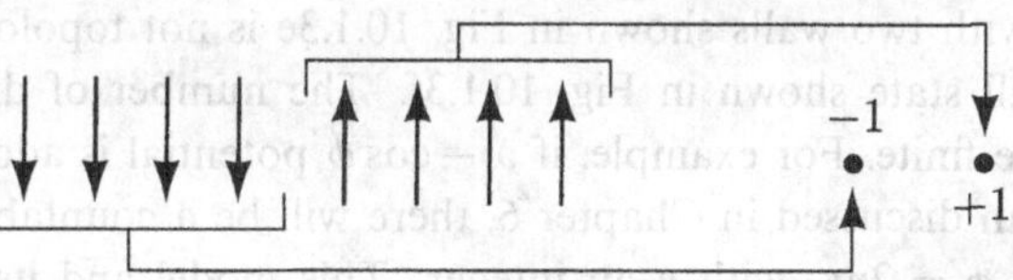

Fig. 10.1.2. Mapping of a one-dimensional Ising model with a single domain wall onto the Ising order parameter space.

10.1 Some simple examples

Consider a linear Ising chain. It has two equivalent ground states, one with spin up and the other with spin down, as shown in Figs. 10.1.1a and b. If a boundary condition is imposed specifying the spins to be up at one end (say at $z = -L/2$) and down at the other end (say at $z = L/2$), then there must be a spin flip somewhere in the chain, as shown in Figs. 10.1.1c and d. This spin flip is a kink that cannot be removed so long as the boundary conditions do not change. Like the topological defects studied in the preceding chapter, the kink can be represented as a map from real space (the lattice of the Ising model) to an order parameter space. The order parameter space for the Ising model consists of two points, $\sigma = \pm 1$. All points to the left of the kink are mapped into the spin-up point, and all points to the right of the kink are mapped into the spin-down point, as shown in Fig. 10.1.2.

A kink can have a positive or a negative sign depending on whether the up spins are on its right or on its left. A state with a positive and a negative kink is topologically equivalent to one of the ground states since the two kinks can annihilate. Thus, the topological character of a state of the one-dimensional Ising model is determined by the direction of spins at $z = \pm\infty$. Since positive and negative kinks can come together and annihilate to produce a uniform ground state, there are only four topologically distinct sectors of the one-dimensional Ising model: the ground states $(+,+)$ and $(-,-)$, and the kink and anti-kink states $(-,+)$ and $(+,-)$. In the paramagnetic state, which is thermodynamically stable at any nonzero temperature, the number of kink-anti-kink pairs is of order the number of spins, and the distinction between the four topological sectors is unimportant. At zero temperature, however, the kink states will have a higher energy than the ground states.

There are only four topologically distinct states in the Ising model because there are only two energetically equivalent ground states. In models with more than two ground states, there are more topologically inequivalent states. For example, the four-state Potts model has four ground states, which we can label 0, 1, 2 and 3. There can be domain walls separating any two ground states. Two walls separating, respectively, 0 and 1 and 1 and 2 can combine to produce a wall separating 0 and 2, but they cannot combine to produce a uniform ground state. The state with two walls shown in Fig. 10.1.3e is not topologically equivalent to the one-wall state shown in Fig. 10.1.3f. The number of distinct ground states need not be finite. For example, if a $-\cos\phi$ potential is added to the elastic *xy*-Hamiltonian discussed in Chapter 6, there will be a countable infinity of ground states with $\phi = 2n\pi$, with n an integer. This model and its generalizations can have topologically distinct states with an extensive number of positive or negative kinks. We will investigate one version of this model in detail in Sec. 10.3.

The point kink of a one-dimensional Ising model becomes a one-dimensional line defect in a two-dimensional Ising model, as depicted in Fig. 10.1.4, and a two-dimensional surface defect in a three-dimensional Ising model. These generalized kinks have spatial dimension one less than the dimension of space, and are generally referred to as domain walls. In the one-dimensional Ising model with Hamiltonian $\mathcal{H} = -J\sum_{<\mathbf{l},\mathbf{l}'>}\sigma_{\mathbf{l}}\sigma_{\mathbf{l}'}$, all spins are aligned in the ground state, and each bond $<\mathbf{l},\mathbf{l}'>$ has an energy $-J$. In a state with a single kink, there is a single broken bond with opposite signs of the spin at its vertices and energy $+J$. Thus, the energy of a kink state relative to the ground state is $+2J$. In higher dimensions, the energy of a domain wall is proportional to the number of broken bonds. Thus, in an Ising model on a d-dimensional lattice of side $L = N_1 a$, the wall energy is proportional to L^{d-1}:

$$E_{\text{kink}} = 2JN_1^{d-1} = \sigma L^{d-1}, \tag{10.1.1}$$

where $\sigma = 2Ja^{-(d-1)}$ is a "surface" tension. The above energy should be compared with the energy that scales as L^{d-2} [Eq. (6.1.3)] of a system with continuous symmetry with similar boundary conditions

In one dimension, a kink can be located at any bond on the lattice. The entropy

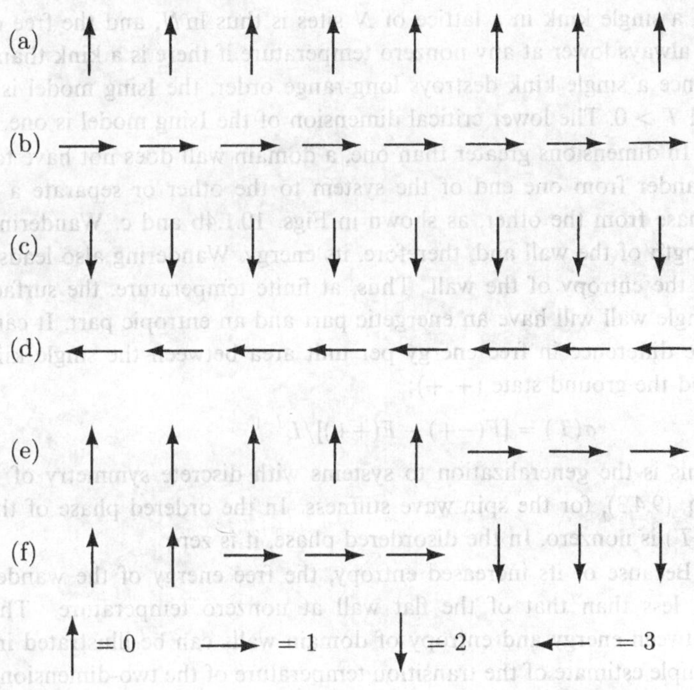

Fig. 10.1.3. (a)–(d) Ground states of the four-state Potts model. (e) State with a single $0-1$ kink. (f) State with a $0-1$ kink and a $1-2$ kink.

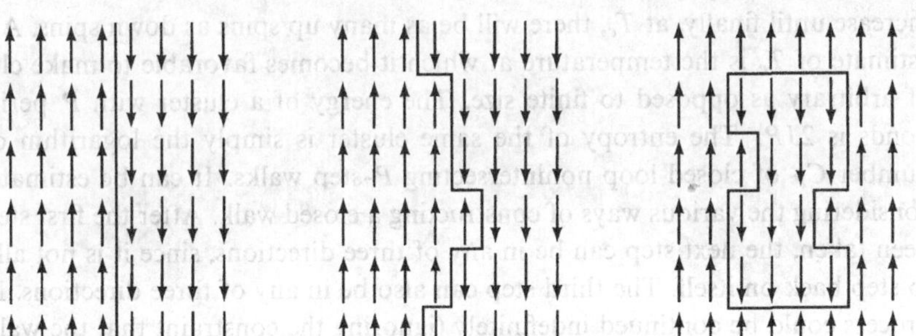

Fig. 10.1.4. Walls in a two-dimensional Ising model. (a) A straight wall. (b) A wandering domain wall. The wandering wall has a greater length (area) than the flat wall and, therefore, greater energy (for positive surface tension). It also has greater entropy so the free energy of the wandering wall at finite temperature is generally less than that of the flat wall. (c) A droplet of down spins separated by a closed domain wall from a sea of up spins.

of a single kink in a lattice of N sites is thus $\ln N$, and the free energy $E - TS$ is always lower at any nonzero temperature if there is a kink than if there is not. Since a single kink destroys long-range order, the Ising model is disordered for all $T > 0$. The lower critical dimension of the Ising model is one.

In dimensions greater than one, a domain wall does not have to be flat. It can wander from one end of the system to the other or separate a droplet of one phase from the other, as shown in Figs. 10.1.4b and c. Wandering increases the length of the wall and, therefore, its energy. Wandering also leads to an increase in the entropy of the wall. Thus, at finite temperature, the surface tension of a single wall will have an energetic part and an entropic part. It can be defined as the difference in free energy per unit area between the single kink state $(-,+)$ and the ground state $(+,+)$:

$$\sigma(T) = [F(-+) - F(++)]/L^{d-1}. \tag{10.1.2}$$

This is the generalization to systems with discrete symmetry of the expression, Eq. (9.4.3), for the spin-wave stiffness. In the ordered phase of the Ising model, $\sigma(T)$ is nonzero. In the disordered phase, it is zero.

Because of its increased entropy, the free energy of the wandering wall may be less than that of the flat wall at nonzero temperature. This competition between energy and entropy of domain walls can be illustrated in the following simple estimate of the transition temperature of the two-dimensional Ising model. At very low temperatures in the ferromagnetic phase, there should be very few flipped spins. There will be small clusters of flipped spins separated by domain walls. As the temperature is increased, the size and number of these clusters will increase until finally, at T_c, there will be as many up spins as down spins. A good estimate of T_c is the temperature at which it becomes favorable to make clusters of arbitrary as opposed to finite size. The energy of a cluster with P perimeter bonds is $2JP$. The entropy of the same cluster is simply the logarithm of the number C_P of closed-loop nonintersecting P-step walks. It can be estimated by considering the various ways of constructing a closed walk. After the first step has been taken, the next step can be in any of three directions, since it is not allowed to step back on itself. The third step can also be in any of three directions. If this process could be continued indefinitely (ignoring the constraint that the walk has to close on itself), C_P would simply be 3^P. There are configurations where there will be only two choices for the fourth step if the walk is not to close after only four steps. A more careful analysis of this effect yields $C_P = (2.639)^P$. The free energy of a single wall of perimeter P is then

$$F = E - TS \sim 2JP - T\ln C_P = [2J - T\ln(2.639)]P. \tag{10.1.3}$$

Thus, for temperatures greater than

$$T_c' = 2J/\ln(2.639), \tag{10.1.4}$$

large perimeter clusters are more favorable than small ones, and a transition to the disordered state takes place. This calculation is over-simplified in that it considers only one cluster of down spins. The estimate of T_c it gives, however, is very close to the exact result $T_c = 2J/\ln(1+\sqrt{2}) = J/\ln(2.414)$ [Eq. (9B.19)]. Note that the reasoning leading to this estimation of T_c is very similar to the simple Kosterlitz-Thouless calculation of the transition temperature of the two-dimensional xy-model presented in Chapter 9.

10.2 Domain walls in mean-field theory

Domain walls are easily generalized to systems with a coarse-grained continuous order parameter field such as those described by the Landau theories of Chapter 3. In this case, the domain wall provides a transition between energetically equivalent minima of the free energy, as shown in Fig. 10.2.1. At one end of the sample, the order parameter ϕ takes on the value ϕ_1 corresponding to one of the minima of the free energy. At the other end of the sample, it takes on the value ϕ_2 corresponding to the other minimum. ϕ changes continuously in passing from one end of the sample to the other, and there is an increase in free energy relative to a spatially uniform equilibrium state both because of the energy cost associated with spatial variation of the order parameter and because the order parameter must take on values for which the local free energy is higher than at its minima. Remember that ϕ can be the order parameter for any system with discrete symmetry. The order parameter profile depicted in Fig. 10.2.1b could, therefore, represent the magnetization near an Ising domain wall, the density profile near a liquid-gas interface, or the relative density profile at the interface separating two coexisting liquid phases. In the latter two cases, the free energy curve shown in Fig. 10.2.1 need not be symmetric about any axis (as discussed in Sec. 4.3).

The field theories of Chapters 4 and 5 can be used to calculate the energy of a flat domain wall. If all spatial variation takes place along the z-direction, the Landau mean-field free energy per unit area relative to the energy of the ground state is

$$\frac{F}{A} = \int_{-\infty}^{\infty} dz \left[\tilde{f}(\phi) + \frac{1}{2}c(d\phi/dz)^2\right] \equiv \int_{-\infty}^{\infty} dz e(z), \tag{10.2.1}$$

where $\tilde{f}(\phi) \equiv f(\phi) - f_0$ is the free energy density measured relative to the ground state and where $A = L^{d-1}$ is the $(d-1)$-dimensional area perpendicular to the z-axis. $e(z)$ is the local energy (strictly speaking free energy) density. The function $f(\phi)$ is depicted schematically in Fig. 10.2.1. It has two equivalent minima at $\phi = \phi_1$ and $\phi = \phi_2$ but, is not necessarily symmetric about any axis. Its value at these minima is $f_0 = f(\phi_1) = f(\phi_2)$. The domain wall is described by an extremum with respect to ϕ of F/A subject to the boundary conditions that $\phi = \phi_1$ at $z = -\infty$ and $\phi = \phi_2$ at $z = \infty$. The value of F/A at ϕ satisfying this extremal solution is simply the surface tension σ of the domain wall.

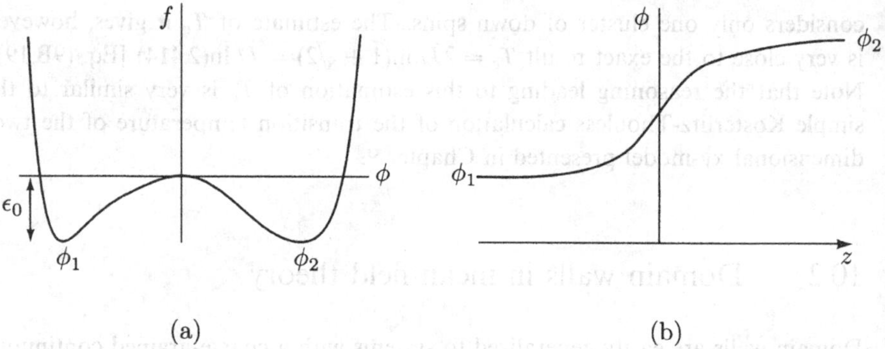

Fig. 10.2.1. (a) Free energy with two energetically equivalent minima at distinct values ϕ_1 and ϕ_2 of the order parameter ϕ. Note that the shape of the free energy curve does not have to be symmetric. (b) Schematic one-dimensional representation of the domain wall as a function of spatial coordinate z.

Before presenting an analytical calculation of domain wall solutions and associated energies, it is instructive to consider a heuristic solution. As discussed above, there are two contributions to the energy of the kink: one coming from the energy density $f(\phi)$ and the other from the gradient term $\frac{1}{2}c(d\phi/dz)^2$. Assume that the kink is localized to a region of width l, as shown in Fig. 10.2.2. Outside this region, the system is in one of its ground states, and $\tilde{f}$ is zero. Within the kink, the order parameter has to change from ϕ_1 to ϕ_2. In this process, $\tilde{f}(\phi)$ takes on all values between 0 and ϵ_0, the height of the barrier separating the two minima. This change occurs over a distance l so that the total free energy cost arising from $\tilde{f}$ is of order $\epsilon_0 l$. The energy associated with the spatial variation of ϕ is of order $\frac{1}{2}c(\phi_1 - \phi_2)^2/l$ so that the total free energy per unit area of the kink is

$$\frac{F}{A} \sim \epsilon_0 l + c\frac{(\Delta\phi)^2}{2l}, \tag{10.2.2}$$

where $\Delta\phi = \phi_2 - \phi_1$. The optimal length of the kink can be obtained by minimizing this equation with respect to l to yield

$$l^* = [c(\Delta\phi)^2/2\epsilon_0]^{1/2} \tag{10.2.3}$$

and

$$\sigma = F/A = 2\epsilon_0 l^*. \tag{10.2.4}$$

Not surprisingly, the surface tension of the kink is of the form (condensation energy / volume) × (length). The scale of the length l^* is set by the length $(c/\epsilon_0)^{1/2}$, which can vary widely depending on the system, temperature and other variables. It is generally of order the correlation length ξ discussed in Chapter 4. In the Ising model near $T = 0$, it is of the order of the lattice spacing, as is the correlation length.

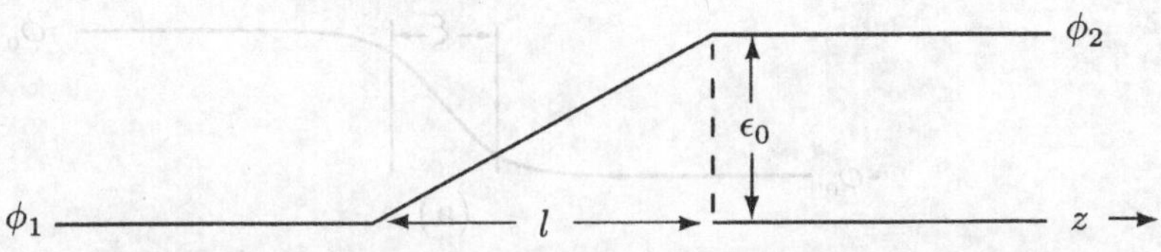

Fig. 10.2.2. Representation of an idealized kink in which all spatial variation takes place in a region of length l.

We will now seek an analytical solution for the kink. In equilibrium, ϕ satisfies the Euler-Lagrange equation

$$\frac{\delta}{\delta\phi}\left(\frac{F}{A}\right) = -c\frac{d^2\phi}{dz^2} + \frac{d\tilde{f}}{d\phi} = 0. \tag{10.2.5}$$

This equation can be solved in the standard way by multiplying both sides by $d\phi/dz$ to yield

$$\frac{d\phi}{dz} = \pm\left(\frac{2}{c}\tilde{f} + B\right)^{1/2}, \tag{10.2.6}$$

where B is a constant of integration. The kink boundary conditions are $d\phi/dz = 0$ at $z = \pm\infty$ and *either* $\phi = \phi_1$ at $z = -\infty$ and $\phi = \phi_2$ at $z = +\infty$ for a positive kink *or* $\phi = \phi_2$ at $z = -\infty$ and $\phi = \phi_1$ at $z = +\infty$ for a negative kink (anti-kink). The positive solution to Eq. (10.2.6) corresponds to the positive kink, and the negative solution corresponds to the negative kink. Since $\tilde{f}$ is zero when $\phi = \phi_1$ or $\phi = \phi_2$, the constant of integration B must be zero in order for $d\phi/dz$ to be zero at $z = \pm\infty$. Integration of Eq. (10.2.6) yields the implicit equation

$$z - z_0 = \pm\int_{\phi(z_0)}^{\phi(z)} \frac{d\phi}{(2\tilde{f}(\phi)/c)^{1/2}} \tag{10.2.7}$$

for $\phi(z)$. This equation clearly satisfies $\phi = \phi(z_0)$ when $z = z_0$. It is useful to choose z_0 to be the position of the center of the kink. This can be done by choosing $\phi(z_0)$ to be the value of ϕ at the kink center. For symmetrical models, this value is usually zero. The surface tension of the kink can be calculated using Eqs. (10.2.1) and (10.2.6) with $B = 0$:

$$\sigma = 2\int_{-\infty}^{\infty} \tilde{f}(\phi(z))dz = 2(c/2)^{1/2}\int_{\phi_1}^{\phi_2} |\, \tilde{f}(\phi)\,|^{1/2}\, d\phi. \tag{10.2.8}$$

This result applies to both positive and negative kinks.

1 The ϕ^4 kink

An analytic solution for the above kink equations can be obtained for the ϕ^4 model discussed in Sec. 4.3 with

$$f = -\frac{1}{2}r\phi^2 + u\phi^4. \tag{10.2.9}$$

(Note that we have set the coefficient of ϕ^2 to $-\frac{1}{2}r$ with $r > 0$.)

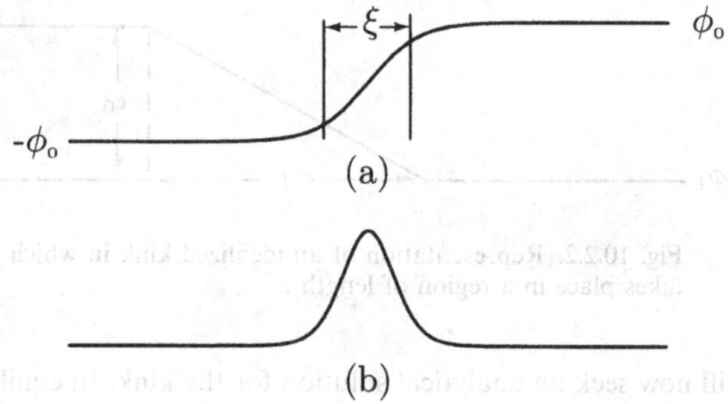

Fig. 10.2.3. (a) $\phi(z)$ and (b) $e(z)$ for a ϕ^4 kink.

There are two minima in f at $\phi = \pm\phi_0 = \pm(r/4u)^{1/2}$ with $f_0 = -r^2/16u = -r\phi_0^2/4$ so that

$$\tilde{f} = \frac{1}{4}r\frac{(\phi^2 - \phi_0^2)^2}{\phi_0^2}. \tag{10.2.10}$$

Using this expression in Eq. (10.2.7), we obtain

$$\begin{aligned} z - z_0 &= \pm \int \frac{d\phi}{\mid 2\tilde{f}/c \mid^{1/2}} = \pm(2c/r)^{1/2} \int \frac{\mid \phi_0 \mid d\phi}{\phi^2 - \phi_0^2} \\ &= \pm(c/2r)^{1/2} \tanh^{-1}(\phi / \mid \phi_0 \mid). \end{aligned} \tag{10.2.11}$$

Thus, the order parameter $\phi(z)$, the energy density $e(z)$, and the surface tension σ are, respectively,

$$\phi(z) = \pm\phi_0 \tanh[(z - z_0)/(\sqrt{2}\xi)], \tag{10.2.12}$$

$$e(z) = 2 \mid f_0 \mid \operatorname{sech}^4[(z - z_0)/(\sqrt{2}\xi)], \tag{10.2.13}$$

and

$$\sigma = (8/3)(\sqrt{2}\xi) \mid f_0 \mid = (2\sqrt{2}/3)(\xi r \phi_0^2), \tag{10.2.14}$$

where $\xi = (c/r)^{1/2}$ is the mean-field correlation length introduced in Sec. 4.3. At the center of the kink at $z = z_0$, ϕ is zero. The energy density is strongly localized in space with a maximum at the center of the kink and falling rapidly to zero for $\mid z - z_0 \mid > \sqrt{2}\xi$. Fig. 10.2.3 depicts both $\phi(z)$ and $e(z)$.

The ϕ^4 model has the same topological character as the Ising model. There are four topologically distinct sectors specified by the values of the order parameter at $z = \pm\infty$. They are the spatially uniform states (ϕ_0, ϕ_0) and $(-\phi_0, -\phi_0)$ and the positive and negative kink states $(-\phi_0, \phi_0)$ and $(\phi_0, -\phi_0)$. A topological charge analogous to a vortex can be defined via

$$Q = \frac{1}{2\phi_0}[\phi(\infty) - \phi(-\infty)] = \frac{1}{2\phi_0}\int_{-\infty}^{\infty} dz \frac{d\phi}{dz}. \tag{10.2.15}$$

Thus, the ground states have charge 0, and the kink and anti-kink states have respective charges of +1 and −1.

2 The sine-Gordon soliton

An analytic kink solution can also be obtained for the periodic potential

$$f = -V_0 \cos m\phi, \tag{10.2.16}$$

where we assume m is an integer. This model describes two-component spins in an m-fold anisotropy field. It is often called the *sine-Gordon* model. Unlike the ϕ^4 model, it has an infinite number of equivalent ground states at $\phi = 2\pi p/m$, where p is an integer. A simple kink (soliton) state in the sine-Gordon model is one in which the order parameter passes from zero to the next minimum ($\phi = 2\pi/m$) in the free energy. Its solution is very similar to that for the angle of the director in a nematic liquid crystal in an external magnetic field and anchoring boundary conditions at $z = 0$ discussed in Sec. 6.2. Following the procedures discussed there, we find

$$\phi_{\pm}(z) = \pm\frac{4}{m}\tan^{-1} e^{zm(V_0/c)^{1/2}}, \tag{10.2.17}$$

where the ϕ_+ describes a soliton and ϕ_- describes an anti-soliton centered at $z = 0$. Solitons centered at z_0 are described by $\phi_{\pm}(z - z_0)$. The energy of a soliton is

$$\sigma = 8\sqrt{cV_0}/m. \tag{10.2.18}$$

Note that the energy of a single kink goes to zero as $m \to \infty$

Because the sine-Gordon model has an infinity of equivalent ground states, there can exist states in which the order parameter increases monotonically undergoing more or less discrete changes of $2\pi/m$ at each kink. The different topological sectors can be indexed by the number of kinks, or equivalently by the charge defined in Eq. (10.2.15). This charge may, as we shall see in the next section, be an extensive quantity.

3 Dynamics

The time-dependent properties of kinks depend on the dynamics assigned to the field $\phi(z, t)$. In one much-studied dynamical model, a kinetic energy proportional to $(\partial\phi/\partial t)^2$ is introduced so that the dynamical Hamiltonian associated with the free energy of Eq. (10.2.1) is

$$\mathscr{H} = \int dz \left[\frac{1}{2}cv_0^{-2}(\partial\phi/\partial t)^2 + \frac{1}{2}c(\partial\phi/\partial z)^2 + \tilde{f}(\phi)\right], \tag{10.2.19}$$

where v_0 is a velocity and cv_0^{-2} is analogous to the mass density in the Lagrangian elasticity theory of Eq. (7.2.1). $\mathscr{H}$ should be interpreted as a Hamiltonian in which $cv_0^{-2}(\partial\phi/\partial t)$ (like $\rho\mathbf{v}$ in Sec. 7.3) is a momentum density. The equation of motion for $\phi(z, t)$ is

$$\frac{1}{v_0^2}\frac{\partial^2\phi}{\partial t^2} - \frac{\partial^2\phi}{\partial z^2} + \frac{1}{c}\frac{d\tilde{f}}{d\phi} = 0. \tag{10.2.20}$$

Both the Hamiltonian and the equation of motion for ϕ are invariant under the Lorentz transformation

$$\begin{aligned} z \to z' &= (1-\beta^2)^{-1/2}(z - vt) \\ v_0 t \to v_0 t' &= (1-\beta^2)^{-1/2}(-\beta z + v_0 t), \end{aligned} \tag{10.2.21}$$

where v is any velocity less than v_0 and $\beta = v/v_0$. Thus, if $\phi(z,t)$ is a solution to the equation of motion, then $\phi'(z,t) = \phi(z',t')$ is also. In particular, if $\phi_s(z)$ is a static solution such as Eqs. (10.2.12) or (10.2.17) to Eq. (10.2.20), then

$$\phi(z,t) = \phi_s[(1-\beta^2)^{-1/2}(z - vt)] \tag{10.2.22}$$

is a time-dependent solution to Eq. (10.2.20). This solution is one in which the center of the kink or soliton moves with a constant velocity v.

The dynamical kink solution just discussed is a moving wave packet in which the energy density and order parameter variations are localized in space. We are familiar with such wave-packet solutions to the wave equation with $\tilde{f} = 0$:

$$\frac{1}{v_0^2}\frac{\partial^2 \phi}{\partial t^2} - \frac{\partial^2 \phi}{\partial z^2} = 0. \tag{10.2.23}$$

Any reasonable smooth function $g(z \pm v_0 t)$ is a solution to this linearized equation. Because the frequency spectrum, $\omega = v_0 q$, is linear in wave number q, a wave packet described by the function $g(z)$ at time $t = 0$ will propagate with velocity $\pm v_0$ and maintain its shape at all future times. In addition, two localized solutions $g_1(z - v_0 t)$ and $g_2(z + v_0 t)$ can be added to produce a third solution $g_3(z,t) = g_1(z - v_0 t) + g_2(z + v_0 t)$. Let $g_1(z - v_0 t)$ and $g_2(z + v_0 t)$ be widely separated wave packets at time $t = 0$. As time increases, the two wave packets will approach each other, creating a distortion described by the function $g_3(z,t)$. At large positive time, the two wave packets will separate and have shapes identical to those they had at $t = 0$. The invariance of the shape of wave packets with respect to time and collisions with other wave packets results from the linearity of the wave equation and the linearity of the dispersion relation $\omega = v_0 q$. Both nonlinearities and dispersion (i.e., deviations from linearity in ω as a function of q) will in general lead to a broadening of wave packets over time. There are, however, special solutions to nonlinear wave equations in which nonlinearities and dispersion work together to produce a wave packet whose shape is constant in time. Such constant shape solutions to nonlinear wave equations are generally called *solitary waves.* The dynamical kink solution of Eq. (10.2.20) is a solitary wave. Most solitary waves are not invariant with respect to collisions. However, some nonlinear equations do have solutions in which well-separated solitary waves collide and emerge at large positive times with precisely the same shape as they had at times well before the collision: the wave packets simply pass through each other. The special solitary-wave solutions are generally called *solitons.* The ϕ^4-field theory has solitary-wave but not soliton solutions. The solitary waves of the sine-Gordon theory, on the other hand, are all solitons. We will leave the study of a two-soliton solution as an exercise

(Problem 10.1). Strictly speaking, the term soliton should only be used for the restricted class of solitary waves just described. It is, however, often used in the literature interchangeably with kink.

10.3 The Frenkel-Kontorowa model

1 Introduction

In Sec. 2.9, we saw that atoms of elements such as xenon adsorbed on a periodic substrate such as graphite form periodic arrays that are either commensurate or incommensurate with the substrate. In the latter case, the reciprocal lattice of the resulting structure has colinear vectors with an irrational magnitude ratio. In this section, we will study in some detail the simplest one-dimensional model for atoms adsorbed on a periodic substrate. This model was originally introduced in the 1930s by Frenkel and Kontorowa (1938) and was subsequently reinvented independently by others, notably Frank and Van der Merwe (1949). It provides a simple and realistic description of commensurate-incommensurate transitions when thermal fluctuations are unimportant as they are near zero temperature. It also provides a starting point for the treatment of the more complex two-dimensional problem at nonzero temperatures to be discussed in the next section.

The adsorbed atoms (adatoms) at positions x_n in the Frenkel-Kontorowa (FK) model are treated as a harmonic chain with equilibrium lattice spacing a. The substrate is a one-dimensional periodic lattice with period b. The interaction between the nth adsorbed atom and the periodic substrate is described by a potential energy $V(x_n)$. The potential energy of the FK model is thus

$$U = \sum_n \left[\frac{1}{2}K(x_{n+1} - x_n - a)^2 + V(x_n)\right]. \tag{10.3.1}$$

To make contact between this model and real adsorbates in equilibrium with their vapor phase, one may assume that the lattice spacing a is a linear function of the adsorbate chemical potential μ, which is determined by the partial pressure of the vapor phase. The potential $V(x)$ is an arbitrary function with period b; it can be approximated, however, without significant loss of physical content by the cosine potential,

$$V(x) = -V_0 \cos(2\pi x/b), \tag{10.3.2}$$

having minima at $x = mb$ for any integer m. The first, or elastic, term in U favors a periodic lattice of adsorbed atoms with $x_n = na$ whereas the second, or potential, term favors *lock-in* to the substrate with each x_n an integral multiple of b. If the potential $V(x)$ is zero, the adsorbate lattice spacing will be independent of b. The resulting structure is called a "floating phase" in which the equilibrium lattice spacing a of the adsorbed lattice can be an arbitrary (including irrational) multiple of the substrate periodicity b. Thus, the floating phase is incommensurate for almost all values of the ratio a/b. In the opposite limit, where the potential is very large, one can expect each atom of the adsorbed lattice to sit in a minimum of the potential V. This leads to a commensurate structure with the average spacing between adsorbed atoms a rational multiple of b. The equilibrium state for general V can be characterized by the average spacing $\tilde{a}$ between adsorbed atoms which is defined via

$$\tilde{a} = \lim_{n\to\infty} \frac{x_n - x_0}{n}. \tag{10.3.3}$$

If $\tilde{a}$ is an irrational multiple of b, the adsorbate lattice is incommensurate with the substrate; otherwise, it is commensurate.

2 Discommensurations

Fig. 10.3.1 depicts schematically the various types of ground states one might expect in the FK model. The floating and commensurate states discussed above are depicted in (a)–(d). The simplest commensurate states are those in which $\tilde{a} = pb$ for an integer p and for which there is exactly one adsorbate atom for every pth substrate minimum. In this case, each adsorbate atom sits at the bottom of a substrate minimum, as depicted in Figs. 10.3.1b and c. More generally, there can be commensurate states with $\tilde{a} = (p/q)b$ with p and q relatively prime integers consisting of periodically repeated unit cells of length pb containing q adsorbate atoms for every p substrate minima. In this case, not every adsorbate atom will sit at the bottom of a substrate minimum. A state with $q = 4$ and $p = 5$ is depicted in Fig. 10.3.1d. In incommensurate states with $\tilde{a}/b$ irrational but near an integer p, one would expect (and indeed we will shortly show) that the best compromise between competing elastic and potential terms in U is realized by placing atoms at the minima of U as often as possible and occasionally stretching (or compressing) the spring between adsorbate atoms so that a minimum of V is missed. More generally, for $\tilde{a}/b$ near p/q, one can expect regions with repeated unit cells of the $\tilde{a}/b = p/q$ commensurate structure separated by regions in which springs are stretched or compressed and in which commensurate registry with the substrate is lost. Configurations in which springs are stretched or compressed correspond precisely to a soliton of the type discussed in the preceding section. In the present model, they are usually called *discommensurations* because they break the commensurate registry of the adsorbate and substrate lattices. Two equivalent ground states of V are separated by a *positive* discommensuration if springs are stretched or by a *negative* discommensuration if springs are compressed, as shown in Figs. 10.3.1e and f. A positive discommensuration leads to a reduction in the density of adsorbate atoms. It is, therefore, often referred to as a *light wall*. A negative discommensuration leads to an increase in adsorbate density and is called a *heavy wall*. The energies of heavy and light walls need not be equal.

All unit cells of the reference commensurate structure to the right of a positive (negative) discommensuration are displaced by one unit cell of the substrate (a distance b) to the right (left). Let $N_{\text{dis}} = \pm \mid N_{\text{dis}} \mid$ be the net number of discommensurations relative to the p/q commensurate structure of N adsorbate atoms. The total length of the adsorbate chain is thus

$$L_N = x_N - x_0 = N(p/q)b \pm \mid N_{\text{dis}} \mid b, \tag{10.3.4}$$

and the average spacing between adatoms [Eq. (10.3.3)] is

$$\tilde{a} = (p/q)b \pm \mathscr{L}^{-1} b, \tag{10.3.5}$$

where $\mathscr{L} = N / \mid N_{\text{dis}} \mid$ is the number of adatoms between discommensurations. The average distance between discommensurations in real space is

$$l = \mathscr{L}\tilde{a} = [\mathscr{L}(p/q) \pm 1]b. \tag{10.3.6}$$

In the thermodynamic limit, $\mathscr{L}$, and thus $\tilde{a}/b$, can take on any, including irrational, values. Note that the addition of one extra unit cell (consisting of q adatoms at fixed length x_N) can be described by an increase (or decrease) in the number of discommensurations by p. For example, if $\tilde{a} < (p/q)b$, then $x_N(N_{\text{dis}}) = x_{N+q}(N'_{\text{dis}})$ implies $N'_{\text{dis}} - N_{\text{dis}} = p$.

3 Devil's staircases and the FK phase diagram

To discuss the dependence of the equilibrium average spacing $\tilde{a}$ on the preferred spacing a, it is convenient to measure all lengths in terms of the substrate periodicity b, and we introduce $\Omega = a/b$ and the *winding number* $\tilde{\Omega} = \tilde{a}/b$, which is a monotonic increasing function of Ω. If the potential strength V_0 is zero, then $\tilde{\Omega} = \Omega$, as shown in Fig. 10.3.2a. If V_0 is not zero, then, as we shall see shortly, there is an interval $\Delta\Omega(p/q)$ about each rational value of $\Omega = p/q$ such that $\tilde{\Omega} = p/q$. Thus, the curve of $\tilde{\Omega}$ versus Ω will have intervals of zero slope of length $\Delta\Omega(p/q)$ centered about every rational value p/q of Ω, as depicted schematically in Figs. 10.3.2b and c. These finite intervals of lock-in to rational values of $\tilde{\Omega}$ exist because the transition from a rational value of $\tilde{\Omega}$ to an irrational value involves the creation of a discommensuration, which costs energy. As Ω increases away from a rational value, the strain energy in the commensurate phase with $\tilde{\Omega} = p/q$ increases. When this energy exceeds the energy required to create a discommensuration, a transition to an incommensurate phase with a nonzero discommensuration density occurs. This *commensurate-incommensurate* (CI) transition is second order in the FK model. The distance between discommensurations in the incommensurate phase is determined, as we shall see, by repulsive interactions between discommensurations, which for the FK model fall off exponentially with distance. The widths $\Delta\Omega(p/q)$ increase with increasing V_0. For V_0 less than a critical value V_c, the total step length $S = \sum \Delta\Omega(p/q)$ for Ω in some interval $I = [\Omega_0, \Omega_0 + S_0]$ is a fraction of the total length of the interval S_0. Thus, the fraction $1 - (S/S_0)$ of the interval I for which $\tilde{\Omega}$ is irrational is nonzero. In this case, the function of $\tilde{\Omega}(\Omega)$ is an *incomplete devil's staircase*, depicted schematically in Fig. 10.3.2b. For $V_0 > V_c$, S/S_0 is

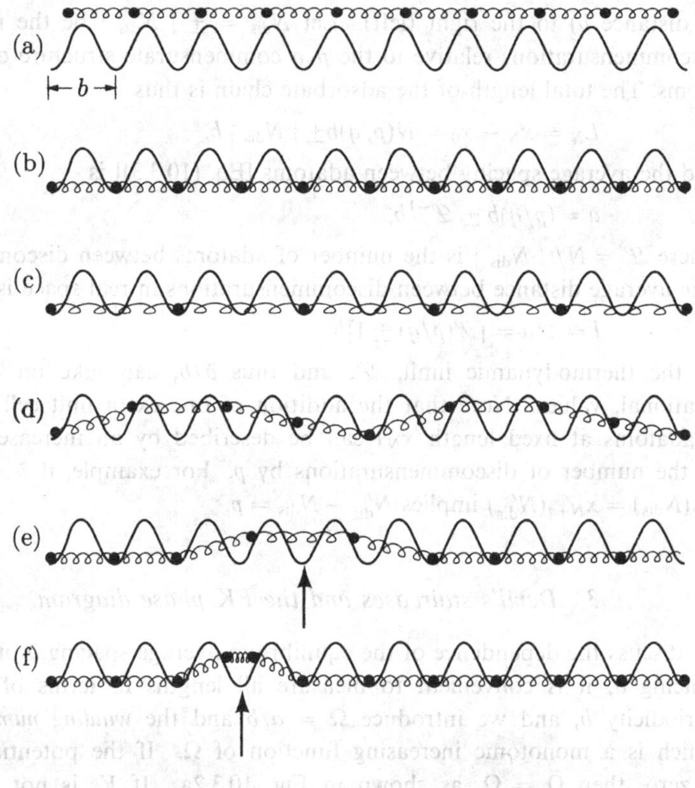

Fig. 10.3.1. (a) The floating phase of the FK model with $\tilde{a} = a$ with a an arbitrary multiple of b. (b) A commensurate structure with $\tilde{a} = b$. (c) A commensurate structure with $\tilde{a} = 2b$. (d) A commensurate structure with $\tilde{a} = (5/4)b$. There are periodically repeated unit cells of length $5b$ containing four adatoms which do not all sit at the bottom of a potential minimum. (e) A portion of an incommensurate structure with $\tilde{a}$ of order but greater than b. There is a regular array of positive discommensurations where springs are stretched relative to the substrate lattice spacing b separating regions where atoms sit in the minima of the potential V. In this figure, there is a single discommensuration marked with an arrow. (f) A portion of an incommensurate structure with negative discommensurations and $\tilde{a} < b$. A single discommensuration (marked with an arrow) is shown.

equal to one, and the interval for which $\tilde{\Omega}$ is irrational becomes a set of measure zero. In this case, the function $\tilde{\Omega}(\Omega)$ is a *complete devil's staircase* (Fig. 10.3.2c). It is a continuous function which has a zero derivative almost everywhere (i.e., except on a set of measure zero). Such functions are called *singular continuous* functions in the mathematical literature. The function $\tilde{\Omega}(\Omega)$ at the transition from an incomplete to a complete devil's staircase is of special interest. The intervals for which $\tilde{\Omega}$ is irrational form a Cantor-set with a non-integral fractal dimension analogous to the fractal dimension of polymers and diffuse objects discussed

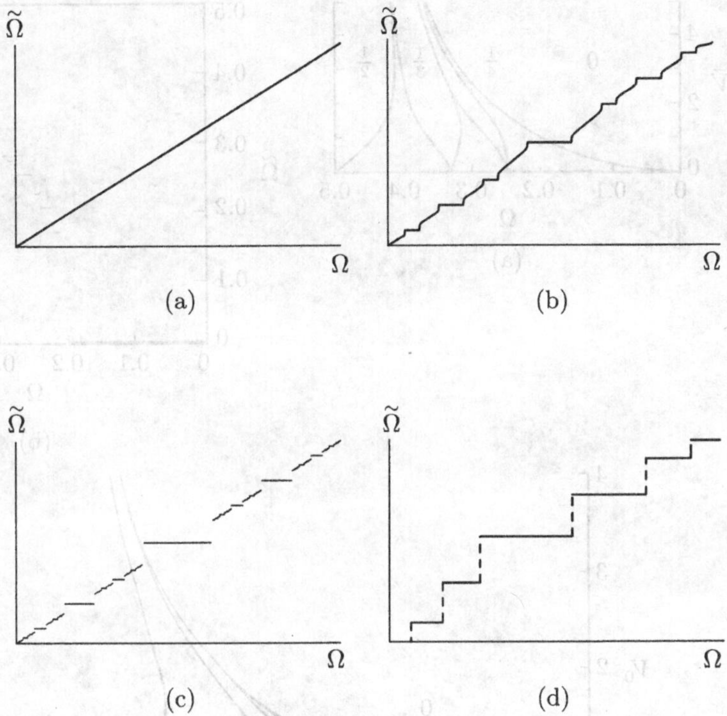

Fig. 10.3.2. The function $\tilde{\Omega}(\Omega)$ for (a) $V_0 = 0$, (b) an incomplete devil's staircase ($0 < V_0 < V_c$), (c) a complete devil's staircase ($V_0 > V_c$), and (d) a harmless staircase.

in Sec. 2.12. In some models, there can be discontinuous or first order jumps between commensurate states leading to a *harmless staircase* (Fig. 10.3.2d).

The numerically calculated phase diagram and the function $\tilde{\Omega}(\Omega)$ for the FK model are shown in Fig. 10.3.3. Also shown is the exact phase diagram for a model in which the cosine potential of the FK model is replaced by a periodic "sawtooth" function.

4 The continuum approximation

A complete determination of the ground states of Eq. (10.3.1) for arbitrary a and b is very complicated, and we will consider in detail only the CI transition near $p = q = 1$ that occurs when $a - b = \delta$ is small (i.e., $\delta / b << 1$) and when the potential is weak (i.e., when $V_0 << Kb^2$). In this limit, the atom index n can be treated as a continuous variable analogous to the indicator variable in the Lagrangian formulation of elasticity (see Sec. 6.5). Since we are interested in the

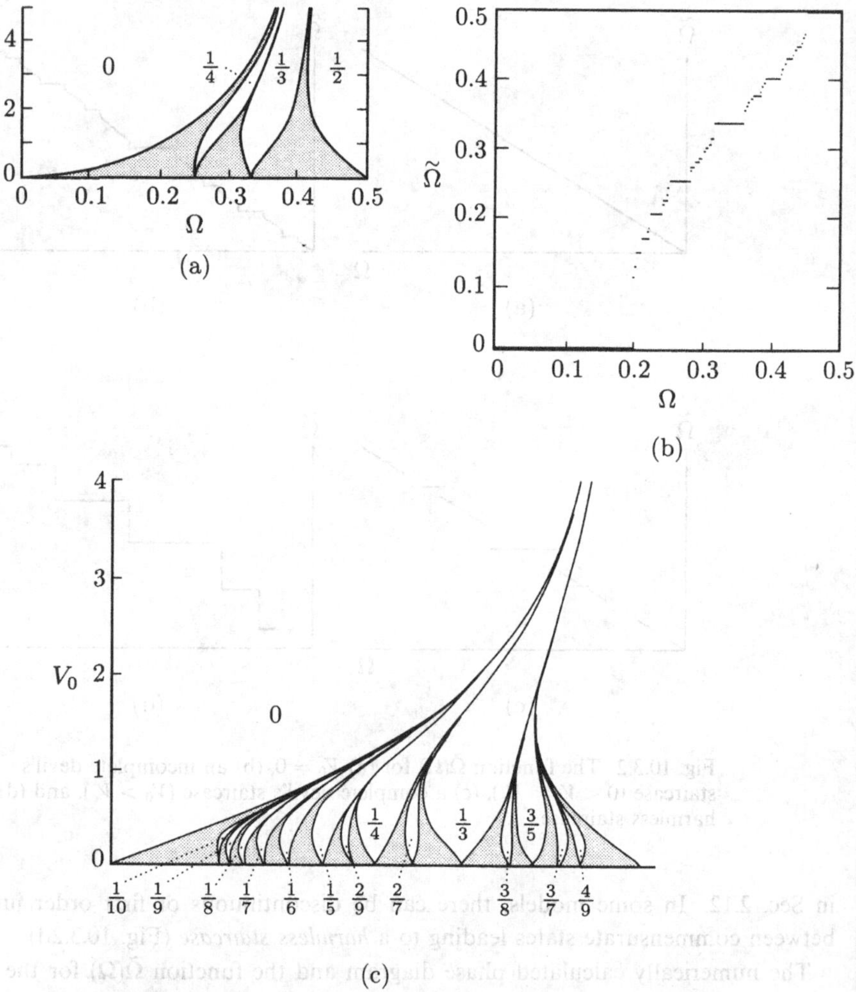

Fig. 10.3.3. (a) Numerically calculated phase diagram for the FK model as a function of the variables $\overline{V} = (2\pi/b)^2(V_0/K)$ and $\Omega = a/b$. The fractions 0, 1/4, 1/3 and 1/2 indicate the value of the winding number $\tilde{\Omega}$ in commensurate phases. The shaded tongues each break up into higher order lock-in regions, such as shown in (c). (b) Winding number $\tilde{\Omega}$ as a function of Ω for the FK model at $\overline{V} = 1$. [W. Chou and R.B. Griffiths, *Phys. Rev. B* **34**, 6219 (1986).] (c) Exact phase diagram for a modified model in which the cosine potential is replaced by $V(x) = (V_0/2)(x - b[(x/b) + (1/2)])$, where $[y]$ is the greatest integer less than or equal to y. This diagram shows many rational lock-in regions. The shaded regions, when expanded, show lock-in to higher order rationals. [S. Aubry, in *Solitons in Condensed Matter Physics*, eds. A.R. Bishop and T. Schneider (Springer-Verlag, New York, 1978).]

transition from the commensurate state with $x_n = nb$, we write x_n in terms of its deviation $\phi(n)$ from nb:

$$x_n = nb + \phi(n). \tag{10.3.7}$$

The potential energy of Eq. (10.3.1) can thus be written as

$$U = \int dn \left[\frac{1}{2}K(d\phi/dn - \delta)^2 + V[\phi(n)]\right]. \tag{10.3.8}$$

Note that, because $V(x)$ is periodic with period b, the potential term is a function of ϕ only. We have taken the atom-index n to be a continuous variable in the above energy. We might alternatively have chosen, in the spirit of Lagrangian elasticity, to index the adsorbate atoms by their position na in their unstretched state. In this formulation, Eq. (10.3.8) would have more the appearance of an elastic medium coupled to an external potential. We will continue to use n rather than na to emphasize that n really is an indexing rather than a distance variable.

The existence of a CI transition as a function of the mismatch δ can be demonstrated by a crude estimate of the energies of the two phases. In the commensurate phase, $\phi = 0$, and the energy U is simply

$$U_{\text{comm}} = N\left[\frac{1}{2}K\delta^2 + V_{\min}\right], \tag{10.3.9}$$

where N is the number of *adsorbate* atoms and $V_{\min}$ is the value of V at its minima ($-V_0$ for the potential of Eq. (10.3.2)). In the incommensurate phase, $d\phi/dn$ is nonzero, and we may estimate it to be equal to δ to minimize the elastic contribution to U. This corresponds to setting $\tilde{a} = a$. If δ is an irrational multiple of b, all values of the argument of V are probed in the limit $N \to \infty$ so that

$$U_{\text{inc}} = \frac{N}{b}\int_0^b V(\phi)du = N\overline{V}, \tag{10.3.10}$$

where $\overline{V}$ is the average of $V(\phi)$ over the interval b. Comparing U_{comm} with U_{inc}, we conclude that there is a first-order transition from the commensurate phase with $\tilde{a} = b$ to the incommensurate phase with $\tilde{a} = a$ when δ equals

$$\delta_c = [(\overline{V} - V_{\min})/K]^{1/2}. \tag{10.3.11}$$

This is clearly only an estimate of δ_c since it compares the commensurate state with an incommensurate state with $x_n = na$. One would expect incommensurate states with discommensurations and an average spacing $\tilde{a}$ less than a to be energetically preferred.

Before proceeding to an analytic solution for the CI transition, it is useful to measure energy relative to the energy of the commensurate state by introducing

$$\begin{aligned}\Delta U = U - U_{\text{comm}} &= \int dn\left[\frac{1}{2}K(d\phi/dn)^2 + \tilde{V}(\phi)\right] - \delta K\int dn(d\phi/dn) \\ &\equiv U' + U_1,\end{aligned} \tag{10.3.12}$$

where $\tilde{V} = V - V_{\min}$, which for the cosine potential of Eq. (10.3.2) becomes

$$\tilde{V}(\phi) = V_0[1 - \cos(2\pi\phi/b)] = 2V_0 \sin^2(\pi\phi/b). \tag{10.3.13}$$

We have divided ΔU in Eq. (10.3.12) into two parts: U', which is harmonic in $(d\phi/dn)$ and is clearly minimized when $d\phi/dn = 0$, and U_1, which is linear in $d\phi/dn$ and depends only on $\phi(n)$ at the boundaries. The latter fact can be seen via

$$U_1 = -\delta K \int_0^N dn(d\phi/dn) = -\delta K[\phi(N) - \phi(0)] \equiv -K\delta N\langle d\phi/dn\rangle, \tag{10.3.14}$$

where

$$\langle d\phi/dn\rangle = \lim_{N\to\infty} [\phi(N) - \phi(0)]/N \tag{10.3.15}$$

is the average rate of growth of ϕ with n. Thus, the final equilibrium phase is determined via the competition between U', which favors a ϕ independent of n, and U_1, which favors nonzero $\langle d\phi/dn\rangle$.

The range in n over which changes in ϕ take place in going from one minimum of the potential to another is determined by the ratio of the elastic constant K to the coefficient of ϕ^2 in the expansion of $\tilde{V}$ around $\phi = 0$, i.e., by

$$\lambda = \left(\frac{1}{K}\frac{d^2\tilde{V}}{d\phi^2}\right)^{-1/2} = \left(\frac{V_0}{K}\right)^{-1/2} \frac{b}{2\pi}, \tag{10.3.16}$$

where the final form applies to the cosine potential. The unitless parameter λ is the analog of a correlation length. Indeed, if we had chosen to index particles by their position na in the unstretched adsorbate lattice, then $a\lambda$ would be precisely a correlation length.

The Euler-Lagrange equations satisfied by u at extrema of ΔU are determined entirely by U' and are not influenced by U_1. These equations are identical in form to Eq. (10.2.6) and can be integrated as in Eq. (10.2.7) to yield

$$\frac{1}{2}K(d\phi/dn)^2 = \tilde{V}(\phi) + \epsilon, \tag{10.3.17}$$

where ϵ is an integration constant with units of energy. In our discussion of domain walls in the preceding section, the integration constant was determined by the boundary conditions that the order parameter be in one minimum of the free energy at $z = -\infty$ and another at $z = +\infty$ with a single kink. Here ϵ defines a continuous family of extremal solutions to U'. The particular solution that minimizes ΔU is determined by minimizing $U' + U_1$ with respect to ϵ. The integration constant ϵ is thus determined by energy minimization rather than by externally imposed boundary conditions.

5 Nature of solutions

Before determining the actual equilibrium solution for a given δ, let us first study the nature of the extremal solutions, Eq. (10.3.17), as a function of the integration constant ϵ. Eq. (10.3.17) can be integrated to yield an implicit equation for $\phi(n)$ as a function of n:

$$J[\phi(n)] \equiv \pm \int_0^{\phi(n)} \frac{d\phi}{g(\epsilon,\phi)} = n - \tilde{u}, \qquad (10.3.18)$$

where

$$g(\epsilon,\phi) = [2(\epsilon + \tilde{V})/K]^{1/2} \qquad (10.3.19)$$

and where the constant of integration $\tilde{u}$ is chosen so that $\phi(\tilde{u}) = 0$. The positive sign in Eq. (10.3.18) corresponds to $\langle d\phi/dn \rangle > 0$ and the negative sign to $\langle d\phi/dn \rangle < 0$. The function $J(\phi)$ is a monotonic single-valued function of ϕ, so that Eq. (10.3.18) can be inverted to yield $\phi(n) = J^{-1}(n - \tilde{u})$. $J(\phi)$ can be expressed after an appropriate change of variables in terms of an elliptic integral of the first kind and the inverse function as the associated amplitude function *am* (see, for example, Abramowitz and Stegun 1965, esp. Chapter 17). The precise form of these relations is not important for our present discussion.

The periodicity of $\tilde{V}$ in the above integral implies that $\phi(n)$ increases by a substrate lattice spacing at regular intervals in n. To see this, we note

$$\begin{aligned} J[\phi(n)+b] &= \pm \int_0^{\phi(n)+b} \frac{d\phi}{g(\epsilon,\phi)} = J[\phi(n)] \pm \mathcal{L} \\ &= n - \tilde{u} \pm \mathcal{L}, \end{aligned} \qquad (10.3.20)$$

where

$$\mathcal{L} = \int_0^b \frac{d\phi}{g(\epsilon,\phi)}. \qquad (10.3.21)$$

From Eq. (10.3.18), we have $J[\phi(n \pm \mathcal{L})] = n - \tilde{u} \pm \mathcal{L}$. Thus, if we apply the inverse function J^{-1} to both sides of Eq. (10.3.20), we obtain

$$\phi(n \pm \mathcal{L}) = \phi(n) + b. \qquad (10.3.22)$$

Therefore, $\phi(n)$ has the form depicted in Fig. 10.3.4 with discommensurations (increasing or decreasing ϕ by b) occurring at intervals of $\mathcal{L}$ separating regions, where $\phi(n)$ is a multiple of b. $\mathcal{L}$ specifies the number of adsorbate atoms between discommensurations. The distance in real space between discommensurations is $\mathcal{L}\tilde{a}$ (not $\mathcal{L}a$). We will, therefore, refer to $\mathcal{L}$ as the distance or separation between discommensurations with the understanding that it must be multiplied by the average adatom spacing $\tilde{a}$ to be a physical distance. $\mathcal{L}$ is a real quantity and can exist according to Eq. (10.3.21) only for $\epsilon > 0$. As $\epsilon \to 0$, $\mathcal{L}$ diverges logarithmically with ϵ^{-1}, as can be seen by

$$\begin{aligned} \mathcal{L} &= 2\int_0^{b/2} \frac{d\phi}{g(\epsilon,\phi)} \\ &\approx \frac{2}{\sqrt{2}} \int_0^{b/2} \frac{d\phi}{[(\epsilon/K) + \lambda^{-2}\phi^2/2]^{1/2}} + \text{const.} \\ &\approx \lambda \ln(\alpha K b^2/\epsilon\lambda^2), \end{aligned} \qquad (10.3.23)$$

where α is a constant that depends on the potential. For the cosine potential, $\alpha = 8/\pi^2$.

The average rate of increase of ϕ is determined by Eqs. (10.3.15) and (10.3.22) to be

$$\langle d\phi/dn \rangle = \pm b/\mathscr{L}. \tag{10.3.24}$$

The average separation between adsorbate atoms is then

$$\tilde{a} = b + \langle d\phi/dn \rangle = (1 \pm \mathscr{L}^{-1})b, \tag{10.3.25}$$

as can be seen with the aid of Eqs. (10.3.3), (10.3.7) and (10.3.24). Note that Eq. (10.3.25) is identical to Eq. (10.3.5) with $p = q = 1$. Eq. (10.3.24) shows that $\langle d\phi/dn \rangle$ can be positive or negative, depending on whether there are positive or negative discommensurations relative to the commensurate state. The average rate of increase (decrease) in ϕ is the increase (decrease) b per discommensuration divided by the distance $\mathscr{L}$ between discommensurations. As in Eq. (10.3.6), the average distance between discommensurations is $l = \mathscr{L}\tilde{a} = (\mathscr{L} \pm 1)b$.

The average linear increase in $\phi(n)$ can be explicitly displayed so that $\phi(n)$ can be expressed as the sum of a linear function of n plus a periodic function $\psi(n-\tilde{u})$ with period $\mathscr{L}$ (as depicted in Fig. 10.3.4b):

$$\phi(n) = (n - \tilde{u})(b/\mathscr{L}) + \psi(n - \tilde{u}), \tag{10.3.26}$$

where the positive sign in Eq. (10.3.24) was chosen. Finally, we can use this equation and Eqs. (10.3.4), (10.3.7), and (10.3.25) to express the actual positions of the adsorbate atoms as

$$x_n = n\tilde{a} - u + \tilde{\psi}(n\tilde{a} - \overline{u}), \tag{10.3.27}$$

where $\tilde{\psi}(x) \equiv \psi(x/\tilde{a})$ is a periodic function with period $l = \mathscr{L}\tilde{a}$ and

$$\overline{u} = \tilde{u}\tilde{a} \quad \text{and} \quad u = \tilde{u}(b/\mathscr{L}) = \overline{u}(b/l). \tag{10.3.28}$$

Eq. (10.3.27) makes it clear that two spatial periods are needed to describe the density of adatoms: the average distance $\tilde{a}$ between atoms and the average spacing $l = \mathscr{L}\tilde{a}$ between discommensurations. The variable $\overline{u}$ describes spatial displacements of discommensurations whereas the variable u describes displacements of adatoms. This relation between u and $\overline{u}$ is valid for arbitrary l.

6 The minimum energy solution

As discussed earlier, the solution $\phi(n)$ that minimizes ΔU is determined by inserting the solution Eq. (10.3.18) into Eq. (10.3.12) for ΔU and then minimizing over the integration constant ϵ. Evaluating ΔU, we obtain

$$\frac{\Delta U}{N} = \frac{1}{\mathscr{L}} \left[\int_0^b d\phi [2K(\epsilon + \tilde{V})]^{1/2} - \epsilon\mathscr{L} \mp Kb\delta \right]. \tag{10.3.29}$$

Note that this energy is independent of $\tilde{u}$. Differentiating this equation with respect to ϵ, we find

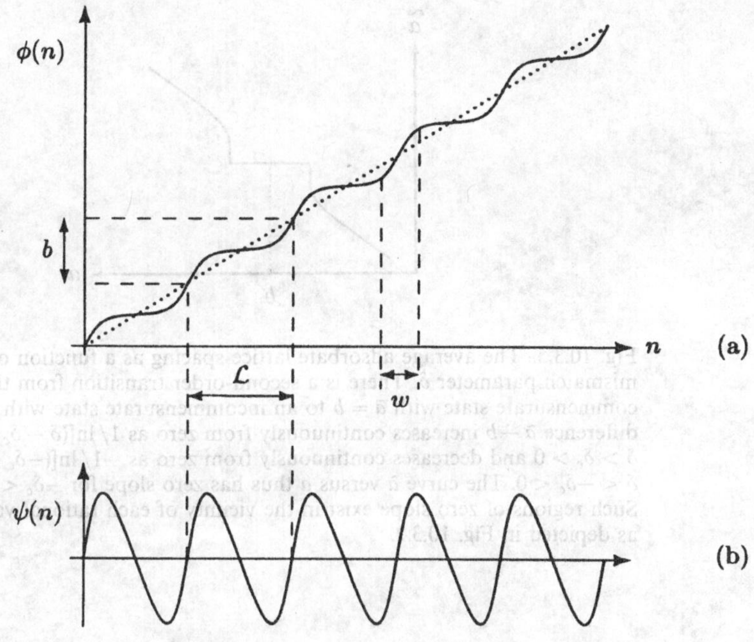

Fig. 10.3.4. (a) The function $\phi(n)$. It consists of a regular periodic array of flat regions separated by discommensurations, where there is rapid change. The width w of the discommensuration is of order λ for ϵ near zero but decreases as ϵ increases. The average slope of $\phi(n)$ is $b/\mathscr{L}$. When the average linear increase with n is subtracted off, there remains a periodic function $\psi(n-\tilde{u})$ with period $\mathscr{L}$ as shown in (b).

$$\frac{\partial}{\partial \epsilon}\left(\frac{\Delta U}{N}\right) = -\frac{1}{\mathscr{L}^2}\left(\int_0^b du[2K(\epsilon+\tilde{V})]^{1/2} \mp Kb\delta\right)\frac{d\mathscr{L}}{d\epsilon}. \tag{10.3.30}$$

This expression is zero if $\mathscr{L}^{-1}=0$ or if

$$\delta = \pm b^{-1}\int_0^b d\phi[2(\epsilon+\tilde{V})/K]^{1/2}. \tag{10.3.31}$$

The first solution corresponds to an infinite separation between discommensurations, and thus to the commensurate phase. This solution with zero ΔU always exists. The second solution only exists if $\epsilon > 0$, and corresponds to the incommensurate phase with a finite separation between discommensurations.

The critical value of δ for which an incommensurate solution first exists is determined by Eq. (10.3.31) with $\epsilon = 0$. Solutions exist for the positive sign of this equation if $\delta > \delta_c$ and for the negative sign if $\delta < -\delta_c$, where

$$\delta_c = b^{-1}\int_0^b d\phi[2\tilde{V}/K]^{1/2} = e_0/bK, \tag{10.3.32}$$

where e_0 is the energy of an isolated discommensuration. For the cosine potential, $e_0 = (4b/\pi)(KV_0)^{1/2}$. For $\delta > \delta_c$, the relation between δ and ϵ is

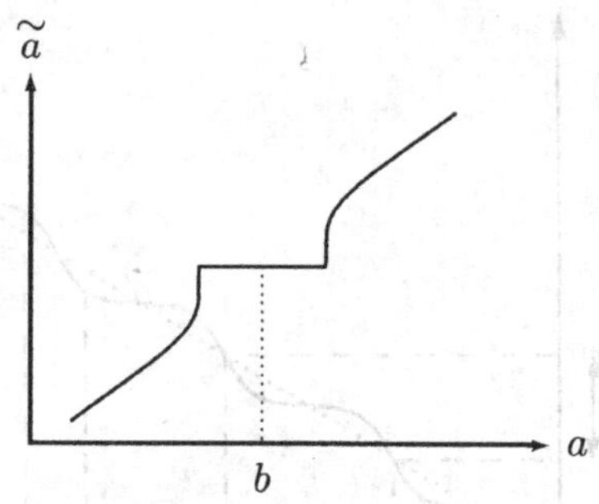

Fig. 10.3.5. The average adsorbate lattice spacing as a function of the mismatch parameter δ. There is a second-order transition from the commensurate state with $\tilde{a} = b$ to an incommensurate state with $\tilde{a} > b$. The difference $\tilde{a} - b$ increases continuously from zero as $1/\ln[(\delta - \delta_c)^{-1}]$ for $\delta > \delta_c > 0$ and decreases continuously from zero as $-1/\ln[(-\delta_c - \delta)^{-1}]$ for $\delta < -\delta_c < 0$. The curve $\tilde{a}$ versus a thus has zero slope for $-\delta_c < \delta < \delta_c$. Such regions of zero slope exist in the vicinity of each rational value of a/b, as depicted in Fig. 10.3.2.

$$| \delta | - \delta_c = b^{-1} \int_0^b d\phi \left[[2(\epsilon + \tilde{V})/K]^{1/2} - (2\tilde{V}/K)^{1/2} \right]. \tag{10.3.33}$$

This equation determines ϵ as a function of δ and can be used in Eq. (10.3.29) to determine the minimum energy as a function of δ. Again, since this solution only exists for $\epsilon > 0$, $| \delta |$ must be greater than δ_c. $| \delta | - \delta_c$ clearly goes to zero as $\epsilon \to 0$. A clean way to determine how $| \delta | - \delta_c$ goes to zero with ϵ is to differentiate Eq. (10.3.33) with respect to ϵ:

$$d(| \delta | - \delta_c)/d\epsilon = b^{-1} K^{-1} \mathscr{L}(\epsilon) \approx (\lambda/bK) \ln(\alpha K b^2/\epsilon\lambda^2). \tag{10.3.34}$$

This equation can be integrated to yield

$$\begin{aligned} | \delta | - \delta_c &= \frac{1}{bK} \int_0^\epsilon \mathscr{L}(\epsilon) \approx \frac{\lambda}{bK} \left[-\epsilon\left(\ln(\alpha K b^2/\epsilon\lambda^2) + 1\right)\right] \\ &\approx \frac{\alpha b}{\lambda} \frac{\mathscr{L}}{\lambda} e^{-\mathscr{L}/\lambda}, \end{aligned} \tag{10.3.35}$$

implying

$$\mathscr{L} \approx \lambda \ln \left[\frac{\alpha b}{\lambda} \frac{1}{| \delta | - \delta_c} \right]. \tag{10.3.36}$$

Thus, the separation between discommensurations diverges logarithmically with $| \delta | - \delta_c$, and the difference between the lattice constant $\tilde{a}$ of the adsorbate lattice and b goes to zero as $1/\ln[(| \delta | - \delta_c)]^{-1}$, as depicted in Fig. 10.3.5. There is, therefore, a second-order CI transition. It should be emphasized, however, that this model is purely mechanical: it has no thermal fluctuations.

7 Repulsive interaction between discommensurations

The energy difference ΔU can be expressed as a function of ϵ, as we have just seen. Alternatively, it can be expressed in terms of the spatial distance $l = \mathscr{L}\tilde{a}$ between discommensurations using Eq. (10.3.35). The latter form is quite instructive and lends itself most readily to generalizations to more complex and higher dimensional problems. The energy ΔU in Eq. (10.3.29) can be expanded near $\epsilon = 0$ with the aid of Eq. (10.3.32) for e_0 and Eq. (10.3.35) for $\int d\epsilon \mathscr{L}(\epsilon)$. The result is

$$\frac{\Delta U}{N\tilde{a}} = \frac{[(e_0 \mp Kb\delta) + \epsilon\lambda]}{\mathscr{L}\tilde{a}} = \frac{\sigma(\delta) + U_0 e^{-l/\lambda\tilde{a}}}{l}, \tag{10.3.37}$$

where $U_0 = \alpha b^2 K \lambda^{-1}$ and

$$\sigma(\delta) = Kb(\delta_c - |\,\delta\,|) \tag{10.3.38}$$

is an effective surface tension for a discommensuration. The interpretation of this formula is as follows: The energy e_0 of an isolated positive or negative discommensuration is positive and contributes an energy $e_0/\mathscr{L}$ to the energy per particle. Discommensurations increase (decrease) $\langle d\phi/dn \rangle$ and give rise to a decrease in the contribution to $\Delta U/N$ coming from U_1 [Eq. (10.3.14)]. The incommensurate phase first becomes favorable relative to the commensurate phase when the positive energy of creating a single discommensuration is just compensated for by the energy gain associated with a nonzero $\langle d\phi/dn \rangle$. When the effective surface tension $\sigma(\delta)$ is positive, discommensurations are energetically unfavorable; when it is negative, they are energetically favorable and tend to proliferate. Neighboring discommensurations repel each other, however, with a potential that dies off exponentially with spatial separation l. The exponential die-off of this potential is a result of the fact that $d\phi/dn$ tends to zero as $\exp(-n/\lambda) = \exp(-l/\lambda\tilde{a})$ away from the center of an isolated discommensuration. The interaction between neighboring discommensurations comes predominantly from the elastic part of ΔU and is approximately $K(d\phi(0)/dn)(d\phi(\mathscr{L})/dn)/2$. The equilibrium separation between discommensurations, and thus the value of $\tilde{a}$, is, therefore, determined by the competition between a negative effective surface tension for an individual discommensuration and the repulsive interaction between neighboring discommensurations.

8 X-ray diffraction

The Fourier transform of the density is easily obtained from the positions given in Eq. (10.3.27):

$$\rho_k = \sum_n e^{ik[n\tilde{a} - u + \tilde{\psi}(n\tilde{a} - \bar{u})]}. \tag{10.3.39}$$

Because $\tilde{\psi}(x)$ has period $\mathscr{L}\tilde{a}$, this function will have delta-function spikes whenever k is an arbitrary linear combination with integer coefficients of $b_1 = 2\pi/\tilde{a}$ and $b_2' = 2\pi/\mathscr{L}\tilde{a}$. Alternatively, the positions of these spikes can be expressed in terms of b_1 and

$$b_2 = b_1 + b_2' = \frac{2\pi}{\tilde{a}}(1 + \mathscr{L}^{-1}) = \frac{2\pi}{b}, \tag{10.3.40}$$

where Eq. (10.3.25) was used to obtain the final form of this expression. Thus, the adsorbate has an incommensurate reciprocal lattice with vectors of the form of Eq. (2.9.2) with primitive translation vectors b_1 and b_2 determined by the average spacing $\tilde{a}$ of the adsorbate direct lattice and the spacing b of the substrate lattice. The amplitude of the mass-density wave at reciprocal lattice vector $G = pb_1 + qb_2$ is easily calculated from Eq. (10.3.39):

$$\rho_G = \frac{N}{\mathscr{L}a} e^{-ipb_1 u} \int_0^{\mathscr{L}\tilde{a}} dx e^{iq(b_2 - b_1)x} e^{iG\tilde{\psi}(x)}. \tag{10.3.41}$$

This expression gives the scattering intensity from the adsorbate atoms only. Scatterers also couple to the substrate, so that, in any real experiment, there will be additional intensity in peaks at reciprocal lattice vectors qb_2 of the substrate. Note that the arbitrary phase u appears only as a factor of pb_1, a reciprocal lattice vector determined by the average adsorbate lattice; it does not appear as a factor of the vectors qb_2 determined by the substrate lattice. In general, one would have expected two phases in an incommensurate system such as this. The substrate lattice is, however, frozen in this model, and there is only a phase associated with translating the adsorbate relative to the substrate. Note that the phase of the mass-density wave that would enter into an Eulerian description of the elasticity, such as discussed in Secs. 6.4 and 6.6, of the incommensurate phase is a multiple $b/\mathscr{L}$ of the discommensuration index $\tilde{u}$.

9 Compressional elastic constants

In equilibrium, for a given value of mismatch, there is a preferred separation between discommensurations in the incommensurate phase. The energy associated with deviations $\delta\mathscr{L} = \mathscr{L}' - \mathscr{L}$ is determined by a compressional elastic constant:

$$\begin{aligned} \tilde{B} &= \mathscr{L}^2 \frac{\partial^2(\Delta U/N)}{\partial \mathscr{L}^2} = \frac{\mathscr{L}}{d\mathscr{L}/d\epsilon} = K \frac{\int_0^b d\phi g^{-1}}{\int_0^b d\phi g^{-3}} \\ &\approx \frac{\mathscr{L}}{\lambda} \frac{\alpha K b^2}{\lambda^2} e^{-\mathscr{L}/\lambda} \approx \frac{Kb(\delta - \delta_c)}{\lambda}. \end{aligned} \tag{10.3.42}$$

To obtain this result, we used Eqs. (10.3.29) and (10.3.30) with $\Delta U(\mathscr{L}) = \Delta U(\epsilon(\mathscr{L}))$ and $\partial^2 \Delta U/\partial \mathscr{L}^2 = (\partial \mathscr{L}/\partial \epsilon)^{-1}[(\partial \mathscr{L}/\partial \epsilon)^{-1} \partial \Delta U/\partial \epsilon]$ evaluated in equilibrium determined by Eq. (10.3.31). Changes in $\mathscr{L}$ can be related to gradients in the discommensuration phase variable $\tilde{u}$ using Eqs. (10.3.26) and (10.3.27):

$$\frac{d\tilde{u}(n)}{dn} = \frac{(\delta\mathscr{L}/\mathscr{L})}{1 + (\delta\mathscr{L}/\mathscr{L})} \approx \frac{\delta\mathscr{L}}{\mathscr{L}}. \tag{10.3.43}$$

The energy of a compressed discommensuration lattice in analogy with the Lagrangian elasticity theory of Chapter 6 can, thus, be written

$$U_{\mathrm{el}} = \frac{1}{2}\int dn\tilde{B}(d\tilde{u}/dn)^2. \tag{10.3.44}$$

As discussed in Eqs. (10.3.39) and (10.3.40), the regular discommensuration lattice gives rise to an incommensurate structure with a reciprocal lattice generated by the vectors $b_1 = 2\pi/\tilde{a}$ and $b_2 = 2\pi/b$. From Eq. (10.3.28), the phase of the mass-density waves generated by the average adsorbate lattice is related to $\tilde{u}$ via $u = \tilde{u}b/\mathscr{L}$. For long-wavelength disturbances, the position in real space of the nth atom is $x_n = n\tilde{a}$, from Eq. (10.3.3), implying $dn = dx/\tilde{a}$. The Eulerian elastic energy of the incommensurate structure with frozen substrate lattice is, therefore,

$$U_{\mathrm{el}} = \frac{1}{2}\int dxB(du/dx)^2, \tag{10.3.45}$$

where

$$B = \tilde{a}(\mathscr{L}/b)^2\tilde{B}. \tag{10.3.46}$$

Note that B tends to zero exponentially with separation between discommensurations, as would be expected from the exponentially decaying interaction between discommensurations; apart from logarithmic corrections, however, it is linear in $|\delta| - \delta_c$.

10 Phasons

There are two reasonable models for the dynamics of the FK model. In the first, dissipative couplings between the adsorbate and the substrate are ignored; in the second, they are not. The energy of an incommensurate phase described by an elastic energy of the form of Eq. (10.3.44) is invariant with respect to arbitrary uniform increments of the displacement variable u, and one expects a zero-frequency Goldstone mode (Sec. 8.1) at wave number $q = 0$ in either model. In commensurate phases, there is no invariance, and modes should either have a gap or be overdamped at $q = 0$. We consider the non-dissipative case first. In this case, the Hamiltonian consists of a kinetic energy and the potential energy of Eq. (10.3.1):

$$\mathscr{H} = \frac{1}{2}\sum_n mv_n^2 + U, \tag{10.3.47}$$

where $v_n = \dot{x}_n \equiv p_n/m$ is the velocity and p_n is the momentum of the nth adsorbate atom of mass m.

In the commensurate phase with $\tilde{a} = (p/q)b$, each adsorbate atom sits in a minimum of the cosine potential, and a harmonic expansion in displacements from the ground state is appropriate:

$$\mathscr{H} = \frac{1}{2}\sum_n m(d\phi_n/dt)^2 + \frac{1}{2}\sum_n \tilde{V}''(0)\phi_n^2 + \frac{1}{2}K\sum_n(\phi_{n+1} - \phi_n - a)^2, \tag{10.3.48}$$

where $\tilde{V}''(0) = V_0(2\pi/b)^2$ for the cosine potential. The frequency spectrum of this harmonic Hamiltonian is

$$\omega^2(q) = \frac{1}{m}[\tilde{V}''(0) + Kq^2]. \tag{10.3.49}$$

There is a gap in the spectrum at $q = 0$ in the $\tilde{a} = a$ commensurate phase. A similar gap also exists for arbitrary p/q, though it is technically more difficult to calculate when $p > 1$ because there are, in general, p branches to the spectrum.

In the incommensurate phase, there should be a gapless Goldstone mode whenever the elastic energy of Eq. (10.3.44) provides a correct description of the long-wavelength statics. To determine the frequency of this mode, we express the kinetic energy of Eq. (10.3.47) in terms of $d\tilde{u}(n)/dt$ rather than v_n with the aid of Eq. (10.3.18):

$$\frac{dJ}{d\phi}\dot{\phi}_n = \pm\frac{\dot{\phi}_n}{g(\epsilon, \phi_n)} = -\frac{d\tilde{u}(n)}{dt}, \tag{10.3.50}$$

where $g(\epsilon, \phi)$ is defined in Eq. (10.3.19). Then, in the continuum limit, the Hamiltonian can be expressed as

$$\mathscr{H} = \frac{1}{2}m\int dn\, g^2(\epsilon, \phi_n)(d\tilde{u}(n)/dt)^2 + U. \tag{10.3.51}$$

The function $g(\epsilon, \phi_n)$ is a periodic function of n with period $\mathscr{L}$. To describe excitations with wavelength long compared to $\mathscr{L}$, we can replace g^2 by its average over $\mathscr{L}$ and U by the elastic energy of Eq. (10.3.44):

$$\mathscr{H} = \frac{1}{2}\int dn m_{\text{eff}}\,[d\tilde{u}(n)/dt]^2 + U_{\text{el}}, \tag{10.3.52}$$

where

$$m_{\text{eff}} = \frac{m}{\mathscr{L}}\int_0^{\mathscr{L}} dng^2(\epsilon, \phi_n) = \frac{m}{\mathscr{L}}\int_0^b d\phi g(\epsilon, \phi). \tag{10.3.53}$$

The Hamiltonian of Eq. (10.3.52) is essentially identical to the elastic Hamiltonian of Sec. 7.3. The equation of motion for $\tilde{u}(n)$ is

$$m_{\text{eff}}\frac{d^2\tilde{u}(n)}{dt^2} = -\tilde{B}\frac{d^2\tilde{u}(n)}{dn^2}. \tag{10.3.54}$$

There is a thus a sound-like mode with $\tilde{u}(n) \sim \exp(iqn)$ with velocity

$$c = \left(\frac{\tilde{B}}{m_{\text{eff}}}\right)^{1/2} = \frac{(K/m)^{1/2}\int_0^b d\phi g^{-1}}{\left[\left(\int_0^b d\phi g^{-3}\right)\left(\int_0^b d\phi g\right)\right]^{1/2}}$$
$$\sim ((|\,\delta\,| - \delta_c)\ln[\delta_c/(|\,\delta\,| - \delta_c)])^{1/2}, \tag{10.3.55}$$

where we used Eq. (10.3.42) for $\tilde{B}$ and Eq. (10.3.21) for $\mathscr{L}$. This gapless mode is called a *phason*. Phasons exist in incommensurate systems because there is an invariance associated with the uniform relative translation of the phases of two mass-density waves with relatively irrational periodicities (in this case those of the adsorbate and the substrate). Note that the phason velocity tends to zero as the commensurate phase is approached.

If dissipative coupling between the adsorbate and the substrate is allowed, a damping term proportional to $(d\phi_n/dt)$ (or to $(d\tilde{u}(n)/dt)$) must be added to the equation of motion. At low frequencies, this term always dominates the inertial term, leading to a long-wavelength diffusive phason with frequency

$$\omega = -i(\tilde{B}/\gamma_{\text{eff}})q^2, \tag{10.3.56}$$

where γ_{eff} is a friction coefficient. Because friction is essentially always present, phasons in incommensurate systems are generally diffusive at long wavelengths provided the phason spectrum is gapless.

11 Pinned phasons

The energy of any incommensurate state is invariant with respect to n-independent increments of $\tilde{u}(n)$. The existence of the gapless phason mode, however, is intimately connected to the existence of an analytic elastic energy of the form of Eq. (10.3.44), which in turn depends on the analyticity of the function $\tilde{\psi}$ [Eq. (10.3.27)] relating x_n to $\tilde{u}_n$. When the potential strength V_0 is sufficiently small, $\tilde{\psi}$ is analytic, and it can be shown that x_n mod b takes on *all* values between 0 and b. This means, in particular, that there will always be at least one atom sitting at the maximum of the periodic potential (x_n mod $b = b/2$), as shown in Fig. 10.3.6a. If the adsorbate is translated relative to the substrate, the atom (or atoms) at the top of potential maximum will move downward into a position of lower potential energy. In the process of translation, however, at least one other atom will have moved to the maximum of the potential. Thus, an incommensurate state can be transformed into an energetically equivalent translated state without passing over any energy barrier, and the gradient expansion of the elastic energy is justified.

As V_0 increases, the tendency of atoms to seek potential minima increases. Also, as we have already discussed, the measure of incommensurate states in any given interval of a decreases. Aubry (1978) has shown that, above the critical strength V_c at which the devil's staircase first becomes complete, $\tilde{\psi}$ becomes a nonanalytic function in the incommensurate states, which now occupy a set of measure zero as a function of a. When $\tilde{\psi}$ ceases to be analytic, x_n mod b ceases to take on all values between 0 and b. There are no atoms sitting at the maximum of the potential wells, as shown in Fig. 10.3.6b. In any given ground state configuration, there will be an atom that is closest to the maximum of the potential. Let e_p be the potential energy required to move this atom to the top of its well. In order to reach another ground state described by a spatially uniform translation of $\tilde{u}_n$, it is necessary to move this atom over the top of its well. Thus, there should be a gap in the phason spectrum of order e_p. Extensive numerical work (Aubry 1978; Peyrard and Aubry 1983; de Seze and Aubry 1984) confirms this hypothesis.

(a)

(b)

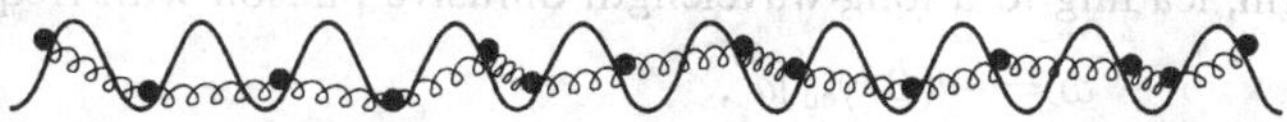

Fig. 10.3.6. (a) Representation of an incommensurate state with unpinned phasons. There is at least one atom at the top of a potential well. Under an infinitesimal translation, this atom will move to a position of lower potential energy, but another atom will move to the top of some other well. (b) Representation of an incommensurate state with pinned phasons. There is no atom at the top of the well. It is necessary to raise at least one atom over an energy barrier in order to reach an energetically equivalent incommensurate state.

12 Extension to two dimensions

The FK model can readily be extended to higher dimensions. The nature of ground state structures depends critically on the rotational symmetry of the substrate. If the substrate has uniaxial symmetry (e.g. a rectangular lattice) with basis vectors $\mathbf{a}_1$ and $\mathbf{a}_2$, then it is in general easier for stretching or compression of the adatom lattice to occur along one direction, say along $\mathbf{a}_1$. In this case, ground state configurations consist of one-dimensional solutions along $\mathbf{a}_1$ repeated along $\mathbf{a}_2$. Thus, for example, there can be a $p \times 1$ commensurate structure in which there is one adatom per substrate unit cell along $\mathbf{a}_2$ but only one adatom per p substrate unit cells along $\mathbf{a}_1$. Point discommensurations in the one-dimensional model become walls parallel to the single *easy direction* $\mathbf{a}_2$, and incommensurate phases are *striped phases*, such as in Fig. 10.3.7a, consisting of a regular array of parallel walls.

If the substrate has square or hexagonal symmetry, there are, respectively, two or three, rather than one, easy directions for walls, and incommensurate states, such as the hexagonal state shown in Fig. 10.3.7b, in which walls cross are possible (Bak *et al.* 1979). Configurations of crossed light and heavy walls relative to a $\sqrt{3} \times \sqrt{3}R30°$ commensurate state of krypton on graphite are shown in Fig. 10.3.8. Each wall crossing costs an energy e_c, which can in principle be calculated from atomic interactions. If e_c is positive, the number of wall crossings will be as small as possible in equilibrium. Thus, in this case, incommensurate ground states will be striped phases with no wall crossings and with lower symmetry than the substrate. If, on the other hand, $e_c < 0$, crossings are favored, and a new calculation is needed to determine the nature of the CI transition. To be concrete, consider a hexagonal substrate and a hexagonal array of walls, as shown in Fig. 10.3.7b. Let l be the length of a hexagonal side. If we assume that the dominant repulsive interaction arises from repulsion from parallel wall

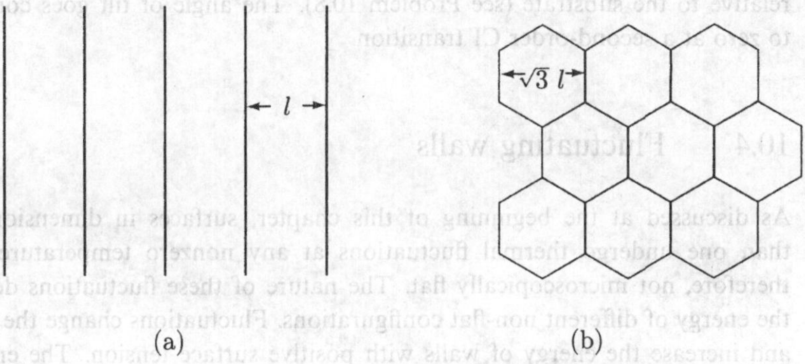

Fig. 10.3.7. (a) A striped incommensurate phase consisting of a linear array of equally spaced domain walls. (b) A hexagonal incommensurate phase in which domain walls follow the edges of a honeycomb lattice.

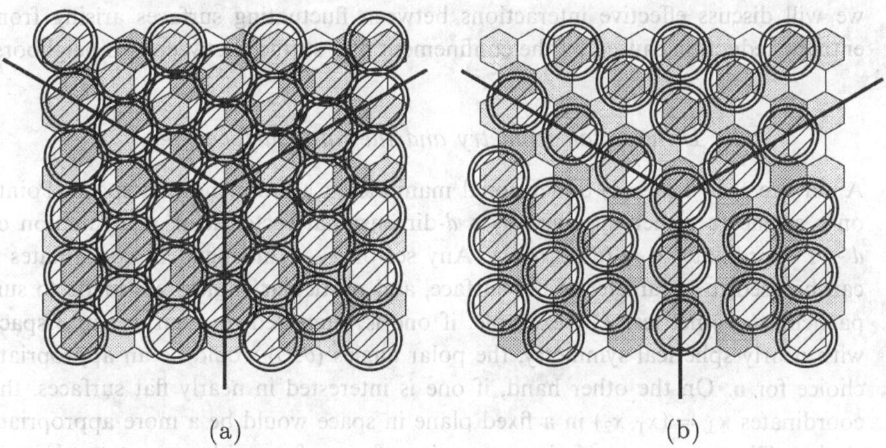

Fig. 10.3.8. (a) Heavy walls and (b) light walls relative to the $\sqrt{3} \times \sqrt{3}R30°$ commensurate state.

segments separated by a distance $\sqrt{3}l$, then the energy per unit area A relative to the commensurate state is

$$\frac{\Delta U}{A} = \frac{6\sigma(\delta)l + 4e_c + U_0 e^{-\sqrt{3}l/\lambda\bar{a}}}{3\sqrt{3}l^2}. \tag{10.3.57}$$

This energy predicts a first-order CI transition. Thus, at $T = 0$, the CI transition on a hexagonal substrate is either first order or it leads to a state of lower than hexagonal symmetry.

Unlike one-dimensional systems, two-dimensional systems have shear as well as longitudinal strain. Transverse strains in general have lower energy than longitudinal strains. This causes the incommensurate adsorbate lattice to tilt relative to the substrate (see Problem 10.8). The angle of tilt goes continuously to zero at a second-order CI transition.

10.4 Fluctuating walls

As discussed at the beginning of this chapter, surfaces in dimensions greater than one undergo thermal fluctuations at any nonzero temperature and are, therefore, not microscopically flat. The nature of these fluctuations depends on the energy of different non-flat configurations. Fluctuations change the total area and increase the energy of walls with positive surface tension. The energy of a wall can also depend on other quantities, such as the local curvature. In this section, we will introduce some rudimentary differential geometry that will allow us to describe spatial configurations of a fluctuating surface. We will then discuss simple properties of fluctuating surfaces in the harmonic approximation. Finally, we will discuss effective interactions between fluctuating surfaces arising from entropy reduction caused by the confinement of a surface between two neighbors.

1 Differential geometry and the total surface area

A surface S is a $(d-1)$-dimensional manifold in a d-dimensional space. Points on a surface are then specified by a d-dimensional vector $\mathbf{R}(\mathbf{u})$ as a function of $d-1$ coordinates $\mathbf{u} = (u^1, ..., u^{d-1})$. Any set of $d-1$ independent coordinates $\mathbf{u}$ can be used to parameterize the surface, and particular choices are made to suit particular geometries. For example, if one is interested in a surface in 3-space with nearly spherical symmetry, the polar angles (θ, ϕ) would be an appropriate choice for $\mathbf{u}$. On the other hand, if one is interested in nearly flat surfaces, the coordinates $\mathbf{x}_\perp = (x_1, x_2)$ in a fixed plane in space would be a more appropriate choice. The invariance of the properties of a surface with respect to changes in parameterization is analogous to gauge invariance in electrodynamics, and a choice of parameterization is often called a *gauge* choice. Since we are most interested in nearly flat surfaces, we will focus on the latter choice called the *Monge* gauge. In this gauge,

$$\mathbf{R}(\mathbf{x}_\perp) = [\mathbf{x}_\perp, h(\mathbf{x}_\perp)], \tag{10.4.1}$$

where $h(\mathbf{x}_\perp)$ specifies the height of the surface above the base plane with coordinates $\mathbf{x}_\perp = (x_1, x_2, ..., x_{d-1})$, as shown in Fig. 10.4.1. In general, the surface can have overhangs, as shown in Fig. 10.4.1b. In this case, the function $h(\mathbf{x}_\perp)$ is multi-valued, and the Monge gauge may not be optimal. It may, of course, be possible to choose a rotated base plane such that there are no overhangs, as shown in Fig. 10.4.1b. In this case, the Monge gauge relative to the rotated base plane is a good one. The Monge gauge is equivalent to specifying the height of the surface above the base plane via the zeroes of the function,

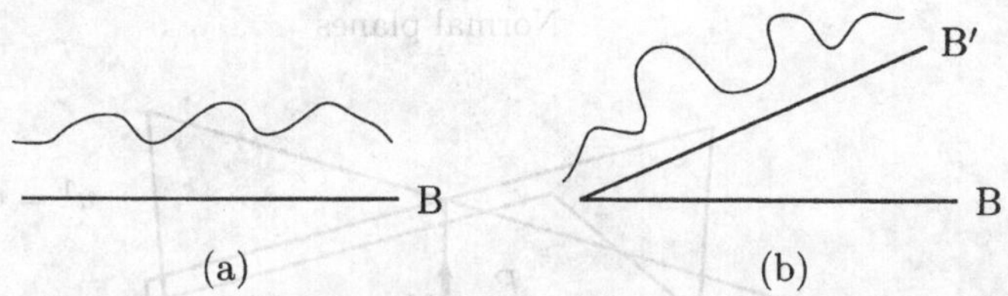

Fig. 10.4.1. (a) A surface without overhangs above a base plane. In the Monge gauge, points on the surface are specified by their height $h(\mathbf{x}_\perp)$ above a coordinate $\mathbf{x}_\perp$ on the base plane. (b) A one-dimensional surface in two dimensions with overhangs relative to base plane B but without overhangs relative to the rotated base plane B'.

$$\phi(\mathbf{x}) = x_d - h(\mathbf{x}_\perp), \tag{10.4.2}$$

where $\mathbf{x} = (\mathbf{x}_\perp, x_d)$ is a d-dimensional vector. Eq. (10.4.2) will prove useful in what follows.

To simplify our discussion, we will, for the moment, restrict our attention to surfaces without overhangs in two and three dimensions. A one-dimensional surface in two dimensions is simply a *planar curve,* which can be parameterized by a single scalar variable u. Then,

$$d\mathbf{R} = \frac{d\mathbf{R}}{du} du \equiv \mathbf{t} du. \tag{10.4.3}$$

The vector $\mathbf{t} = d\mathbf{R}/du$ is tangent to the curve at u. The length of the line segment between u and $u + du$ is $ds = (d\mathbf{R} \cdot d\mathbf{R})^{1/2}$. If u is the arc length s, then $\mathbf{t}$ is a unit vector. In the Monge gauge, $\mathbf{R} = [x, h(x)]$, and

$$d\mathbf{R} = \frac{d\mathbf{R}}{dx} dx = \left(\mathbf{e}_x + \frac{dh}{dx} \mathbf{e}_y \right) dx. \tag{10.4.4}$$

The length of an infinitesimal line segment is ds, and the total length of the curve in the Monge gauge is

$$L = \int_0^L ds = \int_0^{L_B} dx [1 + (dh/dx)^2]^{1/2}, \tag{10.4.5}$$

where L_B is the length of the horizontal base line shown in Fig. 10.4.1.

A two-dimensional surface in three dimensions is parameterized by two variables u^1 and u^2. Infinitesimal displacements along the surface satisfy

$$d\mathbf{R} = \frac{\partial \mathbf{R}}{\partial u^1} du^1 + \frac{\partial \mathbf{R}}{\partial u^2} du^2. \tag{10.4.6}$$

The vectors $\mathbf{t}_1 = \partial \mathbf{R}/\partial u^1$ and $\mathbf{t}_2 = \partial \mathbf{R}/\partial u^2$ are tangent to the surface at (u^1, u^2) but are not necessarily orthogonal, as shown in Fig. 10.4.2. The area of an infinitesimal surface element with sides du^1 and du^2 is, therefore,

$$\begin{aligned} dS &= \left| \frac{\partial \mathbf{R}}{\partial u^1} \times \frac{\partial \mathbf{R}}{\partial u^2} \right| du^1 du^2 \\ &= [1 + (\nabla_\perp h)^2]^{1/2} dx_1 dx_2, \end{aligned} \tag{10.4.7}$$

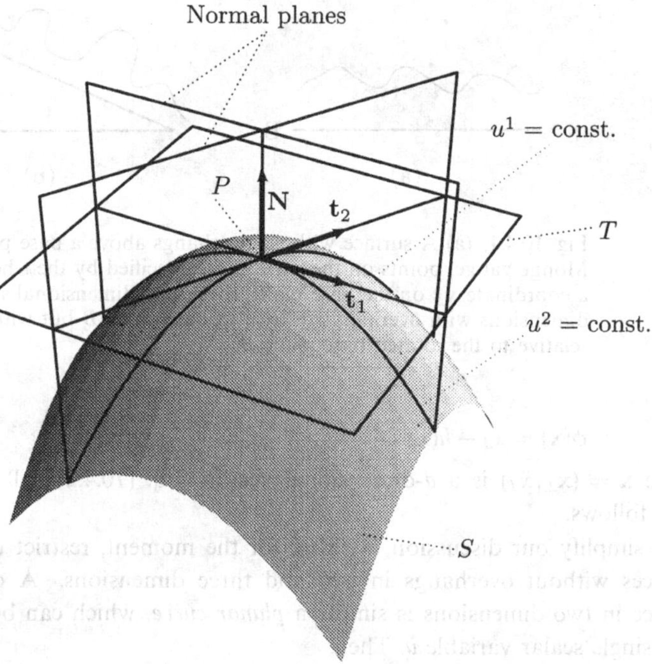

Fig. 10.4.2. Representation of a section of a surface S showing at point P the normal vector $\mathbf{N}$, tangent vectors $\mathbf{t}_1$ and $\mathbf{t}_2$, the tangent plane T, and normal planes perpendicular to T. The intersection of the surface S with the normal planes are called normal sections. The normal sections shown here coincide with the lines u^1 and u^2, which are equal to constants.

where the second form, valid for the Monge gauge with $(u^1, u^1) = (x_1, x_2)$, follows from Eq. (10.4.1) The total area of a surface without overhangs relative to a fixed base plane is, therefore,

$$A = \int dS = \int dx_1 dx_2 [1 + (\nabla_\perp h)^2]^{1/2}, \tag{10.4.8}$$

where the final integral is over the base plane area A_B.

A unit normal $\mathbf{N}$ can be constructed at each point P on a surface. The normal to a planar curve can be obtained by rotating the tangent vector $\partial\mathbf{R}/\partial u$ through 90°. The unit normal to a two-dimensional surface can be constructed from the tangent vectors $\mathbf{t}_1$ and $\mathbf{t}_2$ at P:

$$\mathbf{N} = \frac{\mathbf{t}_1 \times \mathbf{t}_2}{|\,\mathbf{t}_1 \times \mathbf{t}_2\,|}. \tag{10.4.9}$$

The sign of $\mathbf{N}$ is not uniquely determined. If there is nothing to distinguish the two sides of the surface, the choice of the sign of $\mathbf{N}$ is arbitrary. If, however, the two sides are distinguishable, as they are, for example, when the surface is a domain wall separating two coexisting equilibrium phases A and B, or when it is

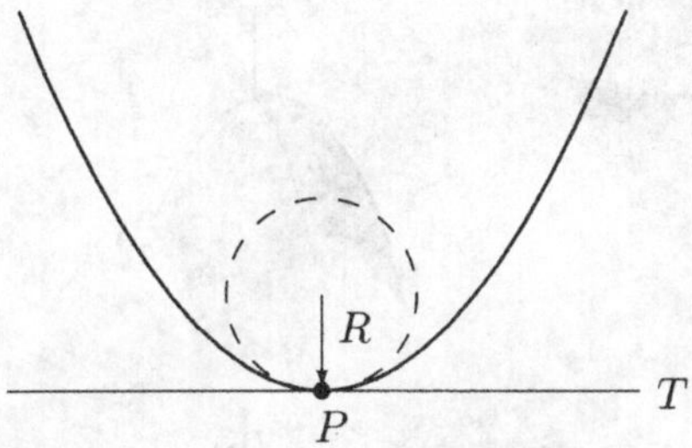

Fig. 10.4.3. A planar curve with a tangent plane T at a point P. The circle tangent to the curve has a radius equal to the local radius of curvature at P.

a surfactant layer separating water and oil in a microemulsion, it is appropriate to choose $\mathbf{N}$ to point always from one phase to the other (say from A to phase B or from water to oil). We will refer to surfaces for which the sign of $\mathbf{N}$ has significance as oriented surfaces. The plane normal to $\mathbf{N}$ at P is tangent to the surface at P and is called the *tangent plane.* An explicit form for $\mathbf{N}$ in the Monge gauge is easily obtained with the aid of Eqs. (10.4.2) and (6.3.6):

$$\mathbf{N} = \frac{\nabla\phi}{|\nabla\phi|} = [1 + (\nabla_\perp h)^2]^{-1/2}(-\nabla_\perp h, 1). \tag{10.4.10}$$

We leave as an exercise the verification that Eqs. (10.4.9) and (10.4.10) are identical.

2 Curvature

Each point on a surface is characterized by a local curvature tensor. For a planar curve in two dimensions, the curvature tensor at point P is a scalar with magnitude equal to the inverse radius of a circle that locally follows the curve at P, as shown in Fig. 10.4.3. Let T be the tangent line to the curve at P, let x be the coordinate along T as measured with respect to P, and let $h(x) = \mathbf{N}\cdot\mathbf{R}$ be the height of the curve above P in the direction of the normal $\mathbf{N}$. For small x, $h(x)$ can be expanded in a power series in x. Because T is tangent to the curve, we have $dh/dx = 0$ and $h(x) = Kx^2/2$. Now imagine a circle of radius R tangent to T at P. If the center of the circle is along $\mathbf{N}$, its height above P as a function of the angle θ measured with respect to $\mathbf{N}$ is $h' = R(1-\cos\theta)$. For small θ, $\theta = x/R$ and $h' = R^{-1}x^2/2$. Thus, the circle and the curve will match to order x^2 if $K = R^{-1}$. If the center of the circle lies along $-\mathbf{N}$, then $h' = -R(1-\cos\theta)$, and $K = -R^{-1}$. Thus, the sign of K is positive if the curve rises towards $\mathbf{N}$ and negative if it falls away from $\mathbf{N}$.

A similar analysis applies to a two-dimensional surface S in three dimensions. Let T be the tangent plane at the point P on S, as shown in Fig. 10.4.2. Planes normal to T at P are called *normal planes.* Each normal plane intersects the surface S in a planar curve called a *normal section.* Each normal section has an

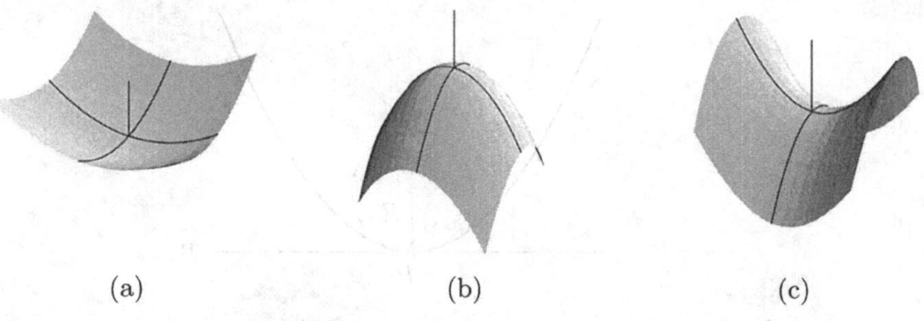

Fig. 10.4.4. Sections of surfaces (a) with two positive radii of curvature, (b) with two negative radii of curvature, and (c) with one positive and one negative radius of curvature. The surface in (a) has positive mean and Gaussian curvature at P; that in (b) has negative mean and positive Gaussian curvature; and that in (c) has negative Gaussian curvature and mean curvature whose sign depends on the relative magnitude of R_1 and R_2.

associated curvature. Let x_1 and x_2 be coordinates in an orthogonal coordinate system on the tangent plane measured with respect to an origin at P. The height $h = \mathbf{N} \cdot \mathbf{R}$ of the surface above P to second order in $\mathbf{x}_\perp = (x_1, x_2)$ is then

$$h(\mathbf{x}_\perp) = \frac{1}{2} K_{ij} x_i x_j, \tag{10.4.11}$$

where K_{ij} are the components of a 2×2 real symmetric tensor $\underline{K}$, which has two real eigenvalues R_1^{-1} and R_2^{-1} and associated orthonormal eigenvectors $\mathbf{e}_1$ and $\mathbf{e}_2$ in T. Therefore, Eq. (10.4.11) can be rewritten as

$$h(\mathbf{x}_\perp) = \frac{1}{2} R_1^{-1} (\mathbf{x}_\perp \cdot \mathbf{e}_1)^2 + \frac{1}{2} R_2^{-1} (\mathbf{x}_\perp \cdot \mathbf{e}_2)^2. \tag{10.4.12}$$

R_1 and R_2 are called the principal radii of curvature of the surface at P and correspond, respectively, to the radii of the circles in the $\mathbf{N} - \mathbf{e}_1$ and the $\mathbf{N} - \mathbf{e}_2$ normal planes that best fit the normal sections in these two planes. The signs of R_1 and R_2 can be positive or negative. If the normal section in the $\mathbf{N} - \mathbf{e}_\alpha$ $(\alpha = 1, 2)$ plane curves towards (away from) $\mathbf{N}$, R_α is positive (negative). If R_1 and R_2 are both positive (negative), all normal sections curve towards (away from) $\mathbf{N}$, as shown in Figs. 10.4.4a and b. If R_1 and R_2 have opposite signs, as they do at a *saddle point*, there will be normal sections that curve towards $\mathbf{N}$ and others that curve away from $\mathbf{N}$, as shown in Fig. 10.4.4c. The curvature of a normal section in a normal plane containing the vector $\mathbf{e}(\gamma) = \cos\gamma \mathbf{e}_1 + \sin\gamma \mathbf{e}_2$ is $R^{-1}(\gamma) = R_1^{-1}\cos^2\gamma + R_2^{-1}\sin^2\gamma$. The extremal values of $R^{-1}(\gamma)$ occur at $\gamma = 0 \bmod (\pi)$ and at $\gamma = (\pi/2) \bmod (\pi)$, where $R^{-1}(\gamma)$ is equal, respectively, to R_1^{-1} and R_2^{-1}. Thus, the principal radii of curvature R_1 and R_2 correspond to the maximum and minimum curvatures of all of the normal sections through P.

Two scalar invariants can be constructed from the tensor $\underline{K}$. They are

$$\mathrm{Tr}\underline{K} = R_1^{-1} + R_2^{-1} \equiv 2H_c \tag{10.4.13}$$

and

$$\det \underline{K} = \frac{1}{R_1 R_2}. \tag{10.4.14}$$

H_c is the average or *mean curvature*, and $\det \underline{K}$ is the *Gaussian curvature*. The mean curvature in the Monge gauge is simply $(\nabla_\perp \cdot \mathbf{N})/2$. This relation is trivially true when the base plane is the tangent plane at P. In this case, to linear order in $\mathbf{x}_\perp$ near P, $\mathbf{N} = (-K_{ij}x_j, 1)$, and $\nabla \cdot \mathbf{N} = \mathrm{Tr}\underline{K}$. It is a straightforward exercise to show that this relation applies to any point on a surface (see Problem 10.6). Thus,

$$\nabla_\perp \cdot \mathbf{N} = \frac{1}{R_1} + \frac{1}{R_2} = \frac{(\nabla_\perp^2 h)(1 + (\nabla_\perp h)^2) - \nabla_{\perp i} h \nabla_{\perp i} \nabla_{\perp j} h \nabla_{\perp j} h}{[1 + (\nabla_\perp h)^2]^{3/2}}. \tag{10.4.15}$$

To linear order in h, the mean curvature is simply $\nabla_\perp^2 h/2$.

3 Energy of a surface

As we indicated at the beginning of this section, the energy of a surface depends in general on its total area and on surface parameters measuring deviations from local flatness, the lowest order of which is the curvature tensor. A phenomenological Hamiltonian for a surface can, therefore, be written as

$$\mathscr{H} = \mathscr{H}_\sigma + \mathscr{H}_c + \mathscr{H}_G, \tag{10.4.16}$$

where

$$\begin{aligned} \mathscr{H}_\sigma &= \int dS \sigma(\mathbf{N}), \\ \mathscr{H}_c &= \frac{1}{2}\kappa \int dS \left(\frac{1}{R_1} + \frac{1}{R_2} - \frac{2}{R_0} \right)^2, \\ \mathscr{H}_G &= \frac{1}{2}\kappa_G \int dS \frac{1}{R_1 R_2}. \end{aligned} \tag{10.4.17}$$

The first term $\mathscr{H}_\sigma$ is the surface tension contribution to $\mathscr{H}$. The second and third terms $\mathscr{H}_c$ and $\mathscr{H}_G$ are, respectively, the mean and Gaussian curvature contributions to $\mathscr{H}$. We discuss $\mathscr{H}_\sigma$ first. When $\sigma(\mathbf{N})$ is independent of the direction of the surface normal $\mathbf{N}$, $\mathscr{H}_\sigma = \sigma A$, where A is the total area of the surface. There are situations in which $\sigma(\mathbf{N})$ can depend on $\mathbf{N}$. For example, the energy of a solid-liquid interface can depend on the direction of $\mathbf{N}$ relative to a crystal axis of the solid. Or the energy of a Frenkel-Kontorowa-like soliton on a two-dimensional adsorbate such as Xe may depend on its direction relative to the periodic substrate. In general, $\sigma(\mathbf{N})$ does not have to be an analytic function of $\mathbf{N}$. Indeed, we will see in the next section how nonanalytic forms for $\sigma(\mathbf{N})$ arise for solid-liquid interfaces and the effect they have on equilibrium crystal shapes. Often, however, $\sigma(\mathbf{N})$ is analytic. If the lowest energy configuration occurs when $\mathbf{N}$ is along a fixed direction $\mathbf{e}$ in space (corresponding, say, to a lattice or substrate direction), then $\sigma(\mathbf{N})$ can be expressed as a function of $\theta = \cos^{-1}(\mathbf{e} \cdot \mathbf{N})$. If $\sigma(\mathbf{N}) = \sigma(\theta)$ is analytic, it can be expanded in a power series about $\theta = 0$:

$$\sigma(\theta) = \sigma + \frac{1}{2}\sigma_1\theta^2 + ..., \qquad (10.4.18)$$

where σ_1 is positive if $\theta = 0$ corresponds to a local minimum of $\sigma(\theta)$.

The second term $\mathcal{H}_c$ measures the energy cost associated with deviations of the mean curvature from a local preferred value R_0. If the lowest energy state is a flat surface, $R_0 = \infty$. Interfaces between coexisting isotropic phases generally have $R_0 = \infty$. Since R_0 has a sign, it can only be nonzero for oriented surfaces. A particularly clear example of a oriented surface is provided by a surfactant layer separating water from oil in microemulsions. A surfactant consists of molecules with a hydrophilic polar head group preferring contact with water over oil and a hydrophobic hydrocarbon tail preferring contact with oil over water. Such a molecule can have a wedge shape, giving rise to a natural surface curvature and a nonzero R_0^{-1}, as shown in Fig. 2.7.13. The curvature rigidity κ has units of energy. In microemulsions, its magnitude is generally of order T at room temperature.

The third, or Gaussian, curvature contribution $\mathcal{H}_G$ is different from the others in that, according to the *Gauss-Bonnet* theorem, it is a constant on a surface of fixed topology. It does not, therefore, affect the fluctuations of a surface of fixed topology. It can, however, become important if a surface is allowed to break up into many disjoint parts or spontaneously to generate handles. The Gauss-Bonnet theorem states that the integral over a surface is a topological invariant:

$$\int dS \frac{1}{R_1 R_2} = 4\pi(1 - g), \qquad (10.4.19)$$

where the *genus* g is an integer equal to the number of handles of the surface. The quantity $\chi = 2(1 - g)$ is called the Euler characteristic of the surface. The proof of the Gauss-Bonnet theorem involves concepts in differential geometry that are beyond the scope of this book. A sphere, and all topologically equivalent closed surfaces that can be mapped continuously onto a sphere, have no handles and have genus $g = 0$. A torus is topologically equivalent to a sphere with a single handle and has genus $g = 1$. Representative examples of higher genus surfaces are shown in Fig. 10.4.5c. The "plumber's nightmare" shown in Fig. 2.7.15 is an example of a surface with an extensive genus proportional to the total area of the surface.

4 Fluctuations in the harmonic approximation

The harmonic approximation to $\mathcal{H}$ in the Monge gauge is obtained by expanding to second order in h. If we restrict our attention to a nearly flat surface of fixed topology, we can ignore $\mathcal{H}_G$, and we obtain

$$\mathcal{H}_{\text{har}} = \sigma A_B + \frac{1}{2}\int d^{(d-1)}x_\perp[\gamma(\nabla_\perp h)^2 + \kappa(\nabla_\perp^2 h)^2], \qquad (10.4.20)$$

where $\gamma = \sigma + \sigma_1$ is the *interfacial stiffness*, which includes a part arising from the θ^2 term [Eq. (10.4.18)] in the anisotropic surface tension. When $\gamma > 0$, the interfacial stiffness determines the nature of the long-wavelength fluctuations, and

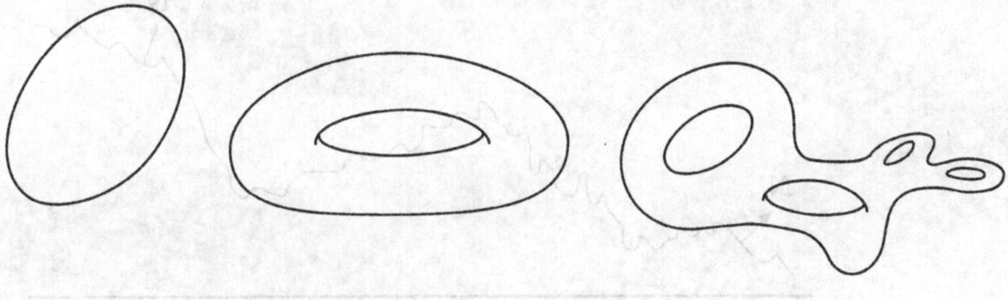

Fig. 10.4.5. Surfaces of (a) genus zero, (b) genus one, and (c) large genus.

the curvature term can be ignored for wave vectors $\mathbf{q}$ with magnitude q less than an appropriately defined cutoff Λ. If the curvature term is dropped, $\mathscr{H}_{\text{har}}$ becomes identical to the xy elastic Hamiltonian discussed in Sec. 6.1. Recall that phase fluctuations in the xy-model become divergent at and below the lower critical dimension $d_L = 2$. The integral in Eq. (10.4.20) is over a $(d-1)$-dimensional surface, so that a surface in three dimensions is at its lower critical dimension, and one in two dimensions is below its lower critical dimension. To be more specific, we have

$$\langle h^2 \rangle = \int \frac{d^{d-1}q}{(2\pi)^{d-1}} \frac{T}{\gamma q^2} = \begin{cases} [T/(2\pi\gamma)]L_B, & d=2; \\ [T/(2\pi\gamma)]\ln L_B, & d=3, \end{cases} \tag{10.4.21}$$

where L_B is the length of the base plane. Similarly,

$$\begin{aligned} \langle [h(\mathbf{x}_\perp) - h(0)]^2 \rangle &= 2\int \frac{d^{d-1}}{(2\pi)^{d-1}} \frac{T}{\gamma q^2}(1 - e^{i\mathbf{q}\cdot\mathbf{x}_\perp}) \\ &= 2 \times \begin{cases} [T/(2\pi\gamma)]|x|, & d=2; \\ [T/(2\pi\gamma)]\ln(|\mathbf{x}_\perp|/a), & d=3. \end{cases} \end{aligned} \tag{10.4.22}$$

Thus, a surface in spatial dimensions less than or equal to three has height fluctuations that diverge with the size of the surface. This implies that the average position of such a surface becomes less well defined as its size increases: the surface lacks *long-range positional order*. On the other hand, the direction of the surface normal remains well defined. The deviation of $\mathbf{N}(\mathbf{x}_\perp)$ from its average direction $\mathbf{e}$ is, from Eq. (10.4.10), $\delta\mathbf{N}(\mathbf{x}_\perp) = \mathbf{N}(\mathbf{x}_\perp) - \mathbf{e} \approx -\boldsymbol{\nabla}_\perp h$, and

$$\langle |\delta\mathbf{N}(\mathbf{x}_\perp)|^2 \rangle \approx \langle (\nabla_\perp h)^2 \rangle = \int \frac{d^{d-1}q}{(2\pi)^{d-1}} \frac{Tq^2}{\gamma q^2} \sim \frac{T}{\gamma}\Lambda^{d-1}. \tag{10.4.23}$$

Thus, fluctuations in $\mathbf{N}(\mathbf{x}_\perp)$ remain finite as the system size diverges. This means that the average surface normal points in the same direction for all parts of an infinite surface. The surface has *long-range orientational order* even though it does not have long-range positional order (see Fig. 10.4.6).

The interfacial tension of some surfaces, such as surfactant layers in a microemulsion, can be effectively zero, so that the curvature term in $\mathscr{H}$ determines

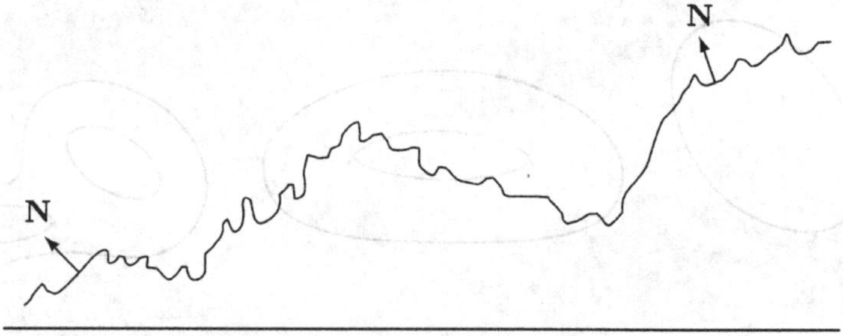

Fig. 10.4.6. Schematic representation of a fluctuating surface with long-range orientational order but no long-range positional order.

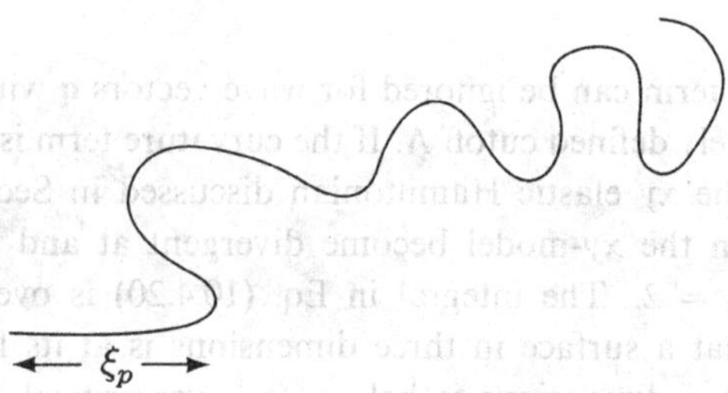

Fig. 10.4.7. A surface not subjected to tension is flat on length scales less than the de Gennes-Taupin length ξ_p and crumpled on larger length scales.

the shape fluctuations. In this case, height and orientation fluctuations are more violent than they are when $\gamma > 0$:

$$\langle h^2 \rangle = \begin{cases} [T/(2\pi\kappa)]L_B^2, & d = 3; \\ [T/(2\pi\kappa)]L_B^3, & d = 2, \end{cases} \tag{10.4.24}$$

and

$$\langle |\,\delta\mathbf{N}\,|^2 \rangle = \begin{cases} [T/(2\pi\kappa)]\ln(L_B/a), & d = 3; \\ [T/(2\pi\kappa)]L_B, & d = 2. \end{cases} \tag{10.4.25}$$

Thus, in this case, not only is there no long-range positional order, there is also no long-range orientational order. The approximation of no overhangs breaks down for base plane lengths larger than the orientational *persistence length*,

$$\xi_p = \begin{cases} ae^{2\pi\kappa/T}, & d = 3; \\ (2\pi\kappa)/T, & d = 2, \end{cases} \tag{10.4.26}$$

at which $\langle |\,\delta\mathbf{N}\,|^2 \rangle$ becomes of order unity. For $L > \xi_p$, orientational order is lost and the surface becomes crumpled, as shown in Fig. 10.4.7. For $d = 2$, ξ_p is a special case of the persistence length of a polymer. For $d = 3$, ξ_p is the de Gennes-Taupin length (de Gennes and Taupin 1982) for fluctuating surfaces.

When a surface has no long-range positional order, it is said to be *rough* in contrast to a crystal surface at zero temperature, which is flat and macroscopically *smooth*. The height fluctuations of a rough surface diverge with the length L_B of its base. It is useful to introduce a critical exponent describing this divergence. Let $l = \langle h^2 \rangle^{1/2}$ be the root mean square height fluctuation. Then

$$(l/b) \sim (L_B/a)^\zeta, \tag{10.4.27}$$

where ζ is called the *wandering exponent, a* is a cutoff length in the base plane, and b is a length determined by a and the rigidity parameters. From our previous discussion, we have

$$\zeta = \begin{cases} (3-d)/2, & \text{for } d \le 3 \text{ and } \gamma > 0; \\ (5-d)/2, & \text{for } d \le 5 \text{ and } \gamma = 0,\ \kappa > 0. \end{cases} \tag{10.4.28}$$

The length b is proportional to $(T/\gamma)a^{3-d}$ when $\gamma > 0$ and to $(T/\kappa)a^{5-d}$ when $\gamma = 0$. Note that $b \to 0$ as $T \to 0$ so that l decreases for a given L_B as T decreases. In general, if $F \sim \int d^D x (\nabla^s h)^2$, then $\langle h^2 \rangle \sim L_B^{2s-D}$, and $\zeta = (2s - D)/2$.

5 Nonlinearities and renormalization in fluid membranes

A membrane in equilibrium has thermally excited height fluctuations or "wiggles" at all wavelengths down to microscopic lengths set by interparticle spacing. A membrane with wiggles is easier to bend than a membrane without wiggles. A simple experiment will verify this. Take two pieces of fairly stiff wire (a paper clip, for example). Bend one piece so that it has several wiggles in it but so that its projected length is the same as that of the other straight, unbent wire. The wire with wiggles is easier to bend than the one without. The short-wavelength wiggles renormalize the long-wavelength bending rigidity. This is fundamentally a nonlinear effect because, in a harmonic system, modes of different wavelengths do not affect each other.

The bending Hamiltonian $\mathcal{H}_c$ [Eq. (10.4.18)] is nonlinear, as can be seen with the aid of Eq. (10.4.15):

$$\begin{aligned} \mathcal{H}_c^\Lambda &= \frac{1}{2}\kappa \int d^2x [1 + (\nabla_\perp h)^2]^{-5/2} [(\nabla_\perp^2 h)(1 + (\nabla_\perp h)^2) - \nabla_{\perp i} h \nabla_{\perp i} \nabla_{\perp j} h \nabla_{\perp j} h]^2 \\ &\approx \frac{1}{2}\kappa \int d^2x \left[(\nabla_\perp^2 h)^2 - \frac{1}{2} (\nabla_\perp^2 h)^2 (\nabla_\perp h)^2 - 2\nabla_\perp^2 h \nabla_{\perp i} h \nabla_{\perp i} \nabla_{\perp j} h \nabla_{\perp j} h \right]. \end{aligned} \tag{10.4.29}$$

The curvature rigidity $\kappa \equiv \kappa(\Lambda^{-1})$ is that at microscopic length scales of order the interparticle spacing $a = 2\pi/\Lambda$. The rigidity at longer length scales or smaller wave number can be calculated by integrating out high wave number degrees of freedom, just as is done in the momentum-shell renormalization group discussed in Sec. 5.8. Let $h(\mathbf{x}) = h^<(\mathbf{x}) + h^>(\mathbf{x})$, where $h^<(\mathbf{x})$ has Fourier components with wave number $\mathbf{q}$, with $0 \le q < \Lambda'$, and $h^>(\mathbf{x})$ has Fourier components with $\Lambda' < q \le \Lambda$, with $\Lambda' < \Lambda$. The Hamiltonian at length scales $2\pi/\Lambda'$,

$$\begin{aligned} \mathcal{H}_c^{\Lambda'} &= -T \ln \int \mathcal{D}h^>(\mathbf{x}) e^{-\mathcal{H}(\Lambda)/T} \\ &= \frac{1}{2}\kappa(\Lambda'^{-1}) \int d^2x (\nabla^2 h^<)^2 + \dots, \end{aligned} \tag{10.4.30}$$

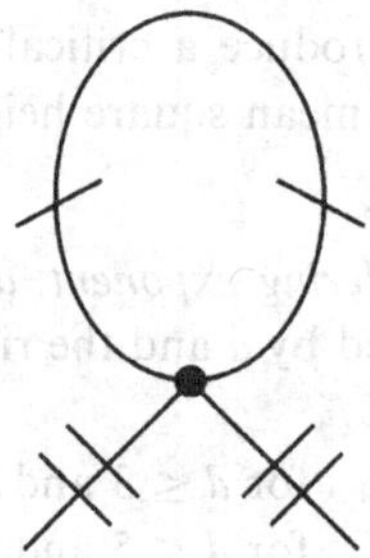

Fig. 10.4.8. One-loop diagram contributing to the renormalization of the bending rigidity κ. Each slash represents one power of the gradient operator. Each of the external lines in this graph has two gradient operators.

can be calculated as a perturbation expansion in T/κ, in analogy with our treatment of the nonlinear sigma model discussed in Sec. 6.7. The contribution to one-loop order to $\kappa(\Lambda')$ is obtained from the diagram in Fig. 10.4.8 with two external ∇^2 legs. The result is (Peliti and Leibler 1985)

$$\kappa(\Lambda'^{-1}) = \kappa(\Lambda^{-1}) - \frac{3T}{4\pi}\ln(\Lambda/\Lambda'), \tag{10.4.31}$$

or equivalently

$$\frac{d\kappa}{dl} = -\frac{3T}{4\pi}. \tag{10.4.32}$$

This equation has no fixed point and runs away to negative values of κ. This is a reflection of the fact that a fluid membrane (like a polymer) is a fractal object at long length scales and cannot be treated as though it were nearly flat with a well-defined bending rigidity. The point at which $\kappa(\Lambda')$ passes through zero defines a length $\xi/a = \Lambda/\Lambda' = e^{4\pi\kappa/3T}$ similar to the de Gennes-Taupin length [Eq. (10.4.26)] that diverges exponentially as $\kappa/T \to \infty$.

We have considered only the renormalization of the bending rigidity. The surface tension and Gaussian curvature also undergo renormalization, the calculation of which is more complicated than of that for κ.

6 Polymerized membranes

A fluid membrane is composed of molecules that freely diffuse; it cannot support a shear. A membrane in which such diffusion is prohibited can be created by forming, via polymerization, a two-dimensional network of connected molecules analogous to a fishnet. Such a fixed-connectivity network is a two-dimensional solid, differing from those we considered in Chapter 6 by its freedom to fluctuate in a third direction. It is characterized by a strain variable measuring distortion from an equilibrium reference state and a strain elastic energy. It can support a shear (at least in the harmonic approximation). Physical examples of fixed connectivity networks include partially polymerized phospholipid vesicles, graphite oxide sheets in an appropriate solvent, and the spectrin network in red blood cells.

The nonlinear coupling of in-plane strain modes and out-of-plane height modes of polymerized membranes leads to nontrivial and surprising properties that are distinct from those of a fluid membrane or a flat two-dimensional solid (Nelson and Peliti 1987; Aronowitz and Lubensky 1988). The long-wavelength bending rigidity is stiffened, rather than softened as in a fluid membrane, and the in-plane bulk and shear moduli are softened by the same effect that favors the use of corrugated steel sheets in some construction projects. It is more difficult to bend a corrugated sheet into a section of a cylinder whose axis is perpendicular to lines of corrugation than it is to bend a flat, uncorrugated sheet. A bend in this direction creates nonzero Gaussian curvature, which is strongly disfavored in a solid that supports shear. It is also easier to bend such a sheet into a cylinder whose axis is parallel to the lines of corrugation than it is to bend a flat sheet. This is the effect leading to softening of the bending rigidity of a fluid membrane. It is, as we shall see, less important than the stiffening effect in polymerized membranes. Finally, it is easier to stretch a corrugated sheet in a direction perpendicular to lines of corrugation than it is to stretch a flat sheet. Fluctuations at finite temperature in a polymerized membrane produce random corrugations in random directions whose effect is to stiffen the bending rigidity and soften the bulk and shear moduli. The wave number-dependent bending rigidity and elastic moduli, respectively, diverge and vanish as q tends to zero with power laws characterized by critical exponents

$$\begin{aligned} \kappa(q) &\sim q^{-\eta_h} \\ \mu(q) &\sim \lambda(q) \sim q^{\eta_u}. \end{aligned} \tag{10.4.33}$$

Rotational invariance implies, as we shall see shortly, that η_h and η_u are not independent but are related via

$$\eta_u = 2(1 - \eta_h). \tag{10.4.34}$$

These power-law singularities lead to nontrivial wandering exponents for height and in-plane displacement correlations:

$$\begin{aligned} \langle |h(\mathbf{x}) - h(0)|^2 \rangle &\sim \int \frac{d^2q}{(2\pi)^2} \frac{1 - e^{i\mathbf{q}\cdot\mathbf{x}}}{q^{4-\eta_h}} \sim |\mathbf{x}|^{2\zeta}, \qquad \zeta = 1 - \frac{\eta_h}{2} \\ \langle |\mathbf{u}(\mathbf{x}) - \mathbf{u}(0)|^2 \rangle &\sim \int \frac{d^2q}{(2\pi)^2} \frac{1 - e^{i\mathbf{q}\cdot\mathbf{x}}}{q^{2+\eta_u}} \sim |\mathbf{x}|^{2\zeta_u}, \qquad \zeta_u = \frac{\eta_u}{2}. \end{aligned} \tag{10.4.35}$$

They also imply that $\langle (\nabla_\perp h)^2 \rangle$ is finite and that the polymerized membrane has long-range orientational order. Fig. 10.4.9 shows typical configurations of a model polymerized membrane obtained from computer simulations. Note that these membranes fluctuate on average about a well-defined plane indicating long-range orientational order. They also have a finite thickness, indicating substantial height fluctuations. The model used in these simulations consisted of spheres connected by unbreakable strings and interacting via a modified Lennard-Jones potential. Simulations generally give $\zeta \approx 0.6$ and $\eta_h \approx 0.8$.

Polymerized membranes have other unusual properties. They have a negative Poisson ratio (Sec. 6.6), i.e., when stretched in one direction, they expand rather than contract in the other. This effect can be seen with a crumpled piece of paper - try it. They also have a nonlinear stress-strain relation with $\langle \nabla u \rangle \sim \sigma^{\phi_\sigma}$, where $\phi_\sigma = (2-\eta_u)/(2+\eta_u)$ (see Problem 10.8).

The Hamiltonian for a polymerized membrane has a strain-elasticity part identical to that of a two-dimensional solid and a bending rigidity part identical to that of a fluid membrane:

$$\mathscr{H} = \frac{1}{2}\int d^2x[\lambda u_{ii}^2 + 2\mu u_{ij}u_{ij}] + \frac{1}{2}\int d^2x\kappa(\nabla^2 h)^2, \tag{10.4.36}$$

where we have displayed only the harmonic part of the bending energy. The strain u_{ij} is the full nonlinear strain

$$\begin{aligned} u_{ij} &= \frac{1}{2}(\partial_i u_j + \partial_i u_j + \partial_i \mathbf{u}\cdot\partial_j\mathbf{u} + \partial_i h\partial_j h) \\ &\approx \frac{1}{2}(\partial_i u_j \partial_j u_i + \partial_i h\partial_j h), \end{aligned} \tag{10.4.37}$$

where the second approximate form applies because fluctuations in $\mathbf{u}$ are much smaller than those in h. In the harmonic approximation, $\mathbf{u}$ and h decouple, and in-plane fluctuations are identical to those for a two-dimensional solid, while out-of-plane fluctuations are identical to those of a fluid membrane. The nonlinear $\partial_i h\partial_j h$ term in u_{ij} couples $\mathbf{u}$ to h. If the $\partial_i\mathbf{u}\cdot\partial_j\mathbf{u}$ term in u_{ij} is neglected, then $\mathscr{H}$ is harmonic in $\mathbf{u}$, and the integral over $\mathbf{u}$ in the partition function can be performed exactly to produce an effective Hamiltonian for h:

$$\mathscr{H}_{\text{eff}} = \frac{1}{2}\kappa\int d^2(\nabla^2 h)^2 + \frac{1}{8}Y_2\int d^2x(P_{ij}^T\partial_i h\partial_j h)^2, \tag{10.4.38}$$

where $P_{ij}^T = \delta_{ij} - (\nabla_i\nabla_j/\nabla^2)$ and $Y_2 = 4\mu(\mu+\lambda)/(2\mu+\lambda)$ is the two-dimensional Young modulus. The nonlinear term arising from coupling to strain is proportional to the shear modulus μ and is not present in a fluid membrane. It has fewer powers of ∇, and is therefore stronger than the leading nonlinear term proportional to $(\nabla h)^2(\nabla^2 h)^2$ arising from the curvature energy alone [Eq. (10.4.29)]. The interaction term is, in fact, a long-range interaction between local Gaussian curvatures, as can be seen by using

$$-\frac{1}{2}\nabla^2(P_{ij}^T\partial_i h\partial_j h) = \det\partial_i h\partial_j h, \tag{10.4.39}$$

which, to lowest order in $\boldsymbol{\nabla}h$, is the Gaussian curvature $S(\mathbf{x})$ [Eq. (10.4.14)]. The interaction term is, therefore, $(1/2)\int d^2xd^2x'S(\mathbf{x})\mathscr{G}(\mathbf{x}-\mathbf{x}')S(\mathbf{x}')$ where, $\mathscr{G}(\mathbf{x}) = \nabla^{-4} \sim |\mathbf{x}|^4\ln|\mathbf{x}|$.

The anomalous elasticity of polymerized membranes can be studied analytically in an ϵ-expansion. To do this, it is necessary to generalize to D-dimensional membranes embedded in a space of dimension $d > D$. In this generalization, $\mathbf{u}$ is a D-dimensional displacement variable, and the scalar h becomes a $(d-D)$-

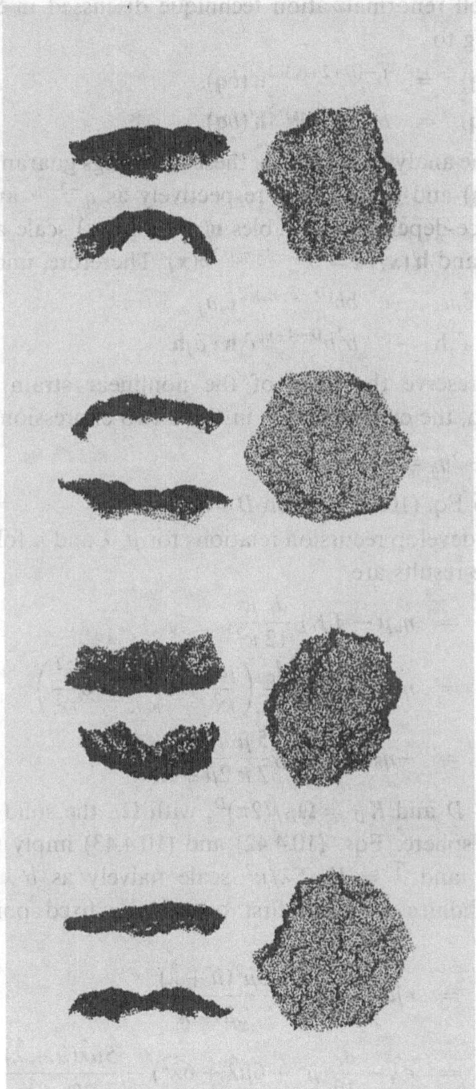

Fig. 10.4.9. Typical configurations of polymerized membranes obtained by computer simulations. [Farid F. Abrahams, W.E. Rudge, and M. Plishke, *Phys. Rev. Lett.* **62**, 1757 (1989).]

dimensional vector $\mathbf{h}$ in the hyperplane perpendicular to $\mathbf{u}$. We now follow the momentum shell renormalization technique discussed in Sec. 5.8. We rescale $\mathbf{u}$ and $\mathbf{h}$ according to

$$\begin{aligned} \mathbf{u}^{<}(\mathbf{q}) &= b^{-(D+2+\eta_u)/2}\mathbf{u}'(b\mathbf{q}), \\ \mathbf{h}^{<}(\mathbf{q}) &= b^{-(D+4-\eta_h)/2}\mathbf{h}'(b\mathbf{q}). \end{aligned} \tag{10.4.40}$$

According to the analysis in Sec. 5.8, these rescalings guarantee that the correlation functions $G_{uu}(\mathbf{q})$ and $G_{hh}(\mathbf{q})$ scale respectively as $q^{-2-\eta_u}$ and $q^{-4+\eta_h}$ at any fixed point. The space-dependent variables $\mathbf{u}(\mathbf{x})$ and $\mathbf{h}(\mathbf{x})$ scale according to $\mathbf{u}'(\mathbf{x}/b) = b^{(D-2-\eta_u)/2}\mathbf{u}(\mathbf{x})$ and $\mathbf{h}'(\mathbf{x}/b) = b^{(D-4+\eta_h)/2}\mathbf{h}(\mathbf{x})$. Therefore, under rescaling, we have

$$\begin{aligned} \partial_i u_j &\rightarrow bb^{(D-2-\eta_u)/2}\partial_i u_j \\ \partial_i \mathbf{h}\cdot\partial_j \mathbf{h} &\rightarrow b^2 b^{D-4+\eta_h}\partial_i\mathbf{h}\cdot\partial_j\mathbf{h}. \end{aligned} \tag{10.4.41}$$

In order to preserve the form of the nonlinear strain [Eq. (10.4.37)] under renormalization, the exponent of b in these two expressions must be equal, or

$$\eta_u + 2\eta_h = 4 - D \equiv \epsilon. \tag{10.4.42}$$

This reduces to Eq. (10.4.34) when $D = 2$.

We can now develop recursion relations for μ, λ and κ following the procedures of Sec. 5.8. The results are

$$\begin{aligned} \frac{d\mu}{dl} &= \eta_u\mu - TK_D\frac{d_c}{12}\frac{\mu^2}{\kappa^2}, \\ \frac{d\lambda}{dl} &= \eta_u\mu - TK_D\frac{d_c}{12}\left(\frac{\mu^2}{\kappa^2} + 6\frac{\mu}{\kappa}\frac{\lambda}{\kappa} + 6\frac{\lambda^2}{\kappa^2}\right), \\ \frac{d\kappa}{dl} &= -\eta_h\kappa + TK_D\frac{5}{2}\frac{\mu}{\kappa}\frac{\mu+\lambda}{2\mu+\lambda}, \end{aligned} \tag{10.4.43}$$

where $d_c = d - D$ and $K_D = \Omega_D/(2\pi)^D$, with Ω_D the solid angle subtended by a D-dimensional sphere. Eqs. (10.4.42) and (10.4.43) imply that the combinations $\hat{\mu} = K_D T\mu/\kappa^2$ and $\hat{\lambda} = K_D T\lambda/\kappa^2$ scale naively as b^{ϵ}. These variables are, therefore, candidates to reach first order in ϵ fixed points. Their recursion relations are

$$\begin{aligned} \frac{d\hat{\mu}}{dl} &= \epsilon\hat{\mu} - \frac{d_c}{12}\hat{\mu}^2 - \frac{5\hat{\mu}^2(\hat{\mu}+\hat{\lambda})}{2\hat{\mu}+\hat{\lambda}}, \\ \frac{d\hat{\lambda}}{dl} &= \epsilon\hat{\lambda} - \frac{d_c}{12}(\hat{\mu}^2 + 6\hat{\mu}\hat{\lambda} + 6\hat{\lambda}^2) - \frac{5\hat{\mu}\hat{\lambda}(\hat{\mu}+\hat{\lambda})}{2\hat{\mu}+\hat{\lambda}}. \end{aligned} \tag{10.4.44}$$

For $\epsilon > 0$, these equations have three fixed points in addition to the Gaussian fixed point. At the globally stable fixed point,

$$\hat{\mu}^* = -3\hat{\lambda}^* = \frac{12\epsilon}{24 + d_c}. \tag{10.4.45}$$

We have not calculated a value for either η_u or η_h. To do this, we use our freedom to fix and overall scale under renormalization. We require that κ appearing in $\mathscr{H}$ remains fixed, just as we required the coefficient of $(\nabla\phi)^2$ to remain constant in our treatment of renormalization of the critical point. (Recall that the κ in $\mathscr{H}$ is not the same thing as the long-wavelength renormalized κ.) This leads to

$$\eta_h = \frac{5}{2}\frac{\hat{\mu}(\hat{\mu}+\hat{\lambda})}{2\hat{\mu}+\hat{\lambda}} = \frac{12\epsilon}{24+d_c},$$

$$\eta_u = \frac{\epsilon}{1+24/d_c}. \tag{10.4.46}$$

Note that these choices for η_h and η_u also fix μ and λ to be constant.

The low-temperature fixed point for polymerized membranes differs from those associated with critical points in that it is globally stable throughout the order phase. It has no relevant direction.

10.5 Arrays of fluctuating walls

There are many examples of physical systems that can be described in terms of arrays of walls. For example, Kr and Xe adsorbed on graphite can form incommensurate phases consisting of regularly spaced discommensurations in either a parallel or hexagonal array, as discussed in Sec. 10.3. Similarly, lamellar phases in microemulsions consist of a stack of regularly spaced surfactant surfaces, as shown in Fig. 2.7.14.

1 *Fluctuating walls and steric entropy*

In the absence of thermal fluctuations, the equilibrium properties of these arrays of walls are generally determined, as in the FK model, by the competition between a negative wall energy (surface tension) and a repulsive interaction potential between walls. We have just seen, however, that isolated walls below their lower critical dimension fluctuate violently with mean-square height fluctuations that diverge with the length of their base planes, and one can ask what effect these fluctuations have on wall arrays. There is always a short-range repulsion between walls that prohibits them from passing through each other. Thus, the amplitude of height fluctuations of a wall confined between two other walls is constrained to be less than the distance between walls. This constraint reduces the phase space available to the wall and leads to a decrease, $\Delta S < 0$, in the entropy of a confined wall relative to an unconfined wall. There is an associated increase $-T\Delta S = T \mid \Delta S \mid$ in the free energy of the wall and thus an effective repulsive interaction between walls. This repulsion is often called *steric repulsion,* and the reduction in entropy is called *steric entropy*. The concept of steric entropy first (de Gennes 1968) arose in connection with polymers confined between two walls. It was introduced in a study of lamellar liquid crystals (Helfrich 1978), and subsequently rediscovered and applied independently to two-dimensional soliton lattices (Pokrovsky and Talapov 1979).

We consider first a parallel array of walls with a preferred orientation relative to a substrate in two dimensions described by Eq. (10.4.20) with $\gamma > 0$ and $\sigma_1 > 0$. Assume the walls are, on average, parallel to the x-axis, and let l be their average separation. Neighboring walls will collide with each other as shown in Fig. 10.5.1. Between collisions, each wall fluctuates as though the other walls were not present, and can, thus, be described by the independent wall Hamiltonian of Eq. (10.4.20). Let L_x and L_y be, respectively, the lengths along the x- and y-axes of the wall array, and let L_B be the distance along the x-axis between collisions. The average separation between layers should be of order the root mean square height fluctuation of a wall of base length L_B, i.e., according to Eq. (10.4.21), $l^2 \sim (T/2\pi\gamma)L_B$. Since a wall behaves freely between collisions, its free energy F_{wall} is simply the independent sum of contributions from free walls with base length L_B, i.e.,

$$F_{\text{wall}} = \sigma L_x + N_w \Delta F_{\text{har}}(L_B), \tag{10.5.1}$$

where $N_w = L_x/L_B$ is the number of independent wall segments and where ΔF_{har} is the free energy of a wall of base length L_B described by the Hamiltonian of Eq. (10.4.20). A good estimate for F_{har} is obtained by replacing $(dh/dx)^2$ by $\langle h^2 \rangle / L_B^2 \sim l^2/L_B^2$ in $\mathcal{H}_{\text{har}}$:

$$\Delta F_{\text{har}} \sim L_B(\gamma l^2/L_B^2) \sim \gamma l^2/L_B. \tag{10.5.2}$$

Then, using $L_B \sim (2\pi\gamma/T)l^2$ [Eq. (10.4.21)], we find

$$F_{\text{wall}} = L_x \left(\sigma + C\frac{T^2}{\gamma l^2} \right), \tag{10.5.3}$$

where C is a numerical constant of order unity. This free energy is often referred to as the *Pokrovsky-Talapov* energy. Calculations based upon a mapping of this problem onto one of interacting fermions in one dimension yield $C = \pi^2/6$ (Okwamoto 1980; Schulz 1980).

The free energy per unit area of a stack of $N = L_y/l$ parallel walls is simply

$$f = \frac{NF_{\text{wall}}}{L_x L_y} = \frac{\sigma}{l} + C\frac{T^2}{\gamma l^3}. \tag{10.5.4}$$

This equation describes the surface tension and steric entropy contributions to the free energy density. Potential energy contributions are, in general, also present. If the walls are FK discommensurations, the repulsive interaction dies off exponentially with l and is always small compared to the steric entropy term at sufficiently large l for $T > 0$. In equilibrium, f is a minimum with respect to variations in l. The interfacial stiffness $\gamma = \sigma + \sigma_1$ must be positive; otherwise, the walls would be unstable with respect to the development of spatial modulations. In order for a minimum with nonzero wall density l^{-1} to exist, however, σ must be negative. Thus, as noted above, σ_1 must be greater than $|\sigma|$. In this case, minimization of f with respect to l yields

$$l = \left(\frac{3CT^2}{|\sigma|\gamma} \right)^{1/2}. \tag{10.5.5}$$

This should be compared with Eq. (10.3.36) for the FK model, which neglects

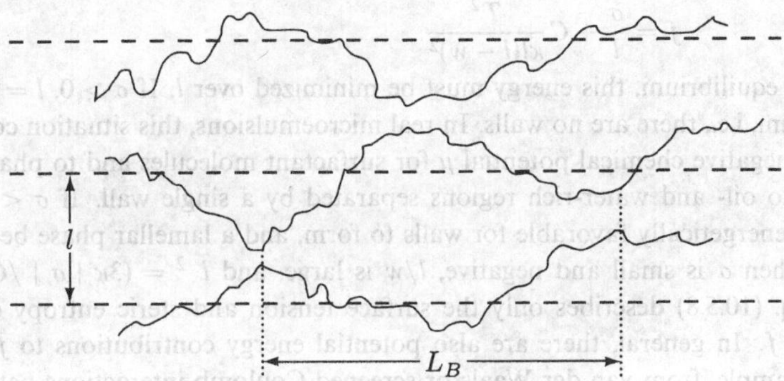

Fig. 10.5.1. Representation of a lattice of fluctuating walls in two dimensions. The average separation between walls is l. Each wall is confined between its two neighbors. The average distance along the x-axis between collisions of a given wall with its neighbors is L_B. Between collisions, each wall behaves like a free wall. The two distances l and L_B are related by the constraint that $l^2 = (T/2\pi\gamma)L_B$.

fluctuation. The surface tension σ is equivalent to $|\,\delta\,| - \delta_c$. At zero temperature, when thermal fluctuations can be ignored, $l^{-1} \sim [\ln(1/\mid\sigma\mid)]^{-1}$. At finite temperatures, however, $l^{-1} \sim T^{-1} \mid \sigma \mid^{1/2}$ rises more slowly from zero when σ becomes negative. (Note that l is the real space distance between walls. Eq. (10.3.36) applies to the average number of atoms $\mathscr{L}$ between discommensurations. From Eq. (10.3.22), $l = \mathscr{L}\tilde{a} = (\mathscr{L} \pm 1)b$, and l and $\mathscr{L}$ differ only by a scale factor when $\mathscr{L} \gg 1$.)

Lamellar phases in microemulsions consisting of stacks of surfactant layers with fluctuations dominated by curvature energy can be stabilized by steric entropy in much the same way as walls in two dimensions with a preferred orientation. It turns out, however, to be of some importance to take into account the finite width w of the surfactant walls. Thus, though the average distance between walls is l, the mean square height fluctuation is only $\langle h^2 \rangle = (l-w)^2$. If the average distance in the xy-plane between collisions is L_B,

$$(l-w)^2 = \frac{T}{2\pi\kappa} L_B^2, \tag{10.5.6}$$

and

$$\begin{aligned} F_{\text{wall}} &= \sigma L_x L_y + N_c L_B^2 \kappa \langle (\nabla_\perp^2 h)^2 \rangle \\ &= L_x L_y \left(\sigma + C \frac{T^2}{\kappa (l-w)^2} \right). \end{aligned} \tag{10.5.7}$$

To a good approximation, the surfactant molecules comprising the walls are incompressible, and increasing the area of walls is equivalent to increasing the number of surfactant molecules. This means that the surface tension σ is equal to the negative of the chemical potential for surfactant molecules times the area per molecule: $\sigma = -\mu a^2$. The energy per unit volume of a stack of walls is

$$f = \frac{\sigma}{l} + C\frac{T^2}{\kappa l(l-w)^2}. \tag{10.5.8}$$

In equilibrium, this energy must be minimized over l. If $\sigma > 0$, $l = \infty$ in equilibrium, i.e., there are no walls. In real microemulsions, this situation corresponds to a negative chemical potential μ for surfactant molecules and to phase separation into oil- and water-rich regions separated by a single wall. If $\sigma < 0$ ($\mu > 0$), it is energetically favorable for walls to form, and a lamellar phase becomes stable. When σ is small and negative, l/w is large, and $l^{-2} = (3\kappa \mid \sigma \mid /CT^2)$. Again, Eq. (10.5.8) describes only the surface tension and steric entropy contributions to f. In general, there are also potential energy contributions to f, arising, for example, from van der Waals or screened Coulomb interactions between walls.

2 Honeycomb lattice of walls

Fluctuations have an even more pronounced effect on hexagonal incommensurate structures, such as that shown in Fig. 10.3.7b, than they do on striped phases. The honeycomb lattice has the very unusual property (Villain 1980) that displacements of vertices where three walls intersect do not change the total length of walls provided no walls cross. A typical configuration of the honeycomb lattice with randomly displaced vertices is shown in Fig. 10.5.2. These configurations, which will be thermally excited at any nonzero temperature, carry considerable entropy. The diameter of any hexagon can change, without shrinking to zero size or colliding with a neighbor, by a factor of order l/b, where l is the average length of a hexagonal side and b is the substrate lattice spacing. Thus, the entropy S associated with fluctuations in the positions of wall intersections is of order $N_h \ln(l/b)$, where N_h is the number of hexagons. At finite temperature, the entropic contribution $-TS$ will dominate the exponential potential term at large l, and Eq. (10.3.57) should be replaced by the free energy density

$$f = \frac{6\sigma(\delta) + 4e_c - BT\ln(l/b),}{3\sqrt{3}l^2} \tag{10.5.9}$$

where B is a constant of order unity and e_c is the wall-crossing energy. This free energy, like the energy of Eq. (10.3.57), predicts a first-order CI transition.

3 Elasticity of sterically stabilized phases

Since equilibrium and stability of sterically stabilized phases are determined by entropic rather than energetic contributions to the free energy, the only important energy scale in these systems is set by the temperature. As a consequence, physical quantities such as elastic constants will tend to zero with temperature. This leads to some interesting experimentally verifiable consequences.

The striped incommensurate phases are essentially two-dimensional smectic liquid crystals with a preferred orientation imposed by the anisotropic substrate.

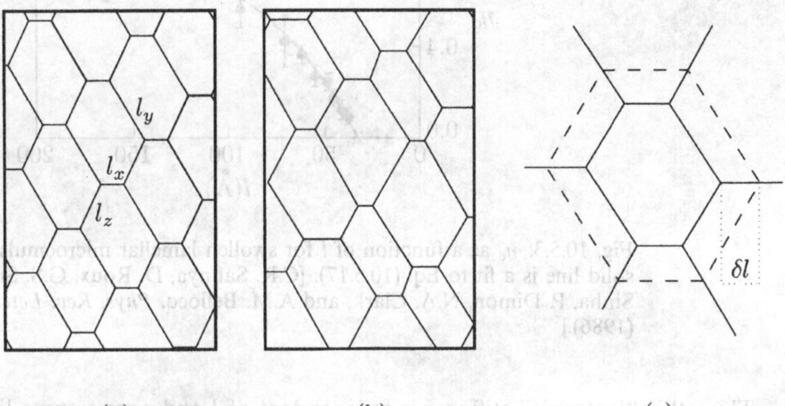

Fig. 10.5.2. (a) A semi-regular honeycomb lattice and (b) a topologically equivalent distorted partner in which each vertex has the same neighbors as in (a). The total length of hexagon edges is the same in (a) and (b), as can be seen by the construction in (c).

The elastic free energy can be expressed in terms of the displacements $\bar{u}(\mathbf{x})$ [see Eqs. (10.3.27) and (10.3.28)] of walls from equilibrium:

$$F_{\text{el}} = \frac{1}{2}\int d^2x[\bar{B}(\partial_y\bar{u})^2 + \bar{K}(\partial_x\bar{u})^2], \tag{10.5.10}$$

where

$$\bar{B} = l^2\frac{\partial^2 f}{\partial l^2} = 6C\frac{T^2}{\gamma l^3} \tag{10.5.11}$$

is the layer compressibility modulus. The constant $\bar{K}$ measures the energy associated with rotating the lattice and is simply

$$\bar{K} = \gamma/l. \tag{10.5.12}$$

Eq. (10.5.10) expresses the elastic energy in terms of the wall displacement variable. Alternatively, it can be expressed in terms of the adatom displacement variable $u = \bar{u}(b/l)$:

$$F_{\text{el}} = \frac{1}{2}\int d^2x[B(\partial_y u)^2 + K(\partial_x u)^2], \tag{10.5.13}$$

where $B = (l/b)^2\bar{B}$ and $K = (l/b)^2\bar{K}$. These free energies are anisotropic versions of the xy elastic free energy discussed extensively in Chapter 6. A simple rescaling of lengths, $x \to x' = (B/K)^{1/4}x$, $y \to y' = (B/K)^{-1/4}y$, leads to the isotropic xy free energy with stiffness

$$\sqrt{BK} = b^{-2}T\sqrt{6C}. \tag{10.5.14}$$

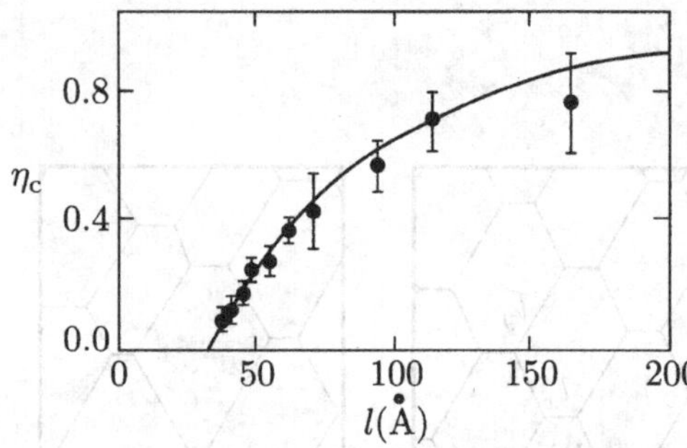

Fig. 10.5.3. η_c as a function of l for swollen lamellar microemulsions. The solid line is a fit to Eq. (10.5.17). [C.R. Safinya, D. Roux, G.S. Smith, S.K. Sinha, P. Dimon, N.A. Clark, and A.M. Bellocq, *Phys. Rev. Lett.* **57**, 2718 (1986).]

Thus, this "isotropic" stiffness is independent of l and goes to zero linearly with temperature.

A sterically stabilized lamellar microemulsion is a smectic liquid crystal which must be described by the long wavelength elastic theory of Eq. (6.3.10). The layer compression modulus is simply

$$B = l^2 \frac{\partial^2 f}{\partial l^2} = 6C \frac{T^2 l}{\kappa(l-w)^4}. \qquad (10.5.15)$$

The bending elastic constant K_1 is clearly proportional to the bending rigidity κ of an individual layer. If the contributions of K_1 arising from thermal fluctuations are ignored (it can in fact be shown that they are small), then

$$K_1 = \kappa / l \qquad (10.5.16)$$

simply by dimensional analysis. An important consequence of Eqs. (10.5.15) and (10.5.16) is that the exponent η_c of Eq. (6.3.16) measuring the power-law decay of order parameter correlations is independent of temperature and independent of l if $l \gg w$:

$$\eta_c = \frac{q_0^2 T}{2\pi\sqrt{BK_1}} = \left(1 - \frac{w}{l}\right)^{-2}. \qquad (10.5.17)$$

Fig. 10.5.3 shows an experimentally measured curve of η_c versus layer-spacing l for a swollen lamellar microemulsion, which provides striking confirmation of Eq. (10.5.17) and the importance of steric entropy in these systems.

4 Dislocations and the CI transition

We have just seen that the striped incommensurate phase is described by an elastic energy identical to that of a two-dimensional *xy*-model. It should, therefore, undergo a Kosterlitz-Thouless transition to a disordered fluid phase. The topological excitations of the striped incommensurate phase are edge dislocations in

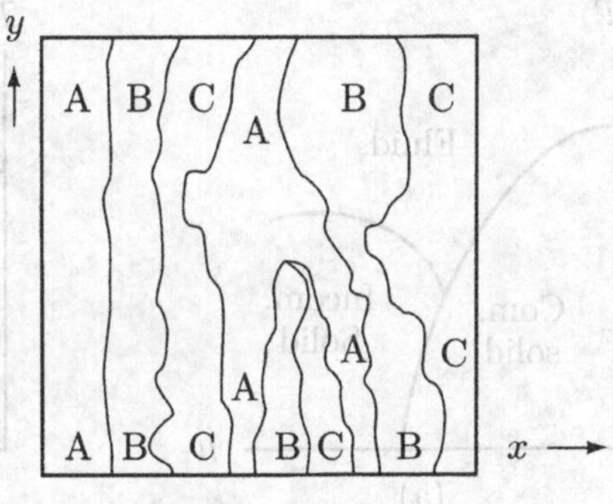

Fig. 10.5.4. Dislocation in a domain wall striped phase near a 3 × 1 commensurate structure. The three possible registries of the adsorbate lattice with the substrate are denoted by *A*, *B* and *C*. A domain wall in which a single extra layer of adatoms is inserted requires the insertion of three domain walls because, in order for *A* regions to connect smoothly from one side of the sample to the other, it is necessary to insert a *B* and a *C* region between two *A* regions as shown.

which the layer displacement variable u undergoes a change in one circuit around a core that leaves the mass-density amplitudes ρ_G unchanged for all reciprocal lattice vectors G. It can be seen from the discussion of X-ray diffraction in Sec. 10.3 and Eq. (10.3.41) that, to satisfy this condition, u must change by an integral multiple of the average adatom spacing $\tilde{a}$ in one circuit around a dislocation core. Thus, the edge dislocations of interest correspond to the insertion or removal of a line of adatoms. In general, one might expect there to be dislocations associated with the second period (namely that of $\tilde{\psi}(n\tilde{a} - \overline{u})$ in Eq. (10.3.39)), the distance $l = \mathscr{L}\tilde{a}$ between discommensurations, appearing in the mass-density wave expansion of Eq. (10.3.39). These, however, can be ignored since they require changes in the phase of the substrate lattice, which we have assumed is frozen. The insertion of one extra line of adatoms in an incommensurate state near a $p \times 1$ structure is equivalent to the insertion or removal of p discommensuration lines, as can be seen from Eq. (10.3.4). To be concrete, consider the case in which $\tilde{a} = (p - \mathscr{L}^{-1})b$ with $\mathscr{L} \gg 1$, and assume there are initially N adatom lines occupying a length $L_N = Npb - N_{\text{dis}}b$ along the y-axis. If one more line is added such that the total length remains fixed, then the number of discommensuration lines must change from N_{dis} to N'_{dis}: $L_{N+1} = (N+1)pb - N'_{\text{dis}}b = L_N$, and $N'_{\text{dis}} = N_{\text{dis}} + p$. This argument gives the correct relation between the number of extra layers and the number of discommensurations. Far from the dislocation core, of course, L_{N+1} will be equal to $(N+1)\tilde{a}$ in equilibrium. A dislocation in an incommensurate state relative to a 3×1 structure is shown in Fig. 10.5.4.

The striped phase is stable with respect to unbinding of dislocations provided the exponent $\eta(T)$ describing the decay of order parameter correlations is less than 1/4 [see Eq. (9.4.38)]. In the xy-model, $\eta(T) = T/2\pi\rho_s$. In the present case,

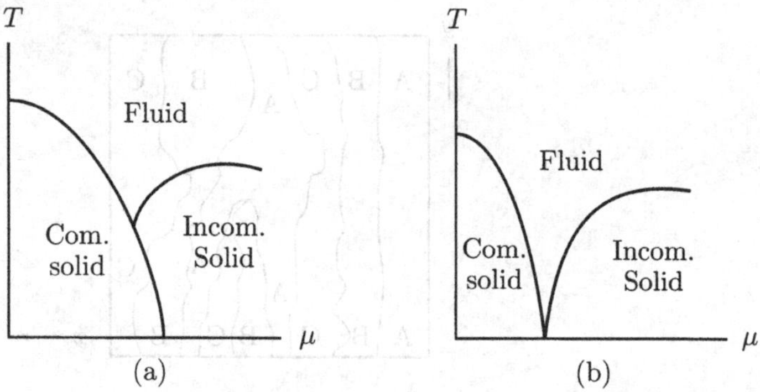

Fig. 10.5.5. Possible phase diagrams in the temperature (T)–chemical potential (μ) plane for a CI transition from a $p \times 1$ commensurate state to an incommensurate state. (a) $p \geq 3$; (b) $p \leq 2$. [S.N. Coppersmith, Daniel S. Fisher, B.I. Halperin, P.A. Lee, and W.F. Brinkman, *Phys. Rev. Lett.* **46**, 549 (1981).]

the elastic variable u has units of length rather than angle, and $\rho_s = (\tilde{a}/2\pi)^2\sqrt{BK}$. Thus, when l is large and steric entropy dominates potential repulsion so that $\sqrt{BK}$ is given by Eq. (10.5.14),

$$\eta(T) = \frac{2\pi}{\tilde{a}^2\sqrt{BK}} = \left(\frac{b}{\tilde{a}}\right)^2 \frac{2\pi}{\sqrt{6C}}. \tag{10.5.18}$$

This quantity is independent of temperature and depends only on $\tilde{a}/b$ and the constant C. As discussed earlier, calculations based on mapping the present problem onto a problem of interacting fermions in one dimension yield $6C = \pi^2$. Thus the condition for stability of a sterically stabilized striped phase with $\tilde{a} \approx pb$ is

$$\eta = 2p^{-2} < 1/4, \tag{10.5.19}$$

or $p^2 > 8$. The separation between discommensurations diverges as the incommensurate-commensurate (IC) transition is approached. Thus, near the CI transition, steric entropy always dominates potential repulsion, and Eq. (10.5.18) applies. We therefore conclude that there can be no CI transition for $p < 3$ at any nonzero T, and that a fluid phase must intervene between a $p \times 1$ commensurate phase and an incommensurate phase at all temperatures greater than zero, as shown in Fig. 10.5.5a. When $p \geq 3$, a CI transition at nonzero temperature is possible, as indicated in Fig. 10.5.5b. As l decreases, exponential potential repulsion cannot be neglected, and elastic constants and $\eta(T)$ acquire nontrivial temperature dependences and a finite transition temperature T_M from the incommensurate to the fluid phase, as shown in Fig. 10.5.5a (see Problem 10.11).

We saw in Sec. 9.5 that a two-dimensional hexagonal solid on an anisotropic substrate undergoes a Kosterlitz-Thouless dislocation unbinding transition to a fluid phase with the rotational symmetry of the substrate when Eq. (9.5.3) is satisfied. Elastic constants for the hexagonal honeycomb lattice can be calculated following procedures similar to those just described by the striped phases. The bulk and shear moduli are proportional to T/l^2 times slowly varying functions of σ, and the left hand side of Eq. (9.5.3) is independent of temperature and less than the critical value necessary for stability against dislocation unbinding. Thus, it is expected that a fluid phase will intervene between the $\sqrt{3} \times \sqrt{3}R30^0$ commensurate and the honeycomb incommensurate structure at all temperatures, provided the zero temperature CI transition is weakly first order (i.e., l is large).

10.6 Roughening and faceting

We saw in the preceding section that the fluid-fluid interface separating coexisting liquid and gas phases is macroscopically rough, i.e., that its height fluctuations diverge with its linear dimension. The solid-fluid interface separating coexisting solid and gas or solid and liquid phases differs from that separating liquid and gas phases because the solid has a periodic density that favors interfaces coinciding with lattice planes. One might expect, and indeed experience generally confirms, that a solid-fluid interface should be macroscopically smooth and should not exhibit divergent height fluctuations. In this section, we will develop models to describe equilibrium solid-fluid interfaces, which predict that the interfaces are smooth at low temperature, but that they can undergo a *roughening transition* to a state in which, like a fluid-fluid interface, they are macroscopically rough. The idea that a crystal interface might roughen goes back to before 1950 (Burton and Cabrera 1949; Burton, Cabrera, and Frank 1951), when it was suggested that the existence of large fluctuations in surface structure might lead to a disappearance of the nucleation barrier to crystal growth. The morphology and growth of a wide class of crystals grown in melts can be understood using these ideas.

1 *The solid-on-solid and discrete Gaussian models*

The atoms of a crystalline solid occupy positions on a regular lattice. At low temperatures, vacancies, interstitials and other defects can be ignored so that every lattice site in the crystal phase is occupied by an atom. The density of the gas phase is essentially zero, and one can define a solid-gas interface as the surface along which occupancy of lattice sites changes from one to zero.

Consider for simplicity the (001) face of a tetragonal lattice with lattice parameters a in the basal plane and b normal to the basal plane. If overhangs are ignored, then the surface configuration is represented by a two-dimensional array of integers specifying the number of atoms in each column perpendicular to the (001) face or, equivalently, by the height of the column relative to the flat $T = 0$ reference surface. Growth or evaporation of the crystal involves the atoms

Fig. 10.6.1. Atoms on a (001) face of a tetragonal crystal. Surface atoms have up to four lateral neighbors. An energy J is associated with each exposed lateral face.

at the tops of their columns. Complex surfaces with random structure can be represented with this column model, as shown in Fig. 10.6.1.

In its lowest energy state, the surface is flat, every column has the same height, and the lateral nearest neighbor sites in the (001) plane of each surface atom are occupied. In excited states, column heights differ, and some atoms will have unoccupied neighbor sites that appear as exposed vertical surfaces, as shown in Fig. 10.6.1. It is natural to assign an energy J to each exposed vertical surface. The number of exposed surfaces between columns $\mathbf{l}$ and $\mathbf{l}'$ is simply $|h_\mathbf{l} - h_{\mathbf{l}'}|/b$, where $h_\mathbf{l}$ is the height of column $\mathbf{l}$. This leads to the absolute solid-on-solid (ASOS) model with Hamiltonian

$$\mathcal{H}_{ASOS} = (J/b) \sum_{<\mathbf{l},\mathbf{l}'>} |h_\mathbf{l} - h_{\mathbf{l}'}|, \tag{10.6.1}$$

where the sum is over nearest neighbor columns $< \mathbf{l}, \mathbf{l}' >$. If a surface with N_x columns along the x-axis and N_y columns along the y-axis is rotated through an angle β about the y-axis (while preserving the orientation of the bulk crystal), there will be $N_x N_y |\tan\beta|$ exposed vertical surfaces and energy

$$e = (J/a^2)|\tan\beta| \tag{10.6.2}$$

per unit area of the reference surface. This expression is nonanalytic in β and is to be compared with the energy proportional to β^2 of a fluid interface. The reference surface itself has $N_x N_y$ exposed horizontal surfaces. If we assign an energy $J_\perp$ to each of these exposed surfaces, then the energy per unit area of the *tilted* surface is

$$\sigma(\mathbf{N}) = (J/a^2)|\sin\beta| + (J_\perp/a^2)|\cos\beta| \tag{10.6.3}$$

since the area of the tilted surface is $N_x N_y a^2/|\cos\beta|$. A crystal with this surface energy is called a Kosel crystal.

At low temperatures, the height differences between neighboring columns will be small, and any model that assigns an energy difference to height differences should provide an accurate description of fluctuations of the surface. These models go under the general heading of *solid-on-solid models* or SOS models. The discrete Gaussian model, which assigns an energy proportional to $(h_\mathbf{l} - h_{\mathbf{l}'})^2$ rather than $|h_\mathbf{l} - h_{\mathbf{l}'}|$, is analytically more tractable than the ASOS model. The Hamiltonian,

$$\mathscr{H}_{DG} = (J/b^2) \sum_{<\mathbf{l},\mathbf{l}'>} (h_\mathbf{l} - h_{\mathbf{l}'})^2, \tag{10.6.4}$$

of the discrete Gaussian model was introduced in Appendix 9B. It is dual to the Villain model, which in turn is in the same universality class as the *xy*-model. Duality, as discussed in Appendix 9B, maps the low-temperature properties of one model onto the high-temperature properties of a dual model. The free energy of a model and its dual will have the same temperature singularities. Since the *xy*-model exhibits a Kosterlitz-Thouless transition, its dual, the discrete Gaussian, model will also. We will rederive this result shortly.

To discuss the long-wavelength statistical mechanics of fluctuating surfaces, it is useful to introduce yet another model that at low temperatures is equivalent to both the ASOS and the discrete Gaussian models. The partition function for the discrete Gaussian model is

$$\begin{aligned} Z_{DG} &= \sum_{h_\mathbf{l}} e^{-\mathscr{H}_{DG}/T} \\ &= \int \mathscr{D}h_\mathbf{l} \prod_\mathbf{l} W(h_\mathbf{l}) e^{-\mathscr{H}_{DG}/T}, \end{aligned} \tag{10.6.5}$$

where

$$W(h) = \sum_{n=-\infty}^{\infty} \delta(h - nb) \tag{10.6.6}$$

is a weighting function that restricts each $h_\mathbf{l}$ to be an integral multiple of the lattice parameter b. Any weight function that favors heights that are integral multiples of b will lead to low-temperature thermodynamic properties that are in the same universality class as the discrete Gaussian model. The simplest such weight function is $W(h) = \exp[-u\cos(2\pi h/b)]$. If we now take the continuum limit, we obtain

$$Z_{SG} = \int \mathscr{D}h(\mathbf{x}) e^{-\mathscr{H}_{SG}/T}, \tag{10.6.7}$$

where

$$\mathscr{H}_{SG} = \mathscr{H}_0 + \mathscr{H}_u \tag{10.6.8}$$

is the two-dimensional sine-Gordon Hamiltonian (which is dual to the two-dimensional Coulomb gas Hamiltonian, Problem 9.10) with

$$\mathscr{H}_0 = \frac{1}{2}\gamma \int d^2x (\nabla h(\mathbf{x}))^2 \tag{10.6.9}$$

and

$$\mathcal{H}_u/T = -u\int \frac{d^2x}{a^2}\cos[2\pi h(\mathbf{x})/b], \tag{10.6.10}$$

where $\gamma = 4J/b^2$. A correlation length

$$\xi = \frac{2\pi}{ab}\left(\frac{\gamma}{uT}\right)^{1/2} \tag{10.6.11}$$

can be introduced [as in Eq. (10.3.16)] by comparing the harmonic parts of Eqs. (10.6.10) and (10.6.11).

2 The roughening transition

The mechanism that leads to a roughening transition is clear from the sine-Gordon Hamiltonian. At low temperatures, the potential u is nonzero and favors smooth surfaces with heights at an integral multiple of b. As temperature increases, fluctuations away from flatness increase and renormalize the long-wavelength value of u. If u renormalizes to zero at some temperature T_R, then, for $T > T_R$, the effective Hamiltonian has only the $\gamma(\nabla h)^2$ term and is thus equivalent to that describing fluid interfaces. Below T_R, when u is nonzero, excitations in which h increases by one step height b are identical to sine-Gordon solitons discussed in Sec. 10.2. These excitations have a finite energy. A surface tilted at an angle θ relative to the reference flat surface has $|\tan\theta| \sim |\theta|$ steps per unit length of the reference surface. Thus, as in the ASOS model, the energy per unit area of the tilted surface is nonanalytic in θ and proportional to $|\theta|$ for small θ.

To derive renormalization equations for γ and u, we proceed in much the same way as we did to derive the Kosterlitz-Thouless recursion relations for the xy-model. The renormalized surface stiffness γ_R is defined via $F(v)-F(0) = \Omega\gamma_R v^2/2$, where $\mathbf{v} = (1/\Omega)\int d^2x\langle\nabla h\rangle$ is the average gradient of h. We therefore define $h(\mathbf{x}) = \mathbf{v}\cdot\mathbf{x} + h'(\mathbf{x})$, where $h'(\mathbf{x})$ is constrained to be zero at the boundaries of the surface. The sine-Gordon Hamiltonian as a function of $\mathbf{v}$ is

$$\mathcal{H}(\mathbf{v}) = \frac{1}{2}\Omega\gamma v^2 + \mathcal{H}_0[h'(\mathbf{x})] + \mathcal{H}_u[h'(\mathbf{x}) + \mathbf{x}\cdot\mathbf{v}]. \tag{10.6.12}$$

Expanding $F(\mathbf{v}) = -T\ln\operatorname{Tr}\exp(-\mathcal{H}(\mathbf{v})/T)$ to second order in v, we obtain

$$\begin{aligned} F(\mathbf{v}) - F(0) &= \frac{1}{2}\gamma\Omega v^2 - T\ln\left\langle e^{-\mathcal{H}_u/T}\right\rangle_0 \\ &= \frac{1}{2}\gamma\Omega v^2 \\ &\quad -T\left[-\langle\mathcal{H}_u\rangle_0/T + \frac{1}{2}(\langle\mathcal{H}_u^2/T^2\rangle_0 - \langle\mathcal{H}_u/T\rangle_0^2)\right] + \cdots, \end{aligned} \tag{10.6.13}$$

where $\langle A\rangle_0$ is the average of A with respect to $\mathcal{H}_0$. Then, differentiating with respect to v_i, we obtain

$$\gamma_R = \gamma + \frac{1}{2\Omega}\operatorname{Tr}\frac{\partial^2\langle\mathcal{H}_u\rangle_0}{\partial v_i\partial v_j} - \frac{1}{4T\Omega}\operatorname{Tr}\frac{\partial^2}{\partial v_i\partial v_j}[\langle\mathcal{H}_u^2\rangle_0 - \langle\mathcal{H}_u\rangle_0^2], \tag{10.6.14}$$

where Tr refers to a sum over the indices i and j. Fluctuations in $h'(\mathbf{x})$ with respect to $\mathcal{H}_0$ are divergent, and $\langle\mathcal{H}_u\rangle_0 = 0$ as $\langle\cos[2\pi h'(\mathbf{x})/b]\rangle_0 = 0$. Thus, the only second-order perturbation term that survives in Eq. (10.6.14) is

$$\langle (\mathscr{H}_u/T)^2\rangle_0 = u^2 \int \frac{d^2x}{a^2}\frac{d^2x'}{a^2}\left\langle \cos\left[\frac{2\pi}{b}(h'(\mathbf{x}_\perp)+\mathbf{v}\cdot\mathbf{x}_\perp)\right] \times \cos\left[\frac{2\pi}{b}(h'(\mathbf{x}'_\perp)+\mathbf{v}\cdot\mathbf{x}'_\perp)\right]\right\rangle, \tag{10.6.15}$$

from which we obtain

$$\frac{1}{\Omega}\frac{\partial^2\langle(\mathscr{H}_u/T)^2\rangle_0}{\partial v_i \partial v_j}\bigg|_{v=0} = u^2\left(\frac{2\pi}{b}\right)^2\frac{1}{\Omega} \times \int \frac{d^2x}{a^2}\frac{d^2x'}{a^2}[(x_ix_j + x'_ix'_j)\langle\cos[2\pi h'(\mathbf{x}_\perp)/b]\cos[2\pi h'(\mathbf{x}'_\perp)/b]\rangle_0 - (x_ix'_j + x'_ix_j)\langle\sin[2\pi h'(\mathbf{x}_\perp)/b]\sin[2\pi h'(\mathbf{x}'_\perp)/b]\rangle_0]. \tag{10.6.16}$$

Then, using the fact that $\langle\cos[(2\pi/b)(h'(\mathbf{x}_\perp)+h'(\mathbf{x}'_\perp))]\rangle = 0$ and expanding the products of sines and cosines in terms of cosines of sum and difference variables, we obtain

$$\gamma_R = \gamma + \frac{1}{8}T\left(\frac{2\pi}{b}\right)^2 u^2 \int \frac{d^2x}{a^2}\left(\frac{r}{a}\right)^2\langle\cos[(2\pi/b)(h(\mathbf{x})-h(0))]\rangle_0. \tag{10.6.17}$$

where $r = |\bar{\mathbf{x}}|$. But

$$\langle\cos[(2\pi/b)(h(\mathbf{x})-h(0))]\rangle_0 = \exp[-2\pi K\ln(r/a)], \tag{10.6.18}$$

where

$$K = \frac{T}{\gamma b^2}. \tag{10.6.19}$$

Defining $K_R = T/(\gamma_R b^2)$, we can rewrite Eq. (10.6.17) as

$$K_R^{-1} = K^{-1} + \pi^3 u^2 \int_a^\infty \frac{dr}{a}\left(\frac{r}{a}\right)^{3-2\pi K}. \tag{10.6.20}$$

The lower cutoff a in this integral, as in the xy-model, occurs because distances between sites on the lattice must be larger than a lattice spacing a. The continuum sine-Gordon model must retain this or some equivalent short-distance cutoff to be well-defined. If u is identified with $2y$, the vortex fugacity, this equation is identical to Eq. (9.4.26) predicting a Kosterlitz-Thouless transition. Note, however, that $K = T/(\gamma b^2)$ in the present case is linear in T, whereas $K = \rho_s/T$ for the xy-model is linear in $1/T$. Thus, the low- and high-temperature phases of the two models are reversed as duality of the xy- and discrete Gaussian models requires.

The properties of the Kosterlitz-Thouless transition are now easily translated to the roughening problem, and we review them here. First, the universal value of $K_c = 2/\pi$ at the roughening temperature leads to

$$T_R = 2\gamma_R(T_R)b^2/\pi. \tag{10.6.21}$$

This result implies that faces with the largest value of b will have the highest transition temperature (assuming γ does not vary significantly with angle). The length b is the distance between equivalent planes; it is a maximum along symmetry directions (see Sec. 2.6), and high-symmetry surfaces, such as the (001) surface of a cubic crystal, are the last to become rough as temperature increases.

The energy per unit length of a step in the flat phase is the sine-Gordon soliton energy [Eq. (10.2.18)], which in the current units is $\epsilon = (4/\pi)(b/a)\sqrt{T\gamma u}$. Dimensional analysis and scaling predict that this quantity must be proportional to T_R/ξ, where $\xi \sim \exp(B/\sqrt{T_R - T})$ is the Kosterlitz-Thouless correlation length [Eq. (9.4.40)]. This result can be obtained from the renormalization group by integrating out degrees of freedom up to length scales ae^{l^*}, where $u(l^*)/K(l^*) \sim 1$:

$$\epsilon = \frac{4T}{\pi a e^{l^*}}\sqrt{\frac{u(l^*)}{K(l^*)}} \sim \frac{T}{\xi}. \tag{10.6.22}$$

The energy of a step thus decreases as ξ^{-1} as the roughening transition is approached from below. As shown in Fig. 10.6.2, steps can wander like the walls in an incommensurate soliton lattice discussed in Sec. 10.4, and there is entropic repulsion between them. A surface making a small angle β relative to the reference surface is called a *vicinal surface.* It is composed of a series of fluctuating steps separated by flat terraces, as shown in Fig. 10.6.2. The average distance between steps is determined by $b/l = |\tan\beta| = |\nabla_\perp h|$. The free energy per unit base plane area of a vicinal surface is, therefore, identical to the Pokrovsky-Talapov free energy of Eq. (10.5.4) with the line tension for a step equal to ϵ (rather than γ):

$$f(\nabla h) = f_0 + \frac{\epsilon}{b}|\nabla_\perp h| + C\frac{T^2}{\epsilon b^3}|\nabla_\perp h|^3, \tag{10.6.23}$$

where f_0 is the free energy per unit area of the flat reference surface. A tilted surface that is in equilibrium at angle β_0 with respect to the base plane, whose energy is given by Eq. (10.6.23), is rough because Eq. (10.6.23) has an analytic expansion in deviations $\delta\beta = \beta - \beta_0$. As β_0 varies, however, there can be commensurate lock-ins to crystallographic directions with energies not described by Eq. (10.6.23).

3 Faceting

A liquid droplet in equilibrium coexistence with its vapor phase will have a spherical shape determined by the requirement that its surface energy σA be a minimum for a fixed volume V. A crystalline solid in equilibrium with its liquid or vapor phase, on the other hand, generally has flat faces, or equivalently *facets,* oriented along symmetry axes of the crystal and sharp edges where differently oriented facets meet. As the triple point is approached along the solid-vapor coexistence line, thermal fluctuations increase, the solid becomes more fluid-like, and one would expect some evolution of the fully faceted low-temperature crystal shape towards a spherical shape, as shown in Fig. 10.6.3. Since crystal faces have a fluid-like energy above their roughening temperature, there should be no

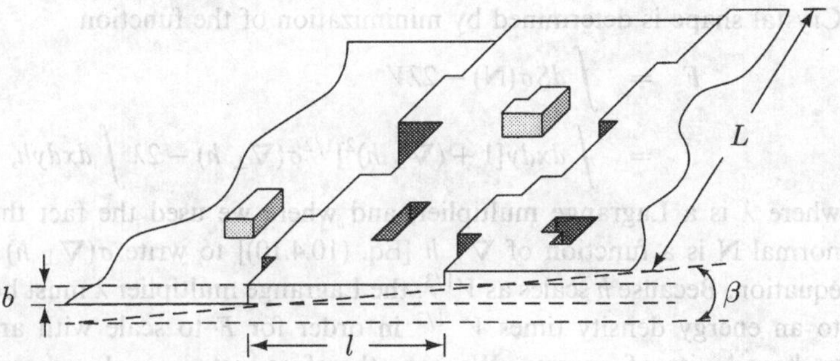

Fig. 10.6.2. Sketch of a vicinal surface at low but nonzero temperature. The steps have a height b and average separation l with $\tan\beta = b/l$. For small β, the steps are widely separated and the total free energy of the interface is a sum of two contributions, one proportional to the total terrace area and the other given by Eq. (10.6.23) arising from the fluctuating and interacting steps.

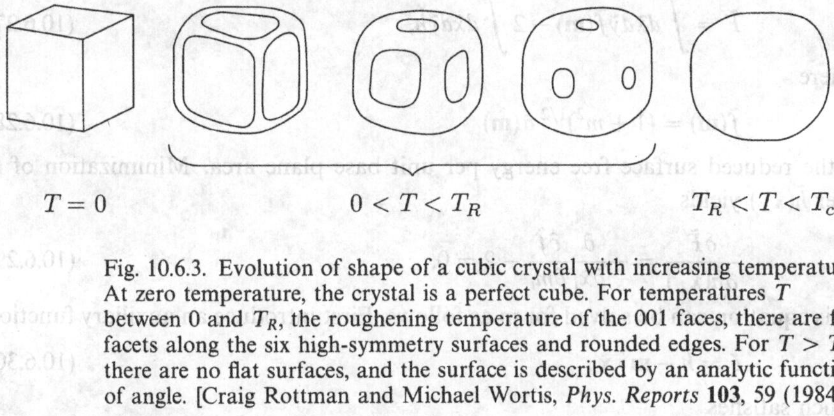

Fig. 10.6.3. Evolution of shape of a cubic crystal with increasing temperature. At zero temperature, the crystal is a perfect cube. For temperatures T between 0 and T_R, the roughening temperature of the 001 faces, there are flat facets along the six high-symmetry surfaces and rounded edges. For $T > T_R$, there are no flat surfaces, and the surface is described by an analytic function of angle. [Craig Rottman and Michael Wortis, *Phys. Reports* **103**, 59 (1984).]

angular discontinuities in the crystal shape above the roughening temperature of the highest symmetry face.

The solid-vapor (or solid-liquid) coexistence curve can be reached by introducing an appropriate amount of matter into a fixed-volume container and varying the temperature. The total volume V of the solid phase along the coexistence curve will be determined thermodynamically by the lever rule. Its shape will be determined by the condition that the total surface energy be a minimum, subject to the constraint that the total volume of the crystal be that determined by the lever rule. The total volume of the crystal is

$$V = \frac{1}{3}\int d\mathbf{S}\cdot\mathbf{r} = \int dxdyh(\mathbf{x}_\perp), \tag{10.6.24}$$

where $\mathbf{r}(\theta,\phi)$ is the radius vector to the surface from the origin and $\mathbf{x}_\perp = (x, y)$.

Crystal shape is determined by minimization of the function

$$F = \int dS\sigma(\mathbf{N}) - 2\lambda V$$
$$= \int dxdy[1+(\nabla_\perp h)^2]^{1/2}\sigma(\nabla_\perp h) - 2\lambda\int dxdyh, \qquad (10.6.25)$$

where λ is a Lagrange multiplier, and where we used the fact that the surface normal $\mathbf{N}$ is a function of $\nabla_\perp h$ [Eq. (10.4.10)] to write $\sigma(\nabla_\perp h)$ in the second equation. Because h scales as $V^{1/3}$, the Lagrange multiplier λ must be proportional to an energy density times $V^{-1/3}$ in order for F to scale with area ($V^{2/3}$) and to have units of energy. We can, therefore, set $\lambda = \gamma L$, where $L = \alpha V^{-1/3}$. The coefficient α, which has a V-independent value in the thermodynamic limit $V \to \infty$, is chosen to fix the total volume after the crystal shape has been determined. Introducing reduced variables

$$\tilde{\mathbf{x}}_\perp = \mathbf{x}_\perp/L, \qquad \tilde{h} = h/L, \qquad \tilde{\sigma} = \sigma/\gamma, \qquad \tilde{F} = F/(\gamma L^2), \qquad (10.6.26)$$

and defining $\mathbf{m} = \nabla_\perp h$, we can write Eq. (10.6.25) as

$$\tilde{F} = \int d\tilde{x}d\tilde{y}\tilde{f}(\mathbf{m}) - 2\int d\tilde{x}d\tilde{y}\tilde{h}, \qquad (10.6.27)$$

where

$$\tilde{f}(\mathbf{m}) = (1+m^2)^{1/2}\,\tilde{\sigma}(\mathbf{m}) \qquad (10.6.28)$$

is the reduced surface free energy per unit base plane area. Minimization of $\tilde{F}$ over $\tilde{h}(\tilde{\mathbf{x}}_\perp)$ yields

$$\frac{\delta\tilde{F}}{\delta\tilde{h}(\tilde{\mathbf{x}}_\perp)} = \frac{\partial}{\partial\tilde{x}_i}\frac{\partial\tilde{f}}{\partial m_i} - 2 = 0. \qquad (10.6.29)$$

This equation can be solved for $\mathbf{m}$ as follows. First introduce an auxiliary function

$$\zeta = h - \mathbf{m}\cdot\tilde{\mathbf{x}}_\perp, \qquad (10.6.30)$$

which satisfies

$$d\zeta = -\tilde{\mathbf{x}}_\perp\cdot d\mathbf{m} \qquad (10.6.31)$$

because, by definition,

$$dh = \mathbf{m}\cdot d\tilde{\mathbf{x}}_\perp. \qquad (10.6.32)$$

Thus, $\zeta = \zeta(\mathbf{m})$ is the Legendre transform of the function $\tilde{h}(\tilde{\mathbf{x}}_\perp)$, and $\tilde{h}(\tilde{\mathbf{x}}_\perp) = [\zeta(\mathbf{m}) + \tilde{\mathbf{x}}_\perp\cdot\mathbf{m}]_{\text{min }\mathbf{m}}$. Next observe that

$$\frac{\partial}{\partial\tilde{x}}\frac{\partial\tilde{f}}{\partial m_x} = \frac{\partial(\partial\tilde{f}/\partial m_x, \tilde{y})}{\partial(\tilde{x},\tilde{y})} = \frac{\partial(\partial\tilde{f}/\partial m_x, -\partial\zeta/\partial m_y)}{\partial(\tilde{x},\tilde{y})},$$
$$\frac{\partial}{\partial\tilde{y}}\frac{\partial\tilde{f}}{\partial m_y} = \frac{\partial(\tilde{x},\partial\tilde{f}/\partial m_y)}{\partial(\tilde{x},\tilde{y})} = \frac{\partial(-\partial\zeta/\partial m_x, \partial\tilde{f}/\partial m_y)}{\partial(\tilde{x},\tilde{y})}, \qquad (10.6.33)$$

and multiply Eq. (10.6.29) by

$$\frac{\partial(\tilde{x},\tilde{y})}{\partial(m_x,m_y)} = \frac{\partial(\partial\zeta/\partial m_x, \partial\zeta/\partial m_y)}{\partial(m_x,m_y)} \qquad (10.6.34)$$

to obtain

$$\frac{\partial(\partial\tilde{f}/\partial m_x, -\partial\zeta/\partial m_y)}{\partial(m_x, m_y)} + \frac{\partial(-\partial\zeta/\partial m_x, \partial\tilde{f}/\partial m_y)}{\partial(m_x, m_y)}$$
$$= -2\frac{\partial(\partial\zeta/\partial m_x, \partial\zeta/\partial m_y)}{\partial(m_x, m_y)}. \tag{10.6.35}$$

This equation has a first integral

$$\tilde{f}(\mathbf{m}) = \zeta(\mathbf{m}). \tag{10.6.36}$$

We have just argued that $\tilde{h}(\tilde{\mathbf{x}}_\perp)$ is the Legendre transform of $\zeta(\mathbf{m})$. But, since $\tilde{f}$ and ζ are equal, this means that the reduced height function $\tilde{h}(\tilde{\mathbf{x}}_\perp)$ is the Legendre transform of the reduced free energy:

$$\tilde{h}(\tilde{\mathbf{x}}_\perp) = g(\tilde{\mathbf{x}}_\perp), \tag{10.6.37}$$

where

$$g(\tilde{\mathbf{x}}_\perp) = [\tilde{f}(\mathbf{m}) + \mathbf{m}\cdot\tilde{\mathbf{x}}_\perp]_{\min \mathbf{m}}. \tag{10.6.38}$$

These observations lead to an interpretation of crystal shape in terms of a thermodynamic phase diagram. The order parameter, which specifies the local orientation of the surface, is $\mathbf{m}$. There can be different thermodynamic phases as a function of $\tilde{\mathbf{x}}_\perp$ and T characterized by different functional dependence of $\mathbf{m}$ on $\tilde{\mathbf{x}}_\perp$. There are phases in which $\mathbf{N}(\mathbf{m})$ is normal to a flat crystal surface; these correspond to crystal facets. Or there can be rough surfaces in which $\mathbf{N}$ changes continuously with angle, as shown in Fig. 10.6.3. These phases can be separated by second-order phase boundaries, across which $\mathbf{m}$ changes continuously, or by first-order phase boundaries, across which $\mathbf{m}$ changes discontinuously. Phase boundaries correspond to singularities in the function $\tilde{h}(\tilde{\mathbf{x}}_\perp, T)$, which can be represented as a phase diagram (analogous to the $\mu - T$ phase diagram of a fluid) in the $\tilde{\mathbf{x}}_\perp - T$ plane. Second-order transitions mark transitions from flat to rough surfaces. First-order boundaries mark discontinuous changes between facets with different crystal orientations.

It is generally more instructive to use polar coordinates (θ, ϕ) relative to the center of the crystal, rather than $(\tilde{x}, \tilde{y})$, as independent variables when representing this phase diagram. A phase diagram in the $T - \phi$ plane in the equatorial plane with $\theta = \pi/2$ for a cubic crystal is shown in Fig. 10.6.4. At $T = 0$, there is a first-order transition between (100) and (010) facets at $\phi = \pi/2$. At all $T > 0$, there are second-order transitions from the (100) and (010) facets to a rough phase corresponding to rounded corners. The width of the flat facets decreases and that of the rough corners increases with increasing temperature until the roughening temperature T_R of the (100) facet is reached. For $T > T_R$, all surfaces are rough, and there are no singularities in $\mathbf{m}(\mathbf{x}_\perp)$. The surface is, however, not spherical because the surface energy, though analytic, still depends on angle.

The curved surface will meet a flat phase with a power-law singularity:

$$\tilde{h}(\tilde{x}, 0) - \tilde{h}_0 \sim (|\tilde{x}| - \tilde{x}_0)^\tau, \tag{10.6.39}$$

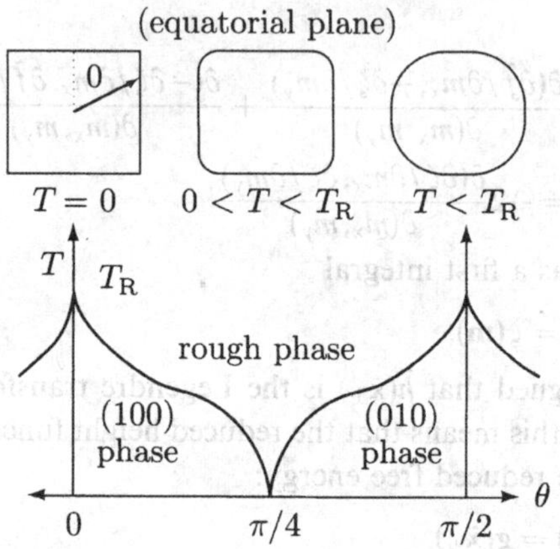

Fig. 10.6.4. Schematic phase diagram for thermal evolution of crystal shape. The full phase diagram is three dimensional. Here, only the evolution in the equatorial plane $\theta = \pi/2$ is shown. The phase boundaries, which are all second order for $T > 0$, specify the position of crystal edges as a function of angle ϕ. The curved boundaries separating the (100) and rough phases belong to the Pokrovsky-Talapov universality class with $\tau = 3/2$. Above the roughening transition at T_R, all surfaces are rough.

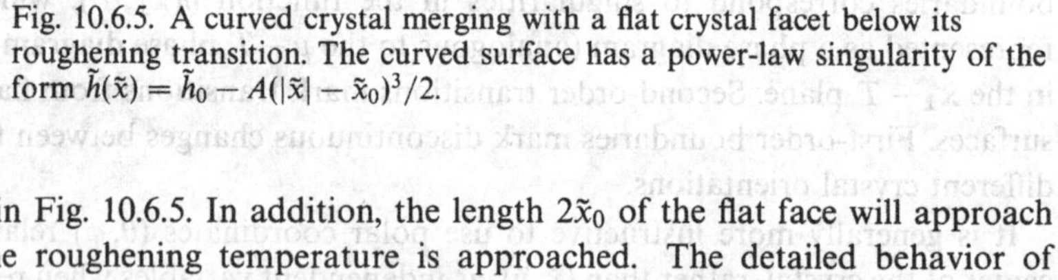

Fig. 10.6.5. A curved crystal merging with a flat crystal facet below its roughening transition. The curved surface has a power-law singularity of the form $\tilde{h}(\tilde{x}) = \tilde{h}_0 - A(|\tilde{x}| - \tilde{x}_0)^3/2$.

as shown in Fig. 10.6.5. In addition, the length $2\tilde{x}_0$ of the flat face will approach zero as the roughening temperature is approached. The detailed behavior of $\tilde{h}(\tilde{x}, 0)$, including the exponent τ, can be calculated near T_R^- using the Pokrovsky-Talapov free energy [Eq. (10.6.23)] with $\nabla_\perp h = \mathbf{m}$ for a vicinal surface and Eqs. (10.6.37) and (10.6.38):

$$\tilde{h}(\tilde{x}, 0) = \begin{cases} \tilde{h}_0, & \text{if } |\tilde{x}| < \tilde{x}_0; \\ \tilde{h}_0 - A(|\tilde{x}| - \tilde{x}_0)^{3/2}, & \text{if } |\tilde{x}| > \tilde{x}_0, \end{cases} \tag{10.6.40}$$

where

$$\begin{aligned} \tilde{x}_0 &= \frac{\epsilon}{\gamma b} \sim \frac{b}{\xi} \sim e^{-B/|T_R - T|^{1/2}} \\ A &= \frac{2}{3^{3/2}} \left(\frac{\epsilon b^3 \gamma}{C T^2} \right)^{1/2} \sim \left(\frac{b}{\xi} \right)^{1/2} \end{aligned} \tag{10.6.41}$$

and $\tilde{h}_0 = f_0/\gamma$, where B is a constant. The final scaling forms in these equations

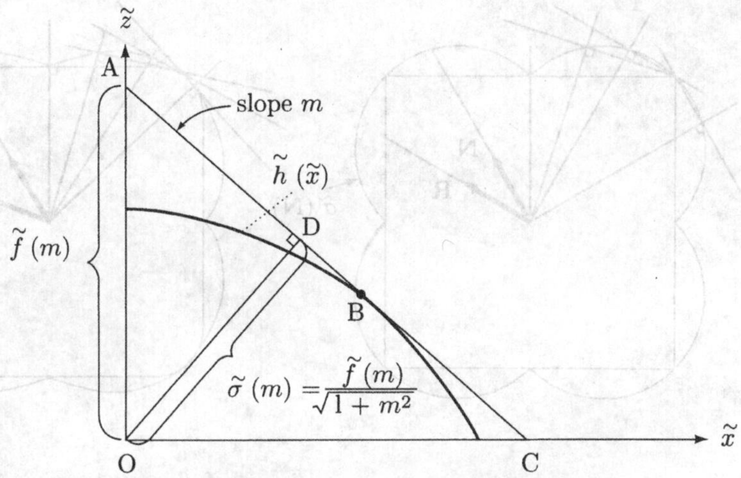

Fig. 10.6.6. Geometric interpretation of the Legendre transform between $\tilde{f}(m)$ and $\tilde{h}(\tilde{x})$ and the Wulff construction. The crystal surface $\tilde{z} = \tilde{h}(\tilde{x})$ is the interior envelope of the family of curves with slope m and $\tilde{z}$-intercept $\tilde{f}(m)$ or, equivalently, the family of lines with normal $\mathbf{N} = (-m, 1)/\sqrt{1+m^2}$ and normal distance to the origin $\tilde{\sigma}(m) = \tilde{f}(m)/\sqrt{1+m^2}$.

were obtained by using $K(l^*) \sim u(l^*) \sim 1$ at the matching point, $a(l^*) = ae^{l^*} = \xi$, and b = constant. Thus, the exponent τ is 3/2, and the length of the flat face approaches zero as ξ^{-1}. The existence of the flat face and singular form of the function $\tilde{h}(\tilde{x}, 0)$ arises from the cusp singularity in $f(\mathbf{m})$. For $T > T_R$, $\tilde{f} = (f_0 + \frac{1}{2}\gamma_R m^2)/\gamma$ and

$$\tilde{h}(\tilde{x}, 0) = \tilde{h}_0 - \frac{1}{2}\frac{\gamma}{\gamma_R}\tilde{x}^2. \tag{10.6.42}$$

The surface above T_R has a curvature $R^{-1} = L^{-1}\gamma/\gamma_R$ that scales with the sample dimension L and tends to a *finite* value at $T = T_R$. The reduced curvature $(L/R)(T_R/\gamma b^2)$ has a universal value of $2/\pi$ at the roughening transition.

We now turn to the geometric interpretation of Eq. (10.6.37) and the Wulff construction, which provides a geometric algorithm for determining the shape of a crystal from the surface free energy $\tilde{\sigma}(\mathbf{N})$ (or $\tilde{f}(\mathbf{m})$). This construction is equivalent to the geometric interpretation of the Legendre transformation. For simplicity, we will treat $\mathbf{m}$ and $\tilde{\mathbf{x}}_\perp$ as scalars m and $\tilde{x}$. Generalization to the vector case is straightforward. The goal is to construct $\tilde{h}(\tilde{x})$ given $\tilde{f}(\tilde{x})$. The crystal surface is described in the $\tilde{z}$-$\tilde{x}$-plane by the curve $\tilde{z} = \tilde{h}(\tilde{x})$, which has slope $m = d\tilde{h}/d\tilde{x}$ at the point $B = (\tilde{x}, \tilde{h}(\tilde{x}))$, as shown in Fig. 10.6.6. The straight line ABC, tangent to the curve $\tilde{z} = \tilde{h}(\tilde{x})$ at B, has slope m and intercepts the $\tilde{z}$ axis at $\tilde{z} = \tilde{h}(\tilde{x}) - m\tilde{x} = \tilde{f}(m)$. Its equation is, therefore,

$$\tilde{z} = m\tilde{x} + \tilde{f}(m). \tag{10.6.43}$$

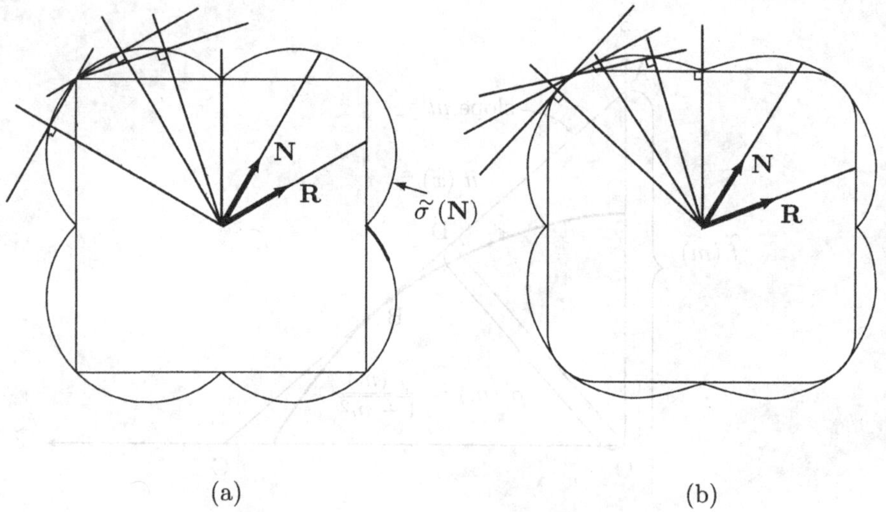

Fig. 10.6.7. (a) The Wulff plot for a Kosel crystal with $\tilde{\sigma}(\mathbf{N}) = |\sin\theta| + |\cos\theta|$. The crystal shape specified by the function $\mathbf{R}(\theta, \phi)$ is the interior envelope of surfaces passing through the Wulff plot with normals $\mathbf{N} = (\sin\theta\cos\phi, \sin\theta\sin\phi, \cos\theta)$. In this case, the equilibirum crystal shape is determined entirely by the cusps in the Wulff plot. All surfaces are flat facets, and there are no curved surfaces. (b) The Wulff plot for what a Kosel crystal becomes when $T > 0$. The flat facets are still determined by cusps, but there are now rounded edges determined by parts of the Wulff plot other than its cusps.

Another point on the crystal surface will have a different slope m and be tangent to another straight line satisfying Eq. (10.6.43) with the new value of m. Thus, the curve $\tilde{z} = \tilde{h}(\tilde{x})$ is the interior envelope of the family of straight lines with slope m and $\tilde{z}$-intercept $\tilde{f}(m)$. This family is determined entirely once the function $f(m)$ is specified.

In application to crystal shapes, the above construction is usually expressed in a slightly modified form. A particular line (e.g., ABC in Fig. 10.6.6) in the family of straight lines is determined by its $\tilde{z}$-intercept $\tilde{f}(m)$ and its slope m. That line is equivalently determined by the vector normal to it and passing through the origin (OD in Fig. 10.6.6). This normal vector points along the unit vector $\mathbf{N} = (-m, 1)/\sqrt{1+m^2}$ and has magnitude $\tilde{\sigma}(\mathbf{N}) = \tilde{f}(m)/\sqrt{1+m^2}$, equal to the surface tension of a surface perpendicular to $\mathbf{N}$. This formulation in terms of the normal rather than the tangent vector applies directly to three dimensions. Thus, the shape of a crystal can be determined as follows. First, construct the surface, called the *Wulff plot*, whose distance from the origin in direction $\mathbf{N}$ specified by the polar coordinates (θ, ϕ) of $\mathbf{N}$ is $\tilde{\sigma}(\mathbf{N})$. The crystal surface $\mathbf{R}(\theta, \phi)$ is the interior envelope of the family of surfaces normal to $\mathbf{N}$ and passing through the Wulff plot. Fig. 10.6.7 shows this construction for a Kosel crystal with the zero-temperature surface energy of Eq. (10.6.3) with $J_\perp = J$ and the same crystal at finite temperature. Note that it is the cusps in the Wulff plot that define flat facets. In the zero-temperature Kosel crystal, all surfaces passing through the curved part of the Wulff plot intersect at one of the corners of the square determined by the cusps, and the equilibrium crystal has no curved surfaces. At finite temperature, curved surfaces develop at the corners.

Bibliography

Kinks and solitons

R. Rajaramam, *Solitons and Instantons* (North-Holland, Amsterdam, 1982).

Frenkel-Kontorowa model and the CI transitions

P. Bak, *Rep. Prog. Phys.* **45**, 587 (1981).

V.L. Pokrovsky and A.L Talopov, *Theory of Incommensurate Crystals,* Soviet Science Reviews (Harwood, Zurich, 1985).

W. Selke, in *Phase Transitions and Critical Phenomena,* Vol. 15, eds. C. Domb and J.L. Lebowitz (Academic Press, New York, 1993).

Differential geometry

B.A. Dubrovin, A.T. Fomenko, and S.P. Novikov, *Modern Geometry - Methods and Applications* (Springer Verlag, New York, 1992).

M. Spivak, *A Comprehensive Introduction to Differential Geometry* (Publish or Perish, Berkeley, 1979).

Fluctuating walls

Michael E. Fisher, *J. Stat. Phys.* **34**, 667 (1984).

R. Lipowsky, *Nature* **349**, 475 (1991).

D.Nelson and S. Weinberg (eds.), *Statistical Mechanics of Membranes and Surfaces* (World Scientific, Singapore, 1989).

Bending energy of fluid membranes

R. Canham, *J. Theor. Biol.* **26**, 61 (1970).

W. Helfrich, *Z. Naturforsch.* **28C**, 693 (1973).

Review of X-ray experiments on lamellar phases composed of fluctuating membranes

C.R. Safinya, in *Phase Transitions in Soft Condensed Matter*, eds. Tormod Riste and David Sherrington (Plenum, New York, 1989).

Roughening transition and the two-dimensional Coulomb gas

S.T. Chui and J.D. Weeks, *Phys. Rev. B* **14**, 4978 (1976).

J.D. Weeks, in *Ordering in Strongly Fluctuating Condensed Matter Systems*, ed. T. Riste (Plenum, New York, 1980), p. 293.

Faceting in equilibrium crystals

Craig Rottman and Michael Wortis, *Phys. Reports* **103**, 59 (1984).

Membranes and Interfaces in Complex Fluids

Samuel A. Safran, *Statistical Mechanics of Membranes and Interfaces* (Addison Wesley, Reading, Mass., 1994).

References

Farid F. Abrahams, W.E. Rudge, and M. Plishke, *Phys. Rev. Lett.* **62**, 1757 (1989).
Milton Abramowitz and Irene A. Stegun, eds., *Handbook of Mathematical Functions* (Dover, New York, 1965).
J.A. Aronovitz and T.C. Lubensky, *Phys. Rev. Lett.* **60**, 2634 (1988).
S. Aubry, in *Solitons in Condensed Matter Physics*, eds. A. Bishop and T. Schnieder (Springer, Berlin, 1978), p. 264.
P. Bak, D. Mukamel, J. Villain, and K. Wentowsaka, *Phys. Rev. B* **19**, 1610 (1979).
W.K. Burton and N. Cabrera, *Disc. Faraday Soc.* **5**, 33 (1949).
W.K. Burton, N. Cabrera, and F.C. Frank, *Phil. Trans. R. Soc.* **243A**, 299 (1951).
W. Chou and R.B. Griffiths, *Phys. Rev. B* **34**, 6219 (1986).
S.N. Coppersmith, Daniel S. Fisher, B.I. Halperin, P.A. Lee, and W.F. Brinkman, *Phys. Rev. Lett.* **46**, 549 (1981).
S. Dieker and R. Pindak, *Phys. Rev. Lett.* **56**, 1819 (1986).
F.C. Frank and J.H. Van der Merwe, *Proc. Roy. Soc. London* **198**, 205 (1949).
Y.I. Frenkel and T. Kontorowa, *Zh. Eksp. Teor. Fiz.* **8**, 1340 (1938).
P.G. de Gennes, *J. Chem. Phys.* **48**, 2257 (1968).
P.G. de Gennes and C. Taupin, *J. Phys. Chem.* **86**, 2294 (1982).
W. Helfrich, *Z. Naturforsch.* **33A**, 205 (1978).
Anthony D. Navaco and John P. Mactague, *Phys. Rev. Lett.* **38**, 1286 (1977).
D.R. Nelson and L. Peliti, *J. Physique* **48**, 1085 (1987).
Y. Okwamoto, *J. Phys. Soc. Japan* **49**, 1645 (1980).
L. Peliti and S. Leibler, *Phys. Rev. Lett.* **54**, 1690 (1985).
David Pettey and T.C. Lubensky, *J. Phys. II France* **3** 1571 (1993).
M. Peyrard and S. Aubry, *J. Phys. C* **16**, 1593 (1983).
V.L. Pokrovsky and A.L. Talapov, *Phys. Rev. Lett.* **42**, 66 (1979).
C.R. Safinya, D. Roux, G.S. Smith, S.K. Sinha, P. Dimon, N.A. Clark, and A.M. Bellocq, *Phys. Rev. Lett.* **57**, 2718 (1986).
H.J. Schulz, *Phys. Rev. B* **22**, 5274 (1980).
J.V. Selinger and D.R. Nelson, *Phys. Rev. A* **39**, 3135 (1989).
J.V. Selinger, Z.-G. Wang, R.F. Bruinsma, and C.M. Knobler, *Phys. Rev. Lett.* **70**, 1139 (1993).
L. de Seze and S. Aubry, *J. Phys. C* **17**, 389 (1984).
J. Villain, in *Ordering in Strongly Fluctuating Condensed Matter Systems*, ed. T. Riste (Plenum, New York, 1980), p. 221.

Problems

10.1 Show that

$$\phi_{SA} = 4\tan^{-1}\left(\frac{\sinh(\beta\bar{t}/\sqrt{1-\beta^2})}{\beta\cosh(\bar{z}/\sqrt{1-\beta^2})}\right)$$

is a solution of the sine-Gordon equation, where $\bar{t} = v_0\sqrt{V_0}/ct$, $\bar{z} = \sqrt{V_0/c}z$, and $\beta = v/v_0$. Then show that

$$\phi_{SA} \to \phi_+\left(\frac{z+z_0+vt}{\sqrt{1-\beta^2}}\right) + \phi_-\left(\frac{z-z_0-vt}{\sqrt{1-\beta^2}}\right), \qquad t \to -\infty,$$

$$\to \phi_+\left(\frac{z-z_0+vt}{\sqrt{1-\beta^2}}\right) + \phi_-\left(\frac{z-z_0-vt}{\sqrt{1-\beta^2}}\right), \qquad t \to +\infty,$$

where ϕ_+ is the soliton and ϕ_- is the anti-soliton solution of Eq. (10.2.17).

This shows that a soliton and an anti-soliton that are well separated at $t = -\infty$ pass through each other and emerge as a soliton-anti-soliton pair at $t = +\infty$. Show that

$$\phi_{SS} = 4\tan^{-1}\left(\frac{\beta\sinh(\bar{z}/\sqrt{1-\beta^2})}{\cosh(\beta\bar{t}/\sqrt{1-\beta^2})}\right)$$

is a solution to the dynamical sine-Gordon equation that reduces to two solitons at $t = \pm\infty$.

10.2 In tilted hexatic films (Sec. 2.8), there are both vector and hexatic order with respective complex order parameters $\psi_1 = |\psi_1|e^{i\theta_1}$ and $\psi_6 = |\psi_6|e^{6i\theta_6}$. There is a preferred relative orientation of θ_1 and θ_6, and the Hamiltonian for the low-temperature phase can be written as

$$\mathcal{H} = \int d^2x\left[\frac{1}{2}K_6|\nabla\theta_6|^2 + \frac{1}{2}K_1|\nabla\theta_1|^2 + K_{16}\nabla\theta_1\cdot\nabla\theta_6 + V(\theta_6-\theta_1)\right],$$

where $V(\theta)$ is a periodic function with period $2\pi/6$.

(a) Show that, by transforming to variables $\theta_- = \theta_6 - \theta_1$ and $\theta_+ = \alpha\theta_6 + (1-\alpha)\theta_1$, $\mathcal{H}$ can be transformed to

$$\mathcal{H} = \int d^2x\left[\frac{1}{2}K_+|\nabla\theta_+|^2 + \frac{1}{2}K_-|\nabla\theta_-|^2 + V(\theta_-)\right],$$

for the proper choice of α. Calculate K_+, K_-, and α as a function of K_1, K_6, and K_{16}. Show that $\alpha \approx 1$, $K_- \approx K_6$, and $K_- \approx K_1$ for $K_6 \gg K_1, K_{16}$.

(b) Let $V(\theta) = -V_0\cos 6\theta$, and calculate $\theta_-(x)$ and the energy per unit length ϵ for a domain wall parallel to the y-axis in which θ_- changes by $2\pi/6$. Discuss the width and energy of the wall for $K_6 \gg K_1, K_{16}$. Interpret this domain wall in terms of the original variables θ_1 and θ_6. A figure would be useful.

(See Selinger and Nelson 1989.)

10.3 Let $\mathbf{c} = (\cos\theta_1, \sin\theta_1)$ be the tilt order parameter for a tilted hexatic film. A coupling of the form

$$\mathcal{H}_c = -\lambda\int d^2x\nabla\cdot\mathbf{c}\cos 6\theta_-$$

is permitted by symmetry and is not included in the simple Hamiltonian of the Problem 10.2. This term favors a nonzero splay, but integrates to the surface and does not contribute to the bulk free energy if $\cos 6\theta_-$ is constant. If $\cos 6\theta_-$ goes to zero inside the sample, as it does at the center of a domain wall of the sort treated in Problem 10.2, then there will be boundaries internal to the sample, and $\mathcal{H}_c$ can contribute to the bulk free energy. This effect favors the formation of modulated phases. Consider the following simple model of a striped phase with one-dimensional modulation in $\mathbf{c}$. There are domain walls of width w, across which θ_- changes linearly by $2\pi/6$, separated by stripes of width l in which $\theta_- = 0 \bmod 2\pi/6$ is a constant. Assume that only θ_- changes in the walls and that $K_6 \gg K_1, K_{16}$. Show that the energy per unit length of this striped phase can be written as

$$\mathcal{H} = \frac{\epsilon - \lambda}{l} + \frac{1}{2}K_6\frac{(\Delta\theta)^2}{l^2},$$

where $\delta\theta = \pi/4$ and $l \gg w$. Show that there is a second-order transition from a uniform to a striped phase when λ is greater than the wall energy ϵ and calculate the stripe width l as a function of $\epsilon - \lambda$. Verify that $w \ll l$ when $K_6 \gg K_1$. Show in a figure the direction of $\mathbf{c}$ and the hexatic order parameter as a function of x.

In chiral films, there can be couplings of the form

$$\mathcal{H}_{c^*} = -\mu \int d^2x \cos 6\theta_-(\mathbf{N}\cdot\nabla\times\mathbf{c}),$$

where $\mathbf{N}$ is the unit normal to the film. Determine the nature of the striped phase when $\mu \neq 0$ and $\lambda = 0$ and when both λ and μ are nonzero (Selinger *et al.* 1993).

10.4 There can be point defects in tilted hexatic films in which θ_1 changes by $2\pi k_1$ and θ_6 changes by $2\pi k_6/6$ in one circuit of the defect, where k_1 and k_6 are integers. Alternatively, θ_+ and θ_- change, respectively, by $2\pi q_+$ and $2\pi q_-$, where $q_+ = (1-\alpha)K_1 + \alpha k_6/6$ and $q_- = (k_6/6) - k_1$. In the tilted hexatic Hamiltonian of Problem 10.2, the variables θ_+ and θ_- decouple completely. The θ_+ part of the Hamiltonian is an xy-model with logarithmic interactions between vortex singularities. The θ_- part is a sine-Gordon Hamiltonian with soliton excitations.

(a) Show that $6q_-$ soliton lines emerge in the lowest energy state from a defect characterized by "charges" q_+ and q_-.

(b) Show that the energy of an arbitrary collection of n defects with charges $q_{+,i}$ and $q_{-,i}$ for $i = 1, ..., n$ in a sample of linear dimension R is

$$E = -\pi K_+ \sum_{i\neq j} q_{+,i}q_{+,j}\ln(r_{ij}/a) + \pi K_+ Q^2 \ln(R/a) + \epsilon L,$$

where $Q = \sum q_{+,i}$, a is the core radius, r_{ij} is the distance between defects i and j, L is the total length of soliton walls, and ϵ is the energy per unit length of a soliton.

(c) Show that the energy of an "N-armed star" with N-soliton walls of length R originating from a defect with $q_+ = 1 - N\alpha/6$ and $q_- = -N\alpha/6$ and terminating with a defect with $q_+ = \alpha/6$ and $q_- = 1/6$ compared to the energy of a single defect with $k_1 = k_6 = 1$ is

$$\begin{aligned} E_N &= -\frac{\pi\alpha N K_+}{36}(12 - \alpha - N\alpha)\ln\frac{R}{a} \\ &\quad -\frac{\pi\alpha^2 K_+}{36}\sum_{n=1}^{N-1}\ln(2\sin\frac{\pi n}{N}) + \epsilon N R. \end{aligned}$$

The five-armed star has been observed in free-standing tilted hexatic films (Dieker and Pindak 1986; Pettey and Lubensky 1993).

10.5 The free energy for a cholesteric liquid crystal in an external magnetic field **H** is

$$F = F_{\text{Frank}} - h\int d^3x\mathbf{n}\cdot\nabla\times\mathbf{n} - \frac{1}{2}\chi_a\int d^3x(\mathbf{H}\cdot\mathbf{n})^2,$$

where F_{Frank} is the Frank free energy for a nematic [Eq. (6.2.3) and Problem 6.7]. Consider a cholesteric with $\mathbf{n} = (\cos\theta(z), \sin\theta(z), 0)$ restricted to the xy-plane and $\mathbf{H} = (H, 0, 0)$ in the x-direction. Show that when $H = 0$, $\theta(z) = k_0 z$, where $k_0 = h/K_2$. Then calculate $\theta(z)$ and the average pitch as a function of H. Determine the critical value of H at which the cholesteric helix unwinds completely and $\mathbf{n} = (1, 0, 0)$.

10.6 Show that $\nabla_\perp\cdot\mathbf{N} = R_1^{-1} + R_2^{-1}$ in the Monge gauge for an arbitrary point on a surface.

10.7 This problem shows in a simple model that there is a preferred orientation of an incommensurate lattice relative to the substrate lattice. Let the substrate be a regular triangular lattice whose smallest reciprocal lattice vectors **G** point to the vertices of a hexagon in reciprocal space. Assume that the adsorbate lattice is, on average, a regular hexagonal lattice with six smallest reciprocal lattice vectors **K** and that $K \approx G$. The adsorbate lattice is described by a two-dimensional elasticity and a coupling with the substrate, which we take to be

$$\begin{aligned} U &= -U_0\sum_{\mathbf{G},\mathbf{l}}\cos[\mathbf{G}\cdot(\mathbf{R}_\mathbf{l}+\mathbf{u}_\mathbf{l})] \\ &\approx -U_0\sum_{\mathbf{G},\mathbf{l}}\cos[\mathbf{G}\cdot\mathbf{R}_\mathbf{l}] + U_0\sum_{\mathbf{G},\mathbf{l}}\mathbf{G}\cdot\mathbf{u}_\mathbf{l}\sin[\mathbf{G}\cdot\mathbf{R}_\mathbf{l}], \end{aligned}$$

where $\mathbf{R}_\mathbf{l}$ is the equilibrium position of the adsorbate atom **l** unmodified by U and $\mathbf{u}_\mathbf{l}$ is the displacement of atom **l** from $\mathbf{R}_\mathbf{l}$. The first term in this expansion is zero in the incommensurate state. The second term provides a linear coupling to $\mathbf{u}_\mathbf{l}$. Calculate $\mathbf{u}_\mathbf{l}$ resulting from coupling with the substrate, and show that its average over all atoms is zero. Show that the change in energy resulting from coupling to the substrate is

$$\Delta E = -\frac{1}{4}U_0^2\sum_{\mathbf{G}} G_i G_{u_iu_j}(\mathbf{K_G}-\mathbf{G})G_j,$$

where $G_{u_iu_j}$ is the strain correlation function of Eq. (6.4.24) and $\mathbf{K}_G$ is the reciprocal lattice vector of the adsorbate such that the difference $\mathbf{K_G} - \mathbf{G}$ is in the first Brillouin zone of the adsorbate lattice. Use this expression to calculate the equilibrium angle θ between the adsorbate and substrate lattices and to show that $\theta \sim |K - G|$ as the commensurate state is approached (Navaco and Mactague 1977).

10.8 This problem concerns polymerized membranes.

(a) Show that the transformation $\mathbf{u} = L_u\tilde{\mathbf{u}}$, $\mathbf{h} = L_h\tilde{\mathbf{h}}$ and $\mathbf{x} = L_x\tilde{\mathbf{x}}$ leads to

$u_{ij} = L_u L_x^{-1} \tilde{u}_{ij}$ if $L_h^2 = L_x L_u$ and

$$\tilde{\mathcal{H}}/T = \frac{1}{2}\int d^D\tilde{x}[2\tilde{u}_{ij}^2 + (\hat{\lambda}/\hat{\mu})\tilde{u}_{ii}^2 + (\tilde{\nabla}^2\tilde{\mathbf{h}})^2]$$

for $L_x = \hat{\mu}^{D-4}$ and $L_h = \kappa/\mu$, where $\hat{\mu} = \mu^2 T/\kappa$. This shows that the initial Hamiltonian can always be rescaled so that μ and κ equal unity, just as the Hamiltonian for a ϕ^4 model can be rescaled so that the coefficient of $(\nabla\phi)^2$ is unity. Furthermore, these coefficients can be fixed to unity under momemtum shell renormalization.

(b) Show that the height correlation function evaluated with respect to $\tilde{\mathcal{H}}$ satisfies

$$\tilde{G}_{hh}(q) = e^{(4-\eta_h)l}\tilde{G}_{hh}(e^l q) = A_h q^{-(4-\eta_h)},$$

where A_h is a constant. Then show that the height correlation function in original variables is

$$G_{hh}(q) = L_h^2 L_x^D \tilde{G}(qL_x) = \frac{T}{\kappa}\frac{A_h}{q^4 (qL_x)^{-\eta_h}}$$

and, thus, that the q-dependent bending rigidity is $\kappa(q) = \kappa(qL_x)^{-\eta_h}$.

(c) Show that the free energy density for a polymerized membrane in the presence of an external stress σ satisfies the scaling relation

$$f(\sigma) = L_x^{-D} e^{-Dl}\tilde{f}(e^{\lambda_\sigma l}\sigma/T\sigma_0),$$

where $\lambda_\sigma = (D+\eta_u)/2$ and $\sigma_0 = (\kappa/T)\hat{\mu}^{2/(4-D)}$. Use this relation to derive the nonlinear stress-strain relation

$$u_{ij} \sim \sigma^{(D-\eta_u)/(D+\eta_u)}.$$

10.9 One way to calculate the steric energy for a confined membrane is to replace the hard-wall confining potential by a harmonic potential. If the wandering exponent for the free membrane is $\zeta = 2s - D > 0$, then the membrane energy is

$$\mathcal{H} = \frac{1}{2}\int d^D x[\kappa(\nabla^s h)^2 + \gamma h^2],$$

where γ is the spring constant for the harmonic restoring force. Let $F_1(T,\gamma,A_B) = -T\ln \mathrm{Tr}e^{-\mathcal{H}/T}$ be the free energy associated with $\mathcal{H}$, where A_B is the D-dimensional "area" in the $\mathbf{x}$-plane. Then $\langle h^2\rangle \equiv l^2 = (1/2)\partial F_1/\partial\gamma$, and a free energy that is a natural function of l rather than γ can be introduced via Legendre transformation: $F_2(T,l,A_B) = F_1 - \frac{1}{2}l^2\gamma$.

(a) Calculate $\delta F_2 = F_2(l) - F_2(l=\infty)$ for general ζ and D and show that it reduces to the Helfrich-Pokrovsky-Talapov energy, const. $\times\, T^2/\kappa l^2$ for $s = (\zeta - D)/2$.

(b) Use the results of (a) to calculate the compressibility B of a smectic phase in three dimensions composed of a stack of two-dimensional membranes with wandering exponent ζ.

10.10 A discotic liquid crystal can be modeled as a hexagonal array of semi-flexible linear chains (polymers). The position of each chain element relative to its straight equilibrium position is $\mathbf{R}(z) = (\mathbf{u}(z), z)$, where z is the coordinate perpendicular to the plane of the hexagonal lattice. Each chain is confined, on average, by its neighbors to occupy, on average, a cylindrical volume with diameter $2l$ equal to the inter-chain distance. If this steric constraint is replaced by a harmonic constraint, the Hamiltonian for an individual chain becomes

$$\mathscr{H} = \frac{1}{2}\int dz\, \left[\kappa(d^2\mathbf{u}(z)/dz^2)^2 + \gamma_{ij}u_i u_j\right].$$

Following the procedure of Problem 10.9, calculate the bulk and shear moduli for this model discotic as a function of l (see Problem 6.6 for the elasticity of a discotic). Also determine the elastic constant K_1.

10.11 If both steric and exponential repulsion between domain walls are kept, Eq. (10.5.4) becomes

$$f = \frac{\sigma}{l} + C\frac{T^2}{\gamma l^3} + \frac{U_0}{l}e^{-l/l_0}.$$

Calculate $b = (l/b)^2\overline{B}$ for this free energy. Then calculate the Kosterlitz-Thouless melting temperature for $p \geq 3$ as a function of σ near $\sigma = 0$ using $\eta(T_c) = 1/4$.

10.12 Consider a two-dimensional square crystal in which there is an energy cost J_1 associated with breaking nearest neighbor bonds and a cost $J_2 = rJ_1$ associated with breaking next nearest neighbor bonds. Show that the surface energy is

$$\sigma(\theta) = J_1[|\cos\theta| + |\sin\theta| + \sqrt{2}r(|\cos(\theta - \pi/4)| + |\sin(\theta - \pi/4)|)].$$

Draw the Wulff plot and determine the equilibrium shape for this crystal for $r = 0.5$ and -0.5.

Glossary

Abrikosov vortex lattice lattice of parallel vortices that forms in type II superconductors in a external field.

absolute solid-on-solid (ASOS) model model for a solid-fluid interface in which the interface is characterized by an integer-valued function $h_{\mathbf{l}}$ specifying the interface height (in units of the crystal lattice parameter) above a flat crystal reference surface with lattice sites $\mathbf{l}$. The Hamiltonian for the ASOS model is $\mathcal{H} = J \sum_{\mathbf{l},\mathbf{l}'} |h_{\mathbf{l}} - h_{\mathbf{l}'}|$.

adatom atom adsorbed on a solid substrate.

amphiphilic molecule molecule with a polar or charged (hydrophilic) head group and a hydrocarbon (hydrophobic) tail. Such molecules tend to segregate at oil-water interfaces. In water, at a concentration above the critical micelle concentration, they self-assemble into structures in which the hydrocarbon tails are shielded from contact with water (micelles, vesicles). Soaps and surfactants are composed of amphiphilic molecules.

anisotropy field field favoring spin order along specific axes in a crystal.

ANNNI model anisotropic next-nearest-neighbor Ising model. An Ising model on a three-dimensional lattice in which there is ferromagnetic exchange between nearest-neighbor sites but antiferromagnetic exchange between next-nearest-neighbor sites in one direction. This model exhibits modulated phases.

atomic form factor Fourier transform of the atomic scattering potential.

basin of attraction region of parameter space flowing to a fixed point of renormalization group (or other) recursion relations.

basis in a crystal the positions of the atoms in a unit cell written as fractions of translation vectors constitute the basis vectors or basis. The crystal structure is defined by specifying the lattice and the basis vectors, which then identify how each atom in the crystal is positioned.

BCC lattice body-centered-cubic lattice. Lattice sites are at the vertices and the center of a simple cubic unit cell. Thirteen elements have BCC structure at room temperature. Even more have BCC phases near their melting points.

bend distortion of the director, $\mathbf{n}$, in a liquid crystal with nonvanishing $\mathbf{n}\times(\nabla\times\mathbf{n})$. Imagine the director following the flow field around the bend in a pipe.

Bernal model random close packed sphere model for a liquid or an amorphous solid.

biaxial nematic nematic liquid crystal with biaxial rather than uniaxial symmetry. The symmetric-traceless-tensor order parameter has two independent eigenvalues.

bicritical point critical point where two second-order lines meet a first-order line.

bipartite lattice lattice such as the BCC lattice that can be decomposed into two distinct interpenetrating sublattices with the property that a site on a given sublattice has nearest neighbors only on the other sublattice.

block-spin variable spin variable replacing a block of spins in a renormalization group decimation on a lattice.

Blume-Emery-Griffiths model lattice model with a three-state order parameter exhibiting a tricritical point. Initially used as a model for the tricritical point in He^3-He^4 mixtures where the superfluid λ line meets the coexistence curve.

bond-angle order orientational order of vectors connecting nearest-neighbor atoms. Crystalline solids have bond-angle order and periodic translational order. Bond-angle order without translation order can be found, e.g., hexatic order in liquid crystals.

bonding orbital singlet configuration of electrons shared between two atoms or molecules giving rise to an attractive interaction.

Bragg peak delta-function peak, at finite wave vector, in the scattering intensity from a periodic solid. The peaks are at reciprocal lattice vectors, and their intensity is proportional to the square of the number of scatterers. Bragg peaks are indicators of the existence of long-range periodic order in crystalline solids.

Bragg scattering scattering from a set of periodic planes in a crystalline solid giving rise to constructive interference and Bragg peaks.

Bragg's law relationship between the scattering angle θ, wavelength λ, and periodic layer spacing d, leading to Bragg scattering: $2d\sin\theta = n\lambda$ for n an integer.

Bragg-Williams theory a formulation of mean-field theory in which the entropy is calculated exactly but in which the internal energy is approximated by replacing microscopic variables by their average value.

Bravais lattice a lattice. Periodic array of points in d dimensions with positions $R_l = l_1\mathbf{a}_1 + l_2\mathbf{a}_2 + \cdots l_d\mathbf{a}_d$, where l_i $(i = 1, \ldots, d)$ is an integer and the vectors $\mathbf{a}_1, \mathbf{a}_2, \ldots, \mathbf{a}_d$ are any set of linearly independent d-dimensional vectors. Equivalent points in unit cells of a periodic lattice lie on a Bravais lattice. One of the 14 three-dimensional or five two-dimensional lattices that can be distinguished as having different point group (reflection and rotation) symmetries.

Brillouin peaks peaks in the density-density correlation function, $S_{nn}(\mathbf{q}, \omega)$, in fluids at $\omega = \pm cq$, where c is the velocity of sound. *Contrast with* Rayleigh peak.

Brillouin zone a Wigner-Seitz cell of a reciprocal lattice, hence the cell of smallest volume enclosed by the planes that are perpendicular bisectors of reciprocal lattice vectors.

broken symmetry term associated with states or thermodynamic phases with a lower symmetry than that of the interaction Hamiltonian. For example, the Ising Hamiltonian is invariant under reversal of all spins, but the all-spins-up or all-spins-down ground states are not and hence are broken-symmetry states.

Brownian motion random erratic motion exhibited by small particles in suspension in a fluid. At long times in d dimensions the motion is diffusive with $\langle \delta R^2(t)\rangle = 2dD_s t$, where $\langle \delta R^2(t)\rangle$ is the mean-square displacement and D_s is the self-diffusion constant ($D_s = k_B T/6\pi\eta a$ for a spherical particle of radius a, in a fluid of viscosity η).

bulk modulus the second derivative (B) of the elastic free energy with respect to isotropic strain. The reciprocal of the compressibility κ. For isotropic solids it is related to the Lamé coefficients by $B = \lambda + 2\mu/d$, where d is the spatial dimension.

bulk viscosity viscosity measuring the change in the isotropic part of the stress tensor resulting from the divergence of the velocity, i.e., $\sigma_{ij} \sim \zeta \nabla \cdot \mathbf{v}\delta_{ij}$, where ζ is the bulk viscosity.

Burgers vector vector specifying the strength of dislocations in periodic and quasi-periodic structures including crystalline solids and smectic liquid crystals. A loop which encloses a dislocation line will contain an extra step (the Burgers vector) corresponding to a direct lattice vector. If the Burgers vector is parallel to the dislocation line, we have a screw dislocation. If it is perpendicular, we have an edge dislocation.

Cahn-Hilliard model purely dissipative dynamical model for a conserved order parameter. The time dependence of a field ϕ is determined by the equation $\partial\phi/\partial t = \lambda\nabla^2(\delta\mathcal{H}/\delta\phi) + \zeta(\mathbf{x}, t)$, where λ is a dissipative coefficient and ζ is a noise. This model is also called model B.

Callen-Welton theorem also known as the fluctuation-dissipation theorem, it relates the imaginary part of the response function for some variable, $\chi''(\omega)$, to the equilibrium fluctuations, $S(\omega)$, in that variable: $\chi''(\omega) = [(1-e^{-\beta\hbar\omega})/2\hbar]S(\omega)$, where β is the inverse temperature and $\hbar$ is Planck's constant. Classically, $\chi''(\omega) = \frac{1}{2}\beta\omega S(\omega)$.

Cantor set take the unit interval, divide it into three equal parts and throw out the middle. This gives the next generation. Repeat this process with each of the remaining parts to obtain successive generations. The limit of inifinite generations gives the original Cantor set with fractal or Hausdorf dimension $d_f = 0.6309$. Now often used to describe any set with $d_f < 1$.

causal property that a disturbance can produce effects only after it occurs.

chiral molecule a molecule that does not have a mirror plane. A molecule with a central atom and four different atoms at the vertices of a tetrahedron is a simple example of a chiral molecule. Chiral liquid crystal molecules, such as cholesterol nonanoate, produce chiral nematic phases.

cholesteric a chiral nematic liquid crystal with an equilibrium director with a helical twist. Often the pitch of the spiral structure (helical twist) is comparable to the wavelength of visible light (microns). The resultant Bragg scattering of visible light is responsible for the colorful appearance of cholesterics.

climb motion of dislocation in a crystal perpendicular to the Burgers vector, **b**, requiring motion of an entire plane of atoms.

clock model lattice spin model in which spins are constrained to point to N equally spaced directions on the unit circle. This model has Z_N symmetry.

close-packed lattice lattice formed by the centers of spheres packed so that they occupy the maximum possible volume fraction, $\phi = 0.7404$. The face-centered-cubic and hexagonal close-packed lattices are close-packed lattices. The random stacking of hexagonal close-packed planes has the same volume fraction but does not form a lattice.

coarse graining replacement of microscopic variables by average variables on an expanded length scale (with an upper wave number cutoff Λ).

codimension difference between the spatial dimension and the dimension of the core of a topological defect; e.g., a line dislocation in a three-dimensional solid has codimension 2.

coexistence simultaneous equilibrium of two or more distinct thermodynamic phases; e.g., water and vapor are in equilibrium on the liquid-gas coexistence line.

coherent and incoherent scattering if the scattering cross-section, b_i, of the particles in a sample varies, there is coherent scattering proportional to $|\langle b\rangle|^2$ and incoherent scattering proportional to $\langle b^2\rangle - |\langle b\rangle|^2$. Coherent scattering provides information about interparticle correlations, whereas incoherent scattering provides information about the motion of individual particles. For neutrons, the scattering cross-section can change for different isotopes or even for the same isotope in different spin states. This is responsible for the large incoherent cross-section of hydrogen as compared to that of deuterium.

columnar discotic phase liquid crystal phase in which plate-like molecules form stacks arranged on a regular two-dimensional lattice. There is, thus, two-dimensional crystalline order but no long-range order in the third dimension.

commensurate lattice lattice that can be divided into two or more sublattices, each of whose basis vectors is a rational multiple of the basis vectors of the other sublattices. *Contrast with* incommensurate lattice.

compressibility modulus κ measuring change of volume V in response to change in pressure p: $\kappa_T = -V^{-1}\partial V/\partial p)_T = n^{-2}\partial n/\partial\mu)_T$, where n is the number density and μ is the chemical potential. Inverse of the bulk modulus B.

Compton scattering scattering of photons by electrons.

conjugate variable the work done in changing an extensive thermodynamic variable is the product of the change in that variable and its conjugate intensive variable. Thus volume and pressure are conjugate variables, as are particle number and chemical potential.

conservative dislocation motion motion of a dislocation along the glide plane, parallel to the Burgers vector, in which there is no net transport of mass.

constitutive relation phenomenological relation between a current and a thermodynamic field. Examples include the relationship between the number current $\mathbf{J}$ and the number density n, $\mathbf{J} = -D\nabla n$, where D is the diffusion constant, and that between the electrical current density and the electric field in a conductor with conductivity σ, $\mathbf{J} = \sigma\mathbf{E}$.

continuous group group, such as the rotation group, whose operations are parametrized by points in a continuous space.

continuous transition phase transition in which the order parameter increases continuously from zero.

conventional unit cell unit cell whose shape most directly reflects the symmetry of the lattice. Thus, the conventional unit cell of all cubic lattices, including BCC and FCC, is a cube. The primitive cells for FCC and BCC lattices are not cubes.

core energy the energy associated with the destruction of order at the core of a topological defect, such as a vortex.

corrections to scaling corrections to the dominant scaling behavior in the vicinity of a critical point arising from irrelevant variables that scale to zero at the critical point.

correlation length characteristic length, ξ, of a correlated region. ξ diverges as the critical temperature, T_c, of a second-order phase transition is approached.

correlation length exponent exponent ν controlling the divergence of the correlation length: $\xi \sim |T - T_c|^{-\nu}$. In mean-field theory, $\nu = \frac{1}{2}$.

covering surface an open surface whose boundary is a specified closed curve, such as a closed vortex or dislocation loop.

creep time-dependent evolution of the strain in a solid subjected to a constant stress. An ideal solid responds elastically to stress (with a time-independent strain). Creep results from the motion of defects, i.e. vacancies, interstitials, dislocations, and grain boundaries.

critical density density at the liquid-gas critical point.

critical dimension the *upper critical dimension* is the spatial dimension below which fluctuations become dominant and mean-field theory breaks down. The *lower critical dimension* is that below which the fluctuations become so large that no transition occurs.

critical endpoint type of critical point where a line of second-order transitions terminates at a line of first-order transitions.

critical exponents exponents controlling the singularities of thermodynamic variables at second-order critical points, e.g. $c_V \sim (T - T_c)^{-\alpha}$, $\phi \sim (T - T_c)^{\beta}$, and $\xi \sim (T - T_c)^{-\nu}$.

critical opalescence enhanced scattering of light and resultant cloudiness of a fluid near its critical point, arising from the development of density fluctuations at length scales comparable to the wavelength of visible light, λ, when $\xi \propto (T - T_c)^{-\nu} \approx \lambda$.

critical point point in a phase diagram characterized by singularities in derivatives of the free energy and related thermodynamics quantities.

critical slowing down slowing down of dynamical processes at a second-order critical point. For example, in the conventional, or Van Hove, theory for a binary mixture, the diffusion coefficient, D, vanishes as the inverse relative concentration susceptibility, χ^{-1}, as the phase separation critical point is approached. More generally, dynamic slowing down is described by a dynamical critical exponent, z, in addition to static exponents (such as γ in χ).

crystallographic point group point group compatible with the symmetry of any periodic crystal lattice composed of regularly repeated identical unit cells. Point groups with five-fold symmetry operations (such as the icosahedral group in three dimensions) are among those that are not crystallographic point groups.

cubatic putative material with "cubic" bond-angle order, resulting, for example, from dislocation melting of a cubic crystal in which the periodic translational order but not the orientational order of the crystal is destroyed.

cubic anisotropy anisotropy of a cubic lattice; it leads to a term $v \sum_i \phi_i^4$ in the Hamiltonian of an O_n field theory.

cubic fixed point fixed point of an O_n field theory resulting from the presence of cubic anisotropy.

Curie spin susceptibility response of noninteracting isolated spins to an external magnet field, $\chi = \mu^2/k_B T$, where μ is the spin's magnetic moment. A Curie law is any linear response proportional to $1/T$.

curvature deviation of a curve or surface in space from local flatness. The curvature at point P on a surface S is characterized by the maximum and minimum radii, R_1 and R_2, of circles in mutually perpendicular planes perpendicular to the plane tangent to S at P, best approximating the curves formed by the intersection of S with these planes. The mean curvature is $(1/R_1 + 1/R_2)/2$, and the Gaussian curvature is $1/(R_1 R_2)$.

cutoff maximum wave number, Λ, of fields in a phenomenological Hamiltonian. In lattice models, $\Lambda = 2\pi/a$, where a is the lattice spacing.

dangerous irrelevant variable irrelevant variable at a critical point that must be retained to provide the correct scaling behavior of some field. Generally, the field in question diverges as some power of the dangerous irrelevant variable. A cubic anisotropy field is a dangerous irrelevant variable for the transverse susceptibility near the Heisenberg critical point below T_c. Above the upper critical dimension, $d_c = 4$, the coefficient of ϕ^4 in a ϕ^4 field theory is a dangerous irrelevant variable for the free energy.

Debye-Hükel theory mean-field theory for mobile charged carriers interacting via the Coulomb potential. This theory is very useful for ions in solution and for unbound vortices. It predicts exponential screening by mobile charges beyond the Debye-Hükel screening length $\kappa^{-1} = (4\pi \sum_q n_q q^2/\epsilon k_B T)^{1/2}$, where there are n_q ions of charge q per unit volume and the dielectric constant is ϵ. The resulting potential has the Yukawa or screened Coulomb form: $\Phi \sim \exp(-\kappa r)/r$.

Debye-Waller Factor Thermal (or quantum) fluctuations reduce the order parameter of a system from its classical $T = 0$ value. In systems such as the xy-model or periodic crystals with a broken continuous symmetry, this reduction is expressed in terms of a factor $\exp(-2W)$, where W is called the Debye-Waller factor. In isotropic systems the fluctuations can be calculated from a d-dimensional sum of modes with wavevector q and energy q^n: $W \propto T \int d^d q/q^n$. W diverges and long-range order is destroyed for d less than or equal to the lower critical dimension d_L. In periodic crystals, the order parameters $\langle n_{\mathbf{G}} \rangle \sim \exp(-W)$ (whose squared amplitudes are proportional to the intensity of Bragg peaks) are averages of complex amplitudes of density waves at wavevectors **G** in the reciprocal lattice, and the Debye-Waller factor is related to the fluctuations in particle displacement u via $W_{\mathbf{G}} \sim G^2\langle u^2\rangle$. Thus the thermal motion of the particles in a crystal leads not to a finite width of the Bragg peak but rather to a decay in amplitude of the higher order peaks.

decimation process of removing degrees of freedom in real-space realizations of the renormalization group. Process of killing every tenth soldier in a Roman legion after a lost battle.

defect imperfection in an ordered structure. There are local defects, such as missing atoms (vacancies) or extra atoms at points other than lattice sites (interstitials) in a crystal, and there are topological defects, such as dislocations characterized by some nonvanishing quantized line (or surface) integral on a loop (surface) enclosing the defect core.

de Gennes-Taupin length length, ξ, beyond which the orientational order of the normal to a fluid membrane is lost: $\xi \sim e^{c\kappa/T}$, where c is a constant and κ is the curvature rigidity of the membrane.

density-density correlation functions the functions

$$\begin{aligned} C_{nn}(\mathbf{x},\mathbf{x}') &= \langle n(\mathbf{x})n(\mathbf{x}')\rangle, \\ S_{nn}(\mathbf{x},\mathbf{x}') &= \langle n(\mathbf{x})n(\mathbf{x}')\rangle - \langle n(\mathbf{x})\rangle\langle n(\mathbf{x}')\rangle, \\ \langle n(\mathbf{x})\rangle\langle n(\mathbf{x}')\rangle g(\mathbf{x},\mathbf{x}') &= C_{nn}(\mathbf{x},\mathbf{x}') - \langle n(\mathbf{x})\rangle\delta(\mathbf{x}-\mathbf{x}') \end{aligned}$$

are all correlation functions of the density $n(\mathbf{x})$. The function $g(\mathbf{x},\mathbf{x}')$ is called the pair-correlation function. Scattering experiments measure the Fourier transform

$$C_{nn}(\mathbf{q}) = S(\mathbf{q}) = V^{-1}\int d^d x d^d x' e^{-i\mathbf{q}\cdot(\mathbf{x}-\mathbf{x}')}C_{nn}(\mathbf{x},\mathbf{x}'),$$

where V is the volume. In fluids,

$$C_{nn}(\mathbf{q}) = \langle n\rangle[1+\langle n\rangle g(q)]$$

has a peak at at $q \approx 2\pi/a$, where a is the average interparticle spacing. In crystals, $C_{nn}(\mathbf{q})$ has Bragg peaks at points in the crystal's reciprocal lattice. For an uncorrelated system, such as an ideal gas, $g(\mathbf{q}) = 1$. From the fluctuation-dissipation theorem, $S_{nn}(\mathbf{q})/T$ is also the susceptibility relating the change in density in response to a change in the chemical potential at wave vector **q**. Thus, $n^{-2}S(q=0)/T$ is the compressibility.

destructive interference addition of two waves whose phases differ by 180°, resulting in a partial or complete cancelation of the amplitude.

devil's staircase a continuous function with flat regions. The average period in the Frenkel-Kontorowa and related models as a function of some control parameter can be a devil's staircase. An *incomplete devil's staircase* is a function with flat regions (with zero derivative) connected by regions with nonzero derivative. A *complete devil's staircase* is a function whose derivative is zero almost everywhere (i.e., except at a countable set of points). It is also called a singular continuous function. A *harmless staircase* is a function with discontinuous jumps between flat regions.

diffuse scattering scattering whose intensity is spread out in wave vector. To be contrasted with Bragg scattering whose intensity is highly concentrated at Bragg peaks. Diffuse scattering can be concentrated in *diffuse rings* (fluids or liquid crystals), in *diffuse sheets* (uncorrelated lines), or *diffuse lines* (uncorrelated sheets).

diffusion term applied to processes controlled by the diffusion equation: $\partial\phi/\partial t = D\nabla^2\phi$, where ϕ is a scalar field and D is the diffusion constant. The diffusion current is $\mathbf{j} = -D\nabla\phi$. The field, ϕ, can, for example, be the temperature or relative concentration of two species. For ideal noninteracting particles in suspension, Brownian motion controls both the mean square displacement of a labeled particle and the relaxation of concentration fluctuations. For more complex situations, the self- (or labeled) diffusion constant, D_s, is defined by $\langle R^2 \rangle = 2dD_s t$, while the cooperative or gradient diffusion, D_c, is defined by $\mathbf{j} = -D_c \nabla n$, where $\mathbf{j}$ is the particle current and n is the density. As repulsive interactions are increased, D_s decreases and D_c increases from the noninteracting value, D_0.

diffusion limited aggregation (DLA) process of forming clusters or aggregates in which diffusing particles have some nonzero probability of irreversibly sticking together once they touch. These clusters are generally fractal, with $d_f \approx 1.75$ in three dimensions.

dilation symmetry invariance with respect to a change of scale. Fractal objects, such as polymers or DLA clusters, have a kind of continuous dilation symmetry (they look the same at different magnifications). Periodic and quasi-crystalline objects have a discrete dilation symmetry.

directed diffusion diffusion with a drift, possibly caused by an external field, in some preferred direction.

direct lattice a lattice of points in coordinate (as opposed to reciprocal) space.

director unit vector specifying the direction of average molecular alignment in liquid crystals, particularly nematic liquid crystals.

disclination orientational topological defect in a crystal or a bond-angle ordered phase, such as a nematic or a hexatic liquid crystal, in which the direction of bond order undergoes a quantized change ($k\pi$ in nematics and $k\pi/3$ in hexatics) in one circuit around the core.

discommensuration soliton-like defect in adsorbed monolayers and in the Frenkel-Kontorowa model separating two regions in which the adatom lattice is commensurate with the substrate lattice. A regular lattice of discommensurations can lead to an incommensurate phase.

discotic liquid crystals liquid crystal composed of disc-shaped molecules. There are nematic and columnar discotic phases. The latter have two-dimensional columnar lattices.

discrete Gaussian model lattice model in which there is an integer-valued function, $h_{\mathbf{l}}$, at each site, $\mathbf{l}$, and a bond energy proportional to $(h_{\mathbf{l}} - h_{\mathbf{l}'})^2$: $\mathscr{H} = J \sum_{<\mathbf{l}\mathbf{l}'>} (h_{\mathbf{l}} - h_{\mathbf{l}'})^2$. The two-dimensional version of this model is used to describe fluid-solid interfaces and is dual to the two-dimensional Coulomb gas.

discrete group group with a countable, usually finite number of elements.

discrete symmetry symmetry in which all symmetry elements are discrete operations such as inversion or rotation through $\pi/2$.

dislocation topological defect in periodic and quasi-periodic solids in which the phase of a mass density wave changes by 2π in one circuit around a core. Alternatively, a defect terminating an extra plane of atoms in the crystal.

dissipation irreversible loss of energy to incoherent degrees of freedom.

dissipationless flow hydrodynamic flow with no irreversible heat loss. *See also* inviscid flow.

domain wall defect separating two distinct but energetically equivalent states in systems with a broken discrete symmetry.

dual lattice each lattice in two dimensions has an associated dual lattice created from the intersections of the perpendicular bisectors of each of its bonds.

Einstein relation equation relating the diffusion constant, D, of a particle to its mobility, $1/\alpha$: $D = k_B T/\alpha$. *See also* Stokes' law.

Euclidean group the symmetry group of isotropic space. It consists of all translations, rotations, and reflections that leave isotropic space invariant.

Euler characteristic $\chi = 2(1 - g)$, where g is the number of handles on a closed surface. The total vorticity of any vector in the tangent plane of a closed surface is equal to χ. For example, a combed hairy sphere ($g = 0$) has two cowlicks.

Euler's equation equation governing the flow of an inviscid (dissipationless) one-component fluid: $\partial \mathbf{v}/\partial t + (\mathbf{v} \cdot \nabla)\mathbf{v} = -(\nabla p)/\rho$, where $\mathbf{v}$ is the velocity, p is the pressure, and ρ is the mass density.

Excluded volume In systems that are dominated by entropy the effect of short-range potentials can be treated in terms of the reduction in available volume produced by a nonzero density of particles. In a hard-sphere gas, the reduced or excluded volume for N particles of volume b is Nb, and the entropy of such a gas confined to a container of volume V is $N \ln(V - Nb)$. The entropic effects are much more interesting in the case of polymers where a random walk becomes a self-avoiding random walk in the presence of monomer excluded volume effects. This changes the dependence of the size of a polymer on the degree of polymerization N from being proportional to $N^{1/2}$ to being proportional to $N^{3/(d+2)}$ where d is the spatial dimension.

FCC lattice face-centered-cubic lattice – lattice in which lattice sites are at the vertices and face centers of a regular cube. Being at once a close-packed structure and of the highest crystalline symmetry (cubic), the FCC lattice is the second most popular solid structure for the elements at room temperature, with 18 takers.

Fermi's golden rule rule for calculating transition probabilities in single scattering events. The transition rate from a quantum state, i, to another state, j, is given by $\tau^{-1} = (2\pi/\hbar)|\langle i|U|j\rangle|^2\rho_f$, where U is the perturbation potential responsible for the transitions and ρ_f is the density of final states.

Fick's law phenomenological relation between the particle current, $\mathbf{j}$, and the gradient of the density, n: $\mathbf{j} = -D_c\nabla n$, where D_c is the cooperative or gradient diffusion constant.

first homotopy group group $\pi_1(\mathcal{M})$ associated with closed loops in an order-parameter space $\mathcal{M}$. Elements of the group correspond to homotopically (*see* homotopy) distinct closed loops in $\mathcal{M}$. Group multiplication rules depend on the topology of $\mathcal{M}$, Two simple cases are $\mathcal{M} = S_1 =$ the unit circle and $\mathcal{M} = S_2 =$ the unit sphere. In S_1, a closed loop is indexed by an integer-valued winding number specifying the number of times S_1 is wrapped. Group multiplication is equivalent to adding winding numbers, and $\pi_1\ (S_1) = Z$, the group of integers under addition. All closed loops in S_2 can be continuously deformed to a point: there is only one homotopy class, and $\pi_1\ (S_2) = 0$.

fixed point a set of recursion relations, such as those used in renormalization group calculations, lead to flows (changes in the variables upon successive iterations) in their parameter space. A point that remains unchanged under application of recursion relations is a fixed point. A fixed point is *stable* if nearby points flow toward it and *unstable* if nearby points flow away from it.

Flory theory for the radius of gyration, R_G, of a polymer. Mean-field theory yielding $R_G \sim N^\nu$, with $\nu = 3/(d+2)$ for a polymer in d dimensions. This result is obtained by minimizing the free energy, $F_G + F_{\text{rep}} \approx (R^2/N) + (N^2/R^d)$, of a polymer composed of N monomers over the polymer radius, R. The entropic contribution, $F_G = R^2/N$, is approximated by that of a Gaussian chain, and the contribution, $F_{\text{rep}} = N^2/R^d$, from monomer-monomer repulsion is estimated to be $\int d^dx\rho^2 \approx N^2/R^d$, where the monomer density, ρ, is N/R^d.

fractal An object whose mass does not scale as R^d where R is its characteristic length and d is the dimension of space. *See* Hausdorff dimension.

Frank free energy elastic energy for a nematic liquid crystal expressed in terms of the director $\mathbf{n}(\mathbf{x})$:

$$F = \frac{1}{2} \int d^3x \left[K_1(\nabla \cdot \mathbf{n})^2 + K_2[\mathbf{n} \cdot (\nabla \times \mathbf{n})]^2 + K_3[\mathbf{n} \times (\nabla \times \mathbf{n})]^2\right]$$

Freedericksz transition transition in a nematic cell from a spatially uniform to a non-uniform state as a function of external magnetic or electric field.

frustration The inability of a model system to satisfy all of its bonds, usually because of topological constraints. A "fully frustrated" system is frustrated on every elementary unit. The classic example is the antiferromagnetic Ising model on a triangular lattice (try it!). The "glass" transition in many disordered systems is blamed on frustration (especially in spin glasses).

functional a function of a function. If $F[f(x_0)]$ depends only on the value $f(x_0)$, then F is a function of $f(x_0)$. If, on the other hand, F depends on the function $f(x)$ at all points in some continuous domain, then F is a functional of f. For example, $F = \int_y^z [(df/dx)^2 - \cos f]dx$ is a functional of f depending on $f(x)$ for all $y < x < z$.

fundamental group *see* first homotopy group.

f-sum rule sum rule for the integral over frequency of the imaginary part of the density-density response function times the frequency:

$$\int \frac{d\omega}{\pi} \omega \chi''(\mathbf{q}, \omega) = \frac{nq^2}{m},$$

where n is the number density and m is the particle mass.

gauge symmetry local, as opposed to, global symmetry. Symmetry that can be applied to any point in space. Quantum electrodynamics (including interaction with matter) has a $U(1)$ gauge symmetry.

Gauss-Bonnet theorem theorem relating the number of handles (genus g) of a closed surface to the integral over the surface of the Gaussian curvature: $\int dS(1/R_1R_2) = 4\pi(1-g)$.

Gaussian critical point critical point at $r = 0$ for a Gaussian model in which the Hamiltonian is quadratic in a field ϕ: $\mathscr{H} = \frac{1}{2}\int d^dx[r\phi^2 + (\nabla\phi)^2]$. The stable fixed point above the upper critical dimension corresponds to this critical point.

Gaussian curvature the product $(1/R_1R_2)$ of the inverse principal radii of curvature, R_1 and R_2, at a point on the surface. *See also* curvature.

Gaussian fluctuations harmonic fluctuations about a local equilibrium state of a field theory.

genus the number of handles on a surface: an integer g that appears in the Gauss-Bonnet theorem. *See also* Euler characteristic.

Gibbs free energy thermodynamic potential $G(T, p, N)$ that is a natural function of temperature T, pressure p, and particle number N.

Gibbs paradox paradox that the entropy of an ideal gas calculated purely classically is not extensive. This paradox is resolved by including a factor, arising from the quantum statistics of particles, of $1/N!$ in integrals over phase space, where N is the number of particles.

Ginzburg criterion criterion that mean-field theory breaks down when the rms fluctuations in the local value of an order parameter exceed the average value of the local order parameter. It states that mean-field theory breaks down in dimension d when $(\xi/\xi_0)^{4-d} > \xi_0^d \Delta c_V$, where ξ is the coherence length, ξ_0 is the bare coherence length, and Δc_V is the mean-field specific heat jump. It can be used to determine the upper critical dimension.

glass a phase of matter with no long-range order but with a nonzero shear rigidity. Usually a disordered material which has a restoring force against shear strain (or its generalizations). Conventionally, any disordered material with a viscosity greater than 10^{13} poise is called a glass.

Glauber model purely dissipative dynamical model for a nonconserved order parameter. The time dependence of a field ϕ is determined by the equation $\partial\phi/\partial t = -\Gamma(\delta\mathcal{H}/\delta\phi)+\zeta(\mathbf{x},t)$, where Γ is a dissipative coefficient and ζ is a noise. This model is also called model A.

global symmetry invariance with respect to operations on all constituents of a system, e.g. rotations of all the spins in a Heisenberg system.

golden mean $(1+\sqrt{5})/2$ (or its inverse); quadratic irrational which solves $\tau^2-\tau-1=0$; "the most irrational number" in that it is least well approximated by rationals, as can be seen by truncating its continued fraction expansion,

$$\tau = 1+\cfrac{1}{1+\cfrac{1}{1+\cfrac{1}{1+\cfrac{1}{1+\ldots}}}}.$$

Greeks described the "most beautiful rectangle", the golden rectangle, as having sides in the ratio of τ. (The default AspectRatio for graphs in Mathematica is τ.) The best rational approximates to τ are given by the ratio of Fibonacci numbers $(F_n = F_{n-1}+F_{n-2},\ F_0=0,\ F_1=1)$.

good solvent the properties of polymers in solution are determined largely by whether their monomers are relatively more attracted to the solvent molecules or to other monomers. Interactions in polymer solutions are characterized by the Flory-Huggins free energy: $F \approx \frac{1}{2}T\phi^2(1-2\chi)$, where ϕ is the monomer volume fraction and χ is the Flory parameter. In a *good solvent*, $1-2\chi>0$, monomers prefer being surrounded by solvent, polymers swell and take on configurations of a self-avoiding random walk with a radius (of gyration) varying as $R_G \sim M^{3(d+2)}$, where M is the molecular weight of the polymer or its polymerization index. In a *θ-solvent*, $1-2\chi=0$, monomers effectively do not interact with the solvent, and polymers behave as ideal chains with $R_G \sim M^{1/2}$. In a *poor solvent*, $1-2\chi<0$, monomers are attracted to each other, and polymers collapse to dense objects that typically precipitate from solution.

grain boundary the boundary between different microcrystallites. Often, crystallite boundaries consist of a low-energy periodic arrangement of dislocations, which orients crystallites at small angles with respect to each other. This is a low-angle grain boundary.

Hansen-Verlet criterion phenomenological criterion, based on computer simulations and experimental observations, that a liquid will condense to a solid phase when $S_{nn}(k_0)/S_{nn}(k=\infty) > 2.7$–$2.9$, where $S_{nn}(k)$ is a density-density correlation function and k_0 is the wave number of the maximum intensity peak.

hard cutoff in calculations where material is considered as continuous, especially in scaling and renormalization, there can be unphysical consequences of allowing the length scales to become too small. A hard cutoff puts atomic dimensions as the smallest wavelength allowable, often as a limit in an integral.

hard spheres particles which intereact with a potential $U(r)=0,\ r>2a;\ U(r)=\infty,\ r<2a$, where a is the particle radius. Dense systems of hard spheres are liquids for $\phi<0.49$, have coexisting liquid ($\phi=0.49$) and FCC solid ($\phi=0.54$) phases for $0.49<\phi<0.54$, and are FCC solids for $\phi > 0.54$, where ϕ is the volume fraction. Random close packing of hard spheres fills space to 63% and represents a reasonable approximation to the structure of glasses and dense liquids. Periodic close packing leads to FCC and HCP structures with a filling fraction of 0.7404.

Hartree approximation a mean-field-like approximation in which ϕ^4 is approximated by $6\phi^2\langle\phi^2\rangle$ and the expectation value, $\langle\phi^2\rangle$, is determined self-consistently. This approximation is exact in the spherical model or the n-vector model when $n = \infty$. A similar approximation is used in treating interacting electron systems.

Hausdorff dimension exponent d_H relating the mass, M, to some characteristic length, R, of an object: $M \sim R^{d_H}$. For a dense object in d dimensions, $M \sim R^d$ and $d_H = d$. For a fractal object, $d_H = d_f < d$, where d_f is the fractal dimension. For an ideal polymer or random walk, $R^2 \sim N \sim M$ and $d_H = 1/2$.

heavy wall a negative discommensuration which tends locally to pile up mass.

hedgehog a topological point defect of three-dimensional spins in a three-dimensional space. Spins in the simplest hedgehog configuration, like the electric field of a point charge, point radially outward from a point.

Heisenberg model model spin Hamiltonian $\mathscr{H} = -J\sum_{\mathbf{l},\mathbf{l}'} \mathbf{s}_\mathbf{l} \cdot \mathbf{s}_{\mathbf{l}'}$. The spin $\mathbf{s}_\mathbf{l}$ can either be quantum mechanical or it can be a classical n-component vector.

Heitler-London theory a model for the electronic structure of small molecules that explicitly correlates electrons so that there are as few per atomic site as possible. *Contrast with* molecular orbital theory.

helicity modulus the elastic constant relating gradients in spin direction θ to a restoring force for an xy-model.

Helmholtz free energy thermodynamic potential $F[T, V, N]$ that is a natural function of the temperature T, the volume V, and the number of particles N.

hexagonal close-packed (HCP) hard spheres at maximum packing fraction (74%) form into hexagonal layers which then stack in an HCP or FCC structure. In an HCP lattice, alternate layers are the same. Each particle has 12 nearest neighbors. This is not a Bravais lattice, requiring a unit cell with two particles (a basis of 2). The layer spacing is $\sqrt{2/3}$ times the distance between sphere centers. This is the most popular structure of the elements, with 24 takers at room temperature.

hexatic phase a structure characterized by long-range six-fold orientational order but no long-range translational order. Such a phase is seen in liquid crystal systems. A scenario for melting in two dimensions involves a hexatic phase as an intermediate stage between liquid and hexagonal crystalline phases.

homogeneous function if $f(x) = b^k f(bx)$, then $f(x)$ is a homogeneous function of degree k. Scaling relations are the result of the homogeneity of thermodynamic functions.

homotopy two mappings, f_0 and f_1, from a closed path in real space to a closed path in order-parameter space are homotopic if they can be continuously deformed into each other. An explicit construction of such a deformation is called a homotopy.

Hubbard-Stratonovich transformation transformation expressing a quadratic form in a field ϕ as integral over an auxiliary field with a Gaussian weight of the exponential of a linear form in the field ϕ:

$$\exp\left(\frac{1}{2}K\phi^2\right) = \frac{1}{(2\pi K)^{1/2}}\int_{-\infty}^{\infty} dy \exp\left(-\frac{1}{2K}y^2 + y\phi\right).$$

Hund's rule electrons on an atom in an unclosed shell occupy the orbitals in such a way as to maximize the total spin. This configuration allows Pauli exclusion to keep like spins apart, and hence reduce their Coulomb repulsion.

hydrodynamic modes the long-wavelength (small q) excitations of a system whose frequency tends to zero with q are called hydrodynamic modes. Hydrodynamic modes are directly related to conservation laws and broken symmetries. In a liquid, there are five hydrodynamic modes associated with conservation of momentum in three directions and conservation of energy and mass. They are the two longitudinal sound modes, two transverse momentum modes, and a thermal diffusion mode. All but the longitudinal sound modes are damped (i.e., nonpropagating). Spin waves in ferromagnets and antiferromagnets are broken-symmetry hydrodynamic modes.

hydrophilic and hydrophobic liking or disliking water and, therefore, soluble or not soluble in water, respectively. Hydrophilic interactions are complicated and controversial. However, naively we know that water consists of dipolar molecules, and we therefore expect that charged or polar molecules can gain some polarization energy in water. On the other hand, organic molecules with low polarizability and no dipoles will disrupt the preferred structure of nearby water molecules and increase their energy. Oils, therefore, separate from water and will not dissolve. *See also* amphiphilic molecule

hyperscaling relation scaling relation involving the spatial dimension d of a system. For example, $\gamma + 2\beta = d\nu$, $\alpha = 2 - d\nu$, and $\beta = (d - 2 + \eta)\nu/2$.

hysteresis history-dependent properties of a system. The thermodynamic properties of a system are completely determined by the temperature, pressure, etc., and are independent of how the system reached these conditions. Therefore, a hysteretic state must be out of equilibrium, or metastable. Hysteresis is often associated with first-order phase transitions in which there is a finite barrier to nucleation of a new phase and a latent heat to be dissipated. It is also associated with glassy systems frozen into nonequilibrium states.

icosahedral symmetry the symmetry of an icosahedran, particularly noteworthy for five-fold rotation axes. It is the point group of highest symmetry, and one which is crystallographically forbidden, i.e., periodic crystals with this symmetry cannot be formed (*but see* quasi crystals). However, the dense packing and high symmetry of icosahedra have led people to look for short-range icosahedral order in liquids and glasses.

icosahedratic the three-dimensional equivalent of hexatic order in two dimensions. Long-range icosahedral orientational (bond-angle) order without long-range translational order.

ideal polymer chain a freely linked chain of monomers with no intermonomer interactions (they can interpenetrate). It should have the same statistical properties as a random walk and hence should have an end-to-end length and a radius of gyration that increases as the square root of the number of monomers ($R_G \sim N^{1/2}$).

incommensurate crystal a structure with long-range periodic order but with two or more periodicities with an irrational ratio. Common examples of incommensurate crystals are magnetic systems in which the magnetic period (e.g. helical order) is irrational with respect to the atomic lattice and systems with density wave instabilities at wave vector q_{DW} related to the Fermi wave vector k_F (in one dimension, $q_{\text{DW}} = 2k_F$), which is often unrelated (hence usually irrationally related) to the underlying lattice.

infrared singularities singularites in integrals over wave number q arising from small q (i.e., long wavelength and thus infrared). These integrals are usually of the form $\int dq q^{-\alpha}$, with $\alpha > 1$. Infrared singularities of the form $\int q^{d-1} q^{-2} dq$ are responsible for the fluctuation destruction of long-range order in systems with a broken continuous symmetry in dimensions d less than or equal to two.

intensive variable a thermodynamic variable that remains unchanged when the system is doubled (or tripled etc.) in size. Examples are pressure, temperature, and chemical potential. *Extensive variables*, such as free energies, entropy, number of particles, and volume can be made intensive by dividing by the volume (to make energy density, etc.).

interfacial stiffness elastic constant relating energy to perpendicular displacement fluctuations of an interface. It contains contributions both from the surface tension (energy per unit area) and the potential, favoring alignment of the interface parallel to some fixed surface.

intermediate function the spatial Fourier transform, $I(\mathbf{q}, t)$, of the density correlation function, $C_{nn}(\mathbf{x}, \mathbf{x}', t, 0)$ – to be contrasted with the Fourier transform, $C_{nn}(\mathbf{q}, \omega)$, with respect to both space and time. This function is often useful in light scattering and neutron scattering.

interstitials atoms or particles that reside at positions between crystalline sites.

inviscid flow fluid flow with no viscosity and hence no dissipation. For incompressible fluids, such flow obeys Euler's equation:

$$\partial \mathbf{v}/\partial t + (\mathbf{v} \cdot \nabla)\mathbf{v} = -(1/\rho)\nabla p.$$

irreducible representation in group theory, a representation of a symmetry operation that cannot be expressed in terms of lower dimensional representations.

irrelevant field a field that successively rescales toward zero as the length scale is increased, i.e., as coarse graining progresses (*but see* dangerous irrelevant variable). Only relevant fields determine universality classes.

Ising model model Hamiltonian $\mathscr{H} = J \sum_{\mathbf{l},\mathbf{l}'} \sigma_{\mathbf{l}} \sigma_{\mathbf{l}'}$ in which the local variables, $\sigma_{\mathbf{l}}$, take on the two values ± 1. This model has Z_2 symmetry.

isobar a constant pressure path. The critical isobar is a path to the critical point (e.g. the liquid-gas critical point) at the critical pressure.

isochore constant density path. The critical isochore is the path to the critical point (e.g. the liquid-gas critical point) at the critical density. It is easily achieved experimentally by enclosing a fixed number of particles in a fixed volume

isotherm constant temperature path. The critical isotherm is a path to the critical point (e.g. the liquid-gas critical point) at the critical temperature.

isotropic fluid a fluid phase whose properties are independent of orientation.

Josephson scaling relation scaling relation, $\rho_s \sim \xi^{2-d}$, expressing the superfluid density (or spin stiffness), ρ_s, in terms of the correlation length, ξ, in dimension d near the critical point.

kink a point defect in a one-dimensional model with discrete symmetry, e.g. a spin-flip in a one-dimensional Ising model with a spin-up chain on one side and a spin-down chain on the other side of the kink.

Kosel crystal crystal whose surface energy is proportional to the number of broken bonds.

Kosterlitz-Thouless transition vortex unbinding transition in two-dimensional systems with xy or $U(1)$ symmetry from a phase with an elastic rigidity (spin stiffness) and power-law correlations to one with no rigidity and exponential correlations.

Kramers' equation stochastic equation for the probability distribution of a particle as a function of its momentum p, position x, and time t.

Kramers-Kronig relation integral relation between the real part, $\chi'(\omega)$, and the imaginary part, $\chi''(\omega)$, of a response function:

$$\chi'(\omega) = \mathscr{P} \int \frac{d\omega}{\pi} \frac{\chi''(\omega')}{\omega' - \omega}, \quad \chi''(\omega) = -\mathscr{P} \int \frac{d\omega}{\pi} \frac{\chi'(\omega')}{\omega' - \omega},$$

where $\mathscr{P}$ means principal part. Results from linear response and causality.

Kubo formula equation relating a transport coefficient, such as Γ in the equation $\mathbf{j} = -\Gamma\nabla\mu$ for the diffusion current in terms of the chemical potential, to a time integral of a correlation function of a current:

$$\Gamma = \frac{\beta}{2}\frac{1}{dV}\int d^dx d^dx' \int_0^\infty dt \left\langle \frac{1}{2}\{j_i(\mathbf{x},t), j_i(\mathbf{x}',0)\}_+ \right\rangle,$$

where $\beta = 1/T$, d is the spatial dimension, V is the volume, and $\{A, B\}_+ = AB + BA$ is the anticommutator of A and B. The diffusion constant, D, is $\Gamma\partial\mu/\partial n$, where n is the density of diffusing species.

Lagrangian elasticity formulation of elasticity in which mass points are indexed by mass points of a reference surface. The elastic free energy is expressed as an integral of local strain energies over the reference surface. To be contrasted with Eulerian free energy, where the strain variable is a phase gradient at a point in space and integrals in the elastic energy are over the volume of material.

λ line a line of second-order phase transitions that meets a line of first-order transitions at a tricritical point. The primary example is the λ (or superfluid transition) line that meets the phase separation line in He^3-He^4 mixtures at low temperature.

Lamé coefficient for an isotropic solid in d dimensions, there are two elastic constants, known as Lamé coefficients, λ and μ, related to the bulk and shear moduli by: $B = \lambda + (2\mu/d)$ and $\mu = \mu$.

lamellar phase a phase with the symmetry of a one-dimensional stack of two-dimensional sheets. In lyotropic liquid crystals, this is the smectic phase. Similar phases occur in microemulsions and block copolymers.

Landau, Lev originator of most of the physics in this book.

Landau mean-field theory phenomenological form of mean-field theory in which the free energy is expanded in a low-order power series in the order parameter.

Landau-Peierls instability Fluctuation destruction of long-range periodic order in one-dimensional solids (such as smectic liquid crystals) in three dimensions. Bragg peaks of an ideal periodic solid are converted to power-law peaks for a lamellar structure.

Landau-Placzek formula equation for the dynamic density-density response function of a one-component fluid in the hydrodynamic limit. This function has both sound-wave (Brillouin) peaks and a thermal diffusion (Rayleigh) peak.

Langevin theory phenomenological theory for dynamical processes in which there is a random force (or noise) with a power spectrum usually chosen to produce thermal equilibrium distributions.

Laplace pressure equilibrium pressure difference, δp, between the inside and outside of a spherical droplet of radius R and surface tension σ: $\delta p = 2\sigma/R$.

lattice a periodic array of points defined by the linear combination with integer coefficients, l_i, of a set of primitive translations vectors, $\mathbf{a}_1, \ldots, \mathbf{a}_d$. The points are located at $\mathbf{R}_l = l_1\mathbf{a}_1 + l_2\mathbf{a}_2 + \ldots + l_d\mathbf{a}_d$.

Laue condition condition that incoming wave vectors, $\mathbf{k}$, which lie on the perpendicular bisectors of reciprocal lattice vectors, $\mathbf{G}$ (i.e., on the Brillouin zone boundary) will be Bragg scattered: $\mathbf{k} - \mathbf{k}' = \mathbf{G}$ or $\mathbf{k}\cdot\mathbf{G}/2 = |\mathbf{G}/2|^2$.

law of rectilinear diameters the linear dependence on temperature, T, of the average density as the liquid-gas critical point is approached along the coexistence curve: $\frac{1}{2}(n_l + n_g) - n_c \propto |T - T_c|$, where n_l and n_g are, respectively, the liquid and gas densities, n_c is the critical density, and T_c is the critical temperature. This behavior is predicted by mean-field theory. Critical fluctuations convert the linear dependence to $|T - T_c|^{1-\alpha}$, where α is the specific heat exponent.

Lennard-Jones potential model potential between neutral atoms incorporating the R^{-6} attractive Van der Waals attraction at large distances and approximating repulsive interactions at short distances by R^{-12}. $U(R) = 4\epsilon[(\sigma/R)^{12} - (\sigma/R)^6]$. Also called the 6-12 potential.

Lifshitz point critical point where a disordered phase, a spatially uniform ordered phase, and a spatially modulated ordered phase meet.

Lifshitz-Slyozov law relation $R \sim t^{1/3}$ between the characteristic length scale, R, and time, t, at long times after a quench below the spinodal line in a system with a conserved order parameter.

Lindemann Criteria When particles in a crystal fluctuate from their equilibrium positions by a distance comparable to the unit cell dimensions, we expect that the crystalline state is no longer well defined. Phenomenologically, solids melt when the rms fluctuations of the constituent particles exceed approximately 20% of the interparticle separation.

lock-in energetic preference for commensurate arrangement of an overlayer on a substrate over a finite range of control parameters.

long-range order (LRO) if you know the value of the order parameter at $\mathbf{x} = 0$, do you know the value at $\mathbf{x} = \infty$? Yes $\rightarrow$ LRO. No $\rightarrow$ no LRO. Equivalently, does the correlation function remain finite as $\mathbf{x} \rightarrow \infty$? If $C_{\phi\phi}(\mathbf{x} \rightarrow \infty) \rightarrow 0$, then no LRO. The existence of (δ function) Bragg peaks implies long-range periodic order.

long-time tail algebraic, rather than exponential, fall-off at long time of correlation functions, particularly the velocity autocorrelation function in a fluid.

Lorentzian peak peak in a function of the form $\Gamma/[x^2 + \Gamma^2]$. Peaks of this form occur in both static and dynamic correlation functions. Being the Fourier transform of exponentials, they characterize short-range correlations over length or time scales of order Γ^{-1}.

Lyotropic - Refers to systems, usually liquid crystals, which exhibit phase changes as a function of concentration rather than temperature as in thermotropic systems. Typical examples are vesicles, micelles, and microemulsions in lamellar and other ordered phases.

magnetic scattering neutrons have a magnetic moment and are, therefore, scattered by magnetic fields. Thus, they are useful probes of magnetic structure such as occurs in paramagnets, ferromagnets, antiferromagnets, etc., or in flux phases of superconductors. With the advent of high photon fluxes in synchrotrons, it is now possible to use the weak scattering of X-rays from magnetic fields (a relativistic effect) to study magnetic structures, but the technique is nowhere near as popular as magnetic neutron scattering.

magnetization the magnetic dipole moment per unit volume. It is the order parameter for a magnetic transition and is the derivative of the energy with respect to magnetic field.

magnon A quantized spin wave. In ferromagnets the long wavelength dispersion is quadratic, $\omega \sim q^2$ while for antiferromagnets it is linear, $\omega \sim |q|$.

mean curvature the average curvature over two perpendicular directions on a surface. The mean radius of curvature is one-half the inverse reciprocal sum of principal radii of curvature, $1/R = \frac{1}{2}[(1/R_1) + (1/R_2)]$. *See also* curvature.

mean-field theory approximate theory, used extensively in the study of phase transitions, in which dynamical variables take on their mean or average values and are then calculated self-consistently. There are no thermodynamic fluctuations in this theory. This is typically the first approximation that one uses to determine the possible thermodynamic phases of a new model Hamiltonian. It has a good chance of giving the correct ground states and topology of phase diagrams. Above an upper critical dimension, mean-field theory yields the correct values of critical exponents.

mean-free path the distance that a particle (or wave) goes (propagates) before undergoing a collision. The transport mean-free path is the distance a particle goes before its direction is randomized. It is weighted (by $(1-\cos\theta) \sim q^2$) as to the effectiveness of the scattering.

memory function the kernel $\gamma(t-t')$ relating the acceleration or force of a (Brownian) particle in suspension to its velocity history: $\partial v/\partial t = -\int_{-\infty}^{\infty} \gamma(t-t')v(t')dt'$. More generally, any time- (or frequency-) dependent transport coefficient.

meniscus the interface between two coexisting fluid phases.

metamagnet a material with antiferromagnetic order in zero external field that undergoes a first-order phase transition to a phase with a non-vanishing ferromagnetic moment in an increasing external field.

metastable state state that is stable with respect to infinitesimal fluctuations but which is not the equilibrium state of the system.

micelle a self-assembled aggregate of surfactant molecules in suspension. A dilute solution of surfactant molecules in water is stable because of entropy of mixing. However, when the surfactant concentration exceeds the critical micelle concentration, the system can lower its free energy via the formation of closed structures whose inner volume is filled with hydrophobic tails shielded from contact with water by hydrophilic heads at the structures' outer surfaces. For surfactants in oil, inverted micelles form with interior volume occupied by polar heads rather than hydrocarbon tails. At low concentrations, micelles tend to be spherical, but at higher concentrations and for different surfactant head to tail dimensions, they can exhibit more complex shapes (e.g. rods, worms, etc.).

Migdal-Kadanoff procedure renomalization procedure in which bonds are cut and moved, leaving sites connected to other sites by only two bonds. The variables at these sites are "decimated" by tracing over their degrees of freedom in the partition function.

mirror plane a plane of reflection symmetry.

mode eigenfunction of linearized equations of motion. A normal mode has a sinusoidal time dependence. For a classical system, the number of normal modes equals the number of degrees of freedom. In systems with translational symmetry, the modes are characterized by a wave vector and a frequency. In fluid systems, the modes are often overdamped.

models A to H phenomenological nonlinear dynamical models describing the time evolution of order parameters and conserved variables that lead to equilibrium probability distributions at long time. Models A to D are purely dissipative models, whereas models E to H have Poisson bracket terms leading to propagating modes. Model A is the Glauber model, and model B is the Cahn-Hilliard model.

modulated phase phase with a spatially modulated order parameter, often associated with a Lifshitz point.

molecular field an internal mean effective field resulting from the interaction of dynamic variables in a system. The most common example is the average effective magnetic field, H_{eff}, acting on a spin, σ, due to the exchange interaction, J, with its neighboring spins, $H_{\text{eff}} = Jz\langle\sigma\rangle/g\mu_B$, where z is the number of nearest neighbors.

molecular orbital approximation the electronic states of a many particle system are often approximated by ignoring the Coulomb interaction among electrons and finding the single particle states in the potential of the bare ions. The states are then occupied by electrons at the lowest energy consistent with the Pauli exclusion principle. The opposite approximation, Heitler-London, assumes strong correlations from the electron repulsions, and hence allows no double occupancy of a site.

Monge gauge parametrization of a surface by its height above some reference frame.

multicritical point a critical point at which two or more second-order, and possibly additional first-order, lines meet. The order of a multicritical point is determined by the number of second-order lines meeting it. Three second-order and one first-order lines meet at a tricritical point.

multiple scattering occurs in any system in which the mean-free path, l, is not much greater than the dimensions of the system, L. Almost all scattering theory (see Chapter 2) applies to the limit $l \gg L$, where there is only single scattering in the sample. The sample is essentially transparent to the probe particle or wave. A scattering probability approaching unity is already much too high to allow meaningful interpretation of scattering data.

Navier-Stokes equations hydrodynamical equations for an incompressible fluid: $\rho(\partial \mathbf{v}/\partial t)+\rho(\mathbf{v}\cdot\nabla)\mathbf{v} = -\nabla p + \eta\nabla^2\mathbf{v}$, with $\nabla\cdot\mathbf{v} = 0$, where ρ is the mass density, $\mathbf{v}$ is the velocity, p is the pressure, and η is the viscosity.

nematic liquid crystal a material composed of anisotropic (rod or disc shaped) particles with long-range orientational but no long-range translational order. The transition from the isotropic to the nematic phase is first order. The term *nematic* comes from the Greek $\nu\epsilon\mu\omega\sigma$ for thread. The most obvious characteristic of early nematics was the threadlike defects, which can be seen connecting other defects on the cover of this book. Most digital liquid crystal displays use nematics.

nematogen a rod-like molecule which tends to form nematic liquid crystals.

Nernst theorem the third law of thermodynamics, stating that entropy tends to zero as temperature approaches absolute zero.

nucleation a first-order transition from one phase to another is characterized by a discontinuous jump in the order parameter, ϕ, and by an energy barrier between the two phases (or values of ϕ). Because of the barrier, there is a surface tension, σ, associated with an interface between the two phases. A droplet (nucleus) of the new equilibrium phase gains bulk free energy but costs surface energy. For the droplet to grow, its radius, R, must exceed a critical radius, $R_c = 2\sigma/\Delta f$, where Δf is the gain in bulk free energy density. The critical nucleus may form from thermodynamic fluctuations (homogeneous nucleation) or on a surface or dust particle (heterogeneous nucleation).

Onsager relation symmetry relation, $\lambda_{AB} = \lambda_{BA}$, for dissipative coefficients, λ_{AB}, relating the current for A to the variable conjugate to B.

***O_n* symmetry** symmetry exhibited by vectors on the surface of the unit sphere in n dimensions. Physical realizations include the vector or Heisenberg models (O_3 symmetry) and the *xy*-model (O_2 symmetry).

Order parameter parameter distinguishing an ordered from a disordered phase. For example, the order parameter for a ferromagnet is the average magnetization $\langle \mathbf{m} \rangle$. $\langle \mathbf{m} \rangle$ is zero in the high-temperature paramagnetic phase where spins are randomly oriented and nonzero in the ferromagnetic phase where spins align on average along a common direction. More generally, the order parameter is the average, $\langle \phi \rangle$, of an operator, ϕ, which is a function of the dynamical variables in the system Hamiltonian $\mathscr{H}$. $\langle \phi \rangle$ is zero in the disordered phase and nonzero in the ordered phase. It is chosen to have values in different equivalent ordered phases reflecting the symmetry of the Hamiltonian $\mathscr{H}$ (i.e., to transform under an irreducible representation of the symmetry group of $\mathscr{H}$).

order-parameter exponent for a second-order transition, the order parameter, $\langle \phi \rangle$, goes to zero as $T \to T_c$ with an exponent β: $\langle \phi \rangle \sim (T_c - T)^{\beta}$. In mean-field theory, $\beta = \frac{1}{2}$; in the three-dimensional Ising model, $\beta \approx \frac{1}{3}$.

Ornstein-Zernicke theory theory for correlations in which the density correlation function is a Lorentzian function, $S_{nn}(\mathbf{q}) = S_{nn}(0)/[1 + q^2\xi^2]$, of the wave number, with a width approaching zero as the critical point is approached.

pair distribution function the normalized probability of finding another particle as a function of distance from the center of a particular particle, statistically averaged over the system. For an ideal gas of pointlike particles, the pair distribution function, $g(\mathbf{x})$, is independent of position and is equal to unity. For hard spheres, it is zero out to twice the sphere radius. It is a maximum at this distance, and then it oscillates and decays with distance until it asymptotes to unity. *See also* density-density correlation functions.

paramagnetic having a positive magnetic susceptibility so that the system energy is decreased upon application of an external field.

paramagnetic phase disordered high-temperature phase of a magnet.

Penrose tiling tiling of a two-dimensional plane with pentagonal symmetry. Two types of tiles are needed to produce this crystallographically disallowed symmetry.

percolation a transition in the connectivity of a system. Imagine a lattice whose bonds (or sites) are randomly occupied with probability p. When p is greater than the percolation threshold, p_c, there is an infinite path of connected bonds (or contiguous sites) from one side of the system to the other. Below p_c, clusters of connected occupied bonds (or contiguous occupied sites) have a characteristic dimension $\xi \propto (p - p_c)^{-\nu}$. At p_c, the infinite cluster is a fractal with fractal dimension $d_F = 1.9$ in two dimensions and $d_F = 2.6$ in three dimensions. In one dimension, $p_c = 1$. For bond percolation, $p_c = 0.5$ for a square lattice and 0.247 for a cubic lattice. In three dimensions, balls on lattices percolate when their volume fraction is ≈ 0.16. In mean-field theory, $p_c = 1/(z - 1)$, where z is the number of nearest neighbors.

persistence length the correlation length for unit tangent vectors to a polymer or unit normals to a surface: the distance over which the object is effectively linear or "flat".

phason the spatial variation of the phase of a density wave in an incommensurate structure is referred to as a phason. It is also the dynamical mode associated with this variation.

phase transition Transition between two equilibrium phases of matter whose signature is a singularity or discontinuity in some observable quantity. First-order transitions are characterized by a discontinuity in a first derivative of a thermodynamic potential. In particular, entropy S, which is the temperature derivative of a free energy ($S = -\partial F/\partial T$) has a discontinuity ΔS leading to a latent heat $L = T\Delta S$. For second-order tansitions, first derivatives of thermodynamic potentials are continuous. *See* order parameter *and* hysteresis.

phonon A quantized sound or elastic wave often with a linear long wavelength dispersion relation, $\omega \sim |q|$.

plumber's nightmare a cubic self-assembled phase of a lyotropic liquid crystal in which there are two continuous, interpenetrating, multiply connected volumes, which are separated by a bilayer membrane. (Imagine trying to clean out one of the volumes with a plumber's snake.) One version of a plumber's nightmare is topologically equivalent to the Fermi surface of copper.

Poisson bracket in a classical system with conjugate coordinate q and momentum p, the Poisson bracket of two functions, A and B, of q and p is $\{A, B\} = (\partial A/\partial p)(\partial B/\partial q) - (\partial A/\partial q)(\partial B/\partial p)$. Poisson bracket relations among continuous fields appear in stochastic equations for continuous fields in which there are nondissipative processes. They lead to propagating, rather than purely dissipative, modes and to mode-mode coupling.

Poisson ratio the Poisson ratio, σ, is the negative ratio of change in width to change in length when a material is pulled along its length: $u_{xx} = -\sigma u_{zz}$. Normally, the width will decrease as the length increases. In an isotropic three-dimensional solid with bulk modulus B and shear modulus μ, thermodynamic stability allows for values of $\sigma = \frac{1}{2}(3B - 2\mu)/(3B + \mu)$ between -2 and $\frac{1}{2}$. If volume is conserved, then $\sigma = \frac{1}{2}$. It is very unusual to find a three-dimensional material with a negative σ, but some have been artificially made. Two-dimensional fluctuating membranes embedded in three dimensions have negative values of σ.

Poisson summation formula formula relating the sum over integers, k, of a function, $g(x)$, evaluated at the integers, k, to the sum over of integers, m, of the Fourier transform of the function: $\sum_k g(k) = \sum_m \int_{-\infty}^{\infty} dx g(x) e^{-2\pi ixm}$.

Pokrovsky-Talapov energy free energy for a one-dimensional wall in two dimensions confined between two other walls. There is a contribution to this energy proportional to T/l^2, where l is the separation between walls, because the confining walls reduce fluctuations and thus entropy.

poor solvent *see* good solvent.

Potts model lattice model in which variables σ_i at sites i on a lattice can take on any of s distinct values. The Hamiltonian is $\mathscr{H} = -J\sum_{<i,j>}[s\delta_{\sigma_i,\sigma_j} - 1]$.

quasi-long-range order (QLRO) type of "order" that exists in systems, such as two-dimensional xy models and three-dimensional smectic liquid crystals, without long-range order but with power-law, rather than exponential, decay of order-parameter correlation functions.

quasicrystal an ordered structure that exhibits (1) long-range incomensurate *see* incommensurate crystal) translation order and (2) long-range orientational order with a crystallographically disallowed point group symmetry. The first condition is often referred to as quasiperiodicity. It implies the existence of a diffraction pattern with Bragg peaks on a dense set of points in reciprocal space. Crystallographically disallowed point groups are those that are incompatible with tiling of space with a single kind of tile. In two dimensions, all point group symmetries with order n not equal to 2, 3, 4, or 6 are disallowed. In three dimensions, icosahedral symmetry is disallowed. Penrose tilings (*see* Penrose tiling) provide an example of a two-dimensional quasicrystal with 5-fold symmetry. Physical examples of quasicrystals include AlMn alloys with icosahedral symmetry and AlCuCo alloys with decagonal symmetry.

radius of gyration for a set of N points at positions $\mathbf{R}_i$, $R_G^2 = \sum_i(\mathbf{R}_i - \langle\mathbf{R}\rangle)^2/N$, where $\langle R\rangle = \sum_i \mathbf{R}_i/N$. A measure of the size of an object (especially a polymer), it is the rms distance of the constituents (monomers) from the center of mass. For an ideal polymer, $R_G = R_{\text{end-to-end}}/\sqrt{6}$ and scales as the square root of the number of monomers.

random walk A series of uncorrelated steps of average step length $\bar{a}$ describe a random path with zero average displacement but characteristic size (as measured by the radius of gyration or the end-to-end length) proportional to $\sqrt{N}\bar{a}$ where N is the number of steps. The Haussdorf dimension of the random walk is 2. The probability of return to the origin for an infinite random walk is 1 in one or two dimensions but is less than 1 in dimension greater than 2. (i.e., for $d > 2$, there is a nonzero probability of escape or equivalently of not returning to the origin)

Rayleigh peak in fluids, the thermal diffusion peak in the density-density correlation function, $S_{nn}(\mathbf{q},\omega)$, as a function of frequency, ω, at fixed wave number $\mathbf{q}$. It is centered at $\omega = 0$, and its width is $D_T q^2$, where D_T is the thermal diffusion coefficient. *Contrast with* Brillouin peaks.

renormalization group a transformation involving thinning of degrees of freedom (coarse graining), coupled with a change in length scale. For example, representing a group of spins as a block spin and then constructing a Hamiltonian on the scale of the block spins.

self-similar a structure that "looks the same" at all length scales. Its correlation functions have no characteristic length and therefore are typically power laws with distance or wave vector. Exhibits dilation symmetry.

smectic from the Greek $\sigma\mu\epsilon\gamma\mu\alpha$, meaning soap, from which the original smectic liquid crystals were made. A smectic phase is a "solid" in one dimension and a fluid in the two orthogonal directions. It is characterized by a mass-density wave with modulations along one spatial direction, and is usually depicted by a cartoon showing equally spaced parallel layers. Thermal fluctuations in the phase of the mass-density wave lead to destruction of long-range positional order in three-dimensional smectics as a result of the Landau-Peierls instability.

solitary wave a time-independent, traveling or oscillating solution to a nonlinear differential equation with a well defined shape.

soliton a solitary wave that remains invariant after collision with another solitary wave.

spin-density wave a static spatial modulation of the spin structure of a system whose wave vectors can either be commensurate or incommensurate with the underlying lattice. The simplest case of a spin-density wave is an antiferromagnet with modulation wave number $q = \pi/a$, where a is a lattice constant.

spin-flop transition an antiferromagnet can be viewed as consisting of antiparallel spin pairs. Due to spin-orbit coupling, spins align along a preferred direction in the crystal, called the easy axis. A magnetic field along this easy axis cannot change the magnetization unless it flips one of the spins. However, if they flop to a configuration perpendicular to the applied field, they can tilt toward the magnetic field and gain energy proportional to $\delta M \cdot H$ and still remain mostly antiparallel to each other, so preserving most of their exchange energy. When the net energy gained is greater than the anisotropy energy, we have the spin-flop transition.

spinodal curve curve separating metastable from unstable regions in the coexistence regions of binary fluids. Decay to equilibrium above the spinodal curve is via droplet nucleation; decay below the spinodal curve is via the formation of initially small amplitude periodic modulations of the order parameter.

spinodal decomposition process of decay towards equilibrium in a locally unstable region of a phase diagram constrained by particle conservation. Typically seen in quenches of binary mixtures.

spin waves the classical normal modes or quantum excitations of a magnetically ordered system. In the long-wavelength limit, they are the hydrodynamic modes related to the symmetry broken by selecting a direction for the spin alignment. The simplest dispersion relations are $\omega \sim q$ for antiferromagnetic spin waves and $\omega \sim q^2$ for ferromagnetic spin waves. Quantized spin waves are known as *magnons*.

spin-wave stiffness elastic constant ρ_s relating free energy density to rotation gradients in an xy- or n-vector model: $f = (\rho_s/2)(\nabla\theta)^2$. In superfluids, which, like the xy-model, have a complex order parameter, ρ_s is the superfluid density.

splay one of three spatial variations of the director, **n**, in a nematic liquid crystal. It is characterized by a finite value of $\nabla \cdot \mathbf{n}$. The usual cartoon describing splay has the director arranged in the same manner as playing cards are held in your hand.

stochastic variable a variable which changes with time such that there is no correlation between different time intervals. A random variable.

Stokes' law says that the drag force on a spherical particle of radius a in a fluid of viscosity η moving at velocity v is $f = 6\pi\eta av$. Equivalently, the mobility is $\alpha^{-1} = (6\pi\eta a)^{-1}$.

strain relative distortion of a solid. Measured by the strain tensor, $u_{ij} = (1/2)(\nabla_i u_j + \nabla_j u_i)$, where $\mathbf{u}$ is the displacement of the solid relative to a reference solid (in Lagrangian coordinates). *Longitudinal strain* has displacement along the gradient; *shear strain* is displacement perpendicular to the gradient. Application of stress to a solid produces strain.

strain rate a gradient of the velocity field measured by the tensor $\gamma_{ij} = \frac{1}{2}(\nabla_i v_j + \nabla_j v_i)$. In a solid, γ_{ij} can be regarded as the strain per unit time or strain-rate tensor. In a fluid phase, application of a stress results in a continuous deformation of the system with time, and stress is linearly related to strain rate by the viscosity (tensor).

stress Force per unit area acting on an element of matter through its bounding surfaces. Normal forces (to the surface) are associated with compression, while transverse forces are associated with shear. The force density is related to the gradient of the (symmetric) stress tensor: $f_i = \nabla_j \sigma_{ij}$. *See also* strain *and* strain rate.

structure factor $S(\mathbf{q}) = V^{-1} \int d^d x d^d x' e^{-i\mathbf{q}\cdot(\mathbf{x}-\mathbf{x}')} C_{nn}(\mathbf{x}, \mathbf{x}')$ is the Fourier transform of the density-density correlation function, $C_{nn}(\mathbf{x}, \mathbf{x}')$, where V is the volume. It contains information about two-body interparticle correlations, and is measured in scattering experiments. *See also* density-density correlation functions.

supercooling for a first-order transition, there is a region between the phase boundary and the spinodal line where the system is in metastable equilibrium, and the free energy prefers the low-temperature phase but there is a barrier to overcome to form a critical nucleus. If there are no inhomogeneities to aid nucleation and the activation energy is sufficiently high, then the high-temperature phase can remain until the sample is supercooled to near the spinodal line. For pure water, the freezing transition can be suppressed to −40°C.

TDGL model time-dependent Ginzburg-Landau model. Dynamic model with purely dissipative dynamics, also called model A or Glauber dynamics.

tetracritical point critical point where four second-order lines meet.

TGB phase twist-grain boundary phase in liquid crystals in which there is a periodic parallel array of twist-grain boundaries, each composed of a regular array of screw dislocations. This phase is the liquid crystal analog of the Abrikosov vortex lattice phase in superconductors. It can have quasicrystalline symmetry.

thermotropic exhibits phase transitions as a function of temperature, as opposed to *lyotropic*, where the transitions occur predominantly as a function of concentration. Usually applied to liquid crystals.

θ-solvent *See under* good solvent

topological defect a defect in an order-parameter field that cannot be eliminated by any continuous distortion of the order parameter. Such a defect is characterized by some integer-valued index, such as the winding number. It generally has a core region, where the order parameter goes to zero, and a far-field region characterized by nonvanishing strain. Examples of topological defects include vortices, dislocations, disclinations, and hedgehogs.

tricritical point critical point where three second-order lines meet a first-order line.

twist distortion of the director, $\mathbf{n}$, in a liquid crystal with nonvanishing $\mathbf{n} \cdot (\nabla \times \mathbf{n})$. The director rotates in a helical fashion as it advances along an axis.

type I and II smectics in a type I smectic, the penetration depth for twist and bend is less than the correlation length for the smectic order parameter. In type II systems, the twist and bend penetration depths are larger than the correlation length.

universality class systems whose properties near a second-order phase transition are controlled by the same renormalization group fixed point are in the same universality class. They have the same relevant critical exponents but may have completely different order parameters and transition temperatures. Typically, universality classes are determined by spatial dimension, symmetry of the order parameter, and range and symmetry of interaction potentials.

upper critical dimension the spatial dimension, d_c, above which fluctuations play a negligible role in the vicinity of a second-order phase transition. Thus, for $d > d_c$, mean-field theory gives the correct values for critical exponents. For the Ising model, $d_c = 4$.

vacancy a pointlike defect in a periodic solid consisting of a missing particle at a crystal site. Point defects in three-dimensional systems have integrable strain fields, and therefore do not destroy long-range order the way that dislocations can.

Van der Waals attraction $1/r^6$ attractive interaction between neutral atoms.

Van der Waals equation of state equation of state for a fluid in which there is a mean-field critical point for a liquid-gas transition:

$$(p + an^2)(n^{-1} - b) = T,$$

where p is the pressure, n is the number density, T is the temperature, and a and b are constants.

Van Hove theory *See under* critical slowing down.

vesicle an object consisting of a closed bilayer membrane with the same phase interior and exterior. Typically a self-assembled bilayer shell with water on either side. Biological cells can be viewed as phospholipid vesicles.

Villain model lattice model with xy symmetry that is dual to the discrete Gaussian model.

viscosity coefficient relating the shear stress to the shear strain rate. It measures the rate of momentum transfer across a transverse velocity gradient.

Volterra construction prescription for creating dislocations and disclinations in crystals by slicing a cylindrical sample along a radius and displacing the sides of the slice by the Burgers vector of the defect and gluing to make the cylinder continuous. For disclinations, slice, rotate, and glue.

Weiner-Kintchine theorem theorem relating the power spectrum, $C_{AA}(\omega)$, of a variable, A, to the temporal Fourier transform of its correlation function:

$$C_{AA}(\omega) = \int_{-\infty}^{\infty} d(t - t')e^{i\omega(t-t')}\langle A(t)A(t')\rangle.$$

Weiss molecular field theory mean-field theory in which the interaction with neighbors is approximated by an average effective field. For an Ising model or a Heisenberg model on a lattice with z nearest neighbors for each site, $H_{\text{eff}} = zJ\langle\sigma\rangle$.

Wigner-Seitz cell unit cell of a lattice that is the interior envelope of all planes that are perpendicular bisectors of bonds connecting a lattice site to all other lattice sites.

work hardening when a material, particularly a metal, is repeatedly plastically deformed, dislocations are introduced and moved. They can multiply and entangle to a degree which prevents further motion. With the dislocations immobile, the yield stress increases.

Wulff plot plot of the surface tension, $\sigma(\theta, \phi)$, of a crystal as a function of the polar angles, θ and ϕ, specifying the direction of the unit normal, **N**, to crystal surfaces. The equilibrium shape of a crystal is the interior envelope of planes perpendicular to **N** and passing through the Wulff plot. The determination of crystal shape in this way is called the Wulff construction.

***xy*-model** model for vectors with O_2 symmetry or complex fields with $U(1)$ symmetry. Applied to spins constrained to lie in a plane, to superfluid and density wave transitions (liquid-solid and smectic transitions).

Yang-Lee edge singular point of the free energy of an Ising model in an external imaginary magnetic field. The upper critical dimension for the critical behavior of this point is 6.

yield stress the stress beyond which a solid material no longer responds elastically. It either flows or deforms plastically, not returning to its original shape when the stress is released. For an ideal model solid, the yield stress is the shear elastic constant divided by 4π. For real materials, it is 10^{-2} to 10^{-3} times smaller because of the presence of dislocations.

Young's modulus the ratio of uniaxial stress to uniaxial strain in a crystal (with orthogonal stresses equal to zero). In terms of Lamé coefficients, $Y = 9B/(3B + \mu)$ in three dimensions.

Z_n symmetry discrete symmetry of the integers mod$n$ under addition. The Ising model has Z_2 symmetry. Clock models with vectors restricted to n equally spaced points on the two-dimensional unit circle have Z_n symmetry.

Index